DevBio Laboratory: *vade mecum*³

An Interactive Guide to Developmental Biology

Mary S. Tyler and Ronald N. Kozlowski

http://labs.devbio.com

Designed to complement the textbook, this unique resource helps you understand the organisms discussed in lecture and prepares you for the laboratory. **DevBio Laboratory: *vade mecum*³** is available online, which allows you the flexibility to use the software from any computer with Internet access. Access to the site is included with each new textbook.

Over 140 interactive videos and 300 labeled photographs take you through the life cycles of model organisms used in developmental biology laboratories. The easy-to-use videos provide you with the concepts, vocabulary, and motivation to enter the laboratory fully prepared. A chapter on zebrafish addresses how to raise the organism and the effects of various teratogens on embryonic development. The site also includes chapters on: the slime mold Dictyostelium discoideum; planarian; sea urchin; the fruit fly Drosophila melanogaster; chick; and amphibian.

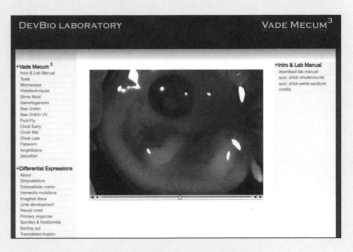

ADDITIONAL FEATURES INCLUDE:

■ **Movie excerpts** from **Differential Expressions**[2]: Short video excerpts about key concepts in development.

■ **PowerPoint® slides** of chick wholemounts and serial sections for self-quizzing and creating tests.

■ **Full video instruction** on histological techniques

■ **Glossary:** Every chapter of the Laboratory Manual includes an extensive glossary.

■ **Laboratory Manual:** The Third Edition of Mary S. Tyler's laboratory manual, **Developmental Biology: A Guide for Experimental Study**, designed for use with the multimedia chapters of **DevBio Laboratory: *vade mecum*³**

■ **Website: www.devbio.net** augments **DevBio Laboratory**, allowing convenient access to recipes from the lab book, a searchable glossary, a module on Laboratory Safety, and more.

■ **Study Questions**

■ **Laboratory Skills Guides**

To access **DevBio Laboratory:** *vade mecum*³ go to **http://labs.devbio.com**, click the "**Register**" link, and follow the instructions to enter your registration code (see left) and create a username and password

Developmental Biology

NINTH EDITION

Developmental Biology NINTH EDITION

SCOTT F. GILBERT

Swarthmore College and The University of Helsinki

 Sinauer Associates, Inc. • *Publishers* • *Sunderland, Massachusetts USA*

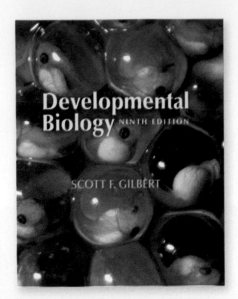

The Cover

Three-day-old embryos of Costa Rican red-eyed tree frogs (*Agalychnis callidryas*). The female frog lays her eggs on a leaf overhanging a pond, so that when the tadpoles hatch (normally in about 7 days), they wiggle out and fall into the pond. The tadpoles' development is exquisitely tuned to their environment, and the larvae place their branching gills near the oxygen-rich egg surface. Tadpoles respond rapidly to the presence of infectious fungi or predaceous snakes, hatching early (often at 5 days) and taking their chances in the pond rather than succumbing to predation (see Chapter 18). Photograph courtesy of Karen Warkentin, Boston University.

Developmental Biology, 9th Edition

Sinauer Associates, Inc.
23 Plumtree Road
Sunderland, MA 01375 USA

FAX 413-549-1118

email: orders@sinauer.com; publish@sinauer.com

www.sinauer.com

Library of Congress Cataloging-in-Publication Data

Gilbert, Scott F., 1949-
  Developmental biology / Scott F. Gilbert. — 9th ed.
      p. cm.
    Includes bibliographical references and index.
    ISBN 978-0-87893-384-6 (casebound)
1. Embryology. 2. Developmental biology. I. Title.

QL955.G48 2010
  571.8—dc22   2010009918

Printed in U.S.A.

5 4 3 2

To Daniel, Sarah, and David

Brief Contents

Contents

PART **II** SPECIFICATION Introducing Cell Commitment and Early Embryonic Development 109

PART III THE STEM CELL CONCEPT Introducing
Organogenesis 323

CHAPTER 11
Paraxial and Intermediate Mesoderm 413

CHAPTER 12
Lateral Plate Mesoderm and the Endoderm 445

PART IV SYSTEMS BIOLOGY Expanding Developmental Biology to Medicine, Ecology, and Evolution 617

An Overview of Plant Development by Susan R. Singer is available at www.devbio.com

Preface

It has become increasingly embarrassing for me to ask students to read the Eighth Edition of this textbook. It's so, well, 2006. Developmental biology has progressed so rapidly in the past four years that my lectures have fundamentally diverged from their reading. My "big" lecture on transcription now focuses on the ability of transcription factors to reprogram cell fates; and my lectures on stem cells and cloning have scrapped the notion of therapeutic cloning altogether, focusing instead on induced pluripotent stem cells. In both instances, we discuss what this means for understanding normal development, as well as what implications these technologies have for the future of medicine. Neither induced pluripotential stem cells nor "transdifferentiation" was established when the last edition of this book was published.

Even my most basic lectures have changed. The lecture on fertilization has to cover the new data on mammalian egg activation. My lectures on sea urchin development—an area of study that has been fundamental to developmental biology for over a century—now include systems theory operations involving double-negative gates and feedforward loops, and my evo-devo talks have led to discussions of mathematical modeling and parasitism. I can't talk about limb development without including the variations seen in dachshunds and bats, and I can't discuss sex determination without using the β-catenin model for mammalian ovary production. None of these areas were covered in earlier editions of my book. So this is really a very new edition. My editor tells me it has close to 700 new references; she only wishes I had deleted at least that many old ones.

Developmental biology is in a state of rapid metamorphosis. And, as in insect and amphibian metamorphosis, some old tissues remain the same, some get substantially remodeled, and some old tissues perish altogether; and all the while, new tissues are forming new structures. I hope that I have gotten these correct, and that the added new material will stand the test of time. I have tried to remodel the retained material into new narratives that are more inclusive of the data, and to appropriately jettison the information that was needed for earlier stages of the book's development but which is no longer needed by undergraduates.

Embryologist John Fallon once wrote me that new data change the story one tells. It is, he said, like putting together a picture puzzle. At first, you think the structure in front of you is a sailboat; but you add another piece, and—no, wait—it's a mountain. Psychologists call these alterations "Gestalt changes," and I think that we are seeing these changes in both our day-to-day interpretations of data and

in the entire field of developmental biology. We are seeing an inversion of relationships within the biological sciences. Genetics is more and more becoming a subset of development. Similarly, the dynamic of evolution is being studied as a question of gene expression as well as gene frequencies. And developmental biology may be on the threshold of changing medicine as much as microbiology did at the turn of the twentieth century.

I began the Preface of the last edition with a quotation from the Grateful Dead, recalling "What a long, strange trip it's been." The epigram for this edition might be Eminem's "Be careful what you wish for." We may achieve biological powers that are "tenfold" what we had hoped to have. And it is axiomatic for this generation that "with great power comes great responsibility."

I hope this Ninth Edition of *Developmental Biology* presents a better way of teaching and learning (and questioning) developmental biology. The introductory section has been streamlined from six chapters to three—one each on developmental anatomy, the mechanisms of gene regulation during differentiation, and cell–cell communication during morphogenesis. Another new feature is the addition of short part-opening "chaplets" that address key concerns in developmental biology. These provide an introduction to the subsequent chapters, placing the forthcoming information into a specific context. Each chapter ends with a guide to web-based resources relevant to that chapter's content, and the Ninth Edition is the first to include an extensive glossary of key terms.

During the writing of this edition, I re-read some of the papers written by the first generation of experimental embryologists, scientists who were experiencing a Gestalt change as important as what we are experiencing today. What impressed me was not necessarily their answers (although some of them were remarkably good even by today's standards); rather, it was their asking the "right" questions. Some of their research did not give us any answers at all. But the results told the next generation of biologists what questions to ask. These embryologists stood in awe of the complexity of the embryo; yet they began to remove, transplant, destroy, and recombine cells in order to find out just how the fertilized egg could give rise to a structured body composed of different cell types. They had faith that these were scientific questions and that science would eventually be able to answer them.

The glory of developmental biology is that we now have interesting answers to many of their questions. But numerous questions that were asked a century ago still lack answers. How does the human brain become organized so that we can think, plan, recall, interpret, hate, and love?

How is the development of plants and insects timed so that the flower opens at the same time when its pollinator has left its cocoon? How does exercise increase muscle mass, and how does our face come to resemble those of our parents more than any one else's? To these questions, we have only very partial answers, but we are on our way.

Developmental biology presents a nascent scientist with a host of fascinating questions that are worth solving. And that's the invitation this book offers. One can enter developmental biology through many portals—genetics, cell biology, embryology, physiology, anatomy—and with many valid motivations. This is a field that needs the help of people with all sorts of competencies and talents. It is an old field that is itself undergoing metamorphic change and emerging as a new field that welcomes newcomers with open arms—full of questions.

Acknowledgments

In addition to the remarkable reviewers listed below, whose candid and thorough criticisms of early chapter drafts made this book so much better, there are some people whose help was absolutely critical. In particular, David McClay and Bill Anderson gave me many suggestions that were outstandingly important in constructing this edition. I also appreciate enormously the cooperation from all those scientists who sent me their photographs, and who even told me about others they had seen. The graphics of this book are truly amazing, and this is due to the community of developmental biologists.

The book's beauty and success is also the result of Andy Sinauer's vision and the hard work of the immensely talented staff he has assembled at Sinauer Associates. David McIntyre's ability to find appropriate photographs from the public and private databases is almost uncanny. Chris Small and Janice Holabird of Sinauer's production department have put together the artwork, the photographs, and the text into a format that is both informative and pleasing. For a book of this size, this is a heroic undertaking. And, more than any other edition of this book (and she has been with it since its inception), this incarnation has been a collaborative effort with my editor Carol Wigg. The book seems to have become a full-time job for both of us.

I especially wish to thank Dr. Hannah Galantino-Homer of the University of Pennsylvania School of Veterinary Medicine, who took it upon her shoulders to compile a glossary for this book. Numerous people have said that they wanted such a glossary for their students, and putting one together is an extremely difficult task, involving an enormous amount of thought and expertise.

This textbook officially entered the "electronic age" in the mid 90s, and its web segment, www.devbio.com, has grown more important with each subsequent edition. With this edition, the *vade mecum*[3] companion that debuted in 2002 is also on the web. Mary Tyler and Ron Kozlowski not only created *vade mecum*[3], with its laboratory sections and its introductions to model animals, they have also produced interviews and filmed the techniques of several developmental biologists. You really have to see these films to realize what a valuable resource these are.

I am blessed by teaching some remarkable students who have not been shy about offering constructive criticism. Their suggestions will, I hope, benefit the next round of students.

And finally, this revising process has taken much longer than expected. I apologize to my wife, Anne Raunio, who has had to put up with me through it all, and to my friends, who may have wondered where I've been. I'll be back.

SCOTT F. GILBERT

Reviewers OF THE NINTH EDITION

It is no longer possible (if it ever was) for one person to comprehend this entire field. As Bob Seger so aptly sings, "I've got so much more to think about … what to leave in, what to leave out." The people who help me leave in and take out the right things are the reviewers. Their expertise in particular areas has become increasingly valuable to me. Their comments were made on early versions of each chapter, and they should not be held accountable for any errors that may appear.

Media and Supplements
to accompany Developmental Biology, Ninth Edition

eBook (ISBN 978-0-87893-412-6)

www.sinauer.com/ebooks

New for the Ninth Edition, *Developmental Biology* is available as an online interactive ebook, at a substantial discount off the list price of the printed textbook. The interactive ebook features a variety of tools and resources that make it flexible for instructors and effective for students. For instructors, the eBook offers an unprecedented opportunity to easily customize the textbook with the addition of notes, Web links, images, documents, and more. Students can readily bookmark pages, highlight text, add their own notes, and customize display of the text. In addition, all of the in-text references to the Companion Website Web topics and to *DevBio Laboratory: vade mecum*[3] are integrated into the ebook as direct links, so the student can easily access a wealth of additional material as they read.

Also available as a CourseSmart eBook (ISBN 978-0-87893-409-6). The CourseSmart eBook reproduces the look of the printed book exactly, and includes convenient tools for searching the text, highlighting, and notes. For more information, please visit **www.coursesmart.com**.

For the Student

Companion Website

www.devbio.com

Available free of charge, this website is intended to supplement and enrich courses in developmental biology. It provides more information for advanced students as well as historical, philosophical, and ethical perspectives on issues in developmental biology. Included are articles, movies, interviews, opinions, Web links, updates, and more. References to specific website topics are included throughout each chapter as well as at the end of each chapter.

DevBio Laboratory: vade mecum[3]: An Interactive Guide to Developmental Biology

http://labs.devbio.com

MARY S. TYLER and RONALD N. KOZLOWSKI

New for Version 3, *DevBio Laboratory: vade mecum*[3] is now online. Access to the program is included with every new copy of the textbook. (See the inside front cover for details.) *DevBio Laboratory: vade mecum*[3] is a rich multimedia learning tool that helps students understand the development of the organisms discussed in lecture and prepares them for laboratory exercises. It also includes excerpts from the *Differential Expressions* series of videos, highlighting some major concepts in developmental biology, famous experiments, and the scientists who performed them.

Developmental Biology: A Guide for Experimental Study, Third Edition

MARY S. TYLER

(Included in *DevBio Laboratory: vade mecum*[3])
This lab manual teaches the student to work as an independent investigator on problems in development and provides extensive background information and instructions for each experiment. It emphasizes the study of living material, intermixing developmental anatomy in an enjoyable balance, and allows the student to make choices in their work.

For the Instructor
(Available to qualified adopters)

Instructor's Resource Library

The *Developmental Biology*, Ninth Edition Instructor's Resource Library includes a rich collection of visual resources for use in preparing lectures and other course materials. The IRL includes:

- All textbook figures (including photos) and tables in JPEG (high and low resolution) and PowerPoint® formats
- A collection of videos illustrating key developmental processes
- Chick embryo cross-sections and chick embryo whole-mounts from *DevBio Laboratory: vade mecum*[3] (PowerPoint® format)
- Video segments from *DevBio Laboratory: vade mecum*[3]
- Instructor's Reference Guide for *Differential Expressions*[2]

Also Available

The following titles are available for purchase separately or, in some cases, bundled with the textbook. Please contact Sinauer Associates for more information.

Ecological Developmental Biology:
Integrating Epigenetics, Medicine, and Evolution
Scott F. Gilbert and David Epel
Paper, 460 pages • ISBN 978-0-87893-299-3

Bioethics and the New Embryology:
Springboards for Debate
Scott F. Gilbert, Anna Tyler, and Emily Zackin
Paper, 261 pages • ISBN 978-0-7167-7345-0

Differential Expressions²:
Key Experiments in Developmental Biology
Mary S. Tyler, Ronald N. Kozlowski, and Scott F. Gilbert
2-DVD Set • UPC 855038001020

A Dozen Eggs:
Time-Lapse Microscopy of Normal Development
Rachel Fink
DVD • ISBN 978-0-87893-329-7

Fly Cycle²
Mary S. Tyler and Ronald N. Kozlowski
DVD • ISBN 978-0-87893-849-0

From Egg to Tadpole:
Early Morphogenesis in Xenopus
Jeremy D. Pickett-Heaps and Julianne Pickett-Heaps
DVD • ISBN 978-0-97752-224-8

PART I

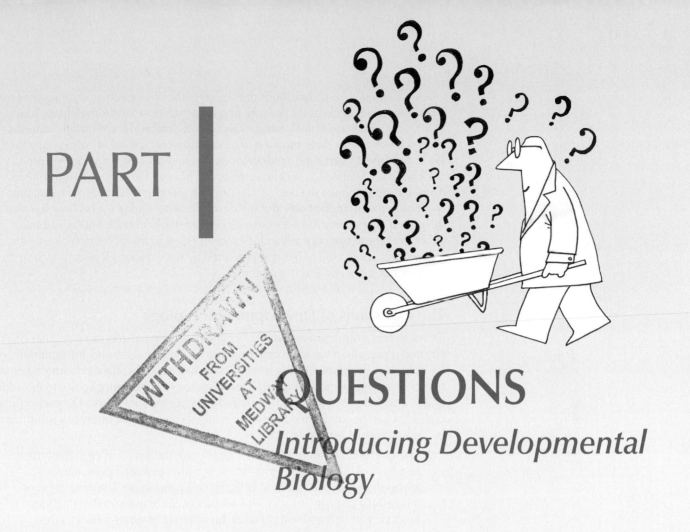

QUESTIONS
Introducing Developmental Biology

Between fertilization and birth, the developing organism is known as an embryo. The concept of an embryo is a staggering one, and forming an embryo is the hardest thing you will ever do. To become an embryo, you had to build yourself from a single cell. You had to respire before you had lungs, digest before you had a gut, build bones when you were pulpy, and form orderly arrays of neurons before you knew how to think. One of the critical differences between you and a machine is that a machine is never required to function until after it is built. Every animal has to function even as it builds itself.

For animals, fungi, and plants, the sole way of getting from egg to adult is by developing an embryo. The embryo mediates between genotype and phenotype, between the inherited genes and the adult organism. Whereas most fields of biology study adult structure and function, developmental biology finds the study of the transient stages leading up to the adult to be more interesting. Developmental biology studies the initiation and construction of organisms rather than their maintenance. It is a science of becoming, a science of process.

This development, this formation of an orderly body from relatively homogeneous material, provokes profound and fundamental questions that *Homo sapiens* have been asking since the dawn of self-awareness: How does the body form with its head always above its shoulders? Why is the heart on the left side of our body? Why do we have five fingers on each hand and not more or fewer? Why can't we regenerate limbs? How do the sexes develop their different anatomies? Why can only females have babies?

Our answers to these questions must respect the complexity of the inquiry and must form a coherent causal network from gene through functional organ. To say that XX mammals are usually females and that XY mammals are usually males does not explain sex determination to a developmental biologist, who wants to know *how* the XX genotype produces a female and *how* the XY genotype produces a male. Similarly, a geneticist might ask how globin genes are transmitted from one generation to the next, and a physiologist might ask about the function of globin proteins in the body. But the developmental biologist asks how it is that the globin genes come to be expressed only in red blood cells, and how these genes become active only at specific times in development. (We don't know the answers yet.) Each field of biology is defined by the questions it asks. *Welcome to a wonderful set of important questions!*

The Questions of Developmental Biology

Development accomplishes two major objectives. First, it generates cellular diversity and order within the individual organism; secondly, it ensures the continuity of life from one generation to the next. Put another way, there are two fundamental questions in developmental biology. How does the fertilized egg give rise to the adult body? And how does that adult body produce yet another body? These two huge questions can be subdivided into seven general categories of questions scrutinized by developmental biologists:

- **The question of differentiation.** A single cell, the fertilized egg, gives rise to hundreds of different cell types—muscle cells, epidermal cells, neurons, lens cells, lymphocytes, blood cells, fat cells, and so on. The generation of this cellular diversity is called *differentiation*. Since every cell of the body (with very few exceptions) contains the same set of genes, how can this identical set of genetic instructions produce different types of cells? How can a single cell, the fertilized egg, generate so many different cell types?*

- **The question of morphogenesis.** How can the cells in our body organize themselves into functional structures? Our differentiated cells are not randomly distributed. Rather, they become organized into intricate tissues and organs. During development, cells divide, migrate, and die; tissues fold and separate. Our fingers are always at the tips of our hands, never in the middle; our eyes are always in our heads, not in our toes or gut. This creation of ordered form is called *morphogenesis*, and it involves coordinating cell growth, cell migration, and cell death.

- **The question of growth.** If each cell in our face were to undergo just one more cell division, we would be considered horribly malformed. If each cell in our arms underwent just one more round of cell division, we could tie our shoelaces without bending over. How do our cells know when to stop dividing? Our arms are generally the same size on both sides of the body. How is cell division so tightly regulated?

- **The question of reproduction.** The sperm and egg are very specialized cells, and only they can transmit the instructions for making an organism

---

*There are more than 210 different cell types recognized in the adult human, but this number has little or no significance. There are many transient cell types that are formed during development but are not seen in the adult. Some of these embryonic cells are transitional stages or precursors of adult cell types. Other embryonic cell types perform particular functions in constructing an organ and then undergo programmed cell death after completing their tasks.

from one generation to the next. How are these germ cells set apart from the cells that are constructing the physical structures of the embryo, and what are the instructions in the nucleus and cytoplasm that allow them to form the next generation?

- **The question of regeneration.** Some organisms can regenerate their entire body. Some salamanders regenerate their eyes and legs, and many reptiles can regenerate their tails. Mammals are generally poor at regeneration, and yet there are some cells in our bodies—*stem cells*—that are able to form new structures even in adults. How do the stem cells retain this capacity, and can we harness it to cure debilitating diseases?

- **The question of evolution.** Evolution involves inherited changes in development. When we say that today's one-toed horse had a five-toed ancestor, we are saying that changes in the development of cartilage and muscles occurred over many generations in the embryos of the horse's ancestors. How do changes in development create new body forms? Which heritable changes are possible, given the constraints imposed by the necessity that the organism survive as it develops?

- **The question of environmental integration.** The development of many (perhaps all) organisms is influenced by cues from the environment that surrounds the embryo or larvae. The sex of many species of turtles, for instance, depends on the temperature the embryo experiences while in the egg. The formation of the reproductive system in some insects depends on bacteria that are transmitted inside the egg. Moreover, certain chemicals in the environment can disrupt normal development, causing malformations in the adult. How is the development of an organism integrated into the larger context of its habitat?

The study of development has become essential for understanding all other areas of biology. Indeed, the questions asked by developmental biologists have also become critical in molecular biology, physiology, cell biology, genetics, anatomy, cancer research, neurobiology, immunology, ecology, and evolutionary biology. In turn, the many advances of molecular biology, along with new techniques of cell imaging, have finally made these questions answerable. This makes developmental biologists extremely happy; for, as the Nobel Prize-winning developmental biologist Hans Spemann stated in 1927,

> We stand in the presence of riddles, but not without the hope of solving them.
> And riddles with the hope of solution—what more can a scientist desire?

So, like the man in the cartoon, I come bearing questions. They are questions bequeathed to us by earlier generations of biologists, philosophers, and parents. They are questions with their own history, questions discussed on an anatomical level by people such as Aristotle, William Harvey, St. Albertus Magnus, and Charles Darwin. More recently, these questions have been addressed on the cellular and molecular levels by men and women throughout the world, each of whom brings to the laboratory his or her own perspectives and training. For there is no one way to become a developmental biologist, and the field has benefitted by having researchers trained in cell biology, genetics, biochemistry, immunology, and even anthropology, engineering, physics, history, and art. You are now invited to become part of a community of question-askers for whom the embryo is a source of both wonder and the most interesting questions in the world.

The next three chapters outline some of the critical framework needed to answer these questions. Chapter 1 discusses *organismal* concepts, including life cycles, the three germ layers that form the organs, and the migration of cells during development. Chapter 2 concentrates on the *genetic* approach to cell differentiation and outlines the principle of differential gene expression (which explains how different proteins can be made in different cells from the same set of inherited genes). Chapter 3 focuses on the *cellular* approach to morphogenesis, showing how communication between cells is critical for their formation into tissues and organs. Thus, you will be introduced to development at the organismal, genetic, and cellular, and much of the textbook thereafter will show how these levels are integrated to produce the remarkable panoply of animal development.

Developmental Anatomy

ACCORDING TO ARISTOTLE, the first embryologist known to history, science begins with wonder: "It is owing to wonder that people began to philosophize, and wonder remains the beginning of knowledge" (Aristotle, *Metaphysics*, ca. 350 BCE). The development of an animal from an egg has been a source of wonder throughout history. The simple procedure of cracking open a chick egg on each successive day of its 3-week incubation period provides a remarkable experience as a thin band of cells is seen to give rise to an entire bird. Aristotle performed this procedure and noted the formation of the major organs. Anyone can wonder at this remarkable—yet commonplace—phenomenon, but it is the scientist seeks to discover how development actually occurs. And rather than dissipating wonder, new understanding increases it.

Multicellular organisms do not spring forth fully formed. Rather, they arise by a relatively slow process of progressive change that we call **development**. In nearly all cases, the development of a multicellular organism begins with a single cell—the fertilized egg, or **zygote**, which divides mitotically to produce all the cells of the body. The study of animal development has traditionally been called **embryology**, after that phase of an organism that exists between fertilization and birth. But development does not stop at birth, or even at adulthood. Most organisms never stop developing. Each day we replace more than a gram of skin cells (the older cells being sloughed off as we move), and our bone marrow sustains the development of millions of new red blood cells every minute of our lives. In addition, some animals can regenerate severed parts, and many species undergo metamorphosis (such as the transformation of a tadpole into a frog, or a caterpillar into a butterfly). Therefore, in recent years it has become customary to speak of **developmental biology** as the discipline that studies embryonic and other developmental processes.

As the introduction to Part I notes, a scientific field is defined by the questions it seeks to answer. Most of the questions in developmental biology have been provided to it by its embryological heritage. We can identify three major approaches to studying embryology:

- Anatomical approaches
- Experimental approaches
- Genetic approaches

Each of these traditions has predominated during a different era. However, although it is true that anatomical approaches gave rise to experimental approaches, and that genetic approaches built on the foundations of the earlier two approaches, all three traditions persist to this day and continue to play a major role in developmental biology. The basis of all research in developmental biology is the changing anatomy of the organism. Today the anatomical approach to

It is a most beautiful thing to study the different changes of life, from the microscopic changes of conception to the more apparent ones of maturity and old age.

FRANKLIN MALL (CA. 1890)

The greatest progressive minds of embryology have not looked for hypotheses; they have looked at embryos.

JANE OPPENHEIMER (1955)

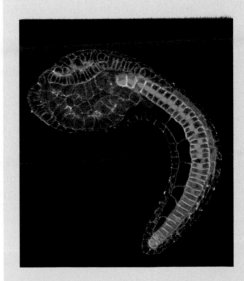

development is continually expanded and enhanced by revolutions in microscopy, computer-aided graphical reconstructions of three-dimensional objects, and methods of applying mathematics to biology. Many of the beautiful photographs in this book reflect this increasingly important component of embryology.

The Cycle of Life

One of the major triumphs of descriptive embryology was the idea of a generalizable animal life cycle. Each animal, whether earthworm or eagle, termite or beagle, passes through similar stages of development. The stages of development between fertilization and hatching are collectively called **embryogenesis**.

Throughout the animal kingdom, an incredible variety of embryonic types exist, but most patterns of embryogenesis are variations on six fundamental processes: fertilization, cleavage, gastrulation, organogenesis, metamorphosis, and gametogenesis.

1. **Fertilization** involves the fusion of the mature sex cells, the sperm and egg, which are collectively called the **gametes**. The fusion of the gamete cells stimulates the egg to begin development and initiates a new individual. The subsequent fusion of the gamete nuclei (both of which have only half the normal number of chromosomes characteristic for the species) gives the embryo its **genome**, the collection of genes that helps instruct the embryo to develop in a manner very similar to that of it parents.

2. **Cleavage** is a series of extremely rapid mitotic divisions that immediately follow fertilization. During cleavage, the enormous volume of zygote cytoplasm is divided into numerous smaller cells called **blastomeres**. By the end of cleavage, the blastomeres have usually formed a sphere, known as a **blastula**.

3. After the rate of mitotic division slows down, the blastomeres undergo dramatic movements and change their positions relative to one another. This series of extensive cell rearrangements is called **gastrulation**, and the embryo is said to be in the **gastrula** stage. As a result of gastrulation, the embryo contains three germ layers that will interact to generate the organs of the body.

4. Once the germ layers are established, the cells interact with one another and rearrange themselves to produce tissues and organs. This process is called **organogenesis**. Many organs contain cells from more than one germ layer, and it is not unusual for the outside of an organ to be derived from one layer and the inside from another. For example, the outer layer of skin (epidermis) comes from the ectoderm, whereas the inner layer (the dermis) comes from the mesoderm. Also during organogenesis, certain cells undergo long migrations from their place of origin to their final location. These migrating cells include the precursors of blood cells, lymph cells, pigment cells, and sex cells.

5. In many species, the organism that hatches from the egg or is born into the world is not sexually mature. Rather, the organism needs to undergo **metamorphosis** to become a sexually mature adult. In most animals, the young organism is a called a **larva**, and it may look significantly different from the adult. In many species, the larval stage is the one that lasts the longest, and is used for feeding or dispersal. In such species, the adult is a brief stage whose sole purpose is to reproduce. In silkworm moths, for instance, the adults do not have mouthparts and cannot feed; the larvae must eat enough so that the adult has the stored energy to survive and mate. Indeed, most female moths mate as soon as they eclose from their pupa, and they fly only once—to lay their eggs. Then they die.

6. In many species, a group of cells is set aside to produce the next generation (rather than forming the current embryo). These cells are the precursors of the gametes. The gametes and their precursor cells are collectively called **germ cells**, and they are set aside for reproductive function. All the other cells of the body are called **somatic cells**. This separation of somatic cells (which give rise to the individual body) and germ cells (which contribute to the formation of a new generation) is often one of the first differentiations to occur during animal development. The germ cells eventually migrate to the gonads, where they differentiate into gametes. The development of gametes, called **gametogenesis**, is usually not completed until the organism has become physically mature. At maturity, the gametes may be released and participate in fertilization to begin a new embryo. The adult organism eventually undergoes senescence and dies, its nutrients often supporting the early embryogenesis of its offspring and its absence allowing less competition. Thus, the cycle of life is renewed.

A Frog's Life

All animal life cycles are modifications of the generalized one described above. Figure 1.1 shows the development of the leopard frog, *Rana pipiens*, and provides a good starting point for a more detailed discussion of a representative life cycle.

Gametogenesis and fertilization

The end of one life cycle and the beginning of the next are often intricately intertwined. Life cycles are often controlled by environmental factors (tadpoles wouldn't survive if they hatched in the fall, when their food is dying), so in most frogs, gametogenesis and fertilization are seasonal events. A combination of photoperiod (hours of daylight) and temperature informs the pituitary gland of the mature female frog that it is spring. The pituitary then secretes hormones that stimulate her ovary to make the hormone estrogen. Estrogen then instructs the liver to make and secrete yolk proteins, which are then transported through the

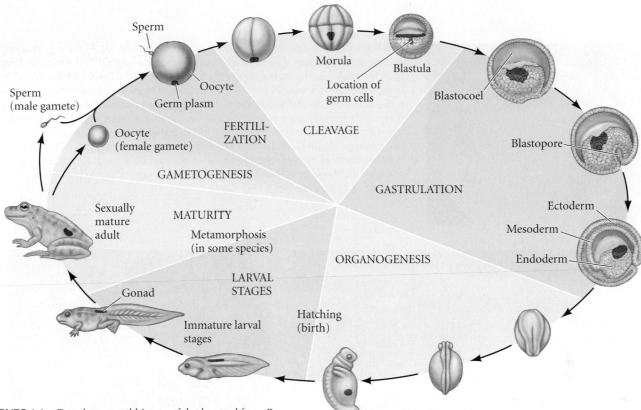

FIGURE 1.1 Developmental history of the leopard frog, *Rana pipiens*. The stages from fertilization through hatching (birth) are known collectively as embryogenesis. The region set aside for producing germ cells is shown in purple. Gametogenesis, which is completed in the sexually mature adult, begins at different times during development, depending on the species. (The sizes of the varicolored wedges shown here are arbitrary and do not correspond to the proportion of the life cycle spent in each stage.)

blood into the enlarging eggs in the ovary. The yolk is transported into the bottom portion of the egg, called the **vegetal hemisphere**, where it will serve as food for the developing embryo (**Figure 1.2A**). The upper half of the egg is called the **animal hemisphere**.* Sperm formation also occurs on a seasonal basis. Male leopard frogs make sperm during the summer, and by the time they begin hibernation in the fall they have produced all the sperm that will be available for the following spring's breeding season.

In most species of frogs, fertilization is external. The male frog grabs the female's back and fertilizes the eggs as the female releases them (**Figure 1.2B**). Some species lay their eggs in pond vegetation, and the egg jelly adheres to the plants and anchors the eggs (**Figure 1.2C**). Other species float their eggs into the center of the pond without any support. So the first important thing to remember about life cycles is that they are often intimately involved with environmental factors.

*The use of the terms *animal* and *vegetal* for the upper and lower hemispheres of the early frog embryo reflect the division rates of the cells. The upper cells divide rapidly and become actively mobile (hence "animated"), while the yolk-filled cells of the lower half were seen as being immobile (hence like plants, or "vegetal").

Fertilization accomplishes several things. First, it allows the haploid nucleus of the egg (the **female pronucleus**) to merge with the haploid nucleus of the sperm (the **male pronucleus**) to form the diploid **zygote nucleus**. Second, fertilization causes the cytoplasm of the egg to move such that different parts of the cytoplasm find themselves in new locations (**Figure 1.2D**). This cytoplasmic migration will be important in determining the three embryonic axes of the frog: anterior-posterior (head-tail), dorsal-ventral (back-belly), and right-left. Third, fertilization activates those molecules necessary to begin cell cleavage and gastrulation (Rugh 1950).

Cleavage and gastrulation

During cleavage, the volume of the frog egg stays the same, but it is divided into tens of thousands of cells (**Figure 1.2E–H**). The cells in the animal hemisphere of the egg divide faster than those in the vegetal hemisphere, and the cells of the vegetal hemisphere become progressively larger the more vegetal the cytoplasm. Meanwhile, a fluid-filled cavity, the **blastocoel**, forms in the animal hemisphere (**Figure 1.2I**). This cavity will be important for allowing cell movements to occur during gastrulation.

Gastrulation in the frog begins at a point on the embryo surface roughly 180 degrees opposite the point of sperm entry with the formation of a dimple, called the **blastopore**. This dimple (which will mark the future dorsal side of the embryo) expands to become a ring, and cells migrating through the blastopore become the mesoderm (**Figure 1.3A–C**). The cells remaining on the outside become the ecto-

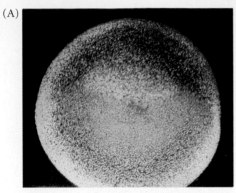

FIGURE 1.2 Early development of the frog *Xenopus laevis*. (A) As the egg matures, it accumulates yolk (here stained yellow and green) in the vegetal cytoplasm. (B) Frogs mate by amplexus, the male grasping the female around the belly and fertilizing the eggs as they are released. (C) A newly laid clutch of eggs. The brown area of each egg is the pigmented animal hemisphere. The white spot in the middle of the pigment is where the egg's nucleus resides. (D) Cytoplasm rearrangement seen during first cleavage. Compare with the initial stage seen in (A). (E) A 2-cell embryo near the end of its first cleavage. (F) An 8-cell embryo. (G) Early blastula. Note that the cells get smaller, but the volume of the egg remains the same. (H) Late blastula. (I) Cross section of a late blastula, showing the blastocoel (cavity). (A–H courtesy of Michael Danilchik and Kimberly Ray; I courtesy of J. Heasman.)

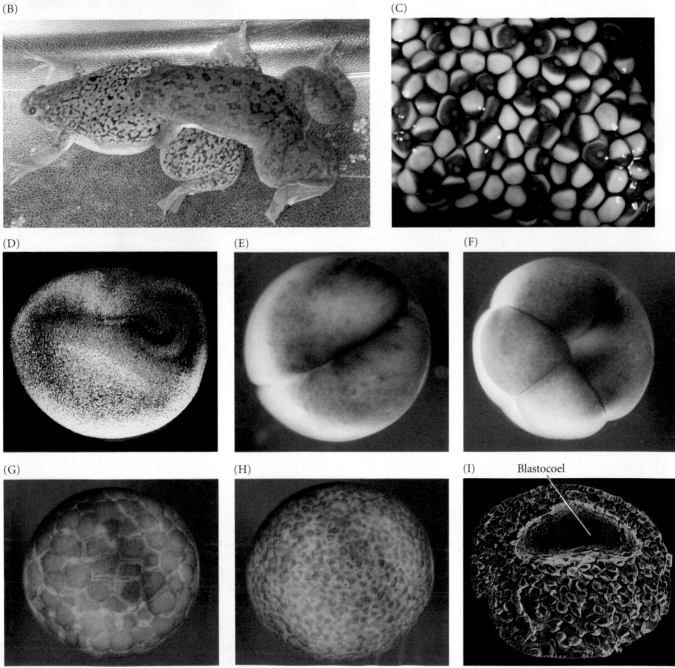

FIGURE 1.3 Continued development of *Xenopus laevis*. (A) Gastrulation begins with an invagination, or slit, in the future dorsal (top) side of the embryo. (B) This slit, the dorsal blastopore lip, as seen from the ventral surface (bottom) of the embryo. (C) The slit becomes a circle, the blastopore. Future mesoderm cells migrate into the interior of the embryo along the blastopore edges, and the ectoderm (future epidermis and nerves) migrates down the outside of the embryo. The remaining part, the yolk-filled endoderm, is eventually encircled. (D) Neural folds begin to form on the dorsal surface. (E) A groove can be seen where the bottom of the neural tube will be. (F) The neural folds come together at the dorsal midline, creating a neural tube. (G) Cross section of the *Xenopus* embryo at the neurula stage. (H) A pre-hatching tadpole, as the protrusions of the forebrain begin to induce eyes to form. (I) A mature tadpole, having swum away from the egg mass and feeding independently. (Courtesy of Michael Danilchik and Kimberly Ray.)

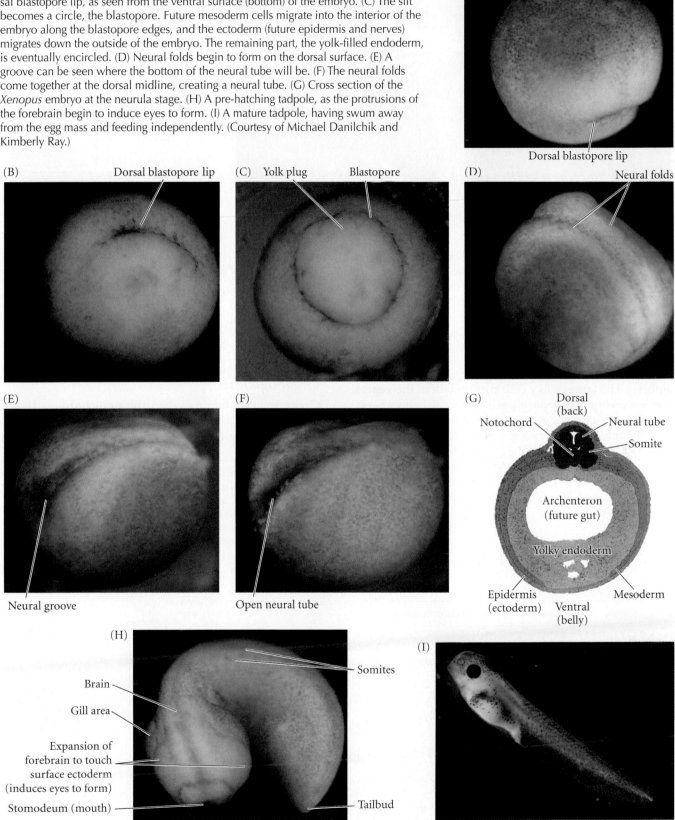

(A)

(B)

(C)

(D)

(E)

(F)

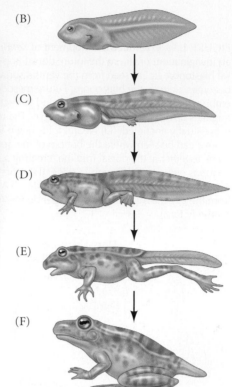

FIGURE 1.4 Metamorphosis of the frog. (A) Huge changes are obvious when one contrasts the tadpole and the adult bullfrog. Note especially the differences in jaw structure and limbs. (B) Premetamorphic tadpole. (C) Prometamorphic tadpole, showing hindlimb growth. (D) Onset of metamorphic climax as forelimbs emerge. (E,F) Climax stages. (A © Patrice Ceisel/Visuals Unlimited.)

derm, and this outer layer expands to enclose the entire embryo. The large, yolky cells that remain in the vegetal hemisphere (until they are encircled by the expanding ectoderm) become the endoderm. Thus, at the end of gastrulation, the ectoderm (precursor of the epidermis, brain, and nerves) is on the outside of the embryo, the endoderm (precursor of the gut and respiratory systems) is on the inside of the embryo, and the mesoderm (precursor of the connective tissue, blood, heart, skeleton, gonads, and kidneys) is between them.

Organogenesis

Organogenesis begins when the **notochord**—a rod of mesodermal cells in the most dorsal portion of the embryo*—signals the ectodermal cells above it that they

---

*The notochord consists of cells such as those mentioned on p. 2 of the Introduction—i.e., cells that are important for constructing the embryo but which, having performed their tasks, die. Although adult vertebrates do not have notochords, this embryonic organ is critical for establishing the fates of the ectodermal cells above it, as we shall see in Chapters 7–9.

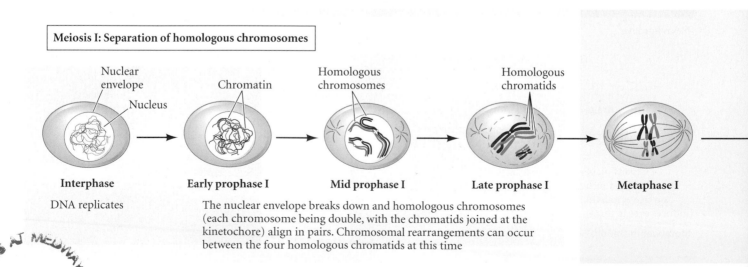

Meiosis I: Separation of homologous chromosomes

Nuclear envelope

Nucleus

Chromatin

Homologous chromosomes

Homologous chromatids

Interphase

DNA replicates

Early prophase I

Mid prophase I

Late prophase I

Metaphase I

The nuclear envelope breaks down and homologous chromosomes (each chromosome being double, with the chromatids joined at the kinetochore) align in pairs. Chromosomal rearrangements can occur between the four homologous chromatids at this time

are not going to become epidermis. Instead, these dorsal ectoderm cells form a tube and become the nervous system. At this stage, the embryo is called a **neurula**. The neural precursor cells elongate, stretch, and fold into the embryo, forming the **neural tube (Figure 1.3D–F)**; the future epidermal cells of the back cover the neural tube.

Once the neural tube has formed, it and the notochord induce changes in their neighbors, and organogenesis continues. The mesodermal tissue adjacent to the neural tube and notochord becomes segmented into **somites (Figure 1.3G,H)**, the precursors of the frog's back muscles, spinal vertebrae, and dermis (the inner portion of the skin). The embryo develops a mouth and an anus, and it elongates into the familiar tadpole structure (**Figure 1.3I**). The neurons make their connections to the muscles and to other neurons, the gills form, and the larva is ready to hatch from its egg jelly. The hatched tadpole will feed for itself as soon as the yolk supplied by its mother is exhausted.

See VADE MECUM
The amphibian life cycle

Metamorphosis and gametogenesis

Metamorphosis of the fully aquatic tadpole larva into an adult frog that can live on land is one of the most striking transformations in all of biology. In amphibians, metamorphosis is initiated by hormones from the tadpole's thyroid gland. (The mechanisms by which thyroid hormones accomplish these changes will be discussed in Chapter 15.) In frogs, almost every organ is subject to modification, and the resulting changes in form are striking and very obvious (**Figure 1.4**). The hindlimbs and forelimbs the adult will use for locomotion differentiate as the tadpole's paddle tail recedes. The cartilaginous tadpole skull is replaced by the predominantly bony skull of the young frog. The horny teeth the tadpole uses to tear up pond plants disappear as the mouth and jaw take a new shape, and the fly-catching tongue muscle of the frog develops. Meanwhile, the tadpole's lengthy intestine—a characteristic of herbivores—shortens to suit the more carnivorous diet of the adult frog. The gills regress and the lungs enlarge. The speed of metamorphosis is carefully keyed to environmental pressures. In temperate regions, for instance, *Rana* metamorphosis must occur before ponds freeze in winter. An adult leopard frog can burrow into the mud and survive the winter; its tadpole cannot.

As metamorphosis ends, the development of the germ cells begins. Gametogenesis can take a long time. In *Rana pipiens*, it takes 3 years for the eggs to mature in the female's ovaries. (Sperm take less time; *Rana* males are often fertile soon after metamorphosis.) To become mature, the germ cells must be competent to complete **meiosis**.

Meiosis (**Figure 1.5**) is one of the most important evolutionary processes characteristic of eukaryotic organisms. It makes fertilization possible and is critical in recombining genes from the two parents. Genetics, development, and evolution throughout the animal kingdom are predicated on meiosis. We will discuss meiosis more thoroughly in Chapter 16, but the most important things to remember about meiosis are:

1. The chromosomes replicate prior to cell division, so that each gene is represented four times.
2. The replicated chromosomes (each called a chromatid) are held together by their kinetochores (centromeres), and the four homologous chromatids pair together.

FIGURE 1.5   Summary of meiosis. The DNA replicates during interphase. During first meiotic prophase, the nuclear envelope breaks down and the homologous chromosomes (each chromosome is double, with its two chromatids joined at the kinetochore) align together. Chromosome rearrangements ("crossing over") can occur at this stage. After the first metaphase, the kinetochore remains unsplit and the pairs of homologous chromosomes are sorted into different cells. During the second meiotic division, the kintochore splits and the sister chromatids are moved into separate cells, each with a haploid set of chromosomes.

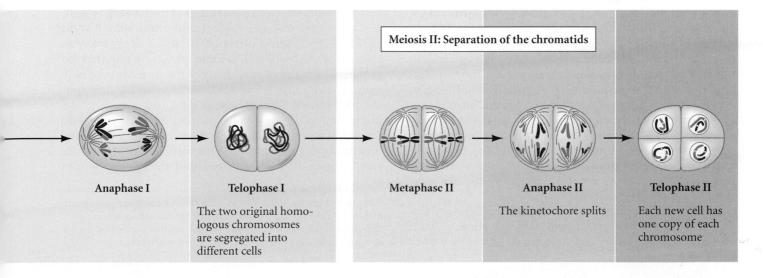

Meiosis II: Separation of the chromatids

Anaphase I	Telophase I	Metaphase II	Anaphase II	Telophase II
	The two original homologous chromosomes are segregated into different cells		The kinetochore splits	Each new cell has one copy of each chromosome

3. The first meiotic division separates the chromatid pairs from one another.
4. The second meiotic division splits the kinetochore such that each chromatid becomes a chromosome.
5. The result is four germ cells, each with a haploid nucleus.

Having undergone meiosis, the mature sperm and egg nuclei can unite in fertilization, restoring the diploid chromosome number and initiating the events that lead to development and the continuation of the circle of life.

"How Are You?"

The fertilized egg has no heart. It has no eye. No limb is found in the zygote. So how did we become what we are? What part of the embryo forms the heart? How do the cells that form the eye's retina migrate the proper distance from the cells that form the lens? How do the tissues that form a bird's wing relate to the tissues that form fish fins or the human hand? What organs are affected by mutations in particular genes? These are the types of questions asked by developmental anatomists.

Several strands weave together to form the anatomical approaches to development. The first strand is **comparative embryology**, the study of how anatomy changes during the development of different organisms. The second strand, based on the first, is **evolutionary embryology**, the study of how changes in development may cause evolutionary change and of how an organism's ancestry may constrain the types of changes that are possible. The third strand of the anatomical approach to developmental biology is **teratology**, the study of birth defects.

Comparative embryology

The first known study of comparative developmental anatomy was undertaken by Aristotle in the fourth century BCE. In *The Generation of Animals* (ca. 350 BCE), he noted some of the variations on the life cycle themes: some animals are born from eggs (**oviparity**, as in birds, frogs, and most invertebrates); some by live birth (**viviparity**, as in placental mammals); and some by producing an egg that hatches inside the body (**ovoviviparity**, as in certain reptiles and sharks). Aristotle also identified the two major cell division patterns by which embryos are formed: the **holoblastic** pattern of cleavage (in which the entire egg is divided into smaller cells, as it is in frogs and mammals) and the **meroblastic** pattern of cleavage (as in chicks, wherein only part of the egg is destined to become the embryo, while the other portion—the yolk—serves as nutrition for the embryo). And should anyone want to know who first figured out the functions of the placenta and the umbilical cord, it was Aristotle.

There was remarkably little progress in embryology for the two thousand years following Aristotle. It was only in 1651 that William Harvey concluded that all animals—even

mammals—originate from eggs. *Ex ovo omnia* ("All from the egg") was the motto on the frontispiece of Harvey's *On the Generation of Living Creatures*, and this precluded the spontaneous generation of animals from mud or excrement. This statement was not made lightly, for Harvey knew that it went against the views of Aristotle, whom Harvey still venerated. (Aristotle had thought that menstrual fluid formed the material of the embryo, while the semen gave it form and animation.) Harvey also was the first to see the **blastoderm** of the chick embryo (the small region of the egg containing the yolk-free cytoplasm that gives rise to the embryo), and he was the first to notice that "islands" of blood tissue form before the heart does. Harvey also suggested that the amniotic fluid might function as a "shock absorber" for the embryo.

As might be expected, embryology remained little but speculation until the invention of the microscope allowed detailed observations. In 1672, Marcello Malpighi published the first microscopic account of chick development. Here, for the first time, the neural groove (precursor of the neural tube), the muscle-forming somites, and the first circulation of the arteries and veins—to and from the yolk—were identified (Figure 1.6).

Epigenesis and preformation

With Malpighi begins one of the great debates in embryology: the controversy over whether the organs of the embryo are formed de novo ("from scratch") at each generation, or whether the organs are already present, in miniature form, within the egg (or sperm). The first view, called **epigenesis**, was supported by Aristotle and Harvey. The second view, called **preformation**, was reinvigorated with Malpighi's support. Malpighi showed that the unincubated* chick egg already had a great deal of structure, and this observation provided him with reasons to question epigenesis. According to the preformationist view, all the organs of the adult were prefigured in miniature within the sperm or (more usually) the egg. Organisms were not seen to be "constructed" but rather "unrolled."

The preformationist hypothesis had the backing of eighteenth-century science, religion, and philosophy (Gould 1977; Roe 1981; Pinto-Correia 1997). First, if all organs were prefigured, embryonic development merely required the growth of existing structures, not the formation of new ones. No extra mysterious force was needed for embryonic development. Second, just as the adult organism was prefigured in the germ cells, another generation already existed in a prefigured state within the germ cells of the first prefigured generation. This corollary, called *emboîtment* (encapsulation), ensured that the species would

---

*As pointed out by Maître-Jan in 1722, the eggs Malpighi examined may technically be called "unincubated," but as they were left sitting in the Bolognese sun in August, they were not unheated. Such eggs would be expected to have developed into chicks.

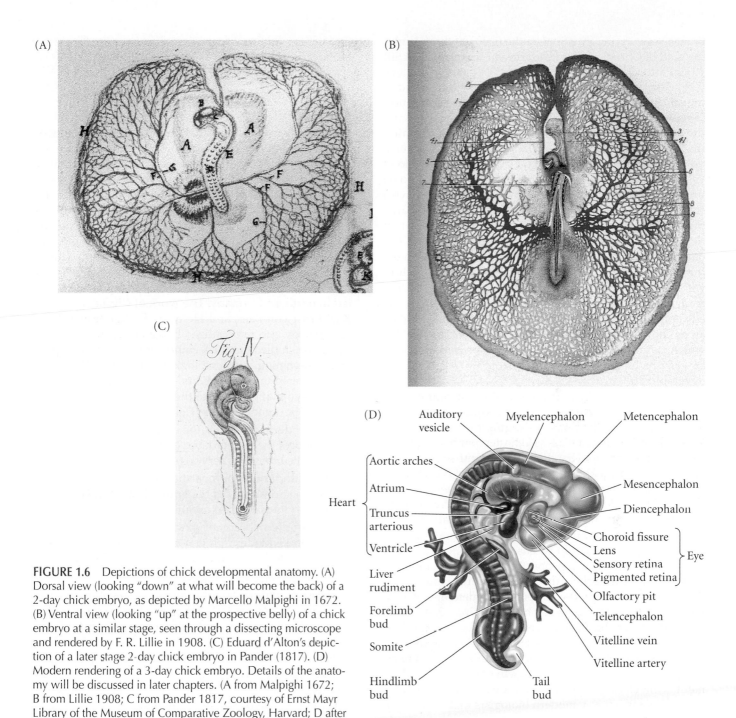

FIGURE 1.6 Depictions of chick developmental anatomy. (A) Dorsal view (looking "down" at what will become the back) of a 2-day chick embryo, as depicted by Marcello Malpighi in 1672. (B) Ventral view (looking "up" at the prospective belly) of a chick embryo at a similar stage, seen through a dissecting microscope and rendered by F. R. Lillie in 1908. (C) Eduard d'Alton's depiction of a later stage 2-day chick embryo in Pander (1817). (D) Modern rendering of a 3-day chick embryo. Details of the anatomy will be discussed in later chapters. (A from Malpighi 1672; B from Lillie 1908; C from Pander 1817, courtesy of Ernst Mayr Library of the Museum of Comparative Zoology, Harvard; D after Carlson 1981.)

remain constant. Although certain microscopists claimed to see fully formed human miniatures within the sperm or egg, the major proponents of this hypothesis—Albrecht von Haller and Charles Bonnet—knew that organ systems develop at different rates, and that structures need not be in the same place in the embryo as they are in the newborn.

The preformationists had no cell theory to provide a lower limit to the size of their preformed organisms (the cell theory arose in the mid-1800s), nor did they view mankind's tenure on Earth as potentially infinite. Rather, said Bonnet (1764), "Nature works as small as it wishes," and the human species existed in that finite time between Creation and Resurrection. This view was in accord with the best science of its time, conforming to the French mathematician-philosopher René Descartes' principle of the infinite divisibility of a mechanical nature initiated, but not interfered with, by God. It also conformed to Enlightenment views of the Deity. The scientist-priest Nicolas Malebranche saw in preformationism the fusion of the rule-giving God of Christianity with Cartesian science (Churchill 1991; Pinto-Correia 1997).*

The embryological case for epigenesis was revived at the same time by Kaspar Friedrich Wolff, a German embryologist working in St. Petersburg. By carefully observing the development of chick embryos, Wolff demonstrated that the embryonic parts develop from tissues that have no counterpart in the adult organism. The heart and blood vessels (which, according to preformationism, had to be present from the beginning to ensure embryonic growth) could be seen to develop anew in each embryo. Similarly, the intestinal tube was seen to arise by the folding of an originally flat tissue. This latter observation was explicitly detailed by Wolff, who proclaimed in 1767 that "When the formation of the intestine in this manner has been duly weighed, almost no doubt can remain, I believe, of the truth of epigenesis." To explain how an organism is created anew each generation, however, Wolff had to postulate an unknown force—the *vis essentialis* ("essential force")—which, acting according to natural laws in the same way as gravity or magnetism, would organize embryonic development.

A reconciliation between preformationism and epigenesis was attempted by the German philosopher Immanuel Kant (1724–1804) and his colleague, biologist Johann Friedrich Blumenbach (1752–1840). Attempting to construct a scientific theory of racial descent, Blumenbach postulated a mechanical, goal-directed force he called *Bildungstrieb* ("developmental force"). Such a force, he said, was not theoretical, but could be shown to exist by experimentation. A hydra, when cut, regenerates its amputated parts by rearranging existing elements (see Chapter 15). Some purposeful organizing force could be observed in operation, and this *Bildungstrieb* was a property of the organism itself, thought to be inherited through the germ cells. Thus, development could proceed through a predetermined force inherent in the matter of the embryo (Cassirer 1950; Lenoir 1980). Moreover, this force was believed to be susceptible to change, as demonstrated by the left-handed variant of snail coiling (where left-coiled snails can produce right-coiled progeny). In this hypothesis, wherein epigenetic development is directed by preformed instructions, we are not far from the view held by modern biologists that most of the

instructions for forming the organism are already present in the fertilized egg.*

Naming the parts: The primary germ layers and early organs

The end of preformationism did not come until the 1820s, when a combination of new staining techniques, improved microscopes, and institutional reforms in German universities created a revolution in descriptive embryology. The new techniques enabled microscopists to document the epigenesis of anatomical structures, and the institutional reforms provided audiences for these reports and students to carry on the work of their teachers. Among the most talented of this new group of microscopically inclined investigators were three friends, born within a year of each other, all of whom came from the Baltic region and studied in northern Germany. The work of Christian Pander, Karl Ernst von Baer, and Heinrich Rathke transformed embryology into a specialized branch of science.

Pander studied the chick embryo for less than two years (before becoming a paleontologist), but in those 15 months, he discovered the **germ layers**[†]—three distinct regions of the embryo that give rise to the differentiated cells types and specific organ systems (**Figure 1.7**).

- The **ectoderm** generates the outer layer of the embryo. It produces the surface layer (epidermis) of the skin and forms the brain and nervous system.
- The **endoderm** becomes the innermost layer of the embryo and produces the epithelium of the digestive tube and its associated organs (including the lungs).
- The **mesoderm** becomes sandwiched between the ectoderm and endoderm. It generates the blood, heart, kidney, gonads, bones, muscles, and connective tissues.

These three layers are found in the embryos of all **triploblastic** ("three-layer") animals. Some phyla, such as the poriferans (sponges) and ctenophores (comb jellies) lack a true mesoderm and are considered **diploblastic** animals.

Pander and Rathke also made observations that weighted the balance in favor of epigenesis. Rathke followed the intricate development of the vertebrate skull, excretory systems, and respiratory systems, showing that these became increasingly complex. He also showed that their complexity took on different trajectories in different classes of vertebrates. For instance, Rathke was the first to identify the

---

*Preformation was a conservative theory, emphasizing the lack of change between generations. Its principal failure was its inability to account for the variations revealed by the limited genetic evidence of the time. It was known, for instance, that matings between white and black parents produced children of intermediate skin color, an impossibility if inheritance and development were solely through either the sperm or the egg. In more controlled experiments, the German botanist Joseph Kölreuter (1766) produced hybrid tobacco plants having the characteristics of both species. Moreover, by mating the hybrid to either the male or female parent, Kölreuter was able to "revert" the hybrid back to one or the other parental type after several generations. Thus, inheritance seemed to arise from a mixture of parental components.

---

*But, as we shall see, not *all* the instructions there. Later in this book, we will see that temperature, diet, predators, symbionts, crowding, and other environmental agents normally regulate gene expression in the embryo and can cause particular phenotypes to occur.

[†]From the same root as *germination*, the Latin *germen* means "sprout" or "bud." The names of the three germ layers are from the Greek: ectoderm from *ektos* ("outside") plus *derma* ("skin"); mesoderm from *mesos* ("middle"); and endoderm from *endon* ("within").

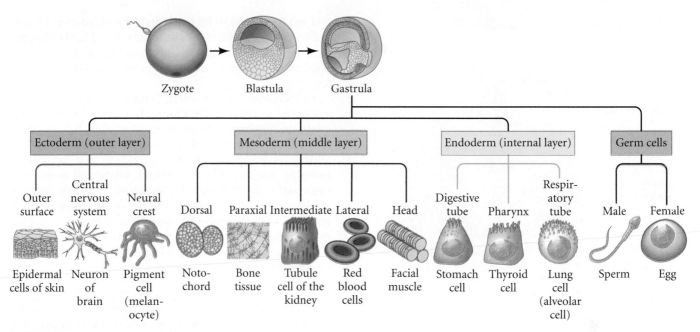

Zygote   Blastula   Gastrula

| Ectoderm (outer layer) | Mesoderm (middle layer) | Endoderm (internal layer) | Germ cells |

Ectoderm (outer layer)
- Outer surface — Epidermal cells of skin
- Central nervous system — Neuron of brain
- Neural crest — Pigment cell (melanocyte)

Mesoderm (middle layer)
- Dorsal — Notochord
- Paraxial — Bone tissue
- Intermediate — Tubule cell of the kidney
- Lateral — Red blood cells
- Head — Facial muscle

Endoderm (internal layer)
- Digestive tube — Stomach cell
- Pharynx — Thyroid cell
- Respiratory tube — Lung cell (alveolar cell)

Germ cells
- Male — Sperm
- Female — Egg

FIGURE 1.7 The dividing cells of the fertilized egg form three distinct embryonic germ layers. Each of the germ layers gives rise to myriad differentiated cell types (only a few representatives are shown here) and distinct organ systems. The germ cells (precursors of the sperm and egg) are set aside early in development and do not arise from any particular germ layer.

pharyngeal arches (Figure 1.8). He showed that these same embryonic structures became gill supports in fish and the jaws and ears (among other things) in mammals. Pander demonstrated that the germ layers did not form their respective organs autonomously (Pander 1817). Rather, each germ layer "is not yet independent enough to indicate what it truly is; it still needs the help of its sister travelers, and therefore, although already designated for different ends, all three influence each other collectively until each has reached an appropriate level." Pander had dis-

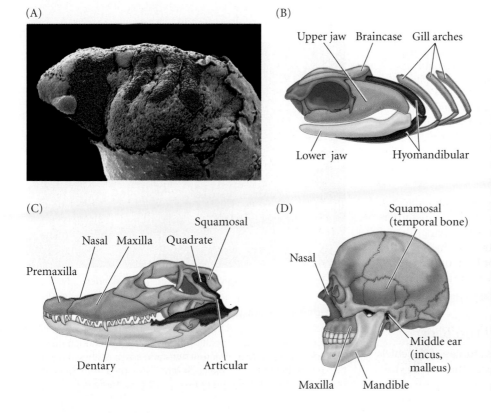

(A)

(B)
Upper jaw   Braincase   Gill arches
Lower jaw   Hyomandibular

(C)
Premaxilla
Nasal   Maxilla   Quadrate   Squamosal
Dentary   Articular

(D)
Nasal
Squamosal (temporal bone)
Maxilla   Mandible   Middle ear (incus, malleus)

FIGURE 1.8 Evolution of pharyngeal arch structures in the vertebrate head. (A) Pharyngeal arches (also called branchial arches) in the embryo of the salamander *Ambystoma mexicanum*. The surface ectoderm has been removed to permit visualization of the arches (highlighted in color) as they form. (B) In adult fish, pharyngeal arch cells form the hyomandibular jaws and gill arches. (C) In amphibians, birds, and reptiles (a crocodile is shown here), these same cells form the quadrate bone of the upper jaw and the articular bone of the lower jaw. (D) In mammals, the quadrate has become internalized and forms the incus of the middle ear. The articular bone retains its contact with the quadrate, becoming the malleus of the middle ear. (A courtesy of P. Falck and L. Olsson; B–D after Zangerl and Williams 1975.)

(A)

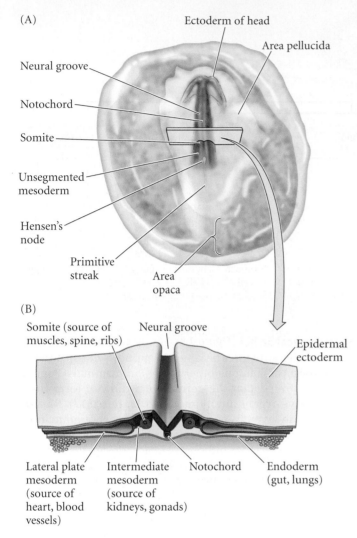

Labels:
- Ectoderm of head
- Area pellucida
- Neural groove
- Notochord
- Somite
- Unsegmented mesoderm
- Hensen's node
- Primitive streak
- Area opaca

(B)

- Somite (source of muscles, spine, ribs)
- Neural groove
- Epidermal ectoderm
- Lateral plate mesoderm (source of heart, blood vessels)
- Intermediate mesoderm (source of kidneys, gonads)
- Notochord
- Endoderm (gut, lungs)

FIGURE 1.9 Notochord in chick development. The notochord separates vertebrate embryos into right and left halves and instructs the ectoderm above it to become the nervous system. (A) Dorsal view of the 24-hour chick embryo. (B) Cross section through the trunk region shows the notochord and developing neural tube. By comparing this illustration and Figure 1.6, you can see the remarkable changes between days 1, 2, and 3 of chick egg incubation. (A after Patten 1951.)

covered the tissue interactions that we now call *induction*. No tissue is able to construct organs by itself; it must interact with other tissues. (We will discuss the principles of induction more thoroughly in Chapter 3.) Thus, Pander showed that preformation could not be true, since the organs come into being through interactions between simpler structures.

The four principles of Karl Ernst von Baer

Karl Ernst von Baer extended Pander's studies of the chick embryo. He discovered the notochord, the rod of dorsalmost mesoderm that separates the embryo into right and left halves and which instructs the ectoderm above it to become the nervous system (**Figure 1.9**). He also discovered the mammalian egg, that long-sought cell that everyone believed existed but no one before von Baer had ever seen.*

In 1828, von Baer reported, "I have two small embryos preserved in alcohol, that I forgot to label. At present I am unable to determine the genus to which they belong. They may be lizards, small birds, or even mammals." **Figure 1.10** allows us to appreciate his quandary. All vertebrate embryos (fish, reptiles, amphibians, birds, and mammals) begin with a basically similar structure. From his detailed study of chick development and his comparison of chick embryos with the embryos of other vertebrates, von Baer derived four generalizations. Now often referred to as "von Baer's laws," they are stated here with some vertebrate examples.

1. *The general features of a large group of animals appear earlier in development than do the specialized features of a smaller group.* All developing vertebrates appear very similar right after gastrulation. It is only later in development that the special features of class, order, and

---

*von Baer could hardly believe that he had at last found what so many others—Harvey, de Graaf, von Haller, Prevost, Dumas, and even Purkinje—had searched for and failed to find. "I recoiled as if struck by lightening ... I had to try to relax a while before I could work up enough courage to look again, as I was afraid I had been deluded by a phantom. Is it not strange that a sight which is expected, and indeed hoped for, should be frightening when it eventually materializes?"

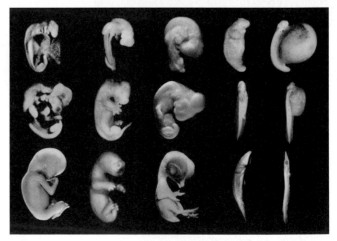

Human   Opossum   Chicken   Salamander   Fish
                            (axolotl)   (gar)

FIGURE 1.10 Similarities and differences among vertebrate embryos as they proceed through development. Each species' embryos begin with a basically similar structure, although they acquire this structure at different ages and sizes. As they develop, the species become less like each other. (Adapted from Richardson et al. 1998; photograph courtesy of M. Richardson.)

finally species emerge. All vertebrate embryos have gill arches, a notochord, a spinal cord, and primitive kidneys.

2. *Less general characters develop from the more general, until finally the most specialized appear.* All vertebrates initially have the same type of skin. Only later does the skin develop fish scales, reptilian scales, bird feathers, or the hair, claws, and nails of mammals. Similarly, the early development of limbs is essentially the same in all vertebrates. Only later do the differences between legs, wings, and arms become apparent.

3. *The embryo of a given species, instead of passing through the adult stages of lower animals, departs more and more from them.** The visceral clefts of embryonic birds and mammals do not resemble the gill slits of adult fish in detail. Rather, they resemble the visceral clefts of *embryonic* fish and other *embryonic* vertebrates. Whereas fish preserve and elaborate these clefts into true gill slits, mammals convert them into structures such as the eustachian tubes (between the ear and mouth).

4. *Therefore, the early embryo of a higher animal is never like a lower animal, but only like its early embryo.* Human embryos never pass through a stage equivalent to an adult fish or bird. Rather, human embryos initially share characteristics in common with fish and avian embryos. Later, the mammalian and other embryos diverge, none of them passing through the stages of the others.

von Baer also recognized that there is a common pattern to all vertebrate development: each of the three germ layers generally gives rise to the same organs, whether the organism itself is a fish, a frog, or a chick.

Comparative embryonic anatomy remains an active field of research today, although it is now done in an evolutionary context. What embryonic interactions, for instance, cover the squirrel's tail with fur but provide scales on the rat's tail? The author's own research concerns how turtles get their shells—a skeletal feature generally composed of 59 bones that no other vertebrate possesses. What is the relationship of these 59 bones to the bones found in alligators and prehistoric marine reptiles? What changes in the "typical" development of the vertebrate skeleton allowed these unique bones to form? Jack Horner and Hans Larsson are looking at the similarities between the developmental anatomy of chick and dinosaur embryos and have found that the embryonic chick, unlike the dinosaur, regresses its tail. They are conducting experiments to block this regression, and actually hope to obtain a chick that more closely resembles its dinosaur ancestors (Horner and Gorman 2009).

---

*von Baer formulated these generalizations prior to Darwin's theory of evolution. "Lower animals" would be those having simpler anatomies.

Keeping Track of Moving Cells: Fate Maps and Cell Lineages

By the late 1800s, the cell had been conclusively demonstrated to be the basis for anatomy and physiology. Embryologists, too, began to base their field on the cell. But unlike those who studied the adult, developmental anatomist found that *cells do not stay still in the embryo.* Indeed, one of the most important conclusions of developmental anatomists is that embryonic cells do not remain in one place, nor do they keep the same shape (Larsen and McLaughlin 1987).

Early embryologists recognized that there are two major types of cells in the embryo: **epithelial cells**, which are tightly connected to one another in sheets or tubes; and **mesenchymal cells**, which are unconnected to one another and operate as independent units. Morphogenesis is brought about through a limited repertoire of variations in cellular processes within these two types of arrangements (Table 1.1):

- *Direction and number of cell divisions.* Think of the faces of two dog breeds—say, a German shepherd and a poodle. The faces are made from the same cell types, but the number and orientation of the cell divisions are different. Think also of the legs of a German shepherd compared with those of a dachshund. The skeleton-forming cells of the dachshund have undergone fewer cell divisions than those of taller dogs (see Figure 1.21).

- *Cell shape changes.* Cell shape change is a critical part of not only of development but also of cancer. In development, change in the shapes of epithelial cells often creates tubes out of sheets (as when the neural tube forms); and a shape change from epithelial to mesenchymal is critical when cells migrate away from the epithelium (as when muscle cells are formed). This same type of epithelial-to-mesenchymal change allows cancer cells to migrate and spread from the primary tumor to new sites.

- *Cell movement.* Cell migration is critical to get cells to their appropriate places. The germ cells have to migrate into the developing gonad, and the primordial heart cells meet in the middle of the vertebrate neck and then migrate to the left part of the chest.

- *Cell growth.* Cells can change in size. This is most apparent in the germ cells: the sperm eliminates most of its cytoplasm and becomes smaller, whereas the developing egg conserves and adds cytoplasm, becoming comparatively huge. Many cells undergo an "asymmetric" cell division that produces one big cell and one small cell, each of which may have a completely different fate.

- *Cell death.* Death is a critical part of life. The cells that in the womb constitute the webbing between our toes and fingers die before we are born. So do the cells of our tails. The orifices of our mouth, anus, and reproductive glands all form through cells dying at particular times and places.

TABLE 1.1 Summary of major morphogenic processes regulated by mesenchymal and epithelial cells

Process	Action	Morphology	Example
MESENCHYMAL CELLS			
Condensation	Mesenchyme becomes epithelium		Cartilage mesenchyme
Cell division	Mitosis produces more cells (hyperplasia)		Limb mesenchyme
Cell death	Cells die		Interdigital mesenchyme
Migration	Cells move at particular times and places		Heart mesenchyme
Matrix secretion and degradation	Synthesis or removal of extracellular layer		Cartilage mesenchyme
Growth	Cells get larger (hypertrophy)		Fat cells
EPITHELIAL CELLS			
Dispersal	Epithelium becomes mesenchyme (entire structure)		Müllerian duct degeneration
Delamination	Epithelium becomes mesenchyme (part of structure)		Chick hypoblast
Shape change or growth	Cells remain attached as morphology is altered		Neurulation
Cell migration (intercalation)	Rows of epithelia merge to form fewer rows		Vertebrate gastrulation
Cell division	Mitosis within row or column		Vertebrate gastrulation
Matrix secretion and degradation	Synthesis or removal of extracellular matrix		Vertebrate organ formation
Migration	Formation of free edges		Chick ectoderm

- *Changes in the composition of the cell membrane or secreted products.* Cell membranes and secreted cell products influence the behavior of neighboring cells. For instance, extracellular matrices secreted by one set of cells will allow the migration of their neighboring cells. Extracellular matrices made by other cell types will *prohibit* the migration of the same set of cells. In this way, "paths and guiderails" are established for migrating cells.

Fate maps

Given such a dynamic situation, one of the most important programs of descriptive embryology became the tracing of **cell lineages**: following individual cells to see what those cells become. In many organisms, resolution of individual cells is not possible, but one can label *groups* of embryonic cells to see what that area becomes in the adult organism. By bringing such studies together, one can construct a **fate map**. These diagrams "map" larval or adult structures onto the region of the embryo from which they arose. Fate maps constitute an important foundation for experimental embryology, providing researchers with information on which portions of the embryo normally become which larval or adult structures. **Figure 1.11** shows fate maps of some vertebrate embryos at the early gastrula stage.

Fate maps can be generated in several ways, and the technology has changed greatly over the past few years. Construction of these maps is an ongoing research pro-gram, and some of the results can be controversial. Recent examples of controversial fate maps include the map for the region of the frog embryo that specifies heart and blood cell precursors (Lane and Sheets 2006), and that for the region of the embryonic turtle that becomes the bones of the plastron (Cebra-Thomas et al. 2007). As we will see later in this book, researchers are currently constructing, refining, and arguing about the fate maps of mammalian embryos.

Direct observation of living embryos

Some embryos have relatively few cells, and the cytoplasm in each of the early blastomeres has a different pigment. In such fortunate cases, it is actually possible to look through the microscope and trace the descendants of a particular cell into the organs they generate. E. G. Conklin patiently followed the fates of each early cell of *Styela partita*, a tunicate (sea squirt) that resides in waters off the coast of Massachusetts (Conklin 1905). The muscle-forming cells of the embryo always had a yellow color, derived from a region of cytoplasm found in the B4.1 blastomere (**Figure 1.12**). Conklin's fate map was confirmed by cell-removal experiments. Removal of the B4.1 cell (which according to Conklin's map should produce all the tail musculature) in fact resulted in a larva with no tail muscles (Reverberi and Minganti 1946).

> **See WEBSITE 1.1 Conklin's art and science**
>
> **See VADE MECUM The compound microscope**

Dye marking

Most embryos are not so accommodating as to have cells of different colors. In the early years of the twentieth century, Vogt (1929) traced the fates of different areas of amphibian eggs by applying **vital dyes** to the region of interest. Vital dyes stain cells but do not kill them. Vogt mixed such dyes with agar and spread the agar on a microscope slide to dry. The ends of the dyed agar were very thin. Vogt cut chips from these ends and placed them on a frog embryo. After the dye stained the cells, he removed the agar chips and could follow the stained cells' movements within the embryo (**Figure 1.13**).

One problem with vital dyes is that as they become more diluted with each cell division, they become difficult to detect. One way around this is to use **fluorescent dyes** that are so intense that once injected into individual cells, they can still be detected in the progeny of these cells many divisions later. Fluorescein-conjugated dextran, for example, can be injected into a single cell of an early embryo, and the descendants of that cell can be seen by examining the embryo under ultraviolet light (**Figure 1.14**).

> **See VADE MECUM**
> **Histotechniques**

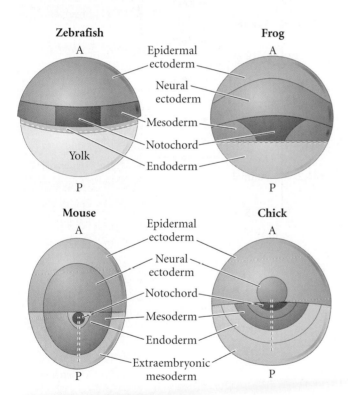

FIGURE 1.11 Fate maps of vertebrates at the early gastrula stage. All are dorsal surface views (looking "down" on the embryo at what will become its back). Despite the different appearances of the adult animals, fate maps of these four vertebrates show numerous similarities among the embryos. The cells that will form the notochord occupy a central dorsal position, while the precursors of the neural system lie immediately anterior to it. The neural ectoderm is surrounded by less dorsal ectoderm, which will form the epidermis of the skin. A indicates the anterior end of the embryo, P the posterior end. The dashed green lines indicate the site of ingression—the path cells will follow as they migrate from the exterior to the interior of the embryo.

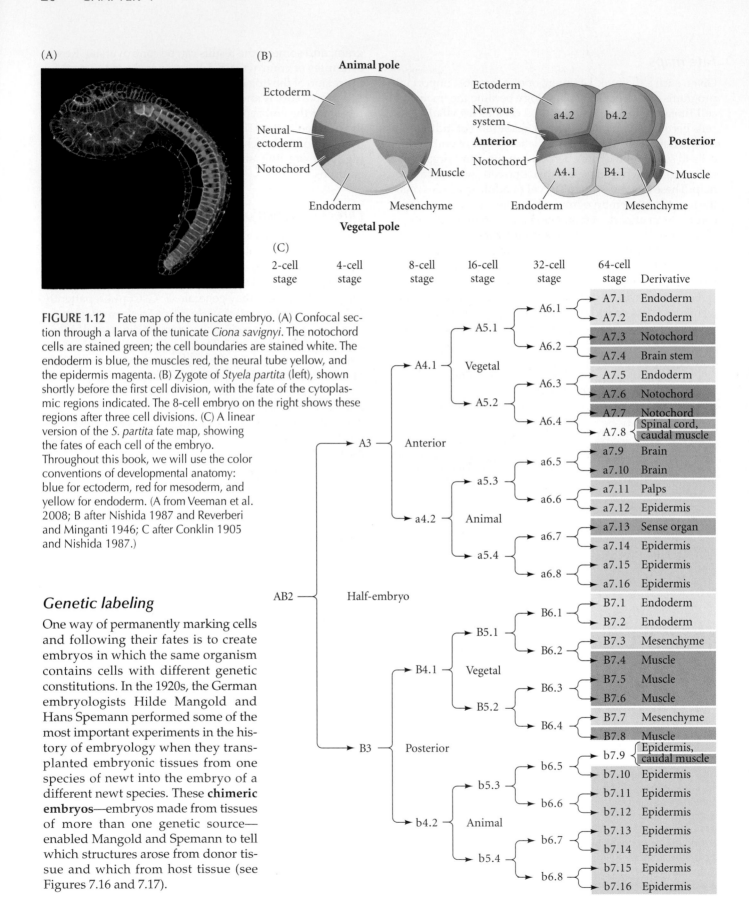

FIGURE 1.12 Fate map of the tunicate embryo. (A) Confocal section through a larva of the tunicate *Ciona savignyi*. The notochord cells are stained green; the cell boundaries are stained white. The endoderm is blue, the muscles red, the neural tube yellow, and the epidermis magenta. (B) Zygote of *Styela partita* (left), shown shortly before the first cell division, with the fate of the cytoplasmic regions indicated. The 8-cell embryo on the right shows these regions after three cell divisions. (C) A linear version of the *S. partita* fate map, showing the fates of each cell of the embryo. Throughout this book, we will use the color conventions of developmental anatomy: blue for ectoderm, red for mesoderm, and yellow for endoderm. (A from Veeman et al. 2008; B after Nishida 1987 and Reverberi and Minganti 1946; C after Conklin 1905 and Nishida 1987.)

Genetic labeling

One way of permanently marking cells and following their fates is to create embryos in which the same organism contains cells with different genetic constitutions. In the 1920s, the German embryologists Hilde Mangold and Hans Spemann performed some of the most important experiments in the history of embryology when they transplanted embryonic tissues from one species of newt into the embryo of a different newt species. These **chimeric embryos**—embryos made from tissues of more than one genetic source—enabled Mangold and Spemann to tell which structures arose from donor tissue and which from host tissue (see Figures 7.16 and 7.17).

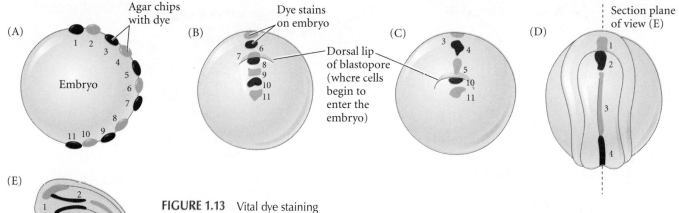

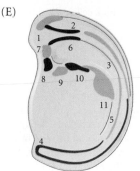

FIGURE 1.13 Vital dye staining of amphibian embryos. (A) Vogt's method for marking specific cells of the embryonic surface with vital dyes. (B–D) Dorsal surface views of stain on successively later embryos. (E) Newt embryo dissected in a medial sagittal section to show the stained cells in the interior. (After Vogt 1929.)

One of the best examples of this technique is the construction of chimeric embryos by grafting quail cells inside a chick embryo. Chicks and quail embryos develop in a similar manner (especially during the early stages development), and the grafted quail cells become integrated into the chick embryo and participate in the construction of the various organs. The substitution of quail cells for chick cells can be performed on an embryo while it is still inside the egg, and the chick that hatches will have quail cells in particular sites, depending on where the graft was placed. Quail cells differ from chick cells in two important ways. First, the quail nucleus has condensed DNA (*heterochromatin*) concentrated around the nucleoli, making quail nuclei easily distinguishable from chick nuclei. Second, cell-specific antigens that are quail-specific can be used to find individual quail cells, even if they are "hidden" within a large population of chick cells. In this way, fine-structure maps of the chick brain and skeletal system have been produced (**Figure 1.15**; Le Douarin 1969; Le Douarin and Teillet 1973).

In addition, the chick-quail chimeras dramatically confirmed the extensive cell migrations taken by neural crest cells during vertebrate development. Mary Rawles (1940) showed that the pigment cells (**melanocytes**) of the chick originate in the **neural crest**, a transient band of cells that

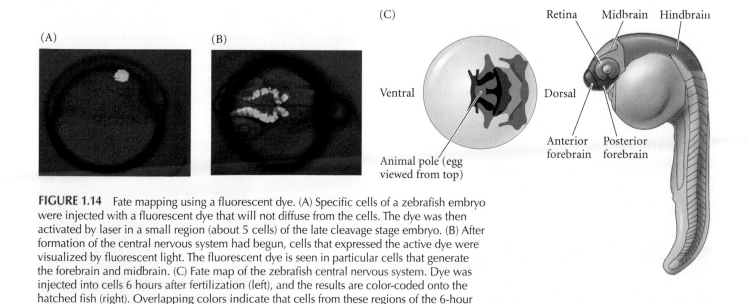

FIGURE 1.14 Fate mapping using a fluorescent dye. (A) Specific cells of a zebrafish embryo were injected with a fluorescent dye that will not diffuse from the cells. The dye was then activated by laser in a small region (about 5 cells) of the late cleavage stage embryo. (B) After formation of the central nervous system had begun, cells that expressed the active dye were visualized by fluorescent light. The fluorescent dye is seen in particular cells that generate the forebrain and midbrain. (C) Fate map of the zebrafish central nervous system. Dye was injected into cells 6 hours after fertilization (left), and the results are color-coded onto the hatched fish (right). Overlapping colors indicate that cells from these regions of the 6-hour embryo contribute to two or more regions. (A,B from Kozlowski et al. 1998, photographs courtesy of E. Weinberg; C after Woo and Fraser 1995.)

(A)

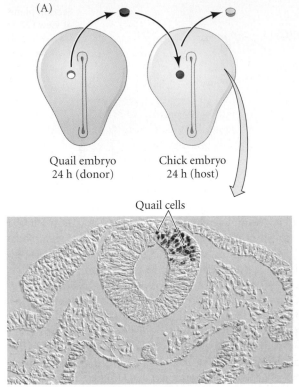

Quail embryo 24 h (donor)   Chick embryo 24 h (host)

Quail cells

Chick embryo with region of quail cells on the neural tube

(B)

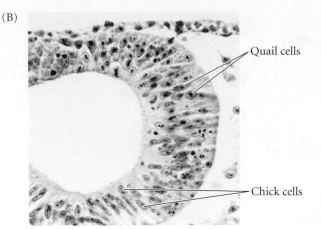

Quail cells

Chick cells

FIGURE 1.15 Genetic markers as cell lineage tracers. (A) Grafting experiment wherein the cells from a particular region of a 1-day quail embryo have been placed into a similar region of a 1-day chick embryo. After several days, the quail cells can be seen by using an antibody to quail-specific proteins. This region of the 3-day embryo produces cells that populate the neural tube. (B) Chick and quail cells can also be distinguished by the heterochromatin of their nuclei. The quail cells have a single large nucleolus (dense purple), distinguishing them from the diffuse nuclei of the chick. (From Darnell and Schoenwolf 1997, courtesy of the authors.)

joins the neural tube to the epidermis. When she transplanted small regions of neural crest-containing tissue from a pigmented strain of chickens into a similar position in an embryo from an unpigmented strain of chickens, the migrating pigment cells entered the epidermis and later entered the feathers (**Figure 1.16**). Ris (1941) used similar techniques to show that while almost all of the external pigment of the chick embryo came from the migrating neural crest cells, the pigment of the retina formed in the retina itself and was not dependent on the migrating neural crest cells. This pattern was confirmed in chick-quail hybrids, in which the quail neural crest cells produced their own pigment and pattern in the chick feathers.

See VADE MECUM
Chick-quail chimeras

Transgenic DNA chimeras

In most animals, it is difficult to meld a chimera from two species. One way of circumventing this problem is to trans-

plant cells from a genetically modified organism. In such a technique, the genetic modification can then be traced only to those cells that express it. One version is to infect the cells of an embryo with a virus whose genes have been altered such that they express the gene for a fluorescently active protein such as GFP.* (This type of altered gene is

*GFP—green fluorescent protein—occurs naturally in certain jellyfish. It emits bright green fluorescence when exposed to ultraviolet light and is widely used as a transgenic label for cells in developmental and other research.

FIGURE 1.16 Chick resulting from transplantation of a trunk neural crest region from an embryo of a pigmented strain of chickens into the same region of an embryo of an unpigmented strain. The neural crest cells that gave rise to the pigment migrated into the wing epidermis and feathers. (From the archives of B. H. Willier.)

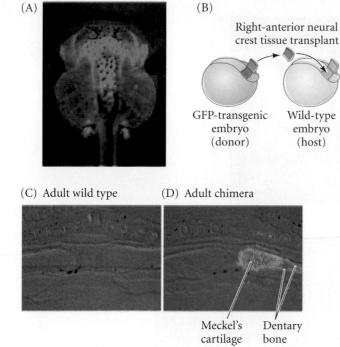

(A)

(B)

Right-anterior neural
crest tissue transplant

GFP-transgenic
embryo
(donor)

Wild-type
embryo
(host)

Chimera

(C) Adult wild type

(D) Adult chimera

Meckel's   Dentary
cartilage   bone

FIGURE 1.17   Fate mapping with transgenic DNA shows that the neural crest is critical in making the bones of the frog jaw. (A) Eggs injected with a virus containing a gene for green fluorescent protein (GFP) on a promoter that is active in all cell types produced tadpoles and frogs that express GFP in every cell (bright green). (B) Frogs labeled in this way were mated, and embryos were obtained that had formed their neural tubes and neural crests. A section from the right-side anterior (head-forming) neural crest was transplanted into the identical region of a wild-type embryo that lacked the *GFP* transgene. (C,D) After transplantation, development through the tadpole stage progressed. Once metamorphosis had occurred and the jawbones formed, the frog was sectioned and viewed under ultraviolet light. (C) In the wild-type, no fluorescence was seen in the jaw. (D) In the transgenic chimera, the right side glowed green wherever neural crest cells had migrated. These areas included Meckel's cartilage, as well as the dentary and other bone structures that formed during metamorphosis. (After Gross et al. 2006; photographs courtesy of J. B. Gross.)

called a *transgene*, because it contains DNA from another species.) When the infected embryonic cells are transplanted into a wild-type host, only the donor cells will express GFP and emit a visible green glow.

Gross and colleagues (2006) used this technique to identify the cells that form the bones of the frog skull. This has been a difficult question, because the tadpole does not have a bony skull; its skull is made of cartilage. So which part of the embryo forms the cells that wait patiently somewhere in the tadpole to become the bony skull of the adult frog? The experimenters first infected the eggs of the frog *Xenopus laevis* with a virus containing an active *GFP* gene. The virus became incorporated into the nuclear DNA of the frog egg, and the viral *GFP* gene was transmitted to

every cell of the *Xenopus* embryo. In this way, Gross and colleagues generated tadpoles and adult frogs in which every cell glowed green when placed under ultraviolet light (Figure 1.17A). They then removed cells from the neural tube and neural crest of a GFP-transgenic embryo and transplanted the cells into a similar region of a normal *Xenopus* embryo (Figure 1.17B). Once the host embryos had metamorphosed into adult frogs, the biologists looked at the lower jaw, a structure whose bones form only after metamorphosis (Figure 1.17C,D). In addition to the cartilage cells that remained there, the dentary (lower jaw) bones glowed bright green, indicating that the cells originally in the transplanted region of neural crest had migrated to form that particular bone in the wild-type host's skull.

Evolutionary Embryology

Charles Darwin's theory of evolution restructured comparative embryology and gave it a new focus. After reading Johannes Müller's summary of von Baer's laws in 1842, Darwin saw that embryonic resemblances would be a strong argument in favor of the genetic connectedness of different animal groups. "Community of embryonic structure reveals community of descent," he would conclude in *On the Origin of Species* in 1859. Darwin's evolutionary interpretation of von Baer's laws established a paradigm that was to be followed for many decades—namely, that relationships between groups can be established by finding common embryonic or larval forms.

Even before Darwin, larval forms had been used for taxonomic classification. J. V. Thompson, for instance, demonstrated in the 1830s that larval barnacles were almost identical to larval shrimp, and therefore (correctly) counted barnacles as arthropods rather than molluscs (Figure 1.18; Winsor 1969). Darwin, an expert on barnacle taxonomy, celebrated this finding: "Even the illustrious Cuvier did not perceive that a barnacle is a crustacean, but a glance at the larva shows this in an unmistakable manner." Alexander Kowalevski (1871) made the similar type of discovery (publicized in Darwin's *The Descent of Man*, 1874) that tunicate larvae have a notochord and pharyngeal pouches, which came from the same germ layers as those same structures in fish and chicks. Thus the tunicate (an invertebrate) was related to the vertebrates, and the two great domains of the animal kingdom—invertebrates and vertebrates— were thereby united through larval structures. Darwin (1874) was thrilled, writing, "Thus, if we may rely on embryology, ever the safest guide in classification, it seems that we have at last gained a clue to the source whence the Vertebrata were derived."

Darwin also noted that embryonic organisms sometimes make structures that are inappropriate for their adult form but that show their relatedness to other animals. He pointed out the existence of eyes in embryonic moles, pelvic

(A) Barnacle

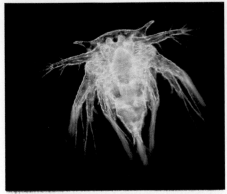

(B) Shrimp

FIGURE 1.18   Larval stages reveal the common ancestry of two crustacean arthropods. (A) Barnacle. (B) Shrimp. Barnacles and shrimp both exhibit a distinctive larval stage (the nauplius) that underscores their common ancestry as crustacean arthropods, even though adult barnacles—once classified as molluscs—are sedentary, differing in body form and lifestyle from the free-swimming adult shrimp. (A © Wim van Egmond/Visuals Unlimited and © Barrie Watts/OSF/Photolibrary.com; B courtesy of U.S. National Oceanic and Atmospheric Administration and © Kim Taylor/Naturepl.com.)

bone rudiments in embryonic snakes, and teeth in baleen whale embryos.

Darwin also argued that adaptations that depart from the "type" and allow an organism to survive in its particular environment develop late in the embryo.* He noted that the differences among species within genera become greater as development persists, as predicted by von Baer's laws. Thus, Darwin recognized two ways of looking at "descent with modification." One could emphasize the

---

*Moreover, as first noted by Weismann (1875), larvae must have their own adaptations to help them survive. The adult viceroy butterfly mimics the monarch butterfly, but the viceroy caterpillar does not resemble the beautiful larva of the monarch. Rather, the viceroy larva escapes detection by resembling bird droppings (Begon et al. 1986).

common descent by pointing out embryonic similarities between two or more groups of animals, or one could emphasize the modifications by showing how development was altered to produce structures that enabled animals to adapt to particular conditions.

Embryonic homologies

One of the most important distinctions made by evolutionary embryologists was the difference between *analogy* and *homology*. Both terms refer to structures that appear to be similar. **Homologous** structures are those organs whose underlying similarity arises from their being derived from a common ancestral structure. For example, the wing of a bird and the forelimb of a human are homologous, both having evolved from the forelimb bones of a common

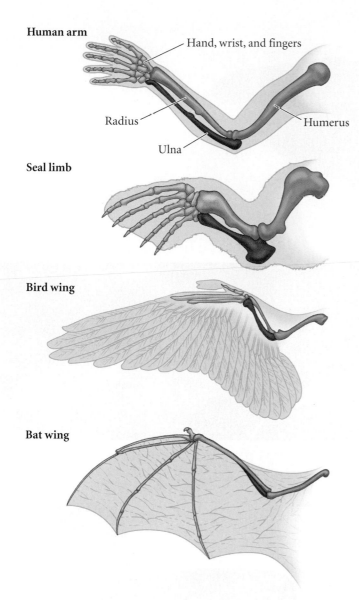

Human arm

Hand, wrist, and fingers

Radius

Ulna

Humerus

Seal limb

Bird wing

Bat wing

FIGURE 1.19 Homologies of structure among a human arm, a seal forelimb, a bird wing, and a bat wing; homologous supporting structures are shown in the same color. All four were derived from a common tetrapod ancestor and thus are homologous as forelimbs. The adaptations of bird and bat forelimbs to flight, however, evolved independently of each other, long after the two lineages diverged from their common ancestor. Therefore, as wings they are not homologous, but analogous.

ancestor. Moreover, their respective parts are homologous (**Figure 1.19**).

Analogous structures are those whose similarity comes from their performing a similar function rather than their arising from a common ancestor. For example, the wing of a butterfly and the wing of a bird are analogous; the two share a common function (and thus both are called wings), but the bird wing and insect wing did not arise from a common ancestral structure that became modified through evolution into bird wings and butterfly wings.*

As we will see in Chapter 19, evolutionary change is based on developmental change. The bat wing, for instance, is made in part by (1) maintaining a rapid growth rate in the cartilage that forms the fingers and (2) preventing the cell death that normally occurs in the webbing between the fingers (**Figure 1.20**). In human development, we start off with webbing between our digits. This webbing is important for creating the anatomical distinctions between our fingers (see Figure 13.26). Once the webbing has serve that function, genetic signals cause its cells to die, leaving us with free digits that can grasp and manipulate objects. Bats, however, use their fingers for flight—a feat accomplished by changing the genes that are activated in the webbing. The genes activated in embryonic bat webbing encode proteins that *prevent* cell death as well as accelerating finger elongation (Cretekos et al. 2005; Sears et al. 2006; Weatherbee et al. 2006). Thus, homologous anatomical structures (in this case, the human hand and the bat wing) can differentiate by altering development.

Changes in development provide the variations needed for evolutionary change. Darwin looked at artificial selection in pigeon and dog breeds, and these examples remain valuable resources for observing selectable variation. For instance, the short legs of dachshunds (**Figure 1.21A**) were selected by breeders who wanted to use these dogs to hunt badgers (German *Dachs*, "badger" + *Hund*, "dog"). The mutation that causes the short legs involves an extra copy of the gene *Fgf4*, which tells the cartilage precursor cells that they have divided enough and can start differentiating. With this extra copy of *Fgf4*, the cartilage cells are told too early that they should stop dividing, so the legs stop growing (Parker et al. 2009). Similarly, long-haired dachshunds (**Figure 1.21B**) differ from their short-haired relatives in having a mutation in the *Fgf5* gene[†] (Cadieu et al. 2009). This gene is involved in hair production and allows each follicle to make a longer hair shaft (Ota et al. 2002). Thus, mutations in genes controlling developmental processes can generate selectable variation.

---

*Homologies must always refer to the level of organization being compared. For instance, bird and bat wings are homologous as forelimbs but not as wings. In other words, they share an underlying structure of forelimb bones because birds and mammals share a common ancestor that possessed such bones. Bats, however, descended from a long line of non-winged mammals, while bird wings evolved independently, from the forelimbs of ancestral reptiles. As we will see, the structure of a bat's wing is markedly different from that of a bird's wing.

[†]The FGF genes will be discussed throughout this book as they regulate construction of numerous organs. Independently acquired mutations in the *Fgf5* gene are also responsible for the long-haired phenotype of Persian cats (Drögemüller et al. 2007; Kehler et al. 2007). However, *Fgf5* is not considered a good candidate to explain the woolliness of mammoths: the sequence of the *Fgf5* gene extracted from the DNA of extinct woolly mammoths appears virtually identical to that of the gene in modern elephants (Roca et al. 2009).

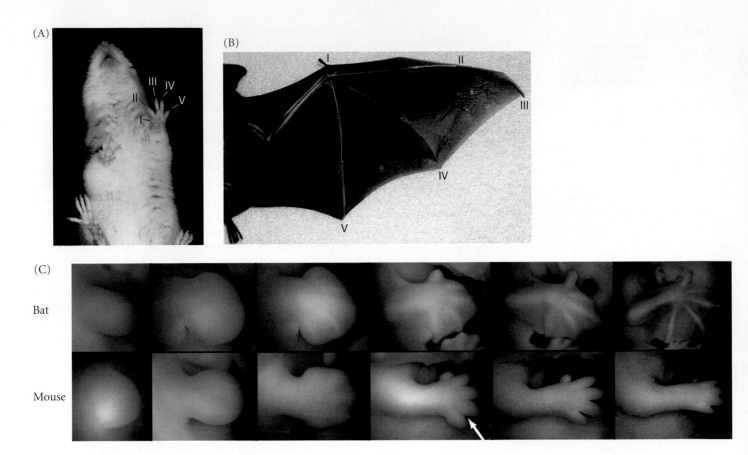

FIGURE 1.20   Development of bat and mouse forelimbs. (A, B) Mouse and bat torsos, showing the elongated fingers and the prominent webbing in the bat wing. A mouse forelimb is shown in the inset, and the digit numbers (I, thumb; V, "pinky") are on both sets. (C) Comparison of mouse and bat forelimb morphogenesis. Both limbs start as webbed appendages, but the webbing between the mouse's digits dies at embryonic day 14 (arrow). The webbing in the bat forelimb does not die, but is sustained as the fingers grow. (A courtesy of David McIntyre; B,C from Cretekos et al. 2008, courtesy of C. J. Cretekos.)

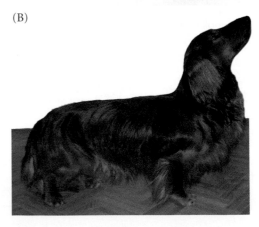

FIGURE 1.21   Selectable variation through mutations of genes that work during developmental. (A) The dachshund has been selected by breeders for its small legs, which enable it to seek badgers in their tunnels. The small legs are a result of premature cessation of cell division in the limb cartilage precursor cells. This premature end to cell division is caused by early activation of the cartilage FGF receptor protein, because the dachshund genome has an extra copy of the *Fgf4* gene. (B) Long-haired dachshunds have an additional mutation, a truncated *Fgf5* gene, which alters the hair follicle cycle, thereby allowing the hair growth beyond the wild-type levels. (A © Alex Potemkin/istockphoto.com; B courtesy of K. Lilleväli.)

Medical Embryology and Teratology

While embryologists could look at embryos to describe the evolution of life and how different animals form their organs, physicians became interested in embryos for more practical reasons. Between 2% and 5% of human infants are born with a readily observable anatomical abnormality (Winter 1996; Thorogood 1997). These abnormalities may include missing limbs, missing or extra digits, cleft palate, eyes that lack certain parts, hearts that lack valves, and so forth. Some birth defects are produced by mutant genes or chromosomes, and some are produced by environmental factors that impede development. Physicians need to know the causes of specific birth defects in order to counsel parents as to the risk of having another malformed infant. In addition, the study of birth defects can tell us how the human body is normally formed. In the absence of experimental data on human embryos, nature's "experiments" sometimes offer important insights into how the human body becomes organized.*

---

*The word *monster*, used frequently in textbooks prior to the mid-twentieth century to describe malformed infants, comes from the Latin *monstrare*, "to show or point out." This is also the root of the English word *demonstrate*. In the 1830s, J. F. Meckel realized that syndromes of congenital anomalies demonstrated certain principles about normal development. Parts of the body that were affected together must have some common developmental origin or mechanism that was being affected. It should also be noted that a condition considered a developmental anomaly in one situation may be considered advantageous in another. The short legs of dachshunds is only one such example.

Genetic malformations and syndromes

Abnormalities caused by genetic events (gene mutations, chromosomal aneuploidies, and translocations) are called **malformations**. Malformations often appear as **syndromes** (Greek, "running together"), in which several abnormalities occur concurrently. For instance, a human malformation called piebaldism, shown in **Figure 1.22A**, is due to a dominant mutation in a gene (*KIT*) on the long arm of chromosome 4 (Spritz et al. 1992). The piebald syndrome includes anemia, sterility, unpigmented regions of the skin and hair, deafness, and the absence of the nerves that cause peristalsis in the gut. The common feature underlying these conditions is that the *KIT* gene encodes a protein that is expressed in the neural crest cells and in the precursors of blood cells and germ cells. The Kit protein enables these cells to proliferate. Without this protein, the neural crest cells—which generate the pigment cells, certain ear cells, and the gut neurons—do not multiply as extensively as they should (resulting in underpigmentation, deafness, and gut malformations), nor do the precursors of the blood cells (resulting in anemia) or the germ cells (resulting in sterility).

Developmental biologists and clinical geneticists often study human syndromes (and determine their causes) by studying animals that display the same syndrome. These are called **animal models** of the disease; the mouse model for piebaldism is shown in **Figure 1.22B**. It has a phenotype very similar to that of the human condition, and it is caused by a mutation in the *Kit* gene of the mouse.*

---

*The mouse *Kit* and human *KIT* genes are considered homologous by their structural similarities and their presumed common ancestry. Human genes are usually italicized and written in all capitals. Mouse genes are italicized, but only the first letter is usually capitalized. Gene products—proteins—are not italicized. If the protein has no standard biochemical or physiological name, it is usually represented with the name of the gene in Roman type, with the first letter capitalized. These rules are frequently bent, however. One is reminded of Cohen's (1982) dictum that "Academicians are more likely to share each other's toothbrush than each other's nomenclature."

(A)   (B)

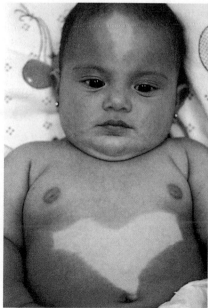

FIGURE 1.22 Developmental anomalies caused by genetic mutation. (A) Piebaldism in a human infant. This genetically produced condition results in sterility, anemia, and underpigmented regions of the skin and hair, along with defective development of gut neurons and the ear. Piebaldism is caused by a mutation in the *KIT* gene. The Kit protein is essential for the proliferation and migration of neural crest cells, germ cell precursors, and blood cell precursors. (B) A piebald mouse with a *Kit* mutation. Mice provide important models for studying human developmental diseases. (Photographs courtesy of R. A. Fleischman.)

(A)

(B)

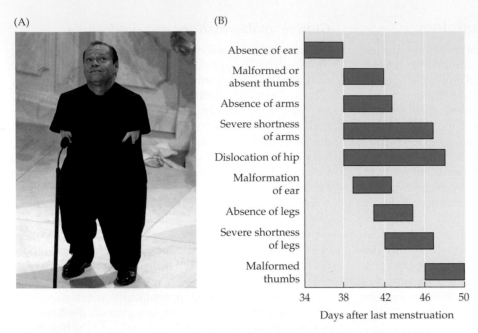

FIGURE 1.23   Developmental anomalies caused by an environmental agent. (A) Pho-comelia, the lack of proper limb development, was the most visible of the birth defects that occurred in many children born in the early 1960s whose mothers took the drug thalidomide during pregnancy. These children are now adults; this photograph is a recent one of Grammy-nominated German singer Thomas Quasthoff. (B) Thalidomide disrupts different structures at different times of human development. (A © AP Photo; B after Nowack 1965.)

Disruptions and teratogens

Abnormalities caused by exogenous agents (certain chemicals or viruses, radiation, or hyperthermia) are called **disruptions**. The agents responsible for these disruptions are called **teratogens** (Greek, "monster-formers"), and the study of how environmental agents disrupt normal development is called *teratology*. Teratogens were brought to the attention of the public in the early 1960s. In 1961, Lenz and McBride independently accumulated evidence that the drug thalidomide, prescribed as a mild sedative to many pregnant women, caused an enormous increase in a previously rare syndrome of congenital anomalies. The most noticeable of these anomalies was *phocomelia*, a condition in which the long bones of the limbs are deficient or absent (**Figure 1.23A**). More than 7000 affected infants were born to women who took thalidomide, and a woman need only have taken one tablet for her child to be born with all four limbs deformed (Lenz 1962, 1966; Toms 1962). Other abnormalities induced by the ingestion of this drug included heart defects, absence of the external ears, and malformed intestines.

Nowack (1965) documented the period of susceptibility during which thalidomide caused these abnormalities (**Figure 1.23B**). The drug was found to be teratogenic only during days 34–50 after the last menstruation (i.e., 20–36 days

postconception). From days 34 to 38, no limb abnormalities are seen, but during this period, thalidomide can cause the absence or deficiency of ear components. Malformations of the upper limbs are seen before those of the lower limbs, because during development the arms form slightly before the legs.

The only animal models for thalidomide are primates, and we still do not know for certain the mechanisms by which this drug causes human developmental disruptions (although it may work by blocking certain molecules from the developing mesoderm, thus preventing blood vessel development). Thalidomide was withdrawn from the market in November 1961. However, the drug is once more beginning to be prescribed (although not to pregnant women) as a potential anti-tumor and anti-autoimmunity drug (Raje and Anderson 1999).

The integration of anatomical information about congenital malformations with our new knowledge of the genes responsible for development has had a revolutionary effect and is currently restructuring medicine. This integration is allowing us to discover the genes responsible for inherited malformations, and it permits us to identify the steps in development that are being disrupted by teratogens. We will see examples of this integration throughout this text, and Chapter 17 will detail some of the remarkable new discoveries in human teratology.

Snapshot Summary: *Developmental Anatomy*

1. The life cycle can be considered a central unit in biology; the adult form need not be paramount. The basic animal life cycle consists of fertilization, cleavage, gastrulation, germ layer formation, organogenesis, metamorphosis, adulthood, and senescence.

2. In gametogenesis, the germ cells (i.e., those cells that will become sperm or eggs) undergo meiosis. Eventually, usually after adulthood is reached, the mature gametes are released to unite during fertilization. The resulting new generation then begins development.

3. Epigenesis happens. New organisms are created de novo each generation from the relatively disordered cytoplasm of the egg.

4. Preformation is not found in the anatomical structures themselves, but in the genetic instructions that instruct their formation. The inheritance of the fertilized egg includes the genetic potentials of the organism. These preformed nuclear instructions include the ability to respond to environmental stimuli in specific ways.

5. The three germ layers give rise to specific organ systems. The ectoderm gives rise to the epidermis, nervous system, and pigment cells; the mesoderm generates the kidneys, gonads, muscles, bones, heart, and blood cells; and the endoderm forms the lining of the digestive tube and the respiratory system.

6. Karl von Baer's principles state that the general features of a large group of animals appear earlier in the embryo than do the specialized features of a smaller group. As each embryo of a given species develops, it diverges from the adult forms of other species. The early embryo of a "higher" animal species is not like the adult of a "lower" animal.

7. Labeling cells with dyes shows that some cells differentiate where they form, whereas others migrate from their original sites and differentiate in their new locations. Migratory cells include neural crest cells and the precursors of germ cells and blood cells.

8. "Community of embryonic structure reveals community of descent" (Charles Darwin, *On the Origin of Species*).

9. Homologous structures in different species are those organs whose similarity is due to sharing a common ancestral structure. Analogous structures are those organs whose similarity comes from serving a similar function (but which are not derived from a common ancestral structure).

10. Congenital anomalies can be caused by genetic factors (mutations, aneuploidies, translocations) or by environmental agents (certain chemicals, certain viruses, radiation).

11. Syndromes consist of sets of developmental abnormalities that "run together."

12. Organs that are linked in developmental syndromes share either a common origin or a common mechanism of formation.

For Further Reading

Complete bibliographical citations for all literature cited in this chapter can be found at the free-access website **www.devbio.com**

Cadieu, E. and 19 others. 2009. Coat variation in the domestic dog is governed by variants in three genes. *Science* 326: 150–153.

Cebra-Thomas, J. A., E. Betters, M. Yin, C. Plafkin, K. McDow and S. F. Gilbert. 2007. Evidence that a late-emerging population of trunk neural crest cells forms the plastron bones in the turtle *Trachemys scripta. Evol. Dev.* 9: 267–277.

Larsen, E. and H. McLaughlin. 1987. The morphogenetic alphabet: Lessons for simple-minded genes. *BioEssays* 7: 130–132.

Le Douarin, N. M. and M.-A. Teillet. 1973. The migration of neural crest cells to the wall of the digestive tract in the avian embryo. *J. Embryol. Exp. Morphol.* 30: 31–48.

Nishida, H. 1987. Cell lineage analysis in ascidian embryos by intracellular injection of a tracer enzyme. III. Up to the tissue-restricted stage. *Dev. Biol.* 121: 526–541.

Pinto-Correia, C. 1997. *The Ovary of Eve: Egg and Sperm and Preformation.* University of Chicago Press, Chicago.

Weatherbee, S. D., R. R. Behringer, J. J. Rasweiler 4th and L. A. Niswander. 2006. Interdigital webbing retention in bat wings illustrates genetic changes underlying amniote limb diversification. *Proc. Natl. Acad. Sci. USA* 103: 15103–15107.

Winter, R. M. 1996. Analyzing human developmental abnormalities. *BioEssays* 18: 965–971.

Woo, K. and S. E. Fraser. 1995. Order and coherence in the fate map of the zebrafish embryo. *Development* 121: 2595–2609.

Go Online

WEBSITE 1.1 Conklin's art and science. The plates from Conklin's remarkable 1905 paper are online. Looking at them, one can see the precision of his observations and how he constructed his fate map of the tunicate embryo.

VADE MECUM

Chick-quail chimeras. We are fortunate to present here a movie made by Dr. Nicole Le Douarin of her chick-quail grafts. You will be able to see how these grafts are actually done.

The compound microscope. The compound microscope has been the critical tool of developmental anatomists. Mastery of microscopic techniques allows one to enter an entire world of form and pattern.

Histotechniques. Most cells must be stained in order to see them; different dyes stain different types of molecules. Instructions on staining cells to observe particular structures (such as the nucleus) are given here.

Developmental Genetics

2

CYTOLOGICAL STUDIES DONE AT THE TURN OF THE TWENTIETH CENTURY established that the chromosomes in each cell of an organism's body are the mitotic descendants of the chromosomes established at fertilization (Wilson 1896; Boveri 1904). In other words, each somatic cell nucleus has the same chromosomes—and therefore the same set of genes—as all the other somatic nuclei. This fundamental concept is called **genomic equivalence**. Given this concept, one of the major questions facing biologists of the early twentieth century was how nuclear genes could direct development when these genes are the same in every cell type (Harrison 1937; Just 1939). If every cell in the body contains the genes for hemoglobin and insulin proteins, why is it that hemoglobin proteins are made only in red blood cells, insulin proteins are made only in certain pancreas cells, and neither protein is made in the kidneys or nervous system?

Based on the embryological evidence for genomic equivalence (as well as on bacterial models of gene regulation), a consensus emerged in the 1960s that the answer to this question lies in **differential gene expression**. The three postulates of differential gene expression are:

- Every cell nucleus contains the complete genome established in the fertilized egg. In molecular terms, the DNAs of all differentiated cells are identical.

- The unused genes in differentiated cells are neither destroyed nor mutated, but retain the potential for being expressed.

- Only a small percentage of the genome is expressed in each cell, and a portion of the RNA synthesized in each cell is specific for that cell type.

Gene expression can be regulated at several levels such that different cell types synthesize different sets of proteins:

- **Differential gene transcription** regulates which of the nuclear genes are transcribed into nuclear RNA.

- **Selective nuclear RNA processing** regulates which of the transcribed RNAs (or which parts of such a nuclear RNA) are able to enter into the cytoplasm and become messenger RNAs.

- **Selective messenger RNA translation** regulates which of the mRNAs in the cytoplasm are translated into proteins.

- **Differential protein modification** regulates which proteins are allowed to remain and/or function in the cell.

Some genes (such as those coding for the globin proteins of hemoglobin) are regulated at all these levels.

But whatever the immediate operations of the genes turn out to be, they most certainly belong to the category of developmental processes and thus belong to the province of embryology.

C. H. WADDINGTON (1956)

We have entered the cell, the mansion of our birth, and have started the inventory of our acquired wealth.

ALBERT CLAUDE (1974)

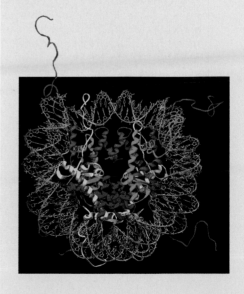

Evidence for Genomic Equivalence

Until the mid-twentieth century, genomic equivalence was not so much proved as it was assumed (because every cell is the mitotic descendant of the fertilized egg. One of the first tasks of developmental genetics was to determine whether every cell of an organism indeed does have the same *genome*—that is, the same set of genes—as every other cell.

Evidence that every cell in the body has the same genome originally came from the analysis of *Drosophila* chromosomes, in which the DNA of certain larval tissues undergoes numerous rounds of DNA replication without separation, such that the structure of the chromosomes can be seen. In these **polytene** (Gr. "many strands") **chromosomes**, no structural differences were seen between cells; but different regions were seen to be "puffed up" at different times and in different cell types, suggesting that these areas were actively making RNA (Beerman 1952).

See WEBSITE 2.1
Does the genome or the cytoplasm direct development?

See WEBSITE 2.2
The origins of developmental genetics

When Giemsa dyes allowed such observations to be made in mammalian chromosomes, it was also found that no chromosomal regions were lost in most cells. These observations, in turn, were confirmed by nucleic acid hybridization studies, which (for instance) found globin genes in pancreatic tissue, which does not make globin proteins.

But the ultimate test of whether the nucleus of a differentiated cell has undergone irreversible functional restriction is to have that nucleus generate every other type of differentiated cell in the body. If each cell's nucleus is identical to the zygote nucleus, then each cell's nucleus should also be capable of directing the entire development of the organism when transplanted into an activated enucleated egg. As early as 1895, the embryologist Yves Delage predicted that "If, without deterioration, the egg nucleus could be replaced by the nucleus of an ordinary embryonic cell, we should probably see this egg developing without changes" (Delage 1895, p. 738).

In 1952, Briggs and King demonstrated that blastula cell nuclei could direct the development of complete tadpoles when transferred into the cytoplasm of an activated enucleated frog egg. This procedure is called **somatic nuclear transfer** or, more commonly, **cloning**. Nuclei from adult frogs, however, were not able to generate adult frogs. However, adult nuclei (from skin cells, for instance) were

---

SIDELIGHTS & SPECULATIONS

The Basic Tools of Developmental Genetics

DNA analysis

Embryologist Theodor Boveri (1904) wrote that to discover the mechanisms of development, it was "not cell nuclei, not even individual chromosomes, but certain parts of certain chromosomes from certain cells that must be isolated and collected in enormous quantities for analysis." This analysis was finally made possible by the techniques of gene cloning, DNA sequencing, Southern blotting, gene knockouts, and enhancer traps. In addition, techniques for showing which enhancers and promoters are methylated and which are unmethylated have become more important, as investigations of differential gene transcription have focused on these elements.

For discussions of these techniques, see Website 2.3.

RNA analysis

Differential gene transcription is critical in development. In order to know the time of gene expression and the place of gene expression, one has to be able to have procedures that actually locate a particular type of messenger RNA. These techniques include northern blots, RT-PCR, in situ hybridization, and array technology. To ascertain the function of these mRNAs, new techniques have been formulated, which include antisense and RNA interference (which destroy messages), Cre-lox analysis (which allows the message to be made or destroyed in particular cell types) and ChIP-on-Chip techniques (which enable one to localize active chromatin).

For discussions of these techniques, see Website 2.4.

Bioinformatics

Modern developmental genetics often involves comparing DNA sequences (especially regulatory units such as enhancers and 3' UTRs) and looking at specific genomes to determine how genes are being regulated. "High-throughput" RNA analysis by micro- and macroarrays enables researchers to compare thousands of mRNAs, and computer-aided synthetic techniques can predict interactions between proteins and mRNAs. Various free websites enable researchers to use the tools that allow such comparisons. Other sites are organism- or organ-specific and are used by researchers studying that particular organ or organism.

For more about these sites and links to them, see Website 2.5

(A)

OOCYTE DONOR
(Scottish blackface strain)

NUCLEAR DONOR
(Finn-Dorset strain)

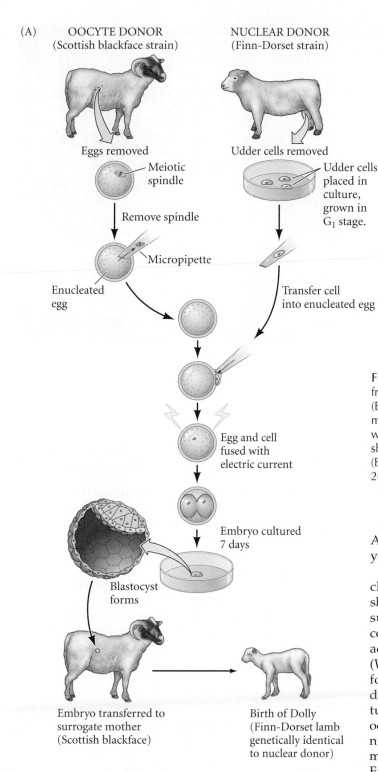

Eggs removed

Meiotic spindle

Remove spindle

Micropipette

Enucleated egg

Udder cells removed

Udder cells placed in culture, grown in G_1 stage.

Transfer cell into enucleated egg

Egg and cell fused with electric current

Embryo cultured 7 days

Blastocyst forms

Embryo transferred to surrogate mother (Scottish blackface)

Birth of Dolly (Finn-Dorset lamb genetically identical to nuclear donor)

(B)

FIGURE 2.1 Cloned mammals have been created using nuclei from adult somatic cells. (A) Procedure used for cloning sheep. (B) Dolly, the adult sheep on the left, was derived by fusing a mammary gland cell nucleus with an enucleated oocyte, which was then implanted in a surrogate mother (of a different breed of sheep) that gave birth to Dolly. Dolly later gave birth to a lamb (Bonnie, at right) by normal reproduction. (A after Wilmut et al. 2000; B, photograph by Roddy Field, © Roslin Institute.)

A nucleus of a skin cell could produce all the cells of a young tadpole.

In 1997, Ian Wilmut announced that a sheep had been cloned from a somatic cell nucleus from an adult female sheep. This was the first time an adult vertebrate had been successfully cloned from another adult. Wilmut and his colleagues had taken cells from the mammary gland of an adult (6-year-old) pregnant ewe and put them into culture (Wilmut et al. 1997; **Figure 2.1A**). The culture medium was formulated to keep the nuclei in these cells at the intact diploid stage (G1) of the cell cycle. This cell-cycle stage turned out to be critical. The researchers then obtained oocytes from a different strain of sheep and removed their nuclei. These oocytes had to be in the second meiotic metaphase (the stage at which they are usually fertilized). Fusion of the donor cell and the enucleated oocyte was accomplished by bringing the two cells together and sending electric pulses through them, destabilizing the cell membranes and allowing the cells to fuse. The same electric pulses that fused the cells activated the egg to begin development. The resulting embryos were eventually transferred into the uteri of pregnant sheep.

able to direct the development of all the organs of the tadpoles (Gurdon et al. 1975). Although the tadpoles all died prior to feeding, their existence showed that a single differentiated cell nucleus still retained incredible potencies.

(A)

(B)

FIGURE 2.2   The kitten "CC" (A) was a clone produced using somatic nuclear transfer from "Rainbow," a female calico cat (B). The two do not appear identical because coat pigmentation pattern in calico cats is affected by the random inactivation of one X chromosome in each somatic cell (see Sidelights & Speculations, p. 50). Their behaviors were also quite different. (Photographs courtesy of College of Veterinary Medicine, Texas A&M University.)

Of the 434 sheep oocytes originally used in this experiment, only one survived: Dolly* (**Figure 2.1B**). DNA analysis confirmed that the nuclei of Dolly's cells were derived from the strain of sheep from which the donor nucleus was taken (Ashworth et al. 1998; Signer et al. 1998). Cloning of adult mammals has been confirmed in guinea pigs, rabbits, rats, mice, dogs, cats, horses, and cows. In 2003, a cloned mule became the first sterile animal to be so reproduced (Woods et al. 2003). Thus it appears that the nuclei of vertebrate adult somatic cells contain all the genes needed to generate an adult organism. No genes necessary for development have been lost or mutated in the somatic cells.[†]

Certain caveats must be applied, however. First, although it appears that all the organs were properly formed in the cloned animals, many of the clones developed debilitating diseases as they matured (Humphreys et al. 2001; Jaenisch and Wilmut 2001; Kolata 2001). As we will shortly see, this problem is due in large part to the differences in methylation between the chromatin of the zygote and the differentiated cell. Second, the phenotype of the cloned animal is sometimes not identical to that of the animal from which the nucleus was derived. There is variability due to random chromosomal events and the effects of environment. The pigmentation of calico cats, for instance, is due to the *random inactivation* of one or the other X chromosome (a genetic mechanism that will be discussed later in this chapter) in each somatic cell of the female cat embryo. Therefore, the coat color pattern of the first cloned cat, a calico named "CC," were different from those of "Rainbow," the adult calico whose cells provided the implanted nucleus that generated "CC" (**Figure 2.2**).

The same genotype gives rise to multiple phenotypes in cloned sheep as well. Wilmut noted that four sheep cloned from blastocyst nuclei from the same embryo "are genetically identical to each other and yet are very different in size and temperament, showing emphatically that an animal's genes do not 'determine' every detail of its physique and personality" (Wilmut et al. 2000, p. 5). Wilmut concludes that for this and other reasons, the "resurrection" of lost loved ones by cloning is not feasible.

See WEBSITE 2.6   Cloning and nuclear equivalence

---

*The creation of Dolly was the result of a combination of scientific and social circumstances. These circumstances involved job security, people with different areas of expertise meeting each other, children's school holidays, international politics, and who sits near whom in a pub. The complex interconnections giving rise to Dolly are told in *The Second Creation* (Wilmut et al. 2000), a book that should be read by anyone who wants to know how contemporary science actually works. As Wilmut acknowledged (p. 36), "The story may seem a bit messy, but that's because life is messy, and science is a slice of life."

[†]Although cloning humans does not seem feasible at present, each cell of the human body (with just a few exceptions, such as lymphocytes) does appear to contain the same genome as every other cell. As we will see in Chapter 17, adding certain activated transcription factors to ordinary skin fibroblasts will convert them into embryonic stem cells that are indeed capable of generating entire embryos, at least in mice.

(A)

(B)

Nucleosome

Histone octamer

H1 histones

DNA "wrap" Linker DNA

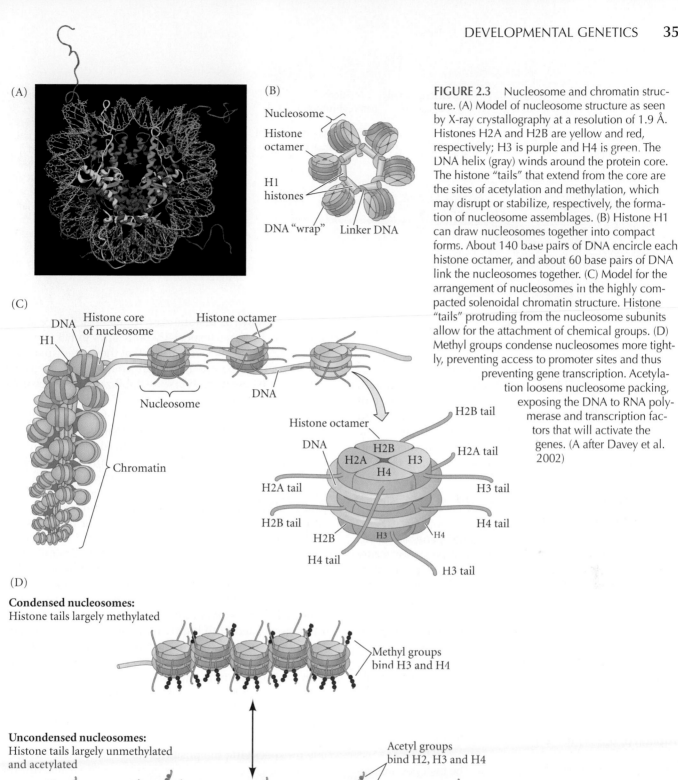

FIGURE 2.3 Nucleosome and chromatin structure. (A) Model of nucleosome structure as seen by X-ray crystallography at a resolution of 1.9 Å. Histones H2A and H2B are yellow and red, respectively; H3 is purple and H4 is green. The DNA helix (gray) winds around the protein core. The histone "tails" that extend from the core are the sites of acetylation and methylation, which may disrupt or stabilize, respectively, the formation of nucleosome assemblages. (B) Histone H1 can draw nucleosomes together into compact forms. About 140 base pairs of DNA encircle each histone octamer, and about 60 base pairs of DNA link the nucleosomes together. (C) Model for the arrangement of nucleosomes in the highly compacted solenoidal chromatin structure. Histone "tails" protruding from the nucleosome subunits allow for the attachment of chemical groups. (D) Methyl groups condense nucleosomes more tightly, preventing access to promoter sites and thus preventing gene transcription. Acetylation loosens nucleosome packing, exposing the DNA to RNA polymerase and transcription factors that will activate the genes. (A after Davey et al. 2002)

(C)

DNA
H1
Histone core of nucleosome
Histone octamer

Nucleosome

DNA

Chromatin

Histone octamer
DNA
H2B
H2A   H3
H4

H2B tail
H2A tail
H3 tail
H4 tail
H3 tail

H2A tail
H2B tail
H2A
H2B
H4 tail
H3
H4

(D)

Condensed nucleosomes:
Histone tails largely methylated

Methyl groups bind H3 and H4

Uncondensed nucleosomes:
Histone tails largely unmethylated and acetylated

Acetyl groups bind H2, H3 and H4

Differential Gene Transcription

So how does the same genome give rise to different cell types? To understand this, one needs to understand the anatomy of the genes. One of the fundamental differences distinguishing most eukaryotic genes from prokaryotic genes is that eukaryotic genes are contained within a complex of DNA and protein called **chromatin**. The protein component constitutes about half the weight of chromatin and is composed largely of **histones**. The **nucleosome** is the basic unit of chromatin structure (**Figure 2.3**). It is composed of an octamer of histone proteins (two molecules

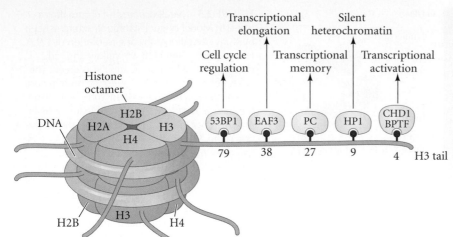

Transcriptional
elongation
Silent
heterochromatin
Cell cycle
regulation
Transcriptional
memory
Transcriptional
activation
Histone
octamer
H2B
H2A   H3
H4
DNA
53BP1   EAF3   PC   HP1   CHD1
BPTF
79   38   27   9   4   H3 tail
H2B   H3   H4

FIGURE 2.4 Histone methylations on histone H3. The tail of histone H3 (its amino-most sequence, at the beginning of the protein) sticks out from the nucleosome and is capable of being methylated or acetylated. Here, lysines can be methylated and recognized by particular proteins. Methylated lysine residues at positions 4, 38, and 79 are associated with gene activation, whereas methylated lysines at positions 9 and 27 are associated with repression. The proteins binding these sites (not shown to scale) are represented above the methyl group. (After Kouzarides and Berger 2007.)

each of histones H2A, H2B, H3, and H4) wrapped with two loops containing approximately 140 base pairs of DNA (Kornberg and Thomas 1974). Histone H1 is bound to the 60 or so base pairs of "linker" DNA between the nucleosomes (Weintraub 1984). There are 14 points of contact between the DNA and the histones (Luger et al. 1997).

Anatomy of the gene: Active and repressed chromatin

Whereas classical geneticists have likened genes to "beads on a string," molecular geneticists liken genes to "string on the beads," an image in which the beads are nucleosomes. Most of the time, the nucleosomes are wound into tight "solenoids" that are stabilized by histone H1 (**Figure 2.3C**). This H1-dependent conformation of nucleosomes inhibits the transcription of genes in somatic cells by packing adjacent nucleosomes together into tight arrays that prevent transcription factors and RNA polymerases from gaining access to the genes (Thoma et al. 1979; Schlissel and Brown 1984). It is generally thought, then, that the "default" condition of chromatin is a repressed state, and that tissue-specific genes become activated by local interruption of this repression (Weintraub 1985).

HISTONES AS AN ACTIVATION SWITCH The histones are critical because they are responsible for maintaining the repression of gene expression. This repression can be locally strengthened (so that it becomes very difficult to transcribe those genes in the nucleosomes) or relieved (so that transcribing them becomes relatively easy) by modifying the histones (**Figure 2.3D**). Repression and activation are controlled to a large extent by modifying the tails of histones H3 and H4 with two small organic groups: methyl (CH_3) and acetyl ($COCH_3$) residues. In general, **histone acetylation**—the addition of negatively charged acetyl groups to histones—neutralizes the basic charge of lysine and loosens the histones. This activates transcription.

Enzymes known as **histone acetyltransferases** place acetyl groups on histones (especially on lysines in H3 and H4), destabilizing the nucleosomes so that they come apart easily. As might be expected, then, enzymes that *remove* acetyl groups—**histone deacetylases**—stabilize the nucleosomes and prevent transcription.

Histone methylation, the addition of methyl groups to histones by **histone methyltransferases**, can either activate or further repress transcription, depending on the amino acid being methylated and the presence of other methyl or acetyl groups in the vicinity (see Strahl and Allis 2000; Cosgrove et al. 2004). For instance, acetylation of the "tails" of H3 and H4 along with methylation of the lysine at position 4 of H3 (i.e., H3K4; remember that K is the abbreviation for lysine) is usually associated with actively transcribed chromatin. In contrast, a combined lack of acetylation of the H3 and H4 tails and methylation of the lysine in the ninth position of H3 (H3K9) is usually associated with highly repressed chromatin (Norma et al. 2001). Indeed, lysine methylations at H3K9, H327, and H4K20 are often associated with highly repressed chromatin. **Figure 2.4** shows a schematic drawing of a nucleosome, with the histone H3 tail having on it some residues whose modification can regulate transcription.

As might be expected, if methyl groups at specific places on the histones repress transcription, then getting rid of these methyl moieties should permit transcription. This has been shown in the activation of those genes responsible for specifying the posterior halves of vertebrate bodies. These genes, called Hox genes, encode transcription factors that are critical in giving cells their identities along the anterior-posterior axis. In early development, Hox genes are repressed by H3K27 trimethylation (the lysine at position 27 having three methyl groups). However, in differentiated cells, a demethylase that is specific for H3K27me3 is recruited to these promoters and enables the gene to be transcribed (Agger et al. 2007; Lan et al. 2007).

The effects of methylation in controlling gene transcription are extensive. So far, we have documented transcriptional regulation by *histone* methylation. Later in this chapter we will discuss the exciting research on the control of transcription by *DNA* methylation.

HISTONE REGULATION OF TRANSCRIPTIONAL ELONGATION
In addition to regulating the initiation of the transcriptional complex (i.e., getting RNA polymerase on the promoter), nucleosomes also appear to regulate the progression of RNA polymerase and the elongation of the mRNA. Indeed, recent evidence suggests that it is relatively common for RNA polymerase to be poised at the promoters, ready to go. For transcription to occur, these nucleosomes need to be modified, and it is possible that the acetylation of histone H3 at positions 9 and 14, coupled with the trimethylation of that histone at position 4, is critical for allowing elongation of the message (Guenther et al. 2007; Li et al. 2007).

Anatomy of the gene: Exons and introns

The second difference between prokaryotic and eukaryotic genes is that eukaryotic genes are not co-linear with their peptide products. Rather, the single nucleic acid strand of eukaryotic mRNA comes from noncontiguous regions on the chromosome. Between **exons**—the regions of DNA that code for a protein*—are intervening sequences called **introns** that have nothing whatsoever to do with the amino acid sequence of the protein. The structure of a typical eukaryotic gene can be illustrated by the human β-globin gene (**Figure 2.5**). This gene, which encodes part of the hemoglobin protein of the red blood cells, consists of the following elements:

- A **promoter region**, which is responsible for the binding of RNA polymerase and for the subsequent initiation of transcription. The promoter region of the human β-globin gene has three distinct units and extends from 95 to 26 base pairs before ("upstream from")[†] the transcription initiation site (i.e., from –95 to –26).

- The **transcription initiation site**, which for human β-globin is ACATTTG. This site is often called the **cap sequence** because it represents the 5' end of the RNA, which will receive a "cap" of modified nucleotides soon after it is transcribed. The specific cap sequence varies among genes.

- The **translation initiation site**, ATG. This codon (which becomes AUG in mRNA) is located 50 base pairs after the transcription initiation site in the human β-globin gene (although this distance differs greatly among different genes). The sequence of 50 base pairs intervening between the initiation points of transcription and trans-

lation is the **5' untranslated region**, often called the **5' UTR** or **leader sequence.** The 5' UTR can determine the rate at which translation is initiated.

- The first exon, which contains 90 base pairs coding for amino acids 1–30 of human β-globin protein.

- An intron containing 130 base pairs with no coding sequences for β-globin. However, the structure of this intron is important in enabling the RNA to be processed into mRNA and exit from the nucleus.

- An exon containing 222 base pairs coding for amino acids 31–104.

- A large intron—850 base pairs—having nothing to do with globin protein structure.

- An exon containing 126 base pairs coding for amino acids 105–146 of the protein.

- A **translation termination codon**, TAA. This codon becomes UAA in the mRNA. The ribosome dissociates at this codon, and the protein is released.

- A **3' untranslated region** (3' UTR) that, although transcribed, is not translated into protein. This region includes the sequence AATAAA, which is needed for **polyadenylation**, the insertion of a "tail" of some 200–300 adenylate residues on the RNA transcript, about 20 bases downstream of the AAUAAA sequence. This polyA tail (1) confers stability on the mRNA, (2) allows the mRNA to exit the nucleus, and (3) permits the mRNA to be translated into protein.

- A **transcription termination sequence**. Transcription continues beyond the AATAAA site for about 1000 nucleotides before being terminated.

The original transcription product is called **nuclear RNA (nRNA),** sometimes called *heterogeneous nuclear RNA* (hnRNA) or *pre-messenger RNA* (pre-mRNA). Nuclear RNA contains the cap sequence, the 5' UTR, exons, introns, and the 3' UTR (**Figure 2.6**). Both ends of these transcripts are modified before these RNAs leave the nucleus. A cap consisting of methylated guanosine is placed on the 5' end of the RNA in opposite polarity to the RNA itself. This means there is no free 5' phosphate group on the nRNA. The 5' cap is necessary for the binding of mRNA to the ribosome and for subsequent translation (Shatkin 1976). The 3' terminus is usually modified in the nucleus by the addition of a polyA tail. The adenylate residues in this tail are put together enzymatically and are added to the transcript; they are not part of the gene sequence. Both the 5' and 3' modifications may protect the mRNA from exonucleases that would otherwise digest it (Sheiness and Darnell 1973; Gedamu and Dixon 1978). The modifications thus stabilize the message and its precursor.

As the nRNA leaves the nucleus, its introns are removed and the remaining exons spliced together. In this way the coding regions of the mRNA—i.e., the exons—are brought together to form a single transcript, and this transcript is translated into a protein. The protein can be further modified to make it functional (see Figure 2.6).

---

*The term *exon* refers to a nucleotide sequence whose RNA "exits" the nucleus. It has taken on the functional definition of a protein-encoding nucleotide sequence. Leader sequences and 3' UTR sequences are also derived from exons, even though they are not translated into protein.

[†]By convention, upstream, downstream, 5', and 3' directions are specified in relation to the RNA. Thus, the promoter is upstream of the gene, near its 5' end.

(A)

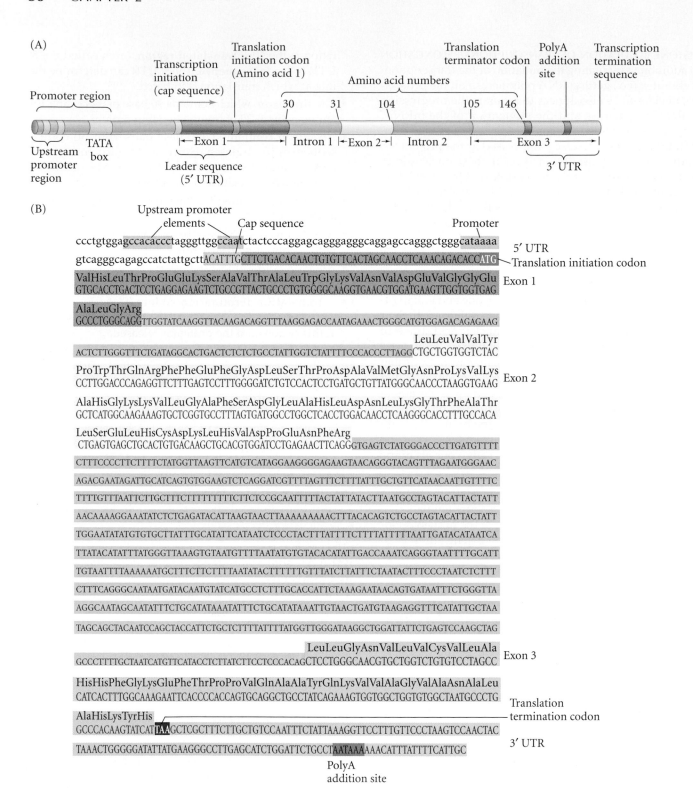

(B)

FIGURE 2.5 Nucleotide sequence of the human β-globin gene. (A) Schematic representation of the locations of the promoter region, transcription initiation site (cap sequence), 5′ untranslated region (leader sequence), exons, introns, and 3′ untranslated region. Exons are shown in color; the numbers flanking them indicate the amino acid positions each exon encodes in β-globin. (B) The nucleotide sequence shown from the 5′ end to the 3′ end of the RNA. The colors correspond to their diagrammatic representation in (A). The promoter sequences are boxed, as are the transla-tion initiation and termination codes ATG and TAA. The large cap-ital letters boxed in color are the bases of the exons, with the amino acids for which they code abbreviated above them. Small-er capital letters indicate the intron bases. The codons after the translation termination site exist in β-globin mRNA but are not translated into proteins. Within this group is the sequence thought to be needed for polyadenylation. By convention, only the RNA-like strand of the DNA double helix is shown. (B after Lawn et al. 1980.)

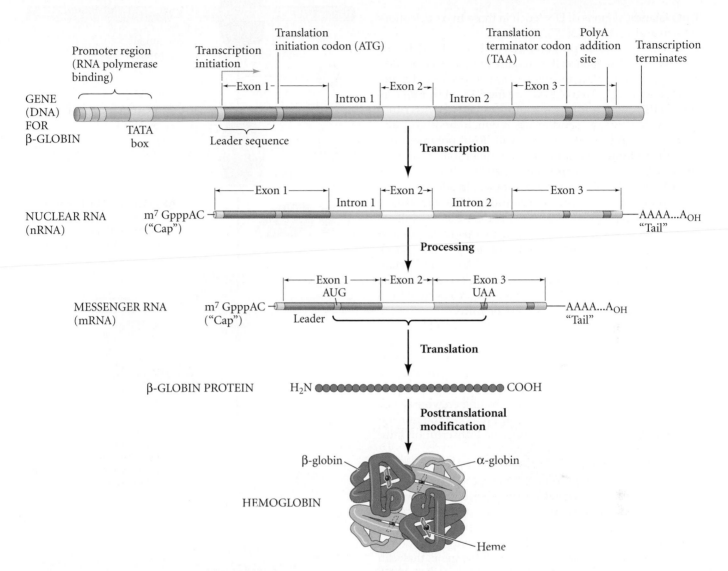

FIGURE 2.6 Summary of steps involved in the production of β-globin and hemoglobin. Transcription of the gene creates a nuclear RNA containing exons and introns, as well as the cap, tail, and 3′ and 5′ untranslated regions. Processing the nuclear RNA into messenger RNA removes the introns. Translation on ribosomes uses the mRNA to encode a protein. The protein is inactive until it is modified and complexed with α-globin and heme to become active hemoglobin (bottom).

Anatomy of the gene: Promoters and enhancers

In addition to the protein-encoding region of the gene, there are regulatory sequences that can be located on either end of the gene (or even within it). These sequences—the promoters and enhancers—are necessary for controlling where and when a particular gene is transcribed.

Promoters are the sites where RNA polymerase binds to the DNA to initiate transcription. Promoters of genes that synthesize messenger RNAs (i.e., genes that encode proteins*) are typically located immediately upstream from the site where the RNA polymerase initiates transcription. Most of these promoters contain the sequence TATA, to which RNA polymerase will be bound. This site, known as the **TATA box**, is usually about 30 base pairs upstream from the site where the first base is transcribed. Since this sequence will appear randomly in the genome at more places than just at promoter sites, other regions flanking it are also important. Many TATA box regions are flanked by

---

*There are several types of RNA that do *not* encode proteins. These include the ribosomal RNAs and transfer RNAs (which are used in protein synthesis) and the small nuclear RNAs (which are used in RNA processing). In addition, there are regulatory RNAs (such as the microRNAs that we will discuss later in this chapter), which are involved in regulating gene expression and are not translated into peptides.

CpG islands, regions of DNA rich in those two nucleotides (Down and Hubbard 2002).

Eukaryotic RNA polymerases will not bind to the "naked" TATA sequence; they require the presence of additional proteins to place the polymerase properly on the promoter (**Figure 2.7**). Two of these are the **TATA-binding protein** (**TBP**), which forms a complex (TFIID) with other proteins to create a "saddle" upon which the RNA polymerase sits; and **TFIIB**, which recruits RNA polymerase to the TBP and positions it in such a manner that it can read the DNA codons (Kostrewa et al. 2009). Other proteins (TFIIA and TFIIH) stabilize the complex. In addition, auxiliary **transcription-associated factors** (**TAFs**) stabilize the RNA polymerase on the promoter and enable it to initiate transcription. These TAFs are bound by **upstream promoter elements** (sometimes called *proximal promoter sites*), which are DNA sequences near the TATA box and usually upstream from it. Eventually, TFIIH will phosphorylate the carboxy terminal of RNA polymerase, releasing it from the saddle so that it can transcribe the mRNA.

An **enhancer** is a DNA sequence that controls the efficiency and rate of transcription from a specific promoter. In other words, enhancers tell where and when a promoter can be used, and how much of the gene product to make. Enhancers bind specific **transcription factors**, proteins that activate the gene by (1) recruiting enzymes (such as histone acetyltransferases) that break up the nucleosomes in the area or (2) stabilizing the transcription initiation complex as described above.

Enhancers can activate only *cis*-linked promoters (i.e., promoters on the same chromosome*); therefore they are sometimes called *cis*-**regulatory elements**. However, because of DNA folding, enhancers can regulate genes at great distances (some as great as a million bases away) from the promoter (Visel et al. 2009). Moreover, enhancers do not need to be on the 5′ (upstream) side of the gene; they can be at the 3′ end, or even in the introns (Maniatis et al. 1987). The human β-globin gene has an enhancer in its 3′ UTR. This enhancer sequence is necessary for the temporal- and tissue-specific expression of the β-globin gene in adult red blood cell precursors (Trudel and Constantini 1987).

One of the principal methods of identifying enhancer sequences is to clone DNA sequences flanking the gene of interest and fuse them to **reporter genes** whose products are both readily identifiable and not usually made in the

---

*Cis- and *trans*-regulatory elements are so named by analogy with *E. coli* genetics and organic chemistry. There, *cis*-elements are regulatory elements that reside on the same strand of DNA (*cis*-, "on the same side as"), while *trans*-elements are those that could be supplied from another chromosome (*trans*-, "on the other side of"). The term *cis*-regulatory elements now refers to those DNA sequences that regulate a gene on the same stretch of DNA (i.e., the promoters and enhancers). *Trans*-regulatory factors are soluble molecules whose genes are located elsewhere in the genome and which bind to the *cis*-regulatory elements. They are usually transcription factors or microRNAs.

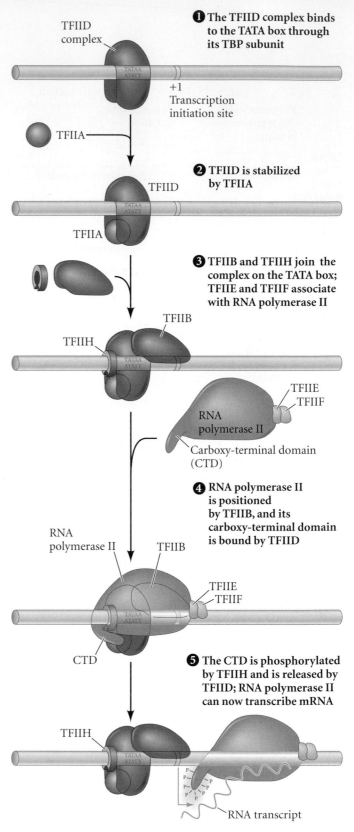

FIGURE 2.7 Formation of the active eukaryotic transcription pre-initiation complex. The diagrams represent the formation of the complex that recruits and stabilizes RNA polymerase onto the promoter. *TF* stands for transcription factor; *II* indicates that the factor was first identified as being needed for RNA polymerase II (the RNA polymerase that transcribes protein-encoding genes); and the letters designate the particular active fraction from the phosphocellulose columns used to purify it.

(A)

(B)

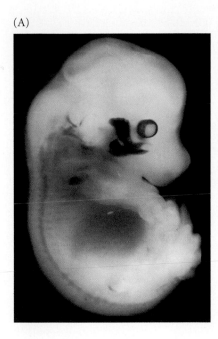

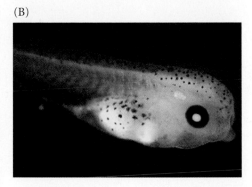

FIGURE 2.8 The genetic elements regulating tissue-specific transcription can be identified by fusing reporter genes to suspected enhancer regions of the genes expressed in particular cell types. (A) The enhancer region of the gene for the muscle-specific protein Myf-5 is fused to a β-galactosidase reporter gene and incorporated into a mouse embryo. When stained for β-galactosidase activity (darkly staining region), the 13.5-day mouse embryo shows that the reporter gene is expressed in the eye muscles, facial muscles, forelimb muscles, neck muscles, and segmented myotomes (which give rise to the back musculature). (B) The *GFP* gene is fused to a lens crystallin gene in *Xenopus tropicalis*. The result is the expression of green fluorescent protein in the tadpole lens. (A courtesy of A. Patapoutian and B. Wold; B from Offield et al. 2000, courtesy of R. Grainger.)

organism being studied. Researchers can insert constructs of possible enhancers and reporter genes into embryos and then monitor the expression of the reporter gene. If the sequence contains an enhancer, the reporter gene should become active at particular times and places. For instance, the *E. coli* gene for β-galactosidase (the *lacZ* gene) can be used as a reporter gene and fused to (1) a promoter that can be activated in any cell and (2) an enhancer that normally directs the expression of a particular mouse gene in muscles. When the resulting transgene is injected into a newly fertilized mouse egg and becomes incorporated into its DNA, β-galactosidase will be expressed in the mouse muscle cells. By staining for the presence of β-galactosidase, the expression pattern of that muscle-specific gene can be seen (**Figure 2.8A**).

Similarly, a sequence flanking a lens crystallin protein in *Xenopus* was shown to be an enhancer. When this sequence was fused to a reporter gene for green fluorescent protein (see Figure 1.17), GFP was expressed only in the lens (**Figure 2.8B**; Offield et al. 2000). GFP reporter genes are very useful because they can be monitored in live embryos and because the changes in gene expression can be seen in single cells.

ENHANCER MODULARITY The enhancer sequences on the DNA are the same in every cell type; what differs is the combination of transcription factor proteins the enhancers bind. Once bound to enhancers, transcription factors are able to enhance or suppress the ability of RNA polymerase to initiate transcription. Enhancers can bind several transcription factors, and it is the specific *combination* of transcription factors present that allows a gene to be active in a particular cell type. That is, the same transcription factor,

in conjunction with different other factors, will activate different promoters in different cells. Moreover, the same gene can have several enhancers, with each enhancer binding transcription factors that enable that same gene to be expressed in different cell types.

Figure 2.9 illustrates this phenomenon for expression of the mouse *Pax6* gene in the cornea and pancreas. The mouse *Pax6* gene (which is expressed in the lens and retina of the eye, in the neural tube, and in the pancreas) has several enhancers (**Figure 2.9A**). The 5′ regulatory regions of the mouse *Pax6* gene were discovered by taking regions from its 5′ flanking sequence and introns and fusing them to a *lacZ* reporter gene. Each of these transgenes was then microinjected into newly fertilized mouse pronuclei, and the resulting embryos were stained for β-galactosidase (**Figure 2.9B**; Kammandel et al. 1998; Williams et al. 1998). Analysis of the results revealed that the enhancer farthest upstream from the promoter contains the regions necessary for *Pax6* expression in the pancreas, while a second enhancer activates *Pax6* expression in surface ectoderm (lens, cornea, and conjunctiva). A third enhancer resides in the leader sequence; it contains the sequences that direct *Pax6* expression in the neural tube. A fourth enhancer sequence, located in an intron shortly downstream of the translation initiation site, determines the expression of *Pax6* in the retina. The *Pax6* gene illustrates the principle of enhancer modularity, wherein having multiple, separate enhancers allows a protein to be expressed in several different tissues while not being expressed at all in others.

COMBINATORIAL ASSOCIATION While enhancers are modular *between* enhancers, there are co-dependent units *within* each enhancer. Enhancers contain regions of DNA that

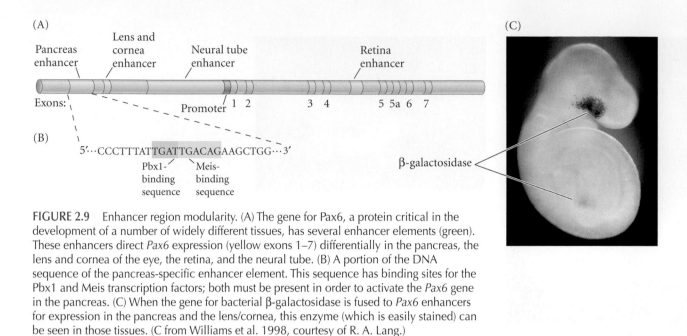

(A)

Pancreas enhancer

Lens and cornea enhancer

Neural tube enhancer

Retina enhancer

Exons:

Promoter   1   2   3   4   5 5a 6   7

(B)

5′···CCCTTTATTGATTGACAGAAGCTGG···3′

Pbx1-binding sequence

Meis-binding sequence

(C)

β-galactosidase

FIGURE 2.9 Enhancer region modularity. (A) The gene for Pax6, a protein critical in the development of a number of widely different tissues, has several enhancer elements (green). These enhancers direct *Pax6* expression (yellow exons 1–7) differentially in the pancreas, the lens and cornea of the eye, the retina, and the neural tube. (B) A portion of the DNA sequence of the pancreas-specific enhancer element. This sequence has binding sites for the Pbx1 and Meis transcription factors; both must be present in order to activate the *Pax6* gene in the pancreas. (C) When the gene for bacterial β-galactosidase is fused to *Pax6* enhancers for expression in the pancreas and the lens/cornea, this enzyme (which is easily stained) can be seen in those tissues. (C from Williams et al. 1998, courtesy of R. A. Lang.)

bind transcription factors, and it is this combination of transcription factors that activates the gene. For instance, the pancreas-specific enhancer of the *Pax6* gene has binding sites for the Pbx1 and Meis transcription factors (see Figure 2.9A). Both need to be present in order for the enhancer to activate *Pax6* in the pancreas cells (Zang et al. 2006).

Moreover, the product of the *Pax6* gene encodes a transcription factor that works in combinatorial partnerships with other transcription factors. Figure 2.10 shows two gene regulatory regions that bind Pax6. The first is that of the chick δ1 lens *crystallin* gene (**Figure 2.10A**; Cvekl and Piatigorsky 1996; Muta et al. 2002). This gene encodes crystallin, a lens protein that is transparent and allows light to reach the retina. A promoter within the *crystallin* gene contains a site for TBP binding, and an upstream promoter element that binds Sp1 (a general transcriptional activator found in all cells). The gene also has an enhancer in its third intron that controls the time and place of crystallin expression. This enhancer has two Pax6-binding sites. The Pax6 protein works with the Sox2 and L-Maf transcription factors to activate the *crystallin* gene only in those head cells that are going to become lens. As we will see in Chapter 10, this involves the cell being head ectoderm (which has Pax6), being in the region of the ectoderm likely to form eyes (L-Maf), and being in contact with the future retinal cells (which induce Sox2 expression; Kamachi et al. 1998).

Meanwhile, another set of regulatory regions that use Pax6 are the enhancers regulating the transcription of the genes for insulin, glucagon, and somatostatin in the pancreas (**Figure 2.10B**). Here, Pax6 is also essential for gene expression, and it works in cooperation with other transcription factors such as Pdx1 (specific for the pancreatic region of the endoderm) and Pbx1 (Andersen et al. 1999; Hussain and Habener 1999). In the absence of Pax6 (as in

the homozygous *small eye* mutation in mice and rats), the endocrine cells of the pancreas do not develop properly and the production of hormones by those cells is deficient (Sander et al. 1997; Zhang et al. 2002).

There are other genes that are activated by Pax6 binding, and one of them is the *Pax6* gene itself. Pax6 protein can bind to a *cis*-regulatory element of the *Pax6* gene (Plaza et al. 1993). This means that once the *Pax6* gene is turned on, it will continue to be expressed, even if the signal that originally activated it is no longer given.

One can see that the genes for specific proteins use numerous transcription factors in various combinations. Thus, *enhancers are modular* (such that the *Pax6* gene is expressed in the eye, pancreas, and nervous system, as shown in Figure 2.9); but *within each cis-regulatory module, transcription factors work in a combinatorial fashion* (such that Pax6, L-Maf, and Sox2 are all needed for the transcription of crystallin in the lens). The combinatorial association of transcription factors on enhancers leads to the spatiotemporal output of any particular gene (see Davidson 2006; Zinzen et al. 2009).

Transcription factor function

Natalie Angier (1992) has written, "A series of new discoveries suggests that DNA is more like a certain type of politician, surrounded by a flock of protein handlers and advisers that must vigorously massage it, twist it, and on occasion, reinvent it before the grand blueprint of the body can make any sense at all." These "handlers and advisers" are the transcription factors. These factors can be grouped together in families based on similarities in structure (Table 2.1). The transcription factors within such a family share a common framework in their DNA-binding sites, and slight

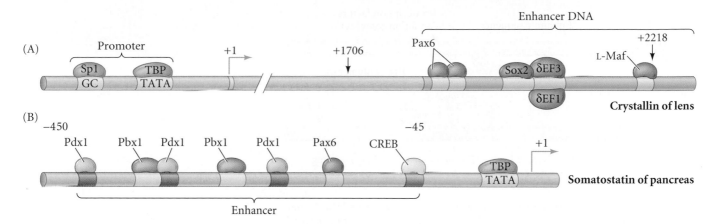

FIGURE 2.10 Modular transcriptional regulatory regions using Pax6 as an activator. (A) Promoter and enhancer of the chick δ1 lens *crystallin* gene. Pax6 interacts with two other transcription factors, Sox2 and L-Maf, to activate this gene. The protein δEF3 binds factors that permit this interaction; δEF1 binds factors that inhibit it. (B) Promoter and enhancer of the rat somatostatin gene. Pax6 activates this gene by cooperating with the Pbx1 and Pdx1 transcription factors. (A after Cvekl and Piatigorsky 1996; B after Andersen et al. 1999.)

differences in the amino acids at the binding site can cause the binding site to recognize different DNA sequences.

As we have already seen, enhancers function by binding transcription factors, and each enhancer can have binding sites for several transcription factors. Transcription factors bind to the enhancer DNA with one part of the protein and use other sites on the protein to interact with one another to recruit histone-modifying enzymes.

For example, the association of the Pax6, Sox2, and L-Maf transcription factors in lens cells recruits a histone acetyltransferase that can transfer acetyl groups to the histones and dissociate the nucleosomes in that area (Yang et al. 2006). Similarly, when MITF, a transcription factor essential for ear development and pigment production, binds to its specific DNA sequence, it also binds a (different) histone acetyltransferase that also facilitates the dissociation of nucleosomes (Ogryzko et al. 1996; Price et al. 1998). And the Pax7 transcription factor that activates muscle-specific genes binds to the enhancer region of these genes within the muscle precursor cells. Pax7 then recruits a histone methyltransferase that methylates the lysine in the fourth position of histone H3 (H3K4), resulting in the trimethylation of this lysine and the activation of transcription (McKinnell et al. 2008). The displacement of nucleosomes along the DNA makes it possible for the transcription fac-

TABLE 2.1 Some major transcription factor families and subfamilies		
Family	**Representative transcription factors**	**Some functions**
Homeodomain:		
Hox	Hoxa1, Hoxb2, etc.	Axis formation
POU	Pit1, Unc-86, Oct-2	Pituitary development; neural fate
LIM	Lim1, Forkhead	Head development
Pax	Pax1, 2, 3, 6, etc.	Neural specification; eye development
Basic helix-loop-helix (bHLH)	MyoD, MITF, daughterless	Muscle and nerve specification; *Drosophila* sex determination; pigmentation
Basic leucine zipper (bZip)	C/EBP, AP1	Liver differentiation; fat cell specification
Zinc finger:		
Standard	WT1, Krüppel, Engrailed	Kidney, gonad, and macrophage development; *Drosophila* segmentation
Nuclear hormone receptors	Glucocorticoid receptor, estrogen receptor, testosterone receptor, retinoic acid receptors	Secondary sex determination; craniofacial development; limb development
Sry-Sox	Sry, SoxD, Sox2	Bend DNA; mammalian primary sex determination; ectoderm differentiation

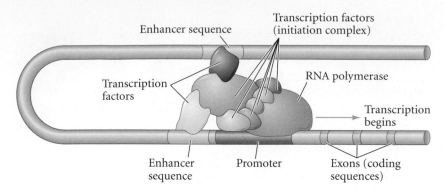

Enhancer sequence

Transcription factors
(initiation complex)

Transcription
factors

RNA polymerase

Transcription
begins

Enhancer
sequence

Promoter

Exons (coding
sequences)

FIGURE 2.11 RNA polymerase is stabilized on the promoter site of the DNA by transcription factors recruited by the enhancers. The TATA sequence at the promoter binds a protein that serves as a "saddle" for RNA polymerase. However, RNA polymerase would not remain bound long enough to initiate transcription were it not for the stabilization by the transcription factors.

tors to find their binding sites (Adkins et al. 2004; Li et al. 2007).

In addition to recruiting nucleosome modifying enzymes, transcription factors can also work by stabilizing the transcription preinitiation complex that enables RNA polymerase to bind to the promoter (**Figure 2.11**). For instance, MyoD, a transcription factor that is critical for muscle cell development (see Chapter 11), stabilizes TFIIB, which supports RNA polymerase at the promoter site (Heller and Bengal 1998). Indeed, MyoD plays several roles in activating gene expression, since it also can bind histone acetyltransferases that initiate nucleosome remodeling and dissociation (Cao et al. 2006).

One of the important consequences of the combinatorial association of transcription factors is **coordinated gene expression**. The simultaneous expression of many cell-specific genes can be explained by the binding of transcription factors by the enhancer elements. For example, many genes that are specifically activated in the lens contain an enhancer that binds Pax6. This means that all the other transcription factors might be assembled at the enhancer, but until Pax6 binds, they cannot activate the gene. Similarly, many of the co-expressed muscle-specific genes contain enhancers that bind the MEF2 transcription factor; and the enhancers on genes encoding pigment-producing enzymes bind MITF (see Davidson 2006).

TRANSCRIPTION FACTOR DOMAINS Transcription factors have three major domains. The first is a **DNA-binding domain** that recognizes a particular DNA sequence in the enhancer. **Figure 2.12** shows a model of such a domain in the Pax6 protein described earlier (see Figure 2.9). The second is a ***trans*-activating domain** that activates or suppresses the transcription of the gene whose promoter or enhancer it has bound. Usually, this *trans*-activating domain enables the transcription factor to interact with the

proteins involved in binding RNA polymerase (such as TFIIB or TFIIE; see Sauer et al. 1995) or with enzymes that modify histones. In addition, there may be a **protein-protein interaction domain** that allows the transcription factor's activity to be modulated by TAFs or other transcription factors.

MITF, a transcription factor essential for ear development and pigment production, has a protein-protein interaction domain that enables it to dimerize with another MITF protein (Ferré-D'Amaré et al. 1993). The resulting homodimer (i.e., two identical protein molecules bound

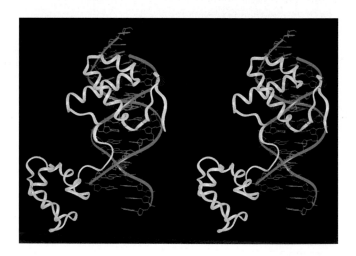

FIGURE 2.12 Stereoscopic model of Pax6 protein binding to its enhancer element in DNA. The DNA-binding region of Pax6 is shown in yellow; the DNA double helix is blue. Red dots indicate the sites of loss-of-function mutations in the *Pax6* gene that give rise to nonfunctional Pax6 proteins. It is worth trying to cross your eyes to see the central three-dimensional figure. (From Xu et al. 1995; photograph courtesy of S. O. Pääbo.)

SIDELIGHTS & *SPECULATIONS*

Reprogramming Cells: Changing Cell Differentiation through Embryonic Transcription Factors

The importance and power of transcription factors were elegantly demonstrated when Zhou and colleagues (2008) used three transcription factors to convert *exocrine* pancreatic cells (which make amylase, chymotrypsin, and other digestive enzymes) into insulin-secreting *endocrine* pancreatic β cells. The researchers infected the pancreases of living 2-month-old mice with harmless viruses containing the genes for three transcription factors: Pdx1, Ngn3, and Mafa.

The Pdx1 protein stimulates the outgrowth of the digestive tube that results in the pancreatic buds. This protein is found throughout the pancreas and is critical in specifying that organ's endocrine cells, as well as in activating genes that encode endocrine proteins (see Figure 2.10B). Ngn3 is a transcription factor found in endocrine, but not exocrine, pancreatic cells. Mafa, a transcription factor regulated by glucose levels, is found only in pancreatic β cells (i.e., those cells that make insulin) and can activate transcription of the insulin gene.

Pdx1, Ngn3, and Mafa activate other transcription factors that work in concert to turn a pancreatic endodermal cell into an insulin-secreting β cell. Zhou and colleagues found that, of all the transcription factor genes tested, these three were the only ones that were crucial for the conversion (**Figure 2.13**). Converted pancreas cells looked identical to normal β cells, and like normal β cells, they were capable of secreting both insulin and a blood vessel-inducing factor. The converted cells retained their new properties for months. Moreover, mice whose normal β cells were destroyed by a particular drug developed diabetes, just like the Type 1 diabetes seen when human adults lose β cells. This diabetes could be cured by injecting the mice with viruses containing the three transcrip-

tion factors. When this was done, about 20% of the exocrine pancreatic cells became β cells and secreted insulin.

This study opens the door to an entire new field of regenerative medicine, illustrating the possibilities of

changing one adult cell type into another by using the transcription factors that had made the new cell type in the embryo.

See VADE MECUM
Transdetermination in *Drosophila*

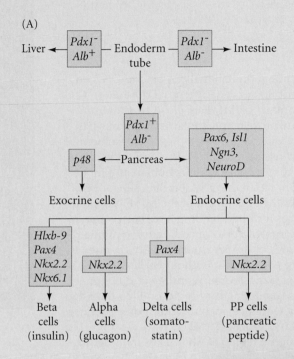

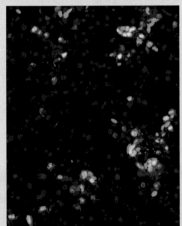

Figure 2.13 Pancreatic lineage and transcription factors. (A) Pdx1 protein is expressed in pancreatic lineages. Several transcription factors, including Ngn3, distinguish the endocrine lineage. Several other transcription factors, including Mafa, are found in the β cells that produce insulin. (B) New pancreatic β cells arise in adult mouse pancreas in vivo after viral delivery of three transcription factors (Ngn3, Pdx1, and Mafa). Virally infected exocrine cells are detected by their expression of nuclear green fluorescent protein. Newly induced β cells are recognized by insulin staining (red). Their overlap produces yellow. The nuclei of all pancreatic cells are stained blue. (B courtesy of D. Melton.)

FIGURE 2.14 Three-dimensional model of the homodimeric transcription factor MITF (one protein shown in red, the other in blue) binding to a promoter element in DNA (white). The amino termini are located at the bottom of the figure and form the DNA-binding domain that recognizes an 11-base-pair sequence of DNA having the core sequence CATGTG. The protein-protein interaction domain is located immediately above. MITF has the basic helix-loop-helix structure found in many transcription factors. The carboxyl end of the molecule is thought to be the *trans*-activating domain that binds the p300/CBP co-activator protein. (From Steingrímsson et al. 1994, courtesy of N. Jenkins.)

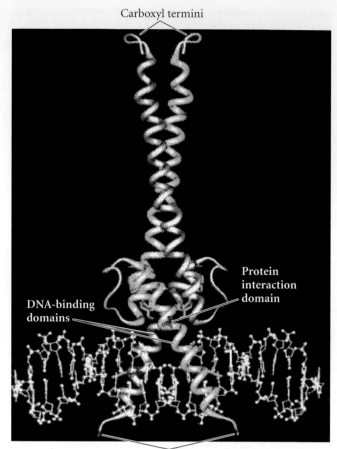

Carboxyl termini

Protein
interaction
domain

DNA-binding
domains

Amino termini

together) is a functional protein that can bind to DNA and activate the transcription of certain genes (**Figure 2.14**). The DNA-binding domain of MITF is close to the amino-terminal end of the protein and contains numerous basic amino acids that make contact with the DNA (Hemesath et al. 1994; Steingrímsson et al. 1994). This assignment was confirmed by the discovery of various human and mouse mutations that map within the DNA-binding site for MITF and which prevent the attachment of the MITF protein to the DNA. Sequences for MITF binding have been found in the regulatory regions of genes encoding three pigment-cell-specific enzymes of the tyrosinase family (Bentley et al. 1994; Yasumoto et al. 1997). Without MITF, these proteins are not synthesized properly, and melanin pigment is not made. These *cis*-regulatory regions all contain the same 11-base-pair sequence, including the core sequence (CATGTG) that is recognized by MITF.

The third functional region of MITF is its *trans*-activating domain. This domain includes a long stretch of amino acids in the center of the protein. When the MITF dimer is bound to its target sequence in the enhancer, the *trans*-activating region is able to bind a TAF, p300/CBP. The p300/CPB protein is a histone acetyltransferase enzyme that can transfer acetyl groups to each histone in the nucleosomes (Ogryzko et al. 1996; Price et al. 1998). Acetylation of the nucleosomes destabilizes them and allows the genes for pigment-forming enzymes to be expressed.

EPIGENETIC MEMORY: KEEPING THE RIGHT GENES ON OR OFF
The modifications of histones can also signal the recruitment of the proteins that can retain the memory of transcriptional state from generation to generation through mitosis. These are the proteins of the Trithorax and Polycomb families. When bound to the nucleosomes of active genes, **Trithorax** proteins keep these genes active, whereas **Polycomb** proteins, which bind to condensed nucleosomes, keep the genes in an inactive state.

The Polycomb proteins fall into two categories that act sequentially in repression. The first set has histone methyltransferase activities that methylate lysines H3K27 and H3K9 to repress gene activity. In many organisms, this repressive state is stabilized by the activity of a second set of Polycomb factors, which bind to the methylated tails of histone 3 and keep the methylation active and also methylate adjacent nucleosomes, thereby forming tightly packed repressive complexes (Grossniklaus and Paro 2007; Margueron et al. 2009).

The Trithorax proteins help retain the memory of activation; they act to counter the effect of the Polycomb proteins. Trithorax proteins can modify the nucleosomes or alter their positions on the chromatin, allowing transcription factors to bind to the DNA previously covered by them. Other Trithorax proteins keep the H3K4 lysine trimethylated (preventing its demethylation into a dimethylated, repressive, state; Tan et al. 2008).

PIONEER TRANSCRIPTION FACTORS: BREAKING THE SILENCE
Finding a promoter is not easy, because the DNA is usually so wound up that the promoter sites are not accessible. Indeed, more than *6 feet* of DNA is packaged into chromosomes of each human cell nucleus (Schones and Zhao 2008).

How can a transcription factor find its binding site, given that the enhancer might be covered by nucleosomes? Several studies have identified certain transcription factors that penetrate repressed chromatin and bind to their enhancer DNA sequences (Cirillo et al. 2002;

FIGURE 2.15 Model for the role of the "pioneer" transcription factor Pbx in aligning the muscle-specific transcription factor MyoD on DNA. (A) Pbx protein recognizes its DNA binding site (TGAT), even within nucleosome-rich chromatin. Pbx probably binds to transcriptional inhibitors. (B) MyoD, complexed with its E12 cofactor, is able to bind to Pbx, replacing the transcriptional inhibitors. MyoD then binds to its recognition element on the DNA. (C) The MyoD/E12 complex can then recruit the histone acetyltransferases and nucleosome remodeling compounds that make the chromatin accessible to other transcription factors (Mef3 and Mef2) and to RNA polymerase. (After Tapscott 2005.)

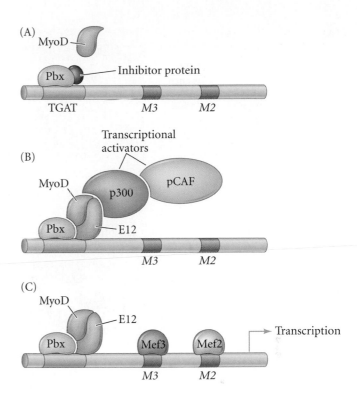

Berkes et al. 2004). They have been called "pioneer" transcription factors, and they appear to be critical in establishing certain cell lineages. One of these transcription factors is FoxA1, which binds to certain enhancers and opens up the chromatin to allow other transcription factors access to the promoter (Lupien et al. 2008). FoxA proteins remain bound to the DNA during mitosis, providing a mechanism to re-establish normal transcription in presumptive liver cells (Zaret et al. 2008). Another pioneer transcription factor is the Pax7 protein mentioned above. It activates muscle-specific gene transcription in a population of muscle stem cells by binding to its DNA recognition sequence and being stabilized there by dimethylated H3K4 on the nucleosomes. It then recruits the histone methyltransferase that converts the dimethylated H3K4 into the trimethylated H3K4 associated with active transcription (McKinnell et al. 2008).

Another pioneer transcription factor in muscle development is Pbx. Members of the Pbx family are made in every cell, and they are able to find their appropriate sites even in highly compacted chromatin. Pbx appears to be used as a "molecular beacon" for another muscle-determining transcription factor, MyoD (mentioned earlier). MyoD is critical for initiating muscle development in the embryo, activating hundreds of genes that are involved with establishing the muscle phenotype. However, MyoD is not able to bind to DNA without the help of Pbx proteins, which bind to DNA elements adjacent to the DNA sequence recognized by MyoD (**Figure 2.15A**). Berkes and colleagues (2004) have shown that MyoD (when complexed with another transcription factor, E12) can bind to the Pbx protein and align itself on its target DNA sequence (**Figure 2.15B**). Once bound there, the E12 protein recruits histone acetyltransferases and nucleosome remodeling complexes to open up the DNA on those genes (**Figure 2.15C**).

SILENCERS Silencers are DNA regulatory elements that actively repress the transcription of a particular gene. They can be viewed as "negative enhancers." For instance, in the mouse, there is a DNA sequence that prevents a promoter's activation in any tissue *except* neurons. This

sequence, given the name **neural restrictive silencer element** (**NRSE**), has been found in several mouse genes whose expression is limited to the nervous system: those encoding synapsin I, sodium channel type II, brain-derived neurotrophic factor, Ng-CAM, and L1. The protein that binds to the NRSE is a zinc finger transcription factor called **neural restrictive silencer factor** (**NRSF**). (It is also called REST). NRSF appears to be expressed in every cell that is *not* a mature neuron (Chong et al. 1995; Schoenherr and Anderson 1995).

To test the hypothesis that the NRSE sequence is necessary for the normal repression of neural genes in non-neural cells, transgenes were made by fusing a β-galactosidase (*lacZ*) gene with part of the *L1* neural cell adhesion gene. (L1 is a protein whose function is critical for brain development, as we will see in later chapters.) In one case, the *L1* gene, from its promoter through the fourth exon, was fused to the *lacZ* sequence. A second transgene was made just like the first, except that the NRSE was deleted from the *L1* promoter. The two transgenes were separately inserted into the pronuclei of fertilized oocytes, and the resulting transgenic mice were analyzed for β-galactosidase expression (Kallunki et al. 1995, 1997). In embryos receiving the complete transgene (which included the NSRE), expression was seen only in the nervous system (**Figure 2.16A**). In mice whose transgene lacked the NRSE sequence, however, expression was seen in the heart, the limb mesenchyme and limb ectoderm, the kidney mesoderm, the ventral body wall, and the cephalic mesenchyme (**Figure 2.16B**).

(A)

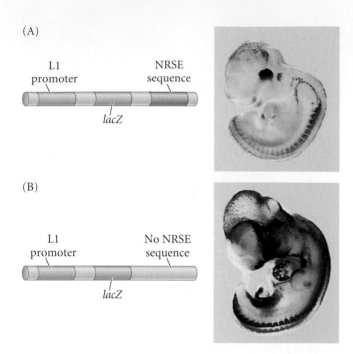

(B)

FIGURE 2.16   Silencers. Analysis of β-galactosidase staining patterns in 11.5-day embryonic mice. (A) Embryo containing a transgene composed of the L1 promoter, a portion of the *L1* gene, and a *lacZ* gene fused to the second exon (which contains the NRSE region). (B) Embryo containing a similar transgene but lacking the NRSE sequence. The dark areas reveal the presence of β-galactosidase (the *lacZ* product). (Photographs from Kallunki et al. 1997.)

DNA Methylation and the Control of Transcription

How does a pattern of gene transcription become stable? How can a lens cell continue to remain a lens cell and not activate muscle-specific genes? How can cells undergo rounds of mitosis and still maintain their differentiated characteristics? The answer appears to be **DNA methylation**. We have already discussed *histone* methylation and

its importance for transcription. Now we look at how the *DNA itself* can be methylated to regulate transcription. Generally speaking, the promoters of inactive genes become methylated at certain cytosine residues, and the resulting methylcytosine stabilizes nucleosomes and prevents transcription factors from binding.

It is often assumed that a gene contains exactly the same nucleotides whether it is active or inactive; that is, a β-globin gene that is activated in a red blood cell precursor has the same nucleotides as the inactive β-globin gene in a fibroblast or retinal cell of the same animal. However, it turns out there is in fact a subtle difference. In 1948, R. D. Hotchkiss discovered a "fifth base" in DNA, 5-methylcytosine. In vertebrates, this base is made enzymatically after DNA is replicated. At this time, about 5% of the cytosines in mammalian DNA are converted to 5-methylcytosine (**Figure 2.17A**). This conversion can occur only when the cytosine residue is followed by a guanosine—in other words, at a CpG sequence (as we will soon see, this restriction is important). Numerous studies have shown that the degree to which the cytosines of a gene are methylated can control the level of the gene's transcription. Cytosine methylation appears to be a major mechanism of transcriptional regulation among vertebrates; however, some other species (*Drosophila* and nematodes among them) do not methylate their DNA.

In vertebrates, the presence of methylated cytosines in a gene's promoter correlates with the repression of transcription from that gene. In developing human and chick red blood cells, for example, the DNA of the globin gene promoters is almost completely unmethylated, whereas the same promoters are highly methylated in cells that do not produce globins. Moreover, the methylation pattern changes during development (**Figure 2.17B**). The cells that produce hemoglobin in the human embryo have unmethylated promoters in the genes encoding the ε-globins ("embryonic globin chains") of embryonic hemoglobin. These promoters become methylated in the fetal tissue, as the genes for fetal-specific γ-globin (rather than the embryonic chains) become activated (van der Ploeg and Flavell 1980; Groudine and Weintraub 1981; Mavilio et al. 1983).

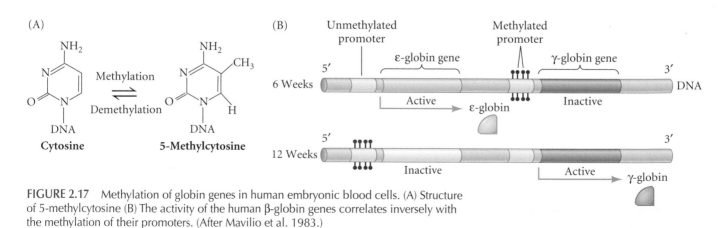

FIGURE 2.17   Methylation of globin genes in human embryonic blood cells. (A) Structure of 5-methylcytosine (B) The activity of the human β-globin genes correlates inversely with the methylation of their promoters. (After Mavilio et al. 1983.)

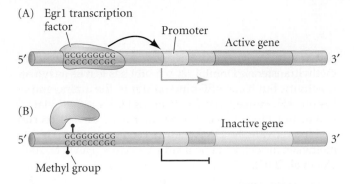

(A) Egr1 transcription factor

5′ ... GCGGGGGCG / CGCCCCCGC ... Promoter — Active gene ... 3′

(B)

5′ ... GCGGGGGCG / CGCCCCCGC ... Inactive gene ... 3′

Methyl group

FIGURE 2.18 DNA methylation can block transcription by preventing transcription factors from binding to the enhancer region. (A) The Egr1 transcription factor can bind to specific DNA sequences such as 5′...GCGGGGGCG...3′, helping to activate transcription of those genes. (B) If the first cytosine residue is methylated, however, Egr1 will not bind and the gene will remain repressed. (After Weaver et al. 2005.)

Similarly, when fetal globin gives way to adult (β) globin, promoters of the fetal (γ) globin genes become methylated.

Mechanisms by which DNA methylation blocks transcription

DNA methylation appears to act in two ways to repress gene expression. First, it can block the binding of transcription factors to enhancers. Several transcription factors can bind to a particular sequence of unmethylated DNA, but they cannot bind to that DNA if one of its cytosines is methylated (**Figure 2.18**). Second, a methylated cytosine can recruit the binding of proteins that facilitate the methylation or deacetylation of histones, thereby stabilizing the nucleosomes. For instance, methylated cytosines in DNA can bind particular proteins such as MeCP2. Once connected to a methylated cytosine, MeCP2 binds to histone deacetylases and histone methyltransferases, which, respectively, remove acetyl groups (**Figure 2.19A**) and add methyl groups (**Figure 2.19B**) on the histones. As a result, the nucleosomes form tight complexes with the DNA and don't allow other transcription factors and RNA polymerases to find the genes. Other proteins, such as HP1 and histone H1, will bind and aggregate methylated histones (Fuks 2005; Rupp and Becker 2005). In this way, repressed chromatin becomes associated with regions where there are methylated cytosines.

Inheritance and stabilization of DNA methylation patterns

Another enzyme recruited to the chromatin by MeCP2 is DNA methyltransferase-3 (Dnmt3). This enzyme methylates previously unmethylated cytosines on the DNA. In this way, a relatively large region can be repressed. The newly established methylation pattern is then transmitted to the next generation by DNA methyltransferase-1 (Dnmt1). This enzyme recognizes methyl cytosines on one

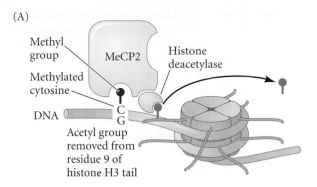

(A)

Methyl group — MeCP2 — Histone deacetylase

Methylated cytosine

DNA

Acetyl group removed from residue 9 of histone H3 tail

(B)

MeCP2 — Histone methyltransferase — Adaptor

DNA

Methyl group added to residue 9 of histone H3 tail

FIGURE 2.19 Modifying nucleosomes through methylated DNA. MeCP2 recognizes the methylated cytosines of DNA. It binds to the DNA and is thereby able to recruit histone deacetylases (which take acetyl groups off the histones) (A) or histone methyltransferases (which add methyl groups to the histones) (B). Both modifications promote the stability of the nucleosome and the tight packing of DNA, thereby repressing gene expression in these regions of DNA methylation. (After Fuks 2003.)

strand of DNA and places methyl groups on the newly synthesized strand opposite it (**Figure 2.20**; see Bird 2002; Burdge et al. 2007). This is why it is necessary for the C to be next to a G in the sequence. Thus, in each cell division, the pattern of DNA methylation can be maintained. The newly synthesized (unmethylated) strand will become

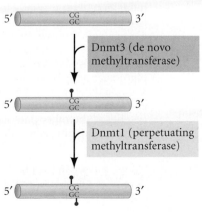

5′ CG / GC 3′

Dnmt3 (de novo methyltransferase)

5′ CG / GC 3′

Dnmt1 (perpetuating methyltransferase)

5′ CG / GC 3′

FIGURE 2.20 Two DNA methyltransferases are critically important in modifying DNA. The "de novo" methyltransferase Dnmt3 can place a methyl group on unmethylated cytosines. The "perpetuating" methyltransferase, Dnmt1, recognizes methylated Cs on one strand and methylates the C on the CG pair on the opposite strand.

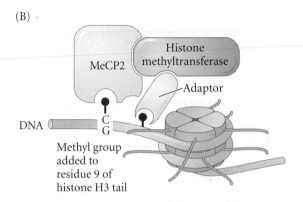

properly methylated when Dnmt1 binds to a methylC on the old CpG sequence and methylates the cytosine of the CpG sequence on the complementary strand. In this way, once the DNA methylation pattern is established in a cell, it can be stably inherited by all the progeny of that cell.

Reinforcement between repressive chromatin and repressive DNA has also been observed. Just as methylated DNA is able to attract proteins that deacetylate histones and attract H1 linker histones (both of which will stabilize nucleosomes), so repressive states of chromatin are able to recruit enzymes that methylate DNA. DNA methylation

patterns during gametogenesis depend in part on the DNA methyltransferase Dnmt3L. It actually has lost its enzymatic activity, but it can still bind avidly to the amino end of histone H3. However, if the lysine at H3K4 is methylated, it will not bind. Once bound, however, it will recruit and/or activate the DNA methyltransferase Dnmt3A2 to methylate the cytosines on nearby CG pairs (Fan et al. 2007; Ooi et al. 2007).

See WEBSITE 2.7
Silencing large blocks of chromatin

SIDELIGHTS & SPECULATIONS

Consequences of DNA Methylation

The control of transcription through DNA methylation has many consequences in addition to cell differentiation. DNA methylation has explained X chromosome inactivation and DNA imprinting. Moreover, as we will see in Chapter 18, improper DNA methylation (when the wrong cytosines are methylated or demethylated) has been associated with aging, cancers, and the poor health of cloned animals.

X chromosome inactivation

In *Drosophila*, nematodes, and mammals, females are characterized as having two X chromosomes per cell, while males are characterized as having a single X chromosome per cell. Unlike the Y chromosome, the X chromosome contains thousands of genes that are essential for cell activity. Yet despite the female's cells having double the number of X chromosomes, male and female cells contain approximately equal amounts of X chromosome-encoded gene products. This equalization phenomenon is called **dosage compensation**, and it can be accomplished in three ways (Migeon 2002). In *Drosophila*, the transcription rate of the male X chromosomes is doubled so that the single male X chromosome makes the same amount of transcript as the two female X chromosomes (Lucchesi and Manning 1987). This is accomplished by acetylation of the nucleosomes throughout the male's X chromosomes, which gives RNA poly-

merase more efficient access to that chromosome's promoters (Akhtar et al. 2000; Smith et al. 2001). In *C. elegans*, both X chromosomes are partially repressed (Chu et al. 2002) so that the male and female* products of the X chromosomes are equalized.

In mammals, dosage compensation occurs through the inactivation of one X chromosome in each female cell. Thus, each mammalian somatic cell, whether male or female, has only one functioning X chromosome. This phenomenon is called **X chromosome inactivation**. The chromatin of the inactive X chromosome is converted into **heterochromatin**—chromatin that remains condensed throughout most of the cell cycle and replicates later than most of the other chromatin (the **euchromatin**) of the nucleus. This was first shown by Mary Lyon (1961), who observed coat color patterns in mice. If a mouse is heterozygous for an autosomal gene controlling hair pigmentation, then it resembles one of its two parents, or has a color intermediate between the two. In either case, the mouse is a single color. But if a female mouse is heterozygous for a pigmentation gene *on the X chromosome*, a different result is seen: patches of one parental color alternate with patches of the other parental color. This also explains why calico and tor-

toiseshell cats[†] are normally female: their coat color alleles (black and orange) are on the X chromosome (Centerwall and Benirscke 1973).

Lyon proposed the following hypothesis to account for these results:

1. Very early in the development of female mammals, both X chromosomes are active. As development proceeds, one X chromosome is inactivated in each cell (**Figure 2.21A**).

2. This inactivation is random. In some cells, the paternally derived X chromosome is inactivated; in other cells, the maternally derived X chromosome is shut down.

3. This process is irreversible. Once a particular X chromosome (either the one derived from the mother or the one derived from the father) has been inactivated in a cell, the same X chromosome is inactivated in all of that cell's progeny (**Figure 2.21B,C**). Because X inactivation happens relatively early in development, an entire region of cells derived from a single cell may all have the same X chromosome inactivated. Thus, all tissues in female mammals are mosaics of two cell types.

[†]Although the terms *calico* and *tortoiseshell* are sometimes used synonymously, tortoiseshell coats are a patchwork of black and orange only; calico cats usually have white patches—i.e., patches with no pigment—as well (see Figure 2.2).

*As we will see in Chapter 5, the "female" is actually a hermaphrodite capable of making both sperm and eggs.

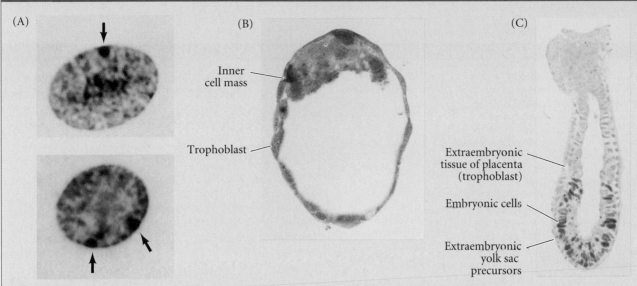

Figure 2.21 X chromosome inactivation in mammals. (A) Inactivated X chromosomes, or *Barr bodies*, in the nuclei of human oral epithelial cells. The top cell is from a normal XX female and has a single Barr body (arrow). In the lower cell, from a female with three X chromosomes, two Barr bodies can be seen. In both cases, only one X chromosome per cell is active. (B,C) The paternally derived X chromosome of this mouse embryo contained a *lacZ* transgene. Those cells in which the chromosome is active make β-galactosidase and stain blue. The other cells are counterstained and appear pink. (B) In the early blastocyst stage (day 4), both X chromosomes are active in all cells. (C) At day 6, random inactivation of one of the chromosomes occurs. Embryonic cells in which the maternal X is active appear pink, while those where the paternal X is active stain blue. In mouse (but not human) trophoblast, the paternally derived X chromosome is preferentially inactivated, so the trophoblast cells are uniformly pink. (A courtesy of M. L. Barr; B,C from Sugimoto et al. 2000, courtesy of N. Takagi.)

The inactivation of the X chromosome is complicated; indeed, it is a bottleneck that many female embryos do not get through (Migeon 2007). The mechanisms of X chromosome inactivation appear to differ between mammalian groups, but these mechanisms converge in that they all inactivate an X chromosome by methylating promoters. In mice and humans, the promoter regions of numerous genes are methylated on the inactive X chromosome and unmethylated on the active X chromosome (Wolf et al. 1984; Keith et al. 1986; Migeon et al. 1991). The memory of this "X inactivation" is transmitted to the progeny of the cells by successive DNA methylation through Dnmt1 (see above).

Genomic Imprinting

The second phenomenon explained by DNA methylation is genomic imprinting. It is usually assumed that the genes one inherits from one's father and the genes one inherits from one's mother are equivalent. In fact, the basis for Mendelian ratios (and the Punnett square analyses used to teach them) is that it does not matter whether the genes came from the sperm or from the egg. But in mammals, there are at least 80 genes for which it *does*

matter.* Here, the chromosomes from the male and the female are not equivalent. In these cases, only the sperm-derived or only the egg-derived allele of the gene is expressed. This means that a severe or lethal condition arises if a mutant allele is derived from one parent, but that the same mutant allele will have no deleterious effects if inherited from the other parent. In some of these cases, the nonfunctioning gene has been rendered inactive by DNA methylation. (This means that a mammal must have both a male parent and a female parent. Unlike sea urchins, flies, and frogs, mammals cannot experience parthenogenesis, or "virgin birth.") The methyl groups are placed on the DNA during spermatogenesis and oogenesis by a series of enzymes that first take the existing methyl groups off the chromatin and then place new sex-specific ones on the DNA (Ciccone et al. 2009).

As described earlier in this chapter, methylated DNA is associated with stable DNA silencing, either (1) by interfering with the binding of gene-activat-

*A list of imprinted mouse genes is maintained at http://www.har.mrc.ac.uk/research/genomic_imprinting/introduction.html

ing transcription factors or (2) by recruiting repressor proteins that stabilize nucleosomes in a restrictive manner along the gene. The presence of a methyl group in the minor groove of DNA can prevent certain transcription factors from binding to the DNA, thereby preventing the gene from being activated (Watt and Molloy 1988).

For example, during early embryonic development in mice, the *Igf2* gene (for insulin-like growth factor) is active only from the father's chromosome 7. The egg-derived *Igf2* gene does not function during embryonic development. This is because the CTCF protein is an inhibitor that can block the promoter from getting activation signals from enhancers. It binds to a region near the *Igf2* gene in females because this region is not methylated. Once bound, it prevents the maternally derived *Igf2* gene from functioning. In the sperm-derived chromosome 7, the region where CTCF would bind is methylated. CTCF cannot bind and the gene is not inhibited from functioning (**Figure 2.22**; Bartolomei et al. 1993; Ferguson-Smith et al. 1993; Bell and Felsenfeld 2000).

In humans, misregulation of *Igf2* methylation causes Beckwith-

(*Continued on next page*)

SIDELIGHTS & SPECULATIONS (Continued)

Figure 2.22 Regulation of the imprinted *Igf2* gene in the mouse. This gene is activated by an enhancer element it shares with the *H19* gene. The differentially methylated region (DMR) is a sequence located between the enhancer and the *Igf2* gene, and is found on both sperm- and egg-derived chromosomes. (A) In the egg-derived chromosome, the DMR is unmethylated. The CTCF insulator protein binds to the DMR and blocks the enhancer signal. (B) In the sperm-derived chromosome, the DMR is methylated. The CTCF insulator protein cannot bind to the methylated sequence, and the signal from the enhancer is able to activate *Igf2* transcription.

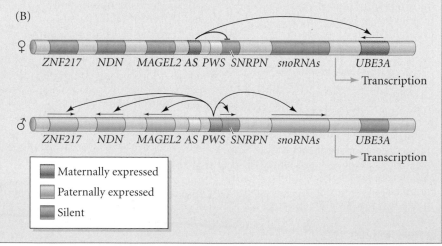

(A) Egg-derived (maternal) chromosome

CTCF insulator protein binds to unmethylated DMR

Enhancer   *H19*   DMR   *Igf2*

Transcription   No transcription

(B) Sperm-derived (paternal) chromosome

Methyl group

Enhancer   *H19*   DMR   *Igf2*

No transcription   Transcription

Wiedemann growth syndrome. Interestingly, although DNA methylation is the mechanism for imprinting this gene in both mice and humans, the mechanisms responsible for the differential *Igf2* methylation between sperm and egg appear to be very different in the two species (Ferguson-Smith et al. 2003; Walter and Paulsen 2003).

Also in humans, the loss of a particular segment of the long arm of chromosome 15 results in different phenotypes, depending on whether the loss is in the male- or the female-derived chromosome (**Figure 2.23A**). If the chromosome with the defective or missing segment comes from the father, the child is born with Prader-Willi syndrome, a disease associated with mild mental retardation, obesity, small gonads, and short stature. If the defective or missing segment comes from the

mother, the child has Angelman syndrome, characterized by severe mental retardation, seizures, lack of speech, and inappropriate laughter (Knoll et al. 1989; Nicholls 1998). The imprinted genes in this region are *SNRPN* and *UBE3A*. In the egg-derived chromosome, *UBE3A* is activated and *SNRPN* is turned off, while in the sperm-derived chromosome, *SNRPN* is activated and *UBE3A* is turned off (**Figure 2.23B**). The expression of either maternal or paternal loci on human chromo-

some 15 also depends on methylation differences at specific regions in the chromosome that regulate these genes (Zeschingk et al. 1997; Ferguson-Smith and Surani 2001; Walter and Paulsen 2003).

Differential methylation is one of the most important mechanisms of epigenetic changes. It provides a reminder that an organism cannot be explained solely by its genes. One needs knowledge of developmental parameters as well as genetic ones.

Figure 2.23 Inheritance patterns for Prader-Willi and Angelman syndromes. (A) A region in the long arm of chromosome 15 contains the genes whose absence causes both these syndromes. However, the two conditions are imprinted in reverse fashion. In Prader-Willi syndrome, the paternal genes are active; in Angelman syndrome, the maternal genes are active. (B) Some of the genes and the "inactivation centers" where methylation occurs on this chromosomal region. In the maternal chromosome, the *AS* inactivation center activates *UBE3A* and suppresses *SNRPN*. Conversely, on the paternal chromosome, the *PWS* inactivation center activates *SNRPN* and several other nearby genes, as well as making antisense RNA to *UBE3A*. (B after Walter and Paulsen 2003.)

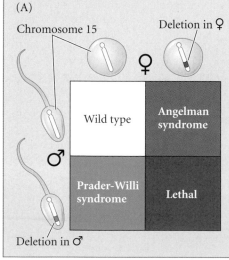

(A)

Chromosome 15

Deletion in ♀

	Wild type	Angelman syndrome
	Prader-Willi syndrome	Lethal

Deletion in ♂

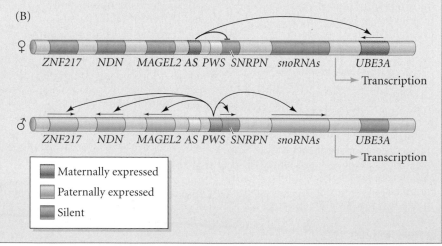

(B)

♀   ZNF217   NDN   MAGEL2   AS   PWS   SNRPN   snoRNAs   UBE3A
    Transcription

♂   ZNF217   NDN   MAGEL2   AS   PWS   SNRPN   snoRNAs   UBE3A
    Transcription

■ Maternally expressed
■ Paternally expressed
■ Silent

FIGURE 2.24 Roles of differential RNA processing during development. By convention, splicing paths are shown by fine V-shaped lines. (A) RNA selection, whereby the same nuclear RNA transcripts are made in two cell types, but the set that becomes cytoplasmic messenger RNA is different. (B) Differential splicing, whereby the same nuclear RNA is spliced into different mRNAs by selectively using different exons.

(A) RNA selection

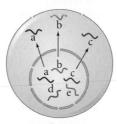

Cell type 1

Cell type 2

(B) Differential splicing

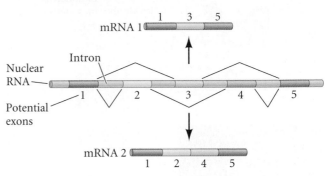

Differential RNA Processing

The regulation of gene expression is not confined to the differential transcription of DNA. Even if a particular RNA transcript is synthesized, there is no guarantee that it will create a functional protein in the cell. To become an active protein, the RNA must be (1) processed into a messenger RNA by the removal of introns, (2) translocated from the nucleus to the cytoplasm, and (3) translated by the protein-synthesizing apparatus. In some cases, the synthesized protein is not in its mature form and must be (4) posttranslationally modified to become active. Regulation during development can occur at any of these steps.

The essence of differentiation is the production of different sets of proteins in different types of cells. In bacteria, differential gene expression can be effected at the levels of transcription, translation, and protein modification. In eukaryotes, however, another possible level of regulation exists—namely, control at the level of RNA processing and transport. There are two major ways in which differential RNA processing can regulate development. The first involves "censorship"—selecting which nuclear transcripts are processed into cytoplasmic messages. Different cells select different nuclear transcripts to be processed and sent to the cytoplasm as messenger RNA. Thus, the same pool of nuclear transcripts can give rise to different populations of cytoplasmic mRNAs in different cell types (**Figure 2.24A**).

The second mode of differential RNA processing is the *splicing* of mRNA precursors into messages that specify different proteins by using different combinations of potential exons. If an mRNA precursor had five potential exons, one cell type might use exons 1, 2, 4, and 5; a different type might use exons 1, 2, and 3; and yet another cell type might use all five (**Figure 2.24B**). Thus a single gene can produce an entire family of proteins.

Control of early development by nuclear RNA selection

In the late 1970s, numerous investigators found that mRNA was not the primary transcript from the genes. Rather, the initial transcript is a nuclear RNA (nRNA). This nRNA is usually many times longer than the corresponding mRNA because nRNA contains introns that get spliced out during the passage from nucleus to cytoplasm (see Figure 2.6). Originally, investigators thought that whatever RNA was transcribed in the nucleus was processed into cytoplasmic mRNA. But studies of sea urchins showed that different cell types could be *transcribing* the same type of nuclear RNA, but *processing* different subsets of this population into mRNA in different types of cells (Kleene and Humphreys 1977, 1985). Wold and her colleagues (1978) showed that sequences present in sea urchin blastula *messenger* RNA, but absent in gastrula and adult tissue mRNA, were nonetheless present in the *nuclear* RNA of the gastrula and adult tissues.

More genes are transcribed in the nucleus than are allowed to become mRNAs in the cytoplasm. This "censoring" of RNA transcripts has been confirmed by probing for the introns and exons of specific genes. Gagnon and his colleagues (1992) performed such an analysis on the transcripts from the *SpecII* and *CyIIIa* genes of the sea urchin *Strongylocentrotus purpuratus*. These genes encode calcium-binding and actin proteins, respectively, which are expressed only in a particular part of the ectoderm of the sea urchin larva. Using probes that bound to an exon (which is included in the mRNA) and to an intron (which is not included in the mRNA), they found that these genes were being transcribed not only in the ectodermal cells, but also in the mesoderm and endoderm. The analysis of the *CyIIIa* gene showed that the concentration of introns was the same in both the gastrula ectoderm and the mesoderm/endoderm samples, suggesting that this gene was being transcribed at the same rate in the nuclei of all cell types, but was made into cytoplasmic mRNA only in ectodermal cells (**Figure 2.25**). The unprocessed nRNA for *CyIIIa* is degraded while still in the nuclei of the endodermal and mesodermal cells.

(A)

(C)

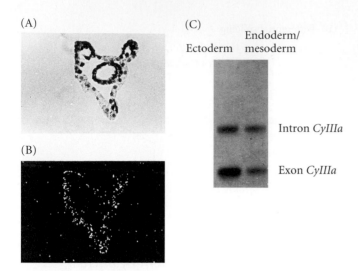

	Ectoderm	Endoderm/ mesoderm
Intron *CyIIIa*		
Exon *CyIIIa*		

(B)

FIGURE 2.25 Regulation of ectoderm-specific gene expression by RNA processing. (A,B) *CyIIIa* mRNA is seen by autoradiography to be present only in the ectoderm. (A) Phase contrast micrograph. (B) In situ hybridization using a probe that binds to a *CyIIIa* exon. (C) The *CyIIIa* nuclear transcript, however, is found in both ectoderm and endoderm/mesoderm. The left lane of the gel represents RNA isolated from the gastrula ectodermal tissue; the right lane represents RNA isolated from endodermal and mesodermal tissues. The upper band is the RNA bound by a probe that binds to an intron sequence (which should be found only in the nucleus) of *CyIIIa*. The lower band represents the RNA bound by a probe complementary to an exon sequence. The presence of the intron indicates that the *CyIIIa* nuclear RNA is being made in both groups of cells, even if the mRNA is seen only in the ectoderm. (From Gagnon et al. 1992, courtesy of R. and L. Angerer.)

Creating families of proteins through differential nRNA splicing

Alternative nRNA splicing is a means of producing a wide variety of proteins from the same gene. The average vertebrate nRNA consists of several relatively short exons (averaging about 140 bases) separated by introns that are usually much longer. Most mammalian nRNAs contain numerous exons. By splicing together different sets of exons, different cells can make different types of mRNAs, and hence, different proteins. Recognizing a sequence of nRNA as either an exon or an intron is a crucial step in gene regulation.

Alternative nRNA splicing is based on the determination of which sequences will be spliced out as introns. This can occur in several ways. Most genes contain "consensus sequences" at the 5′ and 3′ ends of the introns. These sequences are the "splice sites" of the intron. The splicing of nRNA is mediated through complexes known as **spliceosomes** that bind to the splice sites. Spliceosomes are made up of small nuclear RNAs (snRNAs) and proteins called **splicing factors** that bind to splice sites or to

the areas adjacent to them. By their production of specific splicing factors, cells can differ in their ability to recognize a sequence as an intron. That is to say, a sequence that is an *exon* in one cell type may be an *intron* in another (**Figure 2.26A,B**). In other instances, the factors in one cell might recognize different 5′ sites (at the beginning of the intron) or different 3′ sites (at the end of the intron; **Figure 2.26C,D**).

The 5′ splice site is normally recognized by small nuclear RNA U1 (U1 snRNA) and splicing factor 2 (SF2; also known as alternative splicing factor). The choice of alternative 3′ splice sites is often controlled by which splice site can best bind a protein called U2AF. The spliceosome forms when the proteins that accumulate at the 5′ splice site contact those proteins bound to the 3′ splice site. Once the 5′ and 3′ ends are brought together, the intervening intron is excised and the two exons are ligated together.

Researchers estimate that approximately 92% of human genes are alternatively spliced, and that such alternative splicing is a major way by which the rather limited number of genes can create a much larger array of proteins (Wang et al. 2008). The deletion of certain potential exons in some cells but not in others enables one gene to create a family of closely related proteins. Instead of one gene-one polypeptide, one can have one gene-one family of proteins. For instance, alternative RNA splicing enables the gene for α-tropomyosin to encode brain, liver, skeletal muscle, smooth muscle, and fibroblast forms of this protein (Breitbart et al. 1987). The nuclear RNA for α-tropomyosin contains 11 potential exons, but different sets of exons are used in different cells (**Figure 2.27**). Such different proteins encoded by the same gene are called **splicing isoforms** of the protein.

In some instances, alternatively spliced RNAs yield proteins that play similar yet distinguishable roles in the same cell. Different isoforms of the WT1 protein perform different functions in the development of the gonads and kidneys. The isoform without the extra exon functions as a transcription factor during kidney development, whereas the isoform containing the extra exon appears to be involved in splicing different nRNAs and may be critical in testis development (Hammes et al. 2001; Hastie 2001).

The *Bcl-x* gene provides a good example of how alternative nRNA splicing can make a huge difference in a protein's function. If a particular DNA sequence is used as an exon, the "large Bcl-X protein," or Bcl-X$_L$, is made (see Figure 2.26C). This protein inhibits programmed cell death. However, if this sequence is seen as an intron, the "small Bcl-X protein" (Bcl-X$_S$) is made, and this protein *induces* cell death. Many tumors have a higher than normal amount of Bcl-X$_L$.

If you get the impression from this discussion that a gene with dozens of introns could create literally thousands of different, related proteins through differential splicing, you are probably correct. The current champion at making multiple proteins from the same gene is the *Drosophila Dscam1* gene. This gene encodes a membrane receptor protein involved in preventing dendrites from the same neuron

(A) Cassette exon: Type II procollagen

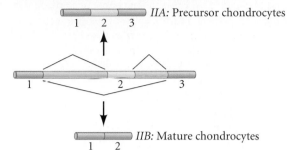

IIA: Precursor chondrocytes

IIB: Mature chondrocytes

(B) Mutually exclusive exons: FgfR2

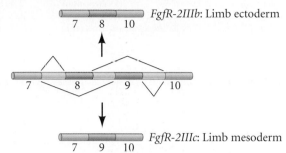

FgfR-2IIIb: Limb ectoderm

FgfR-2IIIc: Limb mesoderm

(C) Alternative 5′ splice site: *Bcl-x*

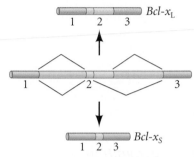

Bcl-x$_L$

Bcl-x$_S$

(D) Alternative 3′ splice site: Chordin

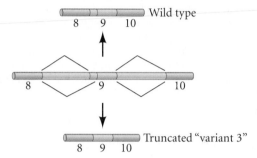

Wild type

Truncated "variant 3"

FIGURE 2.26 Some examples of alternative RNA splicing. Blue and colored portions of the bars represent exons; gray represents introns. Alternative splicing patterns are shown with V-shaped lines. (A) A "cassette" (yellow) that can be used as exon or removed as an intron distinguishes the type II collagen types of chondrocyte precursors and mature chondrocytes (cartilage cells).

(B) Mutually exclusive exons distinguish fibroblast growth factor receptors found in the limb ectoderm from those found in the limb mesoderm. (C) Alternative 5′ splice site selection, such as that used to create the large and small isoforms of the protein Bcl-X. (D) Alternative 3′ splice sites are used to form the normal and truncated forms of chordin. (After McAlinden et al. 2004.)

FIGURE 2.27 Alternative RNA splicing to form a family of rat α-tropomyosin proteins. The α-tropomyosin gene is represented on top. The thin lines represent the sequences that become introns and are spliced out to form the mature mRNAs. Constitutive exons (found in all tropomyosins) are shown in green. Those expressed only in smooth muscle are red; those expressed only in striated muscle are purple. Those that are variously expressed are yellow. Note that in addition to the many possible combinations of exons, two different 3′ ends ("striated" and "general") are possible. (After Breitbart et al. 1987.)

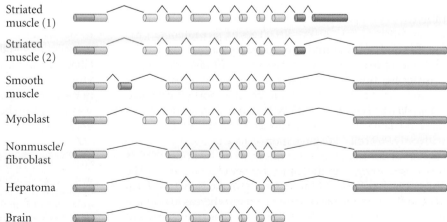

5′ UTR   Striated 3′ UTR   General 3′ UTR

Alternative splicing of mRNA transcripts

Striated muscle (1)

Striated muscle (2)

Smooth muscle

Myoblast

Nonmuscle/ fibroblast

Hepatoma

Brain

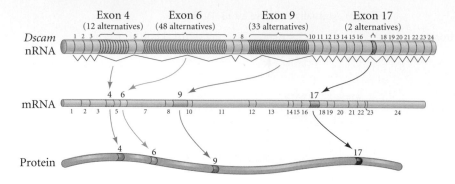

FIGURE 2.28   The *Dscam* gene of *Drosophila* can produce 38,016 different types of proteins by alternative nRNA splicing. The gene contains 24 exons. Exons 4, 6, 9, and 17 are encoded by sets of mutually exclusive possible sequences. Each messenger RNA will contain one of the 12 possible exon 4 sequences, one of the 48 possible exon 6 alternatives, one of the 33 possible exon 9 alternatives, and one of the 2 possible exon 17 sequences. The *Drosophila Dscam* gene is homologous to a DNA sequence on human chromosome 21 that is expressed in the nervous system. Disturbances of this gene in humans may lead to the neurological defects of Down syndrome (After Yamakawa et al. 1998; Saito 2000.)

from binding to one another. *Dscam1* contains 115 exons. However, a dozen different adjacent DNA sequences can be selected to be exon 4. Similarly, more than 30 mutually exclusive adjacent DNA sequences can become exons 6 and 9, respectively (**Figure 2.28**; Schmucker et al. 2000). If all possible combinations of exons are used, this one gene can produce 38,016 different proteins, and random searches for these combinations indicate that a large fraction of them are in fact made. The nRNA of *Dscam1* has been found to be alternatively spliced in different axons, and when two dendrites from the same axon touch each other, they are repelled. This causes the extensive branching of the dendrites. It appears that the thousands of splicing isoforms are needed to ensure that each neuron acquires a unique identity (**Figure 2.29**; Schmucker 2007; Millard and Zipursky 2008; Hattori et al. 2009). The *Drosophila* genome is thought to contain only 14,000 genes, but here is a single gene that encodes three times that number of proteins!

About 92% of human genes are thought to produce multiple types of mRNA. Therefore, even though the human genome may contain 20,000–30,000 genes, its **proteome**—the number and type of proteins encoded by the genome—is far more complex. "Human genes are multitaskers," notes Christopher Burge, one of the scientists who calculated this figure (Ledford 2008). This explains an important paradox. *Homo sapiens* has around 20,000 genes in each nucleus; so does the nematode *Caenorhabditis elegans*, a tubular creature with only 969 cells. We have more cells and cell types in the shaft of a hair than *C. elegans* has in its entire body. What's this worm doing with the same number of genes as us? The answer is that *C. elegans* genes rarely make isoforms. Each gene in the worm makes but one protein, whereas in humans the same number of genes produces an enormous array of different proteins.

Splicing enhancers and recognition factors

The mechanisms of differential RNA processing involve both *cis*-acting sequences on the nRNA and *trans*-acting protein factors that bind to these regions (Black 2003). The *cis*-acting sequences on nRNA are usually close to their potential 5′ or 3′ splice sites. These sequences are called "splicing enhancers," since they promote the assembly of spliceosomes at RNA cleavage sites. Conversely, these same sequences can be "splicing silencers" if they act to exclude exons from an mRNA sequence. These sequences are recognized by *trans*-acting proteins, most of which can recruit spliceosomes to that area. However, some *trans*-acting proteins, such as the polyprimidine tract-binding protein (PTP),* repress spliceosome formation where they bind.

As might be expected, there are some splicing enhancers that appear to be specific for certain tissues. Muscle-specific *cis*-regulatory sequences have been found around those exons characterizing muscle cell messages. These are recognized by certain proteins that are found in the muscle cells early in their development (Ryan and Cooper 1996; Charlet-B et al. 2002). Their presence is able to compete with the PTP that would otherwise prevent the inclusion of the muscle-specific exon into the mature message. In this way, an entire battery of muscle-specific isoforms can be generated.

---

*PTP is involved in making the correct isoform of tropomyosin and may be especially important in determining the mRNA populations of the brain. PTP is also involved in the mutually exclusive use of exon IIIb or IIIc in the mRNA for fibroblast growth factor 2 (see Figure 2.26B; Carstens et al. 2000; Lilleväli et al. 2001; Robinson and Smith 2006).

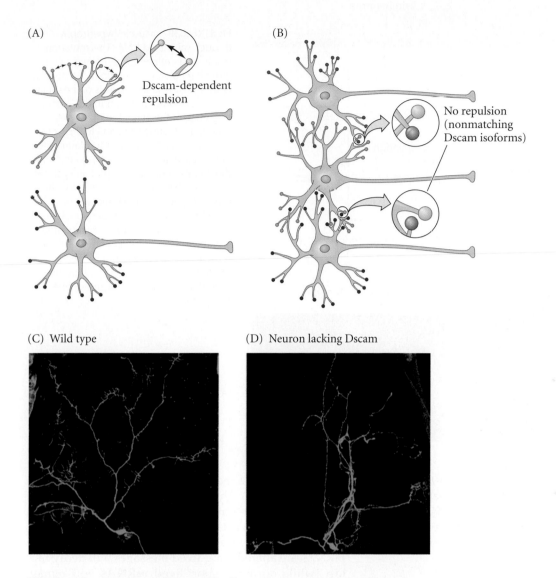

FIGURE 2.29   Dscam protein is specifically required to keep dendrites from the same neuron from adhering to each other. (A) When sister dendrites expressing the same splicing isoform of Dscam touch, the Dscam-Dscam interactions repel them and cause the dendrites to separate. (B) Different neurons express different splicing isoforms that do not interact with one another (and therefore do not trigger repulsion), which allows neurons to interact. (C) Neurons with multiple dendrites normally develop highly branched formations in which none of the branches crosses another. (D) Loss of *Dscam1* in such a neuron abolishes self-repulsion and results in excessive self-crossing and adhesion. (After Schmucker 2007; photographs courtesy of Dietmar Schmucker.)

One might also suspect that mutations of the splicing sites would lead to alternative phenotypes. Most of these splice site mutations lead to nonfunctional proteins and serious diseases. For instance, a single base change at the 5′ end of intron 2 in the human β-globin gene prevents splicing from occurring and generates a nonfunctional mRNA (Baird et al. 1981). This causes the absence of any β-globin from this gene, and thus a severe (and often life-threatening) type of anemia. Similarly, a mutation in the *dystrophin* gene at a par-

ticular splice site causes the skipping of that exon and a severe form of muscular dystrophy (Sironi et al. 2001). In at least one such case, the splice site mutation was not dangerous and actually gave the patient greater strength. In a different case, Schuelke and colleagues (2004) described a family in which individuals in four generations had a splice site mutation in the *myostatin* gene (**Figure 2.30A**). Among the family members were professional athletes and a 4-year-old toddler who was able to hold two 3-kg dumb-

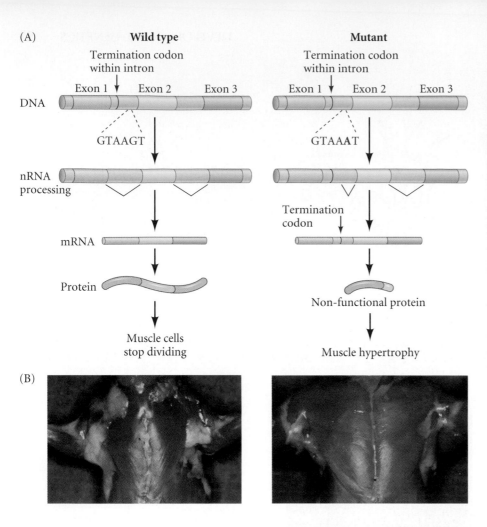

(A)

| Wild type | Mutant |

Termination codon within intron

Termination codon within intron

DNA — Exon 1 | Exon 2 | Exon 3

GTAAGT

GTAAAT

nRNA processing

Termination codon

mRNA

Protein

Non-functional protein

Muscle cells stop dividing

Muscle hypertrophy

(B)

FIGURE 2.30 Muscle hypertrophy through mispliced RNA. This mutation results in a deficiency of the negative growth regulator myostatin in the muscle cells. (A) Molecular analysis of the mutation. There is no mutation in the coding sequence of the gene, but in the first intron, a mutation from a G to an A created a new (and widely used) splicing site. This caused aberrant nRNA splicing and the inclusion of an early protein synthesis termination codon into the mRNA. Thus, proteins made from that message would have been short and nonfunctional. (B) Pectoral musculature of a "mighty mouse" with the mutation (right) compared to the muscles of a wild-type mouse (left). (A after Schuelke et al. 2004; B from McPherron et al. 1997; courtesy of A. C. McPherron.)

bells with his arms fully extended. The *myostatin* gene product is a negative regulator—a factor that tells muscle precursor cells to stop dividing. In mammals with the mutation, the muscles are not told to differentiate until they have undergone many more rounds of cell division, and the result is larger muscles (**Figure 2.30B**).

Control of Gene Expression at the Level of Translation

The splicing of nuclear RNA is intimately connected with its export through the nuclear pores and into the cytoplasm. As the introns are removed, specific proteins bind to the spliceosome and attach the spliceosome-RNA complex to nuclear pores (Luo et al. 2001; Strässer and Hurt 2001). But once the RNA has reached the cytoplasm, there is still no guarantee that it will be translated. The control of gene expression at the level of translation can occur by many means; some of the most important of these are described below.

Differential mRNA longevity

The longer an mRNA persists, the more protein can be translated from it. If a message with a relatively short half-life were selectively stabilized in certain cells at certain times, it would make large amounts of its particular protein only at those times and places.

The stability of a message often depends on the length of its polyA tail. This, in turn, depends largely on sequences in the 3' untranslated region, certain of which allow longer polyA tails than others. If these 3' UTR regions are experimentally traded, the half-lives of the resulting mRNAs are altered: long-lived messages will decay rapidly, while normally short-lived mRNAs will remain around longer (Shaw and Kamen 1986; Wilson and Treisman 1988; Decker and Parker 1995).

In some instances, messenger RNAs are selectively stabilized at specific times in specific cells. The mRNA for casein, the major protein of milk, has a half-life of 1.1 hours in rat mammary gland tissue. However, during periods of lactation, the presence of the hormone prolactin increases this half-life to 28.5 hours (**Figure 2.31**; Guyette et al. 1979). In the development of the nervous system, a group of proteins called HuD proteins stabilizes a group of mRNAs that stop the neuronal precursor cells from dividing and also stabilizes a second group of mRNAs that are critical for these cells to start neuron differentiation (Okano and Darnell 1997; Deschênes-Furry et al. 2006, 2007).

Selective inhibition of mRNA translation: Stored oocyte mRNAs

Some of the most remarkable cases of translational regulation of gene expression occur in the oocyte. The oocyte often makes and stores mRNAs that will be used only after

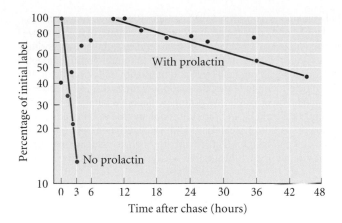

FIGURE 2.31 Degradation of casein mRNA in the presence and absence of prolactin. Cultured rat mammary cells were given radioactive RNA precursors (pulse) and, after a given time, were washed and given nonradioactive precursors (chase). This procedure labeled the casein mRNA synthesized during the pulse time. Casein mRNA was then isolated at different times following the chase and its radioactive label measured. In the absence of prolactin, the labeled (i.e., newly synthesized) casein mRNA decayed rapidly, with a half-life of 1.1 hours. When the same experiment was done in a medium containing prolactin, the half-life was extended to 28.5 hours. (After Guyette et al. 1979.)

fertilization occurs. These messages stay in a dormant state until they are activated by ion signals (discussed in Chapter 4) that spread through the egg during ovulation or fertilization.

Table 2.2 gives a partial list of mRNAs that are stored in the oocyte cytoplasm. Some of these stored mRNAs encode proteins that will be needed during cleavage, when the embryo makes enormous amounts of chromatin, cell membranes, and cytoskeletal components. Some of them encode cyclin proteins that regulate the timing of early cell division (Rosenthal et al. 1980; Standart et al. 1986). Indeed, in many species (including sea urchins and *Drosophila*),

TABLE 2.2 Some mRNAs stored in oocyte cytoplasm and translated at or near fertilization

mRNAs encoding	Function(s)	Organism(s)
Cyclins	Cell division regulation	Sea urchin, clam, starfish, frog
Actin	Cell movement and contraction	Mouse, starfish
Tubulin	Formation of mitotic spindles, cilia, flagella	Clam, mouse
Small subunit of ribonucleotide reductase	DNA synthesis	Sea urchin, clam, starfish
Hypoxanthine phosphoribosyl-transferase	Purine synthesis	Mouse
Vg1	Mesodermal determination(?)	Frog
Histones	Chromatin formation	Sea urchin, frog, clam
Cadherins	Blastomere adhesion	Frog
Metalloproteinases	Implantation in uterus	Mouse
Growth factors	Cell growth; uterine cell growth(?)	Mouse
Sex determination factor FEM-3	Sperm formation	*C. elegans*
PAR gene products	Segregate morphogenetic determinants	*C. elegans*
SKN-1 morphogen	Blastomere fate determination	*C. elegans*
Hunchback morphogen	Anterior fate determination	*Drosophila*
Caudal morphogen	Posterior fate determination	*Drosophila*
Bicoid morphogen	Anterior fate determination	*Drosophila*
Nanos morphogen	Posterior fate determination	*Drosophila*
GLP-1 morphogen	Anterior fate determination	*C. elegans*
Germ cell-less protein	Germ cell determination	*Drosophila*
Oskar protein	Germ cell localization	*Drosophila*
Ornithine transcarbamylase	Urea cycle	Frog
Elongation factor 1a	Protein synthesis	Frog
Ribosomal proteins	Protein synthesis	Frog, *Drosophila*

Compiled from numerous sources.

(A) Circularized mRNA

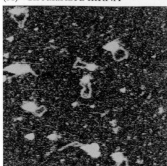

FIGURE 2.32 Translational regulation in oocytes. (A) Messenger RNAs are often found as circles, where the 5′ end and the 3′ end contact one another. Here, a yeast mRNA seen by atomic force microscopy is circularized by eIF4E and eIF4G (5′ end) and the polyA binding protein (3′ end). (B) In *Xenopus* oocytes, the 3′ and 5′ ends of the mRNA are brought together by maskin, a protein that binds to CPEB on the 3′ end and translation initiation factor 4E (eIF4E) on the 5′ end. Maskin blocks the initiation of translation by preventing eIF4E from binding eIF4G. (C) When stimulated by progesterone during ovulation, a kinase phosphorylates CPEB, which can then bind CPSF. CPSF can bind polyA polymerase and initiate growth of the polyA tail. PolyA binding protein (PABP) can bind to this tail and then bind eIF4G in a stable manner. This initiation factor can then bind eIF4E and, through its association with eIF3, position a 40S ribosomal subunit on the mRNA. (A from Wells et al. 1998; B,C after Mendez and Richter 2001.)

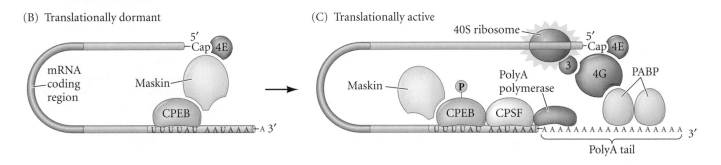

(B) Translationally dormant

(C) Translationally active

maintenance of the normal rate and pattern of early cell divisions does not require a nucleus; rather, it requires continued protein synthesis from stored maternal mRNAs (Wagenaar and Mazia 1978; Edgar et al. 1994). Other stored messages encode proteins that determine the fates of cells. These include the *bicoid*, *caudal*, and *nanos* messages that provide information in the *Drosophila* embryo for the production of its head, thorax, and abdomen.

Most translational regulation in oocytes is negative, as the "default state" of the mRNA is to be available for translation. Therefore, there must be inhibitors preventing the translation of these mRNAs in the oocyte, and these inhibitors must somehow be removed at the appropriate times around fertilization. The 5′ cap and the 3′ untranslated region seem especially important in regulating the accessibility of mRNA to ribosomes. If the 5′ cap is not made or if the 3′ UTR lacks a polyA tail, the message probably will not be translated. The oocytes of many species have "used these ends as means" to regulate the translation of their mRNAs.

It is important to realize that, unlike the usual representations of mRNA, most mRNAs probably form circles, with their 3′ end being brought to their 5′ end (**Figure 2.32A**). The 5′ cap is bound by **eukaryotic initiation factor-4E (eIF4E)**, a protein that is also bound to eIF4A (a helicase that unwinds double-stranded regions of RNA) and **eIF4G**, a scaffold protein that allows the mRNA to bind to the ribosome through its interaction with eIF4E (Wells et al. 1998; Gross et al. 2003). The polyA binding protein, which sits on the polyA tail of the mRNA, also binds to the eIF4G protein. This brings the 3′ end of the message next to the 5′ end and allows the messenger RNA to be recognized by

the ribosome. Thus, the 5′ cap is critical for translation, and some animal's oocytes have used this as a direct means of translational control. For instance, the oocyte of the tobacco hornworm moth makes some of its mRNAs without their methylated 5′ caps. In this state, they cannot be efficiently translated. However, at fertilization, a methyltransferase completes the formation of the caps, and these mRNAs can be translated (Kastern et al. 1982).

In amphibian oocytes, the 5′ and 3′ ends of many mRNAs are brought together by a protein called **maskin** (Stebbins-Boaz et al. 1999; Mendez and Richter 2001). Maskin links the 5′ and 3′ ends into a circle by binding to two other proteins, each at opposite ends of the message. First, it binds to the **cytoplasmic polyadenylation-element-binding protein** (**CPEB**) attached to the UUUUAU sequence in the 3′ UTR; second, maskin also binds to the eIF4E factor that is attached to the cap sequence. In this configuration, the mRNA cannot be translated (**Figure 2.32B**). The binding of eIF4E to maskin is thought to prevent the binding of eIF4E to eIF4G, a critically important translation initiation factor that brings the small ribosomal subunit to the mRNA.

Mendez and Richter (2001) have proposed an intricate scenario to explain how mRNAs bound together by maskin become translated at about the time of fertilization. At ovulation (when the hormone progesterone stimulates the last meiotic divisions of the oocyte and the oocyte is released for fertilization), a kinase activated by progesterone phosphorylates the CPEB protein. The phosphorylated CPEB can now bind to CPSF, the cleavage and polyadenylation specificity factor (Mendez et al. 2000; Hodgman et al. 2001). The bound CPSF protein sits on a

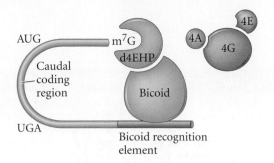

FIGURE 2.33 Protein binding in *Drosophila* oocytes. Bicoid protein binds to a recognition element in the 3' UTR of the *caudal* message. Bicoid can bind to d4EHP, which prevents the binding of eIF4E to the cap structure. Without eIF4E, the eIF4G cannot bind and initiate translation. (After Cho et al. 2005.)

particular sequence of the 3' UTR that has been shown to be critical for polyadenylation, and it complexes with a polymerase that elongates the polyA tail of the mRNA. In oocytes, a message having a short polyA tail is not degraded; however, such messages are not translated.

Once the tail is extended, molecules of the polyA binding protein (PABP) can attach to the growing tail. PABP proteins stabilize eIF4G, allowing it to outcompete maskin for the binding site on the eIF4E protein at the 5' end of the mRNA. The eIF4G protein can then bind eIF3, which can position the small ribosomal subunit onto the mRNA. The small (40S) ribosomal subunit will then find the initiator tRNA, complex with the large ribosomal subunit, and initiate translation (**Figure 2.32C**).

In the *Drosophila* oocyte, Bicoid can act both as a transcription factor (activating genes such as *hunchback*) and also as a translational inhibitor (see Chapter 6). Bicoid represses the translation of *caudal* mRNA, preventing its transcription in the anterior half of the embryo. (The protein made from the *caudal* message is important in activating those genes that specify the cells to be abdomen precursors.) Bicoid inhibits *caudal* mRNA translation by binding to a "bicoid recognition element," a series of nucleotides in the 3' UTR of the *caudal* message (**Figure 2.33**). Once there, Bicoid can bind with and recruit another protein, d4EHP. The d4EHP protein can compete with eIF4E for the cap. Without eIF4E, there is no association with eIF4G and the *caudal* mRNA becomes untranslatable. As a result, the *caudal* message is not translated in the anterior of the embryo (where Bicoid is abundant), but is active in the posterior portion of the embryo.

microRNAs: Specific regulators of mRNA translation and transcription

If proteins can bind to specific nucleic acid sequences to block transcription or translation, you would think that RNA would do the job even better. After all, RNA can be made specifically to bind a particular sequence. Indeed, one of the most efficient means of regulating the translation of a specific message is to make a small RNA complementary to a portion of a particular mRNA. Such a naturally occurring antisense RNA was first seen in *C. elegans*

(Lee et al. 1993; Wightman et al. 1993). Here, the *lin-4* gene was found to encode a 21-nucleotide RNA that bound to multiple sites in the 3' UTR of the *lin-14* mRNA (**Figure 2.34**). The *lin-14* gene encodes a transcription factor, LIN-14, that is important during the first larval phase of *C. elegans* development. It is not needed afterward, and *C. elegans* is able to inhibit synthesis of LIN-14 from these messages by activating the *lin-4* gene. The binding of *lin-*

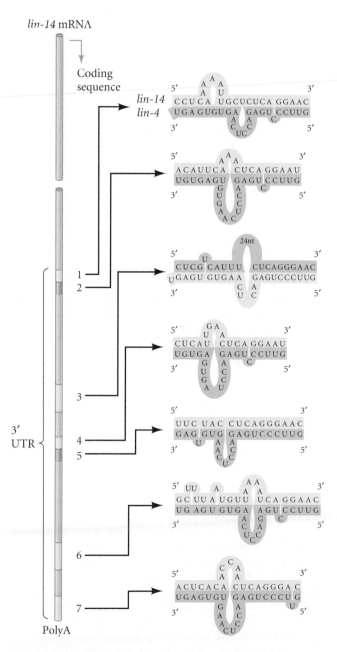

FIGURE 2.34 Hypothetical model of the regulation of *lin-14* mRNA translation by *lin-4* RNAs. The *lin-4* gene does not produce an mRNA. Rather, it produces small RNAs that are complementary to a repeated sequence in the 3' UTR of the *lin-14* mRNA, which bind to it and prevent its translation. (After Wickens and Takayama 1995.)

FIGURE 2.35   Current model for the formation and use of microRNAs. The miRNA gene encodes a pri-miRNA that often has several hairpin regions where the RNA finds nearby complementary bases with which to pair. The pri-miRNA is processed into individual pre-miRNA "hairpins" by the Drosha RNAase, and these are exported from the nucleus. Once in the cytoplasm, another RNAase, Dicer, eliminates the non-base-paired loop. Dicer also acts as a helicase to separate the strands of the double-stranded miRNA. One strand (probably recognized by placement of Dicer) is packaged with proteins into the RNA-induced silencing complex (RISC), which subsequently binds to the 3′ UTRs to effect translational suppression or cleavage, depending (at least in part) on the strength of the complementarity between the miRNA and its target. (After He and Hannon 2004.)

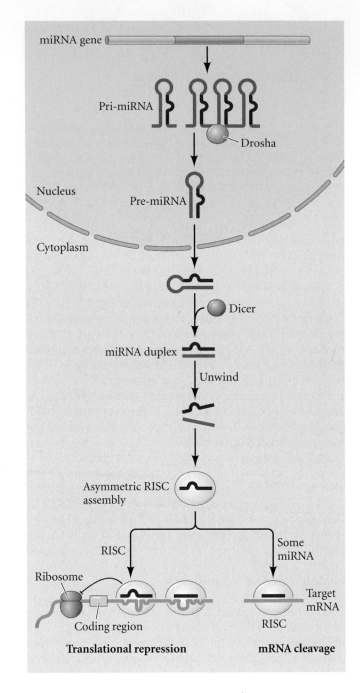

4 transcripts to the *lin-14* mRNA 3′ UTR causes degradation of the *lin-14* message (Bagga et al. 2005).

The *lin-4* RNA is now thought to be the "founding member" of a very large group of **microRNAs (miRNAs)**. These RNAs of about 22 nucleotides are made from longer precursors. These precursors can be in independent transcription units (the *lin-4* gene is far apart from the *lin-14* gene), or they can reside in the introns of other genes (Aravin et al. 2003; Lagos-Quintana et al. 2003). Many of the newly discovered microRNAs have been found in the regions between genes (regions previously considered to contain "junk DNA"). The initial RNA transcript (which may contain several repeats of the miRNA sequence) forms hairpin loops wherein the RNA finds complementary structures within its strand. These stem-loop structures are processed by a set of RNases (Drosha and Dicer) to make single-stranded microRNA (**Figure 2.35**). The microRNA is then packaged with a series of proteins to make an **RNA-induced silencing complex (RISC)**. Such small regulatory RNAs can bind to the 3′ UTR of messages and inhibit their translation. In some cases (especially when the binding of the miRNA to the 3′ UTR is tight), the site is cleaved. More usually, however, several RISCs attach to sites on the 3′ UTR and prevent the message from being translated (see Bartel 2004; He and Hannon 2004).

The abundance of microRNAs and their apparent conservation among flies, nematodes, vertebrates, and even plants suggest that such RNA regulation is a previously unrecognized but potentially very important means of regulating gene expression. This hidden layer of gene regulation parallels the better known protein-level gene control mechanisms, and it may be just as important in regulating cell fate. Recent studies have shown that microRNAs are involved in mammalian heart and blood cell differentiation. During mouse heart development, the microRNA *miR1* can repress the messages encoding the Hand2 transcription factor (Zhao et al. 2005). This transcription factor is critical in the proliferation of ventricle heart muscle cells, and *miR1* may control the balance between ventricle growth and differentiation. The *miR181* miRNA is essential for committing progenitor cells to differentiate into B

lymphocytes, and ectopic expression of *miR181* in mice causes a preponderance of B lymphocytes (**Figure 2.36**; Chen et al. 2004).

MicroRNAs are also used to "clean up" and fine-tune the level of gene products. We mentioned those maternal RNAs that allow early development to occur. How does the embryo get rid of maternal RNAs once they have been used and the embryonic cells are making their own mRNAs? In zebrafish, this cleanup operation is assigned to microRNAs such as *miR430*. This is one of the first genes transcribed by the fish embryonic cells, and there are about 90 copies of this gene in the zebrafish genome. So the level

microRNAs in Transcriptional Gene Regulation

In addition to its role in the translational regulation of gene expression, microRNAs also appear to be able to silence the transcription of certain genes. Such genes are often located in the heterochromatin—that region of the genome where the DNA is tightly coiled and transcription is inhibited by the packed nucleosomes. Volpe and colleagues (2003) discovered that if they deleted the genes in yeast encoding the appropriate RNases or RISC proteins, the heterochromatin around the centromeres became unpacked, the histones in this region lost their inhibitory methylation, and the centromeric heterochromatin started making RNA. Similar phenomena were seen when these proteins were mutated in *Drosophila* (Pal-Bhadra et al. 2004). Indeed, in *Drosophila*, the *Suppressor-of-stellate* gene on the Y chromosome makes a microRNA that represses the transcription of the *stellate* gene on the X chromosome (Gvozdev et al. 2003). This is important for dosage regulation of the X chromosomes in *Drosophila*.

It appears that microRNAs are able to bind to the nuclear RNA as it is being transcribed, and form a complex with the methylating and deacetylating enzymes, thus repressing the gene (Kato et al. 2005; Schramke et al. 2005). If synthetic microRNA made complementary to specific promoters is added to cultured human cells, that microRNA is able to induce that promoter's DNA to become methylated. Lysine 9 on histone H3 also becomes methylated around the promoter, and transcription from that gene stops (Kawasaki and Taira 2004; Morris et al. 2004).

This appears to be the mechanism by which NRSF (see page 47) functions. NRSF prevents gene expression in non-neural cells by repressing microRNAs that would otherwise recruit histone acetyltransferases to activate genes that promote neuron production. In the presence of NRSF, these miRNAs are not present, and so histone deacetylases and methyltransferases are recruited to the chromatin instead. The resulting methylation produces conglomerations of nucleosomes linked together by heterochromatin protein-1 (HP1), thereby stabilizing the conglomerate and preventing transcription of the neuron-promoting genes "hidden" within it (Ooi and Wood 2007; Yoo et al. 2009.) A single silencer protein bound to the DNA can prevent the gene's expression.

Thus, microRNA directed against the 3' end of mRNA may be able to shut down gene expression on the translational level, while microRNA directed at the promoters of genes may be able to block gene expression at the transcriptional level. The therapeutic value of these RNAs in cancer therapy is just beginning to be explored (see Gaur and Rossi 2006).

X chromosome inactivation in mammals is also directed by small noncoding RNAs, albeit not the canonical microRNAs. Although DNA methylation is responsible for keeping one of the two X chromosomes inactive, the choice of which X chromosome to activate arises from the physical interactions between the two X chromosomes and their expressing several small, noncoding RNAs. The mechanisms of differential expression of small RNAs between the two X chromosomes is under intensive investigation (Augui et al. 2007; Migeon 2007; Ogawa et al. 2008).

---

of *miR430* goes up very rapidly. This microRNA has hundreds of targets (about 40% of the maternal RNA types), and when it binds to the 3' UTR of these target mRNAs, these mRNAs lose their polyA tails and are degraded (Giraldez et al. 2006). Slightly later in development, this same microRNA is used in the fish embryo to fine-tune the expression of *Nodal* mRNA (Choi et al. 2007). The consequence of this latter use of *miR430* is the determination of how many cells become committed to the endoderm and how many become committed to be mesoderm.

Although the microRNA is usually about 22 bases long, it recognizes its target primarily through a "seed" region

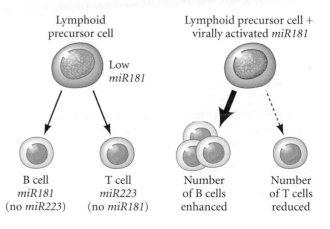

FIGURE 2.36 The lymphoid precursor cell can generate B cells (lymphocytes that make antibodies) or T cells (lymphocytes that kill virally infected cells). This differentiation depends on the organ in which they reside. The regulation of the lineage pathway is controlled in part by levels of the microRNA *miR181*. The lymphocyte precursor cell has little *miR181*. A B cell has high levels of *miR181*, whereas T cells do not appear to have any. If lymphocyte precursor cells are virally transfected with *miR181*, they preferentially generate B cells at the expense of T cells.

FIGURE 2.37 The miRNA complex, including numerous proteins that bind to the miRNA (miRNP), can block translation in several ways. These include (A) blocking the binding of the mRNA to initiation factors or ribosomes; (B) recruiting endonucleases to chew away the polyA tail of the mRNA, thereby causing its destruction; and (C) recruiting protein-digesting enzymes that destroy the nascent protein. (After Filipowicz et al. 2008.)

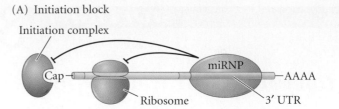

(A) Initiation block

(B) Endonuclease digestion (de-adenylation)

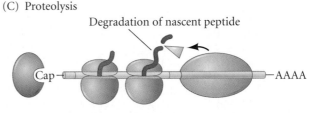

(C) Proteolysis

of about 5 bases in the 5′ end of the microRNA (usually at positions 2–7). This seed region recognizes targets in the 3′ UTR of the message. What happens, then, if an mRNA has a mutated 3′ UTR? Such a mutation appears to have given rise to the Texel sheep, a breed with a large and well-defined musculature that is the dominant meat-producing sheep in Europe. We have already seen that a mutation in the *myostatin* gene that prevents the proper splicing of the nRNA can produce a large-muscled phenotype. Another way of reducing the levels of myostatin involves a mutation in its 3′UTR sequence (see Figure 2.30). Genetic techniques mapped the basis of the sheep's meaty phenotype to the *myostatin* gene. In the Texel breed, there has been a G-to-A transition in the 3′ UTR of the gene for myostatin, creating a target for the *mir1* and *mir206* microRNAs that are abundant in skeletal muscle (Clop et al. 2006). This mutation causes the depletion of myostatin messages and the increase in muscle mass characteristic of these sheep.

The binding of microRNAs to the 3′ UTR can regulate translation in several ways (**Figure 2.37**; Filipowicz et al. 2008). First, they can block initiation of translation, preventing the binding of initiation factors or ribosomes. Second, they can recruit endonucleases that digest the mRNA, usually starting with the polyA tail. In a third mechanism, they allow translation to be initiated, but recruit proteolytic enzymes that digest the protein as it is being made. It is also possible that some microRNAs use more than one method, and it has been proposed (Mathonnet et al. 2007) that the microRNAs may first inhibit translation initiation and then consolidate mRNA silencing by causing the digestion of the message.

Control of RNA expression by cytoplasmic localization

Not only is the time of mRNA translation regulated, but so is the place of RNA expression. A majority of mRNAs (about 70% in *Drosophila* embryos) are localized to specific places in the cell (Lécuyer et al. 2007). Just like the selective repression of mRNA translation, the selective localization of messages is often accomplished through their 3′ UTRs. There are three major mechanisms for the localization of an mRNA (**Figure 2.38**; see Palacios 2007):

• *Diffusion and local anchoring*. Messenger RNAs such as *nanos* diffuse freely in the cytoplasm. However, when they diffuse to the posterior pole of the *Drosophila* oocyte,

they are trapped there by proteins that reside particularly in these regions. These proteins also activate the mRNA, allowing it to be translated.

• *Localized protection*. Messenger RNAs such as those encoding the *Drosophila* heat shock protein hsp83 (which helps protect the embryos from thermal extremes) also float freely in the cytoplasm. Like *nanos* mRNA, *hsp83* accumulates at the posterior pole, but its mechanism for getting there is different. Throughout the embryo, the protein is degraded. However, proteins at the posterior pole protect the *hsp83* mRNA from being destroyed.

• *Active transport along the cytoskeleton*. This is probably the most widely used mechanism for mRNA localization. Here, the 3′ UTR of the mRNA is recognized by proteins that can bind these messages to "motor proteins" that travel along the cytoskeleton to their final destination. These motor proteins are usually ATPases such as dynein or kinesin that split ATP for their motive force. For instance, in *Drosophila* oocytes, the *bicoid* messages (which instruct the formation of the head) are localized to one end of the oocyte. The 3′ UTR of *bicoid* mRNA allows its message to bind to the microtubules through its association with two other proteins (Swallow and Staufen). If the *bicoid* 3′ UTR is attached to some other message, that mRNA will also be bound to the anterior pole of the oocyte (Driever and Nüsslein-Volhard 1988a,b; Ferrandon et al. 1994).

The 3′ UTR of the *bicoid* message binds the Staufen protein that connects it to dynein. Dynein travels along the microtubules in the "minus" direction, that is, toward the site where microtubules begin. In this way, the *bicoid* mRNA is localized to the future anterior part of the oocyte. Other mRNAs, such as the *Oskar* message, in contrast, appear to

(A) Diffusion and local anchoring

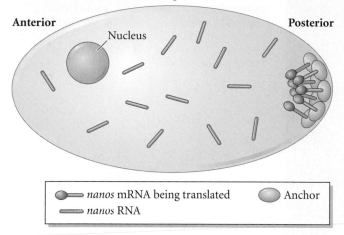

(B) Localized protection

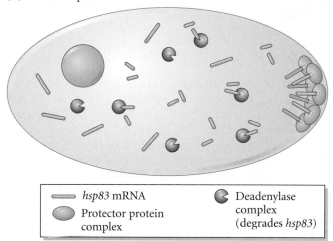

(C) Active transport along cytoskeleton

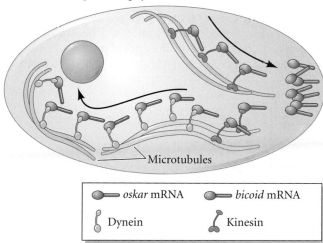

FIGURE 2.38 Localization of mRNAs. (A) Diffusion and local anchoring. *Nanos* mRNA diffuses through the *Drosophila* egg and is bound (in part by the Oskar protein, whose message is described in the text) at the posterior end of the oocyte. This anchoring allows the *nanos* mRNA to be translated. (B) Localized protection. The mRNA for *Drosophila* heat shock protein (hsp83) will be degraded unless it binds to a protector protein (in this case, also at the posterior terminal of the oocyte). (C) Active transport on the cytoskeleton, causing the accumulation of mRNA at a particular site. Here, *bicoid* mRNA is transported to the anterior of the oocyte by dynein and kinesin motor proteins. Meanwhile, *Oskar* mRNA is brought to the posterior pole by transport along microtubules by kinesin ATPases. (After Palacios 2007.)

mRNAs often bind to other cytoskeletal proteins (such as actin microfilaments).

Stored mRNAs in brain cells

One of the most important areas of local translational regulation may be in the brain. The storage of long-term memory requires new protein synthesis, and the local translation of mRNAs in the dendrites of brain neurons has been proposed as a control point for increasing the strength of synaptic connections (Martin 2000; Klann et al. 2004; Wang and Tiedge 2004). The ability to increase the strength of the connections between neurons is critical in forming the original architecture of the brain and also in the ability to learn. Indeed, in recent studies of mice, Kelleher and colleagues (2004) have shown that neuronal activity-dependent memory storage depends on the activation of eIF4E and other components of protein synthesis.

Several mRNAs appear to be transported along the cytoskeleton to the dendrites of neurons (the "receiving portion" of the neuron, where synapse connections are formed with the other neurons). These messages include those mRNAs encoding receptors for neurotransmitters (needed to transmit the signals from one neuron to another); activity-regulated enzymes; and the cytoskeletal components needed to build a synapse (**Figure 2.39**). As we will see in later chapters, one of the proteins responsible for constructing specific synapses is **brain-derived neurotrophic factor**, or **BDNF**. BDNF regulates neural activity and appears to be critical for new synapse formation. Takei and colleagues (2004) have shown that BDNF induces local translation of these neural messages in the dendrites.

Another indication of the importance of dendritic mRNA translation comes from studies of a leading cause of human mental retardation, fragile X syndrome. Fragile X syndrome is caused by loss-of-function mutations in the X-linked *FMR1* gene. The FMR1 protein appears to prevent the translation of several mRNAs that are being transported to the dendrites along microtubules in response to stimulation by glutamic acid (Dictenberg et al. 2008; Wang et al. 2008b). In the absence of functional FMR1, these mRNAs are expressed in the wrong amounts, leading to signaling abnormalities that are believed to cause the prob-

bind to the kinesin motor protein, and it is taken toward the "plus" end of the microtubules, at the tip of their assembly. It is thereby taken to the posterior end of the *Drosophila* oocyte. Once transported to their destinations,

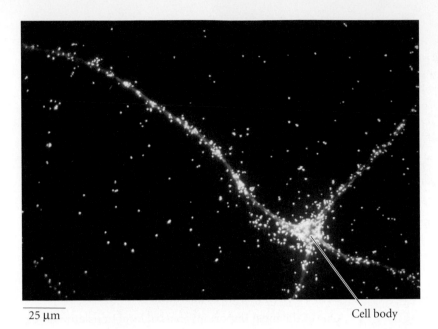

FIGURE 2.39 A brain-specific RNA in a cultured mammalian neuron. BC1 RNA (stained white) appears to be clustered at specific sites in the neuron (stained light blue), especially in the dendrites. (From Wang and Tiedge 2004, courtesy of the authors.)

25 µm

Cell body

lems in cognition and learning. Thus, translational regulation in neurons might be important not only for their initial development but also for their continued ability to learn and change.

Posttranslational regulation of gene expression

When a protein is synthesized, the story is still not over. Once a protein is made, it becomes part of a larger level of organization. For instance, it may become part of the structural framework of the cell, or it may become involved in one of the myriad enzymatic pathways for the synthesis or breakdown of cellular metabolites. In any case, the individual protein is now part of a complex "ecosystem" that integrates it into a relationship with numerous other proteins. Thus, several changes can still take place that determine whether or not the protein will be active.

Some newly synthesized proteins are inactive without the cleaving away of certain inhibitory sections. This is what happens when insulin is made from its larger protein precursor. Some proteins must be "addressed" to their specific intracellular destinations in order to function. Proteins are often sequestered in certain regions of the cell, such as membranes, lysosomes, nuclei, or mitochondria. Some proteins need to assemble with other proteins in order to form a functional unit. The hemoglobin protein, the microtubule, and the ribosome are all examples of numerous proteins joining together to form a functional unit. And some proteins are not active unless they bind an ion (such as Ca^{2+}), or are modified by the covalent addition of a phosphate or acetate group. The importance of this last type of protein modification will become obvious in Chapter 3, since many of the critical proteins in embryonic cells just sit there until some signal activates them.

 Snapshot Summary: *Developmental Genetics*

1. Differential gene expression from genetically identical nuclei creates different cell types. Differential gene expression can occur at the levels of gene transcription, nuclear RNA processing, mRNA translation, and protein modification. Notice that RNA processing and export occur while the RNA is still being transcribed from the gene.

2. Genes are usually repressed, and activating a gene often means inhibiting its repressor. This fact leads to thinking in double and triple negatives: Activation is often the inhibition of the inhibitor; repression is the inhibition of the inhibitor of the inhibitor.

3. Eukaryotic genes contain promoter sequences to which RNA polymerase can bind to initiate transcription. To accomplish this, the eukaryotic RNA polymerases are bound by a series of proteins called transcription-associated factors, or TAFs.

4. Eukaryotic genes expressed in specific cell types contain enhancer sequences that regulate their transcription in time and space.

5. Specific transcription factors can recognize specific sequences of DNA in the promoter and enhancer

regions. These proteins activate or repress transcription from the genes to which they have bound.

6. Enhancers work in a combinatorial fashion. The binding of several transcription factors can act to promote or inhibit transcription from a certain promoter. In some cases transcription is activated only if both factor A and factor B are present; in other cases, transcription is activated if either factor A or factor B is present.

7. A gene encoding a transcription factor can keep itself activated if the transcription factor it encodes also activates its own promoter. Thus, a transcription factor gene can have one set of enhancer sequences to initiate its activation and a second set of enhancer sequences (which bind the encoded transcription factor) to maintain its activation.

8. Often, the same transcription factors that are used during the differentiation of a particular cell type are also used to activate the genes for that cell type's specific products.

9. Enhancers can act as silencers to suppress the transcription of a gene in inappropriate cell types.

10. Transcription factors act in different ways to regulate RNA synthesis. Some transcription factors stabilize RNA polymerase binding to the DNA; some disrupt nucleosomes, increasing the efficiency of transcription.

11. Transcription correlates with a lack of methylation on the promoter and enhancer regions of genes. Methylation differences can account for examples of genomic imprinting, wherein a gene transmitted through the sperm is expressed differently than the same gene transmitted through the egg.

12. Dosage compensation enables the X chromosome-derived products of males (which have one X chromosome per cell in fruit flies and mammals) to equal the X chromosome-derived products of females (which have two X chromosomes per cell). This compensation is accomplished at the level of transcription, either by accelerating transcription from the lone X chromosome in males (*Drosophila*), decreasing the level of transcription from each X chromosome by 50% (*C. elegans*), or by inactivating a large portion of one of the two X chromosomes in females (mammals).

13. Differential nuclear RNA selection can allow certain transcripts to enter the cytoplasm and be translated while preventing other transcripts from leaving the nucleus.

14. Differential RNA splicing can create a family of related proteins by causing different regions of the nRNA to be read as exons or introns. What is an exon in one set of circumstances may be an intron in another.

15. Some messages are translated only at certain times. The oocyte, in particular, uses translational regulation to set aside certain messages that are transcribed during egg development but used only after the egg is fertilized. This activation is often accomplished either by the removal of inhibitory proteins or by the polyadenylation of the message.

16. MicroRNAs can act as translational inhibitors, binding to the 3′ UTR of the RNA.

17. Many mRNAs are localized to particular regions of the oocyte or other cells. This localization appears to be regulated by the 3′ UTR of the mRNA.

For Further Reading

Complete bibliographical citations for all literature cited in this chapter can be found at the free-access website **www.devbio.com**

Clop, A. and 16 others. 2006. A mutation creating a potential illegitimate microRNA target site in the myostatin gene affects muscularity in sheep. *Nature Genet.* 38: 813–818.

Davidson, E. H. 2006. *The Regulatory Genome*. Academic Press, New York.

Migeon, B. R. 2007. *Females Are Mosaics: X Inactivation and Sex Differences in Disease*. Oxford University Press, New York.

Palacios, I. M. 2007. How does an mRNA find its way? Intracellular localization of transcripts. *Sem. Cell Dev. Biol.* 163–170.

Strahl, B. D. and C. D. Allis. 2000. The language of covalent histone modifications. *Nature* 403: 41–45.

Wilmut, I., K. Campbell and C. Tudge. 2001. *The Second Creation: Dolly and the Age of Biological Control*. Harvard University Press, Cambridge, MA.

Zhou, Q., J. Brown, A. Kanarek, J. Rajagopal and D. A. Melton. 2008. In vivo reprogramming of adult pancreatic exocrine cells to β cells. *Nature* 455: 627–632.

Zinzen, R. P., C. Girardot, J. Gagneur, M. Braun and E. E. Furlong. 2009. Combinatorial binding predicts spatiotemporal *cis*-regulatory activity. *Nature* 462: 65–70.

Go Online

WEBSITE 2.1 Does the genome or the cytoplasm direct development? The geneticists versus the embryologists. Geneticists were certain that genes controlled development, whereas embryologists generally favored the cytoplasm. Both sides had excellent evidence for their positions.

WEBSITE 2.2 The origins of developmental genetics. The first hypotheses for differential gene expression came from C. H. Waddington, Salome Gluecksohn-Waelsch, and other scientists who understood both embryology and genetics.

WEBSITE 2.3 Techniques of DNA analysis. The entries of this website describe crucial laboratory skill including gene cloning, DNA sequencing, Southern blotting, "knockouts" of specific genes, enhancer traps, and identification of methylated sites.

WEBSITE 2.4 Techniques of RNA analysis. Techniques described here include northern blots, RT-PCR, in situ hybridization, microarray technology, antisense RNA, interference RNA, Cre-lox analysis, and ChIP-on-Chip.

WEBSITE 2.5 Bioinformatics. This entry provides links to various free websites with tools that enable researchers to compare DNA sequences and specific genomes with the aim of further illuminating the various mechanisms of gene regulation.

WEBSITE 2.6 Cloning and nuclear equivalence. The several entries here address the issues of cloning and whether or not the entire genome is the same in each cell of the body. As it turns out, lymphocytes make new genes during their development and their genomes are not identical.

WEBSITE 2.7 Silencing large blocks of chromatin. The inactivation or the elimination of entire chromosomes is not uncommon among invertebrates and is sometimes used as a mechanism of sex determination. Moreover, among mammals, random X chromosome inactivation may provide females with health benefits—as long as the process occurs flawlessly.

WEBSITE 2.8 So you think you know what a gene is? Different scientists have different definitions, and nature has given us some problematic examples of DNA sequences that may or may not be considered genes.

Vade Mecum

Transdetermination in *Drosophila*. These movies describe Ernst Hadorn's discovery of transdetermination and Walter Gehring's pioneering study of homeotic mutants, changing body parts into eyes through transcription factors.

Cell-Cell Communication in Development

3

THE FORMATION OF ORGANIZED ANIMAL BODIES, or *morphogenesis*, has been one of the great sources of wonder for humankind. Indeed, the "miracle of life" seems just that—inanimate matter becomes organized in such a way that it lives. The twelfth-century rabbi and physician Maimonides (1190) framed the question of morphogenesis beautifully when he noted that the pious men of his day believed that an angel of God had to enter the womb to form the organs of the embryo. This, the men say, is a miracle. How much more powerful a miracle would life be, Maimonides asked, if the Deity had made matter such that it could generate such remarkable order without a matter-molding angel having to intervene in every pregnancy? The problem addressed today is the secular version of Maimonides' question: How can matter alone construct the organized tissues of the embryo?

The idea of angelic intervention remained prevalent in the embryology of the Renaissance. By the eighteenth century, however, scientific advances had allowed the learned to dispense with the necessity of involving heavenly beings in human conception and development, even though the process remained a mystery. In 1782, the Enlightenment essayist Denis Diderot posed the question of morphogenesis in the fevered dream of a noted physicist. This character could imagine that the body was formed from myriad "tiny sensitive bodies" that collected together to form an aggregate, but he could not envision how this aggregate could become an animal.

Diderot's "tiny sensitive bodies" are what we call cells, and we can break his problem into at least five questions that confront modern embryologists who study morphogenesis:

1. *How are separate tissues formed from populations of cells?* For example, how do neural retina cells stick to other neural retina cells rather than becoming part of the pigmented retina or the iris cells next to them? How are the different cell types found in the retina (the three distinct layers of photoreceptors, bipolar neurons, and ganglion cells) arranged such that the retina is functional?
2. *How are organs constructed from tissues?* The retina of the eye forms at a precise distance behind the cornea and the lens. The retina would be useless if it developed behind a bone or in the middle of the kidney. Moreover, neurons from the retina must enter the brain to innervate the regions of the brain cortex that analyze visual information. All these connections must be precisely ordered.
3. *How do organs form in particular locations, and how do migrating cells reach their destinations?* What causes there to be two—and usually only two—kidneys, and how do their ducts form so that they can collect urine made by the filter-

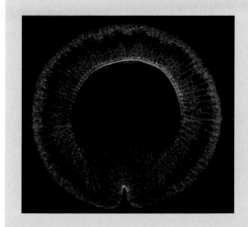

ing tissues of the nephron? Some cells—for instance, the precursors of our pigment cells, germ cells, and blood cells—must travel long distances to reach their final destinations. How are cells instructed to travel along certain routes in our embryonic bodies, and how are they told to stop once they have reached their appropriate destinations?

4. *How do organs and their cells grow, and how is their growth coordinated throughout development?* The cells of all the tissues in the eye must grow in a coordinated fashion if one is to see. Some cells, including most neurons, do not divide after birth. In contrast, the intestine is constantly shedding cells, and new intestinal cells are regenerated each day. The mitotic rate of each tissue must be carefully regulated. If the intestine generated more cells than it sloughed off, it could produce tumorous outgrowths. If it produced fewer cells than it sloughed off, it would soon become nonfunctional. What controls the rate of mitosis in the intestine?

5. *How do organs achieve polarity?* If one were to look at a cross section of the fingers, one would see a certain organized collection of tissues—bone, cartilage, muscle, fat, dermis, epidermis, blood, and neurons. Looking at a cross section of the forearm, one would find the same collection of tissues. But they are arranged very differently. How is it that the same cell types can be arranged in different ways in different parts of the same structure, and that fingers are always at the end of the arm, never in the middle?

Answers to these questions came slowly and are still coming. In the 1850s, Robert Remak (1852, 1855) formulated the cell theory and showed that the fertilized egg divides to produce the myriad "tiny sensitive bodies"—cells—needed to form an embryo. In the mid-twentieth century, E. E.

Just (1939) and Johannes Holtfreter (Townes and Holtfreter 1955) predicted that embryonic cells could have differences in their cell membrane components which would enable the formation of organs. In the late twentieth century, these membrane components—the molecules by which embryonic cells adhere to, migrate over, and induce gene expression in neighboring cells—began to be discovered and described. And presently, these pathways are being modeled to understand how the cell integrates the information from its nucleus and from its surroundings to take its place in the community of cells.

As we discussed in Chapter 1 (see Table 1.1), the cells of an embryo are either epithelial cells or mesenchymal cells. The epithelial cells can form tubes and sheets while remaining adhered to one another, whereas the mesenchymal cells often migrate individually and form extensive *extracellular matrices* that keep the individual cells separate. This chapter will discuss the mechanisms of three behaviors requiring cell-cell communication: cell adhesion, cell migration, and cell signaling.

Cell Adhesion

Differential cell affinity

Many of the answers to our five questions about morphogenesis involve the properties of the cell surface. The cell surface looks pretty much the same in all cell types, and many early investigators thought that the cell surface was not even a living part of the cell. We now know that each type of cell has a different set of proteins in its cell membrane, and that some of these differences are responsible for forming the structure of the tissues and organs during development. Observations of fertilization and early embryonic development made by E. E. Just (1939) suggest-

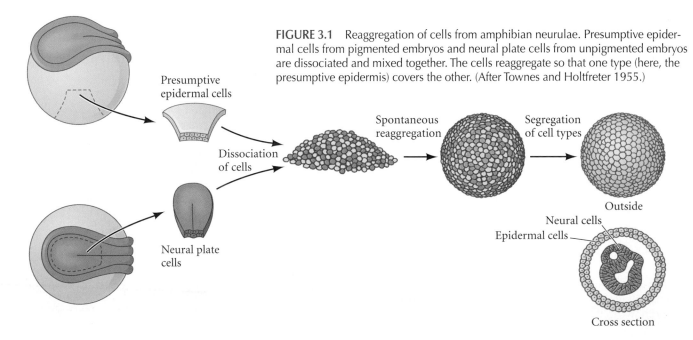

FIGURE 3.1   Reaggregation of cells from amphibian neurulae. Presumptive epidermal cells from pigmented embryos and neural plate cells from unpigmented embryos are dissociated and mixed together. The cells reaggregate so that one type (here, the presumptive epidermis) covers the other. (After Townes and Holtfreter 1955.)

Presumptive epidermal cells

Dissociation of cells

Neural plate cells

Spontaneous reaggregation

Segregation of cell types

Outside

Neural cells

Epidermal cells

Cross section

(A)

Epidermis
+
mesoderm

(B)

Mesoderm
+
endoderm

(C)

Epidermis
+
mesoderm
+
endoderm

(D)

Neural plate
+
epidermis

(E)

Neural plate
+
axial mesoderm
+
epidermis

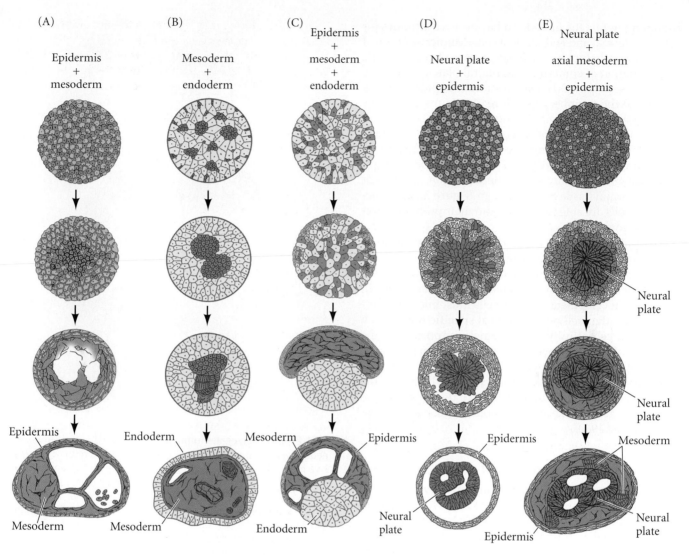

Epidermis

Mesoderm

Endoderm

Mesoderm

Mesoderm

Endoderm

Epidermis

Neural
plate

Neural
plate

Neural
plate

Neural
plate

Mesoderm

Epidermis

Neural
plate

FIGURE 3.2 Sorting out and reconstruction of spatial relationships in aggregates of embryonic amphibian cells. (After Townes and Holtfreter 1955.)

ed that the cell membrane differed among cell types, but the experimental analysis of morphogenesis began with the experiments of Townes and Holtfreter in 1955. Taking advantage of the discovery that amphibian tissues become dissociated into single cells when placed in alkaline solutions, they prepared single-cell suspensions from each of the three germ layers of amphibian embryos soon after the neural tube had formed. Two or more of these single-cell suspensions could be combined in various ways. When the pH of the solution was normalized, the cells adhered to one another, forming aggregates on agar-coated petri dishes. By using embryos from species having cells of different sizes and colors, Townes and Holtfreter were able to follow the behavior of the recombined cells.

The results of their experiments were striking. First, they found that reaggregated cells become spatially segregated. That is, instead of two cell types remaining mixed, each type sorts out into its own region. Thus, when epidermal (ectodermal) and mesodermal cells are brought together in a mixed aggregate, the epidermal cells move to the periphery of the aggregate and the mesodermal cells move

to the inside (**Figure 3.1**). In no case do the recombined cells remain randomly mixed; in most cases, one tissue type completely envelops the other.

Second, the researchers found that the final positions of the reaggregated cells reflect their respective positions in the embryo. The reaggregated mesoderm migrates centrally with respect to the epidermis, adhering to the inner epidermal surface (**Figure 3.2A**). The mesoderm also migrates centrally with respect to the gut or endoderm (**Figure 3.2B**). However, when the three germ layers are mixed together, the endoderm separates from the ectoderm and mesoderm and is then enveloped by them (**Figure 3.2C**). In the final configuration, the ectoderm is on the periphery, the endoderm is internal, and the mesoderm lies in the region between them.

Holtfreter interpreted this finding in terms of **selective affinity**. The inner surface of the ectoderm has a positive affinity for mesodermal cells and a negative affinity for the endoderm, while the mesoderm has positive affinities for both ectodermal and endodermal cells. Mimicry of normal embryonic structure by cell aggregates is also seen in the recombination of epidermis and neural plate cells (**Figure 3.2D**; also see Figure 3.1). The presumptive epidermal cells migrate to the periphery as before; the neural plate cells migrate inward, forming a structure reminiscent of the neural tube. When axial mesoderm (notochord) cells are added to a suspension of presumptive epidermal and presumptive neural cells, cell segregation results in an external epidermal layer, a centrally located neural tissue, and a layer of mesodermal tissue between them (**Figure 3.2E**). Somehow, the cells are able to sort out into their proper embryonic positions.

The third conclusion of Holtfreter and his colleagues was that selective affinities change during development. Such changes should be expected, because embryonic cells do not retain a single stable relationship with other cell types. For development to occur, cells must interact differently with other cell populations at specific times. Such changes in cell affinity are extremely important in the processes of morphogenesis. When tissues from later-stage mammalian and chick embryos were made into single cell suspensions (using the enzyme trypsin, which split the proteins connecting the cells together), the cells reaggregated to form tissuelike arrangements (Moscona 1961; Giudice and Just 1962).

The thermodynamic model of cell interactions

Cells, then, do not sort randomly, but can actively move to create tissue organization. What forces direct cell movement during morphogenesis? In 1964, Malcolm Steinberg proposed the **differential adhesion hypothesis**, a model that sought to explain patterns of cell sorting based on thermodynamic principles. Using cells derived from trypsinized embryonic tissues, Steinberg showed that certain cell types migrate centrally when combined with some cell types, but migrate peripherally when combined with others. **Figure 3.3** illustrates the interactions between pigmented retina cells and neural retina cells. When single-cell suspensions of these two cell types are mixed together, they form aggregates of randomly arranged cells. However, after several hours, no pigmented retina cells are seen on the periphery of the aggregates, and after 2 days, two distinct layers are seen, with the pigmented retina cells lying internal to the neural retina cells. Moreover, such interactions form a hierarchy (Steinberg 1970). If the final position of cell type A is internal to a second cell type B, and the final position of B is internal to a third cell type C, then the final position of A will always be internal to C. For example, pigmented retina cells migrate internally to neural retina cells, and heart cells migrate internally to pigmented retina cells. Therefore, heart cells migrate internally to neural retina cells.

This observation led Steinberg to propose that cells interact so as to form an aggregate with the smallest inter-

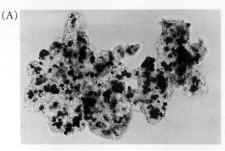

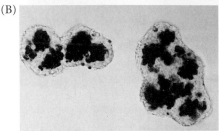

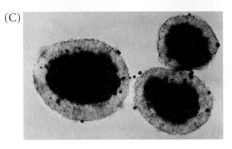

FIGURE 3.3   Aggregates formed by mixing 7-day chick embryo neural retina (unpigmented) cells with pigmented retina cells. (A) Five hours after the single-cell suspensions are mixed, aggregates of randomly distributed cells are seen. (B) At 19 hours, the pigmented retina cells are no longer seen on the periphery. (C) At 2 days, a great majority of the pigmented retina cells are located in a central internal mass, surrounded by the neural retina cells. (The scattered pigmented cells are probably dead cells.) (From Armstrong 1989, courtesy of P. B. Armstrong.)

facial free energy. In other words, the cells rearrange themselves into the most thermodynamically stable pattern. If cell types A and B have different strengths of adhesion, and if the strength of A-A connections is greater than the strength of A-B or B-B connections, sorting will occur, with the A cells becoming central. On the other hand, if the strength of A-A connections is less than or equal to the strength of A-B connections, then the aggregate will remain as a random mix of cells. Finally, if the strength of A-A connections is far greater than the strength of A-B connections—in other words, if A and B cells show essentially no adhesivity toward one another—then A cells and B cells will form separate aggregates. According to this hypothesis, the early embryo can be viewed as existing in an equilibrium state until some change in gene activity changes the cell surface molecules. The movements that result seek to restore the cells to a new equilibrium configuration. All that is required for sorting to occur is that cell types differ in the strengths of their adhesion.

Tissue	Surface tension (dyne/cm)	Equilibrium configuration
Limb bud (green)	20.1	
Pigmented epithelium (red)	12.6	
Heart (yellow)	8.5	
Liver (blue)	4.6	
Neural retina (orange)	1.6	

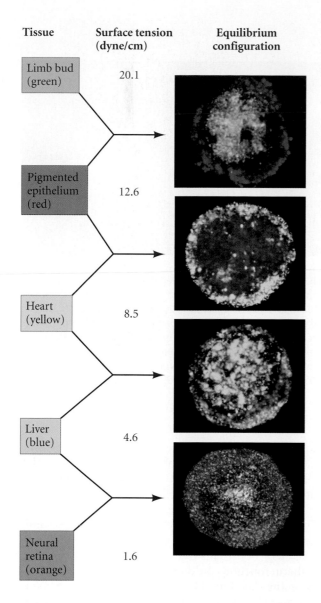

FIGURE 3.4 Hierarchy of cell sorting in order of decreasing surface tensions. The equilibrium configuration reflects the strength of cell cohesion, with the cell types having the greater cell cohesion segregating inside the cells with less cohesion. The images were obtained by sectioning the aggregates and assigning colors to the cell types by computer. The black areas represent cells whose signal was edited out in the program of image optimization. (From Foty et al. 1996, courtesy of M. S. Steinberg and R. A. Foty.)

In 1996, Foty and his colleagues in Steinberg's laboratory demonstrated that this was indeed the case: the cell types that had greater surface cohesion migrated centrally compared with those cells that had less surface tension (**Figure 3.4**; Foty et al. 1996). In the simplest form of this model, all cells could have the same type of "glue" on the cell surface. The amount of this cell surface product, or the cellular architecture that allows the substance to be differentially distributed across the surface, could cause a difference in the number of stable contacts made between cell types. In a more specific version of this model, the thermodynamic differences could be caused by different types of adhesion molecules (see Moscona 1974). When Holtfreter's studies were revisited using modern techniques, Davis and colleagues (1997) found that the tissue surface tensions of the individual germ layers were precisely those required for the sorting patterns observed both in vitro and in vivo.

See VADE MECUM
The differential adhesion hypothesis

Cadherins and cell adhesion

Recent evidence shows that boundaries between tissues can indeed be created by different cell types having both different types and different amounts of cell adhesion molecules. Several classes of molecules can mediate cell adhesion, but the major cell adhesion molecules appear to be the cadherins.

As their name suggests, **cadherins** are *ca*lcium-*d*ependent ad*he*sion molecules. They are critical for establishing and maintaining intercellular connections, and they appear to be crucial to the spatial segregation of cell types and to the organization of animal form (Takeichi 1987). Cadherins are transmembrane proteins that interact with other cadherins on adjacent cells. The cadherins are anchored inside the cell by a complex of proteins called **catenins** (Figure 3.5A), and the cadherin-catenin complex forms the classic adherens junctions that help hold epithelial cells together. Moreover, since the cadherins and the catenins bind to the actin (microfilament) cytoskeleton of the cell, they integrate the epithelial cells into a mechanical unit. Interfering with cadherin function (by univalent antibodies against cadherin or morpholinos against cadherin mRNA) can prevent the formation of tissues and cause the cells to disaggregate (**Figure 3.5B**; Takeichi et al. 1979).

Cadherin proteins perform several related functions. First, their external domains serve to adhere cells together. Second, cadherins link to and help assemble the actin cytoskeleton, thereby providing the mechanical forces for forming tubes. Third, cadherins can serve as signaling molecules that change a cell's gene expression.

In vertebrate embryos, several major cadherin types have been identified. **E-cadherin** is expressed on all early mammalian embryonic cells, even at the zygote stage. Later in development, this molecule is restricted to epithelial tissues of embryos and adults. **P-cadherin** is found predominantly on the placenta, where it helps the placenta stick to the uterus (Nose and Takeichi 1986; Kadokawa et al. 1989). **N-cadherin** becomes highly expressed on the cells of the developing central nervous system (Hatta and Takeichi 1986), and it may play roles in mediating neural signals. **R-cadherin** is critical in retina formation (Babb et al. 2005). A class of cadherins called **protocadherins** (Sano et al. 1993) lack the attachment to the actin skeleton through catenins. Expressing similar protocadherins is an impor-

(A)

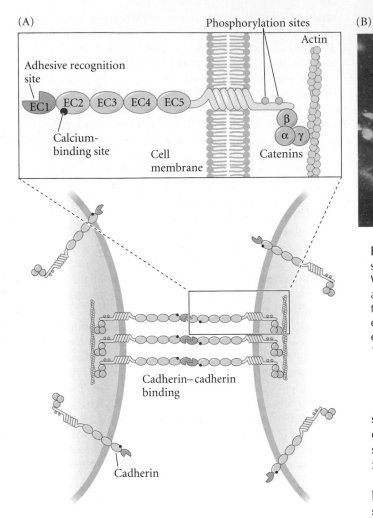

Phosphorylation sites

Actin

Adhesive recognition site

EC1 EC2 EC3 EC4 EC5

Calcium-binding site

Cell membrane

β

α γ

Catenins

Cadherin–cadherin binding

Cadherin

(B)

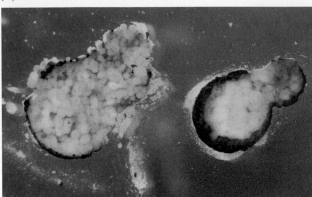

FIGURE 3.5   Cadherin-mediated cell adhesion. (A) Simplified scheme of cadherin linkage to the cytoskeleton via catenins. (B) When an oocyte is injected with an antisense oligonucleotide against a maternally inherited cadherin mRNA (thus preventing the synthesis of the cadherin), the inner cells of the resulting embryo disperse when the animal cap is removed (left). In control embryos (right), the inner cells remain together. (A after Takeichi 1991; B from Heasman et al. 1994, courtesy of J. Heasman.)

tant means of keeping migrating epithelial cells together; and expressing dissimilar protocadherins is an important way of separating tissues (as when the mesoderm forming the notochord separates from the surrounding mesoderm that will form somites).

Differences in cell surface tension and the tendency of cells to bind together depend on the strength of cadherin interactions (Duguay et al. 2003). This strength can be achieved quantitatively (the more cadherins on the apposing cell surfaces, the tighter the adhesion) or qualitatively (some cadherins will bind to different cadherin types, whereas other cadherins will not bind to different types). The ability to sort cells based on the *amount* of cadherin was first shown when Steinberg and Takeichi (1994) collaborated on an experiment using two cell lines that were identical except that they synthesized different amounts of P-cadherin. When these two groups of cells, each expressing a different amount of cadherin, were mixed, the cells that expressed more cadherin had a higher surface cohesion and migrated internally to the lower-expressing group of cells. Foty and Steinberg (2005) demonstrated that this cadherin-dependent sorting directly correlated with the aggregate surface tension (**Figure 3.6**). The surface ten-

sions of these aggregates are linearly related to the amount of cadherin they are expressing on the cell surface. The cell sorting hierarchy is strictly dependent on the cadherin interactions between the cells.

Moreover, the energetic value of cadherin-cadherin binding is remarkably strong—about 3400 kcal/mole, or some 200 times stronger than most metabolic protein-protein interactions. This free energy change associated with cadherin function could be dissipated by depolymerizing the actin skeleton. The underlying actin cytoskeleton appears to be crucial in organizing the cadherins in a manner that allows them to form remarkably stable linkages between cells (Foty and Steinberg 2004).

Qualitative interactions are also important. Duguay and colleagues (2003) showed, for instance, that R-cadherin and β-cadherin do *not* bind well to each other, and in these interactions the type of cadherin expressed becomes important. In another example, the expression of N-cadherin is important in separating the precursors of the neural cells from the precursors of the epidermal cells. All early embryonic cells originally contain E-cadherin, but those cells destined to become the neural tube lose E-cadherin and gain N-cadherin. If epidermal cells are experimentally made to express N-cadherin, or if N-cadherin synthesis is blocked in the prospective neural cells, the border between the nervous system and skin fails to form properly (**Figure 3.7**; Kintner 1993).

The timing of particular developmental events can also depend on cadherin expression. For instance, N-cadherin appears in the mesenchymal cells of the developing chick leg just before these cells condense and form nodules of

(A)

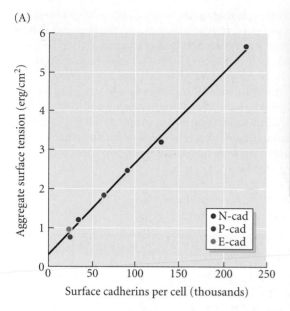

(B)

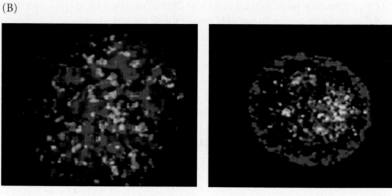

FIGURE 3.6 Importance of the amount of cadherin for correct morphogenesis. (A) Aggregate surface tension correlates with the number of cadherin molecules on the cell membranes. (B) Sorting out of two subclones having different amounts of cadherin on their cell surfaces. The green-stained cells had 2.4 times as many N-cadherin molecules in their membrane as did the other cells. (These cells had no normal cadherin genes being expressed.) At 4 hours of incubation, the cells are randomly distributed, but after 24 hours of incubation, the red cells (with a surface tension of about 2.4 erg/cm^2) have formed an envelope around the more tightly cohering (5.6 erg/cm^2) green cells. (After Foty and Steinberg 2005, photographs courtesy of R. Foty.)

cartilage (which are the precursors of the limb skeleton). N-cadherin is not seen prior to condensation, nor is it seen afterward. If the limbs are injected just prior to condensation with antibodies that block N-cadherin, the mesenchyme cells fail to condense and cartilage fails to form (Oberlander and Tuan 1994). It therefore appears that the signal to begin cartilage formation in the chick limb is the appearance of N-cadherin.

During development, the many cadherins often work with other adhesion systems. For example, one of the most critical times in a mammal's life occurs soon after conception, as the embryo passes from the oviduct and enters the uterus. If development is to continue, the embryo must adhere to and embed itself in the uterine wall. That is why the first differentiation event in mammalian development

FIGURE 3.7 Importance of the types of cadherin for correct morphogenesis. (A–C) Neural and epidermal tissues in a cross section of a mouse embryo showing the domains of E-cadherin expression (B) and N-cadherin expression (C). N-cadherin is critical for separation of presumptive epidermal and neural tissues during organogenesis. (D,E) The neural tube separates cleanly from surface epidermis in wild-type zebrafish embryos (D) but not in mutant embryos where N-cadherin fails to be made (E). In these images, the cell outlines are stained green with antibodies to β-catenin, while the cell interiors are stained blue. (B,C, photographs by K. Shimamura and H. Matsunami, courtesy of M. Takeichi; D,E from Hong and Brewster 2006, courtesy of R. Brewster.)

(A)

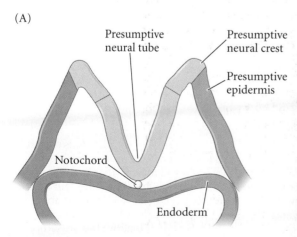

(B) E-cadherin expression

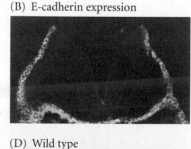

(C) N-cadherin expression

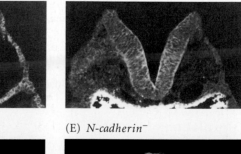

(D) Wild type

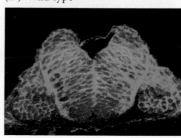

(E) *N-cadherin*⁻

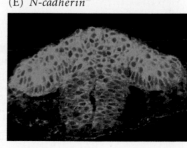

distinguishes the **trophoblast** cells (the outer cells that bind to the uterus) from the **inner cell mass** (those cells that will generate the embryo and eventually the mature organism). This differentiation process occurs as the embryo travels from the upper regions of the oviduct on its way to the uterus.

Trophoblast cells are endowed with several adhesion molecules that anchor the embryo to the uterine wall. First, they contain both E- and P-cadherins (Kadokawa et al. 1989), and these two molecule types recognize similar cad-

herins on the uterine cells. Second, they have receptors (integrin proteins) for the collagen and the heparan sulfate glycoproteins of the uterine wall (Farach et al. 1987; Carson et al. 1988, 1993; Cross et al. 1994). Third, trophoblast cell surfaces have a modified glycosyltransferase enzyme that extends out from the cell membrane and can bind to specific carbohydrate residues on uterine glycoproteins (Dutt et al. 1987). For something as important as the implantation of the mammalian embryo, it is not surprising that several cell adhesion systems appear to work together.

SIDELIGHTS & SPECULATIONS

Shape Change and Epithelial Morphogenesis: "The Force Is Strong in You"

Epithelial cells form sheets and tubes. Their ability to form such structures often depends on cell shape changes that usually involve cadherins and the actin cytoskeleton. The extracellular domains of cadherins bind groups of cells together, while the intracellular domains of the cadherins alter the actin cytoskeleton. The proteins mediating this cadherin-dependent remodeling of the cytoskeleton are usually (1) small GTPases, which convert soluble actin into fibrous actin cables that anchor at the cadherins and (2) non-muscle myosin, which provides the energy for actin contraction. Two examples of cadherin-dependent remodeling of the cytoskeleton are the formation of the neural tube in vertebrates and the internalization of the mesoderm in *Drosophila*. In both cases, the cells (neural ectoderm in frogs, mesoderm in *Drosophila*) are on the outside of the embryo, and it is critical that they migrate to the inside.

Involution of the frog neural tube

In the early frog embryo, each cell's membrane can contain several types of cadherins. Each cell of the gastrula is covered with C-cadherin. However, the presumptive *neural tube* ectoderm cell membranes also contain N-cadherin, concentrated in the apical

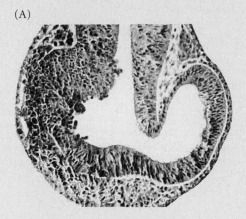 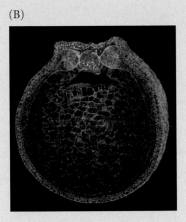

(A)   (B)

Figure 3.8   Importance of cadherin in cell adhesion and morphogenetic movements. (A) Frog gastrula injected with a nonfunctioning N-cadherin gene on one side. The uninjected side (right) develops normally; on the injected side (left), the epidermis and neural tissue fail to separate. (B) Cross section of a *Xenopus* neurula stained for F-actin (green). The red-staining cells of the neural plate (uppermost cells) and notochord (the mesodermal rod beneath them) have in them a morpholino oligonucleotide that prevents N-cadherin mRNA from being translated. The neural plate cells fail to invaginate into the embryo or to form a neural tube because of the loss of the N-cadherin-based actin assembly in their apical cytoplasm. (A from Kintner et al. 1992, courtesy of C. Kintner; B from Nandadasa et al. 2009, courtesy of C. Wylie.)

region of each of these cells; the presumptive *epidermal* cells of the ectoderm contain E-cadherin, which is expressed on the lateral and basal surfaces of the cell. The actin organized in the apical region of the neural cells causes them to change shape and enter the internal region of the embryo as a neural tube. The actin organized on the lateral sides of the epidermal cells enables the movements of the

epidermal (skin) cells over the surface of the embryo. If N-cadherin is experimentally removed from a frog gastrula, the cells still adhere (thanks to the C-cadherin that is still present), but the actin (and the activated myosin that binds to it) fail to assemble apically, so there is no neurulation: the presumptive neural cells do not enter the embryo, and no neural tube forms (**Figure 3.8**).

(A)

Ventral furrow

Pole cells

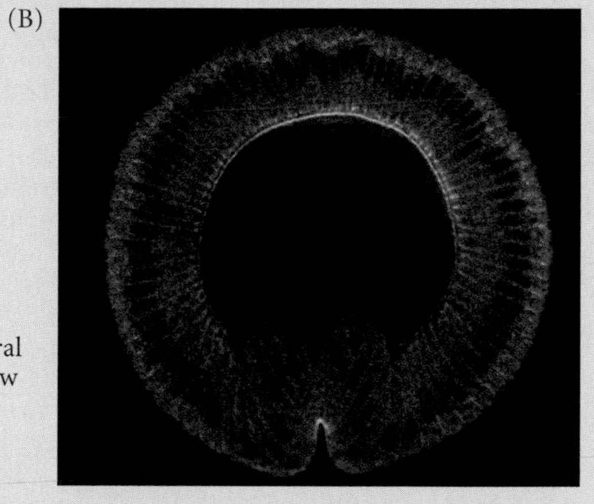

(B)

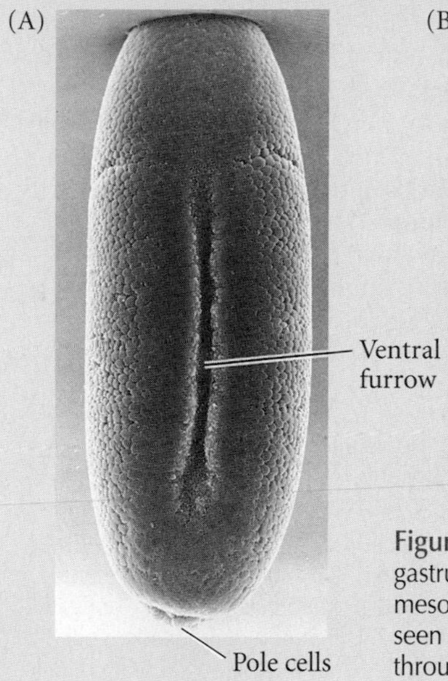

Figure 3.9   Ventral furrow formation during *Drosophila* gastrulation internalizes the cells that will become the mesoderm. (A) Ventral furrow in the *Drosophila* embryo seen by scanning electron microscopy. (B) Cross section through the center of such an embryo, demonstrating cell shape changes and the redistribution of protein products along the dorsal-ventral axis. Apical constrictions of the ventral cells can be observed. (A courtesy of F. R. Turner; B courtesy of V. Kölsch and M. Leptin.)

Drosophila mesoderm formation

In *Drosophila*, the mesoderm is formed from epithelial cells on the ventral side of the embryo. These cells form a furrow and then migrate inside the embryo (**Figure 3.9**). To create this furrow, the cube-shaped cells become wedge-shaped, constricting at their apical surfaces. This transition creates a force that pushes the ventral cells inside the embryo. What creates this force? The apical constriction is brought about by the rearrangement of actin microfilaments and myosin II (a "non-skeletal myosin") to the apical end of the cell (**Figure 3.10**). Actin microfilaments are part of the cytoskeleton and are often found on the periphery of the cell. (Indeed, they are critical for producing the cleavage furrows of cell division.) The instructions for this apical constriction appear to emanate from the *Twist* gene, which

(*Continued on next page*)

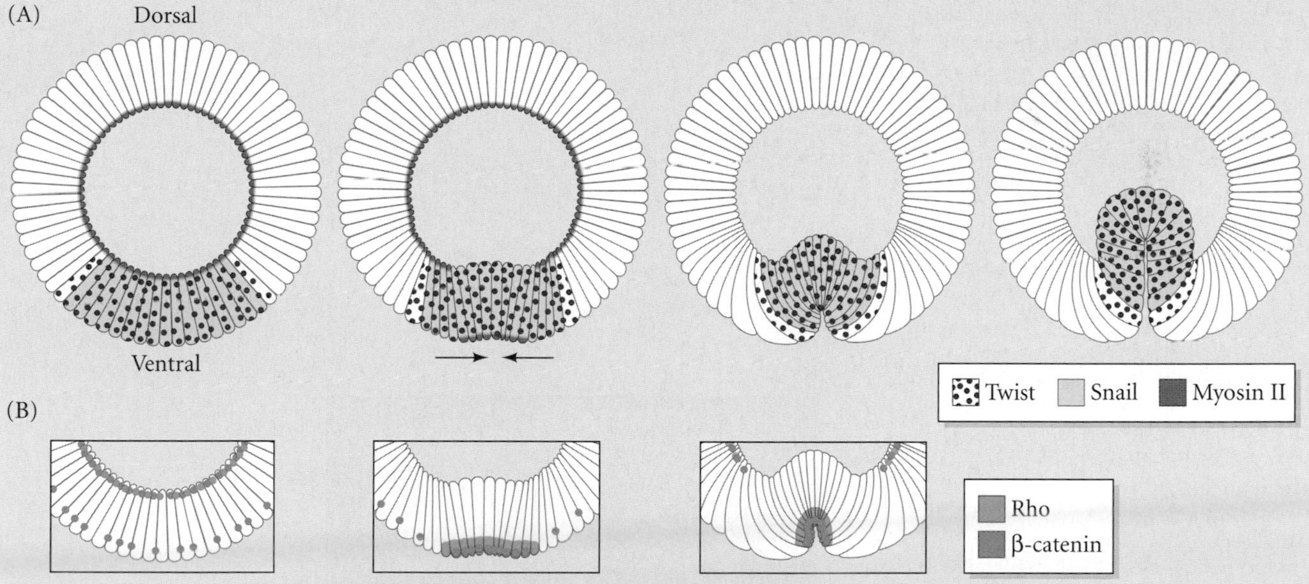

(A)   Dorsal

Ventral

Twist   Snail   Myosin II

(B)

Rho
β-catenin

Figure 3.10   Getting mesodermal cells inside the embryo during *Drosophila* gastrulation by regulation of the cytoskeleton. (A) Schematic representation of ventral furrow formation shown as cross sections through the *Drosophila* embryo, progressing in time from left to right. The ventral cells are defined by the expression of transcription factors Twist and Snail. These cells accumulate myosin II at their apical surfaces. When myosin II interacts with actin already present, the cells begin to constrict apically and thus invaginate. (B) Close-ups of the ventral domain. Before the initiation of ventral furrow formation, Rho (green) and β-catenin (orange) both reside along the basal surface (facing the interior of the embryo) of the ventral cells. β-catenin is also found in a subapical region in all cells. Formation of the ventral furrow begins with the relocalization of Rho and β-catenin, which move from the basal surface to accumulate apically, at the opposite end of the cell. (From Kölsch et al. 2007.)

is only expressed in the nuclei of the ventralmost cells (Kölsch et al. 2007). The Twist protein activates other genes, whose protein products cause the actin cytoskeleton to build up on the apical side of the cell. This build-up is accomplished by the binding of a small GTPase (Rho) and β-catenin to E-cadherin on the apical portion of the cell membrane in these most ventral cells. Once stabilized, the actin-myosin complex in the cell's apical cortex constricts like the drawstring of a purse, causing the cells to change shape, buckle inward, and enter the embryo to form the mesoderm.

External signals: Insect trachea

In the above cases, the instructions for folding come from inside the cell. Instructions for cell shape change can also arise outside the cell. For instance, the tracheal (respiratory) system in *Drosophila* embryos develops from epithelial sacs. The approximately 80 cells in each of these sacs become reorganized into primary, secondary, and tertiary branches without any cell division or cell death (Ghabrial and Krasnow 2006). This reorganization is initiated when nearby cells secrete a protein called Branchless, which acts as a chemoattractant.* Branchless binds to a receptor on the cell membranes of the epithelial cells. The cells receiving the most Branchless protein lead the rest, while the followers (con-

*Chemoattractants are usually diffusible molecules that attract a cell to migrate along an increasing concentration gradient toward the cells secreting the factor. There are also *chemorepulsive* factors that send the migrating cells in an opposite direction. Generally speaking, *chemotactic factors*—soluble factors that cause cells to move in a particular direction—are assumed to be chemoattractive unless otherwise described.

nected to each other by cadherins) receive a signal from the leading cells to form the tracheal tube (**Figure 3.11**). It is the lead cell that will change its shape (by rearranging its actin-myosin cytoskeleton via a small GTPase-mediated process, just like the mesodermal cells) to migrate and to form the secondary branches. During this migration, cadherin proteins are regulated such that the epithelial cells can migrate over one another to form a tube while keeping their integrity as an epithelium (Cela and Llimagas 2006).

But another external force is also at work. The dorsalmost secondary branches of the sac move along a groove that forms between the developing muscles. These tertiary cell migrations cause the trachea to become segmented around the musculature (Franch-Marro and Casanova 2000). In this way, the respiratory tubes are placed close to the larval musculature.

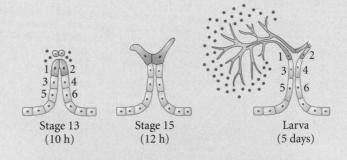

FIGURE 3.11 Tracheal development in *Drosophila*. (A) Diagram of dorsal tracheal branch budding from tracheal epithelium. Nearby cells secrete Branchless protein (Bnl; blue dots), which activates Breathless protein (Btl) on tracheal cells. The activated Btl induces migration of the leader cells and tube formation; the dorsal branch cells are numbered 1–6. Branchless also induces unicellular secondary branches (stage 15). (B) Larval *Drosophila* tracheal system visualized with a fluorescent red antibody. Note the intercalated branching pattern. (A after Ghabrial and Krasnow 2006; B from Casanova 2007.)

(A)

Bnl-secreting cells

Leader cell (Btl activated)

Bnl protein

Tracheal epithelium cells

Stage 11 (6 h)

Stage 12 (8 h)

1 2
3 4
5 6

Stage 13 (10 h)

Stage 15 (12 h)

1 2
3 4
5 6

Larva (5 days)

(B)

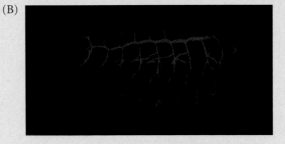

Cell Migration

Cell migration is a common feature of both epithelial and mesenchymal cells (Kurosaka and Kashina 2008). The cells of the embryo move extensively during gastrulation to

form the three germ layers; the neural tube folds into the vertebrate embryo; the mesoderm folds into the fly embryo; and the precursors of the germ cells, blood cells, and pigment cells undergo individual and extensive migrations. In epithelia, the motive force for migration is usual-

(A)   (B)

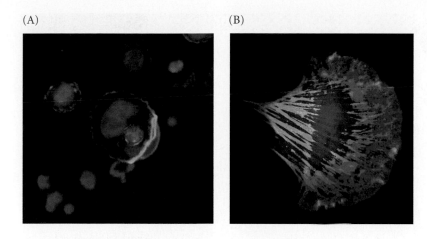

FIGURE 3.12   Cell migration. (A) Polarization of a migrating cell. Cell interiors are stained blue. The actin (stained green with antibodies to β-catenin) becomes redistributed by interstitial flow to the leading edge of the cell. (B) In the lamellipodium of a migrating mesenchymal cell, the ratio of filamentous actin to globular actin is visualized by different colors, blue being the highest filamentous actin level, red being the lowest. (A from Shields et al. 2007, courtesy of M. A. Swartz; B from Cramer et al. 2002, courtesy of L. Cramer.)

ly provided by the cells at the edge of the sheet, and the rest of the cells follow passively. In mesenchymal cell migration, individual cells become polarized and migrate through the extracellular milieu. In both cases, there is a wide-scale reorganization of the actin cytoskeleton (**Figure 3.12A**). The first stage of migration is **polarization**, wherein a cell defines its front and its back. Polarization can be directed by diffusing signals (such as a chemotactic protein) or by signals from the extracellular matrix. These signals reorganize the cytoskeleton so that the cell has a front and a back, and so that the front part of the cell becomes structurally different from the back of the cell (Rodionov and Borisy 1997; Malikov et al. 2005).

The second stage of migration is the protrusion of the cell's leading edge. The mechanical force for this is the polymerization of the actin microfilaments at the cell membrane, creating long parallel bundles (forming filopodia) or broad sheets (forming lamellipodia; **Figure 3.12B**.) The membrane-bound Rho G-proteins activate the WASP-N proteins to nucleate actin and connect it to cadherins and the cell membrane (Co et al. 2007).

The third stage of migration involves the **adhesion** of the cell to its extracellular substrate. The moving cell needs something to push on, and attaches to the surrounding matrix. The key molecules in this process (as we will detail later in this chapter) are *integrin* proteins. Integrins span the cell membrane, connecting the extracellular matrix outside the cell to the actin cytoskeleton on the inside of the cell. These connections of actin to integrin form **focal adhesions** on the cell membrane where the membrane contacts the extracellular matrix. Myosin and its regulators provide the motive force along these actin microfilaments, and they are linked with the lamellipodial actin at the sites of adhesion (Giannone et al. 2007).

The fourth stage of cell migration concerns the release of adhesions in the rear, allowing the cell to migrate in the forward direction. It is probable that stretch-sensitive calcium channels are opened and that the released calcium ions activate proteases that destroy the focal adhesion sites.

Cell Signaling

Induction and competence

From the earliest stages of development through the adult, cell differentiation and behavior (such as adhesion, migration, and cell division) are regulated by signals from one cell being received by another cell. Indeed, these interactions (which are often reciprocal) are what allow organs to be constructed. The development of the vertebrate eye is a classic example used to describe the modus operandi of tissue organization via intercellular interactions.

In the vertebrate eye, light is transmitted through the transparent corneal tissue and focused by the lens tissue (the diameter of which is controlled by muscle tissue), eventually impinging on the tissue of the neural retina. The precise arrangement of tissues in the eye cannot be disturbed without impairing its function. Such coordination in the construction of organs is accomplished by one group of cells changing the behavior of an adjacent set of cells, thereby causing them to change their shape, mitotic rate, or cell fate. This kind of interaction at close range between two or more cells or tissues of different histories and properties is called **induction**.

There are at least two components to every inductive interaction. The first component is the **inducer**: the tissue that produces a signal (or signals) that changes the cellular behavior of the other tissue. Often, this signal is a secreted protein called a *paracrine factor*. **Paracrine factors** are proteins made by a cell or a group of cells that alter the behavior or differentiation of adjacent cells. In contrast to endocrine factors (i.e., hormones), which travel through the blood and exert their effects on cells and tissues far away, paracrine factors are secreted into the extracellular space and influence their close neighbors. (The Branchless protein, mentioned in the Sidelights & Speculations on p. 78, is such a factor). The second component, the **responder**, is the tissue being induced. Cells of the responding tissue must have both a receptor protein for the inducing

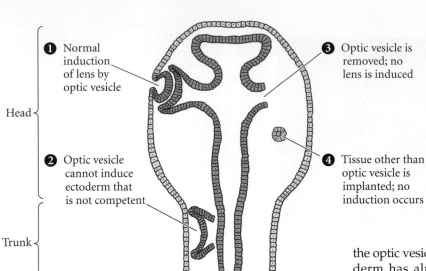

Head

❶ Normal
induction
of lens by
optic vesicle

❷ Optic vesicle
cannot induce
ectoderm that
is not competent

Trunk

❸ Optic vesicle is
removed; no
lens is induced

❹ Tissue other than
optic vesicle is
implanted; no
induction occurs

FIGURE 3.13   Ectodermal competence and the ability to respond to the optic vesicle inducer in *Xenopus*. The optic vesicle is able to induce lens formation in the anterior portion of the ectoderm (1) but not in the presumptive trunk and abdomen (2). If the optic vesicle is removed (3), the surface ectoderm forms either an abnormal lens or no lens at all. (4) Most other tissues are not able to substitute for the optic vesicle.

factor (the receptor for Branchless is the Breathless protein) and the *ability* to respond to the signal. The ability to respond to a specific inductive signal is called **competence** (Waddington 1940).

Even if receptor proteins are present, not every tissue type is competent to respond to an inducer's signal. For instance, if the optic vesicle (the presumptive retina) of a *Xenopus laevis* embryo is placed in an ectopic location underneath the head ectoderm (i.e., in a different part of the head from where the frog's optic vesicle normally occurs), it will induce that ectoderm to form lens tissue. Only the optic vesicle appears to be able to do this; therefore, it is an inducer. However, if the optic vesicle is placed beneath ectoderm in the flank or abdomen of the same organism, that ectoderm will not be able to form lens tissue. Only head ectoderm is *competent* to respond to the signals from the optic vesicle by producing a lens (**Figure 3.13**; Saha et al. 1989; Grainger 1992).

Often, one induction will give a tissue the competence to respond to another inducer. Studies on amphibians suggest that the first inducers of the lens may be the foregut endoderm and heart-forming mesoderm that underlie the lens-forming ectoderm during the early and mid gastrula stages (Jacobson 1963, 1966). The anterior neural plate may produce the next signals, including a signal that promotes the synthesis of Pax6 transcription factor in the anterior ectoderm (**Figure 3.14**; Zygar et al. 1998). Pax6 is important in providing the competence to respond to the inducers from the optic cup (Fujiwara et al. 1994). Thus, although

the optic vesicle appears to be *the* inducer, the anterior ectoderm has already been induced by at least two other tissues. (The situation is like that of the player who kicks the "winning goal" of a soccer match.) The optic vesicle appears to secrete two paracrine factors, one of which may be BMP4 (Furuta and Hogan 1998), a protein that is received by the lens cells and induces the production of the Sox2 transcription factors. The other paracrine factor is Fgf8, a signal that induces the appearance of the L-Maf transcription factor (Ogino and Yasuda 1998; Vogel-Höpker et al. 2000). As we saw in Chapter 2, the combination of Pax6, Sox2, and L-Maf in the ectoderm is needed for the production of the lens and the activation of lens-specific genes such as δ-crystallin.

Cascades of induction: Reciprocal and sequential inductive events

Another feature of induction is the *reciprocal nature* of many inductive interactions. To continue the above example, once the lens has formed, it induces other tissues. One of these responding tissues is the optic vesicle itself; thus the inducer becomes the induced. Under the influence of factors secreted by the lens, the optic vesicle becomes the optic cup and the wall of the optic cup differentiates into two layers, the pigmented retina and the neural retina (**Figure 3.15**; Cvekl and Piatigorsky 1996; Strickler et al. 2007). Such interactions are called **reciprocal inductions**.

At the same time, the lens is inducing the ectoderm above it to become the cornea. Like the lens-forming ectoderm, the cornea-forming ectoderm has achieved a particular competence to respond to inductive signals, in this case the signals from the lens (Meier 1977; Thut et al. 2001). Under the influence of the lens, the corneal ectoderm cells become columnar and secrete multiple layers of collagen. Mesenchymal cells from the neural crest use this collagen matrix to enter the area and secrete a set of proteins (including the enzyme hyaluronidase) that further differentiate the cornea. A third signal, the hormone thyroxine, dehydrates the tissue and makes it transparent (Hay 1980; Bard 1990). Thus, there are sequential inductive events, and multiple causes for each induction.

FIGURE 3.14 Lens induction in amphibians. (A) Additive effects of inducers, as shown by transplantation and extirpation (removal) experiments on the newt *Taricha torosa*. The ability to produce lens tissue is first induced by foregut endoderm, then by cardiac mesoderm, and finally by the optic vesicle. The optic vesicle eventually acquires the ability to induce the lens and retain its differentiation. (B) Sequence of induction postulated by similar experiments performed on embryos of the frog *Xenopus laevis*. Unidentified inducers (possibly from the foregut endoderm and cardiac mesoderm) cause the synthesis of the Otx2 transcription factor in the head ectoderm during the late gastrula stage. As the neural folds rise, inducers from the anterior neural plate (including the region that will form the retina) induce Pax6 expression in the anterior ectoderm that can form lens tissue. Expression of Pax6 protein may constitute the competence of the surface ectoderm to respond to the optic vesicle during the late neurula stage. The optic vesicle secretes factors (probably of the BMP family) that induce the synthesis of the Sox transcription factors and initiate observable lens formation. (A after Jacobson 1966; B after Grainger 1992.)

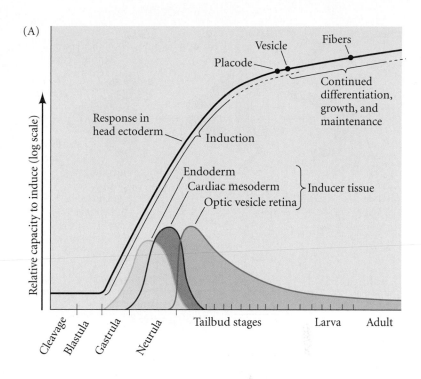

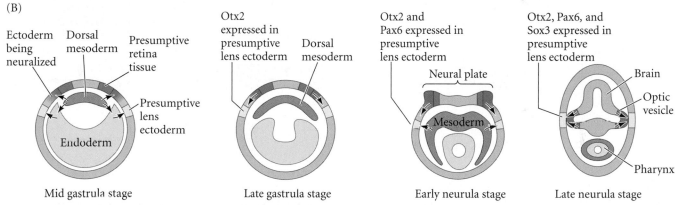

(B)

Mid gastrula stage

Late gastrula stage

Early neurula stage

Late neurula stage

Another principle can be seen in such reciprocal inductions: a structure does not need to be fully differentiated in order to have a function. As we will detail in Chapter 9, the optic vesicle induces the lens placode before it becomes the retina; the lens placode (the prospective lens) reciprocates by inducing the optic vesicle before the lens forms its characteristic fibers. Thus, before a tissue has its "adult" functions, it has critically important transient functions in building the organs of the embryo.

Instructive and permissive interactions

Howard Holtzer (1968) distinguished two major modes of inductive interaction. In **instructive interaction**, a signal from the inducing cell is necessary for initiating new gene expression in the responding cell. Without the inducing cell, the responding cell is not capable of differentiating in that particular way. For example, when the optic vesicle is experimentally placed under a new region of the head ecto-derm and causes that region of the ectoderm to form a lens, that is an instructive interaction.

The second type of inductive interaction is **permissive interaction**. Here, the responding tissue has already been specified, and needs only an environment that allows the expression of these traits. For instance, many tissues need a solid substrate containing *fibronectin* or *laminin* in order to develop. The fibronectin or laminin does not alter the type of cell that is produced, but it enables what has already been determined to be expressed.*

---

*It is easy to distinguish permissive and instructive interactions by an analogy with a more familiar situation. This textbook is made possible by both permissive and instructive interactions. A reviewer can convince me to change the material in the chapters. This is an instructive interaction, as the information expressed in the book is changed from what it would have been. However, the information in the book could not be expressed at all without permissive interactions with the publisher and printer.

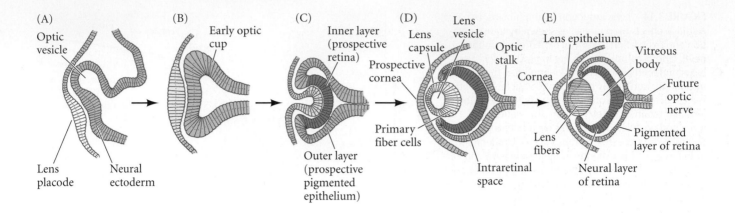

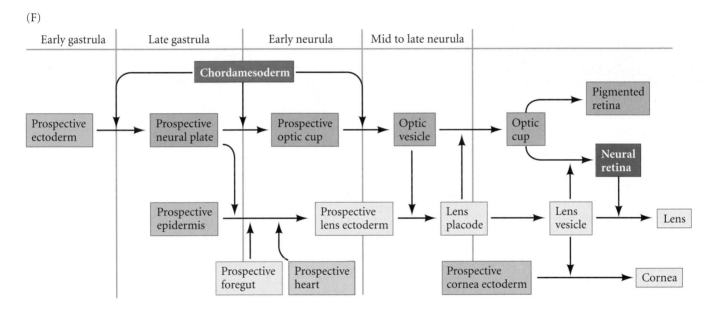

FIGURE 3.15 Schematic diagram of induction of the mouse lens. (A) At embryonic day 9, the optic vesicle extends toward the surface ectoderm from the forebrain. The lens placode (the prospective lens) appears as a local thickening of the surface ectoderm near the optic vesicle. (B) By the middle of day 9, the lens placode has enlarged and the optic vesicle has formed an optic cup. (C) By the middle of day 10, the central portion of the lens-forming ectoderm invaginates while the two layers of the retina become distinguished. (D) By the middle of day 11, the lens vesicle has formed. (E) By day 13, the lens consists of anterior cuboidal epithelial cells and elongating posterior fiber cells. The cornea develops in front of the lens. (F) Summary of some of the inductive interactions during eye development. (A–E after Cvekl and Piatigorsky 1996.)

Epithelial-mesenchymal interactions

Some of the best-studied cases of induction involve the interactions of sheets of epithelial cells with adjacent mesenchymal cells. All organs consist of an epithelium and an associated mesenchyme, so these **epithelial-mesenchymal interactions** are among the most important phenomena in nature. Some examples are listed in **Table 3.1**.

REGIONAL SPECIFICITY OF INDUCTION Using the induction of cutaneous (skin) structures as our examples, we will look at the properties of epithelial-mesenchymal interactions. The first of these properties is the regional specificity of induction. Skin is composed of two main tissues: an outer epidermis (an epithelial tissue derived from ectoderm), and a dermis (a mesenchymal tissue derived from mesoderm). The chick epidermis secretes proteins that signal the underlying dermal cells to form condensations, and the condensed dermal mesenchyme responds by secreting factors that cause the epidermis to form regionally specific cutaneous structures (**Figure 3.16**; Nohno et al. 1995; Ting-Berreth and Chuong 1996). These structures can be the broad feathers of the wing, the narrow feathers of the thigh, or the scales and claws of the feet.

As **Figure 3.17** demonstrates, the dermal mesenchyme is responsible for the regional specificity of induction in the competent epidermal epithelium. Researchers can sep-

(A)

(B)

FIGURE 3.16   Feather induction in the chick. (A) Feather tracts on the dorsum of a 9-day chick embryo. Note that each feather primordium is located between the primordia of adjacent rows. (B) In situ hybridization of a 10-day chick embryo shows Sonic hedgehog expression (dark spots) in the ectoderm of the developing feathers and scales. (A courtesy of P. Sengal; B courtesy of W.-S. Kim and J. F. Fallon.)

arate the embryonic epithelium and mesenchyme from each other and recombine them in different ways (Saunders et al. 1957). The same epithelium develops cutaneous structures according to the region from which the mesenchyme was taken. Here, the mesenchyme plays an instructive role, calling into play different sets of genes in the responding epithelial cells.

GENETIC SPECIFICITY OF INDUCTION   The second property of epithelial-mesenchymal interactions is the genetic specificity of induction. Whereas the mesenchyme may instruct the epithelium as to what sets of genes to activate, the responding epithelium can comply with these instructions only so far as its genome permits. This property was discovered through experiments involving the transplantation of tissues from one species to another.

TABLE 3.1   Some epithelial-mesenchymal interactions

Organ	Mesenchymal component	Epithelial component
Cutaneous structures (hair, feathers, sweat glands, mammary glands)	Epidermis (ectoderm)	Dermis (mesoderm)
Limb	Epidermis (ectoderm)	Mesenchyme (mesoderm)
Gut organs (liver, pancreas, salivary glands)	Epithelium (endoderm)	Mesenchyme (mesoderm)
Foregut and respiratory associated organs (lungs, thymus, thyroid)	Epithelium (endoderm)	Mesenchyme (mesoderm)
Kidney	Ureteric bud epithelium (mesoderm)	Mesenchyme (mesoderm)
Tooth	Jaw epithelium (ectoderm)	Mesenchyme (neural crest)

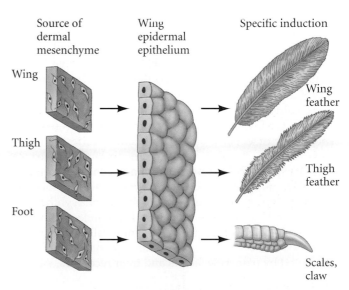

FIGURE 3.17   Regional specificity of induction in the chick. When cells from different regions of the dermis (mesenchyme) are recombined with the epidermis (epithelium), the type of cutaneous structure made by the epidermal epithelium is determined by the original source of the mesenchyme. (After Saunders 1980.)

In one of the most dramatic examples of interspecific induction, Hans Spemann and Oscar Schotté (1932) transplanted flank ectoderm from an early *frog* gastrula to the region of a *newt* gastrula destined to become parts of the mouth. Similarly, they placed presumptive flank ectodermal tissue from a *newt* gastrula into the presumptive oral regions of *frog* embryos. The structures of the mouth region differ greatly between salamander and frog larvae. The salamander larva has club-shaped balancers beneath its mouth, whereas the frog tadpole produces mucus-secreting glands and suckers (**Figure 3.18**). The frog tadpole also has a horny jaw without teeth, whereas the salamander has a set of calcareous teeth in its jaw. The larvae resulting from the transplants were chimeras. The salamander larvae had froglike mouths, and the frog tadpoles had salamander teeth and balancers. In other words, the mesenchymal cells instructed the ectoderm to make a mouth, but the ectoderm responded by making the only kind of mouth it "knew" how to make, no matter how inappropriate.*

Thus the instructions sent by the mesenchymal tissue can cross species barriers. Salamanders respond to frog inducers, and chick tissue responds to mammalian inducers. The response of the epithelium, however, is species-specific. So, whereas organ-type specificity (e.g., feather or claw) is usually controlled by the mesenchyme, species specificity is usually controlled by the responding epithelium. As we will see in Chapter 19, major evolutionary changes in the phenotype can be brought about by changing the response to a particular inducer.

Paracrine Factors: The Inducer Molecules

How are the signals between inducer and responder transmitted? While studying the mechanisms of induction that produce the kidney tubules and teeth, Grobstein (1956) and others (Saxén et al. 1976; Slavkin and Bringas 1976) found that some inductive events could occur despite a filter separating the epithelial and mesenchymal cells. Other inductions, however, were blocked by the filter. The researchers therefore concluded that some of the inductive molecules were soluble factors that could pass through the small pores of the filter, and that other inductive events required physical contact between the epithelial and mesenchymal cells. When cell membrane proteins on one cell surface interact with receptor proteins on adjacent cell surfaces, these events are called **juxtacrine interactions** (since the cell membranes are *juxtaposed*). When proteins synthesized by one cell can diffuse over small distances to induce changes in neighboring cells, the event is called a **paracrine interaction**. A we saw earlier, this type of interaction is mediated by paracrine factors and their receptors. We will

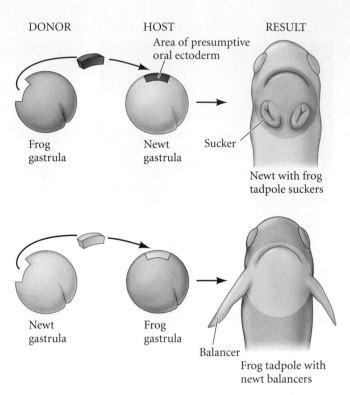

FIGURE 3.18 Genetic specificity of induction in amphibians. Reciprocal transplantation between the presumptive oral ectoderm regions of salamander and frog gastrulae leads to newts with tadpole suckers and tadpoles with newt balancers. (After Hamburgh 1970.)

DONOR HOST RESULT

Area of presumptive oral ectoderm

Frog gastrula

Newt gastrula

Sucker

Newt with frog tadpole suckers

Newt gastrula

Frog gastrula

Balancer

Frog tadpole with newt balancers

consider paracrine interactions first, returning to juxtacrine interactions later in the chapter.

Whereas **endocrine factors*** (hormones) travel through the blood to exert their effects, paracrine factors are secreted into the immediate spaces around the cell producing them. These proteins are the "inducing factors" of the classic experimental embryologists. There is considerable debate as to the distances at which paracrine factors can operate. The proteins Nodal and activin, for instance, can diffuse over many cell diameters and induce different sets of genes at different concentrations (Gurdon et al. 1994, 1995). The Wnt, Vg1, and BMP4 proteins, however, probably work only on their adjacent neighbors (Jones et al. 1996; Reilly and Melton 1996). These factors may induce the expression of other short-range factors from these neighbors, and a cascade of paracrine inductions can be initiated.

---

*Spemann is reported to have put it this way: "The ectoderm says to the inducer, 'you tell me to make a mouth; all right, I'll do so, but I can't make your kind of mouth; I can make my own and I'll do that.'" (Quoted in Harrison 1933.)

---

*Some endocrine factors are active in development; these include the hormones estrogen, testosterone, and thyroxine. They affect many tissues simultaneously and often coordinate development throughout the body (as in metamorphosis or the morphogenesis of sexual phenotypes). These hormones work directly, binding to a dormant transcription factor ("receptor"), thereby activating the transcription factor and allowing it to enter the nucleus and bind to DNA.

In addition to endocrine, paracrine, and juxtacrine interactions, there are also **autocrine** interactions. Autocrine interactions occur when the same cells that secrete paracrine factors also respond to them. In other words, the cell synthesizes a molecule for which it has its own receptor. Although autocrine regulation is not common, it is seen in placental cytotrophoblast cells; these cells synthesize and secrete platelet-derived growth factor, whose receptor is on the cytotrophoblast cell membrane (Goustin et al. 1985). The result is the explosive proliferation of that tissue.

Signal transduction cascades: The response to inducers

The induction of numerous organs is actually effected by a relatively small set of paracrine factors. The embryo inherits a rather compact genetic "tool kit" and uses many of the same proteins to construct the heart, kidneys, teeth, eyes, and other organs. Moreover, the same proteins are used throughout the animal kingdom: the factors active in creating the *Drosophila* eye or heart are very similar to those used in generating mammalian organs. Many of the paracrine factors can be grouped into one of four major families on the basis of their structure:

1. The fibroblast growth factor (FGF) family
2. The Hedgehog family
3. The Wnt family
4. The TGF-β superfamily, encompassing the TGF-β family, the activin family, the bone morphogenetic proteins (BMPs), the Nodal proteins, the Vg1 family, and several other related proteins.

Paracrine factors function by binding to a receptor that initiates a series of enzymatic reactions within the cell. These enzymatic reactions have as their end point either the regulation of transcription factors (such that different genes are expressed in the cells reacting to these paracrine factors) or the regulation of the cytoskeleton (such that the cells responding to the paracrine factors alter their shape or are permitted to migrate). These pathways of responses to the paracrine factor often have several end points and are called **signal transduction cascades**.

The major signal transduction pathways all appear to be variations on a common and rather elegant theme, exemplified in **Figure 3.19**. Each receptor spans the cell membrane and has an extracellular region, a transmembrane region, and a cytoplasmic region. When a ligand (here, the paracrine factor) binds to its receptor's extracellular domain, that ligand induces a conformational change in the receptor's structure. This shape change is transmitted through the membrane and alters the shape of the receptor's cytoplasmic domains. This conformational change in cytoplasmic domains gives those domains enzymatic activity—usually a kinase activity that can use ATP to phosphorylate specific tyrosine residues of particular proteins. Thus, this type of receptor is often called a **receptor tyrosine kinase** (**RTK**). The active receptor can now

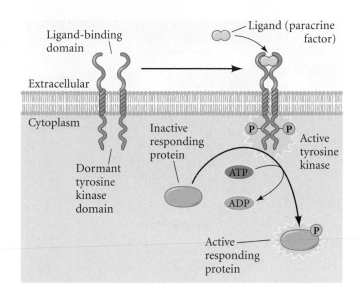

FIGURE 3.19   Structure and function of a receptor tyrosine kinase. The binding of a paracrine factor (such as Fgf8) by the extracellular portion of the receptor protein activates the dormant tyrosine kinase, whose enzyme activity phosphorylates specific tyrosine residues of certain proteins.

catalyze reactions that phosphorylate other proteins, and this phosphorylation in turn activates their latent activities. Eventually, the cascade of phosphorylation activates a dormant transcription factor or a set of cytoskeletal proteins.

Fibroblast growth factors and the RTK pathway

The **fibroblast growth factor** (**FGF**) family of paracrine factors comprises nearly two dozen structurally related members, and the FGF genes can generate hundreds of protein isoforms by varying their RNA splicing or initiation codons in different tissues (Lappi 1995). Fgf1 protein is also known as acidic FGF and appears to be important during regeneration (Yang et al. 2005); Fgf2 is sometimes called basic FGF and is very important in blood vessel formation; and Fgf7 sometimes goes by the name of keratinocyte growth factor and is critical in skin development. Although FGFs can often substitute for one another, the expression patterns of the FGFs and their receptors give them separate functions. In *Drosophila*, Breathless is an FGF protein.

One member of this family, Fgf8, is especially important during limb development and lens induction. Fgf8 is usually made by the optic vesicle that contacts the outer ectoderm of the head (**Figure 3.20**; Vogel-Höpker et al. 2000). After contact with the outer ectoderm occurs, *fgf8* gene expression becomes concentrated in the region of the presumptive neural retina—the tissue directly apposed to the presumptive lens. Moreover, if Fgf8-containing beads are placed adjacent to head ectoderm, this ectopic Fgf8 will

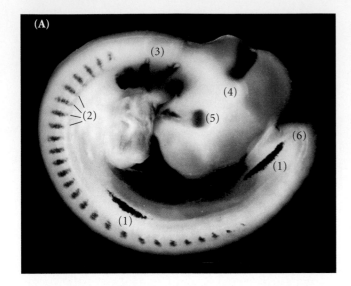

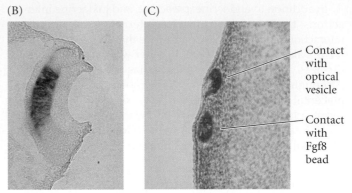

FIGURE 3.20 Fgf8 in the developing chick. (A) *Fgf8* gene expression pattern in the 3-day chick embryo, shown by in situ hybridization. Fgf8 protein (dark areas) is seen in the distalmost limb bud ectoderm (1); in the somitic mesoderm (the segmented blocks of cells along the anterior-posterior axis; 2); in the branchial arches of the neck (3); at the boundary between the midbrain and hindbrain (4); in the developing eye (5); and in the tail (6). (B,C) Fgf8 function in the developing eye. (B) In situ hybridization of *fgf8* in the optic vesicle. The *fgf8* mRNA (purple) is localized to the presumptive neural retina of the optic cup and is in direct contact with the outer ectoderm cells that will become the lens. (C) Ectopic expression of L-Maf in competent ectoderm can be induced by the optic vesicle (above) and by an Fgf8-containing bead (below). (A courtesy of E. Laufer, C.-Y. Yeo, and C. Tabin; B,C courtesy of A. Vogel-Höpker.)

induce this ectoderm to produce ectopic lenses and to express the lens-associated transcription factor L-Maf (see Figure 3.20B). FGFs often work by activating a set of receptor tyrosine kinases called the **fibroblast growth factor receptors (FGFRs)**. The Branchless protein is an FGF receptor in *Drosophila*.

When an FGFR binds an FGF (and only when it binds an FGF), the dormant kinase is activated and phosphorylates certain proteins (including other FGFRs) within the responding cell. These proteins, once activated, can perform new functions. The **RTK pathway** was one of the first signal transduction pathways to unite various areas of developmental biology (**Figure 3.21**). Researchers studying *Drosophila* eyes, nematode vulvae, and human cancers found that they were all studying the same genes. The pathway begins at the cell surface, where an RTK binds its specific ligand. Ligands that bind to RTKs include the fibroblast growth factors, epidermal growth factors, platelet-derived growth factors, and stem cell factor. Each RTK can bind only one or a small set of these ligands. The RTK spans the cell membrane, and when it binds its ligand, it undergoes a conformational change that enables it to dimerize with another RTK. This conformational change activates the latent kinase activity of each RTK, and these receptors phosphorylate each other on particular tyrosine residues (see Figure 3.19). Thus, the binding of the ligand to the receptor causes the autophosphorylation of the cytoplasmic domain of the receptor.

The phosphorylated tyrosine on the receptor is then recognized by an adaptor protein. The adaptor protein serves as a bridge that links the phosphorylated RTK to a powerful intracellular signaling system. While binding to the phosphorylated RTK through one of its cytoplasmic domains, the adaptor protein also activates a **G protein**, such as **Ras**. Normally, the G protein is in an inactive, GDP-bound state. The activated receptor stimulates the adaptor protein to activate the **guanine nucleotide releasing fac-**

tor **(GNRP)**. This protein exchanges a phosphate from a GTP to transform the bound GDP into GTP. The GTP-bound G protein is an active form that transmits the signal to the next molecule. After the signal is delivered, the GTP on the G protein is hydrolyzed back into GDP. This catalysis is greatly stimulated by the complexing of the Ras protein with the **GTPase-activating protein (GAP)**. In this way, the G protein is returned to its inactive state, where it can await further signaling. Without the GAP protein, Ras protein cannot catalyze GTP well, and so remains in its active configuration (Cales et al. 1988; McCormick 1989). Mutations in the *RAS* gene account for a large proportion of cancerous human tumors (Shih and Weinberg 1982), and the mutations of *RAS* that make it oncogenic all inhibit the binding of the GAP protein.

The active Ras G protein associates with a kinase called Raf. The G protein recruits the inactive Raf protein to the cell membrane, where it becomes active (Leevers et al. 1994; Stokoe et al. 1994). The Raf protein is a kinase that activates the MEK protein by phosphorylating it. MEK is itself a kinase, which activates the ERK protein by phosphorylation. In turn, ERK is a kinase that enters the nucleus and phosphorylates certain transcription factors.

The RTK pathway is critical in numerous developmental processes. Moreover, it can be activated by paracrine factors other than those of the FGF family. (And FGF fam-

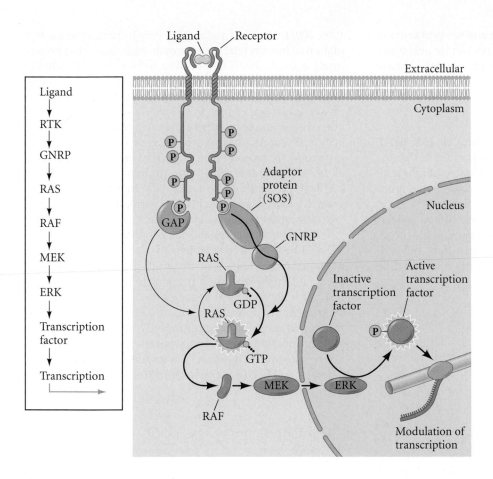

FIGURE 3.21   The widely used RTK signal transduction pathway. The receptor tyrosine kinase is dimerized by the ligand, which causes the autophosphorylation of the receptor. The adaptor protein recognizes the phosphorylated tyrosines on the RTK and activates an intermediate protein, GNRP, which activates the Ras G protein by allowing the phosphorylation of the GDP-bound Ras. At the same time, the GAP protein stimulates the hydrolysis of this phosphate bond, returning Ras to its inactive state. The active Ras activates the Raf protein kinase C (PKC), which in turn phosphorylates a series of kinases. Eventually, the activated kinase ERK alters gene expression in the nucleus of the responding cell by phosphorylating certain transcription factors (which can then enter the nucleus to change the types of genes transcribed) and certain translation factors (which alter the level of protein synthesis). In many cases, this pathway is reinforced by the release of calcium ions. A simplified version of the pathway is shown on the left.

ily proteins can activate other signaling pathways, depending on the receptor.) In the migrating neural crest cells of humans and mice, the pathway is important in activating the microphthalmia transcription factor (MITF) to produce the pigment cells (**Figure 3.22**). MITF, whose mechanism of action was described in Chapter 2, is transcribed in the pigment-forming melanoblast cells that migrate from the neural crest into the skin and in the melanin-forming cells of the pigmented retina. But we have not yet discussed what proteins signal this transcription factor to become active. The clue lies in two mouse mutants whose phenotypes resemble those of mice homozygous for *microphthalmia* mutations. Like *Mitf* mutant mice, homozygous *White* mice and homozygous *Steel* mice are white because

their pigment cells have failed to migrate. Could it be that all three genes (*Mitf*, *Steel*, and *White*) are on the same developmental pathway?

In 1990, several laboratories demonstrated that the *Steel* gene encodes a paracrine protein called **stem cell factor** (see Witte 1990). Stem cell factor binds to and activates the **Kit** receptor tyrosine kinase encoded by the *White* gene (Spritz et al. 1992; Wu et al. 2000; also see Chapter 1). The binding of stem cell factor to the Kit RTK dimerizes the Kit protein, causing it to become phosphorylated. The phosphorylated Kit activates the pathway whereby phosphorylated ERK is able to phosphorylate the MITF transcription factor (Hsu et al. 1997; Hemesath et al. 1998). Only the phosphorylated form of MITF is able to bind the p300/CBP histone acetyltransferase protein that enables it to activate transcription of the genes encoding tyrosinase and other proteins of the melanin-formation pathway (see Figure 3.22; Price et al. 1998).

The JAK-STAT pathway

Fibroblast growth factors can also activate the JAK-STAT cascade. This pathway is extremely important in the differentiation of blood cells, the growth of limbs, and the activation of the casein gene during milk production (**Figure 3.23**; Briscoe et al. 1994; Groner and Gouilleux 1995). Here, the ligand is bound by receptors that are linked to **JAK** (*Ja*nus *k*inase) proteins. The binding of ligand to the receptor phosphorylates the **STAT** (*s*ignal *t*ransducers and *a*ctivators of *t*ranscription) family of transcription factors (Ihle

1996, 2001). The STAT pathway is very important in the regulation of human fetal bone growth. Mutations that prematurely activate the STAT pathway have been implicated in some severe forms of dwarfism, such as the lethal **thanatophoric dysplasia**, wherein the growth plates of the rib and limb bones fail to proliferate. The short-limbed newborn dies because its ribs cannot support breathing. The genetic lesion responsible is in the gene encoding fibroblast growth factor receptor 3,or FgfR3 (**Figure 3.24**; Rousseau et al. 1994; Shiang et al. 1994). This protein is expressed in the cartilage precursor cells—known as **chondrocytes**—in the growth plates of the long bones. Normally, the FgfR3 protein (a receptor tyrosine kinase) is activated by a fibroblast growth factor, and it signals the chondrocytes to stop dividing and begin differentiating into cartilage. This signal is mediated by the Stat1 protein, which is phosphorylated by activated FgfR3 and then translocated into the nucleus. Inside the nucleus, Stat1 activates the genes encoding a cell cycle inhibitor, the p21 protein (Su et al. 1997). Thus, the mutations causing thanatophoric dwarfism result from a gain-of-function phenotype, wherein the mutant FgfR3 is active constitutively—that is, without the need to be activated by an FGF (Deng et al. 1996; Webster and Donoghue

(A)

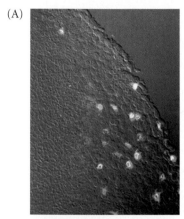

(B)

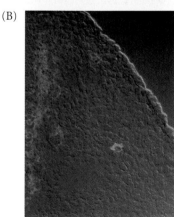

(C)

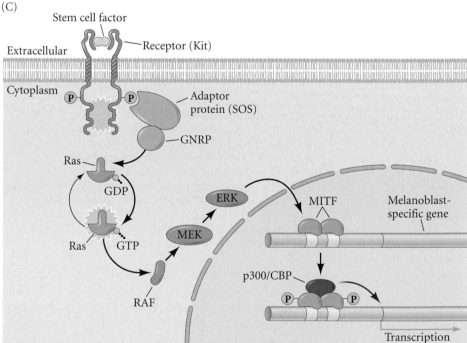

FIGURE 3.22 Activation of MITF transcription factor through the binding of stem cell factor by the Kit RTK protein. The information received at the cell membrane is sent to the nucleus by the RTK signal transduction pathway. (A,B) Demonstration that Kit protein and MITF are present in the same cells. Antibodies to these proteins stain the Kit protein (red) and MITF (green). The overlap is yellow or yellow-green. Both proteins are present in the migrating melanocyte precursor cells (melanoblasts). (A) Migrating melanoblasts can be seen in a wild-type mouse embryo at day 10.5. (B) No melanoblasts are visible in a *microphthalmia* mutant embryo of the same age. The lack of melanoblasts in the mutant is due to the relative absence of MITF. (C) Signal transduction pathway leading from the cell membrane to the nucleus. When the receptor domain of the Kit RTK protein binds the stem cell factor, Kit dimerizes and becomes phosphorylated. This phosphorylation is used to activate the Ras G protein, which activates the chain of kinases that will phosphorylate the MITF protein. Once phosphorylated, MITF can bind the cofactor p300/CBP, acetylate the nucleosome histones, and initiate transcription of the genes needed for melanocyte development. (A,B from Nakayama et al. 1998, courtesy of H. Arnheiter; C after Price et al. 1998.)

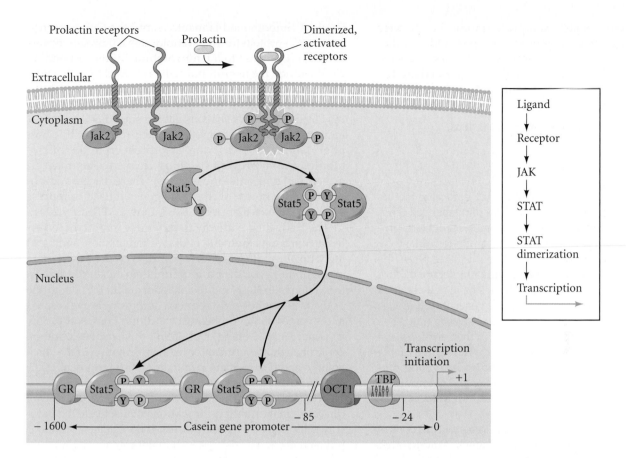

FIGURE 3.23 A STAT pathway: the casein gene activation pathway activated by prolactin. The casein gene is activated during the last (lactogenic) phase of mammary gland development, and its signal is the secretion of the hormone prolactin from the anterior pituitary gland. Prolactin causes the dimerization of prolactin receptors in the mammary duct epithelial cells. A particular JAK protein (Jak2) is "hitched" to the cytoplasmic domain of these receptors. When the receptors bind prolactin and dimerize, the JAK proteins phosphorylate each other and the dimerized receptors, activating the dormant kinase activity of the receptors. The activated receptors add a phosphate group to a tyrosine residue (Y) of a particular STAT protein—in this case, Stat5. This allows Stat5 to dimerize, be translocated into the nucleus, and bind to particular regions of DNA. In combination with other transcription factors (which presumably have been waiting for its arrival), the Stat5 protein activates transcription of the casein gene. GR is the glucocorticoid receptor, OCT1 is a general transcription factor, and TBP is the TATA-binding protein (see Chapter 2) responsible for binding RNA polymerase. A simplified diagram is shown to the right. (For details, see Groner and Gouilleux 1995.)

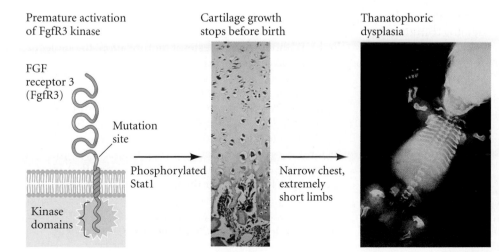

FIGURE 3.24 A mutation in the gene for FgfR3 causes the premature constitutive activation of the STAT pathway and the production of phosphorylated Stat1 protein. This transcription factor activates genes that cause the premature termination of chondrocyte cell division. The result is thanatophoric dysplasia, a condition of failed bone growth that results in the death of the newborn infant because the thoracic cage cannot expand to allow breathing. (After Gilbert-Barness and Opitz 1996.)

1996). The chondrocytes stop proliferating shortly after they are formed, and the bones fail to grow. Other mutations that activate FgfR3 prematurely but to a lesser degree produce **achondroplasic (short-limbed) dwarfism**, the most prevalent of the human dominant syndromes (Legeai-Mallet et al. 2004).

See WEBSITE 3.1 FGF receptor mutations

The Hedgehog family

The proteins of the Hedgehog family of paracrine factors are often used by the embryo to induce particular cell types and to create boundaries between tissues. Hedgehog proteins are processed such that only the amino-terminal two-thirds of the molecule is secreted; once this takes place, the protein must become complexed with a molecule of cholesterol in order to function. Vertebrates have at least three homologues of the *Drosophila hedgehog* gene: *sonic hedgehog* (*shh*), *desert hedgehog* (*dhh*), and *indian hedgehog* (*ihh*). The Desert hedgehog protein is expressed in the Sertoli cells of the testes, and mice homozygous for a null allele of *dhh* exhibit defective spermatogenesis. Indian hedgehog protein is expressed in the gut and cartilage and is important in postnatal bone growth (Bitgood and McMahon 1995; Bitgood et al. 1996).

Sonic hedgehog* has the greatest number of functions of the three vertebrate Hedgehog homologues. Among other important functions, this paracrine factor is responsible for assuring that motor neurons come only from the ventral portion of the neural tube (see Chapter 10), that a portion of each somite forms the vertebrae (see Chapter 12), that the feathers of the chick form in their proper places (see Figure 3.16), and that our pinkies are always our most posterior digits (see Chapter 14). Sonic hedgehog often works with other paracrine factors, such as Wnt and FGF proteins.

THE HEDGEHOG PATHWAY Proteins of the Hedgehog family function by binding to a receptor called Patched. The Patched protein, however, is not a signal transducer. Rather, it is *bound to* a signal transducer, the Smoothened protein. The Patched protein prevents Smoothened from functioning. In the absence of Hedgehog binding to

Patched, Smoothened is inactive, and the Cubitus interruptus (Ci) protein (or the homologous Gli protein in vertebrates) is tethered to the microtubules of the responding cell. While on the microtubules, it is cleaved in such a way that a portion of it enters the nucleus and acts as a transcriptional repressor. When Hedgehog binds to Patched, the Patched protein's shape is altered such that it no longer inhibits Smoothened. Smoothened acts (probably by phosphorylation) to release the Ci protein from the microtubules and to prevent its being cleaved. The intact Ci protein can now enter the nucleus, where it acts as a transcriptional *activator* of the same genes it used to repress (**Figure 3.25**; Aza-Blanc et al. 1997; Lum and Beachy 2004).

The Hedgehog pathway is extremely important in vertebrate limb development, neural differentiation, and facial morphogenesis (McMahon et al. 2003). When mice were made homozygous for a mutant allele of *sonic hedgehog*, they had major limb and facial abnormalities. The midline of the face was severely reduced and a single eye formed in the center of the forehead, a condition known as *cyclopia*[†] (**Figure 3.26**; Chiang et al. 1996). In later development, Sonic hedgehog is critical for feather formation in the chick embryo and hair formation in mammals (Harris et al. 2002; Michino et al. 2003).

While mutations that inactivate the Hedgehog pathway can cause malformations, mutations that activate the pathway ectopically can cause cancers. If the Patched protein is mutated in somatic tissues such that it can no longer inhibit Smoothened, it can cause tumors of the basal cell layer of the epidermis (basal cell carcinomas). Heritable mutations of the *patched* gene cause basal cell nevus syndrome, a rare autosomal dominant condition characterized by both developmental anomalies (fused fingers; rib and facial abnormalities) and multiple malignant tumors such as basal cell carcinoma (Hahn et al. 1996; Johnson et al. 1996).

One remarkable feature of the Hedgehog signal transduction pathway is the importance of cholesterol. First, cholesterol is critical for the catalytic cleavage of Sonic hedgehog protein. Only the amino-terminal portion of the protein is functional and secreted. The cholesterol also binds to the active N-terminus of the Sonic hedgehog protein and allows this paracrine factor to diffuse over a range of a few hundred μm (about 30 cell diameters in the mouse limb). Without this cholesterol modification, the molecule diffuses too quickly and dissipates into the surrounding space. Indeed, Hedgehog proteins probably do not diffuse as single molecules, but they are linked together through their cholesterol-containing regions into lipoprotein packets (Breitling 2007; Guerrero and Chiang 2007). Second, the Patched protein that binds Sonic hedgehog also needs cholesterol in order to function. Some human cyclopia syndromes are caused by mutations in genes that encode either Sonic hedgehog or the enzymes that synthesize cho-

---

*Yes, it is named after the Sega Genesis character. The original *hedgehog* gene was found in *Drosophila*, in which genes are named after their mutant phenotypes—the loss-of-function *hedgehog* mutation causes the fly embryo to be covered with pointy denticles on its cuticle, so it looks like a hedgehog. The vertebrate Hedgehog genes were discovered by searching vertebrate gene libraries (chick, rat, zebrafish) with probes that would find sequences similar to that of the fruit fly *hedgehog* gene. Riddle and colleagues (1993) discovered three genes homologous to *Drosophila hedgehog*. Two were named after existing species of hedgehog; the third was named after the animated character. Two other Hedgehog genes, found only in fish, are named *echidna hedgehog* (after the spiny Australian marsupial mammal) and *Tiggywinkle hedgehog* (after Beatrix Potter's fictional hedgehog).

[†]This pathology, which is named for the one-eyed Cyclops of Homer's *Odyssey*, will be discussed again in Chapter 9.

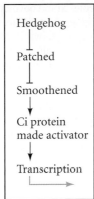

Hedgehog
↓
Patched
⊣
Smoothened
↓
Ci protein
made activator
↓
Transcription

(A)

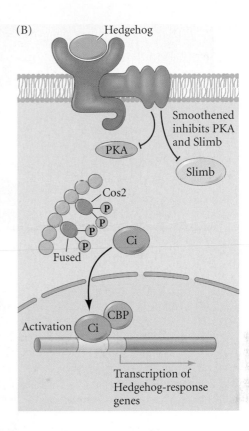

Patched protein

Smoothened protein

Cytoplasm

Patched inhibits Smoothened

Microtubule

Cos2

PKA

Slimb

Ci

Fused

P

Ci

Repressor

No transcription of Hedgehog-responsive genes

(B)

Hedgehog

Smoothened inhibits PKA and Slimb

PKA

Slimb

Cos2

P
P
P
P

Ci

Fused

Activation   Ci   CBP

Transcription of Hedgehog-response genes

FIGURE 3.25 Hedgehog signal transduction pathway. Patched protein in the cell membrane is an inhibitor of the Smoothened protein. (A) In the absence of Hedgehog binding to Patched, the Ci protein is tethered to the microtubules by the Cos2 and Fused proteins. This binding allows the PKA and Slimb proteins to cleave Ci into a transcriptional repressor that blocks the transcription of particular genes. (B) When Hedgehog binds to Patched, its conformation changes, releasing the inhibition of the Smoothened protein. Smoothened then releases Ci from the microtubules (probably by adding more phosphates to the Cos2 and Fused proteins) and inactivates the cleavage proteins PKA and Slimb. The Ci protein enters the nucleus, binds a CBP protein, and acts as a transcriptional activator of particular genes. (After Johnson and Scott 1998.)

lesterol (Kelley et al. 1996; Roessler et al. 1996). Moreover, certain chemicals that induce cyclopia do so by interfering with the cholesterol biosynthetic enzymes (Beachy et al. 1997; Cooper et al. 1998). Two teratogens known to cause cyclopia in vertebrates are jervine and cyclopamine. Both substances are found in the plant *Veratrum californicum*, and both block the synthesis of cholesterol (see Figure 3.26; Keeler and Binns 1968).

See VADE MECUM Cyclopia induced in zebrafish

(A)

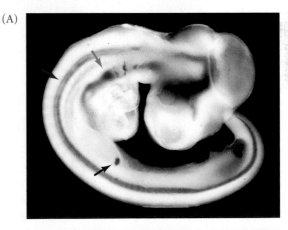

(B)

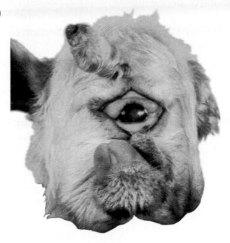

FIGURE 3.26 (A) Sonic hedgehog is shown by in situ hybridization to be expressed in the nervous system (red arrow), gut (blue arrow), and limb bud (black arrow) of a 3-day chick embryo. (B) Head of a cyclopic lamb born of a ewe that ate *Veratrum californicum* early in pregnancy. The cerebral hemispheres fused, resulting in the formation of a single, central eye and no pituitary gland. The jervine alkaloid made by this plant inhibits cholesterol synthesis, which is needed for Hedgehog production and reception. (A courtesy of C. Tabin; B courtesy of L. James and USDA Poisonous Plant Laboratory.)

(A)

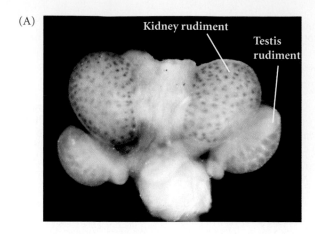

Kidney rudiment

Testis rudiment

FIGURE 3.27 Wnt proteins play several roles in the development of the urogenital organs. Wnt4 is necessary for kidney development and for female sex determination. (A) Whole-mount in situ hybridization of Wnt4 expression in a 14-day mouse embryonic male urogenital rudiment. Expression (dark purple-blue staining) is seen in the mesenchyme that condenses to form the kidney's nephrons. (B) Urogenital rudiment of a wild-type newborn female mouse. (C) Urogenital rudiment of a newborn female mouse with targeted knockout of the *Wnt4* gene shows that the kidney fails to develop. In addition, the ovary starts synthesizing testosterone and becomes surrounded by a modified male duct system. (Courtesy of J. Perasaari and S. Vainio.)

(B)   (C)

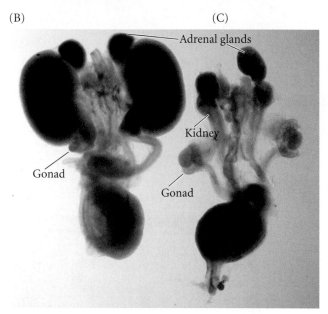

Adrenal glands

Kidney

Gonad

Gonad

The Wnt family

The Wnts are a family of cysteine-rich glycoproteins. There are at least 15 members of this gene family in vertebrates.* Their name is a fusion of the name of the *Drosophila* segment polarity gene *wingless* with the name of one of its vertebrate homologues, *integrated*. While Sonic hedgehog is important in patterning the ventral portion of the somites (causing the cells to become cartilage), Wnt1 appears to be active in inducing the dorsal cells of the somites to become muscle and is involved in the specification of the midbrain cells (see Chapter 11; McMahon and Bradley 1990; Stern et al. 1995). Wnt proteins also are critical in establishing the polarity of insect and vertebrate limbs, promoting the proliferation of stem cells, and in several steps of urogenital system development (**Figure 3.27**).

*A summary of all the Wnt proteins and Wnt signaling components can be found at http://www.stanford.edu/~rnusse/wntwindow.htm

The Wnt proteins were not isolated in their active form until 2003. At that time, Willert and colleagues (2003) discovered that each Wnt protein has a lipid molecule covalently bound to it. These hydrophobic molecules are critical for the activity of the Wnt proteins and probably act to increase their concentration in the cell membrane.

THE "CANONICAL" WNT PATHWAY Members of the Wnt family of paracrine factors interact with transmembrane receptors of the Frizzled family of proteins (Logan and Nusse 2004). In most instances, the binding of Wnt by a Frizzled protein causes Frizzled to activate the Disheveled protein. Once Disheveled is activated, it inhibits the activity of the glycogen synthase kinase 3 (GSK3) enzyme. GSK3, if it were active, would prevent the dissociation of the β-catenin protein from the APC protein, which targets β-catenin for degradation. However, when the Wnt signal is present and GSK3 is inhibited, β-catenin can dissociate from the APC protein and enter the nucleus. Once inside the nucleus, β-catenin binds to a Lef/Tcf transcription factor that is already on the DNA, repressing the genes it has bound. The binding of β-catenin to the Lef/Tcf protein converts the repressor into a transcriptional activator, thereby activating the Wnt-responsive genes (**Figure 3.28A**; Behrens et al. 1996; Cadigan and Nusse 1997).

This model is undoubtedly an oversimplification, because different cells use this pathway in different ways (see McEwen and Peifer 2001). Moreover, its components can have more than one function in the cell. In addition to being part of the Wnt signal transduction cascade, GSK3 is also an enzyme that regulates glycogen metabolism. The β-catenin protein was recognized as being part of the cell adhesion complex on the cell surface before it was also found to be a transcription factor. The APC protein also functions as a tumor suppressor. The transformation of normal adult colon epithelial cells into colon cancer is thought to occur when the *APC* gene is mutated and can no longer keep β-catenin out of the nucleus (Korinek et al. 1997; He et al. 1998). Once in the nucleus, β-catenin can bind with another transcription factor and activate genes for cell division.

One overriding principle is readily evident in both the Wnt pathway and the Hedgehog pathway: *activation is often accomplished by inhibiting an inhibitor.* Thus, in the Wnt

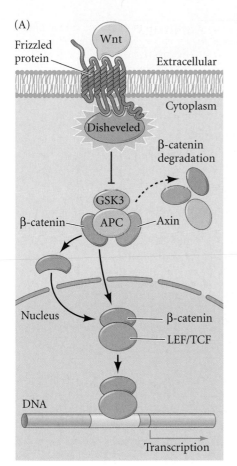

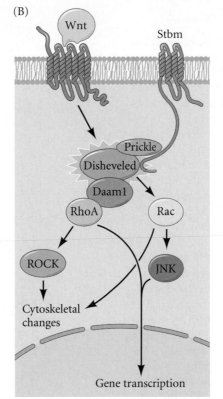

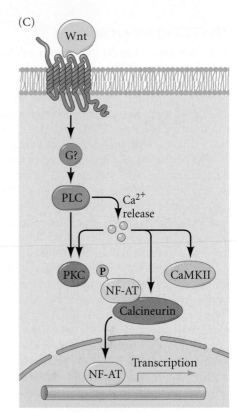

FIGURE 3.28 Wnt signal transduction pathways. (A) The canonical Wnt pathway. The Wnt protein binds to its receptor, a member of the Frizzled family of proteins. In the case of certain Wnt proteins, the Frizzled protein then activates Disheveled, allowing it to become an inhibitor of glycogen synthase kinase 3 (GSK3). GSK3, if it were active, would prevent the dissociation of β-catenin from the APC protein. So by inhibiting GSK3, the Wnt signal frees β-catenin to associate with an LEF or TCF protein and become an active transcription factor. (B) In a pathway that regulates cell morphology, division, and movement, certain Wnt proteins activate Frizzled in a way that causes Frizzled to activate the Disheveled protein, which has been tethered to the plasma membrane (through the Prickle protein). Here, Disheveled activates Rac and RhoA proteins, which coordinate the cytoskeleton and which can also regulate gene expression. (C) In a third pathway, certain Wnt proteins activate Frizzled receptors in a way that releases calcium ions and can cause Ca²⁺-dependent gene expression.

pathway, the GSK3 protein is an inhibitor that is itself repressed by the Wnt signal.

THE "NONCANONICAL" WNT PATHWAYS The pathway described above is often called the "canonical" Wnt pathway because it was the first one to be discovered. However, in addition to sending signals to the nucleus, Wnt can also affect the actin and microtubular cytoskeleton. Here, Wnt activates alternative, "noncanonical," pathways. For instance, when Wnt activates Disheveled, the Disheveled protein can interact with a Rho GTPase. This GTPase can

activate the kinases that phosphorylate cytoskeletal proteins and thereby alter cell shape, cell polarity (where the upper and lower portions of the cell differ), and motility (**Figure 3.28B**; Shulman et al. 1998; Winter et al. 2001). A third Wnt pathway diverges earlier than Disheveled. Here, the Frizzled receptor protein activates a phospholipase (PLC) that synthesizes a compound that releases calcium ions from the endoplasmic reticulum (**Figure 3.28C**). The released calcium can activate enzymes, transcription factors, and translation factors.

It is probable that the Frizzled proteins (of which there are many) can be used to couple different signal transduction cascades to the Wnt signal (see Chen et al. 2005) and that different cells have evolved to use Wnt factors in different ways.

The TGF-β superfamily

There are more than 30 structurally related members of the TGF-β superfamily,* and they regulate some of the most important interactions in development (**Figure 3.29**). The proteins encoded by TGF-β superfamily genes are processed such that the carboxy-terminal region contains the mature peptide. These peptides are dimerized into homodimers (with themselves) or heterodimers (with other TGF-β peptides) and are secreted from the cell. The TGF-β superfamily includes the TGF-β family, the activin family, the bone morphogenetic proteins (BMPs), the Vg1 family, and other proteins, including glial-derived neurotrophic factor (GDNF; necessary for kidney and enteric neuron differentiation) and Müllerian inhibitory factor (which is involved in mammalian sex determination).

TGF-β family members TGF-β1, 2, 3, and 5 are important in regulating the formation of the extracellular matrix between cells and for regulating cell division (both positively and negatively). TGF-β1 increases the amount of extracellular matrix epithelial cells make (both by stimulating collagen and fibronectin synthesis and by inhibiting matrix degradation). TGF-β proteins may be critical in controlling where and when epithelia branch to form the ducts of kidneys, lungs, and salivary glands (Daniel 1989; Hardman et al. 1994; Ritvos et al. 1995). The effects of the individual TGF-β family members are difficult to sort out, because members of the TGF-β family appear to function similarly and can compensate for losses of the others when expressed together.

The members of the BMP family can be distinguished from other members of the TGF-β superfamily by having seven (rather than nine) conserved cysteines in the mature polypeptide. Because they were originally discovered by their ability to induce bone formation, they were given the name **bone morphogenetic proteins**. But it turns out that bone formation is only one of their many functions; the BMPs are extremely multifunctional. [†]They have been found to regulate cell division, apoptosis (programmed cell death), cell migration, and differentiation (Hogan 1996). They include proteins such as BMP4 (which in some tis-

*TGF stands for "*transforming growth factor.*" The designation "superfamily" is often given when each of the different classes of molecules constitutes a "family." The members of a superfamily all have similar structures but are not as close as the molecules within a family are to one another.

[†]One of the many reasons why humans don't seem to need an enormous genome is that the gene products—proteins—involved in our construction and development often have many functions. Many of the proteins we are familiar with in adults (such as hemoglobin, keratins, insulin, and the like) *do* have only one function, which led to the erroneous conclusion that this is the norm. Indeed, the "one-function-per-entity" concept is a longstanding one in science, having been credited to Aristotle. Philosopher John Thorp has called this *monotelism* (Greek, "one end") "Aristotle's worst idea."

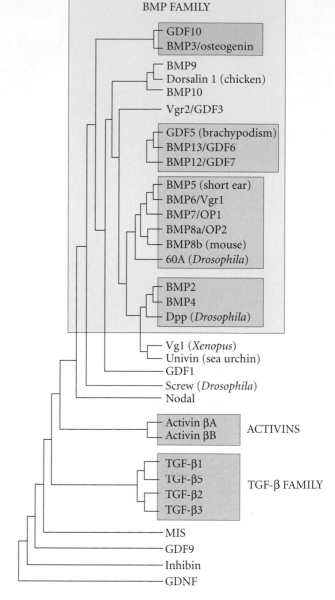

FIGURE 3.29   Relationships among members of the TGF-β superfamily. (After Hogan 1996.)

sues causes bone formation, in other tissues causes cell death, and in other instances specifies the epidermis) and BMP7 (which is important in neural tube polarity, kidney development, and sperm formation). As it (rather oddly) turns out, however, BMP1 is not a member of the BMP family at all; it is a protease.

The *Drosophila* Decapentaplegic (Dpp) protein is homologous to vertebrate BMP4, and human BMP4 can replace Dpp and thus "rescue" *dpp*-deficient flies (Padgett et al. 1993). BMPs are thought to work by diffusion from the cells

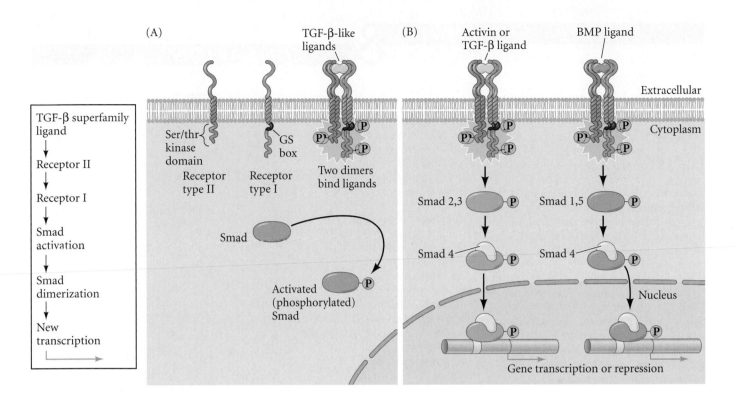

FIGURE 3.30 The Smad pathway activated by TGF-β superfamily ligands. (A) An activation complex is formed by the binding of the ligand by the type I and type II receptors. This allows the type II receptor to phosphorylate the type I receptor on particular serine or threonine residues (of the "GS box"). The phosphorylated type I receptor protein can now phosphorylate the Smad proteins. (B) Those receptors that bind TGF-β family proteins or members of the activin family phosphorylate Smads 2 and 3. Those receptors that bind to BMP family proteins phosphorylate Smads 1 and 5. These Smads can complex with Smad4 to form active transcription factors. A simplified version of the pathway is shown at the left.

producing them. Their range is determined by the amino acids in their N-terminal region, which determine whether the specific BMP will be bound by proteoglycans, thereby restricting its diffusion (Ohkawara et al. 2002).

The Nodal and activin proteins are also members of the TGF-β superfamily. These proteins are extremely important in specifying the different regions of the mesoderm and for distinguishing the left and right sides of the vertebrate body axis.

THE SMAD PATHWAY Members of the TGF-β superfamily activate members of the Smad family of transcription factors (Heldin et al. 1997; Shi and Massague 2003). The TGF-β ligand binds to a type II TGF-β receptor, which allows that receptor to bind to a type I TGF-β receptor. Once the two receptors are in close contact, the type II receptor phos-

phorylates a serine or threonine on the type I receptor, thereby activating it. The activated type I receptor can now phosphorylate the Smad* proteins (**Figure 3.30A**). Smads 1 and 5 are activated by the BMP family of TGF-β factors, while the receptors binding activin, Nodal, and the TGF-β family phosphorylate Smads 2 and 3. These phosphorylated Smads bind to Smad4 and form the transcription factor complex that will enter the nucleus (**Figure 3.30B**).

Other paracrine factors

Although most paracrine factors are members of one of the four families described above, some of these proteins have few or no close relatives. Epidermal growth factor, hepatocyte growth factor, neurotrophins, and stem cell factor are not included among these families, but each of these factors plays important roles during development. In addition, there are numerous paracrine factors involved almost exclusively with developing blood cells: erythropoietin, the cytokines, and the interleukins. These factors will be discussed when we detail blood cell formation in Chapter 12.

---

*Researchers named the Smad proteins by merging the names of the first identified members of this family: the *C. elegans* SMA protein and the *Drosophila* Mad protein.

Cell Death Pathways

"To be, or not to be: that is the question." While we all are poised at life-or-death decisions, this existential dichotomy is exceptionally stark for embryonic cells. **Programmed cell death**, or **apoptosis**,* is a normal part of development (see Baehrecke 2002). In the nematode *C. elegans*, in which we can count the number of cells, exactly 131 cells die according to the normal developmental pattern. All the cells of this nematode are "programmed" to die unless they are actively told not to undergo apoptosis. In humans, as many as 10^{11} cells die in each adult each day and are replaced by other cells. (Indeed, the mass of cells we lose each year through normal cell death is close to our entire body weight!) Within the uterus, we were constantly making and destroying cells, and we generated about three times as many neurons as we eventually ended up with when we were born. Lewis Thomas (1992) has aptly noted,

> By the time I was born, more of me had died than survived. It was no wonder I cannot remember; during that time I went through brain after brain for nine months, finally contriving the one model that could be human, equipped for language.

Apoptosis is necessary not only for the proper spacing and orientation of neurons, but also for generating the middle ear space, the vaginal opening, and the spaces between our fingers and toes (Saunders and Fallon 1966; Roberts and Miller 1998; Rodriguez et al. 1997). Apoptosis prunes unneeded structures (frog tails, male mammary tissue), controls the number of cells in

*The term *apoptosis* (both *p*s are pronounced) comes from the Greek word for the natural process of leaves falling from trees or petals falling from flowers. Apoptosis is an active process that can be subject to evolutionary selection. A second type of cell death, *necrosis*, is a pathological death caused by external factors such as inflammation or toxic injury.

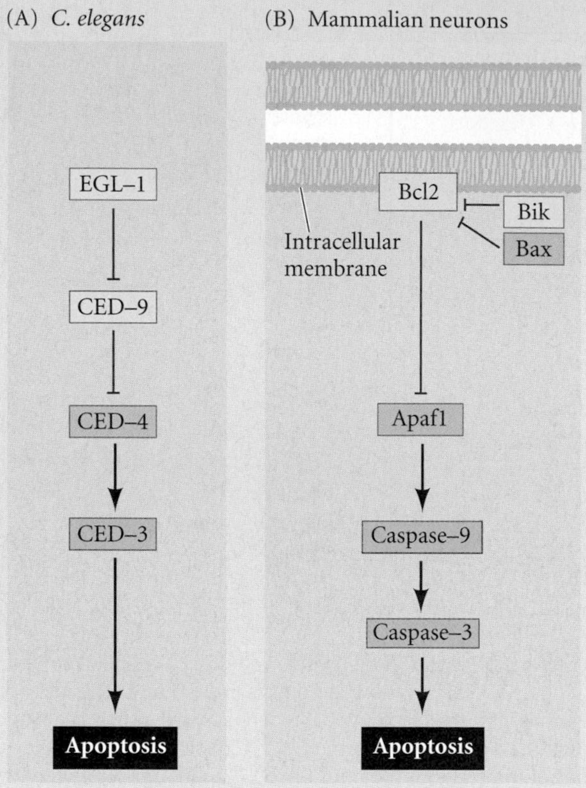

particular tissues (neurons in vertebrates and flies), and sculpts complex organs (palate, retina, digits, and heart).

Different tissues use different signals for apoptosis. One of the signals often used in vertebrates is bone morphogenetic protein 4 (BMP4). Some tissues, such as connective tissue, respond to BMP4 by differentiating into bone. Others, such as the frog gastrula ectoderm, respond to BMP4 by differentiating into skin. Still others, such as neural crest cells and tooth primordia, respond by degrading their DNA and dying. In the developing tooth, for instance, numerous growth and differentiation factors are secreted by the enamel knot. After the cusp has grown, the enamel knot synthesizes BMP4 and shuts itself down by apoptosis (see Chapter 10; Vaahtokari et al. 1996).

In other tissues, the cells are "programmed" to die, and will remain alive only if some growth or differentiation factor is present to "rescue" them. This happens during the development of mammalian red blood cells. The red blood cell precursors in the mouse liver need the hormone erythropoietin in order to survive. If they do not receive it, they undergo apoptosis. The erythropoietin receptor works through the JAK-STAT pathway, activating the Stat5 transcription factor. In this way, the amount of erythropoietin present can determine how many red blood cells enter the circulation.

One of the pathways for apoptosis was largely delineated through genetic studies of *C. elegans*. Indeed, the importance of this pathway was recognized by awarding a Nobel Prize to Sydney Brenner, Bob Horvitz, and Jonathan Sulston in 2002. It was found that the proteins encoded by the *ced-3* and *ced-4* genes were essential for

(A) *C. elegans*

EGL–1 → CED–9 → CED–4 → CED–3 → **Apoptosis**

(B) Mammalian neurons

Intracellular membrane

Bcl2 — Bik / Bax

Apaf1 → Caspase–9 → Caspase–3 → **Apoptosis**

Figure 3.31 Apoptosis pathways in nematodes and mammals. (A) In *C. elegans*, the CED-4 protein is a protease-activating factor that can activate the CED-3 protease. The CED-3 protease initiates the cell destruction events. CED-9 can inhibit CED-4 (and CED-9 can be inhibited upstream by EGL-1). (B) In mammals, a similar pathway exists, and appears to function in a similar manner. In this hypothetical scheme for the regulation of apoptosis in mammalian neurons, Bcl-X_L (a member of the Bcl2 family) binds Apaf1 and prevents it from activating the precursor of caspase-9. The signal for apoptosis allows another protein (here, Bik) to inhibit the binding of Apaf1 to Bcl-X_L. Apaf1 is now able to bind to the caspase-9 precursor and cleave it. Caspase-9 dimerizes and activates caspase-3, which initiates apoptosis. The same colors are used to represent homologous proteins. (After Adams and Cory 1998.)

apoptosis, and that in the cells that did not undergo apoptosis, those genes were turned off by the product of the *ced-9* gene (**Figure 3.31A**; Hengartner et al. 1992). The CED-4 protein is a protease-activating factor that activates the gene for CED-3, a protease that initiates the destruction of the cell. The CED-9 protein can bind to and inactivate CED-4. Mutations that inactivate the gene for CED-9 cause numerous cells that would normally survive to activate their *ced-3* and *ced-4* genes and die, leading to the death of the entire embryo. Conversely, gain-of-function mutations in the *ced-9* gene cause its protein to be made in cells that would normally die, resulting in those cells' survival. Thus, the *ced-9* gene appears to be a binary switch that regulates the choice between life and death on the cellular level. It is possible that every cell in the nematode embryo is poised to die, with those cells that survive being rescued by the activation of the *ced-9* gene.

The CED-3 and CED-4 proteins are at the center of the apoptosis pathway that is common to all animals studied. The trigger for apoptosis can be a developmental cue such as a particular molecule (e.g., BMP4 or glucocorticoids) or the loss of adhesion to a matrix. Either type of cue can activate CED-3 or CED-4 proteins or inactivate CED-9 molecules. In mammals, the homologues of the CED-9 protein are members of the Bcl2 family (which includes Bcl2, Bcl-X, and similar proteins; **Figure 3.31B**). The functional similarities are so strong that if an active human *BCL2* gene is placed in *C. elegans* embryos, it prevents normally occurring cell death (Vaux et al. 1992).

The mammalian homologue of CED-4 is Apaf1 (apoptotic protease activating factor 1), and it participates

in the cytochrome *c*-dependent activation of the mammalian CED-3 homologues, the proteases caspase-9 and caspase-3 (see Figure 3.31B; Shaham and Horvitz 1996; Cecconi et al. 1998; Yoshida et al. 1998). Activation of the caspase proteins results in autodigestion—caspases are strong proteases that digest the cell from within, cleaving cellular proteins and fragmenting the DNA.

While apoptosis-deficient nematodes deficient for CED-4 are viable (despite having 15% more cells than wild-type worms), mice with loss-of-function mutations for either *caspase-3* or *caspase-9* die around birth from massive cell overgrowth in the nervous system (**Figure 3.32**; Kuida et al. 1996, 1998; Jacobson et al. 1997). Mice homozygous for targeted deletions of *Apaf1* have similarly severe craniofacial abnormalities, brain overgrowth, and webbing between their toes.

There are instances where cell death is the normal state unless some

ligand "rescues" the cells. In the chick neural tube, Patched protein (a Hedgehog receptor) will activate caspases. The binding of Sonic hedgehog (from the notochord and ventral neural tube cells) suppresses Patched, and the caspases are not activated to start apoptosis (Thibert et al. 2003). Such "dependence receptors" probably prevent neural cells from proliferating outside the proper tissue, and the loss of such receptors is associated with cancers (Porter and Dhakshinamoorty 2004). Moreover, we will soon see that certain epithelial cells must be attached to the extracellular matrix in order to function. If the cell is removed from the matrix, the apoptosis pathway is activated and the cell dies (Jan et al. 2004). This, too, is probably a mechanism that prevents cancers once cells have lost their adhesion to extracellular matrix proteins.

See WEBSITE 3.2
The uses of apoptosis

(A) *caspase-9*$^{+/+}$ (Wild type)   (B) *caspase-9*$^{-/-}$ (Knockout)

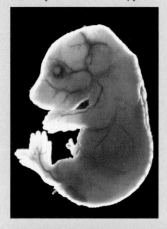   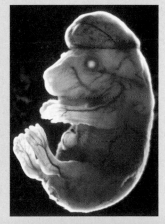

(C) Wild type   (D) Knockout

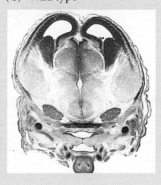   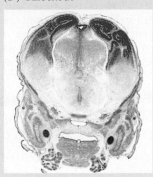

Figure 3.32 Disruption of normal brain development by blocking apoptosis. In mice in which the genes for caspase-9 have been knocked out, normal neural apoptosis fails to occur, and the overproliferation of brain neurons is obvious. (A) 6-day embryonic wild-type mouse. (B) A *caspase-9* knockout mouse of the same age. The enlarged brain protrudes above the face, and the limbs are still webbed. (C,D) This effect is confirmed by cross sections through the forebrain at day 13.5. The knockout exhibits thickened ventricle walls and the near-obliteration of the ventricles. (From Kuida et al. 1998.)

Juxtacrine Signaling

In juxtacrine interactions, proteins from the inducing cell interact with receptor proteins of adjacent responding cells without diffusing from the cell producing it. Two of the most widely used families of juxtacrine factors are the *Notch proteins* (which bind to a family of ligands exemplified by the Delta protein) and the **eph receptors** and their **ephrin ligands**. When the ephrin on one cell binds with the eph receptor on an adjacent cell, signals are sent to each of the two cells (Davy et al. 2004; Davy and Soriano 2005). These signals are often those of either attraction or repulsion, and ephrins are often seen where cells are being told where to migrate or where boundaries are forming. We will see the ephrins and the eph receptors functioning in the formation of blood vessels, neurons, and somites. For the moment, we will look at the Notch proteins and their ligands.

The Notch pathway: Juxtaposed ligands and receptors

While most known regulators of induction are diffusible proteins, some inducing proteins remain bound to the inducing cell surface. In one such pathway, cells expressing the **Delta**, **Jagged**, or **Serrate** proteins in their cell membranes activate neighboring cells that contain **Notch** protein in their cell membranes. Notch extends through the cell membrane, and its external surface contacts Delta, Jagged, or Serrate proteins extending out from an adjacent cell. When complexed to one of these ligands, Notch

undergoes a conformational change that enables a part of its cytoplasmic domain to be cut off by the presenilin-1 protease. The cleaved portion enters the nucleus and binds to a dormant transcription factor of the CSL family. When bound to the Notch protein, the CSL transcription factors activate their target genes (**Figure 3.33**; Lecourtois and Schweisguth 1998; Schroeder et al. 1998; Struhl and Adachi 1998). This activation is thought to involve the recruitment of histone acetyltransferases (Wallberg et al. 2002). Thus, Notch can be considered as a transcription factor tethered to the cell membrane. When the attachment is broken, Notch (or a piece of it) can detach from the cell membrane and enter the nucleus (Kopan 2002).

Notch proteins are involved in the formation of numerous vertebrate organs—kidney, pancreas, and heart—and they are extremely important receptors in the nervous system. In both the vertebrate and *Drosophila* nervous systems, the binding of Delta to Notch tells the receiving cell not to become neural (Chitnis et al. 1995; Wang et al. 1998). In the vertebrate eye, the interactions between Notch and its ligands seem to regulate which cells become optic neurons and which become glial cells (Dorsky et al. 1997; Wang et al. 1998). Notch proteins are also important in the patterning of the nematode vulva. The vulval precursor cell closest to the anchor cell becomes the central vulva cell, and this cell is able to prevent its neighbors from becoming central vulval cells by signaling to them through its Notch homologue, the LIN-12 receptor (Berset et al. 2001).

See WEBSITE 3.3 Notch mutations

(A)

(B)

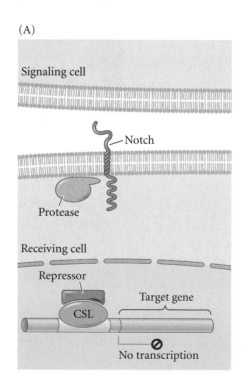
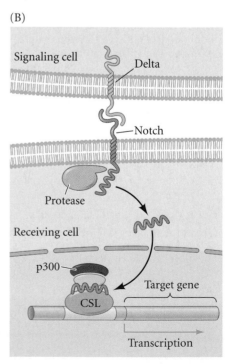

FIGURE 3.33 Mechanism of Notch activity. (A) Prior to Notch signaling, a CSL transcription factor (such as Suppressor of hairless or CBF1) is on the enhancer of Notch-regulated genes. The CSL binds repressors of transcription. (B) Model for the activation of Notch. A ligand (Delta, Jagged, or Serrate protein) on one cell binds to the extracellular domain of the Notch protein on an adjacent cell. This binding causes a shape change in the intracellular domain of Notch, which activates a protease. The protease cleaves Notch and allows the intracellular region of the Notch protein to enter the nucleus and bind the CSL transcription factor. This intercellular region of Notch displaces the repressor proteins and binds activators of transcription, including the histone acetyltransferase p300. The activated CSL can then transcribe its target genes. (After Koziol-Dube, Pers. Comm.)

Juxtacrine Signaling and Cell Patterning

Induction does indeed occur on the cell-to-cell level, and one of the best examples is the formation of the vulva in the nematode worm *Caenorhabditis elegans*. Remarkably, the signal transduction pathways involved turn out to be the same as those used in the formation of retinal receptors in *Drosophila*; only the targeted transcription factors are different. In both cases, an epidermal growth factor-like inducer activates the RTK pathway.

Vulval induction in *C. elegans*

Most *C. elegans* individuals are hermaphrodites. In their early development, they are male and the gonad produces sperm, which is stored for later use. As they grow older, they develop ovaries. The eggs "roll" through the region of sperm storage, are fertilized inside the nematode, and then pass out of the body through the vulva (see Figure 5.43).

The formation of the vulva in *C. elegans* represents a case in which one inductive signal generates a variety of cell types. This organ forms during the larval stage from six cells called the **vulval precursor cells** (**VPCs**). The cell connecting the overlying gonad to the vulval precursor cells is called the **anchor cell** (Figure 3.34). The anchor cell secretes the LIN-3 protein, a paracrine factor (similar to mammalian epidermal growth factor, or EGF) that activates the RTK pathway (Hill and Sternberg 1992). If the anchor cell is

destroyed (or if the *lin-3* gene is mutated), the VPCs will not form a vulva, but instead become part of the hypodermis (skin) (Kimble 1981).

The six VPCs influenced by the anchor cell form an **equivalence group**. Each member of this group is competent to become induced by the anchor cell and can assume any of three fates, depending on its proximity to the anchor cell. The cell directly beneath the anchor cell divides to form the central vulval cells. The two cells flanking that central cell divide to become the lateral vulval cells, while the three cells farther away from the anchor cell generate hypodermal cells. If the anchor cell is destroyed, all six cells of the equivalence group

(*Continued on next page*)

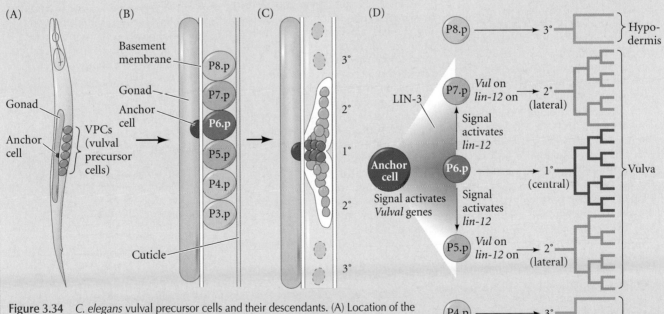

Figure 3.34 *C. elegans* vulval precursor cells and their descendants. (A) Location of the gonad, anchor cell, and VPCs in the second instar larva. (B,C) Relationship of the anchor cell to the six VPCs and their subsequent lineages. Primary lineages result in the central vulval cells; 2° lineages constitute the lateral vulval cells; 3° lineages generate hypodermal cells. (C) Outline of the vulva in the fourth instar larva. The circles represent the positions of the nuclei. (D) Model for the determination of vulval cell lineages in *C. elegans*. The LIN-3 signal from the anchor cell causes the determination of the P6.p cell to generate the central vulval lineage (dark purple). Lower concentrations of LIN-3 cause the P5.p and P7.p cells to form the lateral vulval lineages. The P6.p (central lineage) cell also secretes a short-range juxtacrine signal that induces the neighboring cells to activate the LIN-12 (Notch) protein. This signal prevents the P5.p and P7.p cells from generating the primary, central vulval cell lineage. (After Katz and Sternberg 1996.)

divide once and contribute to the hypodermal tissue. If the three central VPCs are destroyed, the three outer cells, which normally form hypodermis, generate vulval cells instead.

The LIN-3 protein is received by the LET-23 receptor tyrosine kinase on the VPCs, and the signal is transferred to the nucleus through the RTK pathway. The target of the kinase cascade is the LIN-31 protein (Tan et al. 1998). When this protein is phosphorylated in the nucleus, it loses its inhibitory protein partner and is able to function as a transcription factor, promoting vulval cell fates. Two mechanisms coordinate the formation of the vulva through this induction, as shown in Figure 3.34 (Katz and Sternberg 1996; Félix 2007):

1. The LIN-3 protein forms a concentration gradient. Here, the VPC closest to the anchor cell (i.e., the P6.p cell) receives the highest concentration of LIN-3 protein and generates the central vulval cells. The two VPCs adjacent to it (P5.p and P7.p) receive a lower amount of LIN-3 and become the lateral vulval cells. The VPCs farther away from the anchor cell do not receive enough LIN-3 to have an effect, so they become hypodermis (Katz et al. 1995).

2. In addition to forming the central vulval lineage, the VPC closest to the anchor cell also signals laterally to the two adjacent (P5.p and P7.p) cells and instructs them not to generate the central vulval lineages. The P5.p and P7.p cells receive the signal through the LIN-12 (Notch) proteins on their cell membranes. The Notch signal activates a microRNA, *mir-61*, which represses the gene that would specify central vulval fate, as well as promot-

ing those genes that are involved in forming the lateral vulval cells (Sternberg 1988; Yoo et al. 2005). The lateral cells do not instruct the peripheral VPCs to do anything, so they become hypodermis (Koga and Ohshima 1995; Simske and Kim 1995).

Cell-cell interactions and chance in the determination of cell types

The development of the vulva in *C. elegans* offers several examples of induction on the cellular level. We have already discussed the reception of the EGF-like LIN-3 signal by the cells of the equivalence group that forms the vulva. But before this induction occurs, there is an earlier interaction that forms the anchor cell. The formation of the anchor cell is mediated by *lin-12*, the *C. elegans* homologue of the *Notch* gene. In wild-type *C. elegans* hermaphrodites, two adjacent cells, Z1.ppp and Z4.aaa, have the potential to become the anchor cell. They interact in a manner that causes one of them to become the anchor cell while the other one becomes the precursor of the uterine tissue. In loss-of-function *lin-12* mutants, both cells become anchor cells, whereas in gain-of-function mutations, both cells become uterine precursors (Greenwald et al. 1983). Studies using genetic mosaics and cell ablations have shown that this decision is made in the second larval stage, and that the *lin-12* gene needs to function only in that cell destined to become the uterine precursor cell. The presumptive anchor cell does not need it. Seydoux and Greenwald (1989) speculate that these two cells originally synthesize both the signal for uterine differentiation (the LAG-

2 protein, homologous to Delta) and the receptor for this molecule (the LIN-12 protein, homologous to Notch; Wilkinson et al. 1994).

During a particular time in larval development, the cell that, by chance, is secreting more LAG-2 causes its neighbor to cease its production of this differentiation signal and to increase its production of LIN-12 protein. The cell secreting LAG-2 becomes the gonadal anchor cell, while the cell receiving the signal through its LIN-12 protein becomes the ventral uterine precursor cell (**Figure 3.35**). Thus, the two cells are thought to determine each other prior to their respective differentiation events. When the LIN-12 protein is used again during vulva formation, it is activated by the primary vulval lineage to stop the lateral vulval cells from forming the central vulval phenotype (see Figure 3.34). Thus, the anchor cell/ventral uterine precursor decision illustrates two important aspects of determination in two originally equivalent cells. First, the initial difference between the two cells is created by chance. Second, this initial difference is reinforced by feedback.

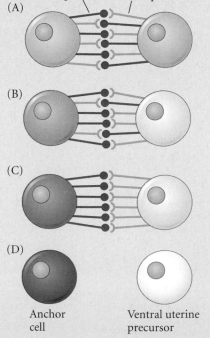

Figure 3.35 Model for the generation of two cell types (anchor cell and ventral uterine precursor) from two equivalent cells (Z1.ppp and Z4.aaa) in *C. elegans*. (A) The cells start off as equivalent, producing fluctuating amounts of signal and receptor (inverted arrow). The *lag-2* gene is thought to encode the signal; the *lin-12* gene is thought to encode the receptor. Reception of the signal turns down LAG-2 (Delta) production and upregulates LIN-12 (Notch). (B) A stochastic (chance) event causes one cell to produce more LAG-2 than the other cell at some particular critical time. This stimulates more LIN-12 production in the neighboring cell. (C) This difference is amplified, since the cell producing more LIN-12 produces less LAG-2. Eventually, just one cell is delivering the LAG-2 signal, and the other cell is receiving it. (D) The signaling cell becomes the anchor cell; the receiving cell becomes the ventral uterine precursor. (After Greenwald and Rubin 1992.)

Maintaining the Differentiated State

Development obviously means more than initiating gene expression. For a cell to become committed to a particular phenotype, gene expression must be maintained. Evolution has resulted in four major pathways for maintaining differentiation once it has been initiated (Figure 3.36):

1. The transcription factor whose gene is activated by a signal transduction cascade can bind to the enhancer of its own gene. In this way, once the transcription factor is made, its synthesis becomes independent of the signal that induced it originally. The MyoD transcription factor in muscle cells is produced in this manner.

2. A cell can stabilize its differentiation by synthesizing proteins that act on chromatin to keep the gene accessible. Such proteins include the Trithorax family discussed in Chapter 2.

3. A cell can maintain its differentiation in an autocrine fashion. If differentiation is dependent on a particular signaling molecule, the cell can make both that signaling molecule and that molecule's receptor. This pro-

duces a "community effect" (Grobstein 1955; Saxén and Wartiovaara 1966; Gurdon 1988), where the capacity to express a developmental potential exists only when a critical cell density of induced cells is present. In other words, once a group of cells has been induced, autocrine factors can sustain that induction and complete their differentiation.* In *Xenopus* muscle development, this community effect is mediated through FGF signaling. Standley and colleagues (2001) have shown (1) that FGF signaling can simulate the community effect in isolated muscle precursor cells; (2) that the muscle precursor cells have the receptors for FGFs at the critical time; and

---

*Community effect is also extremely important in bacterial development. Here it is called "quorum sensing," and it is critical in permitting emergent phenotypes such as light production, biofilm formation, invasiveness, and virulence. These phenotypes are expressed only in groups of bacteria and not in individuals. Each bacterium makes a small amount of a diffusible autocrine inducer that will induce the phenotype only at relatively high concentrations (see Zhu et al. 2002; Podbielski and Kreikemeyer 2004).

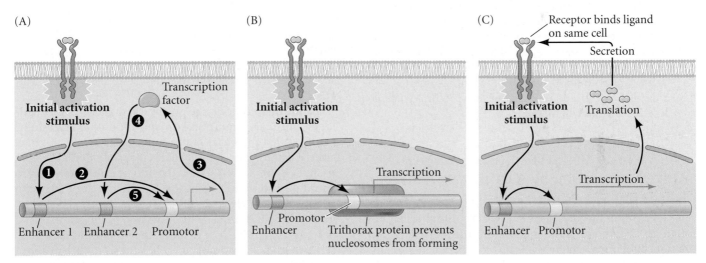

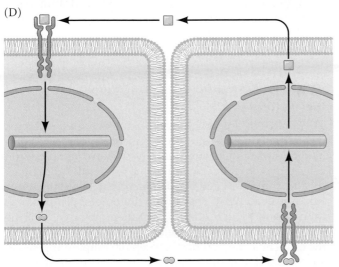

FIGURE 3.36 Four ways of maintaining differentiation after the initial signal has been given. (A) The initial stimulus (1) activates enhancer 1, which stimulates the promoter (2) to transcribe the gene. The gene product (3) is a transcription factor, and one of its targets is enhancer 2 of its own gene (4). This activated enhancer can now stimulate the promoter (5) to make more of this protein. (B) Proteins such as those of the Trithorax group prevent nucleosomes from forming so that the gene remains accessible. (C) Autocrine stimulation of the differentiated state. The cell is stimulated to activate those genes that enable it to synthesize and bind its own stimulatory proteins. (D) Paracrine loop between two cells such that a paracrine factor from one cell stimulates the differentiated state of the second cell, and that differentiated state includes the secretion of a paracrine factor that maintains the first cell's differentiated state.

(3) that the muscle precursor cells express FGFs at this time. It thus appears that part of the developmental program for *Xenopus* muscle cells is to make an FGF protein to which those same cells can also respond (i.e., an autocrine factor). This protein has to be present in sufficiently high density for the continuation of the processes leading to muscle development.

In the ectoderm of sea urchins, the identity of a group of cells that become the ectoderm around the mouth is also coordinated in this manner. Here, a Nodal paracrine factor from one cell causes the phosphorylation of Smad2 transcription factor in the neighboring cell receiving the signal. This activated transcription factor causes the neighboring cell to also make and secrete a Nodal signal. This keeps all the cells in the field actively secreting Nodal (Bolouri and Davidson 2009).

4. A cell may interact with its neighboring cells such that each one stimulates the differentiation of the other, and part of each neighbor's differentiated phenotype is the production of a paracrine factor that stimulates the other's phenotype. This type of I-scratch-your-back-you-scratch-mine strategy is found in the neighboring cells of the developing vertebrate limb and insect segments. This technique can be used to generate two different types of cells next to one another (as it does in insect segments; see Chapter 6). Or this mechanism can be used to generate a single population of cells (if the signal is the same). If this is so, it will also generate a "community effect," like the sea urchin example described for mechanism 3.

The Extracellular Matrix as a Source of Developmental Signals

The **extracellular matrix** is an insoluble network consisting of macromolecules secreted by cells into their immediate environment. These macromolecules form a region of noncellular material in the interstices between the cells. The extracellular matrix is a critical region for much of animal development. Cell adhesion, cell migration, and the formation of epithelial sheets and tubes all depend on the ability of cells to form attachments to extracellular matrices. In some cases, as in the formation of epithelia, these attachments have to be extremely strong. In other instances, as when cells migrate, attachments have to be made, broken, and made again. In some cases, the extracellular matrix merely serves as a permissive substrate to which cells can adhere, or upon which they can migrate. In other cases, it provides the directions for cell movement or the signal for a developmental event. Extracellular matrices are made up of collagen, proteoglycans, and a variety of specialized glycoprotein molecules such as fibronectin and laminin.

Proteoglycans play critically important roles in the delivery of the paracrine factors. These large molecules consist of core proteins (such as syndecan) with covalently attached glycosaminoglycan polysaccharide side chains. Two of the most widespread proteoglycans are heparan sulfate and chondroiton sulfate proteoglycans. Heparan sulfate proteoglycans can bind many members of the TGF-β, Wnt, and FGF families, and they appear to be essential for presenting the paracrine factor in high concentrations to their receptors. In *Drosophila*, *C. elegans*, and mice, mutations that prevent proteoglycan protein or carbohydrate synthesis block normal cell migration, morphogenesis, and differentiation (Garcia-Garcia and Anderson 2003; Hwang et al. 2003; Kirn-Safran et al. 2004).

The large glycoproteins are responsible for organizing the matrix and the cells into an ordered structure. **Fibronectin** is a very large (460 kDa) glycoprotein dimer synthesized by numerous cell types. One function of fibronectin is to serve as a general adhesive molecule, linking cells to one another and to other substrates such as collagen and proteoglycans. Fibronectin has several distinct binding sites, and their interaction with the appropriate molecules results in the proper alignment of cells with their extracellular matrix (**Figure 3.37A**). Fibronectin also has an important role in cell migration, since the "roads" over which certain migrating cells travel are paved with this protein. Fibronectin paths lead germ cells to the gonads and heart cells to the midline of the embryo. If chick embryos are injected with antibodies to fibronectin, the heart-forming cells fail to reach the midline, and two separate hearts develop (Heasman et al. 1981; Linask and Lash 1988).

Laminin (another large glycoprotein) and **type IV collagen** are major components of a type of extracellular matrix called the **basal lamina**. The basal lamina is characteristic of the closely knit sheets that surround epithelial tissue (**Figure 3.37B**). The adhesion of epithelial cells to laminin (upon which they sit) is much greater than the affinity of mesenchymal cells for fibronectin (to which they must bind and release if they are to migrate). Like fibronectin, laminin plays a role in assembling the extracellular matrix, promoting cell adhesion and growth, changing cell shape, and permitting cell migration (Hakamori et al. 1984; Morris et al. 2003).

See VADE MECUM Elements of the ECM

Integrins: Receptors for extracellular matrix molecules

The ability of a cell to bind to adhesive glycoproteins depends on its expressing membrane receptors for the cell-binding sites of these large molecules. The main fibronectin receptors were identified by using antibodies that block the attachment of cells to fibronectin (Chen et al. 1985; Knudsen et al. 1985). The fibronectin receptor complex was found not only to bind fibronectin on the outside of the cell, but also to bind cytoskeletal proteins on the inside of the cell. Thus, the fibronectin receptor complex appears to span the cell membrane and unite two types of matrices.

(A)

(B)

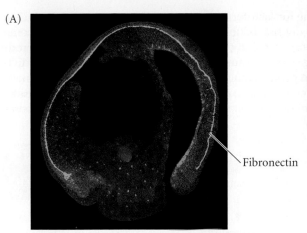

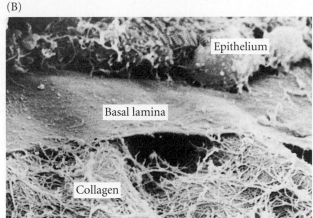

Fibronectin

Epithelium

Basal lamina

Collagen

FIGURE 3.37 Extracellular matrices in the developing embryo. (A) Fluorescent antibodies to fibronectin show fibronectin deposition as a green band in the *Xenopus* embryo during gastrulation. The fibronectin will orient the movements of the mesoderm cells. (B) Fibronectin links together migrating cells, collagen, heparin sulfate proteoglycans, and other extracellular matrix proteins. This scanning electron micrograph shows the extracellular matrix at the junction of the epithelial cells (above) and mesenchymal cells (below). The epithelial cells synthesize a tight, laminin-based basal lamina, while the mesenchymal cells secrete a loose reticular lamina made primarily of collagen. (A courtesy of M. Marsden and D. W. DeSimone; B courtesy of R. L. Trelsted.)

On the outside of the cell, it binds to the fibronectin of the extracellular matrix; on the inside of the cell, it serves as an anchorage site for the actin microfilaments that move the cell (**Figure 3.38**).

Horwitz and co-workers (1986; Tamkun et al. 1986) have called this family of receptor proteins **integrins** because they *integrate* the extracellular and intracellular scaffolds, allowing them to work together. On the extracellular side, integrins bind to the sequence arginine-glycine-aspartate (RGD), found in several adhesive proteins in extracellular matrices, including fibronectin, vitronectin (found in the basal lamina of the eye), and laminin (Ruoslahti and Pierschbacher 1987). On the cytoplasmic side, integrins bind to talin and α-actinin, two proteins that connect to actin microfilaments. This dual binding enables the cell to move by contracting the actin microfilaments against the fixed extracellular matrix.

Bissell and her colleagues (1982; Martins-Green and Bissell 1995) have shown that the extracellular matrix is capable of inducing specific gene expression in developing tis-

sues, especially those of the liver, testis, and mammary gland. In these tissues, the induction of specific transcription factors depends on cell-substrate binding (**Figure 3.39**; Liu et al. 1991; Streuli et al. 1991; Notenboom et al. 1996). Often, the presence of bound integrin prevents the activation of genes that specify apoptosis (Montgomery et al. 1994; Frisch and Ruoslahti 1997). The chondrocytes that produce the cartilage of our vertebrae and limbs can survive and differentiate only if they are surrounded by an extracellular matrix and are joined to that matrix through their integrins (Hirsch et al. 1997). If chondrocytes from the developing chick sternum are incubated with antibodies that block the binding of integrins to the extracellular

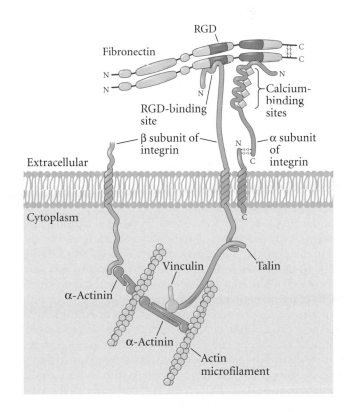

FIGURE 3.38 Simplified diagram of the fibronectin receptor complex. The integrins of the complex are membrane-spanning receptor proteins that bind fibronectin on the outside of the cell while binding cytoskeletal proteins on the inside of the cell. (After Luna and Hitt 1992.)

(A)

(B)

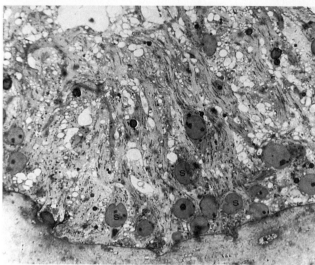

FIGURE 3.39   Role of the extracellular matrix in cell differentiation. Light micrographs of rat testis Sertoli cells grown for 2 weeks on tissue culture plastic dishes (A) and on dishes coated with basal lamina (B). The two photographs were taken at the same magnification, 1200×. (From Hadley et al. 1985, courtesy of M. Dym.)

matrix, they shrivel up and die. Indeed, when the focal adhesions linking the epithelial cell to its extracellular matrix are broken, the caspase-dependent apoptosis pathway is activated and the cells die. Such "death-upon-detachment" is a special type of apoptosis called **anoikis**, and it appears to be a major weapon against cancer (Frisch and Francis 1994; Chiarugi and Giannoni 2008).

While the mechanisms by which bound integrins inhibit apoptosis remain controversial, the extracellular matrix is obviously an important source of signals that can be transduced into the nucleus to produce specific gene expression. Some of the genes induced by matrix attachment are being identified. When plated onto tissue culture plastic, mouse mammary gland cells will divide (**Figure 3.40**). Indeed, the genes for cell division (*c-myc*, *cyclinD1*) are expressed, while the genes for the differentiated products of the mammary gland (casein, lactoferrin, whey acidic protein) are not expressed. If the same cells are plated onto plastic coated with a laminin-containing basement membrane, the cells stop dividing and the differentiated genes of the mammary gland are expressed. This happens only after the integrins of the mammary gland cells bind to the laminin of the extracellular basement membrane. Then the

gene for lactoferrin is expressed, as is the gene for p21, a cell division inhibitor. The *c-myc* and *cyclinD1* genes become silent. Eventually, all the genes for the developmental products of the mammary gland are expressed, and the cell division genes remain turned off. By this time, the mammary gland cells have enveloped themselves in a basal lamina, forming a secretory epithelium reminiscent of the mamma-

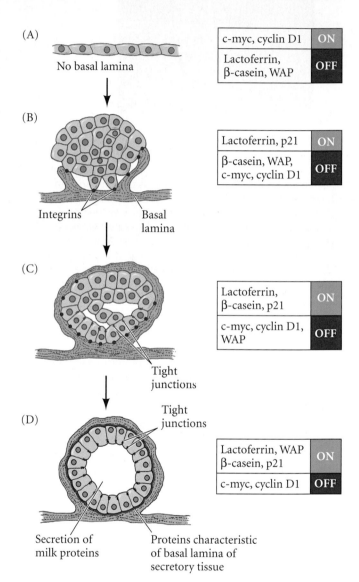

FIGURE 3.40   Basement membrane-directed gene expression in mammary gland tissue. (A) Mouse mammary gland tissue divides when placed on tissue culture plastic. The genes encoding cell division proteins are on, and the genes capable of synthesizing the differentiated products of the mammary gland—lactoferrin, casein, and whey acidic protein (WAP)—are off. (B) When these cells are placed on a basement membrane that contains laminin (basal lamina), the genes for cell division proteins are turned off, while the genes encoding inhibitors of cell division (such as p21) and the gene for lactoferrin are turned on. (C,D) The mammary gland cells wrap the basal lamina around them, forming a secretory epithelium. The genes for casein and WAP are sequentially activated. (After Bissell et al. 2003.)

ry gland tissue. The binding of integrins to laminin is essential for the transcription of the casein gene, and the integrins act in concert with prolactin (see Figure 3.23) to activate that gene's expression (Roskelley et al. 1994; Muschler et al. 1999). Several studies have shown that the binding of integrins to an extracellular matrix can stimulate the RTK pathway. When an integrin on the cell membrane of one cell binds to fibronectin or collagen secreted by a neighboring cell, the integrin can activate the RTK cascade through an adaptor protein-like complex that connects the integrin to the Ras G protein (Wary et al. 1998).

Epithelial-Mesenchymal Transition

One important developmental phenomenon, **epithelial-mesenchymal transition**, or **EMT**, integrates all the processes we have discussed in this chapter. EMT is an orderly series of events whereby epithelial cells are transformed into mesenchymal cells. In this transition, a polarized stationary epithelial cell, which normally interacts with basement membrane through its basal surface, becomes a migratory mesenchymal cell that can invade tissues and form organs in new places (**Figure 3.41A**). EMT is usually initiated when paracrine factors from neighboring cells activate gene expression in the target cells, instructing the target cells to downregulate their cadherins, release

their attachment to laminin and other basement membrane components, rearrange their actin cytoskeleton, and secrete new extracellular matrix molecules characteristic of mesenchymal cells.

Epithelial-mesenchymal transition is critical during development (**Figure 3.41B,C**). Examples of developmental processes in which this transition is active include (1) the formation of neural crest cells from the dorsalmost region of the neural tube; (2) the formation of mesoderm in chick embryos, wherein cells that had been part of an epithelial layer become mesodermal and migrate into the embryo; and (3) the formation of vertebrae precursor cells from the somites, wherein these cells detach from the somite and migrate around the developing spinal cord. EMT is also important in adults, in whom it is needed for wound healing. However, the most critical adult form of EMT is seen in cancer metastasis, wherein cells that have been part of a solid tumor mass leave that tumor to invade other tissues and form secondary tumors elsewhere in the body. It appears that in metastasis, the processes that generated the cellular transition in the embryo have been reactivated, allowing cancer cells to migrate and become invasive. Cadherins are downregulated, the actin cytoskeleton is reorganized, and the cells secrete mesenchymal extracellular matrix while undergoing cell division (Acloque et al. 2009; Kalluri and Weinberg 2009).

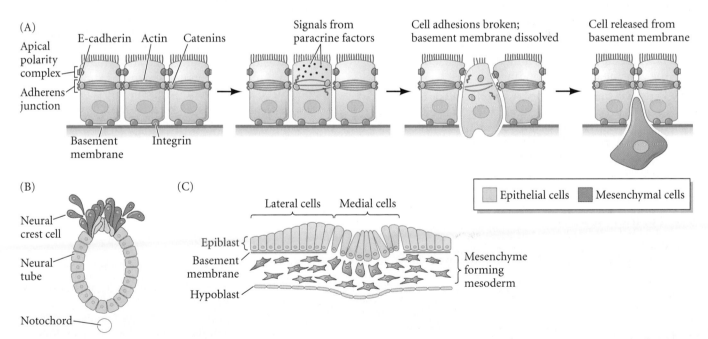

FIGURE 3.41 Epithelial-mesenchymal transition, or EMT. (A) Normal epithelial cells are attached to one another through adherens junctions containing cadherin, catenins, and actin rings. They are attached to the basement membrane through integrins. Paracrine factors can repress the expression of genes that encode these cellular components, causing the cell to lose polarity, lose attachment to the basement membrane, and lose cohesion with other epithelial cells. Cytoskeletal remodeling occurs, as well as the secretion of proteases and extracellular matrix molecules that enable the migration of the newly formed mesenchymal cell. (B,C) EMT is seen in vertebrate embryos during the normal formation of neural crest from the dorsal region of the neural tube (B), and during the formation of the mesoderm by mesenchymal cells delaminating from the epiblast (C).

Snapshot Summary: *Cell-Cell Communication*

1. The sorting out of one cell type from another results from differences in the cell membrane.

2. The membrane structures responsible for cell sorting out are often cadherin proteins that change the surface tension properties of the cells.

3. Cadherin proteins can cause cells to sort out by both quantitative differences (different amounts of cadherin) or qualitative differences (different types of cadherin). Cadherins appear to be critical during certain morphological changes.

4. Migration occurs through changes in the actin cytoskeleton. These changes can be directed by internal instructions (from the nucleus) or by external instructions (from the extracellular matrix or chemoattractant molecules).

5. Inductive interactions involve inducing and responding tissues.

6. The ability to respond to inductive signals depends on the competence of the responding cells.

7. Reciprocal induction occurs when the two interacting tissues are both inducers and are competent to respond to each other's signals.

8. Cascades of inductive events are responsible for organ formation.

9. Regionally specific inductions can generate different structures from the same responding tissue.

10. The specific response to an inducer is determined by the genome of the responding tissue.

11. Paracrine interactions occur when a cell or tissue secretes proteins that induce changes in neighboring cells. Juxtacrine interactions are inductive interactions that take place between the cell membranes of adjacent cells or between a cell membrane and an extracellular matrix secreted by another cell.

12. Paracrine factors are proteins secreted by inducing cells. These factors bind to cell membrane receptors in competent responding cells.

13. Competent cells respond to paracrine factors through signal transduction pathways. Competence is the ability to bind and to respond to inducers, and it is often the result of a prior induction.

14. Signal transduction pathways begin with a paracrine or juxtacrine factor causing a conformational change in its cell membrane receptor. The new shape results in enzymatic activity in the cytoplasmic domain of the receptor protein. This activity allows the receptor to phosphorylate other cytoplasmic proteins. Eventually, a cascade of such reactions activates a transcription factor (or set of factors) that activates or represses specific gene activity.

15. Programmed cell death is one possible response to inductive stimuli. Apoptosis is a critical part of life.

16. The maintenance of the differentiated state can be accomplished by positive feedback loops involving transcription factors, autocrine factors, or paracrine factors.

17. The extracellular matrix is a source of signals for the differentiating cells and plays critical roles in cell migration.

18. Cells can convert from being epithelial to being mesenchymal and vice-versa. The epithlial-mesenchymal transitiion is an series of transformations involved in the dispersion of neural crest cells and the creation of vertebrae from somitic cells. In adults, the EMT is involved in wound healing and cancer metastasis.

For Further Reading

Complete bibliographical citations for all literature cited in this chapter can be found at the free-access website **www.devbio.com**

Ananthakrishnan, R. and A. Ehrlicher. 2007. The forces behind cell movement. *Int. J. Biol. Sci.* 3: 303–317.

Breitling, R. 2007. Greased hedgehogs: New links between hedgehog signaling and cholesterol metabolism. *BioEssays* 29: 1085–1094

Cadigan, K. M. and R. Nusse. 1997. Wnt signaling: A common theme in animal development. *Genes Dev.* 24: 3286–3306.

Casanova, J. 2007. The emergence of shape: notions from the study of the *Drosophila* tracheal system. *EMBO Rep.* 4: 335–339.

Cooper, M. K., J. A. Porter, K. E. Young and P. A. Beachy. 1998. Teratogenmediated inhibition of target tissue response to hedgehog signaling. *Science* 280: 1603–1607.

Engler, A., S. Sen, H. Sweeney and D. Discher. 2006. Matrix elasticity directs

stem cell lineage specification. *Cell* 126: 677–689.

Foty, R. A. and M. S. Steinberg. 2005. The differential adhesion hypothesis: a direct evaluation. *Dev. Biol.* 278: 255–263.

Heldin, C.-H., K. Miyazono and P. ten Dijke. 1997. TGF-β signaling from cell membrane to nucleus through SMAD proteins. *Nature* 390: 465–471.

Lum, L. and P. A. Beachy. 2004. The hedgehog response network: sensors, switches, and routers. *Science* 304: 1755–1759.

Rousseau, F. and 7 others. 1994. Mutations in the gene encoding fibroblast growth factor receptor-3 in achondroplasia. *Nature* 371: 252–254.

Steinberg, M. S. and S. F. Gilbert. 2004. Townes and Holtfreter (1955): Directed movements and selective adhesion of embryonic amphibian cells. *J. Exp. Zool. A: Comp Exp. Biol.* 301: 701–706.

Go Online

WEBSITE 3.1 FGF receptor mutations. Mutations of human FGF receptors have been associated with several skeletal malformation syndromes, including syndromes wherein skull cartilage, rib cartilage, or limb cartilage fail to grow or differentiate.

WEBSITE 3.2 The uses of apoptosis. Apoptosis is used for numerous processes throughout development. This website explores the role of apoptosis in such phenomena as *Drosophila* germ cell development and the eyes of blind cave fish.

WEBSITE 3.3 Notch mutations. Mutations in the genes that encode Notch proteins can cause nervous system abnormalities in humans. Humans have more than one Notch gene and more than one ligand. Their interactions may be critical in neural development. Moreover, the association of Notch with the presenilin protease suggests that disruption of Notch functioning might lead to Alzheimer disease.

Vade Mecum

The differential adhesion hypothesis. These movies show the pioneering work of Townes and Holtfreter and Malcolm Steinberg. These experiments demonstrated the phenomenon of cell sorting and how cell surface adhesion molecules can direct sorting behaviors.

Induced cyclopia in zebrafish. As seen in the segment on zebrafish development, alcohol can act as a teratogen and induce cyclopia in these embryos.

Elements of the ECM. Movies on *Vade Mecum* review the molecular components of the extracellular matrix, how cells are influenced by them, and the work of Elizabeth Hay, who was among the first scientists to show the importance of the ECM to tissue differentiation.

PART II

SPECIFICATION
Introducing Cell Commitment and Early Embryonic Development

The cartoon shows the fundamental phenomenon and mystery of development: a complex and exquisitely ordered body is generated from the relatively unorganized fertilized egg. It is difficult for us in the twenty-first century to realize how totally mysterious embryological development has been. One of America's first embryologists, William Keith Brooks, wrote in 1883 of "the greatest of all wonders of the material universe: the existence, in a simple, unorganized egg, of a power to produce a definite adult animal." Brooks reflected that this property is so complex that "we may fairly ask what hope there is of discovering its solution, of reaching its true meaning, its hidden laws and causes." These hidden laws and causes are now being sought in the way that the cells interpret the genome that is the same in every embryonic cell. How are certain genes activated and repressed in one group of cells to turn them into mesoderm, while a different set of genes is regulated to instruct cells to become endoderm? Moreover, are there any principles or strategies that characterize the origins of different cell types?

Levels of Commitment

The generation of specialized cell types is called *differentiation*. But differentiation is only the last, overt stage of a series of events that commit a particular blastomere to become a particular cell type (**Table II.1**). A red blood cell obviously differs in its protein composition and cell structure from a lens cell or a nerve cell. But these overt changes in cellular biochemistry and function are preceded by a process resulting in the **commitment** of the cell to a certain fate. During these

TABLE II.1 Some differentiated cell types and their major products

Type of cell	Differentiated cell product	Specialized function
Keratinocyte (epidermal cell)	Keratin	Protection against abrasion, desiccation
Erythrocyte (red blood cell)	Hemoglobin	Transport of oxygen
Lens cell	Crystallins	Transmission of light
B lymphocyte	Immunoglobulins	Antibody synthesis
T lymphocyte	Cytokines	Destruction of foreign cells; regulation of immune response
Melanocyte	Melanin	Pigment production
Pancreatic islet cell	Insulin	Regulation of carbohydrate metabolism
Leydig cell (♂)	Testosterone	Male sexual characteristics
Chondrocyte (cartilage cell)	Chondroitin sulfate; type II collagen	Tendons and ligaments
Osteoblast (bone-forming cell)	Bone matrix	Skeletal support
Myocyte (muscle cell)	Muscle actin and myosin	Contraction
Hepatocyte (liver cell)	Serum albumin; numerous enzymes	Production of serum proteins and numerous enzymatic functions
Neurons	Neurotransmitters (acetylcholine, epinephrine, etc.)	Transmission of electric impulses
Tubule cell (♀) of hen oviduct	Ovalbumin	Egg white proteins for nutrition and protection of embryo
Follicle cell (♀) of insect ovary	Chorion proteins	Eggshell proteins for protection of embryo

stages of commitment, the cell might not look differentiated, even though its developmental fate has become restricted.

The process of commitment can be divided into two stages (Harrison 1933; Slack 1991). The first stage is a labile phase called **specification**. The fate of a cell or a tissue is said to be *specified* when it is capable of differentiating autonomously (i.e., by itself) when placed into a Petri dish or test tube—that is, into an environment that is neutral with respect to the developmental pathway. At the stage of specification, cell commitment is still capable of being reversed.

The second stage of commitment is **determination**. A cell or tissue is said to be *determined* when it is capable of differentiating autonomously even when placed into another region of the embryo—a decidedly non-neutral environment. If a cell or tissue type is able to differentiate according to its specified fate even under these circumstances, it is assumed that commitment is irreversible.

There are three major strategies of commitment, and no one embryo uses only one of them. All three strategies are based on mechanisms that apportion certain sets of transcription factors to different cells in the early embryo.

Autonomous Specification

The first mode of commitment is **autonomous specification**. Here, the blastomere inherits a set of transcription factors from the egg cytoplasm, and these transcription factors regulate gene expression, directing the cell into a particular path of development. In other words, the egg cytoplasm is not homogeneous, but rather contains different **morphogenetic determinants** (transcription factors or their mRNAs), which will influence the cell's development. In this type of specification, the cell "knows" what it is to become very early and without interacting with other cells (**Figure II.1**).

Normal development of *Patella*

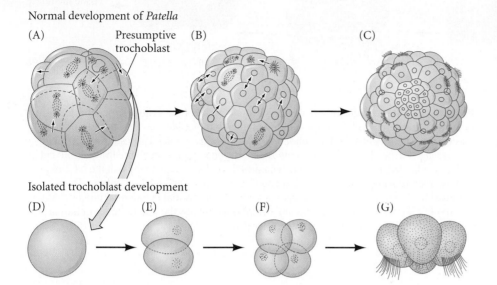

(A)   Presumptive trochoblast   (B)   (C)

Isolated trochoblast development

(D)   (E)   (F)   (G)

FIGURE II.1   Autonomous (mosaic) specification. (A–C) Differentiation of trochoblast (ciliated) cells of the mollusc *Patella*. (A) 16-cell stage seen from the side; the presumptive trochoblast cells are shaded. (B) 48-cell stage. (C) Ciliated larval stage, seen from the animal pole. (D–G) Differentiation of a *Patella* trochoblast cell isolated from the 16-cell stage and cultured in vitro. (E,F) Results of the first and second divisions in culture. (G) Ciliated products of (F). Even in isolated culture, these cells divide and become ciliated at the correct time. (After Wilson 1904.)

For instance, in tunicate (sea squirt) embryos (see Chapter 1), each blastomere will form most of its respective cell types even when separated from the remainder of the embryo (**Figure II.2**). In the 8-cell embryo, the two blastomeres that are going to generate tail muscles contain a yellow-pigmented cytoplasm that has within it a muscle-specific transcription factor called Macho. This transcription factor comes from the egg cytoplasm, and any blastomere that has this factor will become muscle cells, even if that blastomere were to be isolated from the rest of the embryo. Indeed, if Macho-containing cytoplasm is placed into other cells, those cells will form tail muscles (**Figure II.3**; Whittaker 1973; Nishida and Sawada 2001). If the cells normally containing this cytoplasm (the B4.1 blastomeres; see Figure II.2) are removed from the embryo, the embryo will not form tail muscles. Thus, the tail muscles of tunicates are formed autonomously by acquiring a transcription factor from the egg cytoplasm.

When most of the cells of an early embryo are determined by autonomous specification, it gives the appearance that the animal is fully specified this way. This is not the case, and even in tunicate embryos the nervous system arises conditionally by cell interactions. However, embryologists have traditionally called such embryos **mosaic embryos**, since they develop like a mosaic of individually laid tiles—with each cell receiving its instructions independently, without cell-cell interactions.

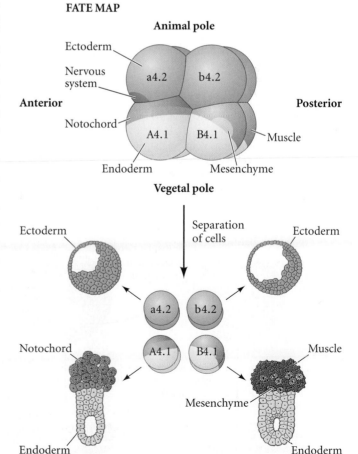

FATE MAP

Animal pole

Ectoderm

Nervous system

Anterior   a4.2   b4.2   Posterior

Notochord

A4.1   B4.1   Muscle

Endoderm   Mesenchyme

Vegetal pole

Separation of cells

Ectoderm   Ectoderm

a4.2   b4.2

Notochord   Muscle

A4.1   B4.1

Mesenchyme

Endoderm   Endoderm

FIGURE II.2   Autonomous specification in the early tunicate embryo. When the four blastomere pairs of the 8-cell embryo are dissociated, each forms the structures it would have formed had it remained in the embryo. The nervous system, however, is conditionally specified. The fate map shows that the left and right sides of the tunicate embryo produce identical cell lineages. Here the yellow crescent muscle-forming material is colored red to conform with its being associated with mesoderm. (After Reverberi and Minganti 1946.)

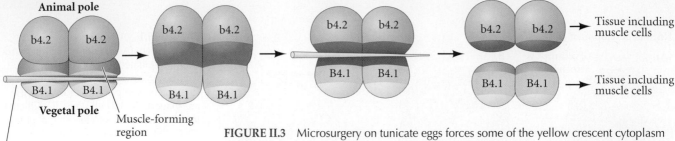

Animal pole

Vegetal pole

Muscle-forming region

Needle pushes muscle-forming cytoplasm into animal cells

Tissue including muscle cells

Tissue including muscle cells

FIGURE II.3 Microsurgery on tunicate eggs forces some of the yellow crescent cytoplasm of the muscle-forming B4.1 blastomeres to enter the b4.2 (epidermis- and nerve-producing) blastomere pair. Pressing the B4.1 blastomeres with a glass needle causes the regression of the cleavage furrow. The furrow re-forms at a more vegetal position where the cells are cut with a needle. The new furrow thereby separates the cells in such a way that the b4.2 blastomeres receive some of the muscle-forming ("yellow crescent") B4.1 cytoplasm. These modified b4.2 cells produce muscle cells as well as their normal ectodermal progeny. As we will see in Chapter 5, the yellow crescent cytoplasm contains the Macho transcription factor that activates the muscle-specific genes. (After Whittaker 1982.)

Conditional Specification

Conditional specification is the ability of cells to achieve their respective fates by interactions with other cells (**Figure II.4**). Here, what a cell becomes is in large measure specified by paracrine factors secreted by its neighbors (see Chapter 3). In one of the ironies of research, conditional specification was demonstrated by attempts to disprove it. In 1888, August Weismann proposed the first testable model of cell specification, the **germ plasm theory**, in which each cell of the embryo would develop autonomously. He boldly proposed that the sperm and egg provided equal chromosomal contributions, both quantitatively and qualitatively, to the new organism. Moreover, he postulated that the chromosomes carried the inherited potentials of this new organism.* However, not all the determinants on the chromosomes were thought to enter every cell of the embryo. Instead of dividing equally, the chromosomes were hypothesized to divide in such a way that different chromosomal deter-

---

*Note that embryologists were thinking in terms of chromosomal mechanisms of inheritance some 15 years before the rediscovery of Mendel's work. Weismann (1892, 1893) also speculated that these nuclear determinants of inheritance functioned by elaborating substances that became active in the cytoplasm!

FIGURE II.4 Conditional specification. (A) What a cell becomes depends on its position in the embryo. Its fate is determined by interactions with neighboring cells. (B) If cells are removed from the embryo, the remaining cells can regulate and compensate for the missing part.

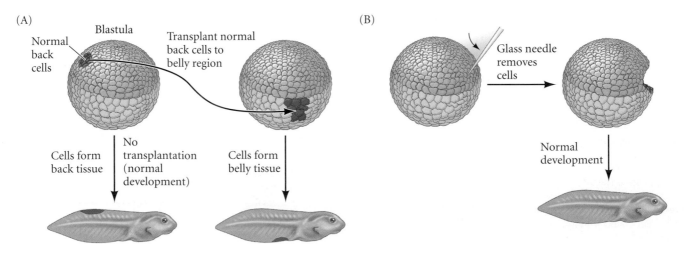

(A)

Blastula

Normal back cells

Transplant normal back cells to belly region

Cells form back tissue

No transplantation (normal development)

Cells form belly tissue

(B)

Glass needle removes cells

Normal development

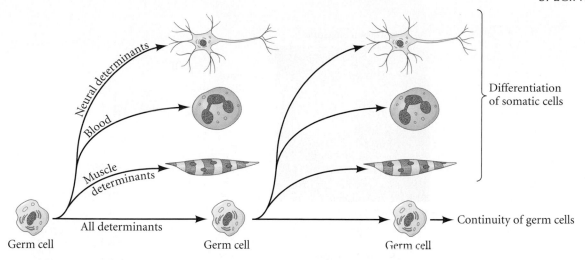

FIGURE II.5 Weismann's germ plasm theory of inheritance. The germ cell (blue) gives rise to the differentiating somatic cells of the body, as well as to new germ cells. Weismann hypothesized that only the germ cells contained all the inherited determinants. The somatic cells were each thought to contain a subset of the determinants, and the types of determinants found in a somatic cell's nucleus would determine its differentiated type. (After Wilson 1896.)

minants entered different cells. Whereas the fertilized egg would carry the full complement of determinants, certain somatic cells would retain the "blood-forming" determinants while others retained the "muscle-forming" determinants, and so forth (**Figure II.5**). Only the nuclei in those cells destined to become germ cells (gametes) were postulated to contain all the different types of determinants. The nuclei of all other cells would have only a subset of the original determinants.

In postulating his germ plasm model, Weismann proposed a hypothesis of development that could be tested immediately. Based on the fate map of the frog embryo, Weismann claimed that when the first cleavage division separated the future right half of the embryo from the future left half, there would be a separation of "right" determinants from "left" determinants in the resulting blastomeres. Wilhelm Roux tested that hypothesis by using a hot needle to kill one of the cells in a 2-cell frog embryo—with the result that only the right or left half of a larva developed (**Figure II.6**). Based on this result, Roux claimed that specification was mosaic, and that all the instructions for normal development were already inside each cell.

Roux's colleague Hans Driesch (1892), however, obtained opposite results. While Roux's studies were *defect* experiments that answered the question of how the remaining blastomeres of an embryo would develop when a subset of blastomeres was destroyed, Driesch (1892) sought to extend this research by performing *isolation* experiments. He separated sea urchin blastomeres from each other by vigorous shaking (or

FIGURE II.6 Roux's attempt to demonstrate autonomous specification. Destroying (but not removing) one cell of a 2-cell frog embryo resulted in the development of only one half of the embryo.

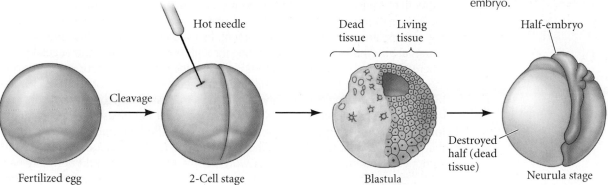

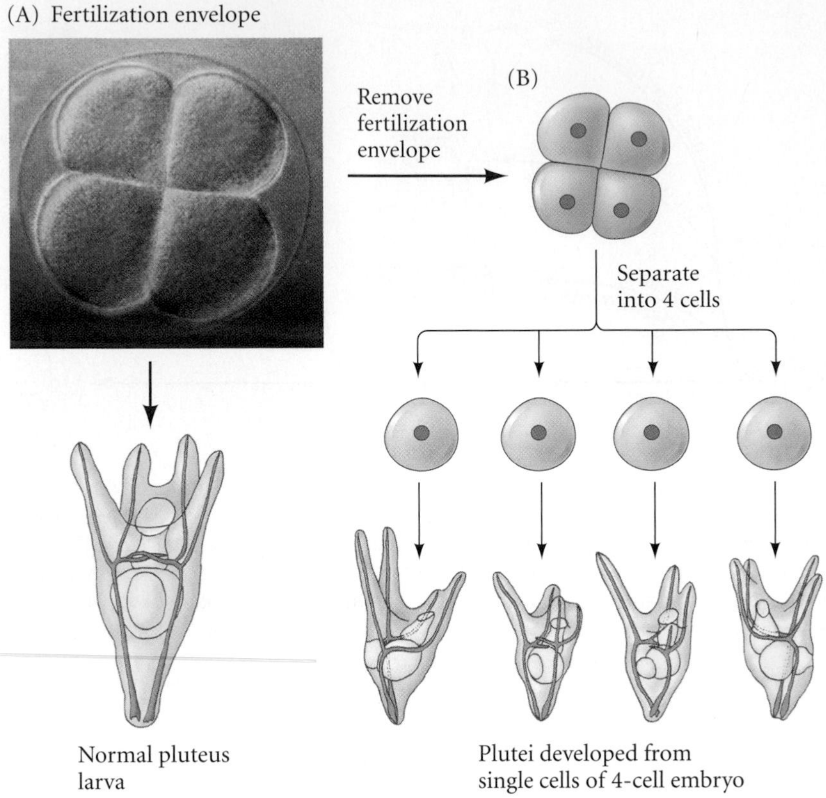

(A) Fertilization envelope

Normal pluteus
larva

Plutei developed from
single cells of 4-cell embryo

FIGURE II.7 Driesch's demonstration of conditional (regulative) specification. (A) An intact 4-cell sea urchin embryo generates a normal pluteus larva. (B) When one removes the 4-cell embryo from its fertilization envelope and isolates each of the four cells, each cell can form a smaller, but normal, pluteus larva. (All larvae are drawn to the same scale.) Note that the four larvae derived in this way are not identical, despite their ability to generate all the necessary cell types. Such variation is also seen in adult sea urchins formed in this way (see Marcus 1979). (Photograph courtesy of G. Watchmaker.)

later, by placing them in calcium-free seawater). To Driesch's surprise, each of the blastomeres from a 2-cell embryo developed into a complete larva. Similarly, when Driesch separated the blastomeres of 4- and 8-cell embryos, some of the isolated cells produced entire pluteus larvae (**Figure II.7**). Here was a result drastically different from the predictions of Weismann and Roux. Rather than self-differentiating into its future embryonic part, each isolated blastomere *regulated* its development to produce a complete organism. These experiments provided the first experimentally observable evidence of conditional specification. In conditional specification, the fate of cells depends on the cells' neighbors. Interactions between cells determine their fates, rather than some cytoplasmic factor that is particular to that type of cell.

Driesch confirmed conditional development in sea urchin embryos by performing an intricate *recombination* experiment. In sea urchin eggs, the first two cleavage planes are normally meridional, passing through both the animal and vegetal poles, whereas the third division is equatorial, dividing the embryo into four upper and four lower cells (**Figure II.8A**). Driesch (1893) changed the direction of the third cleavage by gently compressing early embryos between two glass plates, thus causing the third division to be meridional like the preceding two. After he released the pressure, the fourth division was equatorial. This procedure reshuffled the nuclei, placing nuclei that normally would have been in the region destined to form endoderm into the presumptive ectoderm region. In other words, some nuclei that would normally have produced ventral structures were now found in the dorsal cells (**Figure II.8B**). If seg-

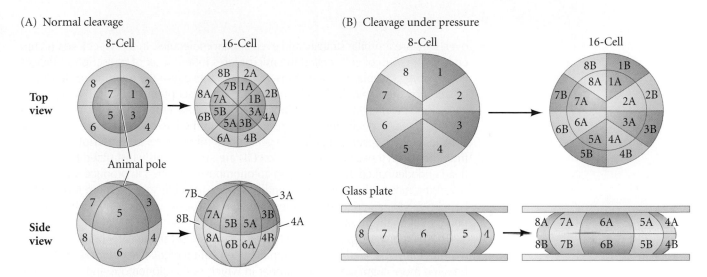

FIGURE II.8 Driesch's pressure-plate experiment for altering the distribution of nuclei. (A) Normal cleavage in 8- to 16-cell sea urchin embryos, seen from the animal pole (upper sequence) and from the side (lower sequence). The nuclei are numbered. (B) Abnormal cleavage planes formed under pressure, as seen from the animal pole and from the side. Some nuclei (such as 6A and 8A) are placed in different regions of the embryo. (After Huxley and de Beer 1934.)

regation of nuclear determinants had occurred (as had been proposed by Weismann and Roux), the resulting embryo should have been strangely disordered. However, Driesch obtained normal larvae from these embryos. He thus concluded that "The relative position of a blastomere within the whole will probably in a general way determine what shall come from it."

The consequences of these experiments were momentous, both for embryology and for Driesch personally.* First, Driesch had demonstrated that the prospective potency of an isolated blastomere (i.e., those cell types it was possible for it to form) is greater than the blastomere's prospective fate (those cell types it would normally give rise to over the unaltered course of its development). According to Weismann and Roux, the prospective potency and the prospective fate of a blastomere should have been identical. Second, Driesch concluded that the sea urchin embryo is a "harmonious equipotential system" because all of its potentially independent parts interacted together to form a single organism. Driesch's experiment implies that cell interaction is critical for normal development. Moreover, if each early blastomere can form all the embryonic cells when isolated, then it follows that, in normal development, the community of cells must prevent it from doing so (Hamburger 1997).

Third, Driesch concluded that the fate of a cell depended solely on its location in the embryo. The interactions between cells determined their fates. We now know (and will see in Chapters 5 and 7) that sea urchin embryos and frog embryos use both autonomous and conditional ways of specifying their early embryonic cells. More-

---

*This idea of nuclear equivalence and the ability of cells to interact eventually caused Driesch to abandon science. Driesch, who thought the embryo was like a machine, could not explain how the embryo could make its missing parts or how a cell could change its fate to become another cell type. Harking back to Aristotle, he invoked a vital force, *entelechy* ("internal goal-directed force") to explain how development proceeds. Essentially, he believed that the embryo was imbued with an internal psyche and the wisdom to accomplish its goals despite the obstacles embryologists placed in its path. However, others, especially Oscar Hertwig (1894), were able to incorporate Driesch's experiments into a more sophisticated experimental embryology, which will be discussed in the introduction to Part IV of this book.

over, they use a similar strategy and even similar molecules. In the 16-cell sea urchin embryo, a group of cells called the micromeres inherit a set of transcription factors from the egg cytoplasm. These transcription factors cause the micromeres to develop *autonomously* into the larval skeleton. These transcription factors also activate the genes for paracrine factors and juxtacrine factors that are employed by the micromeres to *conditionally* specify the cells around them. In the late blastula frog embryo, the cells located opposite the point of sperm entry inherit a set of transcription factors from the egg cytoplasm as well. These cells are autonomously instructed to form the head endodermal cells. They are also autonomously specified to produce and secrete paracrine factors that will conditionally induce the cells around them to form the brain. So the head endoderm is specified autonomously, whereas the brain is specified conditionally, in part, by the head endoderm.

Those embryos (especially vertebrates) wherein most of the early blastomeres are conditionally specified have traditionally been called *regulative embryos*. But as we become more cognizant of the manner in which both autonomous and conditional specification are used in each embryo, the notions of "mosaic" and "regulative" development are appearing less tenable. Indeed, attempts to get rid of these distinctions were begun by no less an embryologist than Edmund B. Wilson (1894, 1904). Wilson (a student of the above-mentioned W. K. Brooks) was one of the first scientists to theorize that chromosomes in the nucleus put forth cell-specifying factors into the cytoplasm. "Mosaic embryos," he wrote, received these factors from the cytoplasm of the egg during cleavage stages, while the nuclei of "regulative embryos" were instructed by other cells to produce these factors later in development.

Morphogen Gradients and Cell Specification

Throughout this book, we will see many instances of cell fate specification that involve morphogen gradients. A **morphogen** (Greek, "form-giver") is a diffusable biochemical molecule that can determine the fate of a cell by its concentration.* Morphogens can be transcription factors produced within cells (as in the *Drosophila* embryos described in the following section). They can also be paracrine factors that are produced in one group of cells and then travel to another population of cells, specifying the target cells differentially according to the concentration of morphogen. Uncommitted cells exposed to high concentrations of the morphogen (nearest its source of production) are specified as one cell type; when the morphogen's concentration drops below a certain threshold, the cells are determined to another fate. When the concentration falls even lower, a cell of the same initial uncommitted type is specified in yet a third manner.

Morphogen gradients provide a very important mechanism for conditional specification. The existence of morphogen gradients as a force in development and regeneration was predicted by Thomas Hunt Morgan (1905, 1906—before he became a geneticist), but it was many years before these gradient models were extended to explain how cells might be placed in specific positions along an embryonic axis (Hörstadius 1939; Toivonen and Saxén 1955; Lawrence 1966; Stumpf 1966; Wolpert 1968, 1969).

Lewis Wolpert illustrated such a gradient of positional information using the "French flag" analogy. Imagine a row of "flag cells," each of which is "multipotential"—capable of differentiating into either a red, a white, or a blue cell. Then imagine a morphogen whose source is on the left-hand edge of the blue stripe and whose sink is at the other end of the flag, on the right-hand edge of the red stripe. A concentration gradient is thus formed, highest at one end of the "flag tissue" and lowest at the other (**Figure II.9**). The specification of the multipotential cells in this tissue is accom-

---

*Although there is overlap in the terminology, a *morphogen* specifies cells in quantitative manner, while a *morphogenetic determinant* specifies cells in a qualitative manner. Morphogens are analog; morphogenetic determinants are digital.

(A)

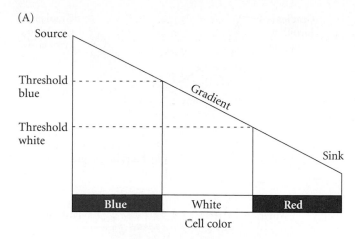

(B)

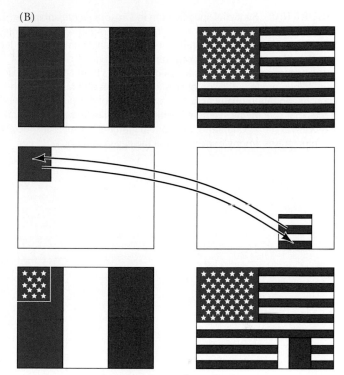

Reciprocal transplants develop according to their final positions in the "donor" flag

FIGURE II.9 The "French flag" analogy for conditional specification. (A) Positional information is delivered by a gradient of a diffusible morphogen extending from a source to a sink. The thresholds indicated on the left are cellular properties that enable the gradient to be interpreted. In this analogy, cells become blue at one concentration of the morphogen. As the concentration declines below a certain threshold, cells become white, and where it falls below another threshold, cells become red. The resulting pattern is that of the French flag. (B) An important feature of this model is that a piece of tissue transplanted from one region of an embryo to another retains its identity of origin, but differentiates according to its new positional instructions. This phenomenon is indicated schematically by reciprocal "grafts" between the flags of the United States and France. (After Wolpert 1978.)

plished by threshold concentrations of the morphogen. Cells sensing the highest concentrations of morphogen become blue. Then there is a threshold of morphogen concentration below which cells become white. As the declining concentration of morphogen falls below another threshold, the cells become red. According to such models (see Crick 1970), the morphogen diffuses from its site of synthesis (source) to its site of degradation (sink), its concentration dropping along the way. This drop in concentration can be due to simple diffusion; to the cells' binding the morphogen and thus "using it up"; or to a combination of a source synthesizing the morphogen and an environment containing an enzyme that degrades it.

Later in the book we will see the concept of morphogen gradients used to model how regions of the vertebrate body axis are established by retinoic acid, Fgf8, and Wnts (Chapters 7 and 8); how the different regions of the mesoderm are specified by bone morphogenetic protein (BMP) from the lateral plate mesoderm (Chapters 11 and 12); and how vertebrate limbs and digits are specified by Sonic hedgehog (Chapter 13).

Syncytial Specification

In addition to autonomous and conditional specification, there is another strategy that uses elements of both. In early embryos of insects, nuclei divide within the egg; but the cell does not divide. In other words, many nuclei are formed within one common cytoplasm. A cytoplasm that contains many nuclei is called a **syncytium**, and the specification of presumptive cells within such a common cytoplasm is called **syncytial specification**. As in the other eggs we have mentioned, the insect egg cytoplasm is not uniform. Nuclei in the anterior part of the cell will be exposed to cyto-

FIGURE II.10 Syncytial specification in *Drosophila melanogaster*. Anterior-posterior specification originates from morphogen gradients in the egg cytoplasm, specifically the morphogenetic transcription factors Bicoid and Caudal. The concentrations and ratios of these two proteins distinguish each position along the axis from any other position. When nuclear division occurs, the amounts and ratios of each morphogen differentially activate transcription of the various nuclear genes that specify the segment identities of the larval and the adult fly. (As we will see in Chapter 6, the Caudal gradient is itself constructed by interactions among constituents of the egg cytoplasm.)

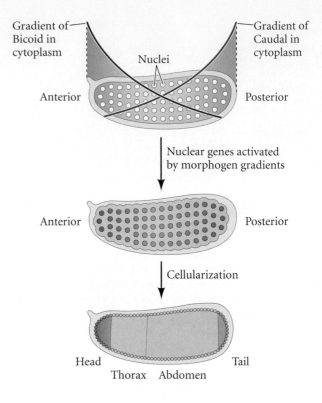

plasmic transcription factors that are not present in the posterior part of the cell, and vice versa. The interactions of nuclei and transcription factors, which eventually result in cell specification, take place in a common cytoplasm.

Each nucleus in *Drosophila* is given positional information (i.e., whether that nucleus is to become part of the anterior, posterior, or midsection of the body) by transcription factors acting as morphogens. These transcription factors are made in specific sites in the embryo, diffuse over long distances, and form concentration gradients where the highest concentration is at the point of synthesis and gets lower as the morphogen diffuses away from its source and degrades over time. The concentration of specific morphogens at any particular site tells the nuclei where they are in relation to the source of the morphogens. As we will detail in Chapter 6, the anteriormost portion of the *Drosophila* embryo produces a morphogen called Bicoid with a concentration that is highest in the anterior and declines toward the posterior. The posteriormost portion of the egg forms a posterior-to-anterior gradient of the morphogen Caudal. Thus, the long axis of the *Drosophila* egg is spanned by opposing morphogen gradients—Bicoid coming from the anterior, and Caudal from the posterior (**Figure II.10**).

Bicoid and Caudal are both transcription factors, and different concentrations and ratios of Bicoid and Caudal proteins activate different sets of genes in the syncytial nuclei. Those nuclei in regions containing high amounts of Bicoid and little Caudal are instructed to activate those genes necessary for producing the head. Nuclei in regions with slightly less Bicoid and a small amount of Caudal are instructed to activate genes that generate the thorax. Nuclei in regions that have little or no Bicoid but plenty of Caudal are instructed to form the abdominal structures (Nüsslein-Volhard et al. 1987). Thus when the syncytial nuclei are eventually incorporated into cells, these cells will have their *general* fate specified. Afterward, the specific fate of each cell will become determined both autonomously (from the transcription factors acquired by the cell's nucleus from the egg cytoplasm) and conditionally (by interactions between the cell and its neighbors).

TABLE II.2   Modes of cell type specification and their characteristics

I. Autonomous specification

Predominates in most invertebrates.

Specification by differential acquisition of certain cytoplasmic molecules present in the egg.

Invariant cleavages produce the same lineages in each embryo of the species. Blastomere fates are generally invariant.

Cell type specification precedes any large-scale embryonic cell migration.

Produces "mosaic" development: cells cannot change fate if a blastomere is lost.

II. Conditional specification

Predominates in vertebrates and a few invertebrates.

Specification by interactions between cells. Relative positions are important.

Variable cleavages produce no invariant fate assignments to cells.

Massive cell rearrangements and migrations precede or accompany specification.

Capacity for "regulative" development; allows cells to acquire different functions.

III. Syncytial specification

Predominates in most insect classes.

Specification of body regions by interactions between cytoplasmic regions prior to cellularization of the blastoderm.

Variable cleavage produces no rigid cell fates for particular nuclei.

After cellularization, both autonomous and conditional specification are seen.

Source: After Davidson 1991.

Summary

Each of the three major strategies of cell specification (summarized in **Table II.2**) offers a different way of providing an embryonic cell with a set of transcription factors that will activate specific genes and cause the cell to differentiate into a particular cell type. In autonomous specification, the transcription factors are already present in different regions of the egg cytoplasm. In conditional specification, the set of transcription factors is determined by paracrine and juxtacrine interactions between neighboring cells. In syncytial specification, there are interactions, not between cells, but between regions of the egg cytoplasm. These regional interactions give each nucleus a different ratio of particular transcription factors, and these ratios determine which genes are on and which are off.

The chapters that follow describe the early development of several organisms. In these chapters, we shall see how the mechanisms of fertilization, cleavage, and gastrulation use the three modes of specification to produce committed cell types and to organize the early embryo.

Fertilization
Beginning a New Organism

FERTILIZATION IS THE PROCESS whereby the sperm and the egg—collectively called the **gametes**—fuse together to begin the creation of a new individual whose genome is derived from both parents. Fertilization accomplishes two separate ends: sex (the combining of genes derived from two parents) and reproduction (the generation of a new organism). Thus, the first function of fertilization is to transmit genes from parent to offspring, and the second is to initiate in the egg cytoplasm those reactions that permit development to proceed.

Although the details of fertilization vary from species to species, conception generally consists of four major events:

1. Contact and recognition between sperm and egg. In most cases, this ensures that the sperm and egg are of the same species.
2. Regulation of sperm entry into the egg. Only one sperm nucleus can ultimately unite with the egg nucleus. This is usually accomplished by allowing only one sperm to enter the egg and actively inhibiting any others from entering.
3. Fusion of the genetic material of sperm and egg.
4. Activation of egg metabolism to start development.

Structure of the Gametes

A complex dialogue exists between egg and sperm. The egg activates the sperm metabolism that is essential for fertilization, and the sperm reciprocates by activating the egg metabolism needed for the onset of development. But before we investigate these aspects of fertilization, we need to consider the structures of the sperm and egg—the two cell types specialized for fertilization.*

Sperm

It is only within the past 135 years that the sperm's role in fertilization has been known. Anton van Leeuwenhoek, the Dutch microscopist who co-discovered sperm in the 1670s, first believed them to be parasitic animals living within the semen (hence the term *spermatozoa*, meaning "seed animals"). Although he originally assumed that they had nothing to do with reproducing the organism in which they were found, he later came to believe that each sperm contained a pre-

---

*Many courses begin with gametogenesis and meiosis. This author believes that meiosis and gametogenesis are the culminating processes in development, and that they cannot be properly appreciated without first understanding somatic organogenesis and differentiation. Also, having gonad formation and gametogenesis in the last lectures completes a circle.

Urge and urge and urge,
Always the procreant urge of the world.
Out of the dimness opposite equals
  advance,
Always substance and increase,
  always sex,
Always a knit of identity, always
  distinction,

WALT WHITMAN (1855)

The final aim of all love intrigues, be they comic or tragic, is really of more importance than all other ends in human life. What it turns upon is nothing less than the composition of the next generation.

A. SCHOPENHAUER
(QUOTED BY C. DARWIN, 1871)

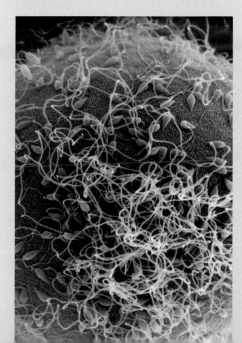

formed embryo. Leeuwenhoek (1685) wrote that sperm were seeds (both *sperma* and *semen* mean "seed"), and that the female merely provided the nutrient soil in which the seeds were planted. In this, he was returning to a notion of procreation promulgated by Aristotle 2000 years earlier.

Try as he might, Leeuwenhoek was continually disappointed in his attempts to find preformed embryos within spermatozoa. Nicolas Hartsoeker, the other co-discoverer of sperm, drew a picture of what he hoped to find: a miniscule human ("homunculus") within the sperm (**Figure 4.1**). This belief that the sperm contained the entire embryonic organism never gained much acceptance, as it implied an enormous waste of potential life. Most investigators regarded the sperm as unimportant.*

See WEBSITE 4.1
Leeuwenhoek and images of homunculi

The first evidence suggesting the importance of sperm in reproduction came from a series of experiments performed by Lazzaro Spallanzani in the late 1700s. Spallanzani induced male toads to ejaculate into taffeta breeches and found toad semen so filtered to be devoid of sperm; such semen did not fertilize eggs. He even showed that semen had to touch the eggs in order to be functional. However, Spallanzani (like many others) felt that the spermatic "animals" were parasites in the fluid; he thought the embryo was contained within the egg and needed spermatic fluid to activate it (see Pinto-Correia 1997).

The combination of better microscopic lenses and the elucidation of the cell theory (that all life is cellular, and all cells come from preexisting cells) led to a new appreciation of sperm function. In 1824, J. L. Prevost and J. B. Dumas claimed that sperm were not parasites, but rather the active agents of fertilization. They noted the universal existence of sperm in sexually mature males and their absence in immature and aged individuals. These observations, coupled with the known absence of sperm in the sterile mule, convinced them that "there exists an intimate relation between their presence in the organs and the fecundating capacity of the animal." They proposed that the sperm entered the egg and contributed materially to the next generation.

These claims were largely disregarded until the 1840s, when A. von Kolliker described the formation of sperm from cells in the adult testes. He ridiculed the idea that the semen could be normal and yet support such an enormous number of parasites. Even so, von Kolliker denied there was any physical contact between sperm and egg. He believed that the sperm excited the egg to develop in much the same way a magnet communicates its presence to iron. It was not until 1876 that Oscar Hertwig and Herman Fol independently demonstrated sperm entry into the egg and

FIGURE 4.1 The human infant preformed in the sperm, as depicted by Nicolas Hartsoeker (1694).

the union of the two cells' nuclei. Hertwig had been seeking an organism suitable for detailed microscopic observations, and he found the Mediterranean sea urchin (*Paracentrotus lividus*) to be perfect for this purpose. Not only was it common throughout the region and sexually mature throughout most of the year, but its eggs were available in large numbers and were transparent even at high magnifications.

When he mixed suspensions of sperm together with egg suspensions, Hertwig repeatedly observed sperm entering the eggs and saw sperm and egg nuclei unite. He also noted that *only one sperm was seen to enter each egg, and that all the nuclei of the resulting embryo were derived mitotically from the nucleus created at fertilization.* Fol made similar observations and also detailed the mechanism of sperm entry. Fertilization was at last recognized as the union of sperm and egg, and the union of sea urchin gametes remains one of the best-studied examples of fertilization.*

See WEBSITE 4.2
The origins of fertilization research

Each sperm cell consists of a **haploid nucleus**, a propulsion system to move the nucleus, and a sac of enzymes that enable the nucleus to enter the egg. In most species, almost all of the cell's cytoplasm is eliminated during sperm maturation, leaving only certain organelles that are modified for spermatic function (**Figure 4.2**). During the course of maturation, the sperm's haploid nucleus becomes very streamlined and its DNA becomes tightly compressed. In front of this compressed haploid nucleus lies the **acrosomal vesicle**, or **acrosome**. The acrosome is derived from

---

*Indeed, sperm was discovered around 1676, while the events of fertilization were not elucidated until 1876. Thus, for nearly 200 years, people had no idea what the sperm actually did. See Pinto-Correia 1997 for details of this remarkable story.

*Hertwig and Fol were actually not the first persons to report fertilization in the sea urchin. At least three other astute observers— Adolphe Dufossé, Karl Ernst von Baer, and Alphonse Derbés— observed sperm-egg contact in 1847. Briggs and Wessel (2006) suggest that the convulsions of Europe during 1848, the low opinion German scientists had of French biology, and the tenuousness of these results (given poor microscopy and the lack of a theory in which to place them) may have confined these papers to obscurity.

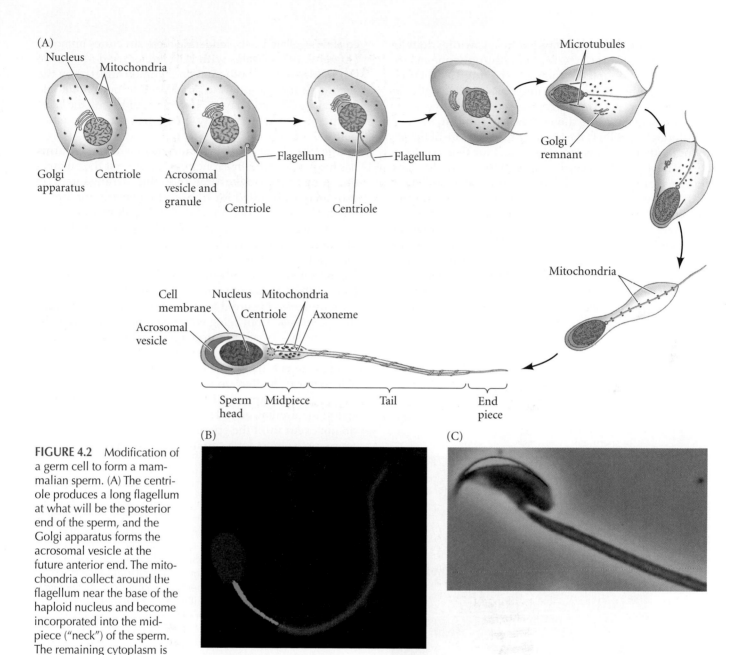

FIGURE 4.2 Modification of a germ cell to form a mammalian sperm. (A) The centriole produces a long flagellum at what will be the posterior end of the sperm, and the Golgi apparatus forms the acrosomal vesicle at the future anterior end. The mitochondria collect around the flagellum near the base of the haploid nucleus and become incorporated into the midpiece ("neck") of the sperm. The remaining cytoplasm is jettisoned, and the nucleus condenses. The size of the mature sperm has been enlarged relative to the other stages. (B) Mature bull sperm. The DNA is stained blue with DAPI, the mitochondria are stained green, and the tubulin of the flagellum is stained red. (C) Acrosome of mouse sperm, stained green by the fusion protein proacrosin-GFP. (A after Clermont and Leblond 1955; B from Sutovsky et al. 1996, courtesy of G. Schatten; C, courtesy of K.-S. Kim and G. L. Gerton.)

the cell's Golgi apparatus and contains enzymes that digest proteins and complex sugars; thus, the acrosome can be considered a modified secretory vesicle. The enzymes stored in the acrosome are used to digest a path through the outer coverings of the egg. In many species, a region of globular actin proteins lies between the sperm nucleus and the acrosomal vesicle. These proteins are used to extend a fingerlike **acrosomal process** from the sperm during the early stages of fertilization. In sea urchins and several other species, recognition between sperm and egg involves molecules on the acrosomal process. Together, the acrosome and nucleus constitute the **sperm head**.

The means by which sperm are propelled vary according to how the species has adapted to environmental conditions. In most species (the major exception is, once again, the nematodes, where the sperm is formed at the sites where fertilization occurs), an individual sperm is able to travel by whipping its **flagellum**. The major motor portion of the flagellum is the **axoneme**, a structure formed by microtubules emanating from the centriole at the base of the sperm nucleus. The core of the axoneme consists of two central microtubules surrounded by a row of 9 doublet microtubules (**Figure 4.3A,B**). Actually, only one microtubule of each doublet is complete, having 13 protofilaments; the other is C-shaped and has only 11 protofilaments (**Figure 4.3C**). The interconnected protofilaments are made exclusively of the dimeric protein **tubulin**.

Although tubulin is the basis for the structure of the flagellum, other proteins are also critical for flagellar function. The force for sperm propulsion is provided by **dynein**, a protein attached to the microtubules (see Figure 4.3B). Dynein is an ATPase, an enzyme that hydrolyzes ATP, converting the released chemical energy into mechanical energy to propel the sperm. This energy allows the active sliding of the outer doublet microtubules, causing the flagellum to bend (Ogawa et al. 1977; Shinyoji et al. 1998). The importance of dynein can be seen in individuals with a genetic syndrome known as the Kartagener triad. These individuals lack functional dynein in all their ciliat-ed and flagellated cells, rendering these structures immotile (Afzelius 1976). Males with Kartagener triad are sterile (immotile sperm). Both men and women affected by the syndrome are susceptible to bronchial infections (immotile respiratory cilia) and have a 50% chance of having the heart on the right side of the body (immotile cilia in the center of the embryo; see Chapter 8).

The "9 + 2" microtubule arrangement with dynein arms has been conserved in axonemes throughout the eukaryotic kingdoms, suggesting that this arrangement is extremely well suited for transmitting energy for movement. The ATP needed to whip the flagellum and propel the sperm comes from rings of mitochondria located in the **midpiece** of the sperm. In many species (notably mammals), a layer of dense fibers has interposed itself between the mitochondrial sheath and the cell membrane. This fiber layer stiffens the sperm tail. Because the thickness of this layer decreases toward the tip, the fibers probably prevent the sperm head from being whipped around too suddenly. Thus, the sperm cell has undergone extensive modification for the transport of its nucleus to the egg.

In mammals, the sperm released during ejaculation are able to move, but they do not yet have the capacity to bind to and fertilize an egg. The final stages of mammalian sperm maturation, cumulatively referred to as *capacitation*, do not occur until the sperm has been inside the female reproductive tract for a certain period of time.

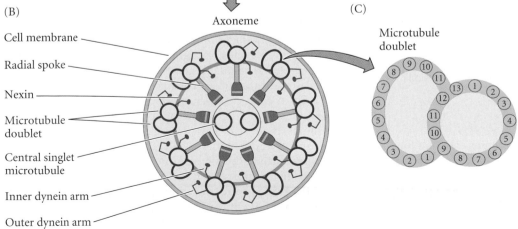

(A) Sperm

Cross section of sperm tail

FIGURE 4.3 Motile apparatus of the sperm. (A) Cross section of the flagellum of a mammalian spermatozoon, showing the central axoneme and the external (course) fibers. (B) Interpretive diagram of the axoneme, showing the "9 + 2" arrangement of the microtubules and other flagellar components. The dynein arms contain the ATPases that provide the energy for flagellar movement. (C) Association of tubulin protofilaments into a microtubule doublet. One portion of the doublet is a fully circular microtubule comprising 13 protofilaments. The second portion of the doublet contains only 11 (occasionally 10) protofilaments. (A © D. M. Phillips/Photo Researchers, Inc.)

(B)

Axoneme

Cell membrane
Radial spoke
Nexin
Microtubule doublet
Central singlet microtubule
Inner dynein arm
Outer dynein arm

(C)

Microtubule doublet

The egg

All the material necessary for the beginning of growth and development must be stored in the egg, or **ovum**. Whereas the sperm eliminates most of its cytoplasm as it matures, the developing egg (called the **oocyte** before it reaches the stage of meiosis at which it is fertilized*) not only conserves the material it has, but actively accumulates more. The meiotic divisions that form the oocyte conserve its cytoplasm rather than giving half of it away (see Figure 16.30); at the same time, the oocyte either synthesizes or absorbs proteins such as yolk that act as food reservoirs for the developing embryo. Birds' eggs are enormous single cells, swollen with accumulated yolk. Even eggs with relatively sparse yolk are large compared with sperm. The volume of a sea urchin egg is about 200 picoliters (2×10^{-4} mm^3), more than 10,000 times the volume of sea urchin sperm (**Figure 4.4**). So, even though sperm and egg have equal haploid nuclear components, the egg also accumulates a remarkable cytoplasmic storehouse during its maturation. This cytoplasmic trove includes the following:

- **Nutritive proteins**. It will be a long time before the embryo is able to feed itself or even obtain food from its mother, so the early embryonic cells need a supply of energy and amino acids. In many species, this is accomplished by accumulating yolk proteins in the egg. Many of these yolk proteins are made in other organs (e.g., liver, fat bodies) and travel through the maternal blood to the oocyte.

- **Ribosomes and tRNA**. The early embryo needs to make many of its own structural proteins and enzymes, and in some species there is a burst of protein synthesis soon after fertilization. Protein synthesis is accomplished by ribosomes and tRNA that exist in the egg. The developing egg has special mechanisms for synthesizing ribosomes; certain amphibian oocytes produce as many as 10^{12} ribosomes during their meiotic prophase.

- **Messenger RNAs**. The oocyte not only accumulates proteins, it also accumulates mRNAs that encode proteins for the early stages of development. It is estimated that sea urchin eggs contain thousands of different types of mRNA that remain repressed until after fertilization (see Chapter 2).

---

*Eggs over easy: The terminology of eggs is confusing. In general, an *egg* is a female gamete capable of binding sperm and being fertilized. An *oocyte* is a developing egg that cannot yet bind sperm or be fertilized (Wessel 2009). The problems in terminology come from the fact that different species of animals have eggs in different stages of meiosis (see Figure 4.5). The human egg, for instance, is in second meiotic metaphase when it binds sperm, whereas the sea urchin egg has finished all of its meiotic divisions when it binds sperm. The contents of the egg also vary greatly from species to species. The synthesis and placement of these materials will be addressed in Chapter 16, when we discuss the differentiation of the germ cells.

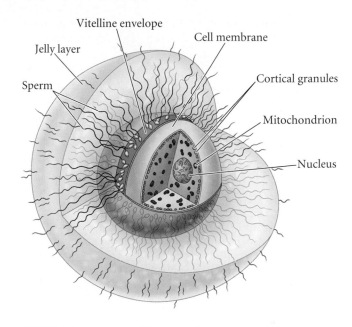

FIGURE 4.4 Structure of the sea urchin egg at fertilization. The drawing shows the relative sizes of egg and sperm. (After Epel 1977.)

- **Morphogenic factors**. Molecules that direct the differentiation of cells into certain cell types are present in the egg. These include transcription factors and paracrine factors. In many species, they are localized in different regions of the egg and become segregated into different cells during cleavage (see Chapter 5).

- **Protective chemicals**. The embryo cannot run away from predators or move to a safer environment, so it must come equipped to deal with threats. Many eggs contain ultraviolet filters and DNA repair enzymes that protect them from sunlight. Some eggs contain molecules that potential predators find distasteful, and the yolk of bird eggs even contains antibodies.

Within this enormous volume of egg cytoplasm resides a large nucleus. In a few species (such as sea urchins), the **female pronucleus** is already haploid at the time of fertilization. In other species (including many worms and most mammals), the egg nucleus is still diploid—the sperm enters before the egg's meiotic divisions are completed (**Figure 4.5**). In these species, the final stages of egg meiosis will take place after the sperm's nuclear material (the **male pronucleus**) is already inside the egg cytoplasm.

See WEBSITE 4.3 The egg and its environment

Enclosing the cytoplasm is the egg **cell membrane**. This membrane must be capable of fusing with the sperm cell membrane and must regulate the flow of certain ions during fertilization. Outside the cell membrane is an extracellular matrix that forms a fibrous mat around the egg and is often involved in sperm-egg recognition (Correia and Carroll 1997). In invertebrates, this structure is usually

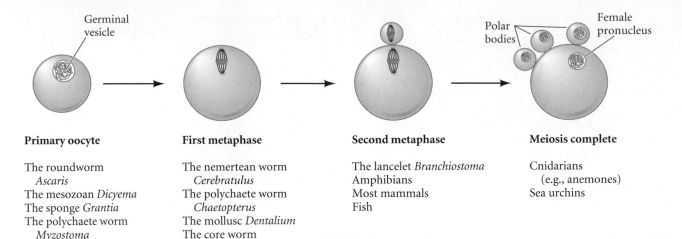

Primary oocyte

The roundworm
  Ascaris
The mesozoan *Dicyema*
The sponge *Grantia*
The polychaete worm
  Myzostoma
The clam worm *Nereis*
The clam *Spisula*
The echiuroid worm
  Urechis
Dogs and foxes

First metaphase

The nemertean worm
  Cerebratulus
The polychaete worm
  Chaetopterus
The mollusc *Dentalium*
The core worm
  Pectinaria
Many insects
Starfish

Second metaphase

The lancelet *Branchiostoma*
Amphibians
Most mammals
Fish

Meiosis complete

Cnidarians
  (e.g., anemones)
Sea urchins

FIGURE 4.5   Stages of egg maturation at the time of sperm entry in different animal species. Note that in most species, sperm entry occurs before the egg nucleus has completed meiosis. The germinal vesicle is the name given to the large diploid nucleus of the primary oocyte. The polar bodies are nonfunctional cells produced by meiosis (see Chapter 16). (After Austin 1965.)

called the **vitelline envelope (Figure 4.6)**. The vitelline envelope contains several different glycoproteins. It is supplemented by extensions of membrane glycoproteins from the cell membrane and by proteinaceous "posts" that adhere the vitelline envelope to the membrane (Mozingo and Chandler 1991). The vitelline envelope is essential for the species-specific binding of sperm.

Many types of eggs also have a layer of **egg jelly** outside the vitelline envelope (see Figure 4.4). This glycoprotein meshwork can have numerous functions, but most commonly it is used either to attract or to activate sperm.

The egg, then, is a cell specialized for receiving sperm and initiating development.

In mammalian eggs, the extracellular envelope is a separate and thick matrix called the **zona pellucida**. The mammalian egg is also surrounded by a layer of cells called the **cumulus (Figure 4.7)**, which is made up of the ovarian follicular cells that were nurturing the egg at the time of its release from the ovary. Mammalian sperm have to get past these cells to fertilize the egg. The innermost layer of cumulus cells, immediately adjacent to the zona pellucida, is called the **corona radiata**.

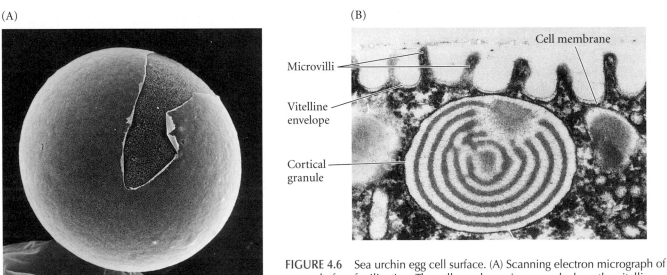

FIGURE 4.6   Sea urchin egg cell surface. (A) Scanning electron micrograph of an egg before fertilization. The cell membrane is exposed where the vitelline envelope has been torn. (B) Transmission electron micrograph of an unfertilized egg, showing microvilli and cell membrane, which are closely covered by the vitelline envelope. A cortical granule lies directly beneath the cell membrane. (From Schroeder 1979, courtesy of T. E. Schroeder.)

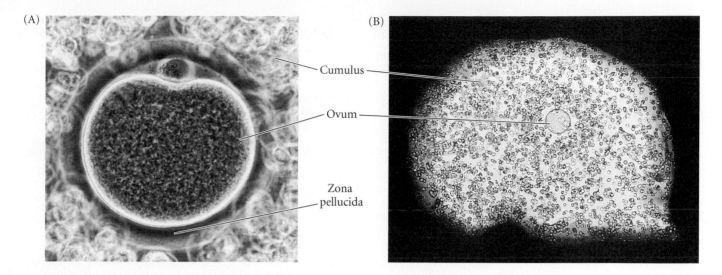

(A)   Cumulus

Ovum

Zona pellucida

(B)

FIGURE 4.7 Mammalian eggs immediately before fertilization. (A) The hamster egg, or ovum, is encased in the zona pellucida. This, in turn, is surrounded by the cells of the cumulus. A polar body cell, produced during meiosis, is also visible within the zona pellucida. (B) At lower magnification, a mouse oocyte is shown surrounded by the cumulus. Colloidal carbon particles (India ink, seen here as the black background) are excluded by the hyaluronidate matrix. (Courtesy of R. Yanagimachi.)

Lying immediately beneath the cell membrane of most eggs is a thin layer (about 5 μm) of gel-like cytoplasm called the **cortex**. The cytoplasm in this region is stiffer than the internal cytoplasm and contains high concentrations of globular actin molecules. During fertilization, these actin molecules polymerize to form long cables of actin known as **microfilaments**. Microfilaments are necessary for cell division. They are also used to extend the egg surface into small projections called **microvilli**, which may aid sperm entry into the cell (see Figure 4.6B; also see Figure 4.16).

Also within the cortex are the **cortical granules** (see Figures 4.4 and 4.6B). These membrane-bound, Golgi-derived structures contain proteolytic enzymes and are thus homologous to the acrosomal vesicle of the sperm. However, whereas a sea urchin sperm contains just one acrosomal vesicle, each sea urchin egg contains approximately 15,000 cortical granules. Moreover, in addition to digestive enzymes, the cortical granules contain mucopolysaccharides, adhesive glycoproteins, and hyalin protein. As we will soon detail, the enzymes and mucopolysaccharides help prevent polyspermy—they prevent additional sperm from entering the egg after the first sperm has entered—and the hyalin and adhesive glycoproteins surround the early embryo and provide support for the cleavage-stage blastomeres.

See VADE MECUM Gametogenesis

Recognition of egg and sperm

The interaction of sperm and egg generally proceeds according to five basic steps (**Figure 4.8**; Vacquier 1998):

1. The chemoattraction of the sperm to the egg by soluble molecules secreted by the egg
2. The exocytosis of the sperm acrosomal vesicle to release its enzymes
3. The binding of the sperm to the extracellular matrix (vitelline envelope or zona pellucida) of the egg
4. The passage of the sperm through this extracellular matrix
5. Fusion of egg and sperm cell membranes

Sometimes steps 2 and 3 can be reversed (as in mammalian fertilization; see Figure 4.8B), and the sperm binds to the extracellular matrix of the egg before releasing the contents of the acrosome. After these five steps are accomplished, the haploid sperm and egg nuclei can meet and the reactions that initiate development can begin. In this chapter, we will focus on the fertilization events of sea urchins, which undergo external fertilization, and mice, which undergo internal fertilization. In subsequent chapters, the variations of fertilization will be described as we study the development of particular organisms.

External Fertilization in Sea Urchins

In many species, the meeting of sperm and egg is not a simple matter. Many marine organisms release their gametes into the environment. That environment may be as small as a tide pool or as large as an ocean (Mead and Epel 1995). Moreover, this environment is shared with other species that may shed their gametes at the same time. Such organisms are faced with two problems: How can sperm and eggs meet in such a dilute concentration, and how can sperm be prevented from attempting to fertilize

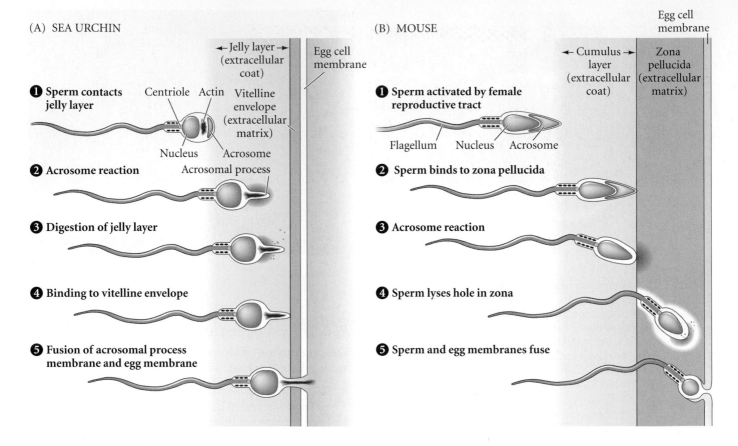

(A) SEA URCHIN

← Jelly layer → (extracellular coat)
Egg cell membrane

❶ **Sperm contacts jelly layer**

Centriole   Actin   Vitelline envelope (extracellular matrix)

Nucleus   Acrosome

❷ **Acrosome reaction**   Acrosomal process

❸ **Digestion of jelly layer**

❹ **Binding to vitelline envelope**

❺ **Fusion of acrosomal process membrane and egg membrane**

(B) MOUSE

Egg cell membrane

← Cumulus → layer (extracellular coat)
Zona pellucida (extracellular matrix)

❶ **Sperm activated by female reproductive tract**

Flagellum   Nucleus   Acrosome

❷ **Sperm binds to zona pellucida**

❸ **Acrosome reaction**

❹ **Sperm lyses hole in zona**

❺ **Sperm and egg membranes fuse**

FIGURE 4.8 Summary of events leading to the fusion of egg and sperm cell membranes in (A) the sea urchin and (B) the mouse. (A) Sea urchin fertilization is external. (1) The sperm is chemotactically attracted to and activated by the egg. (2, 3) Contact with the egg jelly triggers the acrosome reaction, allowing the acrosomal process to form and release proteolytic enzymes. (4) The sperm adheres to the vitelline envelope and lyses a hole in it. (5) The sperm adheres to the egg cell membrane and fuses with it. The sperm pronucleus can now enter the egg cytoplasm. (B) Mammalian fertilization is internal. (1) The contents of the female reproductive tract capacitate, attract, and activate the sperm. (2) The acrosome-intact sperm binds to the zona pellucida, which is thicker than the vitelline envelope of sea urchins. (3) The acrosome reaction occurs on the zona pellucida. (4) The sperm digests a hole in the zona pellucida. (5) The sperm adheres to the egg, and their cell membranes fuse.

eggs of another species? In addition to simply producing enormous numbers of gametes, two major mechanisms have evolved to solve these problems: species-specific sperm attraction and species-specific sperm activation. Here we describe these events as they occur in sea urchins.

Sperm attraction: Action at a distance

Species-specific sperm attraction has been documented in numerous species, including cnidarians, molluscs, echinoderms, amphibians, and urochordates (Miller 1985; Yoshida et al. 1993; Burnett et al. 2008). In many species, sperm

are attracted toward eggs of their species by **chemotaxis**— that is, by following a gradient of a chemical secreted by the egg. In 1978, Miller demonstrated that the eggs of the cnidarian *Orthopyxis caliculata* not only secrete a chemotactic factor but also regulate the timing of its release. Developing oocytes at various stages in their maturation were fixed on microscope slides, and sperm were released at a certain distance from the eggs. Miller found that when sperm were added to oocytes that had not yet completed their second meiotic division, there was no attraction of sperm to eggs. However, after the second meiotic division was finished and the eggs were ready to be fertilized, the sperm migrated toward them. Thus, these oocytes control not only the type of sperm they attract, but also the time at which they attract them.

The mechanisms of chemotaxis differ among species (see Metz 1978; Eisenbach 2004), and the chemotactic molecules are different even in closely related species. In sea urchins, sperm motility is acquired when the sperm are spawned into seawater. As long as sperm cells are in the testes, they cannot move because their internal pH is kept low (about pH 7.2) by the high concentrations of CO_2 in the gonad. However, once spawned into seawater, sperm pH is elevated to about 7.6, resulting in the activation of the dynein ATPase. The splitting of ATP provides the energy for the flagella to wave, and the sperm begin swimming vigorously (Christen et al. 1982).

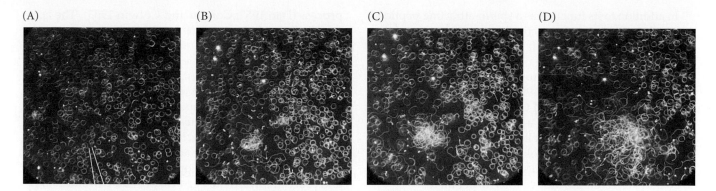

FIGURE 4.9 Sperm chemotaxis in the sea urchin *Arbacia punctulata*. One nanoliter of a 10-n*M* solution of resact is injected into a 20-microliter drop of sperm suspension. (A) A 1-second photographic exposure showing sperm swimming in tight circles before the addition of resact. The position of the injection pipette is shown by the white lines. (B–D) Similar 1-second exposures showing migration of sperm to the center of the resact gradient 20, 40, and 90 seconds after injection. (From Ward et al. 1985, courtesy of V. D. Vacquier.)

But the ability to move does not provide the sperm with direction. In echinoderms, direction is provided by small chemotactic peptides such as resact. **Resact** is a 14-amino acid peptide that has been isolated from the egg jelly of the sea urchin *Arbacia punctulata* (Ward et al. 1985). Resact diffuses readily from the egg jelly into the seawater and has a profound effect at very low concentrations when added to a suspension of *Arbacia* sperm. When a drop of seawater containing *Arbacia* sperm is placed on a microscope slide, the sperm generally swim in circles about 50 μm in diameter. Within seconds after a small amount of resact is injected into the drop, sperm migrate into the region of the

injection and congregate there (**Figure 4.9**). As resact diffuses from the area of injection, more sperm are recruited into the growing cluster. Resact is specific for *A. punctulata* and does not attract sperm of other species. (An analogous compound, speract, has been isolated from the purple sea urchin, *Strongylocentrotus purpuratus*.) *A. punctulata* sperm have receptors in their cell membranes that bind resact (Ramarao and Garbers 1985; Bentley et al. 1986). When the extracellular side of the receptor binds resact, it activates latent guanylyl cyclase activity in the cytoplasmic side of the receptor (**Figure 4.10**). This causes the sperm cell to make more cyclic GMP (cGMP), a compound that activates a calcium channel, allowing the influx of calcium ions (Ca^{2+}) from the seawater into the sperm, thus providing a directional cue (Nishigaki et al. 2000; Wood et al. 2005). Recent studies have demonstrated that the binding of a single resact molecule is able to provide direction for the sperm, which swim up a concentration gradient of this compound until they reach the egg (Kaupp et al. 2003; Kirkman-Brown et al. 2003).

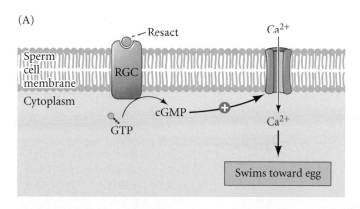

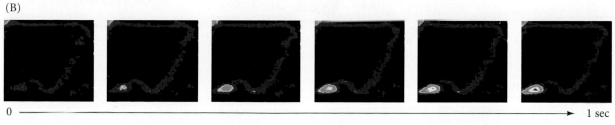

FIGURE 4.10 Model for chemotactic peptides in sea urchin sperm. (A) Resact from *Arbacia* egg jelly binds to its receptor on the sperm. This activates the receptor's guanylyl cyclase (RGC) activity, forming intracellular cGMP in the sperm. The cGMP opens calcium channels in the sperm cell membrane, allowing Ca^{2+} to enter the sperm. The influx of Ca^{2+} activates sperm motility, and the sperm swims up the resact gradient toward the egg. (B) Ca^{2+} levels in different regions of *Strongylocentrotus purpuratus* sperm after exposure to 125 n*M* speract (this species' analog of resact). Red indicates the highest level of Ca^{2+}, blue the lowest. The sperm head reaches its peak Ca^{2+} levels within 1 second. (A after Kirkman-Brown et al. 2003; B from Wood et al. 2003, courtesy of M. Whitaker)

In addition to its function as a sperm-attracting peptide, resact also acts as a **sperm-activating peptide**. One of the major roles of the egg jelly is to increase the motility of sperm, and sperm-activating peptides cause dramatic and immediate increases in mitochondrial respiration and sperm motility (Hardy et al. 1994; Inamdar et al. 2007). The increases in cyclic GMP and Ca^{2+} also activate the mitochondrial ATP-generating apparatus and the dynein ATPase that stimulates flagellar movement in the sperm (Shimomura et al. 1986; Cook and Babcock 1993). Thus, upon meeting resact, *Arbacia* sperm are instructed where to go and are given the motive force to get there.

The acrosome reaction

A second interaction between sperm and egg jelly results in the **acrosome reaction**. In most marine invertebrates, the acrosome reaction has two components: the fusion of the acrosomal vesicle with the sperm cell membrane (an exocytosis that results in the release of the contents of the acrosomal vesicle), and the extension of the acrosomal process (Dan 1952; Colwin and Colwin 1963). The acrosome reaction in sea urchins is initiated by contact of the sperm with the egg jelly. Contact causes the exocytosis of the sperm's acrosomal vesicle and proteolytic enzymes and proteasomes (protein-digesting complexes) that digest a path through the jelly coat to the egg surface (Dan 1967; Franklin 1970). Once the sperm reaches the egg surface, the acrosomal process adheres to the vitelline envelope and tethers the sperm to the egg. It is possible that proteasomes from the acrosome coat the acrosomal process, allowing it to digest the vitelline envelope at the point of attachment and proceed toward the egg (Yokota and Sawada 2007).

In sea urchins, the acrosome reaction is initiated by the interactions of the sperm cell membrane with a specific complex sugar in the egg jelly. These sulfate-containing polysaccharides bind to specific receptors located on the sperm cell membrane directly above the acrosomal vesicle. The egg jelly factors that initiate the acrosome reaction are often highly specific to each species, and egg jelly carbohydrates from one species of sea urchin fail to activate the acrosome reaction even in closely related species (**Figure 4.11**; Hirohashi

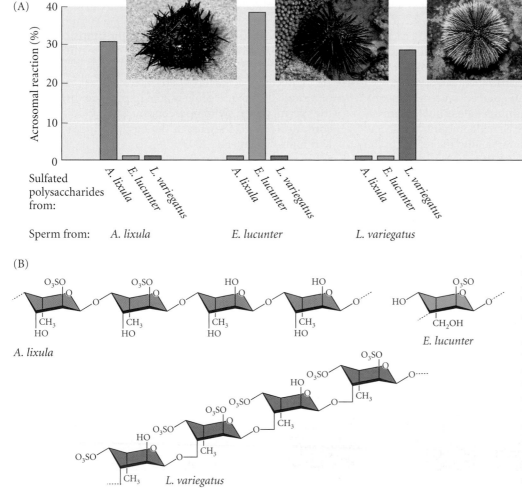

FIGURE 4.11 Species-specific induction of the acrosome reaction by sulfated polysaccharides characterizing the egg jelly coats of three species of sea urchins that co-inhabit the intertidal around Rio de Janeiro. (A) The histograms compare the ability of each polysaccharide to induce the acrosome reaction in the different species of sperm. (B) Chemical structures of the acrosome reaction-inducing sulfated polysaccharides reveal their species-specificity. (After Vilela-Silva et al. 2008; photographs left to right © Interfoto/Alamy; © FLPA/AGE Fotostock; © Water Frame/Alamy.)

and Vacquier 2002; Hirohashi et al. 2002; Vilela-Silva et al. 2008). Thus, the activation of the acrosome reaction constitutes a barrier to interspecies (and thus unviable) fertilizations. This is important when numerous species inhabit the same habitat and when their spawning seasons overlap.

In the sea urchin *Stongylocentrotus purpuratus*, the acrosome reaction is initiated by a repeating polymer of fucose sulfate. When this sulfated carbohydrate binds to its receptor on the sperm, the receptor activates three sperm membrane proteins: (1) a calcium transport channel that allows Ca^{2+} to enter the sperm head; (2) a sodium/hydrogen exchanger that pumps sodium ions (Na^+) into the sperm as it pumps hydrogen ions (H^+) out; and (3) a phospholipase enzyme that makes another second messenger, the phosopholipid **inositol trisphosphate** (**IP3**, of which we will hear much more later in the chapter). IP_3 is able to release Ca^{2+} from *inside* the sperm, probably from within the acrosome itself (Domino and Garbers 1988; Domino et al. 1989; Hirohashi and Vacquier 2003). The elevated Ca^{2+} level in a relatively basic cytoplasm triggers the fusion of the acrosomal membrane with the adjacent sperm cell membrane (**Figure 4.12A–C**), releasing enzymes that can lyse a path through the egg jelly to the vitelline envelope.

The second part of the acrosome reaction involves the extension of the acrosomal process (**Figure 4.12D**). This protrusion arises through the polymerization of globular actin molecules into actin filaments (Tilney et al. 1978). The influx of Ca^{2+} is thought to activate the protein RhoB in the acrosomal region and midpiece of sea urchin sperm (Castellano et al. 1997; de la Sancha 2007). This GTP-binding protein helps organize the actin cytoskeleton in many types of cells, and it is thought to be active in polymerizing actin to make the acrosomal process.

Recognition of the egg's extracellular coat

The sperm's contact with an egg's jelly coat provides the first set of species-specific recognition events (i.e., sperm attraction, activation, and acrosome reaction). Another critical species-specific binding event must occur once the sea urchin sperm has penetrated the jelly and the acrosomal process of the sperm contacts the surface of the egg (**Figure 4.13A**). The acrosomal protein mediating this recognition in sea urchins is called **bindin**. In 1977, Vacquier and co-workers isolated this insoluble, 30,500-Da protein from the acrosome of *Strongylocentrotus purpuratus* and found it to be capable of binding to dejellied eggs of the same species. Further, its interaction with eggs is often species-specific: bindin isolated from the acrosomes of *S. purpuratus* binds to its own dejellied eggs but not to those of *S. franciscanus* (**Figure 4.13B**; Glabe and Vacquier 1977; Glabe and Lennarz 1979). Using immunological techniques, Moy and Vacquier (1979) demonstrated that bindin is located specifically on the acrosomal process—exactly where it should be for sperm-egg recognition (**Figure 4.14**).

See WEBSITE 4.4
The Lillie-Loeb dispute over sperm-egg binding

FIGURE 4.12 Acrosome reaction in sea urchin sperm. (A–C) The portion of the acrosomal membrane lying directly beneath the sperm cell membrane fuses with the cell membrane to release the contents of the acrosomal vesicle. (D) The actin molecules assemble to produce microfilaments, extending the acrosomal process outward. Actual photographs of the acrosome reaction in sea urchin sperm are shown below the diagrams. (After Summers and Hylander 1974; photographs courtesy of G. L. Decker and W. J. Lennarz.)

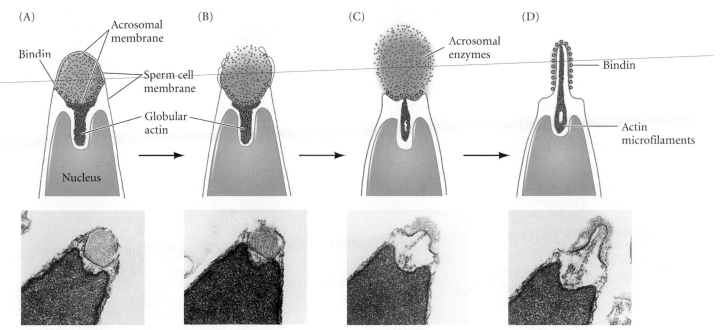

(A) Bindin Acrosomal membrane Sperm cell membrane Globular actin Nucleus

(B)

(C) Acrosomal enzymes

(D) Bindin Actin microfilaments

FIGURE 4.13 Species-specific binding of acrosomal process to egg surface in sea urchins. (A) Actual contact of a sea urchin sperm acrosomal process with an egg microvillus. (B) In vitro model of species-specific binding. The agglutination of dejellied eggs by bindin was measured by adding bindin particles to a plastic well containing a suspension of eggs. After 2–5 minutes of gentle shaking, the wells were photographed. Each bindin bound to and agglutinated only eggs from its own species. (A from Epel 1977, courtesy of F. D. Collins and D. Epel; B based on photographs in Glabe and Vacquier 1977.)

(A)

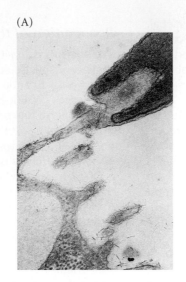

(B)

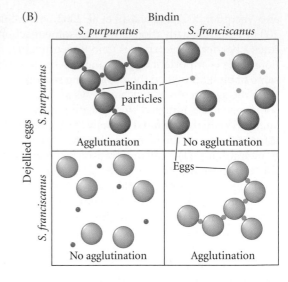

Biochemical studies have shown that the bindins of closely related sea urchin species indeed have different protein sequences. This finding implies the existence of species-specific bindin *receptors* on the egg vitelline envelope. Such receptors were suggested by the experiments of Vacquier and Payne (1973), who saturated sea urchin eggs with sperm. As seen in **Figure 4.15A**, sperm binding does not occur over the entire egg surface. Even at saturating numbers of sperm (approximately 1500), there appears to be room on the ovum for more sperm heads, implying a limiting number of sperm-binding sites. **EBR1**, a 350-kDa glycoprotein that displays the properties expected of a bindin receptor, has been isolated from sea urchin eggs (**Figure 4.15B**; Kamei and Glabe 2003). These bindin receptors are thought to be aggregated into complexes on the vitelline envelope, and hundreds of such complexes may be needed to tether the sperm to the egg. The receptor for sperm bindin on the egg vitelline envelope appears to recognize the protein portion of bindin in a species-specific manner. Closely related species of sea urchins (different species in the same genus) have divergent bindin receptors, and eggs will adhere only to the bindin of their own species* (**Figure 4.15C**). Thus, species-specific recognition of sea urchin gametes can occur at the levels of sperm attrac-

FIGURE 4.14 Localization of bindin on the acrosomal process. (A) Immunochemical technique used to localize bindin. Rabbit antibody was made to the bindin protein, and this antibody was incubated with sperm that had undergone the acrosome reaction. If bindin were present, the rabbit antibody would remain bound to the sperm. After any unbound antibody was washed off, the sperm were treated with swine antibody that had been covalently linked to peroxidase enzymes. The swine antibody bound to the rabbit antibody, placing peroxidase molecules wherever bindin was present. Peroxidase catalyzes the formation of a dark precipitate from diaminobenzidine (DAB) and hydrogen peroxide. Thus, this precipitate formed only where bindin was present. (B) Localization of bindin to the acrosomal process after the acrosome reaction (×33,200). (C) Localization of bindin to the acrosomal process at the junction of the sperm and the egg. (B,C from Moy and Vacquier 1979, courtesy of V. D. Vacquier.)

*Bindin and other gamete recognition proteins are among the fastest evolving proteins known (Metz and Palumbi 1996; Swanson and Vacquier 2002). Even when closely related urchin species have near-identity of every other protein, their bindins may have diverged significantly. The evolution of these gamete recognition proteins shows the hallmarks of sexual selection and coevolved genes. It is thought that coevolution of the genes encoding male and female gamete recognition proteins can lead to reproductive barriers that have the potential to drive speciation by dividing a population into two different mating groups (see Clark et al. 2009).

(A)

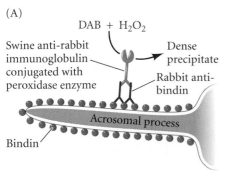

(B) DAB precipitate

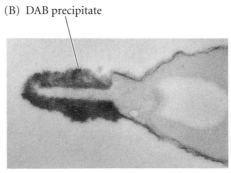

(C) Vitelline envelope of egg   Acrosomal process

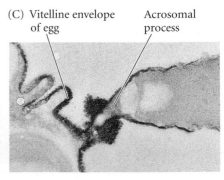

(A)

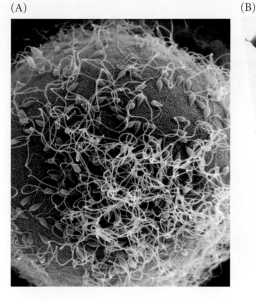

(B)

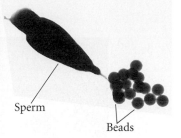

Sperm

Beads

(C)

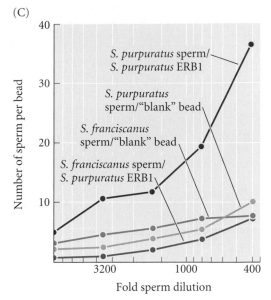

FIGURE 4.15 Bindin receptors on the egg. (A) Scanning electron micrograph of sea urchin sperm bound to the vitelline envelope of an egg. Although this egg is saturated with sperm, there appears to be room on the surface for more sperm, implying the existence of a limited number of bindin receptors. (B) *Strongylocentrotus purpuratus* sperm bind to polystyrene beads that have been coated with purified bindin receptor protein. (C) Species-specific binding of sea urchin sperm to ERB1. *S. purpuratus* sperm bound to beads coated with EBR1 bindin receptor purified from *S. purpuratus* eggs, but *S. franciscanus* sperm did not. Neither sperm bound to uncoated "blank" beads. (A © Mia Tegner/SPL/Photo Researchers, Inc.; B from Foltz et al. 1993; C after Kamei and Glabe 2003.)

tion, sperm activation, the acrosome reaction, and sperm adhesion to the egg surface.

Fusion of the egg and sperm cell membranes

Once the sperm has traveled to the egg and undergone the acrosome reaction, the fusion of the sperm cell membrane with the cell membrane of the egg can begin.

(A)

(B)

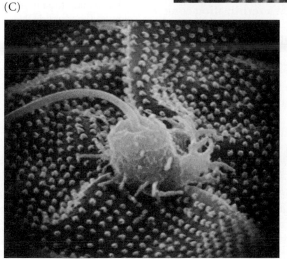

(C)

(D)

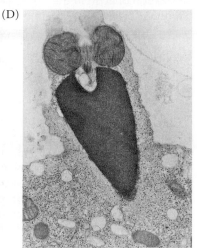

FIGURE 4.16 Scanning electron micrographs of the entry of sperm into sea urchin eggs. (A) Contact of sperm head with egg microvillus through the acrosomal process. (B) Formation of fertilization cone. (C) Internalization of sperm within the egg. (D) Transmission electron micrograph of sperm internalization through the fertilization cone. (A–C from Schatten and Mazia 1976, courtesy of G. Schatten; D courtesy of F. J. Longo.)

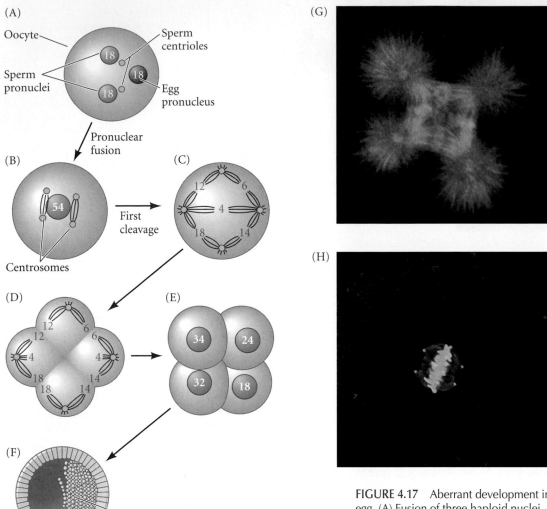

FIGURE 4.17   Aberrant development in a dispermic sea urchin egg. (A) Fusion of three haploid nuclei, each containing 18 chromosomes, and the division of the two sperm centrioles to form four centrosomes (mitotic poles). (B,C) The 54 chromosomes randomly assort on the four spindles. (D) At anaphase of the first division, the duplicated chromosomes are pulled to the four poles. (E) Four cells containing different numbers and types of chromosomes are formed, thereby causing (F) the early death of the embryo. (G) First metaphase of a dispermic sea urchin egg akin to (D). The microtubules are stained green; the DNA stain appears orange. The triploid DNA is being split into four chromosomally unbalanced cells instead of the normal two cells with equal chromosome complements. (H) Human dispermic egg at first mitosis. The four centrioles are stained yellow, while the microtubules of the spindle apparatus (and of the two sperm tails) are stained red. The three sets of chromosomes divided by these four poles are stained blue. (A–F after Boveri 1907; G courtesy of J. Holy; H from Simerly et al. 1999, courtesy of G. Schatten.)

The entry of a sperm into a sea urchin egg is illustrated in **Figure 4.16**. Sperm-egg fusion appears to cause the polymerization of actin in the egg to form a **fertilization cone** (Summers et al. 1975). Homology between the egg and the sperm is again demonstrated, since the sperm's acrosomal process also appears to be formed by the polymerization of actin. The actin from the gametes forms a connection that widens the cytoplasmic bridge between the egg and the sperm. The sperm nucleus and tail pass through this bridge.

In the sea urchin, all regions of the egg cell membrane are capable of fusing with sperm. In several other species, certain regions of the membrane are specialized for sperm recognition and fusion (Vacquier 1979). Fusion is an active process, often mediated by specific "fusogenic" proteins. Indeed, sea urchin sperm bindin plays a second role as a fusogenic protein. In addition to recognizing the egg, bindin contains a long stretch of hydrophobic amino acids near its amino terminus, and this region is able to fuse phospholipid vesicles in vitro (Ulrich et al. 1999; Gage et al. 2004). Under the proper ionic conditions (those in the mature unfertilized egg), bindin can cause the sperm and egg membranes to fuse.

The fast block to polyspermy

As soon as one sperm has entered the egg, the fusibility of the egg membrane—which was so necessary to get the sperm inside the egg—becomes a dangerous liability. In most animals, any sperm that enters the egg can provide a haploid nucleus and a centriole to the egg. In normal **monospermy**, only one sperm enters the egg, and a haploid sperm nucleus and a haploid egg nucleus combine to form the diploid nucleus of the fertilized egg (zygote), thus restoring the chromosome number appropriate for the species. The centriole provided by the sperm divides to form the two poles of the mitotic spindle during cleavage, while the egg-derived centriole is degraded.

The entrance of multiple sperm—**polyspermy**—leads to disastrous consequences in most organisms. In the sea urchin, fertilization by two sperm results in a triploid nucleus, in which each chromosome is represented three times rather than twice. Worse, each sperm's centriole divides to form the two poles of a mitotic apparatus; so instead of a bipolar mitotic spindle separating the chromosomes into two cells, the triploid chromosomes may be divided into as many as four cells. Because there is no mechanism to ensure that each of the four cells receives the proper number and type of chromosomes, the chromosomes are apportioned unequally: some cells receive extra copies of certain chromosomes, while other cells lack them (**Figure 4.17**). Theodor Boveri demonstrated in 1902 that such cells either die or develop abnormally.

Species have evolved ways to prevent the union of more than two haploid nuclei. The most common way is to pre-vent the entry of more than one sperm into the egg. The sea urchin egg has two mechanisms to avoid polyspermy: a fast reaction, accomplished by an electric change in the egg cell membrane, and a slower reaction, caused by the exocytosis of the cortical granules (Just 1919).

The **fast block to polyspermy** is achieved by changing the electric potential of the egg cell membrane. This membrane provides a selective barrier between the egg cytoplasm and the outside environment, so that ion concentrations within the egg differ greatly from those of its surroundings. This concentration difference is especially significant for sodium and potassium ions. Seawater has a particularly high sodium ion (Na^+) concentration, whereas the egg cytoplasm contains relatively little Na^+. The reverse is the case with potassium ions (K^+). These concentration differences are maintained by the cell membrane, which steadfastly inhibits the entry of Na^+ into the oocyte and prevents K^+ from leaking out into the environment. If we insert an electrode into an egg and place a second electrode outside it, we can measure the constant difference in charge across the egg cell membrane. This **resting membrane potential** is generally about 70 mV, usually expressed as –70 mV because the inside of the cell is negatively charged with respect to the exterior.

Within 1–3 seconds after the binding of the first sperm, the membrane potential shifts to a positive level, about +20 mV (Longo et al. 1986). This change is caused by a small influx of Na^+ into the egg (**Figure 4.18A**). Although sperm can fuse with membranes having a resting potential of –70 mV, they cannot fuse with membranes having a positive resting potential, so no more sperm can fuse to the egg. It is not known whether the increased sodium permeability of the egg is due to the *binding* of the first sperm, or to the *fusion* of the first sperm with the egg (Gould and Stephano 1987, 1991; McCulloh and Chambers 1992). However, recent data

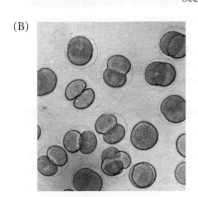

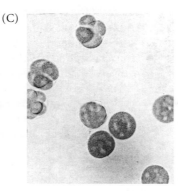

(A)

FIGURE 4.18   Membrane potential of sea urchin eggs before and after fertilization. (A) Before the addition of sperm, the potential difference across the egg cell membrane is about –70 mV. Within 1–3 seconds after the fertilizing sperm contacts the egg, the potential shifts in a positive direction. (B,C) *Lytechinus* eggs photographed during first cleavage. (B) Control eggs developing in 490 m*M* Na^+. (C) Polyspermy in eggs fertilized in similarly high concentrations of sperm in 120 m*M* Na^+ (choline was substituted for sodium). (D) Table showing the rise of polyspermy with decreasing Na^+ concentration. Salt water is about 600 m*M* NaCl. (After Jaffe 1980; B,C courtesy of L. A. Jaffe.)

(D)

Na^+ (m*M*)	Polyspermic eggs (%)
490	22
360	26
120	97
50	100

suggest that the fusogenic region of bindin will not function on a positively charged surface (Rocha et al. 2007).

The importance of Na[+] and the change in resting potential was demonstrated by Laurinda Jaffe and colleagues. They found that polyspermy can be induced if sea urchin eggs are artificially supplied with an electric current that keeps their membrane potential negative. Conversely, fertilization can be prevented entirely by artificially keeping the membrane potential of eggs positive (Jaffe 1976). The fast block to polyspermy can also be circumvented by lowering the concentration of Na[+] in the surrounding water (**Figure 4.18B–D**). If the supply of sodium ions is not sufficient to cause the positive shift in membrane potential, polyspermy occurs (Gould-Somero et al. 1979; Jaffe 1980).

It is not known how the change in membrane potential acts on the sperm to block secondary fertilization. Most likely, the sperm carry a voltage-sensitive component (possibly a positively charged fusogenic protein), and the insertion of this component into the egg cell membrane could be regulated by the electric charge across the membrane (Iwao and Jaffe 1989). An electric block to polyspermy also occurs in frogs* (Cross and Elinson 1980), but probably not in most mammals (Jaffe and Cross 1983).

See WEBSITE 4.5 Blocks to polyspermy

See VADE MECUM E. E. Just

The slow block to polyspermy

The fast block to polyspermy is transient, since the membrane potential of the sea urchin egg remains positive for only about a minute. This brief potential shift is not sufficient to prevent polyspermy permanently, and polyspermy can still occur if the sperm bound to the vitelline envelope are not somehow removed (Carroll and Epel 1975). This sperm removal is accomplished by the **cortical granule reaction**, a slower, mechanical block to polyspermy that becomes active about a minute after the first successful sperm-egg fusion (Just 1919). This reaction—also known as the **slow block to polyspermy**—is found in many animal species, including sea urchins and most mammals.

Directly beneath the sea urchin egg cell membrane are about 15,000 cortical granules, each about 1 μm in diameter (see Figure 4.6B). Upon sperm entry, these cortical granules fuse with the egg cell membrane and release their contents into the space between the cell membrane and the

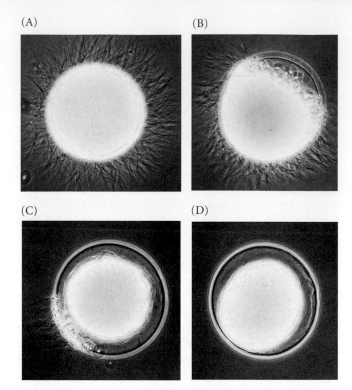

(A) (B)

(C) (D)

FIGURE 4.19 Formation of the fertilization envelope and removal of excess sperm. To create these photographs, sperm were added to sea urchin eggs, and the suspension was then fixed in formaldehyde to prevent further reactions. (A) At 10 seconds after sperm addition, sperm surround the egg. (B,C) At 25 and 35 seconds after insemination, respectively, a fertilization envelope is forming around the egg, starting at the point of sperm entry. (D) The fertilization envelope is complete, and excess sperm have been removed. (From Vacquier and Payne 1973, courtesy of V. D. Vacquier.)

fibrous mat of vitelline envelope proteins. Several proteins are released by this cortical granule exocytosis. One is a trypsin-like protease called **cortical granule serine protease**. This enzyme cleaves the protein posts that connect the vitelline envelope proteins to the cell membrane, and it clips off the bindin receptors and any sperm attached to them (Vacquier et al. 1973; Glabe and Vacquier 1978; Haley and Wessel 1999, 2004).

The components of the cortical granules bind to the vitelline envelope to form a **fertilization envelope** (Wong and Wessel 2004, 2008). This fertilization envelope is elevated from the cell membrane by *glycosaminoglycans* released by the cortical granules. These viscous compounds absorb water to expand the space between the cell membrane and the fertilization envelope, so that the envelope moves radially away from the egg (see Figures 4.19 and 4.20). The fertilization envelope is then stabilized by crosslinking adjacent proteins through egg-specific *peroxidase enzymes* (soluble ovoperoxidase from the cortical granules and Udx1 in the former cortical granule membrane) and a *transglutaminase* released from the cortical granules (Foerder and Shapiro 1977; Wong et al. 2004; Wong and Wessel 2009). This crosslinking allows the egg and early embryo to resist the shear forces of the intertidal waves. As shown in **Figure 4.19**,

---

*One might ask, as did a recent student, how amphibians could have a fast block to polyspermy, since their eggs are fertilized in pond water, which lacks high amounts of sodium ions. It turns out that the ion channels that open in frog egg membranes at fertilization are chloride channels instead of sodium channels as in sea urchin eggs. The concentration of Cl[−] inside the frog egg is much higher than that of pond water. Thus, when chloride channels open at fertilization, the negatively charged chloride ions flow *out* of the cytoplasm, leaving the inside of the egg at a positive potential (see Jaffe and Schlicter 1985; Glahn and Nuccitelli 2003).

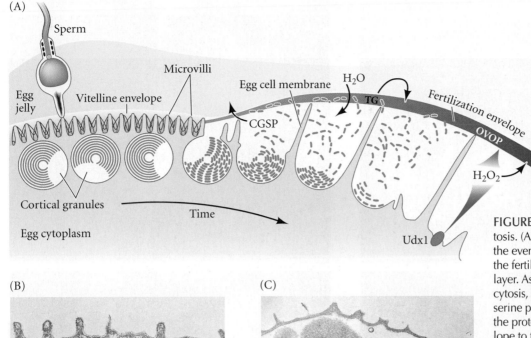

(A)

Sperm

Egg jelly

Vitelline envelope

Microvilli

Egg cell membrane

H₂O

CGSP

TG

Fertilization envelope

OVOP

H₂O₂

Udx1

Cortical granules

Time

Egg cytoplasm

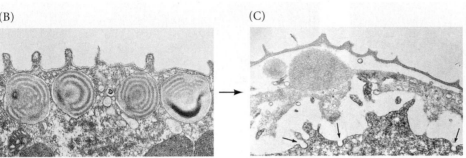

(B)

(C)

FIGURE 4.20 Cortical granule exocytosis. (A) Schematic diagram showing the events leading to the formation of the fertilization envelope and hyaline layer. As cortical granules undergo exocytosis, they release cortical granule serine protease (CGSP), which cleaves the proteins linking the vitelline envelope to the cell membrane. Mucopolysaccharides released by the cortical granules form an osmotic gradient, thereby causing water to enter and swell the space between the vitelline envelope and the cell membrane. Peroxidases (OVOP and Udx1) and transglutaminases (TG) then harden the vitelline envelope, now called the fertilization envelope. Transmission electron micrographs of the cortex of an unfertilized sea urchin egg (B) and the same region of a recently fertilized egg (C). The raised fertilization envelope and the points at which the cortical granules have fused with the egg cell membrane of the egg (arrows) are visible in (C). (A after Wong et al. 2008; B,C from Chandler and Heuser 1979, courtesy of D. E. Chandler.)

the fertilization envelope starts to form at the site of sperm entry and continues its expansion around the egg. This process starts about 20 seconds after sperm attachment and is complete by the end of the first minute of fertilization. As this is happening, a fourth set of cortical granule proteins, including *hyalin*, forms a coating around the egg (Hylander and Summers 1982). The egg extends elongated microvilli whose tips attach to this **hyaline layer**. This layer provides support for the blastomeres during cleavage.

See WEBSITE 4.6
Building the egg's extracellular matrix

See VADE MECUM   Sea urchin fertilization

Calcium as the initiator of the cortical granule reaction

The mechanism of cortical granule exocytosis is similar to that of the exocytosis of the acrosome, and it may involve many of the same molecules.* Upon fertilization, the concentration of free Ca^{2+} in the egg cytoplasm increases greatly. In this high-calcium environment, the cortical granule membranes fuse with the egg cell membrane, releasing their contents (**Figure 4.20**). Once the fusion of the cortical granules begins near the point of sperm entry, a wave of cortical granule exocytosis propagates around the cortex to the opposite side of the egg.

In sea urchins and mammals, the rise in Ca^{2+} concentration responsible for the cortical granule reaction is not due to an influx of calcium into the egg, but rather comes

from within the egg itself. The release of calcium from intracellular storage can be monitored visually using calcium-activated luminescent dyes such as aequorin (like GFP, a protein isolated from luminescent jellyfish) or fluorescent dyes such as fura-2. These dyes emit light when they bind free Ca^{2+}. When a sea urchin egg is injected with dye and then fertilized, a striking wave of calcium release

---

*Exocytotic reactions like the cortical granule reaction and the acrosome reaction are also seen in the release of insulin from pancreatic cells and in the release of neurotransmitters from synaptic terminals. In all cases, there is Ca^{2+}-mediated fusion of the secretory vesicle and the cell membrane. Indeed, the similarity of acrosomal vesicle exocytosis and synaptic vesicle exocytosis may be quite deep. Studies of acrosome reactions in sea urchins and mammals suggest that when the receptors for the sperm-activating ligands bind these molecules, the resulting depolarization of the membrane opens voltage-dependent Ca^{2+} channels in a manner reminiscent of synaptic transmission (González-Martínez et al. 1992; Tulsani and Abou-Haila 2004). The proteins that dock the cortical granules of the egg to the cell membrane also appear to be homologous to those used in the axon terminal. The synaptic granules of the neurons, the acrosomal vesicle of the sperm, and the cortical granules of the egg all appear to use synaptotagmin to bind calcium and initiate fusion of the vesicle with the cell membrane (Bi et al. 1995; Leguia et al. 2006; Roggero et al. 2007).

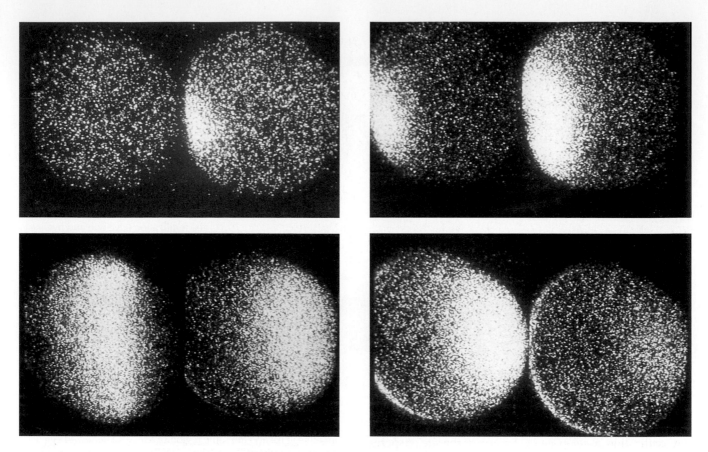

FIGURE 4.21   Wave of Ca^{2+} release across a sea urchin egg during fertilization. The egg is pre-loaded with a dye that fluoresces when it binds Ca^{2+}. When a sperm fuses with the egg, a wave of calcium release is seen, beginning at the site of sperm entry and propagating across the egg. The wave takes 30 seconds to traverse the egg. (Courtesy of G. Schatten.)

propagates across the egg (**Figure 4.21**). Starting at the point of sperm entry, a band of light traverses the cell (Steinhardt et al. 1977; Hafner et al. 1988). The calcium ions do not merely diffuse across the egg from the point of sperm entry. Rather, the release of Ca^{2+} starts at one end of the cell and proceeds actively to the other end, creating a wave of calcium across the egg. The entire release of Ca^{2+} is complete within roughly 30 seconds, and free Ca^{2+} is resequestered shortly after being released. If two sperm enter the egg cytoplasm, Ca^{2+} release can be seen starting at the two separate points of entry on the cell surface (Hafner et al. 1988).

Several experiments have demonstrated that Ca^{2+} ions are directly responsible for propagating the cortical granule reaction, and that these ions are stored within the egg itself. The drug A23187 is a calcium *ionophore* (a compound that allows the diffusion of ions such as Ca^{2+} across lipid membranes, permitting them to traverse otherwise impermeable barriers). Placing unfertilized sea urchin eggs into seawater containing A23187 causes the cortical granule reaction and the elevation of the fertilization envelope.

Moreover, this reaction occurs in the absence of any Ca^{2+} in the surrounding water. Therefore, the A23187 must be stimulating the release of Ca^{2+} already sequestered in organelles within the egg (Chambers et al. 1974; Steinhardt and Epel 1974).

In sea urchins and vertebrates (but not snails and worms), the calcium ions responsible for the cortical granule reaction are stored in the endoplasmic reticulum of the egg (Eisen and Reynolds 1985; Terasaki and Sardet 1991). In sea urchins and frogs, this reticulum is pronounced in the cortex and surrounds the cortical granules (**Figure 4.22**; Gardiner and Grey 1983; Luttmer and Longo 1985). The cortical granules are themselves tethered to the cell membrane by a series of integral membrane proteins that facilitate calcium-mediated exocytosis (Conner et al. 1997; Conner and Wessel 1998). Thus, as soon as Ca^{2+} is released from the endoplasmic reticulum, the cortical granules fuse with the cell membrane above them. Once initiated, the release of calcium is self-propagating. Free calcium is able to release sequestered calcium from its storage sites, thus causing a wave of Ca^{2+} release and cortical granule exocytosis.

(A)

(B)

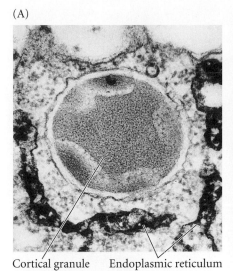

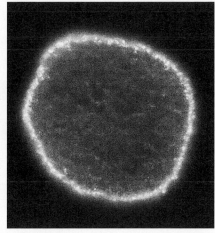

Cortical granule   Endoplasmic reticulum

FIGURE 4.22 Endoplasmic reticulum surrounding cortical granules in sea urchin eggs. (A) The endoplasmic reticulum has been stained to allow visualization by transmission electron microscopy. The cortical granule is seen to be surrounded by dark-stained endoplasmic reticulum. (B) An entire egg stained with fluorescent antibodies to calcium-dependent calcium release channels. The antibodies show these channels in the cortical endoplasmic reticulum. (A from Luttmer and Longo 1985, courtesy of S. Luttmer; B from McPherson et al. 1992, courtesy of F. J. Longo.)

Activation of Egg Metabolism in Sea Urchins

Release of intracellular calcium ions

Although fertilization is often depicted as merely the means to merge two haploid nuclei, it has an equally important role in initiating the processes that begin development. These events happen in the cytoplasm and occur without the involvement of the parental nuclei.* In addition to initiating the slow block to polyspermy (through cortical granule exocytosis), the release of Ca^{2+} that occurs when the sperm enters the egg is critical for activating the egg's metabolism and initiating development. Calcium ions release the inhibitors from maternally stored messages, allowing these mRNAs to be translated; they also release the inhibition of nuclear division, thereby allowing cleavage to occur. Indeed, throughout the animal kingdom, calcium ions are used to activate development during fertilization.

However, the way calcium ions are released varies between species (see Parrington et al. 2007). One way, first proposed by Jacques Loeb (1899, 1902), is that a soluble factor from the sperm is introduced into the egg at the time of cell fusion, and this substance activates the egg by

changing the ionic composition of the cytoplasm (**Figure 4.23A**). This mechanism, as we will see later, probably works in mammals. The other mechanism, proposed by Loeb's rival Frank Lillie (1913), is that the sperm acts like a big hormone, binding to receptors on the egg cell surface and changing their conformation, thus initiating reactions within the cytoplasm that activate the egg (**Figure 4.23B**). This is probably what happens in sea urchins.

IP$_3$: THE RELEASER OF CALCIUM IONS If Ca^{2+} from the egg's endoplasmic reticulum is responsible for the cortical granule reaction and the reactivation of development, what releases Ca^{2+}? Throughout the animal kingdom, it has been found that inositol 1,4,5-trisphosphate (IP_3) is the primary mechanism for releasing Ca^{2+} from intracellular storage.

The IP_3 pathway is shown in **Figure 4.24**. The membrane phospholipid phosphatidylinositol 4,5-bisphosphate (PIP_2) is split by the enzyme **phospholipase C (PLC)** to yield two active compounds: IP_3 and **diacylglycerol (DAG)**. IP_3 is able to release Ca^{2+} into the cytoplasm by opening the calcium channels of the endoplasmic reticulum. DAG activates protein kinase C, which in turn activates a protein that exchanges sodium ions for hydrogen ions, raising the pH of the egg (Nishizuka 1986; Swann and Whitaker 1986). This Na^+/H^+ exchange pump also requires Ca^{2+} for its activity. The result of PLC activation, therefore, is the liberation of Ca^{2+} and the alkalinization of the egg, and both of the compounds this activation creates—IP_3 and DAG—are involved in the initiation of development.

In sea urchin eggs, IP_3 is formed initially at the site of sperm entry and can be detected within seconds of the eggs being fertilized. The inhibition of IP_3 synthesis prevents Ca^{2+} release (Lee and Shen 1998; Carroll et al. 2000), while injected IP_3 can release sequestered Ca^{2+}, leading to cortical granule exocytosis (Whitaker and Irvine 1984; Busa et al. 1985). Moreover, these IP_3-mediated effects can be

---

*In certain salamanders, this developmental function of fertilization has been totally divorced from the genetic function. The silver salamander (*Ambystoma platineum*) is a hybrid subspecies consisting solely of females. Each female produces an egg with an unreduced chromosome number. This egg, however, cannot develop on its own, so the silver salamander mates with a male Jefferson salamander (*A. jeffersonianum*). The sperm from the male Jefferson salamander only stimulates the egg's development; it does not contribute genetic material (Uzzell 1964). For details of this complex mechanism of procreation, see Bogart et al. 1989.

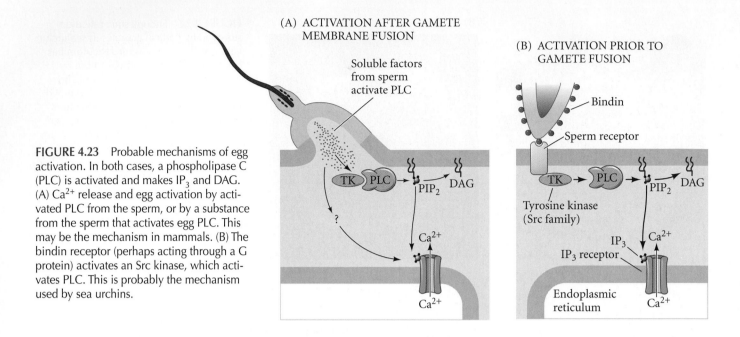

(A) ACTIVATION AFTER GAMETE MEMBRANE FUSION

Soluble factors from sperm activate PLC

TK   PLC

PIP$_2$ DAG

?

Ca^{2+}

Ca^{2+}

(B) ACTIVATION PRIOR TO GAMETE FUSION

Bindin

Sperm receptor

TK   PLC

PIP$_2$ DAG

Tyrosine kinase (Src family)

IP$_3$ Ca^{2+}

IP$_3$ receptor

Endoplasmic reticulum   Ca^{2+}

FIGURE 4.23 Probable mechanisms of egg activation. In both cases, a phospholipase C (PLC) is activated and makes IP$_3$ and DAG. (A) Ca^{2+} release and egg activation by activated PLC from the sperm, or by a substance from the sperm that activates egg PLC. This may be the mechanism in mammals. (B) The bindin receptor (perhaps acting through a G protein) activates an Src kinase, which activates PLC. This is probably the mechanism used by sea urchins.

thwarted by preinjecting the egg with calcium-chelating agents (Turner et al. 1986).

IP$_3$-responsive calcium channels have been found in the egg endoplasmic reticulum. The IP$_3$ formed at the site of sperm entry is thought to bind to the IP$_3$ receptors of these channels, effecting a local release of calcium (Ferris et al. 1989; Furuichi et al. 1989). Once released, Ca^{2+} can diffuse directly, or it can facilitate the release of more Ca^{2+} by binding to calcium-release receptors, also located in the cortical endoplasmic reticulum (McPherson et al. 1992). The binding of Ca^{2+} to these receptors releases more Ca^{2+}, which binds to more receptors, and so on. The resulting wave of calcium release is propagated throughout the cell, starting at the point of sperm entry. The cortical granules, which fuse with the cell membrane in the presence of high calcium concentrations, respond with a wave of exocytosis that follows the calcium wave. Mohri and colleagues (1995) have shown that IP$_3$-released Ca^{2+} is both necessary and sufficient for initiating the wave of calcium release.

PHOSPHOLIPASE C, THE GENERATOR OF IP$_3$ The question then becomes, What activates phospholipase C enzymes?

This question has not been easy to address, since (1) there are numerous types of PLC that (2) can be activated through different pathways, and (3) different species use different mechanisms to activate PLC. Results from studies of sea urchin eggs suggest that the active PLC in echinoderms is a member of the γ (gamma) family of PLCs (Carroll et al. 1997, 1999; Shearer et al. 1999). Inhibitors that specifically block PLCγ inhibit IP$_3$ production as well as Ca^{2+} release. Moreover, these inhibitors can be circumvented by microinjecting IP$_3$ into the egg.

KINASES: A LINK BETWEEN SPERM AND PLCγ The finding that the γ class of PLCs was responsible for generating IP$_3$ during echinoderm fertilization spurred investigators to look at exactly which proteins activated this particular class of phospholipases. Their work soon came to focus on the Src family of protein kinases. Src proteins are found in the cortical cytoplasm of sea urchin and starfish eggs, where they can form a complex with PLCγ. Inhibition of Src protein kinases lowered and delayed the amount of Ca^{2+} released (Kinsey and Shen 2000; Giusti et al. 2003; Townley et al. 2009).

So what activates Src kinase activity? One possibility is heterotrimeric G proteins in the cortex of the egg (**Figure 4.25**). Such G proteins are known to activate Src kinases in mammalian somatic cells, so the cortical G proteins of sea urchin eggs seem like good candidates; blocking these G proteins prevented Ca^{2+} release (Voronina and Wessel 2003, 2004). It is also possible that these G proteins activate PLC directly. Indeed, there may be more than one pathway and more than one way to activate Ca^{2+} release.

Thus, in sea urchins, it is thought that the binding of sperm to the egg (or possibly the fusion of sperm and egg) activates PLCγ through G proteins and Src kinases. The IP$_3$ thus generated opens calcium channels in the nearby cortical endoplasmic reticulum, allowing the initial and local outflow of Ca^{2+}. This first efflux of ions opens calcium-gated calcium release channels, causing a wave of Ca^{2+} that flows across the egg from the point of sperm entry to the opposite side of the egg. In so doing, some of the Ca^{2+} initiates cortical granule exocytosis, fusing the cortical granule with the egg cell membrane. Other Ca^{2+} would be bound by proteins such as calmodulin, which is activated by Ca^{2+} and can regulate numerous functions.

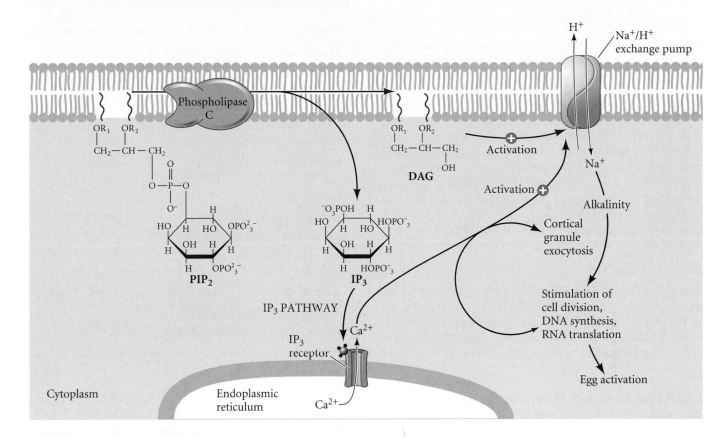

FIGURE 4.24 Roles of inositol phosphates in releasing calcium from the endoplasmic reticulum and the initiation of development. Phospholipase C splits PIP_2 into IP_3 and DAG. IP_3 releases calcium from the endoplasmic reticulum, and DAG, with assistance from the released Ca^{2+}, activates the sodium-hydrogen exchange pump in the membrane.

FIGURE 4.25 G protein involvement in Ca^{2+} entry into sea urchin eggs. (A) Mature sea urchin egg immunologically labeled for the cortical granule protein hyaline (red) and the G protein Gαq (green). (The overlap of signals produces the yellow color.) Gαq is localized to the cortex. (B) A wave of Ca^{2+} appears in the control egg (computer-enhanced to show relative intensities, with red being the highest), but not in the egg injected with an inhibitor of the Gαq protein. (C) Possible model for egg activation by the influx of Ca^{2+}. (After Voronina and Wessel 2003; photographs courtesy of G. M. Wessel.)

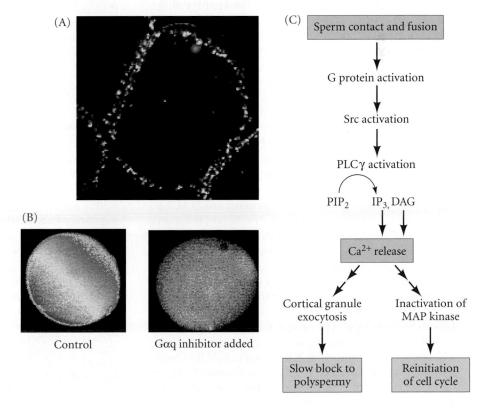

NAADP It is possible that diffusible compounds from the sperm also release calcium directly to the endoplasmic reticulum. There is evidence that *nicotinic acid adenine dinucleotide phosphate* (NAADP), a linear dinucleotide derived from NADP, serves as a sperm-borne Ca^{2+} releaser. NAADP frees stored Ca^{2+} from membrane vesicles during muscle contraction, insulin secretion, and neurotransmitter release (Lee 2001). Upon contact with egg jelly, NAADP concentration in sea uchin sperm increases tenfold, reaching levels that appear to be more than sufficient to release stored Ca^{2+} within the egg (Churchill et al. 2003; Morgan and Galione 2007). Thus, among sea urchins, two pathways may have evolved for the release of stored calcium.

Effects of calcium

The flux of calcium across the egg activates a preprogrammed set of metabolic events. The responses of the egg to the sperm can be divided into "early" responses, which occur within seconds of the cortical granule reaction, and "late" responses, which take place several minutes after fertilization begins (Table 4.1).

EARLY RESPONSES As we have seen, contact or fusion between sea urchin sperm and egg activates the two major blocks to polyspermy: the fast block, initiated by sodium influx into the cell; and the slow block, initiated by the intracellular release of Ca^{2+}.

The same release of Ca^{2+} responsible for the cortical granule reaction is also responsible for the re-entry of the egg into the cell cycle and the reactivation of egg protein synthesis. Ca^{2+} levels in the egg increase from 0.1 to 1 m*M*, and in almost all species, this occurs as a wave or succession of waves that sweep across the egg beginning at the site of sperm-egg fusion (Jaffe 1983; Terasaki and Sardet 1991; Stricker 1999; see Figure 4.21).

Calcium release activates a series of metabolic reactions that initiate embryonic development (Figure 4.26). One of these is the activation of the enzyme NAD⁺ kinase, which converts NAD⁺ to NADP⁺ (Epel et al. 1981). This change may have important consequences for lipid metabolism, since NADP⁺ (but not NAD⁺) can be used as a coenzyme for lipid biosynthesis. Thus the conversion of NAD⁺ to NADP⁺ may be important in the construction of the many new cell membranes required during cleavage. NADP⁺ is also used to make NAADP (see above), which appears to boost calcium release even further. Calcium release also affects oxygen consumption. A burst of oxygen reduction (to hydrogen peroxide) is seen during fertilization, and much of this "respiratory burst" is used to crosslink the fertilization envelope. The enzyme responsible for this reduction of oxygen is also NADPH-dependent (Heinecke and Shapiro 1989; Wong et al. 2004). Lastly, NADPH helps regenerate glutathione and ovothiols, molecules that may be crucial scavengers of free radicals that could otherwise damage the DNA of the egg and early embryo (Mead and Epel 1995).

TABLE 4.1 Events of sea urchin fertilization

Event	Approximate time postinsemination[a]
EARLY RESPONSES	
Sperm-egg binding	0 seconds
Fertilization potential rise (fast block to polyspermy)	within 1 sec
Sperm-egg membrane fusion	within 1 sec
Calcium increase first detected	10 sec
Cortical granule exocytosis (slow block to polyspermy)	15–60 sec
LATE RESPONSES	
Activation of NAD kinase	starts at 1 min
Increase in NADP⁺ and NADPH	starts at 1 min
Increase in O_2 consumption	starts at 1 min
Sperm entry	1–2 min
Acid efflux	1–5 min
Increase in pH (remains high)	1–5 min
Sperm chromatin decondensation	2–12 min
Sperm nucleus migration to egg center	2–12 min
Egg nucleus migration to sperm nucleus	5–10 min
Activation of protein synthesis	starts at 5–10 min
Activation of amino acid transport	starts at 5–10 min
Initiation of DNA synthesis	20–40 min
Mitosis	60–80 min
First cleavage	85–95 min

Main sources: Whitaker and Steinhardt 1985; Mohri et al. 1995.
[a]Approximate times based on data from *S. purpuratus* (15–17°C), *L. pictus* (16–18°C), *A. punctulata* (18–20°C), and *L. variegatus* (22–24°C). The timing of events within the first minute is best known for *Lytechinus variegatus*, so times are listed for that species.

LATE RESPONSES: RESUMPTION OF PROTEIN AND DNA SYNTHESIS The late responses of fertilization include the activation of a new burst of DNA and protein synthesis. The fusion of egg and sperm in sea urchins causes the intracellular pH to increase.* This rise in intracellular pH begins with a second influx of sodium ions, which causes a 1:1 exchange between sodium ions from the seawater and hydrogen ions from the egg. The loss of hydrogen ions causes the pH of the egg to rise (Shen and Steinhardt 1978; Michael and Walt 1999).

It is thought that pH increase and Ca^{2+} elevation act together to stimulate new DNA and protein synthesis

---

*This is due to the production of diacylglycerol, as mentioned above. The sea urchin egg has a Na⁺/H⁺ antiport protein that is regulated by the ionic and cytoskeletal changes at fertilization (Rangel-Mata et al. 2007). Again, variation among species may be prevalent. In the much smaller egg of the mouse, there is no elevation of pH after fertilization, and there is no dramatic increase in protein synthesis immediately following fertilization (Ben-Yosef et al. 1996).

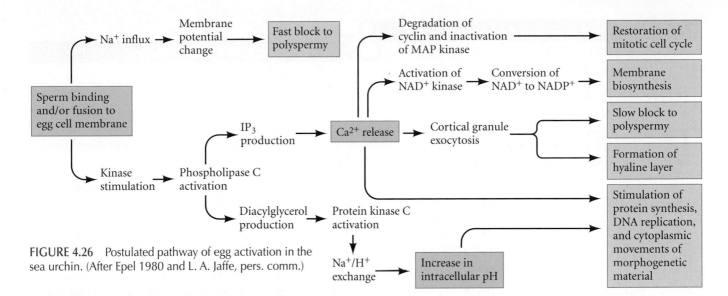

FIGURE 4.26   Postulated pathway of egg activation in the sea urchin. (After Epel 1980 and L. A. Jaffe, pers. comm.)

(Winkler et al. 1980; Whitaker and Steinhardt 1982; Rees et al. 1995). If one experimentally elevates the pH of an unfertilized egg to a level similar to that of a fertilized egg, DNA synthesis and nuclear envelope breakdown ensue, just as if the egg were fertilized (Miller and Epel 1999). Calcium ions are also critical to new DNA synthesis. The wave of free Ca^{2+} inactivates the enzyme MAP kinase, converting it from a phosphorylated (active) to an unphosphorylated (inactive) form, thus removing an inhibition on DNA synthesis (Carroll et al. 2000). DNA and protein synthesis can then resume.

In sea urchins, a burst of protein synthesis usually occurs within several minutes after sperm entry. This protein synthesis does not depend on the synthesis of new messenger RNA; rather, it uses mRNAs already present in the oocyte cytoplasm (**Figure 4.27**; also see Table 2.2). These mRNAs encode proteins such as histones, tubulins, actins, and morphogenetic factors that are used during early development. Such a burst of protein synthesis can be induced by artificially raising the pH of the cytoplasm using ammonium ions (Winkler et al. 1980).

One mechanism for this global rise in the translation of messages stored in the oocyte appears to be the release of inhibitors from the mRNA. In Chapter 2, we discussed

maskin, an inhibitor of translation in the unfertilized amphibian oocyte. In sea urchins, a similar inhibitor binds translation initiation factor eIF4E at the 5′ end of several mRNAs and prevents these mRNAs from being translated. Upon fertilization, however, this inhibitor—the eIF4E-binding protein—becomes phosphorylated and is degraded, thus allowing eIF4E to complex with other translation factors and permit protein synthesis from the stored sea urchin mRNAs (Cormier et al. 2001; Oulhen et al. 2007). One of the mRNAs "freed" by the degradation of eIF4E-binding pro-

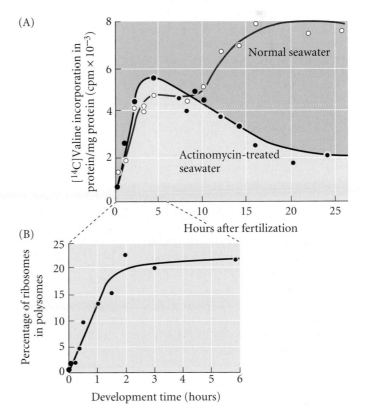

FIGURE 4.27   A burst of protein synthesis at fertilization uses mRNAs stored in the oocyte cytoplasm. (A) Protein synthesis in embryos of the sea urchin *Arbacia punctulata* fertilized in the presence or absence of actinomycin D, an inhibitor of transcription. For the first few hours, protein synthesis occurs with very little new transcription from the zygote or embryo nuclei. A second burst of protein synthesis occurs during the mid-blastula stage, at about 12 hours after fertilization. This burst represents translation of newly transcribed messages and therefore is not seen in embryos growing in actinomycin. (B) The percentage of ribosomes recruited into polysomes increases during the first hours of sea urchin development, especially during the first cell cycle. (A after Gross et al. 1964; B after Humphreys 1971.)

Rules of Evidence: "Find It, Lose It, Move It"

Biology, like all science, does not deal with Facts; rather, it deals with evidence. Several types of evidence will be presented in this book, and they are not equivalent in strength. As an example, we will use the analysis of the role of Ca^{2+} in egg activation.

Correlative evidence

The first, and weakest, type of evidence is **correlative evidence**. Here, we find correlations between two or more events and then make the inference that one event causes the other. For example, upon the meeting of sea urchin sperm and egg, a wave of free Ca^{2+} spreads across the egg (see Figure 4.21), and this wave of Ca^{2+} is thought to activate the egg. This chain of events has been shown in several ways, most convincingly by aequorin fluorescence (Steinhardt et al. 1977; Shimomura 1995; Steinhardt 2006).

Although one might infer that the meeting of egg and sperm caused the Ca^{2+} wave, and that this Ca^{2+} wave caused egg activation, such a correlation of events with one another does not necessarily demonstrate a causal relationship. It is possible the meeting of gametes first caused the flow of Ca^{2+} across the egg and then, separately and by some other mechanism, activated the egg. It is also conceivable that some aspect of egg activation caused the Ca^{2+} release. The correlated occurrence of these events could even be coincidental and have no relationship to one another.* Correla-

*In a tongue-in-cheek letter spoofing such correlative evidence, Sies (1988) demonstrated a remarkably good correlation between the number of storks seen in West Germany from 1965 to 1980 and the number of babies born during those same years. Any cause-and-effect scenario between storks and babies, however, would certainly fly in the face of the evidence presented in this chapter.

tive evidence provides a starting point for many investigations, but one cannot say that one event causes another based solely on correlation.

Functional evidence

The next type of evidence is called **loss-of-function evidence**, also known as **negative inference evidence**. Here, the absence of the postulated cause is associated with the absence of the effect. While stronger than correlative evidence, loss-of-function evidence still does not exclude other explanations. For instance, when calcium chelators such as EDTA were injected into the egg prior to fertilization, released Ca^{2+} failed to activate the egg. This would imply that Ca^{2+} is necessary for egg activation. However, data from such inhibitory studies (including studies from loss-of-function mutations) always leave open the possibility that the inhibitor suppresses more than just the process being studied. For instance, when protease inhibitors caused the failure of mammalian fertilization, it was assumed that these inhibitors were blocking the action of proteases released from the acrosome. As a result, biologists thought that the mammalian acrosome releases soluble proteases that digest the zona. Later experiments, however, demonstrated that the protease inhibitors inhibited the acrosome reaction itself so that the proteases were never released (Llanos et al. 1993). Thus, one couldn't tell whether soluble proteases played any role in mammalian fertilization.

The strongest type of evidence is **gain-of-function evidence**. Here the initiation of the first event causes the second event to happen even in instances where or when neither event usually occurs. Thus, when calcium ionophores (which can shuttle Ca^{2+} across membranes) were added to *unfertilized* eggs, Ca^{2+} was released from intracellular storage

and the egg became activated even without fertilization (Steinhardt and Epel 1974).

Progression of evidence and coherence

Correlative ("find it"), loss-of-function ("lose it"), and gain-of-function ("move it") evidence must consistently support each other to establish and solidify a conclusion. This progression of "find it; lose it; move it" evidence is at the core of nearly all studies of developmental mechanism (Adams 2000). Sometimes it can be found in a single paper, and sometimes, as the case above illustrates, the evidence comes from many laboratories. "Every scientist," writes Fleck (1979), "knows just how little a single experiment can prove or convince." Rather, "an entire system of experiments and controls is needed." Science is a communal endeavor, and it is doubtful that any great discovery is the achievement of a single experiment, or of any individual.

Science also accepts evidence better when it fits into a system of other findings. This is often called **coherence**. For instance, the ability of calcium to activate the egg became a standard part of fertilization physiology when Ca^{2+} was shown to cause both the resumption of cell division and the initiation of translation—two separate components of egg activation. Also, once the sperm was found to activate phopholipase C—the enzyme that synthesizes IP_3—and IP_3 was found to activate intracellular calcium release in numerous cells, the release of Ca^{2+} became understood as being the central element of sea urchin egg activation. It fit into a much wider picture of physiological calcium release, and the mechanisms for its synthesis and its effects all fit together.

tein is the message encoding cyclin B (Salaun et al. 2003, 2004). The cyclin B protein combines with Cdk1 cyclin to create **mitosis-promoting factor** (**MPF**), which is required to initiate cell division.

Thus fertilization activates pathways that target the translational inhibitory protein for degradation, and the newly accessible 5′ end of the mRNA can interact with those proteins that allow the message to be translated. One of these mRNAs encodes a protein critical for cell division. In such a manner, fertilization can initiate mitosis and the sea urchin can begin to form a multicellular organism.

Fusion of genetic material

In sea urchins, the sperm nucleus enters the egg perpendicular to the egg surface. After the sperm and egg cell membranes fuse, the sperm nucleus and its centriole separate from the mitochondria and flagellum. The sperm's mitochondria and the flagellum disintegrate inside the egg, so very few, if any, sperm-derived mitochondria are found in developing or adult organisms. Thus, although each gamete contributes a haploid genome to the zygote, the *mitochondrial* genome is transmitted primarily by the maternal parent. Conversely, in almost all animals studied (the mouse being the major exception), the centrosome needed to produce the mitotic spindle of the subsequent divisions is derived from the sperm centriole (see Figure 4.17; Sluder et al. 1989, 1993).

Fertilization in sea urchin eggs occurs after the second meiotic division, so there is a haploid female pronucleus in the cytoplasm of the egg when the sperm enters. Once inside the egg, the sperm nucleus undergoes a dramatic transformation as it decondenses to form the haploid male pronucleus. First, the nuclear envelope vesiculates into small packets, exposing the compact sperm chromatin to the egg cytoplasm (Longo and Kunkle 1978; Poccia and Collas 1997). Then proteins holding the sperm chromatin in its condensed, inactive state are exchanged for other proteins derived from the egg cytoplasm. This exchange permits the decondensation of the sperm chromatin. Once decondensed, the DNA adheres to the nuclear envelope, where DNA polymerase can initiate replication (Infante et al. 1973).

In sea urchins, sperm chromosome decondensation appears to be initiated by the phosphorylation of the nuclear envelope lamin protein and the phosphorylation of two sperm-specific histones that bind tightly to the DNA. The process begins when sperm comes into contact with a certain glycoprotein in the egg jelly that elevates the level of cAMP-dependent protein kinase activity. These protein kinases phosphorylate several of the basic residues of the sperm-specific histones and thereby interfere with their binding to DNA (Garbers et al. 1980; Porter and Vacquier 1986; Stephens et al. 2002). This loosening is thought to facilitate the replacement of the sperm-specific histones

with other histones that have been stored in the oocyte cytoplasm (Green and Poccia 1985).

After the sea urchin sperm enters the egg cytoplasm, the male pronucleus separates from the tail and rotates 180 degrees so that the sperm centriole is between the sperm pronucleus and the egg pronucleus. The sperm centriole then acts as a microtubule organizing center, extending its own microtubules and integrating them with egg microtubules to form an aster.* Microtubules extend throughout the egg and contact the female pronucleus, at which point the two pronuclei migrate toward each other. Their fusion forms the diploid **zygote nucleus** (**Figure 4.28**). DNA synthesis can begin either in the pronuclear stage (during migration) or after the formation of the zygote nucleus, and depends on the level of Ca^{2+} released earlier in fertilization (Jaffe et al. 2001).

Internal Fertilization in Mammals

It is very difficult to study any interactions between the mammalian sperm and egg that might take place prior to these gametes making contact. One obvious reason for this is that mammalian fertilization occurs inside the oviducts of the female: while it is relatively easy to mimic the conditions surrounding sea urchin fertilization using either natural or artificial seawater, we do not yet know the components of the various natural environments that mammalian sperm encounter as they travel to the egg.

A second reason why it is difficult to study mammalian fertilization is that the sperm population ejaculated into the female is probably very heterogeneous, containing spermatozoa at different stages of maturation. Out of the 280×10^6 human sperm normally ejaculated during coitus, only about 200 reach the vicinity of the egg (Ralt et al. 1991). Thus, since fewer than 1 in 10,000 sperm even gets close to the egg, it is difficult to assay those molecules that might enable the sperm to swim toward the egg and become activated.

Getting the gametes into the oviduct: Translocation and capacitation

The female reproductive tract is not a passive conduit through which sperm race, but a highly specialized set of tissues that actively regulates the transport and maturity of both gametes. Both the male and female gametes use a

---

*When Oscar Hertwig observed this radial array of sperm asters forming in his newly fertilized sea urchin eggs, he called it "the sun in the egg" and thought it was the happy indication of a successful fertilization (Hertwig 1877). More recently, Simerly and co-workers (1999) found that certain types of human male infertility are due to defects in the centriole's ability to form these microtubular asters. This deficiency results in the failure of pronuclear migration and the cessation of further development.

(A)

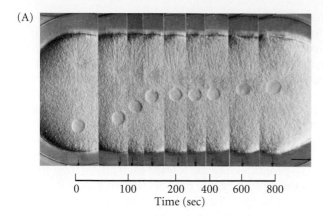

(B)

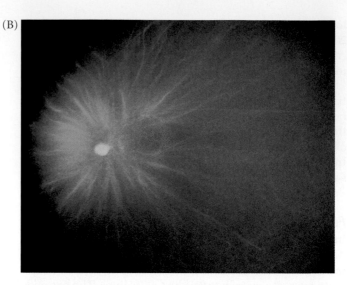

(C)

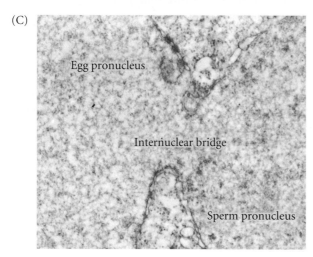

FIGURE 4.28 Nuclear events in the fertilization of the sea urchin. (A) Sequential photographs showing the migration of the egg pronucleus and the sperm pronucleus toward each other in an egg of *Clypeaster japonicus*. The sperm pronucleus is surrounded by its aster of microtubules. (B) The two pronuclei migrate toward each other on these microtubular processes. (The pronuclear DNA is stained blue by Hoechst dye.) The microtubules (stained green with fluorescent antibodies to tubulin) radiate from the centrosome associated with the (smaller) male pronucleus and reach toward the female pronucleus. (C) Fusion of pronuclei in the sea urchin egg. (A from Hamaguchi and Hiramoto 1980, courtesy of the authors; B from Holy and Schatten 1991, courtesy of J. Holy; C courtesy of F. J. Longo.)

combination of small-scale biochemical interactions and large-scale physical propulsion to get to the **ampulla**, the region of the oviduct where fertilization takes place.

TRANSLOCATION A mammalian oocyte just released from the ovary is surrounded by a matrix containing cumulus cells. (Cumulus cells are the cells of the ovarian follicle to which the developing oocyte was attached; see Figures 4.7 and 16.31). If this matrix is experimentally removed or significantly altered, the fimbriae of the oviduct will not "pick up" the oocyte-cumulus complex (see Figure 8.15), nor will the complex be able to enter the oviduct (Talbot et al. 1999). Once it is picked up, a combination of ciliary beating and muscle contractions transport the oocyte-cumulus complex to the appropriate position for its fertilization in the oviduct.

The translocation of sperm from the vagina to the oviduct involves many processes that work at different times and places. Sperm motility (i.e., flagellar action) is probably a minor factor in getting the sperm into the oviduct, although motility is required for mouse sperm to travel through the cervical mucus, and for sperm to

encounter the egg once they are in the oviduct. Sperm are found in the oviducts of mice, hamsters, guinea pigs, cows, and humans within 30 minutes of sperm deposition in the vagina—a time "too short to have been attained by even the most Olympian sperm relying on their own flagellar power" (Storey 1995). Rather, sperm appear to be transported to the oviduct by the muscular activity of the uterus. Recent studies of sperm motility have led to several conclusions, including the following:

1. Uterine muscle contractions are critical in getting the sperm into the oviduct.
2. The region of the oviduct before the ampulla may slow down sperm and release them slowly.
3. Sperm (flagellar) motility is important once sperm arrive within the oviduct; sperm become hyperactive in the vicinity of the oocyte.
4. Sperm may receive directional cues from temperature gradients between the regions of the oviduct and from chemical cues derived from the oocyte or cumulus.
5. During this trek from the vagina to the ampullary region of the oviduct, the sperm matures such that it has the capacity to fertilize the egg when the two finally meet.

CAPACITATION Newly ejaculated mammalian sperm are unable to undergo the acrosome reaction or fertilize an egg until they have resided for some time in the female reproductive tract (Chang 1951; Austin 1952). The set of physiological changes by which sperm become competent to fertilize the egg is called **capacitation**. Sperm that are not capacitated are "held up" in the cumulus matrix and are unable to reach the egg (Austin 1960; Corselli and Talbot 1987). Capacitation can be accomplished in vitro by incubating sperm in a tissue culture medium (such media contain calcium ions, bicarbonate, and serum albumin) or in fluid taken from the oviducts.

Contrary to the opening scenes of the *Look Who's Talking* movies, "the race is not always to the swiftest." Wilcox and colleagues (1995) found that nearly all human pregnancies result from sexual intercourse during a 6-day period ending on the day of ovulation. This means that the fertilizing sperm could have taken as long as 6 days to make the journey. Although some human sperm reach the ampulla of the oviduct within a half-hour after intercourse, "speedy" sperm may have little chance of fertilizing the egg, since they have not undergone capacitation. Eisenbach (1995) has proposed a hypothesis wherein capacitation is a transient event, and sperm are given a relatively brief window of competence during which they can successfully fertilize the egg. As the sperm reach the ampulla, they acquire competence—but they lose it if they stay around too long. By binding and capacitating sperm, the oviduct releases "packets" of capacitated sperm at various intervals, thereby prolonging the time that fertilization can be successful.

The molecular events that take place during capacitation (**Figure 4.29**) have not yet been fully accounted for, but five sets of molecular changes are considered to be important:

1. The sperm cell membrane is altered by the removal of cholesterol by albumin proteins in the female reproductive tract (Cross 1998). The cholesterol efflux from the sperm cell membrane changes the location of "lipid rafts," isolated regions of the cell membrane that often contain receptor proteins. Originally located throughout the sperm cell membrane, lipid rafts now cluster over the anterior sperm head. These lipid microdomains contain proteins that can bind the zona pellucida and participate in the acrosome reaction (Bou Khalil et al. 2006; Gadella et al. 2008).

2. Particular proteins or carbohydrates on the sperm surface are lost during capacitation (Lopez et al. 1985; Wilson and Oliphant 1987). It is possible that these compounds block the recognition sites for the sperm proteins that bind to the zona pellucida. It has been suggested that the unmasking of these sites might be one of the effects of cholesterol depletion (Benoff 1993).

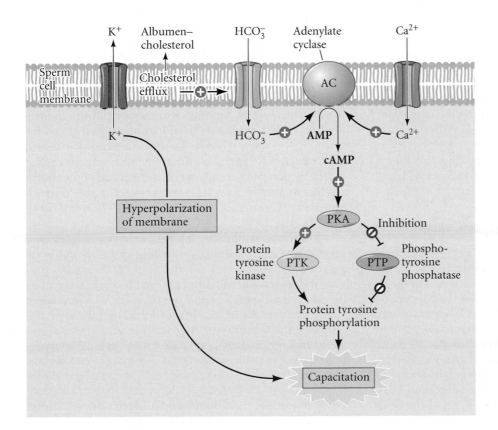

FIGURE 4.29 Hypothetical model for mammalian sperm capacitation. The efflux of potassium (the mechanism of which still remains unknown) results in a change in the resting potential of the sperm cell membrane. The removal of cholesterol by albumin stimulates ion channels that enable calcium and bicarbonate ions to enter the sperm. These ions promote the activity of a sperm-specific soluble adenylate cyclase, which makes cAMP from AMP. The rise in cAMP activates protein kinase A, causing it to activate the protein tyrosine kinases (while inactivating the protein phosphatases). The kinases phosphorylate proteins that are essential for capacitation. (After Visconti and Kopf 1998; Hess et al. 2005.)

3. The membrane potential of the sperm cell membrane becomes more negative as potassium ions leave the sperm. This change in membrane potential may allow calcium channels to be opened and permit calcium to enter the sperm. Calcium and bicarbonate ions may be critical in activating cAMP production and in facilitating the membrane fusion events of the acrosome reaction (Visconti et al. 1995; Arnoult et al. 1999).
4. Protein phosphorylation occurs (Galantino-Homer et al. 1997; Arcelay et al. 2008). In particular, two chaperone (heat-shock) proteins migrate to the surface of the sperm head when they are phosphorylated. Here, they may play an essential role in forming the receptor that binds to the zona pellucida (Asquith et al. 2004, 2005).
5. The outer acrosomal membrane changes and comes into contact with the sperm cell membrane in a way that prepares it for fusion (Tulsiani and Abou-Haila 2004).

It is uncertain whether these events are independent of one another and to what extent each of them contributes to sperm capacitation.

There may be an important connection between sperm translocation and capacitation. Smith (1998) and Suarez (1998) have documented that before entering the ampulla of the oviduct, the uncapacitated sperm bind actively to the membranes of the oviduct cells in the narrow passage (isthmus) preceding it (**Figure 4.30**; see also Figure 8.15). This binding is temporary and appears to be broken when the sperm become capacitated. Moreover, the life span of the sperm is significantly lengthened by this binding, and its capacitation is slowed down. This restriction of sperm entry into the ampulla, the slowing down of capacitation, and the expansion of sperm life span may have important consequences (Töpfer-Petersen et al. 2002; Gwathmey et al. 2003). The binding action may function as a block to polyspermy by preventing many sperm from reaching the egg at the same time; if the oviduct isthmus is excised in cows, a much higher rate of polyspermy results. In addition, slowing the rate of sperm capacitation and extending the active life of sperm may maximize the probability that sperm will still be available to meet the egg in the ampulla.

In the vicinity of the oocyte: Hyperactivation, thermotaxis, and chemotaxis

Different regions of the female reproductive tract may secrete different, regionally specific molecules, and these molecules may influence sperm motility as well as capacitation. During capacitation, sperm become hyperactivated—they swim at higher velocities and generate greater force. This hyperactivation appears to be mediated through the opening of a sperm-specific calcium channel located in the sperm tail (Quill et al. 2003). The asymmetric beating of the flagellum is changed into a rapid synchronous beat with a higher degree of bending. The power of the beat and the direction of sperm head movement are thought to

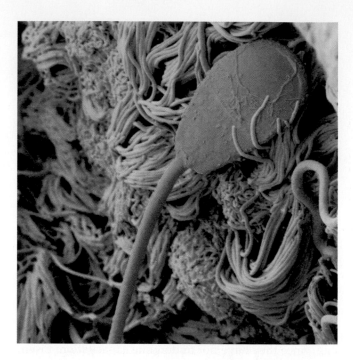

FIGURE 4.30 Scanning electron micrograph (artificially colored) showing bull sperm as it adheres to the membranes of epithelial cells in the oviduct of a cow prior to entering the ampulla. (From Lefebvre et al. 1995, courtesy of S. Suarez.)

release the sperm from their binding with the oviduct epithelial cells. Indeed, only hyperactivated sperm are seen to detach and continue their journey to the egg (Suarez 2008a,b).

Suarez and co-workers (1991) have shown that although this behavior is not conducive to traveling through low-viscosity fluids, it appears to be extremely well suited for linear sperm movement in the viscous fluid that sperm might encounter in the oviduct. Hyperactivation, along with a hyaluronidase enzyme on the outside of the sperm cell membrane, enables the sperm to digest a path through the extracellular matrix of the cumulus cells (Lin et al. 1994; Kimura et al. 2009).

An old joke claims that the reason so many sperm are released at each ejaculation is that none of the male gametes is willing to ask for directions. So what *does* provide the sperm with directions? One cue that can be used by capacitated sperm to find the egg is heat. Mammalian sperm are capable of sensing a thermal gradient of 2°C between the isthmus of the oviduct and the warmer ampullary region. Bahat and colleagues (2003, 2006) measured the temperature difference between these two regions of the oviduct and experimentally demonstrated that rabbit sperm sense the difference and preferentially swim from cooler to warmer sites (thermotaxis). Moreover, only capacitated sperm are able to sense this temperature gradient.

Once in the ampullary region, a second sensing mechanism, chemotaxis, may come into play. It appears that the oocyte and its accompanying cumulus cells secrete molecules that attract the sperm toward the egg during the last stages of sperm migration. Ralt and colleagues (1991) tested this hypothesis using follicular fluid from human follicles whose eggs were being used for in vitro fertilization. When the researchers microinjected a drop of follicular fluid into a larger drop of sperm suspension, some of the sperm changed direction and migrated toward the source of follicular fluid. Microinjection of other solutions did not have this effect. Moreover, these investigations uncovered a fascinating correlation: the fluid from only about half the follicles tested showed a chemotactic effect; in nearly every case, the egg was fertilizable if, and only if, the fluid showed chemotactic ability ($P < 0.0001$; see Eisenbach 1999; Sun et al. 2005). Further research has shown that the ability of human follicular fluid to attract human sperm only occurs if the sperm has been capacitated (Cohen-Dayag et al. 1995; Eisenbach and Tur-Kaspa 1999; Wang et al. 2001).

The identity of these chemotactic compounds is being investigated, but one of them appears to be the hormone **progesterone**. Guidobaldi and colleagues (2008) have shown that progesterone secreted from the cumulus cells surrounding the rabbit oocyte is bound by capacitated sperm and used as a directional cue by the sperm. It is possible that, like certain invertebrate eggs, the human egg secretes a chemotactic factor only when it is capable of being fertilized, and that sperm are attracted to such a compound only when they are capable of fertilizing the egg.

Recognition at the zona pellucida

Before the mammalian sperm can bind to the oocyte, it must first bind to and penetrate the egg's zona pellucida. The zona pellucida in mammals plays a role analogous to that of the vitelline envelope in invertebrates; the zona, however, is a far thicker and denser structure than the vitelline envelope. The binding of sperm to the zona is relatively, but not absolutely, species-specific.

There appear to be several steps in the binding of a hyperactivated, wiggling mouse sperm to the zona pellucida. The mouse zona pellucida is made of three major glycoproteins—**ZP1**, **ZP2**, and **ZP3** (**zona proteins 1, 2,** and **3**)—along with accessory proteins that bind to the zona's integral structure. This glycoprotein matrix, which is synthesized and secreted by the growing oocyte, binds the sperm and, once the sperm is bound, initiates the acrosome reaction (Saling et al. 1979; Florman and Storey 1982; Cherr et al. 1986).

In recent years, a new model for mammalian sperm-zona binding has emerged, emphasizing sequential interactions between several sperm proteins and the components of the zona (**Figure 4.31**). The first step appears to be a relatively weak binding accomplished by the recognition of a sperm protein by a peripheral protein that coats the zona pellucida. This is followed by a somewhat stronger association between the zona and the sperm's SED1 protein. Last, a protein on the sperm (and possibly several other factors) forms strong links with the ZP3 of the zona. This last binding will cause the mouse sperm to undergo its acrosome reaction directly on the zona pellucida.

(A)

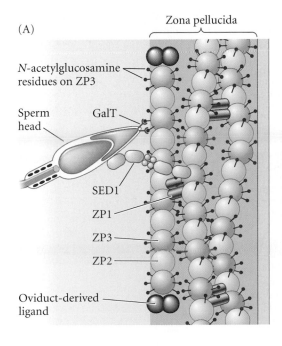

(B)

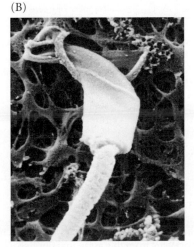

FIGURE 4.31   Sperm-zona binding. (A) Possible model of proteins involved in mouse sperm-egg adhesion. First the sperm binds weakly but specifically to a ligand protein secreted by the oviduct and coating the zona pellucida. The sperm surface protein SED1 (which is localized in the correct area of the sperm head for lateral sperm adhesion) then binds to the ZP complex on the zona. Sperm galactosyltransferase (GalT) crosslinks tightly and specifically to *N*-acetylglucosamine sugar residues on zona protein 3 (ZP3). The clustering of GalT proteins in the sperm cell membrane activates G proteins that open calcium channels and initiate the acrosome reaction. (The diagram is not drawn to scale.) (B) Electron micrograph showing sperm-zona binding in the golden hamster. (A based on data of B. Shur, courtesy of B. Shur; B courtesy of R. Yanagimachi.)

EARLY STAGES OF GAMETE ADHESION Before strong and specific binding of the sperm and the egg's zona pellucida, an initial tethering is accomplished. It appears that sperm initially bind to a 250-kDa protein that is associated with, but not integrally part of, the zona pellucida (Rodeheffer and Shur 2004). This protein can be washed away by preparative techniques (which is probably why it wasn't discovered earlier). Sperm binding is also facilitated by the sperm adhesion protein SED1, which binds to the zona protein complex (Ensslin and Shur 2003). SED1 is found in a discrete domain of the sperm cell membrane, directly overlying the acrosome, and it only binds to the zona of unfertilized oocytes (and not to those of fertilized oocytes). Antibodies against SED1 or solubilized SED1 proteins will inhibit sperm-zona binding. Indeed, the sperm of males whose *SED1* gene has been knocked out are unable to bind to the zona pellucida.

THE FINAL STAGE OF SPERM-ZONA RECOGNITION: BINDING TO ZP3 There are several pieces of evidence demonstrating that ZP3 is the major sperm-binding glycoprotein in the mouse zona pellucida. The binding of mouse sperm to the mouse zona pellucida can be inhibited by first incubating the sperm with solubilized zona glycoproteins. Using this inhibition assay, Bleil and Wassarman (1980, 1986, 1988) found that ZP3 was the active competitor for sperm binding sites (**Figure 4.32A**). This conclusion that sperm bound soluble ZP3 was further supported by the finding that radiolabeled ZP3 (but not ZP1 or ZP2) bound to the heads of mouse sperm with intact acrosomes (**Figure 4.32B**).

The cell membrane overlying the sperm head can bind to thousands of ZP3 glycoproteins in the zona pellucida. Moreover, there appear to be several different proteins on sperm that are capable of binding ZP3 (Wassarman et al. 2001; Buffone et al. 2008). Some of these sperm proteins bind to the serine- and threonine-linked carbohydrate chains of ZP3, and one of the zona-binding proteins, a sperm-surface galactosyltransferase, recognizes carbohydrate residues on ZP3 (Miller et al. 1992; Gong et al. 1995; Lu and Shur 1997). Another set of proteins that bind to the zona are ADAM3 and ADAM2 (also called cyritestin and fertilin β, respectively). These proteins form a complex on the tip of the sperm head, and sperm deficient in either one cannot bind to the zona (Kim et al. 2004; Nishimura et al. 2007).

Although the precise carbohydrate groups on ZP3 to which the proteins of the sperm membrane bind are still undefined, the conclusion that the carbohydrate moieties of ZP3 are critical for sperm attachment to the zona has been confirmed by the finding that if these carbohydrate groups are removed from ZP3, it will not bind sperm as well as intact ZP3 (see Figure 4.32A; Florman and Wassarman 1985; Kopf 1998).

INDUCTION OF THE MOUSE ACROSOME REACTION BY ZP3 ZP3 is the specific glycoprotein in the mouse zona pellucida to which sperm bind. ZP3 also initiates the acrosome reaction after sperm have bound to it. The mouse zona pellucida, unlike the sea urchin vitelline envelope, is a thick structure. By undergoing the acrosome reaction on the zona

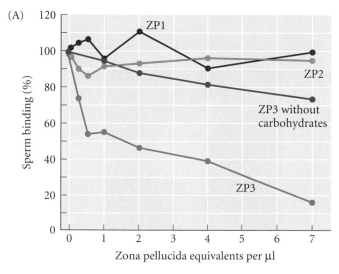
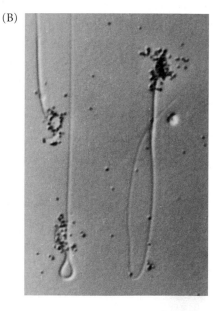

FIGURE 4.32 Mouse ZP3 binds sperm. (A) Inhibition assay showing a specific decrease of mouse sperm binding to zonae pellucidae. It appears from this assay that purified ZP3 (but not ZP1 or ZP2) can bind to sperm and prevent sperm from binding to the zona pellucida. The assay also illustrates the importance of the carbohydrate portion of ZP3 to the binding reaction. (B) Radioactively labeled ZP3 binds to capacitated mouse sperm. (A after Bleil and Wassarman 1980 and Florman and Wassarman 1985; B from Bleil and Wassarman 1986, courtesy of the authors.)

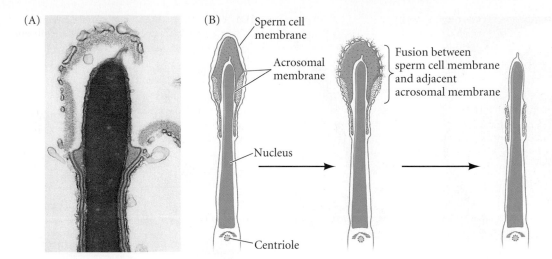

FIGURE 4.33 Acrosome reaction in hamster sperm. (A) Transmission electron micrograph of hamster sperm undergoing the acrosome reaction. The acrosomal membrane can be seen to form vesicles. (B) Interpretive diagram of electron micrographs showing the fusion of the acrosomal and cell membranes in the sperm head. (A from Meizel 1984, courtesy of S. Meizel; B after Yanagimachi and Noda 1970.)

pellucida, the mouse sperm can concentrate its proteolytic enzymes directly at the point of attachment and digest a hole through this extracellular layer (see Figure 4.8B). Indeed, mouse sperm that undergo the acrosome reaction before they reach the zona pellucida are unable to penetrate it (Florman et al. 1998).

ZP3 is able to cause the Ca^{2+}-mediated exocytosis of the acrosomal vesicle. How it does this is still under intensive investigation, although it is known that following ZP3 binding, phospholipase C (which synthesizes IP_3) is activated, the sperm head becomes more alkaline, and membrane Ca^{2+} channels are opened (see Florman et al. 2008). The exact sequence of these events remains undiscovered, but one possibility is that ZP3 initiates these events by crosslinking the sperm cell surface galactosyltransferases, whose active site faces outward and binds to the carbohydrate residues of ZP3 (see Figure 4.31A). This crosslinking activates specific G proteins in the sperm cell membrane, initiating a cascade that opens the membrane's Ca^{2+} channels and results in the calcium-mediated exocytosis of the acrosomal vesicle (**Figure 4.33**; Leyton and Saling 1989; Leyton et al. 1992; Florman et al. 1998; Shi et al. 2001). The source of the Ca^{2+} is also still unknown (Publicover et al. 2007). It is possible that there is a transient IP_3-mediated calcium influx from the outside environment followed by a longer-acting opening of Ca^{2+} channels, releasing Ca^{2+} from the neck of the sperm (see Florman et al. 2008).

TRAVERSING THE ZONA PELLUCIDA The exocytosis of the acrosomal vesicle releases a variety of proteases that lyse the zona pellucida. These enzymes, which may include

proteosomes (Yi et al. 2007a,b), create a hole through which the sperm can travel toward the egg. However, during the acrosome reaction, the anterior portion of the sperm cell membrane (i.e., the region containing the ZP3 binding sites) is shed from the sperm. But if sperm are going to penetrate the zona pellucida, they must somehow retain some adhesion to it. In mammals, there appear to be several adhesion systems that keep the sperm in its channel. First, **zonadhesin** protein from the acrosome may fix the sperm to the point of attachment and provide a pivot point for the sperm to enter into the zona (Hardy and Garbers 1995). Second, certain proteins in the inner acrosomal membrane bind specifically to the ZP2 glycoprotein (Bleil et al. 1988). Whereas acrosome-intact sperm will not bind to ZP2, acrosome-reacted sperm will. Moreover, antibodies against the ZP2 glycoprotein will not prevent the binding of acrosome-intact sperm to the zona but will inhibit the attachment of acrosome-reacted sperm.* Third, in some species, proacrosin, a protein that adheres to the inner acrosomal membrane, binds to sulfated carbohydrate groups on the zona pellucida glycoproteins (Gaboriau et al. 2007).

---

*In guinea pigs, secondary binding to the zona is thought to be mediated by the protein PH-20. Moreover, when this inner acrosomal membrane protein was injected into adult male or female guinea pigs, 100% of them became sterile for several months (Primakoff et al. 1988). The blood sera of these sterile guinea pigs had extremely high concentrations of antibodies to PH-20. The antiserum from guinea pigs sterilized in this manner not only bound specifically to PH-20, but also blocked sperm-zona adhesion in vitro. The contraceptive effect lasted several months, after which fertility was restored. More recently, O'Rand and colleagues (2004) provided reversible immunological contraception by injecting male monkeys with eppin, a sperm-surface protein that interacts with semen components. The antibodies block these interactions, probably slowing down the sperm. These experiments show that the principle of immunological contraception is well founded.

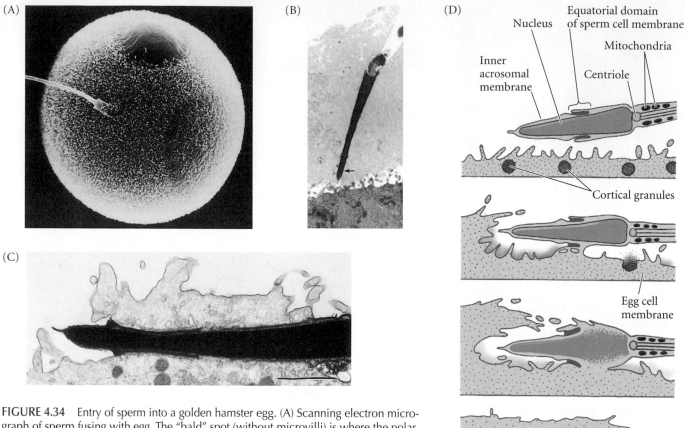

FIGURE 4.34 Entry of sperm into a golden hamster egg. (A) Scanning electron micrograph of sperm fusing with egg. The "bald" spot (without microvilli) is where the polar body has budded off. Sperm do not bind there. (B) Transmission electron micrograph showing the sperm head passing through the zona. (C) Transmission electron micrograph of the sperm fusing parallel to the egg plasma membrane. (D) Diagram of the fusion of the sperm and egg plasma membranes. (A–C from Yanagimachi and Noda 1970 and Yanagimachi 1994, courtesy of R. Yanagimachi.)

Gamete fusion and the prevention of polyspermy

In mammals, the sperm contacts the egg not at its tip (as in the case of sea urchins), but on the side of the sperm head . The acrosome reaction, in addition to expelling the enzymatic contents of the acrosome, also exposes the inner acrosomal membrane to the outside. The junction between this inner acrosomal membrane and the sperm cell membrane is called the **equatorial region**, and this is where membrane fusion between sperm and egg begins (**Figure 4.34**). As in sea urchin gamete fusion, the sperm is bound to regions of the egg where actin polymerizes to extend microvilli to the sperm (Yanagimachi and Noda 1970).

The mechanism of mammalian gamete fusion is still controversial (see Primakoff and Myles 2002; Ikawa et al. 2008). Gene knockout experiments suggest that mammalian gamete fusion may depend on interaction between a sperm protein and integrin-associated CD9 protein on the egg (Le Naour et al. 2000; Miyado et al. 2000; Evans 2001). CD9 protein has been localized to the membranes of the egg microvilli, and female mice with the *CD9* gene knocked out are infertile because their eggs fail to fuse with sperm (Kaji et al. 2002; Runge et al. 2006). This infertility can be reversed by the microinjection of mRNA encoding either mouse or human CD9 protein. It is not known exactly how these proteins facilitate membrane fusion, but CD9 is also known to be critical for the fusion of myocytes (muscle cell precursors) to form striated muscle (Tachibana and Hemler 1999).

On the sperm side of the mammalian fusion process, Inoue and colleagues (2005) have implicated the immunoglobulin-like protein Izumo (named after a Japanese shrine dedicated to marriage). Sperm from mice carrying loss-of-function mutations in the *Izumo* gene are able to bind and penetrate the zona pellucida, but are not able to fuse with the egg cell membrane. Human sperm also contain Izumo protein, and antibodies directed against Izumo prevent sperm-egg fusion in humans as well. There are other candidates for sperm fusion proteins; indeed, there may be several sperm-egg binding systems operating, and each of them may be necessary but not sufficient to insure proper gamete binding and fusion. It is not yet known whether the Izumo on the sperm and the CD9 on the egg bind one another.

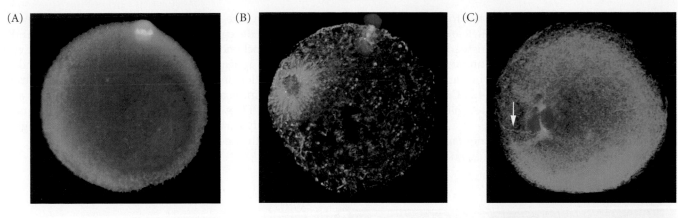

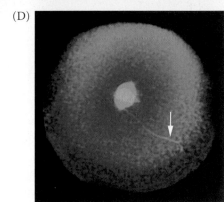

FIGURE 4.35 Pronuclear movements during human fertilization. The microtubules have been stained green, while the DNA is dyed blue. The arrows point to the sperm tail. (A) The mature unfertilized oocyte completes the first meiotic division, budding off a polar body. (B) As the sperm enters the oocyte (left side), microtubules condense around it as the oocyte completes its second meiotic division at the periphery. (C) By 15 hours after fertilization, the two pronuclei have come together, and the centrosome splits to organize a bipolar microtubule array. The sperm tail is still seen (arrow). (D) At prometaphase, chromosomes from the sperm and egg intermix on the metaphase equator and a mitotic spindle initiates the first mitotic division. The sperm tail can still be seen. (From Simerly et al. 1995, courtesy of G. Schatten.)

Polyspermy is a problem for the developing mammal as well as for the sea urchin (see Figure 4.17). In mammals, an electrical "fast" block to polyspermy has not yet been detected; and it may not even be needed, given the limited number of sperm that reach the ovulated egg (Gardner and Evans 2006). However, a *slow* block to polyspermy in mammals occurs by enzymes from the cortical granules modifying the zona pellucida proteins. Released enzymes modify the zona pellucida sperm receptors such that they can no longer bind sperm (Bleil and Wassarman 1980). Cortical granules of mouse eggs have been found to contain *N*-acetylglucosaminidase enzymes capable of cleaving *N*-acetylglucosamine from ZP3 carbohydrate chains. *N*-acetylglucosamine is one of the carbohydrate groups to which sperm can bind. Miller and co-workers (1992, 1993) have demonstrated that when the *N*-acetylglucosamine residues are removed at fertilization, ZP3 will no longer serve as a substrate for the binding of other sperm. ZP2 is clipped by another cortical granule protease and loses its ability to bind sperm as well (Moller and Wassarman 1989). Thus, once one sperm has entered the egg, other sperm can no longer initiate or maintain their binding to the zona pellucida and are rapidly shed.

Fusion of genetic material

As in sea urchins, the mammalian sperm that finally enters the egg carries its genetic contribution in a haploid pronucleus. In mammals, the process of pronuclear migration takes about 12 hours, compared with less than 1 hour in the sea urchin. The mammalian sperm enters almost tangentially to the surface of the egg rather than approaching it perpendicularly, and it fuses with numerous microvilli (see Figure 4.34A). The DNA of the sperm nucleus is bound by basic proteins called *protamines*, which are tightly compacted through disulfide bonds. Glutathione in the egg cytoplasm reduces these disulfide bonds and allows the uncoiling of the sperm chromatin (Calvin and Bedford 1971; Kvist et al. 1980; Perreault et al. 1988).

The mammalian sperm enters the oocyte while the oocyte nucleus is "arrested" in metaphase of its second meiotic division (**Figure 4.35A,B**; see also Figure 4.5). The calcium oscillations brought about by sperm entry inactivate MAP kinase and allow DNA synthesis. But unlike the sea urchin egg, which is already in a haploid state, the mammalian oocyte still has chromosomes in the middle of meiotic metaphase. Oscillations in the level of Ca^{2+} activate another kinase that leads to the proteolysis of cyclin (thus allowing the cell cycle to continue, eventually resulting in a haploid female pronucleus) and securin (the protein holding the metaphase chromosomes together) (Watanabe et al. 1991; Johnson et al. 1998). Mammals appear to undergo several waves of Ca^{2+} release; it is possible that events initiated by the first wave might not go to completion without additional calcium waves (Ducibella et al. 2002).

DNA synthesis occurs separately in the male and female pronuclei. The centrosome (new centriole) accompanying the male pronucleus produces its asters (largely from proteins stored in the oocyte). The microtubules join the two pronuclei and enable them to migrate toward one another other. Upon meeting, the two nuclear envelopes break down (**Figure 4.35C**). However, instead of producing a common zygote nucleus (as in sea urchins), the chromatin condenses into chromosomes that orient themselves on a common mitotic spindle (**Figure 4.35C,D**). Thus, a true diploid nucleus in mammals is first seen not in the zygote, but at the 2-cell stage.

Each sperm brings into the egg not only its pronucleus but also its mitochondria, its centriole, and a small amount of cytoplasm. The sperm mitochondria and their DNA are degraded in the egg cytoplasm, so that all of the new individual's mitochondria are derived from its mother. The egg and embryo appear to get rid of the paternal mitochondria both by dilution and by actively targeting them for destruction (Cummins et al. 1998; Shitara et al. 1998; Schwartz and Vissing 2002). However, in most mammals, the sperm centriole not only survives but serves as the organizing agent for making the new mitotic spindle.

The Nonequivalence of Mammalian Pronuclei

It is generally assumed that males and females carry equivalent haploid genomes. Indeed, one of the fundamental tenets of Mendelian genetics is that genes derived from the sperm are functionally equivalent to those derived from the egg. However, as we saw in Chapter 2, genomic imprinting can occur in mammals such that the sperm-derived genome and the egg-derived genome may be functionally different and play complementary roles during certain stages of development. This *imprinting* is thought to be caused by the different patterns of cytosine methylation on the genome.

The first evidence for nonequivalence came from studies of a human tumor called a **hydatidiform mole**, which resembles placental tissue. A majority of such moles have been shown to arise when a haploid sperm fertilizes an egg in which the female pronucleus is absent. After entering the egg, the sperm chromosomes duplicate themselves, thereby restoring the diploid chromosome number. However, the entire genome is derived from the sperm (Jacobs et al. 1980; Ohama et al. 1981). The cells survive, divide, and have a normal chromosome number, but development is abnormal. Instead of forming an embryo, the egg becomes a mass of placenta-like cells. Normal development does not occur when the entire genome comes from the male parent.

Normal development also does not occur when the genome is derived totally from the egg. The ability to develop an embryo without spermatic contribution is called **parthenogenesis** (Greek, "virgin birth"). The eggs of many invertebrates and some vertebrates are capable of developing normally in the absence of sperm (see Chapter 16). Mammals, however, do not exhibit parthenogenesis. Placing mouse oocytes in a culture medium that artificially activates the oocyte while suppressing the formation of the second polar body produces diploid mouse eggs whose genes are derived exclusively from the oocyte (Kaufman et al. 1977). These eggs divide to form embryos with spinal cords, muscles, skeletons, and organs, including beating hearts. However, development does not continue, and by day 10 or 11 (halfway through the mouse's gestation), these parthenogenetic embryos deteriorate. Neither human nor mouse development can be completed solely with egg-derived chromosomes.

That male and female pronuclei are both needed for normal development was also shown by pronuclear transplantation experiments (Surani and Barton 1983; McGrath and Solter 1984; Surani et al. 1986). Either male or female pronuclei can be removed from recently fertilized mouse eggs

TABLE 4.2 Pronuclear transplantation experiments

Class of reconstructed zygotes	Operation	Number of successful transplants	Number of progeny surviving
Bimaternal		339	0
Bipaternal		328	0
Control		348	18

Source: McGrath and Solter 1984.

and added to other recently fertilized eggs. (The two pronuclei can be distinguished at this stage because the female pronucleus is the one beneath the polar bodies.) Thus, zygotes with two male or two female pronuclei can be constructed. Although these eggs will form diploid cells that undergo normal cleavage, eggs whose genes are derived solely from sperm nuclei or solely from oocyte nuclei do not develop to birth. Control eggs under-going such transplantation (i.e., eggs containing one male and one female pronucleus taken from different zygotes) can develop normally (**Table 4.2**). Thus, for mammalian development to occur, both the sperm-derived and the egg-derived pronuclei are critical.

The importance of DNA methylation in this block to parthenogenesis was demonstrated when Kono and colleagues (2004) generated a female mouse whose genes came exclusively from two oocytes. To accomplish this feat, they had to mutate the DNA methylation system in one of the oocyte genomes to make it more like that of a male mouse, and then they had to perform two rounds of nuclear transfer. "Men," as one reviewer remarked, "do not need to fear becoming redundant any time soon" (Vogel 2004).

Activation of the mammalian egg

As in every other animal studied, a transient rise in cytoplasmic Ca^{2+} is necessary for egg activation in mammals. The sperm induces a series of Ca^{2+} waves that can last for hours, terminating in egg activation (i.e., the resumption of meiosis, cortical granule exocytosis, and the release of the inhibition on maternal mRNAs) and the formation of the male and female pronuclei. And, again as in sea urchins, fertilization triggers intracellular Ca^{2+} release through the production of IP_3 by the enzyme phospholipase C (Swann et al. 2006; Igarashi et al. 2007).

However, the mammalian PLC responsible for egg activation and pronucleus formation may in fact come from the sperm rather than from the egg. Some of the first observations for a sperm-derived PLC came from studies of intracytoplasmic sperm injection (ICSI), an experimental treatment for curing infertility. Here, sperm are directly injected into oocyte cytoplasm, bypassing any interaction with the egg plasma membrane. To the surprise of many biologists (who had assumed that sperm *binding* to some egg receptor protein was critical for egg activation), this treatment worked. The human egg was activated and pronuclei formed. Injecting mouse sperm into mouse eggs will also induce fertilization-like Ca^{2+} oscillations in the egg and lead to complete development (Kimura and Yanagimachi 1995).

It appeared that an activator of Ca^{2+} release was stored in the sperm head. This activator turned out to be a soluble sperm PLC enzyme, **PLCζ** (zeta), that is delivered to the egg by gamete fusion. In mice, expression of PLCζ mRNA in the egg produces Ca^{2+} oscillations, and removing PLCζ from mouse sperm (by antibodies or RNAi) abolishes the sperm's calcium-inducing activity (Saunders et al. 2002; Yoda et al. 2004; Knott et al. 2005). Human sperm that are unsuccessful in ICSI have been shown to have lit-tle or no functional PLCζ. In fact, normal human sperm can activate Ca^{2+} oscillations when injected into mouse eggs, but sperm lacking PLCζ do not (Yoon et al. 2008).

Whereas sea urchin eggs usually are activated as a single wave of Ca^{2+} crosses from the point of sperm entry, the mammalian egg is traversed by numerous waves of Ca^{2+} (Miyazaki et al. 1992; Ajduk et al. 2008; Ducibella and Fissore 2008). In mammals, the calcium ions released by IP_3 bind to a series of proteins including calmodulin-activated protein kinase (which will be important in eliminating the inhibitors of mRNA translation), MAP kinase (which allows the resumption of meiosis), and synaptotagmin (which helps initiate cortical granule fusion). Calcium ions that are not used are pumped back into the endoplasmic reticulum, and additional calcium ions are acquired from outside the cell. If mammalian oocytes are cultured in media without Ca^{2+}, the number and amplitude of oscillations decrease (Igusa and Miyazake 1983; Kline and Kline 1992). The extent (i.e., amplitude, duration, and number) of Ca^{2+} oscillations appears to regulate the timing of egg activation (Ozil et al. 2005; Toth et al. 2006). Thus, cortical granule exocytosis occurs before the resumption of meiosis and the translation of maternal mRNAs.

Coda

Fertilization is not a moment or an event, but a process of carefully orchestrated and coordinated events including the contact and fusion of gametes, the fusion of nuclei, and the activation of development. It is a process whereby two cells, each at the verge of death, unite to create a new organism that will have numerous cell types and organs. It is just the beginning of a series of cell-cell interactions that characterize animal development.

Snapshot Summary: *Fertilization*

1. Fertilization accomplishes two separate activities: sex (the combining of genes derived from two parents), and reproduction (the creation of a new organism).

2. The events of fertilization usually include (1) contact and recognition between sperm and egg; (2) regulation of sperm entry into the egg; (3) fusion of genetic material from the two gametes; and (4) activation of egg metabolism to start development.

3. The sperm head consists of a haploid nucleus and an acrosome. The acrosome is derived from the Golgi apparatus and contains enzymes needed to digest extracellular coats surrounding the egg. The midpiece and neck of the sperm contain mitochondria and the centriole that generates the microtubules of the flagellum. Energy for flagellar motion comes from mitochondrial ATP and a dynein ATPase in the flagellum.

4. The female gamete can be an egg (with a haploid nucleus, as in sea urchins) or an oocyte (in an earlier stage of development, as in mammals). The egg (or oocyte) has a large mass of cytoplasm storing ribosomes and nutritive proteins. Some mRNAs and proteins that will be used as morphogenetic factors are also stored in the egg. Many eggs also contain protective agents needed for survival in their particular environment.

5. Surrounding the egg cell membrane is an extracellular layer often used in sperm recognition. In most animals, this extracellular layer is the vitelline envelope. In mammals, it is the much thicker zona pellucida. Cortical granules lie beneath the egg's cell membrane.

6. Neither the egg nor the sperm is the "active" or "passive" partner. The sperm is activated by the egg, and the egg is activated by the sperm. Both activations involve calcium ions and membrane fusions.

7. In many organisms, eggs secrete diffusible molecules that attract and activate the sperm.

8. Species-specific chemotactic molecules secreted by the egg can attract sperm that are capable of fertilizing it. In sea urchins, the chemotactic peptides resact and speract have been shown to increase sperm motility and provide direction toward an egg of the correct species.

9. The acrosome reaction releases enzymes exocytotically. These proteolytic enzymes digest the egg's protective coating, allowing the sperm to reach and fuse with the egg cell membrane. In sea urchins, this reaction in the sperm is initiated by compounds in the egg jelly. Globular actin polymerizes to extend the acrosomal process. Bindin on the acrosomal process is recognized by a protein complex on the sea urchin egg surface.

10. Fusion between sperm and egg is probably mediated by protein molecules whose hydrophobic groups can merge the sperm and egg cell membranes. In sea urchins, bindin may mediate gamete fusion.

11. Polyspermy results when two or more sperm fertilize an egg. It is usually lethal, since it results in blastomeres with different numbers and types of chromosomes.

12. Many species have two blocks to polyspermy. The fast block is immediate and causes the egg membrane resting potential to rise. Sperm can no longer fuse with the egg. In sea urchins this is mediated by the influx of sodium ions. The slow block, or cortical granule reaction, is physical and is mediated by calcium ions. A wave of Ca^{2+} propagates from the point of sperm entry, causing the cortical granules to fuse with the egg cell membrane. The released contents of these granules cause the vitelline envelope to rise and harden into the fertilization envelope.

13. The fusion of sperm and egg results in the activation of crucial metabolic reactions in the egg. These reactions include reinitiation of the egg's cell cycle and subsequent mitotic division, and the resumption of DNA and protein synthesis.

14. Genetic material is carried in a male and a female pronucleus, which migrate toward each other. In sea urchins, the male and female pronuclei merge and a diploid zygote nucleus is formed. DNA replication occurs after pronuclear fusion.

15. In all species studied, free Ca^{2+}, supported by the alkalization of the egg, activates egg metabolism, protein synthesis, and DNA synthesis. Inositol trisphosphate (IP_3) is responsible for releasing Ca^{2+} from storage in the endoplasmic reticulum. DAG (diacylglycerol) is thought to initiate the rise in egg pH.

16. IP_3 is generated by phospholipases. Different species may use different mechanisms to activate the phospholipases.

17. Mammalian fertilization takes place internally, within the female reproductive tract. The cells and tissues of the female reproductive tract actively regulate the transport and maturity of both the male and female gametes.

18. The translocation of sperm from the vagina to the egg is regulated by the muscular activity of the uterus, by the binding of sperm in the isthmus of the oviduct, and by directional cues from the oocyte (immature egg) and/or the cumulus cells surrounding it.

19. Mammalian sperm must be capacitated in the female reproductive tract before they are capable of fertilizing the egg. Capacitation is the result of biochemical changes in the sperm cell membrane.

20. Capacitated mammalian sperm must penetrate the cumulus and bind to the zona pellucida before undergoing the acrosome reaction. In the mouse, this binding is mediated by ZP3 (zona protein 3) and several sperm proteins that recognize it.

21. ZP3 initiates the mammalian acrosome reaction on the zona pellucida, and the acrosomal enzymes are concentrated there.

22. In mammals, blocks to polyspermy include the modification of the zona proteins by the contents of the cortical granules so that sperm can no longer bind to the zona.

23. The rise in intracellular free Ca^{2+} at fertilization in amphibians and mammals causes the degradation of cyclin and the inactivation of MAP kinase, allowing the second meiotic metaphase to be completed and the formation of the haploid female pronucleus.

24. In mammals, DNA replication takes place as the pronuclei are traveling toward each other. The pronuclear membranes disintegrate as the pronuclei approach each other, and their chromosomes gather around a common metaphase plate.

25. The male and female pronuclei of mammals are not equivalent. If the zygote's genetic material is derived solely from one parent or the other, normal development will not take place. This difference in the male and female genomes is thought to be the result of different methylation patterns on the genes.

For Further Reading

Complete bibliographical citations for all literature cited in this chapter can be found at the free-access website **www.devbio.com**

Bleil, J. D. and P. M. Wassarman. 1980. Mammalian sperm and egg interaction: Identification of a glycoprotein in mouse-egg zonae pellucidae possessing receptor activity for sperm. *Cell* 20: 873–882.

Boveri, T. 1902. On multipolar mitosis as a means of analysis of the cell nucleus. [Translated by S. Gluecksohn-Waelsch.] In B. H. Willier and J. M. Oppenheimer (eds.), *Foundations of Experimental Embryology*. Hafner, New York, 1974.

Briggs, E. and G. M. Wessel. 2006. In the beginning…: Animal fertilization and sea urchin development. *Dev. Biol.* 300: 15–26.

Eisenbach, M. 1999. Mammalian sperm chemotaxis and its association with capacitation. *Dev. Genet.* 25: 87–94.

Florman, H. M. and B. T. Storey. 1982. Mouse gamete interactions: The zona pellucida is the site of the acrosome reaction leading to fertilization in vitro. *Dev. Biol.* 91: 121–130.

Glabe, C. G. and V. D. Vacquier. 1978. Egg surface glycoprotein receptor for sea urchin sperm bindin. *Proc. Natl. Acad. Sci. USA* 75: 881–885.

Jaffe, L. A. 1976. Fast block to polyspermy in sea urchins is electrically mediated. *Nature* 261: 68–71.

Just, E. E. 1919. The fertilization reaction in *Echinarachinus parma*. *Biol. Bull.* 36: 1–10.

Knott, J. G., M. Kurokawa, R. A. Fissore, R. M. Schultz and C. J. Williams. 2005. Transgenic RNA interference reveals role for mouse sperm phospholipase Cζ

in triggering Ca^{2+} oscillations during fertilization. *Biol Reprod.* 72: 992–996.

Parrington, J., L. C. Davis, A. Galione and G. Wessel. 2007. Flipping the switch: How a sperm activates the egg at fertilization. *Dev. Dyn.* 236: 2027–2038.

Swann, K., C. M. Saunders, N. T. Rogers, and F. A. Lai. 2006. PLCζ: A sperm protein that triggers Ca^{2+} oscillations and egg activation in mammals. *Semin. Cell Dev. Biol.* 17: 264–273.

Vacquier, V. D. and G. W. Moy. 1977. Isolation of bindin: The protein responsible for adhesion of sperm to sea urchin eggs. *Proc. Natl. Acad. Sci. USA* 74: 2456–2460.

Go Online

WEBSITE 4.1 Leeuwenhoek and images of homunculi. Scholars in the 1600s thought that either the sperm or the egg carried the rudiments of the adult body. Moreover, these views became distorted by contemporary commentators and later historians.

WEBSITE 4.2 The origins of fertilization research. Studies by Hertwig, Fol, Boveri, and Auerbach investigated fertilization by integrating cytology with genetics. The debates over meiosis and nuclear structure were critical in these investigations of fertilization.

WEBSITE 4.3 The egg and its environment. The laboratory is not where most eggs are found. Eggs have evolved remarkable ways to protect themselves in particular environments.

WEBSITE 4.4 The Lillie-Loeb dispute over sperm-egg binding. In the early 1900s, fertilization research was framed by a dispute between F. R. Lillie and Jacques Loeb, who disagreed over whether the sperm recognized the egg through soluble factors or through cell-cell interactions.

WEBSITE 4.5 Blocks to polyspermy. Theodore Boveri's analysis of polyspermy is a classic of experimental and descriptive biology. E. E. Just's delineation of the fast and slow blocks was a critical paper in embryology. Both papers are reprinted here, along with commentaries.

Vade Mecum

Gametogenesis. Stained sections of testis and ovary illustrate the process of gametogenesis, the streamlining of developing sperm, and the remarkable growth of the egg as it stores nutrients for its long journey. You can see this in movies and labeled photographs that take you at each step deeper into the mammalian gonad.

E. E. Just. Blocks to polyspermy were discovered in the early 1900s by the African American embryologist Ernest Just, who became one of the few embryologists ever to be honored on a postage stamp. The Sea Urchin segment contains videos of Just's work on sea urchin fertilization.

WEBSITE 4.6 Building the egg's extracellular matrix. In sea urchins, the cortical granules secrete not only hyalin but a number of proteins that construct the extracellular matrix of the embryo. This highly coordinated process results in sequential layers.

Sea urchin fertilization. The remarkable reactions that prevent polyspermy in a fertilized sea urchin egg can be seen in the raising of the fertilization envelope. The Sea Urchin segment contains movies of this event shown in real time.

Early Development in Selected Invertebrates

5

"THE GENERAL POLARITY, SYMMETRY, AND PATTERN of the embryo are egg characters which were determined before fertilization," wrote Edwin Grant Conklin in 1920. Indeed, as we will see, the egg cytoplasm plays a major role in determining patterns of cleavage, gastrulation, and cell specification. It does so by interacting with the nuclear genome established at fertilization.

Fertilization gave the organism a new genome and rearranged its cytoplasm. Now the zygote begins the production of a multicellular organism. During *cleavage*, rapid cell divisions divide the cytoplasm of the fertilized egg into numerous cells. These cells undergo dramatic displacements during *gastrulation*, a process whereby they move to different parts of the embryo and acquire new neighbors. During cleavage and gastrulation, the major axes of the embryo are determined and the embryonic cells begin to acquire their respective fates.

While cleavage always precedes gastrulation, axis formation in some species can begin as early as oocyte formation. It can be completed during cleavage (as in *Drosophila*) or extend all the way through gastrulation (as in *Xenopus*). Three body axes must be specified: the anterior-posterior (head-tail) axis; the dorsal-ventral (back-belly) axis; and the left-right axis. Different species specify these axes at different times, using different mechanisms. This chapter will look at the different ways these universal processes take place in some invertebrate embryos.

EARLY DEVELOPMENTAL PROCESSES: AN OVERVIEW

Cleavage

Once fertilization is complete, the development of a multicellular organism proceeds by a process called **cleavage**, a series of mitotic divisions whereby the enormous volume of egg cytoplasm is divided into numerous smaller, nucleated cells. These cleavage-stage cells are called blastomeres. In most species (mammals being the chief exception), both the initial rate of cell division and the placement of the blastomeres with respect to one another are under the control of the proteins and mRNAs stored in the oocyte. Only later do the rates of cell division and the placement of cells come under the control of the newly formed genome.

During the initial phase of development, when cleavage rhythms are controlled by maternal factors, the cytoplasmic volume does not increase. Rather, the zygote cytoplasm is divided into increasingly smaller cells. First the zygote

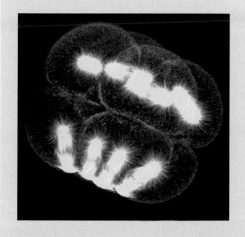

is divided in half, then quarters, then eighths, and so forth. Cleavage occurs very rapidly in most invertebrates, probably as an adaptation to generate a large number of cells quickly and to restore the somatic ratio of nuclear volume to cytoplasmic volume. The embryo often accomplishes this by abolishing the gap periods of the cell cycle (the G_1 and G_2 phases), when growth can occur. A frog egg, for example, can divide into 37,000 cells in just 43 hours. Mitosis in cleavage-stage *Drosophila* embryos occurs every 10 minutes for over 2 hours, and some 50,000 cells form in just 12 hours.

From fertilization to cleavage

As we saw in Chapter 4, fertilization activates protein synthesis, DNA synthesis, and the cell cycle. One of the most important events in this transition from fertilization to cleavage is the activation of **mitosis-promoting factor**, or **MPF**. MPF was first discovered as the major factor responsible for the resumption of meiotic cell divisions in the ovulated frog egg. It continues to play a role after fertilization, regulating the cell cycle of early blastomeres.

Blastomeres generally progress through a biphasic cell cycle consisting of just two steps: M (mitosis) and S (DNA synthesis) (**Figure 5.1**). The MPF activity of early blastomeres is highest during M and undetectable during S. The shift between the M and S phases in blastomeres is driven solely by the gain and loss of MPF activity. When MPF is microinjected into these cells, they enter M. Their nuclear envelope breaks down and their chromatin condenses into chromosomes. After an hour, MPF is degraded and the chromosomes return to S phase (Gerhart et al. 1984; Newport and Kirschner 1984).

What causes this cyclical activity of MPF? Mitosis-promoting factor consists of two subunits. The larger subunit, **cyclin B**, displays the cyclical behavior that is key to mitotic regulation, accumulating during S and being degraded after the cells have reached M (Evans et al. 1983; Swenson et al. 1986). Cyclin B is often encoded by mRNAs stored in the oocyte cytoplasm, and if the translation of this message is specifically inhibited, the cell will not enter mitosis (Minshull et al. 1989).

Cyclin B regulates the small subunit of MPF, the **cyclin-dependent kinase**. This kinase activates mitosis by phosphorylating several target proteins, including histones, the nuclear envelope lamin proteins, and the regulatory subunit of cytoplasmic myosin. It is the actions of this small kinase subunit that bring about chromatin condensation, nuclear envelope depolymerization, and the organization of the mitotic spindle. Without the cyclin B subunit, however, the cyclin-dependent kinase subunit of MPF will not function.

The presence of cyclin B is controlled by several proteins that ensure its periodic synthesis and degradation. In most species studied, the regulators of cyclin B (and thus of MPF) are stored in the egg cytoplasm. Therefore, the cell cycle remains independent of the nuclear genome for a number of cell divisions. These early divisions tend to be rapid and synchronous. However, as the cytoplasmic components are used up, the nucleus begins to synthesize them. In several

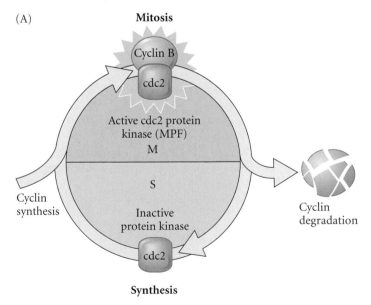

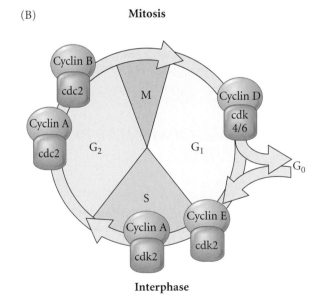

FIGURE 5.1   Cell cycles of somatic cells and early blastomeres. (A) The biphasic cell cycle of early amphibian blastomeres has only two states, S and M. Cyclin B synthesis allows progression to M (mitosis), while degradation of cyclin B allows cells to pass into S (synthesis) phase. (B) The complete cell cycle of a typical somatic cell. Mitosis (M) is followed by an interphase stage. Interphase is subdivided into G_1, S (synthesis), and G_2 phases. Cells that are differentiating are usually taken "out" of the cell cycle and are in an extended G_1 phase called G_0. The cyclins responsible for the progression through the cell cycle and their respective kinases are shown at their point of cell cycle regulation. (B after Nigg 1995.)

(A)

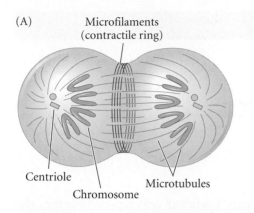

Microfilaments
(contractile ring)

Centriole

Chromosome

Microtubules

(B)

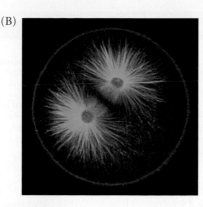

FIGURE 5.2 Role of microtubules and microfilaments in cell division. (A) Diagram of first-cleavage telophase in a sea urchin egg. The chromosomes are being drawn to the centrioles by microtubules, while the cytoplasm is being pinched in by the contraction of microfilaments. (B) Confocal fluorescent image of an echinoderm embryo undergoing first cleavage (early anaphase). The microtubules are stained green, actin microfilaments are stained red. (C) Confocal fluorescent image of sea urchin embryo at the very end of first cleavage. Microtubules are orange; actin proteins (both unpolymerized and in microfilaments) are blue. (B,C courtesy of G. von Dassow and the Center for Cell Dynamics.)

(C)

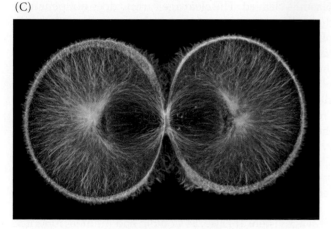

species, the embryo now enters a **mid-blastula transition,** in which several new properties are added to the biphasic cell divisions of the embryo. First, the "gap" stages (G$_1$ and G$_2$) are added to the cell cycle (see Figure 5.1B). *Xenopus* embryos add G$_1$ and G$_2$ phases to the cell cycle shortly after the twelfth cleavage. *Drosophila* adds G$_2$ during cycle 14 and G$_1$ during cycle 17 (Newport and Kirschner 1982; Edgar et al. 1986). Second, the synchronicity of cell division is lost, because different cells synthesize different regulators of MPF. After numerous synchronous rounds of mitosis, the cells begin to "go their own way." Third, new mRNAs are transcribed. Many of these messages encode proteins that will become necessary for gastrulation. In several species, if transcription is blocked cell division will still occur at normal rates and times, but the embryo will not be able to initiate gastrulation. Many of these new messenger RNAs are also used for cell specification. As we will see in sea urchin embryos, the new mRNA expression patterns of the mid-blastula transition map out territories where specific types of cells will later differentiate.

The cytoskeletal mechanisms of mitosis

Cleavage is the result of two coordinated processes. The first of these is **karyokinesis,** the mitotic division of the cell's nucleus. The mechanical agent of karyokinesis is the mitotic spindle, with its microtubules composed of tubu-

lin (the same type of protein that makes up the sperm flagellum). The second process is **cytokinesis:** the division of the cell. The mechanical agent of cytokinesis is a contractile ring of microfilaments made of actin (the same type of protein that extends the egg microvilli and the sperm acrosomal process). Table 5.1 presents a comparison of these agents of cell division. The relationship and coordination between these two systems during cleavage are depicted in **Figure 5.2A,** in which a sea urchin egg is shown undergoing first cleavage. The mitotic spindle and contractile ring are perpendicular to each other, and the spindle is internal to the contractile ring. The contractile ring creates a cleavage furrow, which eventually bisects the plane of mitosis, thereby creating two genetically equivalent blastomeres.

The actin microfilaments are found in the cortex (outer cytoplasm) of the egg rather than in the central cytoplasm. Under the electron microscope, the ring of microfilaments can be seen forming a distinct cortical band 0.1 μm wide (**Figure 5.2B,C**). This contractile ring exists only during cleavage and extends 8–10 μm into the center of the egg. It is responsible for exerting the force that splits the zygote into blastomeres; if the ring is disrupted, cytokinesis stops. Schroeder (1973) likened the contractile ring to an "intracellular purse-string," tightening about the egg as cleavage continues. This tightening of the microfilamentous ring creates the **cleavage furrow.** Microtubules are also seen near the cleavage furrow (in addition to their role in creating the mitotic spindle), since they are needed to bring membrane material to the site of membrane addition (Danilchik et al. 1998).

Although karyokinesis and cytokinesis are usually coordinated, they are sometimes modified by natural or experimental conditions. The placement of the centrioles is critical in orienting the mitotic spindle, and thus the division plane of the blastomeres. Depending on the placement of the centrioles, the blastomeres can separate either into dorsal and ventral daughter cells, anterior and posterior daughter cells, or left and right daughter cells. The spin-

TABLE 5.1 Karyokinesis and cytokinesis

Process	Mechanical agent	Major protein composition	Location	Major disruptive drug
Karyokinesis	Mitotic spindle	Tubulin microtubules	Central cytoplasm	Colchicine, nocodazole[a]
Cytokinesis	Contractile ring	Actin microfilaments	Cortical cytoplasm	Cytochalasin B

[a]Because colchicine has been found to independently inhibit several membrane functions, including osmoregulation and the transport of ions and nucleosides, nocodazole has become the major drug used to inhibit microtubule-mediated processes (see Hardin 1987).

dle can even be at an angle such that one daughter cell is clockwise or counterclockwise to the other. As we will learn Chapter 6, cleavage in insect eggs consists of karyokinesis several times before cytokinesis takes place. In this manner, numerous nuclei exist within the same cell. (The outer membrane of that one big cell will eventually indent to separate the nuclei and form individual cells.)

Patterns of embryonic cleavage

In 1923, embryologist E. B. Wilson reflected on how little we knew about cleavage: "To our limited intelligence, it would seem a simple task to divide a nucleus into equal parts. The cell, manifestly, entertains a very different opinion." Indeed, different organisms undergo cleavage in distinctly different ways. The pattern of embryonic cleavage peculiar to a species is determined by two major parameters: (1) the amount and distribution of yolk protein within the cytoplasm, and (2) factors in the egg cytoplasm that influence the angle of the mitotic spindle and the timing of its formation.

The amount and distribution of yolk determine where cleavage can occur and the relative size of the blastomeres. In general, yolk inhibits cleavage. When one pole of the egg is relatively yolk-free, cellular divisions occur there at a faster rate than at the opposite pole. The yolk-rich pole is referred to as the **vegetal pole**; the yolk concentration in the **animal pole** is relatively low. The zygote nucleus is frequently displaced toward the animal pole. **Figure 5.3** provides a classification of cleavage types and shows the influence of yolk on cleavage symmetry and pattern.

At one extreme are the eggs of sea urchins, mammals, and snails. These eggs have sparse, equally distributed yolk and are thus **isolecithal** (Greek, "equal yolk"). In these species, cleavage is **holoblastic** (Greek holos, "complete"), meaning that the cleavage furrow extends through the entire egg. With little yolk, these embryos must have some other way of obtaining food. Most will generate a voracious larval form, while mammals will obtain their nutrition from the maternal placenta.

At the other extreme are the eggs of insects, fish, reptiles, and birds. Most of their cell volumes are made up of yolk. The yolk must be sufficient to nourish these animals throughout embryonic development. Zygotes containing large accumulations of yolk undergo **meroblastic** cleavage

(Greek meros, "part"), wherein only a portion of the cytoplasm is cleaved. The cleavage furrow does not penetrate the yolky portion of the cytoplasm because the yolk platelets impede membrane formation there. Insect eggs have yolk in the center (i.e., they are **centrolecithal**), and the divisions of the cytoplasm occur only in the rim of cytoplasm, around the periphery of the cell (i.e., **superficial cleavage**). The eggs of birds and fish have only one small area of the egg that is free of yolk (**telolecithal** eggs), and therefore the cell divisions occur only in this small disc of cytoplasm, giving rise to the **discoidal** pattern of cleavage. These are general rules, however, and even closely related species have evolved different patterns of cleavage in different environments.

Yolk is just one factor influencing a species' pattern of cleavage. There are also, as Conklin had intuited, inherited patterns of cell division superimposed on the constraints of the yolk. The importance of this inheritance can readily be seen in isolecithal eggs. In the absence of a large concentration of yolk, holoblastic cleavage takes place, and four major patterns of this cleavage type can be observed: radial, spiral, bilateral, and rotational holoblastic cleavage. We will see examples of all of these cleavage patterns as this chapter takes a more detailed look at four different invertebrate groups.

Gastrulation

The blastula consists of numerous cells, the positions of which were established during cleavage. During **gastrulation**, these cells are given new positions and new neighbors, and the multilayered body plan of the organism is established. The cells that will form the endodermal and mesodermal organs are brought to the inside of the embryo, while the cells that will form the skin and nervous system are spread over its outside surface. Thus, the three germ layers—outer ectoderm, inner endoderm, and interstitial mesoderm—are first produced during gastrulation. In addition, the stage is set for the interactions of these newly positioned tissues.

Gastrulation usually involves some combination of several types of movements (**Figure 5.4**). These movements involve the entire embryo, and cell migrations in one part of the gastrulating embryo must be intimately coordinated with other movements that are taking place simultane-

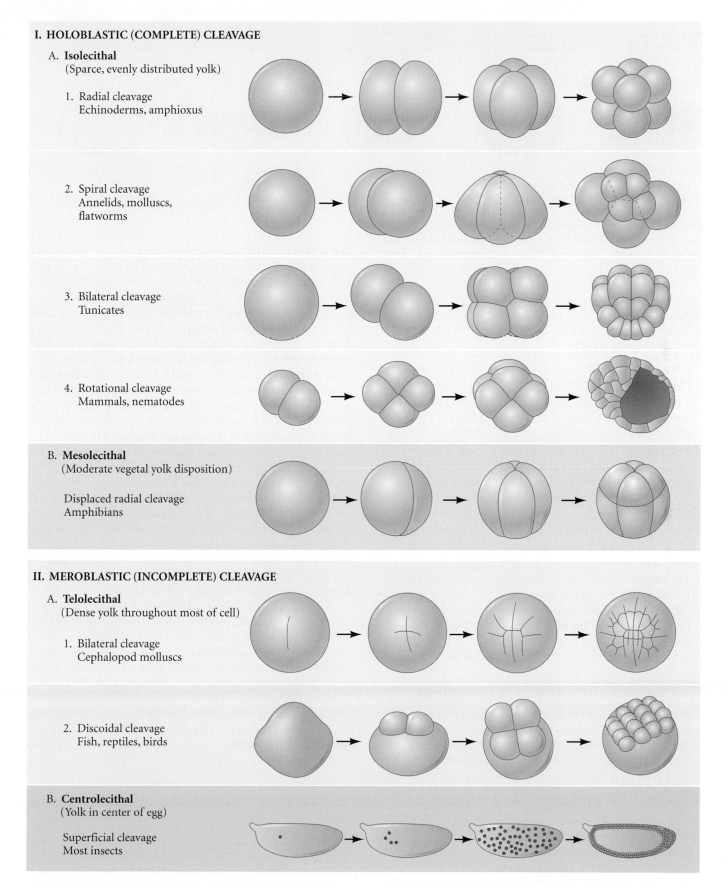

I. HOLOBLASTIC (COMPLETE) CLEAVAGE

A. **Isolecithal**
   (Sparce, evenly distributed yolk)

   1. Radial cleavage
      Echinoderms, amphioxus

   2. Spiral cleavage
      Annelids, molluscs,
      flatworms

   3. Bilateral cleavage
      Tunicates

   4. Rotational cleavage
      Mammals, nematodes

B. **Mesolecithal**
   (Moderate vegetal yolk disposition)

   Displaced radial cleavage
   Amphibians

II. MEROBLASTIC (INCOMPLETE) CLEAVAGE

A. **Telolecithal**
   (Dense yolk throughout most of cell)

   1. Bilateral cleavage
      Cephalopod molluscs

   2. Discoidal cleavage
      Fish, reptiles, birds

B. **Centrolecithal**
   (Yolk in center of egg)

   Superficial cleavage
   Most insects

FIGURE 5.3   Summary of the main patterns of cleavage.

Invagination:
Infolding of cell
sheet into embryo

Example:
Sea urchin
endoderm

Involution:
Inturning of cell sheet
over the basal surface
of an outer layer

Example:
Amphibian
mesoderm

Ingression:
Migration of
individual cells
into the embryo

Example:
Sea urchin mesoderm,
Drosophila neuroblasts

Delamination:
Splitting or migration
of one sheet into
two sheets

Example:
Mammalian and bird
hypoblast formation

Epiboly:
The expansion of
one cell sheet
over other cells

Example: Ectoderm
formation in amphibians,
sea urchins, and tunicates

FIGURE 5.4   Types of cell movements during gastrulation. The gastrulation of any particular organism is an ensemble of several of these movements.

ously. Although patterns of gastrulation vary enormously throughout the animal kingdom, there are only a few basic types of cell movements:

- **Invagination:** The infolding of a region of cells, much like the indenting of a soft rubber ball when it is poked.
- **Involution:** The inturning or inward movement of an expanding outer layer so that it spreads over the internal surface of the remaining external cells.
- **Ingression:** The migration of individual cells from the surface layer into the interior of the embryo. The cells become mesenchymal (i.e., they separate from one another) and migrate independently.
- **Delamination**. The splitting of one cellular sheet into two more or less parallel sheets. While on a cellular basis it resembles ingression, the result is the formation of a new sheet of cells.
- **Epiboly**. The movement of epithelial sheets (usually of ectodermal cells) that spread as a unit (rather than individually) to enclose the deeper layers of the embryo. Epiboly can occur by the cells dividing, by the cells changing their shape, or by several layers of cells intercalating into fewer layers. Often, all three mechanisms are used.

Cell Specification and Axis Formation

Cell fates are specified by cell-cell signaling or by the asymmetric distribution of patterning molecules into particular cells. These unevenly distributed molecules are usually transcription factors that activate or repress the transcription of specific genes in those cells that acquire them. Such asymmetric distributions of patterning molecules begin during cleavage and generally follow one of three mechanisms: (1) the molecules are bound to the egg cytoskeleton and are passively acquired by the cells that obtain this cytoplasm; (2) the molecules are actively transported along the cytoskeleton to one particular cell; or (3) the molecules become associated with a specific centrosome and follow that centrosome into one of the two mitotic sister cells (Lam-

bert and Nagy 2002). Once asymmetry has been established, one cell can specify a neighboring cell by paracrine or juxtacrine interactions at the cell surface (see Chapter 3).

Embryos must develop three crucial axes that are the foundation of the body: the anterior-posterior axis, the dorsal-ventral axis, and the right-left axis (**Figure 5.5**). The **anterior-posterior** (or **anteroposterior**) **axis** is the line extending from head to tail (or mouth to anus in those organisms that lack a head and tail). The **dorsal-ventral (dorsoventral) axis** is the line extending from back (dorsum) to belly (ventrum). For instance, in vertebrates, the neural tube is a dorsal structure. In insects, the neural cord is a ventral structure. The **right-left axis** is a line between the two lateral sides of the body. Although humans (for example) may look symmetrical, recall that in most of us, the heart and liver are in the left half of the body only. Somehow, the embryo knows that some organs belong on one side and other organs go on the other.

In this chapter, we will look at how four invertebrate groups—sea urchins (echinoderms), snails (gastropod molluscs), ascidians (tunicates), and *C. elegans* (a well-studied species of nematode worm)—undergo cleavage, gastrulation, axis specification, and cell fate determination. These four groups have been important **model systems** for developmental biologists. In other words, they are easily maintained in the laboratory, and they have special properties* that allow their mechanisms of development to be readily observed. They also represent a wide variety of cleavage types, patterns of gastrulation, and ways of specifying axes and cell fates.

Despite their differences, the embryos of these four invertebrate groups are all characterized by what Davidson (2001) has called "Type I embryogenesis," which includes:

- Immediate activation of the zygotic genes

---

*These properties include quick generation time, large litters, amenability to genetic and surgical manipulation, and the ability to develop under laboratory conditions. However, this very ability to develop in the laboratory sometimes precludes our asking certain questions concerning the relationship of development to an organism's natural habitat. These questions will be addressed in Chapter 18.

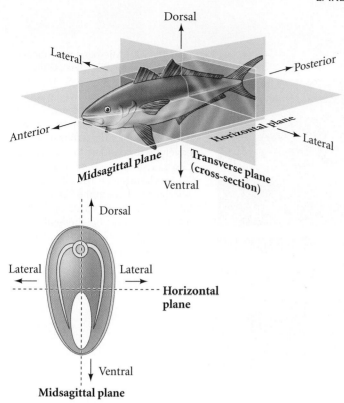

FIGURE 5.5 Axes of a bilaterally symmetrical animal. A single plane, the midsagittal plane, divides the animal into left and right halves. Cross sections are taken along the anterior-posterior axis.

EARLY DEVELOPMENT IN SEA URCHINS

Sea Urchin Cleavage

Sea urchins exhibit radial holoblastic cleavage (**Figures 5.6 and 5.7**). The first seven cleavage divisions are "stereotypic" in that the same pattern is followed in every individual of the same species: the first and second cleavages are both meridional and are perpendicular to each other (that is to say, the cleavage furrows pass through the animal and vegetal poles). The third cleavage is equatorial, perpendicular to the first two cleavage planes, and separates the animal and vegetal hemispheres from each other (see Figure 5.6A, top row, and Figure 5.7A–C). The fourth cleavage, however, is very different from the first three. The four cells of the animal tier divide meridionally into eight blastomeres, each with the same volume. These eight cells are called **mesomeres**. The vegetal tier, however, undergoes an unequal equatorial cleavage (see Figure 5.6B) to produce four large cells—the **macromeres**—and four smaller **micromeres** at the vegetal pole. As the 16-cell embryo cleaves, the eight "animal" mesomeres divide equatorially to produce two tiers, an_1 and an_2, one staggered above the other. The macromeres divide meridionally, forming a tier of eight cells below an_2. Somewhat later, the micromeres divide unequally, producing a cluster of four **small**

- Rapid specification of the blastomeres by the products of the zygotic genes and by maternally active genes
- A relatively small number of cells (a few hundred or less) at the start of gastrulation

FIGURE 5.6 Cleavage in the sea urchin. (A) Planes of cleavage in the first three divisions and the formation of tiers of cells in divisions 3–6. (B) Confocal fluorescence micrograph of the unequal cell division that initiates the 16-cell stage (asterisk in A), highlighting the unequal equatorial cleavage of the vegetal blastomeres to produce the micromeres and macromeres. (B courtesy of G. van Dassow and the Center for Cell Dynamics.)

(A)

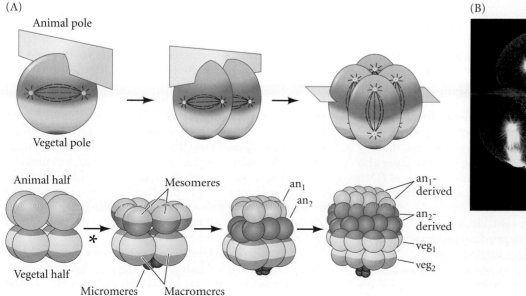

(B)

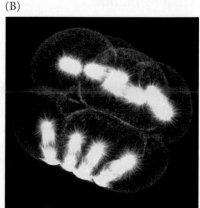

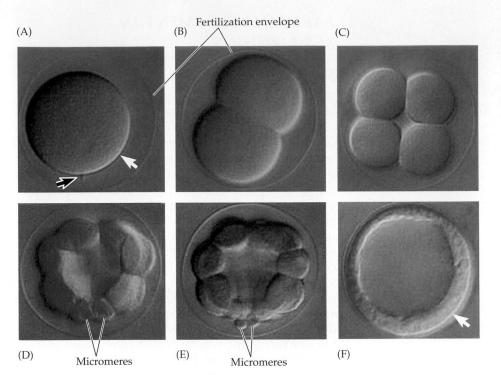

(A)

(B) Fertilization envelope

(C)

(D) Micromeres

(E) Micromeres

(F)

FIGURE 5.7 Micrographs of cleavage in live embryos of the sea urchin *Lytechinus variegatus*, seen from the side. (A) The 1-cell embryo (zygote). The site of sperm entry is marked with a black arrow; a white arrow marks the vegetal pole. The fertilization envelope surrounding the embryo is clearly visible. (B) 2-cell stage. (C) 8-cell stage. (D) 16-cell stage. Micromeres have formed at the vegetal pole. (E) 32-cell stage. (F) The blastula has hatched from the fertilization envelope. The future vegetal plate (arrow) is beginning to thicken. (Courtesy of J. Hardin.)

micromeres at the tip of the vegetal pole, beneath a tier of four **large micromeres**. The small micromeres divide once more, then cease dividing until the larval stage. At the sixth division, the animal hemisphere cells divide meridionally while the vegetal cells divide equatorially; this pattern is reversed in the seventh division (see Figure 5.6A, bottom row). At that time, the embryo is a 120-cell* blastula, in which the cells form a hollow sphere surrounding a central cavity, or **blastocoel** (see Figure 5.7F). From here on, the pattern of divisions becomes less regular.

Blastula formation

By the blastula stage of sea urchin development, all the cells are the same size, the micromeres having slowed down their cell divisions. Every cell is in contact with the proteinaceous fluid of the blastocoel on the inside and with the hyaline layer on the outside. Tight junctions unite the once loosely connected blastomeres into a seamless epithelial sheet that completely encircles the blastocoel. As the cells continue to divide, the blastula remains one cell layer thick, thinning out as it expands. This is accomplished by the adhesion of the blastomeres to the hyaline layer and by an influx of water that expands the blastocoel (Dan 1960; Wolpert and Gustafson 1961; Ettensohn and Ingersoll 1992).

These rapid and invariant cell cleavages last through the ninth or tenth division, depending on the species. By

this time, the fates of the cells have become specified (discussed in the next section), and each cell becomes ciliated on the region of the cell membrane farthest from the blastocoel. This ciliated blastula begins to rotate within the fertilization envelope. Soon afterward, differences are seen in the cells. The cells at the vegetal pole of the blastula begin to thicken, forming a **vegetal plate** (see Figure 5.7F). The cells of the animal hemisphere synthesize and secrete a hatching enzyme that digests the fertilization envelope (Lepage et al. 1992). The embryo is now a free-swimming **hatched blastula**.

Fate maps and the determination of sea urchin blastomeres

The first fate maps of the sea urchin embryo followed the descendants of each of the 16-cell-stage blastomeres. More recent investigations have refined these maps by following the fates of individual cells that have been injected with fluorescent dyes such as diI. These dyes "glow" in the injected cells' progeny for many cell divisions (see Chapter 1). Such studies have shown that by the 60-cell stage, most of the embryonic cell fates are specified, but the cells are not irreversibly committed. In other words, particular blastomeres consistently produce the same cell types in each embryo, but these cells remain pluripotent and can give rise to other cell types if experimentally placed in a different part of the embryo.

A fate map of the 60-cell sea urchin embryo is shown in **Figure 5.8**. The animal half of the embryo consistently gives rise to the ectoderm—the larval skin and its neurons. The

---

*You might have been expecting a 128-cell embryo; but remember that the small micromeres stopped dividing.

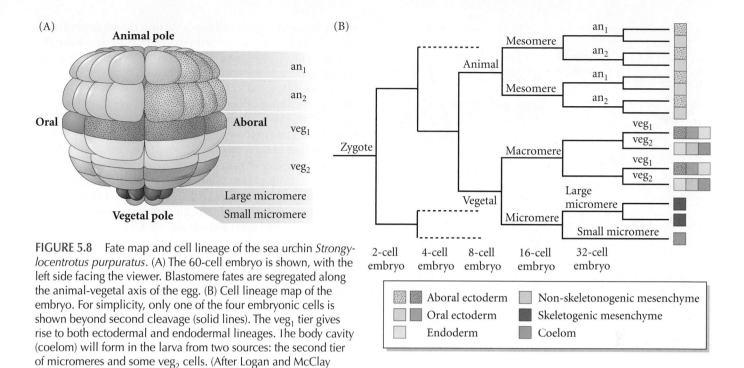

FIGURE 5.8 Fate map and cell lineage of the sea urchin *Strongylocentrotus purpuratus*. (A) The 60-cell embryo is shown, with the left side facing the viewer. Blastomere fates are segregated along the animal-vegetal axis of the egg. (B) Cell lineage map of the embryo. For simplicity, only one of the four embryonic cells is shown beyond second cleavage (solid lines). The veg_1 tier gives rise to both ectodermal and endodermal lineages. The body cavity (coelom) will form in the larva from two sources: the second tier of micromeres and some veg_2 cells. (After Logan and McClay 1999; Wray 1999.)

veg_1 layer produces cells that can enter into either the ectodermal or the endodermal organs. The veg_2 layer gives rise to cells that can populate three different structures—the endoderm, the coelom (internal mesodermal body wall), and the **non-skeletogenic mesenchyme** (sometimes called secondary mesenchyme), which generates pigment cells, immunocytes, and muscle cells. The first tier of micromeres (the large micromeres) produces the **skeletogenic mesenchyme** (also called primary mesenchyme), which forms the larval skeleton. The second-tier micromeres (i.e., the small micromeres) play no role in embryonic development but appear to contribute cells to the larval coelom (Logan and McClay 1997, 1999; Wray 1999).

The fates of the different cell layers are determined in a two-step process. First, the large micromeres become *autonomously* specified. They inherit maternal determinants that had been deposited at the vegetal pole of the egg and which become incorporated into the four micromeres at fourth cleavage. These cells are thus determined to become skeletogenic mesenchyme cells that will leave the blastula epithelium to enter the blastocoel, migrate to particular positions along the blastocoel wall, and then differentiate into the larval skeleton.

Second, the large micromeres produce paracrine and juxtacrine factors that specify the fates of their neighbors. The micromeres are able to produce a signal that tells the cells above them to become endoderm and induces them to invaginate into the embryo. The ability of the micromeres to produce a signal that changes the fates of the embryonic cells is so pronounced that if micromeres are

removed from the embryo and placed on top of an isolated animal cap (the top two animal tiers that usually become ectoderm), the animal cap cells will generate endoderm and a more or less normal larva will develop (**Figure 5.9**; Hörstadius 1939).

These skeletogenic micromeres are the first cells whose fates are determined autonomously. If micromeres are isolated from the 16-cell embryo and placed in test tubes, they will divide the appropriate number of times and produce the skeletal spicules (Okazaki 1975). Thus, the isolated micromeres do not need any other signals to generate their skeletal fates. Moreover, if skeletogenic micromeres are transplanted into the animal region of the blastula, not only will their descendants form skeletal spicules, but the transplanted micromeres will alter the fates of nearby cells by inducing a secondary site for gastrulation. Cells that would normally have produced ectodermal skin cells will be respecified as endoderm and will produce a secondary gut (**Figure 5.10**; Hörstadius 1973; Ransick and Davidson 1993). Therefore, the inducing ability of the micromeres is also established autonomously.

Global regulatory networks and skeletogenic mesenchyme specification

Heredity, according to the embryologist E. B. Wilson, is the transmission from generation to generation of a particular pattern of development. And evolution is the hereditary alteration of such a plan. Even though Wilson was probably the first scientist to write (in 1895) that the hereditary

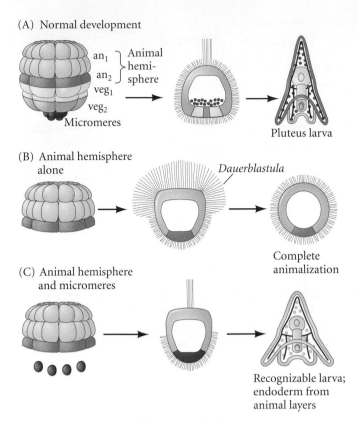

FIGURE 5.9 Ability of micromeres to induce presumptive ectodermal cells to acquire other fates. (A) Normal development of the 64-cell sea urchin embryo, showing the fates of the different layers. (B) An isolated animal hemisphere becomes a ciliated ball of ectodermal cells. (C) When an isolated animal hemisphere is combined with isolated micromeres, a recognizable pluteus larva is formed, with all the endoderm derived from the animal hemisphere. (After Hörstadius 1939.)

instructions were encoded in DNA and transmitted at fertilization through the chromosomes, he had no mechanism for figuring out how the DNA could encode such instructions. Recent studies from the sea urchin developmental biology community, spearheaded by Eric Davidson's laboratory, have started to show how this is done.

Davidson's group has pioneered a network approach to development, where they envision *cis*-regulatory elements (such as enhancers) in a logic circuit connected to each other by transcription factors (see http://sugp.caltech.edu/endomes; Davidson and Levine 2008; Oliveri et al. 2008). The network would receive its first inputs from transcription factors located in the maternal cytoplasm; from then on, the network would self-assemble from (1) the ability of the transcription factors to regulate only certain other factors due to their *cis*-regulatory elements and (2) the ability of these transcription factors to activate paracrine signaling pathways that would activate specific transcription factors in neighboring cells. These studies are trying to show the global regulatory logic by which the genes of the

sea urchin specify the cell types during development, and we will focus on the earliest part of that logic—how the skeletogenic mesenchyme cells receive their developmental fate and their interactive properties.

The skeletogenic mesenchyme cells of the sea urchin embryo are those cells that are specified autonomously to ingress into the blastocoel and become the skeleton of the sea urchin larva. Moreover, as we have seen, they are also the cells that induce their neighbors to become endoderm (gut) and non-skeletogenic mesenchyme (pigment; coelom) cells. By a combination of in situ hybridization, inhibition of normal gene expression, and the addition of active genes to different cell types, the regulatory regions and their interactions have been found. Nearly all the genes of this regulatory network have been identified and found to act as described.

The specification of the micromere lineage (and hence the rest of the embryo) begins inside the undivided egg.

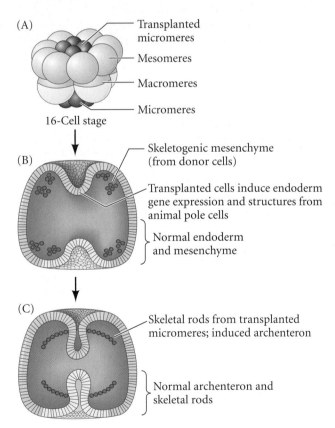

FIGURE 5.10 Ability of micromeres to induce a secondary axis in sea urchin embryos. (A) Micromeres are transplanted from the vegetal pole of a 16-cell embryo into the animal pole of a host 16-cell embryo. (B) The transplanted micromeres invaginate into the blastocoel to create a new set of skeletogenic mesenchyme cells, and they induce the animal cells next to them to become vegetal plate endoderm cells. (C) The transplanted micromeres differentiate into skeletal cables, while the induced animal cap cells form a secondary archenteron. Meanwhile, gastrulation proceeds normally from the original vegetal plate of the host. (After Ransick and Davidson 1993.)

The initial regulatory inputs are the transcription factors Disheveled and β-catenin, both of which are found in the cytoplasm and are inherited by the micromeres as soon as they are formed at the fourth cleavage. During oogenesis, Disheveled becomes located in the vegetal cortex of the egg (**Figure 5.11A**; Weitzel et al. 2004; Leonard and Ettensohn 2007), where it prevents the degradation of β-catenin in the micromere and macromere cells. The β-catenin is therefore allowed to enter into the nucleus, where it combines with the TCF transcription factor to activate gene expression.

Several pieces of evidence suggest that β-catenin specifies the micromeres. First, during normal sea urchin devel-

opment, β-catenin accumulates in the nuclei of those cells fated to become endoderm and mesoderm (**Figure 5.11B**). This accumulation is autonomous and can occur even if the micromere precursors are separated from the rest of the embryo. Second, this nuclear accumulation appears to be responsible for specifying the vegetal half of the embryo. It is possible that the levels of nuclear β-catenin accumulation help determine the mesodermal and endodermal

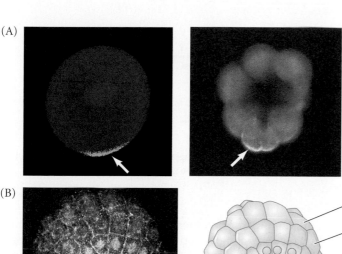

FIGURE 5.11 Role of Disheveled and β-catenin proteins in specifying the vegetal cells of the sea urchin embryo. (A) Localization of Disheveled (arrows) in the vegetal cortex of the sea urchin oocyte before fertilization (left) and in the region of a 16-cell embryo about to become the micromeres (right). (B) During normal development, β-catenin accumulates predominantly in the micromeres and somewhat less in the veg_2 tier cells. (C) In embryos treated with lithium chloride, β-catenin accumulates in the nuclei of all blastula cells (probably by LiCl blocking the GSK3 enzyme of the Wnt pathway), and the animal cells become specified as endoderm and mesoderm. (D) When β-catenin is prevented from entering the nuclei (i.e., it remains in the cytoplasm), the vegetal cell fates are not specified and the entire embryo develops as a ciliated ectodermal ball. (A,B from Weitzel et al. 2004, courtesy of C. Ettensohn; B–D from Logan et al. 1998, courtesy of D. McClay.)

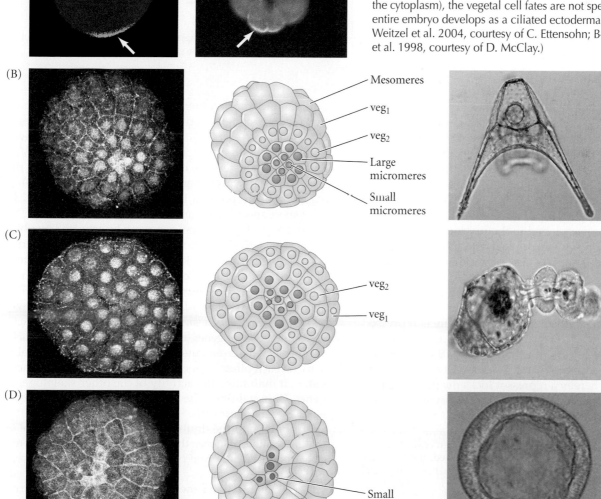

(A)

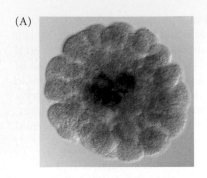

FIGURE 5.12 Simplified, double-negative gated "circuit" for micromere specification. (A) In situ hybridization reveals the accumulation of *Pmar1* mRNA (dark purple) in the micromeres. (B) Otx and β-catenin from the maternal cytoplasm are concentrated in at the vegetal pole of the egg. These transcriptional regulators are inherited by the micromeres and activate the *Pmar1* gene. *Pmar1* encodes a repressor of *HesC*, which in turn encodes a repressor (hence the "double-negative") of several genes involved in micromere specification (e.g., *Alx1*, *Tbr*, and *Ets*). Genes encoding signaling proteins (e.g., *Delta*) are also under the control of HesC. In the micromeres, where activated Pmar1 represses the HesC repressor, the micromere specification and signaling genes are active. In the veg$_2$ cells, *Pmar1* is not activated and the HesC gene product shuts down the skeletogenic genes; however, those cells containing Notch can respond to the Delta signal from the skeletogenic mesenchyme. U, ubiquitous activating transcription factors. (After Oliveri et al. 2008; photograph courtesy of P. Oliveri.)

(B)

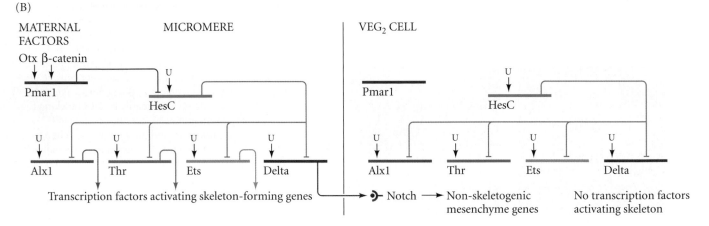

fates of the vegetal cells. Treating sea urchin embryos with lithium chloride causes the accumulation of β-catenin in every cell and transforms the presumptive ectoderm into endoderm. Conversely, experimental procedures that inhibit β-catenin accumulation in the vegetal cell nuclei prevent the formation of endoderm and mesoderm (**Figure 5.11C,D**; Logan et al. 1998; Wikramanayake et al. 1998).

The next regulatory input comes from the Otx transcription factor that is enriched in the micromere cytoplasm. The β-catenin/TCF and Otx interact at the enhancer of the *Pmar1* genes to activate their transcription in the micromeres shortly after their formation (**Figure 5.12A**; Oliveri et al. 2008). Pmar1 protein is a transcriptional repressor of *HesC*, a gene that encodes another repressive transcription factor. When the *Pmar1* gene is repressed, HesC is expressed in every nucleus of the sea urchin embryo *except* those of the micromeres.* In the micromeres, where *Pmar1* is activated, the *HesC* gene is repressed. This mechanism, whereby a repressor locks the genes of specification and these genes can be unlocked by the repressor

of that repressor (in other words, when activation occurs by the repression of a repressor), is called a **double-negative gate** (**Figure 5.12B, 5.13A**). Such a gate allows for tight regulation of fate specification: it promotes the expression of these genes where the input occurs, and it represses the same genes in every other cell type (Oliveri et al. 2008).

The genes repressed by HesC are those involved in micromere specification and differentiation: *Alx1*, *Ets1*, *Tbr*, *Tel*, and *SoxC*. Each of these genes can be activated by ubiquitous transcription factors, but these positive transcription factors cannot work while HesC repressor protein binds to their respective enhancers. When the Pmar1 protein is present, it represses *HesC*, and all these genes become active (Revilla-i-Domingo et al. 2007). The newly activated genes synthesize transcription factors that activate another set of genes, most of which are genes that activate skeletal determinants. These transcription factors also activate each other's genes, so that once one factor is activated, it maintains the activity of the other skeletogenic genes. This stabilizes the regulatory state of the skeletogenic mesenchyme cells.

In contrast to the double-negative gate, the control of the differentiation genes that make the sea urchin skeleton operates on a *feedforward* process (**Figure 5.13B**). Here, regulatory gene A produces a transcription factor that is needed for differentiation gene C and also activates regulatory gene B, which produces a transcription factor also needed for differentiation gene C. The skeletogenic portion of the micromere genetic regulatory network appears to be due

---

*This is an oversimplification of a very complex process. Other transcription factors, such as Blimp, are also needed, and the maternal transcription factor SoxB1 has to be eliminated from the vegetal pole or else it will inhibit the activation of *Pmar1*. In addition, the cytoskeletal processes partitioning the cells and anchoring certain factors are not being considered here. For complete details of the model, see the continually updated website http://sugp.caltech.edu/endomes/.

(A)

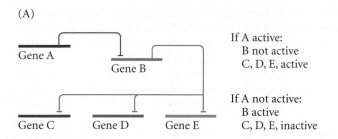

If A active:
B not active
C, D, E, active

If A not active:
B active
C, D, E, inactive

(B)

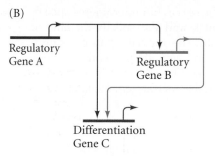

FIGURE 5.13 "Logic circuits" for gene expression. (A) In a double-negative gate, a single gene encodes a repressor of an entire battery of genes. When this repressor gene is repressed, the battery of genes is expressed. (B). In this feedforward circuit, gene product A activates both gene B and gene C, and gene B also activates gene C. Feedforward circuits provide an efficient way of amplifying a signal in one direction.

to the recruitment of a "subroutine" that in most echinoderms (including sea urchins) is used for making the adult skeleton (Gao and Davidson 2008).

The cooption of subroutines by a new lineage is one of the ways evolution occurs. It happens that the genetic regulatory network of the micromeres in sea urchin embryos is very different from that in other echinoderms. Only in the micromeres of sea urchins has the skeletogenic subroutine (which in all other echinoderms is activated late in development) come under the control of the genes that specify cells to the micromere lineage. The most important evolutionary events were those placing the skeletogenic genes *Alx1* and *Ets1* (necessary for adult skeletal development) and *Tbr* (used in later larval skeleton formation) under the regulation of the Pmar1-HesC double-negative gate. This occurred through mutations in the *cis*-regulatory regions of these genes. Thus, the skeletogenic properties that distinguish the sea urchin micromeres appear to have arisen through the co-option of a preexisting skeletogenic regulatory system by the micromere lineage gene regulatory system. When skeletal gene expression was placed under the control of the double-negative gate, it became a property specifically of the micromeres.

Specification of the vegetal cells

The abilities of the primary micromeres go well beyond the autonomous development into skeletal cells. They also produce signals that can induce changes in other tissues. The first of these signals is the paracrine factor Wnt8 (Angerer and Angerer 2000; Wikramanayake et al. 2004). As soon as the micromeres form, maternal β-catenin and Otx activate the *Blimp1* gene, whose product (in conjunction with more β-catenin) activates the gene encoding Wnt8. Wnt8 is then received by the micromeres that made it (autocrine regulation), activating the micromeres' own genes for β-catenin. Since β-catenin activates *Blimp1*, this sets up a positive feedback loop between Blimp1 and Wnt8 that establishes a source of β-catenin for the nuclei of the micromeres. The experiments described in Figure 5.10 demonstrated that the micromeres are able to induce a second embryonic axis when transplanted to the animal hemisphere. However, micromeres from embryos in which β-catenin is prevented from entering the nucleus are unable to induce the animal cells to form endoderm, and no second axis forms (Logan et al. 1998).

The second inducing signal is an as yet unidentified "early signal." This early signal (ES) gene is under the control of the double-negative gate of Pmar1 and HesC (see Figure 5.12B). The early signal instructs the adjacent cells (i.e., cells above the micromeres) to have an endomesodermal fate. It is also responsible for establishing a second axis when micromeres are transplanted to the animal region of the embryo.* If *Pmar1* mRNA is injected into an animal cell, that animal cell will develop into a skeletogenic mesenchyme cell, and the cell adjacent to it will start developing like a macromere endomesodermal cell (Oliveri et al. 2003). It appears that both Wnt8 and the unidentified "early signal" are needed, and in this particular order. If either one is not present, the veg$_2$ cells fail to make non-skeletogenic mesenchyme or endoderm (Ransick and Davidson 1995; Sherwood and McClay 1999; Sweet et al. 1999). The early signal may create competence in the veg$_2$ cells to respond to the Delta signal.

The third signal is the juxtacrine protein Delta, which is also controlled by the double-negative gate. It functions later in development by activating the Notch protein on the adjacent cells. Notch protein on these cells tells the cells below them to become the non-skeletogenic mesenchyme cells. Finally, Wnt8 makes another appearance, a product of the micromeres and endoderm cells. Wnt8 appears to act in an autocrine manner to boost the specification of both the veg$_2$ endoderm cells and the micromeres and to facilitate their separation into two distinct lineages. Wnt8 also acts as a paracrine factor on the endoderm precursor cells to initiate the invagination of the vegetal plate at the start of gastrulation (Davidson et al. 2002; Wikramanayake et al. 2004).

---

*The experiments described in Figure 5.10 demonstrated that the micromeres are able to induce a second embryonic axis when transplanted to the animal hemisphere. However, micromeres from embryos in which β-catenin is prevented from entering the nucleus are unable to induce the animal cells to form endoderm, and no second axis forms (Logan et al. 1998).

In sum, the genes of the sea urchin micromeres specify their cell fates autonomously and also specify the fates of their neighbors conditionally. The original inputs come from the maternal cytoplasm, and activate genes that unlock repressors of a specific cell fate. Once the cytoplasmic factors accomplish their functions, the nuclear genome takes over.

Axis specification

In the sea urchin blastula, the general cell fates (ectoderm, endoderm, skeleton, etc.) line up along the animal-vegetal axis established in the egg cytoplasm prior to fertilization. The animal-vegetal axis also appears to structure the future anterior-posterior axis, with the vegetal region sequestering those maternal components necessary for posterior development (Boveri 1901; Maruyama et al. 1985).

In most sea urchins, the dorsal-ventral and left-right axes are specified after fertilization, but the manner of their specification is not well understood. Lineage tracer dye injected into one blastomere at the 2-cell stage demonstrated that, in nearly all cases, the oral pole of the future oral-aboral (mouth-anus) axis lay 45 degrees clockwise from the first cleavage plane as viewed from the animal pole (Cameron et al. 1989). This oral-aboral axis appears to form through the activation of the *Nodal* gene in the oral (but not in the aboral) ectoderm (Duboc et al. 2004; Flowers et al. 2004). As we will see, *Nodal* (a member of the TGF-β paracrine family; see Chapter 3) is also used by vertebrates to establish the right-left body axis.

The role of *Nodal* was discovered through the classic "find it, lose it, move it" mode of experimentation that we described in Chapter 4 (see p. 144). First, researchers cloned a sea urchin *Nodal* gene and, using in situ hybridization, demonstrated that Nodal protein becomes expressed in the presumptive oral ectoderm at about the 60-cell stage. Nodal then becomes prominent on one side of the blastula and on the presumptive oral side of the gastrula. When the

researchers prevented translation of the *Nodal* message, development was normal until the mesenchyme blastula stage—but the larvae never obtained bilateral symmetry, the archenteron did not bend to one side to form the mouth, and the skeletogenic mesenchyme did not separate into the two sets of spicule-forming skeleton cells. Moreover, genes usually expressed in the oral ectoderm were not expressed. Conversely, when researchers induced ectopic expression of the *Nodal* gene throughout the ectoderm, all of the ectoderm appeared to become oral. Thus, Nodal appears to be crucial in establishing the oral ectoderm.

See WEBSITE 5.1
Sea urchin cell specification

Sea Urchin Gastrulation

The late sea urchin blastula consists of a single layer of about 1,000 cells that form a hollow ball, somewhat flattened at the vegetal end. The blastomeres are derived from different regions of the zygote and have different sizes and properties. **Figures 5.14** and **5.15** illustrate development of the blastula through gastrulation to the pluteus larva stage that is characteristic of sea urchins. The drawings show the fate of each cell layer during gastrulation.

Ingression of the skeletogenic mesenchyme

Shortly after the blastula hatches from its fertilization envelope, the descendants of the large micromeres change their shape, lose their adhesions to their neighboring cells, and then break away from the apical layer to enter the blastocoel as skeletogenic mesenchyme cells (see Figure 5.15, 9–10 hours). The skeletogenic mesenchyme cells then begin extending and contracting long, thin (250 nm in diameter and 25 μm long) processes called **filopodia**. At first the cells appear to move randomly along the inner blastocoel surface, actively making and breaking filopodial connections

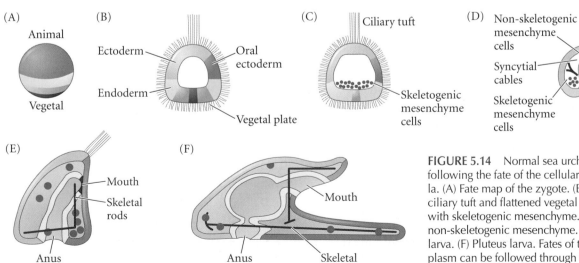

FIGURE 5.14 Normal sea urchin development, following the fate of the cellular layers of the blastula. (A) Fate map of the zygote. (B) Late blastula with ciliary tuft and flattened vegetal plate. (C) Blastula with skeletogenic mesenchyme. (D) Gastrula with non-skeletogenic mesenchyme. (E) Prism-stage larva. (F) Pluteus larva. Fates of the zygote cytoplasm can be followed through the color pattern (see Figure 5.9). (Courtesy of D. McClay.)

to the wall of the blastocoel. Eventually, however, they become localized within the prospective ventrolateral region of the blastocoel. Here they fuse into syncytial cables, which will form the axis of the calcium carbonate spicules of the larval skeletal rods (see Figure 5.14D–F).

THE IMPORTANCE OF EXTRACELLULAR LAMINA INSIDE THE BLASTOCOEL

The ingression of the large micromere descendants into the blastocoel is a result of their losing their affinity for their neighbors and for the hyaline membrane; instead they acquire a strong affinity for a group of proteins that line the blastocoel. This model of mesenchymal migration was first proposed by Gustafson and Wolpert (1967) and was confirmed in 1985, when Rachel Fink and David McClay measured the strength of sea urchin blastomere adhesion to the hyaline layer, to the basal lamina lining the blastocoel, and to other blastomeres.

Originally, all the cells of the blastula are connected on their outer surface to the hyaline layer, and on their inner surface to a basal lamina secreted by the cells. On their lateral surfaces, each cell has another cell for a neighbor. Fink and McClay found that the prospective ectoderm and endoderm cells (descendants of the mesomeres and macromeres, respectively) bind tightly to one another and to the hyaline layer, but adhere only loosely to the basal lamina (Table 5.2). The micromeres originally display a similar pattern of binding. However, the micromere pattern changes at gastrulation. Whereas the other cells retain their tight binding to the hyaline layer and to their neighbors, the skeletogenic mesenchyme precursors lose their affinities for these structures (which drop to about 2% of their original value), while their affinity for components of the basal lamina and extracellular matrix (such as fibronectin) increases a hundredfold. These changes have been correlated with changes in cell surface molecules that occur during this time (Wessel and McClay 1985), and proteins such as fibronectin, integrin, laminin, and cadherins have been shown to be involved in cellular ingression.

These changes in affinity cause the skeletogenic mesenchyme precursors to release their attachments to the external hyaline layer and to their neighboring cells and, drawn in by the basal lamina, to migrate up into the blastocoel (Figure 5.16A). There is a heavy concentration of extracellular material around the ingressing mesenchyme cells (Figure 5.16B,C). Once inside the blastocoel, these cells appear to migrate along the extracellular matrix of the blastocoel wall, extending their filopodia in front of them (Galileo and Morrill 1985; Karp and Solursh 1985; Cherr et al. 1992). Several proteins (including a fibronectin-like protein and a particular sulfated glycoprotein) are necessary to initiate...

9 hr

9.5 hr

10 hr

10.5 hr

11 hr

11.5 hr   Blastopore

12 hr

13 hr

13.5 hr   Syncytial cables

15 hr

17 hr

18 hr   Blastopore
Syncytial cables

FIGURE 5.15   Entire sequence of gastrulation in *Lytechinus variegatus*. Times show the length of development at 25°C. (Courtesy of J. Morrill; pluteus larva courtesy of G. Watchmaker.)

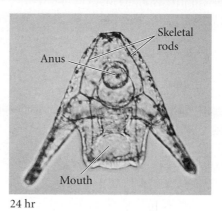

Anus

Skeletal rods

Mouth

24 hr

(A)

Extracellular
matrix fibril   Basal lamina   Skeletogenic
Blastocoel   mesenchyme cell

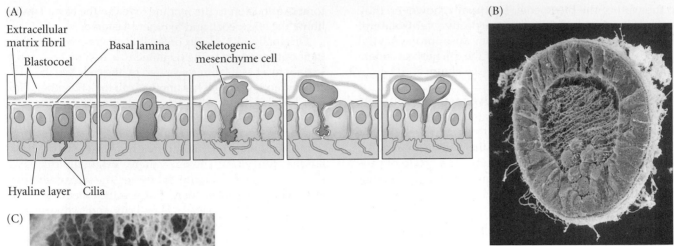

Hyaline layer   Cilia

(B)

(C)

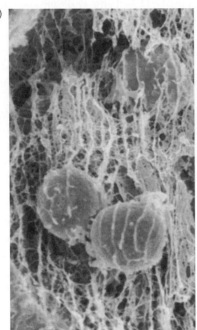

FIGURE 5.16 Ingression of skeletogenic mesenchyme cells. (A) Interpretative depiction of changes in the adhesive affinities of the presumptive skeletogenic mesenchyme cells (pink). These cells lose their affinities for hyalin and for their neighboring blastomeres while gaining an affinity for the proteins of the basal lamina. Nonmesenchymal blastomeres retain their original high affinities for the hyaline layer and neighboring cells. (B) Scanning electron micrograph of skeletogenic mesenchyme cells enmeshed in the extracellular matrix of an early *Strongylocentrotus* gastrula. (C) Gastrula-stage mesenchyme cell migration. The extracellular matrix fibrils of the blastocoel lie parallel to the animal-vegetal axis and are intimately associated with the skeletogenic mesenchyme cells. (B,C from Cherr et al. 1992; courtesy of the authors.)

tiate and maintain this migration (Wessel et al. 1984; Lane and Solursh 1991; Berg et al. 1996).

At two sites near the future ventral side of the larva, many skeletogenic mesenchyme cells cluster together, fuse with one another, and initiate spicule formation (**Figure 5.17**; Hodor and Ettensohn 1998). If a labeled micromere from another embryo is injected into the blastocoel of a gastrulating sea urchin embryo, it migrates to the correct location and contributes to the formation of the embryonic spicules (Ettensohn 1990; Peterson and McClay 2003). It is thought that the necessary positional information is pro-

TABLE 5.2 Affinities of mesenchymal and nonmesenchymal cells to cellular and extracellular components[a]

Cell type	Dislodgment force (in dynes)		
	Hyaline	Gastrula cell monolayers	Basal lamina
16-cell-stage micromeres	5.8×10^{-5}	6.8×10^{-5}	4.8×10^{-7}
Migratory-stage mesenchyme cells	1.2×10^{-7}	1.2×10^{-7}	1.5×10^{-5}
Gastrula ectoderm and endoderm	5.0×10^{-5}	5.0×10^{-5}	5.0×10^{-7}

Source: After Fink and McClay 1985.

[a]Tested cells were allowed to adhere to plates containing hyaline, extracellular basal lamina, or cell monolayers. The plates were inverted and centrifuged at various strengths to dislodge the cells. The dislodgement force is calculated from the centrifugal force needed to remove the test cells from the substrate.

(A)

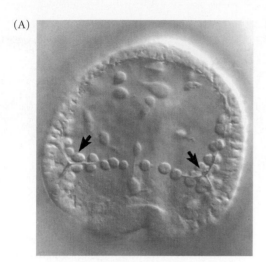

(B)

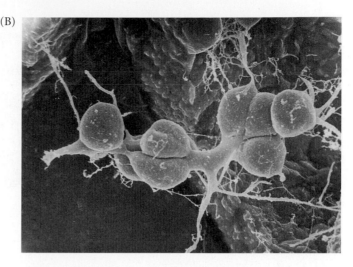

FIGURE 5.17   Formation of syncytial cables by skeletogenic mesenchyme cells of the sea urchin. (A) Skeletogenic mesenchyme cells in the early gastrula align and fuse to lay down the matrix of the calcium carbonate spicule (arrows). (B) Scanning electron micrograph of spicules formed by the fusing of skeletogenic mesenchyme cells into syncytial cables. (A from Ettensohn 1990; B from Morrill and Santos 1985.)

vided by the prospective ectodermal cells and their basal laminae (**Figure 5.18A**; Harkey and Whiteley 1980; Armstrong et al. 1993; Malinda and Ettensohn 1994). Only the skeletogenic mesenchyme cells (and not other cell types or latex beads) are capable of responding to these patterning cues (Ettensohn and McClay 1986). Miller and colleagues (1995) reported that the extremely fine filopodia on the skeletogenic mesenchyme cells appear to explore and sense the blastocoel wall and may be responsible for picking up dorsal-ventral and animal-vegetal patterning cues from the ectoderm (**Figure 5.18B**; Malinda et al. 1995).

Two signals secreted by the blastula wall appear to be critical for this migration. VEGF paracrine factors are emitted from two small regions of the ectoderm where the skeletogenic mesenchyme cells will congregate (Duloquin et al. 2007), and a fibroblast growth factor (FGF) paracrine factor is made in the equatorial belt between endoderm and ectoderm and then becomes defined into the lateral domains where the skeletogenic mesenchyme cells collect (Röttinger et al. 2008). The skeletogenic mesenchyme cells migrate to these points of VEGF and FGF synthesis and arrange themselves in a ring along the animal-vegetal axis.

INTEGRATION OF THE GENETIC REGULATORY NETWORK AND MORPHOGENESIS   The skeletogenic mesenchyme cell receptor for VEGF is under the control of the micromere genet-

(A)

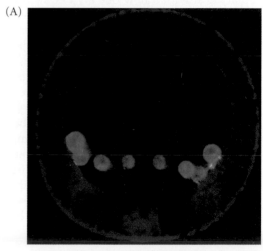

(B)

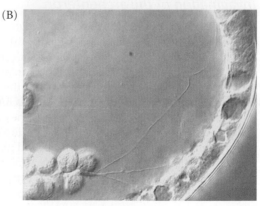

FIGURE 5.18   Localization of skeletogenic mesenchyme cells. (A) Localization of the micromeres to form the calcium carbonate skeleton is determined by the ectodermal cells. The skeletogenic mesenchyme cells are stained green, while β-catenin is stained red. The skeletogenic mesenchyme cells appear to accumulate in those regions characterized by high β-catenin concentrations. (B) Nomarski videomicrograph showing a long, thin filopodium extending from a skeletogenic mesenchyme cell to the ectodermal wall of the gastrula, as well as a shorter filopodium extending inward from the ectoderm. The mesenchymal filopodia extend through the extracellular matrix and directly contact the cell membrane of the ectodermal cells. (B from Miller et al. 1995; photographs courtesy of J. R. Miller and D. McClay.)

(A) (B)

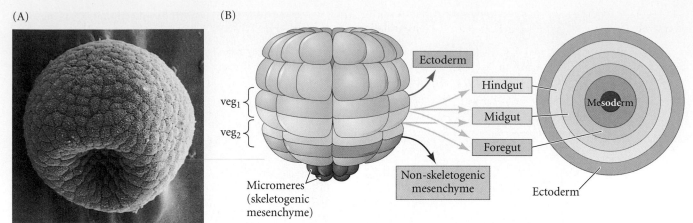

FIGURE 5.19 Invagination of the vegetal plate. (A) Vegetal plate invagination in *Lytechinus variegatus*, seen by scanning electron microscopy of the external surface of the early gastrula. The blastopore is clearly visible. (B) Fate map of the vegetal plate of the sea urchin embryo, looking "upward" at the vegetal surface. The central portion becomes the non-skeletogenic mesenchyme cells, while the concentric layers around it become the foregut, midgut, and hindgut, respectively. The boundary where the endoderm meets the ectoderm marks the anus. The non-skeletogenic mesenchyme and foregut come from the veg_2 layer, the midgut comes from veg_1 and veg_2 cells, and the hindgut (and the ectoderm in contact with it) comes from the veg_1 layer. (A from Morrill and Santos 1985, courtesy of J. B. Morrill; B after Logan and McClay 1999.)

ic regulatory network and thereby connects morphogenesis to cell specification. As we have seen, skeletogenic mesenchyme cells ingress into the blastocoel at a particular time. Moreover, this ingression is a "cell-autonomous" property (i.e., not controlled by neighboring cells), and skeletogenic mesenchyme cells will ingress into the blastocoels even if transplanted into younger or older embryos (Peterson and McClay 2003). The preparation for cell ingression involves the changing of the micromere cell membranes by endocytosing the original membrane and replacing it with a new one (Wu et al. 2007). This new membrane lacks the cadherin necessary for cell-cell adhesion (and presumably altered receptors for hyalin and basal lamina proteins as well). Endocytosis and the downregulation of cadherins are controlled by the transcription factor Snail. And the gene encoding the Snail protein is activated by the Alx1 transcription factor that is regulated by the double-negative gate of the gene regulatory network. The use of Snail protein to downregulate cadherins and prepare the cell for an epithelial-to-mesenchymal transition is a morphogenetic subroutine that exists throughout the animal kingdom. Here, it has been connected to the specification program of the micromere lineage. In this manner, the micromere cells alter their cell membranes and leave the blastula epithelium by ingressing into the blastocoel.

Invagination of the archenteron

FIRST STAGE OF ARCHENTERON INVAGINATION As the skeletogenic mesenchyme cells leave the vegetal region of the spherical embryo, important changes are occurring in the cells that remain there. These cells thicken and flatten to form a vegetal plate, changing the shape of the blastula (see Figure 5.15, 9 hours). The vegetal plate cells remain bound to one another and to the hyaline layer of the egg, and they move to fill the gaps caused by the ingression of the skeletogenic mesenchyme. Moreover, the vegetal plate involutes inward by altering its cell shape, and then invaginates about one-fourth to one-half the way into the blastocoel (**Figure 5.19A**; see also Figure 5.15, 10.5–11.5 hours). Then invagi-

nation suddenly ceases. The invaginated region is called the **archenteron** (primitive gut), and the opening of the archenteron at the vegetal pole is called the **blastopore**.

The movement of the vegetal plate into the blastocoel appears to be initiated by shape changes in the vegetal plate cells and in the extracellular matrix underlying them (see Kominami and Takata 2004 for a review). Actin microfilaments collect in the apical ends of the vegetal cells, causing these ends to constrict, forming bottle-shaped vegetal cells that pucker inward (Kimberly and Hardin 1998; Beane et al. 2006). Destroying these cells with lasers retards gastrulation. In addition, the hyaline layer at the vegetal plate buckles inward due to changes in its composition (Lane et al. 1993).

At the stage when the skeletogenic mesenchyme cells begin ingressing into the blastocoel, the fates of the vegetal plate cells have already been specified (Ruffins and Ettensohn 1996). The non-skeletogenic mesenchyme is the first group of cells to invaginate, forming the tip of the archenteron and leading the way into the blastocoel. The non-skeletogenic mesenchyme will form the pigment cells, the musculature around the gut, and contribute to the coelomic pouches. The endodermal cells adjacent to the micromere-derived mesenchyme become foregut, migrating the farthest distance into the blastocoel. The next layer of endodermal cells becomes midgut, and the last circumferential row to invaginate forms the hindgut and anus (**Figure 5.19B**).

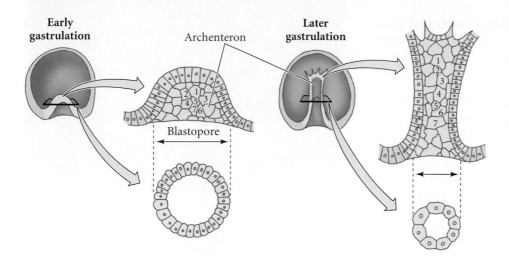

FIGURE 5.20 Cell rearrangement during extension of the archenteron in sea urchin embryos. In this species, the early archenteron has 20 to 30 cells around its circumference. Later in gastrulation, the archenteron has a circumference made by only 6 to 8 cells. (After Hardin 1990.)

SECOND AND THIRD STAGES OF ARCHENTERON INVAGINATION

After a brief pause following the initial invagination, the second phase of archenteron formation begins. The archenteron extends dramatically, sometimes tripling its length. In this process of extension, the wide, short gut rudiment is transformed into a long, thin tube (**Figure 5.20**; see also Figure 5.15, 12 hours). To accomplish this extension, the cells change their shape and start migrating. First, the thickened cells of the vegetal plate become thinner; then these cells rearrange themselves, intercalating between one another, like lanes of traffic merging (Ettensohn 1985; Hardin and Cheng 1986). This phenomenon, where cells intercalate to narrow the tissue and at the same time move it forward, is called **convergent extension** (Martins et al. 1998). In this manner, the archenteron is elongated.

The final phase of archenteron elongation is initiated by the tension provided by non-skeletogenic mesenchyme cells, which form at the tip of the archenteron and remain there. These cells extend filopodia through the blastocoel fluid to contact the inner surface of the blastocoel wall (Dan and Okazaki 1956; Schroeder 1981). The filopodia attach to the wall at the junctions between the blastomeres and then shorten, pulling up the archenteron (**Figure 5.21**; see also Figure 5.15, 12 and 13 hours). Hardin (1988) ablated non-skeletogenic mesenchyme cells of *Lytechinus pictus* gastrulae with a laser, with the result that the archenteron could elongate only to about two-thirds of the normal length. If a few non-skeletogenic mesenchyme cells were left, elongation continued, although at a slower rate. Thus, in this species, the non-skeletogenic mesenchyme cells play an essential role in pulling the archenteron upward to the blastocoel wall during the last phase of invagination.

But can the non-skeletogenic mesenchyme filopodia attach to any part of the blastocoel wall, or is there a specific target in the animal hemisphere that must be present for attachment to occur? Is there a region of the blastocoel wall that is already committed to becoming the ventral side of the larva? Studies by Hardin and McClay (1990) show that

(A)

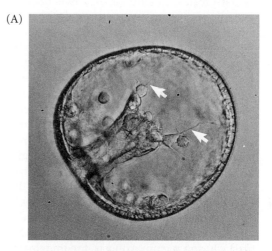

(B)

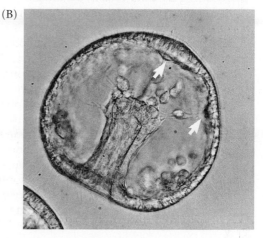

FIGURE 5.21 Mid-gastrula stage of *Lytechinus pictus*, showing filopodial extensions of non-skeletogenic mesenchyme. (A) Non-skeletogenic mesenchyme cells extend filopodia (arrows) from the tip of the archenteron. (B) Filopodial cables connect the blastocoel wall to the archenteron tip. The tension of the cables can be seen as they pull on the blastocoel wall at the point of attachment (arrows). (Courtesy of C. Ettensohn.)

there is a specific target site for the filopodia that differs from other regions of the animal hemisphere. The filopodia extend, touch the blastocoel wall at random sites, and then retract. However, when the filopodia contact a particular region of the wall, they remain attached there, flatten out against this region, and pull the archenteron toward it. When Hardin and McClay poked in the other side of the blastocoel wall so that contacts were made most readily with that region, the filopodia continued to extend and retract after touching it. Only when the filopodia found their target tissue did they cease these movements. If the gastrula was constricted so that the filopodia never reached the target area, the non-skeletogenic mesenchyme cells continued to explore until they eventually moved off the archenteron and found the target as freely migrating cells. There appears, then, to be a target region on what is to become the ventral side of the larva that is recognized by the non-skeletogenic mesenchyme cells, and which positions the archenteron near the region where the mouth will form.

As the top of the archenteron meets the blastocoel wall in the target region, many of the non-skeletogenic mesenchyme cells disperse into the blastocoel, where they proliferate to form the mesodermal organs (see Figure 5.15, 13.5 hours). Where the archenteron contacts the wall, a mouth eventually forms. The mouth fuses with the archenteron to create a continuous digestive tube. Thus, as is characteristic of deuterostomes, the blastopore marks the position of the anus.

As the pluteus larva elongates, the coelomic cavities form from non-skeletogenic mesenchyme and veg$_2$ cells. Under the influence of Nodal protein, the right coelomic sac remains rudimentary. However, the left coelomic sac undergoes extensive development to form many of the structures of the adult sea urchin. The left sac splits into three smaller sacs. An invagination from the ectoderm fuses with the middle sac to form the **imaginal rudiment**. This rudiment develops a fivefold symmetry (**Figure 5.22**), and skeletogenic mesenchyme cells enter the rudiment to synthesize the first skeletal plates of the shell. The left side of the pluteus becomes, in effect, the future oral surface of the adult sea urchin (Bury 1895; Aihara and Amemiya 2001). During metamorphosis, the imaginal rudiment separates from the larva, which then degenerates. While the imaginal rudiment (now called a juvenile) is re-forming its digestive tract and settling on the ocean floor, it is dependent on the nutrition it received from the jettisoned larva.

The echinoderm pattern of gastrulation provides the evolutionary prototype for deuterostome development. In deuterostomes (echinoderms, tunicates, cephalochordates, and vertebrates), the first opening (i.e., the blastopore) becomes the anus while the second opening becomes the mouth (hence, *deutero stoma*, "mouth second"). Moreover, in sea urchins, we see the phenomena of convergent extension and the use of *Nodal* gene expression for the establishment of axes. We will return to the subject of deuterostome development later in this and in succeeding chapters, but

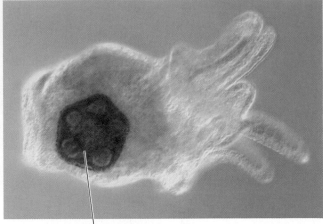

Imaginal rudiment

FIGURE 5.22   The imaginal rudiment growing in the left side of the pluteus larva of a sea urchin. The rudiment will become the adult sea urchin, while the larval stage will be jettisoned. The fivefold symmetry of the rudiment is obvious. (Courtesy of G. Wray.)

we now turn to development in a protostome group, the molluscs.

See VADE MECUM   Sea urchin development

EARLY DEVELOPMENT IN SNAILS

Cleavage in Snail Embryos

Spiral holoblastic cleavage is characteristic of several animal groups, including annelid worms, some flatworms, and most molluscs (see Lambert 2010). It differs from radial cleavage in numerous ways. First, the cleavage planes are not parallel or perpendicular to the animal-vegetal axis of the egg; rather, cleavage is at oblique angles, forming a "spiral" arrangement of daughter blastomeres. Second, the cells touch one another at more places than do those of radially cleaving embryos. In fact, they assume the most thermodynamically stable packing orientation, much like adjacent soap bubbles. Third, spirally cleaving embryos usually undergo fewer divisions before they begin gastrulation, making it possible to follow the fate of each cell of the blastula. When the fates of the individual blastomeres from annelid, flatworm, and mollusc embryos were compared, many of the same cells were seen in the same places, and their general fates were identical (Wilson 1898). Blastulae produced by spiral cleavage have no blastocoel and are called **stereoblastulae**.

Figures 5.23 depicts the cleavage pattern typical of many molluscan embryos. The first two cleavages are nearly meridional, producing four large macromeres (labeled A, B, C, and D). In many species, these four blastomeres are different sizes (D being the largest), a characteristic that allows them to be individually identified. In each successive cleavage, each macromere buds off a small micromere at its animal pole. Each successive quartet of micromeres is dis-

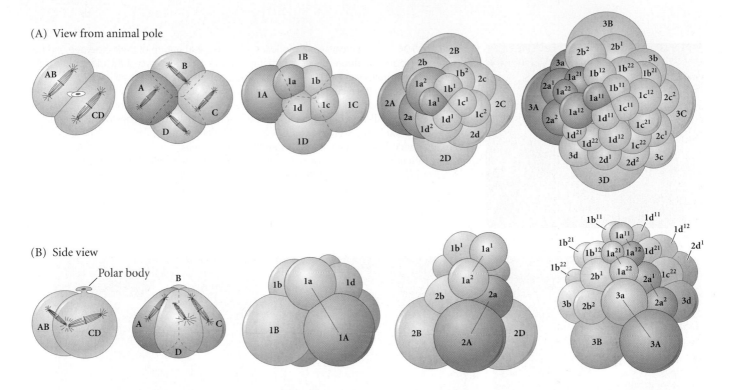

(A) View from animal pole

(B) Side view

FIGURE 5.23 Spiral cleavage of the mollusc *Trochus* viewed from the animal pole (A) and from one side (B). The cells derived from the A blastomere are shown in color. The mitotic spindles, sketched in the early stages, divide the cells unequally and at an angle to the vertical and horizontal axes. Each successive quartet of micromeres (indicated with lowercase letters) is displaced to the right or to the left of its sister macromere (uppercase letters), creating the characteristic spiral pattern.

placed to the right or to the left of its sister macromere, creating the characteristic spiral pattern. Looking down on the embryo from the animal pole, the upper ends of the mitotic spindles appear to alternate clockwise and counterclockwise (**Figure 5.24**). This arrangement causes alternate micromeres to form obliquely to the left and to the right of their macromeres.

At the third cleavage, the A macromere gives rise to two daughter cells, macromere 1A and micromere 1a. The B, C, and D cells behave similarly, producing the first quartet of micromeres. In most species, these micromeres are to the *right* of their macromeres (looking down on the animal pole). At the fourth cleavage, macromere 1A divides to form macromere 2A and micromere 2a, and micromere 1a divides to form two more micromeres, $1a^1$ and $1a^2$ (see Figure 5.23). The micromeres of this second quartet are to the *left* of the macromeres. Further cleavage yields blastomeres 3A and 3a from macromere 2A, and micromere $1a^2$ divides to produce cells $1a^{21}$ and $1a^{22}$. In normal development, the first-quartet micromeres form the head structures, while the second-quartet micromeres form the statocyst (balance organ) and shell. These fates are specified both by cyto-

plasmic localization and by induction (Cather 1967; Clement 1967; Render 1991; Sweet 1998).

The orientation of the cleavage plane to the left or to the right is controlled by cytoplasmic factors in the oocyte. This was discovered by analyzing mutations of snail coiling. Some snails have their coils opening to the right of their shells (**dextral coiling**), whereas the coils of other snails open to the left (**sinistral coiling**). Usually the direction of coiling is the same for all members of a given species, but occasional mutants are found (i.e., in a population of right-coiling snails, a few individuals will be found with coils that open on the left). Crampton (1894) analyzed the embryos of such aberrant snails and found that their early cleavage differed from the norm. The orientation of the cells after the second cleavage was different in the sinistrally coiling snails as a result of a different orientation of the mitotic apparatus (**Figure 5.25**). In some species (such as the pond snail *Physa*, an entirely sinistral species), the sinistrally coiling cleavage patterns are mirror images of the dextrally coiling pattern of the right-handed species. In other instances (such as *Lymnaea*, where about 2% of the snails are lefties), sinistrality is the result of a two-step process: at each division, the initial cleavage is radial; however, as the cleavage furrow forms, the blastomeres shift to the left-hand spiral position (Shibazaki et al. 2004). In Figure 5.25, one can see that the position of the 4d blastomere (which is extremely important, as its progeny will form the mesodermal organs) is different in the two types of spiraling embryos.

See WEBSITE 5.2
Alfred Sturtevant and the genetics of snail coiling

(A)

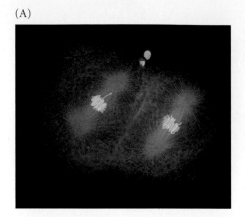

FIGURE 5.24 Spiral cleavage in molluscs. (A) The spiral nature of third cleavage can be seen in the confocal fluorescence micrograph of the 4-cell embryo of the clam *Acila castrenis*. Microtubules stain red, RNA stains green, and the DNA stains yellow. Two cells and a portion of a third cell are visible; a polar body can be seen at the top of the micrograph. (B–E) Cleavage in the mud snail *Ilyanassa obsoleta*. The D blastomere is larger than the others, allowing the identification of each cell. Cleavage is dextral. (B) 8-cell stage. PB, polar body. (C) Mid-fourth cleavage (12-cell embryo). The macromeres have already divided into large and small spirally oriented cells; 1a–d have not divided yet. (D) 32-cell embryo. (A courtesy of G. von Dassow and the Center for Cell Dynamics; B–E from Craig and Morrill 1986, courtesy of the authors.)

(B)

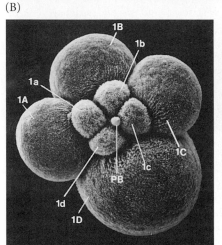

(C)

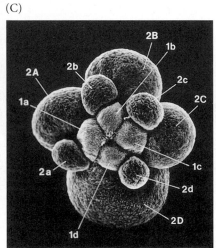

(D)

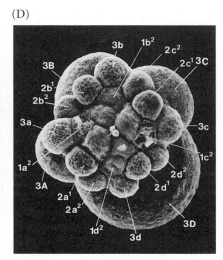

In snails such as *Lymnaea*, the direction of snail shell coiling is controlled by a single pair of genes (Sturtevant 1923; Boycott et al. 1930). In *Lymnaea peregra*, rare mutants exhibiting sinistral coiling were found and mated with wild-type, dextrally coiling snails. These matings showed that the right-coiling allele, *D*, is dominant to the left-coiling allele, *d*. However, the direction of cleavage is determined not by the genotype of the developing snail but by the genotype of the snail's mother. A *dd* female snail can produce only sinistrally coiling offspring, even if the offspring's genotype is *Dd*. A *Dd* individual will coil either left or right, depending on the genotype of its mother. Such matings produce a chart like this:

	Genotype	Phenotype
DD ♀ × dd ♂ →	*Dd*	All right-coiling
DD ♂ × dd ♀ →	*Dd*	All left-coiling
Dd × Dd →	*1DD:2Dd:1dd*	All right-coiling

The genetic factors involved in snail coiling are brought to the embryo by the oocyte cytoplasm. It is the genotype of the ovary in which the oocyte develops that determines which orientation cleavage will take. When Freeman and Lundelius (1982) injected a small amount of cytoplasm from dextrally coiling snails into the eggs of *dd* mothers, the resulting embryos coiled to the right. Cytoplasm from sinistrally coiling snails did not affect right-coiling embryos. These findings confirmed that the wild-type mothers were placing a factor into their eggs that was absent or defective in the *dd* mothers.

Just as in sea urchins (and vertebrates), the right-left axis comes to be defined by the Nodal family of paracrine factors. In the case of snails, Nodal activates genes on the right side of dextrally coiling embryos and on the left side of sinistrally coiling embryos. Changing the direction of cleavage (using glass needles) at the 8-cell stage changes the location of *Nodal* gene expresion (Grande and Patel 2009; Kuroda et al. 2009). Nodal appears to be expressed in the C-quadrant micromere lineages (which give rise to the ectoderm). This signal induces the expression of the gene encoding the Pitx transcription factor (a target of Nodal protein in vertebrate axis formation) in the neighboring D-quadrant blastomeres.

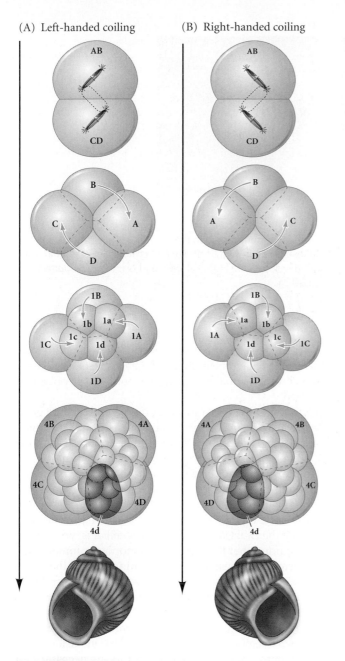

FIGURE 5.25 Looking down on the animal pole of left-coiling (A) and right-coiling (B) snails. The origin of sinistral and dextral coiling can be traced to the orientation of the mitotic spindle at the second cleavage. Left- and right-coiling snails develop as mirror images of each other. (After Morgan 1927.)

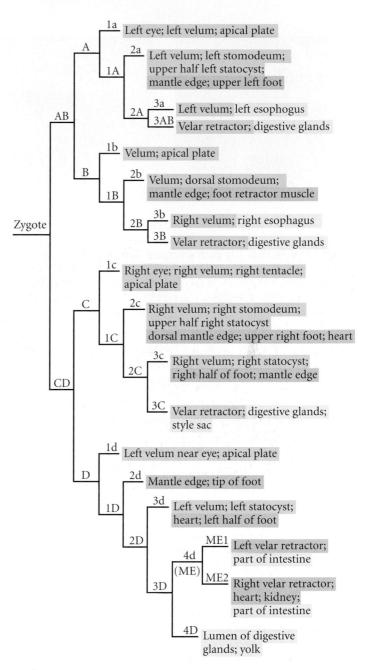

FIGURE 5.26 Fate map of *Ilyanassa obsoleta*. Beads containing Lucifer Yellow were injected into individual blastomeres at the 32-cell stage. When the embryos developed into larvae, their descendants could be identified by their fluorescence. (After Render 1997.)

The snail fate map

The fate maps of *Ilyanassa obsoleta* and *Crepidula fornicata* were constructed by injecting specific micromeres with large polymers conjugated to fluorescent dyes (Render 1997; Hejnol et al. 2007). The fluorescence is maintained over the period of embryogenesis and can be seen in the larval tissue derived from the injected cells. The results of the *Ilyanassa* studies, shown in **Figure 5.26**, indicated that the second-quartet micromeres (2a–d) generally contribute to the shell-forming mantle, the velum, the mouth, and the heart. The third-quartet micromeres (3a–d) generate large

(A) (B) (C)

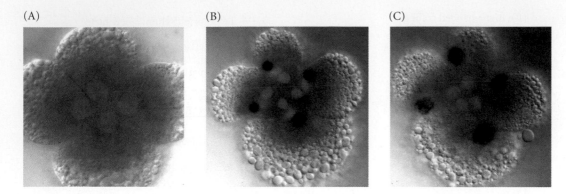

FIGURE 5.27 Association of *decapentaplegic* (*dpp*) mRNA with specific centrosomes of *Ilyanassa*. (A) In situ hybridization of the mRNA for the BMP-like paracrine factor Dpp in the 4-cell snail embryo shows no Dpp accumulation. (B) At prophase of the 4- to 8-cell stage, *dpp* mRNA (black) accumulates at one centrosome of the pair forming the mitotic spindle. (C) As mitosis continues, *dpp* mRNA is seen to attend the centrosome in the macromere rather than the centrosome in the micromere of each cell. The *dpp* message encodes a BMP-like paracrine factor critical to molluscan development. (From Lambert and Nagy 2002, courtesy of L. Nagy.)

regions of the foot, velum, esophagus, and heart. The 4d cell—the mesentoblast—contributes to the larval kidney, heart, retractor muscles, and intestine.

The polar lobe: Cell determination and axis formation

Molluscs provide some of the most impressive examples of both mosaic development—in which the blastomeres are specified autonomously—and of cytoplasmic localization, wherein morphogenetic determinants are placed in a specific region of the oocyte (see the Part II opener). Autonomous specification of early blastomeres is especially prominent in those groups of animals having spiral cleavage, all of which initiate gastrulation at the future anterior end after only a few cell divisions. In molluscs, the mRNAs for some transcription factors and paracrine factors are placed in particular cells by associating them with certain centrosomes (**Figure 5.27**; Lambert and Nagy 2002; Kingsley et al. 2007). This allows the mRNA to enter specifically into one of the two daughter cells. In other cases, the patterning molecules appear to be bound to a certain region of the egg that will form the **polar lobe**.

E. B. Wilson and his student H. E. Crampton observed that certain spirally cleaving embryos (mostly in the mollusc and annelid phyla) extrude a bulb of cytoplasm immediately before first cleavage. This protrusion is the polar lobe. In some species of snails, the region uniting the polar lobe to the rest of the egg becomes a fine tube. The first cleavage splits the zygote asymmetrically, so that the polar lobe is connected only to the CD blastomere (**Figure 5.28A**). In several species, nearly one-third of the total cytoplasmic volume is contained in this anucleate lobe, giving it the appearance of another cell (**Figure 5.28B**). The resulting three-lobed structure is often referred to as the **trefoil-stage** embryo (**Figure 5.28C**). The CD blastomere absorbs the polar lobe mate-

rial but extrudes it again prior to second cleavage. After this division, the polar lobe is attached only to the D blastomere, which absorbs its material. From this point on, no polar lobe is formed.

Crampton (1896) showed that if one removes the polar lobe at the trefoil stage, the remaining cells divide normally. However, instead of a normal trochophore larva,* the result is an incomplete larva, wholly lacking its endoderm (intestine) and mesodermal organs (such as the heart and retractor muscles), as well as some ectodermal organs (such as eyes; **Figure 5.29**). Moreover, Crampton demonstrated that the same type of abnormal larva can be produced by removing the D blastomere from the 4-cell embryo. Crampton concluded that the polar lobe cytoplasm contains the endodermal and mesodermal determinants, and that these determinants give the D blastomere its endomesoderm-forming capacity. Crampton also showed that the localization of the mesodermal determinants is established shortly after fertilization, thereby demonstrating that a specific cytoplasmic region of the egg, destined for inclusion in the D blastomere, contains whatever factors are necessary for the special cleavage rhythms of the D blastomere and for the differentiation of the mesoderm.

Centrifugation studies demonstrated that the morphogenetic determinants sequestered in the polar lobe are probably located in the cytoskeleton or cortex, not in the lobe's diffusible cytoplasm (Clement 1968). Van den Biggelaar (1977) obtained similar results when he removed the cytoplasm from the polar lobe with a micropipette. Cytoplasm from other regions of the cell flowed into the polar lobe, replacing the portion he had removed. The subse-

---

*The trochophore (Greek, *trochos*, "wheel") is a planktonic (free-swimming) larval form found among the molluscs and several other protostome phyla with spiral cleavage, most notably the marine annelid worms.

FIGURE 5.28   Polar lobe formation in certain mollusc embryos. (A) Cleavage. Extrusion and reincorporation of the polar lobe occur twice. (B) Late first division of a scallop embryo, showing the microtubules (red) and the RNA (stained green with propidium iodide). The DNA of the chromosomes is yellow. (C) Section through first-cleavage, or trefoil-stage, embryo of *Dentalium*. The arrow points to the large polar lobe. (A after Wilson 1904; B courtesy of G. von Dassow and the Center for Cell Dynamics; C courtesy of M. R. Dohmen.)

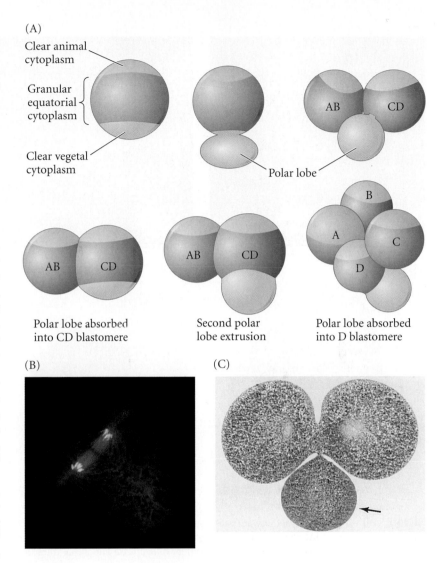

quent development of these embryos was normal. In addition, when he added the diffusible polar lobe cytoplasm to the B blastomere, no duplicated structures were seen (Verdonk and Cather 1983). Therefore, the diffusible part of the polar lobe cytoplasm does not contain the morphogenetic determinants; they probably reside in the nonfluid cortical cytoplasm or on the cytoskeleton.

Clement (1962) also analyzed the further development of the D blastomere in order to observe the further appropriation of these determinants. The development of the D blastomere can be traced in Figure 5.24B–D. This macromere, having received the contents of the polar lobe, is larger than the other three. When one removes the D blastomere or its first or second macromere derivatives (1D or 2D), one obtains an incomplete larva, lacking heart, intestine, velum (the ciliated border of the larva), shell gland, eyes, and foot. This is essentially the same phenotype one gets when one removes the polar lobe. Since the D blastomeres do not directly contribute cells to many of these structures, it appears that the D-quadrant macromeres are involved in inducing other cells to have these fates.

When one removes the 3D blastomere shortly after the division of the 2D cell to form the 3D and 3d blastomeres, the larva produced looks similar to those formed by the removal of the D, 1D, or 2D macromeres.

FIGURE 5.29   Importance of the polar lobe in the development of *Ilyanassa*. (A) Normal trochophore larva. (B) Abnormal larva, typical of those produced when the polar lobe of the D blastomere is removed. (E, eye; F, foot; S, shell; ST, statocyst; V, velum; VC, velar cilia; Y, residual yolk; ES, everted stomodeum; DV, disorganized velum.) (From Newrock and Raff 1975, courtesy of K. Newrock.)

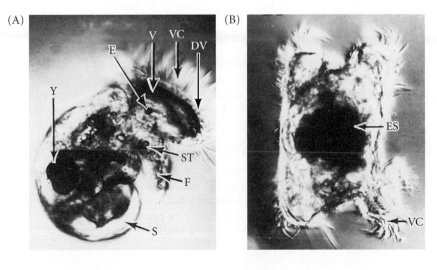

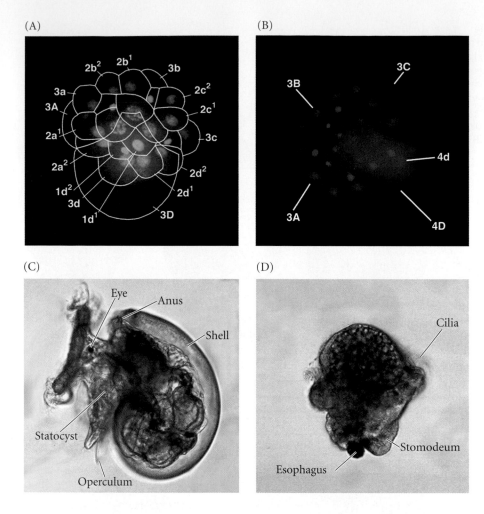

FIGURE 5.30 MAP kinase activity activated by D-quadrant snail blastomeres. (A) The 3D blastomere activates MAP kinase activity in adjacent *Ilyanassa* micromeres.. Activated MAP kinase (blue stain) can be seen in the 3D macromere of *Ilyanassa* and in the micromeres above it (1a–d[1], 1d[2], 2d[1], 2d[2], 3d). The nuclei are counterstained green, and the cell boundaries have been superimposed on the photographic image. Staining was done 30 minutes after the formation of the 3D macromere. (B) The presence of MAP kinase activity in the 4d blastomere (but not 4D) of *Crepidula*. (C) Control larva grown to veliger larval stage. (D) Same age larva treated with MAP kinase inhibitor 15 minutes after 3D blastomere formed. The shell, eye, statocyst, and operculum have not developed. (A,C,D from Lambert and Nagy 2001; B from Henry and Perry 2008, courtesy of the authors.)

However, ablation of the 3D blastomere at a later time produces an almost normal larva, with eyes, foot, velum, and some shell gland, but no heart or intestine (see Figure 5.30). After the 4d cell is given off (by the division of the 3D blastomere), removal of the D derivative (the 4D cell) produces no qualitative difference in development. In fact, all the essential determinants for heart and intestine formation are now in the 4d blastomere, and removal of that cell results in a heartless and gutless larva (Clement 1986). The 4d blastomere is responsible for forming (at its next division) the two **mesentoblasts**, the cells that give rise to both the mesodermal (heart) and endodermal (intestine) organs.

The mesodermal and endodermal determinants of the 3D macromere, then, are transferred to the 4d blastomere, while the inductive ability of the 3D blastomere (to induce eyes and shell gland, for instance) is needed during the time the 3D cell is formed but is not required afterward. The 3D cell appears to activate the MAP kinase signaling pathway in the micromeres above it (Lambert and Nagy 2001). If cells are stained for activated MAP kinase, the stain is seen in those cells that require the signal from the 3D macromere for their normal differentiation (**Figure 5.30**). Removal of 3D prevents MAP kinase signaling, and if the MAP kinase signaling is blocked by specific inhibitors, the

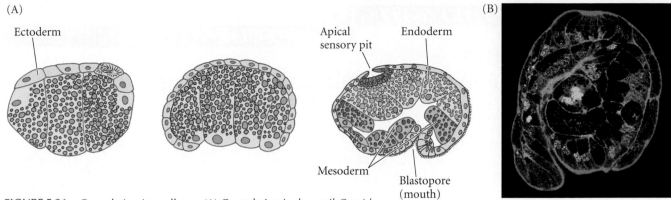

FIGURE 5.31 Gastrulation in molluscs. (A) Gastrulation in the snail *Crepidula*. The ectoderm undergoes epiboly from the animal pole and envelops the other cells of the embryo. (B) Late gastrula of the clam *Acila*, stained for actin microfilaments (orange) and nucleic acid (blue). (A after Conklin 1897; B courtesy of G. von Dassow and the Center for Cell Dynamics.)

resulting larvae look precisely like those formed by the deletion of the D blastomeres (see Figure 5.30). Thus, the 3D macromere appears to activate the MAP kinase cascade in the ectodermal (eye- and shell gland-forming) micromeres above it.* After the 3D divides to produce the 4d and 4D blastomeres, MAP kinase activity persists only in the 4d cell. If the MAP kinase is inhibited then, the structures (such as the heart) produced by the 4d derivatives fail to form (Henry and Perry 2008; Lambert 2008).

In addition to its role in cell differentiation, the material in the polar lobe is also responsible for specifying the dorsal-ventral polarity of the embryo. When polar lobe material is forced to pass into the AB blastomere as well as into the CD blastomere, twin larvae form that are joined at their ventral surfaces (Guerrier et al. 1978; Henry and Martindale 1987).

To summarize, experiments have demonstrated that the nondiffusible polar lobe cytoplasm is extremely important in normal molluscan development for several reasons:

- It contains the determinants for the proper cleavage rhythm and the cleavage orientation of the D blastomere.
- It contains certain determinants (those entering the 4d blastomere and hence leading to the mesentoblasts) for autonomous mesodermal and intestinal differentiation.

- It is responsible for permitting the inductive interactions (through the material entering the 3D blastomere) leading to the formation of the shell gland and eye.
- It contains determinants needed for specifying the dorsal-ventral axis of the embryo.

Although the polar lobe is clearly important in normal snail development, we still do not know the mechanisms for most of its effects. One possible clue has been provided by Atkinson (1987), who observed differentiated cells of the velum, digestive system, and shell gland within lobeless embryos. But even though lobeless embryos can produce these cells, they appear unable to organize them into functional tissues and organs. Tissues of the digestive tract can be found but are not connected; individual muscle cells are scattered around the lobeless larva but are not organized into a functional muscle tissue. Thus, the developmental functions of the polar lobe are probably very complex and may be essential for axis formation.

See WEBSITE 5.3
Modifications of cell fate in spiralian eggs

Gastrulation in Snails

The snail stereoblastula is relatively small, and its cell fates have already been determined by the D series of macromeres. Gastrulation is accomplished primarily by epiboly, wherein the micromeres at the animal cap multiply and "overgrow" the vegetal macromeres (Collier 1997; van den Biggelaar and Dictus 2004). Eventually, the micromeres cover the entire embryo, leaving a small blastopore slit at the vegetal pole (**Figure 5.31**). Molluscs are protostomes, forming their mouth regions from the blastopore; thus this slit will become the mouth.

---

*The MAP kinase cascade is also seen in the 3D blastomere of equally cleaving spiralian embryos, and thus may represent an evolutionarily ancient mechanism for specifying the dorsal-ventral axis among all spirally cleaving taxa (Henry 2002; Lambert and Nagy 2003). In vertebrates, the MAP kinase cascade is the classic response to FGF paracrine factors that activate receptor tyrosine kinases (see Chapter 3).

SIDELIGHTS & SPECULATIONS

Adaptation by Modifying Embryonic Cleavage

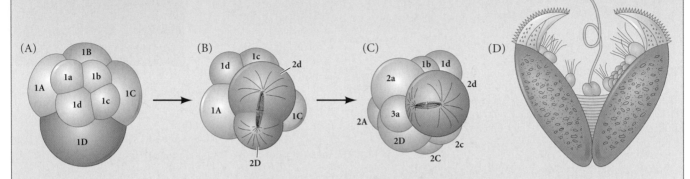

(A)   (B)   (C)   (D)

Figure 5.32   Formation of a glochidium larva by the modification of spiral cleavage. After the 8-cell embryo is formed (A), the placement of the mitotic spindle causes most of the D cytoplasm to enter the 2d blastomere (B). This large 2d blastomere divides (C), eventually giving rise to the large "bear-trap" shell of the larva (D). (After Raff and Kaufman 1983.)

Evolution is often the result of the hereditary alterations of embryonic development. Sometimes we are able to identify a specific modification of embryogenesis that has enabled the organism to survive in an otherwise inhospitable environment. One such modification, discovered by Frank Lillie in 1898, is brought about by an alteration of the typical pattern of molluscan spiral cleavage in the unionid family of clams.

Unlike most clams, *Unio* and its relatives live in swift-flowing streams. Streams create a problem for the dispersal of larvae: because the adults are sedentary, free-swimming larvae would always be carried downstream by the current. *Unio* clams have adapted to this environment via two modifications of their development. The first is an alteration in embryonic cleavage. In typical molluscan cleavage, either all the macromeres are equal in size or the 2D blastomere is the largest cell at that embryonic stage. However, cell division in *Unio* is such that the *2d* blastomere gets the largest amount of cytoplasm (**Figure 5.32**). This cell divides to produce most of the larval structures, including a gland capable of producing a large shell. The resulting larva is called a **glochidium** and resembles a tiny bear trap. Glochidia have sensitive hairs that cause the valves of the shell to

snap shut when they are touched by the gills or fins of a wandering fish. The larvae can thus attach themselves to the fish and "hitchhike" until they are ready to drop off and metamorphose into adult clams. In this manner, they can spread upstream as well as downstream.

In some unionid species, glochidia are released from the female's brood pouch and then wait passively for a fish to swim by. Some other species, such as *Lampsilis altilis*, have increased the chances of their larvae finding a fish by yet another developmental modification. Many clams develop a thin mantle that flaps around the shell and surrounds the brood pouch. In some unionids, the shape of the brood pouch (marsupium) and the undulations of the mantle mimic the shape and swimming behavior of a minnow (Welsh 1969). To make the deception even better, they develop a black "eyespot" on one end and a flaring "tail" on the other (**Figure 5.33**). When a predatory fish is lured within range of this "prey," the clam discharges the glochidia from the brood pouch and

the larvae attach to the fish's gills. Thus, the modification of existing developmental patterns has permitted unionid clams to survive in challenging environments.

Figure 5.33   Phony fish atop the unionid clam *Lampsilis altilis*. The "fish" is actually the brood pouch and mantle of the clam. The "eyes" and flaring "tail" attract predatory fish, and the glochidium larvae attach to the fish's gills. (Courtesy of Wendell R. Haag/USDA Forest Service.)

EARLY DEVELOPMENT IN TUNICATES

Tunicate Cleavage

Members of the tunicate subphylum are fascinating animals for several reasons, but the foremost is that they are invertebrate chordates. As Lemaire (2009) has written, "looking at an adult ascidian, it is difficult, and slightly degrading, to imagine that we are close cousins to these creatures." But even though tunicates such as ascidians lack vertebrae at all stages of their life cycles, a tunicate larva has a notochord and a dorsal nerve cord (making the tunicate a chordate). As larvae, they are free-swimming tadpoles; but when the tadpole undergoes metamorphosis, its nerve cord and notochord degenerate, and it secretes a cellulose tunic (which gave the name "tunicates" to these creatures).

Ascidians are characterized by **bilateral holoblastic cleavage**, a pattern found primarily in tunicates (**Figure 5.34**). The most striking feature of this type of cleavage is that the first cleavage plane establishes the earliest axis of symmetry in the embryo, separating the embryo into its future right and left sides. Each successive division orients itself to this plane of symmetry, and the half-embryo formed on one side of the first cleavage plane is the mirror image of the half-embryo on the other side. The second cleavage is meridional, like the first, but unlike the first division, it does not pass through the center of the egg. Rather, it creates two large anterior cells (the A and a blastomeres) and two smaller posterior cells (blastomeres B and b). Each side now has a large and a small blastomere.

Indeed, from the 8- through the 64-cell stages of tunicate development, every cell division is asymmetrical, such that the posterior blastomeres are always smaller than the anterior cells (Nishida 2005; Sardet et al. 2007). Prior to each of these unequal cleavages, the posterior centrosome in the blastomere migrates toward the **centrosome-attracting body** (**CAB**), a macroscopic subcellular structure. The CAB positions the centrosomes asymmetrically in the cell, resulting in one large and one small cell at each of these three divisions. The CAB also attracts particular mRNAs in such a way that these messengers are placed in the posteriormost (i.e., smaller) cell of each division (Hibino et al. 1998; Nishikata et al. 1999; Patalano et al. 2006). In this way, the CAB integrates cell patterning with cell determination. At the 64-cell stage, a small blastocoel is formed, and gastrulation begins from the vegetal pole.

The tunicate fate map

The fate map and cell lineages of the tunicate *Styela partita* are shown in Figure 1.12. Most of the early tunicate blastomeres are specified autonomously, each cell acquiring a specific type of cytoplasm that will determine its fate. In many tunicates, the different regions of cytoplasm have distinct pigmentation, and the cell fates can easily be seen to correspond to the type of cytoplasm taken up by each cell. These cytoplasmic regions are apportioned to the egg during fertilization. In the unfertilized egg of *Styela partita*, a central gray cytoplasm is enveloped by a cortical layer containing yellow lipid inclusions (**Figure 5.35A**). During meiosis, the breakdown of the nucleus releases a clear substance that accumulates in the animal hemisphere of the egg. Within 5 minutes of sperm entry, the inner clear and cortical yellow cytoplasms contract into the vegetal (lower) hemisphere of the egg (Prodon et al. 2005, 2008; Sardet et al. 2005). As the male pronucleus migrates from the vegetal pole to the equator of the cell along the future posterior side of the embryo, the yellow lipid inclusions migrate with it. This migration forms the **yellow crescent**, extending from the vegetal pole to the equator (**Figure 5.35B–D**);

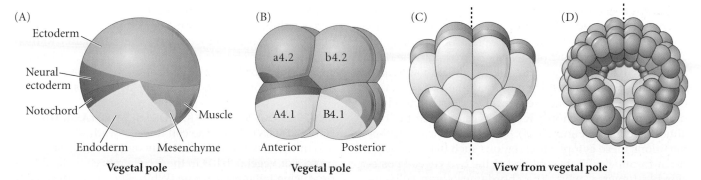

FIGURE 5.34 Bilateral symmetry in the egg of the ascidian tunicate *Styela partita*. (The cell lineages of *Styela* are shown in Figure 1.12C.) (A) Uncleaved egg. The regions of cytoplasm destined to form particular organs are labeled here and coded by color throughout the diagrams. (B) 8-cell embryo, showing the blastomeres and the fates of various cells. The embryo can be viewed as two 4-cell halves; from here on, each division on the right side of the embryo has a mirror-image division on the left. (C,D) Views of later embryos from the vegetal pole. The dashed line shows the plane of bilateral symmetry. (A after Balinsky 1981.)

(A)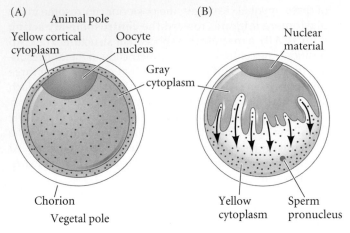

(B)

(C)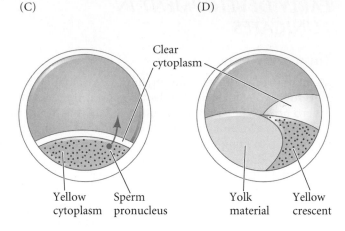

(D)

FIGURE 5.35 Cytoplasmic rearrangement in the fertilized egg of *Styela partita*. (A) Before fertilization, yellow cortical cytoplasm surrounds the gray yolky inner cytoplasm. (B) After sperm entry (in the vegetal hemisphere of the oocyte), the yellow cortical cytoplasm and the clear cytoplasm derived from the breakdown of the oocyte nucleus contract vegetally toward the sperm. (C) As the sperm pronucleus migrates animally toward the newly formed egg pronucleus, the yellow and clear cytoplasms move with it. (D) The final position of the yellow cytoplasm marks the location where cells give rise to tail muscles. (After Conklin 1905.)

this region will produce most of the tail muscles of the tunicate larva. The movement of these cytoplasmic regions depends on microtubules that are generated by the sperm centriole and on a wave of calcium ions that contracts the animal pole cytoplasm (Sawada and Schatten 1989; Speksnijder et al. 1990; Roegiers et al. 1995).

Edwin Conklin (1905) took advantage of the differing coloration of these regions of cytoplasm to follow each of the cells of the tunicate embryo to its fate in the larva (see Figure 1.12C). He found that cells receiving clear cytoplasm become ectoderm; those containing yellow cytoplasm give rise to mesodermal cells; those that incorporate slate gray inclusions become endoderm; and light gray cells become the neural tube and notochord. The cytoplasmic regions are localized bilaterally around the plane of symmetry, so they are bisected by the first cleavage furrow into the right and left halves of the embryo. The second cleavage causes the prospective mesoderm to lie in the two posterior cells, while the prospective neural ectoderm and chordamesoderm (notochord) will be formed from the two anterior cells (see Figure 5.34). The third division further partitions these cytoplasmic regions such that the mesoderm-forming cells are confined to the two vegetal posterior blastomeres, while the chordamesoderm cells are restricted to the two vegetal anterior cells.

See WEBSITE 5.4
The experimental analysis of tunicate cell specification

Autonomous and conditional specification of tunicate blastomeres

The autonomous specification of tunicate blastomeres was one of the first observations in the field of experimental embryology (Chabry 1888). Cohen and Berrill (1936) confirmed Chaby's and Conklin's results, and by counting the number of notochord and muscle cells, they demonstrated that larvae derived from only one of the first two blastomeres had half the expected number of cells. Reverberi and Minganti (1946) extended this analysis in a series of isolation experiments, and they, too, observed the self-differentiation of each isolated blastomere and of the remaining embryo. (The results of one of these experiments are shown in Figure II.2.) When the 8-cell embryo is separated into its four doublets (the right and left sides being equivalent), both mosaic and conditional specification are seen. The animal posterior pair of blastomeres gives rise to the ectoderm, and the vegetal posterior pair produces endoderm, mesenchyme, and muscle tissue—just as expected from the fate map. Autonomous specification is seen in the tunicate gut endoderm, muscle mesoderm, skin ectoderm, and neural cord.

However, organisms are formed by both autonomous and conditional specification. Some organs are not formed by the localization of cytoplasmic determinants. Rather, they are formed by the interactions of cells. Conditional specification (by induction) is seen in the formation of the brain, notochord, heart, and mesenchyme cells.

AUTONOMOUS SPECIFICATION OF THE MYOPLASM: THE YELLOW CRESCENT From the cell lineage studies of Conklin and others, it was known that only one pair of blastomeres (posterior vegetal; B4.1) in the 8-cell embryo is capable of producing tail muscle tissue (Whittaker 1982). These cells contain the yellow crescent cytoplasm. When yellow crescent cytoplasm is transferred from the B4.1 (muscle-forming) blastomere to the b4.2 (ectoderm-forming) blastomere of an 8-cell tunicate embryo, the ectoderm-forming blastomere generates muscle cells as well as its normal ecto-

FIGURE 5.36 Cytoplasmic segregation in the egg of *Boltenia villosa*. The yellow crescent, originally seen in the vegetal pole, becomes segregated into the B4.1 blastomere pair and thence into the muscle cells. (From Swalla 2004, courtesy of B. Swalla, K. Zigler, and M. Baltzley.)

(A)   (B) Yellow crescent   (C)

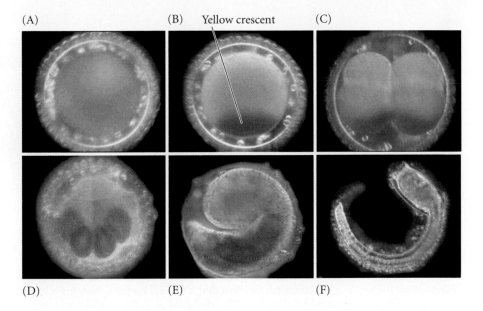

(D)   (E)   (F)

dermal progeny (**Figure 5.36**; see also Figure II.3). Moreover, cytoplasm from the yellow crescent area of the fertilized egg can cause the a4.2 blastomere to express muscle-specific proteins (Nishida 1992a). Conversely, when larval cell nuclei are transplanted into enucleated tunicate egg fragments, the newly formed cells show the structures typical of the egg regions providing the cytoplasm, not of those cells providing the nuclei (Tung et al. 1977). We can conclude, then, that certain determinants present in the egg cytoplasm cause the formation of certain tissues. These morphogenetic determinants appear to work by selectively activating or inactivating specific genes. The determination of the blastomeres and the activation of certain genes are controlled by the spatial localization of the morphogenetic determinants within the egg cytoplasm.

Using RNA hybridization techniques, Nishida and Sawada (2001) found particular mRNAs to be highly enriched in the vegetal hemisphere of the tunicate *Halocynthia roretzi*. One of these RNA messages encodes a zinc-finger transcription factor called **Macho-1**. *Macho-1* mRNA

was found to be concentrated in the vegetal hemisphere of the unfertilized egg and remains present during early fertilization. It appears to migrate with the yellow crescent cytoplasm into the posterior vegetal region of the egg during the second half of the first cell cycle. By the 8-cell stage, *macho-1* mRNA is found only in the B4.1 blastomeres. At the 16- and 32-cell stages, it is seen in those blastomeres that give rise to the muscle cells* (**Figure 5.37**).

When antisense oligonucleotides to deplete *macho-1* mRNA were injected into unfertilized eggs, the resulting larvae lacked all the muscles usually formed by the descendants of the B4.1 blastomere. (They did have the secondary muscles that are generated through the interactions of A4.1 and b4.2 blastomeres.) The tails of these *macho-1*-depleted larvae were severely shortened, but the other regions of the tadpoles appeared structurally and biochem-

---

Macho-1 mRNA is also localized in the cells that become the mesenchyme. However, FGF signals from the endoderm prevent these mesenchyme precursors from developing into muscle cells, as we will see later.

FIGURE 5.37 Autonomous specification by a morphogenetic factor. The *macho-1* mRNA message is localized to the muscle-forming tunicate cytoplasm. In situ hybridization shows the *macho-1* message found first in the vegetal pole cytoplasm (A), then migrating up the presumptive posterior surface of the egg (B) and becoming localized in the B4.1 blastomere (C). (From Nishida and Sawada 2001, courtesy of H. Nishida and N. Satoh.)

(A)   (B)   (C)

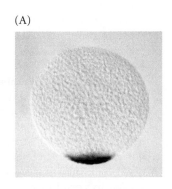 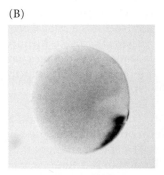 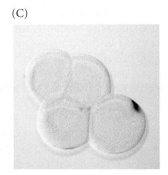

(A)　　　　　　　(B)　　　　　　　(C)

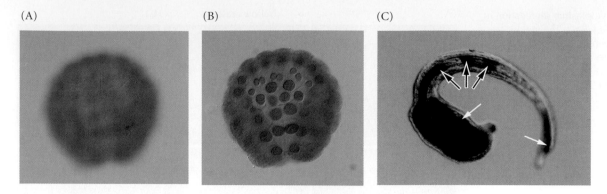

FIGURE 5.38 Antibody staining of β-catenin protein shows its involvement with endoderm formation. (A) No β-catenin is seen in the animal pole nuclei of a 110-cell *Ciona* embryo. (B) In contrast, nuclear β-catenin is readily seen in the nuclei in the vegetal endoderm precursors at the 110-cell stage. (C) When β-catenin is expressed in notochordal precursor cells, those cells will become endoderm and express endodermal markers such as alkaline phosphatase. The white arrows show normal endoderm; the black arrows show notochordal cells that are expressing endodermal enzymes. (From Imai et al. 2000, courtesy of H. Nishida and N. Satoh.)

ically normal. Moreover, B4.1 blastomeres isolated from *macho-1*-depleted embryos failed to produce muscle tissue. Nishida and Sawada then injected *macho-1* mRNA into cells that would not normally form muscle, and found that these ectoderm or endoderm precursors did generate muscle cells when given *macho-1* mRNA.

Macho-1 turns out to be a transcription factor that is required for the activation of several mesodermal genes, including *muscle actin, myosin, tbx6,* and *snail* (Sawada et al. 2005; Yagi et al. 2005a). Of these gene products, only the Tbx6 protein produced muscle differentiation (as Macho-1 did) when expressed in cells ectopically. Macho-1 thus appears to directly activate a set of *tbx6* genes, and Tbx6 proteins activate the rest of muscle development (Yagi et al. 2005b). Thus, the *macho-1* message is found at the right place and at the right time, and these experiments suggest that Macho-1 protein is both necessary and sufficient to promote muscle differentiation in certain ascidian cells.

The Macho-1 and Tbx6 proteins also appear to activate the muscle-specific gene *snail*. Snail protein is important in preventing *Brachyury* (*T*) expression in presumptive muscle cells, and is therefore needed to prevent the muscle precursors from becoming notochord cells. It appears, then, that the Macho-1 transcription factor is a critical component of the tunicate yellow crescent, muscle-forming cytoplasm. Macho-1 activates a transcription factor cascade that promotes muscle differentiation while at the same time inhibiting notochord specification.

AUTONOMOUS SPECIFICATION OF THE ENDODERM: β-CATENIN Presumptive endoderm originates from the vegetal A4.1 and B4.1 blastomeres. The specification of these cells coincides with the localization of β-catenin, a transcription factor discussed earlier in regard to sea urchin

endoderm specification. Inhibition of β-catenin results in the loss of endoderm and its replacement by ectoderm in the ascidian embryo (**Figure 5.38**; Imai et al. 2000). Conversely, increasing β-catenin synthesis causes an increase in the endoderm at the expense of the ectoderm (just as in sea urchins). The β-catenin transcription factor appears to function by activating the synthesis of the homeobox transcription factor Lhx-3. Inhibition of the *lhx-3* message prohibits the differentiation of endoderm (Satou et al. 2001).

CONDITIONAL SPECIFICATION OF THE MESENCHYME AND NOTOCHORD BY THE ENDODERM While most of the muscles are specified autonomously from the yellow crescent cytoplasm, the most posterior muscle cells form through conditional specification by cell interactions with the descendants of the A4.1 and b4.2 blastomeres (Nishida 1987, 1992a,b). Moreover, the notochord, brain, heart, and mesenchyme also form through inductive interactions. In fact, the notochord and mesenchyme appear to be induced by the fibroblast growth factor that is secreted by the endoderm cells (Nakatani et al. 1996; Kim et al. 2000; Imai et al. 2002).

The posterior cells that will become mesenchyme respond differently to the FGF signal due to the presence of Macho-1 in the posterior vegetal cytoplasm (**Figure 5.39**; Kobayashi et al. 2003). Macho-1 prevents notochord induction in the mesenchymal cell precursors by activating the *snail* gene (which will in turn suppress the activation of *Brachyury*). Thus, Macho-1 is not only a muscle-activating determinant, it is also a factor that distinguishes cell response to the FGF signal. These FGF-responding cells do not become muscle, because FGF also activates cascades that block muscle formation—another role that is conserved in vertebrates. As can be seen in Figure 5.40, the

FIGURE 5.39 The two-step process for specifying the marginal cells of the tunicate embryo. The first step involves the acquisition (or nonacquisition) by the cells of the Macho-1 transcription factor. The second step involves the reception (or nonreception) of the FGF signal from the endoderm. (After Kobayashi et al. 2003.)

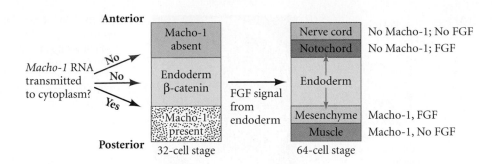

presence of Macho-1 changes the responses to endodermal FGFs, causing the anterior cells to form notochord while the posterior cells become mesenchyme.

Specification of the embryonic axes

The axes of the tunicate larva are among its earliest commitments. Indeed, all of its embryonic axes are determined by the cytoplasm of the zygote *prior* to first cleavage (see Sardet et al. 2007). The first axis to be determined is the dorsal-ventral axis, which is defined by the cap of cytoplasm at the vegetal pole. This vegetal cap is enriched for mitochondria, endoplasmic reticum components, and specific maternal mRNAs (such as *macho-1*) that will be involved in cell specification. This vegetal cap prefigures the future dorsal side of the larva and the site where gastrulation is initiated (Bates and Jeffery 1988). When small regions of vegetal pole cytoplasm were removed from zygotes (between the first and second waves of zygote cytoplasmic movement), the zygotes neither gastrulated nor formed a dorsal-ventral axis.

The anterior-posterior axis is the second axis to appear and is determined during the migration of the oocyte cytoplasm during fertilization. Microtubules originating from the sperm centrosome, followed by cortical actin microfilaments, cause the vegetal cap to become repositioned to what will be the posterior region of the embryo. This can be followed readily, since the yellow crescent forms in the region of the egg that will become the posterior side of the larva (see Figures 5.35 and 5.36). When roughly 10% of the cytoplasm from this posterior vegetal region of the egg was removed after the second wave of cytoplasmic movement, most of the embryos failed to form an anterior-posterior axis. Rather, these embryos developed into radially symmetrical larvae with anterior fates. This posterior vegetal cytoplasm (PVC) is "dominant" to other cytoplasms in that when it was transplanted into the anterior vegetal region of zygotes that had had their own PVC removed, the anterior of the cell became the new posterior, and the axis was reversed (Nishida 1994).

The specification of the left-right axis in tunicates is poorly understood. The first cleavage divides the embryo into its future right and left sides, but we do not yet know how these sides are specified. As in sea urchins and vertebrates, the left-right asymmetry is predicated on the expression of the *Nodal* gene. In tunicates, *Nodal* becomes expressed specifically in the left-side epidermis of the tail-bud-stage embryo (Morokuma et al. 2002; Yoshida and Sauga 2008). But how it becomes expressed there is still not known.

Gastrulation in Tunicates

Tunicates, like sea urchins and vertebrates, are deuterostomes and follow a pattern of gastrulation in which the blastopore becomes the anus. Tunicate gastrulation is characterized by the invagination of the endoderm, the involution of the mesoderm, and the epiboly of the ectoderm. About 4–5 hours after fertilization, the vegetal (endoderm) cells assume a wedge shape, expanding their apical margins and contracting near their vegetal margins (**Figure 5.40**). The A8.1 and B8.1 blastomere pairs appear to lead this invagination into the center of the embryo. The invagination forms a blastopore whose lips will become the mesodermal cells. The presumptive notochord cells are now on the anterior portion of the blastopore lip, while the presumptive tail muscle cells (from the yellow crescent) are on the posterior lip. The lateral lips comprise those cells that will become mesenchyme.

The second step of gastrulation involves the involution of the mesoderm. The presumptive mesoderm cells involute over the lips of the blastopore and, by migrating over the basal surfaces of the ectodermal cells, move inside the embryo. The ectodermal cells then flatten and epibolize over the mesoderm and endoderm, eventually covering the embryo. After gastrulation is complete, the embryo elongates along its anterior-posterior axis. The dorsal ectodermal cells that are the precursors of the neural tube invaginate into the embryo and are enclosed by neural folds. This process forms the neural tube, which will form a brain anteriorly and a spinal chord posteriorly. Meanwhile, the presumptive notochord cells on the right and left sides of the embryo migrate to the midline and interdigitate to form the notochord. The 40 cells of the notochord rearrange themselves from a 4-by-10 sheet of cells into a single row of 40 cells (Jiang et al. 2005). This intercalation and migration of notochord cells is called *convergent extension* (**Figure 5.41**; we also saw this phenomenon in our discussion of the sea urchin archenteron), and it extends the body axis along the anterior-posterior dimension.

(A)

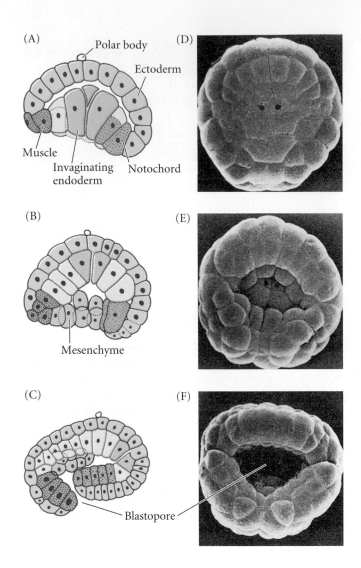

(B)

(C)

(D)

(E)

(F)

Polar body

Ectoderm

Muscle

Invaginating endoderm

Notochord

Mesenchyme

Blastopore

FIGURE 5.40 Gastrulation in the tunicate. Cross sections (A–C) and scanning electron micrographs viewed from the vegetal pole (D–F) illustrate the invagination of the endoderm (A,D), the involution of the mesoderm (B,E), and the epiboly of the ectoderm (C,F). Cell fates are color-coded as in Figure 5.34. (From Satoh 1978 and Jeffery and Swalla 1997, courtesy of N. Satoh.)

the genes responsible for heart differentiation. These two cells migrate to form two regions of cardiac mesoderm on the left and right ventral sides of the tadpole, just anterior to the tail. Like the heart precursor cells of vertebrate embryos, these two cell clusters migrate to meet at the ventral midline of the larva (Davidson and Levine 2003; Satou et al. 2004; Christiaen et al. 2008). After metamorphosis, they will form the functional heart of the adult. During this metamorphosis, the tail and brain degenerate and the tunicate no longer moves.*

THE NEMATODE C. ELEGANS

Our ability to analyze development requires appropriate model organisms. Sea urchins have long been a favorite of embryologists because their gametes are readily obtainable in large numbers, their eggs and embryos are transparent, and fertilization and development can occur under laboratory conditions. But sea urchins are difficult to rear in the laboratory for more than one generation, making their genetics difficult to study.

Geneticists (at least those who have worked with multicellular eukaryotes) have always favored *Drosophila*. The fruit fly's rapid life cycle, its readiness to breed, and the polytene chromosomes of the larva (which allow gene

Indeed, the convergent extension of notochordal precursor cells is characteristic across all the chordate phyla.

The muscle cells of the tail differentiate on either side of the neural tube and notochord (Jeffery and Swalla 1997). This forms the tadpole-like body of the larva. At the 110-cell stage, the B7.5 blastomere pairs express the conserved heart transcription factor Mesp. The anterior daughters of these B7.5 blastomeres respond to FGF signals to activate the cytoskeletal genes responsible for migration as well as

*Such a process, according to neurobiologist Rudolfo Llinás (1987), is "paralleled by some human academics upon obtaining university tenure."

FIGURE 5.41 Convergent extension of the tunicate notochord. The notochord is visualized by a green fluorescent protein (GFP) probe fused to a promoter of the *Brachyury* gene, which is usually expressed in the notochord. The notochordal precursor cells converge and extend the notochord down the length of the animal's tail. (From Deschet et al. 2003, courtesy of the authors.)

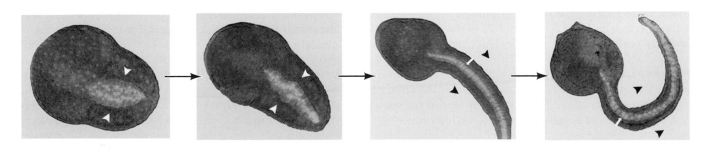

localization) make it superbly suited for hereditary analysis. But *Drosophila* development, as we will detail in Chapter 6, is complex.

A research program spearheaded by Sydney Brenner (1974) was established to identify an organism wherein it might be possible to identify each gene involved in development as well as to trace the lineage of each and every cell. Nematode roundworms seemed like a good group to start with, since embryologists such as Goldschmidt and Boveri had shown that several nematode species have a relatively small number of chromosomes and a small number of cells with invariant cell lineages.

Eventually, Brenner and his colleagues settled on *Caenorhabditis elegans*, a small (1 mm long), free-living (i.e., nonparasitic) soil nematode (**Figure 5.42A**). *C. elegans* has a rapid period of embryogenesis (about 16 hours), which it can accomplish in a petri dish, and relatively few cell types. Moreover, its predominant adult form is hermaphroditic, with each individual producing both eggs and sperm. These roundworms can reproduce either by self-fertilization or by cross-fertilization with the infrequently occurring males.

The body of an adult *C. elegans* hermaphrodite contains exactly 959 somatic cells, and the entire cell lineage has been traced through its transparent cuticle (**Figure 5.42B**; Sulston and Horvitz 1977; Kimble and Hirsh 1979). Furthermore, unlike vertebrate cell lineages, the *C. elegans* lineage is almost entirely invariant from one individual to the next. There is little room for randomness (Sulston et al. 1983). It also has a very compact genome. Although it has about the same number of genes as human beings (*C. elegans* has 18,000–20,000 genes, whereas *H. sapiens* has 20,000–25,000), the nematode has only about 3% the number of nucleotides in its genome (Hodgkin 1998, 2001).* The *C. elegans* genome was the first complete sequence ever obtained for a multicellular organism (*C. elegans* Sequencing Consortium 1999).

C. elegans has the rudiments of nearly all the major types of bodily systems (feeding, nervous, reproductive,

---

*This similarity in gene number was rather surprising, to say the least. "What does a worm want with 20,000 genes?" wrote Jonathan Hodgkin, curator of the *C. elegans* gene map. Humans have trillions of cells, a four-chambered heart, an incredibly regionalized brain, intricate limbs, and remarkable vascular networks. *C. elegans*, on the other hand, has no hands. Nor does the nematode have chambers in its heart, or any head to speak of. Thousands of these organisms would fit under our fingernails (which *C. elegans* also lacks). Hodgkin (2001) notes that human genes and their proteins tend to be more multifunctional than their nematode counterparts. Whereas human developmental proteins often have many functions, each *C. elegans* protein appears to have just a single function. Nematode genes do not have nearly the capacity for producing alternatively spliced RNAs as human genes do. In addition, *C. elegans* may have duplicated many of its genes, thus inflating its gene number. Whatever the case, the mere *number* of genes does not seem to be responsible for the huge physical differences between worms and human beings.

etc.—although there is no skeleton), and it exhibits an aging phenotype before it dies. In addition, *C. elegans* is particularly friendly to molecular biologists. DNA injected into *C. elegans* cells is readily incorporated into their nuclei, and *C. elegans* can take up antisense RNA from its culture medium.

Cleavage and Axis Formation in *C. elegans*

Rotational cleavage of the C. elegans egg

The zygote of *Caenorhabditis* exhibits rotational holoblastic cleavage (**Figure 5.42C,D**). During early cleavage, each asymmetrical division produces one founder cell (denoted AB, E, MS, C, and D) that produces differentiated descendants; and one stem cell (the P1–P4 lineage). In the first cell division, the cleavage furrow is located asymmetrically along the anterior-posterior axis of the egg, closer to what will be the posterior pole. It forms an anterior founder cell (AB) and a posterior stem cell (P1). During the second division, the founder cell (AB) divides equatorially (longitudinally; 90 degrees to the anterior-posterior axis), while the P1 cell divides meridionally (transversely) to produce another founder cell (EMS) and a posterior stem cell (P2). The stem cell lineage always undergoes meridional division to produce (1) an anterior founder cell and (2) a posterior cell that will continue the stem cell lineage.

The descendants of each founder cell can be observed through the transparent cuticle and divide at specific times in ways that are nearly identical from individual to individual. In this way, the exactly 558 cells of the newly hatched larva are generated. Descendant cells are named according to their positions relative to their sister cells. For instance, ABal is the "left-hand" daughter cell of the ABa cell, and ABa is the "anterior" daughter cell of the AB cell.

Anterior-posterior axis formation

The elongated axis of the *C. elegans* egg defines the future anterior-posterior axis of the nematode's body. The decision as to which end will become the anterior and which the posterior seems to reside with the position of the sperm pronucleus. When it enters the oocyte cytoplasm, the centriole associated with the sperm pronucleus initiates cytoplasmic movements that push the male pronucleus to the nearest end of the oblong oocyte. That end becomes the posterior pole (Goldstein and Hird 1996). The integration of cell division, cell specification, and morphogenesis is coordinated by the sperm and several PAR ("partitioning") proteins. The sperm provides a protein, CYK-4, that activates GTPases in the egg cytoplasm. As discussed in Chapter 3, GTPases (G proteins) are often involved in altering cytoskeletal proteins. Here, CYK-4 from the sperm cytoplasm activates the egg actin microfilaments to reposition PAR proteins, pushing them anteriorly (**Figure 5.43**). If females are mated with males whose CYK-4 has been

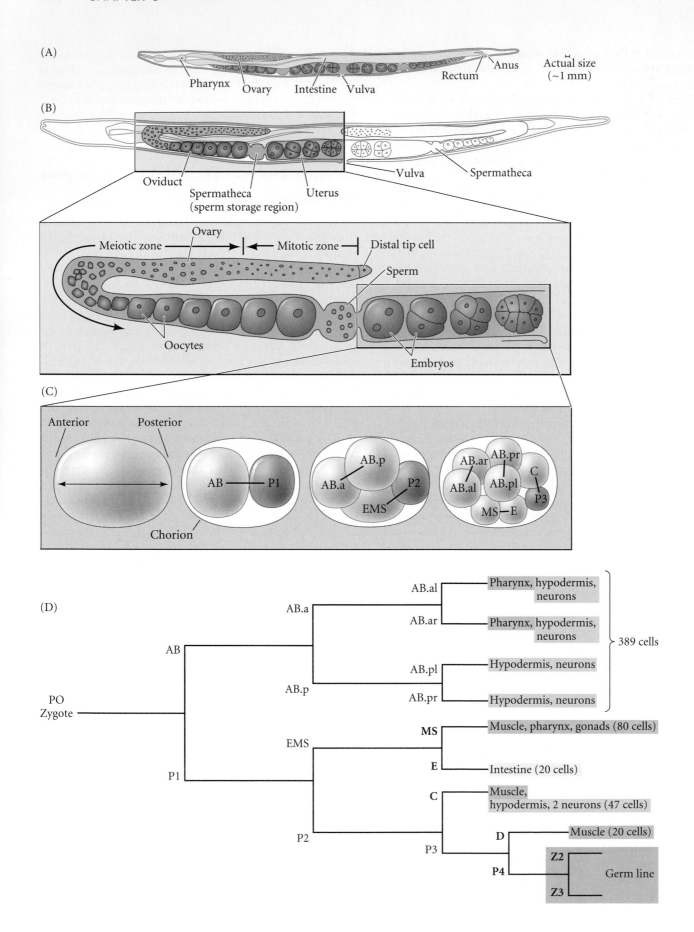

(A)

Pharynx Ovary Intestine Vulva Rectum Anus

Actual size
(~1 mm)

(B)

Oviduct Spermatheca (sperm storage region) Uterus Vulva Spermatheca

Ovary

Meiotic zone Mitotic zone Distal tip cell

Sperm

Oocytes

Embryos

(C)

Anterior Posterior

AB P1

AB.p P2 AB.ar AB.pr C

AB.a AB.al AB.pl P3

EMS MS—E

Chorion

(D)

AB.al — Pharynx, hypodermis, neurons

AB.a

AB.ar — Pharynx, hypodermis, neurons

AB

AB.pl — Hypodermis, neurons

389 cells

AB.p

AB.pr — Hypodermis, neurons

PO
Zygote

MS — Muscle, pharynx, gonads (80 cells)

EMS

E — Intestine (20 cells)

P1

C — Muscle, hypodermis, 2 neurons (47 cells)

D — Muscle (20 cells)

P2

P3

Z2

P4

Germ line

Z3

◀ **FIGURE 5.42**   The nematode *Caenorhabditis elegans*. (A) Side view of adult hermaphrodite. Sperm are stored such that a mature egg must pass through the sperm on its way to the vulva. (B) The gonads. Near the distal end, the germ cells undergo mitosis. As they move farther from the distal tip, they enter meiosis. Early meioses form sperm, which are stored in the spermatheca. Later meioses form eggs, which are fertilized as they roll through the spermatheca. (C) Early development, as the egg is fertilized and moves toward the vulva. The P lineage consists of stem cells that will eventually form the germ cells. (D) Abbreviated cell lineage chart. The germ line segregates into the posterior portion of the most posterior (P) cell. The first three cell divisions produce the AB, C, MS, and E lineages. The number of derived cells (in parentheses) refers to the 558 cells present in the newly hatched larva. Some of these continue to divide to produce the 959 somatic cells of the adult. (After Pines 1992, based on Sulston and Horvitz 1977 and Sulston et al. 1983.)

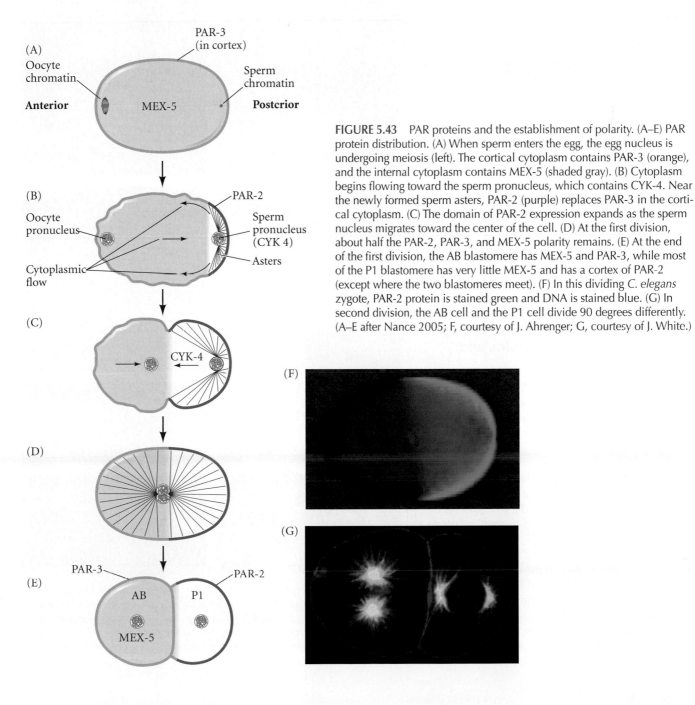

FIGURE 5.43   PAR proteins and the establishment of polarity. (A–E) PAR protein distribution. (A) When sperm enters the egg, the egg nucleus is undergoing meiosis (left). The cortical cytoplasm contains PAR-3 (orange), and the internal cytoplasm contains MEX-5 (shaded gray). (B) Cytoplasm begins flowing toward the sperm pronucleus, which contains CYK-4. Near the newly formed sperm asters, PAR-2 (purple) replaces PAR-3 in the cortical cytoplasm. (C) The domain of PAR-2 expression expands as the sperm nucleus migrates toward the center of the cell. (D) At the first division, about half the PAR-2, PAR-3, and MEX-5 polarity remains. (E) At the end of the first division, the AB blastomere has MEX-5 and PAR-3, while most of the P1 blastomere has very little MEX-5 and has a cortex of PAR-2 (except where the two blastomeres meet). (F) In this dividing *C. elegans* zygote, PAR-2 protein is stained green and DNA is stained blue. (G) In second division, the AB cell and the P1 cell divide 90 degrees differently. (A–E after Nance 2005; F, courtesy of J. Ahrenger; G, courtesy of J. White.)

knocked down with RNAi, many of the result-ing embryos lack polarity and die (Jenkins et al. 2006).

Some of the most important entities localized by the PAR proteins are the **P-granules**, ribonu-cleoprotein complexes that specify the germ cells. The P-granules appear to be a collection of trans-lation regulators. The proteins of these granules include RNA helicases, polyA polymerases, and translation initiation factors (Amiri et al. 2001; Smith et al. 2002; Wang et al. 2002). Using fluo-rescent antibodies to a component of the P-gran-ules, Strome and Wood (1983) discovered that shortly after fertilization, the randomly scattered P-granules move toward the posterior end of the zygote, so that they enter only the blastomere (P1) formed from the posterior cytoplasm (**Fig-ure 5.44**). The P-granules of the P1 cell remain in the posterior of the P1 cell when it divides and are thereby passed to the P2 cell. During the divi-sion of P2 and P3, however, the P-granules become associated with the nucleus that enters the P3 cytoplasm. Eventually, the P-granules will reside in the P4 cell, whose progeny become the sperm and eggs of the adult. The localization of the P-granules requires microfilaments but can occur in the absence of microtubules. Treating the zygote with cytochalasin D (a microfilament inhibitor) prevents the segregation of these gran-ules to the posterior of the cell, whereas demecol-cine (a colchicine-like microtubule inhibitor) fails to stop this movement.

See WEBSITE 5.5 P-granule migration

Formation of the dorsal-ventral and right-left axes

The dorsal-ventral axis of the nematode is seen in the division of the AB cell. As the AB cell divides, it becomes longer than the eggshell is wide. This causes the cells to slide, resulting in one AB daughter cell being anterior and one being posterior (hence their respective names, ABa and ABp; see Figure 5.43). This squeezing also causes the ABp cell to take a position above the EMS cell that results from the division of the P1 blastomere. The ABp cell defines the future dorsal side of the embryo, while the EMS cell—the precursor of the muscle and gut cells—marks the future ventral surface of the embryo. The left-right axis is specified later, at the 12-cell stage, when the MS blastomere (from the division of the EMS cell) contacts half the "granddaughters" of the ABa cell, distinguishing the right side of the body from the left side (Evans et al. 1994).

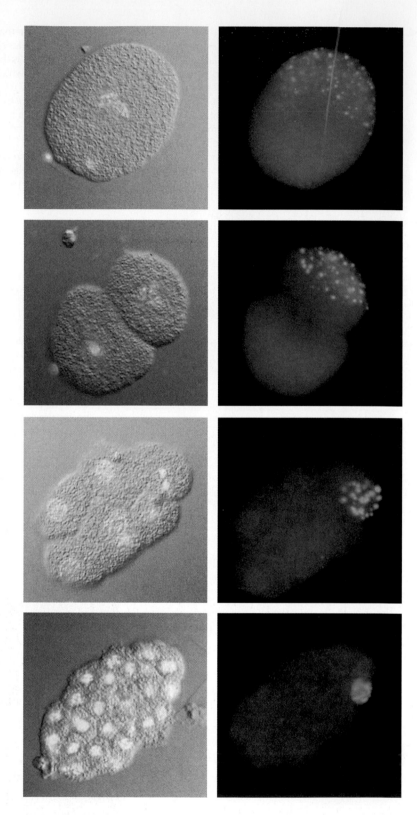

FIGURE 5.44 Segregation of the P-granules into the germ line lineage of the *C. elegans* embryo. The left column shows the cell nuclei (the DNA is stained blue by Hoescht dye); the right column shows the same embryos stained for P-granules. At each successive division, the P-granules enter the P-lineage blas-tomere, whose progeny will become the germ cells. (Courtesy of S. Strome.)

This asymmetric signaling (by Delta in the MS blastomere activating Notch in the cell it contacts) sets the stage for several other inductive events that make the right side of the larva differ from the left (Hutter and Schnabel 1995). Indeed, even the different neuronal fates seen on the left and right sides of the *C. elegans* brain can be traced back to that single change at the 12-cell stage (Poole and Hobert 2006).

Control of blastomere identity

C. elegans demonstrates both the conditional and autonomous modes of cell specification. Both modes can be seen if the first two blastomeres are experimentally separated (Priess and Thomson 1987). The P1 cell develops autonomously without the presence of AB, generating all the cells it would normally make, and the result is the posterior half of an embryo. However, the AB cell in isolation makes only a small fraction of the cell types it would normally make. For instance, the resulting ABa blastomere fails to make the anterior pharyngeal muscles that it would have made in an intact embryo. Therefore, the specification of the AB blastomere is conditional, and it needs to interact with the descendants of the P1 cell in order to develop normally.

AUTONOMOUS SPECIFICATION The determination of the P1 lineages appears to be autonomous, with the cell fates determined by internal cytoplasmic factors rather than by interactions with neighboring cells. The P-granules are localized in a way consistent with a role as a morpho-

genetic determinant, and they act through translational regulation (the exact mechanism is not well understood) to specify germ cells. Meanwhile, the SKN-1, PAL-1, and PIE-1 proteins are thought to encode transcription factors that act intrinsically to determine the fates of cells derived from the four P1-derived somatic founder cells (MS, E, C, and D).

The **SKN-1** protein is a maternally expressed polypeptide that may control the fate of the EMS blastomere, which is the cell that generates the posterior pharynx. After first cleavage, only the posterior blastomere—P1— has the ability to produce pharyngeal cells when isolated. After P1 divides, only EMS is able to generate pharyngeal muscle cells in isolation (Priess and Thomson 1987). Similarly, when the EMS cell divides, only one of its progeny, MS, has the intrinsic ability to generate pharyngeal tissue. These findings suggest that pharyngeal cell fate may be determined autonomously, by maternal factors residing in the cytoplasm that are parceled out to these particular cells.

Bowerman and co-workers (1992a,b, 1993) found maternal effect mutants lacking pharyngeal cells and were able to isolate a mutation in the *skn-1* gene. Embryos from homozygous *skn-1*-deficient mothers lack both pharyngeal mesoderm and endoderm derivatives of EMS (**Figure 5.45**). Instead of making the normal intestinal and pharyngeal structures, these embryos seem to make extra hypodermal (skin) and body wall tissue where their intestine and pharynx should be. In other words, the EMS blastomere appears to be respecified as C. Only those cells destined to form pharynx or intestine are affected by this

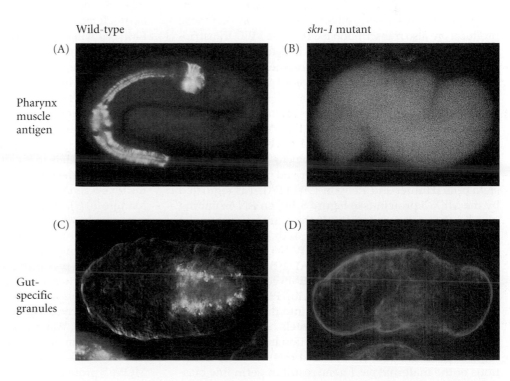

Wild-type   *skn-1* mutant

(A)   (B)

Pharynx muscle antigen

(C)   (D)

Gut-specific granules

FIGURE 5.45 Deficiencies of intestine and pharynx in *skn-1* mutants of *C. elegans*. Embryos derived from wild-type females (A,C) and from females homozygous for mutant *skn-1* (B,D) were tested for the presence of pharyngeal muscles (A,B) and gut-specific granules (C,D). A pharyngeal muscle-specific antibody labels the pharynx musculature of those embryos derived from wild-type females (A), but does not bind to any structure in the embryos from *skn-1* mutant females (B). Similarly, the birefringent gut granules characteristic of embryonic intestines (C) are absent from embryos derived from the *skn-1* mutant females (D). (From Bowerman et al. 1992a, courtesy of B. Bowerman.)

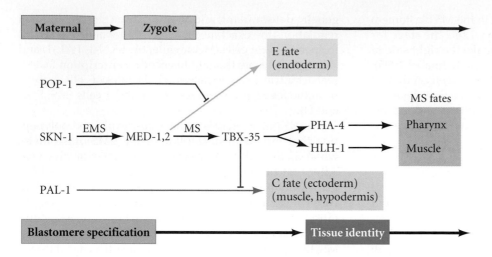

FIGURE 5.46 Model for specification of the MS blastomere. Maternal SKN-1 activates GATA transcription factors MED-1 and MED-2 in the EMS cell. The POP-1 signal prevents these proteins from activating the endodermal transcription factors (such as END-1) and instead activates the *tbx-35* gene. The TBX-35 transcription factor activates mesodermal genes in the MS cell, including *pha-4* in the pharynx lineage and *hlh-1* (which encodes a myogenic transcription factor) in muscles. TBX-35 also inhibits *pal-1* gene expression, thereby preventing the MS cell from acquiring the C-blastomere fates. (After Broitman-Maduro et al. 2006.)

mutation. Moreover, the protein encoded by the *skn-1* gene has a DNA-binding site motif similar to that seen in the bZip family of transcription factors (Blackwell et al. 1994).

SKN-1 is a maternal protein, and it activates the transcription of at least two genes, *med-1* and *med-2*, whose products are also transcription factors. The MED transcription factors appear to specify the entire fate of the EMS cell, since expression of the *med* genes in other cells can cause non-EMS cells to become EMS even if SKN-1 is absent (Maduro et al. 2001).

A second putative transcription factor, **PAL-1**, is also required for the differentiation of the P1 lineage. PAL-1 activity is needed for the normal development of the somatic descendants of the P2 blastomere. Embryos lacking PAL-1 have no somatic cell types derived from the C and D stem cells (Hunter and Kenyon 1996). PAL-1 is regulated by the MEX-3 protein (see Figure 5.44), an RNA-binding protein that appears to inhibit the translation of *pal-1* mRNA. Wherever MEX-3 is expressed, PAL-1 is absent. Thus, in *mex-3*-deficient mutants, PAL-1 is seen in every blastomere. SKN-1 also inhibits PAL-1 (thereby preventing it from becoming active in the EMS cell).

A third putative transcription factor, **PIE-1**, is necessary for germ line fate. PIE-1 is placed into the P blastomeres through the action of the PAR-1 protein, and it appears to inhibit both SKN-1 and PAL-1 function in the P2 and subsequent germ line cells (Hunter and Kenyon 1996). Mutations of the maternal *pie-1* gene result in germ line blas-

tomeres adopting somatic fates, with the P2 cell behaving similarly to a wild-type EMS blastomere. The localization and the genetic properties of PIE-1 suggest that it represses the establishment of somatic cell fate and preserves the totipotency of the germ cell lineage (Mello et al. 1996; Seydoux et al. 1996).

CONDITIONAL SPECIFICATION As mentioned earlier, the *C. elegans* embryo uses both autonomous and conditional modes of specification. Conditional specification can be seen in the development of the endoderm cell lineage. At the 4-cell stage, the EMS cell requires a signal from its neighbor (and sister), the P2 blastomere. Usually, the EMS cell divides into an MS cell (which produces mesodermal muscles) and an E cell (which produces the intestinal endoderm). If the P2 cell is removed at the early 4-cell stage, the EMS cell will divide into two MS cells, and no endoderm will be produced. If the EMS cell is recombined with the P2 blastomere, however, it will form endoderm; it will not do so, however, when combined with ABa, ABp, or both AB derivatives (Goldstein 1992). Specification of the MS cell begins with the maternal SKN-1 protein activating the genes encoding transcription factors such as MED-1 and MED-2. The POP-1 signal blocks the pathway to the E (endodermal) fate in the prospective MS cell to become MS by blocking the ability of MED-1 and MED-2 to activate the *tbx-35* gene (Figure 5.46; Broitman-Maduro et al. 2006; Maduro 2009). Throughout the animal kingdom, TBX proteins are known to be active in mesoderm formation; TBX-35 acts to activate the mesodermal genes in the pharynx (*PHA-4*) and muscles (*HLH-1*) of *C. elegans*.

The P2 cell produces a signal that interacts with the EMS cell and instructs the EMS daughter next to it to become the E cell. This message is transmitted through the Wnt signaling cascade (Figure 5.47; Rocheleau et al. 1997; Thorpe et al. 1997; Walston et al. 2004). The P2 cell produces a *C. elegans* homologue of a Wnt protein, the MOM-2 peptide. The MOM-2 peptide is received in the EMS cell by the MOM-5 protein, a *C. elegans* version of the Wnt receptor

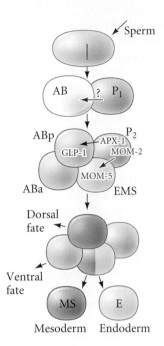

FIGURE 5.47 Cell-cell signaling in the 4-cell embryo of *C. elegans*. The P2 cell produces two signals: (1) the juxtacrine protein APX-1 (Delta), which is bound by GLP-1 (Notch) on the ABp cell, and (2) the paracrine protein MOM-2 (Wnt), which is bound by the MOM-5 (Frizzled) protein on the EMS cell. (After Han 1998.)

protein Frizzled. The result of this signaling cascade is to downregulate the expression of the *pop-1* gene in the EMS daughter destined to become the E cell. In *pop-1*-deficient embryos, both EMS daughter cells become E cells (Lin et al. 1995; Park et al. 2004).

The P2 cell is also critical in giving the signal that distinguishes ABp from its sister, ABa (see Figure 5.47). ABa gives rise to neurons, hypodermis, and the anterior pharynx cells, while ABp makes only neurons and hypodermal cells. However, if one experimentally reverses the positions of these two cells, their fates are similarly reversed and a normal embryo forms. In other words, ABa and ABp are equivalent cells whose fates are determined by their positions in the embryo (Priess and Thomson 1987). Transplantation and genetic studies have shown that ABp becomes different from ABa through its interaction with the P2 cell. In an unperturbed embryo, both ABa and ABp contact the EMS blastomere, but only ABp contacts the P2 cell. If the P2 cell is killed at the early 4-cell stage, the ABp cell does not generate its normal complement of cells (Bowerman et al. 1992a,b). Contact between ABp and P2 is essential for the specification of ABp cell fates, and the ABa cell can be made into an ABp-type cell if it is forced into contact with P2 (Hutter and Schnabel 1994; Mello et al. 1994).

This interaction is mediated by the GLP-1 protein on the ABp cell and the APX-1 (anterior pharynx excess) protein on the P2 blastomere. In embryos whose mothers have mutant *glp-1*, ABp is transformed into an ABa cell (Hutter

and Schnabel 1994; Mello et al. 1994). The GLP-1 protein is a member of a widely conserved family called the Notch proteins, which serve as cell membrane receptors in many cell-cell interactions; it is seen on both the ABa and ABp cells (Evans et al. 1994).* As mentioned earlier in the chapter (in the discussion of sea urchin cleavage), one of the most important ligands for Notch proteins such as GLP-1 is another cell surface protein called Delta. In *C. elegans*, the Delta-like protein is APX-1, and it is found on the P2 cell (Mango et al. 1994a; Mello et al. 1994). This APX-1 signal breaks the symmetry between ABa and ABp, since it stimulates the GLP-1 protein solely on the AB descendant that it touches—namely, the ABp blastomere. In doing this, the P2 cell initiates the dorsal-ventral axis of *C. elegans* and confers on the ABp blastomere a fate different from that of its sister cell.

INTEGRATION OF AUTONOMOUS AND CONDITIONAL SPECIFICATION: DIFFERENTIATION OF THE *C. ELEGANS* PHARYNX It should become apparent from the above discussion that the pharynx is generated by two sets of cells. One group of pharyngeal precursors comes from the EMS cell and is dependent on the maternal *skn-1* gene. The second group of pharyngeal precursors comes from the ABa blastomere and is dependent on GLP-1 signaling from the EMS cell. In both cases, the pharyngeal precursor cells (and only these cells) are instructed to activate the *pha-4* gene (Mango et al. 1994b). The *pha-4* gene encodes a transcription factor that resembles the mammalian HNF3β protein. Microarray studies by Gaudet and Mango (2002) revealed that the PHA-4 transcription factor activates almost all of the pharynx-specific genes. It appears that the PHA-4 transcription factor may be the node that takes the maternal inputs and transforms them into a signal that transcribes the zygotic genes necessary for pharynx development.

Gastrulation in *C. elegans*

Gastrulation in *C. elegans* starts extremely early, just after the generation of the P4 cell in the 24-cell embryo (Skiba and Schierenberg 1992). At this time, the two daughters of the E cell (Ea and Ep) migrate from the ventral side into the center of the embryo. There they divide to form a gut consisting of 20 cells. There is a very small and transient blastocoel prior to the movement of the Ea and Ep cells, and their inward migration creates a tiny blastopore. The next cell to migrate through this blastopore is the P4 cell,

---

*The GLP-1 protein is localized in the ABa and ABp blastomeres, but the maternally encoded *glp-1* mRNA is found throughout the embryo. Evans and colleagues (1994) have postulated that there might be some translational determinant in the AB blastomere that enables the *glp-1* message to be translated in its descendants. The *glp-1* gene is also active in regulating postembryonic cell-cell interactions. It is used later by the distal tip cell of the gonad to control the number of germ cells entering meiosis; hence the name GLP, for "germ line proliferation."

(A)

(B)

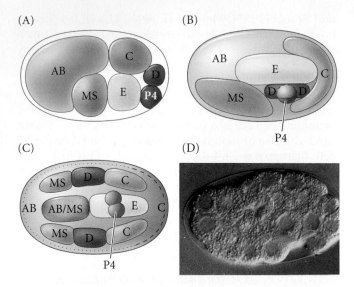

P4

(C)

(D)

P4

FIGURE 5.48   Gastrulation in *C. elegans*. (A) Positions of founder cells and their descendants at the 26-cell stage, at the start of gastrulation. (B) 102-cell stage, after the migration of the E, P4, and D descendants. (C) Positions of the cells near the end of gastrulation. The dotted and dashed lines represent regions of the hypodermis contributed by AB and C, respectively. (D) Early gastrulation, as the two E cells start moving inward. (After Schierenberg 1997; photograph courtesy of E. Schierenberg.

the precursor of the germ cells. It migrates to a position beneath the gut primordium. The mesodermal cells move in next: the descendants of the MS cell migrate inward from the anterior side of the blastopore, and the C- and D-derived muscle precursors enter from the posterior side. These cells flank the gut tube on the left and right sides (**Figure 5.48**; Schierenberg 1997). Finally, about 6 hours after fertilization, the AB-derived cells that contribute to the pharynx are brought inside, while the hypoblast (hypodermal precursor) cells move ventrally by epiboly, eventually closing the blastopore. The two sides of the hypodermis are sealed by E-cadherin on the tips of the leading cells that meet at the ventral midline (Raich et al. 1999).

During the next 6 hours, the cells move and develop into organs, while the ball-shaped embryo stretches out to become a worm with 558 somatic cells (see Priess and Hirsh 1986; Schierenberg 1997). There is evidence (Schnabel et al. 2006) that although these gastrulation move-

ments provide a good "first approximation" of the final form, an additional "cell focusing" is used to move cells into functional arrangements. Here, cells of the same fate sort out along the anterior-posterior axis. Other modeling takes place as well: an additional 115 cells undergo apoptosis (programmed cell death; see Chapter 3). After four molts, the worm is a sexually mature, hermaphroditic adult, containing exactly 959 somatic cells as well as numerous sperm and eggs.

One characteristic of *C. elegans* that distinguishes it from most other well-studied developing organisms is the prevalence of cell fusion. During *C. elegans* gastrulation, about one-third of all the cells fuse together to form syncytial cells containing many nuclei. The 186 cells that comprise the hypodermis (skin) of the nematode fuse into 8 syncytial cells, and cell fusion is also seen in the vulva, uterus, and pharynx. The functions of these fusion events can be determined by observing mutations that prevent syncytia from forming (Shemer and Podbilewicz 2000, 2003). The fusion prevents individual cells from migrating beyond their normal borders. In the vulva, cell fusion prevents hypodermis cells from adopting a vulval fate and making an ectopic (and nonfunctional) vulva.

The *C. elegans* research program integrates genetics and embryology to provide an understanding of the networks that govern cell differentiation and morphogenesis. In addition to providing some remarkable insights into how gene expression can change during development, studies of *C. elegans* have also humbled us by demonstrating how complex these networks are. Even in an organism as "simple" as *C. elegans*, with only a few genes and cell types, the right side of the body is made in a different manner from the left! The identification of the genes mentioned above is just the beginning of our effort to understand the complex interacting systems of development.

Coda

This chapter has described early embryonic development in four invertebrate groups, each of which develops in a different pattern. The largest group of animals on this planet, however, is another invertebrate group—the insects. We probably know more about the development of one particular insect, *Drosophila melanogaster*, than any other organism. The next chapter details the early development of this particularly well-studied creature.

 Snapshot Summary: *Early Invertebrate Development*

1. During cleavage, most cells do not grow. Rather, the volume of the oocyte is cleaved into numerous cells. The major exceptions to this rule are mammals.

2. "Blast" vocabulary: A *blastomere* is a cell derived from cleavage in an early embryo. A *blastula* is an

embryonic structure composed of blastomeres. The cavity within the blastula is the *blastocoel*. If the blastula lacks a blastocoel, it is a *stereoblastula*. (A mammalian blastula is called a *blastocyst*; see Chapter 8.) The invagination where gastrulation begins is the *blastopore*.

3. The blastomere cell cycle is governed by the synthesis and degradation of cyclin B. Cyclin B synthesis promotes the formation of mitosis-promoting factor, and MPF promotes mitosis. Degradation of cyclin B brings the cell back to the S phase. The G phases are added at the mid-blastula transition.

4. The movements of gastrulation include invagination, involution, ingression, delamination, and epiboly.

5. Three axes form the foundations of the body: the anterior-posterior axis (head to tail, or mouth to anus); the dorsal-ventral axis (back to belly); and the right-left axis (the two lateral sides of the body).

6. Body axes are established in different ways in different species. In some, such as sea urchins and tunicates, the axes are established at fertilization through determinants in the egg cytoplasm. In others, such as nematodes and snails, the axes are established by cell interactions later in development.

7. In all four invertebrates described, cleavage is holoblastic. In sea urchins, cleavage is radial; in snails, spiral; in tunicates, bilateral; and in nematodes, rotational.

8. In sea urchins and in tunicates, gastrulation occurs only after hundreds of cells have formed. The blastopore becomes the anus, and the mouth is formed elsewhere; this deuterostome mode of gastrulation is also characteristic of chordates (including vertebrates). In snails and in *C. elegans*, gastrulation occurs when there are relatively few cells, and the blastopore becomes the mouth. This is the protostome mode of gastrulation.

9. Sea urchin cell fates are determined both by autonomous and conditional modes of specification. The micromeres are specified autonomously and become a major signaling center for conditional specification of other lineages. Maternal β-catenin is important for the autonomous specification of the micromeres.

10. Differential cell adhesion is important in regulating sea urchin gastrulation. The micromeres detach first from the vegetal plate and move into the blastocoel. They form the skeletogenic mesenchyme, which becomes the skeletal rods of the pluteus larva. The vegetal plate invaginates to form the endodermal archenteron, with a tip of non-skeletogenic mesenchyme cells. The archenteron elongates by convergent extension and is guided to the future mouth region by the non-skeletogenic mesenchyme.

11. Snails exhibit spiral cleavage and form stereoblastulae, with no blastocoels. The direction of spiral cleavage is regulated by a factor encoded by the mother and placed in the oocyte. Spiral cleavage can be modified by evolution, and adaptations of spiral cleavage have allowed some molluscs to survive in otherwise harsh environments.

12. The polar lobe of certain molluscs contains the morphogenetic determinants for mesoderm and endoderm. These determinants enter the D blastomere.

13. The tunicate fate map is identical on its right and left sides. The yellow cytoplasm contains muscle-forming determinants; these act autonomously. The heart and nervous system of tunicates are formed conditionally, by signaling interactions between blastomeres.

14. The soil nematode *Caenorhabditis elegans* was chosen as a model organism because it has a small number of cells, has a small genome, is easily bred and maintained, has a short life span, can be genetically manipulated, and has a cuticle through which one can see cell movements.

15. In the early divisions of the *C. elegans* zygote, one daughter cell becomes a founder cell (producing differentiated descendants) and the other becomes a stem cell (producing other founder cells and the germ line).

16. Blastomere identity in *C. elegans* is regulated by both autonomous and conditional specification.

For Further Reading

Complete bibliographical citations for all literature cited in this chapter can be found at the free-access website **www.devbio.com**

Lambert, J. D. 2010. Developmental patterns in spiralian embryos. *Curr. Biol.* 20: 272–277.

Lemaire, P. 2009. Unfolding a chordate developmental program, one cell at a time: Invariant lineages, short-range inductions, and evolutionary plasticity in ascidians. *Dev. Biol.* 332: 48–60.

Nishida, H. and K. Sawada. 2001. *Macho-1* encodes a localized mRNA in ascidian eggs that specifies muscle fate during embryogenesis. *Nature* 409: 724–729.

Revilla-i-Domingo, R., P. Oliveri and E. H. Davidson. 2007. A missing link in the sea urchin embryo gene regulatory network: *hesC* and the double-negative specification of micromeres. *Proc. Natl. Acad. Sci. USA* 104: 12383–12388.

Shibazaki, Y., M. Shimizu and R. Kuroda. 2004. Body handedness is directed by genetically determined cytoskeletal dynamics in the early embryo. *Curr. Biol.* 14: 1462–1467.

Sulston, J. E., J. Schierenberg, J. White and N. Thomson. 1983. The embryonic cell lineage of the nematode *Caenorhabditis elegans*. *Dev. Biol.* 100: 64–119.

Wu, S. Y., M. Ferkowicz and D. R. McClay. 2007. Ingression of primary mesenchyme cells of the sea urchin embryo: A precisely timed epithelial mesenchymal transition. *Birth Def. Res. C Embryol. Today* 81: 241–252.

Go Online

WEBSITE 5.1 Sea urchin cell specification. The specification of sea urchin cells was one of the first major research projects in experimental embryology and remains a fascinating area of research. It appears that the initial signaling parses the blastula into domains characterized by the expression of specific transcription factors.

WEBSITE 5.2 Alfred Sturtevant and the genetics of snail coiling. By a masterful thought experiment, Sturtevant demonstrated the power of applying genetics to embryology. To do this, he brought Mendelian genetics into the study of snail coiling.

WEBSITE 5.3 Modifications of cell fate in spiralian eggs. Within the gastropods, differences in the timing of cell fate result in significantly different body plans. Furthermore, in the leeches and nemerteans, the spiralian cleavage pattern has been modified to produce new types of body plans.

WEBSITE 5.4 The experimental analysis of tunicate cell specification. Researchers analyzing tunicate development are using biochemical and molecular probes to find the morphogenetic determinants that are segregated to different regions of the egg cytoplasm. You may also want to look back at **WEBSITE 1.2**, which includes E. G. Conklin's remarkable 1905 fate map of the tunicate embryo. Conklin showed that "all the principal organs of the larva in their definitive positions and proportions are here marked out in the 2-cell stage by distinct kinds of protoplasm." His study of cell lineage has been the basis for all subsequent research on the autonomous specification of tunicates.

WEBSITE 5.5 P-granule migration. Susan Strome's laboratory produced these movies of P-granule migration under natural and experimental conditions. They show P-granule segregation to the P-lineage blastomeres except when perturbed by mutations or chemicals that inhibit microfilament function.

Vade Mecum

Sea urchin development. The Vade Mecum companion site provides an excellent review of sea urchin development, as well as questions on the fundamentals of echinoderm cleavage and gastrulation.

Outside Sites

Stanford University hosts two valuable and freely accessible websites on sea urchin development. A set of interactive tutorials developed in conjunction with the National Science Foundation is found at **http://virtualurchin. stanford.edu/**. Numerous ways of studying sea urchin development in the laboratory are described at **http:// www.stanford.edu/group/Urchin/contents.html**. Other laboratory protocols can be found at **http://www. swarthmore.edu/NatSci/sgilber1/DB_lab/Urchin/ urchin_protocols.html**.

At **http://sugp.caltech.edu/endomes/**, the Davidson Laboratory Sea Urchin Specification Project provides updated systems diagrams showing an hour-by hour account of the specification of the sea urchin cell types during the early cleavage stages. **http://sugp.caltech.edu/endomes/**

The genome of the sea urchin *Strongylocentrotus purpuratus* was sequenced in 2006. The urchin gene database can be accessed at **http://www.hgsc.bcm.tmc.edu/projects/ seaurchin/**.

For materials on tunicate development, Christian Sardet's excellent films and websites can be accessed at **http:// biodev.obs-vlfr.fr/recherche/biomarcell/ascidies/ ascidiemenu.htm**. The Four-Dimensional Ascidian Body Atlas website at **http://chordate.bpni.bio.keio.ac.jp/faba2/ 2.0/top.html** provides confocal images of developing tunicates.

The *C. elegans* community has been especially cognizant of the need for excellent online sources of information. Wormbase, at **http://www.wormbase.org/**, is the one-stop-shopping site for anything Caenorhabitic. The Goldstein laboratory has excellent movies of *C. elegans* development at **http://www.bio.unc.edu/faculty/goldstein/lab/movies. html**. Wormbook, **http://www.wormbook.org/**, is a superb web-based textbook.

The Genetics of Axis Specification in *Drosophila*

6

THANKS LARGELY TO STUDIES spearheaded by Thomas Hunt Morgan's laboratory during the first two decades of the twentieth century, we know more about the genetics of *Drosophila* than that of any other multicellular organism. The reasons have to do with both the flies themselves and with the people who first studied them. *Drosophila* is easy to breed, hardy, prolific, tolerant of diverse conditions, and the polytene chromosomes of its larvae (see Figure 15.17) are readily identified. The progress of *Drosophila* genetics was aided by the relatively free access of every scientist to the mutants and fly breeding techniques of every other researcher. Mutants were considered the property of the entire scientific community, and Morgan's laboratory established a database and exchange network whereby anyone could obtain them.

Undergraduates (starting with Calvin Bridges and Alfred Sturtevant) played important roles in *Drosophila* research, which achieved its original popularity as a source of undergraduate research projects. As historian Robert Kohler noted (1994), "Departments of biology were cash poor but rich in one resource: cheap, eager, renewable student labor." The *Drosophila* genetics program was "designed by young persons to be a young person's game," and the students set the rules for *Drosophila* research: "No trade secrets, no monopolies, no poaching, no ambushes."

But *Drosophila* was a difficult organism on which to study embryology. Although Jack Schultz (originally in Morgan's laboratory) and others following him attempted to relate the genetics of *Drosophila* to its development, the fly embryos proved too complex and intractable to study, being neither large enough to manipulate experimentally nor transparent enough to observe microscopically. It was not until the techniques of molecular biology allowed researchers to identify and manipulate the genes and RNAs of the insect that its genetics could be related to its development. And when that happened, a revolution occurred in the field of biology. This revolution is continuing, in large part because of the ability to generate transgenic flies at high frequency (Pfeiffer et al. 2009). This enables researchers to identify developmental interactions taking place in very small regions of the embryo, and it identifies enhancers that control developmental processes taking place rapidly and in small areas. The merging of our knowledge of the molecular aspects of *Drosophila* genetics with our knowledge of the fly's development built the foundations on which the current sciences of developmental genetics and evolutionary developmental biology are based.

Those of us who are at work on Drosophila *find a particular point to the question. For the genetic material available is all that could be desired, and even embryological experiments can be done.... It is for us to make use of these opportunities. We have a complete story to unravel, because we can work things from both ends at once.*

JACK SCHULTZ (1935)

The chief advantage of Drosophila *initially was one that historians have overlooked: it was an excellent organism for student projects.*

ROBERT E. KOHLER (1994)

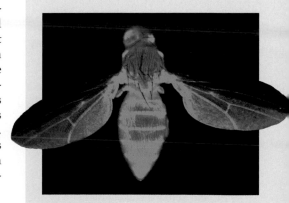

EARLY DROSOPHILA DEVELOPMENT

In Chapter 5 we discussed the specification of early embryonic cells by cytoplasmic determinants stored in the oocyte. The cell membranes that form during cleavage establish the region of cytoplasm incorporated into each new blastomere, and the incorporated morphogenetic determinants then direct differential gene expression in each cell. During *Drosophila* development, however, cellular membranes do not form until after the thirteenth nuclear division. Prior to this time, all the dividing nuclei share a common cytoplasm, and material can diffuse throughout the whole embryo. In these embryos, the specification of cell types along the anterior-posterior and dorsal-ventral axes is accomplished by the interactions of cytoplasmic materials *within* the single multinucleated cell. Moreover, the initiation of the anterior-posterior and dorsal-ventral differences is controlled by the position of the egg within the mother's ovary. Whereas the sperm entry site may fix the axes in ascidians and nematodes, the fly's anterior-posterior and dorsal-ventral axes are specified by interactions between the egg and its surrounding follicle cells.

Fertilization

Drosophila fertilization is not your standard sperm-meets-egg story. First, the sperm enters an egg that is already activated. Egg activation in *Drosophila* is accomplished at ovulation, a few minutes *before* fertilization begins. As the *Drosophila* oocyte squeezes through a narrow orifice, ion channels open, allowing calcium ions to flow into it. The oocyte nucleus then resumes its meiotic divisions and the cytoplasmic mRNAs become translated, even without fertilization (Mahowald et al. 1983; Fitch and Wakimoto 1998; Heifetz et al. 2001; Horner and Wolfner 2008). Second, there

is only one site—the **micropyle**, at the future dorsal anterior region of the embryo—where the sperm can enter the egg. The micropyle is a tunnel in the chorion (eggshell) that allows sperm to pass through it one at a time. The micropyle probably prevents polyspermy in *Drosophila*. There are no cortical granules to block polyspermy, although cortical changes are seen. Third, by the time the sperm enters the egg, the egg already has begun to specify its axes; thus the sperm enters an egg that is already organizing itself as an embryo. Fourth, there is competition between sperm. A sperm tail can be many times longer than the adult fly, and this huge tail is thought to block other sperm from entering the egg. In *Drosophilia melanogaster*, the sperm tail is 1.8 mm—about as long as the adult fly, and some 300 times longer than a human sperm. The entire sperm (huge tail and all) gets incorporated into the oocyte cytoplasm, and the sperm cell membrane does not break down until after it is fully inside the oocyte (Snook and Karr 1998; Clark et al. 1999).

See WEBSITE 6.1
Drosophila fertilization

Cleavage

Most insect eggs undergo **superficial cleavage**, wherein a large mass of centrally located yolk confines cleavage to the cytoplasmic rim of the egg. One of the fascinating features of this cleavage pattern is that cells do not form until after the nuclei have divided several times. Karyokinesis (nuclear division) occurs without cytokinesis (cell division), and the rapid rate of division is accomplished (as it is in sea urchin embryos) by eliminating the gap (G) stages of the cell cycle. Cleavage in the *Drosophila* egg creates a **syncytium**, a single cell with many nuclei residing in a common cytoplasm (**Figure 6.1**). The zygote nucleus under-

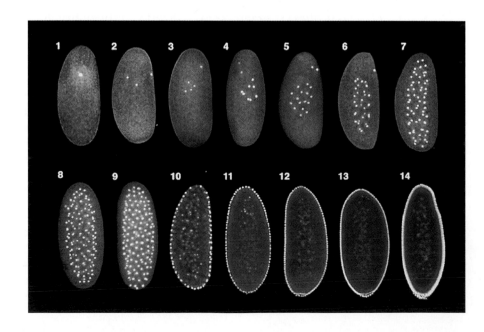

FIGURE 6.1 Laser confocal micrographs of stained chromatin showing superficial cleavage in a *Drosophila* embryo. The future anterior is positioned upward, and the numbers refer to the cell division cycle. The early nuclear divisions occur centrally. Later, the nuclei and their cytoplasmic islands (energids) migrate to the periphery of the cell. This creates the syncytial blastoderm. After cycle 13, the oocyte membranes ingress between the nuclei to form the cellular blastoderm. The pole cells (germ cell precursors) form in the posterior. (Courtesy of D. Daily and W. Sullivan.)

(A)

(B)

(C)

(D)

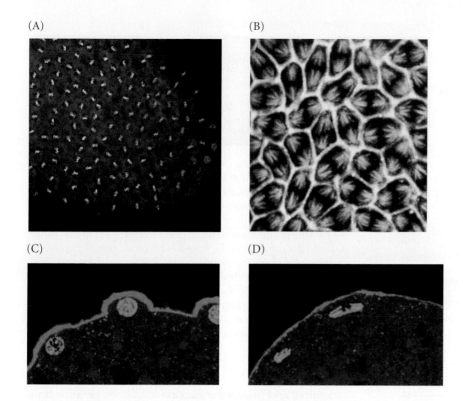

FIGURE 6.2 Nuclear and cell division in *Drosophila*. (A) Nuclear division (but not cell division) can be seen in the single cell of the *Drosophila* embryo using a dye that stains DNA. The first region to cellularize, the pole region, can be seen forming the cells that will eventually become the germ cells (sperm or eggs) of the fly. (B) Confocal fluorescence photomicrographs of nuclei dividing during cellularization of the blastoderm. While there are no cell boundaries, actin (green) can be seen forming regions within which each nucleus divides. The microtubules of the mitotic apparatus are stained red with antibodies to tubulin. (C,D) Cross section of a part of the stage 10 *Drosophila* embryo showing nuclei (green) in the cortex of the syncytial cell, near a layer of actin microfilaments (red). (C) Interphase nuclei. (D) Nuclei in anaphase, dividing parallel to the cortex and enabling the nuclei to stay in the cell periphery. (A from Bonnefoy et al. 2007; B from Sullivan et al. 1993, courtesy of W. Theurkauf and W. Sullivan; C,D from Foe 2000, courtesy of V. Foe.)

goes several mitotic divisions within the central portion of the egg; 256 nuclei are produced by a series of eight nuclear divisions averaging 8 minutes each. During the ninth division cycle, about five nuclei reach the surface of the posterior pole of the embryo. These nuclei become enclosed by cell membranes and generate the **pole cells** that give rise to the gametes of the adult. At cycle 10, the other nuclei migrate to the cortex (periphery) of the egg, where the mitoses continue, albeit at a progressively slower rate (**Figure 6.2**; Foe et al. 2000). During these stages of nuclear division, the embryo is called a **syncytial blastoderm**, since no cell membranes exist other than that of the egg itself.

The nuclei divide within a common cytoplasm, but this does not mean the cytoplasm is itself uniform. Karr and Alberts (1986) have shown that each nucleus within the syncytial blastoderm is contained within its own little territory of cytoskeletal proteins. When the nuclei reach the periphery of the egg during the tenth cleavage cycle, each nucleus becomes surrounded by microtubules and microfilaments. The nuclei and their associated cytoplasmic islands are called **energids**. Following division cycle 13, the oocyte plasma membrane folds inward between the nuclei, eventually partitioning off each somatic nucleus into a single cell. This process creates the **cellular blastoderm**, in which all the cells are arranged in a single-layered jacket around the yolky core of the egg (Turner and Mahowald 1977; Foe and Alberts 1983).

Like any other cell formation, the formation of the cellular blastoderm involves a delicate interplay between microtubules and microfilaments. The membrane movements, the nuclear elongation, and the actin polymerization each

appear to be coordinated by the microtubules (Riparbelli et al. 2007). The first phase of blastoderm cellularization is characterized by the invagination of cell membranes between the nuclei to form furrow canals (**Figure 6.3**). This process can be inhibited by drugs that block microtubules. After the furrow canals have passed the level of the nuclei, the second phase of cellularization occurs. The rate of invagination increases, and the actin-membrane complex begins to constrict at what will be the basal end of the cell (Foe et al. 1993; Schejter and Wieschaus 1993; Mazumdar and Mazumdar 2002). In *Drosophila*, the cellular blastoderm consists of approximately 6000 cells and is formed within 4 hours of fertilization.

The mid-blastula transition

After the nuclei reach the periphery, the time required to complete each of the next four divisions becomes progressively longer. While cycles 1–10 average 8 minutes each, cycle 13—the last cycle in the syncytial blastoderm—takes 25 minutes to complete. Cycle 14, in which the *Drosophila* embryo forms cells (i.e., after 13 divisions), is asynchronous. Some groups of cells complete this cycle in 75 minutes, other groups take 175 minutes (Foe 1989). Zygotic gene transcription (which begins around cycle 11) is greatly enhanced at this stage. This slowdown of nuclear division, cellularization, and concomitant increase in new RNA transcription is often referred to as the **mid-blastula transition** (see Chapter 5). It is at this stage that the maternally provided mRNAs are degraded and hand over control of development to the zygotic genome (Brandt et al. 2006;

(A)

(B)

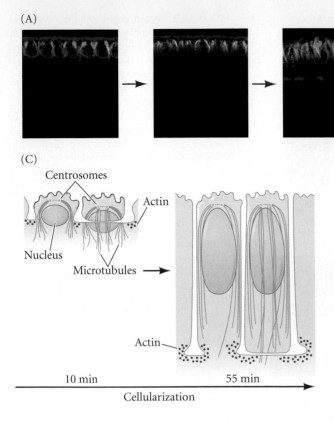

(C)

Centrosomes

Actin

Nucleus

Microtubules

Actin

10 min   55 min

Cellularization

FIGURE 6.3 Formation of the cellular blastoderm in *Drosophila*. Nuclear shape change and cellularization are coordinated through the cytoskeleton. (A) Cellularization and nuclear shape change shown by staining the embryo for microtubles (green), microfilaments (blue), and nuclei (red.) The red stain in the nuclei is due to the presence of the Kugelkern protein, one of the earliest proteins made from the zygotic nuclei. It is essential for nuclear elongation. (B) This embryo was treated with nocadozole to disrupt microtubules. The nuclei fail to elongate, and cellularization is prevented. (C) Diagrammatic representation of cell formation and nuclear elongation. (After Brandt et al. 2006; photographs courtesy of J. Grosshans and A. Brandt.)

De Renzis et al. 2007; Benoit et al. 2009). Such a maternal-to-zygotic transition is seen in the embryos of numerous vertebrate and invertebrate phyla.

In *Drosophila*, the coordination of the mid-blastula transition and the maternal-to-zygotic transition is controlled by several factors, including (1) the ratio of chromatin to cytoplasm; (2) Smaug protein; and (3) cell cycle regulators. The ratio of chromatin to cytoplasm is a consequence of the increasing amount of DNA while the cytoplasm remains constant (Newport and Kirschner 1982; Edgar et al. 1986a). Edgar and his colleagues compared the early development of wild-type *Drosophila* embryos with that of haploid mutants. The haploid *Drosophila* embryos had half the wild-type quantity of chromatin at each cell division. Hence, a haploid embryo at cell division cycle 8 had the same amount of chromatin that a wild-type embryo had at cycle 7. The investigators found that, whereas wild-type embryos formed a cellular blastoderm immediately after the thirteenth division, haploid embryos underwent an extra, fourteenth, division before cellularization. Moreover, the lengths of cycles 11–14 in wild-type embryos corresponded to those of cycles 12–15 in the haploid embryos. Thus, the haploid embryos followed a pattern similar to that of the wild-type embryos—but they lagged by one cell division.

Smaug (yes, it's named after the dragon in *Lord of the Rings*) is an RNA-binding protein often involved in repressing translation. During the mid-blastula transition, however, it targets the maternal mRNAs for destruction (Tadros et al. 2007; Benoit et al. 2009). Maternal Smaug mutants disrupt the slowing down of nuclear division, prevent cellu-

larization, and thwart the increase in zygotic genome transcription. Smaug is encoded by a maternal mRNA, and Smaug protein levels increase during the early cleavage divisions. These levels peak when the zygotic genome begins efficient transcription. Moreover, if Smaug is artificially added to the anterior of an early *Drosophila* embryo, there results a concomitant gradient in the timing of maternal transcript destruction, cleavage cell cycle delays, zygotic gene transcription, cellularization, and gastrulation. Thus, Smaug accumulation appears to regulate the progression from maternal to nuclear control of development and coordinates this progression with the mid-blastula transition.

Cell cycle regulators are critical for introducing the gap stages into the cell cycle and slowing it down. As the maternal replication factors are depleted, the zygotically encoded replication factors take over and regulate the accumulation of cyclins in the cell (Sibon et al. 1997; Royou et al. 2008).

See VADE MECUM *Drosophila* **development**

See WEBSITE 6.2
The early development of other insects

Gastrulation

Gastrulation begins shortly after the mid-blastula transition. The first movements of *Drosophila* gastrulation segregate the presumptive mesoderm, endoderm, and ectoderm. The prospective mesoderm—about 1000 cells constituting the ventral midline of the embryo—folds inward to produce the ventral furrow (**Figure 6.4A**). This furrow eventually pinches off from the surface to become a ventral tube within the embryo. The prospective endoderm invaginates

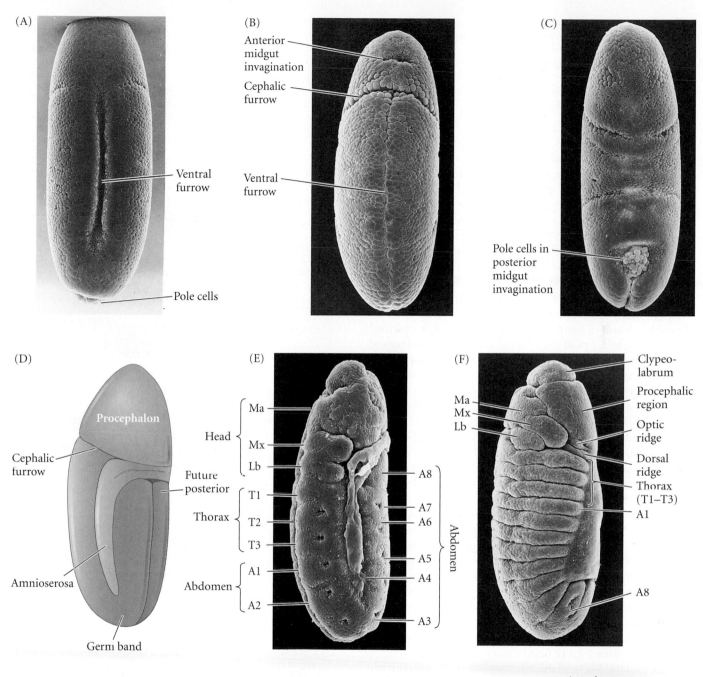

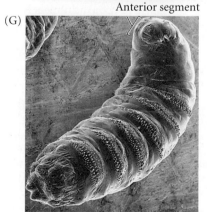

FIGURE 6.4 Gastrulation in *Drosophila*. The anterior points upward in each figure. (A) Ventral furrow beginning to form as cells flanking the ventral midline invaginate. (B) Closing of ventral furrow, with mesodermal cells placed internally and surface ectoderm flanking the ventral midline. (C) Dorsal view of a slightly older embryo, showing the pole cells and posterior endoderm sinking into the embryo. (D) Dorsolateral view of an embryo at fullest germ band extension, just prior to segmentation. The cephalic furrow separates the future head region (procephalon) from the germ band, which will form the thorax and abdomen. (E) Lateral view, showing fullest extension of the germ band and the beginnings of segmentation. Subtle indentations mark the incipient segments along the germ band. Ma, Mx, and Lb correspond to the mandibular, maxillary, and labial head segments; T1–T3 are the thoracic segments; and A1–A8 are the abdominal segments. (F) Germ band reversing direction. The true segments are now visible, as well as the other territories of the dorsal head, such as the clypeolabrum, procephalic region, optic ridge, and dorsal ridge. (G) Newly hatched first-instar larva. (Photographs courtesy of F. R. Turner. D after Campos-Ortega and Hartenstein 1985.)

FIGURE 6.5 Schematic representation of gastrulation in *Drosophila*. Anterior is to the left; dorsal is facing upward. (A,B) Surface and cutaway views showing the fates of the tissues immediately prior to gastrulation. (C) The beginning of gastrulation as the ventral mesoderm invaginates into the embryo. (D) This view corresponds to Figure 6.4A, while (E) corresponds to Figure 6.4B,C. In (E), the neuroectoderm is largely differentiated into the nervous system and the epidermis. (After Campos-Ortega and Hartenstein 1985.)

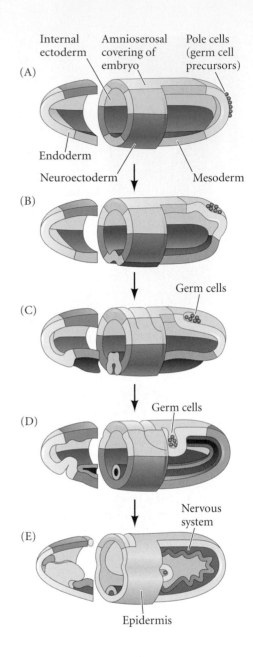

to form two pockets at the anterior and posterior ends of the **ventral furrow**. The pole cells are internalized along with the endoderm (**Figure 6.4B,C**). At this time, the embryo bends to form the **cephalic furrow**.

The ectodermal cells on the surface and the mesoderm undergo convergence and extension, migrating toward the ventral midline to form the **germ band**, a collection of cells along the ventral midline that includes all the cells that will form the trunk of the embryo. The germ band extends posteriorly and, perhaps because of the egg case, wraps around the top (dorsal) surface of the embryo (**Figure 6.4D**). Thus, at the end of germ band formation, the cells destined to form the most posterior larval structures are located immediately behind the future head region (**Figure 6.4E**). At this time, the body segments begin to appear, dividing the ectoderm and mesoderm. The germ band then retracts, placing the presumptive posterior segments at the posterior tip of the embryo (**Figure 6.4F**). At the dorsal surface, the two sides of the epidermis are brought together in a process called **dorsal closure**. The amnioserosa, which had been the most dorsal structure, interacts with the epidermal cells to encourage their migration (reviewed in Panfilio 2007; Heisenberg 2009).

While the germ band is in its extended position, several key morphogenetic processes occur: organogenesis, segmentation, and the segregation of the imaginal discs.* In addition, the nervous system forms from two regions of ventral ectoderm. Neuroblasts differentiate from this neurogenic ectoderm within each segment (and also from the nonsegmented region of the head ectoderm). Therefore, in insects like *Drosophila*, the nervous system is located ventrally, rather than being derived from a dorsal neural tube as in vertebrates (**Figure 6.5**).

The general body plan of *Drosophila* is the same in the embryo, the larva, and the adult, each of which has a distinct head end and a distinct tail end, between which are repeating segmental units (**Figure 6.6**). Three of these segments form the thorax, while another eight segments form the abdomen. Each segment of the adult fly has its own identity. The first thoracic segment, for example, has only

legs; the second thoracic segment has legs and wings; and the third thoracic segment has legs and *halteres* (balancing organs). Thoracic and abdominal segments can also be distinguished from each other by differences in the cuticle of the newly hatched first-instar larvae.

GENES THAT PATTERN THE DROSOPHILA *BODY PLAN*

Most of the genes involved in shaping the larval and adult forms of *Drosophila* were identified in the early 1990s using a powerful "forward genetics" approach. The basic strategy was to randomly mutagenize flies and then screen for mutations that disrupted the normal formation of the body

---

*Imaginal discs are those cells set aside to produce the adult structures. The details of imaginal disc differentiation will be discussed in Chapter 15. For more information on *Drosophila* developmental anatomy, see Bate and Martinez-Arias 1993; Tyler and Schetzer 1996; and Schwalm 1997.

(A)

(B)

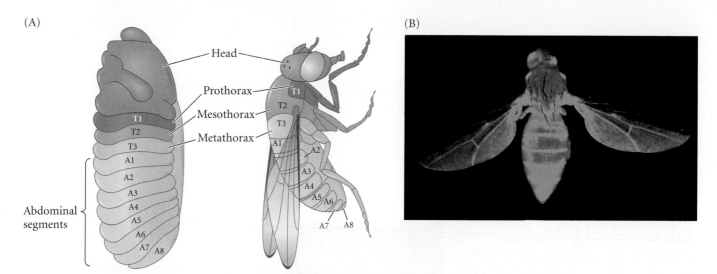

FIGURE 6.6   Comparison of larval (left) and adult (right) segmentation in *Drosophila*. (A) In the adult, the three thoracic segments can be distinguished by their appendages: T1 (prothorax) has legs only; T2 (mesothorax) has wings and legs; T3 (metathorax) has halteres (not visible) and legs. (B) Segments in adult transgenic *Drosophila* in which the gene for green fluorescent protein has been fused to the *cis*-regulatory region of the *engrailed* gene. Thus, GFP is produced in the areas of *engrailed* transcription, which is active at the border of each segment and in the posterior compartment of the wing. (B courtesy of A. Klebes.)

plan. Some of these mutations were quite fantastic, and included embryos and adult flies in which specific body structures were either missing or in the wrong place. These mutant collections were distributed to many different laboratories. The genes involved in the mutant phenotypes were cloned and then characterized with respect to their expression patterns and their functions. This combined effort has led to a molecular understanding of body plan development in *Drosophila* that is unparalleled in all of biology, and in 1995 the work resulted in a Nobel Prize for Edward Lewis, Christiane Nüsslein-Volhard, and Eric Wieschaus.

The rest of this chapter details the genetics of *Drosophila* development as we have come to understand it over the past two decades. First we will examine how the dorsal-ventral and anterior-posterior axes of the embryo are established by interactions between the developing oocyte and its surrounding follicle cells. Next we will see how dorsal-ventral patterning gradients are formed within the embryo, and how these gradients specify different tissue types. The third part of the discussion will examine how segments are formed along the anterior-posterior axis, and how the different segments become specialized. Finally, we will briefly show how the positioning of embryonic tissues along the two primary axes specifies these tissues to become particular organs.

Primary Axis Formation during Oogenesis

The processes of embryogenesis may "officially" begin at fertilization, but many of the molecular events critical for *Drosophila* embryogenesis actually occur during oogenesis. Each oocyte is descended from a single female germ cell—the **oogonium**—which is surrounded by an epithelium of follicle cells. Before oogenesis begins, the oogonium divides four times with incomplete cytokinesis, giving rise to 16 interconnected cells: 15 **nurse cells** and the single oocyte precursor. These 16 cells constitute the **egg chamber** (ovary) in which the oocyte will develop, and the oocyte will be the cell at the posterior end of the egg chamber (see Figure 16.4). As the oocyte precursor develops, numerous mRNAs made in the nurse cells are transported on microtubules through the cellular interconnections into the enlarging oocyte.

Anterior-posterior polarity in the oocyte

The follicular epithelium surrounding the developing oocyte is initially uniform with respect to cell fate, but this uniformity is broken by two signals organized by the oocyte nucleus. Interestingly, both of these signals involve the same gene, *gurken*. The *gurken* message appears to be synthesized in the nurse cells, but it becomes transported specifically to the oocyte nucleus. Here it is localized between the nucleus and the cell membrane and is translated into Gurken protein (Cáceres and Nilson 2005). At this time the oocyte nucleus is very near the posterior tip of the egg chamber, and the Gurken signal is received by the follicle cells at that position through a receptor protein encoded by the ***torpedo*** gene* (Figure 6.7A). This signal results in the "posteriorization" of these follicle cells (Fig-

---

*Molecular analysis has established that *gurken* encodes a homologue of the vertebrate epidermal growth factor (EGF), while *torpedo* encodes a homologue of the vertebrate EGF receptor (Price et al. 1989; Neuman-Silberberg and Schüpbach 1993).

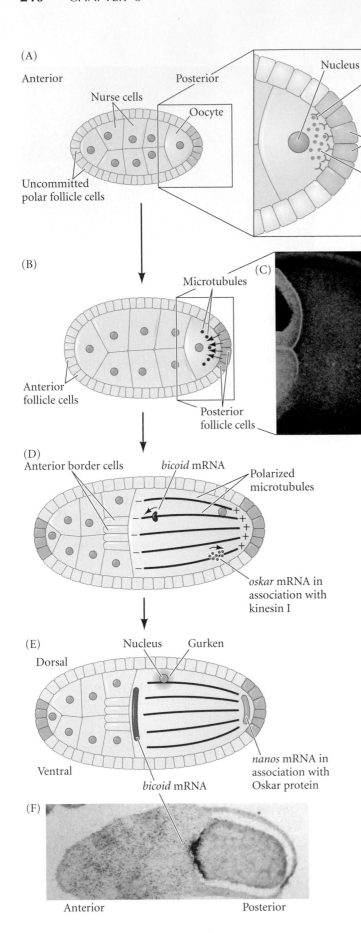

FIGURE 6.7 The anterior-posterior axis is specified during oogenesis. (A) The oocyte moves into the posterior region of the egg chamber, while nurse cells fill the anterior portion. The oocyte nucleus moves toward the terminal follicle cells and synthesizes Gurken protein (green). The terminal follicle cells express Torpedo, the receptor for Gurken. (B) When Gurken binds to Torpedo, the terminal follicle cells differentiate into posterior follicle cells and synthesize a molecule that activates protein kinase A in the egg. Protein kinase A orients the microtubules such that the growing end is at the posterior. (C) The Par-1 protein (green) localizes to the cortical cytoplasm of nurse cells and to the posterior pole of the oocyte. (The Stauffen protein marking the posterior pole is labeled red; the red and green signals combine to fluoresce yellow.) (D) The *bicoid* message binds to dynein, a motor protein associated with the non-growing end of microtubules. Dynein moves the *bicoid* message to the anterior end of the egg. The *oskar* message becomes complexed to kinesin I, a motor protein that moves it toward the growing end of the microtubules at the posterior region, where Oskar can bind the *nanos* message. (E) The nucleus (with its Gurken protein) migrates along the microtubules, inducing the adjacent follicle cells to become the dorsal follicles. (F) Photomicrograph of *bicoid* mRNA (stained black) passing from the nurse cells and localizing to the anterior end of the oocyte during oogenesis. (C courtesy of H. Doerflinger; F from Stephanson et al. 1988, courtesy of the authors.)

ure 6.7B). The posterior follicle cells send a signal back into the oocyte. The identity of this signal is not yet known, but it recruits the par-1 protein to the posterior edge of the oocyte cytoplasm (**Figure 6.7C**; Doerflinger et al. 2006). Par-1 protein organizes microtubules specifically with their

(A)

(B)
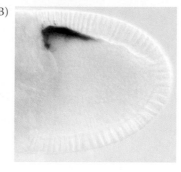

(C)
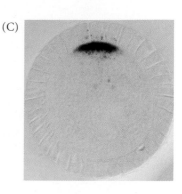

(D)
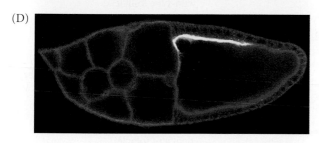

FIGURE 6.8 Expression of the *gurken* message and protein between the oocyte nucleus and the dorsal anterior cell membrane. (A) The *gurken* mRNA is localized between the oocyte nucleus and the dorsal follicle cells of the ovary. Anterior is to the left; dorsal faces upward. (B) The Gurken protein is similarly located (shown here in a younger stage oocyte than A). (C) Cross section of the egg through the region of Gurken protein expression. (D) A more mature oocyte, showing Gurken protein (yellow) across the dorsal region. The actin is stained red, showing cell boundaries. As the oocyte grows, follicle cells migrate across the top of the oocyte, becoming exposed to Gurken. (A from Ray and Schüpbach 1996, courtesy of T. Schüpbach; B,C from Peri et al. 1999, courtesy of S. Roth; D courtesy of C. van Buskirk and T. Schüpbach.)

minus (cap) and plus (growing) ends at the anterior and posterior ends of the oocyte, respectively (Gonzalez-Reyes et al. 1995; Roth et al. 1995; Januschke et al. 2006).

The orientation of the microtubules is critical, because different microtubule motor proteins will transport their mRNA or protein cargoes in different directions. The motor protein kinesin, for instance, is an ATPase that will use the energy of ATP to transport material to the plus end of the microtubule. Dynein, however, is a "minus-directed" motor protein that will transport its cargo the opposite way. One of the messages transported by kinesin along the microtubules to the posterior end of the oocyte is *oskar* mRNA (Zimyanin et al. 2008). The *oskar* mRNA is not able to be translated until it reaches the posterior cortex, at which time it generates the Oskar protein. Oskar protein recruits more par-1 protein, thereby stabilizing the microtubule orientation and allowing more material to be recruited to the posterior pole of the oocyte (Doerflinger et al. 2006; Zimyanin et al. 2007). The posterior pole will thereby have its own distinctive cytoplasm, called **pole plasm**, which contains the determinants for producing the abdomen and the germ cells.

This cytoskeletal rearrangement in the oocyte is accompanied by an increase in oocyte volume, owing to transfer of cytoplasmic components from the nurse cells. These components include maternal messengers such as the *bicoid* and *nanos* mRNAs. These mRNAs are carried by motor proteins along the microtubules to the anterior and posterior ends of the oocyte, respectively (**Figure 6.7D–F**). As we shall soon see, the protein products encoded by *bicoid* and *nanos* are critical for establishing the anterior-posterior polarity of the embryo.

Dorsal-ventral patterning in the oocyte

As oocyte volume increases, the oocyte nucleus moves to an anterior dorsal position where a second major signal-ing event takes place. Here the *gurken* message becomes localized in a crescent between the oocyte nucleus and the oocyte cell membrane, and its protein product forms an anterior-posterior gradient along the dorsal surface of the oocyte (**Figure 6.8**; Neuman-Silberberg and Schüpbach 1993). Since it can diffuse only a short distance, Gurken protein reaches only those follicle cells closest to the oocyte nucleus, and it signals those cells to become the more columnar **dorsal follicle cells** (Montell et al. 1991; Schüpbach et al. 1991; see Figure 6.7E). This establishes the dorsal-ventral polarity in the follicle cell layer that surrounds the growing oocyte.

Maternal deficiencies of either the *gurken* or the *torpedo* gene cause ventralization of the embryo. However, *gurken* is active only in the oocyte, whereas *torpedo* is active only in the somatic follicle cells. This fact was revealed by experiments with germline/somatic chimeras. In one such experiment, Schüpbach (1987) transplanted germ cell precursors from wild-type embryos into embryos whose mothers carried the *torpedo* mutation. Conversely, she transplanted the germ cell precursors from *torpedo* mutants into wild-type embryos (**Figure 6.9**). The wild-type eggs produced mutant, ventralized embryos when they developed in a *torpedo* mutant mother's egg chamber. The *torpedo* mutant eggs were able to produce normal embryos if they developed in a wild-type ovary. Thus, unlike Gurken, the Torpedo protein is needed in the follicle cells, not in the egg itself.

The Gurken-Torpedo signal that specifies dorsalized follicle cells initiates a cascade of gene activities that create

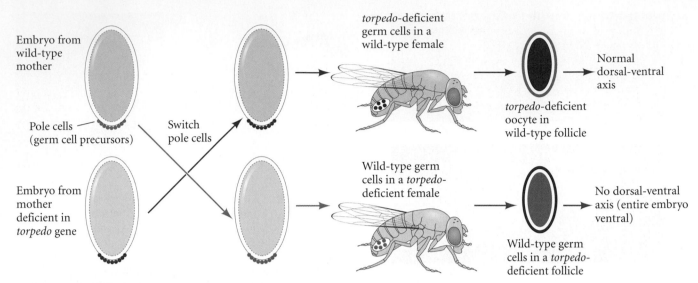

FIGURE 6.9 Germline chimeras made by interchanging pole cells (germ cell precursors) between wild-type embryos and embryos from mothers homozygous for a mutation of the *torpedo* gene. These transplants produced wild-type females whose eggs came from mutant mothers, and *torpedo*-deficient females that laid wild-type eggs. The *torpedo*-deficient eggs produced normal embryos when they developed in the wild-type ovary, whereas the wild-type eggs produced ventralized embryos when they developed in the mutant mother's ovary.

the dorsal-ventral axis of the embryo (**Figure 6.10**). The activated Torpedo receptor protein inhibits the expression of the *pipe* gene. As a result, Pipe protein is made only in the *ventral* follicle cells (Sen et al. 1998; Amiri and Stein 2002). In some as yet unknown way (probably involving sulfation), Pipe activates the Nudel protein, which is secreted to the cell membrane of the neighboring ventral embryonic cells (see Zhang et al. 2009). A few hours later, activated Nudel initiates the activation of three serine proteases that are secreted into the perivitelline fluid (see Figure 6.10B; Hong and Hashimoto 1995). These proteases are the products of the *gastrulation defective* (*gd*), *snake* (*snk*), and *easter* (*ea*) genes. Like most extracellular proteases, these molecules are secreted in an inactive form and are subsequently activated by peptide cleavage. In a complex cascade of events, activated Nudel activates the Gastrulation-defective protease. The Gd protease cleaves the Snake protein, activating the Snake protease, which in turn cleaves the Easter protein. This cleavage activates the Easter protease, which then cleaves the Spätzle protein (Chasan et al. 1992; Hong and Hashimoto 1995; LeMosy et al. 2001).

It is obviously important that the cleavage of these three proteases be limited to the most ventral portion of the embryo. This is accomplished by the secretion of a protease inhibitor from the follicle cells of the ovary (Hashimoto et al. 2003; Ligoxygakis et al. 2003). This inhibitor of Easter

and Snake is found throughout the perivitelline space surrounding the embryo. Indeed, this protein is very similar to the mammalian protease inhibitors that limit blood clotting protease cascades to the area of injury. In this way, the proteolytic cleavage of Easter and Spätzle is strictly limited to the area around the most ventral embryonic cells.

The cleaved Spätzle protein is now able to bind to its receptor in the oocyte cell membrane, the product of the *toll* gene. Toll protein is a maternal product that is evenly

FIGURE 6.10 Generating dorsal-ventral polarity in *Drosophila*. ▶ (A) The nucleus of the oocyte travels to what will become the dorsal side of the embryo. The *gurken* genes of the oocyte synthesize mRNA that becomes localized between the oocyte nucleus and the cell membrane, where it is translated into Gurken protein. The Gurken signal is received by the Torpedo receptor protein made by the follicle cells (see Figure 6.7). Given the short diffusibility of the signal, only the follicle cells closest to the oocyte nucleus (i.e., the dorsal follicle cells) receive the Gurken signal, which causes the follicle cells to take on a characteristic dorsal follicle morphology and inhibits the synthesis of Pipe protein. Therefore, Pipe protein is made only by the *ventral* follicle cells. (B) The ventral region at a slightly later stage of development. Pipe modifies an unknown protein (x) and allows it to be secreted from the ventral follicle cells. Nudel protein interacts with this modified factor to split the product of the *gastrulation defective* gene, which then splits the product of the *snake* gene to create an active enzyme that will split the inactive Easter zymogen into an active Easter protease. The Easter protease splits the Spätzle protein into a form that can bind to the Toll receptor (which is found throughout the embryonic cell membrane). This protease activity of Easter is strictly limited by the protease inhibitor found in the perivitelline space. Thus, only the ventral cells receive the Toll signal. This signal separates the Cactus protein from the Dorsal protein, allowing Dorsal to be translocated into the nuclei and ventralize the cells. (After van Eeden and St. Johnston 1999.)

distributed throughout the cell membrane of the egg (Hashimoto et al. 1988, 1991), but it becomes activated only by binding the Spätzle protein, which is produced only on the ventral side of the egg. Therefore, the Toll receptors on the ventral side of the egg are transducing a signal into the egg, while the Toll receptors on the dorsal side of the egg are not. This localized activation establishes the dorsal-ventral polarity of the oocyte.

Generating the Dorsal-Ventral Pattern in the Embryo

Dorsal, the ventral morphogen

The protein that distinguishes dorsum (back) from ventrum (belly) in the fly embryo is the product of the *dorsal* gene. The mRNA transcript of the mother's *dorsal* gene is

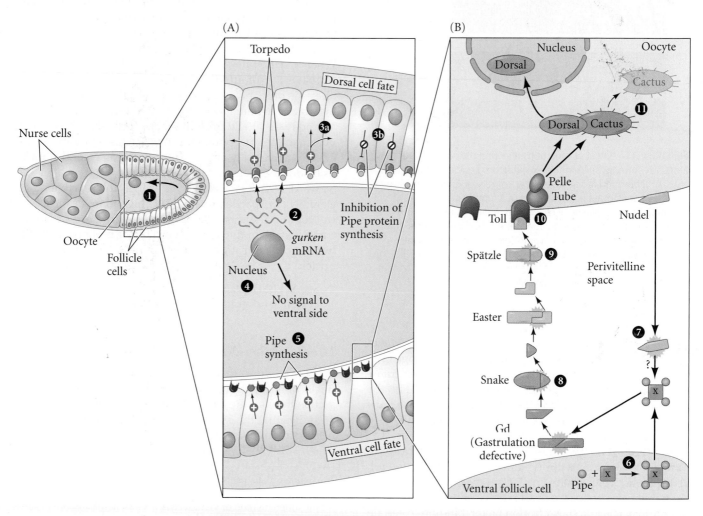

❶ Oocyte nucleus travels to anterior dorsal side of oocyte where it localizes *gurken* mRNA.

❷ *gurken* messages are translated. Gurken is received by Torpedo proteins during mid-oogenesis.

❸ₐ Torpedo signal causes follicle cells to differentiate to a dorsal morphology.

❸♭ Synthesis of Pipe is inhibited in dorsal follicle cells.

❹ Gurken does not diffuse to ventral side.

❺ Ventral follicle cells synthesize Pipe.

❻ In ventral follicle cells, Pipe completes the modification of an unknown factor (x).

❼ Nudel and factor (x) interact to split the Gastrulation-deficient (Gd) protein.

❽ Activated Gd splits the Snake protein, and activated Snake cleaves the Easter protein.

❾ Activated Easter splits Spätzle; activated Spätzle binds to Toll receptor protein.

❿ Toll activation activates Tube and Pelle, which phosphorylate the Cactus protein. Cactus is degraded, releasing it from Dorsal.

⓫ Dorsal protein enters the nucleus and ventralizes the cell.

(A)

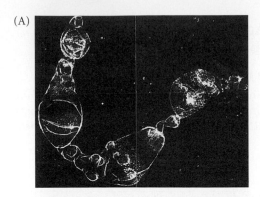

(B)

FIGURE 6.11   Effect of mutations affecting distribution of the Dorsal protein, as seen in the exoskeleton (cuticle) patterns of larvae. (A) Deformed larvae consisting entirely of dorsal cells. Larvae like these developed from the eggs of a female homozygous for a mutation of the *snake* gene, one of the maternal effect genes involved in the signaling cascade that establishes a Dorsal gradient. (B) Larva that developed from *snake* mutant eggs that received injections of mRNA from wild-type eggs. Larvae like this one have a wild-type appearance. (From Anderson and Nüsslein-Volhard 1984, courtesy of C. Nüsslein-Volhard.)

placed in the oocyte by the nurse cells. However, Dorsal protein is not synthesized from this maternal message until about 90 minutes after fertilization. When Dorsal is translated, it is found throughout the embryo, not just on the ventral or dorsal side. How can this protein act as a morphogen if it is located everywhere in the embryo?

In 1989, the surprising answer to this question was found (Roth et al. 1989; Rushlow et al. 1989; Steward 1989). While Dorsal is found throughout the syncytial blastoderm of the early *Drosophila* embryo, it is translocated into nuclei only in the ventral part of the embryo. In the nucleus, Dorsal protein acts as a transcription factor, binding to certain genes to activate or repress their transcription. If Dorsal does not enter the nucleus, the genes responsible for specifying ventral cell types are not transcribed, the genes responsible for specifying dorsal cell types are not repressed, and all the cells of the embryo become specified as dorsal cells.

This model of dorsal-ventral axis formation in *Drosophila* is supported by analyses of maternal effect mutations that give rise to an entirely dorsalized or an entirely ventralized phenotype (**Figure 6.11**; Anderson and Nüsslein-Volhard 1984). In mutants in which all the cells are dorsalized (evident from their dorsal-specific exoskeleton), Dorsal does not enter the nucleus of any cell. Conversely, in

mutants in which all cells have a ventral phenotype, Dorsal protein is found in every cell nucleus.*

Establishing a nuclear Dorsal gradient

So how does the Dorsal protein enter into the nuclei only of the ventral cells? When Dorsal is first produced, it is complexed with a protein called Cactus in the cytoplasm of the syncytial blastoderm. As long as Cactus is bound to it, Dorsal remains in the cytoplasm. Dorsal enters ventral nuclei in response to a signaling pathway that frees it from Cactus (see Figure 6.10B). This separation of Dorsal from Cactus is initiated by the ventral activation of the Toll receptor. When Spätzle binds to and activates the Toll protein, Toll activates a protein kinase called Pelle. Another protein (Tube) is probably necessary for bringing Pelle to the cell membrane, where it can be activated (Galindo et al. 1995). The activated Pelle protein kinase (probably through an intermediate) can phosphorylate Cactus. Once phosphorylated, Cactus is degraded and Dorsal can enter the nucleus (Kidd 1992; Shelton and Wasserman 1993; Whalen and Steward 1993; Reach et al. 1996). Since Toll is activated by a gradient of Spätzle protein that is highest in the most ventral region, there is a corresponding gradient of Dorsal translocation in the ventral cells of the embryo, with the highest concentrations of Dorsal in the most ventral cell nuclei.[†]

> **See WEBSITE 6.3**
> **Evidence for gradients in insect development**

Effects of the Dorsal protein gradient

What does the Dorsal protein do once it is located in the nuclei of the ventral cells? A look at the fate map of a cross section through the *Drosophila* embryo at the division cycle 14 shows that the 16 cells with the highest concentration of Dorsal are those that generate the mesoderm (**Figure**

---

*Remember that a gene in *Drosophila* is usually named after its mutant phenotype. Thus, the product of the *dorsal* gene is necessary for the differentiation of ventral cells. That is, in the absence of *dorsal*, the ventral cells become dorsalized.

[†]Recall that maternal effect mutations (as in the coiling mutant in snails discussed in Chapter 5) involve those genes that are active in the female and provide materials for the oocyte cytoplasm. The process described for the translocation of Dorsal protein into the nucleus is very similar to the process for the translocation of the NF-κB transcription factor into the nucleus of mammalian lymphocytes. In fact, there is substantial homology between NF-κB and Dorsal, between I-B and Cactus, between Toll and the interleukin 1 receptor, between Pelle and an IL-1-associated protein kinase, and between the DNA sequences recognized by Dorsal and by NF-κB (González-Crespo and Levine 1994; Cao et al. 1996). Thus, the biochemical pathway used to specify dorsal-ventral polarity in *Drosophila* appears to be homologous to that used to differentiate lymphocytes in mammals.

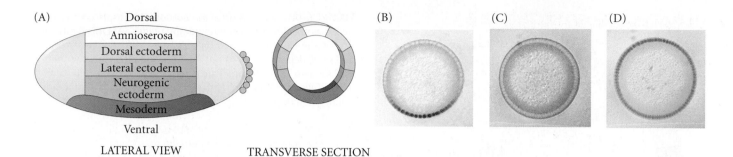

FIGURE 6.12 Specification of cell fate by the gradient of Dorsal protein. The translocation of Dorsal protein into ventral, but not lateral or dorsal, nuclei produces a gradient where the ventral cells with the most Dorsal protein become mesoderm precursors. (A) Fate map of a lateral cross section through the *Drosophila* embryo at division cycle 14. The most ventral part becomes the mesoderm; the next higher portion becomes the neurogenic (ventral) ectoderm. The lateral and dorsal ectoderm can be distinguished in the cuticle, and the dorsalmost region becomes the amnioserosa, the extraembryonic layer that surrounds the embryo. (B–D) Transverse sections of embryos stained with antibody to show the presence of Dorsal protein (dark-stained area). (B) A wild-type embryo, showing Dorsal protein in the ventralmost nuclei. (C) A dorsalized mutant, showing no localization of Dorsal protein in any nucleus. (D) A ventralized mutant, in which Dorsal protein has entered the nucleus of every cell. (A after Rushlow et al. 1989; B–D from Roth et al. 1989, courtesy of the authors.)

6.12). The next cell up from this region generates the specialized glial and neural cells of the midline. The next two cells give rise to the ventrolateral epidermis and ventral nerve cord, while the nine cells above them produce the dorsal epidermis. The most dorsal group of six cells generates the amnioserosal covering of the embryo (Ferguson and Anderson 1991). This fate map is generated by the gradient of Dorsal protein in the nuclei. Large amounts of Dorsal instruct the cells to become mesoderm, while lesser amounts instruct the cells to become glial or ectodermal tissue (Jiang and Levine 1993; Hong et al. 2008).

The first morphogenetic event of *Drosophila* gastrulation is the invagination of the 16 ventralmost cells of the embryo (Figure 6.13). All of the body muscles, fat bodies, and gonads derive from these mesodermal cells (Foe 1989). Dorsal protein specifies these cells to become mesoderm in two ways. First, the protein activates specific genes that create the mesodermal phenotype. Five of the target genes for the Dorsal protein are *twist*, *snail*, *fgf8*, the *fgf8 receptor*, and *rhomboid* (Figure 6.14). These genes are transcribed only in nuclei that have received high concentrations of Dorsal,

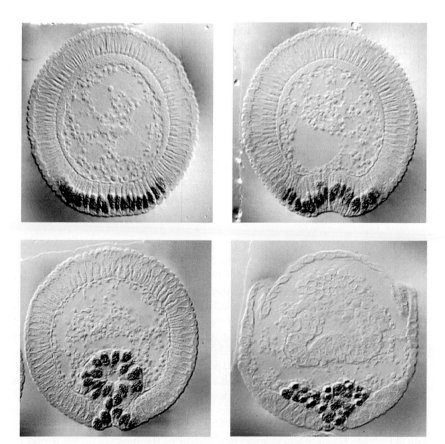

FIGURE 6.13 Gastrulation in *Drosophila*. In this cross section, the mesodermal cells at the ventral portion of the embryo buckle inward, forming the ventral furrow (see Figure 6.4A,B). This furrow becomes a tube that invaginates into the embryo and then flattens and generates the mesodermal organs. The nuclei are stained with antibody to the Twist protein, a marker for the mesoderm. (From Leptin 1991a, courtesy of M. Leptin.)

(A) Dorsal patterning

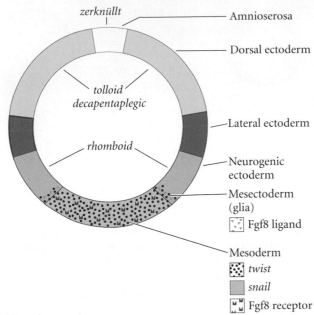

(B) Ventral patterning

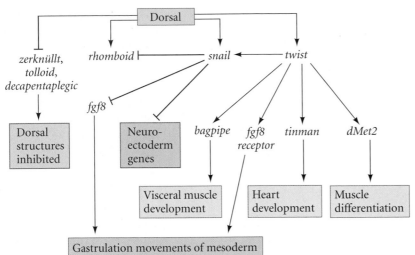

FIGURE 6.14 Subdivision of the *Drosophila* dorsal-ventral axis by the gradient of Dorsal protein in the nuclei. (A) Dorsal protein activates the zygotic genes *rhomboid, twist, fgf8, fgf8 receptor,* and *snail,* depending on its nuclear concentration. The mesoderm forms where Twist and Snail are present, and the glial cells form where Twist and Rhomboid interact. Those cells with Rhomboid, but no Snail or Twist, form the neurogenic ectoderm. The Fgf receptor is expressed in the mesoderm, and the Fgf8 ligands for this receptor are expressed in the mesectoderm (glia and midline central nervous system), adjacent to the mesoderm. The binding of Fgf8 to its receptor triggers the cell movements required for the ingression of the mesoderm. (B) Interactions in the specification of the ventral portion of the *Drosophila* embryo. Dorsal protein inhibits those genes that would give rise to dorsal structures (*tolloid, decapentaplegic,* and *zerknüllt*) while activating the three ventral genes. Snail protein, formed most ventrally, inhibits the transcription of *rhomboid* and prevents ectoderm formation. Twist activates *dMet2* and *bagpipe* (which activate muscle differentiation) as well as *tinman* (heart muscle development). (A after Steward and Govind 1993; B after Furlong et al. 2001 and Leptin and Affolter 2004.)

derm invaginates and brings these ventro-lateral regions together. This mesectoderm gives rise to glial cells and to the midline structures of the central nervous system. Unlike the neurogenic ectoderm adjacent to it, the mesectoderm cells never form typical neuroblasts, never form epidermis, and are not a stem cell population (see Figure 6.14).

The high concentration of Twist protein in the nuclei of the ventralmost cells activates the gene for the Fgf8 receptor (the product of the *heartless* gene) in the presumptive mesoderm (Jiang and Levine 1993; Gryzik and Müller 2004; Strathopoulos et al. 2004). The expression and secretion of Fgf8 by the presumptive neural ectoderm is received by its receptor on the mesoderm cells, causing these mesoderm cells to invaginate into the embryo and flatten against the ectoderm (see Figure 6.13).

Meanwhile, *intermediate* levels of nuclear Dorsal activate transcription of the *Short gastrulation* (*Sog*) gene in two lateral stripes that flank the ventral *twist* expression domain, each 12–14 cells wide (François et al. 1994; Srinivasan et al. 2002). *Sog* encodes a protein that prevents the ectoderm in this region from becoming epidermis and begins the processes of neural differentiation (**Figure 6.15**).

Dorsal protein also determines the mesoderm indirectly. In addition to activating the mesoderm-stimulating genes (*twist* and *snail*), it directly inhibits the dorsalizing genes *zerknüllt* (*zen*) and *decapentaplegic* (*dpp*). Thus, in the same cells, Dorsal can act as an activator of some genes and a repressor of others. Whether Dorsal activates or represses a given gene depends on the structure of the gene's

since their enhancers do not bind Dorsal with a very high affinity (Thisse et al. 1988, 1991; Jiang et al. 1991; Pan et al. 1991). Both Snail and Twist are also needed for the complete mesodermal phenotype and proper gastrulation (Leptin et al. 1991b). The Twist protein activates mesodermal genes, while the Snail protein represses particular non-mesodermal genes that might otherwise be active. The *rhomboid* and *fgf8* genes are interesting because they are activated by Dorsal but repressed by Snail. Thus, *rhomboid* and *fgf8* are not expressed in the most ventral cells (i.e., the mesodermal precursors) but are expressed in the cells adjacent to the mesoderm. These *rhomboid*- and *fgf8*-expressing cells will become the mesectoderm. The mesectoderm tissue is fated to become the ventral midline, once the meso-

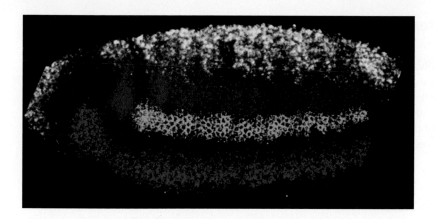

FIGURE 6.15   Dorsal-ventral patterning in *Drosophila*. The readout of the Dorsal gradient can be seen in the anterior region of a whole-mount stained embryo. The expression of the most ventral gene, *ventral nervous system defective* (blue), is from the neurogenic ectoderm. The *intermediate neuroblast defective* gene (green) is expressed in lateral ectoderm. Red represents the *muscle-specific homeobox* gene, expressed in the mesoderm above the intermediate neuroblasts. The dorsalmost tissue expresses *decapentaplegic* (yellow). (From Kosman et al. 2004, courtesy of D. Kosman and E. Bier.)

enhancers. The *zen* enhancer has a silencer region that contains a binding site for Dorsal as well as a second binding site for two other DNA-binding proteins. These two other proteins enable Dorsal to bind a transcriptional repressor protein (Groucho) and bring it to the DNA (Valentine et al. 1998). Mutants of *Dorsal* express *dpp* and *zen* genes throughout the embryo (Rushlow et al. 1987), and embryos deficient in *dpp* and *zen* fail to form dorsal structures (Irish and Gelbart 1987). Thus, in wild-type embryos, the mesodermal precursors express *twist* and *snail* (but not *zen* or

dpp); precursors of the dorsal epidermis and amnioserosa express *zen* and *dpp* (but not *twist* or *snail*). Glial (mesectoderm) precursors express *twist* and *rhomboid*, while the lateral neural ectodermal precursors do not express any of these five genes (Kosman et al. 1991; Ray and Schüpbach 1996). By the cellular responses to the Dorsal protein gradient, the embryo becomes subdivided from the ventral to dorsal regions into mesoderm, neurogenic ectoderm, epidermis (from the lateral and dorsal ectoderm), and amnioserosa (see Figure 6.12A).

SIDELIGHTS & SPECULATIONS

The Left-Right Axis

Very little is known about the formation of the left-right axis in *Drosophila*. Although the fly may look bilaterally symmetric, there are asymmetries in the embryonic hindgut (which loops to the left) and the adult hindgut and gonads. This asymmetry appears to be regulated by microfilaments (Hozumi et al. 2006; Spéder et al. 2006). The mechanism that produces this asymmetry is different from that known to produce left-right asymmetry in vertebrates, which appears to be regulated by microtubules.

If the actin microfilaments are disrupted in the *Drosophila* embryo, many defects occur, and the left-right pattern is randomized. Loss-of-function mutations of certain genes for myosin-1 proteins (which interact with microfilaments) can reverse the insect's left-right asymmetry (**Figure 6.16**).

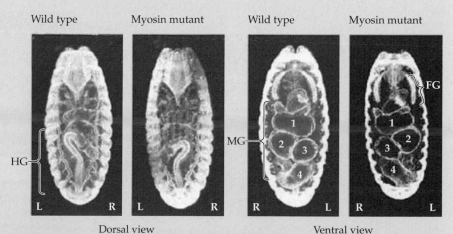

Figure 6.16   Left-right axis formation in *Drosophila* involves the microfilament cytoskeleton. Mutations in the myosin gene *Myo31DF* can reverse the insect's left-right asymmetry. Here the embryonic gut is seen in dorsal and ventral perspectives, showing that in the larva with myosin mutant, the asymmetry of the gut is reversed. HG, hindgut; MG, midgut; FG, foregut. (From Hozumi et al. 2006, courtesy of K. Matsuno.)

Segmentation and the Anterior-Posterior Body Plan

The genetic screens pioneered by Nüsslein-Volhard and Wieschaus identified a hierarchy of genes that establish anterior-posterior polarity, and divide the embryo into a specific number of segments with different identities (Figure 6.17). This hierarchy is initiated by **maternal effect genes** that produce messenger RNAs that are placed in dif-ferent regions of the egg. These messages encode transcriptional and translational regulatory proteins that diffuse through the syncytial blastoderm and activate or repress the expression of certain zygotic genes.

The first such zygotic genes to be expressed are called **gap genes** (because mutations in them cause gaps in the segmentation pattern). These genes are expressed in certain broad (about three segments wide), partially overlapping domains. These gap genes encode transcription fac-

FIGURE 6.17 Generalized model of *Drosophila* anterior-posterior pattern formation. Anterior is to the left; the dorsal surface faces upward. (A) The pattern is established by maternal effect genes that form gradients and regions of morphogenetic proteins. These proteins are transcription factors that activate the gap genes, which define broad territories of the embryo. The gap genes enable the expression of the pair-rule genes, each of which divides the embryo into regions about two segments wide. The segment polarity genes then divide the embryo into segment-sized units along the anterior-posterior axis. Together, the actions of these genes define the spatial domains of the homeotic genes that define the identities of each of the segments. In this way, periodicity is generated from nonperiodicity, and each segment is given a unique identity. (B) Maternal effect genes. The anterior axis is specified by the gradient of Bicoid protein (yellow through red). (C) Gap gene protein expression and overlap. The domain of Hunchback protein (orange) and the domain of Krüppel protein (green) overlap to form a region containing both transcription factors (yellow). (D) Products of the *fushi tarazu* pair-rule gene form seven bands across the blastoderm of the embryo. (E) Products of the segment polarity gene *engrailed*, seen here at the extended germ band stage. (B courtesy of C. Nüsslein-Volhard; C courtesy of C. Rushlow and M. Levine; D courtesy of D. W. Knowles; E courtesy of S. Carroll and S. Paddock.)

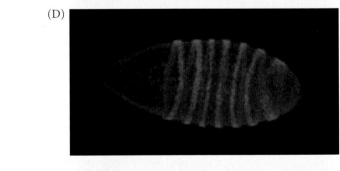

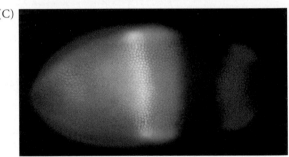

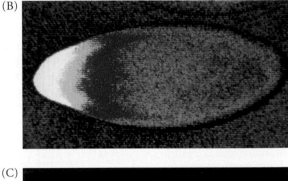

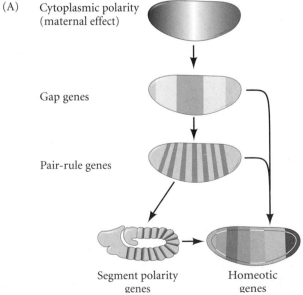

(A) Cytoplasmic polarity (maternal effect)

Gap genes

Pair-rule genes

Segment polarity genes   Homeotic genes

tors, and differing combinations and concentrations of the gap gene proteins regulate the transcription of **pair-rule genes**, which divide the embryo into periodic units. The transcription of the different pair-rule genes results in a striped pattern of seven transverse bands perpendicular to the anterior-posterior axis. The proteins encoded by the pair-rule genes are transcription factors that activate the **segment polarity genes**, whose mRNA and protein products divide the embryo into 14-segment-wide units, establishing the periodicity of the embryo. At the same time, the protein products of the gap, pair-rule, and segment polarity genes interact to regulate another class of genes, the **homeotic selector genes**, whose transcription determines the developmental fate of each segment.

Maternal gradients: Polarity regulation by oocyte cytoplasm

Classical embryological experiments demonstrated that there are at least two "organizing centers" in the insect egg, one in the anterior of the egg and one in the posterior. For instance, Klaus Sander (1975) found that if he ligated the egg early in development, separating the anterior half from the posterior half, one half developed into an anterior embryo and one half developed into a posterior embryo, but neither contained the middle segments of the embryo. The later in development the ligature was made, the fewer middle segments were missing. Thus it appeared that there were indeed morphogenetic gradients emanating from the two poles during cleavage, and that these gradients interacted to produce the positional information determining the identity of each segment.

Moreover, when the RNA in the anterior of insect eggs was destroyed (by either ultraviolet light or RNase), the resulting embryos lacked a head and thorax. Instead, these embryos developed two abdomens and *telsons* (tails) with mirror-image symmetry: telson-abdomen-abdomen-telson (**Figure 6.18**; Kalthoff and Sander 1968; Kandler-Singer and Kalthoff 1976). Sander's laboratory postulated the existence of a gradient at both ends of the egg, and hypothesized that the egg sequesters an mRNA that generates a gradient of anterior-forming material.

The molecular model: Protein gradients in the early embryo

In the late 1980s, the gradient hypothesis was united with a genetic approach to the study of *Drosophila* embryogenesis. If there were gradients, what were the morphogens whose concentrations changed over space? What were the genes that shaped these gradients? And did these morphogens act by activating or inhibiting certain genes in the areas where they were concentrated? Christiane Nüsslein-Volhard led a research program that addressed these questions. The researchers found that one set of genes encoded morphogens for the anterior part of the embryo, another

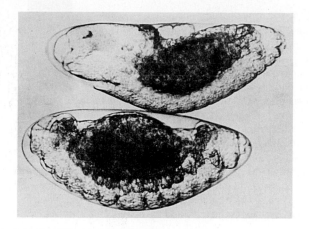

FIGURE 6.18   Normal and irradiated embryos of the midge *Smittia*. The normal embryo (top) shows a head on the left and abdominal segments on the right. The UV-irradiated embryo (bottom) has no head region but has abdominal and tail segments at both ends. (From Kalthoff 1969, courtesy of K. Kalthoff.)

set of genes encoded morphogens responsible for organizing the posterior region of the embryo, and a third set of genes encoded proteins that produced the terminal regions at both ends of the embryo (**Table 6.1**).

Two maternal messenger RNAs, *bicoid* and *nanos*, are most critical to the formation of the anterior-posterior axis. The *bicoid* mRNAs are located near the anterior tip of the unfertilized egg, and *nanos* messages are located at the posterior tip. These distributions occur as a result of the dramatic polarization of the microtubule networks in the developing oocyte (see Figure 6.7). After ovulation and fertilization, the *bicoid* and *nanos* mRNAs are translated into proteins that can diffuse in the syncytial blastoderm, forming gradients that are critical for anterior-posterior patterning (**Figure 6.19**; see also Figure 6.17B).

See WEBSITE 6.4
Christiane Nüsslein-Volhard and the molecular approach to development

BICOID AS THE ANTERIOR MORPHOGEN   That Bicoid was the head morphogen of *Drosophila* was demonstrated by the "find it, lose it, move it" experimentation scheme. Christiane Nüsslein-Volhard, Wolfgang Driever, and their colleagues (Driever and Nüsslein-Volhard 1988; Driever et al. 1990) showed that (1) Bicoid protein was found in a gradient, highest in the anterior (head-forming) region; (2) embryos lacking Bicoid could not form a head; and (3) when *bicoid* mRNA was added to Bicoid-deficient embryos in different places, the place where *bicoid* mRNA was injected became the head. Moreover, the areas around the site of Bicoid injection became the thorax, as expected from a concentration-dependent signal (**Figure 6.20**). When inject-

FIGURE 6.19 Syncytial specification in *Drosophila*. Anterior-posterior specification originates from morphogen gradients in the egg cytoplasm. *Bicoid* mRNA is stabilized in the most anterior portion of the egg, while *Nanos* mRNA is tethered to the posterior end. (The anterior can be recognized by the micropyle on the shell; this structure permits sperm to enter.) When the egg is laid and fertilized, these two mRNAs are translated into proteins. The Bicoid protein forms a gradient that is highest at the anterior end, and the Nanos protein forms a gradient that is highest at the posterior end. These two proteins form a coordinate system based on their ratios. Each position along the axis is thus distinguished from any other position. When the nuclei form, each nucleus is given its positional information by the ratio of these proteins. The proteins forming these gradients activate the transcription of the genes specifying the segmental identities of the larva and the adult fly.

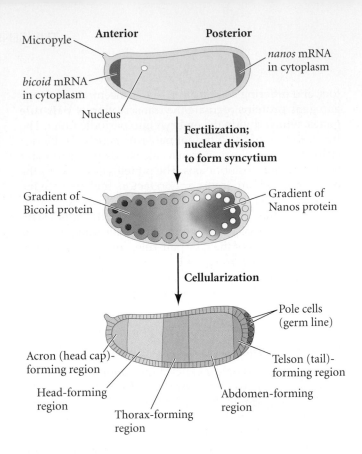

ed into the anterior of *bicoid*-deficient embryos (whose mothers lacked *bicoid* genes), the *bicoid* mRNA "rescued" the embryos and they developed normal anterior-posterior polarity. Moreover, any location in an embryo where the *bicoid* message was injected became the head. If *bicoid* mRNA was injected into the center of an embryo, that mid-

TABLE 6.1 Maternal effect genes that establish the anterior-posterior polarity of the *Drosophila* embryo

Gene	Mutant phenotype	Proposed function and structure
ANTERIOR GROUP		
bicoid (bcd)	Head and thorax deleted, replaced by inverted telson	Graded anterior morphogen; contains homeodomain; represses *caudal* mRNA
exuperantia (exu)	Anterior head structures deleted	Anchors *bicoid* mRNA
swallow (swa)	Anterior head structures deleted	Anchors *bicoid* mRNA
POSTERIOR GROUP		
nanos (nos)	No abdomen	Posterior morphogen; represses *hunchback* mRNA
tudor (tud)	No abdomen, no pole cells	Localization of Nanos protein
oskar (osk)	No abdomen, no pole cells	Localization of Nanos protein
vasa (vas)	No abdomen, no pole cells; oogenesis defective	Localization of Nanos protein
valois (val)	No abdomen, no pole cells; cellularization defective	Stabilization of the Nanos localization complex
pumilio (pum)	No abdomen	Helps Nanos protein bind *hunchback* message
caudal (cad)	No abdomen	Activates posterior terminal genes
TERMINAL GROUP		
torso (tor)	No termini	Possible morphogen for termini
trunk (trk)	No termini	Transmits Torso-like signal to Torso
fs(1)Nasrat[fs(1)N]	No termini; collapsed eggs	Transmits Torso-like signal to Torso
fs(1)polehole[fs(1)ph]	No termini; collapsed eggs	Transmits Torso-like signal to Torso

Source: After Anderson 1989.

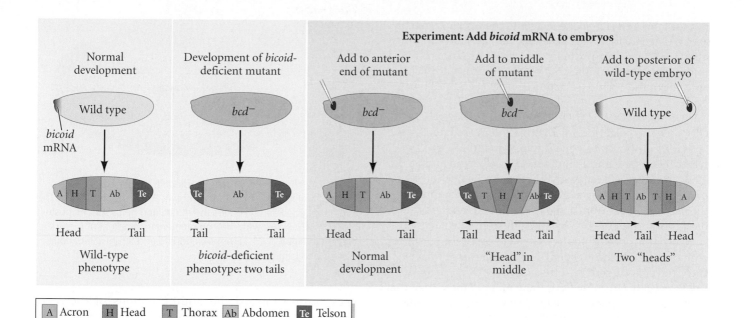

Experiment: Add *bicoid* mRNA to embryos

| Normal development | Development of *bicoid*-deficient mutant | Add to anterior end of mutant | Add to middle of mutant | Add to posterior of wild-type embryo |

Wild-type phenotype | *bicoid*-deficient phenotype: two tails | Normal development | "Head" in middle | Two "heads"

| A | Acron | H | Head | T | Thorax | Ab | Abdomen | Te | Telson |

FIGURE 6.20 Schematic representation of experiments demonstrating that the *bicoid* gene encodes the morphogen responsible for head structures in *Drosophila*. The phenotypes of *bicoid*-deficient and wild-type embryos are shown at left. When *bicoid*-deficient embryos are injected with *bicoid* mRNA, the point of injection forms the head structures. When the posterior pole of an early-cleavage wild-type embryo is injected with *bicoid* mRNA, head structures form at both poles. (After Driever et al. 1990.)

dle region became the head, with the regions on either side of it becoming thorax structures. If a large amount of *bicoid* mRNA was injected into the posterior end of a wild-type embryo (with its own endogenous *bicoid* message in its anterior pole), two heads emerged, one at either end (Driever et al. 1990).

BICOID MRNA LOCALIZATION IN THE ANTERIOR POLE OF THE OOCYTE The 3′ untranslated region (UTR) of *bicoid* mRNA contains sequences that are critical for its localization at the anterior pole (**Figure 6.21**; Ferrandon et al. 1997; Macdonald and Kerr 1998; Spirov et al. 2009). These sequences interact with the

Exuperantia and **Swallow** proteins while the messages are still in the nurse cells of the egg chamber (Schnorrer et al. 2000). Experiments in which fluorescently labeled *bicoid* mRNA was microinjected into living egg chambers of wild-type or mutant flies indicate that Exuperantia must be present in the nurse cells for anterior localization. But having Exuperantia in the oocyte is not sufficient to bring the *bicoid* message into the oocyte (Cha et al. 2001; Reichmann and Ephrussi 2005). The *bicoid*-protein complex is transported out of the nurse cells and into the oocyte via microtubules. The complex seems to ride on a kinesin ATPase that is crit-

(A) mRNA

Anterior   Posterior

(B) Protein

Anterior   Posterior

FIGURE 6.21 *Bicoid* mRNA and protein gradients shown by in situ hybridization and confocal microscopy. (A) *Bicoid* mRNA shows a steep gradient across the anterior portion of the oocyte. (B) When the mRNA is translated, the Bicoid protein gradient can be seen in the anterior nuclei. Anterior is to the left; the dorsal surface is upward. (After Spirov et al., courtesy of S. Baumgartner.)

ical for taking the *bicoid* message into the oocyte (Arn et al. 2003). Once inside the oocyte, the *bicoid*-mRNA complex attaches to dynein proteins that are maintained at the microtubule organizing center (the "minus end") at the anterior of the oocyte (see Figure 6.7; Cha et al. 2001).

NANOS MRNA LOCALIZATION IN THE POSTERIOR POLE OF THE OOCYTE The posterior organizing center is defined by the activities of the *nanos* gene (Lehmann and Nüsslein-Volhard 1991; Wang and Lehmann 1991; Wharton and Struhl 1991). While *bicoid* message is bound to the anchored end of the microtubules by active transport along microtubules, the *nanos* message appears to get "trapped" in the posterior end of the oocyte by passive diffusion. The *nanos* message becomes bound to the cytoskeleton in the posterior region of the egg through its 3′ UTR and its association with the products of several other genes (*oskar, valois, vasa, staufen,* and *tudor*).* If *nanos* (or any other of these maternal effect genes) is absent in the mother, no abdomen forms in the embryo (Lehmann and Nüsslein-Volhard 1986; Schüpbach and Wieschaus 1986). But before the *nanos* message can get "trapped" in the posterior cortex, a *nanos* mRNA-specific trap has to be made; this trap is the Oskar protein (Ephrussi et al. 1991). The *oskar* message and the Staufen protein are transported to the posterior end of the oocyte by the kinesin motor protein (see Figure 6.7). There they become bound to the actin microfilaments of the cortex. Staufen allows the translation of the *oskar* message, and the resulting Oskar protein is capable of binding the *nanos* message (Brendza et al. 2000; Hatchet and Ephrussi 2004).

Most *nanos* mRNA, however, is not trapped. Rather, it is bound in the cytoplasm by the translation inhibitors Smaug and CUP. Smaug binds to the 3′ UTR of *nanos* mRNA and recruits the CUP protein that prevents the association of the message with the ribosome. If the *nanos*-Smaug-CUP complex reaches the posterior pole, however, Oskar can dissociate CUP from Smaug, allowing the mRNA to be bound at the posterior and ready for translation (Forrest et al. 2004; Nelson et al. 2004).

Therefore, at the completion of oogenesis, the *bicoid* message is anchored at the anterior end of the oocyte, and the *nanos* message is tethered to the posterior end (Frigerio et al. 1986; Berleth et al. 1988; Gavis and Lehmann 1992). These mRNAs are dormant until ovulation and fertilization, at which time they are translated. Since the Bicoid and Nanos *protein products* are not bound to the cytoskeleton,

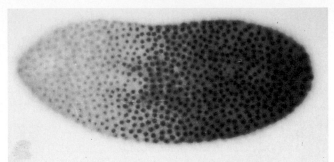

FIGURE 6.22 Caudal protein gradient in the syncytial blastoderm of a wild-type *Drosophila* embryo. Anterior is to the left. The protein (stained darkly) enters the nuclei and helps specify posterior fates. Compare with the complementary gradient of Bicoid protein in Figure 6.21. (From Macdonald and Struhl 1986, courtesy of G. Struhl.)

they diffuse toward the middle regions of the early embryo, creating the two opposing gradients that establish the anterior-posterior polarity of the embryo.

Two other maternally provided mRNAs (*hunchback, hb*; and *caudal, cad*) are critical for patterning the anterior and posterior regions of the body plan, respectively (Lehmann et al. 1987; Wu and Lengyel 1998). These two mRNAs are synthesized by the nurse cells of the ovary and transported to the oocyte, where they are distributed ubiquitously throughout the syncytial blastoderm. But if they are not localized, how do they mediate their localized patterning activities? It turns out that translation of the *hb* and *cad* mRNAs is repressed by the diffusion gradients of Nanos and Bicoid proteins, respectively.

In anterior regions, Bicoid binds to a specific region of *caudal*'s 3′ UTR. Here, it binds d4HEP, a protein that inhibits translation by binding to the 5′ cap of the message, thereby preventing its binding to the ribosome. By recruiting this translational inhibitor, Bicoid thus prevents translation of Caudal in the anterior section of the embryo (**Figure 6.22**; Rivera-Pomar et al. 1996; Chan and Struhl 1997; Cho et al. 2006). This suppression is necessary because if Caudal protein is made in the embryo's anterior, the head and thorax do not form properly. Caudal is critical in specifying the posterior domains of the embryo, activating the genes responsible for the invagination of the hindgut.

At the other end of the embryo, Nanos functions by preventing *hunchback* translation (Tautz 1988). Nanos protein in the posterior of the embryo forms a complex with several other ubiquitous proteins, including Pumilio and Brat. Pumilio appears to direct the complex to the 3′ UTR of the *hunchback* message, and Brat appears to recruit d4HEP, which will inhibit the translation of the *hunchback* message (Cho et al. 2006). The result of these interactions is the creation of four maternal protein gradients in the early embryo (**Figure 6.23**):

- An anterior-to-posterior gradient of Bicoid protein
- An anterior-to-posterior gradient of Hunchback protein

---

*Like the placement of the *bicoid* message, the location of the *nanos* message is determined by its 3′ UTR. If the *bicoid* 3′ UTR is experimentally placed on the protein-encoding region of *nanos* mRNA, the *nanos* message gets placed in the anterior of the egg. When the RNA is translated, the Nanos protein inhibits the translation of *hunchback* and *bicoid* mRNAs, and the embryo forms two abdomens—one in the anterior of the embryo and one in the posterior (Gavis and Lehmann 1992). We will see these proteins again in Chapter 16, since they are critical in forming the germ cells of *Drosophila*.

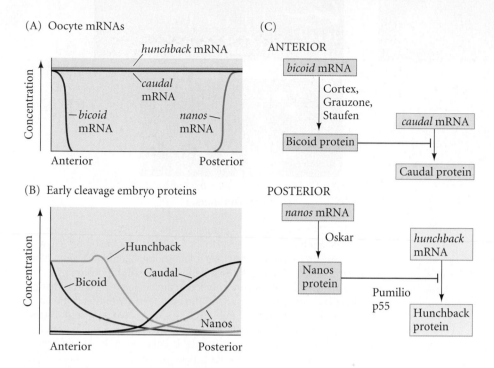

FIGURE 6.23   Model of anterior-posterior pattern generation by *Drosophila* maternal effect genes. (A) The *bicoid, nanos, hunchback,* and *caudal* mRNAs are placed in the oocyte by the ovarian nurse cells. The *bicoid* message is sequestered anteriorly; the *nanos* message is sent to the posterior pole. (B) Upon translation, the Bicoid protein gradient extends from anterior to posterior, while the Nanos protein gradient extends from posterior to anterior. Nanos inhibits the translation of the *hunchback* message (in the posterior), while Bicoid prevents the translation of the *caudal* message (in the anterior). This inhibition results in opposing Caudal and Hunchback gradients. The Hunchback gradient is secondarily strengthened by transcription of the *hunchback* gene in the anterior nuclei (since Bicoid acts as a transcription factor to activate *hunchback* transcription). (C) Parallel interactions whereby translational gene regulation establishes the anterior-posterior patterning of the *Drosophila* embryo. (C after Macdonald and Smibert 1996.)

- A posterior-to-anterior gradient of Nanos protein
- A posterior-to-anterior gradient of Caudal protein

The Bicoid, Hunchback, and Caudal proteins are transcription factors whose relative concentrations can activate or repress particular zygotic genes. The stage is now set for the activation of zygotic genes in those nuclei that were busy dividing while these four protein gradients were being established.

The anterior organizing center: The Bicoid and Hunchback gradients

In *Drosophila*, the phenotype of the *bicoid* mutant provides valuable information about the function of morphogenetic gradients (**Figure 6.24A–C**). Instead of having anterior structures (acron, head, and thorax) followed by abdominal structures and a telson, the structure of the *bicoid* mutant is telson-abdomen-abdomen-telson (**Figure 6.24D**). It would appear that these embryos lack whatever substances are

needed for the formation of anterior structures. Moreover, one could hypothesize that the substance these mutants lack is the one postulated by Sander and Kalthoff to turn on genes for the anterior structures and turn off genes for the telson structures (compare Figures 6.18 and 6.24D).

Several studies support the view that the product of the wild-type *bicoid* gene is the morphogen that controls anterior development. The first type of evidence came from experiments that altered the shape of the Bicoid protein gradient. As we have seen, the *exuperantia* and *swallow* genes are responsible for keeping the *bicoid* message at the anterior pole of the egg. In their absence, the *bicoid* message diffuses farther into the posterior of the egg, and the protein gradient is less steep (Driever and Nüsslein-Volhard 1988a). The phenotype produced by *exuperantia* and *swallow* mutants is similar to that of *bicoid*-deficient embryos but is less severe; these embryos lack their most anterior structures and have an extended mouth and thoracic region. Furthermore, by adding extra copies of the *bicoid* gene, the Bicoid protein gradient can be extended

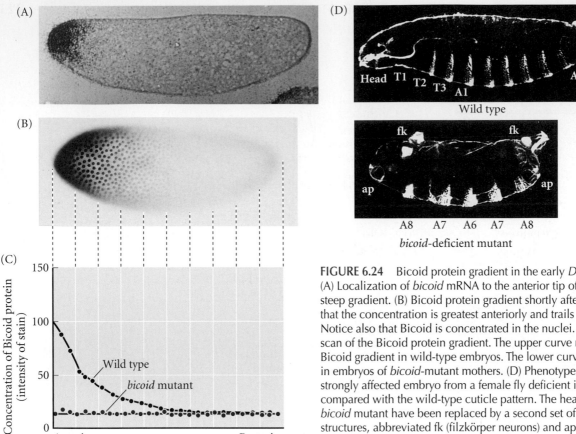

FIGURE 6.24 Bicoid protein gradient in the early *Drosophila* embryo. (A) Localization of *bicoid* mRNA to the anterior tip of the embryo in a steep gradient. (B) Bicoid protein gradient shortly after fertilization. Note that the concentration is greatest anteriorly and trails off posteriorly. Notice also that Bicoid is concentrated in the nuclei. (C) Densitometric scan of the Bicoid protein gradient. The upper curve represents the Bicoid gradient in wild-type embryos. The lower curve represents Bicoid in embryos of *bicoid*-mutant mothers. (D) Phenotype of cuticle from a strongly affected embryo from a female fly deficient in the *bicoid* gene compared with the wild-type cuticle pattern. The head and thorax of the *bicoid* mutant have been replaced by a second set of posterior telson structures, abbreviated fk (filzkörper neurons) and ap (anal plates). (A from Kaufman et al. 1990; B,C from Driever and Nüsslein-Volhard 1988b; D from Driever et al. 1990, courtesy of the authors.)

into more posterior regions, causing anterior structures like the cephalic furrow to be expressed in a more posterior position (Driever and Nüsslein-Volhard 1988a; Struhl et al. 1989). Thus, altering the Bicoid gradient correspondingly alters the fate of specific embryonic regions.

It had been thought that, once the *bicoid* message was translated, the gradient of Bicoid protein would be generated simply by diffusion of the protein; the reality is a bit more complicated. In 2007, Thomas Gregor and his colleagues demonstrated that the speed of diffusion cannot account for the rapid deployment of the Bicoid protein gradient. Shortly thereafter, using highly sensitive confocal microscopy, Weil and colleagues (2008) showed that the anteriorly localized *bicoid* message became dispersed by egg activation (at ovulation), and Spirov and collaborators (2009) showed that the *bicoid* mRNA was transported along microtubules to form a gradient that prefigured the gradient of its protein (see Figures 6.21 and 6.24A,B). The *bicoid* mRNA gradient is established at nuclear cycle 10 (the beginning of the syncytial blastoderm stage), persists through nuclear division 13, and disappears as the mRNA is degraded during the initial stages of cycle 14 (when the blastoderm becomes cellular).

Whether the gradient of Bicoid protein arises from diffusion from a single source or from localized synthesis, it appears to act as a *morphogen*. As described in the Part II opening essay, morphogens are substances that differentially specify the fates of cells by different concentrations. High concentrations of Bicoid produce anterior head struc-

tures. Slightly less Bicoid tells the cells to become jaws. A moderate concentration of Bicoid is responsible for instructing cells to become the thorax, while the abdomen is characterized as lacking Bicoid.

How might a gradient of Bicoid protein control the determination of the anterior-posterior axis? As discussed earlier (see Figure II.9), Bicoid protein acts as a translation inhibitor of *caudal*, and *caudal*'s protein product is critical for the specification of the posterior. However, Bicoid's primary function is to act as a transcription factor that activates the expression of target genes in the anterior part of the embryo.*

The first target of Bicoid to be discovered was the *hunchback* (*hb*) gene. In the late 1980s, two laboratories independently demonstrated that Bicoid binds to and activates *hb* (Driever and Nüsslein-Volhard 1989; Struhl et al. 1989). Bicoid-dependent transcription of *hb* is seen only in the anterior half of the embryo—the region where Bicoid is found. This transcription reinforces the gradient of maternal Hunchback protein produced by Nanos-dependent translational repression. Mutants deficient in maternal and zygotic *hb* genes lack mouthparts and thorax structures. Therefore, both maternal and zygotic Hunchback contribute to the anterior patterning of the embryo.

---

*bicoid appears to be a relatively "new" gene that evolved in the Dipteran (fly) lineage; it has not been found in other insect lineages. The anterior determinant in other insect groups has not yet been determined but appears to have *bicoid*-like properties (Wolff et al. 1998; Lynch and Desplan 2003).

Based on two pieces of evidence, Driever and co-workers (1989) predicted that Bicoid must activate at least one other anterior gene besides *hb*. First, deletions of *hb* produced only some of the defects seen in the *bicoid* mutant phenotype. Second, the *swallow* and *exuperantia* experiments showed that only moderate levels of Bicoid protein are needed to activate thorax formation (i.e., *hunchback* gene expression), but head formation requires higher Bicoid concentrations. Since then, a large number of Bicoid target genes have been identified. These include the head gap genes *buttonhead*, *empty spiracles*, and *orthodenticle*, which are expressed in specific subregions of the anterior part of the embryo (Cohen and Jürgens 1990; Finkelstein and Perrimon 1990; Grossniklaus et al. 1994).

Driever and co-workers (1989) predicted that the promoters of such a head-specific gap gene would have low-affinity binding sites for Bicoid, causing them to be activated only at extremely high concentrations of Bicoid—that is, near the anterior tip of the embryo. In addition to needing high Bicoid levels for activation, transcription of these genes also requires the presence of Hunchback protein (Simpson-Brose et al. 1994; Reinitz et al. 1995). Bicoid and Hunchback act synergistically at the enhancers of these "head genes" to promote their transcription in a feedforward manner (see Figure 5.13).

In the posterior half of the embryo, the Caudal protein gradient also activates a number of zygotic genes, including the gap genes *knirps* (*kni*) and *giant* (*gt*), which are critical for abdominal development (Rivera-Pomar 1995; Schulz and Tautz 1995).

The terminal gene group

In addition to the anterior and posterior morphogens, there is a third set of maternal genes whose proteins generate the unsegmented extremities of the anterior-posterior axis: the **acron** (the terminal portion of the head that includes the brain) and the **telson** (tail). Mutations in these terminal genes result in the loss of the acron and the most anterior head segments as well as the telson and the most posterior abdominal segments (Degelmann et al. 1986; Klingler et al. 1988). A critical gene here appears to be ***torso***, a gene encoding a receptor tyrosine kinase (RTK; see Chapter 3). The embryos of mothers with mutations of *torso* have neither acron nor telson, suggesting that the two termini of the embryo are formed through the same pathway. The *torso* mRNA is synthesized by the ovarian cells, deposited in the oocyte, and translated after fertilization. The transmembrane Torso protein is not spatially restricted to the ends of the egg but is evenly distributed throughout the plasma membrane (Casanova and Struhl 1989). Indeed, a gain-of-function mutation of *torso*, which imparts constitutive activity to the receptor, converts the entire anterior half of the embryo into an acron and the entire posterior half into a telson. Thus, Torso must normally be activated only at the ends of the egg.

Stevens and her colleagues (1990) have shown that this is the case. Torso protein is activated by the follicle cells

only at the two poles of the oocyte. Two pieces of evidence suggest that the activator of Torso is probably the **Torso-like** protein: first, loss-of-function mutations in the *torso-like* gene create a phenotype almost identical to that produced by *torso* mutants; and second, ectopic expression of Torso-like protein activates Torso in the new location. The *torso-like* gene is usually expressed only in the anterior and posterior follicle cells, and secreted Torso-like protein can cross the perivitelline space to activate Torso in the egg membrane (Martin et al. 1994; Furriols et al. 1998). In this manner, Torso-like activates Torso in the anterior and posterior regions of the oocyte membrane.

The end products of the RTK cascade activated by Torso diffuse into the cytoplasm at both ends of the embryo (**Figure 6.25**; Gabay et al. 1997). These kinases are thought to inactivate the Groucho protein, a transcriptional inhibitor

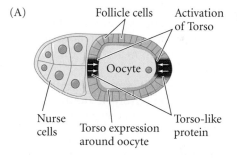

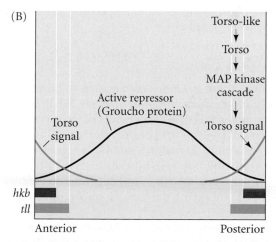

FIGURE 6.25 Formation of the unsegmented extremities by Torso signaling. (A) Torso-like protein is expressed by the follicle cells at the poles of the oocyte. Torso protein is uniformly distributed throughout the plasma membrane of the oocyte. Torso-like activates Torso at the poles (see Casanova et al. 1995). (B) Inactivation of the transcriptional suppression of *huckebein* (*hkb*) and *tailless* (*tll*) genes. The Torso signal antagonizes the Groucho protein. Groucho represses *tailless* and *huckebein* expression. The gradient of Torso is thought to provide the information that allows *tailless* to be expressed farther into the embryo than *huckebein*. (A after Gabay et al. 1997; B after Paroush et al. 1997.)

of the *tailless* and *huckebein* gap genes (Paroush et al. 1997); it is these two gap genes that specify the termini of the embryo. The distinction between the anterior and posterior termini depends on the presence of Bicoid. If *tailless* and *huckebein* act alone, the terminal region differentiates into a telson. However, if Bicoid is also present, the terminal region forms an acron (Pignoni et al. 1992).

Summarizing early anterior-posterior axis specification in Drosophila

The anterior-posterior axis of the *Drosophila* embryo is specified by three sets of genes:

1. **Genes that define the anterior organizing center.** Located at the anterior end of the embryo, the anterior organizing center acts through a gradient of Bicoid protein. Bicoid functions as a *transcription factor* to activate anterior-specific gap genes and as a *translational repressor* to suppress posterior-specific gap genes.
2. **Genes that define the posterior organizing center.** The posterior organizing center is located at the posterior pole. This center acts *translationally* through the Nanos protein to inhibit anterior formation, and *transcriptionally* through the Caudal protein to activate those genes that form the abdomen.
3. **Genes that define the terminal boundary regions.** The boundaries of the acron and telson are defined by the product of the *torso* gene, which is activated at the tips of the embryo.

The next step in development will be to use these gradients of transcription factors to activate specific genes along the anterior-posterior axis.

Segmentation Genes

Cell fate commitment in *Drosophila* appears to have two steps: specification and determination (Slack 1983). Early in development, the fate of a cell depends on cues provided by protein gradients. This specification of cell fate is flexible and can still be altered in response to signals from other cells. Eventually, however, the cells undergo a transition from this loose type of commitment to an irreversible determination. At this point, the fate of a cell becomes cell-intrinsic.*

The transition from specification to determination in *Drosophila* is mediated by **segmentation genes** that divide the early embryo into a repeating series of segmental pri-

---

*Aficionados of information theory will recognize that the process by which the anterior-posterior information in morphogenetic gradients is transferred to discrete and different parasegments represents a transition from analog to digital specification. Specification is analog, determination digital. This process enables the transient information of the gradients in the syncytial blastoderm to be stabilized so that it can be utilized much later in development (Baumgartner and Noll 1990).

(A) Gap: *Krüppel* (as an example)

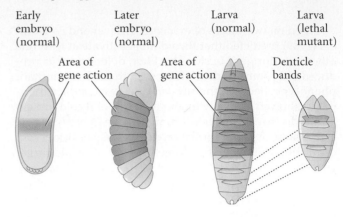

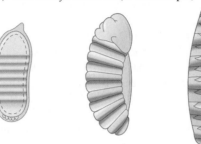
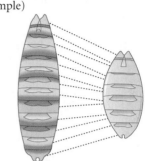

(B) Pair-rule: *fushi tarazu* (as an example)

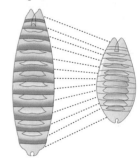

(C) Segment polarity: *engrailed* (as an example)

FIGURE 6.26 Three types of segmentation gene mutations. The left panel shows the early-cleavage embryo, with the region where the particular gene is normally transcribed in wild-type embryos shown in color. These areas are deleted as the mutants (right panel) develop.

mordia along the anterior-posterior axis. Segmentation genes were originally defined by zygotic mutations that disrupted the body plan, and these genes were divided into three groups based on their mutant phenotypes (Nüsslein-Volhard and Wieschaus 1980):

1. **Gap** mutants lacked large regions of the body (several contiguous segments; **Figure 6.26A**).
2. **Pair-rule** mutants lacked portions of every other segment (**Figure 6.26B**).
3. **Segment polarity** mutants showed defects (deletions, duplications, polarity reversals) in every segment (**Figure 6.26C**).

Segments and Parasegments

Mutations in segmentation genes result in *Drosophila* embryos that lack certain segments or parts of segments. However, early researchers found a surprising aspect of these mutations: many of them did not affect actual segments. Rather, they affected the posterior compartment of one segment and the anterior compartment of the immediately posterior segment. These "transegmental" units were named **parasegments** (Figure 6.27A; Martinez-Arias and Lawrence 1985). Moreover, once the means to detect gene expression patterns were available, it was discovered that the expression patterns in the early embryo are delineated by parasegmental boundaries—not by the boundaries of the segments. Thus, the parasegment appears to be the fundamental unit of *embryonic* gene expression.

Although parasegmental organization is also seen in the nerve cord of adult *Drosophila*, it is not seen in the adult epidermis (which is the most obvious manifestation of segmentation), nor is it found in the adult musculature. These adult structures are organized along the segmental pattern. In *Drosophila*, the segmental grooves appear in the epidermis when the germ band is retracted, while the mesoderm becomes segmental later in development.

One can think about the segmental and parasegmental organization schemes as representing different ways of organizing the compartments along the anterior-posterior axis of the embryo. The cells of one compartment do not mix with cells of neighboring compartments, and parasegments and segments are out of phase by one compartment.

Why should there be two modes of metamerism (sequential parts) in flies? Jean Deutsch has proposed that such a twofold way of organizing the body is needed for the coordination of movement. In every group of the Arthropoda—crustaceans, insects, myriapods, and chelicerates (spiders)—the ganglia of the ventral nerve cord are organized by parasegments, but the cuticle

grooves and musculature are segmental. Viewing the segmental border as a movable hinge, this shift in frame by one compartment allows the muscles on both sides of any particular epidermal segment to be coordinated by the same ganglion (**Figure 6.27B**). This in turn allows rapid and coordinated muscle contractions for locomotion.

Therefore, while parts of the body may become secondarily organized according to segments, the parasegment is the basic unit of embryonic construction. A similar situation occurs in vertebrates, where the posterior portion of the anterior somite combines with the anterior portion of the next somite (see Chapter 11).

(A)

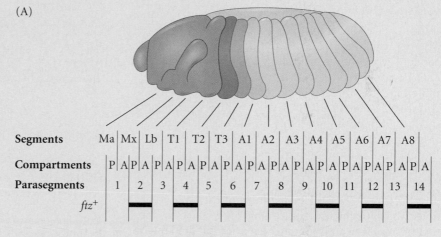

Segments	Ma	Mx	Lb	T1	T2	T3	A1	A2	A3	A4	A5	A6	A7	A8
Compartments	P A	P A	P A	P A	P A	P A	P A	P A	P A	P A	P A	P A	P A	P A
Parasegments	1	2	3	4	5	6	7	8	9	10	11	12	13	14

ftz⁺

(B)

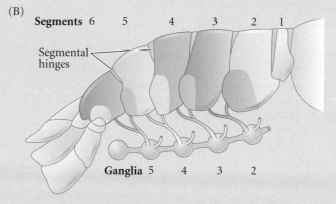

Segments 6   5   4   3   2   1

Segmental hinges

Ganglia 5   4   3   2

Figure 6.27 Overlap and integration of segments and parasegments. (A) Parasegments in the *Drosophila* embryo are shifted one compartment forward in relation to the segments. Ma, Mx, and Lb are the mandibular, maxillary, and labial head segments; T1–T3 are the thoracic segments; and A1–A8 are abdominal segments. Each segment has an anterior (A) and a posterior (P) compartment. Each parasegment (numbered 1–14) consists of the posterior compartment of one segment and the anterior compartment of the segment in the next posterior position. Black bars indicate the boundaries of gene expression observed in the *fushi tarazu* (*ftz*) mutant (see Figure 6.26B). (B) Segments and parasegments integrated in the body of an adult arthropod (the crustacean *Procambarus*). The ventral nerve cord is divided according to parasegments (color). This allows the neurons of the ganglia to regulate the ectodermal scutes and the mesodermal muscles on either side of a segmental hinge. (A after Martinez-Arias and Lawrence 1985; B after Deutsch 2004.)

The gap genes

The gap genes are activated or repressed by the maternal effect genes, and are expressed in one or two broad domains along the anterior-posterior axis. These expression patterns correlate quite well with the regions of the embryo that are missing in gap mutations. For example, the *Krüppel* gene is expressed primarily in parasegments 4–6, in the center of the *Drosophila* embryo (see Figures 6.26A and 6.16C); in the absence of the Krüppel protein, the embryo lacks segments from these and the immediately adjacent regions.

Deletions caused by mutations in three gap genes—*hunchback*, *Krüppel*, and *knirps*—span the entire segmented region of the *Drosophila* embryo. The gap gene *giant* overlaps with these three, and the gap genes *tailless* and *huckebein* are expressed in domains near the anterior and posterior ends of the embryo.

The expression patterns of the gap genes are highly dynamic. These genes usually show low levels of transcriptional activity across the entire embryo that become consolidated into discrete regions of high activity as cleavage continues (Jäckle et al. 1986). The Hunchback gradient is particularly important in establishing the initial gap gene expression patterns. By the end of nuclear division cycle 12, Hunchback is found at high levels across the anterior part of the embryo, and then forms a steep gradient through about 15 nuclei near the middle of the embryo (see Figures 6.16A and 6.22). The last third of the embryo has undetectable Hunchback levels at this time. The transcription patterns of the anterior gap genes are initiated by the different concentrations of the Hunchback and Bicoid proteins. High levels of Bicoid and Hunchback induce the expression of *giant*, while the *Krüppel* transcript appears over the region where Hunchback begins to decline. High levels of Hunchback also prevent the transcription of the posterior gap genes (such as *knirps* and *giant*) in the anterior part of the embryo (Struhl et al. 1992). It is thought that a gradient of the Caudal protein, highest at the posterior pole, is responsible for activating the abdominal gap genes *knirps* and *giant* in the posterior part of the embryo. The *giant* gene thus has two methods for its activation, one for its anterior expression band and one for its posterior expression band (Rivera-Pomar 1995; Schulz and Tautz 1995).

After the initial gap gene expression patterns have been established by the maternal effect gradients and Hunchback, they are stabilized and maintained by repressive interactions between the different gap gene products themselves.* These boundary-forming inhibitions are thought to be directly mediated by the gap gene products, because all four major gap genes (*hunchback*, *giant*, *Krüppel*, and

---

*The interactions between these genes and gene products are facilitated by the fact that these reactions occur within a syncytium, in which the cell membranes have not yet formed.

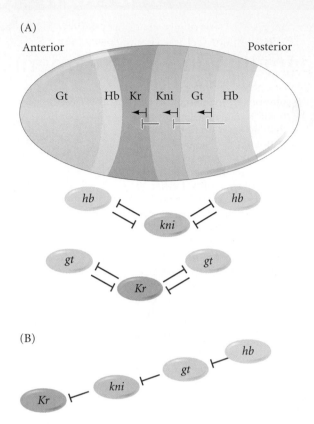

FIGURE 6.28   Expression and regulatory interactions among gap gene products. (A) Schematic expression of the gap genes during the late fourteenth cell cycle. Bars between the domains represent repression of the more anterior domain by the protein posterior to it. Arrows represent the direction in which the domains shift during the cell cycle. (For clarity, overlaps are not shown.) Strong mutual repression (diagrammed below) establishes the basic pattern of gene expression. (B) Asymmetrical repression of gap genes by their posterior neighbors causes an anterior shift in the domains of expression. (After Monk 2004.)

knirps) encode DNA-binding proteins (Knipple et al. 1985; Gaul and Jäckle 1990; Capovilla et al. 1992). The major mechanism involved in this stabilization seems to be strong mutual repression between pairs of *nonadjacent* gap genes (**Figure 6.28A**). Gene misexpression experiments show that Giant and Krüppel are strong mutual repressors, as are Hunchback and Knirps (Kraut and Levine 1991; Clyde et al. 2003). For example, if *hunchback* activity is lacking, the posterior domain of *knirps* expands toward the anterior. Conversely, if *hunchback* is misexpressed in nuclei that normally express *knirps*, strong repression is detected. This system of strong mutual repression results in the precise placement of gap protein domains but permits overlaps between *adjacent* gap genes.

Jaeger and colleagues (2004) used quantified gene expression data to model how stabilization of the gap gene expression patterns occurs during the thirteenth and fourteenth cleavage cycles (at around 71 minutes). Their data suggest that the patterns of gap gene expression were stabilized by three major factors. Two of these were strong mutual inhibitions between Hunchback and Knirps and strong mutual inhibitions between Giant and Krüppel. The data also revealed that these inhibitory interactions are unidirectional, with each protein having a strong effect on the anterior border of the repressed genes. This latter part of the model is important because it may explain the anterior "creeping" of the gap gene transcription patterns (**Figure 6.28B**).

The end result of these repressive interactions is the creation of a precise system of overlapping gap mRNA expression patterns. Each domain serves as a source for diffusion of gap proteins into adjacent embryonic regions. This creates a significant overlap (at least eight nuclei, which accounts for about two segment primordia) between adjacent gap protein domains. This was demonstrated in a striking manner by Stanojević and co-workers (1989). They fixed cellularizing blastoderms (see Figure 6.1), stained Hunchback protein with an antibody carrying a red dye, and simultaneously stained Krüppel protein with an antibody carrying a green dye. Cellularizing regions that contained both proteins bound both antibodies and stained bright yellow (see Figure 6.16C). Krüppel overlaps with Knirps in a similar manner in the posterior region of the embryo (Pankratz et al. 1990).

The pair-rule genes

The first indication of segmentation in the fly embryo comes when the pair-rule genes are expressed during cell division cycle 13, as the cells begin to form at the periphery of the embryo. The transcription patterns of these genes divide the embryo into regions that are precursors of the segmental body plan. As can be seen in **Figure 6.29** (and in Figure 6.16D), one vertical band of nuclei (the cells are just beginning to form) expresses a pair-rule gene, the next

TABLE 6.2 Major genes affecting segmentation pattern in *Drosophila*	
Category	**Gene name**
Gap genes	*Krüppel (Kr)*
	knirps (kni)
	hunchback (hb)
	giant (gt)
	tailless (tll)
	huckebein (hkb)
	buttonhead (btd)
	empty spiracles (ems)
	orthodenticle (otd)
Pair-rule genes (primary)	*hairy (h)*
	even-skipped (eve)
	runt (run)
Pair-rule genes (secondary)	*fushi tarazu (ftz)*
	odd-paired (opa)
	odd-skipped (odd)
	sloppy-paired (slp)
	paired (prd)
Segment polarity genes	*engrailed (en)*
	wingless (wg)
	cubitus interruptusD (ciD)
	hedgehog (hh)
	fused (fu)
	armadillo (arm)
	patched (ptc)
	gooseberry (gsb)
	pangolin (pan)

band of nuclei does not express it, and then the next band expresses it again. The result is a "zebra stripe" pattern along the anterior-posterior axis, dividing the embryo into 15 subunits (Hafen et al. 1984). Eight genes are currently known to be capable of dividing the early embryo in this fashion, and they overlap one another so as to give each cell in the parasegment a specific set of transcription factors. These genes are listed in **Table 6.2**.

The primary pair-rule genes include *hairy*, *even-skipped*, and *runt*, each of which is expressed in seven stripes. All three build their striped patterns from scratch, using distinct enhancers and regulatory mechanisms for each stripe. These enhancers are often modular: control over expression in each stripe is located in a discrete region of the DNA, and these DNA regions often contain binding sites recognized by gap proteins. Thus it is thought that the different concentrations of gap proteins determine whether or not a pair-rule gene is transcribed.

One of the best-studied primary pair-rule gene is *even-skipped* (**Figure 6.30**). Its enhancer is composed of modular units arranged such that each unit regulates a separate stripe or a pair of stripes. For instance, *even-skipped* stripe

FIGURE 6.29 Messenger RNA expression patterns of two pair-rule genes, *even-skipped* (red) and *fushi tarazu* (black) in the *Drosophila* blastoderm. Each gene is expressed as a series of seven stripes. Anterior is to the left, and dorsal is up. (Courtesy of S. Small.)

(A)

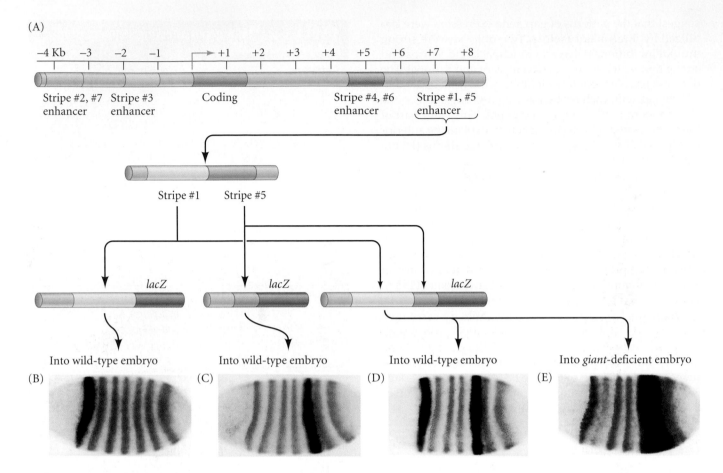

FIGURE 6.30 Specific promoter regions of the *even-skipped* (*eve*) gene control specific transcription bands in the embryo. (A) Partial map of the *eve* promoter, showing the regions responsible for the various stripes. (B–E) A reporter β-galactosidase gene (*lacZ*) was fused to different regions of the *eve* promoter and injected into fly oocytes. The resulting embryos were stained (orange bands) for the presence of Even-skipped protein. (B–D) Wild-type embryos that were injected with *lacZ* transgenes containing the enhancer region specific for stripe 1 (B), stripe 5 (C), or both regions (D). (E) The enhancer region for stripes 1 and 5 was injected into an embryo deficient in *giant*. Here the posterior border of stripe 5 is missing. (After Fujioka et al. 1999 and Sackerson et al. 1999; photographs courtesy of M. Fujioka and J. B. Jaynes.)

2 is controlled by a 500-bp region that is activated by Bicoid and Hunchback and repressed by both Giant and Krüppel proteins (**Figure 6.31**; Small et al. 1991, 1992; Stanojevíc et al. 1991; Janssens et al. 2006). The anterior border is maintained by repressive influences from Giant, while the posterior border is maintained by Krüppel. DNase I footprinting showed that the minimal enhancer region for this stripe contains five binding sites for Bicoid, one for Hunchback, three for Krüppel, and three for Giant. Thus, this region is thought to act as a switch that can directly sense the concentrations of these proteins and make on/off transcriptional decisions.

The importance of these enhancer elements can be shown by both genetic and biochemical means. First, a mutation in a particular enhancer can delete its particular stripe and no other. Second, if a reporter gene (such as *lacZ*, which encodes β-galactosidase) is fused to one of the enhancers, the reporter gene is expressed only in that particular stripe (see Figure 6.30; Fujioka et al. 1999). Third, placement of the stripes can be altered by deleting the gap genes that regulate them. Thus, stripe placement is a result of (1) the modular *cis*-regulatory enhancer elements of the pair-rule genes, and (2) the *trans*-regulatory gap gene and maternal gene proteins that bind to these enhancer sites.

Once initiated by the gap gene proteins, the transcription pattern of the primary pair-rule genes becomes stabilized by interactions among their products (Levine and Harding 1989). The primary pair-rule genes also form the context that allows or inhibits expression of the later-acting secondary pair-rule genes. One such gene is *fushi tarazu* (*ftz*), which means "too few segments" in Japanese (**Figure 6.32**). Early in division cycle 14, *ftz* mRNA and its protein are seen throughout the segmented portion of the embryo. However, as the proteins from the primary pair-rule genes begin to interact with the *ftz* enhancer, the *ftz* gene is repressed in certain bands of nuclei to create interstripe regions. Meanwhile, the Ftz protein interacts with its own

FIGURE 6.31 Hypothesis for formation of the second stripe of transcription from the *even-skipped* gene. The enhancer element for stripe 2 regulation contains binding sequences for several maternal and gap gene proteins. Activators (e.g., Bicoid and Hunchback) are noted above the line; repressors (e.g., Krüppel and Giant) are shown below. Note that nearly every activator site is closely linked to a repressor site, suggesting competitive interactions at these positions. (Moreover, a protein that is a repressor for stripe 2 may be an activator for stripe 5; it depends on which proteins bind next to them.) B, Bicoid; C, Caudal; G, Giant; H, Hunchback; K, Krüppel; N, Knirps; T, Tailless. (After Janssens et al. 2006.)

promoter to stimulate more transcription of *ftz* where it is already present (Edgar et al. 1986b; Karr and Kornberg 1989; Schier and Gehring 1992).

The eight known pair-rule genes are all expressed in striped patterns, but the patterns are not coincident with each other. Rather, each row of nuclei within a parasegment has its own array of pair-rule products that distinguishes it from any other row. These products activate the next level of segmentation genes, the segment polarity genes.

The segment polarity genes

So far our discussion has described interactions between molecules within the syncytial embryo. But once cells form, interactions take place between the cells. These interactions are mediated by the segment polarity genes, and they accomplish two important tasks. First, they reinforce the parasegmental periodicity established by the earlier transcription factors. Second, through this cell-to-cell signaling, cell fates are established within each parasegment.

The segment polarity genes encode proteins that are constituents of the Wingless (Wnt) and Hedgehog signal transduction pathways (see Chapter 3). Mutations in these genes lead to defects in segmentation and in gene expression pattern across each parasegment. The development of the normal pattern relies on the fact that only one row of cells in each parasegment is permitted to express the Hedgehog protein, and only one row of cells in each parasegment is permitted to express the Wingless protein. The key to this pattern is the activation of the *engrailed* gene

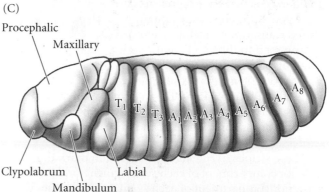

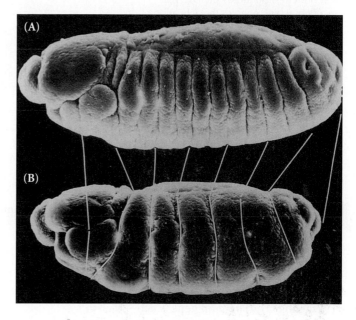

FIGURE 6.32 Defects seen in the *fushi tarazu* mutant. Anterior is to the left; dorsal surface faces upward. (A) Scanning electron micrograph of a wild-type embryo, seen in lateral view. (B) Same stage of a *fushi tarazu* mutant embryo. The white lines connect the homologous portions of the segmented germ band. (C) Diagram of wild-type embryonic segmentation. The shaded areas show the parasegments of the germ band that are missing in the mutant embryo. (D) Transcription pattern of the *fushi tarazu* gene. (After Kaufman et al. 1990; A,B courtesy of T. Kaufman; D courtesy of T. Karr.)

FIGURE 6.33 Model for transcription of the segment polarity genes *engrailed* (*en*) and *wingless* (*wg*). (A) Expression of *wg* and *en* is initiated by pair-rule genes. The *en* gene is expressed in cells that contain high concentrations of either Even-skipped or Fushi tarazu proteins. The *wg* gene is transcribed when neither *eve* or *ftz* genes are active, but when a third gene (probably *sloppy-paired*) is expressed. (B) The continued expression of *wg* and *en* is maintained by interactions between the Engrailed- and Wingless-expressing cells. Wingless protein is secreted and diffuses to the surrounding cells. In those cells competent to express Engrailed (i.e., those having Eve or Ftz proteins), Wingless protein is bound by the Frizzled receptor, which enables the activation of the *en* gene via the Wnt signal transduction pathway. (Armadillo is the *Drosophila* name for β-catenin.) Engrailed protein activates the transcription of the *hedgehog* gene and also activates its own (*en*) gene transcription. Hedgehog protein diffuses from these cells and binds to the Patched receptor protein on neighboring cells. This binding prevents the Patched protein from inhibiting signaling by the Smoothened protein. The Smoothened signal enables the transcription of the *wg* gene and the subsequent secretion of the Wingless protein. For a more complex view, see Sánchez et al. 2008.

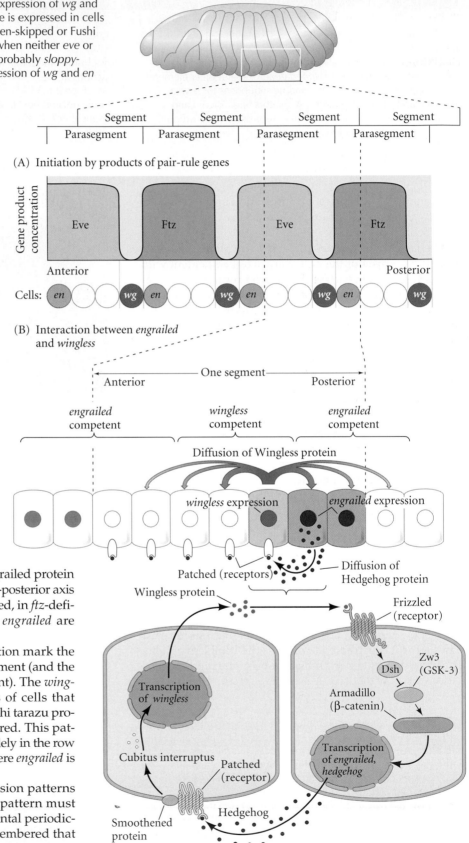

(A) Initiation by products of pair-rule genes

(B) Interaction between *engrailed* and *wingless*

in those cells that are going to express Hedgehog. The *engrailed* gene is activated in cells that have high levels of the Even-skipped, Fushi tarazu, or Paired transcription factors; *engrailed* is repressed in those cells with high levels of Odd-skipped, Runt, or Sloppy-paired proteins. As a result, the Engrailed protein is found in 14 stripes across the anterior-posterior axis of the embryo (see Figure 6.16E). (Indeed, in *ftz*-deficient embryos, only seven bands of *engrailed* are expressed.)

These stripes of *engrailed* transcription mark the anterior compartment of each parasegment (and the posterior compartment of each segment). The *wingless* gene is activated in those bands of cells that receive little or no Even-skipped or Fushi tarazu protein, but which do contain Sloppy-paired. This pattern causes *wingless* to be transcribed solely in the row of cells directly anterior to the cells where *engrailed* is transcribed (**Figure 6.33A**).

Once *wingless* and *engrailed* expression patterns are established in adjacent cells, this pattern must be maintained to retain the parasegmental periodicity of the body plan. It should be remembered that the mRNAs and proteins involved in initiating these patterns are short-lived, and that the patterns must be maintained after their initiators are no longer being synthesized. The maintenance of these patterns is regulated by reciprocal interaction between neighboring cells: cells secreting Hedgehog activate the expression of *wingless* in their neighbors; the Wingless protein is received by the cells that secreted Hedgehog and maintains *hedgehog* expression (**Figure 6.33B**). Wingless protein also acts in an autocrine fashion, maintaining its own expression (Sánchez et al. 2008).

In the cells transcribing the *wingless* gene, wingless mRNA is translocated by its 3′ UTR to the apex of the cell (Simmonds et al. 2001; Wilkie and Davis 2001). At the apex, the *wingless* message is translated and secreted from the cell. The cells expressing *engrailed* can bind this protein because they contain Frizzled, which is the *Drosophila* membrane receptor protein for Wingless (Bhanot et al. 1996). Binding of Wingless to Frizzled activates the Wnt signal transduction pathway, resulting in the continued expression of *engrailed* (Siegfried et al. 1994).

This activation starts another portion of this reciprocal pathway. The Engrailed protein activates the transcription of the *hedgehog* gene in the *engrailed*-expressing cells. Hedgehog protein can bind to its receptor protein (Patched) on neighboring cells. When it binds to the adjacent posterior cells, it stimulates the expression of the *wingless* gene. The result is a reciprocal loop wherein the Engrailed-synthesizing cells secrete the Hedgehog protein, which maintains the expression of the *wingless* gene in the neighboring cells, while the Wingless-secreting cells maintain the expression of the *engrailed* and *hedgehog* genes in their neighbors in turn (Heemskerk et al. 1991; Ingham et al. 1991; Mohler and Vani 1992). In this way, the transcription pattern of these two types of cells is stabilized. This interaction creates a stable boundary, as well as a signaling center from which Hedgehog and Wingless proteins diffuse across the parasegment.

The diffusion of these proteins is thought to provide the gradients by which the cells of the parasegment acquire their identities. This process can be seen in the dorsal epidermis, where the rows of larval cells produce different cuticular structures depending on their position in the segment. The 1° row of cells consists of large, pigmented spikes called denticles. Posterior to these cells, the 2° row produces a smooth epidermal cuticle. The next two cell rows have a 3° fate, making small, thick hairs; they are followed by several rows of cells that adopt the 4° fate, producing fine hairs (**Figure 6.34**).

The fates of the cells can be altered by experimentally increasing or decreasing the levels of Hedgehog or Wingless (Heemskerek and DiNardo 1994; Bokor and DiNardo 1996; Porter et al. 1996). These two proteins thus appear to be necessary for elaborating the entire pattern of cell types across the parasegment. Gradients of Hedgehog and Wingless are interpreted by a second series of protein gradients within the cells. This second set of gradients provides certain cells with the receptors for Hedgehog and (often) with

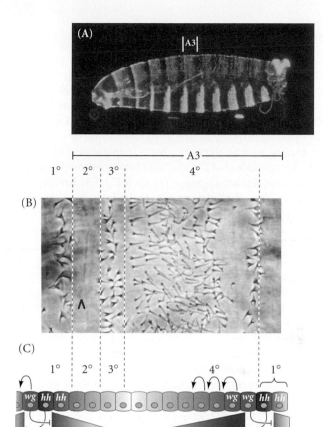

FIGURE 6.34 Cell specification by the Wingless/Hedgehog signaling center. (A) Bright-field photograph of wild-type *Drosophila* embryo, showing the position of the third abdominal segment. Anterior is to the left; the dorsal surface faces upward. (B) Close-up of the dorsal area of the A3 segment, showing the different cuticular structures made by the 1°, 2°, 3°, and 4° rows of cells. (C) Diagram showing a model for the role of Wingless and Hedgehog. Each signal is responsible for about half the pattern. Either each signal acts in a graded manner (shown here as gradients decreasing with distance from their respective sources) to specify the fates of cells at a distance from these sources, or each signal acts locally on the neighboring cells to initiate a cascade of inductions (shown here as sequential arrows). (After Heemskerk and DiNardo 1994; A,B courtesy of the authors.)

the receptor for Wingless (Casal et al. 2002; Lander et al. 2002). The resulting pattern of cell fates also changes the focus of patterning from parasegment to segment. There are now external markers, as the *engrailed*-expressing cells become the most posterior cells of each segment.

See WEBSITE 6.5
 Asymmetrical spread of morphogens

See WEBSITE 6.6
 Getting a head in the fly

The Homeotic Selector Genes

After the parasegmental boundaries are set, the pair-rule and gap genes interact to regulate the homeotic selector genes, which specify the characteristic structures of each segment (Lewis 1978). By the end of the cellular blastoderm stage, each segment primordium has been given an individual identity by its unique constellation of gap, pair-rule, and homeotic gene products (Levine and Harding 1989). Two regions of *Drosophila* chromosome 3 contain most of these homeotic genes (**Figure 6.35**). The first region, known as the **Antennapedia complex**, contains the homeotic genes *labial* (*lab*), *Antennapedia* (*Antp*), *sex combs reduced* (*scr*), *deformed* (*dfd*), and *proboscipedia* (*pb*). The *labial* and *deformed* genes specify the head segments, while *sex combs reduced* and *Antennapedia* contribute to giving the thoracic segments their identities. The *proboscipedia* gene appears to act only in adults, but in its absence, the labial palps of the mouth are transformed into legs (Wakimoto et al. 1984; Kaufman et al. 1990).

The second region of homeotic genes is the **bithorax complex** (Lewis 1978; Maeda and Karch 2009). Three protein-coding genes are found in this complex: *Ultrabithorax* (*Ubx*), which is required for the identity of the third thoracic segment; and the *abdominal A* (*abdA*) and *Abdominal B* (*AbdB*) genes, which are responsible for the segmental identities of the abdominal segments (Sánchez-Herrero et al. 1985). The chromosome region containing both the Antennapedia complex and the bithorax complex is often referred to as the **homeotic complex** (**Hom-C**).

Because the homeotic selector genes are responsible for the specification of fly body parts, mutations in them lead to bizarre phenotypes. In 1894, William Bateson called these organisms **homeotic mutants**, and they have fascinated developmental biologists for decades.* For example, the body of the normal adult fly contains three thoracic segments, each of which produces a pair of legs. The first thoracic segment does not produce any other appendages, but the second thoracic segment produces a pair of wings in addition to its legs. The third thoracic segment produces a pair of wings and a pair of balancers known as **halteres**. In homeotic mutants, these specific segmental identities can be changed. When the *Ultrabithorax* gene is deleted, the third thoracic segment (characterized by halteres) is transformed into another second thoracic segment. The

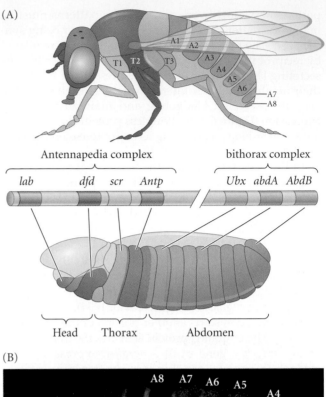

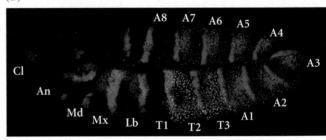

FIGURE 6.35   Homeotic gene expression in *Drosophila*. (A) Expression map of the homeotic genes. In the center are the genes of the Antennapedia and bithorax complexes and their functional domains. Below and above the gene map, the regions of homeotic gene expression (both mRNA and protein) in the blastoderm of the *Drosophila* embryo and the regions that form from them in the adult fly are shown. (B) In situ hybridization for four genes at a slightly later stage (the extended germ band). The *engrailed* (light blue) expression pattern separates the body into segments; *Antennapedia* (green) and *Ultrabithorax* (purple) separate the thoracic and abdominal regions; *Distal-less* (red) shows the placement of jaws and the beginnings of limbs. (A after Kaufman et al. 1990 and Dessain et al. 1992; B courtesy of D. Kosman.)

Homeo, from the Greek, means "similar." *Homeotic mutants* are mutants in which one structure is replaced by another (as where an antenna is replaced by a leg). *Homeotic genes* are those genes whose mutation can cause such transformations; thus, homeotic genes are genes that specify the identity of a particular body segment. The *homeobox* is a conserved DNA sequence of about 180 base pairs that is shared by many homeotic genes. This sequence encodes the 60-amino-acid *homeodomain*, which recognizes specific DNA sequences. The homeodomain is an important region of the transcription factors encoded by homeotic genes. However, not all genes containing homeoboxes are homeotic genes.

result is a fly with four wings (**Figure 6.36**)—an embarrassing situation for a classic dipteran.[†]

Similarly, Antennapedia protein usually specifies the second thoracic segment of the fly. But when flies have a

---

[†]Dipterans (two-winged insects such as flies) are thought to have evolved from four-winged insects; it is possible that this change arose via alterations in the bithorax complex. Chapter 19 includes more speculation on the relationship between the homeotic complex and evolution.

FIGURE 6.36   A four-winged fruit fly constructed by putting together three mutations in *cis*-regulators of the *Ultrabithorax* gene. These mutations effectively transform the third thoracic segment into another second thoracic segment (i.e., halteres into wings). (Courtesy of E. B. Lewis.)

mutation wherein the *Antennapedia* gene is expressed in the head (as well as in the thorax), legs rather than antennae grow out of the head sockets (**Figure 6.37**). This is partly because, in addition to promoting the formation of thoracic structures, the Antennapedia protein binds to and represses the enhancers of at least two genes, *homothorax* and *eyeless*, which encode transcription factors that are critical for antenna and eye formation, respectively (Casares and Mann 1998; Plaza et al. 2001). Therefore, one of Antennapedia's functions is to suppress the genes that would trigger antenna and eye development. In the recessive mutant of *Antennapedia*, the gene fails to be expressed in the second thoracic segment, and antennae sprout in the leg positions (Struhl 1981; Frischer et al. 1986; Schneuwly et al. 1987).

The major homeotic selector genes have been cloned and their expression analyzed by in situ hybridization (Harding et al. 1985; Akam 1987). Transcripts from each gene can be detected in specific regions of the embryo (see Figure 6.35B) and are especially prominent in the central nervous system.

Initiating and maintaining the patterns of homeotic gene expression

The initial domains of homeotic gene expression are influenced by the gap genes and pair-rule genes. For instance, expression of the *abdA* and *AbdB* genes is repressed by the gap gene proteins Hunchback and Krüppel. This inhibition prevents these abdomen-specifying genes from being expressed in the head and thorax (Casares and Sánchez-Herrero 1995). Conversely, the *Antennapedia* gene is activated by particular levels of Hunchback (needing both the maternal and the zygotically transcribed messages), so

Antennapedia is originally transcribed in parasegment 4, specifying the mesothoracic (T2) segment (Wu et al. 2001).

The expression of homeotic genes is a dynamic process. The *Antennapedia* gene, for instance, although initially expressed in presumptive parasegment 4, soon appears in parasegment 5. As the germ band expands, *Antp* expression is seen in the presumptive neural tube as far posterior as parasegment 12. During further development, the domain of *Antp* expression contracts again, and *Antp* transcripts are localized strongly to parasegments 4 and 5. Like that of other homeotic genes, *Antp* expression is negatively regulated by all the homeotic gene products expressed posterior to it (Levine and Harding 1989; González-Reyes and Morata 1990). In other words, each of the bithorax complex genes represses the expression of *Antp*. If the *Ultrabithorax* gene is deleted, *Antp* activity extends through the region that would normally have expressed *Ubx* and stops where the *Abd* region begins. (This allows the third thoracic segment to form wings like the second thoracic segment, as in Figure 6.36.) If the entire bithorax complex is deleted, *Antp* expression extends throughout the abdomen. (Such a larva does not survive, but the cuticle pattern throughout the abdomen is that of the second thoracic segment.)

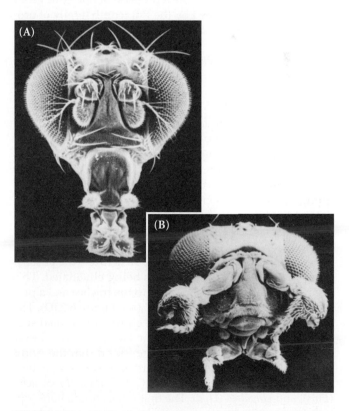

FIGURE 6.37   (A) Head of a wild-type fruit fly. (B) Head of a fly containing the *Antennapedia* mutation that converts antennae into legs. (From Kaufman et al. 1990, courtesy of T. C. Kaufman.)

FIGURE 6.38 Developmental control of posterior spiracle formation through AbdB. The homeotic selector protein AbdB (with the interaction of cofactors) activates the transcription of four genes encoding "intermediate" regulators. The proteins encoded by these genes—Spalt (Sal), Cut (Ct), Empty spiracles (Ems), and Unpaired (Upd)—are necessary and sufficient for specifying posterior spiracle development. They control (directly or indirectly) the local expression of a battery of realisator genes that influence morphogenetic processes such as cell adhesion (cadherins), cell polarity (crumbs), and cytoskeletal organization (G proteins). (After Lohmann 2006; Lovegrove et al. 2006.)

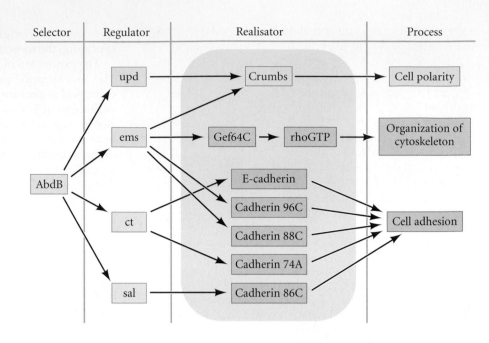

As we have seen, the proteins encoded by the gap and pair-rule genes are transient; however, in order for differentiation to occur, the identities of the segments must be stabilized. So, once the transcription patterns of the homeotic genes have become stabilized, they are "locked" into place by alteration of the chromatin conformation in these genes. The repression of homeotic genes appears to be maintained by the **Polycomb** family of proteins, while the active chromatin conformation appears to be maintained by the **Trithorax** proteins (Ingham and Whittle 1980; McKeon and Brock 1991; Simon et al. 1992).

Realisator genes

Homeotic genes don't do the work alone. In fact, they appear to regulate the action from up in the "executive suite," while the actual business of making an organ is done by other genes on the "factory floor." In this scenario, the homeotic genes work by activating or repressing a group of "realisator genes"—those genes that are the targets of the homeotic gene proteins and that function to form the specified tissue or organ primordia (Garcia-Bellido 1975).

Such a pathway for one simple structure—the posterior spiracle—is well on its way to being elucidated. This organ is a simple tube connecting to the trachea and a protuberance called the "Filzkörper" (see Figure 6.23D). The posterior spiracle is made in the eighth abdominal segment and is under the control of the Hox gene *AbdB*. Lovegrove and colleagues (2006) have found that the AbdB protein controls four genes that are necessary for posterior spiracle formation: *Spalt* (*Sal*), *Cut* (*Ct*), *Empty spiracles* (*Ems*), and *Unpaired* (*Upd*). The first three encode transcription factors; the fourth encodes a paracrine factor. None of them are transcribed without AbdB. Moreover, if these

genes are independently activated in the absence of AbdB, a posterior spiracle will form.

Controlled by AbdB, these four regulator genes in turn control the expression of the realisator genes that control cell structure and function. *Spalt* and *Cut* encode proteins that activate the cadherin genes necessary for cell adhesion and the invagination of the spiracle. *Empty spiracles* and *Unpaired* encode proteins that control the small G proteins (such as Gef64C) that organize the actin cytoskeleton and the cell polarizing proteins that control the elongation of the spiracle (**Figure 6.38**).

Axes and Organ Primordia: The Cartesian Coordinate Model

The anterior-posterior and dorsal-ventral axes of *Drosophila* embryos form a coordinate system that can be used to specify positions within the embryo (**Figure 6.39A**). Theoretically, cells that are initially equivalent in developmental potential can respond to their position by expressing different sets of genes. This type of specification has been demonstrated in the formation of the salivary gland rudiments (Panzer et al. 1992; Bradley et al. 2001; Zhou et al. 2001).

Drosophila salivary glands form only in the strip of cells defined by the activity of the *sex combs reduced* (*scr*) gene along the anterior-posterior axis (parasegment 2). No salivary glands form in *scr*-deficient mutants. Moreover, if *scr* is experimentally expressed throughout the embryo, salivary gland primordia form in a ventrolateral stripe along most of the length of the embryo. The formation of salivary glands along the dorsal-ventral axis is repressed by both Decapentaplegic and Dorsal proteins, which inhibit salivary gland formation both dorsally and ventrally. Thus, the salivary glands form at the intersection of the vertical *scr* expression band (parasegment 2) and the horizontal region in the middle of the embryo's circumference that has neither Decapentaplegic nor Dorsal (**Figure 6.39B**). The cells that form the salivary glands are directed to do so by the intersecting gene activities along the anterior-posterior and dorsal-ventral axes.

A similar situation is seen with tissues that are found in every segment of the fly. Neuroblasts arise from 10 clusters of 4 to 6 cells each that form on each side in every segment in the strip of neural ectoderm at the midline of the embryo (Skeath and Carroll 1992). The cells in each cluster interact (via the Notch pathway discussed in Chapter 3) to generate a single neural cell from each cluster. Skeath and colleagues (1993) have shown that the pattern of neural gene transcription is imposed by a coordinate system. Their expression is repressed by the Decapentaplegic and Snail proteins along the dorsal-ventral axis, while positive enhancement by pair-rule genes along the anterior-posterior axis causes their repetition in each half-segment. It is very likely, then, that the positions of organ primordia in the fly are specified via a two-dimensional coordinate system based on the intersection of the anterior-posterior and dorsal-ventral axes.

Coda

Genetic studies on the *Drosophila* embryo have uncovered numerous genes that are responsible for specification of the anterior-posterior and dorsal-ventral axes. We are far from a complete understanding of *Drosophila* pattern formation, but we are much more aware of its complexity than we were a decade ago. Mutations of *Drosophila* genes have given us our first glimpses of the multiple levels of pattern regulation in a complex organism and have enabled us to isolate these genes and their products. Most importantly, as we will see in forthcoming chapters, the *Drosophila* genes provide clues to a general mechanism of pattern formation that is used throughout the animal kingdom.

(A)

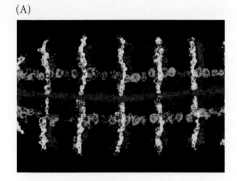

(B)

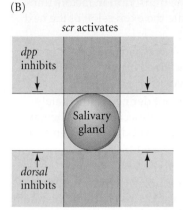

FIGURE 6.39 Cartesian coordinate system mapped out by gene expression patterns. (A) A grid (ventral view, looking "up" at the embryo) formed by the expression of *short-gastrulation* (red), *intermediate neuroblast defective* (green), and *muscle segment homeobox* (magenta) along the dorsal-ventral axis, and by the expression of *wingless* (yellow), and *engrailed* (green) transcripts along the anterior-posterior axis. (B) Coordinates for the expression of genes giving rise to *Drosophila* salivary glands. These genes are activated by the protein product of the *sex combs reduced* (*scr*) homeotic gene in a narrow band along the anterior-posterior axis, and they are inhibited in the regions marked by *decapentaplegic* (*dpp*) and *dorsal* gene products along the dorsal-ventral axis. This pattern allows salivary glands to form in the midline of the embryo in the second parasegment. (A courtesy of D. Kosman; B after Panzer et al. 1992.)

 Snapshot Summary: *Drosophila Development and Axis Specification*

1. *Drosophila* cleavage is superficial. The nuclei divide 13 times before forming cells. Before cell formation, the nuclei reside in a syncytial blastoderm. Each nucleus is surrounded by actin-filled cytoplasm.

2. When the cells form, the *Drosophila* embryo undergoes a mid-blastula transition, wherein the cleavages become asynchronous and new mRNA is made. At

this time, there is a transfer from maternal to zygotic control of development.

3. Gastrulation begins with the invagination of the most ventral region (the presumptive mesoderm), which causes the formation of a ventral furrow. The germ band expands such that the future posterior segments curl just behind the presumptive head.

4. The genes regulating pattern formation in *Drosophila* operate according to certain principles:

   - There are *morphogens*—such as Bicoid and Dorsal—whose gradients determine the specification of different cell types. These morphogens can be transcription factors.

   - There is a *temporal order* wherein different classes of genes are transcribed, and the products of one gene often regulate the expression of another gene.

   - *Boundaries* of gene expression can be created by the interaction between transcription factors and their gene targets. Here, the transcription factors transcribed earlier regulate the expression of the next set of genes.

   - *Translational control* is extremely important in the early embryo, and localized mRNAs are critical in patterning the embryo.

   - *Individual cell fates* are not defined immediately. Rather, there is a stepwise specification wherein a given field is divided and subdivided, eventually regulating individual cell fates.

5. Maternal effect genes are responsible for the initiation of anterior-posterior polarity. *Bicoid* mRNA is bound by its 3′ UTR to the cytoskeleton in the future anterior pole; *nanos* mRNA is sequestered by its 3′ UTR in the future posterior pole. *Hunchback* and *caudal* messages are seen throughout the embryo.

6. Dorsal-ventral polarity is regulated by the entry of Dorsal protein into the nucleus. Dorsal-ventral polarity is initiated when the nucleus moves to the dorsal-anterior of the oocyte and transcribes the *gurken* message, which is then transported to the region above the nucleus and adjacent to the follicle cells.

7. Gurken protein is secreted from the oocyte and binds to its receptor (Torpedo) on the follicle cells. This binding dorsalizes the follicle cells, preventing them from synthesizing Pipe.

8. Pipe protein in the ventral follicle cells modifies an as yet unknown factor that modifies the Nudel protein. This modification allows Nudel to activate a cascade of proteolysis in the space between the ventral follicle cells and the ventral cells of the embryo. As a result of this cascade, the Spätzle protein is activated and binds to the Toll protein on the ventral embryonic cells.

9. The activated Toll protein initiates a cascade that phosphorylates the Cactus protein, which has been bound to Dorsal. Phosphorylated Cactus is degraded, allowing Dorsal to enter the nucleus. Once in the nucleus, Dorsal activates the genes responsible for the ventral cell fates and represses those genes whose proteins would specify dorsal cell fates.

10. Dorsal protein forms a gradient as it enters the various nuclei. Those nuclei at the most ventral surface incorporate the most Dorsal protein and become mesoderm; those more lateral become neurogenic ectoderm.

11. The Bicoid and Hunchback proteins activate the genes responsible for the anterior portion of the fly; Caudal activates genes responsible for posterior development.

12. The unsegmented anterior and posterior extremities are regulated by the activation of the Torso protein at the anterior and posterior poles of the egg.

13. The gap genes respond to concentrations of the maternal effect gene proteins. Their protein products interact with each other such that each gap gene protein defines specific regions of the embryo.

14. The gap gene proteins activate and repress the pair-rule genes. The pair-rule genes have modular promoters such that they become activated in seven "stripes." Their boundaries of transcription are defined by the gap genes. The pair-rule genes form seven bands of transcription along the anterior-posterior axis, each one comprising two parasegments.

15. The pair-rule gene products activate *engrailed* and *wingless* expression in adjacent cells. The *engrailed*-expressing cells form the anterior boundary of each parasegment. These cells form a signaling center that organizes the cuticle formation and segmental structure of the embryo.

16. The homeotic selector genes are found in two complexes on chromosome 3 of *Drosophila*. Together, these regions are called Hom-C, the homeotic gene complex. The genes are arranged in the same order as their transcriptional expression. The Hom-C genes specify the individual segments, and mutations in these genes are capable of transforming one segment into another.

17. The expression of each homeotic selector gene is regulated by the gap and pair-rule genes. Their expression is refined and maintained by interactions whereby their protein products prevent the transcription of neighboring Hom-C genes.

18. The targets of the Hom-C proteins are the realisator genes. These realisator genes are responsible for constructing the specific structure.

19. Organs form at the intersection of dorsal-ventral and anterior-posterior regions of gene expression.

For Further Reading

Complete bibliographical citations for all literature cited in this chapter can be found at the free-access website **www.devbio.com**

Driever, W. and C. Nüsslein-Volhard. 1988a. The Bicoid protein determines position in the *Drosophila* embryo in a concentration-dependent manner. *Cell* 54: 95–104.

Fujioka, M., Y. Emi-Sarker, G. L. Yusibova, T. Goto and J. B. Jaynes. 1999. Analysis of an *even-skipped* rescue transgene reveals both composite and discrete neuronal and early blastoderm enhancers, and multi-stripe positioning by gap gene repressor gradients. *Development* 126: 2527–2538.

Lehmann, R. and C. Nüsslein-Volhard. 1991. The maternal gene *nanos* has a central role in posterior pattern formation of the *Drosophila* embryo. *Development* 112: 679–691.

Leptin, M. 1991. *twist* and *snail* as positive and negative regulators during *Drosophila* mesoderm development. *Genes Dev.* 5: 1568–1576.

Lewis, E. B. 1978. A gene complex controlling segmentation in *Drosophila*. *Nature* 276: 565–570.

Martinez-Arias, A. and P. A. Lawrence. 1985. Parasegments and compartments in the *Drosophila* embryo. *Nature* 313: 639–642.

Panfilio, K. A. 2007. Extraembryonic development in insects and the acrobatics of blastokinesis. *Dev. Biol.* 313: 471–491.

Pankratz, M. J., E. Seifert, N. Gerwin, B. Billi, U. Nauber and H. Jäckle. 1990. Gradients of *Krüppel* and *knirps* gene products direct pair-rule gene stripe patterning in the posterior region of the *Drosophila* embryo. *Cell* 61: 309–317.

Pfeiffer, B. D. and 16 others. 2009. Tools for neuroanatomy and neurogenetics in *Drosophila*. *Proc. Natl. Acad. Sci. USA* 105: 9715–9720.

Roth, S., D. Stein and C. Nüsslein-Volhard. 1989. A gradient of nuclear localization of the dorsal protein determines dorsoventral pattern in the *Drosophila* embryo. *Cell* 59: 1189–1202.

Schüpbach, T. 1987. Germ line and soma cooperate during oogenesis to establish the dorsoventral pattern of egg shell and embryo in *Drosophila melanogaster*. *Cell* 49: 699–707.

Struhl, G. 1981. A homeotic mutation transforming leg to antenna in *Drosophila*. *Nature* 292: 635–638.

Wang, C. and R. Lehman. 1991. Nanos is the localized posterior determinate in *Drosophila*. *Cell* 66: 637–647.

Go Online

WEBSITE 6.1 *Drosophila* fertilization. Fertilization of *Drosophila* can only occur in the region of the oocyte that will become the anterior of the embryo. Moreover, the sperm tail appears to stay in this region.

WEBSITE 6.2 The early development of other insects. *Drosophila* is a highly derived species. There are other insect species that develop in ways very different from the "standard" fruit fly.

WEBSITE 6.3 Evidence for gradients in insect development. The original evidence for gradients in insect development came from studies providing evidence for two "organization centers" in the egg, one located anteriorly and one located posteriorly.

WEBSITE 6.4 Christiane Nüsslein-Volhard and the molecular approach to development. The research that revolutionized developmental biology had to wait for someone to synthesize molecular biology, embryology, and *Drosophila* genetics.

WEBSITE 6.5 Asymmetrical spread of morphogens. It is unlikely that morphogens such as Wingless spread by free diffusion. The asymmetry of Wingless diffusion suggests that neighboring cells play a crucial role in moving this protein.

WEBSITE 6.6 Getting a head in the fly. The segment polarity genes may act differently in the head than in the trunk. Indeed, the formation of the *Drosophila* head may differ significantly from the way the rest of the body is formed.

Vade Mecum

***Drosophila* development.** The Vade Mecum sites have remarkable time-lapse sequences of *Drosophila* development, including cleavage and gastrulation. This segment also provides access to the fly life cycle. The color coding superimposed on the germ layers allows you to readily understand tissue movements.

Outside Sites

"The Interactive Fly," compiled by Thomas Brody, provides an index to the major *Drosophila* websites worldwide. It is hosted by the Society for Developmental Biology (SDB) at **http://www.sdbonline.org/fly/aimain/1aahome.htm**. Two notable entries accessible through the site are "Atlas of Fly Development" by Voker Hartenstein (**http://www.sdbonline.org/fly/atlas/00atlas.htm**) and "Stages in Fly Development: The Movies" (**http://www.sdbonline.org/fly/aimain/2stages.htm**).

Amphibians and Fish

Early Development and Axis Formation

AMPHIBIAN EMBRYOS ONCE DOMINATED the field of experimental embryology. With their large cells and rapid development, salamander and frog embryos were excellently suited for transplantation experiments. However, amphibian embryos fell out of favor during the early days of developmental genetics. Frogs and salamanders undergo a long period of growth before they become fertile, and their chromosomes are often found in several copies, precluding easy mutagenesis.* But new molecular techniques such as in situ hybridization, antisense oligonucleotides, chromatin immunoprecipitation, and dominant negative proteins have allowed researchers to return to the study of amphibian embryos, and to integrate recent molecular analyses of development with earlier experimental findings. The results have been spectacular, revealing new vistas of how vertebrate bodies are patterned and structured.

Also in recent years, the teleost fish *Danio rerio*, commonly known as the zebrafish, has become a model organism for those who study vertebrate development. Zebrafish have large broods, breed all year, are easily maintained, have transparent embryos that develop outside the mother (an important feature for microscopy), and can be raised so that mutants can be readily discovered and propagated in the laboratory. In addition, these fish develop rapidly. At 24 hours after fertilization, the embryo has already formed most of its organ primordia and displays a characteristic tadpole-like form. In addition, the ability to microinject fluorescent dyes into individual cells has allowed scientists to follow individual living cells as an organ develops. Therefore, much of the description of fish development later in this chapter is based on studies of *Danio*.

The analysis of zebrafish mutants has confirmed many of the principles of development originally elucidated in amphibian embryos, and has shown that many of the same molecules are used. Together, the *Xenopus* and zebrafish research communities have established the general rules of vertebrate development, which (as we will see in Chapter 8) are maintained among the amniote vertebrates (reptiles, birds, and mammals). We start here with the amphibians, because it was these organisms that provided us with the "backbone" of questions and tentative conclusions that have enabled vertebrate developmental biology to flourish.

We are standing and walking with parts of our body which could have been used for thinking had they developed in another part of the embryo.

HANS SPEMANN (1943)

Theories come and theories go. The frog remains.

JEAN ROSTAND (1960)

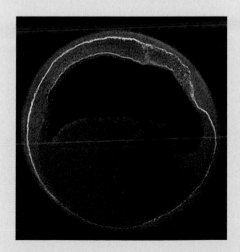

---

*In the 1960s, *Xenopus laevis* replaced *Rana* frogs and the salamanders because it could be induced to mate throughout the year. Unfortunately, *Xenopus laevis* has four copies of each chromosome rather than the more usual two, and it takes 1–2 years to reach sexual maturity. Another *Xenopus* species, *X. tropicalis*, is often studied now, since it has all the advantages of *X. laevis*, plus it is diploid and reaches sexual maturity in a mere 6 months (Gurdon and Hopwood 2000; Hirsch et al. 2002).

EARLY AMPHIBIAN DEVELOPMENT

Fertilization, Cortical Rotation, and Cleavage

Fertilization can occur anywhere in the animal hemisphere of the amphibian embryo. The point of sperm entry is important because it determines the orientation of the dorsal-ventral axis of the larva (tadpole). The point of sperm entry will mark the ventral side of the embryo, while the site 180 degrees opposite the point of sperm entry will mark the dorsal side.*

---

*The axis between the point of sperm entry and the dorsal side approximates, but does not exactly correspond to, the actual ventral-dorsal axis of the amphibian larva (Lane and Smith 1999; Lane and Sheets 2002). However, as the literature in the field has traditionally equated these two axes, we will use the classical terminology here.

The sperm centriole organizes the microtubules of the egg and causes them to arrange themselves in a parallel array in the vegetal cytoplasm, separating the cortical cytoplasm from the yolky internal cytoplasm (**Figure 7.1A,B**). These microtubular tracks allow the cortical cytoplasm to rotate with respect to the inner cytoplasm. Indeed, the arrays are first seen immediately before rotation starts, and they disappear when rotation ceases (Elinson and Rowning 1988; Houliston and Elinson 1991). In the 1-cell embryo, the cortical cytoplasm rotates 30 degrees with respect to the internal cytoplasm (**Figure 7.1C**). In some eggs, this exposes a band of inner gray cytoplasm in the marginal region of the 1-cell embryo, directly opposite the sperm entry point (**Figure 7.1D**; Roux 1887; Ancel and Vintenberger 1948). This region, the **gray crescent**, is where gastrulation will begin. Even in *Xenopus* eggs, which do not expose a gray crescent, cortical rotation occurs and cytoplasmic movements can be seen (Manes and Elinson 1980; Vincent et al. 1986). As we will see later, the site opposite the point

FIGURE 7.1 Reorganization of the cytoplasm and cortical rotation produce the gray crescent in frog eggs. (A,B) Parallel arrays of microtubules (visualized here using fluorescent antibodies to tubulin) form in the vegetal hemisphere of the egg, along the future dorsal-ventral axis. (A) With 50% of the first cell cycle complete, microtubules are present, but they lack polarity. (B) By 70% completion of the cycle, the vegetal shear zone is characterized by a parallel array of microtubules; cortical rotation begins at this time. At the end of rotation, the microtubules will depolymerize. (C) Schematic cross section of cortical rotation. At left, the egg is shown about midway through the first cell cycle. It has radial symmetry around the animal-vegetal axis. The sperm nucleus has entered at one side and is migrating inward. At right, 80% into first cleavage, the cortical cytoplasm has rotated 30 degrees relative to the internal cytoplasm. Gastrulation will begin in the gray crescent—the region opposite the point of sperm entry, where the greatest displacement of cytoplasm occurs. (D) Gray crescent of *Rana pipiens*. Immediately after cortical rotation (left), lighter gray pigmentation is exposed beneath the heavily pigmented cortical cytoplasm. The first cleavage furrow (right) bisects this gray crescent. (A,B from Cha and Gard 1999, courtesy of the authors; C after Gerhart et al. 1989; D courtesy of R. P. Elinson.)

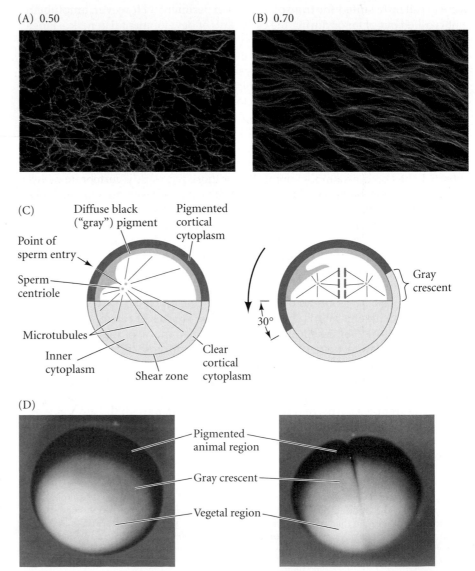

(A) 0.50

(B) 0.70

(C)

Point of sperm entry

Sperm centriole

Microtubules

Inner cytoplasm

Diffuse black ("gray") pigment

Pigmented cortical cytoplasm

Shear zone

Clear cortical cytoplasm

30°

Gray crescent

(D)

Pigmented animal region

Gray crescent

Vegetal region

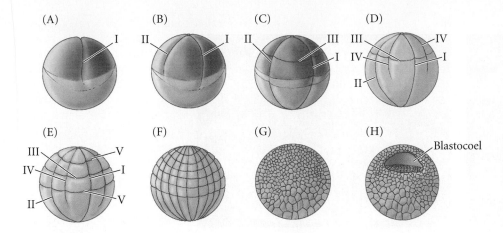

(A) (B) (C) (D)

(E) (F) (G) (H) Blastocoel

FIGURE 7.2 Cleavage of a frog egg. Cleavage furrows, designated by Roman numerals, are numbered in order of appearance. (A,B) Because the vegetal yolk impedes cleavage, the second division begins in the animal region of the egg before the first division has divided the vegetal cytoplasm. (C) The third division is displaced toward the animal pole. (D–H) The vegetal hemisphere ultimately contains larger and fewer blastomeres than the animal half. (H) Cross section through a mid-blastula stage embryo. (After Carlson 1981.)

of sperm entry will be the place where gastrulation begins, and the microtubular array will become extremely important in initiating the dorsal-ventral and anterior-posterior axes of the larva.

Unequal radial holoblastic cleavage

Cleavage in most frog and salamander embryos is radially symmetrical and holoblastic, just like echinoderm cleavage. The amphibian egg, however, contains much more yolk. This yolk, which is concentrated in the vegetal hemisphere, is an impediment to cleavage. Thus, the first division begins at the animal pole and slowly extends down into the vegetal region (**Figure 7.2**; see also Figure 1.2E). In the axolotl salamander, the cleavage furrow extends through the animal hemisphere at a rate close to 1 mm per minute, slowing down to a mere 0.02–0.03 mm per minute as it approaches the vegetal pole (Hara 1977). In many species (especially salamanders and frogs of the genus *Rana*), the first cleavage division bisects the gray crescent.

At the frog egg's first cleavage, one can see the difference in the furrow between the animal and the vegetal

hemispheres (**Figure 7.3A**). While the cleavage furrow is still cleaving the yolky cytoplasm of the vegetal hemisphere, the second cleavage has already started near the animal pole. This cleavage is at right angles to the first one and is also meridional (**Figure 7.3B**). The third cleavage is equatorial. However, because of the vegetally placed yolk, the third cleavage furrow is not actually at the equator but is displaced toward the animal pole (Valles et al. 2002). It divides the amphibian embryo into four small animal blastomeres (micromeres) and four large blastomeres (macromeres) in the vegetal region (**Figure 7.3C**). Despite their unequal sizes, the blastomeres continue to divide at the same rate until the twelfth cell cycle (with only a small delay of the vegetal cleavages). As cleavage progresses, the animal region becomes packed with numerous small cells, while the vegetal region contains a relatively small number of large, yolk-laden macromeres (see Figures 7.2 and 1.2I). An amphibian embryo containing 16 to 64 cells is commonly called a **morula** (plural *morulae*; Latin, "mulberry," whose shape it vaguely resembles). At the 128-cell stage, the blastocoel becomes apparent, and the embryo is considered a blastula.

(A) (B) (C)

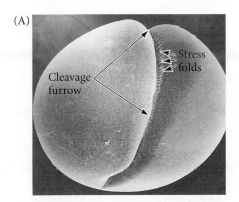
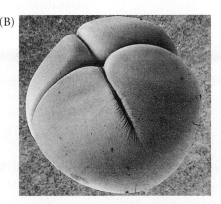
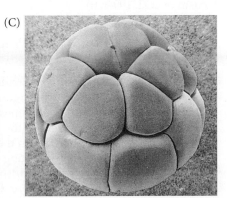

FIGURE 7.3 Scanning electron micrographs of frog egg cleavage. (A) First cleavage. (B) Second cleavage (4 cells). (C) Fourth cleavage (16 cells), showing the size discrepancy between the animal and vegetal cells after the third division. (A from Beams and Kessel 1976, courtesy of the authors; B,C courtesy of L. Biedler.)

(A)

(B)

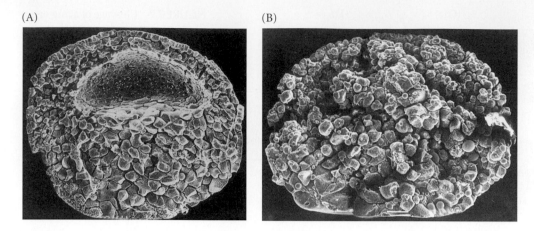

FIGURE 7.4 Depletion of EP-cadherin mRNA in the *Xenopus* oocyte results in the loss of adhesion between blastomeres and the obliteration of the blastocoel. (A) Control embryo. (B) EP-cadherin-depleted embryo. (From Heasman et al. 1994b, courtesy of J. Heasman.)

The amphibian blastocoel serves two major functions. First, it permits cell migration during gastrulation; and second, it prevents the cells beneath it from interacting prematurely with the cells above it. When Nieuwkoop (1973) took embryonic newt cells from the roof of the blastocoel in the animal hemisphere (a region often called the **animal cap**) and placed them next to the yolky vegetal cells from the base of the blastocoel, the animal cap cells differentiated into mesodermal tissue instead of ectoderm. Thus, the blastocoel prevents the contact of the vegetal cells destined to become endoderm with those cells in the ectoderm fated to give rise to the skin and nerves.

Numerous cell adhesion molecules keep the cleaving blastomeres together. One of the most important of these is EP-cadherin. The mRNA for this protein is supplied in the oocyte cytoplasm. If this message is destroyed by antisense oligonucleotides so that no EP-cadherin is made, the adhesion between blastomeres is dramatically reduced, resulting in the obliteration of the blastocoel (**Figure 7.4**; Heasman et al. 1994a,b).

The mid-blastula transition: Preparing for gastrulation

An important precondition for gastrulation is the activation of the genome. In the model organism *Xenopus laevis* (African clawed frog), only a few genes appear to be transcribed during early cleavage. For the most part, nuclear genes are not activated until late in the twelfth cell cycle (Newport and Kirschner 1982a,b; Yang et al. 2002). At that time, the embryo experiences a **mid-blastula transition** (**MBT**; see Chapters 5 and 6); different genes begin to be transcribed in different cells, the cell cycle acquires gap phases, and the blastomeres acquire the capacity to become motile. It is thought that some factor in the egg is being absorbed by the newly made chromatin because (as in

Drosophila) the time of this transition can be changed experimentally by altering the ratio of chromatin to cytoplasm in the cell (Newport and Kirschner 1982a,b).

Some of the events that trigger the mid-blastula transition involve chromatin modification. First, certain promoters are demethylated, allowing transcription of these genes. In *Xenopus* (unlike mammals), high levels of methylated DNA are seen in both the paternally and maternally derived chromosomes. However, during the late blastula stages, there is a loss of methylation on the promoters of genes that are activated at MBT. This demethylation is not seen on promoters that are not activated at MBT, nor is it observed in the coding regions of MBT-activated genes. The methylation of lysine-4 on histone H3 (forming a trimethylated lysine that is associated with active transcription) is also seen on the 5′ ends of many genes during the MBT. It appears, then, that modification of certain promoters and their associated nucleosomes may play a pivotal role in regulating the timing of gene expression at the mid-blastula transition (Stancheva et al. 2002; Akkers et al. 2009).

It is thought that once the chromatin at the promoters has been remodeled, various transcription factors (such as the VegT protein, formed in the vegetal cytoplasm from localized maternal mRNA) bind to the promoters and initiate new transcription. For instance, the vegetal cells (under the direction of the VegT protein) become the endoderm and begin secreting factors that induce the cells above them to become the mesoderm (see Figure 7.23).

See WEBSITE 7.1 Amphibian development movies

Amphibian Gastrulation

The study of amphibian gastrulation is both one of the oldest and one of the newest areas of experimental embryolo-

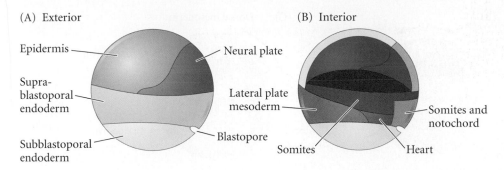

(A) Exterior

Epidermis

Supra-
blastoporal
endoderm

Subblastoporal
endoderm

Neural plate

Blastopore

(B) Interior

Lateral plate
mesoderm

Somites

Somites and
notochord

Heart

FIGURE 7.5   Fate maps of the *Xenopus laevis* blastula exterior (A) and interior (B). Most of the mesodermal derivatives are formed from the interior cells. The placement of ventral mesoderm remains controversial (see Kumano and Smith 2002; Lane and Sheets 2006). (After Lane and Smith 1999; Newman and Kreig 1999.)

gy (see Beetschen 2001; Braukmann and Gilbert 2005). Even though amphibian gastrulation has been studied extensively for the past century, most of our theories concerning its mechanisms have been revised over the past decade. The study of these developmental movements has been complicated by the fact that there is no single way amphibians gastrulate; different species employ different means to achieve the same goal. In recent years, the most intensive investigations have focused on *Xenopus laevis*, so we will concentrate on the mechanisms of gastrulation in that species.

Vegetal rotation and the invagination of the bottle cells

Amphibian blastulae are faced with the same tasks as the invertebrate blastulae we followed in Chapters 5 and 6—namely, to bring inside the embryo those areas destined to form the endodermal organs; to surround the embryo with cells capable of forming the ectoderm; and to place the mesodermal cells in the proper positions between them. Fate mapping by Løvtrup (1975; Landstrom and Løvtrup 1979) and by Keller (1975, 1976) has shown that cells of *Xenopus* blastulae have different fates, depending on whether they are located in the deep or the superficial layers of the embryo (**Figure 7.5**). In *Xenopus*, the mesodermal precursors exist mostly in the deep layer of cells, while the ectoderm and endoderm arise from the superficial layer on the surface of the embryo. Most of the precursors for the notochord and other mesodermal tissues are located beneath the surface in the equatorial (marginal) region of the embryo.*

Gastrulation movements in frog embryos act to position the mesoderm between the outer ectoderm and the inner endoderm. These movements are initiated on the future dorsal side of the embryo, just below the equator, in the region of the gray crescent (**Figure 7.6**)—the region opposite the point of sperm entry. Here the cells invaginate to form a slitlike blastopore (**Figure 7.7**). These cells change their

shape dramatically. The main body of each cell is displaced toward the inside of the embryo while maintaining contact with the outside surface by way of a slender neck. These **bottle cells** line the *archenteron* (primitive gut) as it forms, and as in the gastrulating sea urchin, an invagination of cells initiates formation of the archenteron. However, unlike sea urchins, gastrulation in the frog begins not in the most vegetal region but in the **marginal zone**—the region surrounding the equator of the blastula, where the animal and vegetal hemispheres meet (see Figure 7.6B). Here the endodermal cells are not as large or as yolky as the most vegetal blastomeres.

Working with *Xenopus*, Ray Keller and his students showed that the peculiar shape change of the bottle cells is needed to *initiate* gastrulation; it is the constriction of these cells that forms the blastopore. However, once the subsurface marginal cells are brought into contact with the basal region of the surface blastomeres, they begin to involute on the extracellular matrix secreted onto the basal region of these cells. When such involution movements are underway, bottle cells are no longer essential. When bottle cells are removed after their formation, involution and blastopore formation and closure continue (Keller 1981; Hardin and Keller 1988). Thus, in *Xenopus*, the major factor in the movement of cells into the embryo appears to be the involution of the subsurface cells rather than the superficial marginal cells.

But cell involution is not a passive event. At least 2 hours before the bottle cells form, internal cell rearrangements propel the cells of the dorsal floor of the blastocoel toward

---

*In urodeles (salamanders such as *Triturus* and *Ambystoma*) and in some frog species, many more of the notochord and mesoderm precursor cells are found among the surface cells (Purcell and Keller 1993; Shook et al. 2002). Here, the superficial mesoderm cells move into the deeper layers by forming a primitive streak similar to that of amniote embryos (see Chapter 8). For ways in which the patterning of the mesoderm differs between anurans and urodeles, see Hurtado and De Robertis 2007.

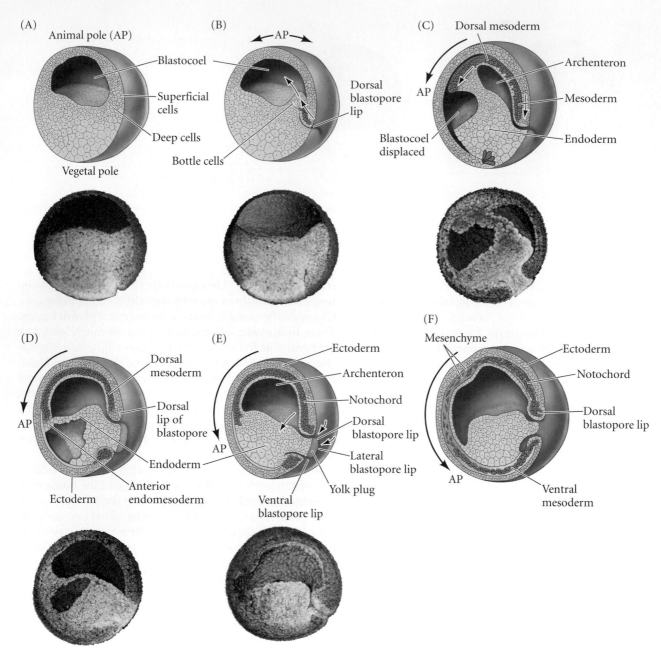

FIGURE 7.6 Cell movements during frog gastrulation. The drawings show meridional sections are cut through the middle of the embryo and positioned so that the vegetal pole is tilted toward the observer and slightly to the left. The major cell movements are indicated by arrows, and the superficial animal hemisphere cells are colored so that their movements can be followed. Below the drawings are corresponding micrographs imaged with a surface imaging microscope (see Ewald et al. 2002). (A,B) Early gastrulation. The bottle cells of the margin move inward to form the dorsal lip of the blastopore, and the mesodermal precursors involute under the roof of the blastocoel. AP marks the position of the animal pole, which will change as gastrulation continues. (C,D) Mid-gastrulation. The archenteron forms and displaces the blastocoel, and cells migrate from the lateral and ventral lips of the blastopore into the embryo. The cells of the animal hemisphere migrate down toward the vegetal region, moving the blastopore to the region near the vegetal pole. (E,F) Toward the end of gastrulation, the blastocoel is obliterated, the embryo becomes surrounded by ectoderm, the endoderm has been internalized, and the mesodermal cells have been positioned between the ectoderm and endoderm. (Drawings after Keller 1986; micrographs courtesy of Andrew Ewald and Scott Fraser.)

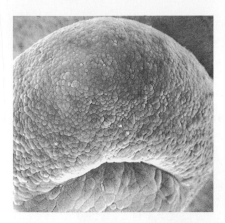

FIGURE 7.7 Surface view of an early dorsal blastopore lip of *Xenopus*. The size difference between the animal and vegetal blastomeres is readily apparent. (Courtesy of C. Phillips.)

the animal cap. This **vegetal rotation** places the prospective pharyngeal endoderm cells adjacent to the blastocoel and immediately above the involuting mesoderm. These cells then migrate along the basal surface of the blastocoel roof, traveling toward the future anterior of the embryo (**Figure 7.8**; Nieuwkoop and Florschütz 1950; Winklbauer and Schürfeld 1999; Ibrahim and Winklbauer 2001). The superficial layer of marginal cells is pulled inward to form the endodermal lining of the archenteron, merely because it is attached to the actively migrating deep cells. Although experimentally removing the bottle cells does not affect the involution of the deep or superficial marginal zone cells into the embryo, removal of the deep **involuting marginal zone (IMZ)** cells stops archenteron formation.

See VADE MECUM Amphibian development

INVOLUTION AT THE BLASTOPORE LIP The next phase of gastrulation involves the involution of the marginal zone cells while the animal cells undergo epiboly and converge at the blastopore (see Figure 7.6C,D). When the migrating marginal cells reach (and become) the **dorsal lip of the blastopore**, they turn inward and travel along the inner surface of the outer animal hemisphere cells (i.e., the blastocoel roof). Thus, the cells constituting the lip of the blastopore are constantly changing. The order of the march into the embryo was determined by the vegetal rotation that abutted the prospective pharyngeal endoderm against the inside of the animal cap tissue. The first cells to compose the dorsal blastopore lip and enter into the embryo are the prospective pharyngeal endoderm of the foregut (including the bottle cells). As these first cells pass into the interior of the embryo, the dorsal blastopore lip becomes composed of cells that involute into the embryo to become the **prechordal plate** (the precursor of the head mesoderm). The next cells involuting into the embryo through the dorsal blastopore lip are the **chordamesoderm** cells. These cells will form the **notochord**, a transient mesodermal rod that plays an important role in inducing and patterning the nervous system. Thus the cells constituting the dorsal blastopore lip are constantly changing as the original cells migrate into the embryo and are replaced by cells migrating down, inward, and upward.

As the new cells enter the embryo, the blastocoel is displaced to the side opposite the dorsal lip of the blastopore. Meanwhile, the lip expands laterally and ventrally as the processes of bottle cell formation and involution continue around the blastopore. The widening blastopore "crescent" develops lateral lips and finally a ventral lip over which additional mesodermal and endodermal precursor cells pass. With the formation of the ventral lip, the blastopore has formed a ring around the large endodermal cells that

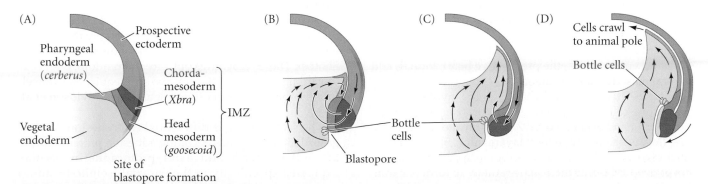

FIGURE 7.8 Early movements of *Xenopus* gastrulation. Orange represents the prospective pharyngeal endoderm (as described by *cerberus* expression). Dark orange represents the prospective head mesoderm (*goosecoid* expression), and the chordamesoderm (*Xbra* expression) is red. (A) At the beginning of gastrulation, the involuting marginal zone (IMZ) forms. (B) Vegetal rotation (arrows) pushes the prospective pharyngeal endoderm to the side of the blastocoel. (C,D) The vegetal endoderm movements push the pharyngeal endoderm forward, driving the mesoderm passively into the embryo and toward the animal pole. The ectoderm (blue) begins epiboly. (After Winklbauer and Schürfeld 1999.)

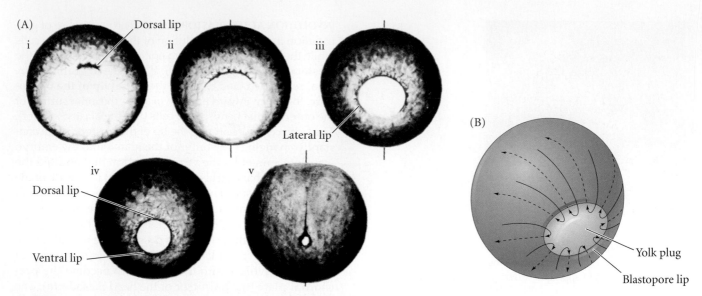

FIGURE 7.9 Epiboly of the ectoderm. (A) Changes in the region around the blastopore as the dorsal, lateral, and ventral lips are formed in succession. When the ventral lip completes the circle, the endoderm becomes progressively internalized. Numbers ii–v correspond to Figure 7.6 B–E, respectively. (B) Summary of epiboly of the ectoderm and involution of the mesodermal cells migrating into the blastopore and then under the surface. The endoderm beneath the blastopore lip (the yolk plug) is not mobile and is enclosed by these movements. (A from Balinsky 1975, courtesy of B. I. Balinsky.)

remain exposed on the vegetal surface (**Figure 7.9**). This remaining patch of endoderm is called the **yolk plug**; it, too, is eventually internalized. At that point, all the endodermal precursors have been brought into the interior of the embryo, the ectoderm has encircled the surface, and the mesoderm has been brought between them.

CONVERGENT EXTENSION OF THE DORSAL MESODERM Involution begins dorsally, led by the pharyngeal endoderm and the head mesoderm. These tissues will migrate most anteriorly beneath the surface ectoderm.* Meanwhile, as the lip of the blastopore expands to have dorsolateral, lateral, and ventral sides, the prospective heart, kidney, and ventral mesodermal precursor cells enter into the embryo.

Figure 7.10 depicts the behavior of the involuting marginal zone cells at successive stages of *Xenopus* gastrulation (Keller and Schoenwolf 1977; Hardin and Keller 1988). The IMZ is originally several layers thick. Shortly before their involution through the blastopore lip, the several layers of deep IMZ cells intercalate radially to form one thin, broad layer. This intercalation further extends the IMZ veg-

---

*The pharyngeal endoderm and head mesoderm cannot be separated experimentally at this stage, so they are sometimes referred to collectively as the head endomesoderm. The notochord is the basic unit of the dorsal mesoderm, but it is thought that the dorsal portion of the somites may have similar properties to the notochord.

etally (**Figure 7.10A**). At the same time, the superficial cells spread out by dividing and flattening. When the deep cells reach the blastopore lip, they involute into the embryo and initiate a second type of intercalation. This intercalation causes a **convergent extension** along the mediolateral axis that integrates several mesodermal streams to form a long, narrow band (**Figure 7.10B**). This movement is reminiscent of traffic on a highway when several lanes must merge to form a single lane (or for that matter, the cell movements of the sea urchin archenteron). The anterior part of this band migrates toward the animal cap. Thus, the mesodermal stream continues to migrate toward the animal pole, and the overlying layer of superficial cells (including the bottle cells) is passively pulled toward the animal pole, thereby forming the endodermal roof of the archenteron (see Figures 7.6 and 7.10A). The radial and mediolateral intercalations of the deep layer of cells appear to be responsible for the continued movement of mesoderm into the embryo.

Several forces appear to drive convergent extension. The first force is a polarized cell cohesion, wherein the involuted mesodermal cells send out protrusions to contact one another. These "reachings out" are not random, but occur toward the midline of the embryo and require an extracellular fibronectin matrix (Goto et al. 2005; Davidson et al. 2006).

The second force is differential cell cohesion. During gastrulation, the genes encoding adhesion proteins **paraxial protocadherin** and **axial protocadherin** become expressed specifically in the paraxial (somite-forming) mesoderm and the notochord, respectively (**Figure 7.11**). An experimental dominant negative form of axial protocadherin prevents the presumptive notochord cells from sorting out from the paraxial mesoderm and blocks normal axis formation. A dominant negative paraxial protocadherin (which is secreted instead of being bound to the cell membrane) prevents convergent extension* (Kim et al.

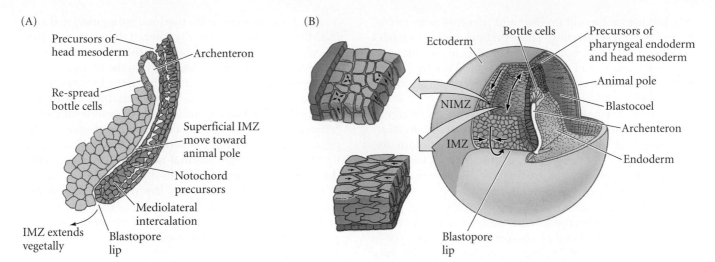

(A)

Precursors of head mesoderm

Archenteron

Re-spread bottle cells

Superficial IMZ move toward animal pole

Notochord precursors

Mediolateral intercalation

Blastopore lip

IMZ extends vegetally

(B)

Ectoderm

Bottle cells

Precursors of pharyngeal endoderm and head mesoderm

Animal pole

Blastocoel

Archenteron

Endoderm

NIMZ

IMZ

Blastopore lip

FIGURE 7.10 *Xenopus* gastrulation continues. (A) The deep marginal cells flatten, and the formerly superficial cells form the wall of the archenteron. (B) Radial intercalation, looking down at the dorsal blastopore lip from the dorsal surface. In the noninvoluting marginal zone (NIMZ) and the upper portion of the IMZ, deep (mesodermal) cells are intercalating radially to make a thin band of flattened cells. This thinning of several layers into a few causes convergent extension (white arrows) toward the blastopore lip. Just above the lip, mediolateral intercalation of the cells produces stresses that pull the IMZ over the lip. After involuting over the lip, mediolateral intercalation continues, elongating and narrowing the axial mesoderm. (After Wilson and Keller 1991; Winklbauer and Schürfeld 1999.)

1998; Kuroda et al. 2002). Moreover, the expression domain of paraxial protocadherin characterizes the trunk mesodermal cells, which undergo convergent extension, distinguishing them from the head mesodermal cells, which do not undergo convergent extension.

The third factor regulating convergent extension is calcium flux. Wallingford and colleagues (2001) found that dramatic waves of calcium ions surge across the dorsal tissues undergoing convergent extension, causing waves of contraction within the tissue. The calcium ions are released from intracellular stores and are required for convergent extension. If the release of calcium ions is blocked, normal cell specification still occurs, but the dorsal mesoderm neither converges nor extends. This calcium is thought to regulate the contraction of the actin microfilaments. This may help explain why the head region does not undergo convergent extension. The *Otx2* gene is the vertebrate homologue of the *Drosophila orthodenticle* gene and encodes a transcription factor expressed in the most anterior region of the embryo. The Otx2 protein is critical in head formation, activating those genes involved in forebrain formation (see Figure 7.34). In addition to specifying the anterior tissues of the embryo, Otx2 prevents the cells expressing it from undergoing convergence and extension by activating the *calponin* gene (Morgan et al. 1999). Calponin pro-

*Dominant negative proteins are mutated forms of the wild-type protein that interfere with the normal functioning of the wild-type protein. Thus, a dominant negative protein will have an effect similar to a loss-of-function mutation in the gene encoding the particular protein.

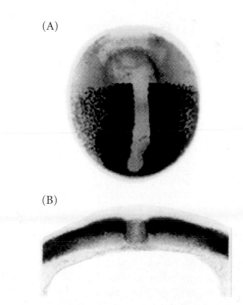

(A)

(B)

FIGURE 7.11 Protocadherin expression separates axial and paraxial mesoderm. (A) The expression pattern of paraxial protocadherin during late gastrulation (dark areas) shows the distinct downregulation in the notochord and the absence of expression in the head region. (B) Double-stained cross section through a late *Xenopus* gastrula shows the separation of notochord (reddish-brown staining for chordin) and the paraxial mesoderm (midnight blue staining for paraxial protocadherin). (Courtesy of E. M. De Robertis.)

tein binds to actin and myosin and prevents actin microfilaments from contracting. These findings support a model wherein regulatory proteins cause changes in the outer surface of the tissue and generate mechanical traction forces that either prevent or encourage cell migration (Beloussov et al. 2006; Davidson et al. 2008).

As mesodermal movement progresses, convergent extension continues to narrow and lengthen the involuting marginal zone. The involuting cells contain the prospective endodermal roof of the archenteron in its superficial layer and the prospective mesodermal cells, including those of the notochord, in its deep region. Toward the end of gastrulation, the centrally located notochord separates from the somitic mesoderm on either side of it, and the notochord elongates separately as its cells continue to intercalate (Wilson and Keller 1991). This may in part be a consequence of the different adhesion molecules in the axial and paraxial mesoderms (see Figure 7.11; Kim et al. 1998). This convergent extension of the mesoderm appears to be autonomous, because the movements of these cells occur even if this region of the embryo is experimentally isolated from the rest of the embryo (Keller 1986).

The dorsal portion of the **noninvoluting marginal zone** (**NIMZ**) extends more rapidly toward the blastopore than the ventral portion does, causing the blastopore lips to move toward the ventral side. While those mesodermal cells entering through the dorsal lip of the blastopore give rise to the central dorsal mesoderm (notochord and somites), the remainder of the body mesoderm (which forms the heart, kidneys, bones, and parts of several other organs) enters through the ventral and lateral blastopore lips to create the **mesodermal mantle**. The endoderm is derived from the superficial cells of the involuting marginal zone that form the lining of the archenteron roof and from the subblastoporal vegetal cells that become the archenteron floor (Keller 1986). The remnant of the blastopore—where the endoderm meets the ectoderm—now becomes the anus.*

---

*As gastrulation expert Ray Keller famously remarked, "Gastrulation is the time when a vertebrate takes its head out of its anus."

SIDELIGHTS & SPECULATIONS

Fibronectin and the Pathways for Mesodermal Migration

How are the involuting cells informed where to go once they enter the inside of the embryo? In many amphibians, it appears that the involuting mesodermal precursors migrate toward the animal pole on a fibronectin lattice secreted by the cells of the blastocoel roof (**Figure 7.12A,B**). Shortly before gastrulation, the presumptive ectoderm of the blastocoel roof secretes an extracellular matrix that contains fibrils of fibronectin (Boucaut et al. 1984; Nakatsuji et al. 1985). The involuting mesoderm appears to travel along these fibronectin fibrils. Confirmation of this hypothesis was obtained by chemically synthesizing a "phony" fibronectin that can compete with the genuine fibronectin of the extracellular matrix. Cells bind to a region of the fibronectin protein that contains a three-amino acid sequence (Arg-Gly-Asp). Boucaut and co-workers injected large amounts of a small peptide containing this sequence into the blastocoels of salamander embryos

shortly before gastrulation began. If fibronectin were essential for cell migration, then cells binding this soluble peptide fragment instead of the real extracellular fibronectin should stop migrating. Unable to find their "road," the mesodermal cells should cease involution. That is precisely what happened (**Figure 7.12C–F**). No migrating cells were seen along the underside of the ectoderm in the experimental embryos. Instead, the mesodermal precursors remained outside the embryos, forming a convoluted cell mass. Other small synthetic peptides (including other fragments of the fibronectin molecule) did not impede migration. Thus the fibronectin-containing extracellular matrix appears to provide both a substrate for adhesion as well as cues for the direction of cell migration. Shi and colleagues (1989) showed that salamander IMZ cells would migrate in the wrong direction if extra fibronectin lattices were placed in their path.

In *Xenopus*, fibronectin is similarly

secreted by the cells lining the blastocoel roof. The result is a band of fibronectin lining the roof, including Brachet's cleft, the part of the blastocoel roof extending vegetally on the dorsal side (see Figure 7.12A,B). The vegetal rotation places the pharyngeal endoderm and involuting mesoderm into contact with these fibronectin fibrils (Winklbauer and Schürfeld 1999). Convergent extension pushes the migrating cells upward toward the animal pole. The fibronectin fibrils are necessary for the head mesodermal cells to flatten and to extend broad (lamelliform) processes in the direction of migration (Winklbauer et al. 1991; Winklbauer and Keller 1996). Studies using inhibitors of fibronectin formation have shown that fibronectin fibrils are necessary for the direction of mesoderm migration, the maintenance of intercalation of animal cap cells, and the initiation of radial intercalation in the marginal zone (Marsden and DeSimone 2001).

The mesodermal cells are thought to adhere to fibronectin through the $\alpha_5\beta_1$ integrin protein (Alfandari et al. 1995). Mesodermal migration can also be arrested by the microinjection of antibodies against either fibronectin or the α_5 subunit of integrin, which serves as part of the fibronectin receptor (D'Arribère et al. 1988, 1990). Alfandari and colleagues (1995) have shown that the α_5 subunit of integrin appears on the mesodermal cells just prior to gastrulation, persists on their surfaces throughout gastrulation, and disappears when gastrulation ends. The integrin coordinates the interaction of the fibronectin on the blastocoel roof with actin microfilaments within the migrating mesendodermal cells. This interaction allows for increased traction and determines the speed of migration (Davidson et al. 2002). It seems, then, that the coordinated synthesis of fibronectin and its receptor signals the times for the mesoderm to begin, continue, and stop migration.

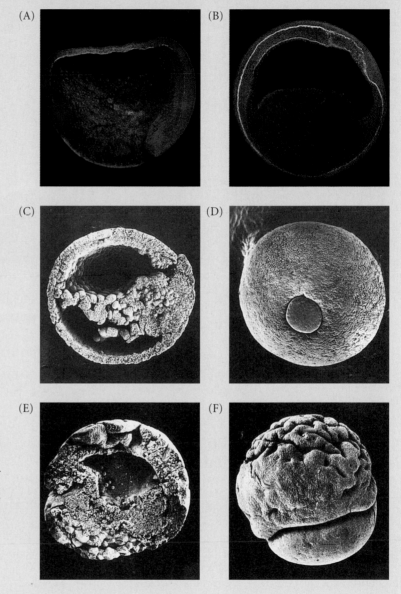

Figure 7.12 Fibronectin and amphibian gastrulation. (A,B) Sagittal section of *Xenopus* embryos at early (A) and late (B) gastrulation. The fibronectin lattice on the blastocoel roof is identified by fluorescent antibody labeling (yellow), while the embryonic cells are counterstained red. (C–F) Scanning electron micrographs of a normal salamander embryo injected with a control solution at the blastula stage (C,D) and an embryo of the same stage injected with the cell-binding fragment of fibronectin (E,F). (C) Section during mid-gastrulation. (D) The yolk plug toward the end of gastrulation. (E,F) The finishing stages of the arrested gastrulation, wherein the mesodermal precursors, having bound the synthetic fibronectin, cannot recognize the normal fibronectin-lined migration route. The archenteron fails to form, and the noninvoluted mesodermal precursors remain on the surface. (A,B from Marsden and DeSimone 2001, photographs courtesy of the authors; C–F from Boucaut et al. 1984, courtesy of J.-C. Boucaut and J.-P. Thiery.)

Epiboly of the prospective ectoderm

During gastrulation, the animal cap and noninvoluting marginal zone cells expand by epiboly to cover the entire embryo (see Figure 7.10B). These cells will form the surface ectoderm. One important mechanism of epiboly in *Xenopus* gastrulation appears to be an increase in cell number (through division) coupled with a concurrent integration of several deep layers into one (**Figure 7.13**; Keller and Schoenwolf 1977; Keller and Danilchik 1988; Saka and Smith 2001). A second mechanism of *Xenopus* epiboly involves the assembly of fibronectin into fibrils by the blastocoel roof (mentioned above). This fibrillar fibronectin is critical in allowing the vegetal migration of the animal cap cells and enclosure of the embryo (Rozario et al. 2009).

See WEBSITE 7.2
Migration of the mesodermal mantle

FIGURE 7.13 Epiboly of the ectoderm is accomplished by cell division and intercalation. (A,B) Cell division in the presumptive ectoderm. Cell division is shown by staining for phosphorylated histone 3, a marker of mitosis. Stained nuclei appear black. In early gastrulae (A; stage 10.5), most cell division occurs in the animal hemisphere presumptive ectoderm. In late gastrulae (B; stage 12), cell division can be seen throughout the ectodermal layer. (Interestingly, the dorsal mesoderm shows no cell division). (C) Scanning electron micrographs of the *Xenopus* blastocoel roof, showing the changes in cell shape and arrangement. Stages 8 and 9 are blastulae; stages 10–11.5 represent progressively later gastrulae. (A,B after Saka and Smith 2001, photographs courtesy of the authors; C from Keller 1980, courtesy of R. E. Keller.)

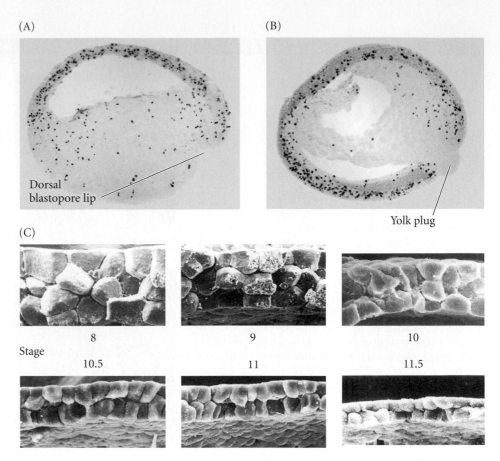

Progressive Determination of the Amphibian Axes

As we have seen, the unfertilized amphibian egg has polarity along the animal-vegetal axis. Thus, the germ layers can be mapped onto the oocyte even before fertilization. The animal hemisphere blastomeres will become the cells of the ectoderm (skin and nerves); the vegetal hemisphere cells will form the cells of the gut and associated organs (endoderm); and the mesodermal cells will form from the internal cytoplasm around the equator. This general fate map is thought to be imposed on the embryo by the vegetal cells. The vegetal cells have two major functions, one being to differentiate into endoderm, the other to induce the cells immediately above them to become mesoderm.

In *Xenopus* oocytes, the mRNA encoding the transcription factor VegT is anchored to the cortex of the vegetal hemisphere and is apportioned to the vegetal cells during cleavage. When VegT transcripts are destroyed in experiments using antisense oligonucleotides, the entire embryo becomes epidermis, with no mesodermal or endodermal components (Zhang et al. 1998; Taverner et al. 2005). VegT activates the zygotic transcription of genes encoding several members of the TGF-β paracrine factor family, including at least six Nodal-related genes and the paracrine factor **Vg1** (see Figure 7.23A). If either Nodal or Vg1 signaling is blocked, there is little or no mesoderm induction (Kofron et al. 1999; Agius et al. 2000; Birsoy et al. 2006).

While animal-vegetal polarity gives an indication as to which cells form each germ layer, this information does not tell us which cells will be anterior, which posterior, which back, and which belly. Rather, the anterior-posterior, dorsal-ventral, and left-right axes are specified by events triggered at fertilization and realized during gastrulation. In *Xenopus* (and in other amphibians), the formation of the anterior-posterior axis is inextricably linked to the formation of the dorsal-ventral axis. Once the dorsal portion of the embryo is established (opposite the point of sperm entry), the movement of the involuting mesoderm establishes the anterior-posterior axis. The first endomesoderm to migrates over the dorsal blastopore lip will induce the ectoderm above it to produce anterior structures, such as the forebrain; mesoderm that involutes later through the dorsal blastopore lip allows the ectoderm to form more posterior structures, such as the hindbrain and spinal cord. This process, whereby the central nervous system forms through interactions with the underlying mesoderm, has been called *primary embryonic induction* and is one of the principal ways that the vertebrate body becomes organized. Indeed, as we will now see, its discoverers called the dorsal blastopore lip and its descendants "the Organizer," and found that this region is different from all the other parts of the embryo.

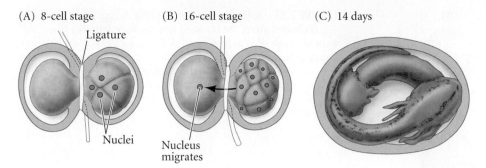

(A) 8-cell stage   (B) 16-cell stage   (C) 14 days

Ligature

Nuclei

Nucleus
migrates

FIGURE 7.14 Spemann's demonstration of nuclear equivalence in newt cleavage. (A) When the fertilized egg of the newt *Triturus taeniatus* was constricted by a ligature, the nucleus was restricted to one half of the embryo. The cleavage on that side of the embryo reached the 8-cell stage, while the other side remained undivided. (B) At the 16-cell stage, a single nucleus entered the as-yet undivided half, and the ligature was further constricted to complete the separation of the two halves. (C) After 14 days, each side had developed into a normal embryo. (After Spemann 1938.)

Hans Spemann: Inductive interactions in regulative development

Amphibian axis formation combines autonomous specification (*mosaic development*) and conditional specification (*regulative development*). The requirement for inductive interactions was demonstrated in Hans Spemann's laboratory at the University of Freiburg (see Hamburger 1988; De Robertis and Aréchaga 2001; Sander and Fässler 2001). Experiments by Spemann and his students framed the questions that experimental embryologists asked for most of the twentieth century and resulted in a Nobel Prize for Spemann in 1935. In recent times, the ongoing saga of discovery in identifying the molecules associated with these inductive processes has provided some of the most exciting moments in contemporary science.

The experiment that began this research program was performed in 1903, when Spemann demonstrated that early newt blastomeres have identical nuclei, each capable of producing an entire larva. His procedure was ingenious: Shortly after fertilizing a newt egg, Spemann used a baby's hair (taken from his infant daughter) to "lasso" the zygote in the plane of the first cleavage. He then partially constricted the egg, causing all the nuclear divisions to remain on one side of the constriction. Eventually—often as late as the 16-cell stage—a nucleus would escape across the constriction into the non-nucleated side. Cleavage then began on this side too, whereupon Spemann tightened the lasso until the two halves were completely separated. Twin larvae developed, one slightly more advanced than the other (**Figure 7.14**). Spemann concluded from this experiment that early amphibian nuclei were genetically identical and that each cell was capable of giving rise to an entire organism.

However, when Spemann performed a similar experiment with the constriction still longitudinal, but perpendicular to the plane of the first cleavage (i.e., separating the future dorsal and ventral regions rather than the right and left sides), he obtained a different result altogether. The nuclei continued to divide on both sides of the constriction, but only one side—the future dorsal side of the embryo—gave rise to a normal larva. The other side produced an unorganized tissue mass of ventral cells, which Spemann called the *Bauchstück*—the belly piece. This tissue mass was a ball of epidermal cells (ectoderm) containing blood and mesenchyme (mesoderm) and gut cells (endoderm), but it contained no dorsal structures such as nervous system, notochord, or somites.

Why should these two experiments give different results? One possibility was that when the egg was divided perpendicular to the first cleavage plane, some *cytoplasmic* substance was not equally distributed into the two halves. Fortunately, the salamander egg was a good place to test that hypothesis. As we saw earlier in this chapter, there are dramatic movements in the cytoplasm following the fertilization of amphibian eggs, and in some amphibians these movements expose a gray, crescent-shaped area of cytoplasm in the region directly opposite the point of sperm entry (see Figure 7.1D). The first cleavage plane normally splits this gray crescent equally between the two blastomeres. If these cells are then separated, two complete larvae develop (**Figure 7.15A**). However, should this cleavage plane be aberrant (either in the rare natural event or in an experiment), the gray crescent material passes into only one of the two blastomeres. Spemann's work revealed that when two blastomeres are separated such that only one of the two cells contains the crescent, only the blastomere containing the gray crescent develops normally (**Figure 7.15B**).

It appeared, then, that *something in the region of the gray crescent was essential for proper embryonic development*. But how did it function? What role did it play in normal development? The most important clue came from fate maps, which showed that the gray crescent region gives rise to those cells that form the dorsal lip of the blastopore. These dorsal lip cells are committed to invaginate into the blastula, initiating gastrulation and the formation of the head endomesoderm and notochord. Because all future amphibian development depends on the interaction of cells that

(A) (B)

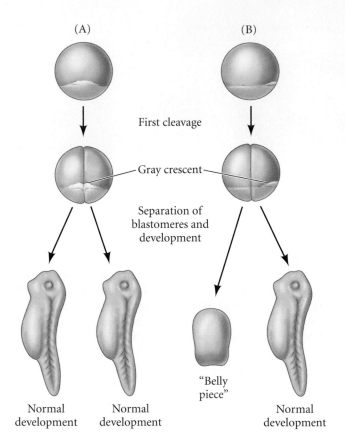

First cleavage

Gray crescent

Separation of
blastomeres and
development

"Belly
piece"

Normal
development

Normal
development

Normal
development

FIGURE 7.15 Asymmetry in the amphibian egg. (A) When the egg is divided along the plane of first cleavage into two blastomeres, each of which gets half of the gray crescent, each experimentally separated cell develops into a normal embryo. (B) When only one of the two blastomeres receives the entire gray crescent, it alone forms a normal embryo. The other blastomere produces a mass of unorganized tissue lacking dorsal structures. (After Spemann 1938.)

are rearranged during gastrulation, Spemann speculated that the importance of the gray crescent material lies in its ability to initiate gastrulation, and that crucial changes in cell potency occur during gastrulation. In 1918, he performed experiments that showed both statements to be true. He found that the cells of the *early* gastrula were uncommitted, but that the fates of *late* gastrula cells were determined.

Spemann's demonstration involved exchanging tissues between the gastrulae of two species of newts whose embryos were differently pigmented—the darkly pigmented *Triturus taeniatus* and the nonpigmented *T. cristatus*. When a region of prospective epidermal cells from an early gastrula of one species was transplanted into an area in an early gastrula of the other species and placed in a region where neural tissue normally formed, the transplanted cells gave rise to neural tissue. When prospective neural tissue from early gastrulae was transplanted to the region fated to become belly skin, the neural tissue became epidermal (**Figure 7.16A; Table 7.1**). Thus, cells of the *early* newt gastrula exhibit conditional (i.e., regulative, or dependent) development because their ultimate fate depends on their location in the embryo.

However, when the same interspecies transplantation experiments were performed on *late* gastrulae, Spemann obtained completely different results. Rather than differentiating in accordance with their new location, the transplanted cells exhibited *autonomous* (independent, or mosaic) development (see the introduction to Part II). Their prospective fate was *determined*, and the cells developed independently of their new embryonic location. Specifically, prospective neural cells now developed into brain tissue even when placed in the region of prospective epidermis (**Figure 7.16B**), and prospective epidermis formed skin even in the region of the prospective neural tube. Within the time separating early and late gastrulation, the potencies of these groups of cells had become restricted to their eventual paths of differentiation. Something was causing

TABLE 7.1 Results of tissue transplantation during early- and late-gastrula stages in the newt

Donor region	Host region	Differentiation of donor tissue	Conclusion
EARLY GASTRULA			
Prospective neurons	Prospective epidermis	Epidermis	Conditional development
Prospective epidermis	Prospective neurons	Neurons	Conditional development
LATE GASTRULA			
Prospective neurons	Prospective epidermis	Neurons	Autonomous development (determined)
Prospective epidermis	Prospective neurons	Epidermis	Autonomous development (determined)

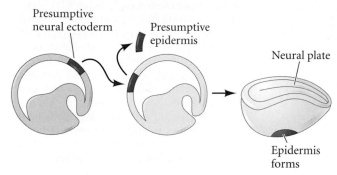

(A) Transplantation in early gastrula

Presumptive
neural ectoderm

Presumptive
epidermis

Neural plate

Epidermis
forms

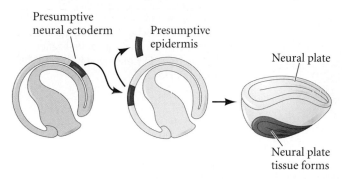

(B) Transplantation in late gastrula

Presumptive
neural ectoderm

Presumptive
epidermis

Neural plate

Neural plate
tissue forms

FIGURE 7.16   Determination of ectoderm during newt gastrulation. Presumptive neural ectoderm from one newt embryo is transplanted into a region in another embryo that normally becomes epidermis. (A) When the tissues are transferred between early gastrulae, the presumptive neural tissue develops into epidermis, and only one neural plate is seen. (B) When the same experiment is performed using late-gastrula tissues, the presumptive neural cells form neural tissue, thereby causing two neural plates to form on the host. (After Saxén and Toivonen 1962.)

them to become committed to epidermal and neural fates. What was happening?

Hans Spemann and Hilde Mangold: Primary embryonic induction

The most spectacular transplantation experiments were published by Spemann and Hilde Mangold in 1924.* They showed that, of all the tissues in the early gastrula, only one has its fate autonomously determined. This self-determining tissue is the dorsal lip of the blastopore—the tissue derived from the gray crescent cytoplasm. When this

---

*Hilde Proescholdt Mangold died in a tragic accident in 1924, when her kitchen's gasoline heater exploded. She was 26 years old, and her paper was just being published. Hers is one of the very few doctoral theses in biology that have directly resulted in the awarding of a Nobel Prize. For more information about Hilde Mangold, her times, and the experiments that identified the organizer, see Hamburger 1984, 1988, and Fässler and Sander 1996.

tissue was transplanted into the presumptive belly skin region of another gastrula, it not only continued to be dorsal blastopore lip but also initiated gastrulation and embryogenesis in the surrounding tissue!

In these experiments, Spemann and Mangold once again used the differently pigmented embryos of *Triturus taeniatus* and *T. cristatus* so they could identify host and donor tissues on the basis of color. When the dorsal lip of an early *T. taeniatus* gastrula was removed and implanted into the region of an early *T. cristatus* gastrula fated to become ventral epidermis (belly skin), the dorsal lip tissue invaginated just as it would normally have done (showing self-determination) and disappeared beneath the vegetal cells (**Figure 7.17A**). The pigmented donor tissue then continued to self-differentiate into the chordamesoderm (notochord) and other mesodermal structures that normally form from the dorsal lip (**Figure 7.17B**). As the donor-derived mesodermal cells moved forward, host cells began to participate in the production of a new embryo, becoming organs that normally they never would have formed. In this secondary embryo, a somite could be seen containing both pigmented (donor) and unpigmented (host) tissue. Even more spectacularly, the dorsal lip cells were able to interact with the host tissues to form a complete neural plate from host ectoderm. Eventually, a secondary embryo formed, conjoined face to face with its host (**Figure 7.17C**). The results of these technically difficult experiments have been confirmed many times and in many amphibian species, including *Xenopus* (**Figure 7.17D**; Capuron 1968; Smith and Slack 1983; Recanzone and Harris 1985).

See WEBSITE 7.3
Spemann, Mangold, and the organizer

Spemann referred to the dorsal lip cells and their derivatives (notochord and head endomesoderm) as the **organizer** because (1) they induced the host's ventral tissues to change their fates to form a neural tube and dorsal mesodermal tissue (such as somites), and (2) they organized host and donor tissues into a secondary embryo with clear anterior-posterior and dorsal-ventral axes. He proposed that during normal development, these cells "organize" the dorsal ectoderm into a neural tube and transform the flanking mesoderm into the anterior-posterior body axis (Spemann 1938). It is now known (thanks largely to Spemann and his students) that the interaction of the chordamesoderm and ectoderm is not sufficient to organize the entire embryo. Rather, it initiates a series of sequential inductive events. Because there are numerous inductions during embryonic development, this key induction—in which the progeny of dorsal lip cells induce the dorsal axis and the neural tube—is traditionally called the **primary embryonic induction**.[†]

[†]This classical term has been a source of confusion because the induction of the neural tube by the notochord is no longer considered the first inductive process in the embryo. We will soon discuss inductive events that precede this "primary" induction.

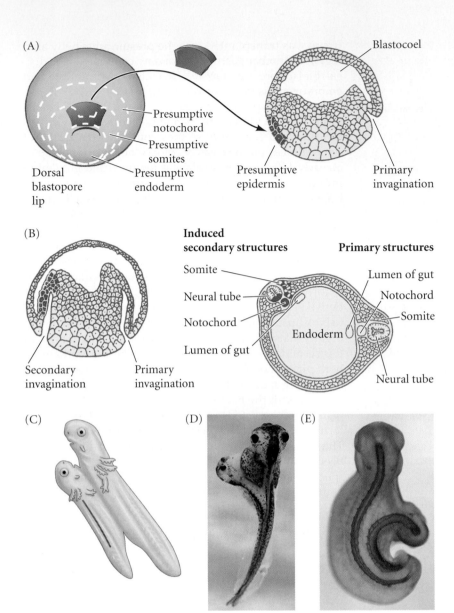

(A)

Blastocoel

Presumptive
notochord

Presumptive
somites

Dorsal
blastopore
lip

Presumptive
endoderm

Presumptive
epidermis

Primary
invagination

(B)

Induced
secondary structures

Primary structures

Somite

Neural tube

Notochord

Lumen of gut

Lumen of gut

Notochord

Somite

Endoderm

Neural tube

Secondary
invagination

Primary
invagination

(C)   (D)   (E)

FIGURE 7.17 Organization of a secondary axis by dorsal blastopore lip tissue. (A–C) Spemann and Mangold's 1924 experiments visualized the process by using differently pigmented newt embryos. (A) Dorsal lip tissue from an early *T. taeniatus* gastrula is transplanted into a *T. cristatus* gastrula in the region that normally becomes ventral epidermis. (B) The donor tissue invaginates and forms a second archenteron, and then a second embryonic axis. Both donor and host tissues are seen in the new neural tube, notochord, and somites. (C) Eventually, a second embryo forms joined to the host. (D) Live twinned *Xenopus* larvae generated by transplanting a dorsal blastopore lip into the ventral region of an early-gastrula host embryo. (E) Similar twinned larvae are seen from below and stained for notochord; the original and secondary notochords can be seen. (A–C after Hamburger 1988; D,E photographs by A. Wills, courtesy of R. Harland.)

Molecular Mechanisms of Amphibian Axis Formation

The experiments of Spemann and Mangold showed that the dorsal lip of the blastopore, along with the dorsal mesoderm and pharyngeal endoderm that form from it, constituted an "organizer" able to instruct the formation of embryonic axes. But the mechanisms by which the organizer itself was constructed and through which it operated remained a mystery. Indeed, it is said that Spemann and Mangold's landmark paper posed more questions than it answered. Among those questions were:

- How did the organizer get its properties? What caused the dorsal blastopore lip to differ from any other region of the embryo?

- What factors were being secreted from the organizer to cause the formation of the neural tube and to create the anterior-posterior, dorsal-ventral, and left-right axes?

- How did the different parts of the neural tube become established, with the most anterior becoming the sensory organs and forebrain, and the most posterior becoming spinal cord?

Spemann and Mangold's description of the organizer was the starting point of one of the first truly international scientific research programs: the search for the organizer molecules. Researchers from Britain, Germany, France, the United States, Belgium, Finland, Japan, and the former Soviet Union all tried to find these remarkable substances (see Gilbert and Saxén 1993). R. G. Harrison referred to the amphibian gastrula as the "new Yukon to which eager miners were now rushing to dig for gold around the blastopore" (see Twitty 1966, p. 39). Unfortunately, their early picks and shovels proved too blunt to uncover the molecules involved. The proteins responsible for induction were present in concentrations too small for biochemical analyses, and the large quantity of yolk and lipids in the amphibian egg further interfered with protein purification (Grunz 1997). The analysis of organizer molecules had to wait until recombinant DNA technologies enabled investigators to make cDNA clones from blastopore lip mRNA and to see which of these clones encoded factors that could dorsalize the embryo. Today, however, we are able to take up each of the above four questions in turn.

How does the organizer form?

Why are the dozen or so initial cells of the organizer positioned opposite the point of sperm entry, and what determines their fate so early? Recent evidence provides an unexpected answer: these cells are in the right place at the right time, at a point where two signals converge. The first signal tells the cells that they are dorsal. The second signal says that they are mesoderm.

See WEBSITE 7.4
The molecular biology of organizer formation

THE DORSAL SIGNAL: β-CATENIN It turns out that one of the reasons the organizer cells are special is that these mesodermal cells reside above a special group of vegetal cells. One of the major clues in determining how the dorsal blastopore lip obtained its properties came from the experiments of Pieter Nieuwkoop (1969, 1973, 1977) and Osamu Nakamura. These studies showed that the organizer receives its properties from the ectoderm beneath it.

Nakamura and Takasaki (1970) showed that the mesoderm arises from the marginal (equatorial) cells at the border between the animal and vegetal poles. The Nakamura and Nieuwkoop laboratories then demonstrated that the properties of this newly formed mesoderm were induced by the vegetal (presumptive endoderm) cells underlying them. Nieuwkoop removed the equatorial cells (i.e., presumptive mesoderm) from a blastula and showed that neither the animal cap (presumptive ectoderm) nor the vegetal cap (presumptive endoderm) produced any mesodermal tissue. However, when the two caps were recombined, the animal cap cells were induced to form mesodermal structures such as notochord, muscles, kidney cells, and blood cells (**Figure 7.18**). The polarity of this induction (i.e., whether the animal cells formed dorsal mesoderm or ventral mesoderm) depended on whether the endodermal (vegetal) fragment was taken from the dorsal or the ventral side: ventral and lateral vegetal cells (those closer to the side of sperm entry) induced ventral (mesenchyme, blood) and intermediate (kidney) mesoderm, while the dorsalmost vegetal cells specified dorsal mesoderm components (somites, notochord)—including those having the properties of the organizer. These dorsalmost vegetal cells of the blastula,

which are capable of inducing the organizer, have been called the **Nieuwkoop center** (Gerhart et al. 1989).

The Nieuwkoop center was demonstrated in the *Xenopus* embryo by transplantation and recombination experiments. First, Gimlich and Gerhart (Gimlich and Gerhart 1984; Gimlich 1985, 1986) performed an experiment analogous to the Spemann and Mangold studies, except that they used early *Xenopus* blastulae rather than newt gastrulae. When they transplanted the dorsalmost vegetal blas-

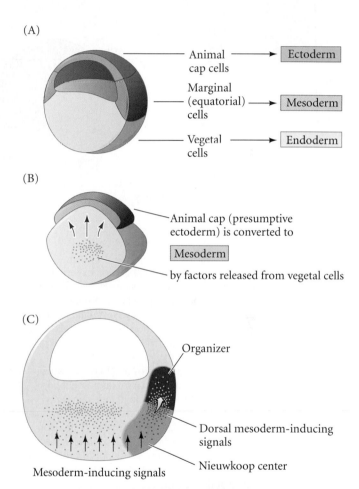

FIGURE 7.18 Summary of experiments by Nieuwkoop and by Nakamura and Takasaki, showing mesodermal induction by vegetal endoderm. (A) Isolated animal cap cells become a mass of ciliated ectoderm, isolated equatorial (marginal zone) cells become mesoderm, and isolated vegetal cells generate gutlike tissue. (B) If animal cap cells are combined with vegetal cap cells, many of the animal cells generate mesodermal tissue. (C) Simplified model for mesoderm induction in *Xenopus*. A ventral signal (probably a complex set of signals from activin-like TGF-β factors and FGFs) is released throughout the vegetal region of the embryo. This signal induces the marginal cells to become mesoderm. On the dorsal side (away from the point of sperm entry), a signal is released by the vegetal cells of the Nieuwkoop center. This dorsal signal induces the formation of the Spemann organizer in the overlying marginal zone cells. The possible identity of this signal will be discussed later in this chapter. (C after De Robertis et al. 1992.)

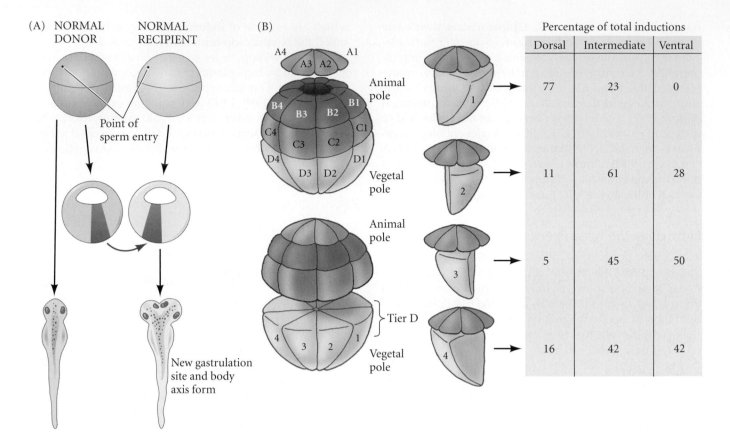

FIGURE 7.19 Transplantation and recombination experiments on *Xenopus* embryos demonstrate that the vegetal cells underlying the prospective dorsal blastopore lip region are responsible for initiating gastrulation. (A) Formation of a new gastrulation site and body axis by the transplantation of the most dorsal vegetal cells of a 64-cell embryo into the ventralmost vegetal region of another embryo. (B) The regional specificity of mesoderm induction demonstrated by recombining blastomeres of 32-cell *Xenopus* embryos. Animal pole cells were labeled with fluorescent polymers so their descendants could be identified, then combined with individual vegetal blastomeres. The inductions resulting from these recombinations are summarized at the right. D1, the dorsalmost vegetal blastomere, was the most likely to induce the animal pole cells to form dorsal mesoderm. These dorsalmost vegetal cells constitute the Nieuwkoop center. (A after Gimlich and Gerhart 1984; B after Dale and Slack 1987.)

tomere from one blastula into the ventral vegetal side of another blastula, two embryonic axes formed (**Figure 7.19A**). Second, Dale and Slack (1987) recombined single vegetal blastomeres from a 32-cell *Xenopus* embryo with the uppermost animal tier of a fluorescently labeled embryo of the same stage. The dorsalmost vegetal cell, as expected, induced the animal pole cells to become dorsal mesoderm. The remaining vegetal cells usually induced the animal cells to produce either intermediate or ventral mesodermal tissues (**Figure 7.19B**). Holowacz and Elinson (1993) found that cortical cytoplasm from the dorsal vegetal cells of the 16-cell *Xenopus* embryo was able to induce

the formation of secondary axes when injected into ventral vegetal cells. Thus, dorsal vegetal cells can induce animal cells to become dorsal mesodermal tissue.

So one important question became, What gives the dorsalmost vegetal cells their special properties? The major candidate for the factor that forms the Nieuwkoop center in these vegetal cells is **β-catenin**, a multifunctional protein that can act as an anchor for cell membrane cadherins (see Chapter 3) or as a nuclear transcription factor (induced by the Wnt pathway). As we saw in Chapter 5, β-catenin is responsible for specifying the micromeres of the sea urchin embryo. β-Catenin is a key player in the formation of the dorsal tissues, and experimental depletion of this molecule results in the lack of dorsal structures (Heasman et al. 1994a). Moreover, injection of exogenous β-catenin into the ventral side of an embryo produces a secondary axis (Funayama et al. 1995; Guger and Gumbiner 1995).

In *Xenopus* embryos, β-catenin is initially synthesized throughout the embryo from maternal mRNA (Yost et al. 1996; Larabell et al. 1997). It begins to accumulate in the dorsal region of the egg during the cytoplasmic movements of fertilization and continues to accumulate preferentially at the dorsal side throughout early cleavage. This accumulation is seen in the nuclei of the dorsal cells and appears to cover both the Nieuwkoop center and organizer regions (**Figure 7.20**; Schneider et al. 1996; Larabell et al. 1997).

If β-catenin is originally found throughout the embryo, how does it become localized specifically to the side oppo-

(A)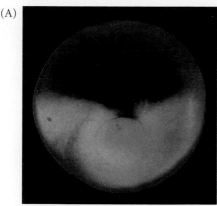

(B)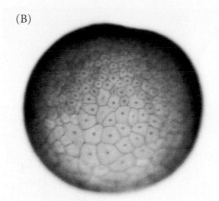

(C)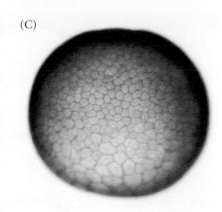

(D)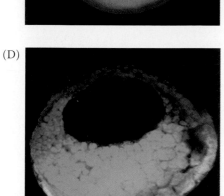

FIGURE 7.20 Role of Wnt pathway proteins in dorsal-ventral axis specification. (A–D) Differential translocation of β-catenin into *Xenopus* blastomere nuclei. (A) Early 2-cell stage, showing β-catenin (orange) predominantly at the dorsal surface. (B) Presumptive dorsal side of a blastula stained for β-catenin shows nuclear localization. (C) Such nuclear localization is not seen on the ventral side of the same embryo. (D) β-Catenin dorsal localization persists through the gastrula stage. (A,D courtesy of R. T. Moon; B,C from Schneider et al. 1996, courtesy of P. Hausen.)

site sperm entry? The answer appears to reside in the translocation of Wnt11 and the **Disheveled (Dsh)** protein from the vegetal pole to the dorsal side of the egg during fertilization. From research done on the Wnt pathway, we have learned that β-catenin is targeted for destruction by **glycogen synthase kinase 3 (GSK3**; see Chapter 3). Indeed, activated GSK3 destroys β-catenin and blocks axis formation when added to the egg, and if endogenous GSK3 is knocked out by a dominant negative form of GSK3 in the ventral cells of the early embryo, a second axis forms (see Figure 7.21F; He et al. 1995; Pierce and Kimelman 1995; Yost et al. 1996).

GSK3 can be inactivated by the **GSK3-binding protein (GBP)** and Disheveled. These two proteins release GSK3 from the degradation complex and prevent it from binding β-catenin and targeting it for destruction. During the first cell cycle, when the microtubules form parallel tracts in the vegetal portion of the egg, GBP travels along the microtubules by binding to kinesin, an ATPase motor protein that travels on microtubules. Kinesin always migrates toward the growing end of the microtubules, and in this case, that means moving to the point opposite sperm entry, i.e., the future dorsal side (**Figure 7.21A–C**). Disheveled, which is originally found in the vegetal pole cortex, grabs onto the GPB, and it too becomes translocated along the microtubular monorail (Miller et al. 1999; Weaver et al. 2003). The cortical rotation is probably important in orient-

ing and straightening the microtubular array and in maintaining the direction of transport when the kinesin complexes occasionally jump the track (Weaver and Kimelman 2004). Once at the site opposite the point of sperm entry, GBP and Dsh are released from the microtubules. Here, on the future dorsal side of the embryo, they inactivate GSK3, allowing β-catenin to accumulate on the dorsal side while ventral β-catenin is degraded (**Figure 7.21D,E**).

But the mere translocation of these proteins to the dorsal side of the embryo does not seem to be sufficient for protecting β-catenin. It appears that a Wnt paracrine factor has to be secreted there to activate the β-catenin protection pathway; this is accomplished by **Wnt11** (see Figure 7.21). If Wnt11 synthesis is suppressed (by the injection of antisense Wnt11 oligonucleotides into the oocytes), the organizer fails to form (Tao et al. 2005). *Wnt11* mRNA is localized to the vegetal cortex during oogenesis and is thought to be translocated to the future dorsal portion of the embryo by the cortical rotation of the egg cytoplasm. Here it is translated into a protein that becomes concentrated in and secreted on the dorsal side of the embryo (Ku and Melton 1993; Schroeder et al. 1999; White and Heasman 2008).

Thus, during first cleavage, GBP, Dsh, and Wnt11 are brought into the future dorsal section of the embryo. Here, GBP and Dsh can *initiate* the inactivation of GSK3 and the consequent protection of β-catenin. The signal from Wnt11 *stabilizes* GBP and Dsh and organizes them to protect β-catenin. The β-catenin transcription factor can associate with other transcription factors to give them new properties. It is known that *Xenopus* β-catenin can combine with a ubiquitous transcription factor known as **Tcf3**, converting the Tcf3 repressor into an activator of transcription. Expression of a mutant form of Tcf3 (one that lacks the β-catenin binding domain) results in embryos without dorsal structures (Molenaar et al. 1996).

(A) Fertilization

Sperm

Egg

Dishevelled
protein (Dsh)

(B) Cortical rotation

V

D

Fast transport
on microtubules

Slow transport by
cortical rotation

Inner
cytoplasm

GBP Dsh

Kinesin

Wnt 11
mRNA

Microtubules

(C) Dorsal enrichment
of Dsh and GBP

V

Wnt 11
Dsh
GBP

D

(D) Dorsal inhibition
of GSK3

V

GSK3

Wnt 11
Dsh, GBP
⊥
GSK3

D

β-catenin
degraded

β-catenin
stable

(E) Dorsal enrichment
of β-catenin

V

D

No β-catenin
in ventral nuclei

β-catenin
in dorsal nuclei

(F)

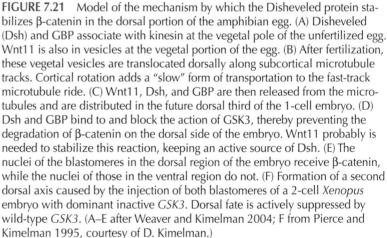

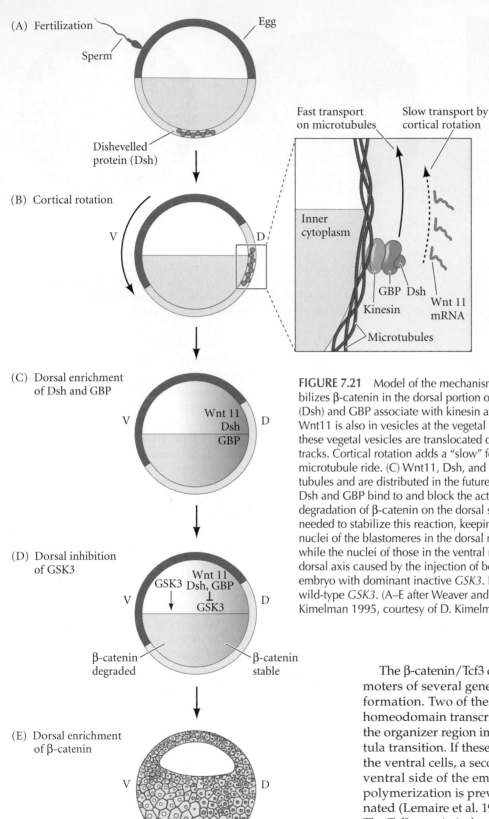

FIGURE 7.21 Model of the mechanism by which the Disheveled protein stabilizes β-catenin in the dorsal portion of the amphibian egg. (A) Disheveled (Dsh) and GBP associate with kinesin at the vegetal pole of the unfertilized egg. Wnt11 is also in vesicles at the vegetal portion of the egg. (B) After fertilization, these vegetal vesicles are translocated dorsally along subcortical microtubule tracks. Cortical rotation adds a "slow" form of transportation to the fast-track microtubule ride. (C) Wnt11, Dsh, and GBP are then released from the microtubules and are distributed in the future dorsal third of the 1-cell embryo. (D) Dsh and GBP bind to and block the action of GSK3, thereby preventing the degradation of β-catenin on the dorsal side of the embryo. Wnt11 probably is needed to stabilize this reaction, keeping an active source of Dsh. (E) The nuclei of the blastomeres in the dorsal region of the embryo receive β-catenin, while the nuclei of those in the ventral region do not. (F) Formation of a second dorsal axis caused by the injection of both blastomeres of a 2-cell *Xenopus* embryo with dominant inactive *GSK3*. Dorsal fate is actively suppressed by wild-type *GSK3*. (A–E after Weaver and Kimelman 2004; F from Pierce and Kimelman 1995, courtesy of D. Kimelman.)

The β-catenin/Tcf3 complex appears to bind to the promoters of several genes whose activity is critical for axis formation. Two of these genes, *twin* and *siamois*, encode homeodomain transcription factors and are expressed in the organizer region immediately following the mid-blastula transition. If these genes are ectopically expressed in the ventral cells, a secondary axis emerges on the former ventral side of the embryo; and if cortical microtubular polymerization is prevented, *siamois* expression is eliminated (Lemaire et al. 1995; Brannon and Kimelman 1996). The Tcf3 protein is thought to inhibit *siamois* and *twin* transcription when it binds to those genes' promoters in the absence of β-catenin. However, when β-catenin binds to Tcf3, the repressor is converted into an activator, and *twin* and *siamois* are activated (**Figure 7.22**).

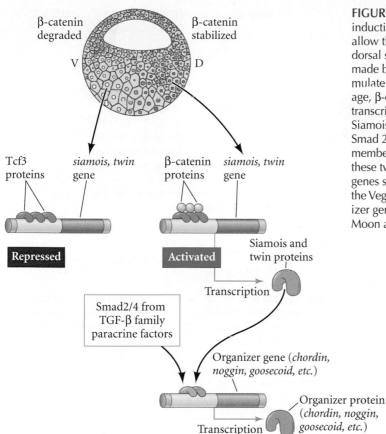

FIGURE 7.22 Summary of events hypothesized to bring about induction of the organizer in the dorsal mesoderm. Microtubules allow the translocation of Disheveled and Wnt11 proteins to the dorsal side of the embryo. Dsh (from the vegetal cortex and newly made by Wnt11) binds GSK3, thereby allowing β-catenin to accumulate in the future dorsal portion of the embryo. During cleavage, β-catenin enters the nuclei and binds with Tcf3 to form a transcription factor that activates genes encoding proteins such as Siamois and Twin. Siamois and Twin interact in the organizer with Smad 2/4 transcription factors activated by vegetal TGF-β family members (Nodal-related proteins, Vg1, activin, etc.). Together, these two sets of transcription factors activate the "organizer" genes such as *chordin*, *noggin*, and *goosecoid*. The presence of the VegT transcription factor in the endoderm prevents the organizer genes from being expressed outside the organizer area. (After Moon and Kimelman 1998.)

Siamois and Twin bind to the enhancers of several genes involved in organizer function (Fan and Sokol 1997; Kessler 1997). These include genes encoding the transcription factors Goosecoid and Xlim1 (which appear to be critical in specifying the dorsal mesoderm) and the paracrine factor antagonists Noggin, Chordin, Frzb, and Cerberus (Laurent et al. 1997; Engleka and Kessler 2001). VegT appears to inhibit the expression of these genes in the vegetal cells (Brannon et al. 1997; Ishibashi et al. 2008). Thus one could expect that if the dorsal side of the embryo contained β-catenin, then β-catenin would allow this region to express Twin and Siamois proteins, and these proteins would initiate formation of the organizer.

THE VEGETAL TGF-β-LIKE SIGNAL Yet another transcription factor also appears to be critical in activating the genes that characterize the organizer cells. This other factor, Smad2/4, is induced in the dorsal mesoderm cells by TGF-β family paracrine factors secreted by the vegetal cells beneath them (Brannon and Kimelman 1996; Engleka and Kessler 2001). TGF-β proteins in the Nieuwkoop center induce the cells in the dorsal marginal zone above them to express Smad2/4 transcription factors, which then bind to the promoter of the organizer genes and cooperate with Twin and Siamois to activate them (see Figure 7.22; Germain et al. 2000).

Two maternal RNAs tethered to the vegetal cortex appear to be crucial for the ability of the vegetal cells to induce the cells above them to be mesodermal. One of these encodes Vg1, a member of the TGF-β superfamily (**Figure 7.23A**). Vg1 is critically important, since embryos whose Vg1 has been depleted lack organizer gene expression and also lack notochords (Birsoy et al. 2006). The other vegetally tethered mRNA encodes VegT, a transcription factor that instructs the endoderm to synthesize and secrete TGF-β family members activin, Derrière, and several Nodal proteins (Latinkic et al. 1997; Smith 2001). These proteins have overlapping functions. Each of them can activate the *Xbra* (*Brachyury*) gene encoding a transcription factor that instructs the cells to become mesoderm. Derrière can induce animal cap cells to become mesoderm over the long-range distances predicted by Nieuwkoop's experiments (White et al. 2002), and activin can induce different types of mesoderm at different concentrations. At moderate concentrations, activin activates the *Xbra* gene, while at higher concentrations it induces organizer genes to become expressed (Green and Smith 1990; Moriya and Asashima 1992; Piepenburg et al. 2004).

Thus the TGF-β-like signal from the endoderm is known to be critical for mesoderm induction; moreover, the *amount* of this signal may control the *type* of mesoderm induced. Since activin, Vg1, Derrier, and Nodal proteins all act

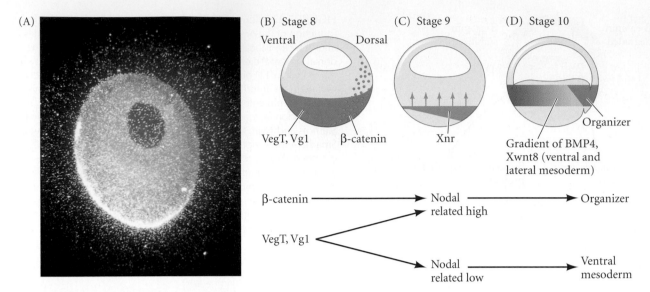

FIGURE 7.23 Vegetal induction of mesoderm. (A) The maternal RNA encoding Vg1 (bright white crescent) is tethered to the vegetal cortex of a *Xenopus* oocyte. The message (along with the maternal VegT message) will be translated at fertilization. Both proteins appear to be crucial for the ability of vegetal cells to induce cells above them to become mesodermal. (B–D) Model for mesoderm induction and organizer formation by the interaction of β-catenin and TGF-β proteins. (B) At late blastula stages, Vg1 and VegT are found in the vegetal hemisphere; β-catenin is located in the dorsal region. (C) β-Catenin acts synergistically with Vg1 and VegT to activate the Nodal-related (*Xnr*) genes. This creates a gradient of Xnr proteins across the endoderm, highest in the dorsal region. (D) The mesoderm is specified by the Xnr gradient. Mesodermal regions with little or no Xnr have high levels of BMP4 and Xwnt8; they become ventral mesoderm. Those having intermediate concentrations of Xnr become lateral mesoderm. Where there is a high concentration of Xnr, *goosecoid* and other dorsal mesodermal genes are activated and the mesodermal tissue becomes the organizer. (A courtesy of D. Melton; B–D after Agius et al. 2000.)

through the same pathway (activating the Smad2/4 transcription factor; see Figure 7.22), the amount of each of these is expected to be additive (Agius et al. 2000). Indeed, this appears to be the case.

During the late blastula stage, several Nodal-related proteins (including Xnr1–6) are expressed in a gradient throughout the endoderm with a low concentration ventrally and a high concentration dorsally (Onuma et al. 2002; Rex et al. 2002; Wright et al. 2005). This gradient is formed by the activation of *Xenopus* Nodal-related gene expression by the synergistic action of VegT with β-catenin. Agius and his colleagues presented a model, shown in **Figure 7.23B–D**, in which the dorsally located β-catenin and the vegetally located Vg1 signals interact to create a gradient of the Nodal-related proteins across the endoderm. Those regions with little Nodal-related protein induce the cells above them to become ventral mesoderm; regions above vegetal cells with some Nodal-related protein become lateral mesoderm; and regions above vegetal cells containing large amounts of these proteins (plus Vg1) become the organizer. Activin, Vg1, and Derriére gradients may behave in the same way. Thus, the initial specification of the mesoderm along the dorsal-ventral axis appears to be accomplished by Nodal-like TGF-β paracrine factors. The region with the highest concentration of these factors may provide the vegetal signal for dorsal mesoderm (organizer) specification, particularly when combined with dorsal β-catenin signal.

In summary, then, the formation of the dorsal mesoderm and the organizer originates through the activation of critical transcription factors by intersecting pathways. The first pathway is the Wnt/β-catenin pathway that activates genes encoding the Siamois and Twin transcription factors. The second pathway is the maternal VegT pathway that activates the expression of Vg1 and other Nodal-related paracrine factors, which in turn activate the Smad2/4 transcription factor in the mesodermal cells above them. The high levels of Smad2/4 and Siamois/Twin transcription factor proteins work within the dorsal mesoderm cells and activate the genes that give these cells their "organizer" properties (review Figures 7.20–7.23).

Functions of the organizer

While the Nieuwkoop center cells remain endodermal, the cells of the organizer become the dorsal mesoderm and migrate underneath the dorsal ectoderm. There, the dorsal mesoderm induces the central nervous system to form. The properties of the organizer tissue can be divided into four major functions:

1. The ability to self-differentiate dorsal mesoderm (prechordal plate, chordamesoderm, etc.)

2. The ability to dorsalize the surrounding mesoderm into paraxial (somite-forming) mesoderm when it would otherwise form ventral mesoderm
3. The ability to dorsalize the ectoderm and induce formation of the neural tube
4. The ability to initiate the movements of gastrulation

It is now thought that the cells of the organizer ultimately contribute to four cell types: pharyngeal endoderm, head mesoderm (prechordal plate), dorsal mesoderm (primarily the notochord), and the dorsal blastopore lip (Keller 1976; Gont et al. 1993). The pharyngeal endoderm and prechordal plate lead the migration of the organizer tissue and induce the forebrain and midbrain. The dorsal mesoderm induces the hindbrain and trunk. The dorsal blastopore lip remaining at the end of gastrulation eventually becomes the chordaneural hinge that induces the tip of the tail.

As we have just seen, the Smad2/4 and β-catenin transcription factors cooperate to activate several genes (Table 7.2). Many of these genes encode secreted proteins that will act to organize the embryo.

See WEBSITE 7.5
Early attempts to locate the organizer molecules

Induction of neural ectoderm and dorsal mesoderm: BMP inhibitors

Evidence from experimental embryology showed that one of the most critical properties of the organizer was its production of soluble factors. The evidence for such diffusible signals from the organizer came from several sources. First, Hans Holtfreter (1933) showed that if the notochord fails to migrate beneath the ectoderm, the ectoderm will not become neural tissue (and will become epidermis). More definitive evidence for the importance of soluble factors came later from the transfilter studies of Finnish investigators (Saxén 1961; Toivonen et al. 1975; Toivonen and Wartiovaara 1976). Here, newt dorsal lip tissue was placed on one side of a filter fine enough so that no processes could fit through the pores, and competent gastrula ectoderm was placed on the other side. After several hours, neural structures were observed in the ectodermal tissue (Figure 7.24). The identities of the factors diffusing from the organizer, however, took another quarter of a century to find.

It turned out that scientists were looking for the wrong mechanism. They were searching for a molecule secreted by the organizer and received by the ectoderm that then converted the ectoderm into neural tissue. However, molecular studies led to a remarkable and non-obvious conclusion: *it is the epidermis that is induced to form, not the neural tissue*. The ectoderm is induced to become epidermal tissue by binding **bone morphogenetic proteins (BMPs)**, while the nervous system forms from that region

TABLE 7.2 Proteins expressed solely or almost exclusively in the organizer (partial list)	
Nuclear proteins	**Secreted proteins**
Twin	Chordin
Siamois	Dickkopf
Xlim1	ADMP
Xnot	Frzb
Otx2	Noggin
XFD1	Follistatin
XANF1	Sonic hedgehog
Goosecoid	Cerberus
HNF3β	Nodal-related proteins (several)

of the ectoderm that is *protected* from epidermal induction by BMP-inhibiting molecules (Hemmati-Brivanlou and Melton 1994, 1997). In other words, (1) the "default fate" of the ectoderm is to become neural tissue; (2) certain parts of the embryo induce the ectoderm to become epidermal tissue by secreting BMPs; and (3) the organizer tissue acts by secreting molecules that block BMPs, thereby allowing the ectoderm "protected" by these BMP inhibitors to become neural tissue.

See WEBSITE 7.6 Competence and bias

Three of the major BMP inhibitors secreted by the organizer are Noggin, chordin, and follistatin. The *noggin, chordin*, and *follistatin* genes are some of the most critical genes activated by Smad2/4 and Siamois/Twin (Carnac et al. 1996; Fan and Sokol 1997; Kessler 1997).

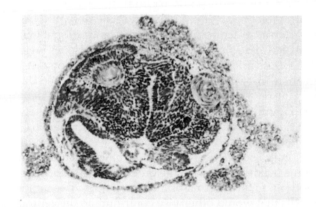

FIGURE 7.24 Neural structures induced in presumptive ectoderm by newt dorsal lip tissue, separated from the ectoderm by a nucleopore filter with an average pore diameter of 0.05 mm. Anterior neural tissues are evident, including some induced eyes. (From Toivonen 1979, courtesy of L. Saxén.)

(A)

(B)

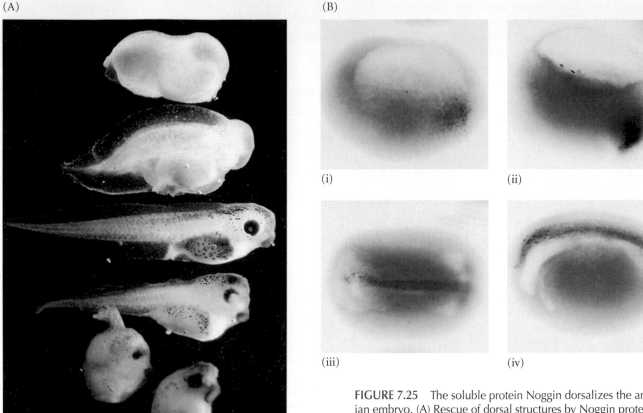

(i)   (ii)

(iii)   (iv)

FIGURE 7.25   The soluble protein Noggin dorsalizes the amphibian embryo. (A) Rescue of dorsal structures by Noggin protein. When *Xenopus* eggs are exposed to ultraviolet radiation, cortical rotation fails to occur, and the embryos lack dorsal structures (top). If such an embryo is injected with *noggin* mRNA, it develops dorsal structures in a dosage-related fashion (top to bottom). If too much *noggin* message is injected, the embryo produces dorsal anterior tissue at the expense of ventral and posterior tissue, becoming little more than a head (bottom). (B) Localization of *noggin* mRNA in the organizer tissue, shown by in situ hybridization. At gastrulation (i), *noggin* mRNA (dark areas) accumulates in the dorsal marginal zone. When cells involute (ii), *noggin* mRNA is seen in the dorsal blastopore lip. During convergent extension (iii), *noggin* is expressed in the precursors of the notochord, prechordal plate, and pharyngeal endoderm, which (iv) extend beneath the ectoderm in the center of the embryo. (Courtesy of R. M. Harland.)

NOGGIN In 1992, Smith and Harland constructed a cDNA plasmid library from dorsalized (lithium chloride-treated) gastrulae. Messenger RNAs synthesized from sets of these plasmids were injected into ventralized embryos (having no neural tube) produced by irradiating early embryos with ultraviolet light. Those plasmid sets whose mRNAs rescued dorsal structures in these embryos were split into smaller sets, and so on, until single-plasmid clones were isolated whose mRNAs were able to restore the dorsal tissue in such embryos. One of these clones contained the gene for the protein **Noggin** (Figure 7.25A). Injection of *noggin* mRNA into 1-cell, UV-irradiated embryos completely rescued dorsal development and allowed the formation of a complete embryo.

Noggin is a secreted protein that is able to accomplish two of the major functions of the organizer: it induces dorsal ectoderm to form neural tissue, and it dorsalizes mesoderm cells that would otherwise contribute to the ventral mesoderm (Smith et al. 1993). Smith and Harland showed that newly transcribed *noggin* mRNA is first localized in the dorsal blastopore lip region and then becomes expressed in the notochord (**Figure 7.25B**). Noggin binds to BMP4 and BMP2 and inhibits their binding to receptors (Zimmerman et al. 1996).

CHORDIN The second organizer protein found was **chordin**. It was isolated from clones of cDNA whose mRNAs were present in dorsalized, but not in ventralized, embryos (Sasai et al. 1994). These clones were tested by injecting them into ventral blastomeres and seeing whether they induced secondary axes. One of the clones capable of inducing a secondary neural tube contained the *chordin* gene. *Chordin* mRNA was found to be localized in the dorsal blastopore lip and later in the notochord (**Figure 7.26**). Of all organizer genes observed, *chordin* is the one most acutely activated by β-catenin (Wesseley et al. 2004). Morpholino antisense oligomers directed against the *chordin* message blocked the ability of an organizer graft to induce a secondary central nervous system (Oelgeschläger et al.

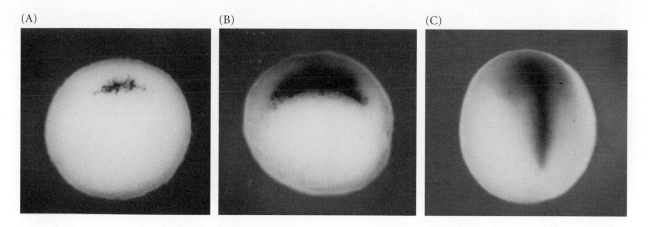

FIGURE 7.26   Localization of *chordin* mRNA. (A) Whole-mount in situ hybridization shows that just prior to gastrulation, *chordin* mRNA (dark area) is expressed in the region that will become the dorsal blastopore lip. (B) As gastrulation begins, *chordin* is expressed at the dorsal blastopore lip. (C) In later stages of gastrulation, the *chordin* message is seen in the organizer tissues. (From Sasai et al. 1994, courtesy of E. De Robertis.)

2003). Like Noggin, chordin binds directly to BMP4 and BMP2 and prevents their complexing with their receptors (Piccolo et al. 1996).

FOLLISTATIN   The mRNA for a third organizer-secreted protein, **follistatin**, is also transcribed in the dorsal blastopore lip and notochord. Follistatin was found in the organizer through an unexpected result of an experiment that was looking for something else. Ali Hemmati-Brivanlou and Douglas Melton (1992, 1994) wanted to see whether the protein activin was required for mesoderm induction. In searching for the mesoderm inducer, they found that follistatin, an inhibitor of both activin and BMPs, caused ectoderm to become neural tissue. They then proposed that under normal conditions, ectoderm becomes neural unless induced to become epidermal by the BMPs. This model was supported by, and explained, certain cell dissociation experiments that had also produced odd results. Three 1989 studies—by Grunz and Tacke, Sato and Sargent, and Godsave and Slack—had shown that when whole embryos or their animal caps were dissociated, they formed neural tissue. This result would be explainable if the "default state" of the ectoderm was not epidermal, but neural, and tissue had to be induced to have an epidermal phenotype. Thus we conclude that *the organizer blocks this epidermalizing induction by inactivating BMPs.*

See WEBSITE 7.7 Specification of the endoderm

Epidermal inducers: The BMPs

In *Xenopus*, the epidermal inducers are the bone morphogenetic protein BMP4 and its close relatives BMP2, BMP7,

and ADMP (anti-dorsalizing morphogenetic protein, a BMP-like paracrine factor). There is an antagonistic relationship between these BMPs and the organizer. If the mRNA for BMP4 is injected into *Xenopus* eggs, all the mesoderm in the embryo becomes ventrolateral mesoderm. Involution is delayed and, when it does occur, has a ventral rather than a dorsal character (Dale et al. 1992; Jones et al. 1992). Conversely, overexpression of a dominant negative BMP4 receptor results in the formation of twinned axes (Graff et al. 1994; Suzuki et al. 1994). In 1995, Wilson and Hemmati-Brivanlou demonstrated that BMP4 induces ectodermal cells to become epidermal. By 1996, several laboratories had demonstrated that Noggin, chordin, and follistatin are all secreted by the organizer, and that each of them prevents BMP from binding to and inducing the ectoderm and mesoderm cells near the organizer (Piccolo et al. 1996; Zimmerman et al. 1996; Iemura et al. 1998).

BMP4 is expressed initially throughout the ectodermal and mesodermal regions of the late blastula. However, during gastrulation, *bmp4* transcripts become restricted to the ventrolateral marginal zone. This is because the Goosecoid protein (and some other transcription factors) are induced by the Siamois/Twin and Smad2/4 interactions in the dorsal (organizer) mesoderm starting at the beginning of gastrulation (Blitz and Cho 1995; Yao and Kessler 2001). These transcription factors repress *bmp4* and *wnt8* transcription (Hemmati-Brivanlou and Thomsen 1995; Northrop et al. 1995; Steinbeisser et al. 1995; Glavic et al. 2001). In the ectoderm, BMPs repress the genes (such as *neurogenin*) involved in forming neural tissue, while activating other genes involved in epidermal specification (Lee et al. 1995). In the mesoderm, it appears that graded levels of BMP4 activate different sets of mesodermal genes, thereby specifying the dorsal, intermediate, and lateral mesodermal tissues (**Figure 7.27**; Gawantka et al. 1995; Hemmati-Brivanlou and Thomsen 1995; Dosch et al. 1997).

In 2005, two important sets of experiments confirmed the default model and the importance of blocking BMPs to specify the nervous system. First, Khokha and colleagues

(A)

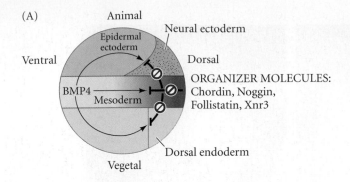

(B)

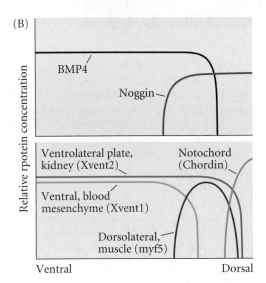

FIGURE 7.27 Model for the action of the organizer. (A) BMP4 (along with certain other molecules) is a powerful ventralizing factor. Organizer proteins such as chordin, Noggin, and follistatin block the action of BMP4; their inhibitory effects can be seen in all three germ layers. (B) BMP4 may elicit the expression of different genes in a concentration-dependent fashion. Thus, in the regions of *noggin* and *chordin* expression, BMP4 is totally prevented from binding, and these tissues become notochord (organizer) tissue. Slightly farther away from the organizer, the *myf5* gene is activated, producing a marker for the dorsolateral muscles. As more and more BMP4 molecules are allowed to bind to the cells, the *Xvent2* (ventrolateral) and *Xvent1* (ventral) genes become expressed. (After Dosch et al. 1997; De Robertis et al. 2000.)

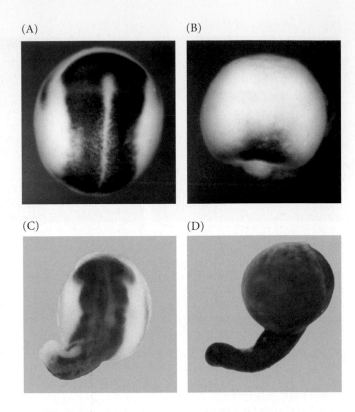

FIGURE 7.28 Control of neural specification by the levels of BMPs. (A,B) Lack of dorsal structures in *Xenopus* embryos whose BMP-inhibitor genes (*chordin*, *noggin*, and *follistatin*) were eliminated by antisense morpholino oligonucleotides. (A) Control embryo with neural folds stained for the expression of the neural gene *Sox2*. (B) Lack of neural tube and *Sox2* expression in an embryo treated with the morpholinos against all three BMP inhibitors. (C,D) Expanded neural development. (C) The neural tube, visualized by *Sox2* staining, is greatly enlarged in embryos treated with antisense morpholinos that destroy BMPs 2, 4, and 7. (D) Complete transformation of the entire ectoderm into neural ectoderm (and loss of the dorsal-ventral axis) by inactivation of ADMP as well as BMPs 2, 4, and 7. (A,B from Khokha et al. 2005, courtesy of R. Harland; C,D from Reversade and De Robertis 2005.)

(2005) used antisense morpholinos to eliminate the three BMP antagonists (i.e., Noggin, chordin, and follistatin) in *Xenopus*. The resulting embryos had catastrophic failure of dorsal development and lacked neural plates and dorsal mesoderm (**Figure 7.28A,B**). Second, Reversade and colleagues blocked BMP activity with antisense morpholinos (Reversade et al. 2005; Reversade and De Robertis 2006).

When they simultaneously blocked the formation of BMPs 2, 4, and 7, the neural tube became greatly expanded, taking over a much larger region of the ectoderm (**Figure 7.28C**). When they did a quadruple inactivation of the three BMPs *and* ADMP, the entire ectoderm became neural—no dorsoventral polarity was apparent (**Figure 7.28D**). Thus the epidermis is instructed by BMP signaling, and the organizer works by blocking that BMP signal from reaching the ectoderm above it.

SIDELIGHTS & SPECULATIONS

BMP4 and Geoffroy's Lobster

The hypothesis that the organizer secretes proteins that block BMPs received further credence from an unexpected source—the emerging field of evolutionary developmental biology (see Chapter 19). Researchers have discovered that the same chordin-BMP4 interaction that instructs the formation of the neural tube in vertebrates also forms neural tissue in fruit flies (Holley et al. 1995; Schmidt et al. 1995; De Robertis and Sasai 1996). The dorsal neural tube of the vertebrate and the ventral neural cord of the fly appear to be generated by the same set of instructions.

The *Drosophila* homologue of the *bmp4* gene is *decapentaplegic* (*dpp*). As discussed in Chapter 6, Dpp protein is responsible for patterning the fly's dorsal-ventral axis; it is present in the dorsal portion of the fly embryo and diffuses ventrally. Dpp is opposed by a protein called Short-gastrulation

(Sog), which is the *Drosophila* homologue of chordin. These insect homologues not only appear to be similar to their vertebrate counterparts, they can actually substitute for each other. When *sog* mRNA is injected into ventral regions of *Xenopus* embryos, it induces the amphibian notochord and neural tube. Injecting *chordin* mRNA into *Drosophila* embryos produces ventral nervous tissue.

Although chordin dorsalizes the *Xenopus* embryo, it ventralizes *Drosophila*. In *Drosophila*, Dpp is made dorsally; in *Xenopus*, BMP4 is made ventrally. In both cases, Sog/chordin helps specify neural tissue by blocking the effects of Dpp/BMP4. In *Drosophila*, Sog interacts with Tolloid and several other proteins to create a gradient of Sog proteins. In *Xenopus*, the homologues of the same proteins act to create a gradient of chordin (see Figure 19.4;

Hawley et al. 1995; Holley et al. 1995; De Robertis et al. 2000).

In 1822, the French anatomist Étienne Geoffroy Saint-Hilaire provoked one of the most heated and critical confrontations in biology when he proposed that the lobster was but a vertebrate upside down. He claimed that the ventral side of the lobster (with its nerve cord) was homologous to the dorsal side of the vertebrate (Appel 1987). It seems that he was correct on the molecular level, if not on the anatomical level. The instructions for producing a nervous system in fact may have evolved only once, and the myriad animal lineages may all have used this same set of instructions—just in different places. The BMP4 (Dpp)/chordin (Sog) interaction is an example of "homologous processes," suggesting a unity of developmental principles among all animals (Gilbert and Bolker 2001).

The Regional Specificity of Neural Induction

One of the most important phenomena in neural induction is the regional specificity of the neural structures that are produced. Forebrain, hindbrain, and spinocaudal regions of the neural tube must be properly organized in an anterior-to-posterior direction. The organizer tissue not only induces the neural tube, it also specifies the regions of the neural tube. This region-specific induction was demonstrated by Hilde Mangold's husband, Otto Mangold, in 1933. He transplanted four successive regions of the archenteron roof of late-gastrula newt embryos into the blastocoels of early-gastrula embryos. The most anterior portion of the archenteron roof (containing head mesoderm) induced balancers and portions of the oral apparatus; the next most anterior section induced the formation of various head structures, including nose, eyes, balancers, and otic vesicles; the third section (including the noto-

chord) induced the hindbrain structures; and the most posterior section induced the formation of dorsal trunk and tail mesoderm* (**Figure 7.29A–D**).

In further experiments, Mangold demonstrated that when dorsal blastopore lips from early salamander gastrulae were transplanted into other early salamander gastrulae, they formed secondary heads. When dorsal lips from later gastrulas were transplanted into early salamander gastrulae, however, they induced the formation of secondary tails (**Figure 7.29E,F**; Mangold 1933). These results show that the first cells of the organizer to enter the embryo induce the formation of brains and heads, while those cells

---

*The induction of dorsal mesoderm—rather than the dorsal ectoderm of the nervous system—by the posterior end of the notochord was confirmed by Bijtel (1931) and Spofford (1945), who showed that the posterior fifth of the neural plate gives rise to tail somites and the posterior portions of the pronephric kidney duct.

FIGURE 7.29 Regional and temporal specificity of induction. (A–D) Regional specificity of structural induction can be demonstrated by implanting different regions (color) of the archenteron roof into early *Triturus* gastrulae. The resulting embryos develop secondary dorsal structures. (A) Head with balancers. (B) Head with balancers, eyes, and forebrain. (C) Posterior part of head, diencephalon, and otic vesicles. (D) Trunk-tail segment. (E,F) Temporal specificity of inducing ability. (E) Young dorsal lips (which will form the anterior portion of the organizer) induce anterior dorsal structures when transplanted into early newt gastrulae. (F) Older dorsal lips transplanted into early newt gastrulae produce more posterior dorsal structures. (A–D after Mangold 1933; E,F after Saxén and Toivonen 1962.)

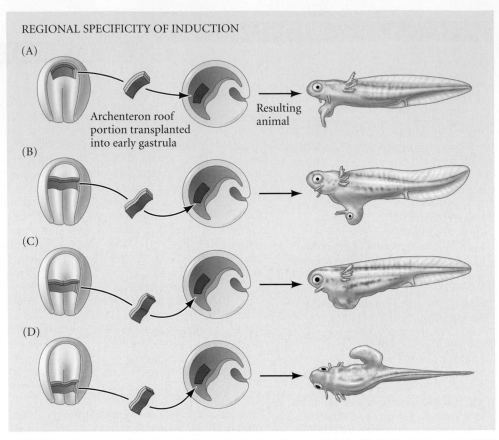

REGIONAL SPECIFICITY OF INDUCTION

(A)

Archenteron roof portion transplanted into early gastrula

Resulting animal

(B)

(C)

(D)

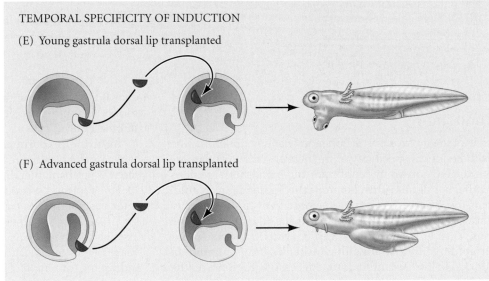

TEMPORAL SPECIFICITY OF INDUCTION

(E) Young gastrula dorsal lip transplanted

(F) Advanced gastrula dorsal lip transplanted

that form the dorsal lip of later-stage embryos induce the cells above them to become spinal cords and tails.

The question then became, What are the molecules being secreted by the organizer in a regional fashion such that the first cells involuting through the blastopore lip (the endomesoderm) induce head structures, while the next portion of involuting mesoderm (notochord) produces trunk and tail structures? **Figure 7.30** shows a possible

model for these inductions, the elements of which we will now describe in detail.

The head inducer: Wnt inhibitors

The most anterior regions of the head and brain are underlain not by notochord but by pharyngeal endoderm and head (prechordal) mesoderm (see Figures 7.6C,D and

(A)

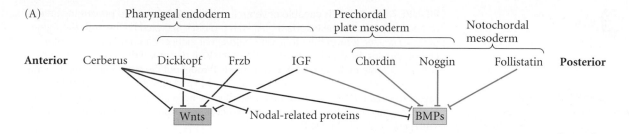

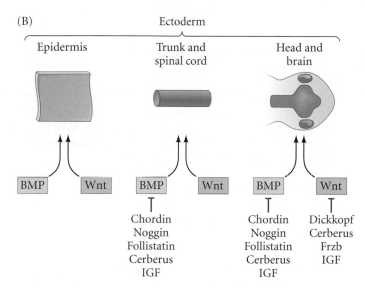

(B)

FIGURE 7.30 Paracrine factor antagonists from the organizer are able to block specific paracrine factors to distinguish head from tail. (A) The pharyngeal endoderm that underlies the head secretes Dickkopf, Frzb, and Cerberus. Dickkopf and Frzb block Wnt proteins; Cerberus blocks Wnts, Nodal-related proteins, and BMPs. The prechordal plate secretes the Wnt-blockers Dickkopf and Frzb, as well as BMP-blockers chordin and Noggin. The notochord contains BMP blockers chordin, Noggin, and follistatin, but it does not secrete Wnt-blockers. IGF from the head endomesoderm probably acts at the junction of the notochord and prechordal mesoderm. (B) Summary of paracrine antagonist function in the ectoderm. Brain formation requires inhibiting both the Wnt and BMP pathways. Spinal cord neurons are produced when Wnt functions without the presence of BMPs. Epidermis is formed when both the Wnt and BMP pathways are operating.

7.30A). This endomesodermal tissue constitutes the leading edge of the dorsal blastopore lip. Recent studies have shown that these cells not only induce the most anterior head structures, but that they do it by blocking the Wnt pathway as well as by blocking BMP4.

CERBERUS In 1996, Bouwmeester and colleagues showed that the induction of the most anterior head structures could be accomplished by a secreted protein called Cerberus.* Unlike the other proteins secreted by the organizer, Cerberus promotes the formation of the cement gland (the most anterior region of tadpole ectoderm), eyes, and olfactory (nasal) placodes. When *cerberus* mRNA was injected into a vegetal ventral *Xenopus* blastomere at the 32-cell stage, ectopic head structures were formed (**Figure 7.31**). These head structures arose from the injected cell as well as from neighboring cells.

The *cerberus* gene is expressed in the pharyngeal endomesoderm cells that arise from the deep cells of the early dorsal lip. Cerberus protein can bind BMPs, Nodal-related proteins, and Xwnt8 (see Figure 7.30; Piccolo et al. 1999). When Cerberus synthesis is blocked, the levels of BMP, Nodal-related proteins, and Wnts all rise in the ante-

rior of the embryo, and the ability of the anterior endomesoderm to induce a head is severely diminished (Silva et al. 2003).

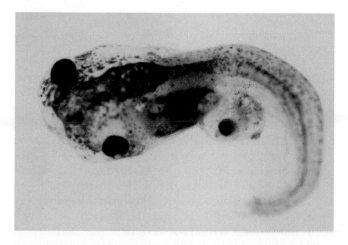

FIGURE 7.31 *Cerberus* mRNA injected into a single D4 (ventral vegetal) blastomere of a 32-cell *Xenopus* embryo induces head structures as well as a duplicated heart and liver. The secondary eye (a single cyclopic eye) and olfactory placode can be readily seen. (From Bouwmeester et al. 1996, courtesy of E. M. De Robertis.)

---

*Cerberus is named after the three-headed dog that guarded the entrance to Hades in Greek mythology.

(A)

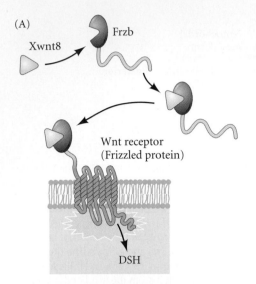

FIGURE 7.32 Xwnt8 is capable of ventralizing the mesoderm and preventing anterior head formation in the ectoderm. (A) Frzb protein is secreted by the anterior region of the organizer. It must bind to Xwnt8 before that inducer can bind to its receptor. Frzb resembles the Wnt-binding domain of the Wnt receptor (Frizzled protein), but Frzb is a soluble molecule. (B) Xwnt8 is made throughout the marginal zone. (C) Double in situ hybridization localizing Frzb (dark stain) and chordin (reddish stain) messages. The *frzb* mRNA is seen to be transcribed in the head endomesoderm of the organizer, but not in the notochord (where *chordin* is expressed). (From Leyns et al. 1997, courtesy of E. M. De Robertis.)

(B)

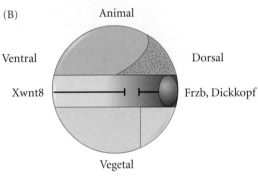

(C)

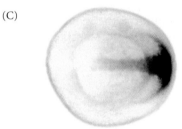

ing embryos to have small, deformed heads with no forebrain (Glinka et al. 1998). Therefore, the induction of trunk structures may be caused by the blockade of BMP signaling from the notochord, while Wnt signals are allowed to proceed. However, to produce a head, both the BMP signal and the Wnt signal must be blocked. This Wnt blockade comes from the endomesoderm, the most anterior portion of the organizer (Glinka et al. 1997).

INSULIN-LIKE GROWTH FACTORS In addition to those proteins that block BMPs and Wnts by physically binding to these paracrine factors, the head region contains yet another set of proteins that prevent BMP and Wnt signals from reaching the nucleus. Pera and colleagues (2001) showed that **insulin-like growth factors (IGFs)** are required for the formation of the anterior neural tube, including the brain and sensory placodes. IGFs accumulate in the dorsal midline and are especially prominent in the anterior neural tube (**Figure 7.33A**). When injected into ventral mesodermal blastomeres, mRNA from IGFs causes the formation of ectopic heads, while blocking the IGF receptors results in the lack of head formation (**Figure 7.33B,C**).

Insulin-like growth factors appear to work by initiating a receptor tyrosine kinase (RTK) signal transduction cascade (see Chapter 3) that interferes with the signal transduction pathways of both BMPs and Wnts (Richard-Parpaillon et al. 2002; Pera et al. 2003).

Trunk patterning: Wnt signals and retinoic acid

Toivonen and Saxén provided evidence for a gradient of a posteriorizing factor that would act to specify the trunk and tail tissues of the amphibian embryo* (Toivonen and

FRZB AND DICKKOPF Shortly after the attributes of Cerberus were demonstrated, two other proteins, Frzb and Dickkopf, were discovered to be synthesized in the involuting endomesoderm. **Frzb** (pronounced "frisbee") is a small, soluble form of Frizzled (the Wnt receptor), and it is capable of binding Wnt proteins in solution (**Figure 7.32**; Leyns et al. 1997; Wang et al. 1997). Frzb is synthesized predominantly in the endomesoderm cells beneath the brain (see Figure 7.32B,C). If embryos are made to synthesize excess Frzb, Wnt signaling fails to occur throughout the embryo; such embryos lack ventral posterior structures and become "all head." The **Dickkopf** protein (German, "thick head," "stubborn") also appears to interact directly with the Wnt receptors, preventing Wnt signaling (Mao et al. 2001, 2002). Injection of antibodies against Dickkopf causes the result-

*The tail inducer was initially thought to be part of the trunk inducer, since transplantation of the late dorsal blastopore lip into the blastocoel often produced larvae with extra tails. However, it appears that tails are normally formed by interactions between the neural plate and the posterior mesoderm during the neurula stage (and thus are generated outside the organizer). Here, Wnt, BMPs, and Nodal signaling all seem to be required (Tucker and Slack 1995; Niehrs 2004). Interestingly, all three of these signaling pathways have to be inactivated if the head is to form.

(A)
(B)

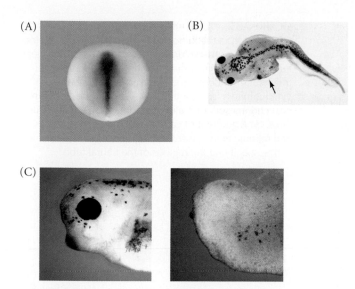

(C)

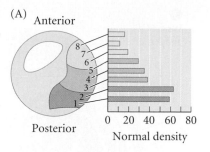

FIGURE 7.33   Insulin-like growth factors enhance anterior neural development. (A) Expression pattern of *Igf3*, showing protein accumulation in the anterior neural tube. (B) An ectopic headlike structure (complete with eyes and cement gland) formed when *Igf2* mRNA was injected into ventral marginal zone blastomeres. (C) Anterior of 3-day control tadpole (left) compared with a tadpole whose 4-cell embryonic blastomeres were injected with an IGF inhibitor. The cement gland and eyes are absent. (From Pera et al. 2001, courtesy of E. M. De Robertis.)

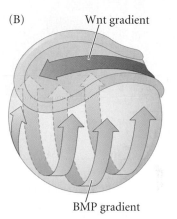

FIGURE 7.34   Wnt signaling pathway and posteriorization of the neural tube. (A) Gradient of β-catenin in the presumptive neural plate during gastrulation. Gastrulating embryos were stained for β-catenin and the density of the stain compared between regions of the ectodermal cells. (B) Double-gradient model whereby a gradient of BMP expression specifies the frog dorsal-ventral axis while a gradient of Wnt proteins specifies the anterior-posterior axis. (After Kiecker and Niehrs 2001; Niehrs 2004.)

Saxén 1955, 1968; reviewed in Saxén 2001). This factor's activity would be highest in the posterior of the embryo and weakened anteriorly. Recent studies have extended this model and have proposed candidates for posteriorizing molecules. The primary protein involved in posteriorizing the neural tube is thought to be a member of the Wnt family of paracrine factors, most likely Xwnt8 (Domingos et al. 2001; Kiecker and Niehrs 2001).

It appears that a gradient of Wnt proteins is necessary for specifying the posterior region of the neural plate (the trunk and tail; Hoppler et al. 1996; Niehrs 2004). In *Xenopus*, an endogenous gradient of Wnt signaling and β-catenin is highest in the posterior and absent in the anterior (**Figure 7.34A**). Moreover, if Xwnt8 is added to developing embryos, spinal cord-like neurons are seen more anteriorly in the embryo, and the most anterior markers of the forebrain are absent. Conversely, suppressing Wnt signaling (by adding Frzb or Dickkopf to the developing embryo) leads to the expression of the anteriormost markers in more posterior neural cells. Therefore, there appear to be two major gradients in the amphibian gastrula—a BMP gradient that specifies the dorsal-ventral axis and a Wnt gradient specifying the anterior-posterior axis (**Figure 7.34B**). It must be remembered, too, that both of these axes are established by the initial axes of Nodal-like TGF-β factors and β-catenin across the vegetal cells. The

basic model of neural induction, then, looks like the diagram in **Figure 7.35**.

While the Wnt proteins probably play a major role in specifying the anterior-posterior axis, they are probably not the only agents involved. Fibroblast growth factors appear to be critical in allowing the cells to respond to the Wnt signal (Holowacz and Sokol 1999; Domingos et al. 2001). Retinoic acid also is seen to have a gradient highest at the posterior end of the neural plate, and RA can also posteriorize the neural tube in a concentration-dependent manner (Cho and De Robertis 1990; Sive and Cheng 1991; Chen et al. 1994). RA signaling appears to be especially important in patterning the hindbrain and appear to interact with Fgf signals to activate the posterior Hox genes (Kolm et al. 1997; Dupé and Lumsden 2001; Shiotsugu et al. 2004).

See WEBSITE 7.8 Regional specification

See VADE MECUM The primary organizer and double gradient hypothesis

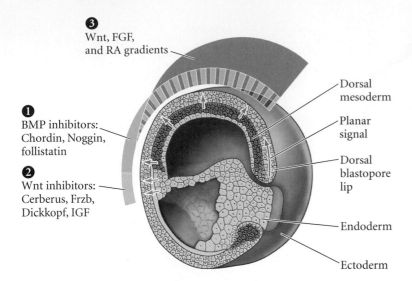

❸ Wnt, FGF, and RA gradients

❶ BMP inhibitors: Chordin, Noggin, follistatin

❷ Wnt inhibitors: Cerberus, Frzb, Dickkopf, IGF

Dorsal mesoderm

Planar signal

Dorsal blastopore lip

Endoderm

Ectoderm

FIGURE 7.35 Model of organizer function and axis specification in the *Xenopus* gastrula. (1) BMP inhibitors from organizer tissue (dorsal mesoderm and pharyngeal mesendoderm) block the formation of epidermis, ventrolateral mesoderm, and ventrolateral endoderm. (2) Wnt inhibitors in the anterior of the organizer (pharyngeal endomesoderm) allow the induction of head structures. (3) A gradient of caudalizing factors (Wnts, Fgfs, and retinoic acid) causes the regional expression of Hox genes, specifying the regions of the neural tube.

Specifying the Left-Right Axis

Although the developing tadpole looks symmetrical from the outside, several internal organs, such as the heart and the gut tube, are not evenly balanced on the right and left sides. In other words, in addition to its dorsal-ventral and anterior-posterior axes, the embryo has a left-right axis. In all vertebrates studied so far, the crucial event in left-right axis formation is the expression of a *nodal* gene in the lateral plate mesoderm on the left side of the embryo. In *Xenopus*, this gene is *Xnr1* (*Xenopus nodal-related 1*). If the expression of this gene is permitted to occur on the right-hand side, the position of the heart (normally found on the left side) and the coiling of the gut are randomized.

But what limits *Xnr1* expression solely to the left-hand side? As in other vertebrates (as we will see in Chapter 8), this concentration of a Nodal protein to the left side is caused by the clockwise rotation of cilia found in the organizer region. In *Xenopus*, these specific cilia are formed during the later stages of gastrulation (after the original specification of the mesoderm) at the dorsal blastopore lip (Schweickert et al. 2007; Blum et al. 2009). That is, they are located in the posterior region of the embryo, at the site where the archenteron is still forming. If rotation of these cilia is blocked, *Xnr1* expression fails to occur in the mesoderm and laterality defects result.

The pathway by which Xnr1 instructs the heart and gut to fold properly is unknown, but one of the key genes activated by Xnr1 signals appears to be **pitx2**. Since it is activated by Xnr1, *pitx2* is normally expressed only on the left side of the embryo. Pitx2 protein persists on the embryo's left side as the heart and gut develop, controlling their respective positions; if Pitx2 is injected into the right side of an embryo, heart placement and gut coiling are randomized (**Figure 7.36**; Ryan et al. 1998). As we will see, the path-

way through which Nodal protein establishes left-right polarity by activating *pitx2* on the left side of the embryo is conserved throughout all vertebrate lineages.

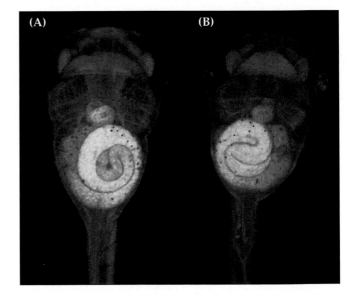

(A)   (B)

FIGURE 7.36 Pitx2 determines the direction of heart looping and gut coiling. (A) Wild-type *Xenopus* tadpole viewed from the ventral side, showing rightward heart looping and counterclockwise gut coiling. (B) If an embryo is injected with Pitx2 so that this protein is present in the mesoderm of both the right and left sides (instead of just the left side), heart looping and gut coiling are random with respect to each other. Sometimes this treatment results in complete reversals, as in this embryo, in which the heart loops to the left and the gut coils in a clockwise manner. (From Ryan et al. 1998, courtesy of J. C. Izpisúa-Belmonte.)

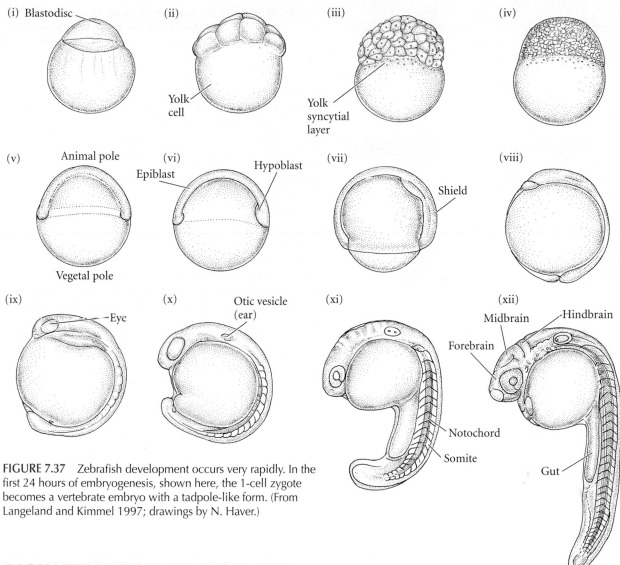

(i) Blastodisc

(ii) Yolk cell

(iii) Yolk syncytial layer

(iv)

(v) Animal pole / Vegetal pole

(vi) Epiblast / Hypoblast

(vii) Shield

(viii)

(ix) Eye

(x) Otic vesicle (ear)

(xi)

(xii) Forebrain / Midbrain / Hindbrain / Notochord / Somite / Gut

FIGURE 7.37 Zebrafish development occurs very rapidly. In the first 24 hours of embryogenesis, shown here, the 1-cell zygote becomes a vertebrate embryo with a tadpole-like form. (From Langeland and Kimmel 1997; drawings by N. Haver.)

EARLY ZEBRAFISH DEVELOPMENT

Xenopus eggs are holoblastic, dividing the entire egg, whereas the yolky zebrafish egg is meroblastic; despite this difference, however, *Xenopus* and zebrafish form their axes and specify their cells in very similar ways. The rapid development, transparency, and breeding ability of zebrafish (*Danio rerio*) have allowed them to confirm many of the principles gleaned from studying *Xenopus*. Zebrafish develop so rapidly that by 24 hours after fertilization, the embryo has formed most of its organ primordia and displays the characteristic tadpole-like form (**Figure 7.37**; see Granato and Nüsslein-Volhard 1996; Langeland and Kimmel 1997).

The zebrafish is the first vertebrate for which intensive mutagenesis has been attempted. By treating parents with mutagens and selectively breeding their progeny, scientists have found thousands of mutations whose normally functioning genes are critical for zebrafish development. The traditional method of genetic screening (modeled after large-scale screens in *Drosophila*) begins when the male parental fish are treated with a chemical mutagen that will cause random mutations in their germ cells (**Figure 7.38**). Each mutagenized male is then crossed to (i.e., mated with) a wild-type female fish to generate F_1 fish. Individuals in the F_1 generation carry the mutations inherited from their father. If the mutation is dominant, it will be expressed in the F_1 generation. If these mutations are recessive, the F_1 fish will not show a mutant phenotype, since the wild-type dominant allele will mask the mutation. The F_1 fish are then crossed to wild-type fish to produce an F_2 generation that includes both males and females who carry the mutant allele. When two F_2 parents carry the same recessive mutation, there is a 25% chance that their offspring will show the mutant phenotype. Since zebrafish development occurs in the open (as opposed to within an opaque shell or inside the mother's body), abnormal developmental stages can be readily observed, and the defects in development can often be traced to changes in a particular group of cells (Driever et al. 1996; Haffter et al. 1996).

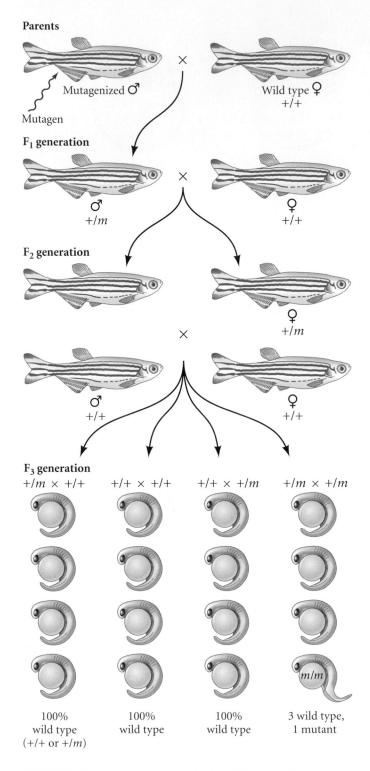

Parents

Mutagenized ♂

Mutagen

Wild type ♀
+/+

F₁ generation

♂
+/m

♀
+/+

F₂ generation

♂
+/+

♀
+/m

♀
+/+

F₃ generation

+/m × +/+   +/+ × +/+   +/+ × +/m   +/m × +/m

100%
wild type
(+/+ or +/m)

100%
wild type

100%
wild type

3 wild type,
1 mutant

(m/m)

FIGURE 7.38 Screening protocol for identifying mutations of zebrafish development. The male parent is mutagenized and mated with a wild-type (+/+) female. If some of the male's sperm carry a recessive mutant allele (m), then some of the F₁ progeny of the mating will inherit that allele. F₁ individuals (here shown as a male carrying the mutant allele m) are then mated with wild-type partners. This creates an F₂ generation wherein some males and some females carry the recessive mutant allele. When the F₂ fish are mated, some of their progeny will show the mutant phenotype. (After Haffter et al. 1996.)

Zebrafish genes are extremely amenable to study. Like *Xenopus* embryos, zebrafish embryos are susceptible to morpholino antisense molecules (Zhong et al. 2001), and researchers can use this method to test whether a particular gene is required for a particular function. Furthermore, the green fluorescent protein (GFP) reporter gene can be fused with specific zebrafish promoters and enhancers and inserted into the fish embryos. The resulting transgenic fish express GFP at the same times and places as they express the proteins controlled by these regulatory sequences. The amazing thing is that one can observe the reporter protein in living transparent embryos* (**Figure 7.39**).

The similarity of developmental programs among all vertebrates and the above-mentioned ability of this fish to be genetically manipulated has given the zebrafish an important role in investigating the genes that operate during human development. When developmental biologists screened zebrafish mutants for cystic kidney disease, they found 12 different genes. Two of these genes were known to cause human cystic kidney disease, and the other 10 were as-yet unknown genes that were found to interact with the first two in a common pathway. Moreover, that pathway, which involves the synthesis of cilia, was not what had been expected. Thus, the zebrafish studies disclosed an important and previously unknown pathway to explore human birth defects (Sun et al. 2004). Zebrafish embryos are also permeable to small molecules placed in

---

*GFP-transgenic zebrafish are the first commercially available transgenic pets, and are marketed as "Glo-Fish."

(A)   (B)

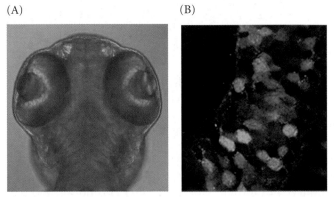

FIGURE 7.39 The gene for green fluorescent protein (GFP) was fused to the regulatory region of a zebrafish *sonic hedgehog* gene. As a result, GFP was synthesized wherever the Hedgehog protein is normally expressed in the fish embryo. (A) In the head of a zebrafish embryo, GFP is seen in the developing retina and nasal placodes. (B) Because GFP is expressed by individual cells, scientists can see precisely which cells make GFP, and thus which cells normally transcribe the gene of interest (in this case, *sonic hedgehog* in the retina). (Photographs courtesy of U. Strahle and C. Neumann.)

the water—a property that allows us to test drugs that may be deleterious to vertebrate development. For instance, zebrafish development can be altered by the addition of ethanol or retinoic acid, both of which produce malformations in the fish that resemble human developmental syndromes known to be caused by these molecules (Blader and Strähle 1998). As one zebrafish researcher joked, "Fish really are just little people with fins" (Bradbury 2004).

Cleavage

The eggs of most bony fish are *telolecithal*, meaning that most of the egg cell is occupied by yolk. Cleavage can take place only in the **blastodisc**, a thin region of yolk-free cytoplasm at the animal pole of the egg. The cell divisions do not completely divide the egg, so this type of cleavage is called **meroblastic** (Greek *meros*, "part"). Since only the blastodisc becomes the embryo, this type of meroblastic cleavage is referred to as **discoidal**.

Scanning electron micrographs show beautifully the incomplete nature of discoidal meroblastic cleavage in fish eggs (**Figure 7.40**). The calcium waves initiated at fertilization stimulate the contraction of the actin cytoskeleton to squeeze non-yolky cytoplasm into the animal pole of the egg. This process converts the spherical egg into a pear-shaped structure with an apical blastodisc (Leung et al. 1998, 2000). In fish, there are many waves of calcium release, and they orchestrate the processes of cell division. The calcium ions coordinate the mitotic apparatus with the actin cytoskeleton; they help propagate cell division across the cell surface; they are needed to deepen the cleavage furrow; and they heal the membrane after the separation of the blastomeres (Lee et al. 2003).

The first cell divisions follow a highly reproducible pattern of meridional and equatorial cleavages. These divisions are rapid, taking about 15 minutes each. The first 12 divisions occur synchronously, forming a mound of cells that sits at the animal pole of a large **yolk cell**. This mound of cells constitutes the **blastoderm**. Initially, all the cells maintain some open connection with one another and with the underlying yolk cell, so that moderately sized (17-kDa) molecules can pass freely from one blastomere to the next (Kimmel and Law 1985).

Maternal effect mutations have shown the importance of oocyte proteins and mRNAs in embryonic polarity, cell

FIGURE 7.40   Discoidal meroblastic cleavage in a zebrafish egg. (A) 1-cell embryo. The mound atop the cytoplasm is the blastodisc. (B) 2-cell embryo. (C) 4-cell embryo. (D) 8-cell embryo, wherein two rows of four cells are formed. (E) 32-cell embryo. (F) 64-cell embryo, wherein the blastodisc can be seen atop the yolk cell. (From Beams and Kessel 1976, courtesy of the authors.)

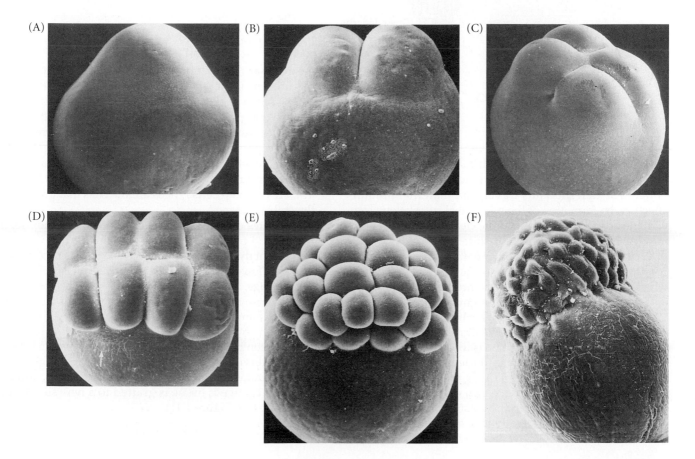

(A)

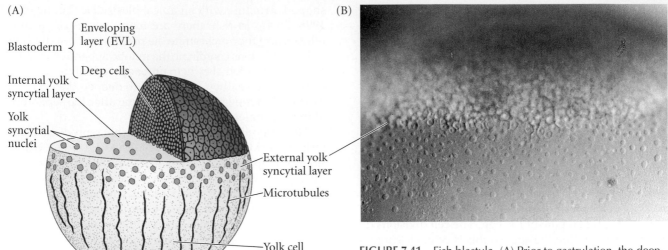

Blastoderm {
Enveloping layer (EVL)
Deep cells

Internal yolk syncytial layer

Yolk syncytial nuclei

External yolk syncytial layer

Microtubules

Yolk cell

(B)

(C)

Animal pole

Nose, eye

Epidermis

Brain

Neural crest

Spinal cord

Ectoderm

Somite muscle

Head

Mesoderm

Ventral

Pronephros

Noto-chord

Dorsal

Blood   Fins   Heart   Muscle

Intestine   Liver   Pharynx

Endoderm

Blastoderm margin

Yolk cell

Vegetal pole

FIGURE 7.41 Fish blastula. (A) Prior to gastrulation, the deep cells are surrounded by the enveloping layer (EVL). The animal surface of the yolk cell is flat and contains the nuclei of the yolk syncytial layer (YSL). Microtubules extend through the yolky cytoplasm and the external YSL. (B) Late blastula-stage embryo of the minnow *Fundulus*, showing the external YSL. The nuclei of these cells were derived from cells at the margin of the blastoderm, which released their nuclei into the yolky cytoplasm. (C) Fate map of the deep cells after cell mixing has stopped. This is a lateral view; for the sake of clarity, not all organ fates are labeled. (A,C after Langeland and Kimmel 1997; B from Trinkaus 1993, courtesy of J. P. Trinkaus.)

division, and cell cleavage (Dosch et al. 2004). As in frogs, the microtubules are important roads upon which morphogenetic determinants travel, and maternal mutants affecting the formation of the microtubal cytoskeleton prevent the normal positioning of the cleavage furrow and of mRNAs in the early embryo (Kishimoto et al. 2004).

As in *Xenopus* and many other embryos, there is a midblastula transition, seen around the tenth cell division, when zygotic gene transcription begins, cell divisions slow, and cell movement becomes evident (Kane and Kimmel 1993). At this time, three distinct cell populations can be distinguished (**Figure 7.41A**). The first of these is the **yolk syncytial layer**, or **YSL**. The YSL is formed at the ninth or tenth cell cycle, when the cells at the vegetal edge of the blastoderm fuse with the underlying yolk cell. This fusion produces a ring of nuclei in the part of the yolk cell cytoplasm that sits just beneath the blastoderm. Later, as the blastoderm expands vegetally to surround the yolk cell, some of the yolk syncytial nuclei will move under the blas-

toderm to form the **internal YSL**, and others will move vegetally, staying ahead of the blastoderm margin, to form the **external YSL** (**Figure 7.41B**). The YSL will be important for directing some of the cell movements of gastrulation.

The second cell population distinguished at the midblastula transition is the **enveloping layer** (**EVL**). It is made up of the most superficial cells from the blastoderm, which form an epithelial sheet a single cell layer thick. The EVL is an extraembryonic protective covering that is sloughed off during later development. Between the EVL and the YSL are the **deep cells**. These are the cells that give rise to the embryo proper.

The fates of the early blastoderm cells are not determined, and cell lineage studies (in which a nondiffusible fluorescent dye is injected into a cell so that its descendants can be followed) show that there is much cell mixing during cleavage. Moreover, any one of these early blastomeres can give rise to an unpredictable variety of tissue descendants (Kimmel and Warga 1987; Helde et al. 1994). A fate map of the blastoderm cells can be made shortly before gastrulation begins. At this time, cells in specific regions of the embryo give rise to certain tissues in a highly predictable manner, although they remain plastic, and cell fates can change if tissue is grafted to a new site (**Figure 7.41C**; see also Figure 1.11).

See VADE MECUM Zebrafish development

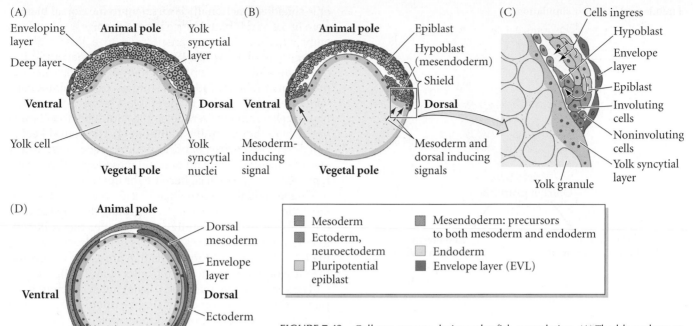

(D)

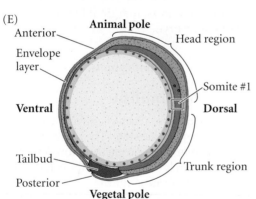

FIGURE 7.42 Cell movements during zebrafish gastrulation. (A) The blastoderm at 30% completion of epiboly (about 4.7 hours). (B) Formation of the hypoblast, either by involution of cells at the margin of the epibolizing blastoderm or by delamination and ingression of cells from the epiblast (6 hours). (C) Close-up of the marginal region. (D) At 90% epiboly (9 hours), mesoderm can be seen surrounding the yolk, between the endoderm and ectoderm. (E) Completion of gastrulation (10.3 hours). (After Driever 1995; Langeland and Kimmel 1997.)

Gastrulation and Formation of the Germ Layers

The first cell movement of fish gastrulation is the epiboly of the blastoderm cells over the yolk. In the initial phase of this movement, the deep cells of the blastoderm move outward to intercalate with the more superficial cells (Warga and Kimmel 1990). Later, this concatenation of cells moves vegetally over the surface of the yolk cell and envelops it completely (Figure 7.42). This downward movement toward the vegetal pole is a result of radial intercalation of the deep cells into the superficial layer. The enveloping layer is tightly joined to the yolk syncytial layer and is dragged along with it. That the vegetal migration of the blastoderm margin is dependent on the epiboly of the YSL can be demonstrated by severing the attachments between

the YSL and the EVL. When this is done, the EVL and the deep cells spring back to the top of the yolk, while the YSL continues its expansion around the yolk cell (Trinkaus 1984, 1992). E-cadherin is critical for these morphogenetic movements and for the adhesion of the EVL to the deep cells (Shimizu et al. 2005).

The expansion of the YSL depends partially on a network of microtubules within it, and radiation or drugs that block the polymerization of tubulin slow epiboly (Strahle and Jesuthasan 1993; Solnica-Krezel and Driever 1994). During epiboly, one side of the blastoderm becomes noticeably thicker than the other. Cell labeling experiments indicate that the thicker side marks the site of the future dorsal surface of the embryo (Schmitz and Campos-Ortega 1994).

After the blastoderm cells have covered about half the zebrafish yolk cell (this occurs earlier in fish species whose eggs have larger yolks), a thickening occurs throughout the margin of the epibolizing blastoderm. This thickening, called the *germ ring*, is composed of a superficial layer, the epiblast, and an inner layer, the hypoblast. The hypoblast in fish (but not in birds and mammals) contains the precursors of the endoderm and mesoderm. The hypoblast forms in a synchronous "wave" of internalization (Keller et al. 2008) that has some characteristics of ingression (especially in the dorsal region; see Carmany-Rampey and Schier 2001) and some elements of involution (especially in the

Endodermal behavior simulation

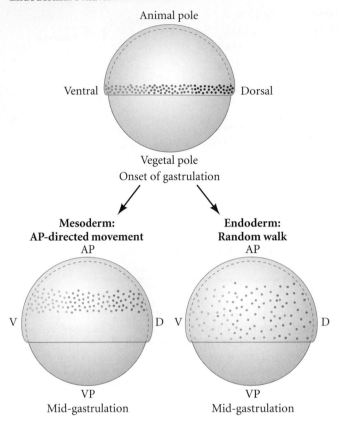

FIGURE 7.43 Cell migration of endodermal and mesodermal precursors. At the beginning of gastrulation, endodermal and mesodermal cells are localized in the hypoblast. Once the hypoblast is internalized, the mesodermal cells migrate vegetally and then toward the animal pole. In contrast, endodermal cells spread through the internal regions of the embryo by a "random walk" over the yolk. (After Pézeron et al. 2008.)

future ventral regions; see Figure 7.42C). Thus, as the cells of the blastoderm undergo epiboly around the yolk, they are also internalizing cells at the blastoderm margin to form the hypoblast. As the hypoblast cells internalize, the future mesoderm cells (the majority of the hypoblast cells) initially migrate vegetally while proliferating to make new mesoderm cells. Later, they alter their direction and procede toward the animal pole. The endodermal precursors, however, appear to move randomly, walking over the yolk (**Figure 7.43**; Pézeron et al. 2008).

Once the hypoblast has formed, cells of the epiblast and hypoblast intercalate on the future dorsal side of the embryo to form a localized thickening, the **embryonic shield** (**Figure 7.44A**). Here, the cells converge and extend anteriorly, eventually narrowing along the dorsal midline (**Figure 7.44B**). This convergent extension in the hypoblast forms the chordamesoderm, the precursor of the notochord (Trinkaus 1992; **Figure 7.44C**). As we will see, the embry-

onic shield is functionally equivalent to the dorsal blastopore lip of amphibians, since it can organize a secondary embryonic axis when transplanted to a host embryo (Oppenheimer 1936; Ho 1992). The cells adjacent to the chordamesoderm—the paraxial mesoderm cells—are the precursors of the mesodermal somites. Concomitant convergence and extension in the epiblast bring presumptive neural cells from all over the epiblast into the dorsal midline, where they form the **neural keel**. The neural keel is usually quite a broad band of neural precursors, and it extends over the axial and paraxial mesoderm. Those cells remaining in the epiblast become the epidermis.

On the ventral side (see Figure 7.44B), the hypoblast ring moves toward the vegetal pole, migrating directly beneath the epiblast that is epibolizing itself over the yolk cell. Eventually, the ring closes at the vegetal pole, completing the internalization of those cells that are going to become the mesoderm and endoderm (Keller et al. 2008).

The zebrafish fate map, then, is not much different from that of the frog. If you can visualize opening a *Xenopus* blastula at the vegetal pole and then stretching that opening into a marginal ring, the resulting fate map closely resembles that of the zebrafish embryo at the stage when half of the yolk has been covered by the blastoderm (see Figure 1.11; Langeland and Kimmel 1997). Meanwhile, the endoderm arises from the most marginal blastomeres of the late blastula-stage embryo.

See WEBSITE 7.9
GFP zebrafish movies and photographs

Axis Formation in Zebrafish

By different mechanisms, the *Xenopus* egg and zebrafish egg have reached the same basic state. They have become multicellular, they have undergone gastrulation, and they have positioned their germ layers such that the ectoderm is on the outside, the endoderm is on the inside, and the mesoderm is between them. As might be expected, zebrafish also form their axes in ways very similar to those of *Xenopus*, and using very similar molecules.

Dorsal-ventral axis formation

The embryonic shield of fish is, as mentioned above, homologous to the dorsal blastopore lip of amphibians, and it is critical in establishing the dorsal-ventral axis. Shield tissue can convert lateral and ventral mesoderm (blood and connective tissue precursors) into dorsal mesoderm (notochord and somites), and it can cause the ectoderm to become neural rather than epidermal. This transformative capacity was shown by transplantation experiments in which the embryonic shield of an early-gastrula embryo was transplanted to the ventral side of another (**Figure 7.45**; Oppenheimer 1936; Koshida et al. 1998). Two axes formed, sharing a common yolk cell. Although the prechordal plate and notochord were derived from the

(A)   Animal pole

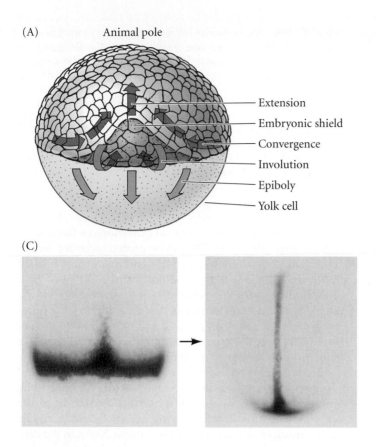

Extension
Embryonic shield
Convergence
Involution
Epiboly
Yolk cell

(B)   Dorsal view

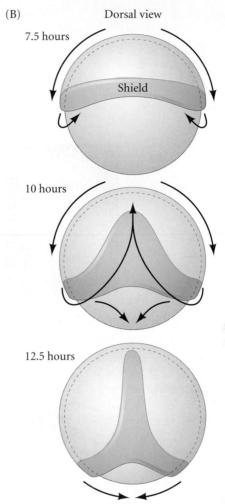

7.5 hours

Shield

10 hours

12.5 hours

(C)

FIGURE 7.44 Convergence and extension in the zebrafish gastrula. (A) Dorsal view of convergence and extension movements during zebrafish gastrulation. Epiboly spreads the blastoderm over the yolk; involution and ingression generate the hypoblast; convergence and extension bring the hypoblast and epiblast cells to the dorsal side to form the embryonic shield. Within the shield, intercalation extends the chordamesoderm toward the animal pole. (B) Model of mesendoderm (hypoblast) formation. Numbers indicate hours after fertilization. On the future dorsal side, the internalized cells undergo convergent extension to form the chordamesoderm (notochord) and the paraxial (somitic) mesoderm adjacent to it. On the ventral side, the hypoblast cells migrate with the epibolizing epiblast toward the vegetal pole, eventually converging there. (C) Convergent extension of the chordamesoderm of the hypoblast cells. These cells are marked by their expression of the *no-tail* gene (dark areas) encoding a T-box transcription factor. (A,C from Langeland and Kimmel 1997, courtesy of the authors; B after Keller et al. 2008.)

donor embryonic shield, the other organs of the secondary axis came from host tissues that would normally form ventral structures. The new axis had been induced by the donor cells.

Like the amphibian blastopore lip, the embryonic shield forms the prechordal plate and the notochord of the developing embryo. The precursors of these two regions are responsible for inducing ectoderm to become neural ecto-

derm. Moreover, the presumptive notochord and prechordal plate appear to do this in a manner very much like the homologous structures in amphibians.* Like amphibians, fish specify the epidermis by using bone morphogenetic proteins (BMPs) and certain Wnt proteins to induce the ectoderm (see Tucker et al. 2008). These BMPs and Wnts are made in the ventral and lateral regions of the embryo. The notochords of both fish and amphibians secrete factors that block this induction, thereby allowing the ectoderm to become neural. In fish, BMP2B induces embryonic cells to acquire ventral and lateral fates, and

---

*Another similarity between the amphibian and fish organizers is that they can be duplicated by rotating the egg and changing the orientation of the microtubules (Fluck et al. 1998). One difference in the axial development of these groups is that in amphibians, the prechordal plate is necessary for inducing the anterior brain to form. In zebrafish, although the prechordal plate appears to be necessary for forming ventral neural structures, the anterior regions of the brain can form in its absence (Schier et al. 1997; Schier and Talbot 1998).

(A)

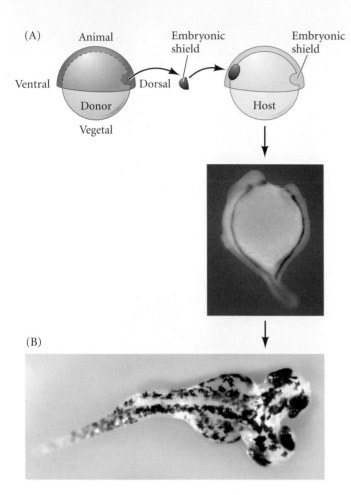

FIGURE 7.45   The embryonic shield as organizer in the fish embryo. (A) A donor embryonic shield (about 100 cells from a stained embryo) is transplanted into a host embryo at the same early-gastrula stage. The result is two embryonic axes joined to the host's yolk cell. In the photograph, both axes have been stained for *sonic hedgehog* mRNA, which is expressed in the ventral midline. (The embryo to the right is the secondary axis.) (B) The same effect can be achieved by activating nuclear catenin in embryos at sites opposite where the embryonic shield will form. (A after Shinya et al. 1999, photograph courtesy of the authors; B courtesy of J. C. Izpisua-Belmonte.)

Wnt8 ventralizes, lateralizes, and posteriorizes the embryonic tissues (see Schier 2001). The protein secreted by the chordamesoderm that binds with and inactivates BMP2B is the chordin homologue called Chordino (**Figure 7.46**; Kishimoto et al. 1997; Schulte-Merker et al. 1997). If the *chordino* gene is mutated, the neural tube fails to form properly; and if *chordino*, *noggin*, and *follistatin* homologues are all inactivated, the neural tube fails to form at all, just like in *Xenopus* (Dal-Pra et al. 2006).

Extracellular BMP antagonists (such as Chordino and Noggin) are not the only important factors for specifying the neural plate. FGFs made in the dorsal side of the embryo also inhibit BMP gene expresson (Furthauer et al. 2004; Tsang et al. 2004; Little and Mullins 2006). In the caudal region of the embryo, FGF signaling is the predominant neural specifier (Kudoh et al. 2004). And as in *Xenopus*, insulin-like growth factors (IGFs) also play a role in the production of the anterior neural plate. Zebrafish IGFs appear to upregulate *chordino* and *goosecoid* while restricting the expression of *bmp2b*. Although IGFs appear to be made throughout the embryo, during gastrulation the IGF receptors are found predominantly in the anterior portion of the embryo (Eivers et al. 2004). Also, Wnt inhibitors appear to play roles in head formation, as antisense mor-

(A)

(B)

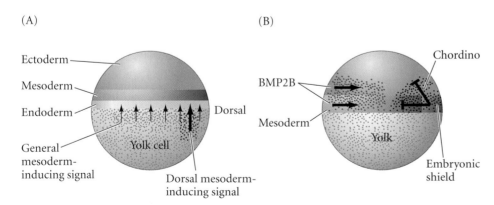

FIGURE 7.46   Axis formation in the zebrafish embryo. (A) Prior to gastrulation, the zebrafish blastoderm is arranged with the presumptive ectoderm near the animal pole, the presumptive mesoderm beneath it, and the presumptive endoderm sitting atop the yolk cell. The yolk syncytial layer (and possibly the endoderm) sends two signals to the presumptive mesoderm. One signal (lighter arrows) induces the mesoderm, while a second signal (heavy arrow) specifically induces an area of mesoderm to become the dorsal mesoderm (embryonic shield). (B) Formation of the dorsal-ventral axis. During gastrulation, the ventral mesoderm secretes BMP2B (arrows) to induce the ventral and lateral mesodermal and epidermal differentiation. The dorsal mesoderm secretes factors (such as Chordino) that block BMP2B and dorsalize the mesoderm and ectoderm (converting the latter into neural tissue). (After Schier and Talbot 1998.)

pholinos that downregulate Wnt3a and Wnt8 lead to the anteriorization of the neural plate at the expense of trunk structures (Shimizu et al. 2005).

The fish Nieuwkoop center

It thus appears that zebrafish have an "organizer" region that dorsalizes the region through BMP and Wnt inhibitors related to similar proteins seen in *Xenopus*. It also turns out that both zebrafish and *Xenopus* use β-catenin- and Nodal-related proteins to form the dorsal mesoderm and enable this mesoderm to express the organizer genes. Both species use BMPs and Wnts to lateralize and posteriorize the embryo, and in both groups the organizer genes encode proteins such as chordin, Noggin, and Dickkopf that inhibit BMPs and Wnts. Furthermore, later in development both zebrafish and *Xenopus* use a particular Wnt protein to posteriorize the ectoderm, forming the trunk neural tube. In some instances, fish and amphibians use these proteins in different ways, but in both groups the result is a structure that is definitely recognizable as a vertebrate embryo.

The zebrafish embryonic shield appears to acquire its organizing ability in much the same way as its amphibian counterparts. In amphibians, the endodermal cells beneath the dorsal blastopore lip (i.e., the Nieuwkoop center) accumulate β-catenin synthesized from maternal messages, enabling the amphibian endoderm to induce the cells above it to become the dorsal blastopore lip. In zebrafish, the nuclei in that part of the yolk syncytial layer lying beneath the cells that will become the embryonic shield similarly accumulate β-catenin. Indeed, the presence of β-catenin distinguishes dorsal YSL from the lateral and ventral YSL regions* (**Figure 7.47A**; Schneider et al. 1996).

---

*Some of the endodermal cells that accumulate β-catenin will become the precursors of the ciliated cells of Kupffer's vesicle (Cooper and D'Amico 1996). As we will discuss in the last section of the chapter, these cells are critical in determining the left-right axis of the embryo.

Inducing β-catenin accumulation on the ventral side of the egg results in dorsalization and a second embryonic axis (Kelly et al. 1995).

β-Catenin in the embryonic shield combines with the zebrafish homologue of Tcf3 to form a transcription factor that activates genes encoding several mesoderm-patterning proteins. One of them is an FGF that represses BMP gene expression on the dorsal side of the embryo (Furthauer et al. 2004; Tsing et al. 2004). Other patterning proteins induced by β-catenin include **Squint** and **Bozozok**, which are very similar to the proteins that pattern the amphibian mesoderm. Squint is a Nodal-like paracrine factor, while Bozozok* is a homeodomain protein similar to the amphibian Nieuwkoop center protein Siamois.

Bozokok protein works in several ways. First, acting alone, it can repress BMP and Wnt genes that would promote ventral functions (Solnica-Krezel and Driever 2001). Second, it suppresses a transcriptional inhibitor (the *vega1* gene), allowing the organizer genes to function (Kawahara et al. 2000). Third, Bozozok and Squint act individually to activate the *chordino* gene, and they act synergistically to activate other organizer genes such as *goosecoid*, *noggin*, and *dickkopf* (**Figure 7.47B**; Sampath et al. 1998; Gritsman et al. 2000; Schier and Talbot 2001). These genes encode the proteins that block BMPs and Wnts and allow the specification of the dorsal mesoderm and neural ectoderm. Thus, the embryonic shield is considered equivalent to the amphibian organizer, and the dorsal part of the yolk cell,

---

Bozozok is Japanese slang for an arrogant youth on a motorcycle. The gene's name was derived from the severe loss-of-function phenotype wherein a single-eyed embryo curves ventrally over the yolk cell (i.e., resembling a rider on a fast motorcycle). However, this gene is also known as *Dharma* (after a famous Buddhist priest) because embryos with gain-of-function—*too much* of this protein, the result of experimentally injecting its mRNA into the embryo— develop huge eyes and head, but no trunk or tail; they thus resemble Japanese Dharma dolls (Yamanaka et al. 1998; Fekany et al. 1999).

(A)

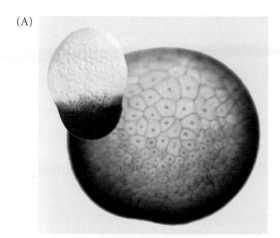

(B)

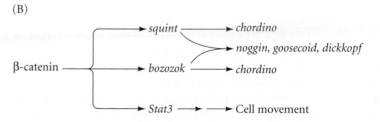

FIGURE 7.47 β-Catenin activates organizer genes in the zebrafish. (A) Nuclear localization of β-catenin marks the dorsal side of the *Xenopus* blastula (larger image) and helps form its Nieuwkoop center beneath the organizer. In the zebrafish late blastula (smaller image), nuclear localization of β-catenin is seen in the yolk syncytial layer nuclei beneath the future embryonic shield. (B) β-Catenin activates *squint* and *bozozok*, whose proteins activate organizer-specific genes as well as the *Stat3* gene whose product is necessary for gastrulation movements. (A courtesy of S. Schneider; B after Schier and Talbot 2001.)

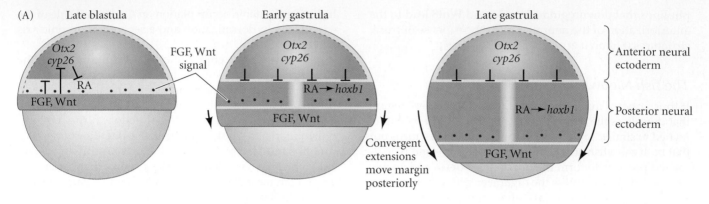

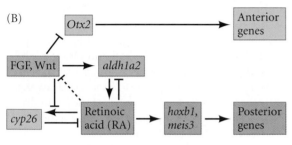

FIGURE 7.48   Model for specification of zebrafish neural ecto-derm by FGF, Wnt, and retinoic acid (RA) signaling. The *cyp26* gene encodes retinoic acid-4-hydroxylase, a cytochrome family enzyme that degrades RA. (A) Temporal sequence of the posterior-ization process. At the late blastula stage, the *cyp26* gene is con-fined to the anterior region by FGFs and Wnts from the margin. After the start of gastrulation, convergent extension takes the mar-gin farther from the anterior. RA accumulates in this region, acti-vating genes associated with the posterior neural ectoderm. (B) Pathway through which a boundary can form between anterior (*cyp26-*, *Otx2*-expressing) and posterior (*hoxb1-*, *meis3*-express-ing) neural ectoderm. In the posterior region, FGFs and/or Wnt suppress anterior genes such as *Otx2*. The FGF/Wnt signal in the posterior also suppresses expression of *cyp26* (which encodes the enzyme that *degrades* RA) and enhances the expression of *aldh1a2* (which encodes the enzyme that *synthesizes* RA), enabling RA to accumulate and activate the posterior genes. (After Kudoh et al. 2002; White et al. 2007.)

together with the dorsal marginal blastomeres (the precur-sors of the Kupffer cells; see below) can be thought of as the Nieuwkoop center of the teleost fish embryo.

Anterior-posterior axis formation

The patterning of the neural ectoderm along the anterior-posterior axis in the zebrafish appears to be the result of the interplay of FGFs, Wnts, and retinoic acid, similar to that seen in *Xenopus*. In fish embryos, there seem to be two separate processes. First a Wnt signal represses the expres-sion of anterior genes; then Wnts, RA, and FGFs are required to activate the posterior genes.

This regulation of anterior-posterior identity appears to be coordinated by **retinoic acid-4-hydroxylase**, an enzyme that degrades RA (Kudoh et al. 2002; Dobbs-McAuliffe et al. 2004). The gene encoding this enzyme, *cyp26*, is expressed specifically in the region of the embryo destined to become the anterior end. Indeed, this gene's expression is first seen during the late blastula stage, and by the time of gastrulation it defines the presumptive anterior neural plate. Retinoic acid-4-hydroxylase prevents the accumula-tion of RA at the embryo's anterior end, blocking the expression of the posterior genes there. This inhibition is reciprocated, since the posteriorly expressed FGFs and Wnts inhibit the expression of the *cyp26* gene, as well as inhibiting the expression of the head-specifying gene *Otx2*. This mutual inhibition creates a border between the zone of posterior gene expression and the zone of anterior gene expression. As epiboly continues, more and more of the body axis is specified to become posterior (**Figure 7.48A**).

Retinoic acid acts as a morphogen, regulating cell prop-erties depending on its concentration. Cells receiving very little RA express anterior genes; cells receiving high levels of RA express posterior genes; and those cells at intermedi-ate levels of RA express genes characteristic of cells between the anterior and posterior regions. This morpho-genesis is extremely important in the hindbrain, where dif-ferent levels of RA specify different types of cells along the anterior-posterior axis (White et al. 2007). FGFs positively regulate RA accumulation both by inhibiting the expres-sion of *cyp26* (which encodes the enzyme that degrades RA) and by enhancing the synthesis of the enzyme that synthesizes RA from vitamin A (**Figure 7.48B**). The result is a gradient of Cyp26, which in turn generates an RA gra-dient that is stable and can be sustained throughout the time the embryo is elongating.

See VADE MECUM: Retinoic acid as a teratogen

Left-right axis formation

In all vertebrates studied, the right and left sides differ both anatomically and developmentally. In fish, the heart is on

(A)

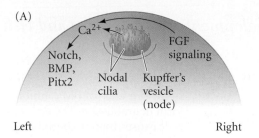

Left   Right

(B)

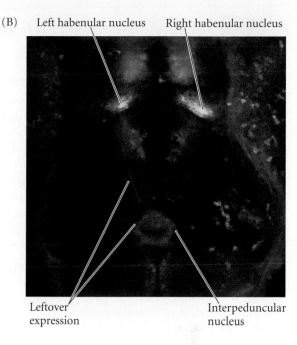

FIGURE 7.49 Left-right asymmetry in the zebrafish embryo. (A) Model for asymmetric gene expression. As in other vertebrates, Nodal-related genes are expressed solely on the embryo's left side: Nodal cilia in Kupffer's vesicle create a current that releases Ca^{2+}, stimulating the Notch and BMP4 pathways and activating the Pitx2 transcription factor in the left-hand mesoderm (blue). On the other hand, FGF expression is seen predominantly on the right-hand side (red). (B) Brain asymmetry in zebrafish. Antibody staining of the Leftover (red) and Right-on (green) proteins in neurons of the habenular nucleus (a behavior-controlling region of the zebrafish forebrain) and the axonal projections to their midbrain target (the interpeduncular nucleus) reveals marked asymmetry. Most Leftover-positive axons emerge from the left habenula to innervate the target. (A after Okada et al. 2005; B from Gamse et al. 2005, courtesy of M. Halpern.)

the left side and there are different structures in the left and right regions of the brain (**Figure 7.49**). Moreover, as in other vertebrates, the cells on the left side of the body are given that information by Notch and Nodal signaling and by the Pitx2 transcription factor, while the cells on the right side of the body are exposed to FGF signaling. The ways the different vertebrate classes accomplish this asymmetry differ, but recent evidence suggests that the currents produced by cilia in the node may be responsible for left-right axis formation in all the vertebrate classes (Okada et al. 2005); when the gene for a dynein subunit of cilia was cloned, it was found to be expressed in the ventral portion of the node (or organizer) in mouse, chick, *Xenopus*, and zebrafish embryos (Essner et al. 2002, 2005). Using extremely rapid (500 frames/sec) photography, Okada and colleagues (2005) showed that the clockwise rotational motion of cilia and the leftward flow of particles were conserved in all these vertebrate groups, despite the different types and shapes of their nodal structures.

In zebrafish, the nodal structure housing the cilia that control left-right asymmetry is a transient fluid-filled organ called **Kupffer's vesicle**. As mentioned earlier, Kupffer's vesicle arises from a group of dorsal cells near the embryonic shield shortly after gastrulation. Essner and colleagues (2002, 2005) were able to inject small beads into Kupffer's vesicle and see their translocation from one side of the vesi-

cle to the other. Blocking ciliary function by preventing the synthesis of dynein or by ablating the precursors of the ciliated cells prevented normal left-right axis formation. The cilia activate the Nodal-related genes on the left side of the embryo. These genes are critically important in initiating the cascades of paracrine factors that are specific to the left side of the body (Rebagliati et al. 1998; Long et al. 2003).

Coda

Researchers analyzing the development of *Xenopus laevis* and *Danio rerio* are finally putting names to the "agents" and "soluble factors" postulated by the early experimental embryologists. We are also delineating the intercellular pathways of paracrine and transcription factors that constitute the first steps in organogenesis. The international research program initiated by Spemann's laboratory in the 1920s is reaching fruition, and this research has revealed layers of complexity beyond anything Spemann could have conceived. Just as Spemann's experiments told us how much we didn't know, the answers we have found to these older questions have generated a whole new set of questions. The ability to perform live imaging of cells, the ability to delete and add new genes to particular cells, and the ability to produce mutant lines of zebrafish have all helped us to understand the mechanisms by which order is generated in the embryo. The remarkable conservation of mechanisms and molecules between model amphibian and fish species, despite the initial differences in their cleavage, shows very strong evolutionary constraints on how this order comes into being.

Snapshot Summary: *Early Development of Amphibians and Fish*

1. Amphibian cleavage is holoblastic, but it is unequal due to the presence of yolk in the vegetal hemisphere.

2. Amphibian gastrulation begins with the invagination of the bottle cells, followed by the coordinated involution of the mesoderm and the epiboly of the ectoderm. Vegetal rotation plays a significant role in directing the involution.

3. The driving forces for ectodermal epiboly and the convergent extension of the mesoderm are the intercalation events in which several tissue layers merge. Fibronectin plays a critical role in enabling the mesodermal cells to migrate into the embryo.

4. The dorsal lip of the blastopore forms the organizer tissue of the amphibian gastrula. This tissue dorsalizes the ectoderm, transforming it into neural tissue, and it transforms ventral mesoderm into lateral and dorsal mesoderm.

5. The organizer consists of pharyngeal endoderm, head mesoderm, notochord, and dorsal blastopore lip tissues. The organizer functions by secreting proteins (Noggin, chordin, and follistatin) that block the BMP signal that would otherwise ventralize the mesoderm and activate the epidermal genes in the ectoderm.

6. The organizer is itself induced by the Nieuwkoop center, located in the dorsalmost vegetal cells. This center is formed by the translocation of the Disheveled protein and Wnt11 to the dorsal side of the egg to stabilize β-catenin in the dorsal cells of the embryo.

7. The Nieuwkoop center is formed by the accumulation of β-catenin, which can complex with Tcf3 to form a transcription factor complex that can activate the transcription of the *siamois* and *twin* genes on the dorsal side of the embryo.

8. The Siamois and Twin proteins collaborate with Smad2/4 transcription factors generated by the TGF-β pathway (Nodal, Vg1, and activin) to activate genes encoding BMP inhibitors. These inhibitors include the secreted factors Noggin, chordin, and follistatin, as well as the transcription factor Goosecoid.

9. In the presence of BMP inhibitors, ectodermal cells form neural tissue. The action of BMP on ectodermal cells causes them to become epidermis.

10. In the head region, an additional set of proteins (Cerberus, Frzb, Dickkopf) blocks the Wnt signal from the ventral and lateral mesoderm.

11. Wnt signaling causes a gradient of β-catenin along the anterior-posterior axis of the neural plate. This graded signaling appears to specify the regionalization of the neural tube.

12. Insulin-like growth factors (IGFs) help transform the neural tube into anterior (forebrain) tissue.

13. The left-right axis appears to be initiated by the activation of a Nodal protein solely on the left side of the embryo. In *Xenopus*, as in other vertebrates, Nodal protein activates expression of *pitx2*, which is critical in distinguishing left-sidedness from right-sidedness in the heart and gut tubes.

14. In fish, cleavage is meroblastic, and the deep cells of the blastoderm form between the yolk syncytial layer and the enveloping layer. These deep cells migrate over the top of the yolk, forming the hypoblast and epiblast layers. On the future dorsal side, these layers intercalate to form the embryonic shield, a structure homologous to the amphibian organizer. Transplantation of the embryonic shield into the ventral side of another embryo will cause a second embryonic axis to form.

15. In both amphibians and fish, neural ectoderm is permitted to form where the BMP-mediated induction of epidermal tissue is prevented. The embryonic shield is the equivalent of the amphibian dorsal blastopore lip, secreting the BMP antagonists. Like the amphibian organizer, the shield receives its abilities by being induced by β-catenin and by underlying endodermal cells expressing Nodal-related paracrine factors.

For Further Reading

Complete bibliographical citations for all literature cited in this chapter can be found at the free-access website **www.devbio.com**

Blum, M., T. Beyer, T. Weber, P. Vick, P. Andre, E. Bitzer and A. Schweickert. 2009. *Xenopus*, an ideal model system to study vertebrate left-right asymmetry. *Dev. Dyn.* 238: 1215–1225.

Cho, K. W. Y., B. Blumberg, H. Steinbeisser and E. De Robertis. 1991. Molecular nature of Spemann's organizer: The role of the *Xenopus* homeobox gene *goosecoid*. *Cell* 67: 1111–1120.

De Robertis, E. M. 2006. Spemann's organizer and self-regulation in amphibian embryos. *Nature Rev. Mol. Cell Biol.* 7: 296–302.

Essner, J. J., J. D. Amack, M. K. Nyholm, E. B. Harris and H. J. Yost. 2005. Kupffer's vesicle is a ciliated organ of asymmetry in the zebrafish embryo that initiates left-right development of the brain, heart and gut. *Development* 132: 1247–1260.

Haffter, P. and 16 others. 1996. The identification of genes with unique and essential functions in the development of the zebrafish, *Danio rerio. Development* 123: 1–36.

Khokha, M. K., J. Yeh, T. C. Grammer and R. M. Harland. 2005. Depletion of three BMP antagonists from Spemann's organizer leads to catastrophic loss of dorsal structures. *Dev. Cell* 8: 401–411.

Langeland, J. and C. B. Kimmel. 1997. The embryology of fish. In S. F. Gilbert and A. M. Raunio (eds.), *Embryology: Constructing the Organism.* Sinauer Associates, Sunderland, MA, pp. 383–407.

Larabell, C. A. and 7 others. 1997. Establishment of the dorsal-ventral axis in *Xenopus* embryos is presaged by early asymmetries in β-catenin which are modulated by the Wnt signaling pathway. *J. Cell Biol.* 136: 1123–1136.

Niehrs, C. 2004. Regionally specific induction by the Spemann-Mangold organizer. *Nature Rev. Genet.* 5: 425–434.

Piccolo, S., E. Agius, L. Leyns, S. Bhattacharyya, H. Grunz, T. Bouwmeester and E. M. DeRobertis. 1999. The head inducer Cerberus is a multifunctional antagonist of Nodal, BMP, and Wnt signals. *Nature* 397: 707–710.

Reversade, B., H. Kuroda, H. Lee, A. Mays, and E. M. De Robertis. 2005. Deletion of BMP2, BMP4, and BMP7 and Spemann organizer signals induces massive brain formation in *Xenopus* embryos. *Development* 132: 3381–3392.

Spemann, H. and H. Mangold. 1924. Induction of embryonic primordia by implantation of organizers from a different species. (Trans. V. Hamburger). In B. H. Willier and J. M. Oppenheimer (eds.), *Foundations of Experimental Embryology.* Hafner, New York, pp. 144–184. Reprinted in *Int. J. Dev. Biol.* 45: 13–38.

White, J. A. and J. Heasman. 2008. Maternal control of pattern formation. *J. Exp. Zool. (MDE)* 310B: 73–84.

Go Online

WEBSITE 7.1 Amphibian development movies. A compilation of amphibian development movies from around the globe.

WEBSITE 7.2 Migration of the mesodermal mantle. Different growth rates coupled with the intercalation of cell layers allows the mesoderm to expand in a tightly coordinated fashion.

WEBSITE 7.3 Spemann, Mangold, and the Organizer. Spemann did not see the importance of this work the first time he and Mangold did it. This website provides a more detailed account of why Spemann and Mangold did this experiment.

WEBSITE 7.4 The molecular biology of organizer formation. FGFs and TGF-β family factors interact to specify both the ventral and the dorsal mesodermal components during the blastula stage. Moreover, the ectoderm protects itself from these powerful signals.

WEBSITE 7.5 Early attempts to locate the organizer molecules. While Spemann did not believe that molecules alone could organize the embryo, his students began a long quest for these factors.

WEBSITE 7.6 Specification of the endoderm. Although the mesoderm is induced, and the differences between neural and epidermal ectoderm are induced, the endoderm appears to be specified autonomously. Recent studies show that TGF-β family signals and the maternally derived VegT transcription factor in the vegetal cells initiate a cascade of events leading to endoderm formation.

WEBSITE 7.7 Regional specification. The research into regional specification has been a fascinating endeavor involving scientists from all over the world. Before molecular biology gave us the tools to uncover morphogenetic proteins, embryologists developed ingenious ways of finding out what those proteins were doing.

WEBSITE 7.8 Competence and bias. The cells destined to become neural ectoderm may have a bias toward that fate that predisposes them not to become epidermis.

WEBSITE 7.9 GFP zebrafish movies and photographs. The ability to photograph and film living embryos expressing the GFP reporter gene driven by promoters of specific genes has opened a new dimension in developmental biology and has allowed us to link gene structure to developmental anatomy.

Vade Mecum

Amphibian development. The events of cleavage and gastrulation are difficult to envision without three-dimensional models. You can see movies of such 3-D models, as well as footage of a living *Xenopus* embryo, in the segments on amphibian development.

Zebrafish development. This full account of zebrafish development includes time-lapse movies of the beautiful and rapid development of this organism.

The primary organizer and double gradient hypothesis. These movies explain Spemann and Mangold's discovery of the primary organizer, Holtfreter's discovery of the "dead organizer," and Lauri Saxén's work on the double gradient hypothesis.

Retinoic acid as a teratogen. One feature within the zebrafish development segment is a visualization of the teratogenic effects of retinoic acid on development.

Outside Sites

XENBASE, A *Xenopus* web resource. Discoveries in the genetics, development, and cell biology of *Xenopus* are updated here (**http://www.xenbase.org**).

ZFIN, the zebrafish model organism database, contains protocols, atlases, and books concerning zebrafish. (**http://zfin.org**)

Birds and Mammals
Early Development and Axis Formation

8

THIS FINAL CHAPTER ON THE PROCESSES of early development extends our survey of vertebrate development to include the **amniotes**—those vertebrates whose embryos form an amnion, or water sac (i.e., the reptiles, birds, and mammals). Birds and reptiles follow a very similar pattern of development (Gilland and Burke 2004; Coolen et al. 2008), and birds are considered by modern taxonomists to be a reptilian clade. Cleavage in bird and reptile eggs, like that of the bony fishes described in the last chapter, is meroblastic, with only a small portion of the egg cytoplasm being used to make the cells of the embryo. In mammals, however, holoblastic cleavage is modified to make a placenta, which enables the embryo to develop inside another organism. Although methods of cleavage and gastrulation differ among the vertebrate classes, certain underlying principles are common to all vertebrates.

EARLY DEVELOPMENT IN BIRDS

Cleavage

Ever since Aristotle first observed and recorded the details of its 3-week-long development, the domestic chicken (*Gallus gallus*) has been a favorite organism for embryological studies. It is accessible year-round and is easily raised. Moreover, at any particular temperature, its developmental stage can be accurately predicted. Thus, large numbers of embryos can be obtained at the same stage. Chick organ formation is accomplished by genes and cell movements similar to those of mammalian organ formation, and the chick is one of the few organisms whose embryos are amenable to both surgical and genetic manipulations (Stern 2005). Thus, the chick embryo has often served as an inexpensive surrogate for human embryos.

Fertilization of the chick egg occurs in the oviduct, before the albumen and shell are secreted to cover it. Like the egg of the zebrafish, the chick egg is telolecithal, with a small disc of cytoplasm—the *blastodisc*—sitting atop a large yolk. Like fish eggs, the yolky eggs of birds undergo *discoidal meroblastic cleavage*. Cleavage occurs only in the blastodisc, which is about 2–3 mm in diameter and is located at the animal pole of the egg. The first cleavage furrow appears centrally in the blastodisc; other cleavages follow to create a single-layered *blastoderm* (**Figure 8.1**). As in the fish embryo, the cleavages do not extend into the yolky cytoplasm, so the early-cleavage cells are continuous with one another and with the yolk at their bases (see Figure 8.1E). Thereafter, equatorial and vertical cleavages divide the blastoderm into a

FIGURE 8.1 Discoidal meroblastic cleavage in a chick egg. Avian eggs include some of the largest cells known to science (inches across), but cleavage takes place in only a small region, the blastodisc. (A–D) Four stages viewed from the animal pole (the future dorsal side of the embryo). (E) An early-cleavage embryo viewed from the side. (After Bellairs et al. 1978.)

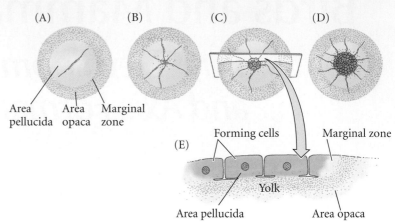

(A)   (B)   (C)   (D)

Area pellucida   Area opaca   Marginal zone

(E)   Forming cells   Marginal zone

Area pellucida   Yolk   Area opaca

tissue 5–6 cell layers thick, and the cells become linked together by tight junctions (Bellairs et al. 1975; Eyal-Giladi 1991).

Between the blastoderm and the yolk of avian eggs is a space called the **subgerminal cavity**, which is created when the blastoderm cells absorb water from the albumen ("egg white") and secrete the fluid between themselves and the yolk (New 1956). At this stage, the deep cells in the center of the blastoderm appear to be shed and die, leaving behind a 1-cell-thick **area pellucida**; this part of the blastoderm forms most of the actual embryo. The peripheral ring of blastoderm cells that have not shed their deep cells constitutes the **area opaca**. Between the area pellucida and the area opaca is a thin layer of cells called the **marginal zone** (or marginal belt*) (Eyal-Giladi 1997; Arendt and Nübler-Jung 1999). Some marginal zone cells become very important in determining cell fate during early chick development.

Gastrulation of the Avian Embryo

The hypoblast

By the time a hen has laid an egg, the blastoderm contains some 20,000 cells. At this time, most of the cells of the area pellucida remain at the surface, forming an "upper layer," called the **epiblast**, while other area pellucida cells have delaminated and migrated individually into the subgerminal cavity to form **hypoblast islands** (sometimes called the *primary hypoblast* or *polyinvagination islands*), an archipelago of disconnected clusters of 5–20 cells each (**Figure 8.2A**). Shortly thereafter, a sheet of cells derived from deep yolky cells at the posterior margin of the blastoderm migrates anteriorly beneath the surface. This sheet of migrating cells pushes the hypoblast islands anteriorly, thereby forming the **secondary hypoblast**, or **endoblast** (**Figure 8.2B–E**; Eyal-Giladi et al. 1992; Bertocchini and Stern

2002; Khaner 2007a,b). The resulting two-layered blastoderm (epiblast and hypoblast) is joined together at the marginal zone of the area opaca, and the space between the layers forms a blastocoel. Thus, although the shape and formation of the avian blastodisc differ from those of the amphibian, fish, or echinoderm blastula, the overall spatial relationships are retained.

However, what the amniotes have evolved is a set of extraembryonic tissues—the hypoblast, the yolk sac, and the amnion. The avian embryo comes entirely from the epiblast; the hypoblast does not contribute any cells to the developing embryo (Rosenquist 1966, 1972). Rather, the hypoblast cells form portions of the external membranes, especially the yolk sac and the stalk linking the yolk mass to the endodermal digestive tube. Hypoblast cells also provide chemical signals that specify the migration of epiblast cells. However, the three germ layers of the embryo proper (plus the amnion) are formed solely from the epiblast (Schoenwolf 1991).

The primitive streak

The major structural characteristic of avian, reptilian, and mammalian gastrulation is the **primitive streak**;* the primitive streak becomes the blastopore lips of amniote embryos. Dye-marking experiments and time-lapse cinemicrography indicate that the primitive streak first arises from a local thickening of the epiblast at the posterior edge of the area pellucida, called **Koller's sickle**, and the epiblast above it (Bachvarova et al. 1998; Lawson and Schoenwolf 2001a,b; Voiculescu et al. 2007).

FORMATION OF THE PRIMITIVE STREAK The streak is first visible as cells accumulate in the middle layer, followed by a thickening of the epiblast at the *posterior marginal zone*, just

---

*Arendt and Nübler-Jung (1999) have argued that the region should be called the marginal *belt* to distinguish it from the marginal *zone* of amphibians. Here we will continue to use the earlier nomenclature.

---

*But, as we saw in the previous chapter, a structure resembling the primitive streak has also been found in certain salamander embryos (Shook et al. 2002).

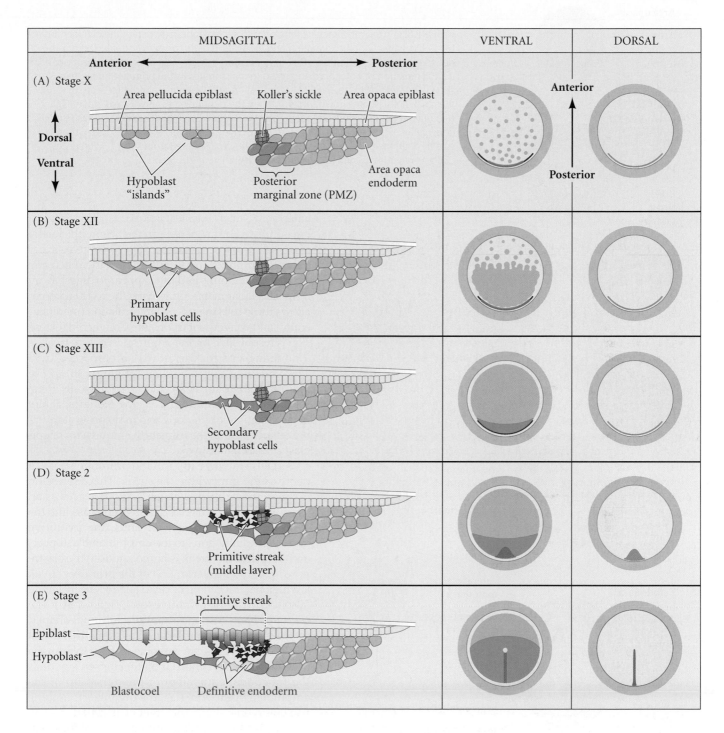

FIGURE 8.2 Formation of the chick blastoderm. The left column depicts a diagrammatic midsagittal section through part of the blastoderm. The middle column depicts the entire embryo viewed from the ventral side. This shows the migration of the primary hypoblast and the secondary hypoblast (endoblast) cells. The right column shows the entire embryo seen from the dorsal side. (A–C) Events prior to laying of the shelled egg, (A) Stage X embryo, where islands of hypoblast cells can be seen, as well as a congregation of hypoblast cells around Koller's sickle. (B) By stage XII, just prior to primitive streak formation, the hypoblast island cells have coalesced to form the primary hypoblast layer, which meets endoblast cells and primitive streak cells at Koller's sickle. (C) By stage XIII, the secondary hypoblast cells migrate anteriorly. (D) By stage 2 (6–7 hours after the egg is laid), the primitive streak cells form a third layer that lies between the hypoblast and epiblast cells. (E) By stage 3 (up to 13 hours post laying), the primitive streak has become a definitive region of the epiblast, with cells migrating through it to become the mesoderm and endoderm. (After Stern 2004.)

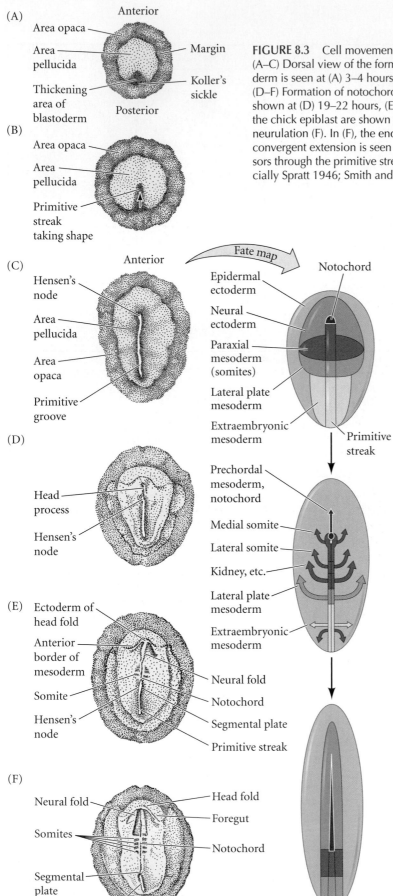

FIGURE 8.3 Cell movements of the primitive streak and fate map of the chick embryo. (A–C) Dorsal view of the formation and elongation of the primitive streak. The blastoderm is seen at (A) 3–4 hours, (B) 7–8 hours, and (C) 15–16 hours after fertilization. (D–F) Formation of notochord and mesodermal somites as the primitive streak regresses, shown at (D) 19–22 hours, (E) 23–24 hours, and (F) the four-somite stage. Fate maps of the chick epiblast are shown for two stages, the definitive primitive streak stage (C) and neurulation (F). In (F), the endoderm has already ingressed beneath the epiblast, and convergent extension is seen in the midline. The movements of the mesodermal precursors through the primitive streak at (C) are shown. (Adapted from several sources, especially Spratt 1946; Smith and Schoenwolf 1998; Stern 2005a,b.)

anterior to Koller's sickle (**Figure 8.3A**). This thickening is initiated by an increase in the height (thickness) of the cells forming the center of the primitive streak. The presumptive streak cells around them become globular and motile, and they appear to digest away the extracellular matrix underlying them. This process allows these cells to undergo intercalation (mediolaterally) and convergent extension. Convergent extension is responsible for the progression of the streak—a doubling in streak length is accompanied by a concomitant halving of its width (**Figure 8.3B**). Those cells that initiate streak formation (i.e., the cells in the midline of the epiblast, overlying Koller's sickle) appear to migrate anteriorly and may constitute a stable cell population that directs the movement of epiblast cells into the streak.

As cells converge to form the primitive streak, a depression forms within the streak. This depression is called the **primitive groove**, and it serves as an opening through which migrating cells pass into the deep layers of the embryo. Thus, the primitive groove is homologous to the amphibian blastopore, and the primitive streak is homologous to the blastopore lips. At the anterior end of the primitive streak is a regional thickening of cells called **Hensen's node** (also known as the **primitive knot**; Figure 8.3C). The center of Hensen's node contains a funnel-shaped depression (sometimes called the **primitive pit**) through which cells can enter the embryo to form the notochord and prechordal plate. Hensen's node is the functional equivalent of the dorsal lip of the amphibian blastopore (i.e., the organizer)* and the fish embryonic shield (Boettger et al. 2001).

The primitive streak defines the axes of the avian embryo. It extends from posterior to anterior; migrating cells enter through its dorsal side and move to its ventral side; and it separates the left portion of the embryo from the right. The axis of the streak is equivalent to the dorsal-ventral axis of amphibians. The anterior end of the streak (Hensen's node) gives rise

---

*Frank M. Balfour proposed the homology of the amphibian blastopore and the chick primitive streak in 1873, while he was still an undergraduate (Hall 2003). August Rauber (1876) provided further evidence for their homology.

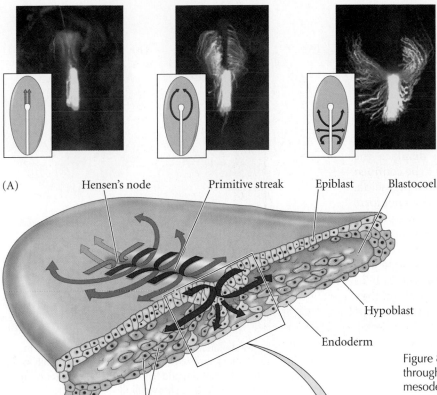

(A)

FIGURE 8.4 Migration of endodermal and mesodermal cells through the primitive streak. (A) Stereogram of a gastrulating chick embryo, showing the relationship of the primitive streak, the migrating cells, and the hypoblast and epiblast of the blastoderm. The lower layer becomes a mosaic of hypoblast and endodermal cells; the hypoblast cells eventually sort out to form a layer beneath the endoderm and contribute to the yolk sac. Above each region of the stereogram are micrographs showing the tracks of GFP-labeled cells at that position in the primitive streak. Cells migrating through Hensen's node travel anteriorly to form the prechordal plate and notochord; those migrating through the next anterior region of the streak travel laterally, but converge near the midline to make notochord and somites; those from the middle of the streak form intermediate mesoderm and lateral plate mesoderm (see the fate map in Figure 8.3). Farther posterior, the cells migrating through the primitive streak make the extraembryonic mesoderm (not shown). (B) This scanning electron micrograph shows epiblast cells passing into the blastocoel and extending their apical ends to become bottle cells. (A after Balinsky 1975, photographs from Yang et al. 2002; B from Solursh and Revel 1978, courtesy of M. Solursh and C. J. Weijer.)

(B)

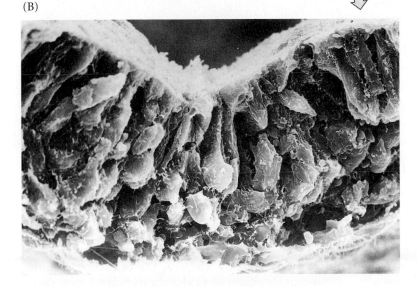

to the prechordal mesoderm, notochord, and medial part of the somites. Cells that ingress through the middle of the streak give rise to the lateral part of the somites and to the heart and kidneys. Cells in the posterior portion of the streak make the lateral plate and extraembryonic mesoderm (Psychoyos and Stern 1996). After the ingression of the mesoderm cells, epiblast cells remaining outside, but close to, the streak will form medial (dorsal) structures such as the neural plate, while those epiblast cells farther from the streak will become epidermis (see Figure 8.3, right-hand panels).

ELONGATION OF THE PRIMITIVE STREAK As cells enter the primitive streak, they undergo an epithelial-to-mesenchymal transformation, and the basal lamina beneath them breaks down. The streak elongates toward the future head region as more anterior cells migrate toward the center of the embryo. Cell division adds to the length produced by convergent extension, and some of the cells from the anterior portion of the epiblast contribute to the formation of Hensen's node (Streit et al. 2000; Lawson and Schoenwolf 2001b).

At the same time, the secondary hypoblast (endoblast) cells continue to migrate anteriorly from the posterior margin of the blastoderm (see Figure 8.2E). The elongation of the primitive streak appears to be coextensive with the anterior migration of these secondary hypoblast cells, and the hypoblast directs the movement of the primitive streak (Waddington 1933; Foley et al. 2000). The streak eventually extends to 60–75% of the length of the area pellucida.

FORMATION OF ENDODERM AND MESODERM As soon as the primitive streak has formed, epiblast cells begin to migrate through it and into the blastocoel (**Figure 8.4**). The streak thus has a continually changing cell population. Cells migrating through the anterior end pass down into the blastocoel and migrate anteriorly, forming the endoderm, head mesoderm, and notochord; cells passing through the more posterior portions of the primitive streak give rise to the majority of mesodermal tissues (Rosenquist et al. 1966; Schoenwolf et al. 1992).

The first cells to migrate through Hensen's node are those destined to become the pharyngeal endoderm of the foregut. Once deep within the embryo, these endodermal cells migrate anteriorly and eventually displace the hypoblast cells, causing the hypoblast cells to be confined to a region in the anterior portion of the area pellucida. This anterior region, the **germinal crescent**, does not form any embryonic structures, but it does contain the precursors of the germ cells, which later migrate through the blood vessels to the gonads (see Chapter 16).

The next cells entering through Hensen's node also move anteriorly, but they do not travel as far ventrally as the presumptive foregut endodermal cells. Rather, they remain between the endoderm and the epiblast to form the **prechordal plate mesoderm** (Psychoyos and Stern 1996). Thus, the head of the avian embryo forms anterior (rostral) to Hensen's node.

The next cells passing through Hensen's node become the **chordamesoderm**. The chordamesoderm has two components: the head process and the notochord. The most anterior part, the **head process**, is formed by central mesoderm cells migrating anteriorly, behind the prechordal plate mesoderm and toward the rostral tip of the embryo (see Figures 8.3 and 8.4). The head process will underlie those cells that will form the forebrain and midbrain. As the primitive streak regresses (see below), the cells deposited by the regressing Hensen's node will become the notochord.

Molecular mechanisms of migration through the primitive streak

FORMATION OF THE PRIMITIVE STREAK The migration of chick epiblast cells to form the primitive streak was first analyzed by Ludwig Gräper, who made time-lapse movies of labeled cells under the microscope in 1926. He wrote that these movements reminded him of the Polonaise, a courtly dance wherein men and women move in parallel rows along the sides of the room, and then a man and a woman at the "posterior end" leave their respective lines to dance forward through the center. The mechanism for this "dance" was revealed by Voiculescu and colleagues (2007), who used a modern version of cinemicrography (specifically, multiphoton time-lapse microscopy) that could identify individual moving cells. They found that cells came down the sides of the epiblast to undergo a medially directed intercalation of cells in the posterior margin where the primitive streak was forming (**Figure 8.5**). And while the movement may look like a dance from far away, "at high power, it looks like a rush hour" (Stern 2007). This rush to the center is mediated by the activation of the Wnt planar cell polarity pathway (see Chapter 3) in the epiblast next to Koller's sickle, at the posterior edge of the embryo. If this pathway is blocked, the mesoderm and

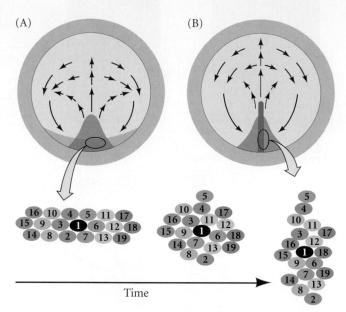

FIGURE 8.5 Mediolateral intercalation in the formation of the primitive streak. Chick embryos at (A) stage XII (immediately prior to primitive streak formation) and (B) stage 2 (shortly after primitive streak formation). Arrows show cell displacement toward the streak and in front of it. The red area represents the streak-forming region; in (A), the original location of this region is shown in green. The circled areas are represented in the lower row. Each colored disc represents an individual cell, and the cells become mediolaterally intercalated as the primitive streak forms. (After Voiculescu et al. 2007.)

endoderm form peripherally instead of centrally. The Wnt pathway in turn appears to be activated by **fibroblast growth factors** (**FGFs**) produced by the hypoblast. If the hypoblast is rotated, the orientation of the primitive streak follows it. Moreover, if FGF signaling is activated in the margin of the epiblast, Wnt signaling will occur there, and the orientation of the primitive streak will change, as if the hypoblast had been placed there. The cell migrations that form the primitive streak thus appear to be regulated by FGFs coming from the hypoblast, which activates the Wnt planar cell polarity pathway in the epiblast.

PREPARING TO MIGRATE The basic rule of amniote cell specification is that germ layer identity (ectoderm, mesoderm, or endoderm) is established before gastrulation, but the specific type of cell is controlled by inductive influences during and after migration through the primitive streak. Chapman and colleagues (2007) transplanted groups of cells from one area of quail embryos to a different area in chick embryos and showed that presumptive mesoderm cells remain mesodermal and will migrate to the middle layer, even if they enter the primitive streak from a new (usually endodermal) site. Similarly, presumptive endodermal cells transplanted to a new position (where presumptive mesodermal cells usually enter the streak) ingress into the lower layer and become epithelial.

(A) *Hoxb4*

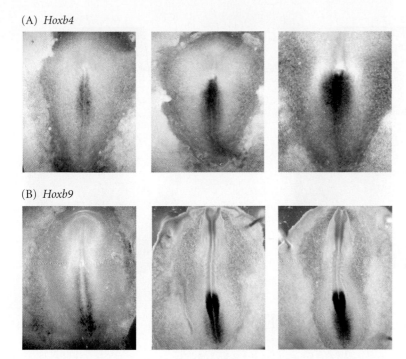

(B) *Hoxb9*

FIGURE 8.6 Hox gene activation begins when the mesodermal precursor cells are still in the epiblast. These genes are activated in an anterior-to-posterior fashion. Migration into the primitive streak is regulated by the Hox gene expression pattern (the most posterior Hox gene having preference). (A) *Hoxb4* expression in stage 4, 5, and 6 chick embryos. (B) *Hoxb9* expression in stage 7, 8, and 8+ chick embryos. Note that the anterior border of expression is more posterior than the anterior border of *Hoxb4* expression. (From Iimura and Pourquié 2006, courtesy of O. Pourquié.)

While they are still in the epiblast, but close to the primitive streak, the mesoderm cells appear to receive instructions that tell them exactly where they are along the anterior-posterior axis (Iimura and Pourquié 2006). At this point, they are induced to activate specific Hox genes. As we will detail later in the chapter, Hox genes are the vertebrate homologues of the homeotic (Hom-C) genes of *Drosophila*. And just like in *Drosophila* embryos, vertebrate Hox genes specify the identity of cells along the anterior-posterior axis of the embryo. In the case of vertebrates, however, there are four gene clusters (*HoxA, HoxB, HoxC,* and *HoxD*) instead of just one, and rather than individual Hox genes appearing at particular segmental levels, there is a nested set of Hox gene expression. For example, the mesodermal precursor cells are patterned along the anterior-posterior axis by the *HoxB* genes, which appear to inform the cells when to leave the epiblast and ingress into the primitive streak. "Anterior" Hox genes (which are identified with lower numbers, e.g., *Hoxb4*) are expressed early and extend farther anterior than genes such as *Hoxb9*, which is expressed later and does not extend as far into the embryo's anterior (**Figure 8.6**). Thus, the more posterior cells have more Hox genes expressed than the more anterior cells.

MIGRATION THROUGH THE PRIMITIVE STREAK The migration of the mesodermal cells through the anterior primitive streak and their condensation to form the chordamesoderm appear to be controlled by FGF signaling. Fgf8 protein is expressed in the primitive streak and repels migrating cells away from the streak. Yang and colleagues (2002) were able to follow the trajectories of these cells as they migrated

through the primitive streak (see Figure 8.4). They were able to deflect these normal trajectories by using beads that released Fgf8. Once inside the embryo, the lateral movement of the mesoderm cells appears to be regulated in part by platelet-derived growth factor (PDGF), a paracrine factor made by the epiblast (Yang et al. 2008).

Meanwhile, cells continue migrating inward through the primitive streak. As they enter the embryo, the cells separate into two layers. The deep layer joins the hypoblast along its midline, displacing the hypoblast cells to the sides. These deep-moving cells give rise to all the endodermal organs of the embryo, as well as to most of the extraembryonic membranes (the hypoblast and peripheral cells of the area opaca form the rest). The second migrating layer spreads to form a loose layer of cells between this endoderm and the epiblast. This middle layer of cells generates the mesodermal portions of the embryo and the mesoderm lining the extraembryonic membranes.

Movement away from the streak appears to be guided by Fgf8-mediated chemorepulsion, and the further movement of the mesodermal precursors appears to be regulated by Wnt proteins. In the more posterior regions, Wnt5a is unopposed and directs the cells to migrate broadly to become the lateral plate mesoderm. In the more anterior regions of the streak, however, Wnt5a is opposed by Wnt3a, which inhibits migration and causes the cells to form paraxial mesoderm. Indeed, the addition of Wnt3a-secreting pellets to the posterior primitive streak suppresses lateral migration and prevents the formation of lateral plate mesoderm (Sweetman et al. 2008). By 22 hours of incubation, most of the presumptive endodermal cells are in the interior of the embryo, although presumptive mesodermal cells continue to migrate inward for a longer time.

Regression of the primitive streak and epiboly of the ectoderm

Now a new phase of development begins. As mesodermal ingression continues, the primitive streak starts to regress, moving Hensen's node from near the center of the area pel-

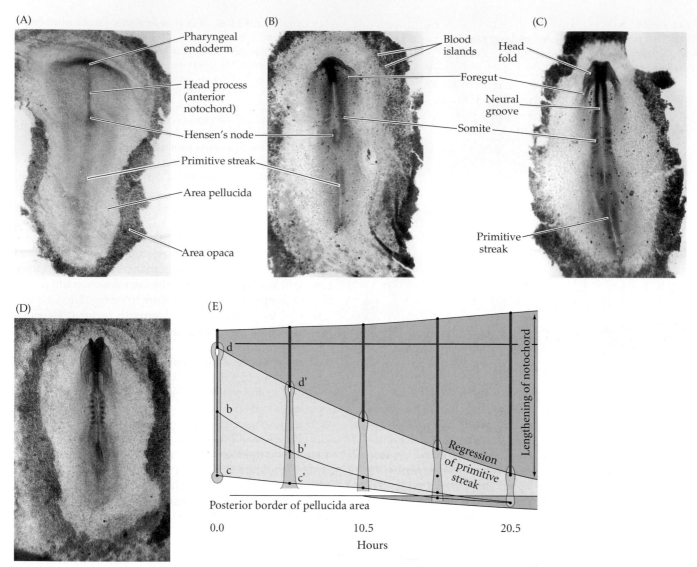

(A)
- Pharyngeal endoderm
- Head process (anterior notochord)
- Hensen's node
- Primitive streak
- Area pellucida
- Area opaca

(B)
- Blood islands
- Foregut
- Somite

(C)
- Head fold
- Neural groove
- Somite
- Primitive streak

(D)

(E)
- d
- d'
- b
- b'
- c
- c'
- Lengthening of notochord
- Regression of primitive streak
- Posterior border of pellucida area
- 0.0   10.5   20.5
- Hours

FIGURE 8.7 Chick gastrulation 24–28 hours after fertilization. (A) The primitive streak at full extension (24 hours). The head process (anterior notochord) can be seen extending from Hensen's node. (B) Two-somite stage (25 hours). Pharyngeal endoderm is seen anteriorly, while the anterior notochord pushes up the head process beneath it. The primitive streak is regressing. (C) Four-somite stage (27 hours). (D) At 28 hours, the primitive streak has regressed to the caudal portion of the embryo. (E) Regression of the primitive streak, leaving the notochord in its wake. Various points of the streak (represented by letters) were followed after it achieved its maximum length. The x axis (time) represents hours after achieving maximum length (the reference line is about 18 hours of incubation). (A–D courtesy of K. Linask; E after Spratt 1947.)

lucida to a more posterior position (Figure 8.7). The regressing streak leaves in its wake the dorsal axis of the embryo, including the notochord. The notochord is laid down in a head-to-tail direction, starting at the level where the ears and hindbrain form and extending caudally to the tail bud. As in the frog, the pharyngeal endoderm and head mesoendoderm will induce the anterior parts of the brain, while the notochord will induce the hindbrain and spinal cord. By this time, all the presumptive endodermal and mesodermal cells have entered the embryo and the epiblast is composed entirely of presumptive ectodermal cells.

While the presumptive mesodermal and endodermal cells are moving inward, the ectodermal precursors proliferate and migrate to surround the yolk by epiboly. The enclosure of the yolk by the ectoderm (again, reminiscent of the epiboly of the amphibian ectoderm) is a Herculean task that takes the greater part of 4 days to complete. It involves the continuous production of new cellular mate-

rial and the migration of the presumptive ectodermal cells along the underside of the vitelline envelope (New 1959; Spratt 1963). Interestingly, only the cells of the outer margin of the area opaca attach firmly to the vitelline envelope. These cells are inherently different from the other blastoderm cells, as they can extend enormous (500 μm) cytoplasmic processes onto the vitelline envelope. These elongated filopodia are believed to be the locomotor apparatus of the marginal cells, by which the marginal cells pull other ectodermal cells around the yolk (Schlesinger 1958). The filopodia bind to fibronectin, a laminar protein that is a component of the chick vitelline envelope. If the contact between the marginal cells and the fibronectin is experimentally broken by adding a soluble polypeptide similar to fibronectin, the filopodia retract and ectodermal migration ceases (Lash et al. 1990).

Thus, as avian gastrulation draws to a close, the ectoderm has surrounded the yolk, the endoderm has replaced the hypoblast, and the mesoderm has positioned itself between these two regions. Although we have identified many of the processes involved in avian gastrulation, we are only beginning to understand the molecular mechanisms by which some of these processes are carried out.

See WEBSITE 8.1 **Epiblast cell heterogeneity**

See VADE MECUM **Chick development**

Axis Specification and the Avian "Organizer"

As a consequence of the sequence in which the head endomesoderm and notochord are established, avian (and mammalian, reptilian, and teleost fish) embryos exhibit a distinct anterior-to-posterior gradient of developmental maturity. While cells of the posterior portions of the embryo are still part of a primitive streak and colonizing the mesoderm, cells at the anterior end are already starting to form organs (see Darnell et al. 1999). For the next several days, the anterior end of the embryo is more advanced in its development (having had a "head start," if you will) than the posterior end. Although the formation of the chick body axes is accomplished during gastrulation, axis specification begins earlier, during the cleavage stage.

The role of gravity and the PMZ

The conversion of the radially symmetrical blastoderm into a bilaterally symmetrical structure appears to be determined by gravity. As the ovum passes through the hen's reproductive tract, it is rotated for about 20 hours in the shell gland.

This spinning, at a rate of 10–12 revolutions per hour, shifts the yolk such that its lighter components (probably containing stored maternal determinants for development) lie beneath one side of the blastoderm. This imbalance tips up one end of the blastoderm, and that end becomes the **posterior marginal zone** (**PMZ**) of the embryo—the part where primitive streak formation begins (**Figure 8.8**; Kochav and Eyal-Giladi 1971; Callebaut et al. 2004).

It is not known what interactions cause this specific portion of the blastoderm to become the PMZ. Early on, the ability to initiate a primitive streak is found throughout the marginal zone; if the blastoderm is separated into parts, each with its own marginal zone, each part will form its own primitive streak (Spratt and Haas 1960). However, once the PMZ has formed, it controls the other regions of the margin. Not only do the cells of the PMZ initiate gastrulation, they also prevent other regions of the margin from forming their own primitive streaks (Eyal-Giladi et al. 1992; Bertocchini et al. 2004).

It now seems apparent that the posterior marginal zone contains cells that act as the equivalent of the amphibian Nieuwkoop center. When placed in the anterior region of the marginal zone, a graft of PMZ tissue (posterior to and not including Koller's sickle) is able to induce a primitive streak and Hensen's node without contributing cells to either structure (Bachvarova et al. 1998; Khaner 1998). Current evidence suggests that the entire marginal zone produces Wnt8c (capable of inducing β-catenin) and that, like the amphibian Nieuwkoop center, the PMZ produces Vg1, a member of the TGF-β family of secreted proteins (Mitrani et al. 1990; Hume and Dodd 1993; Seleiro et al. 1996).

Wnt8c and Vg1 act together to induce expression of Nodal (another secreted TGF-β protein) in the future embryonic epiblast next to Koller's sickle and the PMZ (Skromne and Stern 2002). Thus the pattern appears similar to that of amphibian embryos. Recent studies suggest that Nodal activity is needed to initiate the primitive streak, and that it is the secretion of Cerberus—an antagonist of Nodal—by the primary hypoblast cells that prevents primitive streak formation (Bertocchini and Stern 2002; Bertocchini et al. 2004). As the primary hypoblast cells move away from the PMZ, Cerberus protein is no longer present, allowing Nodal activity and therefore formation of the

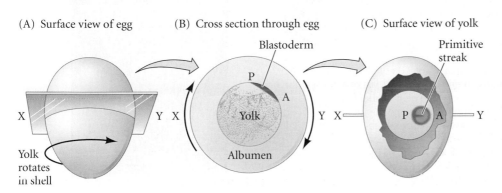

(A) Surface view of egg   (B) Cross section through egg   (C) Surface view of yolk

FIGURE 8.8 Specification of the chick anterior-posterior axis by gravity. (A) Rotation in the shell gland results in (B) the lighter components of the yolk pushing up one side of the blastoderm. (C) This more elevated region becomes the posterior of the embryo. (After Wolpert et al. 1998.)

(A)

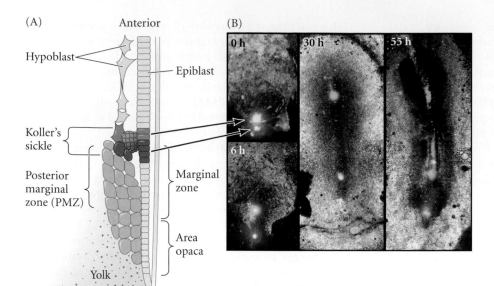

Anterior

Hypoblast

Epiblast

Koller's sickle

Posterior marginal zone (PMZ)

Marginal zone

Area opaca

Yolk

Posterior

(B)

0 h   30 h   55 h

6 h

FIGURE 8.9 Formation of Hensen's node from Koller's sickle. (A) Diagram of posterior end of an early (pre-streak) embryo, showing the cells labeled with fluorescent dyes in the photographs. (B) Just before gastrulation, cells in the anterior end of Koller's sickle (the epiblast and middle layer) were labeled with green dye. Cells of the posterior portion of Koller's sickle were labeled with red dye. As the cells migrated, the anterior cells formed Hensen's node and its notochord derivatives. The posterior cells formed the posterior region of the primitive streak. The time after dye injection is labeled on each photograph. (A after Bachvarova et al. 1998; B courtesy of R. F. Bachvarova.)

primitive streak in the posterior epiblast. Once formed, however, the streak secretes its own Nodal antagonist (the Lefty protein), thereby preventing any further primitive streaks from forming. Eventually, the Cerberus-secreting hypoblast cells are pushed to the future anterior of the embryo, where they contribute to ensuring that neural cells in this region become forebrain rather than more posterior structures of the nervous system.

The chick "organizer"

The "organizer" of the chick embryo forms from cells initially located just anterior to the posterior marginal zone.

The epiblast and middle layer cells in the anterior portion of Koller's sickle become Hensen's node, as described earlier. The posterior portions of Koller's sickle contribute to the posterior portion of the primitive streak (**Figure 8.9**). Hensen's node has long been known to be the avian equivalent of the amphibian dorsal blastopore lip, since (1) it is the region whose cells become the prechordal plate and chordamesoderm, (2) it is the region whose cells can both induce and pattern a second embryonic axis when transplanted into other locations of the gastrula (**Figure 8.10**; Waddington 1933; 1934; Gallera 1966; Nicolet 1970), and (3) it expresses the same marker genes as Spemann's organizer in amphibians and the embryonic shield of teleost fishes, such as the transcription factor Goosecoid (Izpisua-Belmonte et al. 1993). Moreover, Hensen's node can induce

(A)

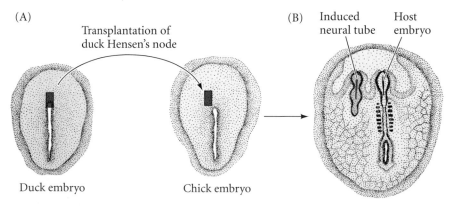

Transplantation of duck Hensen's node

Duck embryo   Chick embryo

(B)   Induced neural tube   Host embryo

(C)

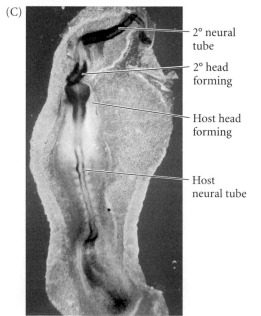

2° neural tube

2° head forming

Host head forming

Host neural tube

FIGURE 8.10 Induction of a new embryo by transplantation of Hensen's node. (A) A Hensen's node from a duck embryo is transplanted into the epiblast of a chick embryo. (B) A secondary embryo is induced (as is evident by the neural tube) from host tissues at the graft site. (C) Graft of Hensen's node from one embryo into the periphery of a host embryo. After further incubation, the host embryo has a neural tube whose regionalization can be seen by in situ hybridization. Probes to *Otx2* (red) recognize the head region, while probes to *Hoxb1* (blue) recognize the trunk neural tube. The donor node has induced the formation of a secondary axis, complete with head and trunk regions. (A,B after Waddington 1933; C from Boettger et al. 2001.)

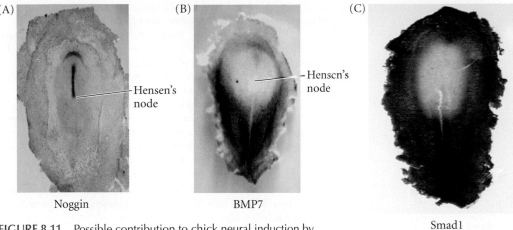

(A) Noggin

(B) BMP7

(C) Smad1

Hensen's node

Hensen's node

FIGURE 8.11 Possible contribution to chick neural induction by the inhibition of BMP signaling. (A) In a neurulating embryo, Noggin protein (purple) is expressed in the notochord and the pharyngeal endoderm. (B) Bmp7 expression (dark purple), which had encompassed the entire epiblast, becomes restricted to the nonneural regions of the ectoderm. (C) Similarly, the product of BMP signaling, the phosphorylated form of Smad1 (recognized by antibodies to the phosphorylated form of the protein; dark brown) is not seen in the neural plate. (From Faure et al. 2002, courtesy of the authors.)

neural tissue when grafted into fish, amphibian, or mammalian embryos (Waddington 1936; Kintner and Dodd 1991; Hatta and Takahashi 1996).

As is the case in all vertebrates, the dorsal mesoderm is able to induce the formation of the central nervous system in the ectoderm overlying it. The cells of Hensen's node and its derivatives act like the amphibian organizer, and they secrete bone morphogenetic protein (BMP) antagonist proteins such as chordin, Noggin, and Nodal. These proteins repress BMP signaling and dorsalize the ectoderm and mesoderm (**Figure 8.11**). However, repression of the BMP signal by these antagonists does not appear to be sufficient for neural induction (see Stern 2005). Fibroblast growth factors synthesized in the hypoblast and in Hensen's node precursor cells just prior to gastrulation appear to be critical for preparing the epiblast to generate neuronal phenotypes. Although FGFs can block BMP signaling, this does not account for the ability of FGFs to induce a transient expression of pre-neural genes in the epiblast (Streit et al. 1998, 2000). These neural genes do not stay active unless they are supported by BMP antagonists (Streit et al. 1998, 2000; Albazerchi and Stern 2006). Thus, FGF signaling inhibits BMPs from inducing the genes that specify ectoderm to become epidermis, and they activate the genes that specify to ectoderm to become neural.

Indeed, fibroblast growth factors play four fundamental roles in cell specification during gastrulation:

- First, as in all vertebrates, FGFs are responsible for specifying the mesoderm. FGFs from the hypoblast (in collaboration with Nodal from the posterior marginal zone) accomplish this specification by activating the *Brachyury* and *Tbx6* genes in the cells passing through the primitive streak (**Figure 8.12**; Sheng et al. 2003).

- Second, FGFs separate mesoderm formation from neurulation. The mechanism by which FGFs help end mesoderm ingression and stabilize the epiblast appears to be due to a gene called *Churchill*. While FGFs are rapidly inducing the mesoderm, they are also slowly inducing activation of *Churchill* in the ectoderm. The Churchill protein (so named because the protein's two zinc fingers extend like the British prime minister's famous "V for Victory" symbol) can activate the Smad-interacting protein SIP1. SIP1 controls the genes whose expression is required for ingression of cells through the primitive streak. Thus, once activated, SIP1 helps prospective neural plate cells remain in the epiblast.

- Third, FGFs help bring about neurulation in the central ectodermal cells. SIP1, probably through its interaction with Smad1, may make the prospective neural plate cells less sensitive to BMP.

- Fourth, FGFs induce *ERNI* and *Sox3*, two pre-neural genes that initiate the signaling cascade leading to the production of neural tissue.

Thus, FGFs appear to be critically important regulators of cell fate in the early chick embryo (Streit et al. 2000; Sheng et al. 2003; Albazerchi and Stern 2006).

Anterior-posterior patterning

The patterning of the definitive anterior-posterior axis occurs differently for the mesoderm and neural ectoderm, but in both cases the process involves timing (the sequential generation of cells from a zone of undifferentiated proliferating cells) and the influence of caudalizing molecules. In the ectoderm, most of the initial neural plate corresponds to the future head region (from forebrain to the level of the future ear vesicle, which lies adjacent to Hensen's node at full primitive streak stage). A small

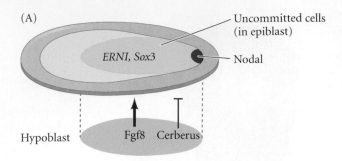

(A)

Uncommitted cells (in epiblast)

ERNI, Sox3

Nodal

Hypoblast

Fgf8 Cerberus

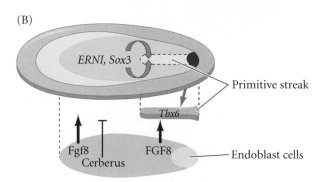

(B)

ERNI, Sox3

Primitive streak

Tbx6

Fgf8
Cerberus

FGF8

Endoblast cells

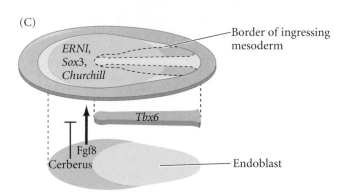

(C)

ERNI,
Sox3,
Churchill

Border of ingressing mesoderm

Tbx6

Fgf8
Cerberus

Endoblast

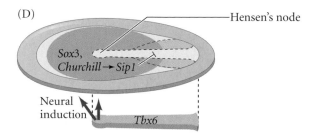

(D)

Hensen's node

Sox3,
Churchill → Sip1

Neural
induction

Tbx6

FIGURE 8.12 Model by which FGFs regulate mesoderm formation and neurulation. (A) Stage XI, where the hypoblast (green) secretes Fgf8, which induces pre-neural genes *ERNI* and *Sox2* (blue) in the epiblast. The cells in this domain, however, remain uncommitted. Nodal, expressed in the posterior epiblast, cannot function; it is inhibited by the Cerberus protein secreted by the hypoblast. (B) At around stage 1, the hypoblast is displaced from the posterior edge by the endoblast (secondary hypoblast; gold), allowing Nodal to function. Nodal plus Fgf8 induces *Brachyury* and *Tbx6* expression to specify the mesoderm and initiate the ingression of mesoderm cells through the primitive streak (red). (C) At stage 4, continued Fgf8 expression activates *Churchill* in the epiblast. (D) By end of stage 4, Churchill protein induces SIP1, which blocks *Brachyury* and *Tbx6*, preventing further ingression of epiblast cells through the streak. The remaining epiblast cells can now become sensitized to neural inducers from Hensen's node (purple). (After Sheng et al. 2003.)

"young" and undifferentiated as it regresses, and that this is antagonized by retinoic acid (RA) activity as cells leave this zone (**Figure 8.13**; Diez del Corral et al. 2003).

In the mesoderm, the anterior-posterior axis is also related to time. The entire length of the notochord at the midline is derived from cells that are present in Hensen's node by the full primitive streak stage. A population of progenitor cells remains in the node; their descendants gradually leave as the node regresses, laying down the chordamesoderm and the ventral midline of the neural tube (the future floor plate of the spinal cord) (Selleck and Stern 1991; Psychoyos and Stern 1996; Tzouanacou et al. 2009). Therefore anterior-posterior identities along the axis from the hindbrain to the tail are specified as a function of the time of emergence from the primitive streak and Hensen's node. It has been proposed that the length of time cells are resident in this region determines which Hox genes are expressed by the cells. This pattern of Hox gene expression can also be under the influence of the FGF and retinoic acid gradients (Gaunt 1992; Wilson et al. 2009).

Left-right axis formation

The vertebrate body has distinct right and left sides. The heart and spleen, for instance, are generally on the left side of the body, while the liver is usually on the right. The distinction between the sides is primarily regulated by two proteins: the paracrine factor Nodal and the transcription factor Pitx2. However, the mechanism by which *nodal* gene expression is activated in the left side of the body differs among the vertebrate classes. The ease with which chick embryos can be manipulated has allowed scientists to elucidate the pathways of left-right axis determination in birds more readily than in other vertebrates.

As the primitive streak reaches its maximum length, transcription of the *sonic hedgehog* gene (*shh*) becomes restricted to the left side of the embryo, controlled by activin and its receptor (**Figure 8.14A**). Activin signaling, along with BMP4, appears to block the expression of Sonic hedgehog and to activate expression of Fgf8 protein on the

region of neural ectoderm just lateral and posterior to the node (sometimes called the caudal lateral epiblast) will give rise to the rest of the nervous system, including the posterior hindbrain and all of the spinal cord. As the primitive streak regresses, this latter region regresses with the node and adds cells to the caudal end of the elongating neural plate. It appears that FGF signaling in the streak and paraxial (future somite) mesoderm keeps this region

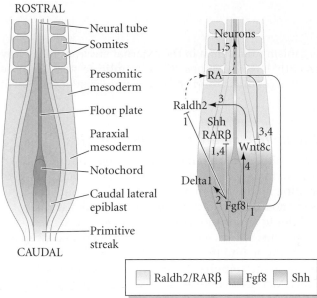

Raldh2/RARβ   Fgf8   Shh

FIGURE 8.13   Signals that regulate axis extension in chick embryos. In the stage 10 chick embryo, Fgf8 inhibits expression of the retinoic acid (RA) synthesizing enzyme Raldh2 in the presomitic mesoderm (1) and the expression of the retinoic acid receptor RARβ in the neural ectoderm (4), thus preventing RA from triggering differentiation in the caudal-lateral epiblast cells (those cells adjacent to the node/streak border and which give rise to lateral and dorsal neural tube) and the caudalmost paraxial mesoderm (1,5). In addition, Fgf8 inhibits Sonic hedgehog (Shh) expression in the neural tube floorplate, controlling the onset of ventral patterning genes (1). FGF signaling is also required for expression of Delta1 in the medial portion of the caudal-lateral epiblast cells (2) and promotes expression of Wnt8c (4). As Fgf8 decays in the caudal paraxial mesoderm, Wnt signaling, most likely provided by Wnt8c, now acts to promote Raldh2 in the adjacent paraxial mesoderm (4). RA produced by Raldh2 activity represses Fgf8 (1) and Wnt8c (3,4). (After Wilson et al. 2009.)

right side of the embryo. Fgf8 blocks expression of the paracrine factor Cerberus on the right-hand side; it may also activate a signaling cascade that instructs the mesoderm to have right-sided capacities (Schlueter and Brand 2009).

Meanwhile, on the left side of the body, Shh protein activates Cerberus (**Figure 8.14B**), which in this case acts with BMP to stimulate the synthesis of Nodal protein (Yu et al. 2008). Nodal activates the *pitx2* gene while repressing *snail*. In addition, Lefty1 in the ventral midline prevents the Cerberus signal from passing to the right side of the embryo (**Figure 8.14C,D**) As in *Xenopus*, Pitx2 is crucial in directing the asymmetry of the embryonic structures. Experimentally induced expression of either Nodal or Pitx2 on the right side

FIGURE 8.14   Model for generating left-right asymmetry in the chick embryo. (A) On the left side of Hensen's node, Sonic hedgehog (Shh) activates Cerberus, which stimulate BMPs to induce the expression of Nodal. In the presence of Nodal, the *pitx2* gene is activated. Pitx2 protein is active in the various organ primordia and specifies which side will be the left. On the right side of the embryo, activin is expressed, along with activin receptor IIa. This activates Fgf8, a protein that blocks expression of the gene for Cerberus. In the absence of Cerberus, Nodal is not activated and thus Pitx2 is not expressed. (B) Whole-mount in situ hybridization of *cerberus* mRNA. This view is from the ventral surface ("from below", so the expression seems to be on the right. Dorsally, the expression pattern would be on the left. (C) Whole-mount in situ hybridization using probes for the chick *nodal* message (stained purple) shows its expression in the lateral plate mesoderm only on the left side of the embryo. This view is from the dorsal side. (D) Similar in situ hybridization, using the probe for *pitx2* at a later stage of development. The embryo is seen from its ventral surface. At this stage, the heart is forming, and *pitx2* expression can be seen on the left side of the heart tube (as well as symmetrically in more anterior tissues. (A after Raya and Izpisua-Belmonte 2004; B from Rodriguez-Esteban et al. 1999, courtesy of J. Izpisúa-Belmonte; C courtesy of C. Stern; D from Logan et al. 1998, courtesy of C. Tabin.)

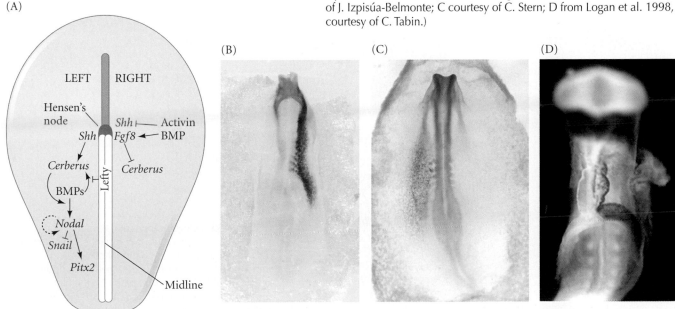

of the chick reverses the asymmetry or cause randomization of asymmetry on the right or left sides* (Levin et al. 1995; Logan et al. 1998; Ryan et al. 1998).

The real mystery is, What processes create the original asymmetry of Shh and Fgf8? One important observation is that the first asymmetry seen during the formation of Hensen's node in chicks involves Fgf8- and Shh-expressing cells rearranging themselves to converge on the right-hand side of the node (Cui et al. 2009; Gros et al. 2009). Therefore, the differences in gene expression can be traced back to differences in cell migration to the right and left sides of the embryo. What establishes this initial asymmetry is still unknown.

EARLY MAMMALIAN DEVELOPMENT

Cleavage

It is not surprising that early mammalian development has been incredibly difficult to study. Mammalian eggs are among the smallest in the animal kingdom, making them hard to manipulate experimentally. The human zygote, for instance, is only 100 μm in diameter—barely visible to the eye and less than one-thousandth the volume of a *Xenopus laevis* egg. Also, mammalian zygotes are not produced in numbers comparable to sea urchin or frog zygotes; a female mammal usually ovulates fewer than 10 eggs at a given time, so it is difficult to obtain enough material for biochemical studies. As a final hurdle, the development of mammalian embryos is accomplished inside another

organism rather than in the external environment. Only recently has it been possible to duplicate some of these internal conditions and observe mammalian development in vitro.

The unique nature of mammalian cleavage

It took time to surmount the difficulties, but our knowledge of mammalian cleavage has turned out to be worth waiting for. Mammalian cleavage is strikingly different from most other patterns of embryonic cell division.

Prior to fertilization, the mammalian oocyte, wrapped in cumulus cells, is released from the ovary and swept by the fimbriae into the oviduct (**Figure 8.15**). Fertilization occurs in the **ampulla** of the oviduct, a region close to the ovary. Meiosis is completed after sperm entry, and the first cleavage begins about a day later (see Figure 4.35). Cleavages in mammalian eggs are among the slowest in the animal kingdom, taking place some 12–24 hours apart. Meanwhile, the cilia in the oviduct push the embryo toward the uterus; the first cleavages occur along this journey.

In addition to the slowness of cell division, several other features distinguish mammalian cleavage, including the unique orientation of mammalian blastomeres with relation to one another. The first cleavage is a normal meridional division; however, in the second cleavage, one of the two blastomeres divides meridionally and the other divides equatorially (**Figure 8.16**). This is called **rotational cleavage** (Gulyas 1975).

Another major difference between mammalian cleavage and that of most other embryos is the marked asynchrony of early cell division. Mammalian blastomeres do not all divide at the same time. Thus, mammalian embryos do not increase exponentially from 2- to 4- to 8-cell stages, but frequently contain odd numbers of cells. And the mam-

---

*In humans, homozygous loss of *PITX2* causes Rieger's syndrome, a condition characterized by asymmetry anomalies. A similar condition is caused by knocking out the *Pitx2* gene in mice (Fu et al. 1999; Lin et al. 1999).

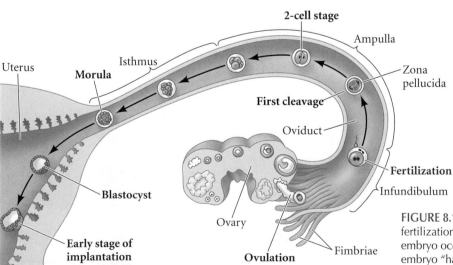

FIGURE 8.15 Development of a human embryo from fertilization to implantation. Compaction of the human embryo occurs on day 4, at the 10-cell stage. The embryo "hatches" from the zona pellucida upon reaching the uterus. During its migration to the uterus, the zona prevents the embryo from prematurely adhering to the oviduct rather than traveling to the uterus.

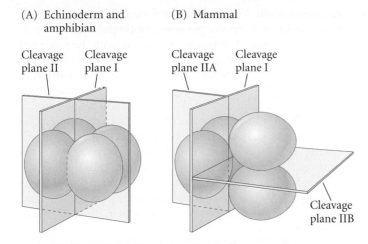

(A) Echinoderm and amphibian

Cleavage plane II   Cleavage plane I

(B) Mammal

Cleavage plane IIA   Cleavage plane I

Cleavage plane IIB

FIGURE 8.16 Comparison of early cleavage in (A) echinoderms and amphibians (radial cleavage) and (B) mammals (rotational cleavage). Nematodes also have a rotational form of cleavage, but they do not form the blastocyst structure characteristic of mammals. (After Gulyas 1975.)

malian genome, unlike the genomes of rapidly developing animals, is activated during early cleavage, and it is the newly formed nuclei (rather than the oocyte cytoplasm) that produce the proteins necessary for cleavage and development. In the mouse and goat, the switch from maternal to zygotic control occurs at the 2-cell stage. In humans, the zygotic genes are first activated between the 4- and 8-cell stages (Piko and Clegg 1982; Braude et al. 1988; Prather 1989).

Most research on mammalian development has focused on the mouse, since mice are relatively easy to breed, have large litters, and can be housed easily in laboratories. Thus, most of the studies discussed here will concern murine development.

Compaction

One of the most crucial differences between mammalian cleavage and all other types involves the phenomenon of **compaction**. Mouse blastomeres through the 8-cell stage form a loose arrangement with plenty of space between them (**Figure 8.17A,B**). Following the third cleavage, however, the blastomeres undergo a spectacular change in their behavior. Cell adhesion proteins such as E-cadherin become expressed, and the blastomeres suddenly huddle together and form a compact ball of cells (**Figure 8.17C,D**; Peyrieras et al. 1983; Fleming et al. 2001). This tightly packed arrangement is stabilized by tight junctions that form between the outside cells of the ball, sealing off the inside of the sphere. The cells within the sphere form gap junctions, thereby enabling small molecules and ions to pass between them.

See WEBSITE 8.2 Mechanisms of compaction and formation of the inner cell mass

See WEBSITE 8.3 Human cleavage and compaction.

The cells of the compacted 8-cell embryo divide to produce a 16-cell **morula** (**Figure 8.17E**). The morula consists of a small group of internal cells surrounded by a larger group of external cells (Barlow et al. 1972). The internal cells are called the **inner cell mass** (**ICM**). They will give rise to the embryo. Most of the descendants of the external cells become the **trophoblast** (trophectoderm) cells. This group of cells produces no embryonic structures. Rather, it forms the tissue of the chorion, the embryonic

FIGURE 8.17 Cleavage of a single mouse embryo in vitro. (A) 2-cell stage. (B) 4-cell stage. (C) Early 8-cell stage. (D) Compacted 8-cell stage. (E) Morula. (F) Blastocyst. (G) Electron micrograph through the center of a mouse blastocyst. (A–F from Mulnard 1967, courtesy of J. G. Mulnard; G from Ducibella et al. 1975, courtesy of T. Ducibella.)

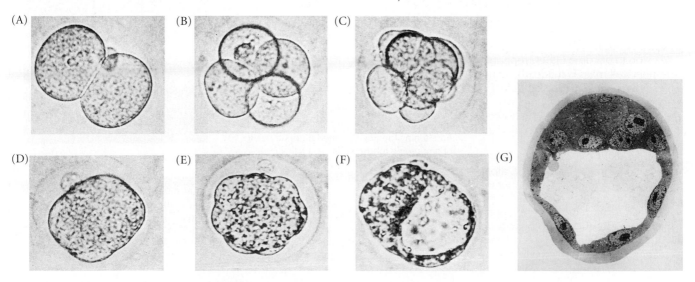

(A)   (B)   (C)

(D)   (E)   (F)   (G)

portion of the placenta. The chorion enables the fetus to get oxygen and nourishment from the mother. It also secretes hormones that cause the mother's uterus to retain the fetus and produces regulators of the immune response so that the mother will not reject the embryo (as she would an organ graft). It is important to remember that the crucial outcome of these first divisions is to generate cells that will stick to the uterus. Thus, formation of the trophectoderm is the first differentiation event in mammalian development. These external cells first adhere to the uterine lining, then digest a path that allows the embryo to lodge itself in the uterine wall.

It is a matter of chance which cells become trophoblast and which become inner cell mass (see Dard et al. 2009) . Indeed, time-lapse photomicroscopy appears to show that the plane of the first division is totally irrelevant to the allocation of blastomeres along the embryonic/abembryonic axis (the axis defined by the position of the inner cell mass within the trophoblast to the area of the trophoblast directly opposite the ICM). The descendants of both of the first two blastomeres were seen to contribute to the trophoblast and to the inner cell mass (Hiiragi and Solter 2004; Motosugi et al. 2005; Kurotaki et al. 2007). The mammalian embryo has no known proteins or other subcellular components

asymmetrically localized in the fertilized egg, and there does not seem to be any predetermined polarity.

The mouse embryo proper is derived from the inner cell mass of the 16-cell stage, supplemented by cells dividing from the outer cells of the morula during the transition to the 32-cell stage (Pedersen et al. 1986; Fleming 1987; Bischoff et al. 2008). The cells of the ICM will give rise to the embryo and its associated yolk sac, allantois, and amnion. By the 64-cell stage, the inner cell mass (approximately 13 cells) and the trophoblast cells have become separate cell layers, with neither contributing cells to the other group (Dyce et al. 1987; Fleming 1987). The inner cell mass actively supports the trophoblast, secreting proteins such as Fgf4 that stimulate the trophoblast cells to divide (Tanaka et al. 1998).

The earliest blastomeres (such as each blastomere of an 8-cell embryo) can form both trophoblast cells and the embryo precursor cells. These very early cells are said to be **totipotent** (Latin, "capable of everything"). The inner cell mass is said to be **pluripotent** (Latin, "capable of many things"). That is, each cell of the ICM can generate any cell type in the body, but because the distinction between ICM and trophoblast has been established, it is thought that ICM cells are not able to form the trophoblast.

SIDELIGHTS & SPECULATIONS

Trophoblast or ICM?

The decision to become either trophoblast or inner cell mass blastomere is the first binary decision in our lives. Prior to blastocyst formation, each early embryonic blastomere expresses both the Cdx2 and the Oct4 transcription factors (Niwa et al. 2005; Dietrch and Hiiragi 2007; Ralston and Rossant 2008). However, once the decision to become either trophoblast or inner cell mass is made, the cells of these two regions express different genes (**Figure 8.18**). Only the trophoblast cells synthesize the homeodomain-containing, caudal-like transcription factor Cdx2, which downregulates Oct4 and Nanog—two more transcription factors that, along with Stat3, characterize the inner cell mass (Strumpf et al. 2005).

The activation of the *Cdx2* gene in the outer cells appears to be regulated by the Yap protein. Yap is a co-factor for the transcription factor Tead4.

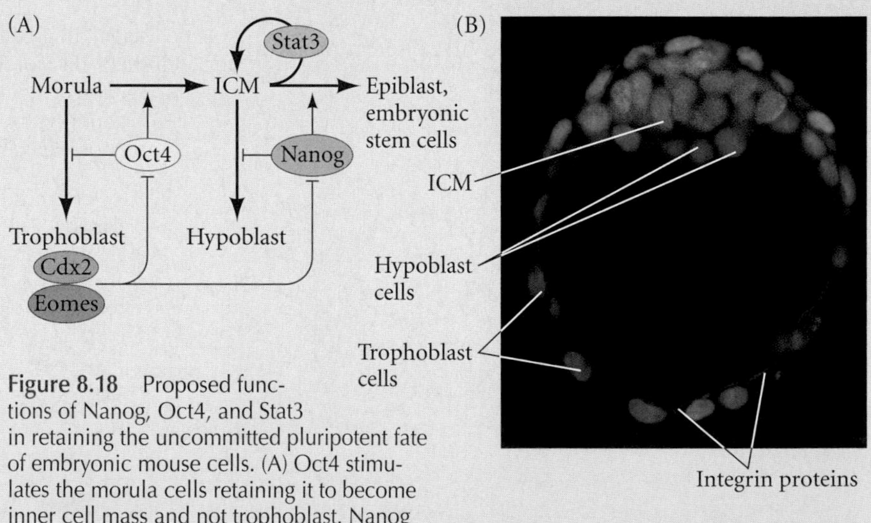

Figure 8.18 Proposed functions of Nanog, Oct4, and Stat3 in retaining the uncommitted pluripotent fate of embryonic mouse cells. (A) Oct4 stimulates the morula cells retaining it to become inner cell mass and not trophoblast. Nanog works at the next differentiation event, preventing the ICM cells from becoming hypoblast and promoting their becoming the pluripotent embryonic epiblast. Stat3 is probably involved in the self-renewal of these pluripotent cells. Cdx2 in the trophoblast prevents *Oct4* and *nanog* expression, thereby stabilizing the trophoblast lineage. (B) Mouse blastocyst in which the Oct4 protein in the ICM is stained orange. (After Mitsui et al. 2003; Strumpf et al. 2005; B courtesy of J. Rossant.)

SIDELIGHTS & SPECULATIONS (Continued)

Tead4 is found in the nuclei of both the inner and outer cell compartments of the blastocyst, but it is activated by Yap only in the outer compartment. That is because Yap protein can enter the nucleus in the outer cells and thereby allow Tead4 to transcribe trophoblast-specifying genes such as *Cdx2* and *eomesodermin*. In contrast, the inner cells, with each of their surfaces surrounded by other cells, activate (probably through Hippo, a cell surface kinase) the Lats protein, a kinase that phosphorylates the Yap protein. Phosphorylated Yap is not able to enter the nucleus and is subsequently degraded (Nishioka et al. 2009; **Figure 8.19**). Therefore, in the inner cells, Tead4 cannot function, and the *Cdx2* gene remains untranscribed.

Cdx2 blocks the expression of Oct4, and Oct4 blocks the expression of Cdx2. In this way, the two lineages become separated. The expression of the three transcription factors characteristic of the inner cell mass—Oct4, Stat3, and Nanog—is critical for the formation of the embryo and for maintaining the pluripotency of the inner cell mass. Oct4 is expressed first, and it is expressed in the morula as well as in the inner cell mass and early epiblast. Oct4 blocks cells from taking on the trophoblastic fate. Later,

Nanog* prevents the ICM blastomeres from becoming hypoblast cells, and stimulates blastomere self-renewal in the epiblast. The activated (phosphorylated) form of Stat3 also stimulates self-renewal of ICM blastomeres (see Figure 8.18; Pesce et al. 2001; Chambers et al. 2003; Mitsui et al. 2003). If the inner cell mass blastomeres are removed in a manner that lets them retain their expression of Nanog, Oct4, and phosphorylated Stat3 proteins, these cells divide and become **embryonic stem cells** (**ES cells**). The pluripotency of ES cells (and the medical uses for them, which will be detailed later in this book) is dependent on their retaining the expression of these three transcription factors.

*The research leading to the discovery of Nanog was partially motivated by the desire to convert normal human somatic cells into stem cell lines. The gene's name derives from the mythical Celtic land of perpetual youth.

(A)

(B)

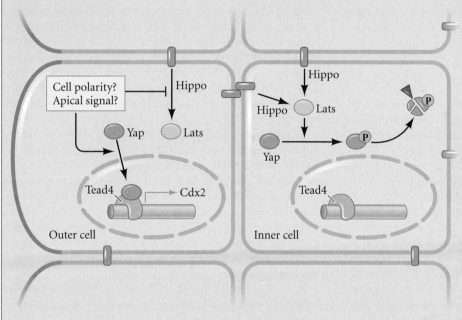

Figure 8.19 Possible pathway initiating the distinction between inner cell mass and trophectoderm blastomeres. (A) The Tead4 transcription factor, when active, promotes transcription of the *Cdx2* gene. Together, the Tead4 and Cdx2 transcription factors activate the genes that specify the outer cells to become trophectoderm. (B) Model for Tead4 activation. In the outer cells, the lack of cells surrounding the embryo sends a signal (as yet unknown) that blocks the Hippo pathway from activating the Lats protein. In the absence of functional Lats, the Yap transcriptional co-factor can bind with Tead4 to activate the *Cdx2* gene. In the inner cells, the Hippo pathway is active and the Lats kinase phosphorylates the Yap transcriptional co-activator. The phosphorylated form of Yap is targeted for degradation and does not enter the nucleus. (After Nishioka et al. 2009.)

Initially, the morula does not have an internal cavity. However, during a process called **cavitation**, the trophoblast cells secrete fluid into the morula to create a blastocoel. The membranes of trophoblast cells contain sodium pumps (an Na^+/K^+-ATPase and an Na^+/H^+ exchanger) that pump Na^+ into the central cavity. The subsequent accumulation of Na^+ draws in water osmotically, thus creating and enlarging the blastocoel (Borland 1977; Ekkert et al. 2004; Kawagishi et al. 2004). Interestingly, this sodium pumping activity appears to be stimulated by the oviduct cells on which the embryo is traveling toward the uterus (Xu et al. 2004).

(A)

(B)

(C)

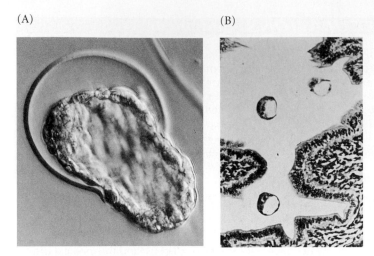

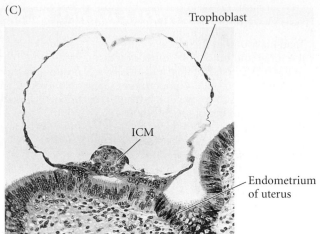

FIGURE 8.20 Hatching from the zona and implantation of the mammalian blastocyst in the uterus. (A) Mouse blastocyst hatching from the zona pellucida. (B) Mouse blastocysts entering the uterus. (C) Initial implantation of a rhesus monkey blastocyst. (A from Mark et al. 1985, courtesy of E. Lacy; B from Rugh 1967; C, Carnegie Institution of Washington, Chester Reather, photographer.)

As the blastocoel expands, the inner cell mass becomes positioned on one side of the ring of trophoblast cells (see Figure 8.17F); the resulting type of blastula, called a **blastocyst**, is another hallmark of mammalian cleavage.*

Escape from the zona pellucida

While the embryo is moving through the oviduct en route to the uterus, the blastocyst expands within the zona pellucida (the extracellular matrix of the egg that was essential for sperm binding during fertilization; see Chapter 4). During this time, the zona pellucida prevents the blastocyst from adhering to the oviduct walls. (If this happens—as it sometimes does in humans—it is called an ectopic, or "tubal," pregnancy, a dangerous condition because an embryo implanted in the oviduct can cause a life-threatening hemorrhage when it begins to grow.) When the embryo reaches the uterus, it must "hatch" from the zona so that it can adhere to the uterine wall.

The mouse blastocyst hatches from the zona pellucida by digesting a small hole in it and squeezing through the hole as the blastocyst expands (**Figure 8.20A**). A trypsin-like protease secreted by the trophoblast seems responsible for hatching the blastocyst from the zona (Perona and

Wassarman 1986; O'Sullivan et al. 2001). Once outside the zona, the blastocyst can make direct contact with the uterus (**Figure 8.20B,C**). The **endometrium**—the epithelial lining of the uterus—"catches" the blastocyst on an extracellular matrix made up of complex sugars, collagen, laminin, fibronectin, cadherins, hyaluronic acid, and heparan sulfate receptors (see Wang and Dey 2006). As in so many intercellular adhesions during development, there appears to be a period of labile attachment, followed by a period of more stable attachment. The first attachment seems to be mediated by L-selectin on the trophoblast cells adhering to sulfated polysaccharides on the uterine cells (Genbacev et al. 2003). These sulfated polysaccharides are synthesized in response to estrogen and progesterone secreted by the corpus luteum (the remnant of the ruptured ovarian follicle).

After the initial binding, several other adhesion systems appear to coordinate their efforts to keep the blastocyst tightly bound to the uterine lining. The trophoblast cells synthesize integrins that bind to the uterine collagen, fibronectin, and laminin, and they synthesize heparan sulfate proteoglycan precisely prior to implantation (see Carson et al. 1993). P-cadherins on the trophoblast and uterine endometrium also help dock the embryo to the uterus (see Chapter 3). Once in contact with the endometrium, Wnt proteins (from the trophoblast, the endometrium, or from both) instruct the trophoblast to secrete a set of proteases, including collagenase, stromelysin, and plasminogen activator. These protein-digesting enzymes digest the extracellular matrix of the uterine tissue, enabling the blastocyst to bury itself within the uterine wall (Strickland et al. 1976; Brenner et al. 1989; Pollheimer et al. 2006).

Mammalian Gastrulation

Birds and mammals are both descendants of reptilian species (albeit different reptilian species). It is not surprising, therefore, that mammalian development parallels that

---

*The interplay of myth and biology certainly comes to the fore when describing mammalian development. Although the mammalian blastocyst was discovered by Rauber in 1881, its first public display was probably in Gustav Klimt's 1908 painting *Danae*, in which blastocyst-like patterns are featured on the heroine's robe as she becomes impregnated by Zeus.

of reptiles and birds. What *is* surprising is that the gastrulation movements of reptilian and avian embryos, which evolved as an adaptation to yolky eggs, are retained in the mammalian embryo even in the absence of large amounts of yolk. The mammalian inner cell mass can be envisioned as sitting atop an imaginary ball of yolk, following instructions that seem more appropriate to its reptilian ancestors.

Modifications for development inside another organism

The mammalian embryo obtains nutrients directly from its mother and does not rely on stored yolk. This adaptation has entailed a dramatic restructuring of the maternal anatomy (such as expansion of the oviduct to form the uterus) as well as the development of a fetal organ capable of absorbing maternal nutrients. This fetal organ—the **chorion**—is derived primarily from embryonic trophoblast cells, supplemented with mesodermal cells derived from the inner cell mass. The chorion forms the fetal portion of the placenta. It also induces the uterine cells to form the maternal portion of the placenta, the decidua. The **decidua** becomes rich in the blood vessels that will provide oxygen and nutrients to the embryo.

The origins of early mammalian tissues are summarized in **Figure 8.21**. The first segregation of cells within the inner cell mass forms two layers. The lower layer is the **hypoblast** (sometimes called the *primitive endoderm* or *visceral endoderm*); the remaining inner cell mass tissue above it is the **epiblast** (**Figure 8.22A,B**). Surprisingly, whether a cell becomes epiblast or hypoblast does not depend on the position of the cell within the ICM. Rather, the blastomeres of the ICM appear to be a mosaic of future epiblast cells (expressing Nanog transcription factor) and hypoblast cells (expressing Gata6 transcription factor) a full day before the layers segregate at day 4.5 (Chazaud et al. 2006). The epiblast and hypoblast form a structure called the **bilaminar germ disc**. The hypoblast cells delaminate from the inner cell mass to line the blastocoel cavity, where they give rise to the **extraembryonic endoderm**, which forms the yolk sac. As in avian embryos, these cells do not produce any part of the newborn organism. The epiblast cell layer is split by small clefts that eventually coalesce to separate the **embryonic epiblast** from the other epiblast cells that line the **amnionic cavity** (**Figure 8.22C,D**). Once the lining of the amnion is completed, the amnionic cavity fills with a secretion called **amnionic fluid**, which serves as a shock absorber for the developing embryo while preventing it from drying out. The embryonic epiblast is thought to contain all the cells that will generate the actual embryo, and it is similar in many ways to the avian epiblast.

By labeling individual cells of the epiblast with horseradish peroxidase, Kirstie Lawson and her colleagues (1991) were able to construct a detailed fate map of the

FIGURE 8.21 Schematic diagram showing the derivation of tissues in human and rhesus monkey embryos. The dashed line indicates a possible dual origin of the extraembryonic mesoderm. (After Luckett 1978; Bianchi et al. 1993.)

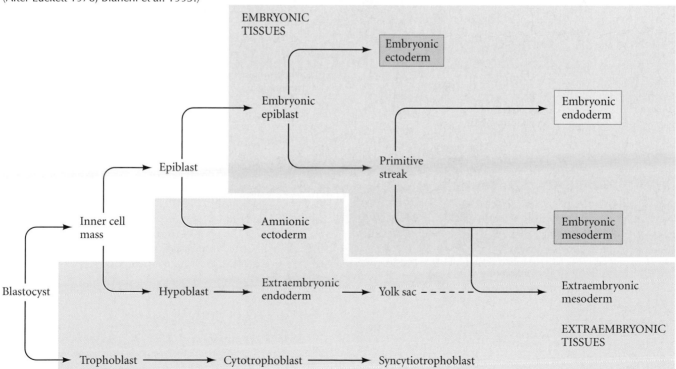

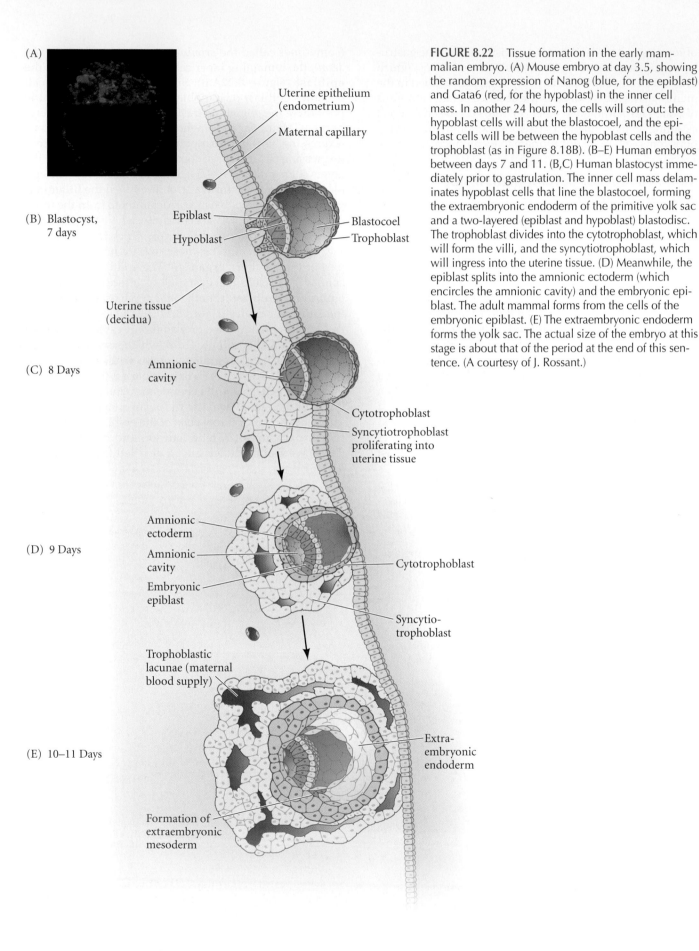

(A)

(B) Blastocyst,
7 days

Uterine epithelium
(endometrium)

Maternal capillary

Epiblast

Hypoblast

Blastocoel

Trophoblast

Uterine tissue
(decidua)

(C) 8 Days

Amnionic
cavity

Cytotrophoblast

Syncytiotrophoblast
proliferating into
uterine tissue

(D) 9 Days

Amnionic
ectoderm

Amnionic
cavity

Embryonic
epiblast

Cytotrophoblast

Syncytio-
trophoblast

Trophoblastic
lacunae (maternal
blood supply)

(E) 10–11 Days

Extra-
embryonic
endoderm

Formation of
extraembryonic
mesoderm

FIGURE 8.22 Tissue formation in the early mammalian embryo. (A) Mouse embryo at day 3.5, showing the random expression of Nanog (blue, for the epiblast) and Gata6 (red, for the hypoblast) in the inner cell mass. In another 24 hours, the cells will sort out: the hypoblast cells will abut the blastocoel, and the epiblast cells will be between the hypoblast cells and the trophoblast (as in Figure 8.18B). (B–E) Human embryos between days 7 and 11. (B,C) Human blastocyst immediately prior to gastrulation. The inner cell mass delaminates hypoblast cells that line the blastocoel, forming the extraembryonic endoderm of the primitive yolk sac and a two-layered (epiblast and hypoblast) blastodisc. The trophoblast divides into the cytotrophoblast, which will form the villi, and the syncytiotrophoblast, which will ingress into the uterine tissue. (D) Meanwhile, the epiblast splits into the amnionic ectoderm (which encircles the amnionic cavity) and the embryonic epiblast. The adult mammal forms from the cells of the embryonic epiblast. (E) The extraembryonic endoderm forms the yolk sac. The actual size of the embryo at this stage is about that of the period at the end of this sentence. (A courtesy of J. Rossant.)

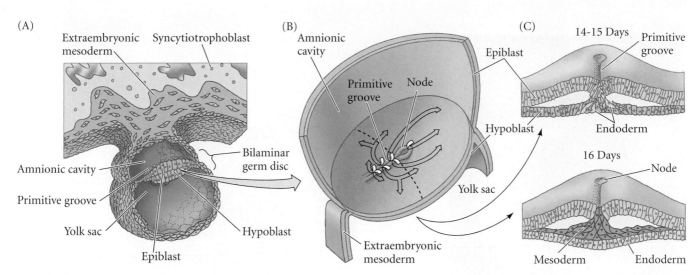

FIGURE 8.23 Amnion structure and cell movements during human gastrulation. (A,B) Human embryo and uterine connections at day 15 of gestation. (A) Sagittal section through the midline. (B) View looking down on the dorsal surface of the embryo. (C) Movements of the epiblast cells through the primitive streak and Hensen's node and underneath the epiblast are superimposed on the dorsal surface view. At days 14 and 15, the ingressing epiblast cells are thought to replace the hypoblast cells (which contribute to the yolk sac lining), while at day 16, the ingressing cells fan out to form the mesodermal layer. (After Larsen 1993.)

mouse epiblast (see Figure 1.11). Gastrulation begins at the posterior end of the embryo, and this is where the cells of the **node*** arise (**Figure 8.23**). Like the chick epiblast cells, the mammalian mesoderm and endoderm migrate through a primitive streak; also like their avian counterparts, the migrating cells of the mammalian epiblast lose E-cadherin, detach from their neighbors, and migrate through the streak as individual cells (Burdsal et al. 1993). Those cells arising from the node give rise to the notochord. However, in contrast to notochord formation in the chick, the cells that form the mouse notochord are thought to become integrated into the endoderm of the primitive gut (Jurand 1974; Sulik et al. 1994). These cells can be seen as a band of small, ciliated cells extending rostrally from the node. They form the notochord by converging medially and "budding" off in a dorsal direction from the roof of the gut.

Cell migration and specification appear to be coordinated by fibroblast growth factors. The cells of the primitive streak appear to be capable of both synthesizing and responding to FGFs (Sun et al. 1999; Ciruna and Rossant 2001). In embryos that are homozygous for the loss of the *fgf8* gene, cells fail to emigrate from the primitive streak, and neither mesoderm nor endoderm are formed. Fgf8 (and per-

haps other FGFs) probably control cell movement into the primitive streak by downregulating the E-cadherin that holds the epiblast cells together. Fgf8 may also control cell specification by regulating *snail*, *Brachyury* (*T*), and *Tbx6*, three genes that are essential (as they are in the chick embryo) for mesodermal migration, specification, and patterning.

The ectodermal precursors are located anterior and lateral to the fully extended primitive streak, as in the chick epiblast; however, in some instances (also as in the chick embryo), a single cell gives rise to descendants in more than one germ layer, or to both embryonic and extraembryonic derivatives. Thus, at the epiblast stage these lineages have not become fully separate from one another. As in avian embryos, the cells migrating into the space between the hypoblast and epiblast layers become coated with hyaluronic acid, which they synthesize as they leave the primitive streak. This substance keeps them separate while they migrate (Solursh and Morriss 1977). It is thought that the replacement of human hypoblast cells by endoderm precursors occurs on days 14–15 of gestation, while the migration of cells forming the mesoderm does not start until day 16 (see Figure 8.23C; Larsen 1993).

Formation of the extraembryonic membranes

While the embryonic epiblast is undergoing cell movements reminiscent of those seen in reptilian or avian gastrulation, the extraembryonic cells are making the placenta, a distinctly mammalian set of tissues that enable the fetus to survive within the maternal uterus. Although the initial trophoblast cells of mice and humans divide like most other cells of the body, they give rise to a population of cells in which nuclear division occurs in the absence of cytokinesis. The original trophoblast cells constitute a layer called the **cytotrophoblast**, whereas the multinucleated cell type forms the **syncytiotrophoblast**. The cytotrophoblast initially adheres to the endometrium through a series of adhesion molecules, as we saw above. Moreover,

---

*In mammalian development, Hensen's node is usually just called "the node," despite the fact that Hensen discovered this structure in rabbit and guinea pig embryos.

(A)

(B)

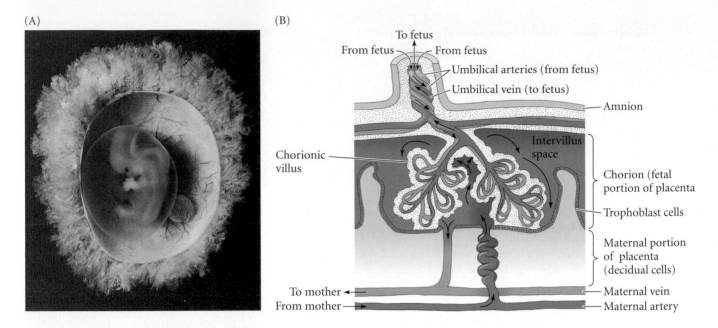

FIGURE 8.24 (A) Human embryo and placenta after 50 days of gestation. The embryo lies within the amnion, and its blood vessels can be seen extending into the chorionic villi. The small sphere to the right of the embryo is the yolk sac. (B) Relationship of the chorionic villi to the maternal blood supply in the primate uterus. In the umbilicus, there are two arteries and a single vein. (A from Carnegie Institution of Washington, courtesy of Chester F. Reather.)

cytotrophoblasts contain proteolytic enzymes that enable them to enter the uterine wall and remodel the uterine blood vessels so that the maternal blood bathes fetal blood vessels. The syncytiotrophoblast tissue is thought to further the progression of the embryo into the uterine wall by digesting uterine tissue. The cytotrophoblast secretes paracrine factors that attract maternal blood vessels and gradually displace their vascular tissue such that the vessels become lined with trophoblast cells (Fisher et al. 1989; Hemberger et al. 2003). Shortly thereafter, mesodermal tissue extends outward from the gastrulating embryo (see Figure 8.22E). Studies of human and rhesus monkey embryos have suggested that the yolk sac (and hence the hypoblast) as well as primitive streak-derived cells contribute this extraembryonic mesoderm (Bianchi et al. 1993).

The extraembryonic mesoderm joins the trophoblastic extensions and gives rise to the blood vessels that carry nutrients from the mother to the embryo. The narrow connecting stalk of extraembryonic mesoderm that links the embryo to the trophoblast eventually forms the vessels of the umbilical cord. The fully developed extraembryonic organ, consisting of trophoblast tissue and blood vessel-containing mesoderm, is the chorion, and it fuses with the uterine wall decidua to create the placenta. Thus, the placenta has both a maternal portion (the uterine endometrium, or decidua, which is modified during pregnancy) and a fetal

component (the chorion). The chorion may be very closely apposed to maternal tissues while still being readily separable from them (as in the contact placenta of the pig), or it may be so intimately integrated with maternal tissues that the two cannot be separated without damage to both the mother and the developing fetus (as in the deciduous placenta of most mammals, including humans).*

Figure 8.24A shows the relationships between the embryonic and extraembryonic tissues of a 6.5-week human embryo. The embryo is seen encased in the amnion and is further shielded by the chorion. The blood vessels extending to and from the chorion are readily observable, as are the villi that project from the outer surface of the chorion. These villi contain the blood vessels and allow the chorion to have a large area exposed to the maternal blood. Although fetal and maternal circulatory systems normally never merge, diffusion of soluble substances can occur through the villi (**Figure 8.24B**). In this manner, the mother provides the fetus with nutrients and oxygen, and the fetus sends its waste products (mainly carbon dioxide and urea) into the maternal circulation. The maternal and fetal blood cells usually do not mix, although a small number of fetal red blood cells are seen in the maternal blood circulation.

See WEBSITE 8.4 Placental functions

---

*There are numerous types of placentas, and the extraembryonic membranes form differently in different orders of mammals (see Cruz and Pedersen 1991). Although mice and humans gastrulate and implant in a similar fashion, their extraembryonic structures are distinctive. It is very risky to extrapolate developmental phenomena from one group of mammals to another. Even Leonardo da Vinci got caught (Renfree 1982). His remarkable drawing of the human fetus inside the placenta is stunning art, but poor science: the placenta is that of a cow.

SIDELIGHTS & SPECULATIONS

Twins and Chimeras

The early cells of the mammalian embryo can replace each other and compensate for a missing cell. This regulative ability was first demonstrated in 1952, when Seidel destroyed one cell of a 2-cell rabbit embryo and the remaining cell produced an entire embryo. Once the inner cell mass (ICM) has become separate from the trophoblast, the ICM cells constitute an equivalence group. In other words, each ICM cell has the same potency (in this case, each cell can give rise to all the cell types of the embryo, but not to the trophoblast), and their fates will be determined by interactions among their descendants. Gardner and Rossant (1976) also showed that if cells of the ICM (but not trophoblast cells) are injected into blas-tocysts, they contribute to the new embryo. Since the ICM blastomeres can generate any cell type in the body, the cells of the blastocyst are referred to as pluripotent.

The regulative capacity of the ICM blastomeres is also seen in humans. Human twins are classified into two major groups: **monozygotic** (**one-egg**, or **identical**) twins and **dizygotic** (**two-egg**, or **fraternal**) twins. Fraternal twins are the result of two separate fertilization events, whereas identical twins

See **WEBSITE 8.5**
Nonidentical monozygotic twins

See **WEBSITE 8.6**
Conjoined twins

are formed from a single embryo whose cells somehow become dissociated from one another. Identical twins may be produced by the separation of early blastomeres, or even by the separation of the inner cell mass into two regions within the same blastocyst.

Identical twins occur in roughly 0.25% of human births. About 33% of identical twins have two complete and separate chorions, indicating that separation occurred before the formation of the trophoblast tissue at day 5 (**Figure 8.25A**). The remaining identical twins share a common chorion, suggesting that the split occurred within

(Continued on next page)

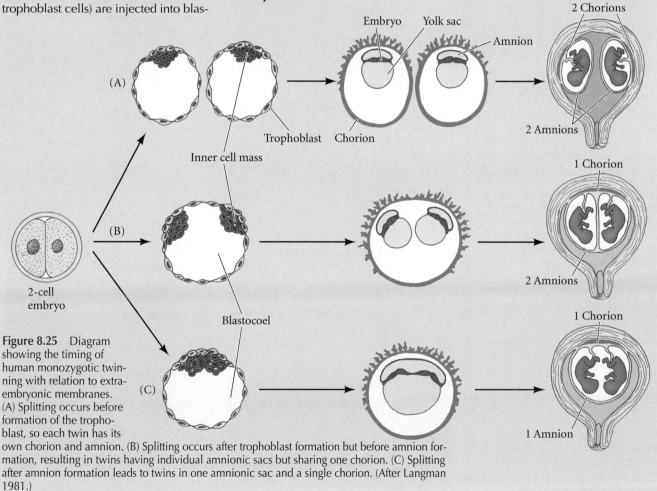

Figure 8.25 Diagram showing the timing of human monozygotic twinning with relation to extra-embryonic membranes. (A) Splitting occurs before formation of the tropho-blast, so each twin has its own chorion and amnion. (B) Splitting occurs after trophoblast formation but before amnion formation, resulting in twins having individual amnionic sacs but sharing one chorion. (C) Splitting after amnion formation leads to twins in one amnionic sac and a single chorion. (After Langman 1981.)

SIDELIGHTS & SPECULATIONS (Continued)

the inner cell mass after the trophoblast formed. By day 9, the human embryo has completed the construction of another extraembryonic layer, the lining of the amnion. This tissue, derived from the ectoderm and mesoderm, forms the amnionic sac, which surrounds the embryo with amnionic fluid and protects it from desiccation and abrupt movement. If the separation of the embryo were to come after the formation of the chorion on day 5 but before the formation of the amnion on day 9, then the resulting

embryos should have one chorion and two amnions (**Figure 8.25B**). This happens in about two-thirds of human identical twins. A small percentage of identical twins are born within a single chorion and amnion (**Figure 8.25C**). This means the division of the embryo came after day 9.

The ability to produce an entire embryo from cells that normally would have contributed to only a portion of the embryo—called *regulation*—was discussed in the Introduction to Part II. Regulation is also seen in the ability of

two or more early embryos to form one chimeric individual rather than twins, triplets, or a multiheaded individual. Chimeric mice can be produced by artificially aggregating two or more early-cleavage (usually 4- or 8-cell) embryos to form a composite embryo. As shown in **Figure 8.26A**, the zonae pellucidae of two genetically different embryos can be artificially removed and the embryos brought together to form a common blastocyst. These blastocysts are then implanted into the uterus of a foster mother. When they are born, the chimeric offspring have some cells from each embryo. This is readily seen when the aggregated blastomeres come from mouse strains that differ in their coat colors. When blastomeres from white and black strains are aggregated, the result is commonly a mouse with black and white bands. Markert and Petters (1978) have shown that three early 8-cell embryos can unite to form a common compacted morula and that the resulting mouse can have the coat colors of the three different strains. Moreover, they showed that each of the three embryos gave rise to precursors of the gametes. When a chimeric (black/brown/white) female mouse was mated to a white-furred (recessive) male, offspring of each of the three colors were produced (**Figure 8.26B**).

(A)

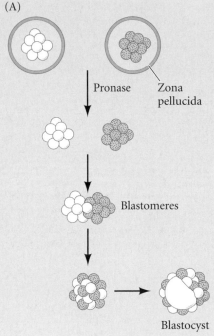

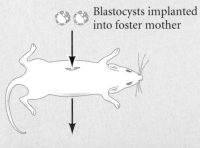

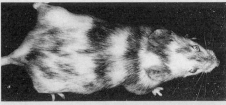

Figure 8.26 Production of chimeric mice. (A) Experimental procedures used to produce chimeric mice. Early 8-cell embryos of genetically distinct mice (here, with coat color differences) are isolated from mouse oviducts and brought together after their zonae are removed by proteolytic enzymes. The cells form a composite blastocyst, which is implanted into the uterus of a foster mother. The photograph shows one of the actual chimeric mice produced in this manner. (B) An adult female chimeric mouse (bottom) produced from the fusion of three 4-cell embryos: one from two white-furred parents, one from two black-furred parents, and one from two brown-furred parents. The resulting mouse has coat colors from all three embryos. Moreover, each embryo contributed germline cells, as is evidenced by the three colors of offspring (above) produced when this chimeric female was mated with recessive (white-furred) males. (A courtesy of B. Mintz; B from Markert and Petters 1978, courtesy of C. Markert.)

(B)

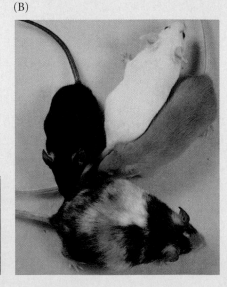

There is even evidence that human embryos can form chimeras (de la Chappelle et al. 1974; Mayr et al. 1979). Some individuals have two genetically different cell types (XX and XY) within the same body, each with its own set of genetically defined characteristics. The simplest explanation for such a phenomenon is that these individuals resulted from the aggregation of two embryos, one male and one female, that were developing at the same time. If this explanation is correct, then two fraternal twins have fused to create the individual* (see Yu et al. 2002).

According to these studies of twin formation and chimeric mice, each blastomere of the inner cell mass should be able to produce any cell of the body. This hypothesis has been confirmed, and it has important consequences for the study of mammalian development. When ICM cells are isolated and grown under certain conditions, they remain undifferentiated and continue to divide in culture (Evans and Kaufman 1981; Martin 1981). Such cells are **embryonic stem (ES) cells**. When ES cells are injected into a mouse blastocyst, they can integrate into the host inner cell mass. The resulting embryo has cells from both the host and the donor tissue. This technique has become extremely important in determining the function of genes during mammalian development.

*There are other explanations, at least for some of the chimeras. Souter and colleagues (2007) have shown that in at least one XX/XY chimera, the maternal alleles were identical and the paternal alleles differed. This would be expected if the egg underwent parthenogenic activation and each of the meiotic cells were fertilized by a different sperm (one bearing an X chromosome, one bearing a Y). The intermingling of the cells would produce the chimera, which in this case was a true hermaphrodite. We still do not know the mechanisms through which human twins and chimeras form.

Mammalian Axis Formation

The formation of the anterior-posterior axis has been studied most extensively in mice. The structure of the mouse epiblast, however, differs from that of humans in that it is cup-shaped rather than disc-shaped. The dorsal surface of the epiblast (the embryonic ectoderm) contacts the amnionic cavity, while the ventral surface of the epiblast contacts the newly formed mesoderm. In this cuplike arrangement, the endoderm covers the surface of the embryo on the "outside" of the cup (Figure 8.27A).

The anterior-posterior axis: Two signaling centers

The mammalian embryo appears to have two signaling centers: one in the node (equivalent to Hensen's node and the trunk portion of the amphibian organizer) and one in the anterior visceral endoderm (AVE; equivalent to the chick hypoblast and similar to the head portion of the amphibian organizer) (**Figure 8.27B**; Beddington and Robertson 1999; Foley et al. 2000). The node (at the "bottom of the cup" in the mouse) appears to be responsible for the creation of all of the body, and the two signaling centers work together to form the anterior region of the embryo (Bachiller et al. 2000). The notochord forms by the dorsal infolding of the small, ciliated cells of the node.

The AVE originates from the visceral endoderm (hypoblast) that migrates forward. As this region migrates, it secretes two antagonists of the Nodal protein, Lefty-1 and Cerberus (Brennan et al. 2001; Perea-Gomez et al. 2001; Yamamoto et al. 2004). (Lefty-1 binds to the Nodal's receptors and blocks Nodal binding, and Cerberus binds to Nodal itself.) While the Nodal proteins in the epiblast acti-vate the expression of posterior genes that are required for mesoderm formation, the AVE creates an anterior region where Nodal cannot act. The AVE also begins expressing the anterior markers Otx2 and Wnt-inhibitor Dickkopf. Studies of mutant mice indicate that the AVE promotes anterior specification by suppressing the formation of the primitive streak, a posterior structure, by Nodal and Wnt proteins, as in the chick (Bertocchini et al. 2002; Perea-Gomez et al. 2002). However, the AVE alone cannot induce neural tissue, as the node can (Tam and Steiner 1999).

Formation of the node is dependent on the trophoblast (Episkopou et al. 2001). As in the chick embryo, the placement of the node and the primitive streak appears to be due to the blocking of Nodal signaling by the Cerberus and Lefty-1 from the AVE (Perea-Gomez et al. 2002). Once formed, the node will secrete Chordin; the head process and notochord will later add Noggin. These two BMP antagonists are not expressed in the AVE. Dickkopf is expressed in both the AVE and in the node, but only the Dickkopf from the node is critical for head development (Mukhopadhyay et al. 2001). While knockouts of either the *chordin* or the *noggin* gene do not affect development, mice missing both genes lack a forebrain, nose, and other facial structures (**Figure 8.28**). It is probable that the AVE functions in the epiblast to restrict Nodal activity, thereby cooperating with the node-produced mesendoderm to promote the head-forming genes to be expressed in the anterior portion of the epiblast.

This regulation can be altered by retinoic acid (RA), which appears to inhibit the migration of the AVE precursors. As a result, more than one body axis can form (**Figure 8.27C,D**; Liao and Collins 2008). This effect has been proposed as a possible cause of conjoined twins.

(A)

(B)

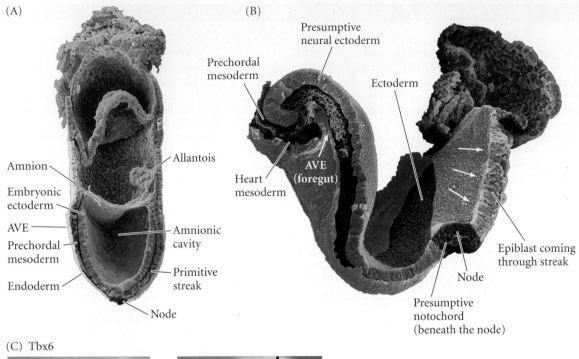

Presumptive
neural ectoderm

Prechordal
mesoderm

Ectoderm

Amnion

Allantois

Heart
mesoderm

AVE
(foregut)

Embryonic
ectoderm

AVE

Prechordal
mesoderm

Amnionic
cavity

Epiblast coming
through streak

Endoderm

Primitive
streak

Node

Presumptive
notochord
(beneath the node)

Node

(C) Tbx6

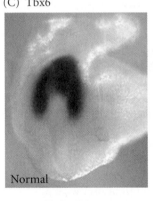
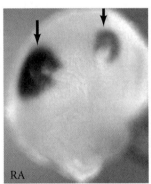

Normal

RA

(D) Sonic hedgehog

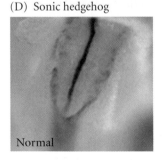
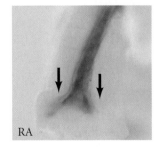

Normal

RA

FIGURE 8.27 Axis and notochord formation in the mouse. (A) In the 7-day mouse embryo, the dorsal surface of the epiblast (embryonic ectoderm) is in contact with the amnionic cavity. The ventral surface of the epiblast contacts the newly formed mesoderm. In this cuplike arrangement, the endoderm covers the surface of the embryo. The node is at the bottom of the cup, and it has generated chordamesoderm. The two signaling centers, the node and the anterior visceral endoderm (AVE), are located on opposite sides of the cup. Eventually, the notochord will link them. The caudal side of the embryo is marked by the presence of the allantois. (B) By embryonic day 8, the AVE lines the foregut, and the prechordal mesoderm is now in contact with the forebrain ectoderm. The node is now farther caudal, largely as a result of the rapid growth of the anterior portion of the embryo. The cells in the midline of the epiblast migrate through the primitive streak (white arrows). (C,D) Retinoic acid appears to inhibit the migration of the AVE precursors. In normal mouse embryos (left column), Tbx6 is expressed in the anterior primitive streak and epiblast surrounding the node (C), while Sonic hedgehog is expressed in the notochord (D). In embryos treated with retinoic acid during early gastrulation (right column), there are often two axes. Here, two areas of Tbx6 are observed, and the notochord is bifurcated (arrows). (A,B courtesy of K. Sulik; C,D from Liao and Collins 2008, courtesy of K. Sulik and G. Schoenwolf.)

Anterior-posterior patterning by FGF and retinoic acid gradients

The head region of the mammalian embryo is devoid of Nodal signaling, and BMPs, FGFs, and Wnts are also inhibited. The posterior region is characterized by Nodal, BMPs, Wnts, FGFs, and retinoic acid. There appears to be a gradient of Wnt, BMP, and FGF proteins that is highest in the posterior and that drops off strongly near the anterior region.

Moreover, in the anterior half of the embryo, starting at the node, there is a high concentration of antagonists that prevent BMPs and Wnts from acting (**Figure 8.29A**). The Fgf8 gradient is created by the decay of mRNA: *fgf8* is expressed at the growing posterior tip of the embryo, but the *fgf8* message is slowly degraded in the newly formed tissues. Thus there is a gradient of *fgf8* mRNA across the posterior of the embryo, which is then converted into an Fgf8 protein gradient (**Figure 8.29B**; Dubrulle and Pourquié 2004).

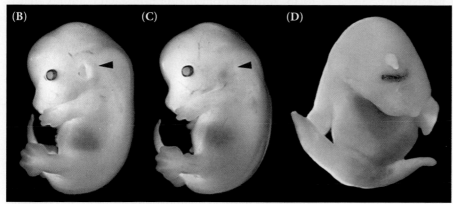

FIGURE 8.28 Expression of BMP antagonists in the mammalian node. (A) Expression of chordin during mouse gastrulation is seen in the anterior primitive streak, node, and axial mesoderm. It is not expressed in the anterior visceral endoderm. (B–D) Phenotypes of 12.5-day embryos. (B) Wild-type embryo. (C) Embryo with the *chordin* gene knocked out has a defective ear but an otherwise normal head. (D) Phenotype of a mouse deficient in both *chordin* and *noggin*. There is no jaw, and there is a single centrally located eye, over which protrudes a large proboscis (nose). (From Bachiller et al. 2000, courtesy of E. M. De Robertis.)

In addition to FGFs, the late gastrula has a gradient of retinoic acid. RA levels are high in the posterior regions and low in the anterior portions of the embryo. This gradient (like that of chick, frog, and fish embryos) appears to be controlled by the expression of RA-synthesizing enzymes in the embryo's posterior and RA-degrading enzymes in the anterior parts of the embryo (Sakai et al. 2001; Oosterveen et al. 2004).

The FGF gradient patterns the posterior portion of the embryo by working through the Cdx family of caudal-related genes (**Figure 8.29C**; Lohnes 2003). The Cdx genes, in turn, integrate the various posteriorization signals and activate particular Hox genes.

(A)

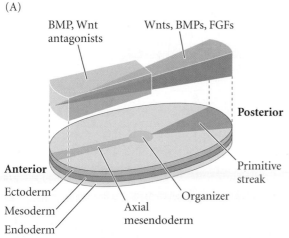

(B)

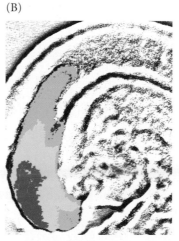

(C)

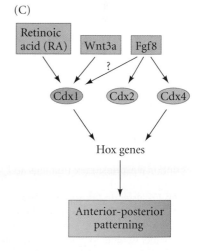

FIGURE 8.29 Anterior-posterior patterning in the mouse embryo. (A) Concentration gradients of BMPs, Wnts, and FGFs in the late gastrula mouse embryo (depicted as a flattened disc). The primitive streak and other posterior tissues are the sources of Wnt and BMP proteins, whereas the organizer and its derivatives (such as the notochord) produce antagonists. Fgf8 is expressed in the posterior tip of the gastrula and continues to be made in the tail bud. Its mRNA decays, creating a gradient across the posterior portion of the embryo. (B) Fgf8 gradient in the tailbud region of a 9-day mouse embryo. The highest amount of Fgf8 (red) is found near the tip. The gradient was determined by in situ hybridization of an Fgf8 probe and staining for increasing amounts of time. (C) Retinoic acid, Wnt3a, and Fgf8 each contribute to posterior patterning, but they are integrated by the Cdx family of proteins that regulates the activity of the Hox genes. (A after Robb and Tam 2004; B from Dubrulle and Pourquié 2004, courtesy of O. Pourquié; C after Lohnes 2003.)

Anterior-posterior patterning: The Hox code hypothesis

In all vertebrates, anterior-posterior polarity becomes specified by the expression of Hox genes. Hox genes are homologous to the homeotic selector genes (Hom-C genes) of the fruit fly (see Chapter 6). The *Drosophila* homeotic gene complex on chromosome 3 contains the *Antennapedia* and *bithorax* clusters of homeotic genes (see Figure 6.35) and can be seen as a single functional unit. (Indeed, in some other insects, such as the flour beetle *Tribolium*, it is a single physical unit.) All of the known mammalian genomes contain four copies of the Hox complex per haploid set, located on four different chromosomes (*Hoxa* through *Hoxd* in the mouse, *HOXA* through *HOXD* in humans; see Boncinelli et al. 1988; McGinnis and Krumlauf 1992; Scott 1992).

The order of these genes on their respective chromosomes is remarkably similar between insects and humans, as is the expression pattern of these genes. Those mammalian genes homologous to the *Drosophila labial, proboscipedia,* and *deformed* genes are expressed anteriorly and early, while those genes homologous to the *Drosophila AbdB* gene are expressed posteriorly and later. As in *Drosophila*, a separate set of genes in mice encodes the transcription factors that regulate head formation. In *Drosophila*, these are the *orthodenticle* and *empty spiracles* genes. In mice, the midbrain and forebrain are made through the expression of genes homologous to these—*Otx2* and *Emx* (see Kurokawa et al. 2004; Simeone 2004).

The mammalian Hox/HOX genes are numbered from 1 to 13, starting from that end of each complex that is expressed most anteriorly. **Figure 8.30** shows the relationships between the *Drosophila* and mouse homeotic gene sets. The equivalent genes in each mouse complex (such as *Hoxa1, Hoxb1,* and *Hoxd1*) are called **paralogues**. It is thought that the four mammalian Hox complexes were formed by chromosome duplications. Because the correspondence between the *Drosophila* Hom-C genes and mouse Hox genes is not one-to-one, it is likely that independent gene duplications and deletions have occurred since these two animal groups diverged (Hunt and Krumlauf 1992). Indeed, the most posterior mouse Hox gene (equivalent to *Drosophila AbdB*) underwent its own set of duplications in some mammalian chromosomes (see Figure 8.30).

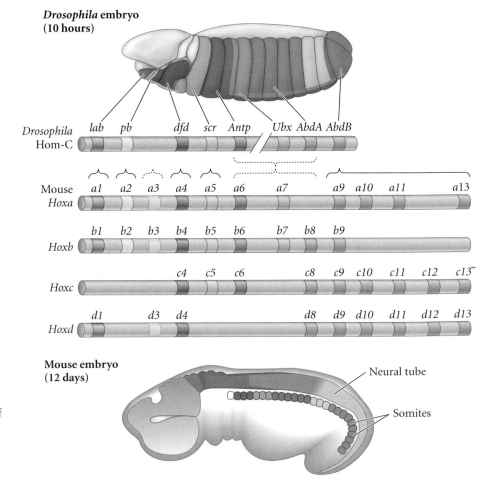

FIGURE 8.30 Evolutionary conservation of homeotic gene organization and transcriptional expression in fruit flies and mice is seen in the similarity between the Hom-C cluster on *Drosophila* chromosome 3 and the four Hox gene clusters in the mouse genome. The mouse genes of the higher numbered paralogous groups are those that are expressed later and more posteriorly. Genes having similar structures occupy the same relative positions on each of the four chromosomes, and paralogous gene groups display similar expression patterns. The comparison of the transcription patterns of the Hom-C and *Hoxb* genes of *Drosophila* and mice are shown above and below the chromosomes, respectively. (After Carroll 1995.)

(A)

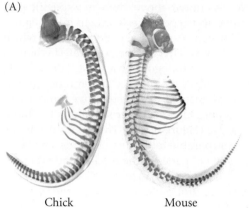

Chick   Mouse

FIGURE 8.31 Schematic representation of the chick and mouse vertebral pattern along the anterior-posterior axis. (A) Axial skeletons stained with Alcian blue at comparable stages of development. The chick has twice as many cervical vertebrae as the mouse. (B) The boundaries of expression of certain Hox gene paralogous groups (*Hox5/6* and *Hox9/10*) have been mapped onto the vertebral type domains. (A from Kmita and Duboule 2003, courtesy of M. Kmita and D. Duboule; B after Burke et al. 1995.)

(B)

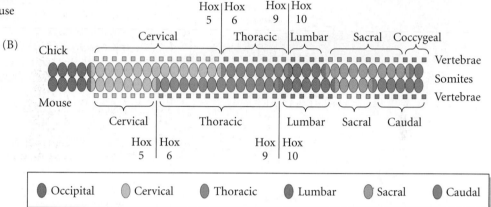

EXPRESSION OF HOX GENES ALONG THE DORSAL AXIS

Hox gene expression can be seen along the mammalian dorsal axis (in the neural tube, neural crest, paraxial mesoderm, and surface ectoderm) from the anterior boundary of the hindbrain through the tail. The regions of expression are not in register, but the 3′ Hox genes (homologous to *labial*, *proboscopedia*, and *deformed* of the fly) are expressed more anteriorly than the 5′ Hox genes (homologous to *Ubx*, *abdA*, and *AbdB*). Thus, one generally finds the genes of paralogous group 4 expressed anteriorly to those of paralogous group 5, and so forth (see Figure 8.30; Wilkinson et al. 1989; Keynes and Lumsden 1990). Mutations in the Hox genes suggest that the level of the body along the anterior-posterior axis is determined primarily by the most posterior Hox gene expressed in that region.

Experimental analysis of the Hox code

The expression patterns of mouse Hox genes suggest a code whereby certain combinations of Hox genes specify a particular region of the anterior-posterior axis (Hunt and Krumlauf 1991). Particular sets of paralogous genes provide segmental identity along the anterior-posterior axis of the body. Evidence for such a code comes from three sources:

- Comparative anatomy, in which the types of vertebrae in different vertebrate species are correlated with the constellation of Hox gene expression

- Gene targeting or "knockout" experiments in which mice are constructed that lack both copies of one or more Hox genes

- Retinoic acid teratogenesis, in which mouse embryos exposed to RA show an atypical pattern of Hox gene expression along the anterior-posterior axis and abnormal differentiation of their axial structures

COMPARATIVE ANATOMY A new type of comparative embryology is emerging based on the comparison of gene expression patterns among species. Gaunt (1994) and Burke and her collaborators (1995) have compared the vertebrae of the mouse and the chick (**Figure 8.31A**). Although the mouse and chick have a similar number of vertebrae, they apportion them differently. Mice (like all mammals, be they giraffes or whales) have 7 cervical (neck) vertebrae. These are followed by 13 thoracic (rib) vertebrae, 6 lumbar (abdominal) vertebrae, 4 sacral (hip) vertebrae, and a variable (20+) number of caudal (tail) vertebrae. The chick, on the other hand, has 14 cervical vertebrae, 7 thoracic vertebrae, 12 or 13 (depending on the strain) lumbosacral vertebrae, and 5 coccygeal (fused tail) vertebrae. The researchers asked, Does the constellation of Hox gene expression correlate with the type of vertebra formed (e.g., cervical or thoracic) or with the relative position of the vertebrae (e.g., number 8 or 9)?

The answer is that the constellation of Hox gene expression predicts the type of vertebra formed. In the mouse,

(A)

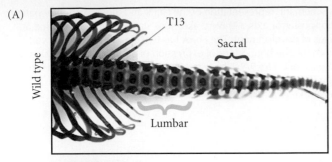

(B)

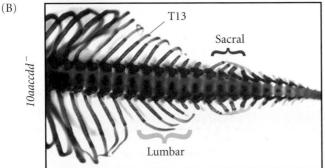

(C)

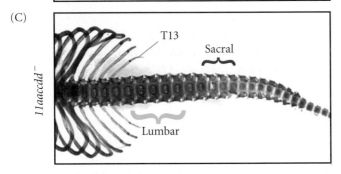

FIGURE 8.32 Axial skeletons of mice in gene knockout experiments. Each photograph is of an 18.5-day embryo, looking upward at the ventral region from the middle of the thorax toward the tail. (A) Wild-type mouse. (B) Complete knockout of *Hox10* paralogues (*Hox10aaccdd*) converts lumbar vertebrae (after the thirteenth thoracic vertebrae) into ribbed thoracic vertebrae. (C) Complete knockout of *Hox11* paralogues (*Hox11aaccdd*) transforms the sacral vertebrae into copies of lumbar vertebrae. (After Wellik and Capecchi 2003, courtesy of M. Capecchi.)

the transition between cervical and thoracic vertebrae is between vertebrae 7 and 8; in the chick, it is between vertebrae 14 and 15 (**Figure 8.31B**). In both cases, the *Hox5* paralogues are expressed in the last cervical vertebra, while the anterior boundary of the *Hox6* paralogues extends to the first thoracic vertebra. Similarly, in both animals, the thoracic-lumbar transition is seen at the boundary between the *Hox9* and *Hox10* paralogous groups. It appears there is a code of differing Hox gene expression along the anterior-posterior axis, and that code determines the type of vertebra formed.

GENE TARGETING As noted above, there is a specific pattern to the number and type of vertebrae in mice (**Figure 8.32A**), and the Hox gene expression pattern dictates which vertebral type will form. This was demonstrated when all six copies of the *Hox10* paralogous group (i.e., *Hoxa10, c10,* and *d10* in Figure 8.30) were knocked out and no lumbar vertebrae developed. Instead, the presumptive lumbar vertebrae formed ribs and other characteristics similar to those of thoracic vertebrae (**Figure 8.32B**). This was a homeotic transformation comparable to those seen in insects; however, the redundancy of genes in the mouse made it much more difficult to produce, because the existence of even one copy of the *Hox10* group genes prevented the transformation (Wellik and Capecchi 2003; Wellik 2009). Similarly, when all six copies of the *Hox11* group were knocked out, the thoracic and lumbar vertebrae were normal, but the sacral vertebrae failed to form and were replaced by lumbar vertebrae (**Figure 8.32C**).

RETINOIC ACID TERATOGENESIS Homeotic changes are also seen when mouse embryos are exposed to teratogenic doses of retinoic acid, a derivative of vitamin A. As we saw earlier, by day 7 of development, an RA gradient has been established that is high in the posterior regions and low in the anterior portions of the embryo. This gradient appears to be controlled by the differential synthesis or degradation of RA in the different parts of the embryo. Hox genes are responsive to retinoic acid either by virtue of having RA receptor sites in their enhancers or by being responsive to *Cdx* (the mammalian homologue of the *Drosophila caudal* gene), which is activated by RA (Conlon and Rossant 1992; Kessel 1992; Sakai et al. 2001; Lohnes 2003).

Exogenous retinoic acid can mimic the RA concentrations normally encountered only by the posterior cells, and high doses of RA can activate Hox genes in more anterior locations along the anterior-posterior axis (Kessel and Gruss 1991; Allan et al. 2001). Thus, when excess retinoic acid is administered to mouse embryos on day 8 of gestation, shifts in Hox gene expression occur such that the last cervical vertebra is turned into a thoracic (ribbed) vertebra. Conversely, impairment of retinoic acid function causes Hox gene expression to become more posterior, and the first thoracic vertebra becomes a copy of the cervical vertebrae (**Figure 8.33**).

The Dorsal-Ventral and Left-Right Axes

The dorsal-ventral axis

Very little is known about the mechanisms of dorsal-ventral axis formation in mammals. After the fifth cell division in the mouse embryo, the blastocyst cavity begins to form, and the inner cell mass resides on one side of this cavity. This axis is probably created by the ellipsoidal shape of the zona pellucida (Kurotaki et al. 2007). In mice and humans, the hypoblast forms on the side of the inner cell mass that

(A) Normal development   (B) Retinoic acid (RA)   (C) Hox genes

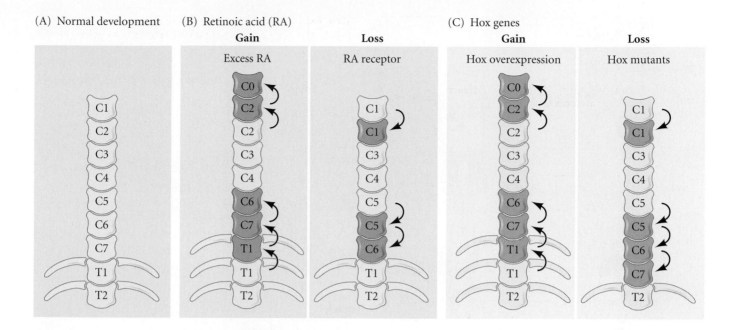

FIGURE 8.33   Effect of retinoic acid on mouse embryos. Addition of exogenous RA leads to differences in Hox gene expression and the transformation of vertebral form. Similarly, loss of RA function (through mutations in the RA receptors) can lead to the opposite type of transformations. (After Houle et al. 2003.)

is exposed to the blastocyst fluid, while the dorsal axis forms from those ICM cells that are in contact with the trophoblast and amnionic cavity. Thus, the dorsal-ventral axis of the embryo is defined, in part, by the embryonic-abembryonic axis of the blastocyst. The embryonic region contains the ICM, while the abembryonic region is that part of the blastocyst opposite the ICM. The first dorsal-ventral polarity is seen at the blastocyst stage, and as development proceeds, the primitive streak maintains this polarity by causing migration ventrally from the dorsal surface of the embryo (Goulding et al. 1993).

The left-right axis

The mammalian body is not symmetrical. Although the human heart begins its formation at the midline of the embryo, it moves to the left side of the chest cavity and loops to the right. The spleen is found solely on the left side of the abdomen, the major lobe of the liver forms on the right side of the abdomen, the large intestine loops right to left as it traverses the abdominal cavity, and the right lung has one more lobe than the left lung (**Figure 8.34**).

Mutations in mice have shown that there are two levels of regulation of the left-right axis: a global level and an organ-specific level. Mutation of the gene *situs inversus viscerum* (*iv*) randomizes the left-right axis for each asymmetrical organ independently (Hummel and Chapman 1959;

Layton 1976). This means that the heart may loop to the left in one homozygous animal but to the right in another. Moreover, the direction of heart looping is not coordinated with the placement of the spleen or stomach. This lack of coordination can cause serious problems, even death. A second gene, *inversion of embryonic turning* (*inv*), causes a more global phenotype. Mice homozygous for an insertion mutation at this locus had all their asymmetrical organs on the wrong side of the body (Yokoyama et al. 1993).* Since all the organs were reversed, this asymmetry did not have dire consequences for the mice.

Several additional asymmetrically expressed genes have recently been discovered, and their influence on one another has enabled scientists to arrange them into a possible pathway. The end of this pathway—the activation of Nodal proteins and the Pitx2 transcription factor on the left side of the lateral plate mesoderm—appears to be the same in mammals as in other vertebrate embryos, although the path leading to this point differs among the species (see Figure 8.14; Collignon et al. 1996; Lowe et al. 1996; Meno et al. 1996). In mammals, the distinction between left and right sides begins in the ciliary cells of the node (**Figure 8.35A**). The cilia cause fluid in the node to flow from right to left. When Nonaka and colleagues (1998) knocked out a mouse gene encoding the ciliary motor protein dynein (see Chapter 4), the nodal cilia did

---

*The *inv* gene was discovered accidentally when Yokoyama and colleagues randomly inserted the transgene for the tyrosinase enzyme into mouse genomes. In one instance, the transgene inserted itself into a region of chromosome 4, knocking out the existing *inv* gene. The resulting homozygous mice had laterality defects.

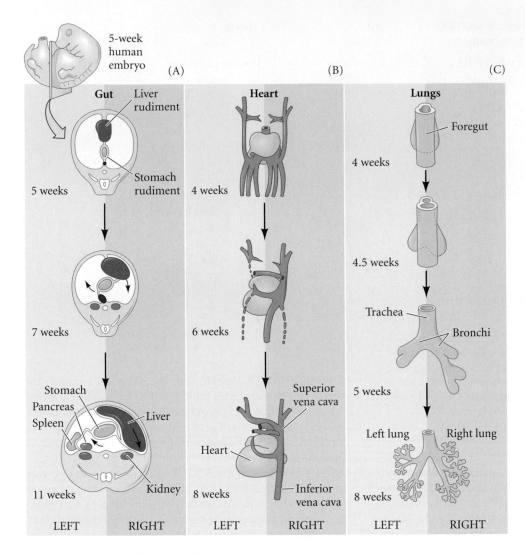

FIGURE 8.34 Left-right asymmetry in the developing human. (A) Abdominal cross sections show that the originally symmetrical organ rudiments acquire asymmetric positions by week 11. The liver moves to the right and the spleen moves to the left. (B) Not only does the heart move to the left side of the body, but the originally symmetrical veins of the heart regress differentially to form the superior and inferior venae cavae, which connect only to the right side of the heart. (C) The right lung branches into three lobes, while the left lung (near the heart) forms only two lobes. In human males, the scrotum also forms asymmetrically. (After Kosaki and Casey 1998.)

not move and the situs (lateral position) of each asymmetrical organ was randomized.

This finding correlated extremely well with other data. First, it had long been known that humans with a dynein deficiency had immotile cilia and a random chance of having their hearts on the left or right side of the body (Afzelius 1976). Second, when the *iv* gene described above was cloned, it was found to encode the ciliary dynein protein (Supp et al. 1997). Third, when Nonaka and colleagues (2002) cultured early mouse embryos under an artificial

flow of medium from left to right, they obtained a reversal of the left-right axis. Moreover, the flow was able to direct the polarity of the left-right axis in *iv*-mutant mice, whose cilia are otherwise immotile.

Why should fluid flow be all-important to establishing left-right asymmetry? The reason may reside in small (around 1 μm), membrane-bound particles called **nodal vesicular parcels** (**NVPs**). These "parcels," which contain Sonic hedgehog protein and retinoic acid, are secreted from the node cells under the influence of FGF signals (**Figure**

(A)

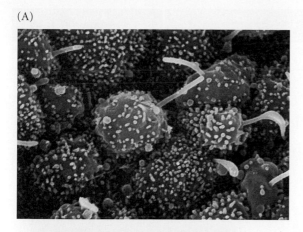

(B)

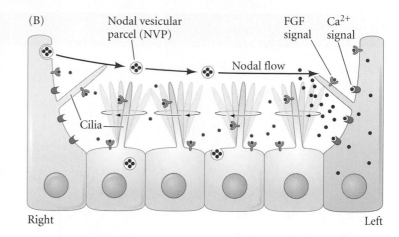

Nodal vesicular parcel (NVP)

FGF signal   Ca²⁺ signal

Nodal flow

Cilia

Right   Left

(C)

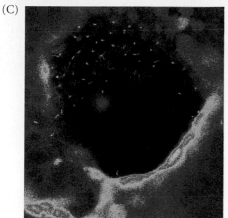

Right   Left

FIGURE 8.35 Laterality axis formation in mammals. (A) Ciliated cells of the mammalian node. (B) Schematic drawing showing the FGF-induced secretion of nodal vesicular parcels from the cells of the node, the movement of the NVPs to the left side, motivated by the ciliary currents, and the rise in Ca²⁺ concentration on the left side of the node. (C) Calcium ions (red, green) concentrated on the left side of the node in mice. (A courtesy of K. Sulik and G. C. Schoenwolf; B after Tanaka et al. 2005; C courtesy of M. Bueckner.)

8.35B). It appears that ciliary flow carries the NVPs to the left side of the body; if FGF signaling is inhibited, the parcels are not secreted and left-right asymmetry fails to become established (Tanaka et al. 2005). Such a method of delivering paracrine factors from one set of cells to another represents a newly discovered mode of signaling.

One of the results of the transport of the NVPs is the rise of calcium ions on the left side of the node (**Figure 8.35C**; Levin 2003; McGrath et al. 2003). It is yet to be discovered how the expression of genes such as *nodal* become placed under the control of these ion fluxes; but we are beginning to understand the differences between right and left.

Coda

Variations on the important themes of development have evolved in the different vertebrate groups (**Figure 8.36**). The major themes of vertebrate gastrulation include:

1. Internalization of the endoderm and mesoderm
2. Epiboly of the ectoderm around the entire embryo
3. Convergence of the internal cells to the midline
4. Extension of the body along the anterior-posterior axis

Although fish, amphibian, avian, and mammalian embryos have different patterns of cleavage and gastrulation, they use many of the same molecules to accomplish the same goals. Each group uses gradients of Nodal proteins to establish polarity along the dorsal-ventral axis. In *Xenopus* and zebrafish, maternal factors induce Nodal proteins in the vegetal hemisphere or marginal zone. In the chick, Nodal expression is induced by Wnt and Vg1 emanating from the posterior marginal zone, while elsewhere Nodal activity is suppressed by the hypoblast. In the mouse, the hypoblast similarly restricts Nodal activity, although the source of its ability to do so remains uncertain.

Each of these vertebrate groups uses BMP inhibitors to specify the dorsal axis; however, they use them in different ways. Similarly, Wnt inhibition and Otx2 expression are important in specifying the anterior regions of the embryo, but different groups of cells may be expressing these proteins. In all cases, the region of the body from the hindbrain to the tail is specified by Hox genes. Finally, the left-right axis is established through the expression of Nodal on the left-hand side of the embryo. Nodal activates Pitx2, leading to the differences between the left and right sides of the embryo. How Nodal becomes expressed on the left side appears to differ among the vertebrate groups. But overall, despite their initial differences in cleavage and gastrulation, the vertebrates have maintained very similar ways of establishing the three body axes.

STAGE	Zebrafish *Danio rerio*	Frog *Xenopus laevis*	Chicken *Gallus gallus*	Mouse *Mus musculus*
Early cleavage				
Late cleavage	NC Yolk cell	Blastocoel NC	Area pellucida Epiblast NC Area opaca Primary hypoblast Secondary hypoblast	Extra-embryonic ectoderm Epiblast Hypoblast
Early gastrula	Epiboly SMO Blastopore	Epiboly Blastocoel Blastopore SMO	Epiblast SMO Hypoblast	SMO
Late gastrula	Forebrain Midbrain Hind-brain Pre-somitic mesoderm Prechordal plate mesoderm Epidermis Gut endoderm Notochord Somite Spinal cord SMO	Prechordal plate mesoderm Forebrain Midbrain Hindbrain Somite Gut endoderm Notochord Spinal cord Post-somitic mesoderm SMO	Epidermis Forebrain Midbrain Hindbrain Somite Spinal cord SMO Prechordal plate mesoderm Notochord Gut endoderm Pre-somitic mesoderm	Prechordal plate mesoderm Forebrain Epidermis Epiblast Mid-brain Hindbrain Gut endoderm Somite SMO Spinal cord
Pharyngula				

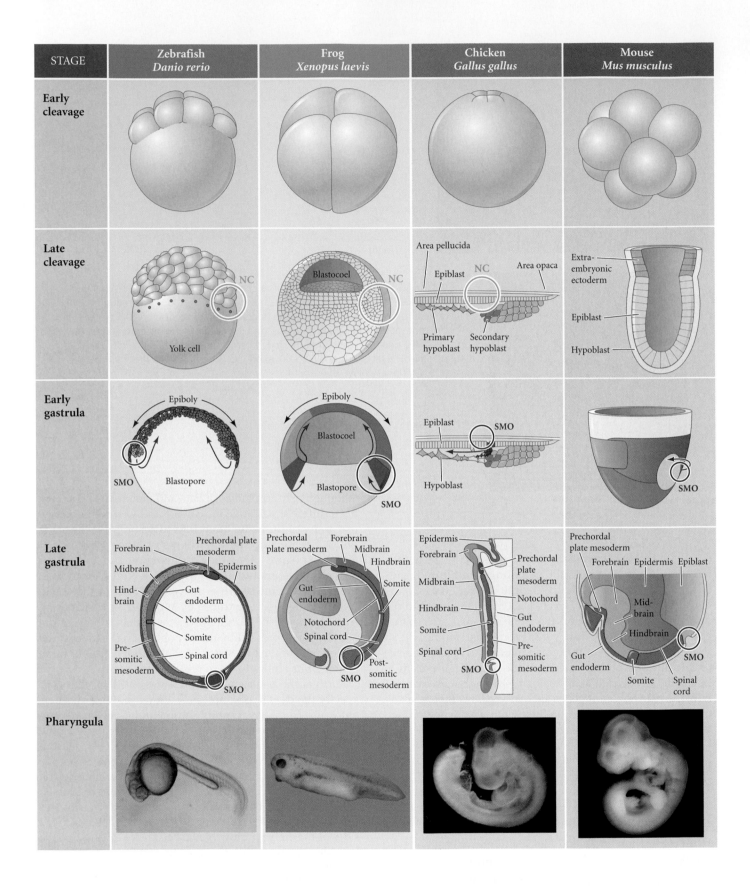

◀ **FIGURE 8.36** Early development of four vertebrate classes. Cleavage differs greatly among the four groups. Zebrafish and chicks have meroblastic discoidal cleavage; frogs have unequal holoblastic cleavage; and mammals have equal holoblastic cleavage. These cleavage patterns form different structures, but there are many conserved features, such as the Nieuwkoop center (NC; green circles). As gastrulation begins, each of the groups has cells equivalent to the Spemann-Mangold organizer (SMO; red circles). The SMO marks the beginning of the blastopore region, and the remainder of the blastopore is indicated by the red arrows extending from the organizer. By the late gastrula stage, the endoderm (yellow) is inside the embryo, the ectoderm (blue, purple) surrounds the embryo, and the mesoderm (red) is between the endoderm and ectoderm. The regionalization of the mesoderm has also begun. The bottom row shows the pharyngula stage that immediately follows gastrulation. This stage, with a pharynx, a central neural tube and notochord flanked by somites, and a sensory cephalic (head) region, characterizes the vertebrates. (After Solnica-Krezel 2005.)

 Snapshot Summary: *Early Development of Birds and Mammals*

1. Reptiles and birds, like fish, undergo discoidal meroblastic cleavage, wherein the early cell divisions do not cut through the yolk of the egg. These early cells form a blastoderm.

2. In each class of vertebrates, neural ectoderm is permitted to form where the BMP-mediated induction of epidermal tissue is prevented.

3. In chick embryos, early cleavage forms an area opaca and an area pellucida. The region between them is the marginal zone. Gastrulation begins in the area pellucida next to the posterior marginal zone, as the hypoblast and primitive streak both start there.

4. The primitive streak is derived from anterior epiblast cells and the central cells of Koller's sickle. As the primitive streak extends rostrally, Hensen's node is formed. Cells migrating out of Hensen's node become prechordal mesendoderm and are followed by the head process and notochord cells.

5. The prechordal plate helps induce formation of the forebrain; the chordamesoderm induces formation of the midbrain, hindbrain, and spinal cord. The first cells migrating laterally through the primitive streak become endoderm, displacing the hypoblast. The mesoderm cells then migrate through the primitive streak. Meanwhile, the surface ectoderm undergoes epiboly around the yolk.

6. In birds, gravity helps determine the position of the primitive streak (which defines the future anterior-posterior axis). The left-right axis is formed by the expression of Nodal protein on the left side of the embryo, which signals Pitx2 expression on the left side of developing organs.

7. Mammals undergo a variation of holoblastic rotational cleavage that is characterized by a slow rate of cell division, a unique cleavage orientation, lack of divisional synchrony, and formation of a blastocyst.

8. The blastocyst forms after the blastomeres undergo compaction. It contains outer cells—the trophoblast cells—that become the chorion, and an inner cell mass that becomes the amnion and the embryo.

9. The chorion forms the fetal portion of the placenta, which functions to provide oxygen and nutrition to the embryo, to provide hormones for the maintenance of pregnancy, and to provide barriers to the mother's immune system.

10. Mammalian gastrulation is not unlike that of birds. There appear to be two signaling centers, one in the node and one in the anterior visceral endoderm. The latter center is critical for head development, while the former is critical in inducing the nervous system and in patterning axial structures caudally from the midbrain.

11. Hox genes pattern the anterior-posterior axis and help specify positions along that axis. If Hox genes are knocked out, segment-specific malformations can arise. Similarly, causing the ectopic expression of Hox genes can alter the body axis.

12. The homology of gene structure and the similarity of expression patterns between *Drosophila* and mammalian Hox genes suggest that this patterning mechanism is extremely ancient.

13. The mammalian left-right axis is specified similarly to that of the chick, but with some significant differences in the roles of certain genes.

14. In amniote gastrulation, the pluripotent epithelium, called the epiblast, produces the mesoderm and endoderm (which migrate through the primitive streak), while the precursors of the ectoderm remain on the surface. By the end of gastrulation, the head and anterior trunk structures are formed. Elongation of the embryo continues through precursor cells in the caudal epiblast surrounding the posteriorized Hensen's node.

For Further Reading

Complete bibliographical citations for all literature cited in this chapter can be found at the free-access website **www.devbio.com**

Beddington, R. S. P. and E. J. Robertson. 1999. Axis development and early asymmetry in mammals. *Cell* 96: 195–209.

Bertocchini, F. and C. D. Stern. 2002. The hypoblast of the chick embryo positions the primitive streak by antagonizing Nodal signaling. *Dev. Cell* 3: 735–744.

Burke, A. C., A. C. Nelson, B. A. Morgan and C. Tabin. 1995. Hox genes and the evolution of vertebrate axial morphology. *Development* 121: 333–346.

Chuai, M. and C. J. Weijer. 2008. The mechanisms underlying prinitive streak formation in the chick embryo. *Curr. Top. Dev. Bio.* 81: 135–156.

Markert, C. L. and R. M. Petters. 1978. Manufactured hexaparental mice show that adults are derived from three embryonic cells. *Science* 202: 56–58.

Sheng, G., M. dos Reis and C. D. Stern. 2003. Churchill, a zinc finger transcriptional activator, regulates the transition between gastrulation and neurulation. *Cell* 115: 603–613.

Strumpf, D., C.-A. Mao, Y. Yamanaka, A. Ralston, K. Chawengsaksophak, F. Beck and J. Rossant. 2005. Cdx2 is required for correct cell fate specification and differentiation of trophectoderm in the mouse blastocyst. *Development* 132: 2093–2102.

Go Online

WEBSITE 8.1 Epiblast cell heterogeneity. Although the early epiblast appears uniform, different cells have different molecules on their cell surfaces. This variability allows some of them to remain in the epiblast while others migrate into the embryo.

WEBSITE 8.2 Mechanisms of compaction and formation of the inner cell mass. What determines whether a cell is to become a trophoblast cell or a member of the inner cell mass? It may just be a matter of chance. However, once the decision is made, different genes are switched on.

WEBSITE 8.3 Human cleavage and compaction. XY blastomeres have a slight growth advantage that may have had profound effects on in vitro fertility operations.

WEBSITE 8.4 Placental functions. Placentas are nutritional, endocrine, and immunological organs. They provide hormones that enable the uterus to retain the pregnancy and also accelerate mammary gland development. Placentas also block the potential immune response of the mother against the developing fetus. Recent studies suggest that the placenta uses several mechanisms to block the mother's immune response.

WEBSITE 8.5 Nonidentical monozygotic twins. Although monozygotic twins have the same genome, random developmental factors or the uterine environment may give them dramatically different phenotypes.

WEBSITE 8.6 Conjoined twins. There are rare events in which more than one set of axes is induced in the same embryo. These events can produce conjoined twins—twins that share some parts of their bodies. The medical and social issues raised by conjoined twins provide a fascinating look at what people throughout history have considered "individuality."

Vade Mecum

Chick development. Viewing these movies of 3-D models of chick cleavage and gastrulation will help you understand these phenomena.

Outside Sites

The University of North Carolina School of Medicine has developed an excellent site for medical embryology. (**http://www.med.unc.edu/embryo_images/**)

The Edinburgh Mouse Atlas Project is a remarkable database for both anatomy and gene expression of Burns's "wee little beastie." (**http://genex.hgu.mrc.ac.uk/**)

The University of New South Wales Embryology Course, constructed by Mark Hill, serves as a reference for human developmental anatomy, with connections to websites for other organisms. (**http://embryology.med.unsw.edu.au/**)

The Visible Embryo shows the progress of human development, using the Carnegie Institution embryos, which have become the standard for staging. (**http://www.visembryo.com/**)

PART III

"REMEMBER WHEN YOU SAID STEM CELL RESEARCH WAS ALL HYPE?"

THE STEM CELL CONCEPT
Introducing Organogenesis

We have now explored the development of the embryo from fertilization through gastrulation. In the next chapters, we will see the how the three germ layers interact with each other to begin **organogenesis**—the processes of organ formation. Early organogenesis is a symphony of interactions between different parts of the embryo, and some of these interactions create privileged sites called *stem cell niches*. These niches provide a milieu of extracellular matrices and paracrine factors that allow cells residing within them to remain relatively undifferentiated. These relatively undifferentiated cells are *stem cells*, and their presence has become central to our vision of organogenesis and critical to the field of modern medicine.

The Stem Cell Concept

A **stem cell** can be defined as a relatively undifferentiated cell that when it divides produces (1) one cell that retains its undifferentiated character; and (2) a second cell that can undergo one or more paths of differentiation (**Figure III.1**). Thus, a stem cell has the potential to renew itself at each division (so that there is always a supply of stem cells) while also producing a daughter cell capable of responding to its environment by differentiating in a particular manner. (This potential is not always realized; in some instances, stem cells divide symmetrically so that both daughter cells remain stem cells.) In many cases, the stem cell remains in the niche while its sister leaves the niche and differentiates.

In some organs, such as the gut, epidermis, and bone marrow, stem cells regularly divide to replace worn-out cells and repair damaged tissues. In other organs, such as the prostate and heart, stem cells divide only under special physiological conditions, usually in response to stress or the need to repair the organ.

(A)

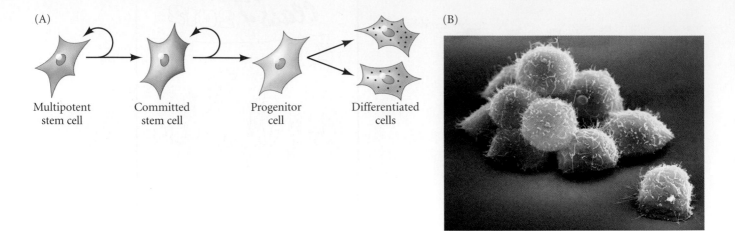

Multipotent
stem cell

Committed
stem cell

Progenitor
cell

Differentiated
cells

(B)

FIGURE III.1 The stem cell concept. (A) Stem cells have two distinct progeny at division: another stem cell and a more committed cell that can differentiate. (B) Hematopoietic stem cells isolated from human bone marrow form all the different blood cell types, as well as producing more hematopoietic stem cells. These multipotent stem cells generate blood cells throughout our entire lives. (B © SPL/Photo Researchers, Inc.)

In the mid-twentieth century, biologists thought that cell specification occurred in the early embryo and that after this stage there was only the growth of the existing parts. This idea is preserved in the classification of human development into *embryonic* stages where specification takes place (i.e., before week 9) and later *fetal* stages that are characterized by growth. However, beginning in the 1960s, biologists studying blood cell development began to ponder a remarkable phenomenon. All of the blood cells—red blood cells (erythrocytes), white blood cells (granulocytes), platelets, and even lymphocytes—are constantly being produced in the bone marrow. Billions of blood cells are destroyed by the spleen each hour, but an equal number of blood cells are generated to replace them.

In an elegant series of experiments, Ernest McCulloch and James Till demonstrated that there was a common generative cell, the "hematopoietic [blood-forming] stem cell," that gave rise to all the different types of blood cells (**Figure III.2**). In a 1961 experiment, Till and McCulloch injected bone marrow cells from a donor mouse into lethally irradiated mice* of the same genetic strain. Some of the individual donor cells produced discrete nodules on the spleens of the host animals (**Figure III.3**), and these nodules contained erythrocytes, granulocytes, and platelet precursor cells. Later experiments in which the donor marrow cells were irradiated to genetically mark each cell with random chromosome breaks confirmed that each of these different cell types in a nodule arose from a single cell, and that there were lymphocytes present in some of these nodules (Becker et al. 1963).

For this "colony-forming cell" to be a true stem cell, however, it had to produce not only the differentiated blood cells but also more colony-forming cells. This was shown to be the case by taking the nodule derived from a single genetically marked colony-forming cell and injecting cells of the nodule into another irradiated mouse. Many spleen colonies emerged, each of them having the same chromosomal arrangement as the original colony (Till et al. 1964; Jurásková and Tkadlécek 1965; Humphries

---

*The mice were "lethally irradiated" in that X-rays were used to destroy their immune and blood cell precursors. The mice survived only if there were stem cells in the transplant that could replace these cells and regenerate blood cells and lymphocytes.

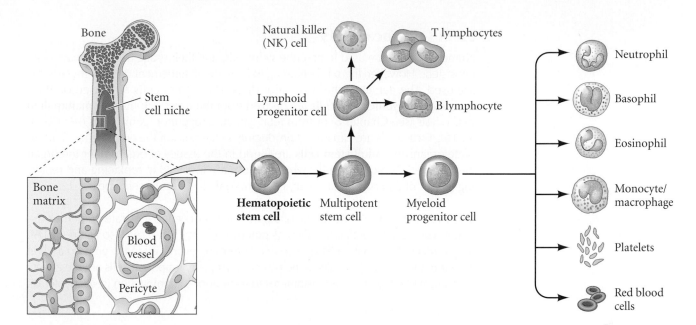

FIGURE III.2   Blood-forming (hematopoietic) stem cells and the bone marrow niche. The hematopoietic stem cell is located in the bone marrow and generates a second stem cell that is capable of becoming either a lymphocyte progenitor cell (which divides to form the cells of the immune system) or a myeloid stem cell (which forms the blood cell precursors). The type of lineage that the cells take is regulated by the niche, which involves contact between the stem cells and the matrices of bone cells, paracrine factors from stromal cells and the pericytes surrounding the blood vessels, and systemic hormones and neural signals. Mesenchymal stem cells also use the bone marrow as a niche. (After http://stemcells.nih.gov/.)

et al. 1979). Thus, we see that a single marrow cell can form numerous different cell types and can also undergo self-renewal. This research on hematopoietic stem cells (which we will discuss in detail in Chapter 12) led to the establishment of the field of bone marrow transplantation.*

---

*Interestingly, the concept that the different blood cell types were continuously generated by a hematopoietic stem cell was first proposed in 1909 by the Russian histologist Alexander Maximov. He is credited with coining the word *Stammzelle* to refer to the regenerative capacities of these cells. Maximov had been a student of Oskar Hertwig, one of the leading German embryologists and one of the originators of the current theory of epigenesis. Returning to St. Petersburg to be a professor of embryology and histology, Maximov's work was cut short by the Russian Revolution. He managed to flee the Soviet Union in 1922, having bribed a guard at the Finnish border with a bottle of laboratory ethanol. Eventually he settled in the United States, where he co-authored the leading textbook of histology with his student William Bloom (Konstantinov 2000).

FIGURE III.3   Spleen colony-forming cells. Hematopoietic stem cells injected into an adult mouse will colonize the spleen and form colonies containing red blood cells, granulocytes (white blood cells), platelet precursors, and lymphocytes. Shown here is the spleen of an irradiated mouse whose immune and blood precursor cells were abolished by radiation. This host mouse was then injected with bone marrow cells from a genetically identical mouse. When the host animal's spleen was removed 12 days later, colonies of donor blood cells were visible on its surface. (From Hall et al. 2003.)

Stem Cell Vocabulary

Numerous terms are used to describe stem cells, and their usage has not always been consistent. However, there is beginning to be general agreement on how these terms are used. The names of the two major divisions of stem cells are based on their sources. **Embryonic stem cells** are derived from the inner cell mass of mammalian blastocysts (see Chapter 8) or from fetal gamete progenitor (germ) cells (see Chapter 16). These cells are capable of producing all the cells of the embryo (i.e., a complete organism). **Adult stem cells** are found in the tissues of organs after the organ has matured. These stem cells, which are usually involved in replacing and repairing tissues of that particular organ, can form only a subset of cell types.

STEM CELL POTENCY The ability of a particular stem cell to generate numerous different types of differentiated cells is its **potency** (**Figure III.4**; see also Figure III.1A). In mammals, **totipotent cells** are capable of forming every cell in the embryo and, in addition, the trophoblast cells of the placenta. The only totipotent cells are the zygote and (probably) the first 4–8 blastomeres to form prior to compaction (see Chapter 8).

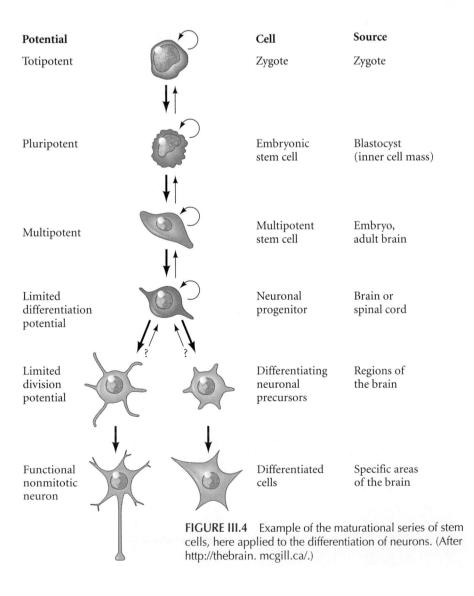

Potential	Cell	Source
Totipotent	Zygote	Zygote
Pluripotent	Embryonic stem cell	Blastocyst (inner cell mass)
Multipotent	Multipotent stem cell	Embryo, adult brain
Limited differentiation potential	Neuronal progenitor	Brain or spinal cord
Limited division potential	Differentiating neuronal precursors	Regions of the brain
Functional nonmitotic neuron	Differentiated cells	Specific areas of the brain

FIGURE III.4 Example of the maturational series of stem cells, here applied to the differentiation of neurons. (After http://thebrain. mcgill.ca/.)

Pluripotent stem cells have the ability to become all the cell types of the embryo *except* trophoblast. Usually these embryonic stem cells are derived from the inner cell mass of the mammalian blastocyst. However, germ cells and germ cell tumors (such as teratocarcinomas; see Chapter 16) can also form pluripotent stem cells.

Multipotent stem cells are stem cells whose commitment is limited to a relatively small subset of all the possible cells of the body. These are usually adult stem cells. The hematopoietic stem cell, for instance, can form the granulocyte, platelet, and red blood cell lineages. Similarly, the mammary stem cell can form all the different cell types of the mammary gland. Some adult stem cells are **unipotent stem cells,** which are found in particular tissues and are involved in regenerating a particular type of cell. Spermatogonia, for example, are stem cells that give rise only to sperm (see Chapter 16). Whereas pluripotent stem cells can produce cells of all three germ layers (as well as producing germ cells), the multipotent and unipotent stem cells are often grouped together as **committed stem cells**, since they have the potential to become relatively few cell types.

PROGENITOR CELLS Although they are related to stem cells, **progenitor cells** are not capable of unlimited self-renewal; they have the capacity to divide only a few times before differentiating (Seaberg and van der Kooy 2003). They are sometimes called *transit-amplifying cells*, since they usually divide while migrating away from the stem cell niche. Both unipotent stem cells and progenitor cells have been called *lineage-restricted cells*, but the stem cells have the capacity for self-renewal, while the progenitor cells do not. Progenitor cells are usually more differentiated than stem cells and have become committed to become a particular type of cell. In many instances, stem cell division generates progeny that become progenitor cells, as is seen in the formation of the blood cells, sperm cells, and the nervous system (see Figures III.1A and III.3).

Adult Stem Cells

Numerous adult organs contain committed stem cells that can give rise to a limited set of cell and tissue types. In addition to the well-known hematopoietic stem cells (see Figure III.2), developmental biologists have discovered epidermal stem cells, neural stem cells, hair stem cells, melanocyte stem cells, muscle stem cells, tooth stem cells, gut stem cells, and germline stem cells. Such cells are not as easy to use as pluripotent embryonic stem cells; they are difficult to isolate, since they often represent fewer than 1 out of every 1000 cells in an organ. In addition, they appear to have a relatively low rate of cell division and do not proliferate readily. However, neither of these facts precludes their usefulness. Each year some 40,000 bone marrow transplant procedures are performed in which hematopoietic stem cells are transferred from one person to another. These multipotential stem cells are rare (about 1 in every 15,000 bone marrow cells), but such transplantation treatment works well for people suffering from red blood cell deficiencies or leukemias.

Techniques to selectively allow the growth and isolation of multipotent stem cells may allow some organ deficiencies to be treated in the same way as these blood cell deficiencies—by administering a source of committed stem cells. In mice, very few (perhaps even one) blood stem cell will reconstitute the mouse's blood and immune systems (Osawa et al. 1996); a single mammary stem cell will generate an entire mammary gland (including epithelium, muscles, and stroma; Shackleton et al. 2006); and a single transplanted prostatic stem cell will produce an entire prostate gland (Leong et al. 2008). Carvey and colleagues have shown that when neural stem cells from the midbrain of adult rats are cultured in a mixture of paracrine factors, they will differentiate into dopaminergic neurons that can ameliorate the rodent version of Parkinson disease (Carvey et al. 2001; see Hall et al. 2007).

Adult Stem Cell Niches

Many tissues and organs contain stem cells that undergo continual renewal; these include the mammalian epidermis, hair follicles, intestinal villi, blood cells, and sperm cells, as well as *Drosophila* intestine, sperm, and egg cells. Such stem cells must maintain the long-term ability to divide, producing some daughter cells that are differentiated and other daughter cells that remain stem cells. The ability of a cell to become an adult stem cell is determined in large part by where it resides. The continuously proliferating stem cells are housed in compartments called **stem cell niches** (Schofield 1978; sometimes called **regulatory microenvironments**). These are particular places in the embryo that allow the controlled proliferation of the stem cells within the niche and the controlled differentiation of the cell progeny that leave the niche.

Stem cell niches regulate stem cell proliferation and differentiation, usually by paracrine factors that are produced by the niche cells (Moore and Lemischka 2006; Jones and Wagers 2008). These factors retain the cells in an uncommitted state. Once the cells leave the niche, the paracrine factors cannot reach them, and the cells begin differentiating. Mouse incisors, for instance, differ from human incisors in that they continue to grow throughout the lifetime of the animal. Each mouse incisor has two stem cell niches; one is on the "inside," facing into the mouth, and one is on the "outside," facing the lips (**Figure III.5**). The stem cells that reside therein are kept in a proliferative and non-differentiated state by an integrated network of paracrine factors, including Fgf3 and activin (which increase the proliferation of stem cells), and their respective inhibitors, BMP4 and follistatin (Wang et al. 2007). Teeth in humans and most other mammals lack stem cell niches and thus do not regenerate.

Similarly, mammalian hair follicles contain a "bulge" that houses the melanocyte stem cells, which provide pigment to the hair (see Figure 9.40). The division of the melanocyte stem cells within this niche is coordinated with hair growth. Moreover, upon cell division, one daughter cell remains in the stem cell niche, retaining its stem cell properties, while the other daughter cell migrates toward the developing hair shaft. The migrating cell is a committed melanocyte progenitor cell, and it will reside in the matrix at the bottom of the hair shaft, producing pigment-forming melanocytes.

The mammalian hematopoietic niche is found in the hollow cavities of trabecular bones (such as the sternum) where the bone marrow resides (see Figure III.2). Here,

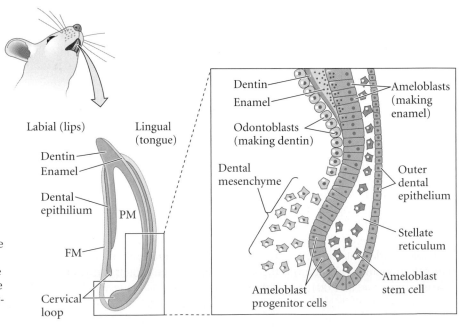

FIGURE III.5   The cervical loop of the mouse incisor is a stem cell niche for the enamel-secreting ameloblast cells. These cells migrate from the base of the stellate reticulum into the enamel layer, allowing the teeth to keep growing. (After Wang et al. 2007.)

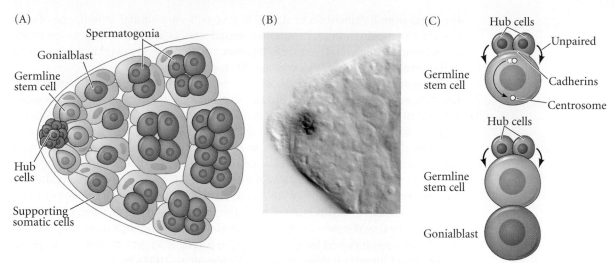

FIGURE III.6 Stem cell niche in *Drosophila* testes. (A) The apical hub consists of about 12 somatic cells, to which are attached 5–9 germ stem cells. The germ stem cells divide asymmetrically to form another germ stem cell (which remains attached to the somatic hub cells) and a gonialblast that will divide to form the sperm precursors (the spermatogonia and the spermatocyte cysts where meiosis is initiated). (B) Reporter β-galactosidase inserted into the gene for Unpaired reveals that this protein is transcribed in the somatic hub cells. (C) Cell division pattern of the germline stem cells, wherein one of the centrosomes remains in the cortical cytoplasm near the site of hub cell adhesion, while the other centrosome migrates to the opposite pole of the germline stem cell. This results in one cell remaining attached to the hub and the other cell detaching from it and differentiating. (After Tulina and Matunis 2001; B courtesy of E. Matunis.)

the stem cells are in close proximity to the bone cells (osteocytes) and the endothelial cells that line the blood vessels. A complex cocktail of paracrine factors, including Wnts, angiopoietin, and stem cell factor, combines with cell surface signals from Notch and integrin to regulate stem cell proliferation and differentiation. Hormonal signals and pressure from the blood vessels, as well as neurotransmitters from adjacent axons, also help regulate hematopoiesis (see Spiegel et al. 2008; Malhotra and Kincade 2009).

In many instances, germ cells are continuously produced in stem cell niches. In *Drosophila* testes, the stem cells for sperm reside in a regulatory microenvironment called the **hub** (**Figure III.6**). The hub consists of about a dozen somatic testes cells and is surrounded by 5–9 germ stem cells. The division of the sperm stem cell is asymmetric, always producing one cell that remains attached to the hub and one unattached cell. The daughter cell attached to the hub is maintained as a stem cell, while the cell that is not touching the hub becomes a gonialblast—a committed cell that will divide to become the precursors of the sperm cells. The somatic cells of the hub create this asymmetric proliferation by secreting the paracrine factor Unpaired onto the cells attached to them. Unpaired protein activates the JAK-STAT pathway in the adjacent germ stem cells to specify their self-renewal. Those cells that are distant from the paracrine factor cannot receive this signal, so they begin their differentiation into the sperm cell lineage (Kiger et al. 2001; Tulina and Matunis 2001).

Physically, this asymmetric division involves the interactions between the sperm stem cells and the somatic cells. In the division of the stem cell, one centrosome remains attached to the cortex at the contact site between the stem cell and the somatic cells. The other centrosome moves to the opposite side, thus establishing a mitotic spindle that will produce one daughter cell attached to the hub and one daughter cell away from it (Yamashita et al. 2003). (We will see a similar inheritance of centrosomes in the division of mammalian neural stem cells.) The cell adhesion molecules linking the hub and stem cells together are probably involved in retaining one of the centrosomes in the region where the two cells touch.

The stem cell niche is a critical component of our phenotype, and it does nothing less than regulate the ratio of cell division to cell differentiation. This means that maintenance of such niches is critical for our health. Too much stem cell differentiation depletes the stem cells and promotes the phenotypes of aging or decay. Too much stem cell division can cause cancers to arise. The "graying" of mammalian hair (in mice as well as humans) can involve a breakdown in the maintenance of the stem cell niche such that both daughters of the dividing stem cells differentiate, thereby leaving a smaller population of stem cells that cannot continue to make pigment-forming melanocytes (Nishimura et al. 2005; Steingrimson et al. 2005). Conversely, myeloproliferative disease—a cancer of blood stem cells and their (non-lymphocytic) derivatives—results when the stem cell niche is unable to provide the signals needed for proper blood cell differentiation (Walkley et al. 2007a,b).

Thus, stem cell niches provide microenvironments that regulate stem cell renewal, survival, and differentiation. Their paracrine factors, cell adhesion molecules, and architecture allow asymmetric cell divisions such that a stem cell divides in a manner that allows one of its daughter cells to have a high probability of leaving the niche and beginning to differentiate according to the new signals it encounters.

Mesenchymal Stem Cells: Multipotent Adult Stem Cells

Adult stem cells are used in the adult body to replace worn-out somatic cells on a regular basis. Our epidermis, our intestinal epithelium, and our blood cells are continually being replaced with cells generated by dividing adult stem cells. Most (if not all) adult stem cells are restricted to forming only a few cell types (Wagers et al. 2002). When hematopoietic stem cells marked with green fluorescent protein were placed in mice, their labeled descendants were found throughout the animals' blood but not in any other tissue. Some adult stem cells, however, appear to have a surprisingly large degree of plasticity. These multipotent cells are called **mesenchymal stem cells**, or **MSCs** (sometimes called **bone marrow-derived stem cells**, or **BMDCs**), and their potency remains a controversial subject.

Originally found in bone marrow (Friedenstein et al. 1968; Caplan 1991), multipotent MSCs have also been found in adult tissue such as fat, muscle, thymus, and dental pulp, as well as in the umbilical cord (see Kuhn and Tuan 2010). Indeed, the finding that human umbilical cords and deciduous ("baby") teeth contain MSCs (Gronthos et al. 2000; Hirata et al. 2004; Traggiai et al. 2004; Perry et al. 2008) has led some physicians to propose that parents freeze cells from their child's umbilical cord or teeth so that these cells will be available for transplantation later in life. However, the crucial test for pluripotency—the ability of a mouse stem cell to generate cells of all germ layers when inserted into a blastocyst—has not yet been achieved.

Mesenchymal stem cells *are* able to give rise to numerous bone, cartilage, muscle, and fat lineages (Pittenger et al. 1999; Dezawa et al. 2005). The differentiation of MSCs is predicated on both paracrine factors and cell matrix molecules in the stem cell niche. Certain cell matrix components, especially laminin, have been implicated in keeping MSCs in a state of undifferentiated "stemness" (Kuhn and Tuan 2010). Certain paracrine factors appear to direct development into specific lineages. In one study (Ng et al. 2008), platelet-derived growth factor was critical for chondrogenesis and fat formation, TGF-β signaling was important for chondrogenesis, and FGF signaling was crucial for the differentiation of MSCs into bone cells.

In addition to paracrine factors, the repertoire of cell types from mesenchymal stem cells may also be enhanced by the surfaces on which the stem cells reside. Human

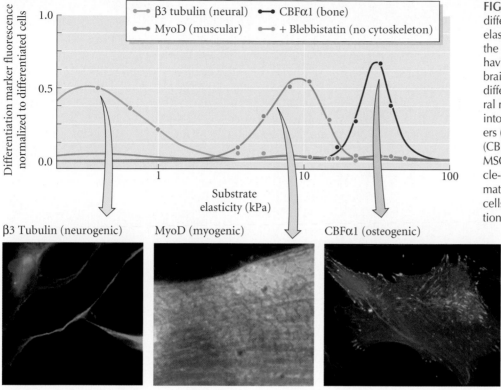

FIGURE III.7 Mesenchymal stem cell differentiation is influenced by the elasticity of the matrices upon which the cells sit. (On collage-coated gels having elasticity similar to that of the brain (about 01–1 kPa), Human MSCs differentiated into cells containing neural markers (such as β3 tubulin) but not into cells containing muscle cell markers (MyoD) or bone cell markers (CBFα1). As the gels became stiffer, the MSCs generated cells exhibiting muscle-specific proteins, and even stiffer matrices elicited the differentiation of cells with bone markers. Differentiation of the MSC on any matrix could be abolished with blebbistatin, which inhibits microfilament assembly at the cell membrane. (After Engler et al. 2006; photographs courtesy of J. Shields.)

MSCs can differentiate according to the elasticity of the surface on which they are placed. If placed on soft matrices of collagen-coated polyacrilamide, these stem cells differentiate into neurons. A moderately elastic matrix of the same materials causes the same stem cells to become muscle cells, while harder matrices cause the MSCs to produce bone cells (**Figure III.7**; Engler et al. 2006). It is not yet known whether this range of potency is found normally in the body.

Mesenchymal stem cells have recently been linked to normal growth and repair conditions in the human body. Indeed, one premature aging syndrome, Hutchinson-Gilford progeria, appears to be caused by the inability of MSCs to differentiate into certain cell types, such as fat cells (Scaffidi and Misteli 2008). These findings lead to speculation that the loss of MSC ability to differentiate may be a component of the normal aging syndrome. Moreover, MSCs may work in ways other than differentiating into needed cell types. They may produce paracrine factors that aid other, more specific stem cells to divide and repair tissues (Gnecchi et al. 2009).

A New Perspective on Organogenesis

The ability to create, isolate, and manipulate stem cells offers a vision of regenerative medicine, wherein a patient can have his or her diseased organs regrown and replaced by one's own stem cells. We will detail the medical possibilities of stem cell therapy in Chapter 17. But beyond their potential medical uses, stem cells tell us a great many facts about how the body is constructed and how it maintains its structure. Organs often form by the regulation of stem cells, and we will see that the skin, hair, blood, and parts of the nervous system routinely use stem cells in their construction. Stem cells certainly give credence to the view that "development never ends," and offer fascinating (if not frightening) potential ways to modify development.

The Emergence of the Ectoderm

Central Nervous System and Epidermis

9

For the real amazement, if you wish to be amazed, is this process. You start out as a single cell derived from the coupling of a sperm and an egg; this divides in two, then four, then eight, and so on, and at a certain stage there emerges a single cell which has as all its progeny the human brain. The mere existence of such a cell should be one of the great astonishments of the earth. People ought to be walking around all day, all through their waking hours calling to each other in endless wonderment, talking of nothing except that cell.

LEWIS THOMAS (1979)

"WHAT IS PERHAPS THE MOST INTRIGUING question of all is whether the brain is powerful enough to solve the problem of its own creation." So Gregor Eichele (1992) ended a review of research on mammalian brain development. The construction of an organ that perceives, thinks, loves, hates, remembers, changes, deceives itself, and coordinates all of our conscious and unconscious bodily processes is undoubtedly the most challenging of all developmental enigmas. A combination of genetic, cellular, and organismal approaches is now giving us a very preliminary understanding of how the basic anatomy of the brain becomes ordered.

The fates of the vertebrate ectoderm are shown in **Figure 9.1**. In the past two chapters, we have seen how the ectoderm is instructed to form the vertebrate nervous system and epidermis. A portion of the dorsal ectoderm is specified to become neural ectoderm, and its cells become distinguishable by their columnar appearance. This region of the embryo is called the **neural plate**. The process by which the neural plate tissue forms a **neural tube**—the rudiment of the central nervous system—is called **neurulation**, and an embryo undergoing such changes is called a **neurula** (**Figure 9.2**). The neural tube forms the brain anteriorly and the spinal cord posteriorly. This chapter will look at the processes by which the neural tube and the epidermis arise and acquire their distinctive patterns.

Establishing the Neural Cells

This chapter follows the story of the committed neural and epidermal precursor cells. Neural cells become specified through their interactions with other cells. There are at least four stages through which the pluripotent cells of the epiblast or blastula become neural precursor cells, or **neuroblasts** (see Figure III.4; Wilson and Edlund 2001):

- **Competence**, wherein multipotent cells can become neuroblasts if they are exposed to the appropriate combination of signals.
- **Specification**, wherein cells have received the appropriate signals to become neuroblasts, but progression along the neural differentiation pathway can still be repressed by other signals.
- **Commitment** (**determination**), wherein neuroblasts enter the neural differentiation pathway and will become neurons even in the presence of inhibitory signals.
- **Differentiation**, wherein the neuroblasts leave the mitotic cycle and express those genes characteristic of neurons.

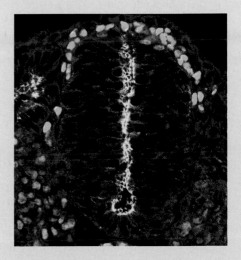

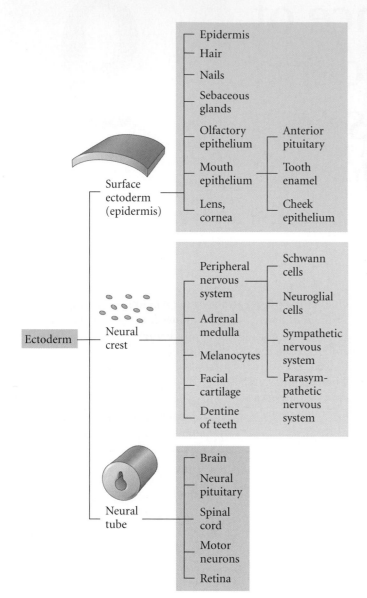

FIGURE 9.1   Major derivatives of the ectoderm germ layer. The ectoderm is divided into three major domains: the surface ectoderm (primarily epidermis), the neural crest (peripheral neurons, pigment, facial cartilage), and the neural tube (brain and spinal cord).

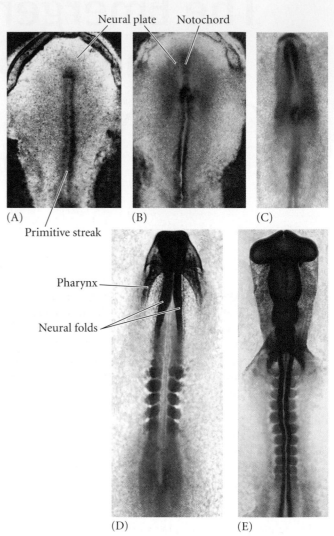

FIGURE 9.2   Gastrulation and neurulation in a chick embryo (dorsal view). (A) Flat neural plate. (B) Flat neural plate with underlying notochord (head process). (C) Neural groove. (D) Neural folds begin closing at the dorsalmost region, forming the incipient neural tube. (E) Neural tube, showing the three brain regions and the spinal cord. The neural tube remains open at the anterior end, and the optic bulges (which become the retinas) have extended to the lateral margins of the head. (F) 24-hour chick embryo, as in (D). The cephalic (head) region has undergone neurulation, while the caudal (tail) regions are still undergoing gastrulation. (A–E courtesy of G. C. Schoenwolf; F after Patten 1971.)

CONSTRUCTING THE CENTRAL NERVOUS SYSTEM

Formation of the Neural Tube

There are two major ways of converting the neural plate into a neural tube. In **primary neurulation**, the cells surrounding the neural plate direct the neural plate cells to proliferate, invaginate, and pinch off from the surface to form a hollow tube. In **secondary neurulation**, the neural tube arises from the coalescence of mesenchyme cells into a solid cord that subsequently forms cavities that coalesce to create a hollow tube. In general, the anterior portion of the neural tube is made by primary neurulation, while the posterior portion of the neural tube is made by secondary neurulation. The complete neural tube forms by joining these two separately formed tubes together (Harrington et al. 2009).

(F)

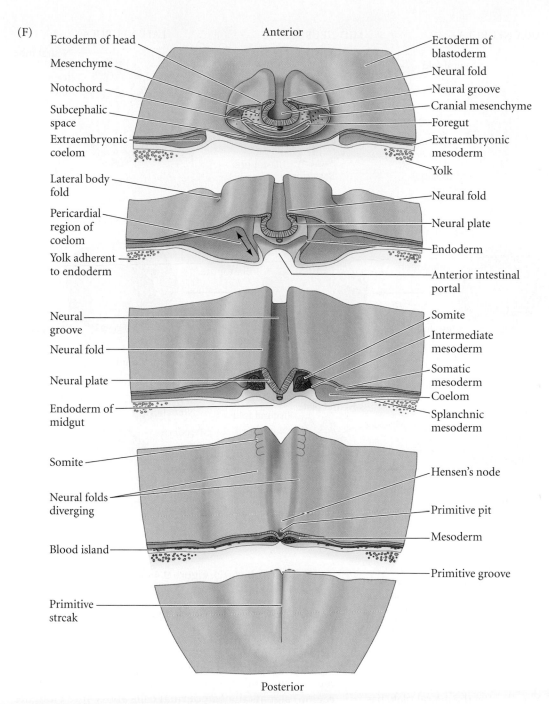

Anterior

Ectoderm of head
Mesenchyme
Notochord
Subcephalic space
Extraembryonic coelom

Ectoderm of blastoderm
Neural fold
Neural groove
Cranial mesenchyme
Foregut
Extraembryonic mesoderm
Yolk

Lateral body fold
Pericardial region of coelom
Yolk adherent to endoderm

Neural fold
Neural plate
Endoderm
Anterior intestinal portal

Neural groove
Neural fold
Neural plate
Endoderm of midgut

Somite
Intermediate mesoderm
Somatic mesoderm
Coelom
Splanchnic mesoderm

Somite
Neural folds diverging
Blood island

Hensen's node
Primitive pit
Mesoderm
Primitive groove

Primitive streak

Posterior

In birds, the neural tube anterior to the twenty-eighth somite pair (i.e., everything anterior to the hindlimbs) is made by primary neurulation (Pasteels 1937; Catala et al. 1996). In mammals, secondary neurulation begins at the level of the sacral vertebrae of the tail (Schoenwolf 1984; Nievelstein et al. 1993). In amphibians such as *Xenopus*, only the tail neural tube is derived from secondary neurulation (Gont et al. 1993); the same pattern occurs in fish (whose neural tubes were formerly thought to be formed solely by secondary neurulation) (Lowery and Sive 2004).

See WEBSITE 9.1
Homologous specification of the neural tissue

Primary neurulation

The events of primary neurulation divide the original ectoderm into three sets of cells: (1) the internally positioned neural tube, which will form the brain and spinal cord; (2) the externally positioned epidermis of the skin; and (3) the neural crest cells (see Figure 9.1). The neural crest cells form in the region that connects the neural tube and epidermis, but they migrate to new locations where they will generate the peripheral neurons and glia, the pigment cells of the skin, and several other cell types (see Chapter 10).

The process of primary neurulation appears to be similar in all vertebrates; **Figure 9.3** illustrates the process in

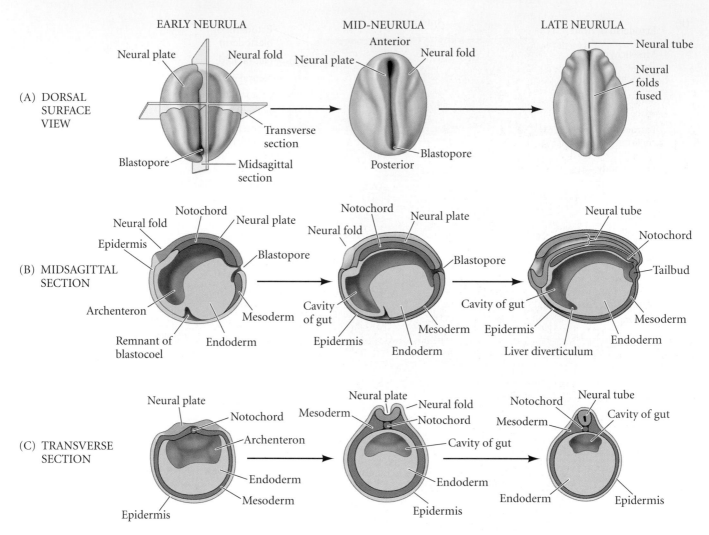

EARLY NEURULA   MID-NEURULA   LATE NEURULA

(A) DORSAL SURFACE VIEW

Neural plate   Neural fold   Neural plate   Anterior   Neural fold   Neural tube

Neural folds fused

Transverse section

Blastopore   Midsagittal section

Blastopore

Posterior

(B) MIDSAGITTAL SECTION

Notochord   Notochord   Neural plate   Neural tube

Neural fold   Neural plate   Neural fold   Notochord

Epidermis   Blastopore   Blastopore   Tailbud

Archenteron   Mesoderm   Cavity of gut   Cavity of gut   Mesoderm

Remnant of blastocoel   Endoderm   Epidermis   Endoderm   Epidermis   Endoderm

Liver diverticulum

(C) TRANSVERSE SECTION

Neural plate   Neural plate   Notochord   Neural tube

Notochord   Mesoderm   Neural fold   Mesoderm   Cavity of gut

Archenteron   Notochord

Cavity of gut

Endoderm   Endoderm   Endoderm   Epidermis

Mesoderm

Epidermis   Epidermis

FIGURE 9.3   Three views of neurulation in an amphibian embryo, showing early (left), middle (center), and late (right) neurulae in each case. (A) Looking down on the dorsal surface of the whole embryo. (B) Midsagittal section through the medial plane of the embryo. (C) Transverse section through the center of the embryo. (After Balinsky 1975.)

amphibians (Gallera 1971). Shortly after the neural plate has formed, its edges thicken and move upward to form the **neural folds**, while a U-shaped **neural groove** appears in the center of the plate, dividing the future right and left sides of the embryo. The neural folds migrate toward the midline of the embryo, eventually fusing to form the neural tube beneath the overlying ectoderm. The cells at the dorsalmost portion of the neural tube become the neural crest cells.

Primary neurulation can be divided into four distinct but spatially and temporally overlapping stages: (1) *formation* and *folding* of the neural plate; (2) *shaping* and *elevation* of the neural plate; (3) *convergence* of the neural folds, creating a neural groove; and (4) *closure* of the neural groove to form the neural tube (Smith and Schoenwolf 1997; Colas

and Schoenwolf 2001). **Figure 9.4** illustrates these stages in the chick embryo.

See VADE MECUM Chick neurulation

FORMATION AND SHAPING OF THE NEURAL PLATE The process of neurulation begins when the underlying dorsal mesoderm (and the pharyngeal endoderm in the head region) signals the ectodermal cells above it to elongate into columnar neural plate cells (Smith and Schoenwolf 1989; Keller et al. 1992). Their elongated shape distinguishes the cells of the prospective neural plate from the flatter pre-epidermal cells surrounding them. As much as half of the ectoderm is included in the neural plate.

The neural plate is shaped by the movements of the epidermal and neural plate regions. The neural plate lengthens along the anterior-posterior axis and narrows by convergent extension, intercalating several layers of cells into a few layers. These convergence and extension movements are critical for the shaping the neural plate, and mutations disturbing these movements can block neural tube closure (Ueno and Greene 2003).

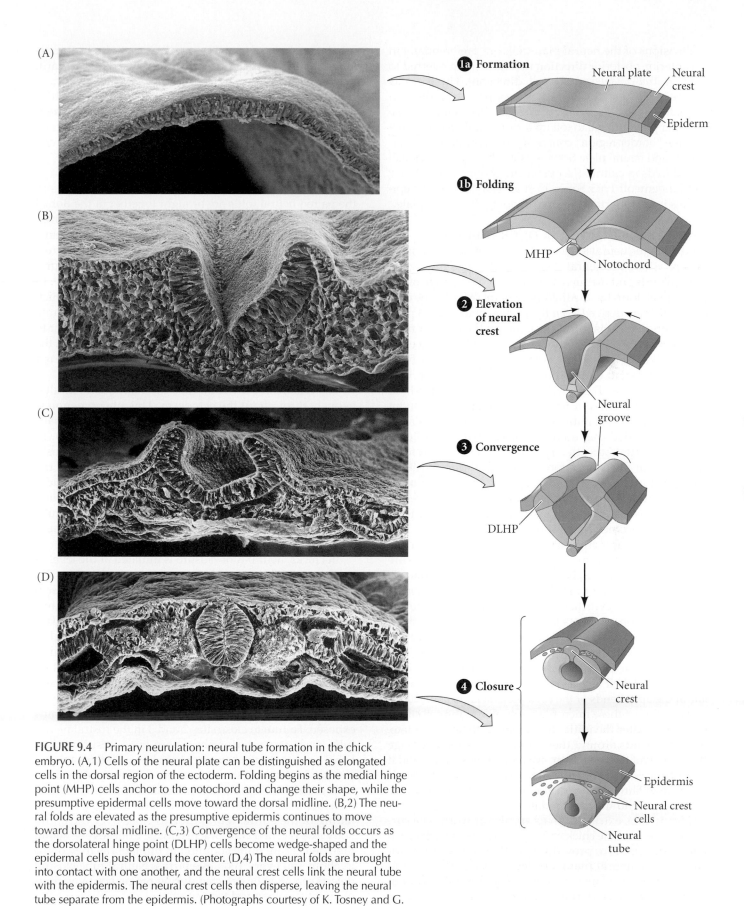

FIGURE 9.4   Primary neurulation: neural tube formation in the chick embryo. (A,1) Cells of the neural plate can be distinguished as elongated cells in the dorsal region of the ectoderm. Folding begins as the medial hinge point (MHP) cells anchor to the notochord and change their shape, while the presumptive epidermal cells move toward the dorsal midline. (B,2) The neural folds are elevated as the presumptive epidermis continues to move toward the dorsal midline. (C,3) Convergence of the neural folds occurs as the dorsolateral hinge point (DLHP) cells become wedge-shaped and the epidermal cells push toward the center. (D,4) The neural folds are brought into contact with one another, and the neural crest cells link the neural tube with the epidermis. The neural crest cells then disperse, leaving the neural tube separate from the epidermis. (Photographs courtesy of K. Tosney and G. Schoenwolf; drawings after Smith and Schoenwolf 1997.)

Divisions of the neural plate cells are preferentially in the anterior-posterior direction (in chick, often referred to as the *rostral-caudal*, or beak-tail, direction). These events occur even if the involved tissues are separated. If the neural plate is isolated, its cells converge and extend to make a thinner plate, but fail to roll up into a neural tube. However, if the "border region" containing both presumptive epidermis and neural plate tissue is isolated, it will form small neural folds in culture (Jacobson and Moury 1995; Moury and Schoenwolf 1995; Sausedo et al. 1997). Hence the epidermis is an important factor in shaping the neural plate.

BENDING AND CONVERGENCE OF THE NEURAL PLATE The bending of the neural plate involves the formation of hinge regions where the neural plate contacts surrounding tissues. In birds and mammals, the cells at the midline of the neural plate form the **medial hinge point** (**MHP**). The cells of the MHP are derived from the portion of the neural plate just anterior to Hensen's node and from the anterior midline of Hensen's node (Schoenwolf 1991a,b; Catala et al. 1996). MHP cells become anchored to the notochord beneath them and form a hinge, which forms a furrow at the dorsal midline. The notochord induces the MHP cells to decrease in height and to become wedge-shaped (van Straaten et al. 1988; Smith and Schoenwolf 1989); cells lateral to the MHP do not undergo such a change (see Figure 9.4B,C). Shortly thereafter, two other hinge regions form furrows near the connection of the neural plate with the remainder of the ectoderm. These regions are the **dorsolateral hinge points** (**DLHPs**), and they are anchored to the surface ectoderm. The DLHP cells also increase their height and become wedge-shaped.

Cell wedging is intimately linked to changes in cell shape. In the DLHPs, microtubules and microfilaments are both involved in these changes. Colchicine, an inhibitor of microtubule polymerization, inhibits the elongation of these cells, while cytochalasin B, an inhibitor of microfilament formation, prevents the apical constriction of these cells, thereby inhibiting wedge formation (Karfunkel 1972; Burnside 1973; Nagele and Lee 1987). As the neuroepithelial cells enlarge (by means of the microtubules), the microfilaments and their associated regulatory proteins accumulate at the apical ends of the cells (Zolessi and Arruti 2001). Constriction of these microfilaments allows the cell shape to change. After the initial furrowing of the neural plate, the plate bends around these hinge regions. Each hinge acts as a pivot that directs the rotation of the cells around it (Smith and Schoenwolf 1991). In *Xenopus*, the actin-binding protein Shroom is critical in initiating this apical constriction to bend the neural plate (Haigo et al. 2003).

Meanwhile, extrinsic forces are also at work. The surface ectoderm of the chick embryo pushes toward the midline of the embryo, providing another motive force for bending the neural plate (see Figure 9.4C; Alvarez and Schoenwolf 1992; Lawson et al. 2001). This movement of the presumptive epidermis and the anchoring of the neural plate to the underlying mesoderm may also be important for ensuring that the neural tube invaginates inward, or into the embryo and not outward. If small pieces of neural plate are isolated from the rest of the embryo (including the mesoderm), they tend to roll inside out (Schoenwolf 1991a). The pushing of the presumptive epidermis toward the center and the furrowing of the neural tube create the neural folds.

CLOSURE OF THE NEURAL TUBE The neural tube closes as the paired neural folds are brought together at the dorsal midline. The folds adhere to each other, and the cells from the two folds merge. In some species, the cells at this junction form the neural crest cells. In birds, the neural crest cells do not migrate from the dorsal region of the neural tube until after the neural tube has closed at that site. In mammals, however, the cranial neural crest cells (which form facial and neck structures; see Chapter 10) migrate while the neural folds are still being elevated (i.e., prior to neural tube closure), whereas in the spinal cord region, the neural crest cells do not migrate until closure has occurred (Nichols 1981; Erickson and Weston 1983).

The closure of the neural tube does not occur simultaneously throughout the ectoderm. This phenomenon is best seen in those vertebrates (such as birds and mammals) whose body axis is elongated prior to neurulation. In amniotes, induction occurs in an anterior-to-posterior fashion; so in the 24-hour chick embryo, neurulation in the **cephalic** (head) region is well advanced, while the **caudal** (tail) region of the embryo is still undergoing gastrulation (see Figure 9.2D-F). The two open ends of the neural tube are called the **anterior neuropore** and the **posterior neuropore**.

In chicks, neural tube closure is initiated at the level of the future midbrain and "zips up" in both directions. By contrast, in mammals neural tube closure is initiated at several places along the anterior-posterior axis. In humans, there are probably three sites of neural tube closure (**Figure 9.5A–C**; Nakatsu et al. 2000; O'Rahilly and Muller 2002). Failure of closure in different regions of the neural tube causes different neural tube defects. Failure to close the posterior neuropore around day 27 of development results in a condition called **spina bifida**, the severity of which depends on how much of the spinal cord remains exposed. Failure to close sites 2 and 3 in the rostral neural tube keeps the anterior neuropore open, resulting in a lethal condition called **anencephaly** in which the forebrain remains in contact with the amnionic fluid and subsequently degenerates (**Figure 9.5D**). The fetal forebrain ceases development, and the vault of the skull fails to form. The failure of the entire neural tube to close over the entire body axis is called **craniorachischisis**. Collectively, neural tube defects are quite common in humans, as they occur in about 1 in every 1000 live births. Neural tube closure defects can often be detected during pregnancy by various physical and chemical tests.

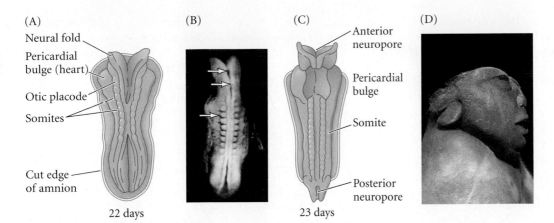

(A) Neural fold
Pericardial bulge (heart)
Otic placode
Somites
Cut edge of amnion
22 days

(B)

(C) Anterior neuropore
Pericardial bulge
Somite
Posterior neuropore
23 days

(D)

FIGURE 9.5 Neurulation in the human embryo. (A) Dorsal view of a 22-day (8-somite) human embryo initiating neurulation. Both anterior and posterior neuropores are open to the amnionic fluid. (B) 10-somite human embryo showing the three major sites of neural tube closure (arrows). (C) Dorsal view of a 23-day neurulating human embryo with only its neuropores open. (D) Photograph of a stillborn infant with anencephaly. (B from Nakatsu et al. 2000; D courtesy of National March of Dimes.)

The neural tube eventually forms a closed cylinder that separates from the surface ectoderm. This separation appears to be mediated by the expression of different cell adhesion molecules. Although the cells that will become the neural tube originally express E-cadherin, they stop producing this protein as the neural tube forms, and instead synthesize N-cadherin and N-CAM (**Figure 9.6A**). As a result, the surface ectoderm and neural tube tissues no longer adhere to each other. If the surface ectoderm is experimentally made to express N-cadherin (by injecting N-cadherin mRNA into one cell of a 2-cell *Xenopus* embryo), the separation of the neural tube from the pre-

sumptive epidermis is dramatically impeded (**Figure 9.6B**; Detrick et al. 1990; Fujimori et al. 1990).

Human neural tube closure is the result of a complex interplay between genetic and environmental factors (Cabrera et al. 2005; Fournier-Thibault et al. 2009). The *Pax3, Sonic hedgehog,* and *openbrain* genes are essential for the formation of the mammalian neural tube, but dietary factors, such as cholesterol and folate (also known as folic acid or vitamin B_9), also appear to be critical. It has been estimated that more than half of all human neural tube birth defects could be prevented by a pregnant woman's taking supplemental folate. Therefore, the U.S. Public Health Service recommends that women of childbearing age take 0.4 milligrams of folate daily (Milunsky et al. 1989; Centers for Disease Control 1992; Czeizel and Dudas 1992).

While the mechanism by which folate facilitates neural tube closure is not understood, recent studies have demonstrated expression of a folate receptor protein on the dorsalmost regions of the mouse neural tube immediately prior to fusion (**Figure 9.7**; Saitsu et al. 2003). Rothenberg and colleagues (2004) showed that most women who deliv-

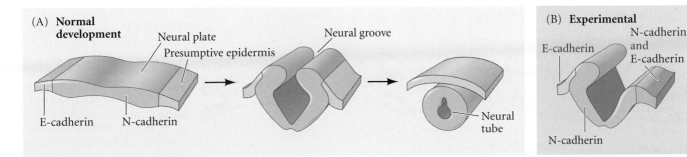

(A) **Normal development** Neural plate / Presumptive epidermis / Neural groove / E-cadherin / N-cadherin / Neural tube

(B) **Experimental** N-cadherin and E-cadherin / E-cadherin / N-cadherin

FIGURE 9.6 Expression of N- and E-cadherin adhesion proteins during neurulation in *Xenopus*. (A) Normal development. In the neural plate stage, N-cadherin is seen in the neural plate, while E-cadherin is seen on the presumptive epidermis. Eventually, the N-cadherin-bearing neural cells separate from the E-cadherin-containing epidermal cells. (The neural crest cells express neither N- nor E-cadherin, and they disperse.) (B) No separation of the neural tube occurs when one side of the frog embryo is injected with N-cadherin mRNA, so that N-cadherin is expressed in the epidermal cells as well as in the presumptive neural tube.

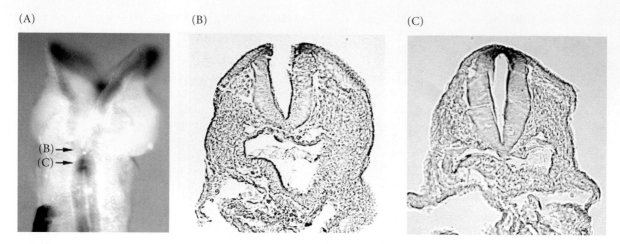

FIGURE 9.7 Folate-binding protein in the neural folds as neural tube closure occurs. (A) 10-somite mouse embryo stained for folate-binding protein mRNA. (B,C) Sections through embryo (A) at the two arrows, showing the folate-binding protein (dark blue) at the points of neural tube closure. (From Saitsu et al. 2003; photographs courtesy of K. Shiota.)

ered babies with neural tube defects had antibodies against this receptor protein, whereas such antibodies were seldom found in women whose babies did not have open neural tubes. Mice with mutations in the gene encoding the folate receptor were born with a high incidence of neural tube abnormalities. The incidence of neural tube defects decreased if pregnant mice were fed folate supplements (Piedrahita et al. 1999; Spiegelstein et al. 2004).

Folate deficiency appears to be only one risk factor, however. Women in low socioeconomic groups appear to have a higher incidence of babies with neural tube defects, even when vitamin use is taken into account (Little and Elwood 1992a; Wasserman et al. 1998). Moreover, there appears to be a seasonal variation in the incidence of neu-

ral tube defects (Little and Elwood 1992b). Although this phenomenon remains mysterious, one possible explanation may be contaminated crops. Marasas and colleagues (2004) documented that a fungal contaminant of corn produces a teratogen (fumonisin) that causes neural tube failure by perturbing the function of several lipids and proteins—including the folate receptor protein. This fungus has been found in regions where neural tube defects are prevalent. In mice, the teratogenic effects of fumonisin can be reduced by the administration of folate supplements during pregnancy.

Secondary neurulation

Secondary neurulation involves the production of mesenchyme cells from the prospective ectoderm and endoderm, followed by the condensation of these cells into a **medullary cord** beneath the surface ectoderm (**Figure 9.8A,B**). After this mesenchymal-to-epithelial transition, the central portion of this cord undergoes cavitation to form

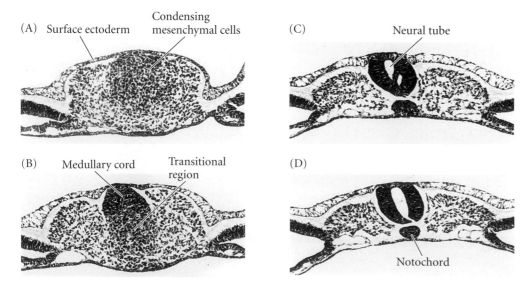

FIGURE 9.8 Secondary neurulation in the caudal region of a 25-somite chick embryo. (A) Mesenchymal cells condense to form the medullary cord at the most caudal end of the chick tailbud. (B) The medullary cord at a slightly more anterior position in the tailbud. (C) The neural tube is cavitating and the notochord forming; note the presence of separate lumens. (D) The lumens coalesce to form the central canal of the neural tube. (From Catala et al. 1995; photographs courtesy of N. M. Le Douarin.)

several hollow spaces, or *lumens* (**Figure 9.8C**); the lumens then coalesce into a single central cavity (**Figure 9.8D**; Schoenwolf and Delongo 1980). We have seen that after Hensen's node has migrated to the posterior end of the embryo, the caudal region of the epiblast contains a precursor cell population that gives rise to both neural ectoderm and mesoderm as the embryo's trunk elongates (Tzouanacou et al. 2009). The ectodermal cells that invaginate (and which will form the secondary neural tube) express the *Pax2* gene, while the mesodermal cells adjacent to them (which will form somites) do not express *Pax2* (Shimokita and Takahashi 2010). Presumably, these precursor cells provide the material for both the neural tube and mesoderm.

In human and chick embryos, there appears to be a transitional region at the junction of the anterior ("primary") and posterior ("secondary") neural tubes. In human embryos, coalescing cavities are seen in the transitional region, but the neural tube also forms by the bending of neural plate cells. Some posterior neural tube anomalies result when the two regions of the neural tube fail to coalesce (Saitsu et al. 2007). Given the prevalence of human posterior spinal cord malformations, further understanding of the mechanisms of secondary neurulation may have important clinical implications.

BUILDING THE BRAIN

Differentiation of the Neural Tube

Differentiation of the neural tube into the various regions of the central nervous system (i.e., the brain and spinal cord) occurs simultaneously in three different ways. On the gross anatomical level, the neural tube and its lumen bulge and constrict to form the chambers of the brain and spinal cord. At the tissue level, the cell populations in the wall of the neural tube rearrange themselves to form the different functional regions of the brain and spinal cord. Finally, on the cellular level, the neuroepithelial cells themselves differentiate into the numerous types of nerve cells (**neurons**) and supportive cells (**glia**) present in the body.

The early development of most vertebrate brains is similar, but because the human brain may be the most organized piece of matter in the solar system and is arguably the most interesting organ in the animal kingdom, we will concentrate on the development that is supposed to make *Homo* sapient.

The anterior-posterior axis

The early mammalian neural tube is a straight structure. However, even before the posterior portion of the tube has formed, the most anterior portion of the tube is undergoing drastic changes. In the anterior region, the neural tube balloons into the three primary vesicles: the forebrain (**prosencephalon**), midbrain (**mesencephalon**), and hindbrain (**rhombencephalon**) (**Figure 9.9**). By the time the posterior end of the neural tube closes, secondary bulges—the *optic vesicles*, which will become retina—have extended laterally from each side of the developing forebrain.

The prosencephalon becomes subdivided into the anterior **telencephalon** and the more caudal **diencephalon** (see Figure 9.9). The telencephalon will eventually form the **cerebral hemispheres**. The diencephalon will form the

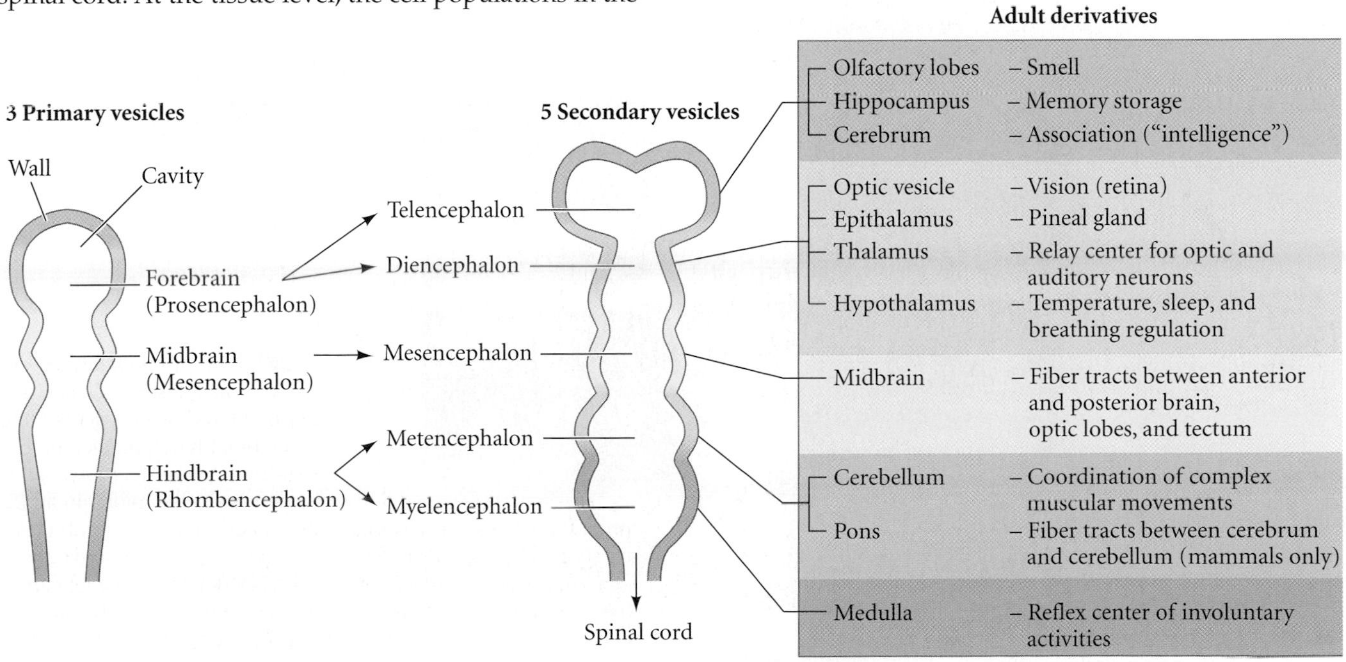

FIGURE 9.9 Early human brain development. The three primary brain vesicles are subdivided as development continues. At the right is a list of the adult derivatives formed by the walls and cavities of the brain. (After Moore and Persaud 1993.)

3 Primary vesicles

Wall
Cavity

Forebrain (Prosencephalon)
Midbrain (Mesencephalon)
Hindbrain (Rhombencephalon)

5 Secondary vesicles

Telencephalon
Diencephalon
Mesencephalon
Metencephalon
Myelencephalon

Spinal cord

Adult derivatives

Olfactory lobes	– Smell
Hippocampus	– Memory storage
Cerebrum	– Association ("intelligence")
Optic vesicle	– Vision (retina)
Epithalamus	– Pineal gland
Thalamus	– Relay center for optic and auditory neurons
Hypothalamus	– Temperature, sleep, and breathing regulation
Midbrain	– Fiber tracts between anterior and posterior brain, optic lobes, and tectum
Cerebellum	– Coordination of complex muscular movements
Pons	– Fiber tracts between cerebrum and cerebellum (mammals only)
Medulla	– Reflex center of involuntary activities

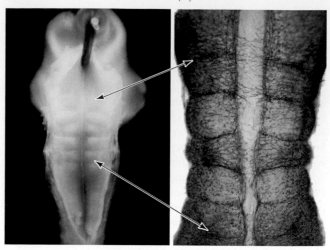

FIGURE 9.10 Rhombomeres of the chick hindbrain. (A) Hindbrain of a 3-day chick embryo. The roof plate has been removed so that the segmented morphology of the neural epithelium can be seen. The r1/r2 boundary is at the upper arrow, and the r6/r7 boundary is at the lower arrow. (B) A chick hindbrain at a similar stage stained with antibody to a neurofilament subunit. The rhombomere boundaries are emphasized because they serve as channels for neurons crossing from one side of the brain to the other. (From Lumsden 2004, courtesy of A. Lumsden.)

optic vesicles—the future retina—as well as the **thalamic** and **hypothalamic** brain regions, which receive neural input from the retina. Indeed, the retina itself is a derivative of the diencephalon. The mesencephalon does not become subdivided, and its lumen eventually becomes the **cerebral aqueduct**. The rhombencephalon becomes subdivided into a posterior **myelencephalon** and a more anterior **metencephalon**. The myelencephalon eventually becomes the **medulla oblongata**, whose neurons generate the nerve centers responsible for pain relay to the head and neck, auditory connections, tongue movement, and heartbeat, as well as respiratory and gastrointestinal movements. The metencephalon gives rise to the **cerebellum**, the part of the brain responsible for coordinating movements, posture, and balance.

The rhombencephalon develops a segmental pattern that specifies the places where certain nerves originate. Periodic swellings called **rhombomeres** divide the rhombencephalon into smaller compartments. The rhombomeres represent separate "territories" in that the cells within each rhombomere mix freely within it, but not with cells from adjacent rhombomeres (Guthrie and Lumsden 1991; Lumsden 2004). Moreover, each rhombomere has a

different fate. The neural crest cells above each rhombomere will form **ganglia**—clusters of neuronal cell bodies whose *axons* form a nerve. The generation of the cranial nerves from the rhombomeres has been studied most extensively in the chick, in which the first neurons appear in the even-numbered rhombomeres, r2, r4, and r6 (**Figure 9.10**; Lumsden and Keynes 1989). Neurons originating from r2 ganglia form the fifth (trigeminal) cranial nerve; those from r4 form the seventh (facial) and eighth (vestibuloacoustic) cranial nerves; and those from r6 form the ninth (glossopharyngeal) cranial nerve.

See WEBSITE 9.2 Specifying the brain boundaries

The zebrafish neural tube follows the same basic differentiation pattern as the mammalian neural tube (although the eyes form earlier). When red fluorescent dye is injected into the zebrafish hindbrain as that vesicle is forming, the dye soon diffuses into the midbrain and forebrain as well (**Figure 9.11**). This diffusion is dependent on the inflation

Inject red dye into hindbrain ventricle (18 h postfertilization)

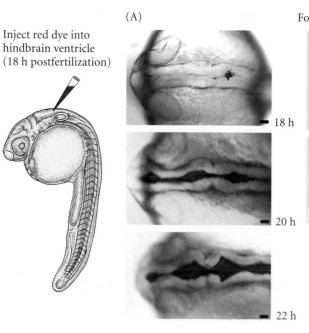

(A)

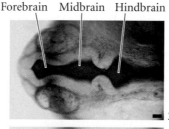

Forebrain Midbrain Hindbrain

(B) *Snakehead* mutant

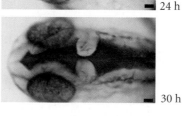

FIGURE 9.11 Brain ventricle formation in zebrafish. (A) When a fluorescent red dye is injected into the incipient hindbrain ventricle of 18-hour zebrafish embryos, the dye diffuses to the midbrain and forebrain regions. As development proceeds, the brain ventricles are shaped by regional differences in cell proliferation and adhesion. (B) In the *snakehead* mutant, the Na$^+$/K$^+$ ATPase is deficient. The brain ventricles fail to form, and dye injected into the presumptive hindbrain does not diffuse. (After Lowery and Sive 2005; photographs courtesy of H. Sive.)

of the lumen by the Na⁺/K⁺ ATPase, which sets up an osmotic gradient causing water to fill the ventricles* (Lowery and Sive 2005). (A similar mechanism is used to fill the mammalian blastocoel.) Further shaping of the ventricles depends on the secretion of the cerebrospinal fluid and on regional cell proliferation and cell adhesion. Both the midbrain and hindbrain ventricles are shaped by lateral hinge regions where the epithelium bends sharply.

The ballooning of the early embryonic brain is remarkable in its rate and extent, as well as in the fact that brain expansion is primarily the result of an increase in cavity size, not tissue growth. In the chick embryo, brain volume expands 30-fold between days 3 and 5 of development. This rapid expansion is thought to be caused by positive cerebrospinal fluid pressure exerted against the walls of the neural tube. It might be expected that this fluid pressure would be dissipated by the spinal cord, but this does not appear to happen. Rather, as the neural folds close in the region between the presumptive brain and spinal cord, the surrounding dorsal tissues push inward to constrict the neural tube at the base of the brain (**Figure 9.12**; Schoenwolf and Desmond 1984; Desmond and Schoenwolf 1986; Desmond and Levitan 2002). This occlusion (which also occurs in the human embryo) separates the presumptive brain region from the future spinal cord (Desmond 1982). If the fluid pressure in the anterior portion of an occluded

---

*The *vesicles* of the developing brain become the *ventricles* of the more mature brain.

neural tube is experimentally decreased, the chick brain enlarges at a much slower rate and contains many fewer cells than normal. The occluded region of the neural tube reopens after the initial rapid enlargement of the brain ventricles.

The anterior-posterior patterning of the hindbrain and spinal cord is controlled by a series of genes that include the Hox gene complexes. These genes were discussed more fully in Chapter 8 and will be revisited in Chapter 19.

The dorsal-ventral axis

The neural tube is polarized along its dorsal-ventral axis. In the spinal cord, for instance, the dorsal region is the place where the spinal neurons receive input from sensory neurons, while the ventral region is where the motor neurons reside. In the middle are numerous interneurons that relay information between the sensory and motor neurons.

The dorsal-ventral polarity of the neural tube is induced by signals coming from its immediate environment. The ventral pattern is imposed by the notochord, while the dorsal pattern is induced by the overlying epidermis. Specification of the axis is initiated by two major paracrine factors: Sonic hedgehog protein, originating from the notochord, and TGF-β proteins, originating in the dorsal ectoderm (**Figure 9.13**). In both cases, these factors induce a second signaling center within the neural tube itself. Sonic hedgehog is secreted from the notochord and induces the medial hinge point cells to become the **floor plate** of the neural tube. These floor plate cells also secrete

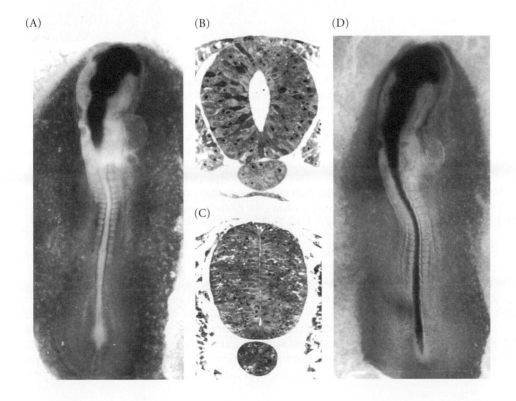

(A) (B) (D) (C)

FIGURE 9.12 Occlusion of the neural tube allows expansion of the future brain region. (A) Dye injected into the anterior portion of a 3-day chick neural tube fills the brain region but does not pass into the spinal region. (B,C) Sections of the chick neural tube at the base of the brain before occlusion (B) and during occlusion (C). (D) Reopening of the occlusion after initial brain enlargement allows dye to pass from the brain region into the spinal cord region. (Courtesy of M. Desmond.)

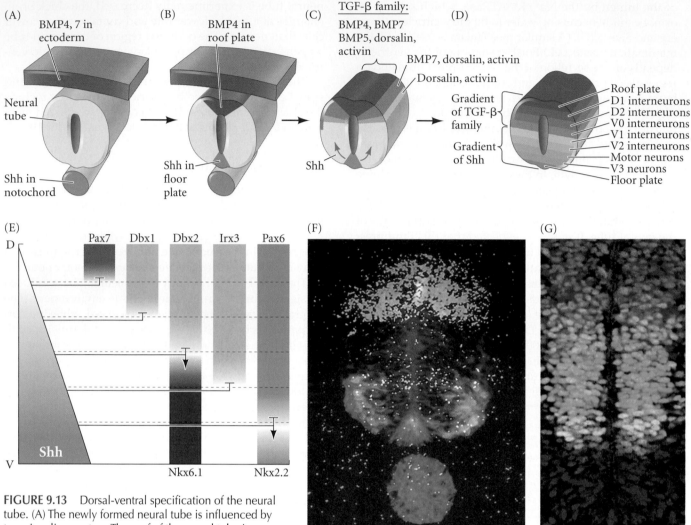

FIGURE 9.13 Dorsal-ventral specification of the neural tube. (A) The newly formed neural tube is influenced by two signaling centers. The roof of the neural tube is exposed to BMP4 and BMP7 from the epidermis, and the floor of the neural tube is exposed to Sonic hedgehog protein (Shh) from the notochord. (B) Secondary signaling centers are established in the neural tube. BMP4 is expressed and secreted from the roof plate cells; Sonic hedgehog is expressed and secreted from the floor plate cells. (C) BMP4 establishes a nested cascade of TGF-β factors, spreading ventrally into the neural tube from the roof plate. Sonic hedgehog diffuses dorsally as a gradient from the floor plate cells. (D) The neurons of the spinal cord are given their identities by their exposure to these gradients of paracrine factors. The amounts and types of paracrine factors present cause different transcription factors to be activated in the nuclei of these cells, depending on their position in the neural tube. (E) Sonic hedgehog is confined to the ventral region of the neural tube by the TGF-β factors, and the gradient of Sonic hedgehog specifies the ventral neural tube by activating and inhibiting the synthesis of particular transcription factors. (F) Chick neural tube, showing areas of Sonic hedgehog (green) and the expression domain of the TGF-β-family protein dorsalin (blue). Motor neurons induced by a particular concentration of Sonic hedgehog are stained orange/yellow. (G) In situ hybridization for three transcription factors: Pax7 (blue, characteristic of the dorsal neural tube cells), Pax6 (green), and Nkx6.1 (red). Where Nkx6.1 and Pax6 overlap (yellow), the motor neurons become specified. (E, F courtesy of T. M. Jessell; F from Jessell 2000; G courtesy of J. Briscoe.)

Sonic hedgehog, and this paracrine factor from the floor plate cells forms a gradient that is highest at the most ventral portion of the neural tube (Roelink et al. 1995; Briscoe et al. 1999).

The dorsal fates of the neural tube are established by proteins of the TGF-β superfamily, especially bone morphogenetic proteins (BMPs) 4 and 7, dorsalin, and activin

(Liem et al. 1995, 1997, 2000). Initially, BMP4 and BMP7 are found in the epidermis. Just as the notochord establishes a secondary signaling center—the floor plate cells—on the ventral side of the neural tube, the epidermis establishes a secondary signaling center by inducing BMP4 expression in the **roof plate** cells of the neural tube. The BMP4 protein from the roof plate induces a cascade of TGF-β pro-

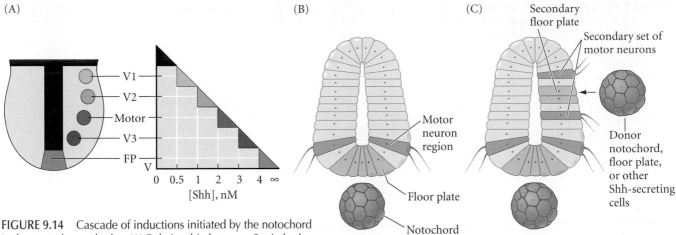

(A)

(B)

(C)

Secondary floor plate

Secondary set of motor neurons

Motor neuron region

Floor plate

Notochord

Donor notochord, floor plate, or other Shh-secreting cells

FIGURE 9.14 Cascade of inductions initiated by the notochord in the ventral neural tube. (A) Relationship between Sonic hedgehog concentrations, the generation of particular neuronal types in vitro, and distance from the notochord. (B) Two cell types in the newly formed neural tube. Those closest to the notochord become the floor plate neurons; motor neurons emerge on the ventrolateral sides. (C) If a second notochord, floor plate, or any other Sonic hedgehog-secreting cell is placed adjacent to the neural tube, it induces a second set of floor plate neurons, as well as two other sets of motor neurons. (A after Briscoe et al. 1999; B,C after Placzek et al. 1990.)

teins in adjacent cells (see Figure 9.13C). Dorsal sets of cells are thus exposed to higher concentrations of TGF-β proteins and at earlier times, when compared with the more ventral neural cells. The importance of the TGF-β superfamily factors in patterning the dorsal portion of the neural tube was demonstrated by the phenotypes of zebrafish mutants. Those mutants deficient in certain BMPs lacked dorsal and intermediate types of neurons (Nguyen et al. 2000).

The paracrine factors interact to instruct the synthesis of different transcription factors along the dorsal-ventral axis of the neural tube. In the case of the ventral neural tube, a gradient of Sonic hedgehog specifies different cell identities as a function of its concentration. Cells adjacent to the floor plate that receive high concentrations of Sonic hedgehog (and hardly any TGF-β signal) synthesize the Nkx6.1 and Nkx2.2 transcription factors and become the ventral (V3) neurons. The cells dorsal to these, exposed to slightly less Sonic hedgehog and slightly more TGF-β factors, produce Nkx6.1 and Pax6 transcription factors, and they become the motor neurons. The next two groups of cells, receiving progressively less Sonic hedgehog, become the V2 and V1 interneurons (**Figure 9.14A**; Lee and Pfaff 2001; Muhr et al. 2001).

The importance of Sonic hedgehog in inducing and patterning the ventral portion of the neural tube can be shown experimentally. If notochord fragments are taken from one embryo and transplanted to the lateral side of a host neural tube, the host neural tube will form another set of floor plate cells at its sides (**Figure 9.14B,C**). These ectopic floor plate cells then induce the formation of motor neurons on either side of them. The same results can be obtained if the notochord fragments are replaced by pellets of cultured cells secreting Sonic hedgehog (Echelard et al. 1993). Moreover, if a piece of notochord is removed from an embryo, the neural tube adjacent to the deleted region will have no floor plate cells (Placzek et al. 1990).

By combining information about the anterior-posterior axis with information establishing the dorsal-ventral axis, neurons can become uniquely specified for certain regions. For instance, one would expect the motor neurons innervating the forelimb to recognize different targets than the motor neurons innervating hindlimb targets. The Hox genes (and especially the HoxC paralogue group) play important roles in determining motor neuron targets (Dasen et al. 2003, 2005).

See WEBSITE 9.3 Constructing the pituitary gland

Differentiation of Neurons in the Brain

The human brain consists of more than 10^{11} neurons associated with over 10^{12} glial cells. The neuroepithelial cells of the neural tube give rise to three main types of cells. First, they become the *ventricular (ependymal) cells* that remain integral components of the neural tube lining and that secrete the cerebrospinal fluid. Second, they generate the *precursors of the neurons* that conduct electric potentials and coordinate our bodily functions, our thoughts, and our sensations of the world. Third, they give rise to the *precursors of the glial cells* that aid in constructing the nervous system, provide insulation around the neurons, and that may be important in memory storage. As we have seen, the differentiation of these precursor cells is believed to be largely determined by the environment they enter (Rakic and Goldman 1982). At least in some cases, a given neuroepithelial cell can presumably give rise to both neurons and glia (Turner and Cepko 1987).

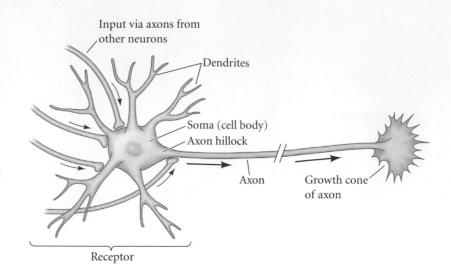

FIGURE 9.15 Diagram of a motor neuron. Electric impulses (red arrows) are received by the dendrites, and the stimulated neuron transmits impulses through its axon to its target tissue. The axon (which may be 2–3 feet long) is a cellular process through which the neuron sends its signals. The growth cone of the axon is both a locomotor and a sensory apparatus that actively explores the environment and picks up directional cues telling it where to go. Eventually the growth cone will form a connection, or synapse, with the axon's target tissue.

The brain contains a wide variety of neuronal and glial types (as is evident from a comparison of the relatively small *granule cell* neuron with the enormous *Purkinje neuron*). The fine, branching extensions of the neuron that are used to pick up electric impulses from other cells are called **dendrites** (Figure 9.15). Some neurons develop only a few dendrites, whereas other cells (such as the Purkinje neurons) develop extensive *dendritic arbors*. Very few dendrites are found on cortical neurons at birth, and one of the amazing events of the first year of human life is the increase in the number of these receptive processes. During this year, each cortical neuron develops enough dendritic surface to accommodate as many as 100,000 connections (**synapses**) with other neurons. The average cortical neuron connects with 10,000 other neural cells. This pattern of synapses enables the human cortex to function as the center for learning, reasoning, and memory; to develop the capacity for symbolic expression; and to produce voluntary responses to interpreted stimuli.

Another important feature of a developing neuron is its **axon** (sometimes called a **neurite**). Whereas dendrites are often numerous and do not extend far from the neuronal cell body, or **soma**, axons may extend 2–3 feet (see Figure 9.15). The pain receptors on your big toe, for example, must transmit their messages all the way to your spinal cord. One of the fundamental concepts of neurobiology is that the axon is a continuous extension of the nerve cell body. At the beginning of the twentieth century, there were many competing theories of axon formation. Theodor Schwann, one of the founders of the cell theory, believed that numerous neural cells linked themselves together in a chain to form an axon. Viktor Hensen, the discoverer of the embryonic node, thought that the axon formed around preexisting cytoplasmic threads between the cells. Wilhelm His (1886) and Santiago Ramón y Cajal (1890) postulated that the axon was an outgrowth (albeit an extremely large one) of the neuronal soma.

In 1907, Ross Harrison demonstrated the validity of the outgrowth theory in an elegant experiment that founded both the science of developmental neurobiology and the technique of tissue culture. Harrison isolated a portion of

the neural tube from a 3-mm frog tadpole. (At this stage, shortly after the closure of the neural tube, there is no visible differentiation of axons.) He placed this neuroblast-containing tissue in a drop of frog lymph on a coverslip and inverted the coverslip over a depression slide so he could watch what was happening within this "hanging drop." What Harrison saw was the emergence of axons as outgrowths from the neuroblasts, elongating at about 56 μm per hour.

Nerve outgrowth is led by the tip of the axon, called the **growth cone** (Figure 9.16A). The growth cone does not proceed in a straight line but rather "feels" its way along the substrate. The growth cone moves by the elongation and contraction of pointed filopodia called **microspikes** (Figure 9.16B). These microspikes contain microfilaments, which are oriented parallel to the long axis of the axon. (This mechanism is similar to that seen in the filopodial microfilaments of secondary mesenchyme cells in echinoderms; see Chapter 5.) Treating neurons with cytochalasin B destroys the actin microspikes, inhibiting their further advance (Yamada et al. 1971; Forscher and Smith 1988). Within the axon itself, structural support is provided by microtubules, and the axon will retract if the neuron is placed in a solution of colchicine (an inhibitor of microtubule polymerization). Thus the developing neuron displays the same mechanisms we noted in the dorsolateral hinge points of the neural tube—namely, elongation by microtubules and apical shape changes by microfilaments.

As in most migrating cells, the exploratory microspikes of the growth cone attach to the substrate and exert a force that pulls the rest of the cell forward. Axons will not grow if the growth cone fails to advance (Lamoureux et al. 1989). In addition to their structural role in axonal migration, the microspikes also have a sensory function. Fanning out in front of the growth cone, each microspike samples the microenvironment and sends signals back to the soma (Davenport et al. 1993). As we will see in Chapter 10, microspikes are the fundamental organelles involved in neuronal pathfinding.

Neurons transmit electric impulses from one region of the body to another. These impulses usually go from the

(A)

(B)

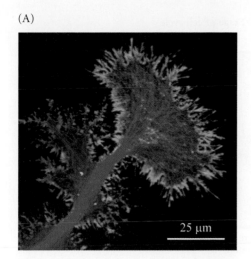

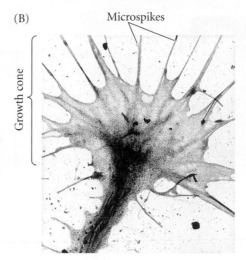

Microspikes

Growth cone

FIGURE 9.16   Axon growth cones. (A) Growth cone of the hawkmoth *Manduca sexta* during axon extension and pathfinding. The actin in the filopodia is stained green with fluorescent phalloidin, while the microtubules are stained red with a fluorescent antibody to tubulin. (B) Actin microspikes in an axon growth cone, seen by transmission electron microscopy. (A courtesy of R. B. Levin and R. Luedemanan; B from Letourneau 1979.)

25 μm

dendrites into the soma, where they are focused into the axon. To prevent dispersal of the electric signal and to facilitate its conduction to the target cell, the axon is insulated at intervals by glial cells. Within the central nervous system, axons are insulated at intervals by processes that originate from a type of glial cell called an **oligodendrocyte**. The oligodendrocyte wraps itself around the developing axon, then produces a specialized cell membrane called a **myelin sheath**. In the peripheral nervous system, myelination is accomplished by a glial cell type called the **Schwann cell (Figure 9.17)**. Transplantation experiments have shown that the axon, and not the glial cell, controls

the thickness of the myelin sheath. Mikhailov and colleagues (2004) have demonstrated that sheath diameter is regulated by the amount of neuregulin-1 secreted by the axon.

The myelin sheath is essential for proper neural function, and demyelination of nerve fibers is associated with convulsions, paralysis, and certain debilitating afflictions such as multiple sclerosis. There are also mouse mutants where subsets of neurons are poorly myelinated. In the *trembler* mutant, the Schwann cells are unable to produce

FIGURE 9.17   Myelination in the central and peripheral nervous systems. (A) In the peripheral nervous system, Schwann cells wrap themselves around the axon; in the central nervous system, myelination is accomplished by the processes of oligodendrocytes. (B) The mechanism of this wrapping entails the production of an enormous membrane complex. (C) Micrograph of an axon enveloped by the myelin membrane of a Schwann cell. (C courtesy of C. S. Raine.)

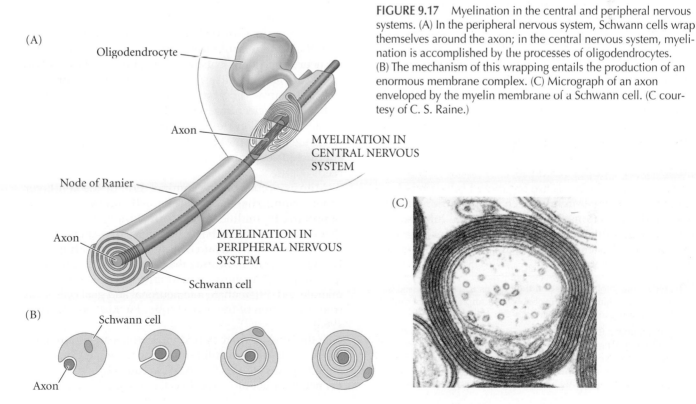

(A)

Oligodendrocyte

Axon

MYELINATION IN CENTRAL NERVOUS SYSTEM

Node of Ranier

Axon

MYELINATION IN PERIPHERAL NERVOUS SYSTEM

Schwann cell

(B)

Schwann cell

Axon

(C)

(A)

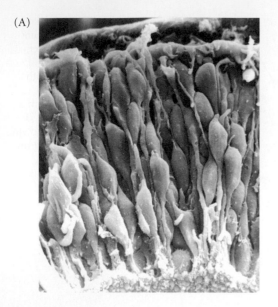

(B)

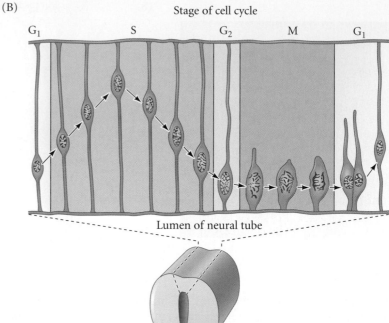

Lumen of neural tube

a particular protein component such that myelination is deficient in the peripheral nervous system but normal in the central nervous system. Conversely, in the mouse mutant *jimpy*, the central nervous system is deficient in myelin, while the peripheral nerves are unaffected (Sidman et al. 1964; Henry and Sidman 1988).

Axons are specialized for secreting specific chemical **neurotransmitters** across the small gap (the **synaptic cleft**) that separates the axon of a neuron from the surface of its target cell. Some neurons develop the ability to synthesize and secrete acetylcholine, while others develop the enzymatic pathways for making and secreting epinephrine, norepinephrine, octopamine, serotonin, γ-aminobutyric acid (GABA), or dopamine, among other neurotransmitters. Each neuron must activate those genes responsible for making the enzymes that can synthesize its neurotransmitter. Thus, neuronal development involves both structural and molecular differentiation.*

Tissue Architecture of the Central Nervous System

The neurons of the brain are organized into layers (**cortices**) and clusters (**nuclei**†), each having different functions and connections. The original neural tube is composed of a **germinal neuroepithelium**—a layer of rapidly dividing neural stem cells one cell layer thick. Sauer and

---

*The regeneration of neurons and their axons will be discussed in Chapter 15. The glial cells are probably very important in permitting or preventing axon regeneration.

†In neuroanatomy, the term *nucleus* refers to an anatomically discrete collection of neurons within the brain that typically serves a specific function. Note that this is a completely distinct structure from the "cell nucleus" (which, indeed, is a term that appears later in this same paragraph).

FIGURE 9.18   Neural stem cells in the germinal epithelium. (A) Scanning electron micrograph of a newly formed chick neural tube, showing cells at different stages of their cell cycles. (B) Schematic section of a chick embryo neural tube, showing the position of the nucleus in a neuroepithelial cell as a function of the cell cycle. Mitotic cells are found near the inner surface of the neural tube, adjacent to the lumen. (A courtesy of K. Tosney; B after Sauer 1935.)

colleagues (1935) showed that the cells of the germinal neuroepithelium are continuous from the luminal surface of the neural tube to the outside surface, but that the *cell nuclei* are at different heights, giving the superficial impression that the neural tube has numerous cell layers (**Figure 9.18A**). The nuclei move within their cells as they progress through the cell cycle. DNA synthesis (S phase) occurs while the nucleus is at the outside edge of the neural tube, and the nucleus migrates luminally as the cell cycle proceeds (**Figure 9.18B**).

Mitosis occurs on the luminal side of the cell layer. When mammalian neural tube cells are labeled with radioactive thymidine during early development, 100% of them incorporate this base into their DNA (Fujita 1964). Shortly thereafter certain cells stop incorporating these DNA precursors, indicating that they are no longer participating in DNA synthesis and mitosis. These cells then migrate and differentiate into neuronal and glial cells away from the lumen of the neural tube (Fujita 1966; Jacobson 1968).

If the labeled progeny of dividing cells are found in the outer cortex in the adult brain, then those neurons must have migrated to their cortical positions from the germinal neuroepithelium. When a cell of the germinal neuroep-

Stage of cell cycle

G₁   S   G₂   M   G₁

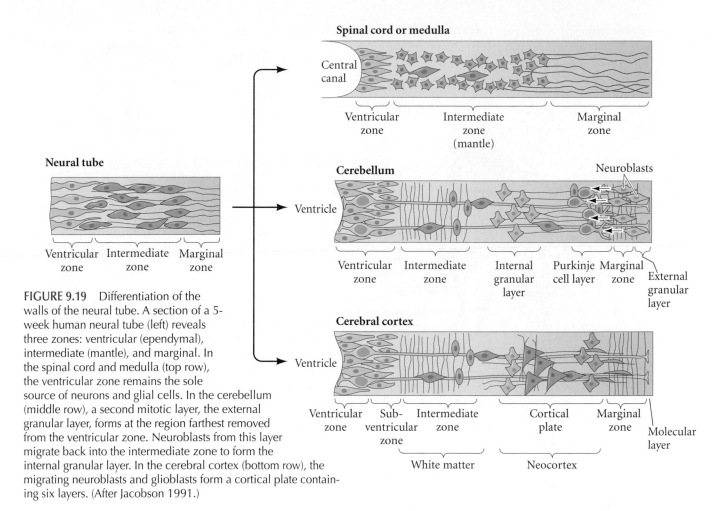

FIGURE 9.19 Differentiation of the walls of the neural tube. A section of a 5-week human neural tube (left) reveals three zones: ventricular (ependymal), intermediate (mantle), and marginal. In the spinal cord and medulla (top row), the ventricular zone remains the sole source of neurons and glial cells. In the cerebellum (middle row), a second mitotic layer, the external granular layer, forms at the region farthest removed from the ventricular zone. Neuroblasts from this layer migrate back into the intermediate zone to form the internal granular layer. In the cerebral cortex (bottom row), the migrating neuroblasts and glioblasts form a cortical plate containing six layers. (After Jacobson 1991.)

ithelium is ready to generate neurons (instead of more neural stem cells), the plane of cell division shifts. Instead of having both cells attached to the luminal surface, one of the two daughter cells remains in the epithelium while the other becomes detached. The cell connected to the luminal surface usually remains a stem cell, while the other cell migrates and differentiates (Chenn and McConnell 1995; Hollyday 2001). The time of this vertical division is the last time the migrating cell will divide, and is called that neuron's *birthday*. Different types of neurons and glial cells have birthdays at different times. Labeling cells at different times during development shows that the cells with the earliest birthdays migrate the shortest distances; those with later birthdays migrate to form the more superficial regions of the brain cortex. Subsequent differentiation depends on the positions the neurons occupy once outside the germinal neuroepithelium (Letourneau 1977; Jacobson 1991).

Spinal cord and medulla organization

As the cells adjacent to the lumen continue to divide, the migrating cells form a second layer around the original neural tube. This layer becomes progressively thicker as more cells are added to it from the germinal neuroepithelium. This new layer is called the **mantle** (or **intermediate**) **zone**, and the germinal epithelium is now called the **ventricular zone** (and, later, the **ependyma**) (Figure 9.19). The mantle zone cells differentiate into both neurons and glia. The neurons make connections among themselves and send forth axons away from the lumen, thereby creating a cell-poor **marginal zone**. Eventually glial cells cover many of the axons in the marginal zone in myelin sheaths, giving them a whitish appearance. Hence, the axonal marginal layer is often called **white matter**, while the mantle zone, containing the neuronal cell bodies, is referred to as **gray matter**.

In the spinal cord and medulla, this basic three-zone pattern of ventricular (ependymal), mantle, and marginal layers is retained throughout development. When viewed in cross section, the gray matter (mantle) gradually becomes a butterfly-shaped structure surrounded by white matter, and both become encased in connective tissue. As the neural tube matures, a longitudinal groove—the **sulcus limitans**—divides it into dorsal and ventral halves. The dorsal portion receives input from sensory neurons, whereas the ventral portion is involved in effecting various motor functions (**Figure 9.20**).

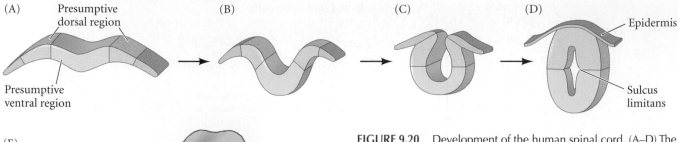

(A) Presumptive dorsal region

Presumptive ventral region

(B)

(C)

(D) Epidermis

Sulcus limitans

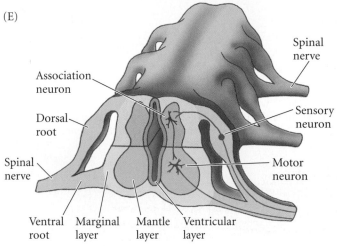

(E)

Spinal nerve

Association neuron

Dorsal root

Sensory neuron

Spinal nerve

Motor neuron

Ventral root

Marginal layer

Mantle layer

Ventricular layer

FIGURE 9.20   Development of the human spinal cord. (A–D) The neural tube is functionally divided into dorsal and ventral regions, separated by the sulcus limitans. As cells from the adjacent somites form the spinal vertebrae, the neural tube differentiates into the ventricular (ependymal), mantle, and marginal zones, as well as the roof and floor plates. (E) A segment of the spinal cord with its sensory (dorsal) and motor (ventral) roots. (After Larsen 1993.)

Cerebellar organization

In the brain, cell migration, differential neuronal proliferation, and selective cell death produce modifications of the three-zone pattern seen in Figure 9.19. In the cerebellum, some neuronal precursors enter the marginal zone to form nuclei (clusters of neurons; see second footnote, p. 348). Each nucleus works as a functional unit, serving as a relay station between the outer layers of the cerebellum and other parts of the brain. Other neuronal precursors migrate away from the germinal neuroepithelium. These cerebellar neuroblasts migrate to the outer surface of the developing cerebellum and form a new germinal zone, the **external granular layer** (see Figure 9.19), near the outer boundary of the neural tube.

At the outer boundary of the external granular layer, which is 1–2 cells thick, neuroblasts proliferate and come into contact with cells that secrete BMP factors. The BMPs specify the postmitotic products of these neuroblasts to become a type of neuron called the **granule cells** (Alder et al. 1999). Granule cells migrate back toward the ventricular (ependymal) zone, where they produce a region called the **internal granular layer** (see Figure 9.19). Meanwhile, the original ventricular zone of the cerebellum generates a wide variety of neurons and glial cells, including the distinctive and large **Purkinje neurons**, the major cell type of the cerebellum (**Figure 9.21**). Purkinje neurons secrete Sonic hedgehog, which sustains the division of granule cell precursors in the external granular layer (Wallace 1999). Each Purkinje neuron has an enormous **dendritic arbor** that spreads like a tree above a bulblike cell body (see Figure

9.21B). A typical Purkinje neuron may form as many as 100,000 synapses with other neurons—more connections than any other type of neuron studied. Each Purkinje neuron also sends out a slender axon, which connects to neurons in the deep cerebellar nuclei.

Purkinje neurons are critical in the electrical pathway of the cerebellum. All electric impulses eventually regulate the activity of these neurons, which are the only output neurons of the cerebellar cortex. For this to happen, the proper cells must differentiate at the appropriate place and time. How is this accomplished?

One mechanism thought to be important for positioning young neurons in the developing mammalian brain is **glial guidance** (Rakic 1972; Hatten 1990). Throughout the cortex, neurons are seen to ride a "glial monorail" to their respective destinations. In the cerebellum, the granule cell precursors travel on the long processes of the **Bergmann glia** (see Figure 9.21B; Rakic and Sidman 1973; Rakic 1975). As **Figure 9.22** illustrates, this neuron-glia interaction is a complex and fascinating series of events, involving reciprocal recognition between glia and neuroblasts (Hatten 1990; Komuro and Rakic 1992). The neuron maintains its adhesion to the glial cell through a number of proteins, one of them an adhesion protein called **astrotactin**. If the astrotactin on a neuron is masked by antibodies to that protein, the neuron will fail to adhere to the glial processes (Edmondson et al. 1988; Fishell and Hatten 1991). Mice deficient in astrotactin have slow neuronal migration rates, abnormal Purkinje cell development, and problems coordinating their balance (Adams et al. 2002). The direction of this migration appears to be regulated by a complex series of events orchestrated by brain-derived neurophic factor (BDNF), a paracrine factor that is made by the internal granular layer (Zhou et al. 2007).

Much insight into the mechanisms of spatial ordering in the brain has come from the analysis of neurological mutations in mice. More than 30 mutations are known to affect the arrangement of cerebellar neurons. Many of these

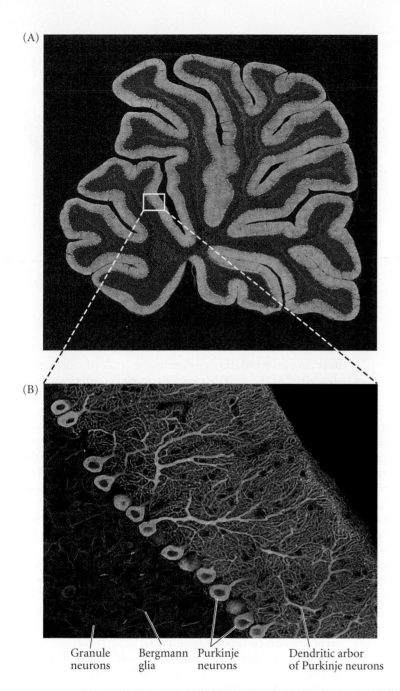

FIGURE 9.21 Cerebellar organization. (A) Sagittal section of a fluorescently labeled rat cerebellum photographed using dual-photon confocal microscopy. (B) A vast enlargement of one area of (A). Purkinje neurons are light blue with bright green processes, Bergmann glia are red, and granule cells are dark blue. This close-up illustrates the highly structured organization of neurons and glial cells. (Courtesy of T. Deerinck and M. Ellisman, University of California, San Diego.)

Granule neurons   Bergmann glia   Purkinje neurons   Dendritic arbor of Purkinje neurons

distinct ways. First, like the cerebellum, it is organized vertically into layers that interact with one another. Certain neuroblasts from the mantle zone migrate on glial processes through the white matter to generate a second zone of neurons at the outer surface of the brain. This new layer of gray matter will become the **neocortex**. The specification of the neocortex is accomplished largely through the Lhx2 transcription factor, which activates numerous other cerebral genes. In *Lhx2*-deficient mice, the cerebral cortex fails to form (Mangale et al. 2008; Chou et al. 2009). The neocortex eventually stratifies into six layers of neuronal cell bodies; the adult forms of these layers are not completed until the middle of childhood. Each layer of the neocortex differs from the others in its functional properties, the types of neurons found there, and the sets of connections they make. For instance, neurons in layer 4 receive their major input from the thalamus (a region that forms from the diencephalon), whereas neurons in layer 6 send their major output back to the thalamus.

In addition to the six vertical layers, the cerebral cortex is organized horizontally into more than 40 regions that regulate anatomically and functionally distinct processes. For instance, neurons in layer 6 of the visual cortex project axons to the lateral geniculate nucleus of the thalamus, which is involved in vision (see Chapter 10), while layer 6 neurons of the auditory cortex (located more anteriorly than the visual cortex) project axons to the medial geniculate nucleus of the thalamus, which functions in hearing. One of the major questions in developmental neurobiology is whether the different functional regions of the cerebral cortex are already specified in the ventricular region, or if specification is accomplished much later by the synaptic connections between the regions. Evidence that specification is early (and that there might be some "proto-map" of the cerebral cortex) is suggested by certain human mutations that destroy the layering and functional abilities in only one part of the cortex, leaving the other regions intact (Piao et al. 2004).

Indeed, most of the neuroblasts generated in the ventricular zone migrate outward along radial glial processes to form the **cortical plate** at the outer surface of the brain. As in the rest of the brain, those neurons with the earliest

cerebellar mutants have been found because the phenotype of such mutants—namely, the inability to maintain balance while walking—can be easily recognized. For obvious reasons, these mutations are given names such as *weaver*, *reeler*, and *staggerer*.

See WEBSITE 9.4
Cerebellar mutations of the mouse

Cerebral organization

The three-zone arrangement of the neural tube is also modified in the cerebrum. The cerebrum is organized in two

FIGURE 9.22 Neuron-glia interaction in the mouse. (A) Diagram of a cortical neuron migrating on a glial cell process. (B) Sequential photographs of a neuron migrating on a cerebellar glial process. The leading process has several filopodial extensions. The neuron can reach speeds around 40 μm per hour as it travels. (A after Rakic 1975; B from Hatten 1990, photograph courtesy of M. Hatten.)

(A)

Leading process
of neuron

Migrating
neuron

Process of
glial cell

(B)

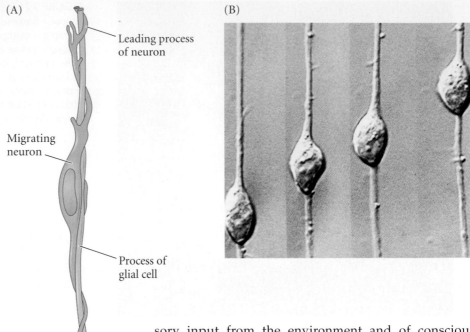

birthdays form the layer closest to the ventricle. Subsequent neurons travel greater distances to form the more superficial layers of the cortex. This process forms an "inside-out" gradient of development (Rakic 1974). McConnell and Kaznowski (1991) have shown that the determination of laminar identity (i.e., which layer a cell migrates to) is made during the final cell division. Newly generated neuronal precursors transplanted after this last division from young brains (where they would form layer 6) into older brains whose migratory neurons are forming layer 2 are committed to their fate, and migrate only to layer 6. However, if these cells are transplanted prior to their final division (during mid-S phase), they are uncommitted and can migrate to layer 2 (**Figure 9.23**). The fates of neuronal precursors from older brains are more fixed. While the neuronal precursor cells formed early in development have the potential to become any neuron (at layers 2 or 6, for instance), later precursor cells give rise only to upper-level (layer 2) neurons (Frantz and McConnell 1996). Once the cells arrive at their final destination, it is thought that they produce particular adhesion molecules that organize them together as brain nuclei (Matsunami and Takeichi 1995).

According to Gaiano (2008), "the construction of the mammalian neocortex is perhaps the most complex biological process that occurs in nature. A pool of seemingly homogeneous stem cells first undergoes proliferative expansion and diversification and then initiates the production of successive waves of neurons. As these neurons are generated, they take up residence in the nascent cortical plate where they integrate into the developing neocortical circuitry. The spatial and temporal coordination of neuronal generation, migration, and differentiation is tightly regulated and of paramount importance to the creation of a mature brain capable of processing and reacting to sen-

sory input from the environment and of conscious thought." And—as if this weren't enough—after neurogenesis subsides, the neuronal stem cells start making glia!

See WEBSITE 9.5 Constructing the cerebral cortex

STEM CELLS AND PRECURSOR CELLS The stem cells of the neocortex are originally generated by symmetrical division of the neural tube epithelium before neurogenesis (neuron formation) actually occurs. Initially, the proliferative layer of the mouse cortex forms the ventricular zone (VZ). Shortly thereafter (around day 13), it divides to give rise to a subventricular zone (SVB) directly outside it. Together, these zones form the germinal strata that generate the neuroblasts (neuronal precursor cells) that migrate into the cortical plate and form the layers of neurons. The VZ will form the lower (deeper) layers of neurons, while the SVB will give rise to those cells that form the upper layer of neurons (Frantz et al. 1994).

There are three major progenitor cells in the germinal strata: radial glial cells (RGCs), short neural precursors (SNPs), and intermediate progenitor cells (IPCs; **Figure 9.24**). RGCs (which aren't really glia but tend to look like them) are thought to be stem cells, and they are found in the VZ. At each division, they generate another VZ cell and a more committed cell type. Interestingly, at each division, the cell receiving the "old" centriole (which contains different proteins than the newly made centriole) stays in the VZ, while the cell receiving the "young" centriole leaves to differentiate (Wang et al. 2009). The more committed cells can be either neuroblasts (which divide to generate neurons) or IPCs, which migrate to the SVZ, where they generate neuroblasts. The SNPs also appear to generate neuroblasts, but from the VZ. A single stem cell in the ventricular layer can give rise to neurons (and glial cells) in any of the cortical layers (Walsh and Cepko 1988).

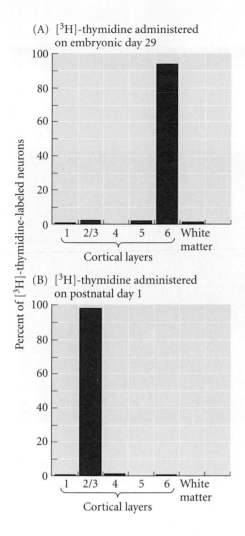

(A) [³H]-thymidine administered on embryonic day 29

(B) [³H]-thymidine administered on postnatal day 1

FIGURE 9.23   Determination of cortical laminar identity in the ferret cerebrum. (A) "Early" neuronal precursors (birthdays on embryonic day 29) migrate to layer 6. (B) "Late" neuronal precursors (birthdays on postnatal day 1) migrate farther, into layers 2 and 3. (C) When early neuronal precursors (dark blue) are transplanted into older ventricular zones after their last mitotic S phase, the neurons they form migrate to layer 6. (D) If these precursors are transplanted before or during their last S phase, however, they migrate (with the host neurons) to layer 2. (After McConnell and Kaznowski 1991.)

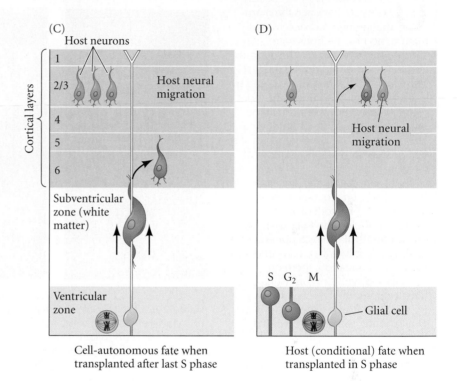

Cell-autonomous fate when transplanted after last S phase

Host (conditional) fate when transplanted in S phase

FIGURE 9.24   Cortical neurons are generated from three types of neural precursor cells: radial glia cells, short neural precursors, and intermediate progenitor cells. RGCs and SNPs divide at the apical (luminal) surface of the ventricular layer. SNPs are committed neural precursors. IPCs divide away from the luminal surface in the ventricular and subventricular zones. Most IPCs undergo neurogenic divisions, with a small fraction undergoing symmetrical proliferative divisions (dotted circular arrow). Through asymmetrical divisions, RGCs give rise to IPCs that migrate to the subventricular layer. The ventricular zone generates lower-layer neurons; the subventricular zone generates upper-layer neurons. (After Dehay and Kennedy 2007.)

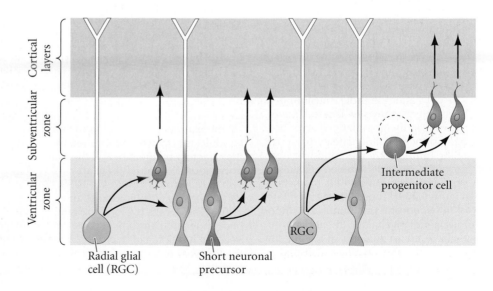

SIDELIGHTS & SPECULATIONS

Adult Neural Stem Cells

Ontil recently, it was generally believed that once the mammalian nervous system was mature, no new neurons were "born"—in other words, the neurons formed in utero and during the first few years of life were all we could ever expect to have. The good news from recent studies, however, is that the adult brain is capable of producing new neurons, and environmental stimulation can increase the number of these new neurons.

In these experiments, researchers injected adult mice, rats, or marmosets with bromodeoxyuridine (BrdU), a nucleoside that resembles thymidine. BrdU is incorporated into a cell's DNA only if the cell is undergoing DNA replication; therefore, any cell labeled with BrdU must have been undergoing DNA synthesis during the time it was exposed to BrdU. This labeling technique revealed that *thousands* of new neurons are being made each day in adult mice. Moreover, these new brain cells integrated with other cells of the brain, had normal neuronal morphology, and exhibited action potentials (**Figure 9.25A**; van Praag et al. 2002).

Injecting humans with BrdU is usually unethical, since large doses of BrdU are often lethal. However, in certain cancer patients, the progress of chemotherapy is monitored by transfusing the patient with a small amount of BrdU. Gage and colleagues (see Erikkson et al. 1998) took postmortem samples from the brains of five such patients who died between 16 and 781 days after the BrdU infusion. In all five subjects, they saw labeled (new) neurons in the granular cell layer of the hippocampal dentate gyrus (a part of the brain where memories may be formed). The BrdU-labeled cells also stained for neuron-specific markers (**Figure 9.25B**). Thus, although the rate of new neuron formation in adulthood may be relatively low, the human brain is not an anatomical fait accompli at birth, or even after childhood.

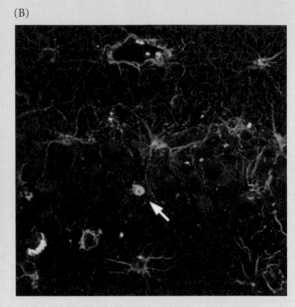

Figure 9.25 Evidence of adult neural stem cells. The green staining, which indicates newly divided cells, is from a fluorescent antibody against BrdU (a thymidine analogue that is taken up only during the S phase of the cell cycle). (A) Newly generated adult mouse neurons (green cells) have a normal morphology and receive synaptic inputs. The red spots are synaptophysin, a protein found on the dendrites at the synapses of axons from other neurons. (B) Newly generated neuron (arrow) in the adult human brain. This cell is located in the dentate gyrus of the hippocampus. The red fluorescence is from an antibody that stains only neural cells. Yellow indicates the overlap of red and green. Glial cells are stained purple. (A from van Praag et al. 2002; B from Erikksson et al. 1998, photograph courtesy of F. H. Gage.)

The existence of neural stem cells in adults is now well established for the olfactory epithelium and the hippocampus (Kempermann et al. 1997a,b; Kornack and Rakic 1999; van Praag et al. 1999; Kato et al. 2001). These cells respond to Sonic hedgehog and can proliferate to become multiple cell types for at least the first year of a mouse's life (Ahn and Joyner 2005). It appears that the stem cells producing these neurons are located in the ependyma (the former ventricular zone, in which the embryonic neural stem cells once resided) or in the subventricular zone (SVZ) adjacent to it (Doetsch et al. 1999; Johansson et al. 1999; Cassidy and Frisén 2001). These adult neural stem cells represent only about 0.3% of the ventricle wall cell population, but they can be distinguished from more differentiated cells by their cell surface proteins* (Rietze et al. 2001). In the adult mouse, thousands of new neuroblasts are generated each day, migrating from the lateral SVZ to the olfactory bulb, where they differentiate into several different types of neurons. Recent evidence suggests that

*These neural stem cells may have particular physiological roles as well. During pregnancy, prolactin stimulates the production of neuronal progenitor cells in the subventricular zone of the adult mouse forebrain. These progenitor cells migrate to produce olfactory neurons that may be important for maternal behavior of rearing offspring (Shingo et al. 2003).

these stem cells are not multipotent (becoming specified only when they reach the olfactory bulb) but instead are a population of heterogenous neuroblasts that are already committed to becoming certain neuronal types (Merkle et al. 2007). These adult neural stem cells proliferate in response to exercise, learning, and stress (Zhang et al. 2008).

The existence of adult neural stem cells in the cortex is more controversial (see Gould 2007). Some investiga-

tors (Gould et al. 1999a,b; Magavi et al. 2000) claim to have identified them; other scientists (see Rakic 2002) question the existence of these cortical neural stem cells. The mechanisms by which neural stem cells are kept in a state of ready quiescence well into adulthood are currently being explored. Before they become neurons, neural stem cells are characterized by the expression of the NRSE translational inhibitor that prevents neuronal differentiation by binding to

a silencer region of DNA (see Chapter 2). When neural stem cells begin to differentiate, they synthesize a small, double-stranded RNA that has the same sequence as the silencer and which might bind NRSE and thereby permit neuronal differentiation (Kuwabara et al. 2004). The use of cultured neuronal stem cells to regenerate or repair parts of the brain will be considered in Chapter 17.

CORTICAL CELL MIGRATION The first cortical neurons to be generated migrate out of the germinal zone to form the transient **preplate** (Kawauchi and Hoshino 2008). Subsequently generated neurons migrate into the preplate and separate it into two layers: the Cajal-Retzius layer and the subplate. The Cajal-Retzius layer becomes and remains the most superficial layer of the neocortex, and its cells express the cell surface glycoprotein Reelin. The subplate remains the deepest layer through which the successive waves of neuroblasts travel to form the cortical plate. The Reelin-producing cells of the Cajal-Retzius layer are critical in the separation of the preplate. In Reelin-deficient mice, the preplate fails to split, and the neurons produced by the germinal layers pile up behind the previously generated neurons (instead of migrating through them). By activating the Notch pathway, Reelin on the surface of the Cajal-Retzius cells allows the neuronal stem cell to produce a long fiber that extends through the cortical plate (Hashimoto-Torii et al. 2008; Nomura et al. 2008). This (and the fact that it produced some proteins thought to be glial-specific) caused the neural stem cell to be called the *radial glial cell*. The process from this cell becomes critical for the migration of the neural cells produced by the germinal zones.

We also know that there are mutations that specifically affect the microtubular cytoskeleton of the migrating neuroblasts. Mutations in the *DISC1* gene prevent neuronal migration in the cortex by interfering with microtubule assembly; humans with such mutations have been seen to suffer from mental dysfunctions, among them autism, bipolar diso, and schizophrenia* (Kamiya et al. 2005, 2008).

We still do not know the nature of the information given to the cell as it becomes committed. However, Hanashina and her colleagues (2004) have shown that there are several genetic switches that get "thrown" at these division

times. One of these switches is the gene encoding the transcription factor Foxg1. When the mouse neuronal progenitor cells divide to form the first layer of cortical neurons, *Foxg1* is not expressed in the progenitor cells or in the first-formed neurons. However, later, when the progenitor cells generate those neurons destined for layers 4 and 5, they express this gene. If the *Foxg1* gene is conditionally knocked out of this lineage, the neural precursor cells continually give rise to layer 1 neurons. Therefore, it seems that the Foxg1 transcription factor is required to suppress the "layer 1" neural fate.

Neither the vertical nor the horizontal organization of the cerebral cortex is clonally specified—that is, none of the functional units form from the progeny of a single cell. Rather, the developing cortex forms from the mixing of cells derived from numerous stem cells. The early regionalization of the neocortex is thought to be organized by paracrine factors secreted by the epidermis and neural crest cells at the margins of the developing brain (Rakic et al. 2009). The paracrine factors induce the expression of transcription factors in the specific brain regions, which then mediate the survival, differentiation, proliferation, and migration of the newly generated neurons.

For instance, Fgf8 protein is secreted by the anterior neural ridge and is important for specifying the telencephalon (**Figure 9.26**). If Fgf8 is overexpressed in the ridge, specification of the telencephalon is extended caudally, whereas if Fgf8 is ectopically added to the caudal region of the cortex, part of that caudal region will become anterior (Fukuchi-Shimogori and Grove 2001, 2005). Sonic hedgehog is secreted by the medial ganglionic eminence and helps form the ventral neurons of the cortex, including those of the substantia nigra (whose absence causes Parkinson disease).

NEURONS TO GLIA The stem cells of the vertebrate cortex make neurons first and then make glial cells. In mice, neurons are formed from embryonic day 12 through embryonic day 18. Then, at embryonic day 18, the same precursor cells generate glia. When cortical progenitor cells (cortical

---

*DISC1 mutations are not the cause of most cases of these diseases, but they are seen in the rare cases where these behavioral diseases are transmitted as Mendelian traits (see Marx 2007).

FIGURE 9.26 Building the proto-map of the cerebral cortex. (A) Day 13 mouse embryo at the point of the formation of the two telencephalic vesicles. Fgf8 is secreted from the anterior neural ridge and the commissural plate; Wnts and BMPs are secreted from the cortical hem; and sonic hedgehog (Shh) is secreted from the medial ganglionic eminence. These result in a graded expression of transcription factors that translates into a molecular regionalization of the germinal zone. (B) Fgf signaling affects the identity of cortical regions. Ectopic expression of Fgf8 in the caudal pole of the neocortex generates a partial duplication of the S1 barrel-field (the receptive field for whisker responses) usually seen in the anterior. (After Dehay and Kennedy 2007.)

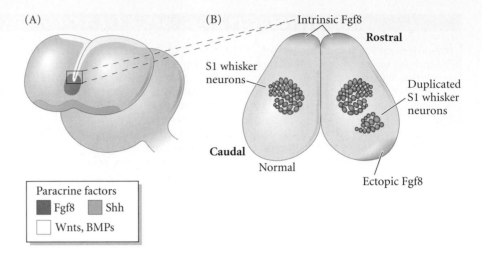

stem cells such as RGCs and their more committed descendants) are cultured on embryonic day 12, they differentiate into neurons for the first few days. After multiple days of culture, the progenitors switch and start making glia (Götz and Barde 2005). Retroviral tracers and time-lapse microcinematography have shown that individual precursor cells make both neurons and glia (Reid et al. 1995; Qian et al. 2000; Shen et al. 2006).

Both internal and external factors regulate this transition. Cortical precursor cells form neurons when cultured on embryonic cortical slices, but they become glia when placed on older cortical slices (Morrow et al. 2001). One of the main factors involved in this environmental regulation is CT-1, a paracrine factor that activates the JAK-STAT pathway. The STAT transcription factors activate glial-specific genes. Moreover, this transition must be done in the absence of neurotrophic factors, as these activate the RTK-MAPK pathway, resulting in neural-specific gene expression (see Miller and Gauthier 2007).

SIDELIGHTS & SPECULATIONS

The Unique Development of the Human Brain

There are many differences between humans and chimpanzees. These include our hairless, sweaty skin and striding bipedal posture. However, the most striking and significant differences occur in brain development. The enormous growth and asymmetry of our neocortex and our ability to reason, remember, plan for the future, and learn language and cultural skills make humans unique in the animal kingdom (Varki et al. 2008). The development of the human neocortex is strikingly plastic and is an almost constant work in progress. Several developmental phenomena have been identified that distinguish the development of the human brain from that of other species, including other primates:

1. The retention of the fetal neuronal growth rate after birth.
2. The activity of human-specific RNA genes.
3. The high activity of transcription.
4. Human-specific alleles of developmental regulatory genes.
5. The continuation of brain maturation into adulthood.

Fetal neuronal growth rate after birth

If there is one important developmental trait that distinguishes humans from the rest of the animal kingdom, it is our retention of the fetal neuronal growth rate. Both human and ape brains have a high growth rate before birth. After birth, however, this rate slows greatly in

the apes, whereas human brain growth continues at a rapid rate for about 2 years (**Figure 9.27A**; Martin 1990). During early postnatal development, we add approximately 250,000 neurons per minute (Purves and Lichtman 1985). The ratio of brain weight to body weight at birth is similar for great apes and humans, but by adulthood the ratio for humans is literally "off the chart" when compared with that of other primates (**Figure 9.27B**; Bogin 1997).

At the cellular level, we find that no fewer than 30,000 synapses per second per cm² of cortex are formed during the first few years of human life. It is thought that this increase in neuron number may (1) generate new modules (addressable sites) that can acquire new functions; (2) store new

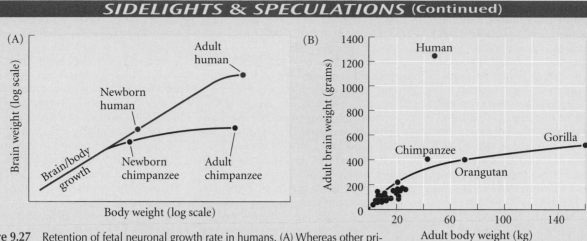

Figure 9.27 Retention of fetal neuronal growth rate in humans. (A) Whereas other primates, such as chimpanzees, complete neuronal proliferation around the time of birth, the neurons of newborn humans continue to proliferate at the same rate as the fetal brain neurons. (B) The human brain/body weight ratio (encephalization index) is about 3.5 times higher than that of apes. (After Bogin 1997.)

memories for use in thinking and forecasting; and (3) enable learning by interconnecting among themselves and with prenatally generated neurons (Rose 1998; Barinaga 2003). It is in this early postnatal stage that intervention is thought to be able to raise IQ (reviewed in Wickelgren 1999).

The prodigious rate of continued neuron production has important consequences. Portmann (1941), Montagu (1962), and Gould (1977) have each made the claim that we are essentially "extrauterine fetuses" for the first year of life. If one follows the charts of ape maturity, actual human gestation is 21 months. Our "premature" birth" is an evolutionary compromise between maternal pelvic width, the fetal head circumference, and lung maturity. The mechanism for retaining the fetal neuronal growth rate has been called *hypermorphosis*—the phyletic extension of development beyond its ancestral state (Vrba 1996; Vinicius and Lahr 2003).

See WEBSITE 9.6 Neuronal growth and the invention of childhood

Genes for neuronal growth
Yet humans and chimpanzees have remarkably similar genomes. When the protein-encoding DNAs are compared, we are around 99% identical. However, the protein-coding regions comprise only around 2% of the

genome. In 1975, King and Wilson concluded from their studies of human and chimpanzee proteins that "The organismal differences between chimpanzees and humans would then result chiefly from genetic changes in a few regulatory systems, while amino acid substitutions in general would rarely be a key factor in major adaptive shifts." This was one of the first suggestions that evolution can occur through changes in developmental regulatory genes. When the total genomes are compared, humans and chimpanzees differ at about 4% of their sequences, and most of these are in the noncoding regions (see Varki et al. 2008).

Although there are some brain growth genes (e.g., *ASPM* and *microcephalin*) whose sequences differ between humans and apes, these differences have not been correlated with the huge growth of human brains. Rather, the critical differences may reside in the DNA sequences that encode noncoding RNAs. These RNAs are highly expressed in the developing brain and may regulate the transcription or translation of neuronal transcription factors. Computer analysis comparing different mammalian genomes may have found such genes (Pollard et al. 2006a,b; Prabhakar et al. 2006). First, these studies identified a relatively small group of noncoding DNA regions that were conserved among non-human mammals. This represents around 2% of the genome, and

it was assumed that if these regions were conserved throughout mammalian evolution, they were important. Second, they compared these sequences to their human homologues to see if any of these regions have changed between humans and other mammals. This study identified 49 such regions where the sequence is highly conserved among mammals but has diverged rapidly between humans and chimps. The most rapid divergence is seen in the sequence *HAR1* (*human accelerated region-1*), where 18 sequence changes were seen between chimps and humans when one would expect, based on mammalian evolution, to see only one or no substitutions. *HAR1* is expressed in the developing brains of humans and apes, especially in the *Reelin*-expressing Cajal-Retzius neurons that are known to be responsible for directing neuronal migration during the formation of the six-layered neocortex. Research is ongoing to discover the function of *HAR1* and the other HAR genes that are in the conserved noncoding region of the genome.

High transcriptional activity
In the 1970s, A. C. Wilson suggested that the difference between humans and chimpanzees might reside in the *amount* of proteins made from their genes (see Gibbons 1998), and there is now evidence supporting this hypothesis. Using microarrays to study global patterns of gene expression, several

(Continued on next page)

recent investigations have found that, although the quantities and types of genes expressed in human and chimp livers and blood are indeed extremely similar, human *brains* produce more than five times as much mRNA as chimpanzee brains (Enard et al. 2002a; Preuss et al. 2004). In humans, transcription of some genes (such as *SPTLC1*, a gene whose defect causes sensory nerve damage) was elevated 18-fold over the same genes' expression in the chimp cortex, while other genes (such as *DDX17*, whose product is involved in RNA processing) are expressed 10 times less in human than in chimp cortices.

Speech, language, and the FOXP2 gene

Spoken language is a uniquely human trait and is presumed to be the prerequisite for the evolution of cultures. Speech entails the fine-scale control of the larynx (voice box) and mouth. Individuals who are heterozygous for mutations at the *FOXP2* locus have severe problems with language articulation and with forming sentences (Vargha-Khadem et al. 1995; Lai et al. 2001). This observation has provided genetic anthropologists with an interesting gene to study. Enard and colleagues (2002b) have shown that, although the *FOXP2* gene is conserved throughout most of mammalian evolution, it has a unique form in humans, having accumulated at least two amino acid-changing mutations just since our divergence from the com-

mon ancestor of humans and chimpanzees. These differences are significant, since human and chimpanzee forms of the Foxp2 protein differentially regulate more than 100 genes (Konopka et al. 2009).

In the mouse, the *Foxp2* gene is expressed in the developing brain, but its major site of expression is the lung (Shu et al. 2001). In humans, *FOXP2* is predominantly expressed in those brain regions that coordinate speech (i.e., the caudate nucleus and inferior olive nuclei); these sites are abnormal in patients with *FOXP2* deficiency (Lai et al. 2003). In birds, the Foxp2 protein is associated with song learning, and the experimental downregulation of *Foxp2* expression in certain areas of the brain prevents young male birds from imitating their species-specific song (Teramitsu and White 2006; Haesler et al. 2007). Although it is not certain that *FOXP2* is the most critical gene for human language acquisition, it seems to be very important for allowing the orofacial movements and grammar characteristic of human speech.

Teenage brains: Wired and unchained

Until recently, most scientists thought that after the initial growth of neurons during fetal development and early childhood, rapid neural proliferation ceased. However, magnetic resonance imaging (MRI) studies have shown that the brain keeps developing until around puberty, and that not all areas of the brain mature simultaneously

(Giedd et al. 1999; Sowell et al. 1999). Soon after puberty, neuronal growth ceases and pruning occurs. The time of this "pruning" correlates with the time when language acquisition becomes difficult (which may be why children learn language more readily than adults). There is also a wave of myelin production ("white matter" from the glial cells that surround neuronal axons) at this time. Myelination is critical for proper neural functioning, and the greatest differences between brains in early puberty and those in early adulthood involve the frontal cortex (**Figure 9.28**; Sowell et al. 1999; Gogtay et al. 2004).

These differences in brain development may explain the extreme responses teenagers have to certain stimuli, as well as their ability to learn certain tasks. In tests using functional MRI to scan subjects' brains while emotion-charged pictures flashed on a computer screen, the brains of young teens showed activity in the amygdala, which mediates fear and strong emotions. When older teens were shown the same pictures, most of their brain activity was centered in the frontal lobe, an area involved in more reasoned perceptions (Baird et al. 1999; Luna et al. 2001). The teenage brain is a complicated and dynamic entity, and (as any parent knows) one that is not easily understood. But if one survives these years, the resulting adult brain is usually capable of making reasoned decisions, even in the onslaught of emotional situations.

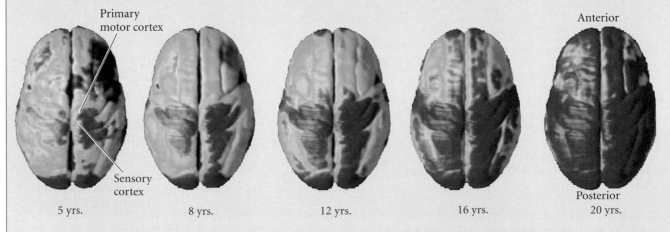

Figure 9.28 Dorsal view of the human brain showing the progression of myelination ("white matter") over the cortical surface during adolescence. (Images courtesy of N. Gogtay.)

Primary motor cortex

Sensory cortex

5 yrs.　8 yrs.　12 yrs.　16 yrs.　20 yrs.

Anterior

Posterior

Low　Myelination (percent "white matter")　High

DEVELOPMENT OF THE VERTEBRATE EYE

An individual gains knowledge of its environment through its sensory organs. The major sensory organs of the head develop from interactions of the neural tube with a series of epidermal thickenings called the **cranial ectodermal placodes** (discussed in more detail in Chapter 10). Most of these placodes form neurons and sensory epithelia. The two **olfactory placodes** form the nasal epithelium (smell receptors) as well as the ganglia for the olfactory nerves. Similarly, the two **otic placodes** invaginate to form the inner ear labyrinth, whose neurons form the acoustic ganglia that enable us to hear. In this section, we will focus on the development of the eye from the *lens placode*.

The lens placode does not form neurons; rather, it forms the transparent lens that allows light to impinge on the retina. The interactions between the cells of the lens placode and the presumptive retina structure the eye via a cascade of reciprocal changes that enable the construction of an intricately complex organ.

The Dynamics of Optic Development

We first discussed the interactive induction of the vertebrate eye in Chapter 3. At gastrulation, the involuting endoderm and mesoderm interact with the adjacent prospective head ectoderm to give the head ectoderm a lens-forming bias (Saha et al. 1989). But not all parts of the head ectoderm eventually form lenses, and the lens must have a precise spatial relationship with the retina. The activation of the head ectoderm's latent lens-forming ability and the positioning of the lens in relation to the retina are accomplished by the **optic vesicle (Figure 9.29)**. The optic vesicle extends from the diencephalon, and where it meets the head ectoderm, it induces the formation of a **lens placode**, which then invaginates to form the lens. This invagination is accomplished by the cells of the lens placode extending adhesive filopodia to contact the optic vesicle. As the optic vesicle bends to form the optic cup, the presumptive lens cells are brought inside the embryo (Chauhan et al. 2009). As the optic vesicle becomes the **optic cup**, its two layers differentiate in different ways. The cells of the outer layer produce melanin pigment (being one of the few tissues other than the neural crest cells that can form this pigment) and ultimately become the **pigmented retina**. The cells of the inner layer proliferate rapidly and generate a variety of glia, ganglion cells, interneurons, and light-sensitive photoreceptor neurons. Collectively, these cells constitute the **neural retina**. The retinal ganglion cells are neurons whose axons send electric impulses to the brain. Their axons meet at the base of the eye and travel down the optic stalk, which is then called the **optic nerve**.

How is a specific region of neural ectoderm informed that it will become the optic vesicle? It appears that a group of transcription factors—Six3, Pax6, and Rx1—are expressed together in the most anterior tip of the neural plate. This single domain will later split into the bilateral regions that form the optic vesicles (Adelmann 1929; Mathers et al. 1995; Chiang et al. 1996). Again, we see the similarities between the *Drosophila* and vertebrate nervous systems, for these three proteins are also necessary for the formation of the *Drosophila* eye.

Formation of the Eye Field

Eye formation has been studied in several organisms, but historically the amphibian eye has received the most attention. (Indeed, this was this organ that Spemann began studying in fin de siècle Germany; see Saha 1991.) Using

(A) 4-mm embryo

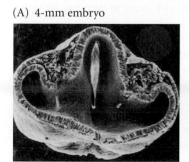

(B) 4.5-mm embryo

Lens placode

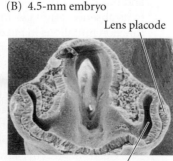

Optic vesicle

(C) 5-mm embryo

Lens vesicle

Optic cup

(D) 7-mm embryo

Retina   Lens

Cornea

FIGURE 9.29 Development of the vertebrate eye (see also Figure 3.15). (A) The optic vesicle evaginates from the brain and contacts the overlying ectoderm, inducing a lens placode. (B,C) The overlying ectoderm differentiates into lens cells as the optic vesicle folds in on itself, and the lens placode becomes the lens vesicle. (C) The optic vesicle becomes the neural and pigmented retina as the lens is internalized. (D) The lens vesicle induces the overlying ectoderm to become the cornea. (A–C from Hilfer and Yang 1980, courtesy of S. R. Hilfer; D courtesy of K. Tosney.)

FIGURE 9.30 Dynamic formation of the eye field in the anterior neural plate. (A) Formation of the eye field. Light blue represents the neural plate; moderate blue indicates the area of Otx2 expression (forebrain); and dark blue indicates the region of the eye field as it forms in the forebrain. (B) Dynamic expression of transcription factors leading to specification of the eye field. Prior to stage 10, Noggin inhibits *ET* expression but promotes expression of *Otx2*. Otx2 protein then represses the inhibition of *ET* by Noggin signaling. The ET transcription factor activates *Rx1*, which encodes a transcription factor that blocks *Otx2* and promotes *Pax6* expression. Pax6 protein initiates the cascade of gene expression constituting the eye field (at right). (C) Location of the transcription factors in the nascent eye field of stage 12.5 (early neurula) and stage 15 (mid-neurula) *Xenopus* embryos, showing a concentric organization of transcription factors having domains of decreasing size: Six3 > Pax6 > Rx1 > Lhx2 > ET. (After Zuber et al. 2003.)

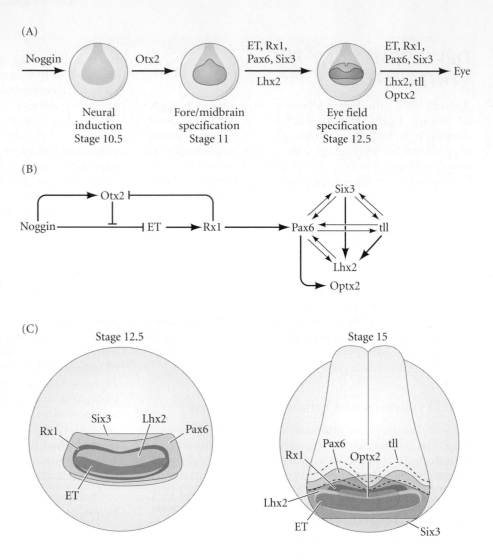

in situ hybridization and injection of transcription factor mRNA into developing *Xenopus* embryos, Zuber and colleagues (2003) were able to outline the mechanisms by which the amphibian eye field becomes specified.

Formation of the eye field begins with the specification of the neural tube. The anterior portion of the neural tube, where both BMP and Wnt pathways are inhibited, is specified by the expression of the *Otx2* gene. Noggin is especially important in blocking BMPs, as this not only allows *Otx2* expression but also inhibits expression of the transcription factor ET, one of the first proteins expressed in the eye field. However, once Otx2 protein accumulates, it blocks Noggin's ability to inhibit the *ET* gene, so ET protein is produced.

One of the genes controlled by ET is *Rx1* (also called *Rx*), whose product helps specify the retina. Rx1 is a transcription factor that acts first by inhibiting *Otx2*, and second by activating *Pax6*, the major gene in forming the eye field in the anterior neural plate (**Figure 9.30**; Zuber et al. 2003). As discussed in Chapters 2 and 3, the Pax6 protein is especially important in the specification of the lens and retina. Indeed, it appears to be a common denominator for photoreceptive cells in all phyla. When the mouse *Pax6* gene is

inserted into the *Drosophila* genome and activated randomly, *Drosophila* eyes form (see Figure 19.3; Halder et al. 1995).

Pax6 is also expressed in the mouse forebrain, hindbrain, and nasal placodes, but the eyes seem to be most sensitive to its absence. Humans and mice heterozygous for loss-of-function mutations in *Pax6* have small eyes, while homozygotic mice and humans (and *Drosophila*) lack eyes altogether (Jordan et al. 1992; Glaser et al. 1994; Quiring et al. 1994). In both flies and vertebrates, Pax6 protein initiates a cascade of transcription factors that mutually activate one another to constitute the eye field. These genes have overlapping functions that are critical for forming the retina and producing the compounds that will induce the lens (see Gestri et al. 2005; Tétreault et al. 2009).

In vertebrates, separation of the single eye field into two bilateral fields depends on the secretion of Sonic hedgehog (Shh; **Figure 9.31**). If the mouse *Shh* gene is mutated, or if the processing of this protein is inhibited, the single median eye field does not split. The result is **cyclopia**—a single eye in the center of the face, usually below the nose (see Figure 9.31C and Figure 3.26; Chiang et al. 1996; Kelley et al. 1996; Roessler et al. 1996). Shh from the prechordal plate suppresses *Pax6* expression in the center of the

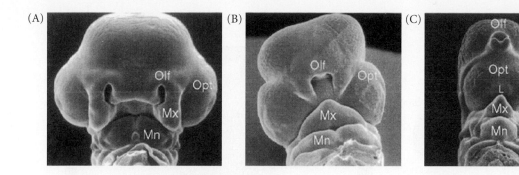

(A) (B) (C)

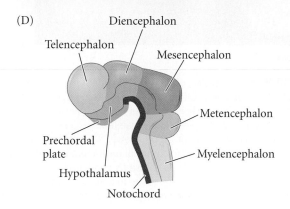

(D)

Telencephalon
Diencephalon
Mesencephalon
Metencephalon
Myelencephalon
Prechordal plate
Hypothalamus
Notochord

FIGURE 9.31 Sonic hedgehog separates the eye field into bilateral fields. Jervine, an alkaloid found in certain plants, inhibits endogenous *Shh* signaling. (A) Scanning electron micrograph showing the external facial features of a normal mouse embryo. (B) Mouse embryos exposed to 10 µM jervine had variable loss of midline tissue and resulting fusion of the paired, lateral olfactory processes (Olf), optic vesicles (Opt), and maxillary (Mx) and mandibular (Mn) processes of the jaw. (C) Complete fusion of the mouse optic vesicles and lenses (L) resulted in cyclopia. (D) Drawing showing the location of the prechordal plate (the source of Shh) in the 12-day mouse embryo. (A–C from Cooper et al. 1998, courtesy of P. A. Beachy.)

embryo, dividing the field in two. The phenomenon of human cyclopia also involves loss-of-function mutations that prevent Shh from functioning.

Conversely, if too much Shh is synthesized by the prechordal plate, *Pax6* is suppressed in too large an area and the eyes fail to form at all. This phenomenon explains why cavefish are blind. Yamamoto and colleagues (2004) demonstrated that the difference between the surface population of the Mexican tetra fish *Astyanax mexicanus* and eyeless cave-dwelling populations of the same species is the amount of Shh secreted from the prechordal plate. Elevated Shh was probably selected in cave-dwelling species because it resulted in heightened oral sensing and larger jaws (Yamamoto et al. 2009). However, Shh also downregulates *Pax6*, resulting in the disruption of optic cup development, apoptosis of lens cells, and arrested eye development (**Figure 9.32**). This sequence can be verified experimentally: injecting *Shh* mRNA into one side of the surface of fish embryos blocks eye development on that side only.

FIGURE 9.32 Surface-dwelling (A) and cave-dwelling (B) Mexican tetras (*Astyanax mexicanus*). The eye fails to form in the population that has lived in caves for over 10,000 years (top right). Two genes that respond to Shh proteins, *Ptc2* and *Pax2*, are expressed in broader domains in the cavefish embryos than in those of surface dwellers (center). The embryonic optic vesicles (bottom) of surface-dwelling fish are normal size and have small domains of *Pax2* expression (specifying the optic stalk). The optic vesicles of the cave-dwelling fishes' embryos (where *Pax6* is usually expressed) are much smaller, and the *Pax2*-expressing region has grown at the expense of the *Pax6* region. (From Yamamoto et al. 2004; photographs courtesy of W. Jeffery.)

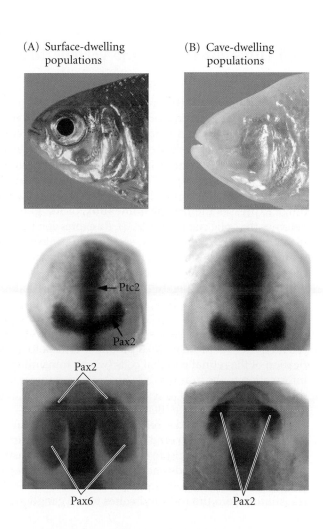

(A) Surface-dwelling populations   (B) Cave-dwelling populations

(A)

Xrx1 in eye field

Pineal gland

(B)

Retina

Ventral
hypothalamus

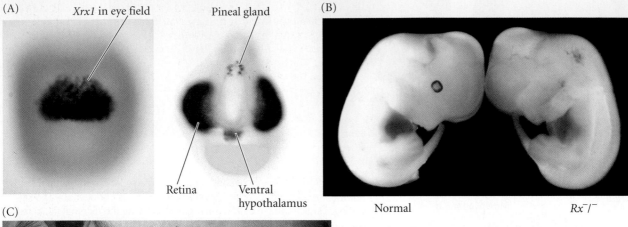

(C)

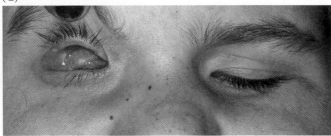

FIGURE 9.33 Expression of *Rx* genes in vertebrate retina development. (A) Expression pattern of the *Xenopus Xrx1* gene in the single eye field of the early neurula (left) and in the two developing retinas (as well as in the pineal, an organ that has a presumptive retina-like set of photoreceptors) of a newly hatched tadpole (right). (B) Eye development in a normal mouse embryo (left) and lack of eyes in a mouse embryo whose *Rx* gene has been knocked out (right). (C) Absence of ocular tissue in a human patient with mutant *Rx* genes. (From Bailey et al. 2004, courtesy of M. Jamrich.)

Normal

Rx⁻/⁻

Cell Differentiation in the Vertebrate Eye

Neural retina differentiation

As in other regions of the brain, the dorsal-ventral polarity of the eye depends on gradients of BMP signals from the dorsal region and hedgehog proteins from the ventral domains. The most ventral cells express Pax2 protein and become the optic stalk; the most dorsal region responds to BMPs by expressing the MITF transcription factor and generates the pigmented epithelium. The central region of the bulge expresses the **retinal homeobox**, or ***Rx*, gene** and becomes the optic cup (**Figure 9.33**). *Rx* expression is upregulated by Otx2 protein, which specifies the anterior head region and encodes a transcription factor that activates numerous genes; among these genes are *Pax6* and *Six3*, both of which help specify the retinal neurons (Bailey et al. 2004; Voronina et al. 2004; see Figure 9.30). While initial expression of *Pax6* and *Six3* is not dependent on *Rx* expression, their continued expression in the retinal progenitor cells requires Rx protein. Pax6 protein is critical for specification of the retinal ganglial cells (which transmit visual information to the brain), while Six3 is necessary for coordinating the number of times retinal precursor cells divide before they differentiate (Bene et al. 2004).

Like the cerebral and cerebellar cortices, the neural retina develops into a layered array of different neuronal types. These layers include the light- and color-sensitive photoreceptor cells (**rods** and **cones**); the cell bodies of the ganglion cells; and **bipolar interneurons** that transmit electric stimuli from the rods and cones to the ganglion cells

(A)   (B)

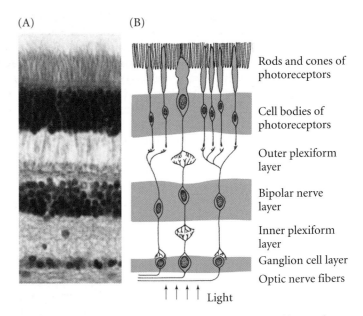

Rods and cones of photoreceptors

Cell bodies of photoreceptors

Outer plexiform layer

Bipolar nerve layer

Inner plexiform layer

Ganglion cell layer

Optic nerve fibers

↑ ↑ ↑ ↑ Light

FIGURE 9.34 Retinal neurons sort out into functional layers during development. (A) The three layers of neurons in the adult retina and the synapses between them. (B) A functional depiction of the major neuronal pathway in the retina. Light traverses the layers until it is received by the photoreceptors. The axons of the photoreceptors synapse with bipolar neurons, which transmit electric signals to the ganglion cells. The axons of the ganglion cells join to form the optic nerve, which enters the brain. (A courtesy of G. Grunwald.)

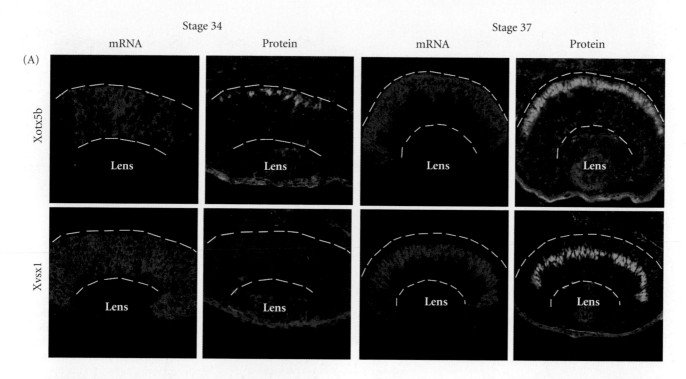

(A)

Stage 34
mRNA   Protein

Stage 37
mRNA   Protein

Xotx5b

Xvsx1

(B) MicroRNA *miR222*

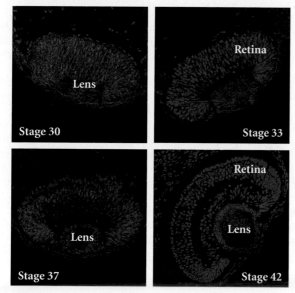

FIGURE 9.35 Timing of retinal neurogenesis by the translation of mRNAs encoding transcription factors Xotx5b and Xvsx1. (A) Comparison of *Xotx5b* (photoreceptor) and *Xvsx1* (bipolar neuron) mRNAs and proteins at stages 34 (early retinal neurogenesis) and 37 (late retinal neurogenesis). The mRNAs for both proteins are synthesized, but at stage 34 only the *Xotx5b* message has been translated. At stage 37, both messages have been translated. (B) The diminishing expression of *miR222* (pink) from stage 30 through stage 42. (From Decembrini et al. 2006, 2009.)

(Figure 9.34). In addition, the retina contains numerous **Müller glial cells** that maintain its integrity, **amacrine neurons** (which lack large axons), and **horizontal neurons** that transmit electric impulses in the plane of the retina.

The neuroblasts of the retina are competent to generate all seven retinal cell types. For instance, if one injects an individual retinal neuroblast with a genetic marker, that marker will be seen in a strip that can include all the different cell types of the retina (Turner and Cepko 1987; Yang 2004). In amphibians, the type of neuron produced from the multipotent retinal stem cell appears to be dependent on the timing of gene translation, not the location of gene transcription. Photoreceptor neurons, for instance, are specified by expression of the *Xotx5b* gene, while expression of *Xotx2* and *Xvsx1* are critical for specifying the bipolar neurons. Interestingly, these three genes are *transcribed* in all retinal cells, but they are *translated* differently (**Figure 9.35A**). Those neurons whose birthday is at stage 30 translate the *Xotx5b* mRNA and become photoreceptors, while those neurons forming later, whose birthdays are at stage 35, translate the *Xotx2* and *Xvsx1* messages and become bipolar interneurons (Decembrini et al. 2006, 2009).

This time-dependent regulation of translation is mediated by four microRNAs: *miR129*, *miR155*, *miR214*, and *miR222*. These microRNAs are highly expressed in early retinal progenitor cells and bind to the 3′ UTRs of the *Xotx2* and *Xvsx1* mRNAs, inhibiting their translation and thus preventing early-differentiating retinal cells from becoming bipolar neurons. Expression of these microRNAs diminishes with time (**Figure 9.35B**; Decembrini et al. 2009), so that later-forming neurons are no longer blocked from acquiring bipolar interneuron characteristics.

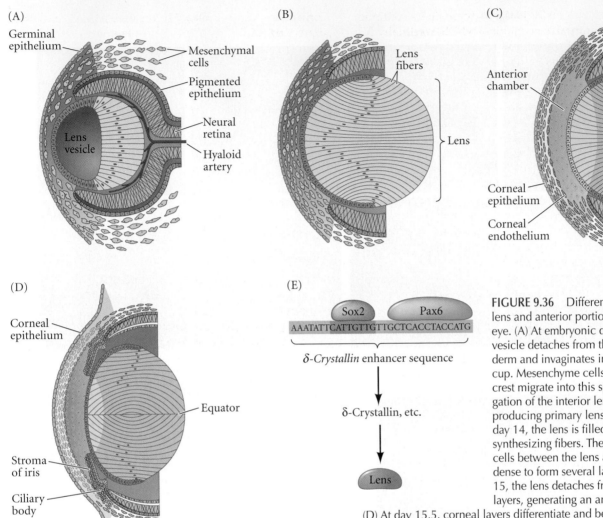

(A)
Germinal epithelium
Mesenchymal cells
Pigmented epithelium
Neural retina
Hyaloid artery
Lens vesicle

(B)
Lens fibers
Lens

(C)
Anterior chamber
Corneal epithelium
Corneal endothelium

(D)
Corneal epithelium
Equator
Stroma of iris
Ciliary body

(E)
Sox2   Pax6
AAATATTCATTGTTGTTGCTCACCTACCATG

δ-*Crystallin* enhancer sequence

↓

δ-Crystallin, etc.

↓

Lens

FIGURE 9.36 Differentiation of the lens and anterior portion of the mouse eye. (A) At embryonic day 13, the lens vesicle detaches from the surface ectoderm and invaginates into the optic cup. Mesenchyme cells from the neural crest migrate into this space. The elongation of the interior lens cells begins, producing primary lens fibers. (B) At day 14, the lens is filled with crystallin-synthesizing fibers. The mesenchyme cells between the lens and surface condense to form several layers. (C) At day 15, the lens detaches from the corneal layers, generating an anterior cavity. (D) At day 15.5, corneal layers differentiate and begin to become transparent. The anterior edge of the optic cup enlarges to form a non-neural region containing the iris muscles and the ciliary body. New lens cells are derived from the anterior lens epithelium. As the lens grows, the nuclei of the primary lens cells degenerate and new lens fibers grow from the epithelium on the lateral sides. (E) Close binding of the Sox2 and Pax6 transcription factors on a small region of the δ-*crystallin* enhancer. (A–D after Cvekl and Tamm 2004; E after Kondoh et al. 2004.)

Not all the cells of the optic cup become neural tissue. The tips of the optic cup on either side of the lens develop into a pigmented ring of muscular tissue called the **iris**. The iris muscles control the size of the pupil (and give an individual his or her characteristic eye color). At the junction between the neural retina and the iris, the optic cup forms the **ciliary body**. This tissue secretes the **aqueous humor**, a fluid needed for the nutrition of the lens and for forming the pressure needed to stabilize the curvature of the eye and the constant distance between the lens and the cornea.

Lens and cornea differentiation

We have been focusing on the retina, but the eye can't focus on the retina unless it has a lens and a cornea. The differentiation of the lens tissue into a transparent membrane capable of directing light onto the retina involves changes in cell structure and shape as well as the synthesis of transparent, lens-specific proteins called **crystallins**.

Shortly after the lens vesicle has detached from the surface ectoderm, mesenchymal cells from the neural crest migrate into the space between the lens and the surface epithelium. These cells condense to form several flat layers of cells, eventually becoming the corneal precursor cells (**Figure 9.36A,B**; Cvekl and Tamm 2004). As these cells mature, they dehydrate and form tight junctions among the cells, forming the cornea. Intraocular fluid pressure (from the aqueous humor) is necessary for the correct curvature of the cornea, allowing light to be focused on the retina (Coulombre 1956, 1965). Intraocular pressure is sustained by a ring of scleral bones (derived from the neural crest), which acts as an inelastic restraint. As invagination proceeds, the cells at the inner portion of the lens vesicle elongate and, under the influence of the neural retina, become the lens fibers (Piatigorsky 1981). As these fibers continue to grow, they synthesize crystallins, which eventually fill up the cell and cause the extrusion of the nucleus.

The crystallin-containing fibers eventually fill the space between the two layers of the lens vesicle (**Figure 9.36C**). The anterior cells of the lens vesicle constitute a germinal epithelium, which continues dividing. These dividing cells move toward the equator of the vesicle, and as they pass through the equatorial region, they, too, begin to elongate (**Figure 9.36D**). Thus, the lens contains three regions: an anterior zone of dividing epithelial cells, an equatorial zone of cellular elongation, and a posterior and central zone of crystallin-containing fiber cells. This arrangement persists throughout the lifetime of the animal as fibers are continuously being laid down. In the adult chicken, the process of differentiation from an epithelial cell to a lens fiber takes 2 years (Papaconstantinou 1967).

The optic vesicle induces the expression of many transcription factors in the presumptive lens ectoderm. In *Xenopus*, for instance, contact is needed between the optic vesicle and the presumptive lens ectoderm. When this contact is made, Delta proteins on the optic vesicle activate the Notch receptors on the presumptive lens ectoderm (Ogino et al. 2008). The Notch intracellular domain binds to one of the enhancer elements of the *Lens1* gene, and in the presence of the Otx2 transcription factor (which is expressed throughout the entire head region), this gene is activated. The Lens1 protein is itself a transcription factor that is essential for epithelial cell proliferation (making and growing the lens placode) and eventually for closing the lens vesicle. In this interaction we see a principle that is observed throughout development—namely, that some transcription factors (such as Otx2) specify a particular field and provide competence for cells to respond to a more specific induction (such as Notch) within the field.

The optic vesicle also provides paracrine factors needed for inducing lens-specific transcription factors. As mentioned in Chapters 2 and 3, the regulation of *crystallin* genes is under the control of Pax6, Sox2, and L-Maf (**Figure 9.36E**). Like Otx2, Pax6 appears in the head ectoderm before the lens is formed, and Sox2 (formerly called δEF2) is induced in the lens placode by BMP4 secreted from the optic vesicle. Co-expression of Pax6 and Sox2 in the same cells initiates lens differentiation, and activation of the crystallin genes. L-Maf is induced by Fgf8, secreted by the optic vesicle, and it appears later than Sox2. L-Maf is needed for the maintenance of crystallin gene expression and the completion of lens fiber differentiation (Kondoh et al. 2004; Reza et al. 2007).

Relatively little is known about the development of the cornea. As the developing lens invaginates, it stimulates the overlying ectoderm to start differentiating into the cornea. This overlying ectoderm is induced to secrete layers of collagen into which neural crest cells migrate and make new cell layers and to secrete a corneal-specific extracellular matrix (Meier and Hay 1974; Johnston et al. 1979; Kanakubo et al. 2006). Pax6 is especially important in

corneal development. In addition, while Pax6 is downregulated in most adult mammalian tissues, it remains expressed in the adult cornea, where it is important for wound healing (Ramaesh et al. 2006). Repair and regeneration are critical to the cornea, since, like the epidermis, it is exposed to the outside world. Moreover, as in the epidermis, a layer of basal cells continually renews the corneal cells throughout the life of the individual. Long-lived stem cells found at the edge of the cornea contribute to corneal repair and can regenerate the cornea in humans (Cotsarelis et al. 1989; Tsai et al. 2000; Majo et al. 2008).

See WEBSITE 9.7 **Why babies don't see well**

THE EPIDERMIS AND ITS CUTANEOUS APPENDAGES

The skin is the largest organ in our bodies. A tough, elastic, water-impermeable membrane, skin protects our body against dehydration, injury, and infection. Moreover, it is constantly renewable, able to take assaults and regenerate itself. This property of regeneration is due to a population of epidermal stem cells that last the lifetime of our bodies. Mammalian skin has three major components: (1) a stratified epidermis; (2) an underlying dermis composed of loosely packed fibroblasts; and (3) neural crest-derived melanocytes that reside in the basal epidermis and hair follicles. It is the melanocytes (discussed in detail in Chapter 10) that provide the skin's pigmentation. A subcutaneous ("below the skin") fat layer is present beneath the dermis.

Origin of the Epidermis

The epidermis originates from the ectodermal* cells covering the embryo after neurulation. As detailed in Chapter 7, this surface ectoderm is induced to form epidermis rather than neural tissue by the actions of BMPs. The BMPs promote epidermal specification and at the same time induce transcription factors that block the neural pathway (see Bakkers et al. 2002).

Originally, the epidermis is only one cell layer thick, but in most vertebrates it soon becomes a two-layered structure. Then the outer layer gives rise to the **periderm**, a temporary covering that is shed once the inner layer differentiates to form a true epidermis. The inner layer, called the **basal layer** or **stratum germinativum**, contains epidermal stem cells attached to a basement membrane that they help to make (**Figure 9.37**). Lechler and Fuchs (2005) have shown that, like the neural stem cells of the ependymal layer, the

---

*Some clarifying vocabulary: *Epidermis* is the outer layer of skin. The *ectoderm* is the germ layer that forms the epidermis, neural tube, and neural crest. *Epithelial* refers to a sheet of cells that are held together tightly (as opposed to the loosely connected *mesenchymal* cells). Epithelia can be produced by any germ layer. The epidermis and the neural tube both happen to be ectodermal epithelia; the lining of the gut is an *endodermal* epithelium.

epidermal stem cells divide asymmetrically: the daughter cell that remains attached to the basal lamina remains a stem cell, while the cell that leaves the basal layer migrates outward and starts differentiating, making the keratins characteristic of skin and joining them into dense intermediate filaments. It is likely that all the basal layer cells of the adult mouse skin have stem cell properties (Clayton et al. 2007). These differentiated epidermal cells, the **keratinocytes**, are bound tightly together and produce a water-impermeable seal of lipid and protein.

Cell division from the basal layer produces younger cells and pushes the older cells to the border of the skin. After the synthesis of the differentiated products, the cells cease transcriptional and metabolic activities. As they reach the surface, the cells are dead, flattened sacs of keratin protein, and their nuclei are pushed to one edge of the cell. These cells constitute the **cornified layer** (**stratum corneum**). Throughout life, the dead keratinocytes of the cornified layer are shed—humans lose about 1.5 grams of these cells every day*—and are replaced by new cells. In

---

*Most of this skin becomes "house dust" on furniture and floors. If you doubt this, burn some dust; it will smell like singed skin.

mice, the journey from the basal layer to the sloughed cell takes about 2 weeks. Human epidermis turns over a bit more slowly. The proliferative ability of the basal layer is remarkable in that it can supply the cellular material to continuously replace 1–2 m² of skin for several decades throughout adult life.

Several factors stimulate development of the epidermis. BMPs help initiate epidermal production by inducing the p63 transcription factor in the basal layer. This transcription factor's multiple roles may depend in part on different splicing isoforms of p63 that are expressed in the epidermis. The p63 protein is required for keratinocyte proliferation and differentiation (Truong and Khavari 2007); it also appears to stimulate the production of the Notch ligand Jagged. Jagged is a juxtacrine protein in the basal cells that activates the Notch protein on the cells above them, activating the keratinocyte differentiation pathway and preventing further cell divisions (see Mack et al. 2005; Blanpain and Fuchs 2009).

The Cutaneous Appendages

The epidermis and dermis interact at specific sites to create the sweat glands and the **cutaneous appendages**: hairs, scales, or feathers (depending on the species). The formation of these appendages requires a series of reciprocal inductive interactions between the dermal mesenchyme and the ectodermal epithelium, resulting in the formation of epidermal thickenings called **placodes** that are the base precursors of hair follicles. Epidermal cells in the regions capable of forming these placodes secrete Wnt protein among themselves, and Wnt signaling is critical for the initiation of follicle development (see Reddy et al. 2001; Andl et al. 2002).

In mammals, the first indication that a hair follicle placode will form at a particular place is an aggregation of cells in the basal layer of the epidermis (**Figure 9.38A,B**). This aggregation is directed by the underlying dermal fibroblast cells and occurs at different times and different places in the embryo. The basal epidermal cells elongate, divide, and sink into the dermis. The dermal fibroblasts respond to this ingression of epidermal cells by forming a small node (the **dermal papilla**) beneath the hair germ (**Figure 9.38C**). The dermal papilla then pushes up on the basal stem cells and stimulates them to divide more rapidly. The basal cells respond by producing postmitotic cells that will differentiate into the keratinized hair shaft (**Figure**

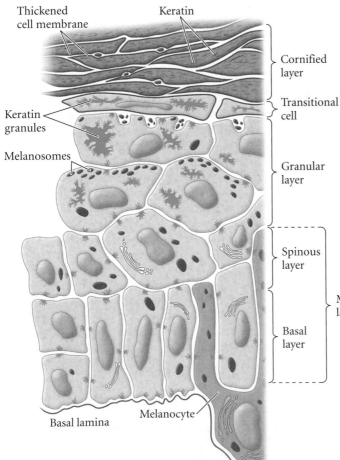

Thickened cell membrane

Keratin

Keratin granules

Melanosomes

Cornified layer

Transitional cell

Granular layer

Spinous layer

Malpighian layer

Basal layer

Melanocyte

Basal lamina

FIGURE 9.37 Layers of the human epidermis. The basal cells are mitotically active, whereas the fully keratinized cells characteristic of external skin are dead and are continually shed. The keratinocytes obtain their pigment through the transfer of melanosomes from the processes of melanocytes that reside in the basal layer. (After Montagna and Parakkal 1974.)

(A)

Basal keratinocytes of epidermis

Dermal fibroblasts

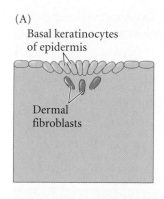

(B)

Epidermis

Placode

Dermal papilla fibroblasts

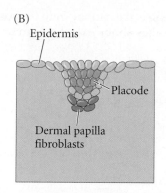

(C)

Inner root sheath   Epidermis

Melanin in early hair shaft

Dermal papilla

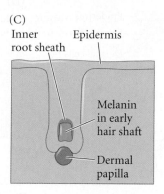

(D)

Epidermis

Sebocytes

Bulge

Inner root sheath

Melanocyte

Dermal papilla

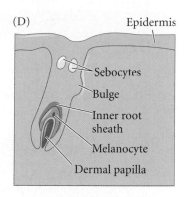

(E)

Sebaceous gland

Bulge

Hair shaft

Inner root sheath

Outer root sheath

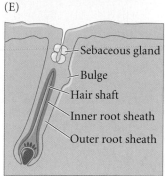

(F)

Hair shaft

Sebaceous gland

Bulge

Melanocytes

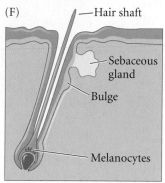

FIGURE 9.38   Early development of the hair follicle and hair shaft. (A) Signals initiate local proliferation of the basal keratinocytes in the epidermis. (B) Proliferation of epidermal cells results in formation of the hair follicle placode, which signals the dermal mesenchymal cells to aggregate beneath it into a dermal papilla. (C) The papilla signals the proliferation of the hair germ, making it into a primitive hair shaft (or "hair peg"). (D) The hair shaft engulfs the dermal papilla and forms the inner hair root directly above the papilla. Sebaceous cells (sebocytes) and the bulge appear as melanin granules enter into the cortex. (E) Sebaceous glands form, and the hair canal is made. The hair shaft differentiates the inner root sheath of epidermal cells. (F) The sebaceous gland is localized on the lateral wall of the follicle, while the hair shaft extends into the hair canal and out past the skin. (A–F after Philpott and Paus 1998.)

See WEBSITE 9.8
Developmental genetics of hair formation

9.38D,F; see Hardy 1992; Miller et al. 1993; Duverger and Morasso 2009).

Wnt signaling is critical in placode formation. Forced expression of the Wnt-inhibitor Dickkopf, or deletion of epithelial β-catenin, preclude placode formation, while mutation of epithelial β-catenin to a constitutively active form results in adoption of hairlike fate by the entire surface ectoderm. This indicates that ectodermal β-catenin signaling determines hair follicle versus epidermal fate (Gat et al. 1998; Andl et al. 2002; Närhi et al. 2008; Zhang et al. 2008, 2009). Patterning of placode fate within the surface ectoderm is thought to rely on a reaction-diffusion mechanism based on competition between placode promoting-factors (such as Wnts) and secreted inhibitors (including the Dickkopf family of Wnt inhibitors and several BMP family members; Jiang et al. 2004; Mou et al. 2006; Sick et al. 2006; Bazzi et al. 2007).

In one such regulatory loop, Wnt activates expression of its own inhibitor, Dkk4, in placodes. As Dkk4 is thought to be more diffusible than Wnt ligands, it may suppress placode fate in cells adjacent to the placode, thereby contributing to patterning (**Figure 9.39**; Sick et al. 2006; Bazzi et al. 2007). Expression of Dkk4 is additionally enhanced by another signaling pathway that is crucial for placode formation: the ectodysplasin (EDA) pathway (Fliniaux et al. 2008; Zhang et al. 2009; see Sidelights & Speculations, p. 369).

Interestingly, signaling interactions similar to those that take place in the embryo can occur during healing of large wounds in adult mouse skin, resulting in regeneration of follicles within the wound. The epidermis appears to revert back to an embryonic state and uses the same Wnt signaling pathway to establish new hair follicles (Ito et al. 2007). This regenerative process is inhibited by forced expression of Dkk1 and enhanced by overexpression of Wnt7a (indicating that it requires Wnt activity).

Once placodes are established, signals from each placode induce clumping of underlying dermal fibroblasts, forming a dermal condensate (see Figure 9.38B). In response to a "second dermal message" from the dermal condensate, the epithelial placode cells proliferate and invade the dermis, eventually surrounding the dermal condensate, which develops into the hair follicle dermal papilla, an important signaling center in the mature hair follicle. Further proliferation and differentiation of the epithelial cells result in the formation of the inner root sheath and hair shaft of the mature follicle, processes that are likely to require lateral communication between epithelial cells (see Hardy 1992; Andl et al. 2002). As the hair follicle matures, an epithelial swelling begins to form on the periphery of the hair germ and eventually develops into the **sebaceous gland** (see Figure 9.38E,F). Sebaceous glands produce an oily secretion known as **sebum** that functions to lubricate the hair follicles and skin.

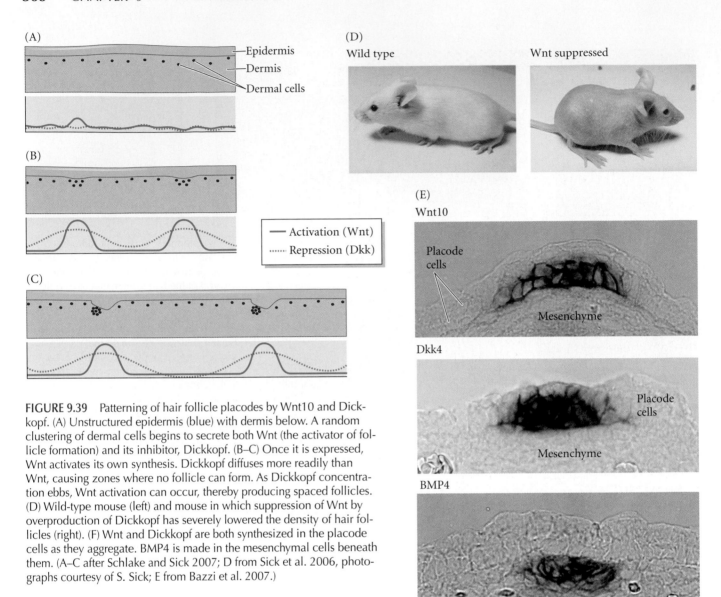

FIGURE 9.39 Patterning of hair follicle placodes by Wnt10 and Dickkopf. (A) Unstructured epidermis (blue) with dermis below. A random clustering of dermal cells begins to secrete both Wnt (the activator of follicle formation) and its inhibitor, Dickkopf. (B–C) Once it is expressed, Wnt activates its own synthesis. Dickkopf diffuses more readily than Wnt, causing zones where no follicle can form. As Dickkopf concentration ebbs, Wnt activation can occur, thereby producing spaced follicles. (D) Wild-type mouse (left) and mouse in which suppression of Wnt by overproduction of Dickkopf has severely lowered the density of hair follicles (right). (F) Wnt and Dickkopf are both synthesized in the placode cells as they aggregate. BMP4 is made in the mesenchymal cells beneath them. (A–C after Schlake and Sick 2007; D from Sick et al. 2006, photographs courtesy of S. Sick; E from Bazzi et al. 2007.)

The first hairs in the human embryo are of a thin, closely spaced type called **lanugo**. This type of hair is usually shed before birth and is replaced (at least partially by new follicles) by the short and silky **vellus hair**. Vellus hair remains on many parts of the human body that are usually considered hairless, such as the forehead and eyelids. In other areas of the body, vellus hair gives way to longer and thicker "terminal" hair. During a person's life, some of the follicles that produced vellus hair can later form terminal hair and still later revert to vellus hair production. Hair follicles in the armpits, for instance, grow vellus hair until adolescence, when they begin producing terminal shafts in response to androgens. Conversely, in male pattern baldness, scalp hair follicles revert to producing unpigmented and very fine vellus hair (Montagna and Parakkal 1974).*

Hair is one structure that mammals are able to regenerate. Throughout life, hair follicles undergo cycles of growth (known as **anagen**), regression (**catagen**), rest (**telogen**),

and regrowth. Hair length is determined by the amount of time the hair follicle spends in the anagen phase. Human scalp hair can spend several years in anagen, whereas arm hair grows for only 6–12 weeks in each cycle.

The ability of hair follicles to regenerate depends on the existence of a population of epithelial stem cells that forms

---

*One of the great mysteries is why testosterone induces hair growth on the chin to procede from thin vellus hair to full beard growth, but causes the hair on the male scalp to become vellus (as in male pattern balding). As we will see in Chapter 10, head dermis is derived from two sources. The "balding" follicles of the frontal-parietal scalp probably come from the neural crest, while the non-balding follicles of the jaw and of the occipital-temporal region of the scalp (the fringe of the scalp that usually retains its full hair potential) are probably derived from the mesoderm. These different dermal populations have different abilities to respond to testosterone (see Randall 2003).

EDAR Syndromes

Vertebrates are classified by their cutaneous appendages. Hair and mammary glands are the basis for classifying organisms as mammals, and birds are the only extant lineage with feathers. Fish are recognized by their variety of scales. Amazingly, hair, feathers, mammary glands, fish scales, and teeth all form through reciprocal interactions of the epidermis and its underlying mesenchymal cells. (As we will see in Chapter 10, the mesenchyme beneath the tooth placodes is derived from neural crest, not meso-

derm.) One pathway that connects all cutaneous appendages is the ectodysplasin (EDA) cascade (**Figure 9.40**). This pathway seems to be specific for cutaneous appendage formation, and mutations involving the components of the EDA pathway often cause syndromes involving two or more of these appendage types. When EDA binds to one of its receptors (EDAR or EDARADD), it separates the NF-κB transcription factor from its inhibitor and allows it to activate or repress particular genes.

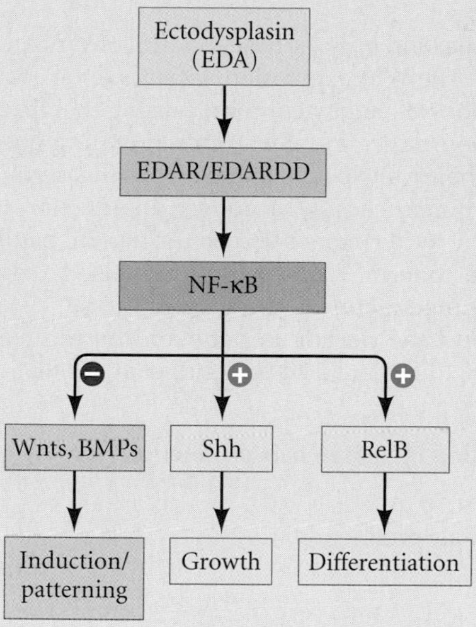

EDA signaling directly or indirectly regulates the Wnt, BMP, and Shh signaling pathways and is thought to regulate both the patterning of hair follicles and the differentiation of the follicles. Indeed, the thick hair associated with Asian populations may be due to a variant of the *EDAR* gene that predominates in Japan and China and is essentially absent in native European populations (Fujimoto et al. 2008). People lacking *EDA* have a syndrome that includes sparse hair, absent eyelashes and eyebrows, the lack of most teeth, and the inability to sweat (**Figure 9.41**). Often they also lack mammary glands (which are modified sweat glands; see Gritli-Linde et al. 2007). The syndromes caused by mutations in the *EDA*, *EDAR*, and *EDARADD* genes are essentially identical and are referred to as **anhidrotic ectodermal dysplasia** (Mikkola 2007).

Ectodermal dysplasia also occurs in other vertebrates. In zebrafish, the EDA pathway is criti-

Figure 9.41 Facial anomalies of anhidrotic ectodermal dysplasia, caused by mutation of an *EDA* gene. Manifestations of this condition include thin hair and lack of eyebrows, eyelids, and teeth. (Photograph © The Orange County Register/ZUMA Press.)

cal in forming the placodes that generate the body scales and the pharyngeal teeth (Harris et al. 2007), and the chick feather placodes are similarly regulated by the EDA pathway (Houghton et al. 2005; Drew et al. 2007). In stickleback fish, the armorless phenotype seen in many freshwater lakes is the result of independent mutations of the *EDA* gene (Colosimo et al. 2005). Thus the ectodysplasin pathway appears to regulate many types of cutaneous appendages in many species.

Ectodysplasin (EDA)

↓

EDAR/EDARDD

↓

NF-κB

⊖ ⊕ ⊕

Wnts, BMPs — Shh — RelB

↓ ↓ ↓

Induction/patterning — Growth — Differentiation

Figure 9.40 Outline of the ectodysplasin (EDA) pathway and some of the genes regulated by it.

in the permanent bulge region of the follicle late in embryogenesis. When Philipp Stöhr drew the histology of the human hair for his 1903 textbook, he showed this bulge (*Wulst*) as the attachment site for the arrector pili muscles (which give a person "goosebumps" when they contract). However, research carried out during the 1990s suggested that the bulge in fact houses populations of at least two remarkable adult stem cells: the multipotent **follicular stem cell**, which gives rise to the hair shaft, sheath, and seba-

ceous gland (Cotsarelis et al. 1990; Morris and Potten 1999; Taylor et al. 2000), and the **melanocyte stem cell**, which gives rise to the pigment of the skin and hair (see Chapter 10; Nishimura et al. 2002). The bulge appears to be a niche that allows cells to retain the quality of "stemness."

The mesenchymal-epithelial interactions at the start of the hair cycle direct the epidermal stem cells to migrate downward through the outer root sheath to the dermal papilla and produce the seven concentric columns of cells

(A)

(B)

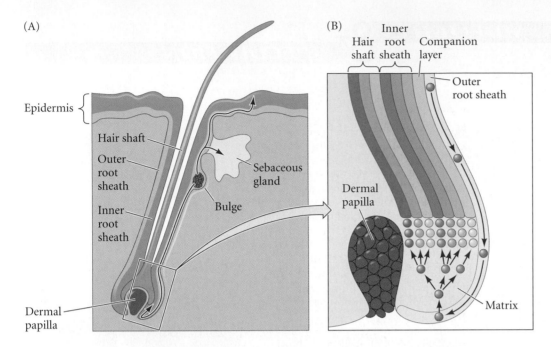

FIGURE 9.42 Model of follicle stem cell migration and differentiation. (A) During the normal hair cycle, bulge stem cells are periodically activated to form a new hair follicle. As the hair follicle grows, the more committed progeny of the bulge cells migrate through the outer root sheath, becoming a highly proliferative population of cells that will form a new hair. (B) As the bulge stem cells' progeny colonize the base of the follicle, they subsequently differentiate along one of seven hair lineages. Differentiation occurs at the base of the follicle. (After Fuchs 2007; Blanpain and Fuchs 2009).

that form the hair shaft (**Figure 9.42**). This migration and differentiation appear to be regulated by a complicated reaction-diffusion scheme wherein Wnt paracrine factors in the placode activate BMP paracrine factors as well as Wnt inhibitors in the condensed mesoderm beneath them.

This causes a cyclic (and out-of-phase) pattern of Wnt and BMP expression. The Wnt expression appears to be critical in activating the follicular stem cells, causing their proliferation and their migration. But if this signaling goes unopposed, the hair follicle fails to grow into the dermis and hair development ceases. Moreover, in addition to being regulated by local factors between the dermal papilla and placodal ectoderm, Wnt signaling can also be regulated by long-range factors coming from the subcutaneous fat pad and the dermis in between the follicles (Närhi et al. 2008; Plikus et al. 2008; Rendl et al. 2008).

See WEBSITE 9.9
Normal variation in human hair production

See WEBSITE 9.10
Mutations of hair production

 ## Snapshot Summary: *Neural and Epidermal Ectoderm*

1. The neural tube forms from the shaping and folding of the neural plate. In primary neurulation, the surface ectoderm folds into a tube that separates from the surface. In secondary neurulation, the ectoderm forms a cord, then forms a cavity within the cord.

2. Primary neurulation is regulated both by intrinsic and extrinsic forces. Intrinsic wedging occurs within cells of the hinge regions, bending the neural plate. Extrinsic forces include the migration of the surface ectoderm toward the center of the embryo.

3. Neural tube closure is also the result of extrinsic and intrinsic forces. In humans, congenital anomalies can

result if the neural tube fails to close. Folate is important in mediating neural tube closure.

4. The neural crest cells arise at the borders of the neural tube and surface ectoderm. They become located between the neural tube and surface ectoderm, and they migrate away from this region to become peripheral neural, glial, and pigment cells.

5. There is a gradient of maturity in many embryos (especially those of amniotes) so that the anterior develops earlier than the posterior.

6. The brain forms three primary vesicles: prosencephalon (forebrain), mesencephalon (midbrain),

and rhombencephalon (hindbrain). The pros-encephalon and rhombencephalon become subdivided.

7. The brain expands as fluid secretion puts positive pressure on the vesicles.

8. The dorsal-ventral patterning of the neural tube is accomplished by proteins of the TGF-β superfamily secreted from the surface ectoderm and the roof plate of the neural tube, and by Sonic hedgehog protein secreted from the notochord and floor plate cells. Gradients of these proteins trigger the synthesis of particular transcription factors that specify the neuroepithelium.

9. Dendrites receive signals from other neurons, while axons transmit signals to other neurons. The gap between cells where signals are transferred from one neuron to another (through the release of neurotransmitters) is called a synapse.

10. Axons grow from the nerve cell body, or soma. They are led by the growth cone.

11. Neural stem cells have been observed in the adult human brain. We now believe that humans can continue making neurons throughout life, although at nowhere near the fetal rate.

12. The neurons of the brain are organized into cortices (layers) and nuclei (clusters).

13. New neurons are formed by the division of neural stem cells in the wall of the neural tube (called the ventricular zone). The resulting neural precursors, or neuroblasts, can migrate away from the ventricular zone and form a new layer, called the mantle zone (gray matter). Neurons forming later have to migrate through the existing layers. This process forms the cortical layers.

14. In the cerebellum, migrating neurons form a second germinal zone, called the external granular layer. Other neurons migrate out of the ventricular zone on the processes of glial cells.

15. The cerebral cortex in humans, called the neocortex, has six layers. Cell fates are often fixed as they undergo their last division. Neurons derived from the same stem cell may end up in different functional regions of the brain.

16. Human brains appear to differ from those of other primates by their retention of the fetal neuronal growth rate during early childhood, the high transcriptional activity, and the presence of human-specific alleles of developmental regulatory genes

17. The vertebrate retina forms from an optic vesicle that extends from the brain. Pax6 plays a major role in eye formation, and the downregulation of Pax6 by Sonic hedgehog (Shh) in the center of the brain splits the eye-forming region of the brain in half. If Shh is not expressed there, a single medial eye results.

18. The photoreceptor cells of the retina gather light and transmit an electric impulse through interneurons to the retinal ganglion cells. The axons of the retinal ganglion cells form the optic nerve.

19. The lens and cornea form from the surface ectoderm. Both must become transparent.

20. The basal layer of the surface ectoderm becomes the germinal layer of the skin. Epidermal stem cells divide to produce differentiated keratinocytes and more stem cells.

21. The follicular stem cells, which regenerate hair follicles during periods of cyclical growth, reside in the bulge of the hair follicle.

For Further Reading

Complete bibliographical citations for all literature cited in this chapter can be found at the free-access website **www.devbio.com**

Bailey, T. J., H. El-Hodiri, L. Zhang, R. Shah, P. H. Mathers and M. Jamrich. 2004. Regulation of vertebrate eye development by *Rx* genes. *Int. J. Dev. Biol.* 48: 761–770.

Catala, M., M.-A. Teillet, E. M. De Robertis and N. M. Le Douarin. 1996. A spinal cord fate map in the avian embryo: While regressing, Hensen's node lays down the notochord and floor plate thus joining the spinal cord lateral walls. *Development* 122: 2599–2610.

Colas, J.-F. and G. C. Schoenwolf. 2001. Towards a cellular and molecular understanding of neurulation. *Dev. Dynam.* 221: 117–145.

Cooper, M. K., J. A. Porter, K. E. Young and P. A. Beachy. 1998. Teratogen-mediated inhibition of target tissue response to Shh signaling. *Science* 280: 1603–1607.

Erikksson, P. S., E. Perfiliea, T. Björn-Erikksson, A.-M. Alborn, C. Nordberg, D. A. Peterson and F. H. Gage. 1998. Neurogenesis in the adult human hippocampus. *Nature Med.* 4: 1313–1317.

Gogtay, N. and 11 others. 2004. Dynamic mapping of human cortical development during childhood through early adulthood. *Proc. Nat. Acad. Sci. USA* 101: 8174–8179.

Halder, G., P. Callaerts and W. J. Gehring. 1995. Induction of ectopic eyes by targeted expression of the *eyeless* gene in *Drosophila. Science* 267: 1788–1792.

Hanashina, C., S. C. Li, L. Shen, E. Lai and G. Fishell. 2004. Foxg1 suppresses early cortical cell fate. *Science* 303: 56–59.

Hatten, M. E. 1990. Riding the glial monorail: A common mechanism for glial-guided neuronal migration in different regions of the mammalian brain. *Trends Neurosci.* 13: 179–184.

Jessell, T. M. 2000. Neuronal specification in the spinal cord: Inductive signals and transcriptional codes. *Nature Rev. Genet.* 1: 20–29.

Lawson, A., H. Anderson and G. C. Schoenwolf. 2001. Cellular mechanisms of neural fold formation and morphogenesis in the chick embryo. *Anat. Rec.* 262: 153–168.

Milunsky, A., H. Jick, S. S. Jick, C. L. Bruell, D. S. Maclaughlen, K. J. Rothman and W. Willett. 1989. Multivitamin folic acid supplementation in early pregnancy reduces the prevalence of neural tube defects. *J. Am. Med. Assoc.* 262: 2847–2852.

Pollard, K. S. and 15 others. 2006. An RNA gene expressed during cortical development evolved rapidly in humans. *Nature* 443: 167–172.

Sick, S., S. Reinker, J. Timmer and T. Schlake. 2006. WNT and DKK determine hair follicle spacing through a reaction-diffusion mechanism. *Science.* 314: 1447–1450.

Turner, D. L. and C. L. Cepko. 1987. A common progenitor for neurons and glia persists in rat retina late in development. *Nature* 328: 131–136.

Varki, A., D. H. Geschwind and E. E. Eichler. 2008. Explaining human uniqueness: Genome interactions with environment, behaviour, and culture. *Nature Rev. Genet.* 9: 749–763.

Go Online

WEBSITE 9.1 Homologous specification of the neural tissue. The insect nervous system develops in a manner very different from that of the vertebrate nervous system. However, the genetic instructions for forming the central neural system and specifying its regions appear to be homologous. In *Drosophila*, neural fate often depends on the number of cell divisions preceding neuronal differentiation.

WEBSITE 9.2 Specifying the brain boundaries. The Pax transcription factors and the paracrine factor FGF8 are critical in establishing the boundaries of the forebrain, midbrain, and hindbrain.

WEBSITE 9.3 Constructing the pituitary gland. The developmental genetics of the pituitary have shown that (as in the trunk neural tube) paracrine signals elicit the expression of overlapping sets of transcription factors. These factors regulate the different cell types of the gland responsible for producing growth hormone, prolactin, gonadotropins, and other hormones.

WEBSITE 9.4 Cerebellar mutations of the mouse. The mouse mutations affecting cerebellar function have given us remarkable insights into the ways in which the cerebellum is constructed. The *reeler* mutation, in particular, has been extremely important in our knowledge of how cerebellar neurons migrate.

WEBSITE 9.5 Constructing the cerebral cortex. Three genes have recently been shown to be necessary for the proper lamination of the mammalian brain. They appear to be important for cortical neural migration, and when mutated in humans can produce profound mental retardation.

WEBSITE 9.6 Neuronal growth and the invention of childhood. An interesting hypothesis claims that the caloric requirements of this brain growth necessitated a new stage of the human life cycle—childhood—during which the child is actively fed by adults.

WEBSITE 9.7 Why babies don't see well. The retinal photoreceptors are not fully developed at birth. As the child grows older, the density of photoreceptors increases, allowing far better discrimination and nearly 350 times the light-absorbing capacity that is present at birth.

WEBSITE 9.8 Developmental genetics of hair formation. A complex cascade of proteins in the Wnt and FGF signaling pathways regulate the epithelial-mesenchymal interactions of hair formation. Several mutations in humans and mice show how sensitive that formation is to paracrine factors.

WEBSITE 9.9 Normal variation in human hair production. The human hair has a complex life cycle. Moreover, some hairs (such as those of our eyelashes) grow short while other hairs (such as those of our scalp) grow long. The pattern of hair size and thickness (or lack thereof) is determined by paracrine and endocrine factors.

WEBSITE 9.10 Mutations of hair production. In addition to normal variation, there are also inherited mutations that interfere with normal hair development. Some people are born without the ability to grow hair, while others develop hair over their entire bodies. These genetic conditions give us insights into the mechanisms of normal hair growth.

Vade Mecum

Chick neurulation. By 33 hours of incubation, neurulation in the chick embryo is well underway. Both whole-mounts and a complete set of serial cross sections through a 33-hour chick embryo are included in the "Chick-Mid" segment so you can see this amazing event. The serial sections can be displayed either as a continuum in movie format or individually, along with labels and color-coding that designates germ layers.

Neural Crest Cells and Axonal Specificity

10

THIS CHAPTER CONTINUES our discussion of ectodermal development, focusing here on neural crest cells and axonal growth cones (the motile tips that guide axons to their destinations). Both share the property of having to migrate far from their source of origin to specific places in the embryo; both must recognize and respond to signals that guide them along specific routes to their final destination. Recent research has revealed that neural crest cells and axonal growth cones recognize many of the same signals.

THE NEURAL CREST

Although it is derived from the ectoderm, the neural crest is so important that it has sometimes been called the "fourth germ layer" (see Hall 2009). It has even been said, somewhat hyperbolically, that "the only interesting thing about vertebrates is the neural crest" (quoted in Thorogood 1989). Certainly, the emergence of the neural crest is one of the pivotal events of animal evolution, as it led to the jaws, face, skull, and sensory ganglia of the vertebrates (Northcutt and Gans 1983). The neural crest is a transient structure. Adults do not have a neural crest, nor do later-stage vertebrate embryos. Rather, the cells of the neural crest undergo an epithelial-to-mesenchymal transition from the dorsal neural tube, after which they migrate extensively to generate a prodigious number of differentiated cell types (Figure 10.1A; Table 10.1).

The neural crest is a population of multipotent progenitor cells that can produce tissues as diverse as (1) the neurons and glial cells of the sensory, sympathetic, and parasympathetic nervous systems; (2) the epinephrine-producing (medulla) cells of the adrenal gland; (3) the pigment-containing cells of the epidermis; and (4) many of the skeletal and connective tissue components of the head. It remains uncertain whether the majority of the individual cells that leave the neural crest are multipotent or whether most are already restricted to certain fates. Bronner-Fraser and Fraser (1988, 1989) provided evidence that many individual trunk neural crest cells are multipotent as they leave the crest. They injected fluorescent dextran molecules into individual chick neural crest cells while the cells were still above the neural tube, and then looked to see what types of cells their descendants became after migration. The progeny of a single neural crest cell could become sensory neurons, **melanocytes** (pigment-forming cells), adrenomedullary cells, and glia (**Figure 10.1B**). Henion and Weston (1997) also found that the initial avian trunk neural crest population was a heterogeneous mixture of precursor cells, and that nearly half of the cells that emerge from the neural crest are restricted to generate a single cell type. Recent studies of the *cranial* region of the chick neural crest provide evidence that a large majority of

Thus, beyond all questions of quantity there lie questions of pattern which are essential for understanding Nature.

ALFRED NORTH WHITEHEAD (1934)

Like the entomologist in search of brightly colored butterflies, my attention hunted, in the garden of the gray matter, cells with delicate and elegant forms, the mysterious butterflies of the soul.

SANTIAGO RAMÓN Y CAJAL (1937)

FIGURE 10.1 Neural crest cell migration. (A) The neural crest is a transient structure dorsal to the neural tube. Neural crest cells (stained blue in this micrograph) undergo epithelial-to-mesenchymal transition and migrate ventrally. (B) In two separate experiments, single neural crest cells were injected with fluorescent dextran shortly before migration was initiated; Two days later, neural crest-derived tissues contain dextran-labeled cells descended from the injected precursor. The figure summarizes data from two different experiments. (A courtesy of J. Briscoe; B after Lumsden 1988a.)

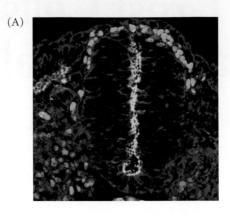

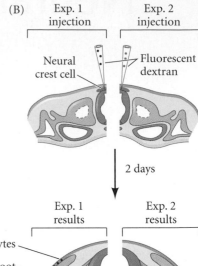

early migrating neural crest cells can generate nearly all the numerous cells types (Calloni et al. 2009).

Reviewing the results of numerous laboratories as well as their own studies, Nicole Le Douarin and others proposed a model whereby an original multipotent neural crest cell divides and progressively loses its developmental potentials (**Figure 10.2**; see Creuzet et al. 2004; Martinez-Morales et al. 2007; Le Douarin et al. 2008). Thus, the cells of the neural crest appear to differ greatly in their developmental potential and commitments. While some exhibit multipotency, we still do not know if the precursor cells

can give rise to other precursor cells. In other words, we don't know whether the neural crest multipotent precursor cell is or is not a true stem cell.

Specification of Neural Crest Cells

The specification of the neural crest at the neural plate-epidermis boundary is a multistep process (see Huang and Saint-Jeannet 2004; Meulemans and Bronner-Fraser 2004). The first step appears to be the location of the neural plate border. In amphibians, this border appears to be specified by intermediate concentrations of BMPs. Indeed, in the 1940s Raven and Kloos (1945) showed that while the presumptive notochord could induce both the amphibian neural plate and neural crest tissue (presumably blocking nearly all BMPs), the somite mesoderm and lateral plate mesoderm could induce only the neural crest. In chick embryos, this specification occurs during gastrulation, when the borders between the neural and non-neural ectoderm are still forming (Basch et al. 2006; Schmidt et al. 2007; Ezin et al. 2009). Here, **neural plate inductive signals** (especially BMPs and Wnts) secreted from the ventral ectoderm and paraxial mesoderm interact to specify the boundaries.

In the anterior region, the timing of BMP and Wnt signal expression is critical for discriminating between neural

TABLE 10.1 Some derivatives of the neural crest	
Derivative	**Cell type or structure derived**
Peripheral nervous system (PNS)	Neurons, including sensory ganglia, sympathetic and parasympathetic ganglia, and plexuses Neuroglial cells Schwann cells
Endocrine and paraendocrine derivatives	Adrenal medulla Calcitonin-secreting cells Carotid body type I cells
Pigment cells	Epidermal pigment cells
Facial cartilage and bone	Facial and anterior ventral skull cartilage and bones
Connective tissue	Corneal endothelium and stroma Tooth papillae Dermis, smooth muscle, and adipose tissue of skin, head, and neck Connective tissue of salivary, lachrymal, thymus, thyroid, and pituitary glands Connective tissue and smooth muscle in arteries of aortic arch origin

Source: After Jacobson 1991, based on multiple sources.

Neural crest specifiers

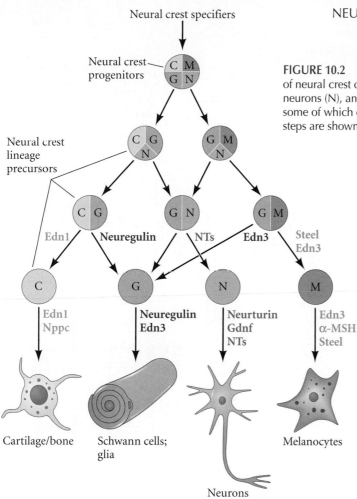

Neural crest progenitors

Neural crest lineage precursors

Edn1   **Neuregulin**   NTs   **Edn3**   Steel Edn3

Edn1 Nppc

Neuregulin Edn3

Neurturin Gdnf NTs

Edn3 α-MSH Steel

Cartilage/bone   Schwann cells; glia   Neurons   Melanocytes

FIGURE 10.2 Model for neural crest lineage segregation and the heterogeneity of neural crest cells. The committed precursors of cartilage/bone (C), glia (G), neurons (N), and melanocytes (M) are derived from intermediate progenitors, some of which could act as stem cells. The paracrine factors regulating these steps are shown in color type. (After Martinez-Morales et al. 2007.)

plate, epidermis, placode,* and neural crest cell tissues (**Figure 10.3A**). As we saw in Chapters 7 and 8, if both BMP and Wnt signaling are continuous, the fate of the ectoderm is epidermal; but if BMP antagonists (e.g., Noggin or FGFs) block BMP signaling, the ectoderm becomes neural. Studies by Patthey and colleagues (2008, 2009) have shown that if Wnts induce BMPs and then Wnt signaling is turned off, the cells become committed to be anterior placodes, whereas if the Wnt signaling induces BMPs but stays on, the cells become neural crest.

The neural plate induces in these border cells a set of transcription factors called **neural plate border specifiers** (**Figure 10.3B**). These factors, including Distalless-5, Pax3, and Pax7, collectively prevent the border region from becoming either neural plate or epidermis. The border-specifying transcription factors then induce a second set of transcription factors, the **neural crest specifiers**, in those cells that are to become the neural crest. These include genes encoding the transcription factors FoxD3, Sox9, Id, Twist, and Snail.

---

*The *placodes*, which form structures such as the lens of the eye and sensory organ neurons, will be discussed later in the chapter.

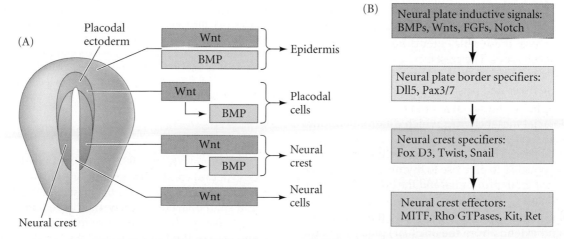

(A)

Placodal ectoderm

Neural crest

| Wnt |
| BMP |
→ Epidermis

| Wnt |
→ | BMP |
→ Placodal cells

| Wnt |
→ | BMP |
→ Neural crest

| Wnt |
→ Neural cells

(B)

Neural plate inductive signals: BMPs, Wnts, FGFs, Notch

↓

Neural plate border specifiers: Dll5, Pax3/7

↓

Neural crest specifiers: Fox D3, Twist, Snail

↓

Neural crest effectors: MITF, Rho GTPases, Kit, Ret

FIGURE 10.3 Specification of neural crest cells. (A) The neural plate is bordered by neural crest anteriorly and caudally, and by placodal ectoderm anteriorly. If the ectodermal cells receive both BMP and Wnt for an extended period of time, they become epidermis. If Wnt induces BMPs and is then downregulated, the cells become placodal border cells (expressing the placode specifier genes *Six1*, *Six4*, and *Eya2*). If Wnt induces BMP but remains active, the border cells become neural crest (expressing neural crest specifier genes *Pax7*, *Snail2*, and *Sox9*). If they receive Wnt only (because the BMP signal is blocked by Noggin or FGF), the ectodermal cells become neural. (B) Stages in the specification of neural crest cells. (A after Patthey et al. 2009; B after Nikitina and Bronner-Fraser 2009.)

When FoxD3, Snail, and Sox9 are experimentally expressed in the lateral neural tube, these lateral neuroepithelial cells become neural crest-like, undergo epithelial-mesenchymal transition, and delaminate from the neuroepithelium. Sox9 and Snail, together, are sufficient to induce such a transition in neuroepithelial cells. Sox9 is also required for the survival of trunk neural crest cells after delamination, since in the absence of Sox9, neural crest cells undergo apoptosis as soon as they delaminate. FoxD3 may play many roles. It is needed for the expression of the cell-surface proteins needed for cell migration, and also appears to be critical for the specification of ectodermal cells as neural crest. Inhibiting expression of the *FoxD3* gene inhibits neural crest differentiation. Conversely, when *FoxD3* is expressed ectopically by electroporating the active gene into neural plate cells, those neural plate cells express proteins characteristic of the neural crest (Nieto et al. 1994; Kos et al. 2001; Soo et al. 2002; Cheung et al. 2005; Taneyhill et al. 2007; Teng et al. 2008).

Neural crest specifiers activate the transcription of those genes that give the neural crest cells their migratory properties and some of their differentiated properties. These **neural crest effectors** include some transcription factors (such as MITF in the melanocyte lineage that forms pigment cells); small G proteins (such as Rho GTPases) that allow cells to change shape and migrate; and cell-surface receptors (such as receptor tyrosine kinases Ret and **Kit**) that allow the neural crest cells to respond to patterning and inducing proteins in their environments.

See VADE MECUM
Nicole Le Douarin and the neural crest

Regionalization of the Neural Crest

The neural crest is a transient structure, as its cells undergo epithelial-to-mesenchymal transition to disperse throughout the body. But at different levels along the anterior-posterior axis, these cells enter different tissues and form different cell types. The crest can be divided into four main (but overlapping) anatomical regions, each with characteristic derivatives and functions (**Figure 10.4**):

• **Cranial (cephalic) neural crest cells** migrate to produce the craniofacial mesenchyme, which differentiates into the cartilage, bone, cranial neurons, glia, and connective tissues of the face. These cells also enter the pharyngeal arches and pouches to give rise to thymic cells, the odontoblasts of the tooth primordia, and the bones of the middle ear and jaw.*

• The **cardiac neural crest** is a subregion of the cranial neural crest and extends from the otic (ear) placodes to the

---

*The pharyngeal (branchial) arches (see Figure 1.8A) are outpocketings of the head and neck region into which cranial neural crest cells migrate. The pharyngeal pouches form between these arches and become the thyroid, parathyroid, and thymus.

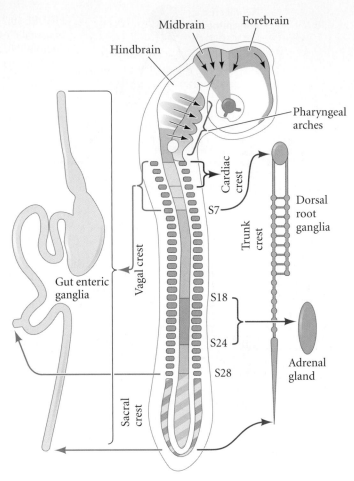

FIGURE 10.4 Regions of the chick neural crest. The cranial neural crest migrates into the pharyngeal arches and the face to form the bones and cartilage of the face and neck. It also produces the cranial nerves. The vagal neural crest (near somites 1–7) and the sacral neural crest (posterior to somite 28) form the parasympathetic nerves of the gut. The cardiac neural crest cells arise near somites 1–3; they are critical in making the division between the aorta and the pulmonary artery. Neural crest cells of the trunk (about somite 6 through the tail) make sympathetic neurons and pigment cells (melanocytes), and a subset of these (at the level of somites 18–24) forms the medulla portion of the adrenal gland. (After Le Douarin 1982.)

third somites (Kirby 1987; Kirby and Waldo 1990). Cardiac neural crest cells develop into melanocytes, neurons, cartilage, and connective tissue (of the third, fourth, and sixth pharyngeal arches). This region of the neural crest also produces the entire muscular-connective tissue wall of the large arteries (the "outflow tracts") as they arise from the heart, as well as contributing to the septum that separates pulmonary circulation from the aorta (Le Lièvre and Le Douarin 1975).

• **Trunk neural crest** cells take one of two major pathways. One migratory pathway takes trunk neural crest cells ventrolaterally through the anterior half of each somitic sclerotome. (**Sclerotomes**, derived from somites, are blocks of mesodermal cells that will differentiate into the vertebral cartilage of the spine.) Those trunk neural crest cells that remain in the sclerotomes form the **dorsal root ganglia** containing the sensory neurons. Those that continue more ventrally form the sympathetic ganglia, the adrenal medulla, and the nerve clusters surrounding the aorta. The second major migratory path for trunk neural crest cells proceeds dorsolaterally, allowing the precursors of melanocytes to move through the dermis from the dorsum to the belly (Harris and Erickson 2007).

- The **vagal** and **sacral neural crest** cells generate the **parasympathetic (enteric) ganglia** of the gut (Le Douarin and Teillet 1973; Pomeranz et al. 1991). The vagal (neck) neural crest overlaps the cranial/trunk crest boundary, lying opposite chick somites 1–7, while the sacral neural crest lies posterior to somite 28. Failure of neural crest cell migration from these regions to the colon results in the absence of enteric ganglia and thus to the absence of peristaltic movement in the bowels.

The trunk neural crest and the cranial neural crest cells are not equivalent. Cranial crest cells can form cartilage and bone, whereas trunk neural crest cannot. When trunk neural crest cells are transplanted into the head region, they can migrate to the sites of cartilage and cornea formation, but they make neither cartilage nor cornea (Noden 1978; Nakamura and Ayer-Le Lievre 1982; Lwigale et al. 2004). However, both cranial and trunk neural crest cells can generate neurons, melanocytes, and glia. The cranial neural crest cells that normally migrate into the eye region to become cartilage cells can form sensory ganglion neurons, adrenomedullary cells, glia, and Schwann cells if the cranial region is transplanted into the trunk region (Noden 1978; Schweizer et al. 1983).

The inability of the trunk neural crest to form skeleton is most likely due to the expression of Hox genes in the trunk neural crest. If Hox genes are expressed in the cranial neural crest, these cells fail to make skeletal tissue; if trunk crest cells lose Hox gene expression, they can form skeleton. Moreover, if transplanted into the trunk region, cranial crest cells participate in forming trunk cartilage that normally does not arise from neural crest components. This ability to form bone may have been a primitive property of the neural crest and may have been critical for forming the bony armor found in several extinct fish species (Smith and Hall 1993). In other words, the trunk crest has apparently lost the ability to form bone, rather than the cranial crest acquiring this ability. McGonnell and Graham (2002) have shown that bone-forming capacity may still be latent in the trunk neural crest: if cultured in certain hormones and vitamins, the trunk cells become capable of forming bone and cartilage when placed into the head region. Moreover, Abzhanov and colleagues (2003) have shown that the trunk crest cells can act like cranial crest cells (and

make skeletal tissue) if the trunk cells are cultured in conditions that cause them to lose the expression of their Hox genes.

So, even though the cells of the cranial neural crest and trunk neural crest are multipotent (a cranial crest cell can form neurons, cartilage, bone, and muscles; a trunk neural crest cell can form glia, pigment cells, and neurons), they have different repertoires of cell types that they can generate.

Trunk Neural Crest

Migration pathways of trunk neural crest cells

Cells migrating from the trunk-level neural crest follow one of two major pathways (**Figure 10.5A**). Many cells that leave early follow a **ventral pathway** away from the neural tube. Fate mapping experiments show that these cells become sensory (dorsal root) and autonomic neurons, adrenomedullary cells, and Schwann cells (Weston 1963; Le Douarin and Teillet 1974). In birds and mammals (but not fish and frogs),* these cells migrate ventrally through the anterior, but not the posterior, section of the sclerotomes (**Figure 10.5B–D**; Rickmann et al. 1985; Bronner-Fraser 1986; Loring and Erickson 1987; Teillet et al. 1987).

Trunk crest cells that emigrate via the second pathway—the **dorsolateral pathway**—become melanocytes, the melanin-forming pigment cells. These cells travel between the epidermis and the dermis, entering the ectoderm through minute holes in the basal lamina (which they themselves may create). Once in the ectoderm, they colonize the skin and hair follicles (Mayer 1973; Erickson et al. 1992). The dorsolateral pathway was demonstrated in a series of classic experiments by Mary Rawles (1948), who transplanted the neural tube and neural crest from a pigmented strain of chickens into the neural tube of an albino chick embryo (see Figure 1.16).

By transplanting quail neural tubes and crests into chick embryos, Teillet and co-workers (1987) were able to mark neural crest cells both genetically and immunologically. The antibody marker recognized and labeled neural crest cells of both species; the genetic marker enabled the investigators to distinguish between quail and chick cells. These studies showed that neural crest cells initially located opposite the posterior region of a somite migrate anteriorly or posteriorly along the neural tube, and then enter the anterior region of their own or an adjacent somite. These cells join with the neural crest cells that initially were opposite the anterior portion of the somite, and they form the same structures. Thus, each dorsal root ganglion is composed of three neural crest cell populations: one from the

---

*In the migration of fish neural crest cells, the sclerotome is not important; rather, the myotome appears to guide the migration of the crest cells ventrally (Morin-Kensicki and Eisen 1997).

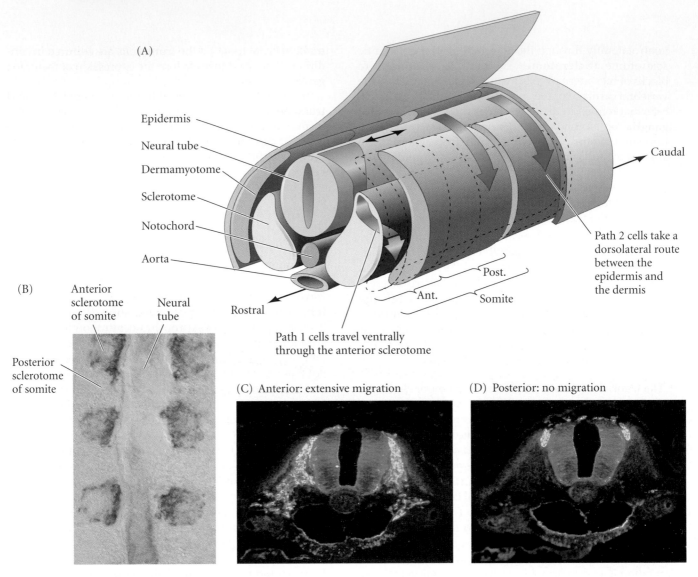

FIGURE 10.5   Neural crest cell migration in the trunk of the chick embryo. (A) Schematic diagram of trunk neural crest cell migration. Cells taking path 1 (the ventral pathway) travel ventrally through the anterior of the sclerotome (that portion of the somite that generates vertebral cartilage). Those cells initially opposite the posterior portion of a sclerotome migrate along the neural tube until they come to an anterior region. These cells contribute to the sympathetic and parasympathetic ganglia as well as to the adrenomedullary cells and dorsal root ganglia. Other trunk neural crest cells enter path 2 (the dorsolateral pathway) somewhat later. These cells travel along a dorsolateral route beneath the ectoderm and become pigment-producing melanocytes. (Migration pathways are shown on only one side of the embryo.) (B) These fluorescence photomicrographs of longitudinal sections of a 2-day chick embryo are stained red with antibody to HNK-1, which selectively recognizes neural crest cells. Extensive staining is seen in the anterior, but not in the posterior, half of each sclerotome. (C,D) Cross sections through these areas, showing extensive migration through the anterior portion of the sclerotome (C), but no migration through the posterior portion (D). Here, the antibodies to HNK-1 are stained green. (B from Wang and Anderson 1997; C,D from Bronner-Fraser 1986, courtesy of the authors.)

neural crest opposite the anterior portion of the somite, and one each from the two neural crest regions opposite the posterior portions of its own and the neighboring somites.

The mechanisms of trunk neural crest migration

Any analysis of migration (be it of birds, butterflies, or neural crest cells) has to ask these questions:

1. What signals initiate migration?
2. When does the migratory agent become competent to respond to these signals?
3. How do the migratory agents know the route to travel?
4. What signals indicate that the destination has been reached?

The timing of neural crest cell migration is controlled by the neural tube's environment. The trigger for the epithelial-mesenchymal transition (EMT) appears to be the acti-

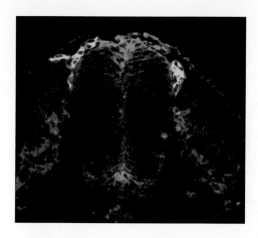

FIGURE 10.6 All migrating neural crest cells are stained red by antibody to HNK-1. The RhoB protein (green stain) is expressed in cells as they leave the neural crest. Cells expressing both HNK-1 and RhoB appear yellow. (From Liu and Jessell 1998, courtesy of T. M. Jessell.)

vation of the Wnt genes by BMPs. The BMPs (which can be produced by the dorsal region of the neural tube; see Chapter 9) are held in check by Noggin protein produced in the somites (see Chapter 11). When the somites cease making Noggin, the neural tube BMPs can function, and they activate EMT in the dorsal neural tube cells (Burstyn-Cohen et al. 2004). The Wnt and BMP signals combine to induce expression of proteins from the Snail and Rho families in those cells destined to become neural crest (**Figure 10.6**; Nieto et al. 1994; Mancilla and Mayor 1996; LaBonne and Bronner-Fraser 1998; Liu and Jessell 1998). If either Snail2 or Rho is inactivated or inhibited, the crest cells fail to emigrate from the neural tube.

Rho and Snail proteins are involved in the pushes and pulls that enable neural crest cell migration. RhoB is thought to be involved in establishing the cytoskeletal conditions for migration by promoting actin polymerization into microfilaments and the attachment of these microfilaments to the cell membrane (Hall 1998). It is possible that (at least in some species) RhoB is activated by noncanonical Wnt signaling by Wnt11 (De Calisto et al. 2005). However, the crest cells cannot leave the neural tube as long as they are tightly connected to one another. One of the functions of the Snail2 protein is to activate factors that dissociate the cadherins that bind these cells together. Originally found on the surface of the neural crest cells, cadherin 6B and E-cadherin adhesion proteins are downregulated at the time of cell migration. Moreover, after migration, the migrating neural crest cells re-express cadherins as they aggregate to form the dorsal root and sympathetic ganglia (Takeichi 1988; Akitaya and Bronner-Fraser 1992; Coles et al. 2007).

The pushing out of the neural crest from the dorsal neural tube appears to be accomplished by their fellow neural crest cells (Carmona-Fontaine et al. 2008). When migrating neural crest cells meet, they stop migration in that direction and extend protrusions from the other side of the cell. This **contact inhibition** of locomotion results in the forward migration of the leading edge of the cells. The contact of the two neural crest cells (but not of a neural crest cell with another cell type) causes the reassortment of noncanonical Wnt pathway proteins to the sites where the cells touch. Here, these proteins activate the RhoA protein that disaggregates the cytoskeleton of the lamellipodiae responsible for migration, and allows them to form on the opposite side of the cell.

The ventral pathway

The choice between the dorsolateral versus the ventral trunk pathway is made at the dorsal neural tube shortly after neural crest cell specification (Harris and Erickson 2007). The first migrating cells are inhibited from entering the dorsolateral pathway by chondroitin sulfate proteoglycans, ephrins, Slit proteins, and probably several other molecules. Because they are so inhibited, these cells migrate ventrally and there give rise to the neurons and glial cells of the peripheral nervous system.

The next choice concerns whether the ventrally migrating cells migrate *between* the somites (to form the sympathetic ganglia of the aorta) or *through* the somites (Schwarz et al. 2009). In the mouse embryo, the first few neural crest cells that form go between the somites, but this pathway is soon blocked by **semaphorin-3F**, a protein that repels neural crest cells; thus, most neural crest cells traveling ventrally migrate through the somites. These cells migrate through the sclerotome (the portion of the somite that gives rise to the cartilage of the spine) and are directed to their destinations by proteins of the extracellular matrices (Newgreen and Gooday 1985; Newgreen et al. 1986).

The extracellular matrices of the sclerotome differ in the anterior and posterior region of each somite, and only the extracellular matrix of the *anterior* sclerotome allows neural crest cell migration. Like the extracellular matrix molecules that prevented neural crest cells from migrating dorsolaterally, the extracellular matrix of the posterior portion of each sclerotome contains proteins that actively exclude neural crest cells. Two such proteins are the **ephrins** and semaphorin-3F (**Figure 10.7A**). The ephrin on the posterior sclerotome is recognized by its receptor, Eph, on the neural crest cells. Similarly, semaphorin-3F on the posterior sclerotome cells is recognized by its receptor, neuropilin-2, on the migrating neural crest cells. When neural crest cells are plated on a culture dish containing stripes of immobilized cell membrane proteins alternately with and without ephrins, the cells leave the ephrin-containing stripes and move along the stripes that lack ephrin (**Figure 10.7B**; Krull et al. 1997; Wang and Anderson 1997; Davy and Soriano 2007). Mutant mice deficient in either semaphorin-3F or neuropilin-2 have severe neural crest migration abnormalities throughout the trunk. This patterning of neural crest cell migration generates the overall segmental character of the peripheral nervous system, reflected in

(A)   Ephrin   Neural crest cells

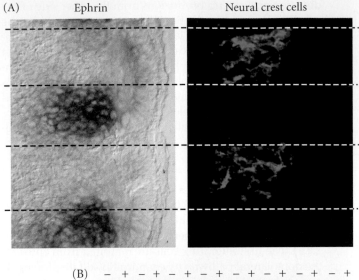

(B) − + − + − + − + − + − + − +

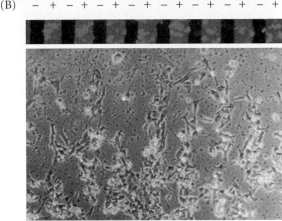

(C) ← Anterior   Posterior →

Motor axons

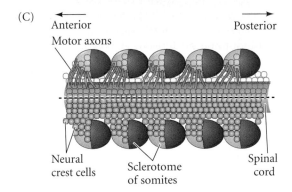

Neural   Sclerotome   Spinal
crest cells   of somites   cord

FIGURE 10.7 Segmental restriction of neural crest cells and motor neurons by the ephrin proteins of the sclerotome. (A) Negative correlation between regions of ephrin in the sclerotome (dark blue stain, left) and the presence of neural crest cells (green HNK-1 stain, right). (B) When neural crest cells are plated on fibronectin-containing matrices with alternating stripes of ephrin, they bind to those regions lacking ephrin. (C) Composite scheme showing the migration of spinal cord neural crest cells and motor neurons through the ephrin-deficient anterior regions of the sclerotomes. (For clarity, the neural crest cells and motor neurons are each depicted on only one side of the spinal cord.) (A,B from Krull et al. 1997; C after O'Leary and Wilkinson 1999.)

terior portion of the sclerotome (Tucker et al. 1999). Thrombospondin is a good substrate for neural crest cell adhesion and migration, and it may cooperate with fibronectin and laminin to promote neural crest cell migration. Therefore, most of the ventrally migrating neural crest cells travel through the anterior region of each somite and are excluded from the posterior regions.

CELL DIFFERENTIATION IN THE VENTRAL PATHWAY The neural crest cells entering the somites differentiate to become two major types of neurons, depending on their location. Those cells that differentiate within the sclerotome become the dorsal root ganglia. These contain the sensory neurons that relay information into the central nervous system.* Those migrating through the anterior portion of the sclerotome and continuing ventrally until they reach the dorsal aorta (but stopping before they enter the gut) become the autonomic ganglia (Vogel and Weston 1990). It is likely that, as they begin to migrate ventrally, the neural crest cells produce progeny that express different receptors. Those migrating neural crest cells that have neurotrophin and Wnt receptors respond to Wnt and neurotrophin-3 from the dorsal neural tube and differentiate into the glia and neurons of the dorsal root ganglia (Weston 1963). Within the dorsal root ganglia, those cells having more Notch will become the glia, while those cells having more Delta (the ligand for Notch) will become the neurons (Wakamatsu et al. 2000; Harris and Erickson 2007).

The cells that continue migrating lack the Wnt and neurotropin receptors. They become the parasympathetic (cholinergic, "rest and digest") and sympathetic (adrenergic, "flight or fight") portions of the autonomic nervous system, the adrenal medulla, and the enteric neurons of the gut.

*These are *afferent* neurons, since they carry information *to* the spinal cord and brain. *Efferent* neurons carry information *away* from the central nervous system; these are the motor neurons generated in the ventral region of the neural tube (as discussed in Chapter 9).

the positioning of the dorsal root ganglia and other neural crest-derived structures.

Both anterior and posterior sclerotome matrices include fibronectin, laminin, tenascin, various collagen molecules, and proteoglycans. Chick trunk neural crest cells begin to express the α4β1 integrin protein, which binds to several of the extracellular matrix proteins, and this allows the neural crest cells to leave the dorsal neural tube (Testaz and Duband 2001). Thrombospondin, another extracellular matrix molecule, is found in the anterior but not the pos-

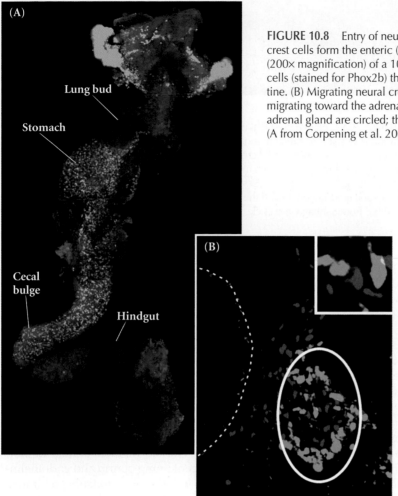

FIGURE 10.8   Entry of neural crest cells into the gut and adrenal gland. (A) Neural crest cells form the enteric (gut) ganglia necessary for peristalsis. Confocal image (200× magnification) of a 10.5-day mouse gut showing the migration of neural crest cells (stained for Phox2b) through the foregut and into the cecal bulge of the intestine. (B) Migrating neural crest cells (stained red for the Sox8 transcription factor) migrating toward the adrenal cortical cells (stained green for SF1). The limits of the adrenal gland are circled; the dorsal aorta boundary is shown by a dotted line. (A from Corpening et al. 2008; B from Reiprich et al. 2008.)

GOING FOR THE GUT   The next choice involves which neural crest cells are able to colonize the gut and which cannot. This distinction involves both extracellular matrix components and soluble paracrine factors. Neural crest cells from the vagal and sacral regions form the enteric ganglia of the gut tube and control intestinal peristalsis. Cells from the vagal neural crest, once past the somites, enter into the foregut and spread to most of the digestive tube, while the sacral neural crest cells colonize the hindgut (**Figure 10.8**). Various inhibitory extracellular matrix proteins (including the Slit proteins) block this more ventral migration in most parts of the embryo, but these inhibitory proteins are absent around the vagal and sacral crests, allowing these neural crest cells to reach the gut tissue. Once in the developing gut, these crest cells are attracted to the digestive tube by **glial-derived neurotrophic factor (GDNF)**, a paracrine factor produced by the gut mesenchyme (Young et al. 2001; Natarajan et al. 2002). GDNF from the gut mesenchyme binds to its receptor, Ret, on the neural crest cells. The vagal neural crest cells have more Ret in their cell membranes than do the sacral cells, and this makes them more invasive (Delalande et al. 2008).

If either GDNF or Ret is deficient in mice or humans, the pup or child suffers from Hirschsprung disease, a syndrome wherein the intestine cannot properly void its solid wastes. In humans, this is most often due to the failure of the vagal neural crest cells to complete their colonization of the foregut and midgut, thus leaving a section of the lower intestine without the ability to undergo peristalsis. By combining the experimental analysis of crest cell migration with mathematical modeling, Landman and colleagues (2007) modeled the migration of the vagal crest cells and explained the genetic deficiencies that cause Hirschsprung disease. In this model, the vagal crest cells normally do not migrate in a directed manner once they are in the anterior portion of the gut. Rather, they proliferate until all the niches in that region of the intestine are saturated, after which the migrating front moves posteriorly. Meanwhile, the gut itself continues to elongate. Whether or not the colonization is complete depends on the initial

These cell lineages may each arise from a mutipotent neural crest progenitor cell, and the restriction of fate into these three lineages may come relatively late. Sieber-Blum (1989) has shown that many individual neural crest cells have the ability to form clones containing numerous cell types. Moreover, BMPs from the aorta appear to convert neural crest cells into the sympathetic and adrenal lineages, whereas glucocorticoids from the adrenal cortex block neuron formation, directing the neural crest cells near them to become adrenomedullary cells (Unsicker et al. 1978; Doupe et al. 1985; Anderson and Axel 1986; Vogel and Weston 1990). Moreover, when chick vagal and thoracic neural crests are reciprocally transplanted, the former thoracic crest produces the cholinergic neurons of the parasympathetic ganglia, and the former vagal crest forms adrenergic neurons in the sympathetic ganglia (Le Douarin et al. 1975). Kahn and co-workers (1980) found that premigratory neural crest cells from both the thoracic and the vagal regions contain enzymes for synthesizing both acetylcholine and norepinephrine. Thus there is good evidence that, although some neural crest cells are committed soon after their formation, the differentiation of the ventrally migrating neural crest cells depends on the pathway they follow and their final location.

(A)

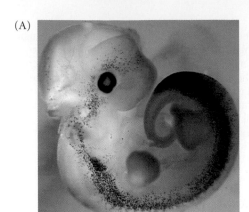

(B)

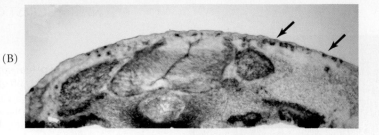

FIGURE 10.9   Neural crest cell migration in the dorsolateral pathway through the skin. (A) Wholemount in situ hybridization of day 11 mouse embryo staining neural crest-derived melanoblasts (purple). (B) Stage 18 chick embryo seen in cross-section through the trunk. Melanoblasts (arrows) can be seen moving through the dermis, from the neural crest region toward the periphery. (A from Baxter and Pavan 2003; B from Santiago and Erickson 2002, courtesy of C. Erickson.)

number of vagal crest cells entering the anterior gut and the ratio of cell motility to gut growth. These results were not obvious, and they show the power of combining experimental and mathematical approaches to development.

THE ADRENAL GLAND   Chemotactic factors are also important for forming the adrenal gland, an organ generated from two distinct cell types. The outer portion, or cortex, of the adrenal gland secretes hormones such as cortisol, while the inner portion—the medulla—secretes hormones such as epinephrine. The cells of the medulla are derived from the trunk neural crest, while those of the adrenal cortex are derived from intermediate mesoderm (see Chapter 11). The neural crest cells are directed into the medulla-forming region by chemotactic factors from those tissues of the intermediate mesoderm that form the adrenal cortex. If other regions of the intermediate mesoderm are converted into adrenal cortex (by electroporating the SF1 transcription factor into these cells), the crest cells will migrate to those ectopic places (Saito and Takahashi 2005; Huber et al. 2008). This migration appears to be directed by a gradient of BMP coming from the adrenal cortical precursors. If Noggin is overexpressed near the adrenal-forming region, trunk neural crest cells are not induced to migrate there. Thus, both extracellular matrices and soluble chemotactic factors influence the migration of these neural crest cells.

The dorsolateral migration pathway

It appears that the cells that take the dorsolateral pathway have already become specified as melanoblasts—pigment cell progenitors—and they are led along the dorsolateral route by chemotactic factors and cell matrix glycoproteins (**Figure 10.9**). In the chick (but not in the mouse), the first neural crest cells to migrate enter the ventral pathway, while cells that migrate later enter the dorsolateral pathway (see Harris and Erickson 2007). These late-migrating cells remain above the neural tube in what is often called the "staging area," and it is these cells that become specified as melanoblasts (Weston and Butler 1966; Tosney 2004). The switch between glial/neural precursor and melanoblast precursor seems to be controlled by the FoxD3 transcription

factor. If FoxD3 is present, it represses expression of MITF, a transcription factor necessary for melanoblast specification and pigment production. If *FoxD3* gene expression is downregulated, MITF is expressed and the cells become melanoblasts. MITF is involved in three signaling cascades. The first cascade activates those genes responsible for pigment production; the second allows these neural crest cells to travel along the dorsolateral pathway into the skin; and the third prevents apoptosis in the migrating cells (Kos et al. 2001; McGill et al. 2002; Thomas and Erickson 2009).

Once specified, the melanoblasts in the staging area upregulate the ephrin receptor (Eph B2) and the endothelin receptor (EDNRB2). This allows them to migrate along extracellular matrices containing ephrin and endothelin-3 (Harris et al. 2008). Indeed, the melanocyte lineage migrates on exactly those same molecules that *repelled* the glial/neural lineage of crest cells. Ephrin expressed along the dorsolateral migration pathway stimulates the migration of melanocyte precursor cells. Ephrin activates its receptor, Eph B2, on the neural crest cell membrane, and this Eph signaling appears to be critical for promoting this migration. Disruption of Eph signaling in late-migrating neural crest cells prevents their dorsolateral migration (Santiago and Erickson 2002).

In mice (but not in chicks), the Kit receptor protein is critical in causing the committed melanoblast precursors to migrate on the dorsolateral pathway. This protein is found on those mouse neural crest cells that also express MITF—that is, the presumptive neuroblasts. Kit protein binds **stem cell factor** (**SCF**), which is made by the dermal cells. When bound to SCF, Kit prevents apoptosis and stimulates cell division among the melanoblast precursors. SCF is critical for dorsolateral migration. If SCF is experimentally secreted from tissues (such as cheek epithelium or the footpads) that do not usually synthesize this protein (and do not usually have melanocytes), neural crest cells will enter those regions and become melanocytes (Kunisada et al. 1998; Wilson et al. 2004).

In vertebrates, all of the pigment cells (except those of the pigmented retina) are derived from the neural crest. As they travel through the dermis, they multiply. In certain mutations, the neural crest cells do not form enough pigment cell precursors to totally encircle the body (see Figure 1.22A). Eventually, the melanoblasts enter the epidermis, where they rapidly enter the developing hair follicles or feather primordia and take up residence at the base of the follicle bulge. In mice, all of the melanoblasts go into the hair follicle (the mouse epidermis is transparent), whereas in birds and humans, both the epidermis and its cutaneous appendage (hair, feather) become pigmented. In the feather or hair follicle, the melanoblasts become **melanocyte stem cells** (Mayer 1973; Nishimura et al. 2002). In the hair shaft, they reside in the bulge, along with the hair stem cells. A portion of these cells migrate outside the bulge at the beginning of each hair development cycle to differentiate into mature melanocytes and provide pigmentation to the hair shaft. Here, the epithelial cells secrete Fgf2, which stimulates the melanocytes to transfer packets of pigment into the epithelial cells (Weiner et al. 2007). Nishimura and colleagues (2005) have documented that the reason the hair of mice and humans grays with age is that melanocyte stem cells become depleted from the bulge.

Thus, the differentiation of the trunk neural crest is accomplished by autonomous factors (such as the Hox genes distinguishing trunk and cranial neural crest cells, or MITF committing cells to a melanocyte lineage), by specific conditions of the environment (such as the adrenal cortex inducing adjacent neural crest cells into adrenomedullary cells), or by a combination of the two (as when cells migrating through the sclerotome respond to Wnt signals depending on their types of receptors.) The fate of an individual neural crest cell is determined both by its starting position (anterior-posterior along the neural tube) and by its migratory path.

Cranial Neural Crest

The head, comprising the face and the skull, is the most anatomically sophisticated portion of the vertebrate body. It is the evolutionary novelty that separates the vertebrates from the other deuterostomes—echinoderms, tunicates, and lancelets (Northcutt and Gans 1983; Wilkie and Morriss-Kay 2001). The head is largely the product of the cranial neural crest, and the evolution of jaws, teeth, and facial cartilage occurs through changes in the placement of these cells (see Chapter 19).

TABLE 10.2 Some derivatives of the pharyngeal arches

Pharyngeal arch	Skeletal elements (neural crest plus mesoderm)	Arches, arteries (mesoderm)	Muscles (mesoderm)	Cranial nerves (neural tube)
1	Incus and malleus (from neural crest); mandible, maxilla, and temporal bone regions (from neural crest)	Maxillary branch of the carotid artery (to the ear, nose, and jaw)	Jaw muscles; floor of mouth; muscles of the ear and soft palate	Maxillary and mandibular divisions of trigeminal nerve (V)
2	Stapes bone of the middle ear; styloid process of temporal bone; part of hyoid bone of neck (all from neural crest cartilage)	Arteries to the ear region: cortico-tympanic artery (adult); stapedial artery (embryo)	Muscles of facial expression; jaw and upper neck muscles	Facial nerve (VII)
3	Lower rim and greater horns of hyoid bone (from neural crest)	Common carotid artery; root of internal carotid	Stylopharyngeus (to elevate the pharynx)	Glossopharyngeal nerve (IX)
4	Laryngeal cartilages (from lateral plate mesoderm)	Arch of aorta; right subclavian artery; original spouts of pulmonary arteries	Constrictors of pharynx and vocal cords	Superior laryngeal branch of vagus nerve (X)
6*	Laryngeal cartilages (from lateral plate mesoderm)	Ductus arteriosus; roots of definitive pulmonary arteries	Intrinsic muscles of larynx	Recurrent laryngeal branch of vagus nerve (X)

Source: Based on Larsen 1993.
*The fifth arch degenerates in humans

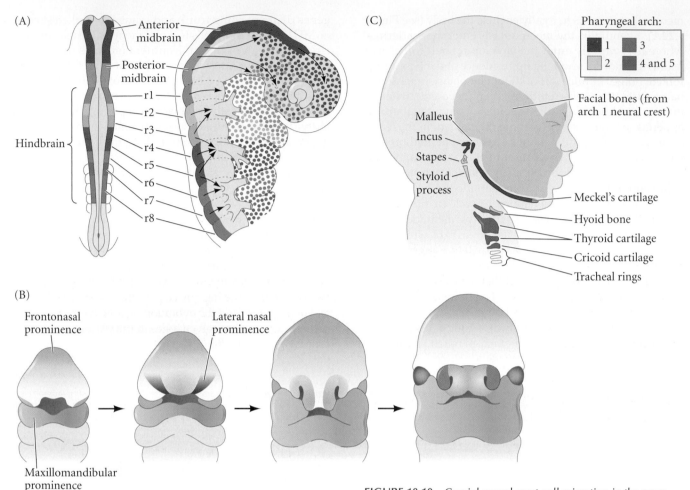

(A) Anterior midbrain, Posterior midbrain, Hindbrain, r1, r2, r3, r4, r5, r6, r7, r8

(B) Frontonasal prominence, Lateral nasal prominence, Maxillomandibular prominence

(C) Pharyngeal arch: 1, 2, 3, 4 and 5. Malleus, Incus, Stapes, Styloid process, Facial bones (from arch 1 neural crest), Meckel's cartilage, Hyoid bone, Thyroid cartilage, Cricoid cartilage, Tracheal rings

FIGURE 10.10   Cranial neural crest cell migration in the mammalian head. (A) Migrational pathways from the cranial neural crest into the pharyngeal arches and frontonasal process. (B) Continued migration of the cranial neural crest to produce the human face. The frontonasal prominence contributes to the forehead, nose, philtrum of the upper lip (the area between the lip and nose), and primary palate. The lateral nasal prominence generates the sides of the nose. The maxillomandibular prominences give rise to the lower jaw, much of the upper jaw, and the sides of the middle and lower regions of the face. (C) Structures formed in the human face by the mesenchymal cells of the neural crest. The cartilaginous elements of the pharyngeal arches are indicated by colors, and the darker pink region indicates the facial skeleton produced by anterior regions of the cranial neural crest. (A after Le Douarin 2004; B after Helms et al. 2005; C after Carlson 1999.)

The cranial neural crest is a mixed population of cells in different stages of commitment, and about 10% of the population comprises multipotent progenitor cells that can differentiate to become neurons, glia, melanocytes, muscle cells, cartilage, and bone (Calloni et al. 2009). The migration of these cells is directed by an underlying segmentation of the hindbrain. As mentioned in Chapter 9, the hindbrain is segmented along the anterior-posterior axis into compartments called rhombomeres. The cranial neural crest cells migrate ventrally from those regions anterior to rhombomere 8 into the pharyngeal arches and the *frontonasal process*; the final destination of these crest cells will determine their eventual fate (**Figure 10.10A; Table 10.2**).

The cranial crest cells follow one of three major streams:

1. Neural crest cells from the midbrain and rhombomeres 1 and 2 of the hindbrain migrate to the first pharyngeal arch (the mandibular arch), forming the jawbones as well as the incus and malleus bones of the middle ear. These cells will also differentiate into the neurons of the trigeminal ganglion—the cranial nerve that innervates the teeth and jaw—and will contribute to the ciliary ganglion that innervates the ciliary muscle of the eye. These neural crest

cells are also pulled by the expanding epidermis to form the **frontonasal process**, which forms the forehead, the middle of the nose, and the primary palate. Thus, the cranial neural crest cells generate the facial skeleton (**Figure 10.10B,C**; Le Douarin and Kalcheim 1999).

2. Neural crest cells from rhombomere 4 populate the second pharyngeal arch, forming the hyoid cartilage of the neck as well as the stapes bone of the middle ear. These cells will also form the neurons of the facial nerve.

3. Neural crest cells from rhombomeres 6–8 migrate into the third and fourth pharyngeal arches and pouches to form the hyoid cartilages and the thymus, parathyroid, and thyroid glands (Serbedzija et al. 1992; Creuzet et al. 2005). They also go to the region of the developing heart, where they help construct the outflow tracts (i.e., the aorta and pulmonary artery). If the neural crest is removed from those regions, these structures fail to form (Bockman and Kirby 1984). Some of these cells migrate caudally to the clavicle (collarbone), where they settle at the sites that will be used for the attachment of certain neck muscles (McGonnell et al. 2001).

Observations of labeled neural crest cells from chick hindbrain, wherein individually marked cells were followed by cameras focusing through a Teflon membrane window in the egg, found that the migrating cells were "kept in line" by interactions with their environment (Kulesa and Fraser 2000). In frog embryos, there is evidence that the separate streams are kept apart by ephrins. Blocking the activity of the Eph receptors causes cells from the different streams to mix together (Smith et al. 1997; Helbling et al. 1998). Neural crest cells from rhombomeres 3 and 5 do not migrate; rather, they undergo apoptosis. If this cell death is prevented, the r3 and r5 neural crest cells enter the first and third pharyngeal arches, respectively, and create muscle attachment sites that are not normally present (Ellies et al. 2002).

Intramembranous ossification: Neural crest-derived head skeleton

Bones form in two major ways. The bones of the trunk and some of the bones in the back of the skull are made from the mesoderm. This type of bone formation, where mesodermal mesenchyme becomes cartilage and the cartilage is replaced by bone, is called *endochondral ossification*; we will talk about it in detail in Chapter 11. The other type of bone formation, where mesenchyme forms bones directly, is called **intramembranous ossification** (Figure 10.11).

There are three main types of intramembranous bone: **sesamoid bone** (such as the patella), which forms as a result of mechanical stress; **periostal bone**, which adds thickness to long bones; and **dermal bone**, which forms in the dermis of the skin. The first two types of intramembranous ossification are characteristic of mesodermal cells. Many of the dermal bones of the skull and face, however, originate from cranial neural crest-derived mesenchymal cells.

The pathway from neural crest to intramembranous bone begins when cranial neural crest cells, under the influence of BMPs from the head epidermis, proliferate and condense into compact nodules (**Figure 10.12**). High levels of BMPs induce these nodules to become cartilage, while lower levels of BMPs induce them to become pre-osteoblast progenitor cells that express the Runx2 transcription factor and the mRNA for collagens II and IX. Later, these cells downregulate Runx2 and begin expressing the *osteopontin* gene, giving them a phenotype similar to a developing chondrocyte (cartilage cell); thus, this stage is called a **chon-**

(A) Osteoblasts   Osteocytes

Mesenchyme

Spicule of bone

FIGURE 10.11 Intramembranous ossification. (A) Mouse mesenchyme cells condense and change shape to produce osteoblasts, which deposit osteoid matrix. Osteoblasts then become arrayed along the calcified region of the matrix. Osteoblasts that are embedded within the calcified matrix become osteocytes. (B) Bone formation in chick embryo heads (days 9-13) as revealed with Alizarin red (which stains bone matrix). (A from Komori et al. 1997; B from Abzhanov et al. 2007, courtesy of P. Abzhanov.)

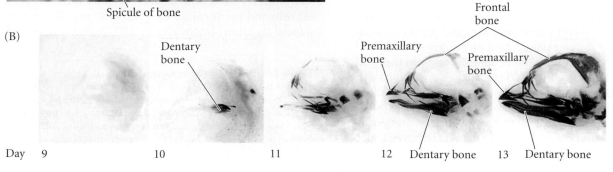

(B)

Day 9   10   11   12 Dentary bone 13 Dentary bone

Dentary bone   Premaxillary bone Frontal bone Premaxillary bone

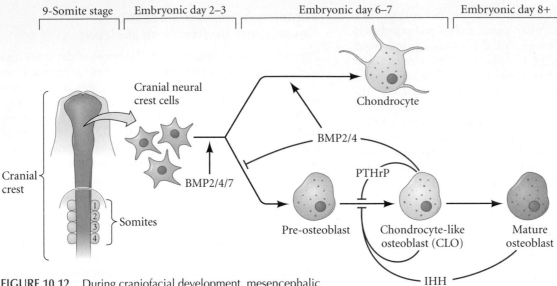

FIGURE 10.12 During craniofacial development, mesencephalic cranial neural crest cells migrate to become the mesenchyme of the future face and much of the skull. Cells of the early cranial skeletogenic condensations depend on BMP2/4/7 activities to form pre-osteoblastic progenitors (which eventually become bone), whereas high levels of BMP2 and/or BMP4 alone induce a chondrogenic (cartilage) fate. Differentiation into the chondrocyte-like osteoblasts is negatively regulated by both Indian hedgehog (IHH) and parathyroid hormone-related protein (PTHrP) activities that are probably autocrine. (After Abzhanov et al. 2007.)

drocyte-like osteoblast. Under the influence of Indian hedgehog (which it secretes and receives in an autocrine fashion), the chondrocyte-like osteoblast becomes a mature **osteoblast**—a committed bone precursor cell (Abzhanov et al. 2007). The osteoblasts secrete a collagen-proteoglycan **osteoid matrix** that is able to bind calcium. Osteoblasts that become embedded in the calcified matrix become **osteocytes** (bone cells). As calcification proceeds, bony spicules radiate out from the region where ossification began. Furthermore, the entire region of calcified spicules becomes surrounded by compact mesenchymal cells that form the **periosteum** (a membrane that surrounds bone). The cells on the inner surface of the periosteum also become osteoblasts and deposit matrix parallel to the existing spicules. In this manner, many layers of bone are formed.

The vertebrate skull, or **cranium**, is composed of the **neurocranium** (skull vault and base) and the **viscerocranium** (jaws and other pharyngeal arch derivatives). Skull bones are derived from both the neural crest and the head mesoderm (Le Lièvre 1978; Noden 1978; Evans and Noden 2006). While the neural crest origin of the viscerocranium has been well documented, the contributions of cranial neural crest cells to the skull vault are more controversial. In 2002, Jiang and colleagues constructed transgenic mice that expressed β-galactosidase only in their cranial neural

crest cells.* When the embryonic mice were stained for β-galactosidase, the cells forming most of the skull—the nasal, frontal, alisphenoid, and squamosal bones—turned blue; the parietal bone did not (**Figure 10.13**). Although the specifics may vary among the vertebrate groups, in general the front of the head is derived from the neural crest, while the back of the skull is derived from a combination of neural crest-derived and mesodermal bones. The neural crest contribution to facial muscle mixes with the cells of the cranial mesoderm, such that facial muscles probably also have dual origins (Grenier et al. 2009).

Given that the neural crest forms our facial skeleton, it follows that even small variations in the rate and direction of cranial neural crest cell divisions will determine what we look like. Moreover, since we look more like our biological parents than our friends do (at least, we hope this is true), such small variations must be hereditary. The regulation of our facial features is probably coordinated in large part by numerous paracrine growth factors. Wnt signaling causes the protrusion of the frontonasal and maxillary prominences, giving shape to the face (Brugmann et al. 2006). FGFs from the pharyngeal endoderm are responsible for the attraction of the cranial neural crest cells into the arches as well as patterning the skeletal elements within the arches. Fgf8 is both a survival factor for the cranial crest cells and is critical for the proliferation cells forming the facial skeleton (Trocovic et al. 2003, 2005; Creuzet et al.

---

*These experiments were done using the Cre-Lox technique. The mice were heterozygous for both (1) a β-galactosidase allele that could be expressed only when Cre-recombinase was activated in that cell, and (2) a Cre-recombinase allele fused to a *Wnt1* gene promoter. Thus, the gene for β-galactosidase was activated (blue stain) only in those cells expressing Wnt1—a protein that is activated in the cranial neural crest and in certain brain cells.

(A)

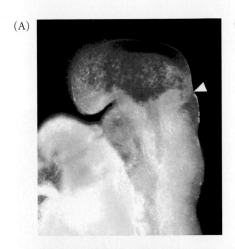

(B)

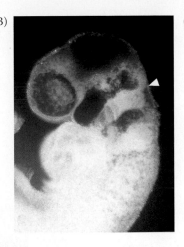

(C)

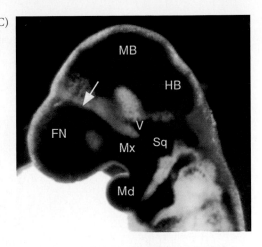

(D)

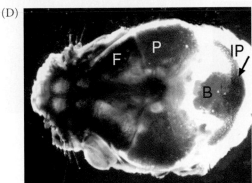

FN, frontal nasal process
MB, midbrain
HB, hindbrain
Mx, maxilla

Md, mandible
Sq, squamosal bone
F, frontal bone
P, parietal

B, mineralized bone
 (non-neural crest)
IP, interparietal bone
V, trigeminal ganglion

FIGURE 10.13   Cranial neural crest cells in embryonic mice, stained for β-galactosidase expression. (A–C) Cranial neural crest migration from day 6 to day 9.5, shown by the dark blue staining neural crest cells. (D) Dorsal view of 17.5-day embryonic mouse showing lack of staining in the parietal bone. (From Jiang et al. 2002, courtesy of G. Morriss-Kay.)

2004, 2005). The FGFs work in concert with BMPs, sometimes activating them and sometimes repressing them (Lee et al. 2001; Holleville et al. 2003).

FGFs also work in concert with Sonic hedgehog (Shh; Haworth et al. 2008). We saw in Chapter 9 that Shh is critical for the proper growth of the facial midline (see Figure 9.31), and Shh is also crucial for shaping the neural crest derivatives of the head. The epithelia (both neural and epidermal) of the dorsal part of the frontonasal process secrete Fgf8, while the ventral epithelia of the frontonasal process secrete Shh (see Figure 9.26). The crest-derived mesenchyme between the epithelia receives both signals. Where these signals meet is where a bird's beak cartilage grows out. If the region of frontnasal process containing the Fgf/Shh boundary is inverted in the chick, the beak forms in reverse direction (Hu et al. 2003; Abzhanov and Tabin 2004). The interactions among these (and other) paracrine factors to pattern the face and skull have become an important area of research in developmental biology.

Coordination of face and brain growth

It is a generalization in clinical genetics that "the face reflects the brain." While this is not always the case, physicians are aware that children with facial anomalies may have brain malformations as well. The coordination between facial form and brain growth was highlighted in studies by Le Douarin and her colleagues (2007). First they found that the region of cranial neural crest that forms the facial skeleton is also critical for the growth of the anterior brain. When that region of chick neural crest was removed, not only did the bird's face fail to form, but the telencephalon failed to grow as well (Figure 10.14A,B). Next they found that the forebrain development could be rescued by adding Fgf8-containing beads to the anterior neural ridge (the neural folds of the anterior neuropore; Figure 10.14C). But this was strange. Cranial neural crest cells do not make or secrete Fgf8; the anterior neural ridge usually does (Figure 10.14D; see Figure 9.26). It seemed that removing the cranial neural crest cells prevented the anterior neural ridge from making the Fgf8 necessary for forebrain proliferation.

Looking at the effects of activated genes added to the anterior neural ridge region, Le Douarin and colleagues hypothesized that the BMP4 from the surface ectoderm was capable of blocking Fgf8. The cranial neural crest cells secreted Noggin and Gremlin, two extracellular proteins that bind to and inactivate BMP4. This allows the synthesis of Fgf8 in the anterior neural ridge and the development of the forebrain structures. Thus, the cranial neural crest cells not only provide the cells that build the facial skeleton and connective tissues, they also regulate the pro-

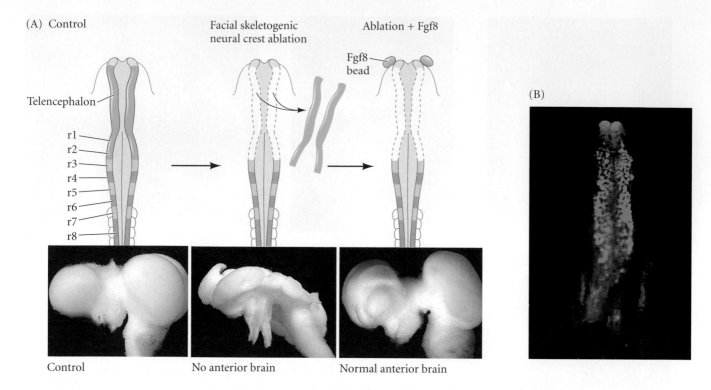

(A) Control

Facial skeletogenic
neural crest ablation

Ablation + Fgf8

Fgf8
bead

Telencephalon

r1
r2
r3
r4
r5
r6
r7
r8

(B)

Control

No anterior brain

Normal anterior brain

FIGURE 10.14   The cranial neural crest that forms the facial skeleton is also critical for the growth of the anterior region of the brain. (A) Removal of the facial skeleton-forming neural crest cells from a 6-somite stage chick embryo stops the telencephalon from forming, as well as inhibiting formation of the facial skeleton. However, telencephalon development could be rescued by adding Fgf8-containing beads to the anterior neural ridge. (B) Embryo stained with HNK-1 (which labels neural crest cells green). Fgf8 appears pink in this micrograph. (After Creuzet et al. 2006, 2009; photographs courtesy of N. Le Douarin.)

duction of Fgf8 in the anterior neural ridge, thereby allowing development of the midbrain and forebrain.

Tooth formation

The cranial neural crest cells of the first pharyngeal arch also form the interior, dentin-secreting odontoblasts of the teeth. The jaw epithelium becomes the outer, enamel-secreting ameloblasts. The mouse tooth is specified by either Fgf8 or BMP4. Fgf8 induces the expression of the Barx1 transcription factor and molds the tooth bud to form molars, whereas BMP4 induces the expression of the Msx1 and Msx2 transcription factors that mold the tooth to form incisors (Mina and Kollar 1987; Lumsden 1988b; Tucker et al. 1998).

The signaling center of the tooth is the **enamel knot**, a group of cells induced in the epithelium by the neural crest-derived mesenchyme (Jernvall et al. 1994; Vaahtokari et al. 1996a,b; Tummers and Thesleff 2009). The enamel knot secretes a cocktail of paracrine factors, including Shh, BMPs 2, 4, and 7, and Fgf4. Shh and Fgf4 induce the pro-

liferation of cells to form a cusp, while the BMPs inhibit the formation of new enamel knots. As will be shown in Chapter 19, these proteins act through a reaction-diffusion mechanism to pattern the cusps of the tooth.

Cardiac Neural Crest

The heart originally forms in the neck region, directly beneath the pharyngeal arches, so it should not be surprising that it acquires cells from the neural crest. The caudal region of the cranial neural crest is sometimes called the cardiac neural crest, since its cells (and only these particular neural crest cells) generate the endothelium of the aortic arch arteries and the septum between the aorta and the pulmonary artery (**Figure 10.15**; Kirby 1989; Waldo et al. 1998). Cardiac crest cells also enter pharyngeal arches 3, 4, and 6 to become portions of other neck structures such as the thyroid, parathyroid, and thymus glands. These cells are often referred to as the circumpharyngeal crest (Kuratani and Kirby 1991, 1992). It is also likely that the carotid body, which monitors oxygen in the blood and reg-

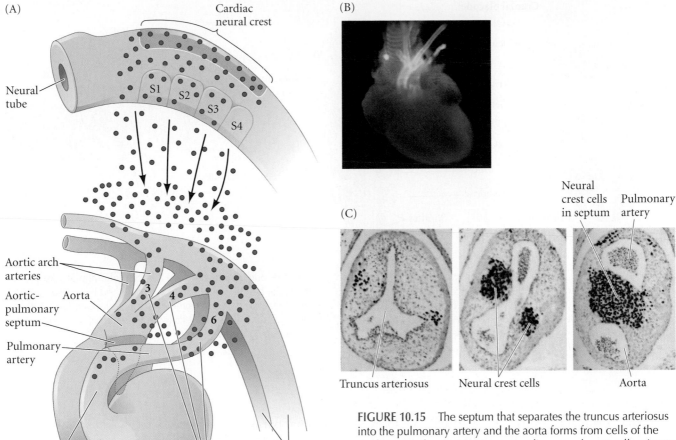

(A)

Cardiac neural crest

Neural tube

S1 S2 S3 S4

Aortic arch arteries

Aortic-pulmonary septum

Aorta

3 4

6

Pulmonary artery

Conotruncus

Pharyngeal arches

Right and left dorsal aorta

(B)

(C)

Neural crest cells in septum

Pulmonary artery

Truncus arteriosus

Neural crest cells

Aorta

FIGURE 10.15 The septum that separates the truncus arteriosus into the pulmonary artery and the aorta forms from cells of the cardiac neural crest. (A) Human cardiac neural crest cells migrate to pharyngeal arches 3, 4, and 6 during the fifth week of gestation and enter the truncus arteriosus to generate the septum. (B) In a transgenic mouse where the fluorescent green protein is expressed only in cells having the Pax3 cardiac neural crest marker, the outflow regions of the heart become labeled. (C) Quail cardiac crest cells were transplanted into the analogous region of a chick embryo and the embryos were allowed to develop. Quail cardiac neural crest cells are visualized by a quail-specific antibody, which stains them darkly. In the heart, these cells can be seen separating the truncus arteriosus into the pulmonary artery and the aorta. (A after Hutson and Kirby 2007; B from Stoller and Epstein 2005; C from Waldo et al. 1998, courtesy of K. Waldo and M. Kirby.)

ulates respiration accordingly, is derived from the cardiac neural crest (see Pardal et al. 2007).

In mice, cardiac neural crest cells are peculiar in that they express the transcription factor Pax3. Mutations of *Pax3* result in fewer cardiac neural crest cells, which in turn leads to persistent truncus arteriosus (failure of the aorta and pulmonary artery to separate), as well as to defects in the thymus, thyroid, and parathyroid glands (Conway et al. 1997, 2000). The path from the dorsal neural tube to the heart appears to involve the coordination between the attractive cues provided by semaphorin-3C and the repulsive signals provided by semaphorin-6 (Toyofuku et al. 2008).

Congenital heart defects in humans and mice often occur along with defects in the parathyroid, thyroid, or thymus glands. It would not be surprising to find that all these problems are linked to defects in the migration of cells from the neural crest (Hutson and Kirby 2007).

See WEBSITE 10.1
Communication between migrating neural crest cells

Cranial Placodes

In addition to forming the cranial neural crest cells, the anterior borders between the epidermal and neural ectoderm also form the **cranial placodes**, which are local and transient thickenings of the ectoderm in the head and neck. (The cranial neural crest and the cranial placodes may have originated from the same cell population during early vertebrate evolution; see Northcutt and Gans 1983; Baker and Bronner-Fraser 1997.) These sensory placodes have neurogenic potential; they give rise to the neurons that form the distal portions of those ganglia associated with hearing, balance, taste, and smell (**Figure 10.16A**; also see Figure

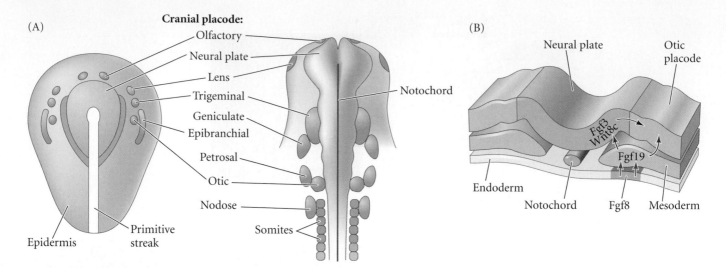

FIGURE 10.16   Cranial placodes form sensory neurons. (A) Fate map of the cranial placodes in the developing chick embryo at the neural plate (left) and 8-somite (right) stages. (B) Induction of the otic (inner ear) placode in the chick embryo. A portion of the pharyngeal endoderm secretes Fgf8, which induces the mesoderm overlying it to secrete Fgf19. Fgf19 is received by both the prospective otic placode and the adjacent neural plate. Fgf19 instructs the neural plate to secrete Wnt8c and Fgf3, two paracrine factors that work synergistically to induce *Pax2* and other genes that allow the cells to produce the otic placode and become sensory cells.

10.3). The proximal neurons of these ganglia are formed from neural crest cells (Baker and Bronner-Fraser 2001). For example, the olfactory placode gives rise to the sensory neurons involved in smell, as well as to migratory neurons that will travel into the brain and secrete gonadotropin releasing hormone. The otic placode gives rise to the sensory epithelium of the ear and to neurons that help form the cochlear-vestibular ganglion. The lens placode is the only placode that does not form neurons.

The placodes are induced by the neighboring tissue, and there is evidence that the different placodes are each a small portion of what had earlier been a common "pre-placodal" territory (Streit 2004; Schlosser 2005). Histological evidence shows that the anterior neural plate is surrounded by a single thickening during the early neurula stages, and the cranial placodes may arise from a common set of inductive interactions between the pharyngeal endoderm and head mesoderm (Platt 1896; Jacobson 1966). Jacobson (1963) also showed that the pre-placodal cells adjacent to the anterior neural tube are competent to give rise to any placode. This columnar pre-placodal epithelium contains the transcription factors Six1, Six4, and Eya. These proteins are maintained in all the placodes and are downregulated in the interplacodal regions (Bhattacharyya et al. 2004; Schlosser and Ahrens 2004). Later, specific interactions define the individual placodes. For instance, the chick otic placode, which develops into the sensory cells of the inner ear, is induced by a combination of FGF and Wnt signals (Ladher et al. 2000, 2005). Here, Fgf19 from the underlying cranial paraxial mesoderm is received by both the presumptive otic vesicle and the adjacent neural plate. Fgf19 induces the neural plate to secrete Wnt8c and Fgf3, which in turn act synergistically to induce formation of the otic placode. The localization of Fgf19 to the specific region of the mesoderm is controlled by Fgf8 secreted in the endodermal region beneath it (**Figure 10.16B**).

The epibranchial placodes form dorsally to where the pharyngeal pouches contact the epidermis. These structures split to form the geniculate, petrosal, and nodose placodes which give rise to the sensory neurons of the facial, glossopharyngeal, and vagal nerves, respectively. The connections made by these placodal neurons are critical, in that they enable taste and other pharyngeal sensations to be appreciated. But how do these neurons find their way into the hindbrain? Late-migrating cranial neural crest cells do not travel ventrally to enter the pharyngeal arches; rather, they migrate dorsally to generate glial cells (Weston and Butler 1966; Baker et al. 1997). These glia form tracks that guide neurons from the epibranchial placodes to the hindbrain. Indeed, if the hindbrain is removed before the neural crest cells emigrate, neurons leaving the placodes enter a crest-free environment and fail to migrate into the hindbrain (Begbie and Graham 2001). Therefore, glial cells from the late-migrating cranial neural crest cells are critical in organizing the innervation of the hindbrain.

See WEBSITE 10.2 Kallmann syndrome

Cranial Neural Crest Cell Migration and Specification

Regional specificity of cranial neural crest migration

There appear to be two "waves" of cranial neural crest cell emigration. The first wave consists of a multipotent population of neural crest-derived mesenchymal cells that migrate to form the cartilage and bones of the head and neck, as well as the stromal tissue of pharyngeal organs such as the thyroid and thymus (see Figure 10.10). Hox genes are critical for specifying these regions. The first stream of cells (down to r2) that form the jaw and face do not express any Hox genes, while the other two streams express the first four Hox paralogue genes. If Hox gene expression is experimentally induced in the anterior neural fold, no facial skeleton is formed.

In *Xenopus*, the first paralogous Hox group (*Hoxa1, Hoxb1, Hoxd1*) is responsible for the migration of neural crest cells into the pharyngeal arches. When antisense morpholinos knock out the expression of all these genes, no migration of neural crest cells extends the arches, and the gill cartilages are totally absent (McNulty et al. 2005).

In the mouse, different Hox genes may play different roles in this specification. For instance, *Hoxa2* appears to be essential for the proper development of the second arch derivatives (**Figure 10.17**). When *Hoxa2* is knocked out from mouse embryos, the neural crest cells of the second pharyngeal arch generate Meckel's cartilage—the jaw cartilage characteristic of the first pharyngeal arch (Gendron-Maguire et al. 1993; Rijli et al. 1993). Conversely, if *Hoxa2* is expressed ectopically in the neural crest cells of the first arch, the crest cells of the first arch form cartilage characteristic of the second arch (Grammatopolous et al. 2000; Pasqualetti et al. 2000).

There are reciprocal interactions by which the neural crest cells and the pharyngeal arch regions mutually specify each other. First, the pharyngeal endoderm dictates the differentia-

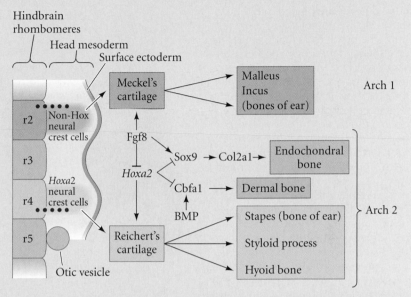

Figure 10.17   The influence of mesoderm and ectoderm on the axial identity of cranial neural crest cells and the role of *Hoxa2* in regulating second-arch morphogenesis. Neural crest cells (small black circles) from rhombomere 2 do not express Hox genes as they enter pharyngeal arch 1. *Hoxa2* is expressed in cranial crest cells migrating into pharyngeal arch 2, but not into arch 1, and it plays a major role in imposing second-arch identity on second-arch structures. The diagram indicates the emerging pathways by which *Hoxa2* interacts with ectodermally secreted Fgf8 and BMPs to regulate morphogenesis in the two arches. (After Trainor and Krumlauf 2001.)

tion of the neural crest cells. When *individual* mouse or zebrafish cranial crest cells are transplanted from one region of the hindbrain to another, they take on the characteristic Hox gene expression pattern of their host (surrounding pharyngeal) region and lose their original gene expression pattern. However, if *groups* of cranial neural crest cells are so transplanted, they tend to keep their original identity (Trainor and Krumlauf 2000; Schilling 2001). Couly and colleagues (2002) found that removing a specific region of pharyngeal endoderm caused the loss of that particular skeletal structure formed by the neural crest. Moreover, if an extra piece of first ("mandibular") arch pharyngeal endoderm was grafted into the existing pharynx of a chick embryo, neural crest cells would enter this graft and

make an extra jaw (**Figure 10.18A**). Reciprocally, the neural crest cells instruct gene expression in the surrounding tissues. This influences the growth patterns of the arch—for example, whether a bird's face has a narrow beak or a broad, ducklike bill (**Figure 10.18B**; Noden 1991; Schneider and Helms 2003).

Why don't birds have teeth?

One characteristic of the jaws of all birds is that they lack teeth. The ancestors of birds and crocodiles had teeth, and modern crocodiles have teeth, but birds lack these structures. Instead, bird beaks have a keratinized epithelium.

Transplantation studies have shown that the oral epithelium of birds retains the ability to form teeth:

(Continued on next page)

SIDELIGHTS & SPECULATIONS (Continued)

transplanting mouse neural crest-derived tooth mesenchyme allows the chick oral epithelium to form toothlike structures (Kollar and Fisher 1980; Mitsiadis et al. 2003). Indeed, if BMP-coated beads are placed into the Fgf8-expressing region of chick jaw epithelium, the beads induce the neural crest-derived mesenchyme to synthesize Msx1 and Msx2 transcription factors and form toothlike structures in the chick jaw (Chen et al. 2000). Harris and colleagues (2006) demonstrated that the *talpid²* mutation of the chick forms teeth, and that these teeth look just like the first set of crocodile teeth. The reason teeth can form in this mutant is that the boundary between the oral and non-oral has changed. In those animals having teeth, the oral ectoderm signaling center (formed by the interaction of Fgf8, BMP4, and Sonic hedgehog) overlies the neural crest mesenchyme that is capable of forming teeth. In birds, the oral ectoderm signaling center is not positioned over the competent mesenchyme, and so the signal is not received. In the *talpid²* mutant, the oral/non-oral ectoderm boundary has again shifted, so that the oral ectoderm signaling center once more meets the competent mesenchyme. Obviously, such mutants don't occur often. In fact, they are "as rare as hen's teeth."

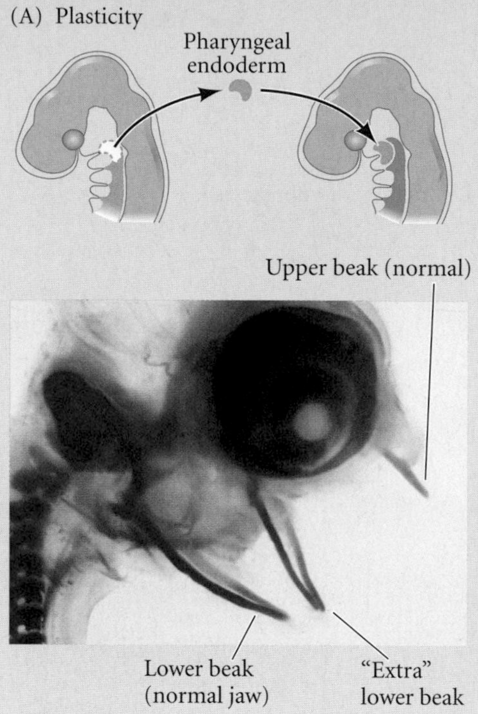

(A) Plasticity

Pharyngeal endoderm

Upper beak (normal)

Lower beak (normal jaw)　　"Extra" lower beak

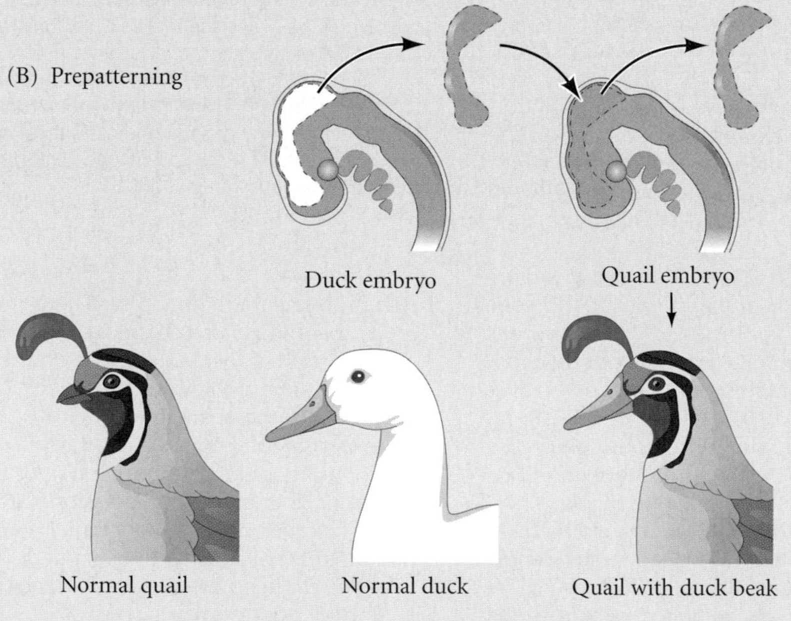

(B) Prepatterning

Duck embryo　　Quail embryo

Normal quail　　Normal duck　　Quail with duck beak

Figure 10.18 Plasticity and prepatterning of the neural crest both play roles in beak morphology. (A) Plasticity of crest cells is shown when a piece of chick endoderm at the level of the mandibular arch is transplanted to a host chick embryo. An extra lower beak develops, as seen in the photograph. The chick has two lower beaks, one from the host and one from neural crest cells developing a jaw in the transplanted pharyngeal endoderm. (B) Prepatterning in duck and quail beaks. A quail has a narrow beak, while a duck's beak is broad and flat. When duck cranial neural crest is transplanted into a quail embryo, the quail develops a duck beak. (The reverse experiment also works, giving the duck a quail's beak.) (After Santagati and Rijli 2003; photograph courtesy of N. M. Le Douarin.)

NEURONAL SPECIFICATION AND AXONAL SPECIFICITY

Not only do neuronal precursor cells and neural crest cells migrate to their place of function, but so do the axons extending from the cell bodies of neurons. Unlike most cells, whose parts all stay in the same place, the neuron can produce axons that may extend for meters. As we saw in Chapter 9, the axon has its own locomotory apparatus, which resides in the axonal growth cone. The growth cone has been called "a neural crest cell on a leash" because, like neural crest cells, it migrates and senses the environment. Moreover, it can respond to the same types of signals that

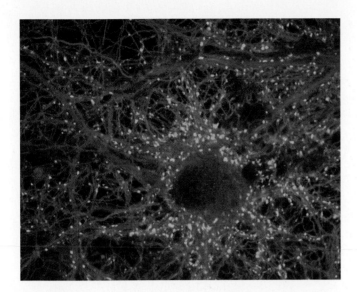

FIGURE 10.19 Connections of axons to a cultured rat hippocampal neuron. The neuron has been stained red with fluorescent antibodies to tubulin. The neuron appears to be outlined by the synaptic protein synapsin (green), which is present in the terminals of axons that contact it. (Photograph courtesy of R. Fitzsimmons and PerkinElmer Life Sciences.)

migrating cells can sense. The cues for axonal migration, moreover, may be even more specific than those used to guide specific cell types to particular areas. Each of the 10^{11} neurons in the human brain has the potential to interact specifically with thousands of other neurons (Figure 10.19). A large neuron (such as a Purkinje cell or motor neuron) can receive input from more than 10^5 other neurons (Gershon et al. 1985). Understanding the generation of this stunningly ordered complexity is one of the greatest challenges to modern science.

Goodman and Doe (1993) list eight stages of neurogenesis:

1. Induction and patterning of a neuron-forming (neurogenic) region
2. Birth and migration of neurons and glia
3. Specification of cell fates
4. Guidance of axonal growth cones to specific targets
5. Formation of synaptic connections
6. Binding of trophic factors for survival and differentiation
7. Competitive rearrangement of functional synapses
8. Continued synaptic plasticity during the organism's lifetime

The first two of these processes were described in Chapter 9. Here, we continue our investigation of the processes of neural development.

See WEBSITE 10.3
The evolution of developmental neurobiology

The Generation of Neuronal Diversity

Neurons are specified in a hierarchical manner. The first decision is whether the ectodermal epithelium will become neural epithelium, epidermal epithelium, or undergo epithelial-mesenchymal transition and become neural crest. If the choice is neural epithelium, the next decision (as we saw in Chapter 9) involves becoming either a glial cell or a neuron. If it is to become a neuron, the next decision is, what type of neuron? Will it become a motor neuron, a sensory neuron, a commissural neuron, or some other type? After this fate is determined, still another decision gives the neuron a specific target. To illustrate this process of progressive specification, we will focus on the motor neurons of vertebrates.

Vertebrates form a dorsal neural tube by blocking a BMP signal, and the specification of neural (as opposed to glial or epidermal) fate is accomplished through the Notch-Delta pathway (see Chapter 3). The specification of the *type* of neuron appears to be controlled by the position of the neuronal precursor within the neural tube and by its birthday. As described in Chapter 9, neurons at the ventrolateral margin of the vertebrate neural tube become motor neurons, while interneurons are derived from cells in the dorsal region of the tube. Since the grafting of floor plate or notochord cells (which secrete Sonic hedgehog) to lateral areas of the neural tube can re-specify dorsolateral cells as motor neurons, the specification of neuron type is probably a function of the cell's position relative to the floor plate. Ericson and colleagues (1996) have shown that two periods of Shh signaling are needed to specify the motor neurons: an early period wherein the cells of the ventrolateral margin are instructed to become ventral neurons, and a later period (which includes the S phase of its last cell division) that instructs a ventral neuron to become a motor neuron rather than an interneuron. The first decision is probably regulated by the secretion of Shh from the notochord, while the second is more likely regulated by Shh from the floor plate cells. Sonic hedgehog appears to specify motor neurons by inducing certain transcription factors at different concentrations (Ericson et al. 1992; Tanabe et al. 1998; see Figure 9.14).

The next decision involves target specificity. If a cell is to become a neuron, and specifically a motor neuron, will that motor neuron be one that innervates the thigh, the forelimb, or the tongue? The anterior-posterior specification of the neural tube is regulated primarily by Hox genes from the hindbrain through the spinal cord, and by specific head genes (such as *Otx*) in the brain (Dasen et al. 2005). Within a region of the body, motor neuron specificity is regulated by the cell's age when it last divides. As discussed in Chapter 9, a neuron's birthday determines which layer of the cortex it will enter. As younger cells migrate to the periphery, they must pass through neurons that differentiated earlier in development. As younger motor neurons migrate through the region of older motor neurons in

(A)

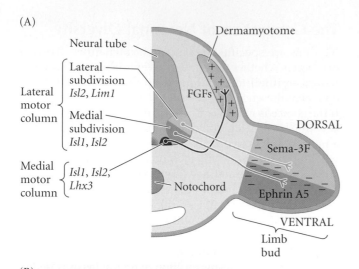

Lateral motor column
- Lateral subdivision *Isl2, Lim1*
- Medial subdivision *Isl1, Isl2*

Medial motor column
- *Isl1, Isl2, Lhx3*

Neural tube

Dermamyotome

FGFs

DORSAL

Sema-3F

Notochord

Ephrin A5

VENTRAL

Limb bud

FIGURE 10.20 Motor neuron organization and Lim specification in the spinal cord innervating the chick hindlimb. (A) Neurons in each of three different columns express specific sets of LIM family genes (including *Isl1* and *Isl2*), and neurons within each column make similar pathfinding decisions. Neurons of the medial motor column are attracted to the axial muscles by FGFs secreted by the dermamyotome. Neurons of the lateral motor column send axons to the limb musculature. Where these columns are subdivided, medial subdivisions project to ventral positions because they are repelled by semaphorin-3F in the dorsal limb bud; and lateral subdivisions send axons to dorsal regions of the limb bud, as they are repelled by ephrin A5 synthesized in the ventral half. (B) Motor neurons in different regions of the chick spinal cord express different transcription factors (visualized here using various stains), giving them different cell-surface receptors that affect axonal migration. (A after Polleux et al. 2007; B courtesy of J. S. Dasen.)

(B)

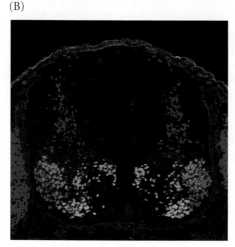

the intermediate zone, they express new transcription factors as a result of a retinoic acid (or other retinoid) signal secreted by the early-born motor neurons (Sockanathan and Jessell 1998). These transcription factors are encoded by the Lim genes and are structurally related to those encoded by the Hox genes.

As a result of their differing birthdays and migration patterns, motor neurons form three major groups: the columns of Terni (CT), and the lateral and medial motor columns (LMC and MMC). The cell bodies of the motor neurons projecting to a single muscle are "pooled" in a longitudinal column of the spinal cord. This pooling is performed by different cadherins that become expressed on these different populations of cells (Landmesser 1978; Hollyday 1980; Price et al. 2002). The pools are grouped into the CT, LMC, and MMC, and neurons in similar places have similar targets. For instance, in the chick hindlimb, muscles are innervated by the LMC axons, with lateral neurons entering the dorsal musculature, while the motor neurons of the MMC innervate ventral limb musculature (**Figure 10.20**; Tosney et al. 1995; Polleux et al. 2007). This

arrangement of motor neurons is consistent throughout the vertebrates.

The targets of motor neurons are specified before their axons extend into the periphery. This was shown by Lance-Jones and Landmesser (1980), who reversed segments of the chick spinal cord so that the motor neurons were placed in new locations. The axons went to their original targets, not to the ones expected from their new positions (**Figure 10.21**). The molecular basis for this target specificity resides in the members of the Hox and Lim protein families that are induced during neuronal development (Tsushida et al. 1994; Sharma et al. 2000; Price and Briscoe 2004). For instance, all motor neurons express Islet1 and (slightly later) Islet2. If no other Lim protein is expressed, the neurons project to the ventral limb muscles. This is because the axons (just like the trunk neural crest cells) synthesize neuropilin-2, the receptor for the chemorepellant semaphorin-3F, which is made in the dorsal part of the limb bud. However, if Lim1 protein is also synthesized, the motor neurons project dorsally to the dorsal limb muscles because Lim1 induces the expression of Eph A4, the receptor for the chemorepellent protein ephrin A5, which is made in the ventral part of the limb bud. Thus, the innervation of the limb by motor neurons depends on repulsive signals. The motor neurons entering the axial muscles of the body wall, however, are brought there by chemoattraction. (Indeed, these axons make an abrupt turn to get to the developing musculature.) This is because these motor neurons express Lhx3, which induces the expression of a receptor for fibroblast growth factors such as those secreted by the dermamyotome (the somatic region that contains muscle precursor cells).

Pattern Generation in the Nervous System

Vertebrate brain function depends not only on the differentiation and positioning of the neurons, but also on the specific connections these cells make among themselves and their

(A) 2–5 days

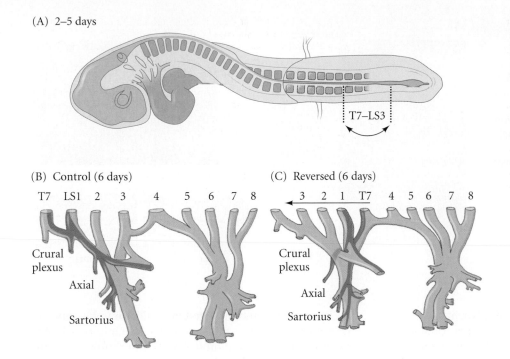

T7–LS3

(B) Control (6 days)

T7   LS1   2   3   4   5   6   7   8

Crural
plexus

Axial

Sartorius

(C) Reversed (6 days)

3   2   1   T7   4   5   6   7   8

Crural
plexus

Axial

Sartorius

FIGURE 10.21 Compensation for small dislocations of axonal initiation position in the chick embryo. (A) A length of spinal cord comprising segments T7–LS3 (seventh thoracic to third lumbosacral segments) is reversed in a 2.5-day embryo. (B) Normal pattern of axon projection to the muscles at 6 days. (C) Projection of axons from the reversed segment at 6 days. The ectopically placed neurons eventually found their proper neural pathways and innervated the appropriate muscles. (After Lance-Jones and Landmesser 1980.)

peripheral targets. Nerves from a sensory organ such as the eye or nose must connect to specific neurons in the brain that can interpret stimuli from that organ, and axons from the central nervous system must cross large expanses of tissue before innervating their target tissue. How does the neuronal axon "know" how to traverse numerous potential target cells to make its specific connection?

Ross G. Harrison (1910) first suggested that the specificity of axonal growth is due to **pioneer nerve fibers**, axons that go ahead of other axons and serve as guides for them.* This observation simplified, but did not solve, the problem of how neurons form appropriate patterns of interconnection. Harrison also noted, however, that axons must grow on a solid substrate, and he speculated that differences among embryonic surfaces might allow axons to travel in certain specified directions. The final connections would occur by complementary interactions on the target cell surface:

> That it must be a sort of a surface reaction between each kind of nerve fiber and the particular structure to be innervated seems clear from the fact that sensory and motor fibers, though running close together in the same bundle, nevertheless form proper peripheral connections,

the one with the epidermis and the other with the muscle. … The foregoing facts suggest that there may be a certain analogy here with the union of egg and sperm cell.

Research on the specificity of neuronal connections has focused on two major systems: motor neurons, whose axons travel from the spinal cord to a specific muscle; and the optic system, wherein axons originating in the retina find their way back into the brain. In both cases, the specificity of axonal connections is seen to unfold in three steps (Goodman and Shatz 1993):

1. **Pathway selection**, wherein the axons travel along a route that leads them to a particular region of the embryo.
2. **Target selection**, wherein the axons, once they reach the correct area, recognize and bind to a set of cells with which they may form stable connections.
3. **Address selection**, wherein the initial patterns are refined such that each axon binds to a small subset (sometimes only one) of its possible targets.

The first two processes are independent of neuronal activity. The third process involves interactions between several active neurons and converts the overlapping projections into a fine-tuned pattern of connections.

It has been known since the 1930s that the axons of motor neurons can find their appropriate muscles even if the neural activity of the axons is blocked. Twitty (who was Harrison's student) and his colleagues found that embryos of the newt *Taricha torosa* secreted a toxin, tetrodotoxin, that blocked neural transmission in other species. By grafting pieces of *T. torosa* embryos onto embryos of other salamander species, they were able to paralyze the host embryos

---

*The growth cones of pioneer neurons migrate to their target tissue while embryonic distances are still short and the intervening embryonic tissue is still relatively uncomplicated. Later in development, other neurons bind to pioneer neurons and thereby enter the target tissue. Klose and Bentley (1989) have shown that in some cases, pioneer neurons die after the "follow-up" neurons reach their destination. Yet if the pioneer neurons are prevented from differentiating, the other axons do not reach their target tissue.

FIGURE 10.22 Repulsion of dorsal root ganglion growth cones. (A) Motor axons migrating through the rostral (anterior), but not the caudal (posterior), compartments of each sclerotome. (B) In vitro assay, wherein ephrin stripes were placed on a background surface of laminin. Motor axons grew only where the ephrin was absent. (C) Inhibition of growth cones by ephrin after 10 minutes of incubation. The left-hand photograph shows a control axon subjected to a similar (but not inhibitory) compound; the axon on the right was exposed to an ephrin found in the posterior somite. (From Wang and Anderson 1997, courtesy of the authors.)

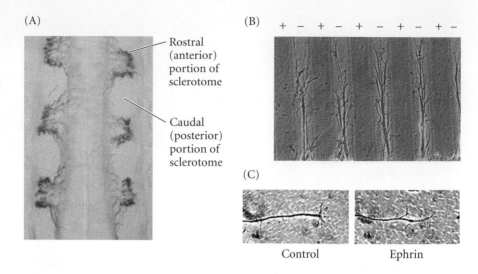

(A)

Rostral (anterior) portion of sclerotome

Caudal (posterior) portion of sclerotome

(B)

(C)

Control   Ephrin

for days while development occurred. Normal neuronal connections were made, even though no neuronal activity could occur. At about the time the tadpoles were ready to feed, the toxin wore off, and the young salamanders swam and fed normally (Twitty and Johnson 1934; Twitty 1937). More recent experiments using zebrafish mutants with nonfunctional neurotransmitter receptors similarly demonstrated that motor neurons can establish their normal patterns of innervation in the absence of neuronal activity (Westerfield et al. 1990). But the question remains: How are the axons instructed where to go?

Cell adhesion and contact guidance by attraction and repulsion

The initial pathway an axonal growth cone follows is determined by the environment the growth cone experiences. Some of the substrates the growth cone encounters permit it to adhere to them, and thus promote axon migration. Other substrates cause the growth cone to retract, preventing its axon from growing in that direction. Growth cones prefer to migrate on surfaces that are more adhesive than their surroundings, and a track of adhesive molecules (such as laminin) can direct them to their targets (Letourneau 1979; Akers et al. 1981; Gundersen 1987).

In addition to general extracellular matrix cues, there are protein guidance cues specific to certain groups of neurons. Axons in the developing nervous system respond to attractive and repulsive signals of the ephrin, semaphorin, netrin, and Split protein families. We have already seen that neural crest cells are patterned by their recognition of ephrin, and that what is an attractive cue to one set of cells (such as the presumptive melanocytes going through the dermis) can be a repulsive signal to another set of cells (such as the presumptive sympathetic ganglia). We will see that whether a guidance signal is attractive or repulsive can depend on the type of cell receiving that signal and on the time when that cell receives it.

Two of the membrane protein families involved in neural patterning are the ephrins and the semaphorins. Just as neural crest cells are inhibited from migrating across the posterior portion of a sclerotome, the axons from the dorsal root ganglia and motor neurons also pass only through the anterior portion of each sclerotome and avoid migrating through the posterior portion (**Figure 10.22A**; also see Figure 10.7). Davies and colleagues (1990) showed that membranes isolated from the posterior portion of a somite cause the growth cones of these neurons to collapse (**Figure 10.22B,C**). These growth cones contain Eph receptors and neuropilin receptors that are responsive to ephrins and semaphorins on the posterior sclerotome cells (Wang and Anderson 1997; Krull et al. 1999; Kuan et al. 2004). Thus the same signals that pattern neural crest cell migration also pattern the spinal neuronal outgrowths.

Found throughout the animal kingdom, the semaphorins often guide growth cones by selective repulsion. They are especially important in forcing "turns" when an axon must change direction. Semaphorin-1 is a transmembrane protein that is expressed in a band of epithelial cells in the developing insect limb. This protein appears to inhibit the growth cones of the Ti1 sensory neurons from moving forward, thus causing them to turn (**Figure 10.23**; Kolodkin et al. 1992, 1993). In *Drosophila*, semaphorin-2 is secreted by a single large thoracic muscle. In this way, the thoracic muscle prevents itself from being innervated by inappropriate axons (Matthes et al. 1995).

The proteins of the semaphorin-3 family, also known as collapsins, are found in mammals and birds. These secreted proteins collapse the growth cones of axons originating in the dorsal root ganglia (Luo et al. 1993). There are several types of neurons in the dorsal root ganglia whose axons enter the dorsal spinal cord. Most of these axons are prevented from traveling farther and entering the ventral spinal cord; however, a subset of them does travel ventrally through the other neural cells (**Figure 10.24**). These particular axons are not inhibited by semaphorin-3, whereas

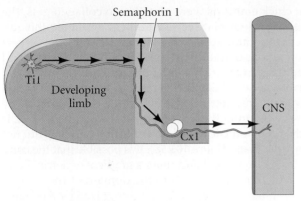

Semaphorin 1

Ti1

Developing
limb

Cx1

CNS

Ventral nerve cord

FIGURE 10.23 Action of semaphorin-1 in the developing grasshopper limb. The axon of sensory neuron Ti1 projects toward the central nervous system. (The arrows represent sequential steps en route.) When it reaches a band of semaphorin-1-expressing epithelial cells, the axon reorients its growth cone and extends ventrally along the distal boundary of the semaphorin-1-expressing cells. When its filopodia connect to the Cx1 pair of cells, the growth cone crosses the boundary and projects into the CNS. When semaphorin-1 is blocked by antibodies, the growth cone searches randomly for the Cx1 cells. (After Kolodkin et al. 1993.)

those of the other neurons are (Messersmith et al. 1995). This finding suggests that semaphorin-3 patterns sensory projections from the dorsal root ganglia by selectively repelling certain axons so that they terminate dorsally. A similar scheme is seen in the brain, where semaphorin made in one region of the brain is used to prevent the entry of neurons made in another region (Marín et al. 2001).

As any psychologist knows, the line between attraction and repulsion is often thin. At the base of both phenomena is some sort of recognition event. This is also the case with

neurons. When the receptor protein Eph A7 recognizes ephrin A5 in the mouse neural tube, the result is attraction rather than repulsion. The interaction between these two proteins is critical for the closure of the neural tube. The Eph A7 and ephrin A5 proteins cause *adhesion* of the neural plate cells, and deletion of either gene results in a condition resembling anencephaly. The change from repulsion to adhesion is caused by alternative RNA processing. By using a different splice site, the mouse neural plate produces Eph A7 lacking the tyrosine kinase domain that transmits the repulsive signal (Holmberg et al. 2000). The result is that the cells recognize each other through these proteins, and no repulsion occurs.

(A)

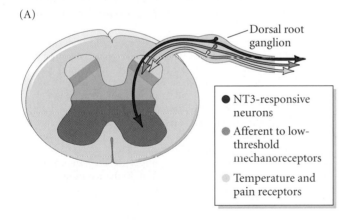

Dorsal root
ganglion

● NT3-responsive
neurons

● Afferent to low-
threshold
mechanoreceptors

● Temperature and
pain receptors

FIGURE 10.24 Semaphorin-3 as a selective inhibitor of axonal projections into the ventral spinal cord. (A) Trajectory of axons in relation to semaphorin-3 expression in the spinal cord of a 14-day embryonic rat. Neurotrophin 3-responsive neurons can travel to the ventral region of the spinal cord, but the afferent axons for the mechanoreceptors and for temperature and pain receptor neurons terminate dorsally. (B) Transgenic chick fibroblast cells that secrete semaphorin-3 inhibit the outgrowth of mechanoreceptor axons. These axons are growing in medium treated with NGF, which stimulates their growth, but are still inhibited from growing toward the source of semaphorin-3. (C) Neurons that are responsive to NT3 for growth are not inhibited from extending toward the source of semaphorin-3 when grown with NT3. (A after Marx 1995; B,C from Messersmith et al. 1995, courtesy of A. Kolodkin.)

(B)

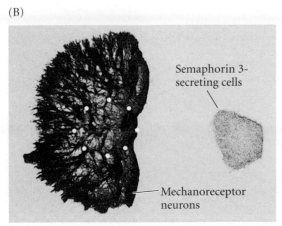

Semaphorin 3-
secreting cells

Mechanoreceptor
neurons

(C)

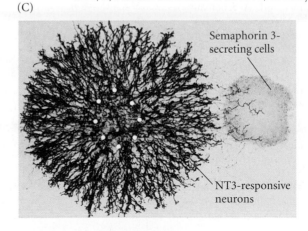

Semaphorin 3-
secreting cells

NT3-responsive
neurons

In a different way, semaphorins can also be attractants. Semaphorin-3A is a classic chemorepellant for the axons coming from pyramidal neurons in the mammalian cortex. However, it is a chemoattractant for the dendrites of the same cells. In this way, a target can "reach out" to the dendrites of these cells without attracting their axons as well (Polleux et al. 2000).

See WEBSITE 10.4 The pathways of motor neurons

Guidance by diffusible molecules

NETRINS AND THEIR RECEPTORS The idea that chemotactic cues guide axons in the developing nervous system was first proposed by Santiago Ramón y Cajal (1892). He suggested that the commissural neurons of the spinal cord might be told by diffusible factors to send axons from their dorsal positions to the ventral floor plate. Commissural neurons are interneurons that cross the ventral midline to coordinate right and left motor activities. Thus, they somehow must migrate to (and through) the ventral midline. The axons of these neurons begin growing ventrally down the side of the neural tube. However, about two-thirds of the way down, their direction changes, and they project through the ventrolateral (motor) neuron area of the neural tube toward the floor plate cells (Figure 10.25A).

There appear to be two systems involved in attracting the commissural neurons to the ventral midline. The first is the Sonic hedgehog protein that is made in the floor plate (see Figure 9.14; Charron et al. 2003). If Shh is inhibited by cyclopamine or conditionally knocked out of the floor plate cells, the commissural axons have difficulty getting to the ventral midline. However, a gradient of Shh does not provide a full explanation of the migration. Some other factor is also involved. In 1994, Serafini and colleagues developed an assay that allowed them to screen for the presence of a presumptive diffusible molecule that might be guiding the commissural neurons. When dorsal spinal cord explants from chick embryos were plated onto collagen gels, the presence of floor plate cells near them promoted the outgrowth of commissural axons. Serafini and his co-workers took fractions of embryonic chick brain homogenate and tested them to see if any of the proteins therein mimicked explant activity. This research resulted in the identification of two proteins, **netrin-1** and **netrin-2**. Netrin-1 is made by and secreted from the floor plate cells, whereas netrin-2 is synthesized in the lower region of the spinal cord, but not in the floor plate (Figure 10.25B). It is possible that the commissural neurons first encounter a gradient of netrin-2 and Shh, which brings them into the domain of the steeper netrin-1 gradient. The netrins are recognized by receptors, DCC and DSCAM, found in the axon growth cones (Liu et al. 2009).

Although they are soluble molecules, both netrins become associated with the extracellular matrix.* Such associations can play important roles and may change the effect of the netrin from attractive to repulsive. The growth cones of *Xenopus* retinal neurons, for example, are attracted to netrin-1 and are guided to the head of the optic nerve by this diffusible factor. Once there, however, the combination of netrin-1 and laminin *prevents* the axons from leaving the optic nerve. It appears that the laminin of the extracellular matrix surrounding the optic nerve converts the netrin from an attractive molecule to a repulsive one (Höpker et al. 1999).

The structures of the netrin proteins have numerous regions of homology with UNC-6, a protein implicated in

---

*The binding of a soluble factor to the extracellular matrix makes for an interesting ambiguity between *chemotaxis* and migrating on preferred substrates (*haptotaxis*). Nature doesn't necessarily conform to the categories we create. There is also some confusion between the terms *neurotropic* and *neurotrophic*. Neuro*tropic* (Latin, *tropicus*, "a turning movement") means that something attracts the neuron. Neuro*trophic* (Greek, *trophikos*, "nursing") refers to a factor's ability to keep the neuron alive, usually by supplying growth factors. Since many agents have both properties, they are alternatively called *neurotropins* and *neurotrophins*. In the recent literature, *neurotrophin* appears to be more widely used.

FIGURE 10.25 Trajectory of commissural axons in the rat spinal cord. (A) Schematic drawing of a model wherein commissural neurons first experience a gradient of Sonic hedgehog and netrin-2, and then a steeper gradient of netrin-1. The commissural axons are chemotactically guided ventrally down the lateral margin of the spinal cord toward the floor plate. Upon reaching the floor plate, contact guidance from the floor plate cells causes the axons to change direction. (B) Autoradiographic localization of netrin-1 mRNA by in situ hybridization of antisense RNA to the hindbrain of a young rat embryo. Netrin-1 mRNA (dark area) is concentrated in the floor plate neurons. (B from Kennedy et al. 1994, courtesy of M. Tessier-Lavigne.)

(A)

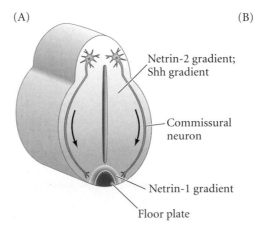

Netrin-2 gradient;
Shh gradient

Commissural neuron

Netrin-1 gradient

Floor plate

(B)

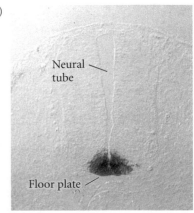

Neural tube

Floor plate

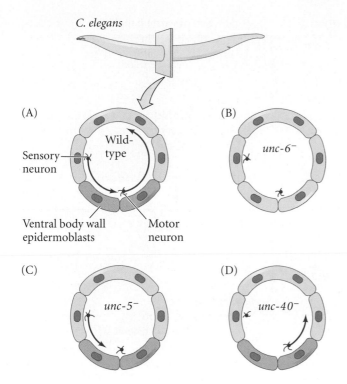

C. elegans

(A)

Sensory neuron

Wild-type

Ventral body wall epidermoblasts

Motor neuron

(B)

unc-6⁻

(C)

unc-5⁻

(D)

unc-40⁻

FIGURE 10.26 UNC expression and function in axonal guidance. (A) In the body of the wild-type *C. elegans* embryo, sensory neurons project ventrally and motor neurons project dorsally. The ventral body wall epidermoblasts expressing UNC-6 are darkly shaded. (B) In the *unc-6* mutant embryo, neither of these migrations occurs. (C) The *unc-5* loss-of-function mutation affects only the dorsal movements of the motor neurons. (D) The *unc-40* loss-of-function mutation affects only the ventral migration of the sensory growth cones. (After Goodman 1994.)

directing the circumferential migration of axons around the body wall of *Caenorhabditis elegans*. In the wild-type nematode, UNC-6 induces axons from certain centrally located sensory neurons to move ventrally while inducing ventrally placed motor neurons to extend axons dorsally. In *unc-6* loss-of-function mutations, neither of these axonal movements occurs (Hedgecock et al. 1990; Ishii et al. 1992; Hamelin et al. 1993). Mutations of the *unc-40* gene disrupt ventral (but not dorsal) axonal migration, whereas mutations of the *unc-5* gene prevent only dorsal migration (**Figure 10.26**). Genetic and biochemical evidence suggests that UNC-5 and UNC-40 are portions of the UNC-6 receptor complex, and that UNC-5 can convert a UNC-40-mediated attraction into a repulsion (Leonardo et al. 1997; Hong et al. 1999; Chang 2004).

There is reciprocity in science, and just as research on vertebrate netrin genes led to the discovery of their *C. elegans* homologues, research on the nematode *unc-5* gene led to the discovery of the gene encoding the human netrin receptor. This turns out to be a gene whose mutation in mice causes a disease called rostral cerebellar malformation (Ackerman et al. 1997; Leonardo et al. 1997).

See WEBSITE 10.5
Genetic control of neuroblast migration in *C. elegans*

SLIT AND ROBO Diffusible proteins can also provide guidance by repulsion. One important chemorepulsive molecule is the Slit protein. In *Drosophila*, Slit is secreted by the neural cells in the midline, and it acts to prevent most neurons from crossing the midline from either side. The

growth cones of *Drosophila* neurons express a Roundabout (Robo) protein, which is the receptor for Slit. In this way, most *Drosophila* neurons are prevented from migrating across the midline. However, the commissural neurons that traverse the embryo from side to side avoid this repulsion by downregulating Robo protein as they approach the midline. Once the growth cone is across the middle of the embryo, the neurons re-express Robo and become sensitive again to the midline inhibitory actions of Slit (**Figure 10.27A**; Brose et al. 1999; Kidd et al. 1999; Orgogozo et al. 2004).

In vertebrates, the Slit/Robo system cooperates with the netrin/DCC system to permit the commissural neurons to cross the midline; there are several vertebrate Robo and Slit proteins, and they do different tasks (Mambetisaeva et al. 2005). As the axon extends toward its target in the developing brain, the neuron is kept from crossing the midline by Slit, which binds to Robo1 (**Figure 10.27B**). Expressed Robo1 protein prevents DCC from binding to the netrin proteins. When the axon gets near the midline, however, Robo3 is expressed and blocks Robo1. Robo1 is no longer able to bind Slit or to block DCC. This enables netrin to bind DCC and turns the axon growth cone toward the midline. The axon grows through the midline, but after crossing it, Robo3 is downregulated while Robo1 is upregulated. This allows Slit to act as a chemorepellent, forcing the growth cone away from the midline. DCC is once again blocked, preventing the axon from going back (Woods 2004). Mutations in the human *ROBO3* gene disrupt the normal crossing of axons from one side of the brain's medulla to the other (Jen et al. 2004). Among other problems, people with this mutation are unable to coordinate their eye movements.

See WEBSITE 10.6 The early evidence for chemotaxis

Target selection

In some cases, nerves in the same ganglion may have several different targets. How do the different neurons know where to go? It seems they use the same strategy as the neural crest cells. In some cases, different neurons in the same ganglion have different receptors and can therefore respond to certain cues and not others. For instance, some neurons in the superior cervical ganglia (the biggest ganglia in the neck) go toward the carotid artery, while other

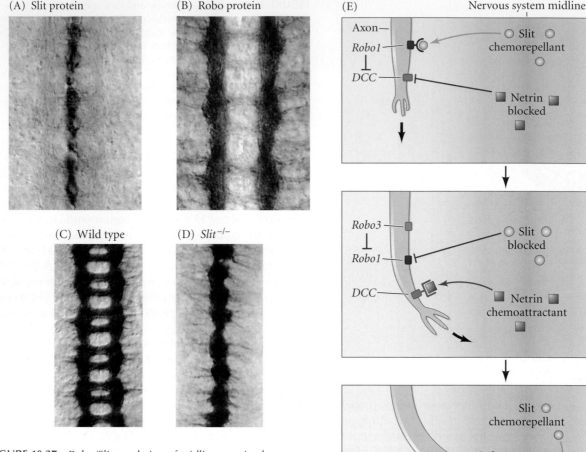

(A) Slit protein (B) Robo protein

(C) Wild type (D) *Slit*$^{-/-}$

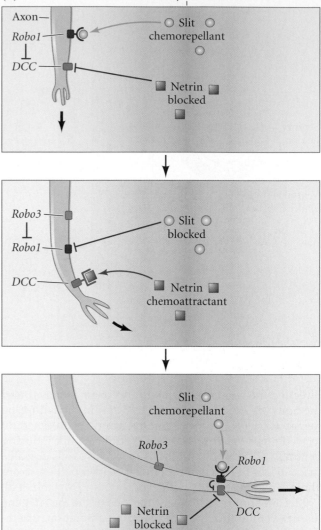

(E) Nervous system midline

FIGURE 10.27 Robo/Slit regulation of midline crossing by neurons. (A–D) Robo and Slit in the *Drosophila* central nervous system. (A) Antibody staining reveals Slit protein in the midline neurons. (B) Robo protein appears along the neurons of the longitudinal tracts of the CNS axon scaffold. (C) Wild-type CNS axon scaffold shows the ladderlike arrangement of neurons crossing the midline. (D) Staining of the CNS axon scaffold with antibodies to all CNS neurons in a *Slit* loss-of-function mutant shows axons entering but failing to leave the midline (instead of running alongside it). (E) Regulation by Slit and Robo in vertebrates. Neurons are prevented from crossing the midline by Slit, which activates the *Robo1* gene. Robo1 then blocks DCC from binding to the netrin proteins. When the axon gets near the midline, Robo3 is expressed and blocks these functions of Robo1, thus enabling netrin to bind to DCC and turning the axon growth cone toward the midline. After the cone crosses the midline, Robo1 is upregulated and Robo3 is downregulated. This allows Slit to act as a chemorepellent, forcing the growth cone away from the midline. (A–D from Kidd et al. 1999, courtesy of C. S. Goodman; E after Woods 2004.)

neurons from these ganglia do not. It appears that those axons extending to the carotid artery follow the blood vessels that lead there. These blood vessels secrete small peptides called **endothelins**. In addition to their adult roles constricting blood vessels, endothelins appear to have an embryonic role as well, as they are able to direct the extension of certain sympathetic axons that have endothelin receptors (Makita et al. 2008). Similarly, the trigeminal ganglion has three peripheral axon bundles that innervate the eye regions, the upper jaw, and the lower jaw (**Figure 10.28**); these include the neurons dentists "put to sleep" with novocaine while filling cavities or extracting teeth. BMP4 from the target organs causes the differential growth and differentiation of these neurons, but intrinsic differences in transcription factors enable them to respond differently to this signal and allow their axons to migrate in their particular ways (Hodge et al. 2007).

Once a neuron reaches a group of cells in which lie its potential targets, it is responsive to various proteins produced by the target cells. In addition to the proteins already mentioned, some target cells produce a set of chemotactic

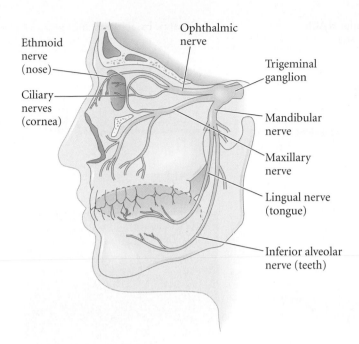

Ethmoid nerve (nose)

Ciliary nerves (cornea)

Ophthalmic nerve

Trigeminal ganglion

Mandibular nerve

Maxillary nerve

Lingual nerve (tongue)

Inferior alveolar nerve (teeth)

FIGURE 10.28 The trigeminal ganglion has three main branches: the ophthalmic (to the eyes), the maxillary (to the upper jaw), and the mandibular (to the lower jaw). The growth of these nerves is regulated by BMP4 from the target tissues combined with differential receptors for the BMPs on the neurons.

interneuron axons in the developing cerebrum and to initiate the elongation and fasciculation of long-range axons from the cerebral cortex (Harkany et al. 2008; Mulder et al. 2008).

The attachment of an axon to its target can be either "digital" or "analogue." In "analogue" mode, different axons recognize the same molecule on the target, but the *amount* of the molecule on the target appears to be critical to the connections that form. This may be the case in the attachment of retinal neurons to the tectum in the fish brain (Gosse et al. 2008). In other cases, there may be extremely molecule-specific ("digital") binding such that certain connections are neuron-specific. This may be the case for retinal neurons in *Drosophila*. Given the complexity of neural connections, it is probable that both qualitative and quantitative cures are used. Growth cones do not rely on a single type of molecule to recognize their target, but integrate the simultaneously presented attractive and repulsive cues, selecting their targets based on the combined input of these multiple signals (Winberg et al. 1998).

factors collectively called **neurotrophins**. These proteins include **nerve growth factor (NGF)**, **brain-derived neurotrophic factor (BDNF)**, and **neurotrophins 3 and 4/5 (NT3, NT4/5)**. These proteins are released from potential target tissues and work at short ranges as either chemotactic factors or chemorepulsive factors (Paves and Saarma 1997). Each can promote and attract the growth of some axons to its source while inhibiting other axons. For instance, sensory neurons from the rat dorsal root ganglia are attracted to sources of NT3 (**Figure 10.29**) but are inhibited by BDNF. These factors are probably transported from the axonal growth cone to the soma of the neuron. For instance, NGF derived from the hippocampus of the brain binds to receptors on the axons of basal forebrain neurons and is endocytosed into these neurons. It is then transported back to the nerve cell body, where it stimulates gene expression. Increased amounts of the gene *App* (a gene on chromosome 21 encoding ameloid precursor protein) are seen in people with Down syndrome and Alzheimer's disease, and increased App locks the retrograde transport of NGF from the axon to the cell soma (Salehi et al. 2006).

Another group of soluble molecules that appear to steer developing axons are **endocannabinoids**, which normally bind to cannibinol receptors during development.* Endocannabinoids appear to regulate the migration of

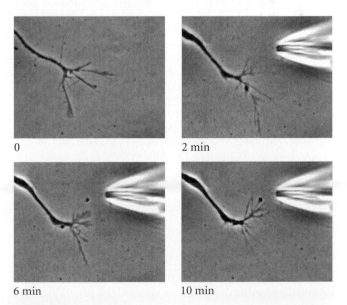

0

2 min

6 min

10 min

FIGURE 10.29 Embryonic axon from a rat dorsal root ganglion turning in response to a source of NT3. The photographs document the growth cone's turn over a 10-minute period. The same growth cone was insensitive to other neurotrophins. (From Paves and Saarma 1997, courtesy of M. Saarma.)

*Exogenous cannabinoids, such as those in marijuana, can also bind to these receptors. While marijuana and its cannabinoids are not considered classic teratogens, the importance of endocannaboids in brain development has led to concern over whether marijuana smoking during pregnancy can cause subtle and long-lasting alterations in the fetal brain (Campololongo et al. 2009; Jutras-Aswad et al. 2009).

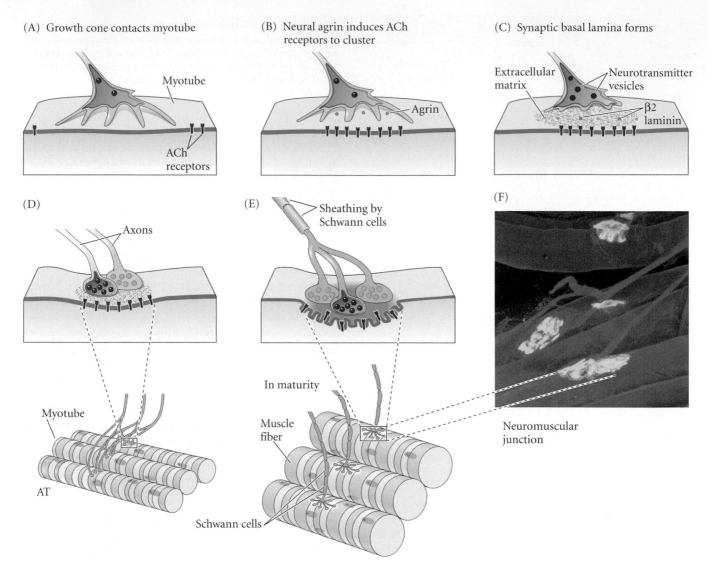

(A) Growth cone contacts myotube

Myotube

ACh receptors

(B) Neural agrin induces ACh receptors to cluster

Agrin

(C) Synaptic basal lamina forms

Extracellular matrix

Neurotransmitter vesicles

β2 laminin

(D)

Axons

Myotube

AT

(E) Sheathing by Schwann cells

In maturity

Muscle fiber

Schwann cells

(F)

Neuromuscular junction

FIGURE 10.30 Differentiation of a motor neuron synapse with a muscle in mammals. (A) A growth cone approaches a developing muscle cell. (B) The axon stops and forms an unspecialized contact on the muscle surface. Agrin, a protein released by the neuron, causes acetylcholine (ACh) receptors to cluster near the axon. (C) Neurotransmitter vesicles enter the axon terminal, and an extracellular matrix connects the axon terminal to the muscle cell as the synapse widens. This matrix contains a nerve-specific laminin. (D) Other axons converge on the same synaptic site. The wider view (below) shows muscle innervation by several axons (seen in mammals at birth). (E) All but one of the axons are eliminated. The remaining axon can branch to form a complex neuromuscular junction with the muscle fiber. Each axon terminal is sheathed by a Schwann cell process, and folds form in the muscle cell membrane. The overview shows muscle innervation several weeks after birth. (F) Whole mount view of mature neuromuscular junction in a mouse. (A–E after Hall and Sanes 1993; Purves 1994; Hall 1995; F courtesy of M. A. Ruegg.)

Forming the synapse: Activity-dependent development

When an axon contacts its target (usually either a muscle or another neuron), it forms a specialized junction called a **synapse**. Neurotransmitters from the axon terminal are released at these synapses to depolarize or hyperpolarize the membrane of the target cell across the synaptic cleft.

The construction of a synapse involves several steps (Burden 1998). When motor neurons in the spinal cord extend axons to muscles, growth cones that contact newly formed muscle cells migrate over their surfaces. When a growth cone first adheres to the cell membrane of a muscle fiber, no specializations can be seen in either membrane. However, the axon terminal soon begins to accumulate neurotransmitter-containing synaptic vesicles, the membranes of both cells thicken at the region of contact, and the synaptic cleft between the cells fills with an extracellular matrix that includes a specific form of laminin (**Figure 10.30A–C**). This muscle-derived laminin specifically binds the growth cones of motor neurons and may act as a "stop signal" for axonal growth (Martin et al. 1995; Noakes et al.

(A) Sympathetic   (B) Dorsal root   (C) Nodose (taste)

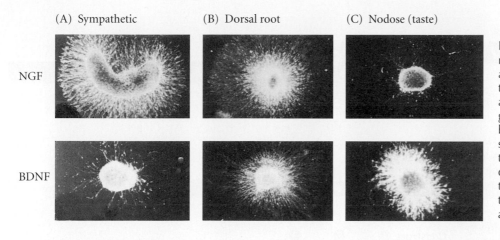

NGF

BDNF

FIGURE 10.31 Effects of NGF (top row) and BDNF (bottom row) on axonal outgrowths from (A) sympathetic ganglia, (B) dorsal root ganglia, and (C) nodose (taste perception) ganglia. While both NGF and BDNF had a mild stimulatory effect on dorsal root ganglia axonal outgrowth, the sympathetic ganglia responded dramatically to NGF and hardly at all to BDNF; the converse was true of the nodose ganglia. (From Ibáñez et al. 1991.)

1995). In at least some neuron-to-neuron synapses, the synapse is stabilized by N-cadherin. The activity of the synapse releases N-cadherin from storage vesicles in the growth cone (Tanaka et al. 2000).

In muscles, after the first contact is made, growth cones from other axons converge at the site to form additional synapses. During mammalian development, all muscles that have been studied are innervated by at least two axons. However, this *polyneuronal innervation* is transient; soon after birth, all but one of the axon branches are retracted (**Figure 10.30D–F**). This "address selection" is based on competition between the axons (Purves and Lichtman 1980; Thompson 1983; Colman et al. 1997). When one of the motor neurons is active, it suppresses the synapses of the other neurons, possibly through a nitric oxide-dependent mechanism (Dan and Poo 1992; Wang et al. 1995). Eventually, the less active synapses are eliminated. The remaining axon terminal expands and is ensheathed by a Schwann cell (see Figure 10.30E).

Differential survival after innervation: Neurotrophic factors

One of the most puzzling phenomena in the development of the nervous system is neuronal cell death. In many parts of the vertebrate central and peripheral nervous systems, more than half the neurons die during the normal course of development (see Chapter 3). Moreover, there do not seem to be great similarities in apoptosis patterns across species. For example, about 80% of a cat's retinal ganglion cells die, while in the chick retina, this figure is only 40%. In fish and amphibian retinas, no ganglion cells appear to die (Patterson 1992).

The apoptotic death of a neuron is not caused by any obvious defect in the neuron itself. Indeed, these neurons have differentiated and have successfully extended axons to their targets. Rather, it appears that the target tissue regulates the number of axons innervating it by limiting the supply of a neurotrophin. In addition to their roles as chemotrophic factors described in the previous section, neurotrophins regulate the survival of different subsets of neurons (**Figure 10.31**). NGF, for example, is necessary for the survival of sympathetic and sensory neurons. Treating

mouse embryos with anti-NGF antibodies reduces the number of trigeminal sympathetic and dorsal root ganglion neurons to 20% of their control numbers (Levi-Montalcini and Booker 1960; Pearson et al. 1983). Furthermore, removal of these neurons' target tissues causes the death of the neurons that would have innervated them, and there is a good correlation between the amount of NGF secreted and the survival of the neurons that innervate these tissues (Korsching and Thoenen 1983; Harper and Davies 1990).

BDNF does not affect sympathetic or sensory neurons, but it can rescue fetal motor neurons in vivo from normally occurring cell death, and from induced cell death following the removal of their target tissue. The results of these in vitro studies have been corroborated by gene knockout experiments, wherein the deletion of particular neurotrophic factors causes the loss of only certain subsets of neurons (Crowley et al. 1994; Jones et al. 1994).

Neurotrophic factors are produced continuously in adults, and loss of these factors may produce debilitating diseases. BDNF is required for the survival of a particular subset of neurons in the striatum (a region of the brain involved in modulating the intensity of coordinated muscle activity such as movement, balance, and walking) and enables these neurons to differentiate and synthesize the receptor for dopamine. BDNF in this region of the brain is upregulated by Huntingtin, a protein that is mutated in Huntington disease. Patients with Huntington disease have decreased production of BDNF, which leads to the death of striatal neurons (Guillin et al. 2001; Zuccato et al. 2001). The result is a series of cognitive abnormalities, involuntary muscle movements, and eventual death. Two other neurotrophins—glial-derived neurotrophic factor (GDNF, which we discussed earlier in terms of neural crest migration) and conserved dopamine neurotrophic factor (CDNF)—enhance the survival of another group of neurons, the midbrain dopaminergic neurons, whose destruction characterizes Parkinson disease (Lin et al. 1993; Lindholm et al. 2007). These neurons send axons to the striatum, whose ability to respond to their dopamine signals is dependent on BDNF. GDNF and CDNF can prevent the death of these neurons in adult brains (see Lindsay 1995) and are being considered as possible therapies for Parkinson disease (Zurn et al. 2001).

The Brainbow

One of the difficulties encountered by developmental biologists is trying to follow any one of the thousands of axons from its source to its specific destination. (Think about how difficult it is to trace wires from a computer to a surge protector; then multiply by several hundred thousand.) But now there is a way that scientists can color-code axons. Livet and colleagues (2007) made a transgene for genes encoding four fluorescent proteins, each fluorescing at a different wavelength (orange, yellow, magenta, and blue). After each of these genes is a 3' UTR, and in front of each gene is a variant cite encoding a specific lox recombination site (**Figure 10.32A**). At the beginning of the transgene is a constitutively active promoter and three variant lox recombination sites. When Cre is introduced into the cell, there is random recombination that can cause the promoter to be brought next to any of the fluorescent protein genes. Moreover, if more than one transgene enters the neuron, combinatorial expression can yield nearly 100 different colors (**Figure 10.32B**). This ability to uniquely label and trace individual axons through the embryo should greatly facilitate studying the development of complex neural circuitry.

FIGURE 10.32 Making a "brainbow." (A) Experimentally manipulated mice contain a region of DNA constructed such that lox sites at the promoter randomly find homologous lox sites adjacent to one of four genes, each encoding a different fluorescent protein. Depending on the rearrangement induced by the Cre protein, different fluorescent proteins are expressed in different neurons. (B) The result is that adjacent neurons fluoresce different colors, allowing researchers to trace axonal routes and connections through the embryo. (A after Livet et al. 2007; B courtesy of J. Lichtman.)

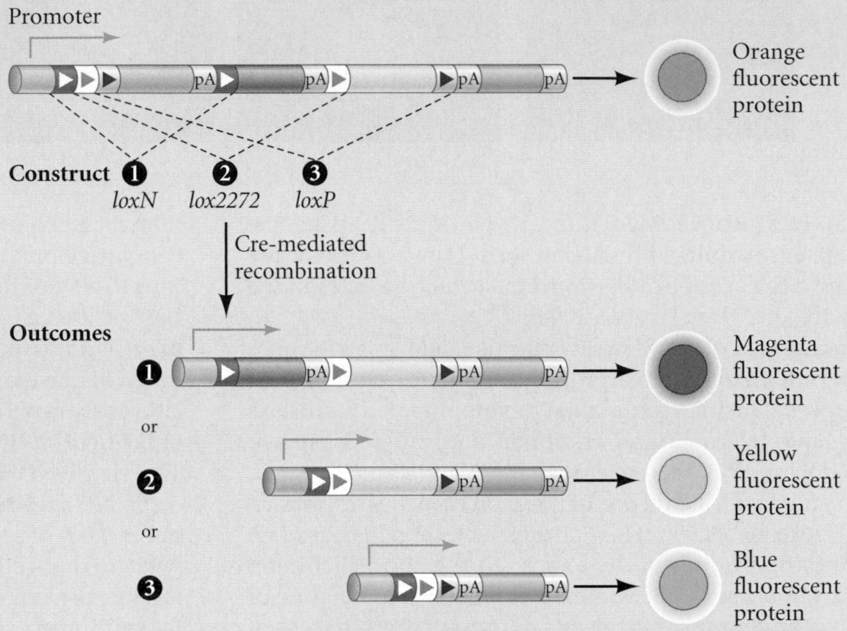

Paths to glory: The travels of the retinal ganglion axons

Nearly all the mechanisms for neuronal specification and axon specificity mentioned in this chapter can be seen in the ways individual retinal neurons send axons to the vision-processing areas of the brain. Even when retinal neurons are transplanted far away from the eye, they are able to find these brain areas (Harris 1986). The ability of the brain to guide the axons of translocated neurons to their appropriate target sites implies that the guidance cues are not distributed solely along the normal pathway but

(A)

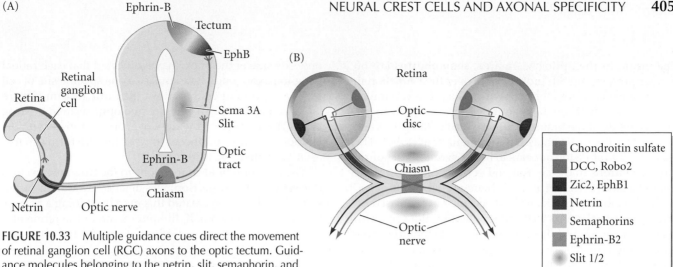

(B)

FIGURE 10.33 Multiple guidance cues direct the movement of retinal ganglion cell (RGC) axons to the optic tectum. Guidance molecules belonging to the netrin, slit, semaphorin, and ephrin families are expressed in discrete regions at several sites along the pathway to direct the RGC growth cones. RGC axons are repelled from the retinal periphery by chondroitin sulfate. At the optic disc, the axons exit the retina and enter the optic nerve, guided by netrin/DCC-mediated attraction. Once in the optic nerve, the axons are kept within the pathway by inhibitory interactions. Slit proteins in the optic chiasm create zones of inhibition. Zic2-expressing ganglia in the ventrotemporal retina project EphB1-expressing axons, which are repelled at the chiasm by ephrin-B2, thus terminating at ipsilateral (same-side) targets. Neurons from the medial portions of the retina do not express EphB1 and proceed to the opposite (contralateral) side. (A) Cross section. (B) Dorsal view. Not all cues are shown. (A after van Horck et al. 2004; B after Harada et al. 2007.)

exist throughout the embryonic brain. Guiding an axon from a nerve cell body to its destination across the embryo is a complex process, and several different types of cues may be used simultaneously to ensure that the correct connections are established. Although we are looking here at non-mammalian vertebrates, the processes we describe apply to mammals as well.

GROWTH OF THE RETINAL GANGLION AXON TO THE OPTIC NERVE The first steps in getting retinal ganglion cell (RGC) axons to their specific regions of the optic tectum take place within the retina. As the RGCs differentiate, their position in the inner margin of the retina is determined by cadherin molecules (N-cadherin and retina-specific R-cadherin) on their cell membranes (Matsunaga et al. 1988; van Horck et al. 2004). The RGC axons grow along the inner surface of the retina toward the optic disc (the head of the optic nerve). The mature human optic nerve will contain over a million retinal ganglion axons.

The adhesion and growth of the retinal ganglion axons along the inner surface of the retina may be governed by its laminin-containing basal lamina. However, simple adhesion to laminin cannot explain the directionality of this growth. Other factors also play a role. The embryonic lens and the periphery of the retina secrete inhibitory factors (probably chondroitin sulfate proteoglycans) that repel the ganglial cell axons, thereby preventing them from traveling in the wrong direction (**Figure 10.33**; Hynes and Lan-

der 1992; Ohta et al. 1999). Moreover, N-CAM may also be especially important here, since the directional migration of the retinal ganglion growth cones depends on the N-CAM-expressing glial endfeet at the inner retinal surface (Stier and Schlosshauer 1995). The secretion of netrin-1 by the cells of the optic disc (where the axons are assembled to form the optic nerve) plays a role in this migration as well. Mice lacking *netrin-1* genes (or the genes for the netrin receptor found in the retinal ganglion axons) have poorly formed optic nerves, as many of the axons fail to leave the eye and grow randomly around the disc (Deiner et al. 1997). The role of netrin may change in different parts of the eye. At the entrance to the optic nerve, netrin-1 is co-expressed with laminin on the surface of the retina. Laminin converts netrin from having an attractive signal to having a repulsive signal. This repulsion might "push" the growth cone away from the retinal surface and into the head of the optic nerve, where netrin is expressed without laminin (Mann et al. 2004).

Upon their arrival at the optic nerve, the migrating axons fasciculate (bundle) with axons already present there. N-CAM and L1 cell adhesion molecules are critical to this fasciculation, and antibodies against L1 or N-CAM cause the axons to enter the optic nerve in a disorderly fashion, which in turn causes them to emerge at the wrong positions in the tectum (Thanos et al. 1984; Brittis et al. 1995; Yin et al. 1995).

GROWTH OF THE RETINAL GANGLION AXON THROUGH THE OPTIC CHIASM When the axons enter the optic nerve, they grow on glial cells toward the midbrain. In non-mammalian vertebrates, the axons will go to a portion of the brain called the optic tectum. (Mammalian axons go to the lateral geniculate nuclei). After leaving the eye, the retinal axons appear to grow on surfaces of netrin, surrounded on all sides by semaphorins keeping them on track by providing repulsive cues (see Harada et al. 2007). Upon entering the brain, mammalian retinal ganglion axons reach the optic chiasm, where they have to "decide" if they are to continue straight or if they are to turn 90 degrees and enter the other side of the brain. In the optic chasm area, semaphorins are no longer present, but Slit proteins take over their function in establishing corridors through which the axons can travel. It appears that those axons not destined to cross to the opposite side of the brain are repulsed from doing so when they enter the optic chiasm (Godement et al. 1990). The basis of this repulsion appears to be the synthesis of ephrin on the neurons in the chiasm and ephrin receptors on the retinal axons (Cheng et al. 1995; Marcus et al. 2000).

In the mouse eye, the Eph B1 receptor is expressed on those temporal axons that are repelled by the optic chiasm's ephrin B2, and those axons project to the side of the tectum on the same side as their eye (the ipsilateral projections); Eph B1 is nearly absent on axons that are allowed to cross over. Mice lacking the *EphB1* gene show hardly any ipsilateral projections. This pattern of *EphB1* expression appears to be regulated by the Zic2 transcription factor found on those retinal axons that do form ipsilateral projections (Herrera et al. 2003; Williams et al. 2003; Pak et al. 2004).

Ephrin appears to play a similar role in the retinotectal mapping in the frog. In the developing frog, the ventral regions express the Eph B receptor, while the dorsal axons do not. Before metamorphosis, both axons cross the optic chiasm. However, when the frog nervous system is being remodeled during metamorphosis, the chiasm expresses ephrin B, which causes a subpopulation of ventral cells to be repulsed and project to the same side rather than cross the chiasm (Mann et al. 2002). (This allows the frog to have binocular vision, which is very good if one is trying to catch flies with one's tongue.)

Laminin appears to promote crossing of the optic chiasm. On their way to the optic tectum, the axons of non-mammalian vertebrates travel on a pathway (the optic tract) over glial cells whose surfaces are coated with laminin. Very few areas of the brain contain laminin, and the laminin in this pathway exists only when the optic nerve fibers are growing on it (Cohen et al. 1987).

TARGET SELECTION When the axons come to the end of the laminin-lined optic tract, they spread out and find their specific targets in the optic tectum. Studies on frogs and fish (in which retinal neurons from each eye project to the opposite side of the brain) have indicated that each retinal ganglion axon sends its impulse to one specific site (a cell or small group of cells) within the optic tectum (**Figure 10.34A**; Sperry 1951). There are two optic tecta in the frog brain. The axons from the right eye form synapses with the left optic tectum, and those from the left eye form synapses in the right optic tectum.

The map of retinal connections to the frog optic tectum (the **retinotectal projection**) was detailed by Marcus Jacobson (1967). Jacobson created this map by shining a narrow beam of light on a small, limited region of the retina and noting, by means of a recording electrode in the tectum,

(A)

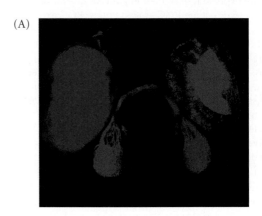

(B)

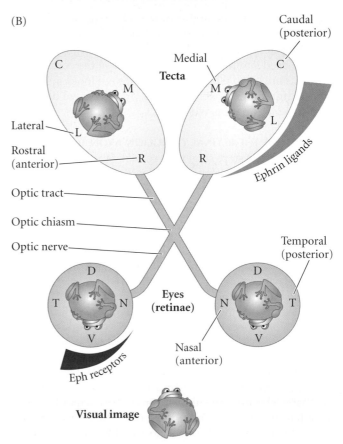

FIGURE 10.34 Retinotectal projections. (A) Confocal micrograph of axons entering the tectum of a 5-day zebrafish embryo. Fluorescent dyes were injected into the eyes of zebrafish embryos mounted in agarose. The dyes diffused down the axons and into the tectum, showing the retinal axons from the right eye going to the left tectum and vice versa. (B) Map of the normal retinotectal projection in adult *Xenopus*. The right eye innervates the left tectum, and the left eye innervates the right tectum. The dorsal portion of the retina (D) innervates the lateral (L) regions of the tectum. The nasal (anterior) region of the retina projects to the caudal (C) region of the tectum. (A courtesy of M. Wilson; B after Holt 2002, courtesy of C. Holt.)

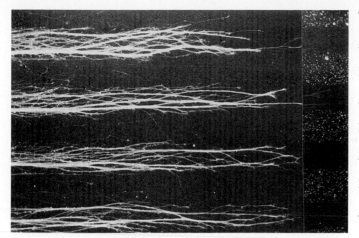

Tectal membranes

Anterior

Posterior

Anterior

Posterior

Anterior

Posterior

Anterior

FIGURE 10.35   Differential repulsion of temporal retinal ganglion axons on tectal membranes. Alternating stripes of anterior and posterior tectal membranes were absorbed onto filter paper. When axons from temporal (posterior) retinal ganglion cells were grown on such alternating carpets, they preferentially extended axons on the anterior tectal membranes. (From Walter et al. 1987.)

which tectal cells were being stimulated. The retinotectal projection of *Xenopus laevis* is shown in **Figure 10.34B**. Light illuminating the ventral part of the retina stimulates cells on the lateral surface of the tectum. Similarly, light focused on the temporal (posterior) part of the retina stimulates cells in the caudal portion of the tectum. These studies demonstrate a point-for-point correspondence between the cells of the retina and the cells of the tectum. When a group of retinal cells is activated, a very small and specific group of tectal cells is stimulated. Furthermore, the points form a continuum; in other words, adjacent points on the retina project onto adjacent points on the tectum. This arrangement enables the frog to see an unbroken image. This intricate specificity caused Sperry (1965) to put forward the *chemoaffinity hypothesis*:

> The complicated nerve fiber circuits of the brain grow, assemble, and organize themselves through the use of intricate chemical codes under genetic control. Early in development, the nerve cells, numbering in the millions, acquire and retain thereafter, individual identification tags, chemical in nature, by which they can be distinguished and recognized from one another.

Current theories do not propose a point-for-point specificity between each axon and the neuron that it contacts. Rather, evidence now demonstrates that gradients of adhesivity (especially those involving repulsion) play a role in defining the territories that the axons enter, and that activity-driven competition between these neurons determines the final connection of each axon.

ADHESIVE SPECIFICITIES IN DIFFERENT REGIONS OF THE OPTIC TECTUM   There is good evidence that retinal ganglion cells can distinguish between regions of the optic tectum. Cells taken from the ventral half of the chick neural retina preferentially adhere to dorsal (medial) halves of the tectum, and vice versa (Gottlieb et al. 1976; Roth and Marchase 1976; Halfter et al. 1981). Retinal ganglion cells are specified along the dorsal-ventral axis by a gradient of transcrip-

tion factors. Dorsal retinal cells are characterized by high concentrations of Tbx5 transcription factor, while ventral cells have high levels of Pax2. These transcription factors are induced by paracrine factors (BMP4 and retinoic acid, respectively) from nearby tissues (Koshiba-Takeuchi et al. 2000). Misexpression of Tbx5 in the early chick retina results in marked abnormalities of the retinotectal projection. Therefore, the retinal ganglion cells are specified according to their location.

One gradient that has been identified functionally is a gradient of repulsion that is highest in the posterior tectum and weakest in the anterior tectum. Bonhoeffer and colleagues (Walter et al. 1987; Baier and Bonhoeffer 1992) prepared a "carpet" of tectal membranes with alternating "stripes" of membrane derived from the posterior and the anterior tecta. They then let cells from the nasal (anterior) or temporal (posterior) regions of the retina extend axons into the carpet. The nasal ganglion cells extended axons equally well on both the anterior and posterior tectal membranes. The neurons from the temporal side of the retina, however, extended axons only on the anterior tectal membranes (**Figure 10.35**). When the growth cone of a temporal retinal ganglion axon contacted a posterior tectal cell membrane, the growth cone's filopodia withdrew, and the cone collapsed and retracted (Cox et al. 1990).

The basis for this specificity appears to be two sets of gradients along the tectum and retina. The first gradient set consists of ephrin proteins and their receptors. In the optic tectum, ephrin proteins (especially ephrins A2 and A5) are found in gradients that are highest in the posterior (caudal) tectum and decline anteriorly (rostrally) (**Figure 10.36A**). Moreover, cloned ephrin proteins have the ability to repulse axons, and ectopically expressed ephrin will prohibit axons from the temporal (but not from the nasal) regions of the retina from projecting to where it is expressed (Drescher et al. 1995; Nakamoto et al. 1996). The complementary Eph receptors have been found on chick retinal ganglion cells, expressed in a temporal-to-nasal gradient along the retinal ganglion axons (Cheng et al. 1995).

This gradient appears to be due to a spatially and temporally regulated expression of retinoic acid (Sen et al. 2005).

Ephrin appears to be a remarkably pliable molecule. Concentration differences in ephrin A in the tectum can account for the smooth topographic map (wherein the position of neurons in the retina maps continuously onto the targets). Hansen and colleagues (2004) have shown that ephrin A can be an attractive as well as a repulsive signal for retinal axons. Moreover, their quantitative assay for axon growth showed that the origin of the axon determined whether it was attracted or repulsed by ephrins. Axon growth is promoted by low ephrin A concentrations that are anterior to the proper target and inhibited by higher concentrations posterior to the correct target (**Figure 10.36B**). Each axon is thus led to the appropriate place and then told to go no further. At that equilibrium point, there would be no growth and no inhibition, and the synapses with the target tectal neurons could be made.

The second set of gradients parallels the ephrins and Ephs. The tectum has a gradient of Wnt3 that is highest at the medial region and lowest laterally (like the ephrin gradient). In the retina, a gradient of Wnt receptor is highest ventrally (like the Eph proteins). The two sets of gradients are both required to specify the position of the axon in the tectum (Schmitt et al. 2006).

The Development of Behaviors: Constancy and Plasticity

One of the most fascinating aspects of developmental neurobiology is the correlation of certain neuronal connections with certain behaviors. Although this could be the subject of another (large) textbook, two remarkable aspects of this phenomenon need to be mentioned here. First, there are cases in which complex behavioral patterns are inherently present in the "circuitry" of the brain at birth. The heartbeat of a 19-day chick embryo quickens when it hears a distress call, and no other call will evoke this response (Gottlieb 1965). Furthermore, a newly hatched chick will immediately seek shelter if presented with the shadow of a hawk. An actual hawk is not needed; the shadow cast by a paper silhouette will suffice. The shadow of a different bird will not elicit the same response (Tinbergen 1951). There appear, then, to be certain neuronal connections that lead to "hard-wired" behaviors in vertebrates.

Synapses also can form as the result of neural activity, and activity-dependent synapse formation appears to be involved in the final stages of retinal projection to the brain. In frog, bird, and rodent embryos treated with tetrodotoxin, axons will grow normally to their respective territories and will make synapses with the tectal neurons. However, the retinotectal map is a coarse one, lacking fine resolu-

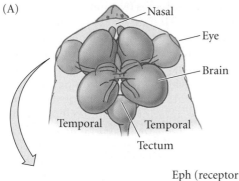

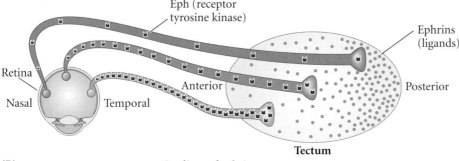

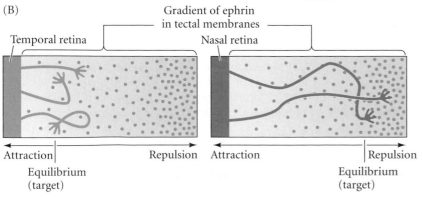

FIGURE 10.36 Differential retinotectal adhesion is guided by gradients of Eph receptors and their ligands. (A) Representation of the dual gradients of Eph receptor tyrosine kinase in the retina and its ligands (ephrin A2 and A5) in the tectum. (B) Experiment showing that temporal, but not nasal, retinal ganglion axons respond to a gradient of ephrin ligand in tectal membranes by turning away or slowing down. An equilibrium of attractive and repulsive forces inherent in the gradient may lead specific axons to their targets. (After Barinaga 1995; Hansen et al. 2004.)

tion. Just as in the final specification of motor neuron synapses, neural activity is needed for the point-for-point retinal projection onto the tectal neurons (Harris 1984; Fawcett and O'Leary 1985; Kobayashi et al. 1990). The fine-tuning of the retinotectal map involves the NMDA receptor, a protein on the tectal neurons. When the NMDA receptor is inhibited, the fine-scale resolution is not obtained (Debski et al. 1990; Dunfield and Haas 2009). It appears that NMDA may be coordinating the interaction between nitric oxide (NO) and BDNF (Wu et al. 1994; Ernst et al. 1999; Cogen and Cohen-Cory 2000). Nitric oxide is involved in the elimination of mistargeted retinal axons, while BDNF may stabilize retinal axon connections. It appears that NO induces growth cone collapse and retraction of developing retinal axons, whereas BDNF protects growth cones and axons from the effects of NO (Ernst et al. 2000). Exposure to both BDNF and NO, but not to either factor alone, stabilized growth cones and axons.

Activity-dependent synapse formation is extremely important during the development of the mammalian visual system, and it will be discussed in Chapter 18. The Nobel Prize-winning research of Hubel and Wiesel (1962, 1963) demonstrated that there is competition between the retinal neurons of each eye for targets in the cortex, and that their connections must be strengthened by experience. The nervous system continues to develop in adult life, and the pattern of neuronal connections is a product of both inherited patterning and patterning produced by experience.

As Dale Purves (1994) concluded in his analysis of brain development:

> Although a vast majority of this construction must arise from developmental programs laid down during the evolution of each species, neural activity can modulate and instruct this process, thus storing the wealth of idiosyncratic information that each of us acquires through individual experience and practice.

Snapshot Summary: *Neural Crest Cells and Axonal Specificity*

1. The neural crest is a transitory structure. Its cells migrate to become numerous different cell types.

2. Trunk neural crest cells can migrate dorsolaterally into the ectoderm, where they become melanocytes. They can also migrate ventrally, to become sympathetic and parasympathetic neurons and adrenomedullary cells.

3. Cranial neural crest cells enter the pharyngeal arches to become the cartilage of the jaw and the bones of the middle ear. They also form the bones of the frontonasal process, the papillae of the teeth, and the cranial nerves.

4. Cardiac neural crest enters the heart and forms the septum (separating wall) between the pulmonary artery and aorta.

5. The formation of the neural crest depends on interactions between the prospective epidermis and the neural plate. Paracrine factors from these regions induce the formation of transcription factors that enable neural crest cells to emigrate.

6. The path a neural crest cell takes depends on the extracellular matrix it meets.

7. Trunk neural crest cells will migrate through the anterior portion of each sclerotome, but not through the posterior portion of a sclerotome. Semaphorin and ephrin proteins expressed in the posterior portion of each sclerotome can prevent neural crest cell migration.

8. Some neural crest cells appear to be capable of forming a large repertoire of cell types. Other neural crest cells may be restricted even before they migrate. The final destination of the neural crest cell can sometimes change its specification.

9. The fates of the cranial neural crest cells are influenced by Hox genes. They can acquire their Hox gene expression pattern through interaction with neighboring cells.

10. Motor neurons are specified according to their position in the neural tube. The Lim family of transcription factors plays an important role in this specification.

11. The targets of motor neurons are specified before their axons extend into the periphery.

12. The growth cone is the locomotor organelle of the neuron, and it senses environmental cues. Axons can find their targets without neuronal activity.

13. Some proteins are generally permissive to neuron adhesion and provide substrates on which axons can migrate. Other substances prohibit migration.

14. Some growth cones recognize molecules that are present in very specific areas and are guided by these molecules to their respective targets.

15. Some neurons are "kept in line" by repulsive molecules. If the neurons wander off the path to their target, these molecules send them back. Some molecules, such as the semaphorins, are selectively repulsive to particular sets of neurons.

16. Some neurons sense gradients of a protein and are brought to their target by following these gradients. The netrins may work in this fashion.

17. Target selection can be brought about by neurotrophins, proteins that are made by the target tissue and that stimulate the particular set of axons able to innervate it. In some cases, the target makes only enough of these factors to support a single axon.

18. Address selection is activity-dependent. An active neuron can suppress synapse formation by other neurons on the same target.

19. Retinal ganglion cells in frogs and chicks send axons that bind to specific regions of the optic tectum. This process is mediated by numerous interactions, and target selection appears to be mediated through ephrins.

20. Some behaviors appear to be innate ("hard-wired"), while others are learned. Experience can strengthen certain neural connections.

For Further Reading

Complete bibliographical citations for all literature cited in this chapter can be found at the free-access website **www.devbio.com**

Bronner-Fraser, M. and S. E. Fraser. 1988. Cell lineage analysis reveals multipotency of some avian neural crest cells. *Nature* 335: 161–164.

Hall, B. K. 2000. The neural crest as a fourth germ layer and vertebrates as quadroblastic, not triploblastic. *Evol. Dev.* 2: 3–5.

Harada, T., C. Harada and L. F. Parada. 2007. Molecular regulation of visual system development: More than meets the eye. *Genes Dev.* 21: 367–378.

Harris, M. L. and C. A. Erickson. 2007. Lineage specification in neural crest cell pathfinding. *Dev. Dyn.* 236: 1–19.

Jiang, X., S. Iseki, R. E. Maxson, H. M. Sucov and G. Morriss-Kay. 2002. Tissue origins and interactions in the mammalian skull vault. *Dev. Biol.* 241: 106–116.

Le Douarin, N. M. 2004. The avian embryo as a model to study the development of the neural crest: A long and still ongoing study. *Mech. Dev.* 121: 1089–1102.

Livet, J., T. A. Weissman, H. Kang, R. W. Draft, J. Lu, R. A. Bennis, J. R. Sanes and J. W. Lichtman. 2007. Transgenic strategies for combinatorial expression of fluorescent proteins in the nervous system. *Nature* 450: 56–62.

Mambetisaeva, E., T. W. Andrews, L. Camurri, A. Annan and V. Sundaresan. 2005. Robo family of proteins exhibit differential expression in mouse spinal cord and Robo-Slit interaction is required for midline crossing in vertebrate spinal cord. *Dev. Dynam.* 233: 41–51.

Polleux, F., G. Ince-Dunn and A. Ghosh. 2007. Transcriptional regulation of vertebrate axon guidance and synapse formation. *Nature Rev. Neurosci.* 8: 331–340.

Schlosser, G. 2005. Evolutionary origins of vertebrate placodes: Insights from developmental studies and from comparisons with other deuterostomes. *J. Exp. Zool.* 304B: 347–399.

Teillet, M.-A., C. Kalcheim and N. M. Le Douarin. 1987. Formation of the dorsal root ganglia in the avian embryo: Segmental origin and migratory behavior of neural crest progenitor cells. *Dev. Biol.* 120: 329–347.

Tosney, K. W. 2004. Long-distance cue from emerging dermis stimulates neural crest melanoblast migration. *Dev. Dynam.* 229: 99–108.

Vaahtokari, A., T. Aberg, J. Jernvall, S. Keränen and I. Thesleff. 1996. The enamel knot as a signalling center in the developing mouse tooth. *Mech. Dev.* 54: 39–43.

Waldo, K., S. Miyagawa-Tomita, D. Kumiski and M. L. Kirby. 1998. Cardiac neural crest cells provide new insight into septation of the cardiac outflow tract: Aortic sac to ventricular septal closure. *Dev. Biol.* 196: 129–144.

Walter, J., S. Henke-Fahle and F. Bonhoeffer. 1987. Avoidance of posterior tectal membranes by temporal retinal axons. *Development* 101: 909–913.

Go Online

WEBSITE 10.1 Communication between migrating neural crest cells. Recent research has shown that neural crest cells might cooperate with one another as they migrate. There may be subtle communication between these cells through their gap junctional complexes, and this communication may be important for heart development.

WEBSITE 10.2 Kallmann syndrome. Some infertile men have no sense of smell. The relationship between sense of smell and male fertility was elusive until the gene for Kallmann syndrome was identified. The gene produces a protein that is necessary for the proper migration of both olfactory axons and hormone-secreting neurons from the olfactory placode.

WEBSITE 10.3 The evolution of developmental neurobiology. Santiago Ramón y Cajal, Viktor Hamburger, and Rita Levi-Montalcini helped bring order to the study of neural development by identifying some of the important questions that still occupy us today.

WEBSITE 10.4 The pathways of motor neurons. To innervate the limb musculature, a motor axon extends over hundreds of cells in a complex and changing environment. Recent research has discovered several paths and several barriers that help guide these axons to their appropriate destinations.

WEBSITE 10.5 Genetic control of neuroblast migration in *C. elegans*. The homeotic gene *mab-5* controls the direction in which certain neurons migrate in the nematode. The expression of this gene can alter which way a neuron travels.

WEBSITE 10.6 The early evidence for chemotaxis. Before molecular techniques, investigators used transplantation experiments and ingenuity to reveal evidence that chemotactic molecules were being released by target tissues.

Vade Mecum

Nicole Le Douarin and the importance of the neural crest. The segment on Dr. Le Douarin's work shows original footage of the experimental techniques and results of her work on neural crest cell regionalization, migration, and differentiation.

Paraxial and Intermediate Mesoderm

11

IN CHAPTERS 9 AND 10 we followed the various tissues formed by the vertebrate ectoderm. In this chapter and the next, we will follow the development of the mesodermal and endodermal germ layers. We will see that the endoderm forms the lining of the digestive and respiratory tubes, with their associated organs. The mesoderm generates all the organs between the ectodermal wall and the endodermal tissues, as well as helping the ectoderm and endoderm to form their own tissues.

The trunk mesoderm of a neurula-stage embryo can be subdivided into four regions (Figure 11.1A):

1. The central region of trunk mesoderm is the **chordamesoderm**. This tissue forms the notochord, a transient organ whose major functions include inducing the formation of the neural tube and establishing the anterior-posterior body axis. The formation of the notochord on the future dorsal side of the embryo was discussed in Chapters 7 and 8.
2. Flanking the notochord on both sides is the **paraxial**, or **somitic, mesoderm**. The tissues developing from this region will be located in the back of the embryo, surrounding the spinal cord. The cells in this region will form somites—blocks of mesodermal cells on either side of the neural tube—which will produce muscle and many of the connective tissues of the back (dermis, muscle, and the vertebral and rib cartilage).
3. The **intermediate mesoderm** forms the urogenital system, consisting of the kidneys, the gonads, and their associated ducts. The outer (cortical) portion of the adrenal gland also derives from this region.
4. Farthest away from the notochord, the **lateral plate mesoderm** gives rise to the heart, blood vessels, and blood cells of the circulatory system, as well as to the lining of the body cavities and to all the mesodermal components of the limbs except the muscles. It also helps form a series of extraembryonic membranes that are important for transporting nutrients to the embryo.

These four subdivisions are thought to be specified along the mediolateral (center-to-side) axis by increasing amounts of BMPs (Pourquié et al. 1996; Tonegawa et al. 1997). The more lateral mesoderm of the chick embryo expresses higher levels of BMP4 than do the midline areas, and one can change the identity of the mesodermal tissue by altering BMP expression. While it is not known how this patterning is accomplished, it is thought that the different BMP concentrations may cause differential expression of the Forkhead (Fox) family of transcription factors. *Foxf1* is transcribed in those regions that will become the lateral plate and extraembryonic mesoderm, whereas *Foxc1* and *Foxc2* are expressed in the paraxial mesoderm that will form the somites (Wilm et al. 2004). If *Foxc1* and *Foxc2* are both deleted

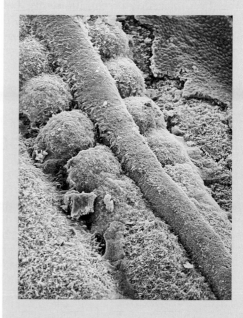

(A)

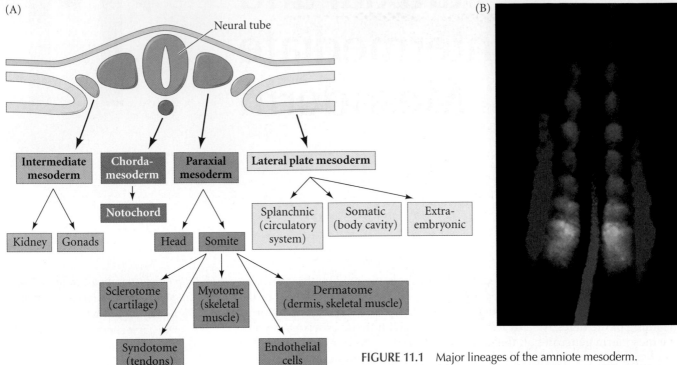

(B)

FIGURE 11.1 Major lineages of the amniote mesoderm. (A) Schematic of the mesodermal compartments of the amniote embryo. (B) Staining for the medial mesodermal compartments in the trunk of a 12-somite chick embryo (about 33 hours). In situ hybridization was performed with probes binding to *chordin* mRNA (blue) in the notochord, *paraxis* mRNA (green) in the somites, and *Pax2* mRNA (red) in the intermediate mesoderm. (B from Denkers et al. 2004, courtesy of T. J. Mauch.)

from the mouse genome, the paraxial mesoderm is respecified as intermediate mesoderm and initiates the expression of *Pax2*, which encodes a major transcription factor of the intermediate mesoderm (**Figure 11.1B**).

Anterior to the trunk mesoderm is a fifth mesodermal region, the **head mesoderm**, consisting of the unsegmented paraxial mesoderm and prechordal mesoderm. This region provides the head mesenchyme that forms much of the connective tissues and musculature of the face and eyes (Evans and Noden 2006). The muscles derived from the head mesoderm form differently than those formed from the somites. Not only do they have their own set of transcription factors, but the head and trunk muscles are affected by different types of muscular dystrophies (Emery 2002; Bothe and Dietrich 2006; Harel et al. 2009).

PARAXIAL MESODERM: THE SOMITES AND THEIR DERIVATIVES

One of the major tasks of gastrulation is to create a mesodermal layer between the endoderm and the ectoderm. As seen in **Figure 11.2**, the formation of mesodermal and endodermal tissues is not subsequent to neural tube formation but occurs synchronously. The notochord extends beneath the neural tube, from the base of the head into the tail. On either side of the neural tube lie thick bands of mesodermal cells. These bands of paraxial mesoderm are referred to either as the **segmental plate** (in chick embryos) or the

unsegmented mesoderm (in other vertebrate embryos). As the primitive streak regresses and the neural folds begin to gather at the center of the embryo, the cells of the paraxial mesoderm will form **somites**. The paraxial mesoderm appears to be specified by the antagonism of BMP signaling by the Noggin protein. Noggin is usually synthesized by the early segmental plate mesoderm, and if Noggin-expressing cells are placed into the presumptive lateral plate mesoderm, the lateral plate tissue will be respecified into somite-forming paraxial mesoderm (**Figure 11.3**; Tonegawa and Takahashi 1998).

The mature somites contain three major compartments: the **sclerotome**, which forms the vertebrae and rib cartilage; the **myotome**, which forms the musculature of the back, rib cage, and ventral body wall; and the **dermamyotome**, which contains skeletal muscle progenitor cells, (including those that migrate into the limbs) and the cells that generate the dermis of the back. In addition, other, smaller compartments are formed from these three. The *syndetome* arising from the most dorsal sclerotome cells generates the tendons, while the most internal cells of the sclerotome (sometimes called the *arthrotome*) become the vertebral joints, the intervertebral discs, and the proximal

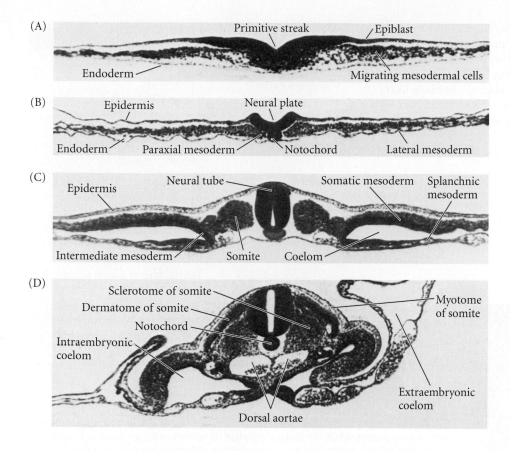

FIGURE 11.2   Gastrulation and neurulation in the chick embryo, focusing on the mesodermal component. (A) Primitive streak region, showing migrating mesodermal and endodermal precursors. (B) Formation of the notochord and paraxial mesoderm. (C,D) Differentiation of the somites, coelom, and the two aortae (which will eventually fuse). A–C, 24-hour embryos; D, 48-hour embryo.

portion of the ribs (Mittapalli et al. 2005; Christ et al. 2007). Moreover, an as-yet unnamed group of cells in the posterior sclerotome generates vascular cells of the dorsal aorta and intervertebral blood vessels (Table 11.1; Pardanaud et al. 1996; Sato et al. 2008).

See VADE MECUM
Mesoderm in the vertebrate embryo

Formation of the Somites

The important components of **somitogenesis** (somite formation) are (1) periodicity, (2) fissure formation (separation of the somites), (3) epithelialization, (4) specification, and (5) differentiation. The first somites appear in the anterior portion of the trunk, and new somites "bud off" from the rostral end of the presomitic mesoderm at regular intervals (**Figure 11.4**). Somite formation begins as paraxial mesoderm cells become organized into whorls of cells, sometimes called **somitomeres** (Meier 1979).

FIGURE 11.3   Specification of somites. Placing Noggin-secreting cells into a prospective region of chick lateral plate mesoderm will respecify that mesoderm into somite-forming paraxial mesoderm. Induced somites (bracketed) were detected by in situ hybridization with Pax3. (From Tonegawa and Takahashi 1998, courtesy of Y. Takahashi.)

The somitomeres become compacted and split apart as fissures separate them into discrete, immature somites. The mesenchymal cells making up the immature somite now change, with the outer cells joining into an epithelium while the inner cells remain mesenchymal. Because individual embryos in any species can develop at slightly different rates (as when chick embryos are incubated at slightly different temperatures), the number of somites present is usually the best indicator of how far development has proceeded.

TABLE 11.1 Derivatives of the somite

Traditional view	Current view
DERMAMYOTOME	
Myotome forms skeletal muscles	Lateral edges generate primary myotome that forms muscle
Dermatome forms back dermis	Central region forms muscle, muscle stem cells, dermis, brown fat cells
SCLEROTOME	
Forms vertebral and rib cartilage	Forms vertebral and rib cartilage
	Dorsal region forms tendons (syndetome)
	Medial region forms blood vessels and meninges
	Central mesenchymal region forms joints (arthrotome)
	Forms smooth muscle cells of dorsal aorta

(A)

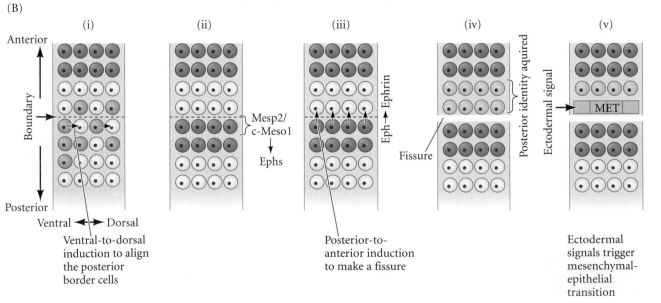

FIGURE 11.4 Formation of new somites. (A) Neural tube and somites seen by scanning electron microscopy. When the surface ectoderm is peeled away, well-formed somites are revealed, as well as paraxial mesoderm (bottom right) that has not yet separated into distinct somites. A rounding of the paraxial mesoderm into a somitomere is seen at the lower left, and neural crest cells can be seen migrating ventrally from the roof of the neural tube. (B) Sequential molecular and cellular events in somitogenesis. (i) At the boundary (determined by Notch signaling), a ventral-to-dorsal signal aligns the posterior border cells (i.e., those cells immediately posterior to the presumptive border. (ii, iii) Mesp2/c-Meso1 induces Ephs in the posterior border cells, and the Ephs induce ephrin in the cells across the border. This creates the fissure. (iv) As the fissure forms, a separate signal aligns the cells that will form the posterior boundary of the somite. (v) Ectodermal signals act on GTPases to coordinate the transition from mesenchymal to epithelial cell, completing the somite. (A courtesy of K. W. Tosney; B after Takahashi and Sato 2008.)

Periodicity of somite formation

Somite formation depends on a "clock and wave" mechanism, in which an oscillating signal (the "clock") is provided by the Notch and Wnt pathways, and a rostral-to-caudal gradient provides a moving "wave" of an FGF that sets the somite boundaries. In the chick embryo, a new somite is formed about every 90 minutes. In mouse embryos, this time frame is more variable (Tam 1981). However, somites appear at exactly the same time on both sides of the embryo, and the clock for somite formation is set when the cells first enter the presomitic mesoderm. If the presomitic mesoderm is inverted such that the caudal end is rostral and the rostral end faces the tail, somite formation will start from the caudal end and proceed rostrally. Even if isolated from the rest of the body, the presomitic mesoderm will segment at the appropriate time and in the right direction (Palmeirim et al. 1997).

Moreover, the number of somites is set at the initial stages of presomitic mesoderm formation. When *Xenopus* or mouse embryos are experimentally or genetically reduced in size, the number of somites remains the same

(Tam 1981). The total number of somites formed is characteristic of a species (about 50 in chicks, 65 in mice, and as many as 500 in some snakes).

Where somites form: The Notch pathway

Although we do not completely understand the mechanisms controlling the temporal periodicity of somite formation, one of the key agents in determining where somites form is the Notch signaling pathway (see Aulehla and Pourquié 2008). When a small group of cells from a region constituting the posterior border at the presumptive somite boundary is transplanted into a region of unsegmented mesoderm that would not ordinarily be part of the boundary area, a new boundary is created. The transplanted boundary cells instruct the cells anterior to them to epithelialize and separate. Nonboundary cells will not induce border formation when transplanted to a nonborder area. However, these nonboundary cells can acquire boundary-forming ability if an activated Notch protein is electroporated into those cells (**Figure 11.5A–C**; Sato et al. 2002). Morimoto and colleagues (2005) have been able to

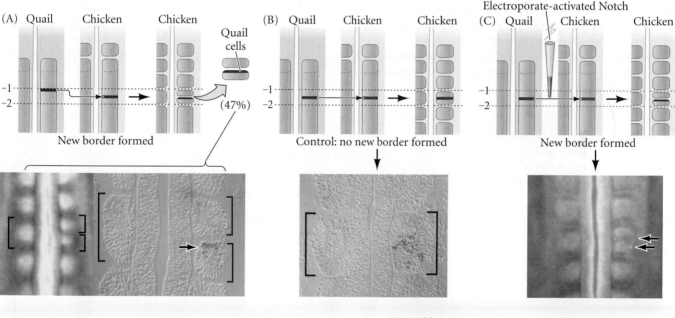

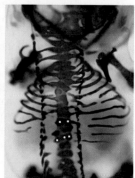

FIGURE 11.5 Notch signaling and somite formation. (A) Transplantation of a prospective somite boundary region into a nonboundary region creates a new boundary and a new somite. The transplanted quail cells can be identified by staining for a quail-specific protein. (B) Transplantation of nonboundary cells into a nonboundary region does not create a new boundary or a new somite. (C) Transplantation of a nonboundary region that has had Notch activated will cause a new somite boundary to occur. (D) Dorsal views of a control mouse and its littermate (E) with the *Delta-like3* gene (the gene encoding a Notch ligand) knocked out. The *Dll3* mutant has several ossification centers (white squares) in rows instead of in a column, and its ribs are malformed. (A–C after Sato et al. 2002, photographs courtesy of Y. Takahashi; D from Dunwoodie et al. 2002, courtesy of S. Dunwoodie.)

visualize the endogenous level of Notch activity in mouse embryos, and have shown that it oscillates in a segmentally defined pattern. The somite boundaries were formed at the interface between the Notch-expressing and Notch-nonexpressing areas.

Notch has also been implicated in somite fissioning by mutations. Segmentation defects have been found in mice that are mutant for important components of the Notch pathway. These include the Notch protein itself, as well as its ligands Delta-like1 and Delta-like3 (Dll1 and Dll3). Mutations affecting Notch signaling have been shown to be responsible for aberrant vertebral formation in mice and humans. In humans, individuals with spondylocostal dysplasia have numerous vertebral and rib defects that have been linked to mutations of the *Delta-like3* gene. Mice with knockouts of *Dll3* have a phenotype similar to that of the human syndrome (**Figure 11.5D**; Bulman et al. 2000; Dunwoodie et al. 2002).

Moreover, Notch signaling follows a remarkable wave-like pattern wherein the *Notch* gene becomes highly expressed in the posterior region of the forming somite, just anterior to the fissure. *Notch* genes are transcribed in a cyclic fashion and function as an autonomous segmentation "clock" (Palmeirim et al. 1997; Jouve et al. 2000, 2002). If Notch signaling determines the placement of somite formation, then Notch protein must control a cascade of gene expression that ultimately separates the tissues.

This wave can be illustrated by following the expression pattern of *hairy1*, a segmentation gene regulated by Notch activity. It is expressed in the presomitic segmental plate in a cyclic, wavelike manner, cresting every 90 minutes in the chick embryo (**Figure 11.6**). The caudal domain of the *hairy1* expression pattern rises anteriorly and then recedes like a wave, leaving a band of expression at what will become the posterior half of the somite. The caudal boundary of this domain is exactly where the transplantation experiments showed Notch expression to be important.

One of the most critical genes in somite formation is *Mesp2/c-Meso1* (the first name refers to the mouse homologue, the second to the chicken homologue). This gene is activated by Notch, and its protein product, a transcription factor, initiates the reactions that *suppress* Notch activity. This activation-suppression cycle causes Mesp2/c-Meso1 expression to oscillate in time and space. Wherever it is expressed, that site is the most anterior group of cells in the next somite, and the boundary forms immediately anterior to those cells (see Figure 11.4B). Mesp2/c-Meso1 induces Eph A4 (one of the compounds whose repulsive interaction separates the somites) in the rostral half of the somite (Saga et al. 1997; Watanabe et al. 2005). In the caudal (posterior) half of the somite, Mesp2/c-Meso1 induces the expression of the transcription factor Uncx4.1 (Takahashi et al. 2000; Saga 2007). In this way, the somite boundary is determined, and the somite is given anterior/posterior polarity at the same time. This, as we saw in Chapter 10, is critical for the migration patterns of neural crest cells and neurons.

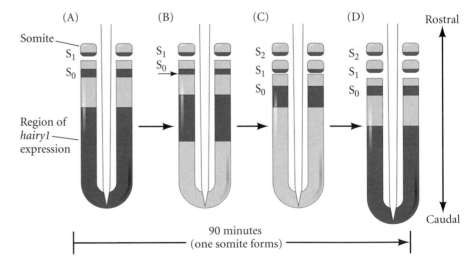

FIGURE 11.6 Somite formation correlates with the wavelike expression of the *hairy1* gene in the chick. (A) In the posterior portion of a chick embryo somite, S_1 has just budded off the presomitic mesoderm. Expression of the *hairy1* gene (purple) is seen in the caudal half of this somite, as well as in the posterior portion of the presomitic mesoderm and in a thin band that will form the caudal half of the next somite (S_0). (B) A caudal fissure (small arrow) begins to separate the new somite from the presomitic mesoderm. The posterior region of *hairy1* expression extends anteriorly. (C) The newly formed somite is now referred to as S_1; it retains the expression of *hairy1* in its caudal half, as the posterior domain of *hairy1* expression moves farther anteriorly and shortens. The former S_1 somite, now called S_2, undergoes differentiation. (D) The formation of somite S_1 is complete, and the anterior region of what had been the posterior *hairy1* expression pattern is now the anterior expression pattern. It will become the caudal domain of the next somite. The entire process takes 90 minutes.

SIDELIGHTS & SPECULATIONS

Coordinating Waves and Clocks in Somite Formation

Acyclic activation of Notch appears to be critical for forming the somites, but what controls Notch activation? The predominant model of somite formation is the clock-and-wavefront model, first proposed by Cooke and Zeeman (1976). In zebrafish, this been found to be relatively simple: the clock involves a negative feedback loop of the Notch signaling pathway. One of the proteins activated by the Notch protein is also able to inhibit *Notch*, which would establish such a negative feedback loop. When the inhibitor is degraded, *Notch* would become active again. Such a cycle would create a "clock" whereby the *Notch* gene would be turned on and off by a protein it itself induces. These off-and-on oscillations could provide the molecular basis for the periodicity of somite segmentation (Holley and Nüsslein-Volhard 2000; Jiang et al. 2000; Dale et al. 2003). More recent reports indicate that the Mesp2/c-Meso1 protein may be such a regulator of *Notch* (Morimoto et al. 2005). The output from this protein controls the ephrins that mediate the separation of the block of cells that form the somite.

The wave appears to be the gradient of Fgf8, which moves caudally as more cells are added to the posterior.

As long as the unsegmented paraxial mesenchyme is in a region of relatively high Fgf8, the clock will not function. This appears to be due to the repression of Delta, the protein that is the major ligand of Notch. The binding of Fgf8 to its receptor enables the expression of the Her13.2 protein, which is necessary to inhibit the transcription of Delta (see Dequéant and Pourquié 2008). Interestingly, the Fgf8 signal can be perturbed by the right-left laterality signals being given at the same time, and the presence of retinoic acid is needed to insulate the somite wavefront from the laterality signals of Nodal proteins. If retinoic acid is not made, somitogenesis becomes asymmetric (Kawakami et al. 2005; Brend and Holley 2009).

In chicks and mice, the situation appears to be far more complicated. The Notch pathway still provides the clock, but the clock appears to be sensitive not only to Fgf but also to the Wnt signaling pathway, which also exhibits a posterior-to-anterior gradient and the retinoic acid gradient that extends in the opposite direction. The details of how the clock and wavefront are coordinated in amniotes have not been elucidated and are an extremely active area of research. It may be that the Fgf8 wavefront is

merely providing the needed amount of β-catenin that can be localized into the nucleus by Wnt signaling. Alternatively, Fgf signaling may provide another clock whose oscillations form a segment when they coincide with both a Wnt clock and a Notch clock (**Figure 11.7**). This latter model has been proposed by Goldbeter and Pourquié (2008), who show that this might have the best fit mathematically.

One way to unravel the complex interactions of signaling molecules in amniotes may be to look at embryos that have extreme numbers of somites. Snake embryos, for instance, can have hundreds of somites. The pattern of gene expression in corn snake embryos indicates that they have the same Fgf8 and Wnt waves and Notch clockwork as chicks and mice; but the snake somite clock is four times faster relative to the embryo's growth rate, rapidly segregating off blocks of presomitic mesenchyme into more numerous, albeit smaller, somites (Gomez et al. 2008). If we can elucidate the alterations that are involved in regulating the speed of this clock, we might be able to figure out how the signaling pathways become integrated to give a single output at specific temporal and spatial intervals.

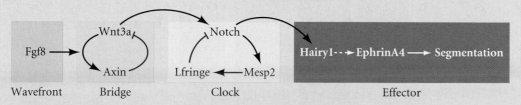

Figure 11.7 Hypothetical pathway for regulation of the clock through which an Fgf8 gradient regulates a Wnt oscillating clock, which in turn controls a Notch clock that can inhibit its own activity in a negative feedback loop. Different species might use different molecules in such a scheme.

Separation of somites from the unsegmented mesoderm

Two proteins whose roles appear to be critical for fissure formation and somite separation are the Eph tyrosine kinases and their ligands, the ephrin proteins. We saw in Chapter 10 that the Eph tyrosine kinase receptors and their ephrin ligands are able to elicit cell-cell repulsion between the posterior somite and migrating neural crest cells. The separation of the somite from the presomitic mesoderm occurs at the ephrin B2/Eph A4 border. In the zebrafish, the boundary between the most recently separated somite and the presomitic mesoderm forms between ephrin B2 in the posterior of the somite and Eph A4 in the most anterior portion of the presomitic mesoderm (**Figure 11.8**; Durbin et al. 1998). Eph A4 is restricted to the boundary area in chick embryos as well. Interfering with this signaling (by injecting embryos with mRNA encoding dominant negative Ephs) leads to the formation of abnormal somite boundaries.

In addition to the posterior-to-anterior induction of the fissure (from Eph proteins to ephrin proteins on their neighboring cells), a second signal originates from the ventral posterior cells of the somite, putting all the cells in register so that the cut is clean from the ventral to the doral aspects of the somite (Sato and Takahashi 2005).

Epithelialization of the somites

Several studies in the chick have shown that epithelialization occurs immediately after somitic fission occurs. As seen in Figure 11.4A, the cells of the newly formed somite are randomly organized as a mesenchymal mass. These cells have to be compacted into an outer epithelium and an internal mesenchyme (**Figure 11.9**). Ectodermal signals appear to cause the peripheral somitic cells to undergo mesenchymal-to-epithelial transition by lowering the Cdc42 levels in these cells. Low Cdc42 levels alter the cytoskeleton, allowing epithelial cells to form a box around the remaining mesenchymal cells, which have a higher level of Cdc42. Another small GTPase, Rac1, must be at a certain level that allows it to activate Paraxis, another transcription factor involved in epithelialization (Burgess et al. 1995; Barnes et al. 1997; Nakaya et al. 2004).

The epithelialization of each somite is stabilized by synthesis of the extracellular matrix protein fibronectin and the adhesion protein N-cadherin (Lash and Yamada 1986; Hatta et al. 1987; Saga et al. 1997; Linask et al. 1998). N-cadherin links the adjoining cells into an epithelium, while the fibronectin matrix acts alongside the Ephrin and Eph to promote the separation of the somites from each other (Martins et al. 2009).

Somite specification along the anterior-posterior axis

Although all somites look identical, they will form different structures. For instance, the somites that form the cervical vertebrae of the neck and lumbar vertebrae of the abdomen are not capable of forming ribs; ribs are generated only by the somites forming the thoracic vertebrae. Moreover, specification of the thoracic vertebrae occurs

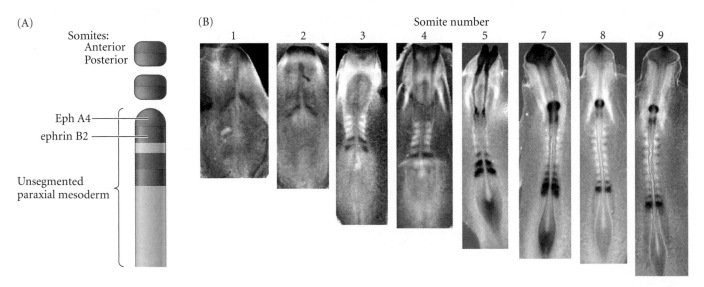

FIGURE 11.8 Ephrin and its receptor constitute a possible fissure site for somite formation. (A) Expression pattern of the receptor tyrosine kinase Eph A4 (blue) and its ligand, ephrin B2 (red), as somites develop. The somite boundary forms at the junction between the region of ephrin expression on the posterior of the last somite formed and the region of Eph A4 expression on the anterior of the next somite to form. In the presomitic mesoderm, the pattern is created anew as each somite buds off. The posteriormost region of the next somite to form does not express ephrin until that somite is ready to separate. (B) In situ hybridization showing Eph A4 (dark blue) expression as new somites form in the chick embryo. (A after Durbin et al. 1998; B courtesy of J. Kastner.)

(A) Epithelial cells

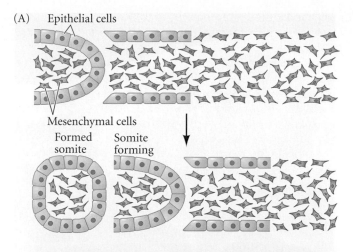

Mesenchymal cells

Formed somite   Somite forming

(B)

FIGURE 11.9 Epithelialization and de-epithelialization in somites of a chick embryo. (A) Changes in cell shape from mesenchymal (pink) to epithelial (gray) cells when a somite forms from presomitic mesenchyme. A formed somite is surrounded by epithelial cells, with mesenchymal cells remaining inside. In chickens, epithelialization occurs first at the posterior edge of the somite, with the anterior edge becoming epithelial later. (B) Changes in cell polarity as somites form are revealed by staining that visualizes F-actin accumulation (red). (After Nakaya et al. 2004; B courtesy of Y. Takahashi.)

very early in development. The segmental plate mesoderm is determined by its position along the anterior-posterior axis before somitogenesis. If one isolates the region of chick segmental plate that will give rise to a thoracic somite and transplants this mesoderm into the cervical (neck) region of a younger embryo, the host embryo will develop ribs in its neck—but only on the side where the thoracic mesoderm has been transplanted (**Figure 11.10**; Kieny et al. 1972; Nowicki and Burke 2000).

The somites are specified according to the Hox genes they express (see Chapter 8). These Hox genes are active in the segmental plate mesoderm before it becomes organized into somites (Carapuço et al. 2005). Mice that are homozygous for a loss-of-function mutation of *Hoxc8* convert a lumbar vertebra into an extra thoracic vertebra, complete with ribs (see Figure 8.32). The Hox genes are activated concomitantly with somite formation, and the embryo appears to "count somites" in setting the expression boundaries of the Hox genes. If Fgf8 levels are manipulated to create extra (albeit smaller) somites, the appropriate Hox gene expression will be activated in the appropriately numbered somite, even if it is in a different position along the anterior-posterior axis. Moreover, when mutations affect the autonomous segmentation clock, they also affect the activation of the appropriate Hox genes (Dubrulle et al. 2001; Zakany et al. 2001). Once established, each somite retains its pattern of Hox gene expression, even if that somite is transplanted into another region of the embryo (Nowicki and Burke 2000). The regulation of the Hox genes by the segmentation clock should allow coordination between the formation and the specification of the new segments.

Differentiation of the somites

In contrast to the early commitment of the presomitic segmental plate mesoderm along the anterior-posterior axis, the commitment of the cells *within* a somite occurs relative-

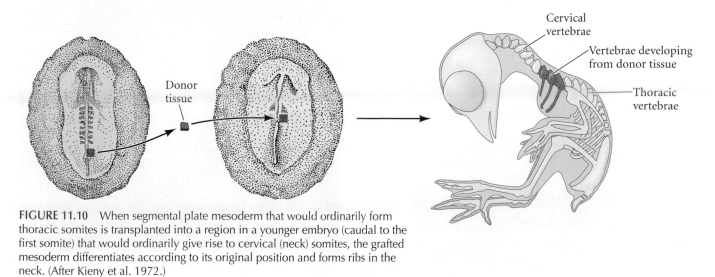

Donor tissue

Cervical vertebrae

Vertebrae developing from donor tissue

Thoracic vertebrae

FIGURE 11.10 When segmental plate mesoderm that would ordinarily form thoracic somites is transplanted into a region in a younger embryo (caudal to the first somite) that would ordinarily give rise to cervical (neck) somites, the grafted mesoderm differentiates according to its original position and forms ribs in the neck. (After Kieny et al. 1972.)

(A) 2-day embryo

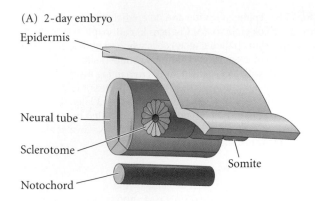

Epidermis

Neural tube

Sclerotome

Notochord

Somite

(B) 3-day embryo

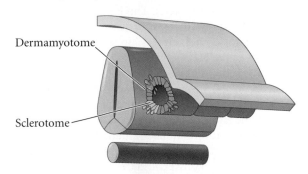

Dermamyotome

Sclerotome

(C) 4-day embryo

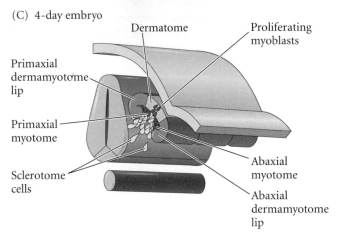

Dermatome

Proliferating myoblasts

Primaxial dermamyotome lip

Primaxial myotome

Sclerotome cells

Abaxial myotome

Abaxial dermamyotome lip

(D) Late 4-day embryo

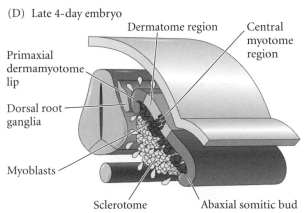

Dermatome region

Central myotome region

Primaxial dermamyotome lip

Dorsal root ganglia

Myoblasts

Sclerotome

Abaxial somitic bud

(E)

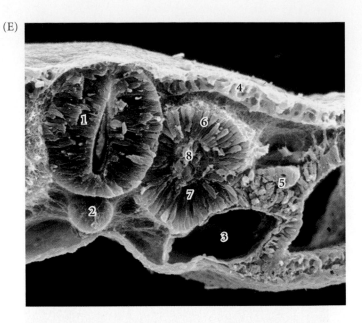

FIGURE 11.11 Transverse section through the trunk of a chick embryo on days 2–4. (A) In the 2-day somite, the sclerotome cells can be distinguished from the rest of the somite. (B) On day 3, the sclerotome cells lose their adhesion to one another and migrate toward the neural tube. (C) On day 4, the remaining cells divide. The medial cells form a primaxial myotome beneath the dermamyotome, while the lateral cells form an abaxial myotome. (D) A layer of muscle cell precursors (the myotome) forms beneath the epithelial dermamyotome. (E,F) Scanning electron micrographs correspond to (A) and (D), respectively; 1, neural tube; 2, notochord; 3, dorsal aorta; 4, surface ectoderm; 5, intermediate mesoderm; 6, dorsal half of somite; 7, ventral half of somite; 8, somitocoel/arthrotome; 9, central sclerotome; 10, ventral sclerotome; 11, lateral sclerotome; 12, dorsal sclerotome;. 13, dermamyotome. (A,B after Langman 1981; C,D after Ordahl 1993; E,F from Christ et al. 2007, courtesy of H. J. Jacob and B. Christ.)

(F)

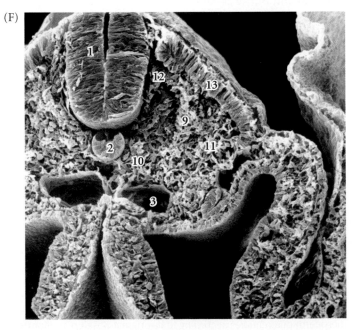

of its cells can become any of the somite-derived structures. These structures include:

- The cartilage of the vertebrae and ribs
- The muscles of the rib cage, limbs, abdominal wall, back, and tongue
- The tendons that connect the muscles to the bones
- The dermis of the dorsal skin
- Vascular cells that contribute to the formation of the aorta and the intervertebral blood vessels
- The cellular sheaths, or *meninges*, of the spinal cord that protect the central nervous system

Thus, the somite contains a population of multipotent cells whose specification depends on their location within the somite. Paracrine factors from the surrounding tissues (neural tube, notochord, epidermis, and intermediate mesoderm) will have profound influences on the regions of the somite adjacent to them. As the somite matures, its various regions become committed to forming only certain cell types. The ventral-medial cells of the somite (those cells closest to the neural tube and notochord) undergo mitosis, lose their round epithelial characteristics, and become mesenchymal cells again. The portion of the somite that gives rise to these cells is the sclerotome, and these mesenchymal cells ultimately become the cartilage cells (chondrocytes) of the vertebrae and a major part of each rib (**Figure 11.11A,B**; see also Figure 11.2).

The remaining epithelial portion of the somite is the dermamyotome. Fate mapping with chick-quail chimeras (Ordahl and Le Douarin 1992; Brand-Saberi et al. 1996; Kato and Aoyama 1998) has revealed that the dermamyotome is arranged into three regions (**Figure 11.11C,D**). The cells in the two lateral portions of this epithelium (i.e., the dorsomedial and ventrolateral lips closest to and farthest from the neural tube, respectively) are the myotomes and will form muscle cells. The central region, the dermatome, will form back dermis and several other derivatives. In the lateral myotomes, muscle precursor cells will migrate beneath the dermamyotome to produce a lower layer of muscle precursor cells, the **myoblasts**. Those myoblasts in the myotome closest to the neural tube form the centrally located **primaxial muscles**,* which include the intercostal musculature between the

ribs and the deep muscles of the back; those myoblasts formed in the region farthest from the neural tube produce the **abaxial muscles** of the body wall, limbs, and tongue (**Figure 11.12**). The boundary between the primaxial and abaxial muscles and between the somite-derived and lateral plate-derived dermis is called the **lateral somitic frontier** (Christ and Ordahl 1995; Burke and Nowicki 2003; Nowicki et al. 2003.) Various transcription factors distinguish the primaxial and abaxial muscles.

The central portion of the dermamyotome has traditionally been called the **dermatome**, since its major prod-

---

*The terms *primaxial* and *abaxial* are used here to designate the muscles from the medial and lateral portions of the somite, respectively. The terms *epaxial* and *hypaxial* are commonly used, but these terms are derived from secondary modifications of the adult anatomy (the hypaxial muscles being innervated by the ventral regions of the spinal cord) rather than from the somitic myotome lineages.

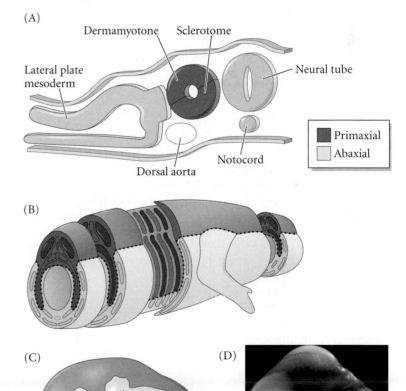

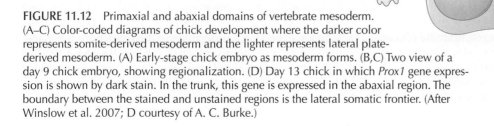

FIGURE 11.12 Primaxial and abaxial domains of vertebrate mesoderm. (A–C) Color-coded diagrams of chick development where the darker color represents somite-derived mesoderm and the lighter represents lateral plate-derived mesoderm. (A) Early-stage chick embryo as mesoderm forms. (B,C) Two view of a day 9 chick embryo, showing regionalization. (D) Day 13 chick in which *Prox1* gene expression is shown by dark stain. In the trunk, this gene is expressed in the abaxial region. The boundary between the stained and unstained regions is the lateral somatic frontier. (After Winslow et al. 2007; D courtesy of A. C. Burke.)

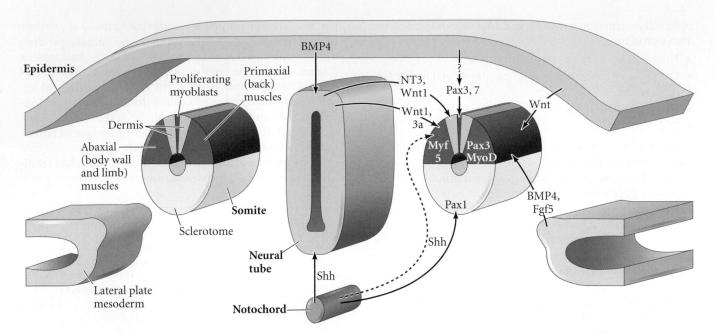

FIGURE 11.13   Model of major postulated interactions in the patterning of the somite. A combination of Wnts (probably Wnt1 and Wnt3a) is induced by BMP4 in the dorsal neural tube. These Wnt proteins, in combination with low concentrations of Sonic hedgehog from the notochord and floor plate, induce the primaxial myotome, which synthesizes the myogenic transcription factor Myf5. High concentrations of Shh induce Pax1 expression in those cells fated to become the sclerotome. Certain concentrations of neurotrophin-3 (NT3) from the dorsal neural tube appear to specify the dermatome, while Wnt proteins from the epidermis, in conjunction with BMP4 and Fgf5 from the lateral plate mesoderm, are thought to induce the primaxial myotome. (After Cossu et al. 1996b.)

uct is the precursors of the dermis of the back. (The dermis of the ventral portion of the body is derived from the lateral plate, and the dermis of the head and neck comes, at least in part, from the cranial neural crest.) However, recent studies have shown that this central region of the dermamyotome also gives rise to a third population of muscle cells (Gros et al. 2005; Relaix et al. 2005). Therefore, some researchers (Christ and Ordahl 1995; Christ et al. 2007) prefer to retain the term *dermamyotome* (or *central dermamyotome*) for this epithelial region. Soon, however, this part of the somite also undergoes an epithelial-to-mesenchymal transition (EMT). FGF signals from the myotome activate the transcription of the *Snail* gene, whose product is a well-known regulator of EMT (see Chapters 3 and 10; Delfini et al. 2009). During this EMT, the mitotic spindles of these epithelial cells are realigned so that cell division takes place along the dorsal-ventral axis. One daughter cell enters the ventral region and becomes part of the myotome, while the other daughter cell locates dorsally to become a precursor of the dermis. The N-cadherin holding the cells together is downregulated and the cells go their separate ways, with the remaining N-cadherin found only on the cells entering the myotome (Ben-Yair and Kalcheim 2005).

These muscle precursor cells that delaminate from the epithelial plate to join the primary myotome cells remain undifferentiated, and they proliferate rapidly to account for most of the myoblast cells. While most of these progenitor cells differentiate to form muscles, some remain undifferentiated and surround the mature muscle cells. These undifferentiated cells become the *satellite cells* responsible for postnatal muscle growth and muscle repair.

Determination of the sclerotome

Like the proverbial piece of real estate, the destiny of a somitic region depends on three things: location, location, and location. As shown in **Figure 11.13**, the locations of the somitic regions place them close to different signaling centers such as the notochord (source of Sonic hedgehog and Noggin), neural tube (source of Wnts and BMPs), and surface epithelium (also a source of Wnts and BMPs).

The specification of the somite is accomplished by the interaction of several tissues. The ventromedial portion of the somite is induced to become the sclerotome by paracrine factors (especially Noggin and Sonic hedgehog) secreted from the notochord (Fan and Tessier-Lavigne 1994; Johnson et al. 1994). If portions of the notochord are transplanted next to other regions of the somite, those regions will also become sclerotome cells. Paracrine factors induce the presumptive sclerotome cells to express the transcription factor Pax1, which is required for their epithelial-to-mesenchymal transition and subsequent differentiation into cartilage (Smith and Tuan 1996). In this EMT, the epithelial cells lose N-cadherin expression and become motile (Sosic et al. 1997). Sclerotome cells also express I-mf, an inhibitor of the myogenic (muscle-forming) family of transcription factors (Chen et al. 1996).

The sclerotome contains several regions, each of which becomes specified according to its location. While most sclerotome cells become the prescursors of the vertebral and rib cartilage, the medial sclerotome cells closest to the neural tube generate the meninges (coverings) of the spinal cord as well as giving rise to blood vessels that will provide the spinal cord with nutrients and oxygen (Halata et al. 1990; Nimmagadda et al. 2007). The cells in the center of the somite (which remain mesenchymal) also contribute to the sclerotome, becoming the vertebral joints, the intervertebral discs, and the portions of the ribs closest to the vertebrae (Mittapalli et al. 2005; Christ et al. 2007). This region of the somite has been called the **arthrotome**.

The notochord, with its secretion of Sonic hedgehog, is critical for sclerotome formation. Moreover, we will also see that it produces compounds that direct the migration of sclerotome cells to the center of the embryo to form the vertebrae. But what happens to the notochord, that central mesodermal structure that induced the nervous system and caused the sclerotome to form? After it has provided the axial integrity of the early embryo and has induced the formation of the dorsal neural tube, most of it degenerates by apoptosis. This apoptosis is probably signaled by mechanical forces. Wherever the sclerotome cells have formed a vertebral body, the notochordal cells die. However, between the vertebrae, the notochordal cells form part of the intervertebral discs, the nuclei pulposi (Aszódi et al. 1998; Guehring et al. 2009). These are the spinal discs that "slip" in certain back injuries.

Determination of the central dermamyotome

The central dermamyotome generates muscle precursors as well as the dermal cells that constitute the connective tissue layer of the *dorsal* skin. The dermis of the *ventral* and *lateral* sides of the body is derived from the lateral plate mesoderm that forms the body wall. There is a sharp demarcation between the somite- and lateral plate-derived dermis. This corresponds to the lateral somitic frontier (see Figure 11.12; Nowicki et al. 2003; Shearman and Burke 2009), a boundary that may have medical importance for the spread of skin diseases (such as viruses that cause rashes only in the chest and belly but not in the back).

The maintenance of the central dermamyotome depends on Wnt6 coming from the epidermis (Christ et al. 2007), and its epithelial-to-mesenchymal transition appears to be regulated by two factors secreted by the neural tube: neurotrophin-3 (NT3) and Wnt1. Antibodies that block the activities of NT3 prevent the conversion of epithelial dermatome into the loose dermal mesenchyme that migrates

beneath the epidermis (Brill et al. 1995). Removing or rotating the neural tube prevents this dermis from forming (Takahashi et al. 1992; Olivera-Martinez et al. 2002). The Wnt signals from the epidermis promote the differentiation of the dorsally migrating central dermamyotome cells into dermis (Atit et al. 2006).

But muscle precursor cells and dermal cells are not the only derivatives of the central dermamyotome. Atit and her colleagues (2006) have shown that **brown adipose cells** ("brown fat") are also somite-derived and appear to come from the central dermamyotome. Brown fat plays active roles in energy utilization by burning fat (unlike the better known adipose tissue, or "white fat," which stores fat). Tseng and colleagues (2008) have found that skeletal muscle and brown fat cells share the same somitic precursor that originally expresses *bHLH proteins*. In brown fat precursor cells, the transcription factor PRDM16 is induced (probably by BMP7); PRDM16 appears to be critical for the conversion of myoblasts to brown fat cells, as it activates a battery of genes that are specific for the fat-burning metabolism of brown adipocytes (Kajimura et al. 2009).

Determination of the myotome

All the skeletal musculature in the vertebrate body (with the exception of the head muscles) comes from the dermamyotome of the somite. In similar ways, the myotome is induced in two different places by at least two distinct signals. Studies using transplantation and knockout mice indicate that the primaxial myoblasts coming from the medial portion of the somite are induced by factors from the neural tube—probably Wnt1 and Wnt3a from the dorsal region and low levels of Sonic hedgehog from the ventral region (Münsterberg et al. 1995; Stern et al. 1995; Borycki et al. 2000). The abaxial myoblasts coming from the lateral edge of the somite are probably specified by a combination of Wnt proteins from the epidermis and signals from the lateral plate mesoderm (see Figure 11.13; Cossu et al. 1996a; Pourquié et al. 1996; Dietrich et al. 1998). These proteins, which probably include Scatter factor (a protein that induces epithelial-to-mesenchymal transition), cause the myoblasts to migrate away from the dorsal region and delay their differentiation until they are in a more ventral position. These myoblasts are also the cells that form the muscles of the limb (Chapter 13).

In addition to positive signals, there are inhibitory signals preventing the positive signals from affecting an inappropriate group of cells. For example, Sonic hedgehog and Noggin both inhibit BMP4 (Marcelle et al. 1997). Indeed, one model of myotome determination holds that the two conditions needed to produce muscle precursors in the somite are (1) the presence of Wnt signals and (2) the absence of BMPs (Reshef et al. 1998). Recent studies (Gerhart et al. 2006) have traced the development of these BMP-inhibiting centers to cells that arise prior to somite formation. They found a small population of surface epiblast cells

(A) Control   (B) Ablated

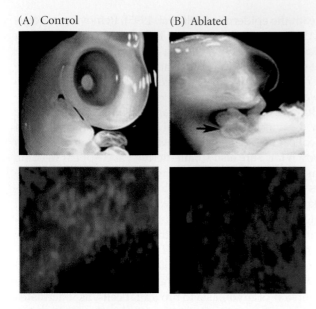

FIGURE 11.14 Ablating Noggin-secreting epiblast cells results in severe muscle defects. Noggin-secreting epiblast cells were ablated in Stage 2 chick embryos using antibodies against G8. (A) The control embryo has normal morphology and abundant staining of myosin (lower photograph) in the muscles. (B) Embryos whose Noggin-secreting epiblast cells are ablated have severe eye defects, severely reduced somatic musculature, and the herniation of abdominal organs through the thin abdominal wall. Severely reduced musculature (sparse myosin in lower photograph) is characteristic of these embryos. (From Gerhart et al. 2006, courtesy of J. Gerhart and M. George-Weinstein.)

that express the mRNA for MyoD but do not translate this message into protein. These particular cells migrate to become paraxial mesoderm and specifically sort out to the dorsomedial and ventrolateral lips of the dermamyotome. There, they synthesize and secrete the BMP inhibitor Noggin to promote the differentiation of myoblasts. If these particular cells are removed from the epiblast, there is a decrease in the skeletal musculature throughout the body, and the ventral body wall is so weak that the heart and abdominal organs often are herniated through it (**Figure 11.14**). (This defect can be prevented by implanting Noggin-releasing beads into the somites lacking these cells.)

See WEBSITE 11.1
Calling the competence of the somite into question

See WEBSITE 11.2 Cranial paraxial mesoderm

Myogenesis: The Generation of Muscle

Myogenic bHLH proteins

What do these Wnt signals activate in the absence of BMPs? As we have seen, muscle cells come from two cell lineages in the somite, the primaxial and the abaxial. In both lineages, paracrine factors instruct cells to become

muscles by inducing them to synthesize the Myf5 and MyoD proteins (see Figure 11.13; Maroto et al. 1997; Tajbakhsh et al. 1997; Pownall et al. 2002). MyoD and Myf5 belong to a family of transcription factors called the **bHLH** (basic helix-loop-helix) **proteins** (sometimes also referred to as **myogenic regulatory factors**, or **MRFs**). The proteins of this family all bind to similar sites on the DNA and activate muscle-specific genes. The mechanism of induction of bHLHs differs slightly between the primaxial and abaxial lineages and between different vertebrate classes. In the lateral portion of the mouse dermamyotome, which forms the abaxial muscles, factors from the surrounding environment induce the Pax3 transcription factor. In the absence of other inhibitory transcription factors (such as those found in the sclerotome cells), Pax3 activates the *myoD* and *myf5* genes (Buckingham et al. 2006). In the medial region of the dermamyotome, which forms the primaxial (epaxial) muscles, MyoD is induced by the Myf5 protein.

In the formation of skeletal muscles, MyoD establishes a temporal cascade of gene activation. First, it can bind directly to certain regulatory regions to activate gene expression. For instance, the MyoD protein appears to directly activate the muscle-specific creatine phosphokinase gene by binding to the DNA immediately upstream from it (Lassar et al. 1989). There are also two MyoD-binding sites on the DNA adjacent to the genes encoding a subunit of the chicken muscle acetylcholine receptor (Piette et al. 1990). MyoD also directly activates its own gene. Therefore, once the *myoD* gene is activated, its protein product binds to the DNA immediately upstream of *myoD* and keeps this gene active.

Second, MyoD can activate other genes whose products act as cofactors for MyoDs binding to a later group of enhancers. For instance, MyoD activates the *p38* gene (which is not muscle-specific) and the *Mef2* gene. The p38 protein facilitates the binding of MyoD and Mef2 to a new set of enhancers, activating a second set of muscle-specific genes (Penn et al. 2004).

Although Pax3 is found in several other cell types, the myogenic bHLH proteins are specific to muscle cells. Any cell making a myogenic bHLH transcription factor such as MyoD or Myf5 becomes committed to forming a muscle cell. Transfection of genes encoding any of these myogenic proteins into a wide range of cultured cells converts those cells into muscles (Thayer et al. 1989; Weintraub et al. 1989).

Specification of muscle progenitor cells

As any athlete or sports fan knows, adult muscles are capable of limited regeneration following injury. The new myofibers come from sets of stem cells or progenitor cells that reside alongside the adult muscle fibers. One type of putative stem cell, the **satellite cell**, is found within the basal lamina of mature myofibers. Satellite cells respond to injury or exercise by proliferating into myogenic cells that fuse and form new muscle fibers; these cells may be stem cells with the capacity to generate daughter cells for renewal or differentiation. Lineage tracing using chick-quail chimeras indicates that these are somite-derived muscle progenitor cells (see Figure 11.15) that have not fused and remain potentially available throughout adult life (Armand et al. 1983).

In 2005, the source of mouse and chick satellite cells was determined to be the central part of the dermamyotome (Ben-Yair and Kalcheim 2005; Gros et al. 2005; Kassar-Duchossoy et al. 2005; Relaix et al. 2005). While the myoblast-forming cells of the dermamyotome form at the lips and express Myf5 and MyoD, the cells that enter into the myotome from the central region usually express Pax3 and Pax7 and do not initially express the bHLH transcription factors. The combination of Pax3 and Pax7 appears to inhibit MyoD expression and muscle differentiation in these cells, and Pax7 protects the satellite cells against apoptosis (Olguin and Olwin 2004; Kassar-Duchossoy et al. 2005; Buckingham et al. 2006).

Injury or exercise causes the satellite cells to enter the cell cycle, produce the bHLH transcription factors, and fuse with the existing muscle fibers (see Gilbert and Epel 2009). Recent experimentation has shown that these satellite cells are not a homogeneous population but contain both stem cells and progenitor cells. The satellite cells that express Pax7 but not Myf5 ($Pax7^+/Myf5^-$ cells) appear to be stem cells that can divide asynchronously to produce two types of cells: another $Pax7^+/Myf5^-$ stem cell and a $Pax7^+/Myf5^+$ satellite cell. This latter cell differentiates into muscle. The $Pax7^+/Myf5^-$ cells, when transplanted into other muscles, contribute to the stem cell population there (Kuang et al. 2007). Muscle stem cell research is a controversial field, and the cell types responsible for muscle regeneration and repair are being intensely explored for therapeutic purposes (Darabi et al. 2008; Buckingham and Vincent 2009).

Myoblast fusion

The myotome cells producing the myogenic bHLH proteins are the myoblasts—committed muscle cell precursors. Experiments with chimeric mice and cultured myoblasts showed that these cells align and fuse to form the multinucleated myotubes characteristic of muscle tissue. Thus, the multinucleated myotube cells are the product of several myoblasts joining together and dissolving the cell membranes between them (Konigsberg 1963; Mintz and Baker 1967; Richardson et al. 2008). By the time a mouse is born, it has the adult number of myofibers, and these multinucleated myofibers grow during the first week by the fusion of singly nucleated myoblasts (Ontell et al. 1988). After the first week, muscle cells can still continue to grow by the fusion of satellite cells into existing myofibers and by an increase in contractile proteins within the myofibers.

Muscle cell fusion begins when the myoblasts leave the cell cycle. As long as particular growth factors (particularly FGFs) are present, myoblasts will proliferate without differentiating. When these factors are depleted, the myoblasts stop dividing, secrete fibronectin onto their extracellular matrix, and bind to it through $\alpha 5\beta 1$ integrin, their major fibronectin receptor (Menko and Boettiger 1987; Boettiger et al. 1995). If this adhesion is experimentally blocked, no further muscle development ensues, so it appears that the signal from the integrin-fibronectin attachment is critical for instructing myoblasts to differentiate into muscle cells (**Figure 11.15**).

The second step is the alignment of the myoblasts into chains. This step is mediated by cell membrane glycoproteins, including several cadherins (Knudsen 1985; Knudsen et al. 1990). Recognition and alignment between cells take place only if the cells are myoblasts. Fusion can occur even between chick and rat myoblasts in culture (Yaffe and Feldman 1965); the identity of the species is not critical. The internal cytoplasm is also rearranged in preparation for the fusion, with actin regulating the regions of contact between the cells (Duan and Gallagher 2009).

The third step is the cell fusion event itself. As in most membrane fusions, calcium ions are critical, and fusion can be activated by calcium ionophores, such as A23187, that carry Ca^{2+} across cell membranes (Shainberg et al. 1969; David et al. 1981). Fusion appears to be mediated by a set of metalloproteinases called **meltrins**. These proteins were discovered during a search for myoblast proteins that would be homologous to fertilin, a protein implicated in sperm-egg membrane fusion. Yagami-Hiromasa and colleagues (1995) found that one of these, meltrin-α, is expressed in myoblasts at about the same time that fusion begins, and that antisense RNA to the meltrin-α message inhibited fusion when added to myoblasts. As the myoblasts become capable of fusing, another myogenic bHLH protein—**myogenin**—becomes active. Myogenin binds to the regulatory region of several muscle-specific genes and activates their expression. Thus, while MyoD and Myf5 are active in the lineage specification of muscle cells, myogenin appears to mediate muscle cell differentiation (Bergstrom and Tapscott 2001).

The last step of cell fusion involves the re-sealing ("healing") of the apposed membranes. This is accomplished by proteins such as myoferlin and dysferlin, which appear to stabilize phospholipids (Doherty et al. 2005). These proteins are similar to those that reseal the membranes at axon nerve synapses after membrane vesicle fusion releases neurotransmitters.

After the original fusion of myoblasts to become a myotube, the myotube secretes interleukin-4 (IL4). IL4 is a paracrine factor that was originally identified as being an important signaling molecule in the adult immune system, and until 2003 it was not known to play a role in the embryo. However, Horsely and colleagues (2003) found that IL4 secreted by the new myotubes recruits other myoblasts to fuse with the tube, thereby forming the mature myotube (see Figure 11.15).

The number of muscle fibers in the embryo and the growth of these fibers after birth appear to be negatively regulated by **myostatin**, a member of the TGF-β family of

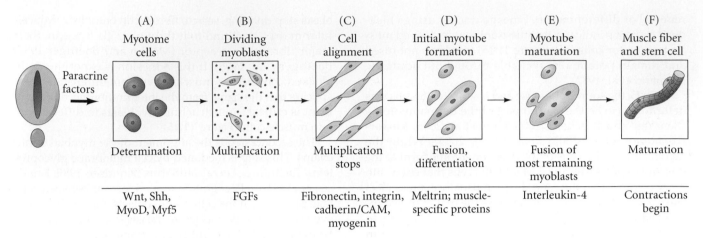

(A) Myotome cells	(B) Dividing myoblasts	(C) Cell alignment	(D) Initial myotube formation	(E) Myotube maturation	(F) Muscle fiber and stem cell
Determination	Multiplication	Multiplication stops	Fusion, differentiation	Fusion of most remaining myoblasts	Maturation
Wnt, Shh, MyoD, Myf5	FGFs	Fibronectin, integrin, cadherin/CAM, myogenin	Meltrin; muscle-specific proteins	Interleukin-4	Contractions begin

(Paracrine factors →)

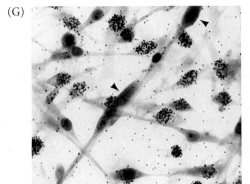

FIGURE 11.15 Conversion of myoblasts into muscles in culture. (A) Determination of myotome cells by paracrine factors. (B) Committed myoblasts divide in the presence of growth factors (primarily FGFs) but show no obvious muscle-specific proteins. (C–E) When the growth factors are used up, the myoblasts cease dividing, align, and fuse into myotubes. (F) The myotubes become organized into muscle fibers that spontaneously contract. (G) Autoradiograph showing DNA synthesis in myoblasts and the exit of fusing cells from the cell cycle. Phospholipase C can "freeze" myoblasts after they have aligned with other myoblasts, but before their membranes fuse. Cultured myoblasts were treated with phospholipase C and then exposed to radioactive thymidine. Unattached myoblasts continued to divide and thus incorporated the radioactive thymidine into their DNA. Aligned (but not yet fused) cells (arrowheads) did not incorporate the label. (A–F after Wolpert 1998; G from Nameroff and Munar 1976, courtesy of M. Nameroff.)

paracrine factors (McPherron et al. 1997; Lee 2004). Myostatin is made by developing and adult skeletal muscle and most probably works in an autocrine fashion. As mentioned in Chapter 2, *myostatin* loss-of-function mutations allow both hyperplasia (more fibers) and hypertrophy (larger fibers) of the muscle. These changes give rise to Herculean phenotypes in dogs,* cattle, mice, and humans (see Figure 2.30).

See WEBSITE 11.3 Muscle formation

Osteogenesis: The Development of Bones

Three distinct lineages generate the skeleton. The somites generate the axial (vertebral) skeleton, the lateral plate mesoderm generates the limb skeleton, and the cranial neural crest gives rise to the pharyngeal arch and craniofacial bones and cartilage. There are two major modes of bone formation, or **osteogenesis**, and both involve the transformation of preexisting mesenchymal tissue into bone tissue. The direct conversion of mesenchymal tissue into bone is called **intramembranous ossification** and was discussed

in Chapter 10. In other cases, the mesenchymal cells differentiate into cartilage, which is later replaced by bone. The process by which a cartilage intermediate is formed and then replaced by bone cells is called **endochondral ossification**. Endochondral ossification is seen predominantly in the vertebral column, ribs, pelvis, and limbs.

Endochondral ossification

Endochondral ossification involves the formation of cartilage tissue from aggregated mesenchymal cells and the subsequent replacement of cartilage tissue by bone (Horton 1990). This is the type of bone formation characteristic of the vertebrae, ribs, and limbs. The vertebrae and ribs form from the somites, while the limb bones (to be discussed in Chapter 13) form from the lateral plate mesoderm.

The process of endochondral ossification can be divided into five stages. First, the mesenchymal cells commit to becoming cartilage cells (**Figure 11.16A**). This commitment is stimulated by Sonic hedgehog, which induces nearby sclerotome cells to express the Pax1 transcription factor (Cserjesi et al. 1995; Sosic et al. 1997). Pax1 initiates a cascade that is dependent on external paracrine factors and internal transcription factors.

During the second phase of endochondral ossification, the committed mesenchyme cells condense into compact nodules and differentiate into chondrocytes, or cartilage

---

*A loss-of-function mutation in the *myostatin* gene has found its way into whippets bred for dog racing. In these dogs, the homozygous loss-of-function condition is not advantageous, but heterozygotes have more muscle power and are significantly overrepresented among the top racers (Mosher et al. 2007).

cells (**Figure 11.16B**). BMPs appear to be critical in this stage. They are responsible for inducing the expression of the adhesion molecules N-cadherin and N-CAM and the transcription factor Sox9. N-cadherin appears to be important in the initiation of these condensations, and N-CAM seems to be critical for maintaining them (Oberlender and Tuan 1994; Hall and Miyake 1995). Sox9 activates other transcription factors as well as the genes encoding collagen 2 and agrican, which are critical in cartilage function. In humans, mutations of the *SOX9* gene cause camptomelic dysplasia, a rare disorder of skeletal development that results in deformities of most of the bones of the body. Most affected babies die from respiratory failure due to poorly formed tracheal and rib cartilage (Wright et al. 1995).

During the third phase of endochondral ossification, the chondrocytes proliferate rapidly to form the cartilage model for the bone (**Figure 11.16C**). As they divide, the chondrocytes secrete a cartilage-specific extracellular matrix. In the fourth phase, the chondrocytes stop dividing and increase their volume dramatically, becoming **hypertrophic chondrocytes** (**Figure 11.16D**). This step appears to be mediated by the transcription factor Runx2 (also called Cbfa1), which is necessary for the development of both intramembranous and endochondral bone (see Figure 10.11). Runx2 is itself regulated by histone deacetylase-4 (HDAC4), a form of chromatin restructuring enzyme that is expressed solely in the prehypertrophic cartilage. If HDAC4 is overexpressed in the cartilaginous ribs or limbs, ossification is seriously delayed; if the *HDAC4* gene is knocked out of the mouse genome, the limbs and ribs ossify prematurely (Vega et al. 2004).

These large chondrocytes alter the matrix they produce (by adding collagen X and more fibronectin) to enable it to become mineralized (calcified) by calcium phosphate. They also secrete the angiogenesis factor VEGF, which can transform mesodermal mesenchyme cells into blood vessels (see Chapter 12; Gerber et al. 1999; Haigh et al. 2000). A number of events lead to the hypertrophy and mineralization (calcification) of the chondrocytes, including an initial switch from aerobic to anaerobic respiration that alters chondrocyte cell metabolism and mitochondrial energy potential (Shapiro et al. 1982). Hypertrophic chondrocytes secrete numerous small, membrane-bound vesicles into the extracellular matrix. These vesicles contain enzymes that are active in the generation of calcium carbonate crystals, called *hydroxyapatite*, which mineralize the cartilaginous matrix (Wu et al. 1997). The hypertrophic chondrocytes, with their metabolism and mitochondrial membranes altered, then die by apoptosis (Hatori et al. 1995; Rajpurohit et al. 1999).

In the fifth phase, the blood vessels induced by VEGF invade the cartilage model (**Figure 11.16E–G**). As the hypertrophic chondrocytes die, the cells that surround the cartilage model differentiate into osteoblasts. The replacement of chondrocytes by bone cells depends on the mineralization of the extracellular matrix. This remodeling releases VEGF, and more blood vessels are made around the dying cartilage. The blood vessels bring in both osteoblasts and

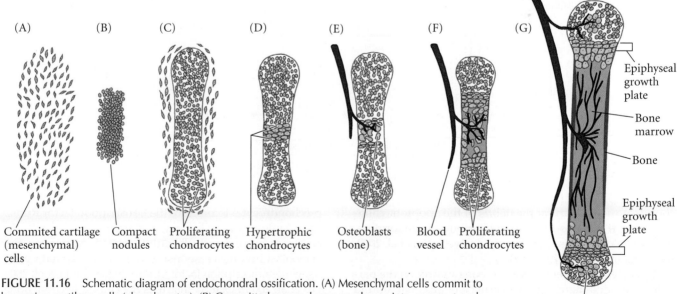

Commited cartilage (mesenchymal) cells   Compact nodules   Proliferating chondrocytes   Hypertrophic chondrocytes   Osteoblasts (bone)   Blood vessel   Proliferating chondrocytes   Epiphyseal growth plate — Bone marrow — Bone — Epiphyseal growth plate — Secondary ossification center

FIGURE 11.16 Schematic diagram of endochondral ossification. (A) Mesenchymal cells commit to becoming cartilage cells (chondrocytes). (B) Committed mesenchyme condenses into compact nodules. (C) Nodules differentiate into chondrocytes and proliferate to form the cartilage model of bone. (D) Chondrocytes undergo hypertrophy and apoptosis while they change and mineralize their extracellular matrix. (E) Apoptosis of chondrocytes allows blood vessels to enter. (F) Blood vessels bring in osteoblasts, which bind to the degenerating cartilaginous matrix and deposit bone matrix. (G) Bone formation and growth consist of ordered arrays of proliferating, hypertrophic, and mineralizing chondrocytes. Secondary ossification centers also form as blood vessels enter near the tips of the bone. (After Horton 1990.)

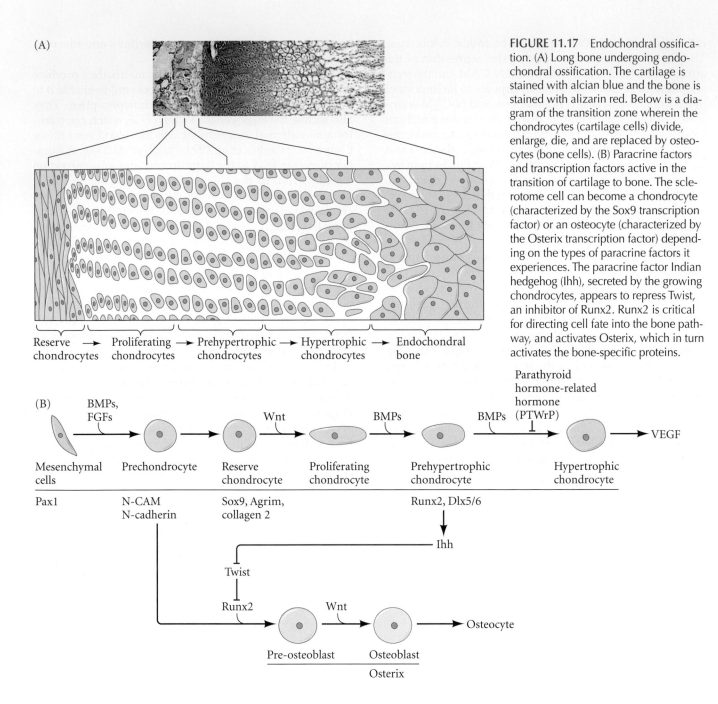

(A)

Reserve → Proliferating → Prehypertrophic → Hypertrophic → Endochondral
chondrocytes chondrocytes   chondrocytes   chondrocytes   bone

(B)

| Mesenchymal cells | Prechondrocyte | Reserve chondrocyte | Proliferating chondrocyte | Prehypertrophic chondrocyte | Hypertrophic chondrocyte |
| Pax1 | N-CAM N-cadherin | Sox9, Agrim, collagen 2 | | Runx2, Dlx5/6 | |

FIGURE 11.17 Endochondral ossification. (A) Long bone undergoing endochondral ossification. The cartilage is stained with alcian blue and the bone is stained with alizarin red. Below is a diagram of the transition zone wherein the chondrocytes (cartilage cells) divide, enlarge, die, and are replaced by osteocytes (bone cells). (B) Paracrine factors and transcription factors active in the transition of cartilage to bone. The sclerotome cell can become a chondrocyte (characterized by the Sox9 transcription factor) or an osteocyte (characterized by the Osterix transcription factor) depending on the types of paracrine factors it experiences. The paracrine factor Indian hedgehog (Ihh), secreted by the growing chondrocytes, appears to repress Twist, an inhibitor of Runx2. Runx2 is critical for directing cell fate into the bone pathway, and activates Osterix, which in turn activates the bone-specific proteins.

chondroclasts (which eat the debris of the apoptotic chondrocytes). If the blood vessels are inhibited from forming, bone development is significantly delayed (Yin et al. 2002; see Karsenty and Wagner 2002).

The osteoblasts begin forming bone matrix on the partially degraded matrix and construct a bone collar around the calcified cartilage matrix (Bruder and Caplan 1989; Hatori et al. 1995; St-Jacques et al. 1999). It is thought that the osteoblasts are derived from same sclerotomal precursors as the chondrocytes (**Figure 11.17**). The osteoblasts form when Indian hedgehog (secreted by the prehypertrophic chondrocytes) causes a relatively immature cell (probably

prechondrocyte) to produce the transcription factor Runx2. Runx2 allows the cell to make bone matrix but keeps the cell from becoming fully differentiated. Moreover, this osteoblast becomes responsive to Wnt signals that upregulate the transcription factor Osterix (Nakashima et al. 2002; Hu et al. 2005). Osterix instructs the cells to become bone. New bone material is added peripherally from the internal surface of the **periosteum**, a fibrous sheath containing connective tissue, capillaries, and bone progenitor cells and that covers the developing bone. At the same time, there is a hollowing out of the internal region of the bone to form the bone marrow cavity. This destruction of bone tissue is

(A)   (B)

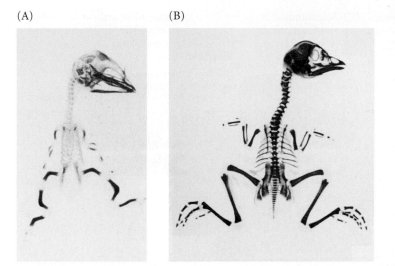

FIGURE 11.18   Skeletal mineralization in 19-day chick embryos that developed (A) in shell-less culture and (B) inside an egg during normal incubation. The embryos were fixed and stained with alizarin red to show the calcified bone matrix. (From Tuan and Lynch 1983, courtesy of R. Tuan.)

removed from their shells at day 3 and grown in plastic wrap for the duration of their development, much of the cartilaginous skeleton fails to mature into bony tissue (**Figure 11.18**; Tuan and Lynch 1983).

See WEBSITE 11.4 Paracrine factors, their receptors, and human bone growth

carried out by **osteoclasts**, multinucleated cells that enter the bone through the blood vessels (Kahn and Simmons 1975; Manolagas and Jilka 1995). Osteoclasts are not derived from the somite; rather, they are derived from a blood cell lineage (in the lateral plate mesoderm) and come from the same precursors as macrophage blood cells (Ash et al. 1980; Blair et al. 1986).

The importance of the mineralized extracellular matrix for bone differentiation is clearly illustrated in the developing skeleton of the chick embryo, which uses the calcium carbonate of the egg's shell as its calcium source. During development, the circulatory system of the chick embryo translocates about 120 mg of calcium from the shell to the skeleton (Tuan 1987). When chick embryos are

Vertebrae formation

The notochord appears to induce its surrounding mesenchyme cells to secrete epimorphin, and this epimorphin attracts sclerotome cells to the region around the notochord and neural tube, where they begin to condense and differentiate into cartilage (Oka et al. 2006). However, before the sclerotome cells form a vertebra, they must split into a rostral (anterior) and a caudal segment (**Figure 11.19**). As the motor neurons from the neural tube grow laterally to innervate the newly forming muscles, the rostral segment of each sclerotome recombines with the caudal segment of the next anterior sclerotome to form the vertebral rudiment (Remak 1850; Aoyama and Asamoto 2000; Morin-Kensicki et al. 2002). As we will see in our discussion of tendons, this **resegmentation** enables the muscles to coordinate the movement of the skeleton, permitting the body to move

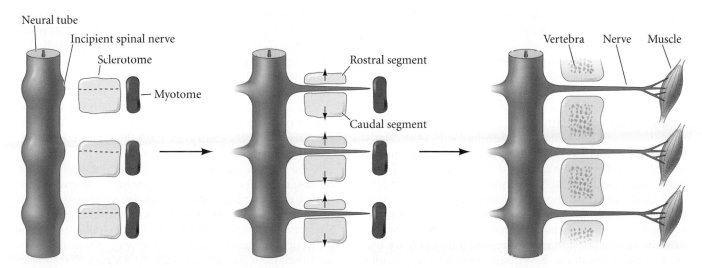

FIGURE 11.19   Respecification of the sclerotome to form each vertebra. Each sclerotome splits into a rostral and caudal segment. As the spinal neurons grow outward to innervate the muscles from the myotome, the rostral segment of each sclerotome combines with the caudal segment of the next anterior sclerotome to form a vertebral rudiment. (After Larson 1998.)

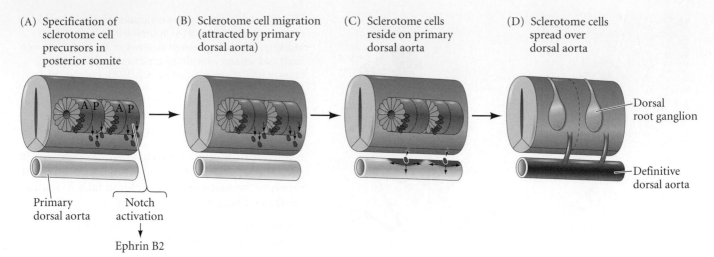

(A) Specification of sclerotome cell precursors in posterior somite

(B) Sclerotome cell migration (attracted by primary dorsal aorta)

(C) Sclerotome cells reside on primary dorsal aorta

(D) Sclerotome cells spread over dorsal aorta

Dorsal root ganglion

Definitive dorsal aorta

Primary dorsal aorta

Notch activation

Ephrin B2

FIGURE 11.20 Model showing contribution of somitic cells to the dorsal aorta. (A) At an early stage, the primary dorsal aorta is of lateral plate origin (pink). A subpopulation of sclerotome cells becomes specified by Notch in the posterior half of somites as endothelial precursors. (B) Chemoattractants made in the primary dorsal aorta cause these cells to migrate through the somite to the aorta. (C) The sclerotome cells take up residence in the dorsal region of the vessel. (D) These cells then spread along both the anterior-posterior and dorsal-ventral axes, ultimately occupying the entire region of the aorta. The primary aortic endothelial cells become blood cell precursors. (After Sato et al. 2008.)

laterally. The resegmentation of somites to allow coordinated movement is reminiscent of the strategy used by insects when constructing segments out of parasegments (see Chapter 6). The bending and twisting movements of the spine are permitted by the intervertebral (synovial) joints that form from the arthrotome region of the sclerotome. Removal of these sclerotome cells leads to the failure of synovial joints to form and to the fusion of adjacent vertebrae (Mittapalli et al. 2005).

Dorsal Aorta Formation

While most of the circulatory system of the early amniote embryo is directed outside the embryo (to obtain nutrients from the yolk or placenta), the intraembryonic circulatory system begins with the formation of the dorsal aorta. The dorsal aorta is composed of two cell layers: an internal lining of endothelial cells surrounded concentrically by a layer of smooth muscle cells. While these two layers of blood vessels are usually derived from the lateral plate mesoderm (and this will be an important subject of Chapter 12), the posterior sclerotome provides the endothelial cells and smooth muscle cells for the dorsal aorta and intervertebral blood vessels (Pardanaud et al. 1996; Wiegreffe et al. 2007). The presumptive endothelial cells are formed by Notch signaling, and they are then instructed to migrate ventrally by a chemoattractant that has been made by the primary dorsal aorta, a transitory structure made by the lateral plate mesoderm. Eventually, the endothelial cells from the sclerotome replace the cells of the primary dorsal aorta (which will become part of the blood stem cell population) (**Figure 11.20**; Pouget et al. 2008; Sato et al. 2008).

Tendon Formation: The Syndetome

The most dorsal part of the sclerotome will become the fourth compartment of the somite, the **syndetome** (Greek *syn*, "connected"). Since the tendons connect muscles to bones, it is not surprising that the syndetome is derived from the most dorsal portion of the sclerotome—that is, from sclerotome cells adjacent to the muscle-forming myotome. The tendon-forming cells of the syndetome can be visualized by their expression of the *scleraxis* gene (**Figure 11.21**; Schweitzer et al. 2001; Brent et al. 2003). Because there is no obvious morphological distinction between the sclerotome and syndetome cells (they are both mesenchymal), our knowledge of this somitic compartment had to wait until we had molecular markers (Pax1 for the sclerotome, scleraxis for the syndetome) that could distinguish them and allow one to follow their cells' fates.

The syndetome is made from the myotome's secretion of Fgf8 onto the immediately subjacent row of sclerotome cells (**Figure 11.22A**; Brent et al. 2003; Brent and Tabin 2004). Other transcription factors limit the expression of scleraxis to the anterior and posterior portions of the syndetome, causing two stripes of scleraxis expression. Meanwhile, the developing cartilage cells, under the influence of Sonic hedgehog from the notochord and floorplate, synthesize Sox5 and Sox6—transcription factors that block scleraxis transcription (**Figure 11.22B**). In this way, the cartilage protects itself from any spread of the Fgf8 signal. The tendons then associate with the muscles directly above them and with the skeleton (including the ribs) on either side of them (**Figure 11.22C**; Brent et al. 2005).

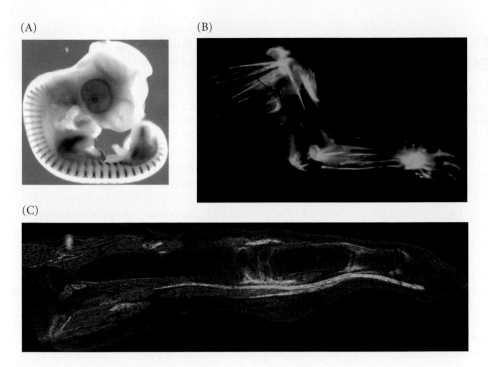

FIGURE 11.21 Scleraxis is expressed in the progenitors of the tendons. (A) In situ hybridization showing scleraxis pattern in the developing chick embryo. (B) Areas of scleraxis expression in the tendons of a newborn mouse forelimb (ventral view). The *GFP* gene had been fused on the scleraxis enhancer. (C) Wrist and digit of a newborn mouse, showing scleraxis in the tendons (green) connecting muscles (red) to the digit and wrist. (A from Schweitzer et al. 2001; all photographs courtesy of R. Schweitzer.)

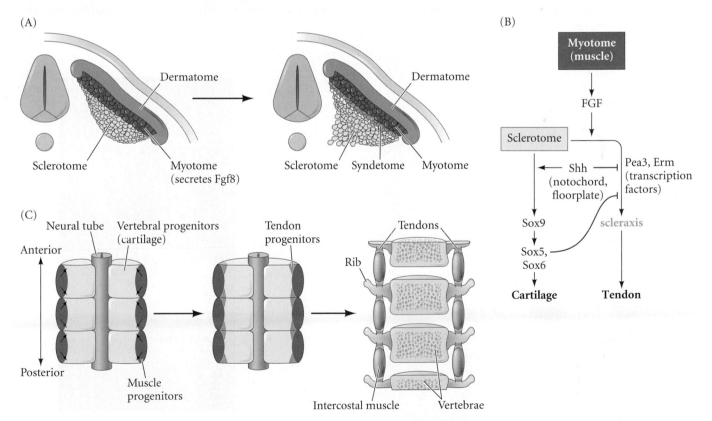

FIGURE 11.22 Induction of scleraxis in the chick sclerotome by Fgf8 from the myotome. (A) The dermatome, myotome, and sclerotome are established before the tendon precursors are specified. Tendon precursors (syndetome) are specified in the dorsalmost tier of sclerotome cells by Fgf8 received from the myotome. (B) Pathway by which Fgf8 signals from the muscle precursor cells induce the subjacent sclerotome cells to become tendons. (C) Syndetome cells migrate (arrows) along the developing vertebrae. They differentiate into tendons that connect the ribs to the intercostal muscles beloved by "spare-ribs" devotees. (A,C after Brent et al. 2003.)

INTERMEDIATE MESODERM: THE UROGENITAL SYSTEM

The intermediate mesoderm generates the urogenital system—the kidneys, the gonads, and their respective duct systems. Reserving the gonads for our discussion of sex determination in Chapter 14, we will concentrate here on the development of the mammalian kidney.

The Progression of Kidney Types

Homer Smith noted in 1953 that "our kidneys constitute the major foundation of our philosophical freedom. Only because they work the way they do has it become possible for us to have bone, muscles, glands, and brains." While this statement may smack of hyperbole, the human kidney is an incredibly intricate organ whose importance cannot be overestimated. Its functional unit, the **nephron**, contains more than 10,000 cells and at least 12 different cell types, each cell type having a specific function and being located in a particular place in relation to the others along the length of the nephron.

Mammalian kidney development progresses through three major stages. The first two stages are transient; only the third and last persists as a functional kidney. Early in

development (day 22 in humans; day 8 in mice), the **pronephric duct** arises in the intermediate mesoderm just ventral to the anterior somites. The cells of this duct migrate caudally, and the anterior region of the duct induces the adjacent mesenchyme to form the **pronephros**, or tubules of the initial kidney (**Figure 11.23A**). The pronephric tubules form functioning kidneys in fish and in amphibian larvae, but they are not believed to be active in amniotes. In mammals, the pronephric tubules and the anterior portion of the pronephric duct degenerate, but the more caudal portions of the pronephric duct persist and serve as the central component of the excretory system throughout development (Toivonen 1945; Saxén 1987). This remaining duct is often referred to as the **nephric**, or **Wolffian, duct**.

As the pronephric tubules degenerate, the middle portion of the nephric duct induces a new set of kidney tubules in the adjacent mesenchyme. This set of tubules constitutes the **mesonephros**, or mesonephric kidney (**Figure 11.23B**; Sainio and Raatikainen-Ahokas 1999). In some mammalian species, the mesonephros functions briefly in urine filtration, but in mice and rats it does not function as a working kidney. In humans, about 30 mesonephric tubules form, beginning around day 25. As more tubules are induced caudally, the anterior mesonephric tubules

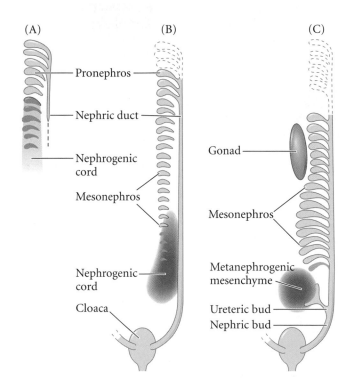

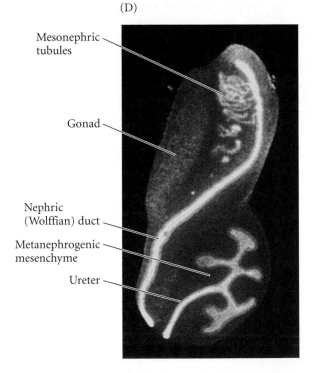

FIGURE 11.23   General scheme of development in the vertebrate kidney. (A) The original tubules, constituting the pronephros, are induced from the nephrogenic mesenchyme by the pronephric duct as it migrates caudally. (B) As the pronephros degenerates, the mesonephric tubules form. (C) The final mammalian kidney, the metanephros, is induced by the ureteric bud, which branches

from the nephric duct. (D) Intermediate mesoderm of a 13-day mouse embryo showing initiation of the metanephric kidney (bottom) while the mesonephros is still apparent. The duct tissue is stained with a fluorescent antibody to a cytokeratin found in the pronephric duct and its derivatives. (A–C after Saxén 1987; D courtesy of S. Vainio.)

(A) (B) (C)

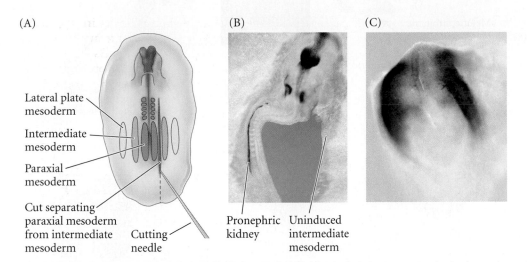

Lateral plate mesoderm

Intermediate mesoderm

Paraxial mesoderm

Cut separating paraxial mesoderm from intermediate mesoderm

Cutting needle

Pronephric kidney

Uninduced intermediate mesoderm

FIGURE 11.24 Signals from the paraxial mesoderm induce pronephros formation in the intermediate mesoderm of the chick embryo. (A) The paraxial mesoderm was surgically separated from the intermediate mesoderm on the right side of the body. (B) As a result, a pronephric kidney (Pax2-staining duct) developed only on the left side. (C) Lim1 expression in an 8-day mouse embryo, showing the prospective intermediate mesoderm. (A,B after Mauch et al. 2000; B courtesy of T. J. Mauch and G. C. Schoenwolf; C courtesy of K. Sainio and M. Hytönen.)

begin to regress through apoptosis (although in mice, the anterior tubules remain while the posterior ones regress; **Figure 11.23C,D**). While it remains unknown whether the human mesonephros actually filters blood and makes urine, it does provide important developmental functions during its brief existence. First, as we will see in Chapter 12, it is one of the main sources of the hematopoietic stem cells necessary for blood cell development (Medvinsky and Dzierzak 1996; Wintour et al. 1996). Second, in male mammals, some of the mesonephric tubules persist to become the tubes that transport the sperm from the testes to the urethra (the epididymis and vas deferens; see Chapter 14).

The permanent kidney of amniotes, the **metanephros**, is generated by some of the same components as the earlier, transient kidney types (see Figure 11.23C). It is thought to originate through a complex set of interactions between epithelial and mesenchymal components of the intermediate mesoderm. In the first steps, the **metanephrogenic mesenchyme** is committed and forms in the posterior regions of the intermediate mesoderm, where it induces the formation of a branch from each of the paired nephric ducts. These epithelial branches are called the **ureteric buds**. These buds eventually separate from the nephric duct to become the collecting ducts and ureters that take the urine to the bladder. When the ureteric buds emerge from the nephric duct, they enter the metanephrogenic mesenchyme. The ureteric buds induce this mesenchymal tissue to condense around them and differentiate into the nephrons of the mammalian kidney. As this mesenchyme differentiates, it tells the ureteric bud to branch and grow.

Specification of the Intermediate Mesoderm: Pax2/8 and Lim1

The intermediate mesoderm of the chick embryo acquires its ability to form kidneys through its interactions with the paraxial mesoderm. While its bias to become intermediate mesoderm is probably established through the BMP gradient mentioned earlier, specification appears to become stabilized through signals from the paraxial mesoderm. Mauch and her colleagues (2000) showed that signals from the paraxial mesoderm induced primitive kidney formation in the intermediate mesoderm of the chick embryo. They cut developing embryos such that the intermediate mesoderm could not contact the paraxial mesoderm on one side of the body. That side of the body (where contact with the paraxial mesoderm was abolished) did not form kidneys, but the undisturbed side was able to form kidneys (**Figure 11.24A,B**). The paraxial mesoderm appears to be both necessary and sufficient for inducing kidney-forming ability in the intermediate mesoderm, since co-culturing lateral plate mesoderm with paraxial mesoderm causes pronephric tubules to form in the lateral plate mesoderm, and no other cell type can accomplish this.

These interactions induce the expression of a set of homeodomain transcription factors—including Lim1, Pax2, and Pax8—that cause the intermediate mesoderm to form the kidney (**Figure 11.24C**; Karavanov et al. 1998; Kobayashi et al. 2005). In *Xenopus*, the Pax8 and Lim1 proteins have overlapping boundaries, and kidney development originates from cells expressing both. Ectopic co-expression of the *Pax8* and *Lim1* genes will produce kidneys in other tissues as well (Carroll and Vize 1999).

In the chick embryo, Pax2 and Lim1 are expressed in the intermediate mesoderm, starting at the level of the sixth somite (i.e., only in the trunk, not in the head); if Pax2 is experimentally expressed in the presomitic mesoderm, it converts that paraxial mesoderm into intermediate meso-

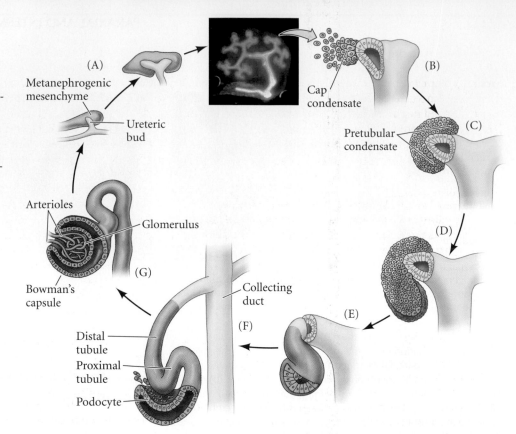

FIGURE 11.25 Reciprocal induction in the development of the mammalian kidney. (A) As the ureteric bud enters the metanephrogenic mesenchyme, the mesenchyme induces the bud to branch. (B–G) At the tips of the branches, the epithelium induces the mesenchyme to aggregate and cavitate to form the renal tubules and glomeruli (where the blood from the arteriole is filtered). When the mesenchyme has condensed into an epithelium, it digests the basal lamina of the ureteric bud cells that induced it and connects to the ureteric bud epithelium. A portion of the aggregated mesenchyme (the pretubular condensate) becomes the nephron (renal tubules and Bowman's capsule), while the ureteric bud becomes the collecting duct for the urine. (After Saxén 1987; Sariola 2002.)

derm, expresses Lim1, and forms kidneys (Mauch et al. 2000; Suetsugu et al. 2005). Similarly, in mouse embryos with knockouts of both the *Pax2* and *Pax8* genes, the mesenchyme-to-epithelial transition necessary to form the kidney duct fails, the cells undergo apoptosis, and no nephric structures form (Bouchard et al. 2002). Moreover, in the mouse, Lim1 and Pax2 proteins appear to induce one another.

Lim1 plays several roles in the formation of the mouse kidney. First it is needed for converting the intermediate mesenchyme into the kidney duct (Tsang et al. 2000), and later it is required for the formation of the ureteric bud and the tubular structure both in mesonephric and metanephric mesenchyme (Shawlot and Behringer 1995; Karavanov et al. 1998; Kobayashi et al. 2005).

The anterior border of the Lim1- and Pax2-expressing cells appears to be established by the cells above a certain region losing their competence to respond to the activin secreted by the neural tube. This competence is established by the transcription factor Hoxb4, which is not expressed in the anteriormost region of the intermediate mesoderm. The anterior boundary of Hoxb4 is established by a retinoic acid gradient, and adding activin locally will allow the kidney to extend anteriorly (Barak et al. 2005; Preger-Ben Noon et al. 2009).

Reciprocal Interactions of Developing Kidney Tissues

The two intermediate mesodermal tissues—the ureteric bud and the metanephrogenic mesenchyme—interact and reciprocally induce each other to form the kidney (**Figure** 11.25). The metanephrogenic mesenchyme causes the ureteric bud to elongate and branch. The tips of these branches induce the loose mesenchyme cells to form epithelial aggregates. Each aggregated nodule of about 20 cells proliferates and differentiates into the intricate structure of a renal nephron. Each nodule first elongates into a "comma" shape, then forms a characteristic S-shaped tube. Soon afterward, the cells of this epithelial structure begin to differentiate into regionally specific cell types, including the capsule cells, the podocytes, and the proximal and distal tubule cells. While this transformation is happening, the epithelializing nodules break down the basal lamina of the ureteric bud ducts and fuse with them.* This fusion creates a connection between the ureteric bud and the newly formed tubule, allowing material to pass from one into the other (Bard et al. 2001). These tubules derived from the mesenchyme form the secretory nephrons of the functioning kidney, and the branched ureteric bud gives rise to the renal collecting ducts and to the ureter, which drains the urine from the kidney.

Clifford Grobstein (1955, 1956) documented this reciprocal induction in vitro. He separated the ureteric bud from the metanephrogenic mesenchyme and cultured them either individually or together. In the absence of mesenchyme, the ureteric bud does not branch. In the absence

---

*The intricate coordination of nephron development with the blood capillaries that the nephrons filter is accomplished by the secretion of VEGF from the podocytes. VEGF, as we will see in Chapter 12, is a powerful inducer of blood vessels, causing endothelial cells from the dorsal aorta to form the capillary loops of the glomerular filtration apparatus (Aitkenhead et al. 1998; Klanke et al. 2002).

of the ureteric bud, the mesenchyme soon dies. But when they are placed together, the ureteric bud grows and branches, and nephrons form throughout the mesenchyme. This has been confirmed by experiments using GFP-labeled proteins to monitor cell division and branching (**Figure 11.26**; Srinvas et al. 1999.)

Mechanisms of reciprocal induction

The induction of the metanephros can be viewed as a dialogue between the ureteric bud and the metanephrogenic mesenchyme. As the dialogue continues, both tissues are altered. We will eavesdrop on this dialogue more intently than we have done for other organs. This is because the kidney has become a model for organogenesis. There are many reasons for this. First, early kidney development has only two major components. Second, the identity and roles of many of the paracrine and transcription factors produced during this dialogue have been discovered from studies of knockout mice. Third, the absence of many of these transcription factors is associated with serious pathologies characterized by absent or rudimentary kidneys. While there are several simultaneous dialogues between different groups of kidney cells, there appear to be at least nine critical sets of signals operating in the reciprocal induction of the metanephros.

STEP 1: FORMATION OF THE METANEPHROGENIC MESENCHYME: FOXES, HOXES, AND WT1   Only the metanephrogenic mesenchyme has the competence to respond to the ureteric bud and form kidney tubules; indeed, this mesenchyme cannot become any tissue other than nephrons. If induced by non-nephric tissues (such as embryonic salivary gland or neural tube tissue), metanephric mesenchyme responds by forming kidney tubules and no other structures (Saxén 1970; Sariola et al. 1982).

The positional specification of the metanephrogenic mesenchyme is positively regulated by Hoxb4 and negatively regulated by the transcription factors Foxc1 and Foxc2. *Foxc1⁻/⁻* double-mutant mice have an expanded metanephric area that induces extra ureters and kidneys (Kume et al. 2000). Next, the permanent kidney-forming metanephrogenic mesenchyme is specified by genes of the Hox11 paralogue group. When *Hox11* genes are knocked

out of mouse embryos, mesenchymal differentiation is arrested and the mesenchyme cannot induce the ureteric bud to form (Patterson et al. 2001; Wellik et al. 2002). The competence to respond to ureteric bud inducers is regulated by a tumor suppressor gene, *WT1*. Without *WT1*, the metanephric mesenchyme cells remain uninduced and die (Kreidberg et al. 1993). In situ hybridization shows that, normally, *WT1* is first expressed in the intermediate mesoderm prior to kidney formation and is then expressed in the developing kidney, gonad, and mesothelium (Pritchard-Jones et al. 1990; van Heyningen et al. 1990; Armstrong et al. 1992). Although the metanephrogenic mesenchyme appears homogeneous, it may contain both mesodermally derived tissue and some cells of neural crest origin (Le Douarin and Tiellet 1974; Sariola et al. 1989; Sainio et al. 1994).

STEP 2: THE METANEPHROGENIC MESENCHYME SECRETES GDNF TO INDUCE AND DIRECT THE URETERIC BUD   So what are all these factors doing in the metanephrogenic mesenchyme? It seems that they are setting the stage for the secretion of a paracrine factor that can induce the ureteric buds to emerge. This second signal in kidney development is **glial-derived neurotrophic factor (GDNF)**.* GDNF is expressed through a complex network initiated by Pax2 and Hox11 (Xu et al. 1999; Wellik et al. 2002). Pax2 and Hox11 (in concert with other transcription factors that permit this interaction) activate GDNF expression in the metanephrogenic mesenchyme (see Brodbeck and Englert 2004).

If GDNF were secreted throughout the metanephrogenic mesenchyme, numerous epithelial buds would sprout from the nephric duct; thus GDNF expression must be limited to the posterior region of this mesenchymal tissue. This restriction is accomplished by the proteins Sprouty1 and Robo2 (Robo2 being the same protein that helps deflect axon growth cones). If the Robo2 protein is mutated, both the anterior and posterior metanephrogenic

---

*This is the same compound that we saw in Chapter 10 as critical for the induction of dopaminergic neurons in the mammalian brain and for the entry of neural crest cells into the gut. GDNF is one busy protein; we will meet it again when we discuss sperm cell production in Chapter 16.

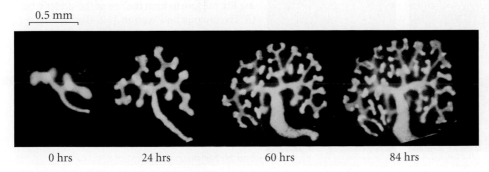

0.5 mm

0 hrs   24 hrs   60 hrs   84 hrs

FIGURE 11.26   Kidney induction observed in vitro. (A) A kidney rudiment from an 11.5-day mouse embryo was placed into culture. This transgenic mouse had a *GFP* gene fused to a Hoxb7 promoter, so it expressed green fluorescent protein in the nephric (Wolffian) duct and in the ureteric buds. Since GFP can be photographed in living tissues, the kidney could be followed as it developed. (Srinivas et al. 1999, courtesy of F. Costantini.)

mesenchyme express GDNF, and the nephric duct sends out an anterior and a posterior ureteric bud (Grieshammer et al. 2004). Sprouty1 is an inhibitor of FGF signaling. In mice whose *Sprouty1* gene is knocked out, GDNF is produced throughout the metanephrogenic mesenchyme, initiating numerous ureteric buds (Basson et al. 2005).

The receptors for GDNF (the Ret tyrosine kinase receptor and the GFRα1 co-receptor) are synthesized in the nephric ducts and later become concentrated in the growing ureteric buds (Schuchardt et al. 1996). Indeed, during formation of the ureter, those cells expressing the Ret receptor migrate to positions closest to the source of GDNF (Figure 11.27A; Chi et al. 2009). Whether this is a sorting-out phenomenon or whether the cells are attracted to the GDNF is not yet known. Mice whose *gdnf* or GDNF receptor genes are knocked out die soon after birth from renal agenesis—lack of kidneys (Figure 11.27B–D; Moore et al. 1996; Pichel et al. 1996; Sánchez et al. 1996). The ability of nephric duct cells to proliferate appears to be suppressed by activin, and the major mechanism of GDNF action may be to locally suppress the activities of inhibitory activin. When activin was experimentally inhibited, GDNF induced numerous ureteric buds (Maeshima et al. 2006).

STEP 3: THE URETERIC BUD SECRETES FGF2 AND BMP7 TO PREVENT MESENCHYMAL APOPTOSIS The third signal in kidney development is sent from the ureteric bud to the metanephrogenic mesenchyme, and it alters the fate of the mesenchyme cells. If left uninduced by the ureteric bud, the mesenchyme cells undergo apoptosis (Grobstein 1955; Koseki et al. 1992). However, if induced by the ureteric bud, the mesenchyme cells are rescued from the precipice of death and are converted into proliferating stem cells (Bard and Ross 1991; Bard et al. 1996). The factors secreted from the ureteric bud include Fgf2 and BMP7. Fgf2 has three modes of action in that it inhibits apoptosis, promotes the condensation of mesenchyme cells, and maintains the synthesis of WT1 (Perantoni et al. 1995). BMP7 has similar effects (Dudley et al. 1995; Luo et al. 1995).

STEP 4: WNT9B AND WNT6 FROM THE URETERIC BUD INDUCE MESENCHYME CELLS TO AGGREGATE The ureteric bud causes dramatic changes in the behavior of the metanephrogenic mesenchyme cells, converting them into an epithelium. The newly induced mesenchyme synthesizes E-cadherin, which causes the mesenchyme cells to clump together. These aggregated nodes of mesenchyme now

(A)

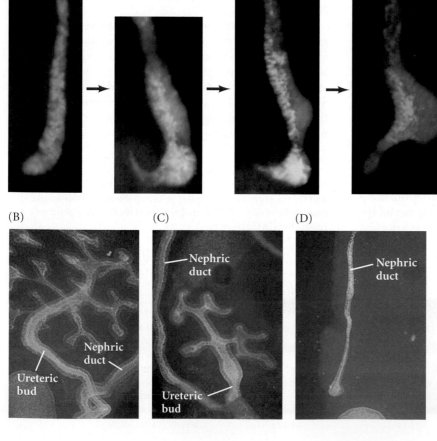

(B)　　　　(C)　　　　(D)

Nephric duct

Nephric duct

Nephric duct

Ureteric bud

Ureteric bud

FIGURE 11.27 Ureteric bud growth is dependent on GDNF and its receptors. (A) When mice are constructed from Ret-deficient cells (green) and Ret-expressing cells (blue), the cells expressing Ret migrate to form the tips of the ureteric bud. (B) The ureteric bud from an 11.5-day wild-type mouse kidney cultured for 72 hours has a characteristic branching pattern. (C) In embryonic mice heterozygous for a mutation of the gene encoding GDNF, both the size of the ureteric bud and the number and length of its branches are reduced. (D) In mouse embryos missing both copies of the *gdnf* gene, the ureteric bud does not form. (A from Chi et al. 2009, courtesy of F. Costantini; B–D from Pichel et al. 1996, courtesy of J. G. Pichel and H. Sariola.)

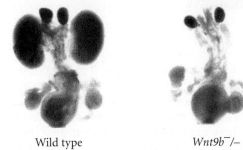

(A) Wnt9b   Wnt11

Wolffian (nephric) duct

Ureteric bud

(B)

Wild type   *Wnt9b⁻/–*

FIGURE 11.28 Wnts are critical for kidney development. (A) In the 11-day mouse kidney, Wnt9b is found on the stalk of the ureteric bud, while Wnt11 is found at the tips. Wnt9b induces the metanephrogenic mesenchyme to condense; Wnt11 will partition the metanephrogenic mesoderm to induce branching of the ureteric bud. Borders of the bud are indicated by a dashed line. (B) A wild-type 18.5-day male mouse (left) has normal kidneys, adrenal glands, and ureters. In a mouse deficient for *Wnt9b* (right), the kidneys are absent. (From Carroll et al. 2005.)

synthesize an epithelial basal lamina containing type IV collagen and laminin. At the same time, the mesenchyme cells synthesize receptors for laminin, allowing the aggregated cells to become an epithelium (Ekblom et al. 1994; Müller et al. 1997). The cytoskeleton also changes from one characteristic of mesenchyme cells to one typical of epithelial cells (Ekblom et al. 1983; Lehtonen et al. 1985).

The transition from mesenchymal to epithelial organization may be mediated by several molecules, including the expression of Pax2 in the newly induced mesenchyme cells. When antisense RNA to *Pax2* prevents the translation of the *Pax2* mRNA that is transcribed as a response to induction, the mesenchyme cells of cultured kidney rudiments fail to condense (Rothenpieler and Dressler 1993). Thus, Pax2 may play several roles during kidney formation.

In the mouse, Wnt6 and Wnt9b from the lateral sides of the ureteric bud (but not from the tip) are critical for transforming the metanephrogenic mesenchyme cells into tubular epithelium. The mesenchyme has receptors for these Wnts. Wnt6 appears to promote the condensation of mesenchyme in an FGF-independent way (Itäranta et al. 2002), and Wnt9b induces Wnt4 in the mesenchyme. As we will see, Wnt4 is very important for the formation of the nephron, and mice deficient in Wnt9b lack kidneys* (**Figure 11.28**; see also Figure 3.27).

STEP 5: WNT4 CONVERTS AGGREGATED MESENCHYME CELLS INTO A NEPHRON Once induced, and after it has started to condense, the mesenchyme begins to secrete Wnt4, which acts in an autocrine fashion to complete the transition from mesenchymal mass to epithelium (Stark et al. 1994; Kispert et al. 1998). Wnt4 expression is found in the condensing mesenchyme cells, in the resulting S-shaped tubules, and in the region where the newly epithelialized cells fuse with the ureteric bud tips. In mice lacking the *Wnt4* gene, the mesenchyme becomes condensed but does not form epithelia. Therefore, the ureteric bud induces the changes in the metanephrogenic mesenchyme by secreting FGFs, Wnt9, and Wnt6; but these changes are mediated by the effects of the mesenchyme's secretion of Wnt4 on itself.

One molecule that may be involved in the transition from aggregated mesenchyme to nephrons is the Lim1 homeodomain transcription factor (Karavanov et al. 1998; Kobayashi et al. 2005). This protein is found in the mesenchyme cells after they have condensed around the ureteric bud, and its expression persists in the developing nephron (**Figure 11.29**). Two other proteins that may be critical for the conversion of the aggregated cells into a nephron are polycystins 1 and 2. These proteins are the

*Wnt9b appears to be critical for inducing mesenchyme-to-epithelium transitions throughout the intermediate mesoderm. This includes the formation of the nephric (Wolffian) and Müllerian ducts, as well as the kidney tubules. In addition to having no kidneys, Wnt9b-deficient mice have no uterus (if they are female) and no vas deferens (if male). In rats, leukemia inhibitory factor (LIF) appears to substitute for Wnt6 and Wnt9 (Barasch et al. 1999; Carroll et al. 2005).

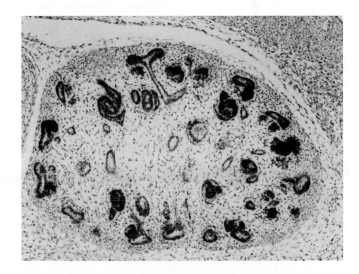

FIGURE 11.29 Lim1 expression (dark stain) in a 19-day embryonic mouse kidney. In situ hybridization shows high levels of Lim1 protein in the newly epithelialized comma-shaped and S-shaped bodies that will become nephrons. Compare with earlier Lim1 expression shown in Figure 11.24C. (From Karavanov et al. 1998, courtesy of A. A. Karavanov.)

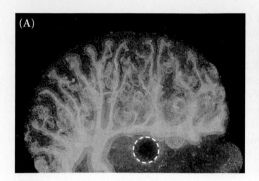

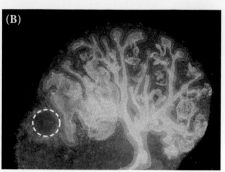

FIGURE 11.30 Effect of GDNF on branching of the ureteric epithelium. The ureteric bud and its branches are stained orange (with antibodies to cytokeratin 18), while the nephrons are stained green (with antibodies to nephron brush border antigens). (A) 13-day embryonic mouse kidney cultured 2 days with a control bead (circle) has a normal branching pattern. (B) A similar kidney cultured 2 days with a GDNF-soaked bead shows a distorted pattern, as new branches are induced in the vicinity of the bead. (From Sainio et al. 1997, courtesy of K. Sainio.)

products of the genes whose loss-of-function alleles give rise to human polycystic kidney disease. Mice deficient in these genes have abnormal, swollen nephrons (Ward et al. 1996; van Adelsberg et al. 1997).

STEP 6: SIGNALS FROM THE MESENCHYME INDUCE THE BRANCHING OF THE URETERIC BUD Recent evidence has implicated several paracrine factors in the branching of the ureteric bud, and these factors probably work as pushes and pulls. Some factors may preserve the extracellular matrix surrounding the epithelium, thereby preventing branching from taking place (the "push"), while other factors may cause the digestion of this extracellular matrix, permitting branching to occur (the "pull").

The first protein regulating ureteric bud branching is GDNF (Sainio et al. 1997). GDNF from the mesenchyme not only induces the initial ureteric bud from the nephric duct, but it can also induce secondary buds from the ureteric bud once the bud has entered the mesenchyme (**Figure 11.30**). GDNF from the mesenchyme promotes cell division in the Ret-expressing cells at the tip of the ureteric bud (Shakya et al. 2005). GDNF also appears to induce Wnt11 synthesis in these responsive cells at the tip of the bud (see Figure 11.28A), and Wnt11 reciprocates by regulating GDNF levels. Wnt11 may also be important in inducing the differentiation of the metanephrogenic mesenchyme (Kuure et al. 2007). The cooperation between the GDNF/Ret pathway and the Wnt pathway appears to coordinate the balance between branching and metanephrogenic cell division such that continued metanephric development is ensured (Majumdar et al. 2003).

The second candidate branch-regulating molecule is transforming growth factor-β1 (TGF-β1). When exogenous

TGF-β1 is added to cultured kidneys, it prevents the epithelium from branching (**Figure 11.31A,B**; Ritvos et al. 1995). TGF-β1 is known to promote the synthesis of extracellular matrix proteins and to inhibit the metalloproteinases that can digest these matrices (Penttinen et al. 1988; Nakamura et al. 1990). Thus, it is possible that TGF-β1 stabilizes branches once they form.

A third molecule that may be important in epithelial branching is BMP4 (Miyazaki et al. 2000). BMP4 is found in the mesenchymal cells surrounding the nephric duct, and BMP4 receptors are found in the epithelial tissue of the duct. Because BMPs antagonize branching signals, BMP4 restricts the branching of the duct to the appropriate sites. When the BMP4 signaling cascade is activated ectopically in embryonic mouse kidney rudiments, it severely distorts the normal branching pattern (**Figure 11.31C**).

The fourth molecule involved in branching is collagen XVIII, which is part of the extracellular matrix induced by the mesenchyme. It may provide the specificity for the branching pattern (Lin et al. 2001). Collagen XVIII is found on the branches of the kidney epithelium but not at the tips; in the developing lung, the reciprocal pattern is seen. This pattern is generated in part by GDNF, which downregulates collagen XVIII expression in the tips of the ureteric bud branches. When ureteric duct epithelium is incubated in lung mesenchyme, the collagen XVIII expression pattern seen is typical of that of the lung, and the branching pattern resembles that of the lung epithelium (**Figure 11.31D**).

STEPS 7 AND 8: DIFFERENTIATION OF THE NEPHRON AND GROWTH OF THE URETERIC BUD The interactions we have described to this point create a *cap condensate* of metanephrogenic mesenchyme cells that covers the tips of the ureteric bud branches. The cap condensate maintains GDNF production, ensuring the outward growth of the ureteric branches. Moreover, the mesenchyme of this cap condensate is a multipotent, self-renewing population that can make all the different cell types of the nephron (Mugford et al. 2009). As such, it is not unlike the somite cells, which also condense into a population of self-renewing multipotent progenitor cells.

(A)

(B)

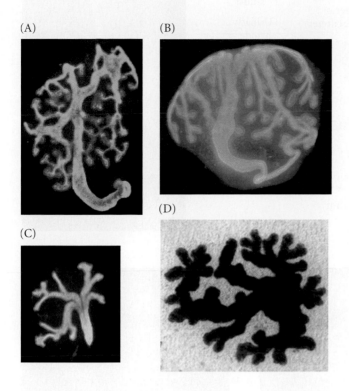

(C)

(D)

FIGURE 11.31 Signaling molecules and branching of the ureteric epithelium. (A) An 11-day embryonic mouse kidney cultured for 4 days in control medium has a normal branching pattern. (B) An 11-day mouse kidney cultured in TGF-β1 shows no branching until reaching the periphery of the mesenchyme, and the branches formed are elongated. (C) An 11-day mouse kidney cultured in activin (which activates the same receptor as BMP4) shows a marked distortion of branching. (D) Epithelial branching of kidney cells grown in lung mesenchyme takes on an appearance similar to that of lung epithelium. (A–C from Ritvos et al. 1995; D from Lin et al. 2001. Photographs courtesy of Y. Lin and S. Vainio.)

The transcription factor Foxb2 is synthesized in the cap cells. When the *Foxb2* gene is knocked out in mouse embryos, the resulting kidney lacks a branched ureteric tree (it branches only 3–4 times instead of the normal 7–8, resulting in an 8- to 16-fold reduction in the number of branches). In addition, the aggregates do not differentiate, and thus no nephrons form (Hatini et al. 1996).

The cap cells are also able to convert vitamin A to retinoic acid, and this RA functions to retain the expression of Ret (one of the GDNF receptors) in the ureteric bud (Batourina et al. 2002). The cap cells also secrete Fgf7, a growth factor whose receptor is found on the ureteric bud. Fgf7 is critical for maintaining ureteric epithelial growth and ensuring an appropriate number of nephrons in the kidney (Qiao et al. 1999; Chi et al. 2004).

STEP 9: INSERTING THE URETER INTO THE BLADDER The branching epithelium becomes the collecting system of the kidney. This epithelium collects the filtered urine from the nephron and secretes anti-diuretic hormone for the resorption of water (a process that, not so incidentally, makes life on land possible). The rest of the ureteric bud becomes the ureter, the tube that carries urine into the bladder. The junction between the ureter and bladder is extremely important, and hydronephrosis, a birth defect involving renal filtration, occurs when this junction is so tightly formed that urine cannot enter the bladder. The ureter is made into a watertight connecting duct by the condensation of mesenchymal cells around it (but not around the collecting ducts). These mesenchymal cells become smooth muscle cells capable of wavelike contractions (peristalsis) that

allow the urine to move into the bladder; these cells also secrete BMP4 (Cebrian et al. 2004). BMP4 upregulates genes for uroplakin, a protein that causes differentiation of this region of the ureteric bud into the ureter. BMP inhibitors protect the region of the ureteric bud that forms the collecting ducts from this differentiation.

The bladder develops from a portion of the cloaca (**Figure 11.32A,B**). The cloaca* is an endodermally lined chamber at the caudal end of the embryo that will become the waste receptacle for both the intestine and the kidney. Amphibians, reptiles, and birds still have this organ and use it to void both liquid and solid wastes. In mammals, the cloaca becomes divided by a septum into the urogenital sinus and the rectum. Part of the urogenital sinus becomes the bladder, while another part becomes the urethra (which will carry the urine out of the body). The ureteric bud originally empties into the bladder via the nephric (Wolffian) duct (which grows toward the bladder by an as-yet-unknown mechanism). Once at the bladder, the urogenital sinus cells of the bladder wrap themselves around both the ureter and the nephric duct. Then the nephric ducts migrate ventrally, opening into the urethra rather than into the bladder and the nephric duct (Figure 11.32C–F). The caudal end of the nephric duct appears to undergo apoptosis, allowing the ureter to separate from the nephric duct. Expansion of the bladder then moves the ureter to its final position at the neck of the bladder (Batourina et al. 2002; Mendelsohn 2009). In females, the entire nephric duct (often referred to as the Wolffian duct) degenerates, while the Müllerian duct opens into the vagina (see Chapter 14). In males, however, the nephric duct also forms the sperm outflow track, so males expel sperm and urine through the same opening.

The Hox13 paralogue group appears to be important in specifying the distal ureter. In human *HOXA13* defects, there are abnormalities of the cloaca, the male and female reproductive tracts, and the urethra (Pinsky 1974; Mortlock and Innis 1997). Because *HOXA13* is also involved in digit specification (see Chapter 13), the person's fingers

---

*The term *cloaca* is Latin for "sewer"—a bad joke on the part of the early European anatomists.

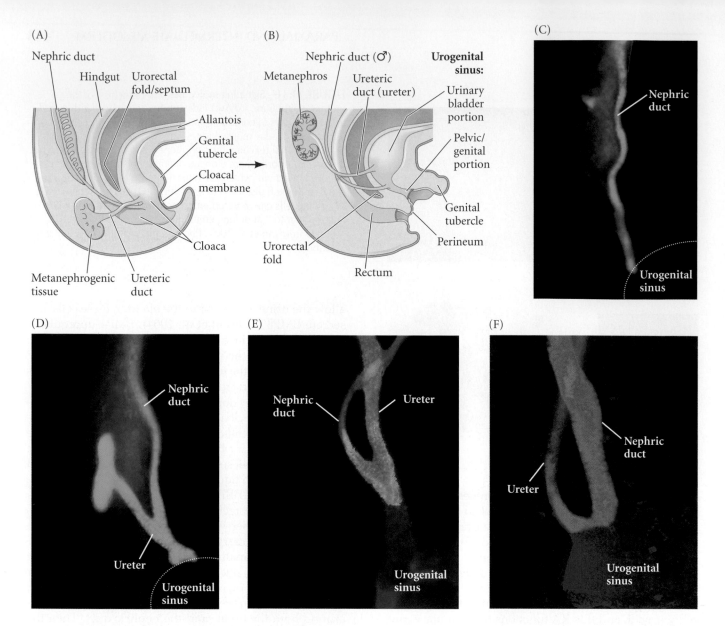

FIGURE 11.32 Development of the bladder and its connection to the kidney via the ureter. (A) The cloaca originates as an endodermal collecting area that opens into the allantois. (B) The urogenital septum divides the cloaca into the future rectum and the urogenital sinus. The bladder forms from the anterior portion of that sinus, and the urethra develops from the posterior region of the sinus. The space between the rectal opening and the urinary opening is the perineum. (C–F) Insertion of the ureter into the embryonic mouse bladder. (C) Day10 mouse urogenital tract. The nephric duct is stained with GFP attached to a Hoxb7 enhancer. (D) Urogenital tract from a day-11 embryo, after ureteric bud outgrowth. (E) Whole-mount urogenital tract from a day12 embryo. The ducts are stained green and the urogenital sinus red. (F) The ureter separates from the nephric duct and forms a separate opening into the bladder. (A,B after Cochard 2002; C–F from Batourina et al. 2002, courtesy of C. Mendelsohn.)

and toes are also malformed; thus the syndrome has been called *hand-foot-genital syndrome*.

FURTHER STEPS The next stages of kidney development now occur in the separate groups. The glomerulus, for instance, forms from the epithelialized descendents of the metanephrogenic mesenchyme located distally, while the more proximal descendents, closer to the collecting ducts, become the distal and proximal convoluted tubules, and so forth. The developing kidney epitomizes the reciprocal interactions needed to form an organ. It also shows us that we have only begun to understand how organs form.

Coda

The central mesodermal compartments—the notochord, paraxial, and intermediate mesoderm—form the core of our musculoskeletal and reproductive systems. We have seen how reciprocal induction functions both in the relationship between tendons and muscles and in kidney formation. We have also seen how some cells (such as the somite cells) are multipotent and can generate numerous cell types, while others (such as the muscle satellite cells) can produce but one type. Probably the most important phenomenon of the central mesodermal compartments is the rich tapestry of integrated co-construction of organs, which emerges from interactions *within* the mesodermal lineages and *between* the mesoderm and its surrounding ectoderm and endoderm. We now proceed to the lateral mesodermal compartments and their interactions among themselves and with the endoderm to provide the basis for circulation and nutrient absorption in the developing body.

 Snapshot Summary: *Paraxial and Intermediate Mesoderm*

Paraxial and Intermediate Mesoderm

1. The paraxial mesoderm forms blocks of tissue called somites. Somites give rise to three major divisions: the sclerotome, the myotome, and the central dermamyotome.

2. Somites are formed from the segmental plate (unsegmented mesoderm) by the interactions of several proteins. The Notch pathway is extremely important in this process, and Eph receptor systems may be involved in the separation of the somites from the unsegmented paraxial mesoderm. N-cadherin, fibronectin, and Rac1 appear to be important in causing these cells to become epithelial.

3. The sclerotome of the somite forms the vertebral cartilage. In thoracic vertebrae, the sclerotome cells also form the ribs. The intervertebral joints as well as the meninges and dorsal aortic cells also come from the sclerotome.

4. The primaxial myotome forms the back musculature. The abaxial myotome forms the muscles of the body wall, limb, diaphragm, and tongue.

5. The central demamyotome forms the dermis of the back, as well as forming precursors of muscle and brown fat cells.

6. The somite regions are specified by paracrine factors secreted by neighboring tissues. The sclerotome is specified to a large degree by Sonic hedgehog, which is secreted by the notochord and floor plate cells. The two myotome regions are specified by different factors, and in both instances, myogenic bHLH transcription factors are induced in the cells that will become muscles.

7. To form muscles, the myoblasts stop dividing, align themselves into myotubes, and fuse.

8. The major lineages that form the skeleton are the somites (axial skeleton), lateral plate mesoderm (appendages), and neural crest and head mesoderm (skull and face).

9. There are two major types of ossification. In intramembranous ossification, which occurs primarily in the skull and facial bones, mesenchyme is converted directly into bone. In endochondral ossification, mesenchyme cells become cartilage. These cartilaginous models are later replaced by bone cells.

10. The replacement of cartilage by bone during endochondral ossification depends on the mineralization of the cartilage matrix.

11. Osteoclasts continually remodel bone throughout a person's lifetime. The hollowing out of bone for the bone marrow is accomplished by osteoclasts.

12. Tendons are formed through the conversion of the dorsalmost layer of sclerotome cells into syndetome cells by FGFs secreted by the myotome.

13. The intermediate mesoderm is specified through interactions with the paraxial mesoderm. It generates the kidneys and gonads. This specification requires Pax2, Pax6, and Lim1.

14. The metanephric kidney of mammals is formed by the reciprocal interactions of the metanephrogenic mesenchyme and a branch of the nephric duct called the ureteric bud.

15. The metanephrogenic mesenchyme becomes competent to form nephrons by expressing WT1, and it starts to secrete GDNF. GDNF is secreted by the mesoderm and induces the formation of the ureteric bud.

16. The ureteric bud secretes Fgf2 and BMP7 to prevent apoptosis in the metanephrogenic mesenchyme. Without these factors, this kidney-forming mesenchyme dies.

17. The ureteric bud secretes Wnt9 and Wnt6, which induce the competent metanephrogenic mesenchyme to form epithelial tubules. As they form these tubules, the cells secrete Wnt4, which promotes and maintains their epithelialization.

18. The condensing mesenchyme secretes paracrine factors that mediate the branching of the ureteric bud. These factors include GDNF, BMP4 and TGF-β1. The branching also depends on the extracellular matrix of the epithelium.

For Further Reading

Complete bibliographical citations for all literature cited in this chapter can be found at the free-access website **www.devbio.com**

Brent, A. E., R. Schweitzer and C. J. Tabin. 2003. A somitic compartment of tendon precursors. *Cell* 113: 235–248.

Christ, B., R. Huang and M. Scaal. 2007. Amniote somite derivatives. *Dev. Dyn.* 236: 2382–2396.

Gerber, H. P., T. H. Vu,. A. M. Ryan, J. Kowalski and Z. Werb. 1999. VEGF couples hypertrophic cartilage remodeling, ossification, and angiogenesis during endochondral bone formation. *Nature Med.* 5: 623–628.

Kuang, S., K. Kuroda, F. Le Grand and M. A. Rudnicki. 2007. Asymmetric self-renewal and commitment of satellite stem cells in muscle. *Cell* 129: 999–1010.

Mauch, T. J., G. Yang, M. Wright, D. Smith and G. C. Schoenwolf. 2000. Signals from trunk paraxial mesoderm induce pronephros formation in chick intermediate mesoderm. *Dev. Biol.* 220: 62–75.

Nowicki, J. L., R. Takimoto, and A. C. Burke. 2003. The lateral somitic frontier: Dorso-ventral aspects of anterior-posterior regionalization in the embryo. *Mech. Dev.* 120: 227–240.

Ordahl, C. P. and N. Le Douarin. 1992. Two myogenic lineages within the developing somite. *Development* 114: 339–353.

Palmeirim, I., D. Henrique, D. Ish-Horowicz and O. Pourquié. 1997. Avian hairy gene expression identifies a molecular clock linked to vertebrate segmentation and somitogenesis. *Cell* 91: 639–648.

Pichel, J. G. and 11 others. 1996. Defects in enteric innervation and kidney development in mice lacking GDNF. *Nature* 382: 73–76.

Sato, Y., K. Yasuda and Y. Takahashi. 2002. Morphological boundary forms by a novel inductive event mediated by Lunatic-Fringe and Notch during somitic segmentation. *Development* 129: 3633–3644.

Go Online

WEBSITE 11.1 Calling the competence of the somite into question. When the *tbx6* gene was knocked out from mice, the resulting embryos had three neural tubes in the posterior of their bodies. Without the *tbx6* gene, the somitic tissue responded to the notochord and epidermal signals as if it were neural ectoderm.

WEBSITE 11.2 Cranial paraxial mesoderm. Most of the head musculature does not come from somites. Rather, it comes from the cranial paraxial (prechordal plate) mesoderm. These cells originate adjacent to the sides of the brain, and they migrate to their respective destinations.

WEBSITE 11.3 Muscle formation. Research on chimeric mice has shown that skeletal muscle becomes multinucle-ate by the fusion of cells, while heart muscle becomes multinucleate by nuclear divisions within a cell.

WEBSITE 11.4 Paracrine factors, their receptors, and human bone growth. Mutations in the genes encoding paracrine factors and their receptors cause numerous skeletal anomalies in humans and mice. The FGF and Hedgehog pathways are especially important.

Vade Mecum

Mesoderm in the vertebrate embryo. The organization of the mesoderm in the neurula stage is similar for all vertebrates. You can see this organization by viewing serial sections of the chick embryo in the Chick-Mid segment.

Lateral Plate Mesoderm and the Endoderm

12

IN THE CHAOS OF THE ENGLISH CIVIL WARS, William Harvey, physician to the King and discoverer of the blastoderm, was comforted by viewing the heart as the undisputed ruler of the body, through whose divinely ordained powers the lawful growth of the organism was assured. Later embryologists saw the heart as more of a servant than a ruler, the chamberlain of the household who assured that nutrients reached the apically located brain and the peripherally located muscles. In either metaphor, the heart, circulation, and digestive system were known to be absolutely critical for development. As Harvey persuasively argued in 1651, the chick embryo must form its own blood without any help from the hen, and this blood is crucial in embryonic growth. How this happened was a mystery to him. "What artificer," he wrote, could create blood "when there is yet no liver in being?" The nutrition provided by the egg was also paramount to Harvey. His conclusion about the nutritive value of the yolk and albumen was "The egge is, as it were, an exposed womb; wherein there is a substance concluded as the Representative and Substitute, or Vicar of the breasts."

This chapter outlines the mechanisms by which the circulatory system, the respiratory system, and the digestive system emerge in the amniote embryo. We will first discuss the formation of the heart and then proceed to the mechanisms whereby the blood vessels and blood cells develop. At the end of the chapter, we will briefly follow the development of the gut tube and its associated organs. As we will see, the lateral plate mesoderm and the endoderm interact to create the circulatory and the digestive organs.*

The Heart of Creatures is the Foundation of Life, the Prince of All, the Sun of the Microcosm, on which all Vegetation doth depend, from whence all Vigor and Strength doth flow.

WILLIAM HARVEY (1628)

Blut is ein ganz besonderer Saft. [Blood is a very special juice.]

WOLFGANG GOETHE (1805)

LATERAL PLATE MESODERM

On the peripheral sides of the two bands of intermediate mesoderm resides the **lateral plate mesoderm** (see Figures 11.1 and 11.2). Each plate splits horizontally into two layers. The dorsal layer is the **somatic (parietal) mesoderm**, which underlies the ectoderm and, together with the ectoderm, forms the **somatopleure**. The ventral layer is the **splanchnic (visceral) mesoderm**, which overlies the endoderm and, together with the endoderm, forms the **splanchnopleure** (Figure

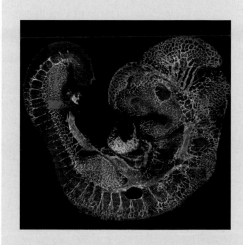

---

*Some scientists think that the mesoderm and endoderm were originally a single germ layer, the "mesendoderm," that accomplished all these functions. Recall from Chapter 5 that what in vertebrates is a broad territory of embryonic cells is actually the progeny of a single "mesentoblast" in many invertebrates. The signals that regulate the mesentoblast and the entire mesodermal and endodermal territory in vertebrates may be very similar (Maduro et al. 2001; Rodaway and Patient 2001).

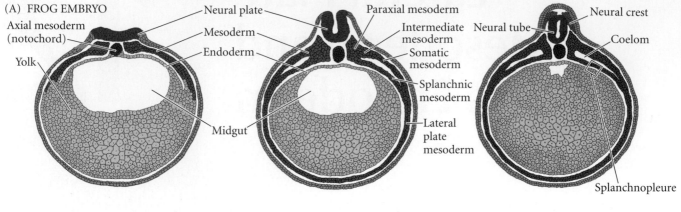

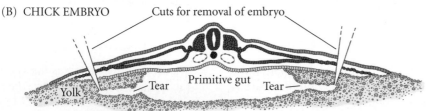

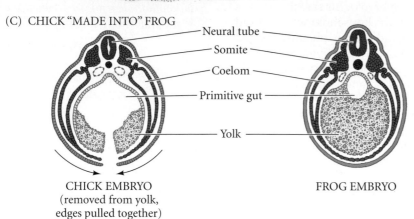

FIGURE 12.1 Mesodermal development in frog and chick embryos. (A) Neurula-stage frog embryos, showing progressive development of the mesoderm and coelom. (B) Transverse section of a chick embryo. (C) When the chick embryo is separated from its enormous yolk mass, it resembles the amphibian neurula at a similar stage. (A after Rugh 1951; B,C after Patten 1951.)

12.1A). The space between these two layers becomes the body cavity—the **coelom**—which stretches from the future neck region to the posterior of the body. During later development, the right-side and left-side coeloms fuse, and folds of tissue extend from the somatic mesoderm, dividing the coelom into separate cavities. In mammals, the coelom is subdivided into the **pleural**, **pericardial**, and **peritoneal cavities**, enveloping the thorax, heart, and abdomen, respectively. The mechanism for creating the linings of these body cavities from the lateral plate mesoderm has changed little throughout vertebrate evolution, and the development of the chick mesoderm can be compared with similar stages of frog embryos (**Figure 12.1B,C**).

See WEBSITE 12.1 Coelom formation

Heart Development

Consisting of a heart, blood cells, and an intricate system of blood vessels, the circulatory system provides nourish-ment to the developing vertebrate embryo. The circulatory system is the first functional unit in the developing embryo, and the heart is the first functional organ. The vertebrate heart arises from two regions of splanchnic mesoderm—one on each side of the body—that interact with adjacent tissue to become specified for heart development.

Specification of heart tissue

In the early amniote gastrula, the heart progenitor cells (about 50 of them in mice) are located in two small patches, one each on the epiblast close to the rostral portion of the primitive streak. These cells migrate together through the streak and form two groups of cells in the lateral plate mesoderm, at the level of the node (Tam et al. 1997; Colas et al. 2000). As can be seen in **Figure 12.2**, the general specification of this **cardiogenic mesoderm**, or **heart field**, has already been started during this migration. Labeling experiments by Stalberg and DeHaan (1969) and Abu-Issa and

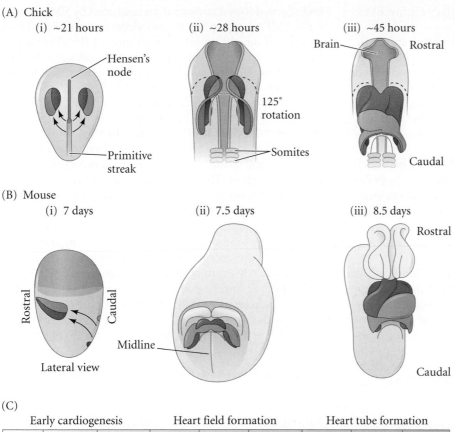

(A) Chick

(i) ~21 hours — Hensen's node, Primitive streak

(ii) ~28 hours — 125° rotation, Somites

(iii) ~45 hours — Brain, Rostral, Caudal

(B) Mouse

(i) 7 days — Rostral, Caudal, Lateral view

(ii) 7.5 days — Midline

(iii) 8.5 days — Rostral, Caudal

FIGURE 12.2 Overview of heart development. (A,B) Schematic of heart development in the chick and mouse. Progenitors of the ascending region, or outflow tract, of the heart tube (right ventricle, conus and truncus arteriosus) are shown in red; those of the descending region, or inflow tract (atria, left ventricle), are in blue. Except for mouse view (i), the embryos are shown from the ventral side. In both initial views (i), the myocardial progenitor cells migrate into the lateral plate mesoderm (arrows) and undergo specification. In the chick, these regions are separated; in the mouse, they are connected in the midline. (C) Developmental times of the events shown in (A) and (B), showing the four developmental processes taking place. (D) Some of the major genes associated with these processes. (After Abu-Issa and Kirby 2007.)

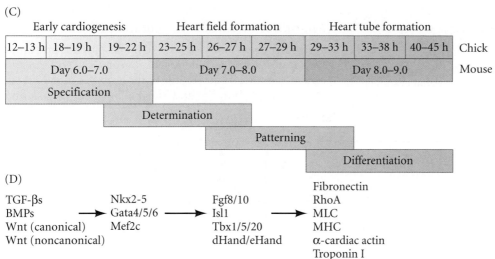

(C)

	Early cardiogenesis			Heart field formation			Heart tube formation			
12–13 h	18–19 h	19–22 h	23–25 h	26–27 h	27–29 h	29–33 h	33–38 h	40–45 h	Chick
Day 6.0–7.0			Day 7.0–8.0			Day 8.0–9.0			Mouse

Specification
Determination
Patterning
Differentiation

(D)

TGF-βs
BMPs → Nkx2-5 → Fgf8/10 → Fibronectin
Wnt (canonical) Gata4/5/6 Isl1 RhoA
Wnt (noncanonical) Mef2c Tbx1/5/20 MLC
  dHand/eHand MHC
  α-cardiac actin
  Troponin I

Kirby (2008) have shown that the cardiac cells of the heart field migrate such that the medial-lateral arrangement of these early cells will become the anterior-posterior (rostral-caudal) axis of the developing heart tube. The progenitors of the inflow tract (the left ventricle and the atria, which receive the blood) are located most laterally, while the outflow tract precursors (the **conus arteriosus** and **truncus arteriosus**, which become the base of the aorta and pulmonary arteries, and the right ventricle) are located medially, closest to the primitive streak. All the cells of the heart—the **cardiomyocytes** that form the muscular layers, the **endocardium** that forms the internal layer, the *endocar-*

dial cushions of the valves, the **epicardium** that forms the coronary blood vessels that feed the heart, and the **Purkinje fibers***** that coordinate the heartbeat—are generated from these two clusters (Mikawa 1999; van Wijk et al. 2009). These cells will be supplemented by cells recruited to specific places, such as the cardiac neural crest cells that make the septum (which separates the aorta from the pulmonary

*Note that these specialized myocardial nerve fibers are not the same thing as the Purkinje *neurons* of the cerebellum mentioned in Chapter 9. Both were named for the nineteenth-century Czech anatomist and histologist Jan Purkinje.

trunk) and portions of the outflow tract (see Figure 10.15; Porras and Brown 2008).

The heart cells on either side of the primitive streak are specified but not yet determined. This is to say, if development is normal, these cells will follow identities dictated by their locations. However, if at this stage the cells are transplanted to another part of the heart field, they will take on a new identity rather than retaining their old one. One can also see that each side of the heart field can be subdivided into smaller units (see Figure 12.2). This anatomical compartmentation of the heart field is based on which cells will contribute to the inflow tract and which to the outflow tract. The inflow tract (left ventricle and the atria) is called the *primary heart field*, while the cells forming the outflow tract (right ventricle, conus and truncus arteriosus) comprise the *anterior heart field** (see Abu-Issa and Kirby 2007).

The specification of the cardiogenic mesoderm cells is induced by the endoderm adjacent to the heart through the BMP and FGF signaling pathways. The heart does not form if the anterior endoderm is removed, and the posterior endoderm cannot induce heart cells to form. Thus, isolated mesoderm from this region will form heart muscle when combined with anterior, but not posterior, endoderm (Nascone and Mercola 1995; Schultheiss et al. 1995). The

endodermal signal appears to be mediated by BMPs, especially BMP2. BMPs from the endoderm promote both heart and blood development. Endodermal BMPs also induce Fgf8 synthesis in the endoderm directly beneath the cardiogenic mesoderm, and Fgf8 appears to be critical for the expression of cardiac proteins (Alsan and Schultheiss 2002).

Inhibitory signals prevent heart formation where it should not occur. The notochord secretes Noggin and chordin, blocking BMP signaling in the center of the embryo, and Wnt proteins from the neural tube, especially Wnt3a and Wnt8, *inhibit* heart formation but *promote* blood formation. The anterior endoderm, however, produces Wnt inhibitors such as Cerberus, Dickkopf, and Crescent, which prevent Wnts from binding to their receptors. In this way, cardiac precursor cells are specified in those places where BMPs (lateral mesoderm and endoderm) and Wnt antagonists (anterior endoderm) coincide (**Figure 12.3**; Marvin et al. 2001; Schneider and Mercola 2001; Tzahor and Lassar 2001). In order to prevent the heart cells from being respecified, one of the first proteins made in the cardiac field cells is Mesp1, a transcription factor that activates the *Dickkopf* gene in these cells (David et al. 2008).

See VADE MECUM Early heart development

---

*Nature doesn't always correspond to such simple distinctions, however, and it turns out that some of the cells in the primary heart field share a progenitor cell with cells in the anterior heart field. So another arrangement looks at the clonal populations of cells and assigns them to either the **first heart field** or the **second heart field** based on their lineage.

Migration of cardiac precursor cells

When the chick embryo is 18–20 hours old, the presumptive heart cells move anteriorly between the ectoderm and endoderm toward the middle of the embryo, remaining in close contact with the endodermal surface (Linask and

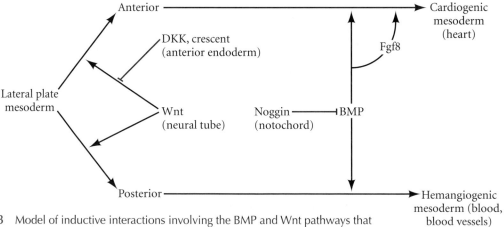

FIGURE 12.3 Model of inductive interactions involving the BMP and Wnt pathways that form the boundaries of the cardiogenic mesoderm. Wnt signals from the neural tube instruct lateral plate mesoderm to become precursors of the blood and blood vessels. In the anterior portion of the body, however, Wnt inhibitors (Dickkopf, Crescent) from the pharyngeal endoderm prevent Wnt from functioning, allowing later signals (BMP, Fgf8) to convert lateral plate mesoderm into cardiogenic mesoderm. BMP signals will also be important for the differentiation of hemangiogenic (blood, blood vessel) mesoderm. In the center of the embryo, Noggin signals from the notochord block BMPs. Thus the cardiac and blood-forming fields do not form in the center of the embryo.

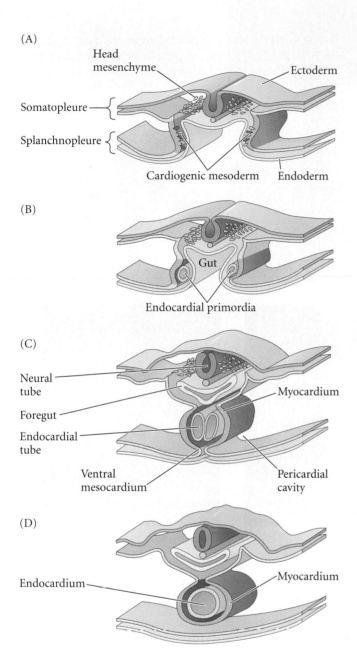

(A)

Head
mesenchyme

Ectoderm

Somatopleure

Splanchnopleure

Cardiogenic mesoderm   Endoderm

(B)

Gut

Endocardial primordia

(C)

Neural
tube

Myocardium

Foregut

Endocardial
tube

Ventral
mesocardium

Pericardial
cavity

(D)

Endocardium

Myocardium

FIGURE 12.4 Formation of chick heart from splanchnic lateral plate mesoderm. The endocardium forms the inner lining of the heart, the myocardium forms the heart muscles, and the epicardium will eventually cover the heart. Transverse sections through the heart-forming region of the chick embryo are shown at (A) 25 hours, (B) 26 hours, (C) 28 hours, and (D) 29 hours. (After Carlson 1981.)

to the anterior-posterior axis. This rotation repositions the cells so that those of the ascending (outflow tract) portion of the heart tube (right ventricle, conus and truncus arteriosus) are brought rostrally, while the descending (inflow tract) portions of the heart tube (atria, atrioventricular canal, left ventricle) are brought caudally (see Figure 12.2A, view ii). In the mouse and human, the bilateral fields are brought together to form one horseshoe-shaped field sometimes called the *cardiac crescent* (see Figure 12.2B, view ii).

In the chick, the fields are brought together around the 7-somite stage, when the foregut is formed by the inward folding of the splanchnopleure (**Figure 12.4**; see also Figure 12.2A, views i and ii). This movement places the two cardiac tubes together. The two endocardial tubes lie within the common tube for a short time, but eventually these two tubes also fuse. The bilateral origin of the heart can be demonstrated by surgically preventing the merger of the lateral plate mesoderm (Gräper 1907; DeHaan 1959). This manipulation results in a condition called **cardia bifida**, in which two separate hearts form, one on each side of the body (**Figure 12.5A**).

In the zebrafish, the heart precursor cells migrate actively from the lateral edges toward the midline. Several mutations affecting endoderm differentiation disrupt this process, indicating that, as in the chick, the endoderm is critical for cardiac precursor specification and migration. The *faust* gene, which encodes the GATA5 protein, is expressed in the endoderm and is required for the migration of cardiac precursor cells to the midline and also for their division and specification. It appears to be important in the pathway leading to activation of the *Nkx2-5* gene in the cardiac precursor cells (Reiter et al. 1999). Another particularly interesting zebrafish mutation is *miles apart*. Its phenotype is limited to cardiac precursor migration to the midline and resembles the cardia bifida seen in experimentally manipulated chick embryos (**Figure 12.5B,C**). The *miles apart* gene encodes a receptor for a cell surface sphingolipid molecule, and it is expressed in the endoderm on either side of the midline (Kupperman et al. 2000).

In mice, cardia bifida can also be produced by mutations of genes that are expressed in the endoderm. One of these genes, *Foxp4*, encodes a transcription factor expressed in the early foregut cells along the pathway the cardiac precursors travel toward the midline. In these mutants, each heart primordia develops separately, and the embryonic mouse contains two hearts, one on each side of the body (**Figure 12.5D,E**; Li et al. 2004).

Lash 1986). When these cells reach the lateral walls of the anterior gut tube, migration ceases. The directionality of this migration appears to be provided by the foregut endoderm. If the cardiac region endoderm is rotated with respect to the rest of the embryo, migration of the cardiogenic mesoderm cells is reversed. It is thought that the endodermal component responsible for this movement is an anterior-to-posterior concentration gradient of fibronectin. Antibodies against fibronectin stop the migration, while antibodies against other extracellular matrix components do not (Linask and Lash 1988).

In the chick, the two cardiac fields do not meet until after there is a 120-degree rotation of the field with respect

(A)

(B)

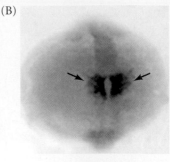

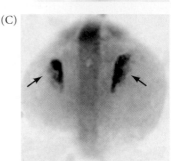

(C)

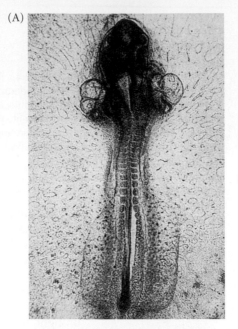

FIGURE 12.5 Migration of heart primordia. (A) Cardia bifida (two hearts) in a chick embryo, induced by surgically cutting the ventral midline, thereby preventing the two heart primordia from fusing. (B) Wild-type zebrafish and (C) *miles apart* mutant, stained with probes for the cardiac myosin light chain. There is a lack of migration in the *miles apart* mutant. (D) Mouse heart stained with antisense RNA probe to ventricular myosin shows fusion of the heart primordia in a wild-type 12.5-day embryo. (E) Cardia bifida in a *Foxp4*-deficient mouse embryo. Interestingly, each of these hearts has ventricles and atria, and they both loop and form all four chambers with normal left-right asymmetry. (A courtesy of R. L. DeHaan; B,C from Kupperman et al. 2000, courtesy of Y. R. Didier; D,E from Li et al. 2004, courtesy of E. E. Morrisey.)

(D)

(E)

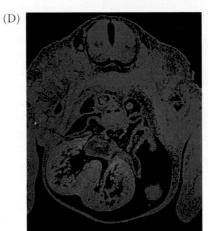

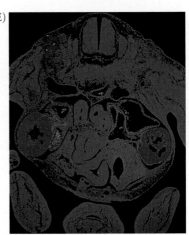

Determination of anterior and posterior cardiac domains

As the cardiac precursor cells migrate, the posterior region becomes exposed to increasingly higher concentrations of retinoic acid (RA) produced by the posterior mesoderm (see Figure 8.13). RA is critical in specifying the posterior cardiac precursor cells into becoming the inflow, or "venous," portions of the heart—the sinus venosus and atria. Originally, these fates are not fixed, as transplantation or rotation experiments show that these precursor cells can regulate and differentiate in accordance with a new environment. But once the posterior cardiac precursors enter the realm of active RA synthesis, they express the gene for retinaldehyde dehydrogenase; they then can produce their own RA, and their posterior fate becomes committed (**Figure 12.6A,B**; Simões-Costa et al. 2005). This ability of retinoic acid to specify and commit heart precursor to become atria explains its teratogenic effects on heart development, wherein exposure of vertebrate embryos to RA can cause expansion of atrial tissues at the expense of ventricular tissues (Stainier and Fishman 1992; Hochgreb et al. 2003).

Interestingly, this relationship between heart development and retinoic acid appears to be conserved, even in embryos that form their hearts in very different ways. In tunicates, the heart is a single-layered U-shaped tube. At the 110-cell stage, the heart lineage is represented by a pair of mesodermal cells near the vegetal pole, the B7.5 blastomeres. These cells give rise to the anteriormost tail cells and the heart precursor cells (trunk ventral cells). When the B7.5 blastomeres split into the heart lineage and the anterior tail lineage, the heart precursor cells express the tunicate homologue of the *Nkx2-5* gene and migrate ventrally into the "head" of the developing tadpole (Davidson and Levine 2003; Simões-Costa et al. 2005). The ante-

rior tail cells do not migrate, but they express retinaldehyde dehydrogenase and initiate a retinoid acid gradient (**Figure 12.6C**).

Heart cell differentiation

One of the most important new discoveries of cardiac development has been the demonstration that all the different cells of the heart—the ventricular myocytes, the atrial myocytes, the smooth muscles that generate the venous and arterial vasculature, the endothelial lining of the heart and valves, and the epicardium that forms an envelope for the heart—are derived from the same progenitor cell type (Kattman et al. 2006; Moretti et al. 2006; Wu et al. 2006). The cardiac fields contain multipotent progenitor cells. Indeed, there appears to be an early progenitor cell population that bears the responsibility for forming the entire circulatory system. Under one set of influences, its descendents become **hemangioblasts**, those cells that form blood vessels and blood cells; under other conditions (i.e., the conditions in the cardiac fields), they form the **multipotent cardiac precursor cells** (Figure 12.7; Anton et al. 2007). Several investigators have proposed slightly different pathways for gen-

(A) Chick, stage 8

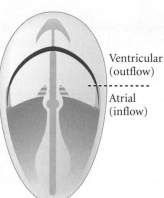

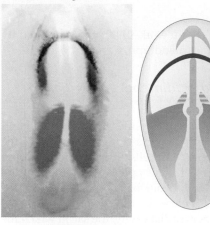

Ventricular (outflow)

Atrial (inflow)

(B) Mouse, 8 days

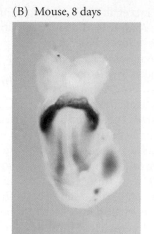

FIGURE 12.6 Double in situ hybridization for the expression of *RADH2* (orange), which encodes the retinoic acid-synthesizing enzyme retinaldehyde dehydrogenase-2; and *Tbx5* (purples), a marker for the early heart fields. In the developmental stages seen here, the heart precursor cells are exposed to progressively increasing amounts of retinoic acid. (A) Chick, stage 8 (26–29 hours). (B) Mouse, 8 days. (C) *Ciona* (a tunicate) at the larval stage. (From Simões-Costa et al. 2005, courtesy of J. Xavier-Neto.)

(C) *Ciona*, larval stage

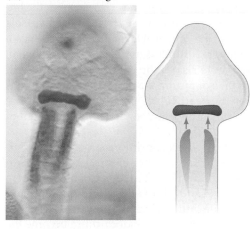

erating these cells, but the differences may be caused by differences in assaying and culturing procedures, and it would not be surprising were we to find that each of these pathways is actually used in different situations.

INITIAL CELL DIFFERENTIATION Several genes are expressed very early during heart development (see Figure 12.2C). The **GATA4** transcription factor is first seen in the precardiac cells of chicks and mice as these cells emerge from the primitive streak. GATA4 expression is retained in all the cells making up the heart fields on both sides of the embryo. This transcription factor is necessary for the activation of numerous heart-specific genes, and it also activates expression of the gene for N-cadherin, a protein that is critical for the fusion of the two heart rudiments into one tube (H. Zhang et al. 2003).

The BMP pathway is critical in inducing the synthesis of the **Nkx2-5** transcription factor in the migrating cardiogenic

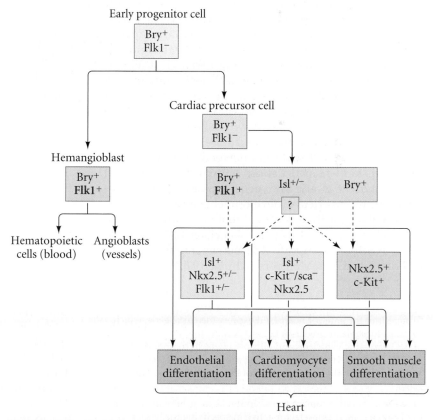

FIGURE 12.7 Model for early cardiovascular lineages. The splanchnic mesoderm gives rise to two lineages that have Flk1 (a VEGF receptor) on their cell membranes. The earlier population gives rise to the hemangioblasts (precursors to blood cells and blood vessels), while the later population gives rise to the cardiac (heart) progenitor cells. This latter population in turn gives rise to a variety of cell types whose relationships are still obscure. However, all the cell types of the heart can be traced back to the cardiac progenitor cells. (After Anton 2007.)

mesoderm (Komuro and Izumo 1993; Lints et al. 1993; Sugi and Lough 1994; Schultheiss et al. 1995; Andrée et al. 1998). Nkx2-5 is crucial in instructing the mesoderm to become heart tissue, and it also activates the synthesis of numerous cardiac transcription factors* (especially members of the T-box, GATA, and Mef2 families). GATA4 can also induce Nkx2-5, so activating one will activate the other and convert mesoderm into cardiac precursor cells. Working together, these transcription factors activate the expression of genes encoding cardiac muscle-specific proteins (such as cardiac actin, atrial natriuretic factor, and the α-myosin heavy chains) (Sepulveda et al. 1998; Kawamura et al. 2005). The Nkx2-5 transcription factor can also work as a repressor, and if the *Nkx2-5* gene is specifically knocked out in those cells destined to become ventricles, these chambers express BMP10, causing massive overgrowth of the ventricles such that the ventricular chambers fill with muscle cells (Pashmforoush et al. 2004). Nkx2-5 also can downregulate BMPs, and in early heart cell development, it limits the number of heart cell precursors there can be in the cardiac fields (Prall et al. 2007).

The Tbx transcription factors are also critical in heart development (Plageman and Yutzey 2004). The **Tbx5** gene is expressed in the cardiac fields on both sides of the chick and mouse embryos (see Figure 12.6). In these early cells, Tbx5 acts with GATA4 and Nkx2-5 to activate numerous genes involved in heart specification.† Later, Tbx5 protein expression becomes restricted to the atria and left ventricle. The ventricular septum (the wall separating the left and right ventricles) is formed at the boundary between those cells expressing Tbx5 and those that do not. Tbx5 works antagonistically to Tbx20, which becomes expressed in the right ventricle. When the Tbx5 expression domain is ectopically expanded, the location of the ventricular septum shifts to this new location. Moreover, a conditional knockout of the mouse *Tbx5* gene—specifically inactivating it during ventricular development—leads to the formation of a lizardlike ventrical that lacks any septation (Takeuchi et al. 2003; Koshiba-Takeuchi et al. 2009). Thus, Tbx5 is extremely important in the separation of the left and right ventricles. Mutations in human *TBX5* cause Holt-

Oram syndrome (Bruneau et al. 1999), characterized by abnormalities of the heart and upper limbs.

Fusion of the heart rudiments and initial heartbeats

Cell differentiation occurs independently in the two heart-forming primordia. As they migrate toward each other, the ventral splanchnic mesoderm cells of the primordia begin to express N-cadherin on their apices, sort out from the somatic mesoderm cells, and join together to form an epithelial layer. This joining of the somatic mesoderm will lead to the formation of the pericardial cavity, the sac in which the heart is formed (Linask 1992). A small population of splanchnic mesoderm then downregulates N-cadherin and delaminates from the epithelium to form the endocardium, the lining of the heart that is continuous with the blood vessels (see Figure 12.4C,D). The epithelial layer of splanchnic mesoderm forms the **myocardium** (Manasek 1968; Linask et al. 1997), which will give rise to the heart muscles that will pump for the lifetime of the organism. The endocardial cells produce many of the heart valves, secrete the proteins that regulate myocardial growth, and regulate the placement of nervous tissue in the heart.

The fusion of the two heart primordia occurs at about 29 hours in chick development (see Figure 12.4C,D) and at 3 weeks in human gestation. The myocardia unite to form a single tube. The two endocardia lie within this common tube for a short while, but eventually they also fuse. The unfused posterior portions of the endocardium become the openings of the **vitelline veins** into the heart (see Figure 12.8A). These veins will carry nutrients from the yolk sac into the **sinus venosus**, the posterior region where the two major veins fuse. The blood then passes through a valve-like flap into the atrial region of the heart. Contractions of the truncus arteriosus speed the blood into the aorta. In addition to the heart cells derived from the paired heart fields, cells from the rostral paraxial mesoderm also appear to form portions of the truncus (de la Cruz and Markwald 1998; Abu-Issa et al. 2004; Harel et al. 2009).

Pulsations of the chick heart begin while the paired primordia are still fusing. Heart muscle cells develop an inherent ability to contract, and isolated heart cells from 7-day rat or chick embryos will continue to beat when placed in petri dishes (Harary and Farley 1963; DeHaan 1967; Imanaka-Yoshida et al. 1998). The pulsations are made possible by the appearance of the sodium-calcium exchange pump in the muscle cell membrane; inhibiting this channel's function prevents the heartbeat from starting (Wakimoto et al. 2000; Linask et al. 2001). Eventually, the rhythmicity of the heartbeat becomes coordinated by the sinus venosus. The electric impulses generated here initiate waves of muscle contraction through the tubular heart. In this way, the heart can pump blood even before its intricate system of valves has been completed. Studies of mutations of cardiac cell calcium channels have implicated these channels in the

---

*The cells of the anterior heart field express the *Foxh1* gene, which commits these heart precursor cells to become the "arterial" portion of the heart—the ventricles and outflow tracts. In combination with the Nkx2-5 transcription factor, Foxh1 activates those genes (such as *Mef2c*) that begin the differentiation of the ventricles (von Both et al. 2004).

†The Nkx2-5 homeodomain transcription factor is homologous to the Tinman transcription factor active in specifying the heart tube of *Drosophila*. Moreover, neither Tinman nor Nkx2-5 alone is sufficient to complete heart development in their respective organisms. Mice lacking the *Nkx2-5* gene start heart tube formation, but the tube fails to thicken or to loop (Lyons et al. 1995). Humans with a mutation in one of their *NKX2-5* genes have congenital heart malformations (Schott et al. 1998).

pacemaker function (Rottbauer et al. 2001; Zhang et al. 2002).

Looping and formation of heart chambers

In 3-day chick embryos and 4-week human embryos, the heart is a two-chambered tube, with one atrium to receive blood and one ventricle to pump blood. In the chick embryo, the unaided eye can see the remarkable cycle of blood entering the lower chamber and being pumped out through the aorta. Looping of the heart thus converts the original anterior-posterior polarity of the heart tube into the right-left polarity seen in the adult organism. When this looping is completed, the portion of the heart tube destined to become the atria lies anterior to the portion that will become the ventricles (Figure 12.8).

The direction of heart looping is dependent on the left-right patterning proteins (Nodal and Pitx2) discussed in Chapter 8. Within the heart primordium, Nkx2-5 regulates the Hand1 and Hand2 transcription factors. Both Hand proteins appear to be synthesized throughout the early heart tube, but as looping commences, Hand1 becomes

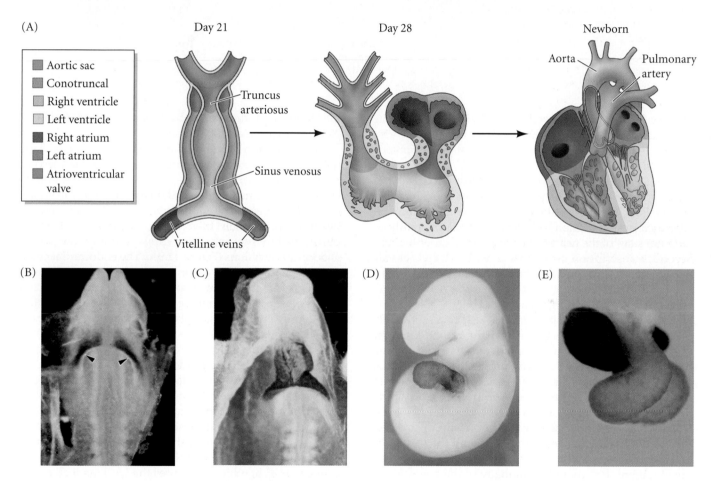

FIGURE 12.8 Cardiac looping and chamber formation. (A) Schematic diagram of cardiac morphogenesis in humans. On day 21, the heart is a single-chambered tube. Regional specification of the tube is shown by the different colors. By day 28, cardiac looping has occurred, placing the presumptive atria anterior to the presumptive ventricles. (B,C) *Xin* expression in the fusion of left and right heart primordia of a chick. The cells fated to form the myocardium are shown by staining for the *Xin* message, whose protein product is essential for the looping of the heart tube. (B) Stage-9 chick neurula, in which expression of *Xin* (purple) is seen in the two symmetrical heart-forming fields (arrows). (C) Stage-10 chick embryo, showing fusion of the two heart-forming regions prior to looping. (D,E) Specification of the atria and ventricles occurs even before heart looping. The atria and ventricles of the mouse embryo have separate types of myosin proteins, which allows them to be differentially stained. In these photographs, atrial myosin is stained blue and ventricular myosin is stained orange. (D) In the tubular heart (prior to looping), the two myosins (and their respective stains) overlap at the atrioventricular channel joining the future regions of the heart. (E) After looping, the blue stain is seen in the definitive atria and inflow tract, while the orange stain is seen in the ventricles. The unstained region above the ventricles is the truncus arteriosus. Derived primarily from the neural crest, the truncus arteriosus becomes separated into the aorta and pulmonary arteries. (A after Srivastava and Olson 2000; B,C from Wang et al. 1999, courtesy of J. J.-C. Lin; D,E from Xavier-Neto et al. 1999, courtesy of N. Rosenthal.)

FIGURE 12.9 Formation of the chambers and valves of the heart. (A) Diagrammatic cross section of the human heart at 4.5 weeks. The atrial and ventricular septa are growing toward the endocardial cushions. (B) Cross section of the human heart during the third month of gestation. Blood can cross from the right side of the heart to the left side through openings in the primary and secondary atrial septa. (After Larsen 1993.)

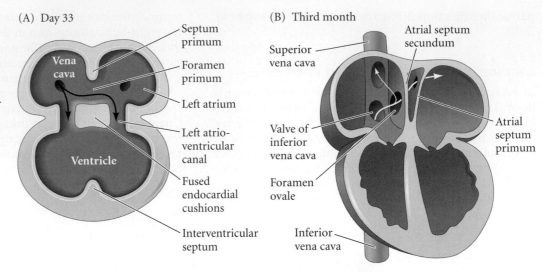

(A) Day 33

Septum primum
Foramen primum
Left atrium
Left atrioventricular canal
Fused endocardial cushions
Interventricular septum

Vena cava
Ventricle

(B) Third month

Atrial septum secundum
Superior vena cava
Valve of inferior vena cava
Foramen ovale
Inferior vena cava
Atrial septum primum

restricted to the future left ventricle and Hand2 to the right. Without these proteins, looping is abnormal, since the ventricles fail to form properly* (Srivastava et al. 1995; Biben and Harvey 1997).

Cytoskeletal proteins and their adhesion to the extracellular matrices are very important in these turning events. First, the myosin-like matrix protein **flectin** regulates the physical tension of the heart tissues differently on different sides of the heart (Tsuda et al. 1996; Lu et al. 2008). Second, transcription factors Nkx2-5 and Mef2C activate the *Xin* gene (Chinese for "heart"), whose product binds to actin microfilaments and mediates cytoskeletal changes essential for heart looping (Wang et al. 1999; Grosskurth et al. 2008). Third, metalloproteinases—especially metalloproteinase-2 (MMP2)—are critical for remodeling the cytoskeleton. If *MMP2* gene expression is blocked by antibodies or pharmaceutical inhibitors, the extracellular matrix fails to change, the asymmetric cell divisions (which cause the left side to grow faster than the right) fail to occur, and heart looping stops (Linask et al. 2005).

Differential cell division is also important in structuring the right and left sides of the heart, and the regulation of differential growth remains a major question in heart development. It is thought that many of the genes found in one particular area of the heart may be responsible for differential growth. For instance, *Tbx5* is expressed in the left ventricle and atrium, while *Wnt11* is expressed in the right side of the heart. A series of regulators (including Tbx2 and Tbx20) probably control growth by activating or interfering with these chamber-specific factors (Cai et al. 2005; Stennard et al. 2005).

The formation of the heart valves—four leaflike flaps that must open and shut without failure once each second for the duration of our life—is just starting to be understood. In mammals, **endocardial cushions** form from the endocardium and divide the tube into right and left atrioventricular channels. Meanwhile, the primitive atrium is partitioned by two **septa** that grow ventrally toward the endocardial cushions (**Figure 12.9A**). The endocardial cushions serve as a valve during early heart development (Lamers et al. 1988), but as the heart enlarges, specialized valves develop to prevent the return of blood into the atria and to prevent the mixing of bloods from the two sides of the heart (**Figure 12.9B**). These valves begin to form when cells from the myocardium produce a factor that causes cells from the adjacent endocardium to detach and enter the hyaluronate-rich "cardiac jelly" extracellular matrix between the two layers (Markwald et al. 1977; Potts et al. 1991). In mammals, BMP2 appears to be necessary for inducing this epithelial-mesenchymal transition and for forming the endocardial cushions from cardiac myocytes (Ma et al. 2005; Rivera-Feliciano and Tabin 2006). BMP2 induces TGF-β, which promotes the EMT, as well as the enzymes that synthesize the hyaluronic acid that separates the cells and becomes a major part of the cardiac jelly (Shirai et al. 2009).

The septa between the atria, however, have openings in them, so blood can still cross from one side of the heart into the other. This crossing of blood is needed for the survival of the fetus before circulation to functional lungs has been established. Upon the first breath, the septal openings close and the left and right circulatory loops are established. With the formation of the septa (which usually occurs in the seventh week of human development), the heart is a four-chambered structure with the pulmonary artery connected to the right ventricle and the aorta connected to the left ventricle.

---

*Zebrafish, with only one ventricle, have only one type of Hand protein. When the gene encoding this protein is mutated, the entire ventricular portion of the heart fails to form (Srivastava and Olson 2000). Nongenetic agents are also critical in normal zebrafish heart formation. In the absence of high-pressure blood flow, heart looping, chamber formation, and valve development are impaired (Hove et al. 2003).

Redirecting Blood Flow in the Newborn Mammal

Although the developing mammalian fetus shares with the adult the need to get oxygen and nutrients to its tissues, the physiology of the fetus differs dramatically from that of the adult. Chief among the differences is the fetus's lack of functional lungs and intestines. All of its oxygen and nutrients must come from the placenta. This observation raises two questions. First, how does the fetus obtain oxygen and nutrients from maternal blood? And second, how is blood circulation redirected to the lungs once the umbilical cord is cut and breathing becomes necessary?

Human embryonic circulation

The human embryonic circulatory system is a modification of that used in other amniotes, such as birds and reptiles. The circulatory system to and from the chick embryo and yolk sac is shown in **Figure 12.10A**. Blood pumped through the dorsal aorta passes over the aortic arches and down into the embryo. Some of this blood leaves the embryo through the vitelline arteries and enters the yolk sac. Nutrients and oxygen are absorbed from the yolk, and the blood returns through the vitelline veins to re-enter the heart through the sinus venosus.

Mammalian embryos obtain food and oxygen from the placenta. Thus, although the embryo has vessels homologous to the vitelline veins, the main supply of food and oxygen comes from the umbilical vein, which unites the embryo with the placenta

Figure 12.10 Embryonic circulatory systems. (A) Circulatory system of a 2-day chick embryo. The sinus terminalis is the outer limit of the circulatory system and the site of blood cell generation. (B) Circulatory system of a 4-week human embryo. Although at this stage all the major blood vessels are paired left and right, only the right vessels are shown here. In both views, arteries are shown in red, veins in blue. (A adapted from Popoff 1894; B after Carlson 1981.)

(Figure 12.10B). This vein, which brings oxygenated, food-laden blood into the embryo, is derived from what would be the right vitelline vein in birds. The umbilical artery, carrying wastes to the placenta, is derived from what would have become the allantoic artery of the chick. It extends from

the caudal portion of the aorta and proceeds along the allantois and then out to the placenta.

Fetal hemoglobin

The solution to the fetus's problem of getting oxygen from its mother's blood

(Continued on next page)

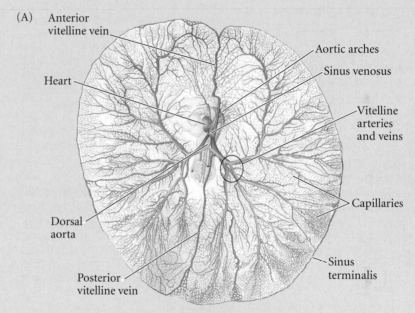

(A) Anterior vitelline vein · Heart · Dorsal aorta · Posterior vitelline vein · Aortic arches · Sinus venosus · Vitelline arteries and veins · Capillaries · Sinus terminalis

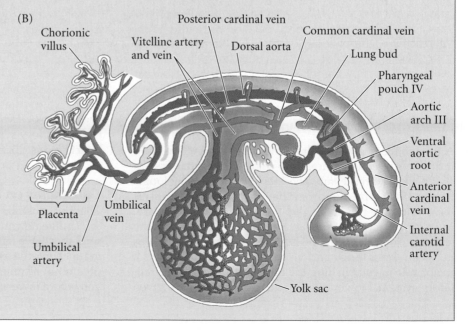

(B) Chorionic villus · Vitelline artery and vein · Posterior cardinal vein · Dorsal aorta · Common cardinal vein · Lung bud · Pharyngeal pouch IV · Aortic arch III · Ventral aortic root · Anterior cardinal vein · Internal carotid artery · Placenta · Umbilical vein · Umbilical artery · Yolk sac

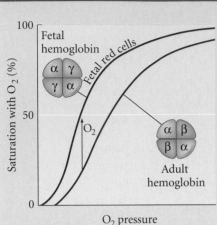

Figure 12.11 Adult and fetal hemoglobin molecules differ in their globin subunits. The fetal γ-chain binds diphosphoglycerate less avidly than does the adult β-chain. Consequently, fetal hemoglobin can bind oxygen more efficiently than can adult hemoglobin. In the placenta, there is a net flow (arrow) of oxygen from the mother's blood (which gives up oxygen to the tissues at lower oxygen pressures) to the fetal blood (which at the same pressure is still taking up oxygen).

involves the development of a specialized fetal hemoglobin. The hemoglobin in fetal red blood cells differs slightly from that in adult corpuscles. Two of the four peptides—the alpha (α) chains—that make up fetal and adult hemoglobin chains are identical, but adult hemoglobin has two beta (β) chains, while the fetus has two gamma (γ) chains (**Figure 12.11**). The β-chains bind the natural regulator diphosphoglycerate, which assists in the unloading of oxygen. The γ-chain proteins do not bind diphosphoglycerate as well, and therefore have a higher affinity for oxygen. In the low-oxygen environment of the placenta, oxygen is released from adult hemoglobin. In this same environment, fetal hemoglobin does not release oxygen, but binds it. This small difference in oxygen affinity mediates the transfer of oxygen from the mother to the fetus. In the fetus, the

myoglobin of the fetal muscles has an even higher affinity for oxygen, so oxygen molecules pass from fetal hemoglobin to the fetal muscles. Fetal hemoglobin is not deleterious to the newborn, and in humans, the replacement of fetal hemoglobin-containing blood cells with adult hemoglobin-containing blood cells is not complete until about 6 months after birth. (The molecular basis for this switch in globins is discussed in Chapter 2.)

From fetal to newborn circulation

Once the fetus is no longer obtaining its oxygen from its mother, how does it restructure its circulation to get oxygen from its own lungs? During fetal development, an opening—the **ductus arteriosus**—diverts blood from the pulmonary artery into the aorta (and thus to the placenta). Because blood does not return from the pulmonary vein in the fetus, the developing mammal has to have some other way of getting

blood into the left ventricle to be pumped. This is accomplished by the **foramen ovale**, an opening in the septum separating the right and left atria. Blood can enter the right atrium, pass through the foramen into the left atrium, and then enter the left ventricle (**Figure 12.12**). When the first breath is drawn, blood pressure in the left side of the heart increases. This pressure closes the septa over the foramen ovale, thereby separating the pulmonary and systemic circulations. Moreover, the decrease in prostaglandins experienced by the newborn cause the muscles surrounding the ductus arteriosus to close that opening as well (Nguyen et al. 1997). Thus, when breathing begins, the respiratory circulation is shunted from the placenta to the lungs.

In some infants, the septa fail to close and the foramen ovale is left open. Indeed, atrial and ventricular septum defects are among the most common congenital anomalies. Usually the atrial opening is so small that there are no physical symptoms, and the foramen eventually closes. If it does not close completely, however, and the secondary septum fails to form, the atrial septal opening may cause enlargement of the right side of the heart, which can lead to heart failure in early adulthood. This fine-tuning of septal growth is controlled by miR-1-2, a microRNA that regulates translation of several proteins involved in cardiac muscle growth and electrical conduction (Zhao et al. 2007).

Formation of Blood Vessels

Although the heart is the first functional organ of the body, it does not even begin to pump until the vascular system of the embryo has established its first circulatory loops. Rather than sprouting from the heart, the blood vessels form independently, linking up to the heart soon afterward. Everyone's circulatory system is different, since the genome cannot encode the intricate series of connections between the arteries and veins. Indeed, chance plays a major role in establishing the microanatomy of the circulatory system. However, all circulatory systems in a given

species look very much alike, because the development of the circulatory system is severely constrained by physiological, physical, and evolutionary parameters.

Constraints on the construction of blood vessels

The first constraint on vascular development is *physiological*. Unlike new machines, which do not need to function until they have left the assembly line, new organisms have to function even as they develop. The embryonic cells must obtain nourishment before there is an intestine, use oxygen before there are lungs, and excrete wastes before there

SIDELIGHTS & SPECULATIONS (Continued)

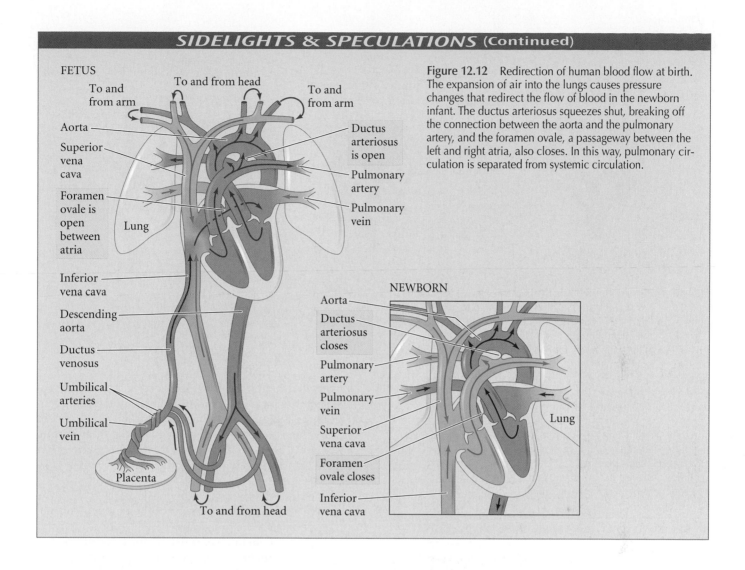

Figure 12.12 Redirection of human blood flow at birth. The expansion of air into the lungs causes pressure changes that redirect the flow of blood in the newborn infant. The ductus arteriosus squeezes shut, breaking off the connection between the aorta and the pulmonary artery, and the foramen ovale, a passageway between the left and right atria, also closes. In this way, pulmonary circulation is separated from systemic circulation.

are kidneys. All these functions are mediated through the embryonic circulatory system. Therefore, the circulatory physiology of the developing embryo must differ from that of the adult organism. Food is absorbed not through the intestine, but from either the yolk or the placenta, and respiration is conducted not through the gills or lungs, but through the chorionic or allantoic membranes. The major embryonic blood vessels must be constructed to serve these extraembryonic structures.

The second constraint is *evolutionary*. The mammalian embryo extends blood vessels to the yolk sac even though there is no yolk inside (see Figure 12.10). Moreover, the blood leaving the heart via the truncus arteriosus passes through vessels that loop over the foregut to reach the dorsal aorta. Six pairs of these aortic arches loop over the pharynx (**Figure 12.13**). In primitive fish, these arches persist and enable the gills to oxygenate the blood through the gills. In adult birds and mammals, in which lungs oxygenate the blood, such a system makes little sense—but all

six pairs of aortic arches are formed in mammalian and avian embryos before the system eventually becomes simplified into a single aortic arch. Thus, even though our physiology does not require such a structure, our embryonic condition reflects our evolutionary history.

The third set of constraints is *physical*. According to the laws of fluid movement, the most effective transport of fluids is performed by large tubes. As the radius of a blood vessel gets smaller, resistance to flow increases as r^{-4} (Poiseuille's law). A blood vessel that is half as wide as another has a resistance to flow 16 times greater. However, diffusion of nutrients can take place only when blood flows slowly and has access to cell membranes. So here is a paradox: the constraints of diffusion mandate that vessels be small, while the laws of hydraulics mandate that vessels be large. This paradox has been solved by the evolution of circulatory systems with a hierarchy of vessel sizes (LaBarbera 1990). In dogs, for example, blood in the large vessels (aorta and vena cava) flows over 100 times faster than it

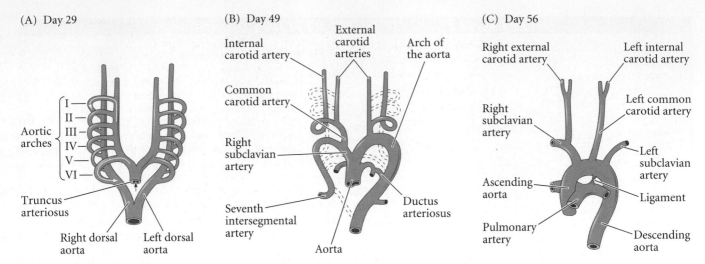

(A) Day 29

Aortic arches { I — II — III — IV — V — VI — }

Truncus arteriosus

Right dorsal aorta   Left dorsal aorta

(B) Day 49

Internal carotid artery

External carotid arteries

Arch of the aorta

Common carotid artery

Right subclavian artery

Seventh intersegmental artery

Ductus arteriosus

Aorta

(C) Day 56

Right external carotid artery

Left internal carotid artery

Right subclavian artery

Left common carotid artery

Ascending aorta

Left subclavian artery

Pulmonary artery

Ligament

Descending aorta

FIGURE 12.13 Aortic arches of the human embryo. (A) Originally, the truncus arteriosus pumps blood into the aorta, which branches on either side of the foregut. The six aortic arches take blood from the truncus arteriosus and allow it to flow into the dorsal aorta. (B) As development proceeds, arches begin to disintegrate or become modified (the dotted lines indicate degenerating structures). The first two pairs of arches completely disappear. The third arches become the internal carotid arteries. The right fourth arch becomes the right subclavian artery, and the left fourth arch becomes the major arch of the aorta. The fifth arches disintegrate. The right sixth arch disappears, but the left sixth arch gives rise to the pulmonary arteries and the ductus arteriosus (which will start closing on the first breath). (C) Eventually, the remaining arches are modified and the adult arterial system is formed. However, numerous variations of this scheme are found in the human population.

does in the capillaries. With a system of large vessels specialized for transport and small vessels specialized for diffusion (where the blood spends most of its time), nutrients and oxygen can reach the individual cells of the growing organism. This hierarchy is seen very early in development; it is already well established in the 3-day chick embryo.

But this is not the entire story. If fluid under constant pressure moves directly from a large-diameter tube into a small-diameter tube (as in a hose nozzle), the fluid velocity increases. The evolutionary solution to this problem was the emergence of many smaller vessels branching out from a larger one, making the collective cross-sectional area of all the smaller vessels greater than that of the larger vessel. Circulatory systems show a relationship (known as Murray's law) in which the cube of the radius of the parent vessel approximates the sum of the cubes of the radii of the smaller vessels. Computer models of blood vessel formation must take into account not only gene expression patterns but also the fluid dynamics of blood flow, if they are to show the branching and anastomosing of the arteries and veins (Gödde and Kurz 2001). The construction of any circulatory system negotiates among all of these physical, physiological, and evolutionary constraints.

Vasculogenesis: The initial formation of blood vessels

The development of blood vessels occurs by two temporally separate processes: **vasculogenesis** and **angiogenesis** (Figure 12.14). During vasculogenesis, a network of blood vessels is created de novo from the lateral plate mesoderm. During angiogenesis, this primary network is remodeled and pruned into a distinct capillary bed, arteries, and veins.

In the first phase of vasculogenesis, cells leaving the primitive streak in the posterior of the embryo become hemangioblasts,* the precursors of both blood cells and blood vessels. Labeling zebrafish embryos with fluorescent probes to make single-cell fate maps confirms that hemangioblasts are the common progenitor for both the hematopoietic (blood) and endothelial (vascular) lineages in zebrafish. This population of bipotential progenitor cells is found only in the ventral mesoderm, the region that had been known to produce these two cell types. The pathway whereby ventral mesoderm cells differentiate into hemangioblasts appears to be induced by the *Cdx4* gene, while the determination of whether the hemangioblast becomes a blood cell precursor or a blood vessel precursor is regulated by the Notch signaling pathway. Notch signaling increases the conversion of hemangioblasts into blood cell precursors, while reduced amounts of Notch cause hemangioblasts to become endothelial (Vogeli et al. 2006; Hart et al. 2007; Lee et al. 2009).

---

*The prefixes *hem-* and *hemato-* refer to blood (as in hemoglobin). Similarly, the prefix *angio-* refers to blood vessels. The suffix *-blast* denotes a rapidly dividing cell, usually a stem cell. The suffixes *-poesis* and *-poietic* refer to generation or formation (*poeisis* is also the root of the word *poetry*). Thus, hematopoietic stem cells are those cells that generate the different types of blood cells. The Latin suffix *-genesis* (as in *angiogenesis*) means the same as the Greek *-poiesis*.

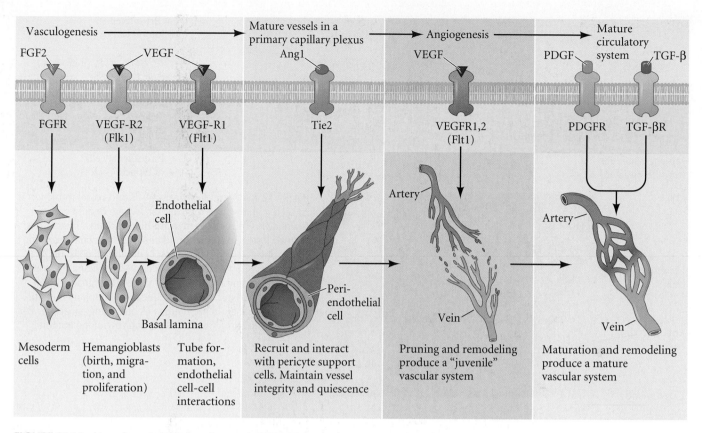

FIGURE 12.14 Vasculogenesis and angiogenesis. Vasculogenesis involves the formation of blood islands and the construction of capillary networks from them. Angiogenesis involves the formation of new blood vessels by remodeling and building on older ones. Angiogenesis finishes the circulatory connections begun by vasculogenesis. The major paracrine factors involved in each step are shown at the top of the diagram, and their receptors (on the vessel-forming cells) are shown beneath them. (After Hanahan 1997 and Risau 1997.)

In amniotes, however, there is much more controversy about hemangioblasts, and there may be more than one pathway by which to generate blood vessels and blood cells (Ueno and Weissman 2006; Weng et al. 2007). Hemangioblasts in the splanchnic mesoderm condense into aggregations that are often called **blood islands*** (Shalaby et al. 1997; Huber et al. 2004). It is generally thought that the inner cells of these blood islands become blood progenitor cells, while the outer cells become **angioblasts**, the progenitors of blood vessels. In the second phase of vasculogenesis, the angioblasts multiply and differentiate into **endothelial cells**, which form the lining of the blood vessels. In the third phase, the endothelial cells form tubes and connect to form the **primary capillary plexus**, a network of capillaries.

SITES OF VASCULOGENESIS In amniotes, formation of the primary vascular networks occurs in two distinct and independent regions. First, **extraembryonic vasculogenesis** occurs in the blood islands of the yolk sac. These are the blood islands formed by the hemangioblasts, and they give rise to the early vasculature needed to feed the embryo and also to a red blood cell population that functions in the early embryo (**Figure 12.15A**). Second, **intraembryonic vasculogenesis** forms the dorsal aorta, and vessels from this large vessel connect with capillary networks that form from mesodermal cells within each organ.

The aggregation of endothelial-forming cells in the yolk sac is a critical step in amniote development, for the blood islands that line the yolk sac produce the veins that bring nutrients to the embryo and transport gases to and from the sites of respiratory exchange (**Figure 12.15B**). In birds, these vessels are called the vitelline veins; in mammals, they are often called the **omphalomesenteric (umbilical) veins**. In the chick, blood islands are first seen in the area opaca, when the primitive streak is at its fullest extent (Pardanaud et al. 1987). They form cords of hemangioblasts, which soon become hollowed out and become the flat endothelial cells lining the vessels (while the central cells give rise to blood cells). As the blood islands grow, they

---

*Again, the endoderm plays a major role in lateral plate mesoderm specification. Here, the visceral endoderm of the splanchnopleure interacts with the yolk sac mesoderm to induce the blood islands. The endoderm is probably secreting Indian hedgehog, a paracrine factor that activates BMP4 expression in the mesoderm. BMP4 expression feeds back on the mesoderm itself, causing it to form hemangioblasts (Baron 2001).

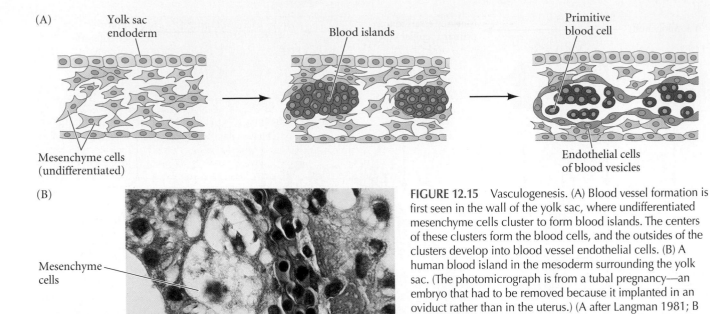

(A) Yolk sac endoderm | Blood islands | Primitive blood cell

Mesenchyme cells (undifferentiated)

Endothelial cells of blood vesicles

(B)

Mesenchyme cells

Yolk sac endoderm

Blood island

FIGURE 12.15 Vasculogenesis. (A) Blood vessel formation is first seen in the wall of the yolk sac, where undifferentiated mesenchyme cells cluster to form blood islands. The centers of these clusters form the blood cells, and the outsides of the clusters develop into blood vessel endothelial cells. (B) A human blood island in the mesoderm surrounding the yolk sac. (The photomicrograph is from a tubal pregnancy—an embryo that had to be removed because it implanted in an oviduct rather than in the uterus.) (A after Langman 1981; B from Katayama and Kayano 1999, courtesy of the authors.)

eventually merge to form the capillary network draining into the two vitelline veins, which bring food and blood cells to the newly formed heart.

In some vertebrates, intraembryonic vasculogenesis occurs in a manner that was totally unexpected. When Kamei and colleagues (2006) took high-resolution time-lapse movies of zebrafish endothelial formation in culture and in transparent embryos, they discovered that the cells get together in groups and form the lumen of the blood vessel by the fusion of intracellular vacuoles. Fluid-filled vacuoles, formed by endocytosis, coalesce within the cells to form larger fluid-filled vacuoles. These larger vacuoles then fuse with the cell membrane at the site where the cells meet. The result is a fluid-filled lumen between the cells (**Figure 12.16A**). It is also possible that the fluid-filled vacuoles fuse within a single cell to create a lumen made from different regions of the same cell (**Figure 12.16B**). This latter mechanism of forming vascular tubes through intracellular endocytosis followed by exocytosis to form a lumen has also been seen in the formation of human umbilical cord veins in vitro, as well as in the formation of tubes in *Drosophila* and *C. elegans*. So this may be a standard way to form blood vessels.

The intraembryonic vascular networks usually arise from individual angioblast progenitor cells in the mesoderm surrounding a developing organ. These cells do not appear to be associated with blood cell formation (Noden 1989; Pardanaud et al. 1989; Risau 1995). It is important to realize that these intraembryonic capillary networks arise in or around the organ itself and are not extensions of larg-er vessels. Indeed, in some cases the developing organ produces paracrine factors that induce blood vessels to form only in its own mesenchyme (Auerbach et al. 1985; LeCouter et al. 2001). This allows each capillary network to have its own specific properties. For instance, the capillary network that forms in the brain is modified by Wnt proteins to produce the extracellular matrices of the blood-brain barrier and to express the glucose transporter proteins that enable the brain to consume 25% of the body's oxygen (Stenman et al. 2008). In limbs, the chondrogenic nodules that form the bones produce VEGF paracrine factors to generate blood vessels in the surrounding mesenchyme (Eshkar-Oren et al. 2009).

GROWTH FACTORS AND VASCULOGENESIS Three growth factors may be responsible for initiating vasculogenesis (see Figure 12.14). One of these, **basic fibroblast growth factor (Fgf2)**, is required for the generation of hemangioblasts from the splanchnic mesoderm. When cells from quail blastodiscs are dissociated in culture, they do not form blood islands or endothelial cells. However, when these cells are cultured in Fgf2, blood islands emerge and form endothelial cells (Flamme and Risau 1992). Fgf2 is synthesized in the chick embryonic chorioallantoic membrane and is responsible for the vascularization of this tissue (Ribatti et al. 1995).

The second family of proteins involved in vasculogenesis is the **vascular endothelial growth factors (VEGFs)**. This family includes several VEGFs, as well as placental growth factor (PlGF), which directs the expansive growth

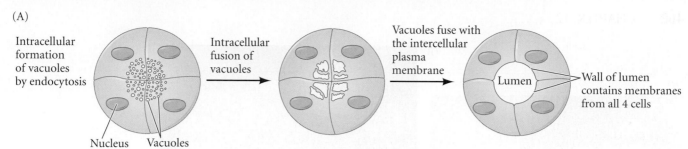

(A)

Intracellular formation of vacuoles by endocytosis

Intracellular fusion of vacuoles

Vacuoles fuse with the intercellular plasma membrane

Lumen

Wall of lumen contains membranes from all 4 cells

Nucleus   Vacuoles

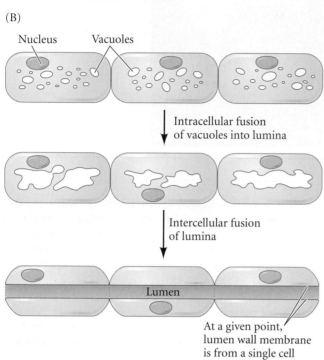

(B)

Nucleus   Vacuoles

Intracellular fusion of vacuoles into lumina

Intercellular fusion of lumina

Lumen

At a given point, lumen wall membrane is from a single cell

FIGURE 12.16 The lumen, or central space, in the vascular tubes is formed by the fusion of intracellular vacuoles. (A) In this scebairui, vacuoles form within the endothelial cells, apparently by endocytosis. These vaculoes merge with other vacuoles to form larger vacuoles. Large vacuoles then fuse with the cell membrane at the point where the cells come together, forming the lumen. (B) Alternatively, the vacuoles may form intracellular lumina within each cell; these individual lumina then fuse such that in any single portion of the final lumen, the lining is made up of membranes from the same cell. (A after Mostov and Martin-Belmonte 2006; B after Kamei et al. 2006.)

of blood vessels in the placenta. Each VEGF appears to enable the differentiation of the angioblasts and their multiplication to form endothelial tubes. The most important VEGF in normal development, VEGF-A, is secreted by the mesenchymal cells near the blood islands, and hemangioblasts and angioblasts have receptors for this VEGF* (Millauer et al. 1993). If mouse embryos lack the genes encoding either VEGF-A or its major receptor (the Flk1 receptor tyrosine kinase), yolk sac blood islands fail to appear, and vasculogenesis fails to take place (**Figure 12.17A**; Ferrara et al. 1996). Mice lacking genes for this

VEGF receptor have blood islands and differentiated endothelial cells, but these cells are not organized into blood vessels (Fong et al. 1995; Shalaby et al. 1995). As we saw in Chapter 11, VEGF-A is also important in forming blood vessels to the developing bone and kidney.

A third set of proteins, the **angiopoietins**, mediate the interaction between the endothelial cells and the **pericytes**—smooth muscle-like cells the endothelial cells recruit to cover them. Mutations of either the angiopoietins or their receptor protein, Tie2, lead to malformed blood vessels deficient in the smooth muscles that normally surround them (Davis et al. 1996; Suri et al. 1996; Vikkula et al. 1996; Moyon 2001).

Angiogenesis: Sprouting of blood vessels and remodeling of vascular beds

After an initial phase of vasculogenesis, angiogenesis begins. By this process, the primary capillary networks are remodeled and veins and arteries are made (see Figure 12.14). The critical factor for angiogenesis is VEGF-A (Adams and Alitalo 2007). In many cases, an organ will secrete VEGF-A in order to induce the migration of endothelial cells from existing blood vessels into the organ and to cause them to form capillary networks there. Other factors, including hypoxia (low oxygen levels), can also induce the secretion of VEGF-A and induce blood vessel formation.

During angiogenesis, some endothelial cells in the existing blood vessel can respond to the VEGF signal and begin "sprouting" to form a new vessel. These cells are known as the **tip cells**, and they differ from the other vessel cells. (If all the endothelial cells responded equally, then the original blood vessel would fall apart.) The tip cells express the Notch ligand Delta-like-4 (Dll4) on their cell surfaces. This protein activates Notch signaling in the adjacent cells, preventing them from responding to VEGF-A (Noguera-Troise et al. 2006; Ridgway et al. 2006; Hellström et al. 2007). If the expression of Dll4 is experimentally reduced, tip cells form along a large portion of the blood vessel in response to VEGF-A.

---

*VEGF needs to be regulated very carefully in adults, and recent studies indicate that it can be affected by diet. The consumption of green tea has been associated with lower incidences of human cancer and the inhibition of tumor cell growth in laboratory animals. Cao and Cao (1999) have shown that green tea and one of its components, epigallocatechin-3-gallate (EGCG), prevent angiogenesis by inhibiting VEGF. Moreover, in mice given green tea instead of water (at levels similar to humans after drinking 2–3 cups of tea), the ability of VEGF to stimulate new blood vessel formation was reduced by more than 50%. The drinking of moderate amounts of red wine has been correlated with reduced coronary disease. Red wine has been shown to reduce VEGF production in adults, and it appears to do so by inhibiting endothelin-1, a compound that induces VEGF and that is crucial for the formation of atherosclerotic plaques (Corder et al. 2001; Spinella et al. 2002).

(A)

(B)

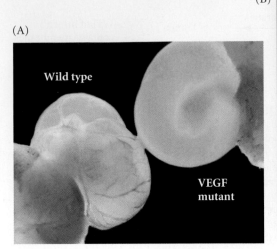

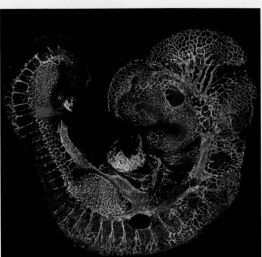

FIGURE 12.17   VEGF and its receptors in mouse embryos. (A) Yolk sacs of a wild-type mouse and a littermate heterozygous for a loss-of-function mutation of VEGF-A. The mutant embryo lacks blood vessels in its yolk sac and dies. (B) In a 9.5-day mouse embryo, VEGFR-3 (red), a VEGF receptor found on tip cells, is found at the angiogenic front of the capillaries (stained green). (A from Tammela et al. 2008, courtesy of the authors; B from Ferrara and Alitalo 1999, courtesy of K. Alitalo.)

The tip cells produce filopodia that are densely packed with VEGFR-2 (VEGF receptor-2) on their cell surfaces. They also express another VEGF receptor, VEGFR-3, and blocking VEGFR-3 greatly suppresses sprouting (**Figure 12.17B**; Tammela et al. 2008). These receptors enable the tip cell to extend toward the source of VEGF, and when the cell divides, the division is along the gradient of VEGFs. Indeed, the filopodia of the tip cells act just like the filopodia of neural crest cells and neural growth cones, and they respond to similar cues (Carmeliet and Tessier-Lavigne 2005; Eichmann et al. 2005). Semaphorins, netrins, neuropilins, and split proteins have roles in directing the sprouting tip cells to the source of VEGF.

Arterial and venous differentiation

Arteries and veins differ substantially from one another even though they are made from the same endothelial precursor cells. Arteries have an extensive coating of smooth muscle and a rich and elastic extracellular matrix. Veins have less extensive musculature and are characterized by valves that direct the flow of blood. A key to our understanding of the mechanism by which veins and arteries form was the discovery that the primary capillary plexus in mice actually contains two types of endothelial cells. The precursors of the arteries contain ephrin B2 in their cell membranes, and the precursors of the veins contain one of the receptors for this molecule, Eph B4 tyrosine kinase, in their cell membranes (Wang et al. 1998). If ephrin B2 is knocked out in mice, vasculogenesis occurs but angiogenesis does not. It is thought that during angiogenesis Eph

B4 interacts with its ligand, ephrin B2, in two ways. First, at the borders of the venous and arterial capillaries, it ensures that arterial capillaries connect only to venous ones. Second, in nonborder areas, it ensures that the fusion of capillaries to make larger vessels occurs only between the same type of vessel (**Figure 12.18**). As we saw before, the same proteins involved in neural patterning are involved in endothelial patterning.

In zebrafish, the separation of arterial cells and venous cells occurs very early in development. The angioblasts develop in the posterior part of the lateral mesoderm, and they migrate to the midline of the embryo, where they coalesce to form the aorta (artery) and the cardinal vein beneath it (**Figure 12.19**). Zhong and colleagues (2001) followed individual angioblasts and found that, contrary to expectations, all the progeny of a single angioblast formed either veins or arteries, never both. In other words, each angioblast was already specified as to whether it would form aorta or cardinal vein. This specification appears to be controlled by the Notch signaling pathway* (Lawson et al. 2001, 2002). Repression of Notch signaling resulted in the loss of ephrin B2-expressing arteries and their replacement by veins. Conversely, activation of Notch signaling suppressed venous development, causing more arterial cells to form. Activation of the Notch proteins in the membranes of the presumptive angioblasts causes the activation of the transcription factor **Gridlock**. Gridlock in turn activates ephrin B2 and other arterial markers, while those angioblasts with low amounts of Gridlock became Eph B4-expressing vein cells.

Weinstein and Lawson (2003) speculate that vascular beds are formed in a two-step process. First, new arteries form in response to VEGF. Second, these arteries then induce neighboring angioblasts (possibly through the ephrin-Eph interactions) to form the venous vessels that

---

*The coordinated use of Notch and Eph signaling pathways is also used to regulate the production of neuroblasts and somites.

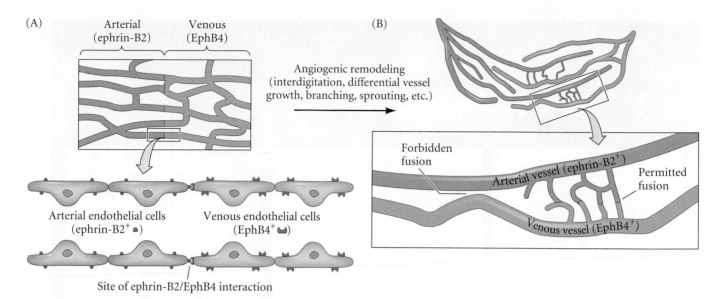

FIGURE 12.18 Roles of ephrin and Eph receptors during angiogenesis. (A) Primary capillary plexus produced by vasculogenesis. The arterial and venous endothelial cells have sorted themselves out by the presence of ephrin B2 or Eph B4 in their respective cell membranes. (B) A maturing vascular network wherein the ephrin-Eph interaction mediates the joining of small branches (future capillaries) and may prevent fusion laterally. (After Yancopopoulos et al. 1998.)

will provide the return for the arterial blood (**Figure 12.20A**). This speculation fits well with the detailed observations of chick vascular development done (and exquisitely drawn) by Popoff (1894) and Isida (1956). These researchers found that the vitelline arteries appeared first within the capillary network and that these capillaries

appeared to induce veins on either side of them (**Figure 12.20B–D**).

Organ-specific capillary formation

As mentioned above for the cases of the brain and the placenta, several organs induce vasculogenesis and angiogenesis in their own mesenchyme. One of the main inducers of VEGF proteins is hypoxia (low oxygen). The HIF-1α transcription factor that activates the *VEGF-A* gene (among others) is functional only at lower oxygen levels (Cramer et al. 2004). The competence of the mesenchyme cells to respond to this signal is governed by their extracellular matrices. Some extracellular matrices can stress the cell membranes, activating certain GTPases. These GTPases can activate transcription factors (such as GATA2) that acti-

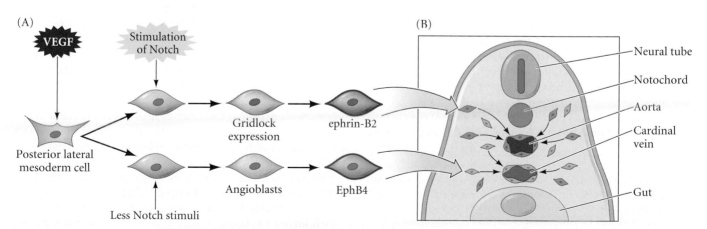

FIGURE 12.19 Blood vessel specification in the zebrafish embryo. (A) Angioblasts experiencing activation of Notch upregulate the Gridlock transcription factor. These cells express ephrin B2 and become aorta cells. Those angioblasts experiencing significantly less Notch activation do not express Gridlock, and they

become Eph B4-expressing cells of the cardinal vein. (B) Once committed to forming veins or arteries, the cells migrate toward the midline of the embryo and contribute to forming the aorta or cardinal vein.

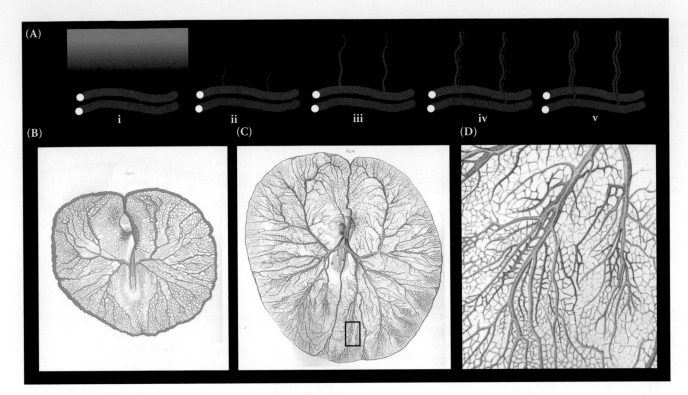

FIGURE 12.20 Blood vessel formation in the chick blastoderm. (A) In response to VEGF (green gradient), endothelial cells are induced to become arteries (red), and these arteries induce veins (blue) to form adjacent to them. New arterial vessels sprout from the arteries and then induce venous vessels adjacent to them. (B) In the chick embryo, a complex branched venous network emerges in the vascular region, with venous drainage at the periphery, via the marginal vein. (C) At later stages, collateral veins emerge adjacent to the arteries. (D) Higher magnification of the boxed region of (C). (After Weinstein and Lawson 2003; B–D modified from Popoff 1894, courtesy of N. D. Weinstein.)

vate the genes encoding VEGF receptors, while inactivating the transcription factors that inhibit VEGF receptors* (Mammoto et al. 2009).

The kidney vasculature is mainly derived from the sprouting of endothelial cells from the dorsal aorta during the initial steps of nephrogenesis. The developing nephrons secrete VEGF, thus allowing the blood vessels to enter the developing kidney and form the capillary loops of the glomerular apparatus (Kitamoto et al. 2002). Another striking example of organ-specific angiogenesis is produced by the developing peripheral nerves. Anatomists have known for decades that blood vessels follow peripheral nerves (see Greenberg and Jin 2005). Their proximity allows the nerves

to obtain oxygen and allows hormones in the blood to regulate vasoconstriction and vasodilation.

The mechanism allowing the nerves and blood vessels to become adjacent is a reciprocal induction: the nerves secrete an angiogenesis factor, and the blood vessels secrete a nerve growth factor. Mukouyama and colleagues (2002, 2005) have demonstrated that arteries become associated with peripheral neurons, although veins do not (**Figure 12.21**). Moreover, peripheral neurons induce arteries to form near them. If the peripheral neurons in the skin fail to form (because of mutations that specifically target peripheral neurons), the arteries likewise fail to form properly. If other mutations cause the peripheral neurons to form haphazardly, the arteries will follow them. This property is due to the secretion of VEGF by the neurons and their associated Schwann cells, which is necessary for arterial formation. Thus, the peripheral nerves appear to provide a template for organ-specific angiogenesis through their ability to secrete VEGF. This interaction is not a one-way street; in some instances, the blood vessels are formed in an area first, before the neurons enter. In those cases, the vascular smooth muscle cells can secrete a compound (most likely GDNF) that allows the neuron to grow along-

---

*This mechanical induction of VEGF receptors explains one of the great conundrums of developmental anatomy: why the aortic arches are asymmetrical. The sixth aortic arch develops only on the left side, whereas the right aortic arch degenerates (see Figure 12.13). The Nodal-induced Pitx2 signal (see Chapter 8) causes rotation of the outflow tract, producing an asymmetric blood flow into the left arch. The left side gets the blood, and the shear force from the blood flow activates the *VEGFR-2* gene on that side only. Without VEGFR-2, the endothelial cells on the right side degenerate. Thus, the sixth aortic arch forms only on the left side (Yashiro et al. 2007).

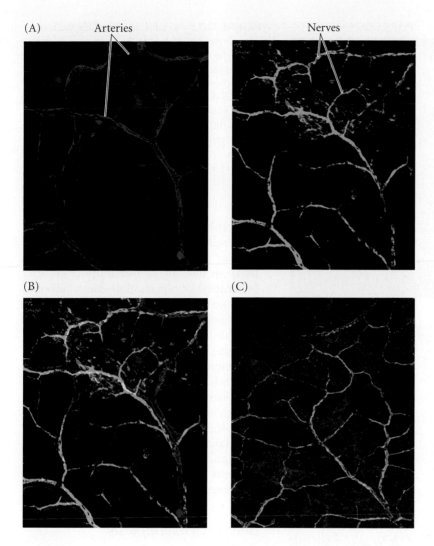

(A) Arteries   Nerves

(B)   (C)

FIGURE 12.21 Arteries are specifically aligned with peripheral nerves in mouse limb skin. (A) Antibody staining of arteries (red; left) and nerves (green; right). (B) Placing the photographs together reveals that the arteries and nerves coincide. (C) Doing the same operation with stained nerves and veins reveals that the veins and nerves do not follow one another. (From Mukouyama et al. 2002, courtesy of Y. Mukouyama.)

side it. In this way, neurons can reach their destinations by following the blood vessels (Honma et al. 2002).

Anti-angiogenesis in normal and abnormal development

Like any powerful process in development, angiogenesis has to be powerfully regulated. The blood vessel formation has to be told to cease, and some tissues have to prevent blood vessel formation. For example, the cornea of most mammals is avascular. This absence of blood vessels allows the transparency of the cornea and optical acuity. Ambati and colleagues (2006) have shown that the cornea

secretes a soluble form of the VEGF receptor that "traps" VEGF and prevents angiogenesis in the cornea.*

Soluble VEGF receptor also appears to be part of the normal mechanisms for regulating the increased formation of vasculature in the uterus during pregnancy. However, if too much soluble VEGF receptor is produced during pregnancy, there can be a dramatic reduction of normal angiogenesis. The spiral arteries that supply the fetus with nutrition fail to form, and the capillary bed of the kidneys is also reduced. This is thought to be a major cause of **preeclampsia**, a condition of pregnant women characterized by hypertension and poor renal filtration (both of which are kidney problems) and fetal distress. Preeclampsia is the leading cause of premature infants and a major cause of maternal and fetal deaths (Levine et al. 2006; Mutter and Karumanchi et al. 2008).

As we will see in Chapter 17, abnormal blood vessel formation occurs in solid tumors and in the retina of patients with diabetes. This vascularization results in the growth and spread of tumor cells and blindness, respectively. By targeting the VEGF receptors and the Notch pathway involved in regulating them, researchers are seeking ways to block angiogenesis and prevent cancer cells and the retina from becoming vascularized.

The lymphatic vessels

In addition to the blood vessels, there is a second circulatory system, the **lymphatic vasculature**. The lymphatic vasculature forms a separate system of vessels which is essential for draining fluid and transporting lymphocytes. The development of the lymphatic system commences when a subset of endothelial cells from the jugular vein (in the neck) sprout to form the lymphatic sacs. After the formation of these sacs, the peripheral lymphatic vessels are generated by further sprouting (Sabin 1902; van der Putte 1975). Commitment to the lymphatic lineage appears to be mediated through the **Prox1** transcription factor, which downregulates blood vessel-specific genes and upregulates genes involved in forming lymphatic vessels (Wigle and Oliver 1999; Wigle

---

*Then what about the manatee, the only animal known to have a vascularized cornea? It turns out that this is one of the exceptions that prove the rule: the cornea of the manatee does not express the soluble VEGF receptor. The manatee's closest relatives (dugongs and elephants) do express it, and their corneas are avascular (Ambati et al. 2006). This morphological distinction among related taxa provides further evidence of the importance of soluble VEGF in preventing corneal vascularization.

(A)   Wild type   VEGF-C-deficient

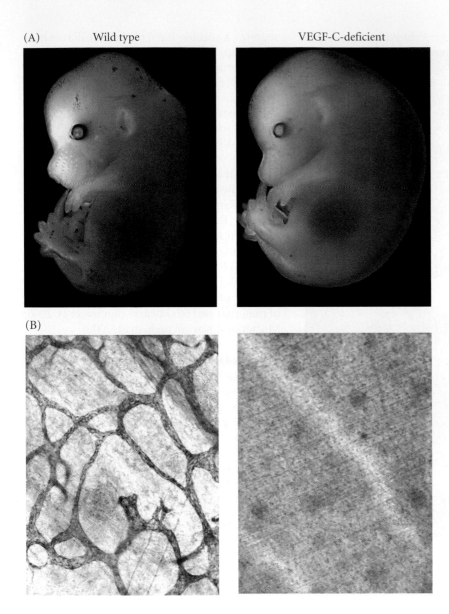

(B)

FIGURE 12.22 VEGF-C is critical for the formation of lymphatic vessels. (A) Compared with the wild-type control, a 15.5-day mouse embryo heterozygous for a *VEGF-C* deficiency suffers from severe edema (bloating with excess fluid). (B) 16.5-day mouse embryos stained for lymphatic vasculature. The lack of lymphatics in the skin of the *VEGF-C* mutant (right) is obvious when compared with that of the wild-type embryo (left). (From Karkkainen et al. 2004, courtesy of K. Alitalo.)

sive proliferation, creating both more stem cells (self-renewal) and more differentiated cellular progeny (see Figures III.1A and III.4). In the case of **hematopoiesis**—the generation of blood cells—stem cells can divide either to produce more stem cells or to replenish differentiated cells; they can also regulate to form more cells when body equilibrium is stressed by injury or environmental factors. The critical stem cell in hematopoiesis is the *pluripotential hematopoietic stem cell*. Often referred to simply as the **hematopoietic stem cell** (**HSC**), this cell type is capable of producing all the blood cells and lymphocytes of the body. It achieves this by generating a series of intermediate progenitor cells whose potency is restricted to certain lineages.

Sites of hematopoiesis

Earlier in this chapter, we mentioned that there are hemangioblasts in early zebrafish embryos, and that these cells generate both the blood vessels and blood cells of the early embryo. We also mentioned that there is considerably more controversy concerning the existence of such hemangioblasts in amniote embryos.

et al. 2002; Françoise et al. 2008). One of the genes upregulated by Prox1 is *VEGFR-3*, which encodes the receptor for the paracrine factor VEGF-C. As important as VEGF-A is for blood vessel development, VEGF-C is equally necessary for proper lymphatic development (**Figure 12.22**; Karkkainen et al. 2004; Alitalo et al. 2005). VEGF-C produced in the area of the jugular vein attracts Prox1-positive endothelial cells out from the vein and then promotes their proliferation and development into the lymphatic sacs (see Adams and Alitalo 2007; Hosking and Makinen 2007).

Hematopoiesis: The Stem Cell Concept

Each day we lose and replace about 10^{11} red blood cells. As red blood cells are killed in the spleen, their replacements come from populations of stem cells. As mentioned in the Part III introduction, a stem cell is capable of exten-

The original sites of hematopoiesis are associated with the blood islands in the ventral mesoderm surrounding the yolk sac—that is, the splanchnopleure. In chick embryos, the first blood cells are seen in those blood islands that form in the posterior marginal zone near the site of hypoblast initiation. Although the hematopoietic stem cells of these blood islands seem capable of generating all blood cell (but not lymphocyte) lineages, they usually produce red blood cells (Moore and Metcalf 1970; Rampon and Huber 2003). In non-amniote vertebrates, this splanchnopleure is also the source of the hematopoietic stem cells, and BMPs are crucial in inducing the blood-forming cells in all vertebrates studied. In *Xenopus*, the ventral mesoderm forms a large blood island that is the first site of hematopoiesis. Ectopic BMP2 and BMP4 can induce blood and blood vessel formation in *Xenopus*, and interference with BMP signaling prevents blood formation (Maeno et al. 1994; Hemmati-Brivanlou and Thomsen 1995). In the zebrafish, the *swirl* mutation, which prevents BMP2 signaling, also abolishes ventral mesoderm and blood cell production (Mullins et al. 1996). As mentioned above, BMP4 is critical in the formation of the blood islands in the mammalian extraembryonic mesoderm.

Early studies suggested that in amniotes, hematopoietic stem cells originated in the yolk sac and then subsequently colonized the bone marrow and fetal liver to generate the adult HSC population (Wilt 1974; Azar and Eyal-Giladi 1979; Kubai and Auerbach 1983). These latter sites would be the places where the definitive hematopoiesis would occur throughout the lifetime of the animal (after the yolk sac disappeared). However, later transplantation studies seemed to show that there are two distinct populations of hematopoietic stem cells: a yolk sac-derived stem cell that produces the blood for the embryos, and a definitive hematopoietic stem cell, originating in the lateral plate splanchnopleure near the aorta, and whose progeny serve the fetus and adult (Dieterlen-Lièvre and Martin 1981; Medvinsky et al. 1993). This mesenchymal area is often called the **aorta-gonad-mesonephros region**, or **AGM**. Recent studies, using less invasive techniques of genetic marking, have returned scientific attention to the hypothesis that the original hematopoietic stem cells

of the yolk sac may generate the definitive hematopoietic stem cells of the bone marrow and liver. However, there is still a great deal of controversy and discussion about this.

In 2009, several laboratories proposed a new mechanism for blood cell production. This new hypothesis was based on the discovery of a new cell type, the **hemogenic endothelial cell**. Recall that in our discussion of somites, we noted that the sclerotome produces angioblasts that migrate to the dorsal aorta and replace most of the primary dorsal aorta cells. These primary endothelial cells of the dorsal aorta, especially those in the ventral area, then give rise to blood-forming stem cells (**Figure 12.23A**). This blood vessel-derived hematopoietic stem cell may be an important source of adult blood stem cells. By analyzing the types of cells made by the blood vessel endothelium, researchers were able to isolate the hemogenic endothelial cells and showed that they produce the hematopoietic stem cells that migrate to the liver and bone marrow (Eilken et al. 2009; Lancrin et al. 2009). Furthermore, the transition from endothelial cell to hematopoietic stem cell was seen to be mediated by the activation of the Runx1 transcription factor. In mice lacking this factor, the blood stem cells failed to form in the yolk sac, umbilical arteries, dorsal aorta, and placental vessels (Chen et al. 2009).

This set of observations dovetails and interacts with a second set of studies (Adamo et al. 2009; North et al. 2009) show-

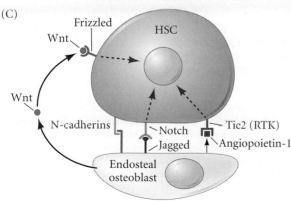

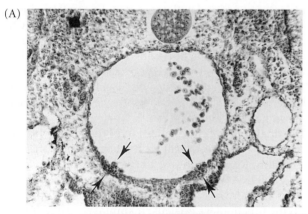

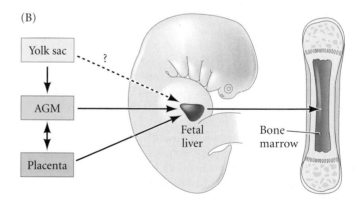

FIGURE 12.23 Sources of blood cells to adult bone marrow. (A) Section through the aorta of a 3-day chick embryo, showing the cells (arrows) that give rise to hematopoietic stem cells in the AGM (aorta-gonad-mesonephros region) of the chick. (B) In mammals, the yolk sac, the AGM, and the placenta each probably contribute stem cells to the fetal liver. Stem cells from the fetal liver then populate the bone marrow as the hematopoietic niche in the bone marrow is constructed. There is controversy as to whether the stem cells from the yolk sac populate the AGM or the liver, or if a new set of stem cells is formed. (C) Schematic diagram of an interaction in the stem cell niche by which bone marrow (endosteal) osteoblasts maintain HSCs by activating the Wnt, Notch, and receptor tyrosine kinase (RTK) pathways (red dashed arrows). (A from Dieterlen-Lièvre and Martin 1981, photograph courtesy of F. Dieterlen-Lièvre; B after Ottersbach and Dzierak 2005.)

ing that blood flow causes shear forces that activate the *Runx1* gene in the ventral endothelium of the dorsal aorta.* The shear forces appear to elevate levels of nitrous oxide (NO) in the endothelium, and NO in turn activates *Runx1* and other genes known to be critical for blood cell formation.

Chen and colleagues (2009) claim that the hematopoietic stem cells arise from hemogenic endothelial cells that originate separately in the blood islands, the dorsal aorta, the placental arteries and veins, and the umbilical vessels; and it is possible that they each contribute hematopoietic stem cells to the liver and bone marrow (Figure 12.23B). Indeed, Samokhvalov and colleagues (2007) have shown that the same hematopoietic stem cells that had formed red blood cells in the mouse yolk sac later generated stem cells that colonized the fetal umbilical cord and AGM. This multi-origin model of hematopoietic stem cells is a testable hypothesis, and the extent to which these various hemogenic endothelial populations colonize the adult stem cell niches in the liver and bone marrow should soon be known.

Committed stem cells and their fates

The bone marrow HSC is a remarkable cell, in that it is the common precursor of red blood cells (erythrocytes), white blood cells (granulocytes, neutrophils, and platelets), and lymphocytes. When transplanted into inbred, irradiated mice (who are genetically identical to the donor cells and whose own stem cells have been eliminated by radiation), HSCs can repopulate the mouse with all the blood and lymphoid cell types. It is estimated that only about 1 in every 10,000 blood cells is a pluripotential HSC (Berardi et al. 1995).

The HSC (sometimes called the CFU-M,L) appears to be dependent on the transcription factor SCL. Mice lacking SCL die from the absence of all blood and lymphocyte lineages. SCL is thought to specify blood cell fate in mesoderm cells, and it continues to be expressed in the HSCs (Porcher et al. 1996; Robb et al. 1996). The pluripotent HSC is also dependent on osteoblasts in the bone marrow. Osteoblasts that have finished making bone can still have a function: **endosteal osteoblasts** that line the bone marrow are responsible for providing the niche that attracts HSCs, prevents apoptosis, and keeps the HSCs in a state of plasticity. These osteoblasts bind HSCs (probably through N-cadherin) and provide several other signals (Calvi et al. 2003; J. Zhang et al. 2003; Chan et al. 2009). One signal is provided by the Jagged protein, which activates Notch protein on the HSC surface (**Figure**

12.23C). A second signal comes from angiopoietin-1 on the osteoblasts, which activates the receptor tyrosine kinase Tie2 on the surface of the HSC (Arai et al. 2004). A third signal is from the Wnt pathway, localizing β-catenin into the nucleus. Wnt proteins are made by osteoblasts, and are probably made by differentiated osteocytes as well. This Wnt pathway seems critical for the self-renewal of the HSC (Reya et al. 2003).

HSCs give rise to **lineage-restricted stem cells** that produce blood cells and lymphocytes. While there are disputes about the exact lineage and time of commitment to certain cell fates (see Adolffson et al. 2005; Hock and Orkin 2005), **Figure 12.24** is a simplified depiction of one plausible model. An HSC can give rise to the blood cell precursor (common myeloid precursor cell, or CMP; sometimes called the CFU-S) or to the lymphocyte stem cell (CLP). These cells may also be stem cells, although this is not certain. The CMPs produce the megakaryocyte/erythroid precursor cell (MEP), which can generate either the red blood cell (erythrocyte) lineage or the platelet lineage. The CMP can also give rise to the granulocyte/monocyte precursor cell (GMP), which generates the basophils, eosinophils, neutrophils, and monocytes. Eventually, these cells produce **progenitor cells** that can divide but produce only *one* type of cell in addition to renewing themselves. For instance, the **erythroid progenitor cell** (BFU-E) is a committed stem cell that can form only red blood cells. Its immediate progeny is capable of responding to the hormone **erythropoietin** to produce the first recognizable differentiated member of the erythrocyte lineage, the **proerythroblast**, a red blood cell precursor. This cell begins to make globin (Krantz and Goldwasser 1965). As the proerythroblast matures, it becomes an **erythroblast**, synthesizing enormous amounts of hemoglobin. The mammalian erythroblast eventually expels its nucleus and becomes a **reticulocyte**. Although reticulocytes, lacking a nucleus, can no longer synthesize globin mRNA, they can translate *existing* messages into globins.

The final stage of differentiation of the erythroid lineage is the **erythrocyte**, or mature red blood cell. In this cell, no division, RNA synthesis, or protein synthesis takes place. The DNA of the erythrocyte condenses and translates no further messages. Amphibians, fish, and birds retain the functionless nucleus; mammals extrude it from the cell.* The erythrocyte then leaves the bone marrow and

---

*Shear stress from blood flow is a major player in development. Recall that it is also required for normal heart development (Mironov et al. 2005) and for the correct patterning of blood vessels (Lucitti et al. 2007; Yashiro et al. 2007). It is also needed for the fragmentation of the platelet precursor cell—the megakaryocyte—into platelets. The megakaryocyte in the bone marrow inserts small processes into the blood vessels surrounding the stem cell niche, and the shear force there fragments these processes into platelets (Junt et al. 2007).

---

*In 1846, the young Joseph Leidy (then an assistant coroner, later the most famous American biologist of his day) was probably the first person to use a microscope to solve a murder mystery. A man accused of killing a Philadelphia farmer had blood on his clothes and hatchet. The suspect claimed innocence, saying that the blood was from chickens he had just slaughtered. Examining the blood cells under his microscope, Leidy found no nuclei in these erythrocytes. Moreover, he found that even if he let chick erythrocytes remain outside the body for hours, they did not lose their nuclei. Thus, he concluded that the blood stains could not have been chicken blood. The suspect subsequently confessed (Warren 1998).

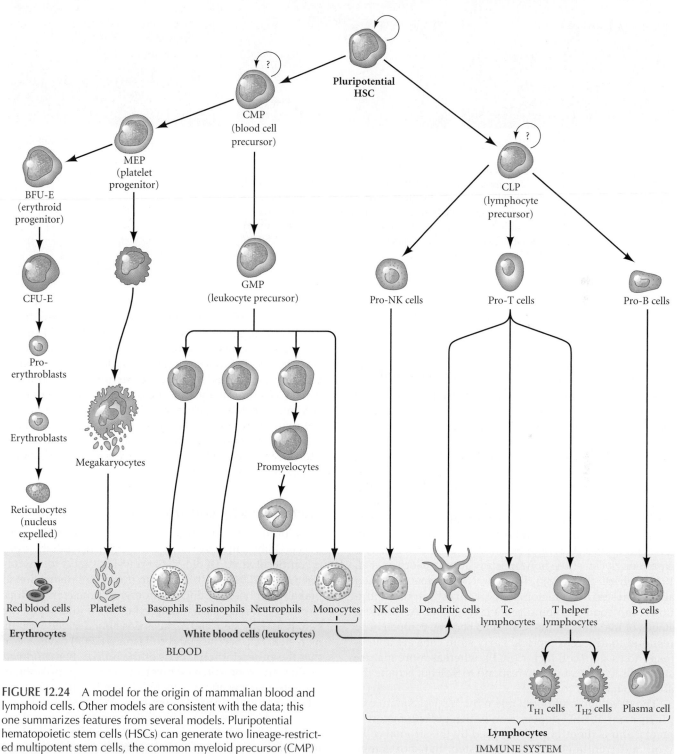

FIGURE 12.24 A model for the origin of mammalian blood and lymphoid cells. Other models are consistent with the data; this one summarizes features from several models. Pluripotential hematopoietic stem cells (HSCs) can generate two lineage-restricted multipotent stem cells, the common myeloid precursor (CMP) of all blood cells, and the common lymphoid precursor (CLP) of the immune system. Whether or not these are actually stem cells is still uncertain. The CMP can generate the precursors of red blood cells and white blood cells. (After Kluger et al. 2004.)

(A)

(B)

(C)

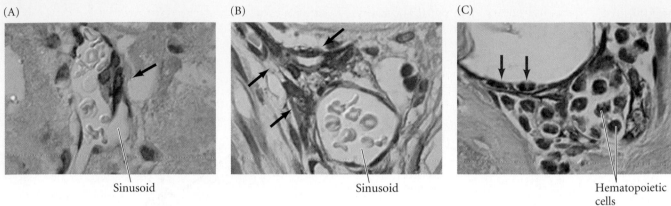

Sinusoid

Sinusoid

Hematopoietic
cells

FIGURE 12.25   Development of human subendothelial cells implanted into mice.(A) At 4 weeks, small mononuclear cells (stained brown with antibodies to the subendothelial cell marker CD146) associate with blood vessels (arrow). (B) At 7 weeks, the elongated CD146$^+$ cells are found over and around sinusoids in the bone marrow (arrows). (C) At 8 weeks, processes from CD146$^+$ cells establish contacts with hematopoietic cells (as in the human bone marrow). The red arrows show hematopoietic cells between endothelial and CD146$^+$ cells. (After Sacchetti et al. 2007, courtesy of P. Bianco.)

enters the circulation, where it delivers oxygen to the body tissues.

Hematopoietic inductive microenvironments

Different paracrine factors are important in causing hematopoietic stem cells to differentiate along the particular pathways illustrated in Figure 12.24. The paracrine factors involved in blood cell and lymphocyte formation are called **cytokines**. Cytokines can be made by several cell types, but they are collected and concentrated by the extracellular matrix of the stromal (mesenchymal) cells at the sites of hematopoiesis (Hunt et al. 1987; Whitlock et al. 1987). For instance, granulocyte-macrophage colony-stimulating factor (GM-CSF) and the multilineage growth factor **interleukin 3** (**IL3**) both bind to the heparan sulfate glycosaminoglycan of the bone marrow stroma (Gordon et al. 1987; Roberts et al. 1988). The extracellular matrix is then able to present these factors to the stem cells in concentrations high enough to bind to their receptors. At different stages of maturation, the stem cells become competent to respond to different factors. Early stem cells are able to respond to stem cell factor (SCF), whereas more mature stem cells lose the ability to respond to SCF but acquire the ability to respond to IL3.

The developmental path taken by the descendant of a pluripotential HSC depends on which growth factors it meets, and is therefore determined by the stromal cells. Wolf and Trentin (1968) demonstrated that short-range interactions between stromal cells and stem cells determine the developmental fates of the stem cells' progeny. These investigators placed plugs of bone marrow in a spleen and then injected stem cells into it. Those CMPs that came to reside in the spleen formed colonies that were predominantly erythroid, whereas those that came to reside in the bone marrow formed colonies that were predominantly granulocytic. In fact, colonies that straddled the borders of the two tissue types were predominantly erythroid in the spleen and granulocytic in the marrow. Such regions of determination are referred to as **hematopoietic inductive microenvironments** (**HIMs**). As expected, the HIMs induce different sets of transcription factors in these cells, and these transcription factors specify the fate of the particular cells (see Kluger et al. 2004).

Stem cell niche construction

Ian Wilmut, who directed the cloning program that produced Dolly, famously remarked that "Although the story is complicated, it is biology, not physics: that is to say, it is not *weird*" (2001, p. 17). This is an important distinction to make when one is trying to understand the complexities of blood cell formation. Because here's another layer: the hematopoietic stem cell niche is possibly derived from its own stem cell. Sacchetti and colleagues (2007) have shown that the **subendothelial cells** of the human hematopoietic microenvironment, a relatively minor cell population in the bone marrow and having no known function, are actually hematopoietic niche stem cells. When such cells are implanted into the skin of immunosuppressed mice, these cells divide and differentiate into small hematopoietic microenvironments, complete with miniature bones and more subendothelial cells (**Figure 12.25**). Mouse hematopoietic stem cells migrate into these regions and start forming blood. Moreover, when those new subendothelial cells are transferred to yet other mice, they too develop a new hematopoietic microenvironment. One of the proteins secreted by these cells is angiopoietin-1, a paracrine factor known to be involved in vascular remodeling and the formation of the stem hematopoietic microenvironment (Suri et al. 1996; Arai et al. 2004). Thus, this hematopoietic stem cell not only differentiates into the osteoblasts that create the hematopoietic microenvironment; it also produces a new population of bone marrow stem cells that can produce the paracrine factors necessary for remodeling the vasculature to integrate the blood vessels with the microenvironment.

We ask a lot of our circulatory system. We require a flawless flow of blood through the valves each second of our lives; we demand fine-tuned coordination between our brain, heart, bone marrow, and hormones such that the cardiac muscle contractions adapt to our physiological needs; and we demand that the production of our blood cells—cells made by precursors that formed in our embryo—be so precise that we get neither cancer nor anemia. Given all this, it is not surprising that blood cell differentiation, heart development, and vessel formation are now among the most important fields of study in medical science. As we will see in Chapter 17, controlling blood cell differentiation and stem cell proliferation is at the root of leukemia research, and regulating angiogenesis holds promise for preventing tumor formation. Congenital heart defects are among the most prevalent types of birth defects, and cardiovascular disease is the most common cause of death in industrialized nations. The questions of cardiogenesis, angiogenesis, and hematopoiesis that engaged Aristotle and Harvey still excite major research programs today.

ENDODERM

The first of the embryonic endoderm's two major functions is to induce the formation of several mesodermal organs. As we have seen in this and earlier chapters, the endoderm is critical for instructing the formation of the notochord, the heart, the blood vessels, and even the mesodermal germ layer. The second function is to construct the linings of two tubes within the vertebrate body. The **digestive tube** extends the length of the body. Buds from the digestive tube form the liver, gallbladder, and pancreas. The **respiratory tube** forms as an outgrowth of the digestive tube and eventually bifurcates into the two lungs. The region of the digestive tube anterior to the point where the respiratory tube branches off is the **pharynx**. Epithelial outpockets of the pharynx give rise to the tonsils, to the thyroid, thymus, and parathyroid glands, and eventually to the respiratory tube itself.

In 1769, Caspar Friedrich Wolff, an embryologist in St. Petersburg, demonstrated that the gut tube formed by the curving of an initially flat sheet. His *De Formatione Intestinorum* was the first microscopic evidence for *epigenesis*—

the view that the embryo constructed itself "from scratch" and that no small, preformed individual resided within the sperm or egg (see Chapter 1). In addition to confirming this, modern work has shown that gut tube development begins at two sites that migrate toward each other and fuse in the center (Lawson et al. 1986; Franklin et al. 2008). In the foregut, cells from the lateral portions of the anterior endoderm move ventrally to form the tube of the **anterior intestinal portal** (**AIP**); the **caudal intestinal portal** (**CIP**) forms from the posterior endoderm. The AIP and CIP migrate toward each other and come together to form the midgut (**Figure 12.26**).

There is an ectodermal entrance at either end of the gut tube. At first, the oral end is blocked by a region of ectoderm called the **oral plate**, or **stomodeum**. Eventually (at about 22 days in human embryos), the stomodeum breaks, creating the oral opening of the digestive tube. The opening itself is lined by ectodermal cells. This arrangement creates an interesting situation, because the oral plate ectoderm is in contact with the brain ectoderm, which has curved around toward the ventral portion of the embryo. These two ectodermal regions interact with each other, with the roof of the oral region forming Rathke's pouch and becoming the glandular portion of the pituitary gland. The neural tissue on the floor of the diencephalon gives rise to the infundibulum, which becomes the neural portion of the pituitary. Thus, the pituitary gland has a dual origin, which is reflected in its adult functions. There is a similar meeting of endoderm and ectoderm at the anus; this is called the **anorectal junction**.

The Pharynx

The anterior endodermal portion of the digestive and respiratory tubes begins in the pharynx. Here, the mammalian embryo produces four pairs of **pharyngeal pouches**. Between these pouches are the four **pharyngeal arches** (**Figure 12.27**). The first pair of pharyngeal pouches become the auditory cavities of the middle ear and the associated eustachian tubes. The second pair of pouches give rise to the walls of the tonsils. The thymus is derived from the third pair of pharyngeal pouches; it will direct the differentiation of T lymphocytes during later stages of development. One pair of parathyroid glands is also derived from the third pair of pharyngeal pouches, while the other pair is derived from the fourth pharyngeal pair. In addition to these paired pouches, a small, central diverticulum is formed between the second pharyngeal pouches on the floor of the pharynx. This pocket of endoderm and mesenchyme will bud off from the pharynx and migrate down the neck to become the thyroid gland. The respiratory tube sprouts from the pharyngeal floor (between the fourth pair of pharyngeal pouches) to form the lungs, as we will see below.

The pharynx is where the endoderm meets the ectoderm, and the endoderm plays a critical role in determin-

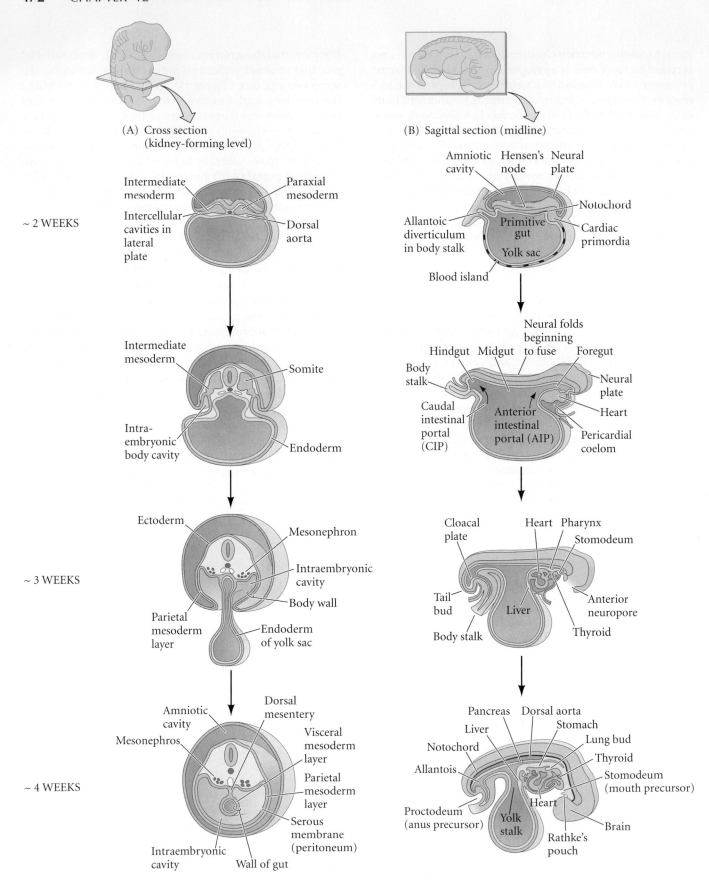

(A) Cross section (kidney-forming level)

(B) Sagittal section (midline)

~ 2 WEEKS

Intermediate mesoderm
Paraxial mesoderm
Intercellular cavities in lateral plate
Dorsal aorta

Amniotic cavity   Hensen's node   Neural plate
Allantoic diverticulum in body stalk
Primitive gut
Yolk sac
Blood island
Notochord
Cardiac primordia

Intermediate mesoderm
Somite
Intra-embryonic body cavity
Endoderm

Neural folds beginning to fuse
Hindgut   Midgut   Foregut
Body stalk
Neural plate
Caudal intestinal portal (CIP)
Anterior intestinal portal (AIP)
Heart
Pericardial coelom

~ 3 WEEKS

Ectoderm
Mesonephron
Intraembryonic cavity
Body wall
Parietal mesoderm layer
Endoderm of yolk sac

Cloacal plate
Heart   Pharynx
Stomodeum
Tail bud
Anterior neuropore
Body stalk
Liver
Thyroid

~ 4 WEEKS

Amniotic cavity
Dorsal mesentery
Mesonephros
Visceral mesoderm layer
Parietal mesoderm layer
Serous membrane (peritoneum)
Intraembryonic cavity
Wall of gut

Pancreas   Dorsal aorta
Liver   Stomach
Notochord   Lung bud
Allantois   Thyroid
Stomodeum (mouth precursor)
Proctodeum (anus precursor)
Heart
Yolk stalk
Rathke's pouch
Brain

FIGURE 12.26 Endodermal folding during early human development. (A) Cross sections through the kidney-forming region. (B) Sagittal sections through the embryo's midline. (After Sadler 2009.)

ing which pouches develop. Sonic hedgehog from the endoderm appears to act as a survival factor, preventing apoptosis of the neural crest cells (Moore-Scott and Manley 2005). In zebrafish, genetic analysis combined with transplantation studies has shown that FGFs (mainly Fgf3 and Fgf8) from the ectoderm and mesoderm also are important not only for the migration and survival of neural crest cells, but also for the formation of the pouches themselves. Mice deficient in both *Fgf8* and *Fgf3* genes lack all the pharyngeal pouches, even when endoderm is present. Instead of migrating laterally and ventrally to form pouches, the endoderm remains in the anterior pharynx and does not spread out (Crump et al. 2004).

The Digestive Tube and Its Derivatives

Posterior to the pharynx, the digestive tube constricts to form the esophagus, which is followed in sequence by the stomach, small intestine, and large intestine. The endodermal cells generate only the lining of the digestive tube and its glands; mesenchyme cells from the splanchnic portion of the lateral plate mesoderm will surround the tube to provide the muscles for peristalsis.

As **Figure 12.28A** shows, the stomach develops as a dilated region of the gut close to the pharynx. The intestines develop more caudally, and the connection between the intestine and yolk sac is eventually severed. The intestine originally ends in the endodermal cloaca, but after the cloaca separates into the bladder and rectal regions (see

Chapter 11), the intestine joins with the rectum. At the caudal end of the rectum, a depression forms where the endoderm meets the overlying ectoderm, and a thin **cloacal membrane** separates the two tissues. When the cloacal membrane eventually ruptures, the resultant opening becomes the anus.

Specification of the gut tissue

The digestive tube proceeds from the pharynx to the anus, differentiating along the way into the esophagus, stomach, duodenum, and intestines, and putting out branches that become (among other things) the thyroid, thymus, pancreas, and liver. What tells the endodermal tube to become these tissues at particular places? Why do we never see a mouth opening directly into a stomach? There appear to be two major ways to specify the gut tube. First, there is a global specification by a retinoic acid gradient that leads to expression of regionally specific transcription factors in the endoderm. Second, signals from the lateral plate-derived mesenchymal cells help specify the endodermal regions.

The gut appears to be regionally specified at a very early stage. Indeed, in the chick, the endoderm appears to be regionally specified even before it forms a tube. The gradients of retinoic acid and FGF signals mentioned earlier in this chapter pattern the endoderm. Studies on chick embryos using beads containing either retinoic acid or inhibitors of its synthesis (Bayha et al. 2009) show that the pharynx can develop only in areas containing little or no

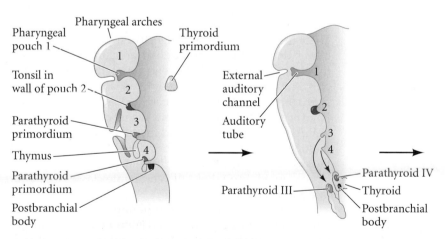
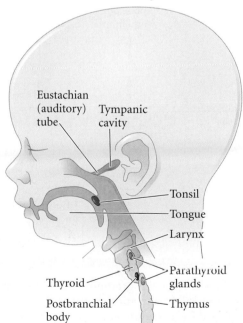

FIGURE 12.27 Formation of glandular primordia from the pharyngeal pouches. The end of each of the first pharyngeal pouches becomes the tympanic cavity of the middle ear and the eustachian tube. The second pouches receive aggregates of lymphoid tissue and become the tonsils. The dorsal portion of the third pharyngeal pouches forms part of the parathyroid gland, while the ventral portion forms the thymus. Both migrate caudally and meet with the tissue from the fourth pharyngeal pouches to form the rest of the parathyroid and the postbranchial body. The thyroid, which had originated in the midline of the pharynx, also migrates caudally into the neck region. (After Carlson 1981.)

(A)

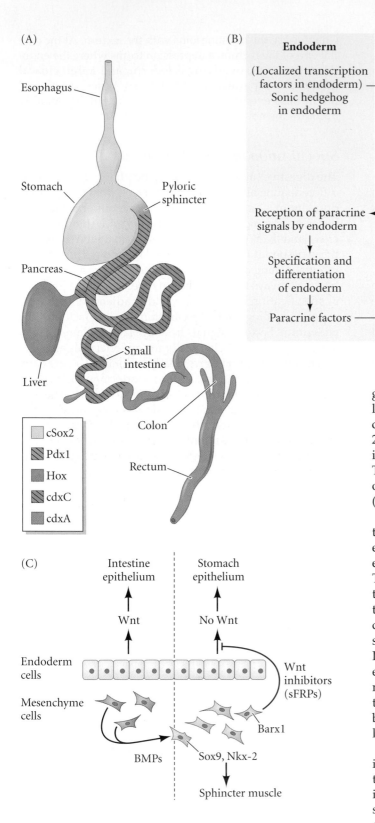

Esophagus

Stomach

Pyloric sphincter

Pancreas

Liver

Small intestine

Colon

Rectum

	cSox2
	Pdx1
	Hox
	cdxC
	cdxA

(B)

Endoderm	Mesoderm
(Localized transcription factors in endoderm) Sonic hedgehog in endoderm →	Reception of Sonic hedgehog protein in mesoderm ↓ Hox gene expression in mesoderm ↓ Specification of mesoderm
Reception of paracrine signals by endoderm ← ↓ Specification and differentiation of endoderm ↓ Paracrine factors →	BMPs, FGFs made by mesoderm Differentiation of mesoderm

FIGURE 12.28 Regional specification of the gut endoderm and splanchnic mesoderm through reciprocal interactions. (A) Regional transcription factors of the (mature) chick gut endoderm. These factors are seen prior to interactions with the mesoderm, but they are not stabilized. (B) Possible course of interactions between the endoderm and the splanchnic mesoderm. (C) Suggested mechanisms by which mesenchymal cells may induce gut endoderm to become either intestine or stomach. (A after Grapin-Botton et al. 2001.)

(C)

Intestine epithelium

Stomach epithelium

Wnt

No Wnt

Endoderm cells

Wnt inhibitors (sFRPs)

Mesenchyme cells

Barx1

BMPs

Sox9, Nkx-2

Sphincter muscle

gene that is essential for the duodenum, pancreas, and liver, as well as the *cSox2* gene that is expressed in the precursors of the stomach and esophagus (Matsushita et al. 2002, 2008). Still higher concentrations of RA directly induce the posterior gene *CdxA* in the intestinal endoderm. This patterning of the axis by retinoic acid may represent one of the oldest developmental patterning mechanisms (see Figure 12.28A; Simões-Costa et al. 2008).

In addition to this direct mechanism of cell specification, the endodermal epithelium is able to respond differently to signals from different regionally specific mesenchymes that are derived from the splanchnic mesoderm. These responses enable the digestive tube and respiratory tube to develop their very different structures. As the digestive tube meets different mesenchymes, the mesenchyme cells instruct the endoderm to differentiate into esophagus, stomach, small intestine, and colon (Okada 1960; Gumpel-Pinot et al. 1978; Fukumachi and Takayama 1980; Kedinger et al. 1990). The endoderm and the splanchnic lateral plate mesoderm undergo a complicated set of interactions, and the signals for generating the different gut tissues appear to be conserved throughout the vertebrate classes (see Wallace and Pack 2003).

The stabilization of the gut boundaries results from interaction with the mesoderm (**Figure 12.28B**). As the gut tube begins to form at the anterior and posterior ends, it induces the splanchnic mesoderm to become regionally specific. Roberts and colleagues (1995, 1998) have implicated Sonic hedgehog (Shh) in this specification. Shh is secreted in different concentrations at different sites, and its target appears to be the mesoderm surrounding the gut tube. The secretion of Shh by the hindgut endoderm induces a nested pattern of "posterior" Hox gene expres-

RA, while the RA gradient patterns the pharyngeal arch endoderm in a graded manner. This is probably accomplished by activating and repressing particular sets of transcription factor genes. Higher levels of RA specify the *Pdx1*

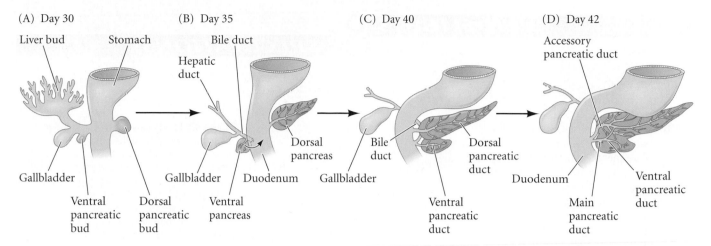

(A) Day 30

Liver bud   Stomach

Gallbladder

Ventral
pancreatic
bud

Dorsal
pancreatic
bud

(B) Day 35

Bile duct

Hepatic
duct

Gallbladder   Duodenum

Dorsal
pancreas

Ventral
pancreas

(C) Day 40

Bile
duct   Gallbladder

Dorsal
pancreatic
duct

Ventral
pancreatic
duct

(D) Day 42

Accessory
pancreatic duct

Duodenum

Main
pancreatic
duct

Ventral
pancreatic
duct

FIGURE 12.29 Pancreatic development in humans. (A) At 30 days, the ventral pancreatic bud is close to the liver primordium. (B) By 35 days, it begins migrating posteriorly, and (C) comes into contact with the dorsal pancreatic bud during the sixth week of development. (D) In most individuals, the dorsal pancreatic bud loses its duct into the duodenum; however, in about 10% of the population, the dual duct system persists.

sion in the mesoderm. As in the vertebrae (see Chapter 8), the anterior borders of Hox gene expression delineate the morphological boundaries of the regions that will form the cloaca, large intestine, cecum, mid-cecum (at the midgut-hindgut border), and posterior portion of the midgut (Roberts et al. 1995; Yokouchi et al. 1995). When Hox-expressing viruses cause the misexpression of these Hox genes in the mesoderm, the mesodermal cells alter the differentiation of the adjacent endoderm (Roberts et al. 1998). The Hox genes are thought to specify the mesoderm so that it can interact with the endodermal tube and specify its regions. Once the boundaries of the transcription factors are established, differentiation can begin. The regional differentiation of the mesoderm (into smooth muscle types) and the regional differentiation of the endoderm (into different functional units such as the stomach, duodenum, small intestine, etc.) are synchronized.

The molecular signals by which the mesenchyme influences the gut tube are just becoming known (**Figure 12.28C**). For instance, the mesenchyme lining the posterior (intestine-forming) region of the endodermal tube secretes Wnt proteins. These Wnt proteins suppress genes such as *Hhex* and prevent the stomach, liver, and pancreas from forming (Bossard and Zaret 2000; McLin et al. 2007). In the anterior regions of the gut tube (which form the thymus, pancreas, stomach, and liver), Wnt signaling must be blocked. In the stomach-forming region, the mesenchyme lining the gut tube expresses the transcription factor Barx1, which activates production of Frzb-like Wnt-blocking proteins (sFRP1 and sFRP2). These Wnt antagonists block Wnt signaling in the vicinity of the stomach but not around the intestine. (Indeed, *Barx1*-deficient mice do not develop stomachs and express intestinal markers in that tissue; Kim et al. 2005.)

In addition to FGF and Wnt signaling, BMPs also act in regional specification. The intestinal mesenchyme secretes BMP4, which instructs the mesoderm anterior to it to express the Sox9 and Nkx2-5 transcription factors (see Figure 12.28C). These factors tell the mesoderm to become the muscles of the pyloric sphincter rather than the smooth

muscle that normally lines the stomach and intestine (Theodosiou and Tabin 2005). Thus, a signal from the small intestine instructs the mesoderm of the adjacent stomach to tell its mesoderm and endoderm how to differentiate. Such communication allows the endoderm to form a tube with very exact regional compartments.

Liver, pancreas, and gallbladder

The endoderm also forms the lining of three accessory organs that develop immediately caudal to the stomach: the liver, pancreas, and gall bladder. The **hepatic diverticulum** is a bud of endoderm that extends out from the foregut into the surrounding mesenchyme. The endoderm of this bud comes from two populations of cells—a lateral group that exclusively forms liver cells, and ventral-medial endoderm cells that form several midgut regions, including the liver (Tremblay and Zaret 2005). The mesenchyme induces this endoderm to proliferate, branch, and form the glandular epithelium of the liver. A portion of the hepatic diverticulum (the region closest to the digestive tube) continues to function as the drainage duct of the liver, and a branch from this duct produces the gallbladder (**Figure 12.29**).

The pancreas develops from the fusion of distinct dorsal and ventral diverticula. As they grow, they come closer together and eventually fuse. In humans, only the ventral duct survives to carry digestive enzymes into the intestine. In other species (such as the dog), both dorsal and ventral ducts empty into the intestine.

Specification of Liver and Pancreas

There is an intimate relationship between the splanchnic lateral plate mesoderm and the foregut endoderm. Just as the foregut endoderm is critical in specifying the cardiogenic mesoderm, the blood vessel endothelial cells induce the endodermal tube to produce the liver primordium and the pancreatic rudiments.

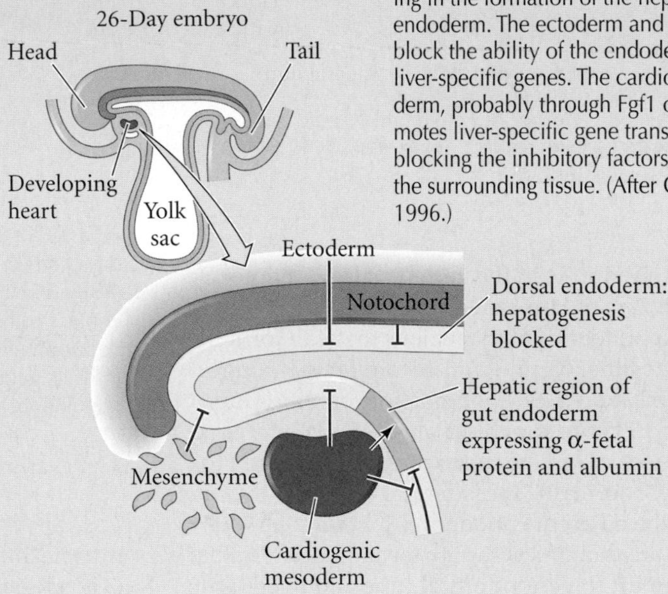

Liver formation

The expression of liver-specific genes (such as the genes for α-fetoprotein and albumin) can occur in any region of the gut tube that is exposed to cardiogenic mesoderm. However, this induction can occur only if the notochord is removed. If the notochord is placed by the portion of the endoderm normally induced by the cardiogenic mesoderm to become liver, the endoderm will not form liver (hepatic) tissue. Therefore, the developing heart appears to induce the liver to form, while the presence of the notochord inhibits liver formation (**Figure 12.30**). This induction is probably due to FGFs secreted by the developing heart cells (Le Douarin 1975; Gualdi et al. 1996; Jung et al. 1999).

However, Matsumoto and colleagues (2001) found that the heart is not the only mesodermal derivative needed to form the liver. Blood vessel endothelial cells are also critical. If endothelial cells are not present in the area around the hepatic region of the gut tube, the liver buds fail to form. This induction occurs even before the endothelial cells have formed tubes, so it does not have anything to do with getting nutrients or oxygen into the region. Thus, the heart endothelial cells have a developmental function in addition to their circulatory roles: they induce the formation of the liver bud by secreting FGFs.

But in order to respond to the FGF signal, the endoderm has to become competent. This competence is given to the foregut endoderm by the **forkhead transcription factors**. Forkhead

Figure 12.30 Positive and negative signaling in the formation of the hepatic (liver) endoderm. The ectoderm and the notochord block the ability of the endoderm to express liver-specific genes. The cardiogenic mesoderm, probably through Fgf1 or Fgf2, promotes liver-specific gene transcription by blocking the inhibitory factors induced by the surrounding tissue. (After Gualdi et al. 1996.)

transcription factors Foxa1 and Foxa2 are required to open the chromatin surrounding the liver-specific genes. These proteins displace nucleosomes from the regulatory regions surrounding these genes and are required before the FGF signal is given (Lee et al. 2005). Mouse embryos lacking *Foxa1* and *Foxa2* expression in their endoderm fail to produce a liver bud or to express liver-specific enzymes.

Once the signal is given, other forkhead transcription factors, such as HNF4α, become critical. HNF4α is essential for the morphological and biochemical differentiation of the hepatic bud into liver tissue (Parviz et al. 2003). When conditional (floxed) mutants of HNF4α were made such that this factor was absent only in the developing liver, neither the tissue architecture, cellular structure, or liver-specific enzymes formed in the liver bud cells. Meanwhile, Odom and colleagues (2004) found that forkhead transcription factors were also critical for the differentiation of the endocrine islands of the pancreas.

HNF4α bound to the regulatory regions of almost half the actively transcribed pancreas-specific genes in these tissues, including those involved in insulin secretion. A link between HNF4α mutations and late-onset type 2 diabetes has been observed (see Kulkarni and Kahn 2004), confirming the importance of this transcription factor in pancreatic, as well as hepatic, development.

Pancreas formation

The formation of the pancreas may be the flip side of liver formation. Whereas the heart cells promote and the notochord prevents liver formation, the notochord may actively promote pancreas formation, while the heart may block the pancreas from forming. It seems that this particular region of the digestive tube has the ability to become either pancreas or liver. One set of conditions (presence of heart, absence of notochord) induces the liver, while another set of conditions (presence of notochord, absence of heart) causes the pancreas to form.

SIDELIGHTS & SPECULATIONS (Continued)

The notochord activates pancreas development by repressing *shh* expression in the endoderm (Apelqvist et al. 1997; Hebrok et al. 1998). (This was a surprising finding, since we saw in Chapter 10 that the notochord is a source of Shh protein and an inducer of further *shh* gene expression in ectodermal tissues.) Sonic hedgehog is expressed throughout the gut endoderm, *except* in the region that will form the pancreas. The notochord in this region of the embryo secretes Fgf2 and activin, which are able to down-regulate *shh* expression. If *shh* is experimentally expressed in this region, the tissue reverts to being intestinal (Jonnson et al. 1994; Ahlgren et al. 1996; Offield et al. 1996).

The lack of Shh in this region of the gut seems to enable it to respond to signals coming from the blood vessel endothelium. Indeed, pancreatic development is initiated at precisely those three locations where the foregut endoderm contacts the endothelium of the major blood vessels. It is at these points—where the endodermal tube meets the aorta and the vitelline veins—that the transcription factors Pdx1 and Ptf1a are expressed (**Figure 12.31A–C**; Lammert et al. 2001; Yoshitomi and Zaret 2004;). If the blood vessels are removed from this area, the Pdx1 and Ptf1a expression regions fail to form, and the pancreatic endoderm fails to bud; if more blood vessels form in this area, more of the endodermal tube becomes pancreatic tissues .

The association of the pancreatic tissues with blood vessels is critical in the formation of the insulin-secreting cells of the pancreas. Pdx1 appears to act in concert with other such transcription factors to form the endocrine cells of the pancreas, the islets of Langerhans (Odom et al. 2004; Burlison et al. 2008; Dong et al. 2008). The exocrine cells (which produce digestive enzymes such as chymotrypsin) and the endocrine cells appear to have the same progenitor (Fishman and Melton 2002), and the level of Ptf1a appears to regulate the proportion of cells in these lineages. The exocrine pancreatic cells have higher amounts of Ptf1a (Dong et al. 2008). The islet

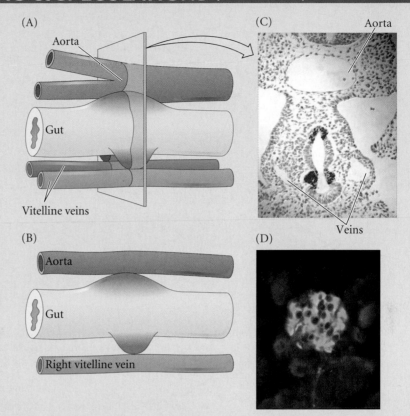

Figure 12.31   Induction of *pdx1* gene expression in the gut epithelium. (A) In the chick embryo, *pdx1* (purple) is expressed in the gut tube and is induced by contact with the aorta and vitelline veins. The regions of *pdx1* gene expression create the dorsal and ventral rudiments of the pancreas. (B) In the mouse embryo, only the right vitelline vein survives, and it contacts the gut endothelium. *Pdx1* gene expression is seen only on this side, and only one ventral pancreatic bud emerges. (C) In situ hybridization of *pdx1* mRNA in a section through the region of contact between the blood vessels and the gut tube of a mouse embryo. The regions of *pdx1* expression show as deep blue. (D) Blood vessels (stained red) direct islets (stained green with antibodies to insulin) of chick embryo to differentiate. The nuclei are stained deep blue.

cells secrete VEGF to attract blood vessels, and these vessels surround the developing islet (**Figure 12.31D**).

Pdx1 is exceptionally important in pancreatic development. This transcription factor elicits budding from the gut epithelium, represses the expression of genes that are characteristic of other regions of the digestive tube, maintains the repression of *shh*, initiates (but does not complete) islet cell differentiation, and is necessary (but not sufficient) for insulin gene expression (Grapin-Botton et al. 2001). Using an inducible Cre-Lox system to mark the progeny of cells expressing either NGN3 or Pdx1, Gu

and colleagues (2002) demonstrated that the Ngn3-expressing cells are specifically islet cell progenitors, whereas the cells expressing Pdx1 give rise to all three types of pancreatic tissue (exocrine, endocrine, and duct cells) (**Figure 12.32**; see Chapter 2). Moreover, Horb and colleagues (2003) have shown that Pdx1 can respecify developing liver tissue into pancreas. When *Xenopus* tadpoles were given a *pdx1* gene attached to a promoter active in liver cells, Pdx1 was made in the liver, and the liver was converted into a pancreas containing both exocrine and endocrine

(Continued on next page)

SIDELIGHTS & SPECULATIONS (Continued)

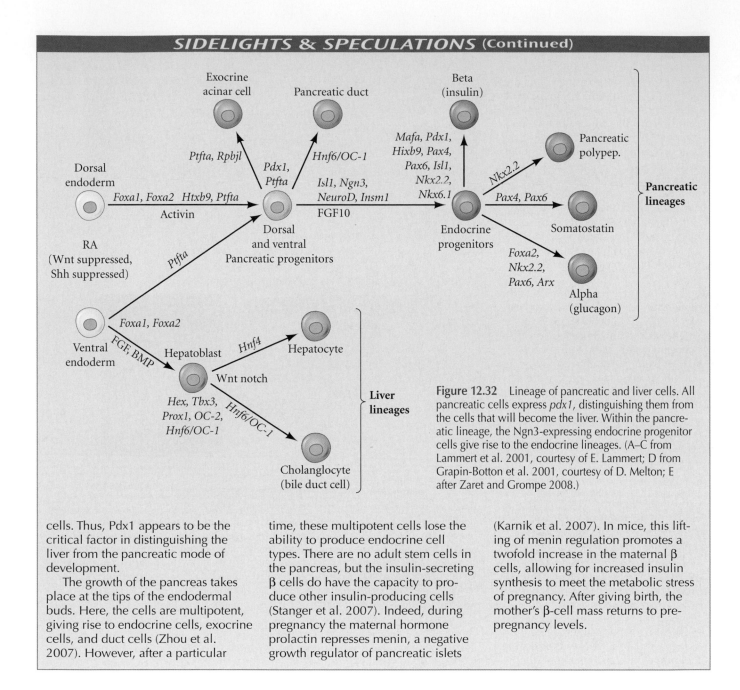

Figure 12.32 Lineage of pancreatic and liver cells. All pancreatic cells express *pdx1*, distinguishing them from the cells that will become the liver. Within the pancreatic lineage, the Ngn3-expressing endocrine progenitor cells give rise to the endocrine lineages. (A–C from Lammert et al. 2001, courtesy of E. Lammert; D from Grapin-Botton et al. 2001, courtesy of D. Melton; E after Zaret and Grompe 2008.)

cells. Thus, Pdx1 appears to be the critical factor in distinguishing the liver from the pancreatic mode of development.

The growth of the pancreas takes place at the tips of the endodermal buds. Here, the cells are multipotent, giving rise to endocrine cells, exocrine cells, and duct cells (Zhou et al. 2007). However, after a particular

time, these multipotent cells lose the ability to produce endocrine cell types. There are no adult stem cells in the pancreas, but the insulin-secreting β cells do have the capacity to produce other insulin-producing cells (Stanger et al. 2007). Indeed, during pregnancy the maternal hormone prolactin represses menin, a negative growth regulator of pancreatic islets

(Karnik et al. 2007). In mice, this lifting of menin regulation promotes a twofold increase in the maternal β cells, allowing for increased insulin synthesis to meet the metabolic stress of pregnancy. After giving birth, the mother's β-cell mass returns to pre-pregnancy levels.

The Respiratory Tube

The lungs are a derivative of the digestive tube, even though they serve no role in digestion. In the center of the pharyngeal floor, between the fourth pair of pharyngeal pouches, the **laryngotracheal groove** extends ventrally (**Figure 12.33**). This groove then bifurcates into the branches that form the paired bronchi and lungs. The laryngotracheal endoderm becomes the lining of the trachea, the two bronchi, and the air sacs (alveoli) of the lungs. Sometimes this separation is not complete and a baby is born with a connection between the two tubes. This digestive and respiratory condition is called a **tracheal-esophageal fistula**,

and must be surgically repaired so the baby can breathe and swallow properly.

The production of the laryngotracheal groove is correlated with the appearance of retinoic acid in the ventral mesoderm, and is probably induced by the same wave of RA that induces the posterior region of the heart. If RA is blocked, the foregut will not produce the lung bud (Desai et al. 2004). Retinoic acid probably induces the formation of Fgf10 by activating Tbx4 in the splanchnic mesoderm adjacent to the ventral foregut (Sakiyama et al. 2003).

Wnt signaling is also required to form the lungs from the digestive tract. Two Wnts—Wnt2 in the earliest stages of lung development, and the related protein Wnt2b—

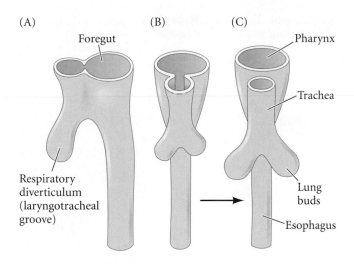

FIGURE 12.33 Partitioning of the foregut into the esophagus and respiratory diverticulum during the third and fourth weeks of human gestation. (A,B) Lateral and ventral views, end of week 3. (C) Ventral view, week 4. (After Langman 1981.)

cause the accumulation of β-catenin in the region of the gut tube that will become the lung and trachea. Without these signals, the separation from the gut tube of the trachea and its development into the lungs fail to happen (**Figure 12.33A,B**; Goss et al. 2009). Conversely, extra lungs can form if β-catenin is expressed ectopically in the gut tube (Harris-Johnson et al. 2009).

As in the digestive tube, the regional specificity of the mesenchyme determines the differentiation of the developing respiratory tube. In the developing mammal, the respiratory epithelium responds in two distinct fashions. In the region of the neck, it grows straight, forming the trachea. After entering the thorax, it branches, forming the two bronchi and then the lungs. The respiratory epithelium of an embryonic mouse can be isolated soon after it has split into two bronchi, and the two sides can be treated differently. **Figure 12.34C** shows the result when the right bronchial epithelium was allowed to retain its lung mesenchyme while the left bronchus was surrounded with tracheal mesenchyme (Wessells 1970). The right bronchus proliferated and branched under the influence of the lung

FIGURE 12.34 Wnt signaling is critical for separation of the trachea and early differentiation of the lung. (A) In normal mice, the trachea separates from the gut tube between days 11 and 12. Lungs are visible 3 days later. (B) In mutant mice that lack both *Wnt2* and *Wnt2b*, the trachea does not separate from the gut tube and no lungs develop. (C) After embryonic mouse lung epithelium had branched into two bronchi, the entire rudiment was excised and cultured. The right bronchus was left untouched while the tip of the left bronchus was covered with tracheal mesenchyme. The tip of the right bronchus formed the branches characteristic of the lung, whereas hardly any branching has occurred in the left bronchus. (A,B from Goss et al. 2009, courtesy of E. E. Morrisey; C from Wessells 1970, courtesy of N. Wessells.)

(A) Wild type

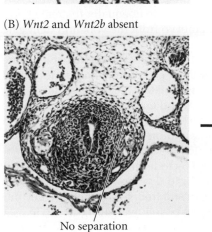

(B) *Wnt2* and *Wnt2b* absent

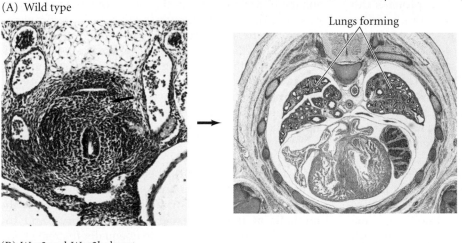

Lungs forming

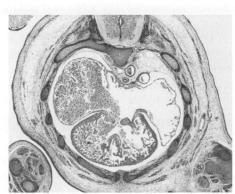

No separation

(C)

Left　　　　　Right

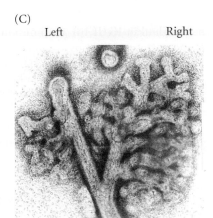

FIGURE 12.35 The immune system relays a signal from the embryonic lung. Surfactant protein-A (SP-A) activates macrophages in the amnionic fluid to migrate into the uterine muscles, where the macrophages secrete IL1β. IL1β stimulates production of cyclooxygenase-2, an enzyme that in turn triggers the production of the prostaglandin hormones responsible for initiating uterine muscle contractions and birth.

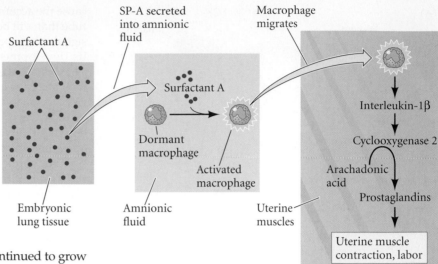

mesenchyme, whereas the left bronchus continued to grow in an unbranched manner. Moreover, the differentiation of the respiratory epithelia into trachea cells or lung cells depends on the mesenchyme it encounters (Shannon et al. 1998).

See WEBSITE 12.2 Induction of the lung

The lungs are among the last of the mammalian organs to fully differentiate. The lungs must be able to draw in oxygen at the newborn's first breath. To accomplish this, the alveolar cells secrete a surfactant into the fluid bathing the lungs. This surfactant, consisting of specific proteins and phospholipids such as sphingomyelin and lecithin, is secreted very late in gestation, and it usually reaches physiologically useful levels at about week 34 of human gestation. The surfactant enables the alveolar cells to touch one another without sticking together. Thus, infants born prematurely often have difficulty breathing and have to be placed on respirators until their surfactant-producing cells mature.

Mammalian birth occurs very soon after lung maturation. New evidence suggests that the embryonic lung may actually signal the mother to start delivery. Condon and colleagues (2004) have shown that surfactant-A—one of the final products produced by the embryonic mouse lung—activates macrophages in the amnionic fluid. These macrophages migrate from the amnion into the uterine muscle, where they produce immune system proteins such as interleukin 1β (IL1β). IL1β initiates the contractions of labor, both by activating cyclooxygenase-2 (which stimulates production of the prostaglandins that contract the uterine muscle cells) and by antagonizing the progesterone receptor (**Figure 12.35**). Surfactant-stimulated macrophages injected into the uteri of female mice induce early labor.* Thus the signal for birth may be transmitted to the mother via her immune system.

---

*IL1β is also produced by macrophages when they attack bacterial infections, which may explain how uterine infections can cause premature labor.

The Extraembryonic Membranes

In reptiles, birds, and mammals, embryonic development took a new evolutionary direction—the amniote egg. This remarkable adaptation, which allowed development to take place on dry land, evolved in the reptile lineage, freeing them to explore niches that were not intimately linked to water. This evolutionary adaptation is so significant and characteristic that reptiles, birds, and mammals are grouped together as the amniote vertebrates, or **amniotes**.

To cope with the challenges of terrestrial development, the amniote embryo produces four sets of extraembryonic membranes to mediate between it and the environment. The evolution of internal development and of the placenta displaced the hard-shelled egg in mammals, but the basic pattern of extraembryonic membranes remains the same.

In developing amniotes, there initially is no distinction between the embryonic and extraembryonic domains. However, as the body of the embryo takes shape, the epithelia at the border between the embryo and the extraembryonic domain divide unequally to create body folds that isolate the embryo from the yolk and delineate which areas are to be embryonic and which extraembryonic (Miller et al. 1994, 1999). These membranous folds are formed by the extension of ectodermal and endodermal epithelium underlain with lateral plate mesoderm. The combination of ectoderm and mesoderm, often referred to as the somatopleure (see Figure 12.4A), forms the amnion and chorion; the combination of endoderm and mesoderm—the splanchnopleure—forms the yolk sac and allantois. The endodermal or ectodermal tissue supplies functioning epithelial cells, and the mesoderm generates the essential blood supply to and from the epithelium. The formation of these folds can be followed in **Figure 12.36**.

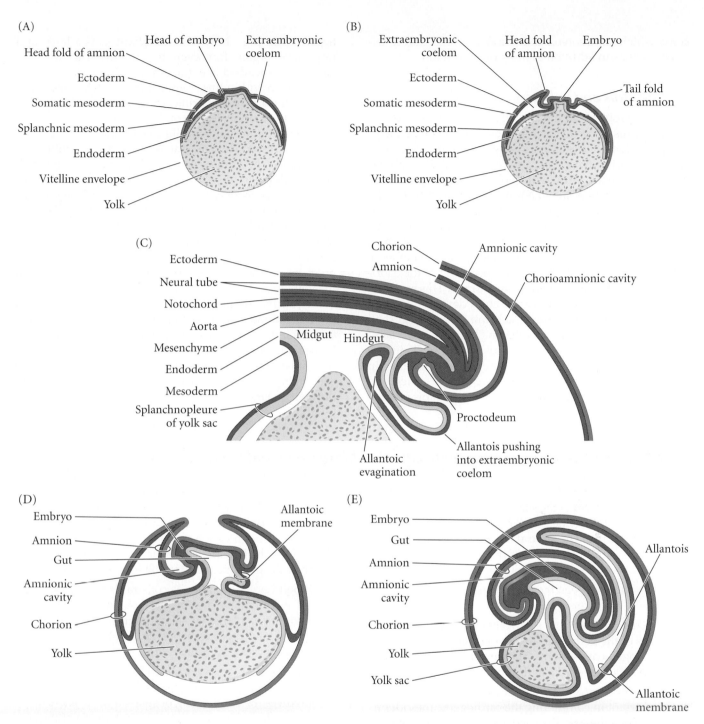

FIGURE 12.36 Schematic drawings of the extraembryonic membranes of the chick. The embryo is cut longitudinally, and the albumen and shell coatings are not shown. (A) 2-day embryo. (B) 3-day embryo. (C) Detailed schematic diagram of the caudal (hind) region of the chick embryo, showing the formation of the allantois. (D) 5-day embryo. (E) 9-day embryo. (After Carlson 1981.)

The amnion and chorion

The first problem a land-dwelling egg faces is desiccation. Embryonic cells would quickly dry out outside an aqueous environment. Such an environment is supplied by the **amnion**. The cells of this membrane secrete **amnionic fluid**; thus, embryogenesis still occurs in water.

The second problem of a terrestrial egg is gas exchange. This exchange is provided for by the **chorion**, the outermost extraembryonic membrane. In birds and reptiles, this membrane adheres to the shell, allowing the exchange of gases between the egg and the environment. In mammals, as we have seen, the chorion has developed into the **pla-**

centa which has evolved endocrine, immune, and nutritive functions in addition to those of respiration.

The allantois and yolk sac

The third problem for a terrestrial egg is waste disposal. The **allantois** stores urinary wastes and also helps mediate gas exchange. The allantois originates from the caudal end of the primitive streak, and in mammals it combines with visceral endoderm (of the yolk sac) to elongate into a chamber (Downs et al. 2009). In reptiles and birds, the allantois becomes a large sac, as there is no other way to keep the toxic by-products of metabolism away from the developing embryo. In some amniote species, such as chickens, the mesodermal layer of the allantoic membrane reaches and fuses with the mesodermal layer of the chorion to create the **chorioallantoic membrane**. This extremely vascular envelope is crucial for chick development and is responsible for transporting calcium from the eggshell into the embryo for bone production (Tuan 1987). In mammals, the size of the allantois depends on how well nitrogenous wastes can be removed by the chorionic placenta. In humans (in which nitrogenous wastes are efficiently removed via the maternal circulation), the allantois is a ves-

tigial sac. In pigs, however, the allantois is a large and important organ. But even in humans, the allantois becomes ensheathed in extraembryonic mesoderm. Blood vessels form within this connecting cord, which becomes the **umbilical cord** that brings the embryonic blood circulation to the uterine vessels of the mother.

And finally, the land-dwelling egg must solve the problem of nutrition. The **yolk sac** is the first extraembryonic membrane to be formed, as it mediates nutrition in developing birds and reptiles. It is derived from splanchnopleural cells that grow over the yolk to enclose it. The yolk sac is connected to the midgut by an open tube, the **yolk duct**, so that the walls of the yolk sac and the walls of the gut are continuous. The blood vessels within the mesoderm of the splanchnopleure transport nutrients from the yolk into the body, for yolk is not taken directly into the body through the yolk duct. Rather, endodermal cells digest the protein in the yolk into soluble amino acids that can then be passed on to the blood vessels within the yolk sac. Other nutrients, including vitamins, ions, and fatty acids, are stored in the yolk sac and transported by the yolk sac blood vessels into the embryonic circulation. In these ways, the four extraembryonic membranes enable the amniote embryo to develop on land.

 ## Snapshot Summary: *Lateral Plate Mesoderm and the Endoderm*

1. The lateral plate mesoderm splits into two layers. The dorsal layer is the somatic (parietal) mesoderm, which underlies the ectoderm and forms the somatopleure. The ventral layer is the splanchnic (visceral) mesoderm, which overlies the endoderm and forms the splanchnopleure.

2. The space between the two layers of lateral plate mesoderm forms the body cavity, or coelom.

3. The heart arises from splanchnic mesoderm on both sides of the body. This region of cells is called the cardiogenic mesoderm. The cardiogenic mesoderm is specified by BMPs in the absence of Wnt signals.

4. The Nkx2-5 and GATA transcription factors are important in committing the cardiogenic mesoderm to become heart cells. These cardiac precursor cells migrate from the sides to the midline of the embryo, in the neck region.

5. The cardiogenic mesoderm forms the endocardium (which is continuous with the blood vessels) and the myocardium (the muscular component of the heart).

6. The endocardial tubes form separately and then fuse. The looping of the heart transforms the original anterior-posterior polarity of the heart tube into a right-left polarity.

7. In mammals, fetal circulation differs dramatically from adult circulation. When the infant takes its first breath, changes in air pressure close the foramen ovale through which blood had passed from the right to the left atrium. At that time, the lungs, rather than the placenta, become the source of oxygen.

8. Blood vessel formation is constrained by physiological, evolutionary, and physical parameters. The subdividing of a large vessel into numerous smaller ones allows rapid transport of the blood to regions of gas and nutrient diffusion.

9. Blood vessels are constructed by two processes, vasculogenesis and angiogenesis. Vasculogenesis involves the condensing of splanchnic mesoderm cells to form blood islands. The outer cells of these islands become endothelial (blood vessel) cells. Angiogenesis involves the remodeling of existing blood vessels.

10. Numerous paracrine factors are essential in blood vessel formation. Fgf2 is needed for specifying the angioblasts. VEGF-A is essential for the differentiation of the angioblasts. Angiopoietins allow the smooth muscle cells (and smooth muscle-like pericytes) to cover the vessels. Ephrin ligands and Eph receptor tyrosine kinases are critical for capillary bed formation.

11. The pluripotential hematopoietic stem cell (HSC) generates other pluripotent stem cells, as well as lineage-restricted stem cells. It gives rise to both blood cells and lymphocytes.

12. In mammals, HSCs are thought to originate from hemogenic endothelial cells that characterize the blood islands, the dorsal aorta, and the placental vessels.

13. The common myeloid precursor (CMP) is a blood stem cell that can generate the more committed stem cells for the different blood lineages. Hematopoietic inductive microenvironments (HIMs) determine the blood cell differentiation.

14. The endoderm constructs the digestive tube and the respiratory tube.

15. Four pairs of pharyngeal pouches become the endodermal lining of the eustacian tubes, the tonsils, the thymus, and the parathyroid glands. The thyroid also forms in this region of endoderm.

16. The gut tissue forms by reciprocal interactions between the endoderm and the mesoderm. Sonic hedgehog from the endoderm appears to play a role in inducing a nested pattern of Hox gene expression in the mesoderm surrounding the gut. The regionalized mesoderm then instructs the endodermal tube to become the different organs of the digestive tract.

17. The endoderm helps specify the splanchnic mesoderm; the splanchnic mesoderm, especially the heart and the blood vessels, helps specify the endoderm.

18. The pancreas forms in a region of endoderm that lacks *sonic hedgehog* expression. The Pdx1 and Ptf1a transcription factors are expressed in this region.

19. The endocrine and exocrine cells of the pancreas have a common origin. The Ngn3 transcription factor probably decides endocrine fate.

20. The respiratory tube is derived as an outpocketing of the digestive tube. The regional specificity of the mesenchyme it meets determines whether the tube remains straight (as in the trachea) or branches (as in the bronchi and alveoli).

21. The chorion and amnion are made by the somatopleure. In birds and reptiles, the chorion abuts the shell and allows for gas exchange. The amnion in birds, reptiles, and mammals bathes the embryo in amnionic fluid.

22. The yolk sac and allantois are derived from the splanchnopleure. The yolk sac (in birds and reptiles) allows yolk nutrients to pass into the blood. The allantois collects nitrogenous wastes.

For Further Reading

Complete bibliographical citations for all literature cited in this chapter can be found at the free-access website **www.devbio.com**

Bruneau, B. G., M. Logan, N. Davis, T. Levi, C. J. Tabin, J. G. Seidman and C. E. Seidman. 1999. Chamber-specific cardiac expression of Tbx5 and heart defects in Holt-Oram syndrome. *Dev. Biol.* 211: 100–108.

Calvi, L. M. and 12 others. 2003. Osteoblastic cells regulate the haematopoietic stem cell niche. *Nature* 425: 841–846.

Fishman, M. P. and D. A. Melton. 2002. Pancreatic lineage analysis using a retroviral vector in embryonic mice demonstrates a common progenitor for endocrine and exocrine cells. *Int. J. Dev. Biol.* 46: 201–207.

Gothert, J. R., S. E. Gustin, M. A. Hall, B. Gottgens, D. J. Izon and C. G. Begley. 2005. In vivo fate-tracing studies using the Scl stem cell enhancer: Embryonic hematopoietic stem cells significantly contribute to adult hematopoiesis. *Blood* 105: 2724–2732.

Jung, J., M. Goldfarb and K. S. Zaret. 1999. Initiation of mammalian liver development from endoderm by fibroblast growth factors. *Science* 284: 1998–2003.

Karkkainenen, M.J. and 11 others. 2004. Vascular endothelial growth factor C is required for sprouting of the first lymphatic vessels from embryonic aveins. *Nature Immunol.* 5: 74–80.

Kim, B.-M., G. Buchner, I. Miletich, P. T. Sharpe and R. A. Shivdasani. 2005. The stomach mesenchymal transcription factor Barx1 specifies gastric epithelial identity through inhibition of transient Wnt signaling. *Dev. Cell* 8: 611–622.

Odom, D. T. and 12 others. 2004. Control of pancreas and liver gene expression by HNF transcription factors. *Science* 303: 1378–1381.

Ottersbach, K. and E. Dzierak. 2005. The murine placenta contains hematopoietic stem cells within the vascular labyrinth region. *Dev. Cell* 8: 377–387.

Simões-Costa, M. S. and 8 others. 2005. The evolutionary origin of heart chambers. *Dev. Biol.* 277: 1–15.

Wang, H. U., Z.-F. Chen and D. J. Anderson. 1998. Molecular distinction and angiogenic interaction between embryonic arteries and veins revealed by ephrin-B2 and its receptor Eph-B4. *Cell* 93: 741–753.

Go Online

WEBSITE 12.1 Coelom formation. Coelom formation is readily visualized by animations. The animation presented here shows the expansion of the mesoderm during chick development.

WEBSITE 12.2 Induction of the lung. The induction of the lung involves interplay between FGFs and Shh. However, it appears to be different from the induction of either the pancreas or the liver.

Vade Mecum

Early heart development. The vertebrate heart begins to function early in its development. You can see this in movies of the living chick embryo at early stages when the heart is little more than a looped tube.

Development of the Tetrapod Limb

13

CONSIDER YOUR LIMB. First, consider its polarity. It has fingers or toes at one end, a humerus or femur at the other. You won't find anyone with fingers in the middle of their arm. Consider also the differences between your hands and your feet. The differences are subtle but obvious. If your fingers were replaced by toes, you would know it. But also consider how *similar* the bones of your feet are to the bones of your hand; it's easy to see that they share a common pattern. The bones of any tetrapod limb, be it arm or leg, wing or flipper, consist of a proximal **stylopod** (humerus/femur) adjacent to the body wall; a **zeugopod** (radius-ulna/tibia-fibula) in the middle region; and a distal **autopod** (carpals-fingers/tarsals-toes) (Figure 13.1). Last, consider the growth of your limbs. Both hands are remarkably similar in size; so are both your feet. So after about 20 years of growth, each of your feet turns out, independently, to be the same length.

These commonplace phenomena present fascinating questions to the developmental biologist. How can growth be so precisely regulated? How is it that we have four limbs and not six or eight? How is it that the fingers form at one end of the limb and nowhere else? How is it that the little finger develops at one edge of the limb and the thumb at the other? How does the forelimb grow differently than the hindlimb?

All of these questions are really about pattern formation. **Pattern formation** is the set of processes by which embryonic cells form ordered spatial arrangements of differentiated tissues. Pattern formation is one of the most dramatic properties of a developing organism, and one that has provoked awe in scientists and laypeople alike. It is one thing to differentiate the chondroblasts and osteoblasts that synthesize the cartilage and bone matrices, respectively; it is another thing to produce those cells in a temporal-spatial orientation that generates a functional bone. It is still another thing to make that bone a humerus and not a pelvis or a femur. The tissues of the finger—bone, cartilage, blood vessels, nerves, dermis, and epidermis—are the same in the toe and thigh; it is their arrangement that differs. The ability of limb cells to sense their relative positions and to differentiate with regard to those positions has been—and still is—the subject of intense study, experimentation, and debate.

The vertebrate limb is an extremely complex organ with an asymmetrical arrangement of parts. The positional information needed to construct a limb has to function in a three-dimensional* coordinate system:

---

*Actually, it is a *four*-dimensional system, in which time is the fourth axis. Developmental biologists get used to seeing nature in four dimensions.

> *What can be more curious than that the hand of a man, formed for grasping, that of a mole for digging, the leg of a horse, the paddle of the porpoise, and the wing of the bat should all be constructed on the same pattern and should include similar bones, and in the same relative positions?*
>
> CHARLES DARWIN (1859)

> *If only you had realized how easy it would have been to create a limb when you were an inch-long embryo— all you would have needed were the right words. Just in case you ever manage to turn back time, those right words are 'fibroblast growth factors.'*
>
> DEBRA NIEHOFF (2005)

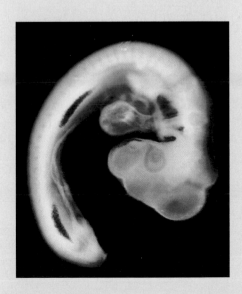

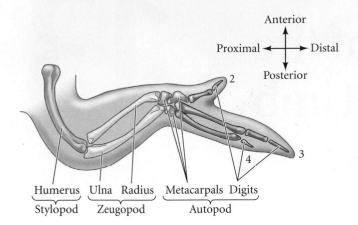

Anterior

Proximal ← → Distal

Posterior

Humerus Ulna Radius Metacarpals Digits
Stylopod Zeugopod   Autopod

FIGURE 13.1 Skeletal pattern of the chick wing. According to convention, the digits are numbered 2, 3, and 4. The cartilage condensations forming the digits appear similar to those forming digits 2, 3, and 4 of mice and humans; however, new evidence (discussed later in the chapter) suggests that the correct designation may be 1, 2, and 3.

- The first dimension is the *proximal-distal axis* ("close-far"; shoulder-finger or hip-toe). The bones of the limb are formed by endochondral ossification. They are initially cartilaginous, but eventually most of the cartilage is replaced by bone. Somehow the limb cells develop differently at early stages of development (when they make the stylopod) than at later stages (when they make the autopod).

- The second dimension is the *anterior-posterior axis* (thumb-pinkie). Our little fingers or toes mark the *posterior* side, while our thumbs or big toes are at the *anterior* end. In humans, it is obvious that each hand develops as a mirror image of the other. One can imagine other arrangements to exist—such as the thumb developing on the left side of both hands—but these patterns do not occur.

- Finally, limbs have a *dorsal-ventral axis*: our palms (ventral) are readily distinguishable from our knuckles (dorsal).

During the past decade, proteins have been identified that play critical roles in the formation of each of the three limb axes. The proximal-distal axis (shoulder-finger; hip-toe) appears to be regulated by proteins of the fibroblast growth factor (FGF) family. The anterior-posterior axis (thumb-pinkie) seems to be regulated by Sonic hedgehog. The dorsal-ventral axis (knuckle-palm) is regulated, at least in part, by Wnt7a. The interactions of these proteins mutually support one another and determine the differentiation of cell types.

The fundamental problem of morphogenesis—how specific structures arise in particular places—is exemplified in limb development. Since the limbs, unlike the heart or brain, are not essential for embryonic or fetal life, one can

experimentally remove or transplant parts of the developing limb, or create limb-specific mutants, without interfering with the vital processes of the organism. Such experiments have shown that the basic "morphogenetic rules" for forming a limb appear to be the same in all tetrapods. Grafted pieces of reptile or mammalian limb buds can direct the formation of chick limbs, and regions of frog limb buds can direct the patterning of salamander limbs (Fallon and Crosby 1977; Sessions et al. 1989; see Hinchliffe 1991). Moreover, as will be detailed in Chapter 15, *regenerating* salamander limbs appear to follow the same rules as developing limbs (Muneoka and Bryant 1982). But what are these morphogenetic rules?

Formation of the Limb Bud

Specification of the limb fields

Limbs do not form just anywhere along the body axis. Rather, there are discrete positions where limbs are generated. The mesodermal cells that give rise to a vertebrate limb can be identified by (1) removing certain groups of cells and observing that a limb does not develop in their absence (Detwiler 1918; Harrison 1918); (2) transplanting groups of cells to a new location and observing that they form a limb in this new place (Hertwig 1925); and (3) marking groups of cells with dyes or radioactive precursors and observing that their descendants partake in limb development (Rosenquist 1971).

Figure 13.2 shows the prospective forelimb areas in the tailbud stage of a salamander and the 10-somite stage of a mouse. The fate-mapping studies on salamanders, pioneered by Ross Granville Harrison's laboratory (see Harrison 1918, 1969), showed that the center of this disc of cells normally gives rise to the limb itself. Adjacent to it are the

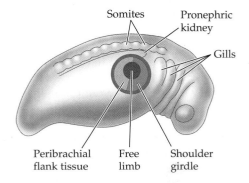

Somites   Pronephric kidney

Gills

Peribrachial   Free   Shoulder
flank tissue   limb   girdle

FIGURE 13.2 Prospective forelimb fields of the salamander and mouse. The forelimb field of the salamander *Ambystoma maculatum*. The central area contains cells destined to form the limb per se (the free limb). The cells surrounding the free limb give rise to the peribrachial flank tissue and the shoulder girdle. The ring of cells outside these regions usually is not included in the limb, but can form a limb if the more central tissues are extirpated. (After Stocum and Fallon 1982.)

(A)
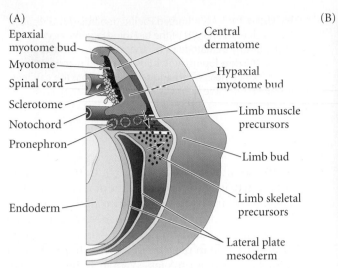

(B)
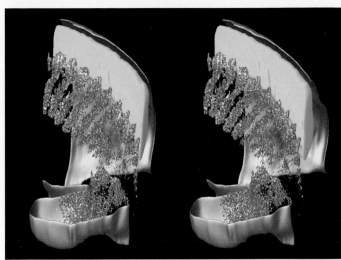

Epaxial myotome bud
Myotome
Spinal cord
Sclerotome
Notochord
Pronephron
Endoderm

Central dermatome
Hypaxial myotome bud
Limb muscle precursors
Limb bud
Limb skeletal precursors
Lateral plate mesoderm

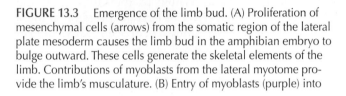

FIGURE 13.3 Emergence of the limb bud. (A) Proliferation of mesenchymal cells (arrows) from the somatic region of the lateral plate mesoderm causes the limb bud in the amphibian embryo to bulge outward. These cells generate the skeletal elements of the limb. Contributions of myoblasts from the lateral myotome provide the limb's musculature. (B) Entry of myoblasts (purple) into the limb bud. This computer stereogram was created from sections of an in situ hybridization to the *myf5* mRNA found in developing muscle cells. If you can cross your eyes (or try focusing "past" the page, looking through it to your toes), the three-dimensionality of the stereogram will become apparent. (B courtesy of J. Streicher and G. Müller.)

cells that will form the peribrachial flank tissue and the shoulder girdle. However, if all these cells are extirpated from the embryo, a limb will still form, albeit somewhat later, from an additional ring of cells that surrounds this area (and which would not normally form a limb). If this last ring of cells is included in the extirpated tissue, no limb will develop. This larger region, representing all the cells in the area capable of forming a limb, is called the **limb field**.

Limb development begins when mesenchyme cells proliferate from the somatic layer of the limb field lateral plate mesoderm (the limb *skeletal* precursor cells) and from the somites (the limb *muscle* precursor cells) at the same level. These mesenchymal cells accumulate under the ectodermal tissue to create a circular bulge called a **limb bud** (Figure 13.3A).

When it first forms, the limb field has the ability to regulate for lost or added parts. In the tailbud stage of *Ambystoma*, any half of the limb disc is able to generate an entire limb when grafted to a new site (Harrison 1918). This potential can also be shown by splitting the limb disc vertically into two or more segments and placing thin barriers between the segments to prevent their reunion. When this is done, each segment develops into a full limb. Thus, like an early sea urchin embryo, the limb field represents a "harmonious equipotential system" wherein a cell can be instructed to form any part of the limb.

In land vertebrates, there are only four limb buds per embryo, and they are always opposite each other with respect to the midline. Although the limbs of different vertebrates differ with respect to the somite level at which they

arise, their position is constant with respect to the level of Hox gene expression along the anterior-posterior axis (see Chapter 8). For instance, in fish (in which the pectoral and pelvic fins correspond to the anterior and posterior limbs, respectively), amphibians, birds, and mammals, the forelimb buds are found at the most anterior expression region of *Hoxc6*, the position of the first thoracic vertebra* (Oliver et al. 1988; Molven et al. 1990; Burke et al. 1995). The lateral plate mesoderm in the four limb fields is also special in that it induces myoblasts to migrate out from the somites and enter the limb bud to become the limb musculature (Figure 13.3B). No other regions of the lateral plate mesoderm can do that (Hayashi and Ozawa 1995).

The regulative ability of the limb bud has recently been highlighted by a remarkable experiment of nature. In numerous ponds in the United States, multilegged frogs and salamanders have been found (Figure 13.4). The presence of these extra appendages has been linked to the infestation of the larval abdomen by parasitic trematode worms. The eggs of these worms apparently split the developing tadpole limb buds in several places, and the limb bud fragments develop as multiple limbs (Sessions and Ruth 1990; Sessions et al. 1999).

*Interestingly, Hox gene expression in at least some snakes (such as *Python*) creates a pattern in which each somite is specified to become a thoracic (ribbed) vertebra. The patterns of Hox gene expression associated with limb-forming regions are not seen (Cohn and Tickle 1999; see Chapter 19).

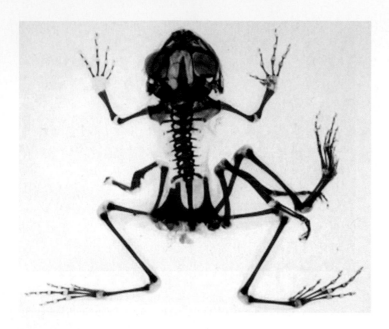

FIGURE 13.4 Multilimbed Pacific tree frog (*Hyla regilla*), the result of infestation of the tadpole-stage developing limb buds by trematode cysts. The parasitic cysts apparently split the developing limb buds in several places, resulting in extra limbs. In this adult frog's skeleton, the cartilage is stained blue; the bones are stained red. (Courtesy of S. Sessions.)

by the somites (Zhao et al. 2009). In this FGF-free zone, Fgf10 is produced throughout the lateral plate mesoderm. Immediately prior to limb formation, however, it becomes restricted to those regions of the lateral plate mesoderm where the limbs will form. This restriction in placement appears to result from the actions of Wnt proteins—Wnt2b in the chick forelimb region and Wnt8c in the chick hindlimb region—which stabilize Fgf10 expression at these sites (**Figure 13.6**; Ohuchi et al. 1997; Kawakami et al. 2001).

Induction of the early limb bud: Wnt proteins and fibroblast growth factors

Molecular studies on the earliest stages of limb formation have shown that the signal for limb bud formation comes from the lateral plate mesoderm cells that will become the prospective limb skeleton. These cells secrete the paracrine factor **Fgf10**, which is capable of initiating the limb-forming interactions between ectoderm and mesoderm. If beads containing Fgf10 are placed ectopically beneath the flank ectoderm, extra limbs emerge (**Figure 13.5**; Ohuchi et al. 1997; Sekine et al. 1999).

After Fgf8 signaling has helped establish the anterior-posterior axis, it is eliminated in the areas capable of forming limbs. This elimination is due to retinoic acid secreted

Specification of forelimb or hindlimb

The mechanisms by which instructions from the Hox code enable the production of Fgf10 is being investigated. Interestingly, the question of how Hox gene expression causes Fgf10 expression may be intimately linked to the question of how the forelimbs and hindlimbs are distinguished. One clue came from specific gene expression patterns. The gene encoding the **Tbx5** transcription factor in mice is transcribed in the anterior lateral plate mesoderm and in the forelimbs,

---

*Tbx stands for T-box. The T (*Brachyury*) gene and its relatives have a sequence that encodes this specific DNA-binding domain. Humans heterozygous for the *TBX5* gene have Holt-Oram syndrome, characterized by abnormalities of the heart and upper limbs (Basson et al. 1996; Li et al. 1996).

(A)

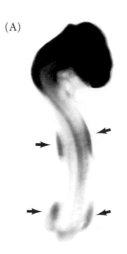

(B)

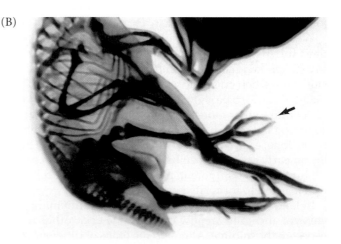

FIGURE 13.5 Fgf10 expression and action in the developing chick limb. (A) Fgf10 becomes expressed in the lateral plate mesoderm in precisely those positions (arrows) where limbs normally form. (B) When transgenic cells that secrete Fgf10 are placed in the flanks of a chick embryo, the Fgf10 can cause the formation of an ectopic limb (arrow). (From Ohuchi et al. 1997, courtesy of S. Noji.)

while the genes encoding the transcription factors **Tbx4** and **Pitx1** are expressed in the posterior lateral plate mesoderm and in the hindlimbs* (Chapman et al. 1996; Gibson-Brown et al. 1996; Takeuchi et al. 1999). Could these three transcription factors be causally involved in directing forelimb versus hindlimb specificity, and could they help bridge the gap between Hox and Fgf10 expression?

In 1998 and 1999, several laboratories (Logan et al. 1998; Ohuchi et al. 1998; Rodriguez-Esteban et al. 1999; Takeuchi et al. 1999, among others) provided gain-of-function evidence that Tbx4 and Tbx5 help specify hindlimbs and forelimbs, respectively. First, if FGF-secreting beads were used to induce an ectopic limb between the chick hindlimb and forelimb buds, the type of limb produced was determined by the Tbx protein expressed. Limb buds induced by placing FGF beads close to the hindlimb (opposite somite 25) expressed *Tbx4* and became hindlimbs. Buds induced close to the forelimb (opposite somite 17) expressed *Tbx5* and developed as forelimbs (wings). Limb buds induced in the center of the flank tissue expressed *Tbx5* in the anterior portion of the limb and *Tbx4* in the posterior portion of the limb. These limbs developed as chimeric structures, with the anterior resembling a forelimb and the posterior resembling a hindlimb (**Figure 13.7**). Moreover, when a chick embryo was made to express *Tbx4* throughout the flank tissue (by infecting the tissue with a virus that expressed

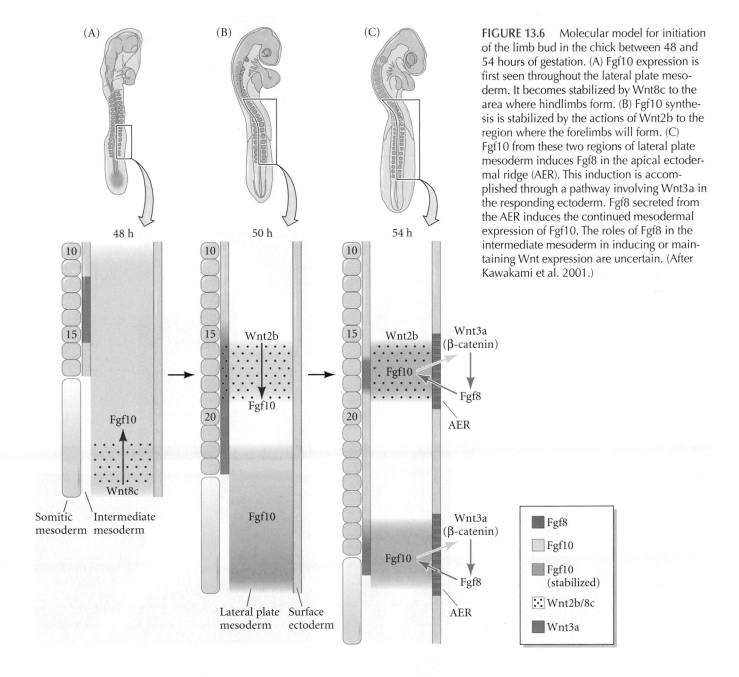

FIGURE 13.6 Molecular model for initiation of the limb bud in the chick between 48 and 54 hours of gestation. (A) Fgf10 expression is first seen throughout the lateral plate mesoderm. It becomes stabilized by Wnt8c to the area where hindlimbs form. (B) Fgf10 synthesis is stabilized by the actions of Wnt2b to the region where the forelimbs will form. (C) Fgf10 from these two regions of lateral plate mesoderm induces Fgf8 in the apical ectodermal ridge (AER). This induction is accomplished through a pathway involving Wnt3a in the responding ectoderm. Fgf8 secreted from the AER induces the continued mesodermal expression of Fgf10. The roles of Fgf8 in the intermediate mesoderm in inducing or maintaining Wnt expression are uncertain. (After Kawakami et al. 2001.)

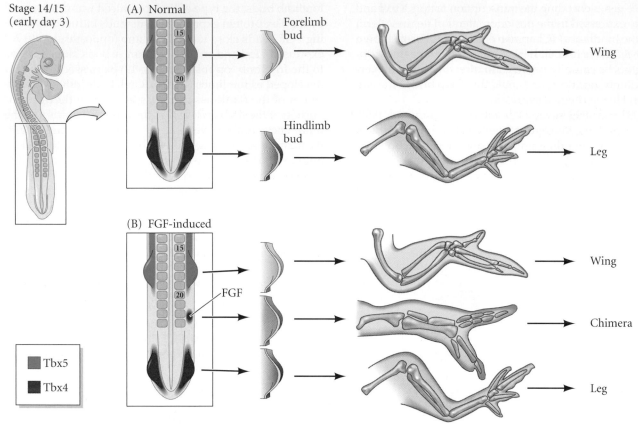

Stage 14/15
(early day 3)

(A) Normal

Forelimb
bud → Wing

Hindlimb
bud → Leg

(B) FGF-induced

Tbx5
Tbx4

FGF

Wing

Chimera

Leg

FIGURE 13.7 Specification of limb type in the chick by Tbx4 and Tbx5. (A) In situ hybridizations show that during normal chick development, Tbx5 (blue) is found in the anterior lateral plate mesoderm, while Tbx4 (red) is found in the posterior lateral plate mesoderm. Tbx5-containing limb buds produce wings, while Tbx4-containing limb buds generate legs. (B) If a new limb bud is induced with an FGF-secreting bead, the type of limb formed depends on which Tbx gene is expressed in the limb bud. If placed between the regions of *Tbx4* and *Tbx5* expression, the bead will induce the expression of *Tbx4* posteriorly and *Tbx5* anteriorly. The resulting limb bud will also express *Tbx5* anteriorly and *Tbx4* posteriorly and will generate a chimeric limb. (C) Expression of *Tbx5* in the forelimb (w, wing) buds and in the anterior portion of a limb bud induced by an FGF-secreting bead. (The somite level can be determined by staining for *Mrf4* mRNA, which is localized to the myotomes.) (D) Expression of *Tbx4* in the hindlimb (le, leg) buds and in the posterior portion of an FGF-induced limb bud. (E,F) A chimeric limb induced by an FGF bead contains anterior wing structures and posterior leg structures. (F) is at a later developmental stage, after feathers form. (A,B after Ohuchi et al. 1998, Ohuchi and Noji 1999; C–F courtesy of S. Noji.)

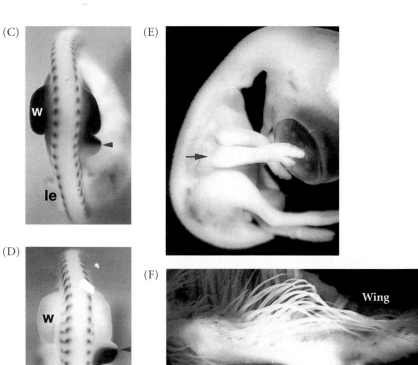

Tbx4), limbs induced in the anterior region of the flank often became legs instead of wings. Thus, Tbx4 and Tbx5 appear to be critical in instructing the limbs to become hindlimbs and forelimbs, respectively.

However, the Tbx genes are not the complete story of limb specification. In mice, at least, Tbx4 and Tbx5 are responsible for the outgrowth of the limbs but not their specification. When either *Tbx5* or *Tbx4* is knocked out, the lateral plate mesoderm fails to make Fgf10 in the forelimb or hindlimb regions. Therefore, it seems likely that Hox proteins may activate Tbx genes and that Tbx proteins activate Fgf10. Moreover, Tbx4 can be replaced by Tbx5 without changing the limb from forelimb to hindlimb. Rather in mice, it appears that Pitx1 is critical in constructing the hindlimbs. Indeed, misexpression of *Pitx1* in the mouse forelimb results in the transformation and translocation of the muscles, bones, and tendons, such that the forelimb acquires a hindlimb-like morphology (Minguillon et al. 2005; DeLaurier et al. 2006). Tbx4 expressed in the forelimb will not do this. Moreover, Pitx1 protein activates hindlimb-specific genes in the forelimb, including *Hoxc10* and *Tbx4*. Ongoing research seeks to discover the mechanisms by which the forelimbs and hindlimbs become distinct from one another.

Generating the Proximal-Distal Axis of the Limb

The apical ectodermal ridge

When mesenchyme cells enter the limb field, they secrete Fgf10 that induces the overlying ectoderm to form a structure called the **apical ectodermal ridge**, or **AER** (Figure 13.8). The AER runs along the distal margin of the limb

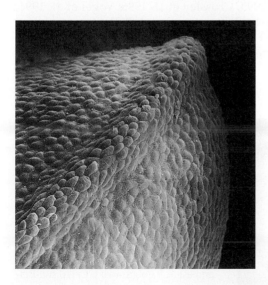

FIGURE 13.8 Scanning electron micrograph of an early chick forelimb bud, with the apical ectodermal ridge in the foreground. (Courtesy of K. W. Tosney.)

bud and will become a major signaling center for the developing limb (Saunders 1948; Kieny 1960; Saunders and Reuss 1974; Fernandez-Teran and Ros 2008). Its roles include (1) maintaining the mesenchyme beneath it in a plastic, proliferating state that enables the linear (proximal-distal, shoulder-finger) growth of the limb; (2) maintaining the expression of those molecules that generate the anterior-posterior (thumb-pinkie) axis; and (3) interacting with the proteins specifying the anterior-posterior and dorsal-ventral (knuckle-palm) axes so that each cell is given instructions on how to differentiate.

The proximal-distal growth and differentiation of the limb bud are made possible by a series of interactions between the AER and the limb bud mesenchyme directly (200 μm) beneath it. This distal mesenchyme is called the **progress zone** (**PZ**) mesenchyme (and sometimes the **undifferentiated zone**), since its proliferative activity extends the limb bud (Harrison 1918; Saunders 1948; Tabin and Wolpert 2007). These interactions were demonstrated by the results of several experiments on chick embryos (Figure 13.9):

1. If the AER is removed at any time during limb development, further development of distal limb skeletal elements ceases.
2. If an extra AER is grafted onto an existing limb bud, supernumerary structures are formed, usually toward the distal end of the limb.
3. If leg mesenchyme is placed directly beneath the wing AER, distal hindlimb structures (toes) develop at the end of the limb. (However, if this mesenchyme is placed farther from the AER, the hindlimb [leg] mesenchyme becomes integrated into wing structures.)
4. If limb mesenchyme is replaced by nonlimb mesenchyme beneath the AER, the AER regresses and limb development ceases.

Thus, although the mesenchyme cells induce and sustain the AER and determine the type of limb to be formed, the AER is responsible for the sustained outgrowth and development of the limb (Zwilling 1955; Saunders et al. 1957; Saunders 1972; Krabbenhoft and Fallon 1989). The AER keeps the mesenchyme cells directly beneath it in a state of mitotic proliferation and prevents them from forming cartilage. Hurle and co-workers (1989) found that if they cut away a small portion of the AER in a region that would normally fall between the digits of the chick leg, an extra digit emerged at that position.*

See WEBSITE 13.1 Induction of the AER

---

*When referring to the hand, one has an orderly set of names to specify each digit (*digitus pollicis, d. indicis, d. medius, d. annularis,* and *d. minimus,* respectively, from thumb to little finger). No such nomenclature exists for the pedal digits, but the plan proposed by Phillips (1991) has much merit. The pedal digits, from hallux to small toe, would be named *porcellus fori, p. domi, p. carnivorus, p. non voratus,* and *p. plorans domi,* respectively.

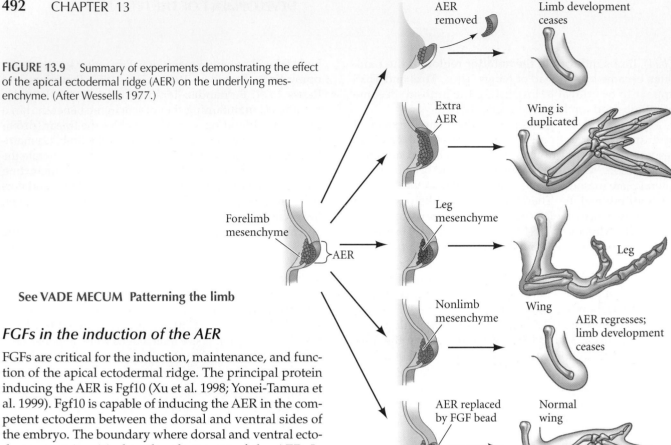

FIGURE 13.9 Summary of experiments demonstrating the effect of the apical ectodermal ridge (AER) on the underlying mesenchyme. (After Wessells 1977.)

See VADE MECUM **Patterning the limb**

FGFs in the induction of the AER

FGFs are critical for the induction, maintenance, and function of the apical ectodermal ridge. The principal protein inducing the AER is Fgf10 (Xu et al. 1998; Yonei-Tamura et al. 1999). Fgf10 is capable of inducing the AER in the competent ectoderm between the dorsal and ventral sides of the embryo. The boundary where dorsal and ventral ectoderm meet is critical to the placement of the AER. In mutants in which the limb bud is dorsalized and there is no dorsal-ventral junction (as in the chick mutant *limbless*), the AER fails to form and limb development ceases (Carrington and Fallon 1988; Laufer et al. 1997a; Rodriguez-Esteban et al. 1997; Tanaka et al. 1997). Fgf10 from the lateral plate mesoderm acts by inducing a Wnt protein (Wnt3a in chicks; Wnt3 in humans and mice) in the prospective limb bud ectoderm. The Wnt protein acts through the canonical β-catenin pathway to induce Fgf8 expression and form the AER (**Figure 13.10**; see Figure 13.6; also see Fernandez-Teran and Ros 2008).

Fgf8 stimulates mitosis in the mesenchyme cells beneath it and causes these cells to keep expressing Fgf10. Thus, a positive feedback loop is established: Fgf10 in the mesenchyme induces Fgf8 in the AER, and Fgf8 in the AER maintains Fgf10 expression in the mesenchyme. Each FGF activates the synthesis of the other (see Figure 13.6; Mahmood et al. 1995; Crossley et al. 1996; Vogel et al. 1996; Ohuchi et al. 1997; Kawakami et al. 2001). The continued expression of these FGFs maintains mitosis in the mesenchyme beneath the AER.

(A)   (B)   (C)

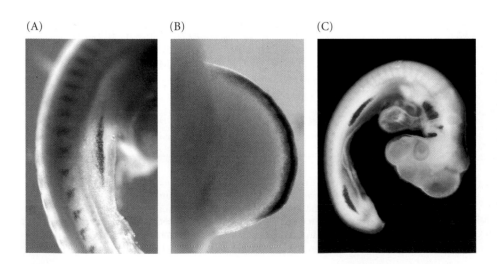

FIGURE 13.10 Fgf8 in the apical ectodermal ridge. (A) In situ hybridization showing expression of *fgf8* message (stained dark purple) in the ectoderm as the limb bud begins to form. (B) Expression of *fgf8* RNA in the AER, the source of mitotic signals to the underlying mesoderm. (C) In the normal 3-day chick embryo, *fgf8* is expressed in the AER of both the forelimb and hindlimb buds. It is also expressed in several other places in the embryo, including the pharyngeal arches. (A,B courtesy of J. C. Izpisúa-Belmonte; C courtesy of A. López-Martínez and J. F. Fallon.)

(A)

FIGURE 13.11 The AER is necessary for wing development. (A) Skeleton of a normal chick wing (dorsal view). (B) Dorsal views of skeletal patterns after removal of the entire AER from the right wing bud of chick embryos at various stages. (From Iten 1982, courtesy of L. Iten.)

(B)

Specifying the limb mesoderm: Determining the proximal-distal polarity of the limb

In 1948, John Saunders made a simple and profound observation: if the AER is removed from an early-stage wing bud, only a humerus forms. If the AER is removed slightly later, humerus, radius, and ulna form (**Figure 13.11**; Saunders 1948; Iten 1982; Rowe and Fallon 1982). Explaining how this happens has not been easy. First it had to be determined whether the positional information for proximal-distal polarity resided in the AER or in the progress zone mesenchyme. Through a series of reciprocal transplantations, this specificity was found to reside in the mesenchyme. If the *AER* had provided the positional information—somehow instructing the undifferentiated mesoderm beneath it as to what structures to make—then older AERs combined with younger mesoderm should have produced limbs with deletions in the middle, while younger AERs combined with older mesoderm should have produced duplications of structures. This was not found to be the case; rather, normal limbs formed in both experiments (Rubin and Saunders 1972). But when the entire progress zone (including both the mesoderm and the AER) from an early embryo was placed on the limb bud of a later-stage embryo, new proximal structures were produced beyond those already present. Conversely, when old progress zones were added to young limb buds, distal structures developed immediately, so that digits were seen to emerge from the humerus without an intervening ulna and radius (**Figure 13.12**; Summerbell and Lewis 1975).

But how does the mesenchyme specify the proximal-distal axis? Three models have been proposed to account for this phenomenon. One model emphasizes the dimen-

(A)

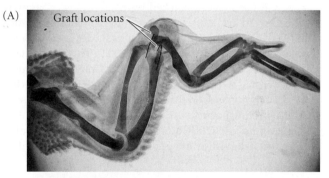

Graft locations

(B)

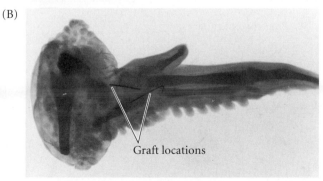

Graft locations

FIGURE 13.12 Control of proximal-distal specification by the progress zone mesenchyme. (A) An extra set of ulna and radius formed when an early wing-bud progress zone was transplanted to a late wing bud that had already formed ulna and radius. (B) Lack of intermediate structures seen when a late wing-bud progress zone was transplanted to an early wing bud. (From Summerbell and Lewis 1975, courtesy of D. Summerbell.)

sion of *time*, one model emphasizes the dimension of *space*, and one model emphasizes generation of *periodicity* .

The **progress zone model** postulates that each mesoderm cell is specified by the amount of time it spends dividing in the progress zone. The longer a cell spends in the PZ, the more mitoses it achieves, and the more distal its specification becomes. Since the PZ remains the same size, cells must exit the PZ constantly as the limb grows; once they leave, they can form cartilage. Thus, the first cells leaving the PZ become the stylopod, and the last cells to leave it become the autopod (**Figure 13.13A**; Summerbell 1974; Wolpert et al. 1979). According to this model, removing the AER means that the progress zone cells can no longer divide and be further specified, so only the proximal structures form.

Although the progress zone model has been the prevailing means of explaining limb polarity, it does not explain recent data from experiments in which the progress zone mesenchyme was prevented from dividing in mice lacking both *fgf8* and *fgf4* (both of which are expressed in the AER). In the limbs of these mice, the proximal elements were altogether missing, while the distal elements were present and often normal. This result could not be explained by a model wherein the specification of distal elements depended on the number of mitoses in the progress zone (Sun et al. 2002). Moreover, when individual early PZ cells were marked with a dye, the progeny of each cell were limited to a particular segment of the limb. The tip of the early limb bud provided cells only to the autopod, for instance (Dudley et al. 2002). This finding suggested that specification was early and that the time spent in the progress zone did not specify cell position. In addition, cell division is seen throughout the limb bud, not solely in the progress zone.

Gail Martin and Cliff Tabin have proposed an **early allocation and progenitor expansion model** as an alternative

to the progress zone model. Here, the cells of the entire early limb bud are already specified; subsequent cell divisions simply expand these cell populations (**Figure 13.13B**). The effects of AER removal are explained by the observation of Rowe et al. (1982) that when the AER is taken off the limb bud, there is apoptosis for approximately 200 μm. If one removes the AER of an early limb bud, before the mesenchymal cell populations have expanded, the cells specified to be zeugopod and autopod die, and only the cells already specified to be stylopod remain. In later embryonic stages, the limb bud grows and the 200-μm region of apoptosis merely deletes the autopod. Research is presently underway to distinguish the contributions of these mechanisms to anterior-posterior axis specification in the limb.

In recent years, the proponents of both models have gotten together and proclaimed that neither model fits the data generated over the past decade (Tabin and Wolpert 2007), and that new models must demonstrate how the progenitors of each region of the limb are generated and how they become specified to a particular segment. To this end, we turn to molecular and physical analyses of limb polarity.

A reaction-diffusion model for limb specification

In contrast to models wherein the proximal-distal axis is specified by paracrine-induced transcription factors, the activator-inhibitor model of Newman and Bhat (2007) portrays these factors as secondarily stabilizing patterns that have been initiated by a reaction-diffusion mechanism (**Figure 13.14**). According to this model, the limb bud becomes sequentially patterned into three areas. In the first area, basically the progress zone, cells respond to FGFs from the AER that suppress the synthesis of fibronectin and there-

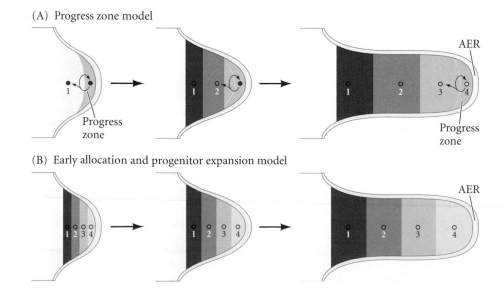

(A) Progress zone model

(B) Early allocation and progenitor expansion model

FIGURE 13.13 Two models for mesodermal specification of the proximal-distal axis of the limb. (A) Progress zone model, wherein the length of time a mesodermal cell remains in the progress zone specifies its position. (B) Early allocation and progenitor expansion model, wherein the territories of the limb bud are established very early and the cells grow in each of these areas.

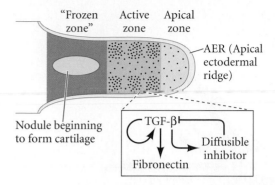

FIGURE 13.14 Reaction-diffusion model for proximal-distal limb specification. The AER divides the limb into three domains. The apical domain (basically the progress zone) is a region where FGF signals from the AER prevent the expression of fibronectin, an extracellular matrix protein that assembles the mesenchymal cells into nodules. The second domain, outside the influence of the AER, is where the nodules actively form according to a reaction-diffusion system (shown on the bottom of the figure). Here each cell secretes and can respond to the paracrine factors of the TGF-β family. TGF-β stimulates its own synthesis as well as that of fibronectin. TGF-β also stimulates the synthesis of a diffusible inhibitor of its own synthesis, thereby limiting the synthesis of fibronectin according to the geometry of the limb bud. (i.e., how many "waves" of activator will be allowed.) The fibronectin-aggregated nodules can now differentiate into cartilage, "freezing" the configuration. This constitutes the third domain.

by prevent condensation of the mesenchyme cells into cartilaginous nodules. In the area just beyond the reach of the AER signals, each mesenchymal cell becomes responsive to TGF-β signals produced by other mesenchymal cells and by themselves. This is the region of patterning activity that directs the formation of cartilaginous nodules. It is thought that the TGF-β acts (1) as a positive inducer of itself, (2) as a positive inducer of the extracellular matrix protein fibronectin (which induces cartilaginous nodule formation), and (3) as an inducer of its own soluble inhibitor (which has not yet been identified).

This unidentified inhibitor diffuses more rapidly than TGF-β and prevents TGF-β synthesis as long as the inhibitor's concentration is above a certain level. When its concentration falls below that level, TGF-β (and its inhibitor) are synthesized. This signaling creates a periodic pattern of regions alternately containing and lacking TGF-β and fibronectin. As the nodules form and coalesce, they become refractile to the signals and are "frozen" in form.

Also according to this model, the form of each bone depends on the physical shape of the bud. At different sizes of the limb, different numbers of cartilaginous units can form bones. First a single condensation can fit (humerus), then two (ulna and radius), then several (wrist, digits). In this reaction-diffusion hypothesis, the aggregations of cartilage actively recruit more cells from the surrounding area

and laterally inhibit the formation of other foci of condensation. The number of foci, then, depends on the geometry of the tissue and the strength of the lateral inhibition. If the inhibition remains the same, tissue volume must increase in order to get two foci forming where one had been allowed before. This model predicts that slight size changes of the distal limb bud alter the number of digits. This is indeed found to be the case, and it may be a very simple way of gaining or losing digits during evolution. According to this model, these standing waves of synthesis and inhibition would constitute the original pattern of the limb. Such patterning of the limb into the regions of stylopod, zeugopod, and autopod by reaction-diffusion mechanisms rather than by transcription factors has been demonstrated as being possible by computer modeling (see Forgacs and Newman 2005). Since no single model currently explains all the experimental phenomena, the generation of the proximodistal axis of the limb remains under intense investigation.

Specification of the Anterior-Posterior Axis

The zone of polarizing activity

The specification of the anterior-posterior axis of the limb is the earliest restriction in limb bud cell potency from the pluripotent condition. In the chick, this axis is specified shortly before a limb bud is recognizable. Hamburger (1938) showed that as early as the 16-somite stage, prospective wing mesoderm transplanted to the flank area develops into a limb with the anterior-posterior and dorsal-ventral polarities of the donor graft, not those of the host tissue.

Several experiments (Saunders and Gasseling 1968; Tickle et al. 1975) suggest that the anterior-posterior axis is specified by a small block of mesodermal tissue near the posterior junction of the young limb bud and the body wall. When tissue from this region is taken from a young limb bud and transplanted to a position on the anterior side of another limb bud, the number of digits of the resulting wing is doubled (**Figure 13.15**). Moreover, the structures of the extra set of digits are mirror images of the normally produced structures. The polarity has been maintained, but the information is now coming from both an anterior and a posterior direction. Thus this region of the mesoderm has been called the **zone of polarizing activity** (**ZPA**).

SONIC HEDGEHOG DEFINES THE ZPA The search for the molecule(s) conferring this polarizing activity on the ZPA became one of the most intensive quests in developmental biology. In 1993, Riddle and colleagues showed by in situ hybridization that *sonic hedgehog* (*shh*), a vertebrate homologue of the *Drosophila hedgehog* gene, was expressed specifically in that region of the limb bud known to be the ZPA (**Figure 13.16A**).

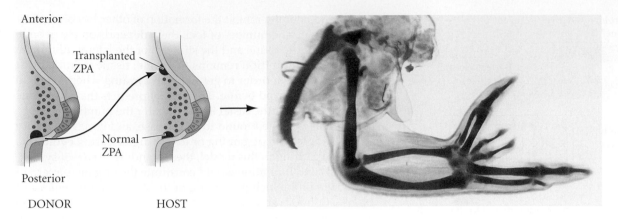

Anterior

Transplanted ZPA

Normal ZPA

Posterior

DONOR   HOST

FIGURE 13.15 When a ZPA is grafted to anterior limb bud mesoderm, duplicated digits emerge as a mirror image of the normal digits. (From Honig and Summerbell 1985, photograph courtesy of D. Summerbell.)

As evidence that this association between the ZPA and *sonic hedgehog* was more than just a correlation, Riddle and co-workers (1993) demonstrated that the secretion of Sonic hedgehog protein is sufficient for polarizing activity. They transfected embryonic chick fibroblasts (which normally would never synthesize Shh) with a viral vector containing the *shh* gene (**Figure 13.16B**). The gene became expressed, translated, and secreted in these fibroblasts, which were then inserted under the anterior ectoderm of an early chick limb bud. Mirror-image digit duplications like those induced by ZPA transplants were the result. More recently, beads containing Sonic hedgehog protein were shown to cause the same duplications (López-Martínez et al. 1995; Yang et al. 1997). Thus, Sonic hedgehog appears to be the active agent of the ZPA.

FIGURE 13.16 Sonic hedgehog protein is expressed in the ZPA. (A) In situ hybridization showing the sites of *sonic hedgehog* expression (arrows) in the posterior mesoderm of the chick limb buds. These are precisely the regions that transplantation experiments defined as the ZPA. (B) Assay for polarizing activity of Sonic hedgehog protein. The *shh* gene was inserted adjacent to an active promoter of a chicken virus, and the recombinant virus was placed into cultured chick embryo fibroblast cells. The virally transfected cells were pelleted and implanted in the anterior margin of a limb bud of a chick embryo. The resulting limbs produced mirror-image digits, showing that the secreted protein had polarizing activity. (A courtesy of R. D. Riddle; B after Riddle et al. 1993.)

(B) Transfect *shh*-expressing virus and allow viral spread

Infectable strain of chick embryo fibroblast cells

Centrifuge cells

Cells containing the *shh* gene

Implant in anterior portion of limb bud (stage 19–23 embryo)

Anterior

Cell pellet secreting Shh

Resistant strain of host embryo

Posterior

ZPA

3

4

Mirror image duplication from graft

2

2

Normal development

4

3

(A)

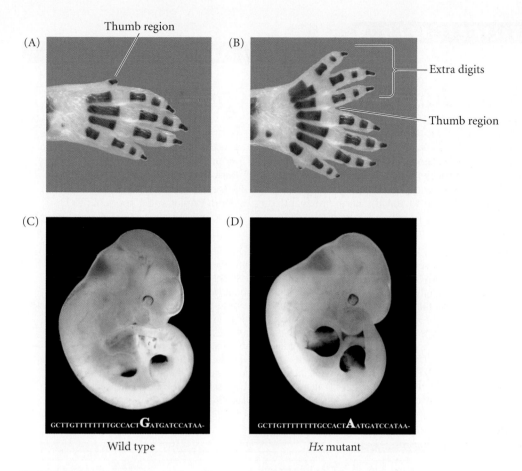

FIGURE 13.17   Ectopic expression of mouse *sonic hedgehog* in the anterior limb causes extra digit formation. (A) Wild-type mouse paw. The bones are stained with Alizarin red. (B) *Hx* (hemimelic extra toes) mutant mouse paws, showing the extra digits associated with the anterior ("thumb") region. (The small extra nodule of posterior bone is peculiar to the *Hx* phenotype on the genetic background used and is not seen on other genetic backgrounds.) (C) Reporter constructs from wild-type *shh* limb enhancer direct transcription solely to the posterior part of each limb bud. (D) Reporter constructs from the *Hx* mutant direct transcription to both the anterior and posterior regions of each limb bud. The wild-type and mutant *shh* limb-specific enhancer region DNA sequences are shown below and highlight the single G-to-A nucleotide substitution that differentiates the two. (From Maas and Fallon 2005, courtesy of B. Robert, Y. Lallemand, S. A. Maas, and J. F. Fallon.)

This fact was confirmed by a remarkable gain-of-function mutation. The *hemimelic extra-toes* (*Hx*) mutant of mice has extra digits on the thumb side of the paws (**Figure 13.17A,B**). This phenotype is associated with a single base-pair difference in the limb-specific *shh* enhancer, a highly conserved region located a long distance (about a million base pairs) upstream from the *shh* gene itself (Lettice et al. 2003; Sagai et al. 2005). Maas and Fallon (2005) made a reporter construct by fusing a β-galactosidase gene to this "long-range" limb enhancer region from both wild-type and *Hx*-mutant genes. They injected these reporter constructs into the pronuclei of newly fertilized mouse eggs to obtain transgenic embryos. In the transgenic mouse embryos carrying the reporter gene with wild-type limb enhancer, staining for β-galactosidase activity revealed a single patch of expression in the posterior mesoderm of each limb bud (**Figure 13.17C**). However, the mice carrying the mutant *Hx* reporter construct showed β-galactosidase activity in *both* the anterior and posterior regions of the limb bud (**Figure 13.17D**). It appears that this enhancer has both positive and negative functions, and that in the anterior region of the limb bud, some factor represses the ability of this enhancer to activate *shh* transcription. The inhibitors probably cannot bind to the mutated enhancer, and thus *shh* expression is seen in both the anterior and posterior regions of the limb bud. This anterior *shh* expression, in turn, causes extra digits to develop in the mutant mice. Similar mutations in the long-range limb enhancer of SHH give polydactylous phenotyes in humans (Gurnett et al. 2007).

Hox Gene Changes during Limb Development

We have seen that the Hox genes play a role in specifying the location where the limbs will form. Now we will see that they play a second role in specifying whether a particular mesenchymal cell will become stylopod, zeugopod, or autopod, and a third role in specifying both the ZPA and the identity of the digits.

Hox specification code for limb skeleton identity

The 5′ (*AbdB*-like) portions (paralogues 9–13) of the HoxA and HoxD gene complexes appear to be active in the forelimb buds of mice. Based on the expression patterns of these genes, and

on naturally occurring and gene knockout mutations, Mario Capecchi's laboratory (Davis et al. 1995) proposed a model wherein these Hox genes specify the identity of a limb region (**Figure 13.18A,B**). Here, *Hox9* and *Hox10* paralogues specify the stylo-

pod, *Hox11* paralogues specify the zeugopod, and *Hox12* and *Hox13* paralogues specify the autopod. This scenario has been confirmed by numerous recent experiments. For instance, when Wellik and Capecchi (2003) knocked out all six *Hox10* paralogues

Figure 13.18 Deletion of limb bone elements by the deletion of paralogous Hox genes. (A) 5′ Hox gene patterning of the forelimb. *Hox9* and *Hox10* paralogues specify the humerus, and *Hox10* paralogues are expressed to a lesser extent in the zeugopod. *Hox11* paralogues are chiefly responsible for patterning the zeugopod; *Hox12* paralogues function in the wrist, with a little patterning of the autopod; and the *Hox13* paralogue group functions in the autopod. (B) Similar but somewhat differing pattern is seen in the hindlimb. (C) Forelimb of a wild-type mouse (left) and of a double-mutant mouse that lacks functional *Hoxa11* and *Hoxd11* genes (right). The ulna and radius are absent in the mutant. (D) Human polysyndactyly ("many fingers joined together") syndrome resulting from a homozygous mutation at the *HOXD13* loci. This syndrome includes malformations of the urogenital system, which also expresses *HOXD13*. (A,B after Wellik and Capecchi 2003; C from Davis et al. 1995, courtesy of M. Capecchi; D from Muragaki et al. 1996, courtesy of B. Olsen.)

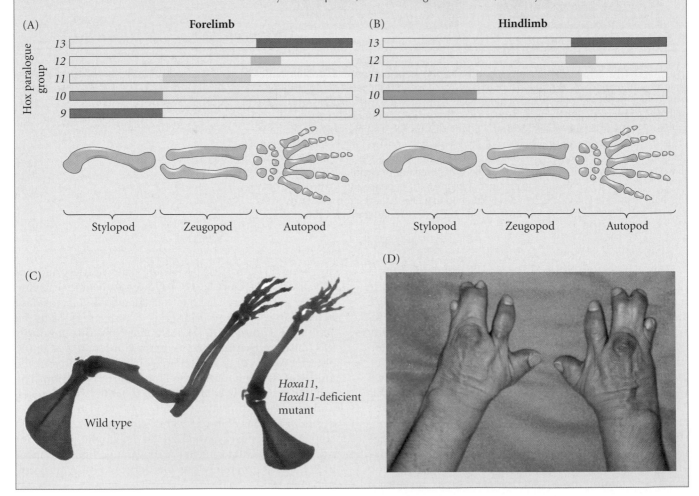

(A) **Forelimb**
Hox paralogue group: 13, 12, 11, 10, 9
Stylopod | Zeugopod | Autopod

(B) **Hindlimb**
13, 12, 11, 10, 9
Stylopod | Zeugopod | Autopod

(C) Wild type | *Hoxa11, Hoxd11*-deficient mutant

(D)

(*Hox10aaccdd*) from mouse embryos, the resulting mice not only had severe axial skeletal defects, they also had no femur or patella. (These mice did have humeruses, because the *Hox9* paralogues are expressed in the forelimb stylopod but not in the hindlimb stylopod.) When all six *Hox11* paralogues were knocked out, the resulting hindlimbs had femurs but neither tibias nor fibulas (and the forelimbs lacked the ulnas and radii). Thus the *Hox11* knockouts got rid of the zeugopods (**Figure 13.18C**). Similarly, knocking out all four *Hoxa13* and *Hoxd13* loci resulted in loss of the autopod (Fromental-Ramain et al. 1996). Humans homozygous for a *HOXD13* mutation show abnormalities of the hands and feet wherein the digits fuse (**Figure 13.18D**), and humans with homozygous mutant alleles of *HOXA13* also have deformities of their autopods (Muragaki et al. 1996; Mortlock and Innis 1997). In both mice and humans, the autopod (the most distal portion of the limb) is affected by the loss of function of the most 5′ Hox genes.

Hox specification of the ZPA

The limb bud has an anterior-posterior polarity as soon as it is formed. The Gli3 transcription repressor is prominent in the anterior region of the limb bud, while the Hand2 transcription activator is present in the posterior mesoderm. These factors are expressed in the mesenchyme immediately before the *sonic hedgehog* gene is expressed (**Figure 13.19A**). In mice lacking *Hand2*, *shh* is not expressed in the limb buds and the limbs are severely truncated. Conversely, ectopic expression of Hand2 protein in the anterior region of the mouse limb bud generates a second ZPA and duplications of the digits (Charité et al. 2000). In the mouse limb bud, Tbx2, which is made in the dermis below the non-AER ectoderm (it is inhibited by Fgf8 from the AER), works with Hand2 and FGFs from the AER to establish the domain of Shh in the posterior mesoderm (i.e., the intersection of Hand2 and Tbx2 domains close to the AER; Nissim et al. 2007). These transcription factors probably operate on the "long-range" *shh* enhancer to activate the

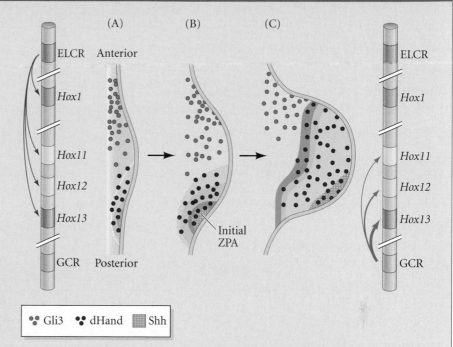

Gli3   dHand   Shh

Figure 13.19 Hox gene expression changes during formation of the tetrapod limb. (A) At the early limb bud stage (day 9 in chicks), the Gli3 repressor transcription factor is in the anterior region, while the Hand2 activating transcription factor is in the posterior. A nested set of HoxD transcription factors arises in the posterior (under the control of the ELCR ehancer), with Hoxd13 being the most limited, followed by Hoxd12 and Hoxd11. The remainder of the limb bud has *Hoxd1* expression. (B) A short time later, Sonic hedgehog protein is seen in the region of cells expressing all of these Hox genes at the posterior of the limb bud mesenchyme. This is the initial ZPA. (C) Under the influence of Sonic hedgehog, a new enhancer (GCR) is activated, and it *reverses* the Hox expression domains. Now, *Hoxd13* is expressed throughout the distal mesenchyme of the limb bud. The zone of *Hoxd12* transcription falls a little more distally, and the domain of *Hoxd11* transcription is a bit more circumscribed still. (After Deschamps 2004.)

shh gene and limit its expression to the posterior mesenchyme (Lettice et al. 2003; Sagai et al. 2005; Maas and Fallon 2005).

In addition, there is a polar arrangement of Hox genes in the nascent limb bud. Within the domain established by Hand2, *Hoxd13* is expressed in the posterior region of the bud, surrounded by a layer of *Hoxd12*-expressing cells and another layer of *Hoxd11*-expressing cells (**Figure 13.19B**). The center and anterior regions of the limb bud (not expressing Hand2) express *Hoxd11*. The expression of *shh* reverses the Hox gene expression pattern. Under a new enhancer, the domain of Hox gene expression expands, and the *Hoxd13* gene is expressed throughout the distal region of the limb bud (**Figure**

13.19C). Closer to the source of Shh, the *Hox12* gene is also transcribed, and the region immediately surrounding the zone of Shh-secreting cells expresses *Hoxd11*. Thus, *Hoxd13* is expressed alone in the anteriormost portion of the distal limb bud, while the region expressing Sonic hedgehog has at least three 5′ *Hoxd* genes expressed (Zákány et al. 2004, 2007).

Moreover, transplantation of either the ZPA or other Shh-secreting cells to the anterior margin of the limb bud at this stage leads to the formation of mirror-image patterns of *Hoxd* gene expression and results in mirror-image digit patterns (Izpisúa-Belmonte et al. 1991; Nohno et al. 1991; Riddle et al. 1993). In normal embryos, the *shh*

(Continued on next page)

SIDELIGHTS & SPECULATIONS (Continued)

gene is kept active by the FGFs of the limb bud, while FGF genes in the AER are kept active by Shh protein. Thus, once the AER and ZPA are established, they mutually support one another (Figure 13.20).

But what genes are these Hox proteins regulating? Some clues come from the analysis of mutations of the Hox13 series of genes. In humans (as mentioned above), portions of the autopod appear to stick together rather than separate. Ectopic expression of the chicken *Hoxa13* gene (which is usually expressed in the distal ends of developing chick limbs) appears to make the cells expressing it "stickier." This property might cause the cartilaginous nodules to condense in specific ways (Yokouchi et al. 1995; Newman 1996).

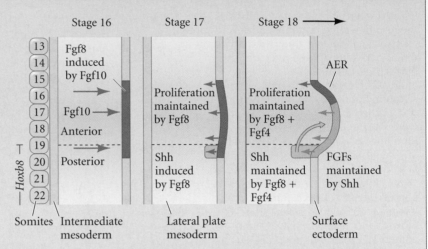

Figure 13.20   Feedback between the AER and ZPA in the forelimb bud. Stage 16 (about 54 hours gestation) represents the same stage as (C) in Figure 13.6. The other two stages are about 5 hours apart. At stage 17, the newly induced AER secretes Fgf8 into the underlying mesenchyme. The mesenchyme expressing *Hoxb8* and *Hand2* is induced to express *shh*, thereby forming the ZPA in the posterior region of the forelimb bud. At stage 18, Shh protein maintains FGF expression in the AER, while the FGFs from the AER maintain *shh* gene expression.

Specifying digit identity by Sonic hedgehog

How does Sonic hedgehog specify the identities of the digits? Surprisingly, when scientists were able to perform fine-scale fate-mapping experiments on the Shh-secreting cells of the ZPA (using recombinase to express a label only in those cells expressing *shh*), they found that *shh*-expressing cells do not undergo apoptosis in the way that the AER does after it finishes its job. Rather, the Shh-secreting cells become the bone and muscle of the posterior limb (Ahn and Joyner 2004; Harfe et al. 2004). Indeed, digits 5 and 4 (and part of digit 3) of the mouse hindlimb are formed from these Shh-secreting cells (Figure 13.21).

It seems that specification of the digits is dependent on the amount of time Shh is expressed and only a little bit by the concentration of Shh that other cells receive (see Tabin and McMahon 2008). The difference between digits 4 and 5 is that the cells of the more posterior digit 5 express Shh longer and are exposed to Shh (in an autocrine manner) for a longer time. Digit 3 has some cells that secrete Shh for a shorter period of time than those of digit 4, and they also depend on Shh secretion by diffusion from the ZPA (since this digit is lost when Shh is modified such that it cannot diffuse away from cells). Digit 2 is dependent entirely on Shh diffusion for its specification, and digit 1 is specified independently of Shh. Indeed, in a naturally occurring chick mutant that lacks *shh* expression in the limb, the only digit that forms is digit 1. Furthermore, when the genes for Shh and Gli3 are conditionally knocked

out of the mouse limb, the resulting limbs have numerous digits, but the digits have no obvious specificity (Litingtung et al. 2002; Ros et al. 2003; Scherz et al. 2007). Vargas and Fallon (2005) propose that digit 1 is specified by Hoxd13 in the absence of Hoxd12. Forced expression of Hoxd12 throughout the digital primordia leads to the transformation of digit 1 into a more posterior digit (Knezevic et al. 1997).

The mechanism by which Sonic hedgehog works to establish a digit's identity may involve the BMP pathway. Shh initiates and sustains a gradient of BMP proteins across the limb bud, and there is evidence that BMP concentration also can specify the digits (Laufer et al. 1994; Kawakami et al. 1996; Drossopoulou et al. 2000). Digit identity is not specified directly in each digit primordium, however. Rather, the identity of each digit is determined by the *interdigital* mesoderm. In other words, the identity of each digit is specified by the webbing between the digits—that region of mesenchyme that will shortly undergo apoptosis.

The interdigital tissue specifies the identity of the digit forming anteriorly to it (i.e., toward the thumb or big toe). Thus, when Dahn and Fallon (2000) removed the webbing between the cartilaginous condensations forming chick hindlimb digits 2 and 3, the second digit was changed into a copy of digit 1. Similarly, when the webbing on the other side of digit 3 was removed, the third digit formed a copy of digit 2 (Figure 13.22A–C). Moreover, the positional value of the webbing could be altered by changing the BMP level. When beads containing BMP antagonists such as Noggin

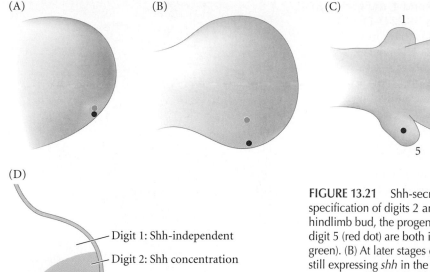

(A)

(B)

(C)

(D)

- Digit 1: Shh-independent
- Digit 2: Shh concentration
- Digit 3: Shh time of expression and concentration
- Digits 4–5: Shh time of expression

☐ Shh diffusion ■ Shh descendants

FIGURE 13.21 Shh-secreting cells form digits 4 and 5 and contribute to the specification of digits 2 and 3 in the mouse limb. (A) In the early mouse hindlimb bud, the progenitors of digit 4 (green dot) and the progenitors of digit 5 (red dot) are both in the ZPA and express *sonic hedgehog* (light green). (B) At later stages of limb development, the cells forming digit 5 are still expressing *shh* in the ZPA, but the cells forming digit 4 no longer do. (C) When the digits form, the cells in digit 5 will have seen high levels of Shh protein for a longer time than the cells in digit 4. (D) Schematic by which digits 4 and 5 are specified by the amount of time they were exposed to Shh in an autocrine fashion; digit 3 is specified by the amount of time the cells were exposed to Shh both in an autocrine and paracrine fashion. Digit 2 is specified by the concentration of Shh its cells received by paracrine diffusion, and digit 1 is specified independently of Shh. (After Harfe et al. 2004.)

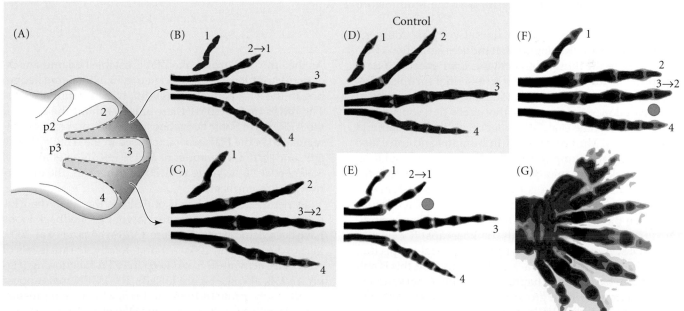

FIGURE 13.22 Regulation of digit identity by BMP concentrations in the interdigital space anterior to the digit and by Gli3. (A) Scheme for removal of interdigital (ID) regions. The results are shown in (B) and (C), respectively. (B) Removal of ID region 2 between digit primordia 2 (p2) and 3 (p3) causes digit 2 to change to the structure of digit 1. (C) Removing ID region 3 (between digit primordia 3 and 4) causes digit 3 to form the structures of digit 2. (D) Control digits and their ID spaces. (E,F) The same transformations as in (B) and (C) can be obtained by adding beads containing the BMP inhibitor Noggin to the ID regions. (E) When a Noggin-containing bead (green dot) is placed in ID region 2, digit 2 is transformed into a copy of digit 1. (F) When the Noggin bead is placed in ID region 3, digit 3 is transformed into a copy of digit 2. (G) The forelimb of a mouse homozygous for deletions of both *gli3* and *shh* is characterized by extra digits of no specific type. (After Dahn and Fallon 2000; Litingtung et al. 2002; B–G, photographs courtesy of R. D. Dahn and J. F. Fallon.)

were placed in the webbing between digits 3 and 4, digit 3 was anteriorly transformed into digit 2 (Figure 13.22 D–F). Each digit has a characteristic array of nodules that form the digit skeleton, and Suzuki and colleagues (2008) have shown that the different levels of BMP signaling in the interdigital webbing regulate the recruitment of progress zone mesenchymal cells into the nodules that make the digits.

Generation of the Dorsal-Ventral Axis

The third axis of the limb distinguishes the dorsal half of the limb (knuckles, nails) from the ventral half (pads, soles). In 1974, MacCabe and co-workers demonstrated that the dorsal-ventral polarity of the limb bud is determined by the ectoderm encasing it. If the ectoderm is rotated 180 degrees with respect to the limb bud mesenchyme, the dorsal-ventral axis is partially reversed; the distal elements (digits) are "upside-down." This suggests that the late specification of the dorsal-ventral axis of the limb is regulated by its ectodermal component.

One molecule that appears to be particularly important in specifying dorsal-ventral polarity is **Wnt7a**. The *Wnt7a* gene is expressed in the dorsal (but not the ventral) ectoderm of chick and mouse limb buds (Dealy 1993; Parr et al. 1993). When Parr and McMahon (1995) knocked out the *Wnt7a* gene, the resulting mouse embryos had ventral footpads on both surfaces of their paws, showing that Wnt7a is needed for the dorsal patterning of the limb (**Figure 13.23**).

Wnt7a is the first known dorsal-ventral axis gene expressed in limb development. It induces activation of the *Lim1* gene in the dorsal mesenchyme. *Lim1* encodes a transcription factor that appears to be essential for specifying dorsal cell fates in the limb. If Lim1 protein is expressed in the ventral mesenchyme cells, those cells develop a dorsal phenotype (Riddle et al. 1995; Vogel et al. 1995; Altabef and Tickle 2002). Mutants of *Lim1* in humans and mice also show this gene's importance for specifying dorsal limb fates. Knockouts in mice produce a syndrome in which the dorsal limb phenotype is lacking, and loss-of-function mutations in humans produce the nail-patella syndrome, in which the dorsal sides of the limbs have been ventralized (e.g., no nails on the digits, no kneecaps) (Chen et al. 1998; Dreyer et al. 1998). Lim1 protein probably specifies the cells to differentiate in a dorsal manner, and this is critical, as we have seen (Chapter 10), for the innervation of motor neurons, whose growth cones recognize inhibitory factors made differentially in the dorsal and ventral compartments of the limb bud.

Coordinating the Three Axes

The mechanisms specifying the three axes of the tetrapod limb are interrelated and coordinated. At first, when the limb bud is relatively small, an initial positive feedback loop between Fgf10 and Fgf8 is established (**Figure 13.24A**).

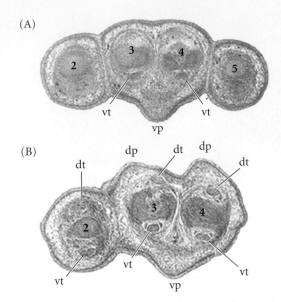

FIGURE 13.23 Dorsal-to-ventral transformations of limb regions in mice deficient for both *Wnt7a* genes. (A) Histological section (stained with hemotoxylin and eosin) of wild-type 15.5-day embryonic mouse forelimb paw. The ventral tendons (vt) and ventral footpads (vp) are readily seen. Numbers indicate digit identity. (B) Same section through a mutant embryo deficient in *Wnt7a*. Ventral tendons and footpads are duplicated on what would normally be the dorsal surface of the paw—that is, the dorsal tendons (dt) and dorsal footpads (dp). (From Parr and McMahon 1995; photographs courtesy of the authors.)

As the limb bud grows, the ZPA is established and another regulatory loop is created (**Figure 13.24B**). Sonic hedgehog in the ZPA activates Gremlin, which inhibits BMPs. The BMPs (mostly BMP4) are able to inhibit the FGFs of the AER. Thus, Sonic hedgehog in the ZPA prevents downregulation of the FGF genes expressed in the AER: *fgf4*, *fgf8*, *fgf9*, and *fgf17* (Niswander et al. 1994; Zúñiga et al. 1999; Scherz et al. 2004; Vokes et al. 2008). After this, the interactivity gets complicated.

Depending on the levels of FGFs in the AER, the ZPA can be either activated or shut down; two feedback loops have been demonstrated (**Figure 13.25A**; Bénazet et al. 2009; Verheyden and Sun 2008). At first, relatively low levels of AER FGFs activate Shh and keep the ZPA functioning (**Figure 13.25B**; also see Figure 13.24B). The FGF signals appear to repress the proteins Etv4 and Etv5, which are repressors of Sonic hedgehog transcription (Mao et al. 2009; Zhang et al. 2009). Thus, the AER and ZPA mutually support each other through the positive loop of Sonic hedgehog and FGFs (Todt and Fallon 1987; Laufer et al. 1994; Niswander et al. 1994). In this phase of limb development, levels of Gremlin (a powerful BMP antagonist) are high, and the positive FGF/Shh loop sustains the limb growth. However, as FGF levels rise (**Figure 13.25C**), they *block* Gremlin expression in the distal mesenchyme. This repression of

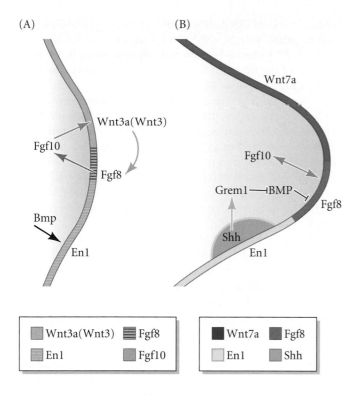

FIGURE 13.24 Early interactions between the AER and limb bud mesenchyme. (A) In the limb bud, Fgf10 from mesenchyme generated by the lateral plate mesoderm activates a Wnt (Wnt3a in chicks; Wnt3 in mice and humans) in the ectoderm. Wnt activates the β-catenin pathway, which induces synthesis of Fgf8 in the region near the AER. Fgf8 activates Fgf10, causing a positive feedback loop. (B) As the limb bud grows, Sonic hedgehog in the posterior mesenchyme creates a new signaling center that induces posterior-anterior polarity, and it also activates Gremlin (Grem1) to prevent BMPs from blocking FGF synthesis in the AER. (After Fernandez-Teran and Ros 2008.)

lier lacked not only dorsal limb structures but also posterior digits, suggesting that Wnt7a is also needed for the anterior-posterior axis (Parr and McMahon 1995). Yang and Niswander (1995) made a similar set of observations in chick embryos. These investigators removed the dorsal ectoderm from developing limbs and found that this resulted in the loss of posterior skeletal elements from the limbs. The reason these limbs lacked posterior digits was that *shh* expression was greatly reduced. Viral-induced expression of *Wnt7a* was able to replace the dorsal ectoderm and restore *shh* expression and posterior phenotypes. These findings showed that the synthesis of Sonic hedgehog is stimulated by the combination of Fgf4 and Wnt7a proteins. Conversely, overactive Wnt signaling in the dorsal ectoderm causes an overgrowth of the AER and extra digits, indicating that the proximal-distal patterning is not independent of dorsal-ventral patterning either (Adamska et al. 2004).

Thus, at the end of limb patterning, BMPs are responsible for simultaneously shutting down the AER, indirectly shutting down the ZPA, and inhibiting the Wnt7a signal along the dorsal-ventral axis (Pizette et al. 2001). The

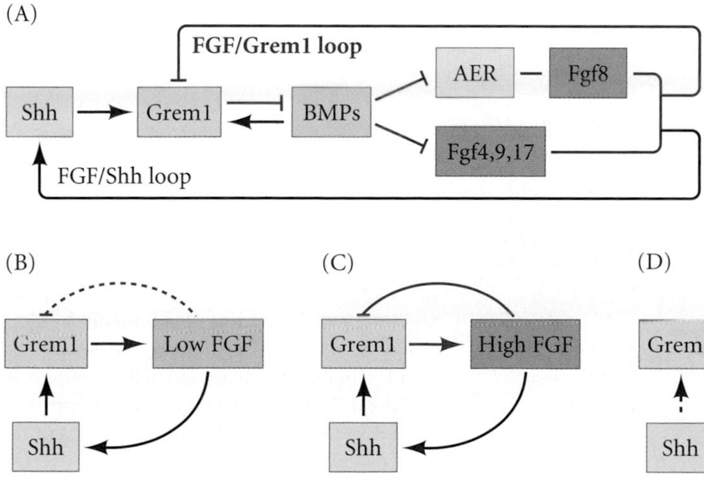

Gremlin synthesis activates the inhibitory feedback loop and creates a zone of cells in the distal mesenchyme that do not secrete Gremlin.

As long as the Gremlin signal can diffuse to the AER, FGFs will be made and the AER maintained. It is thought that as the limb bud expands, the Gremlin signal eventually fails to reach the ectoderm. At that time, the BMPs abrogate FGF synthesis, the AER collapses, and the ZPA (with no FGFs to support it) is terminated. The embryonic phase of limb development ends (**Figure 13.25D**).

The dorsal-ventral axis is also coordinated with these other two. Indeed, the *Wnt7a*-deficient mice described ear-

FIGURE 13.25 Later coordination of AER and ZPA signaling. This is one of several models to explain the coordinated termination of limb outgrowth. (A) The two feedback loops linking the AER and ZPA. In the positive feedback loop (bottom), FGFs from the AER (specifically, Fgf4, 9, and 17) activate Shh transcription, stabilizing the ZPA. Reciprocally, in the inhibitory loop (top), Shh from the ZPA activates Gremlin (Grem1), which blocks the BMPs, thereby preventing the BMP-mediated inactivation of the FGFs in the AER. Thus, in stage (B), there is a mutual accelerated synthesis of both Shh (ZPA) and FGFs (AER). However, as FGF concentration climbs (C), it eventually reaches a threshold where it will inhibit Gremlin. This allows the BMP to begin repressing the AER FGFs. Eventually, as more cells multiply in the area not expressing Gremlin, the Gremlin signal near the AER is too weak to prevent the BMPs from repressing the FGFs. At that point (D), the AER disappears, removing the signal that stabilizes the ZPA. The ZPA then disappears too. (After Verheyden and Sun 2008.)

BMP signal eliminates growth and patterning along all three axes. When exogenous BMP is applied to the AER, the elongated epithelium of the AER reverts to a cuboidal epithelium and ceases to produce FGFs; and when BMPs are inhibited by Noggin, the AER continues to persist days after it would normally have regressed (Gañan et al. 1998; Pizette and Niswander 1999).

Cell Death and the Formation of Digits and Joints

Sculpting the autopod

Cell death plays a role in sculpting the tetrapod limb. Indeed, cell death is essential if our joints are to form and if our fingers are to become separate (Zaleske 1985; Zuzarte-Luis and Hurle 2005). The death (or lack of death) of specific cells in the vertebrate limb is genetically programmed and has been selected for over the course of evolution. The difference between a chicken's foot and the webbed foot of a duck is the presence or absence of cell death between the digits (**Figure 13.26**). Saunders and co-workers have shown that after a certain stage, chick cells between the digit cartilage are destined to die, and will do so even if transplanted to another region of the embryo or placed in culture (Saunders et al. 1962; Saunders and Fallon 1966). Before that time, however, transplantation to a duck limb will save them. Between the time when the cell's death is determined and when death actually takes

place, levels of DNA, RNA, and protein synthesis in the cell decrease dramatically (Pollak and Fallon 1976).

In addition to the **interdigital necrotic zone**, three other regions of the limb are "sculpted" by cell death. The ulna and radius are separated from each other by an **interior necrotic zone**, and two other regions, the **anterior** and **posterior necrotic zones**, further shape the end of the limb (see Figure 13.26B; Saunders and Fallon 1966). Although these zones are referred to as "necrotic," this term is a holdover from the days when no distinction was made between necrotic cell death and apoptotic cell death (see Chapter 3). These cells die by apoptosis, and the death of the interdigital tissue is associated with the fragmentation of their DNA (Mori et al. 1995).

The signal for apoptosis in the autopod is probably provided by the BMP proteins. BMP2, BMP4, and BMP7 are each expressed in the interdigital mesenchyme, and blocking BMP signaling (by infecting progress zone cells with retroviruses carrying dominant negative BMP receptors) prevents interdigital apoptosis (Zou and Niswander 1996; Yokouchi et al. 1996). Since these BMPs are expressed throughout the progress zone mesenchyme, it is thought that cell death would be the "default" state unless there were active suppression of the BMPs. This suppression may come from the Noggin protein, which is made in the developing cartilage of the digits and in the perichondrial cells surrounding it (Capdevila and Johnson 1998; Merino et al. 1998). If Noggin is expressed throughout the limb bud, no apoptosis is seen.

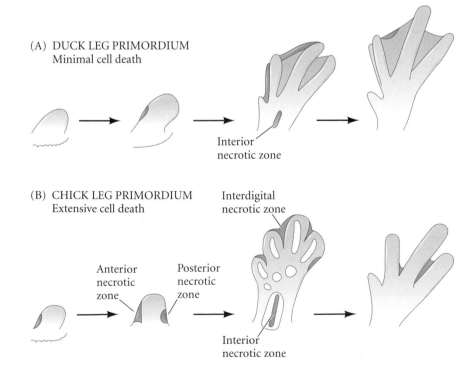

(A) DUCK LEG PRIMORDIUM
Minimal cell death

Interior necrotic zone

(B) CHICK LEG PRIMORDIUM
Extensive cell death

Interdigital necrotic zone

Anterior necrotic zone

Posterior necrotic zone

Interior necrotic zone

FIGURE 13.26 Patterns of cell death in leg primordia of (A) duck and (B) chick embryos. Shading indicates areas of cell death. In the duck, the regions of cell death are very small, whereas there are extensive regions of cell death in the interdigital tissue of the chick leg. (After Saunders and Fallon 1966.)

Limb Development and Evolution

The importance of development to the study of evolution was articulated by C. H. Waddington. He wrote that when we say that a modern horse has one toe and that it came from an ancestor with five toes, what we are saying is that during the evolution of the horse lineage, developmental changes occurred in the cartilage deposition of the embryonic horses' feet. These changes were selected over time to the point of today's *Equus* species. While we still do not know much about the details of evolutionary change in horse limbs, we have found several remarkable cases of limb evolution caused by developmental changes.*

*Earlier in this chapter, it was noted that developmental biologists get used to thinking in four dimensions. Evolutionary developmental biologists have to think in *five* dimensions: the three standard dimensions of space, the dimension of developmental time (hours or days), and the dimension of evolutionary time (millions of years).

Web-footed friends

We can start with the sculpting of the autopod. The regulation of BMPs is critical in creating the webbed feet of ducks (Laufer et al. 1997b; Merino et al. 1999). The interdigital regions of duck feet exhibit the same pattern of BMP expression as the webbing of chick feet. However, whereas the interdigital regions of the chick feet appear to undergo BMP-mediated apoptosis, developing duck feet synthesize the BMP inhibitor Gremlin and block this regional cell death (**Figure 13.27**). Moreover, the webbing of chick feet can be preserved if Gremlin-soaked beads are placed in the interdigital regions (**Figure 13.28**). Thus the evolution of web-footed birds probably involved the inhibition of BMP-mediated apoptosis in the interdigital regions. In Chapter 19, we will find that the bat embryo uses a similar mechanism to acquire its wings.

Dinosaurs and chicken fingers

Next, there is that postulated connection between birds and dinosaurs. Are birds really the descendants of dinosaurs? Although Thomas Huxley (1868, 1870) proposed in the late nineteenth century that the birds descended from dinosaurs, it was J. H. Ostrom's 1969 description of the dinosaur *Deinonychus antirrhopus* and its similarities to the fossils of the first

(Continued on next page)

Figure 13.27   Autopods of chicken feet (upper row) and duck feet (lower row) are shown at similar stages. Both show BMP4 expression (dark blue) in the interdigital webbing; BMP4 induces apoptosis. The duck foot (but not the chicken foot) expresses the BMP4-inhibitory protein Gremlin (dark brown; arrows) in the interdigital webbing. Thus the chicken foot undergoes interdigital apoptosis (as seen by neutral red dye accumulation in the dying cells), but the duck foot does not. (Courtesy of J. Hurle and E. Laufer.)

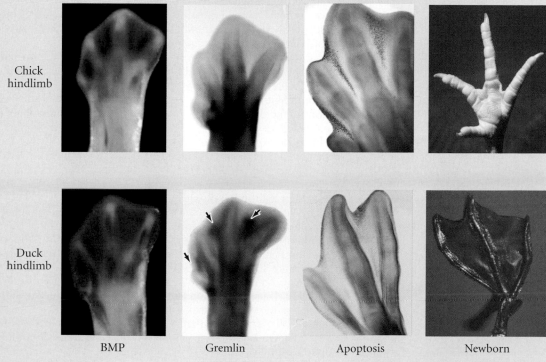

Chick hindlimb

Duck hindlimb

BMP   Gremlin   Apoptosis   Newborn

known bird, *Archaeopteryx*, that was critical in making the dinosaur-to-bird hypothesis acceptable. Ostrom listed 22 similarities between *Deinonychus* and *Archaeopteryx*, similarities found in no other groups and linking birds and dinosaurs.

Without its feathers, *Archaeopteryx* looks exactly like a small coelurosaur (such as Jurassic Park's *Velociraptor*). Indeed, one specimen of *Archaeopteryx* was misidentified as a coelurosaur for over 100 years until its feathers were noticed by Peter Wellnhofer (1993). Gauthier's cladistic work in the mid-1980s (see Gauthier 1986) provided systematic support for the theory that birds are the descendants of coelurosaurian dinosaurs. Unlike any other reptiles, both birds and theropod dinosaurs (of which the coelurosaurs are a group) have a 3-fingered grasping hand and a 4-toed foot supported by three main toes. Thus Padian and Chiappe (1998a) conclude that "in fact, living birds are nothing less than small, feathered, short-tailed theropod dinosaurs."

However, whereas paleontologists were nearly unanimous in their appraisal that birds are the direct descendants of dinosaurs, some developmental biologists harbored serious doubts. Fossil evidence unambiguously identified the theropod-like birds as having wing digits 1, 2, and 3 (Padian and Chiappe 1998b),* but embryological evidence suggested that the wing digits of current birds are 2, 3, and 4. For instance, Burke and Feduccia (1997) found that digit primordia in the fingers of early and present-day birds correspond to the index, middle, and ring fingers (2-3-4). Moreover, the arrangement of cartilaginous condensations is the one expected for the 2-3-4 pattern, not the 1-2-3 pattern. This would mean that the similarity of dinosaur and bird digits is based on

*For a detailed analysis of these transition forms, see http://www.ucmp.berkeley.edu/diapsids/avians.html. Further speculations, based on the single fossil of a newly discovered Jurassic theropod, maintain that dinosaur digits may also have been 2-4 (see Xu et al. 2009). At least, the fossil's wrist appears 2-4; the digits appear 1-3.

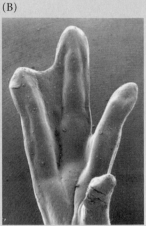

Figure 13.28 Inhibition of cell death by inhibiting BMPs. (A) Control chick limb shows extensive apoptosis in the space between the digits, leading to the absence of webbing. (B) When beads soaked with Gremlin protein are placed in the interdigital mesoderm, the webbing persists and generates a ducklike pattern. (From Merino et al. 1999, courtesy of E. Hurle.)

independent selection for three digits (convergent evolution) and is not based on shared ancestry. This developmental critique of the bird-dinosaur link has been made by other scientists studying chick limb development (see Galis 2005; Welten et al. 2005). They point out that bird feet have reversed toes used for perching on branches (something dinosaurs never developed), and that theropods had a characteristic joint in their lower jaw for grasping prey (something never found in birds). Alan Feduccia has called the notion that birds arose directly from dinosaurs a "delusional fantasy by which one can vicariously study dinosaurs at the backyard bird feeder" (Feduccia 1997).

But a study by Vargas and Fallon (2005a,b) suggests that embryologists have been wrong in their assessment of bird digits. Although the condensations of the digits *look* like those expected for digits 2, 3, and 4, the Hox gene expression patterns suggest that the actual digits are indeed 1, 2, and 3, just as in the theropod dinosaurs. Fallon and Vargas claim that digit 1 (thumb/hallux) is uniquely characterized (at least in the chicken hindlimb and the mouse forelimb and hindlimb) by *Hoxd13* expression in the absence of *Hoxd12* expression (see Figure 13.19). All other digit primordia express both *Hoxd12* and *Hoxd13*. Thus, Vargas and Fallon propose that the wing digits of chickens are actually 1-2-3, and that avian digit arrangement is further proof rather than a rebuttal of the idea that birds are the descendants of dinosaurs.

Recent studies show some support for this "frameshift" model. The Hox expression of alligator digits also suggests that in the bird wing, digits 1, 2, and 3 develop from the embryological positions expected of digits 2, 3, and 4 (Vargas et al. 2008). Furthermore, Vargas and Wagner (2009) showed that if cyclopamine is used to inhibit Shh signaling in chick limbs, the digit morphologies that usually develop from positions 2 and 3 shift such that they develop from positions 3 and 4. Their findings link the developmental analysis of limb development to paleontological studies by Wagner and Gauthier (1999), who hypothesized that when digit 4 was lost in birds, a "homeotic frameshift" occurred, causing digits 2, 3, and 4 to develop from the embryological positions formerly giving rise to digits 2, 3, and 4. If this developmental genetic observation is confirmed, then the "greatest challenge to the theropod-bird link" (Zhou 2004) will have been eliminated.

Developmental biologist Richard Hinchliffe (1994, 1997) sees the argument in a larger context. While evolutionary biologists have reached consensus that birds and dinosaurs evolved from the same class of prehistoric creatures, he says, "the only question we are arguing about is whether [birds] derived very late in time from a specific group of theropod dinosaurs, the so-called raptors, or are they derived from a common-stem ancestor with dinosaurs."

Forming the joints

The function first ascribed to BMPs was the formation, not the destruction, of bone and cartilage tissue. In the developing limb, BMPs induce the mesenchymal cells either to undergo apoptosis or to become cartilage-producing chondrocytes—depending on the stage of development. The same BMPs can induce death or differentiation, depending on the age of the target cell. This "context dependency" of signal action is a critical concept in developmental biology. It is also critical for the formation of joints. Macias and colleagues (1997) have shown that during early limb bud stages (before cartilage condensation), beads secreting BMP2 or BMP7 cause apoptosis. Two days later, the same beads cause the limb bud cells to form cartilage.

In the normally developing limb, BMPs use both of these properties to form joints. BMP7 is made in the perichondrial cells surrounding the condensing chondrocytes and promotes further cartilage formation (**Figure 13.29A**). Two other BMP proteins, BMP2 and GDF5, are expressed at the regions between the bones, where joints will form (**Figure 13.29B**; Macias et al. 1997; Brunet et al. 1998). Mouse mutations have suggested that the function of these proteins in joint formation is critical. Mutations of *Gdf5* produce brachypodia, a condition characterized by a lack of limb joints (Storm and Kingsley 1999). In mice homozygous for loss-of-function alleles of *noggin*, no joints form either. It appears that the BMP7 in these *noggin*-defective embryos is able to recruit nearly all the surrounding mesenchyme into the digits (**Figure 13.29C**).

Wnt proteins and blood vessels also appear to be critical in joint formation. The conversion of mesenchyme cells into nodules of cartilage-forming tissue establishes where the bone boundaries are. The mesenchyme will not form such nodules in the presence of blood vessels, and one of the first indications of cartilage formation is the regression of blood vessels in the region wherein the nodule will form (Yin and Pacifici 2001). Wnt proteins are critical in sustaining transcription of *Gdf5*, and β-catenin produced by the Wnts is able to suppress the *Sox9* and *collagen-2* genes that characterize pre-cartilage cells (Hartmann and Tabin 2001; Tufan and Tuan 2001).

Joints are not merely just "absences" of bone. Rather, they are complex structures that incorporate a lubrication system, an immune system, and a ligament system, all joined to the proper articulation of the skeleton. One of the critical elements of joint formation that allows this differentiation is muscle contraction. In normal joint formation, the cells that will form the joint lose their chondrocyte characteristics (such as the expression of collagen-2 and Sox9), and instead begin to express GDF5, Wnt4, and Wnt9a. These cells will form the articulate cartilage and the synovium (which secretes the lubricating synovial fluid) (Pacifici et al. 2005; Koyama et al. 2008). Kahn and colleagues (2009) have shown that the movement of the bones is necessary for maintaining this commitment to form joints. In mutant mice, where the muscles do not form or are para-

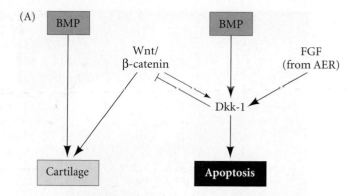

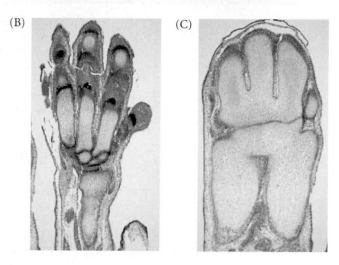

FIGURE 13.29 Possible involvement of BMPs in stabilizing cartilage and apoptosis. (A) Model for the dual role of BMP signals in limb mesodermal cells. BMP can be received in the presence of FGFs (to produce apoptosis) or Wnts (to induce bone). When FGFs from the AER are present, Dickkopf (Dkk) is activated. This protein mediates apoptosis and at the same time inhibits Wnt from aiding in skeleton formation. (B,C) The effects of Noggin. (B) 16.5-day autopod from a wild-type mouse, showing GDF5 expression (dark blue) at the joints. (C) 16.5-day *noggin*-deficient mutant mouse autopod, showing neither joints nor GDF5 expression. Presumably, in the absence of Noggin, BMP7 was able to convert nearly all the mesenchyme into cartilage. (A after Grotewold and Rüther 2002; B,C from Brunet et al. 1998, courtesy of A. P. McMahon.)

lyzed, the joint cells revert back to a cartilaginous phenotype.

The discovery of the fossil *Tiktaalik roseae*, the "fish with fingers," highlights the importance of joint development in limb evolution. Fish fins, including those of some of the most primitive species, develop using the same three Hox gene expression phases as tetrapods use to form their limbs (Davis et al. 2007; Ahn and Ho 2008). What may have allowed the independent modification of fin bones into limb bones are the joints. The joints of *Tiktaalik*'s pectoral fins are very similar to those of amphibians and indicate that *Tiktaalik* had mobile wrists and a substrate-supported

(A)

(B)

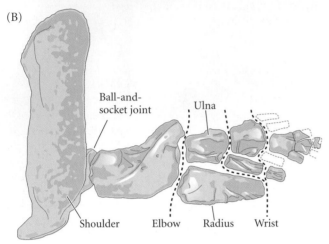

Ball-and-socket joint

Ulna

Shoulder   Elbow   Radius   Wrist

(C)

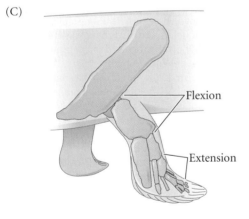

Flexion

Extension

FIGURE 13.30 *Tiktaalik*, a fish with wrists and fingers, lived in shallow waters about 375 million years ago. (A) This reconstruction shows *Tiktaalik*'s fishlike gills, fins, scales, and (lack of) neck. The external nostrils on its snout, however, indicate that it could breath air. (B) Fossilized *Tiktaalik* bones reveal the beginnings of digits, wrists, elbows, and shoulders, and suggest that this amphibian-like fish could propel itself on stream bottoms and perhaps live on land for small durations. The joints of the fin included a ball-and-socket joint in the shoulder and a planar joint that allowed the wrist to bend. Other joints allowed the animal to get purchase on its substrate. (C) Resistant contact with a substrate would allow flexion at the proximal joints (shoulder and elbow) and extension at the distal ones (wrist and digits). (A, model and photograph © Tyler Keillor; B,C after Shubin et al. 2006.)

stance in which the elbow and shoulder could flex (Figure 13.30; Shubin et al. 2006; Shubin 2008). In addition, the presence of wristlike structures and the loss of dermal scales in these regions suggest that this Devonian fish was able to propel itself on moist substrates. So *Tiktaalik* is thought to be a transition between fish and amphibians—a "fishapod" (as one of its discoverers, Neil Shubin, called it), "capable of doing push-ups."

Continued Limb Growth: Epiphyseal Plates

If all of our cartilage were turned into bone before birth, we could not grow any larger, and our bones would be only as large as the original cartilaginous model. However, as the ossification front nears the ends of the cartilage model, the chondrocytes near the ossification front proliferate prior to undergoing hypertrophy, pushing out the cartilaginous ends of the bone. In the long bones of many mammals (including humans), endochondral ossification spreads outward in both directions from the center of the bone. These cartilaginous areas at the ends of the long bones are called **epiphyseal growth plates**. As we saw in Chapter 11, these plates contain three regions: a region of chondrocyte proliferation, a region of mature chondrocytes, and a region of hypertrophic chondrocytes (see Figure 11.17; Chen et al. 1995). As the inner cartilage hypertrophies and the ossification front extends farther outward, the remaining cartilage in the epiphyseal growth plate proliferates. As long as the epiphyseal growth plates are able to produce chondrocytes, the bone continues to grow.

Fibroblast growth factor receptors: Dwarfism

Recent discoveries of human and mouse mutations resulting in abnormal skeletal development have provided remarkable insights into how the differentiation, proliferation, and patterning of chondrocytes are regulated.

The proliferation of the epiphyseal growth plate cells and facial cartilage can be halted by the presence of fibroblast growth factors, which appear to instruct the cartilage precursors to differentiate rather than divide (Deng et al. 1996; Webster and Donoghue 1996). In humans, mutations of the receptors for FGFs can cause these receptors to become activated prematurely. Such mutations give rise to the major types of human dwarfism. **Achondroplasia** is a dominant condition caused by mutations in the transmembrane region of FGF receptor 3 (FgfR3). Roughly 95% of achondroplastic dwarfs have the same mutation of FgfR3, a base-pair substitution that converts glycine to arginine at position 380 in the transmembrane region of the protein. In addition, mutations in the extracellular portion of the FgfR3 protein or in the tyrosine kinase intracellular domain may result in thanatophoric dysplasia (see Figure 3.24), a lethal form of dwarfism that resembles homozygous achondroplasia (Bellus et al. 1995; Tavormina et al. 1995).

As mentioned in Chapter 1, dachshunds have an achondroplastic mutation, but its cause is slightly different than that of the human form. Dachshunds have an extra copy of the *fgf4* gene, which is also expressed in the developing limb (see Figure 1.21). This causes Fgf4 to be made in excess, activating FgfR3 and accelerating the pathway that stops the growth of chondroblasts and hastens their differentiation. The same extra *fgf4* copy has been found in other short-limbed dogs such as corgis and basset hounds (Parker et al. 2009).

Growth hormone and estrogen receptors

The pubertal growth spurt and subsequent cessation of growth are induced by growth hormone; but growth hormone can be induced by sex hormones (Kaplan and Grumbach 1990), and estrogen appears to be important for bone growth in both men and women. Estrogen receptors are found on the cells that regulate growth hormone production, as well as in all the cells of the human growth plate. Thus the effects of estrogen on growth may involve both the regulation of growth hormone and more local effects on the growth plate itself (Juul 2001).

Relatively low levels of estrogen stimulate skeletal growth in men and women, while the higher levels that occur at the end of puberty induce apoptosis in the hypertrophic chondrocytes and simulate the invasion of bone-forming osteoblasts into the growth plate. Without any further cartilage formation, these bones stop growing, a process known as **growth plate closure**. Different alleles of the estrogen receptors may play important roles in the genetic variations in growth rates and adult height (Lehrer et al. 1994; Lorentzon et al. 1999).

Estrogen is an important factor in regulating bone growth, even in men. There have been several documented cases of men who do not produce estrogen (Smith 1994; Juul 2001). These men continue to grow even in adulthood and approach 7 feet in height. Their epiphyseal growth plates do not mature and thus remain full of chondroblasts. In most cases, this deficiency is due to a loss-of-function mutation in the gene encoding aromatase, the enzyme that converts testosterone into estrogen; in such instances, therapy with estrogens is able to stop the excessive growth.

Parathyroid hormone-related peptide and Indian hedgehog

Parathyroid hormone-related peptide (PTHrP) maintains cell division of chondrocytes (by activating cyclin D1 synthesis) and prevents their hypertrophy. In humans, loss-of-function mutations in the protein encoding PTHrP result in severe growth defects due to the lack of limb growth (Provot and Schipiani 2005). PTHrP is stimulated in the cartilage by Indian hedgehog. Indian hedgehog also stimulates BMP production, coordinating the rates of cell division and matrix deposition (Vortkamp et al. 1996).

Coda

FGF proteins generate the proximal-distal axis of the vertebrate limb; Shh and BMPs generate the anterior-posterior axis; and Wnts appear to mediate formation of the dorsal-ventral axis. Thus all the major paracrine factor families act in coordination to build the limb. While many of the "executives" of the limb bud formation have been identified, we still remain largely ignorant about how the orders of these paracrine factors and transcription factors are carried out. Niswander (2002) writes:

> There is a very large gap in our understanding of how the activity of Shh, BMP, FGF, and Wnt genes influences, for example, where the cartilaginous condensation will form, how the elements are sculpted, how the number of phalangeal elements are specified, and where the tendon/muscle will insert.

Limb development is an exciting meeting place for developmental biology, evolutionary biology, and medicine. Within the next decade, we can expect to know the bases for numerous congenital diseases of limb formation, and perhaps we will understand how limbs are modified into flippers, wings, hands, and legs.

 ## Snapshot Summary: *The Tetrapod Limb*

1. The positions where limbs emerge from the body axis depend upon Hox gene expression.

2. The proximal-distal axis of the developing limb is determined by the induction of the ectoderm at the dorsal-ventral boundary by Fgf10 from the mesenchyme. This induction forms the apical ectodermal ridge (AER). The AER secretes Fgf8, which keeps the underlying mesenchyme proliferative and undifferentiated. This area of mesenchyme is called the progress zone.

3. As the limb grows outward, the stylopod forms first, then the zeugopod, and the autopod is formed last. Each phase of limb development is characterized by a specific pattern of Hox gene expression. The evolution of the autopod involved a reversal of Hox gene expression that distinguishes fish fins from tetrapod limbs.

4. The anterior-posterior axis is defined by the expression of Sonic hedgehog in the posterior mesoderm of the limb bud. This region is called the zone of polar-

izing activity (ZPA). If ZPA or Sonic hedgehog-secreting cells or beads are placed in the anterior margin of a limb bud, they establish a second, mirror-image pattern of Hox gene expression and a corresponding mirror-image duplication of the digits.

5. The ZPA is maintained by the interaction of FGFs from the AER with mesenchyme made competent to express Sonic hedgehog by its expression of Hand2 and particular Hox genes. Sonic hedgehog acts, probably in an indirect manner, and probably through the Gli factors, to change the expression of the Hox genes in the limb bud.

6. The dorsal-ventral axis is formed in part by the expression of Wnt7a in the dorsal portion of the limb ectoderm. Wnt7a also maintains the expression level of Sonic hedgehog in the ZPA and of Fgf4 in the posterior AER. Fgf4 and Sonic hedgehog reciprocally maintain each other's expression.

7. The levels of FGFs in the AER can either support or inhibit the production of Shh by the ZPA. As the limb bud grows and more FGFs are produced in the AER, inhibition of Shh occurs. This causes the lowering of FGFs, and eventually proximodistal outgrowth ceases.

8. Cell death in the limb is necessary for the formation of digits and joints. It is mediated by BMPs. Differences between the unwebbed chicken foot and the webbed duck foot can be explained by differences in the expression of Gremlin, a protein that antagonizes BMPs.

9. The BMPs are involved both in inducing apoptosis and in differentiating the mesenchymal cells into cartilage. The effects of BMPs can be regulated by the Noggin and Gremlin proteins. This is critical in forming the joints between the bones of the limb and regulating proximal-distal outgrowth.

10. The ends of the long bones of humans and other mammals contain cartilaginous regions called epiphyseal growth plates. The cartilage in these regions proliferates so that the bone grows larger. Eventually, the cartilage is replaced by bone and growth stops.

For Further Reading

Complete bibliographical citations for all literature cited in this chapter can be found at the free-access website **www.devbio.com**

Mahmood, R. and 9 others. 1995. A role for FGF-8 in the initiation and maintenance of vertebrate limb outgrowth. *Curr. Biol.* 5: 797–806.

Merino, R, J. Rodriguez-Leon, D. Macias, Y. Gañan, A. N. Economides and J. M. Hurle. 1999. The BMP antagonist Gremlin regulates outgrowth, chondrogenesis and programmed cell death in the developing limb. *Development* 126: 5515–5522.

Niswander, L., S. Jeffrey, G. R. Martin and C. Tickle. 1994. A positive feedback loop coordinates growth and patterning in the vertebrate limb. *Nature* 371: 609–612.

Riddle, R. D., R. L. Johnson, E. Laufer and C. Tabin. 1993. Sonic hedgehog mediates the polarizing activity of the ZPA. *Cell* 75: 1401–1416.

Scherz, P. J., E. McGlinn, S. Nissim and C. J. Tabin. 2007. Extended exposure to Sonic hedgehog is required for patterning the posterior digits of the vertebrate limb. *Dev. Biol.* 308: 343–354.

Sekine, K. and 10 others. 1999. Fgf10 is essential for limb and lung formation. *Nature Genet.* 21: 138–141.

Tabin, C. and L. Wolpert. 2007. Rethinking the proximodistal axis of the vertebrate limb in the molecular era. *Genes Dev.* 21: 1433–1442.

Verheyden, J. M. and X. Sun. 2008. An Fgf–Gremlin inhibitory feedback loop triggers termination of limb bud outgrowth. *Nature* 454: 638–641.

Vortkamp, A, K. Lee, B. Lanske, G. V. Segre, H. M. Kronenberg and C. J. Tabin. 1996. Regulation of rate of cartilage differentiation by Indian hedgehog and PTH-related protein. *Science* 273: 613–622.

Wellik, D. M. and M. R. Capecchi. 2003. *Hox10* and *Hox11* genes are required to globally pattern the mammalian skeleton. *Science* 301: 363–367.

Go Online

WEBSITE 13.1 Induction of the AER. The induction of the AER is a complex event involving the interaction between the dorsal and ventral compartments of the ectoderm. The Notch pathway may be critical in this process. Misexpression of the genes in this pathway can result in the absence or duplication of limbs.

Vade Mecum

Patterning the limb. An interview with John Saunders contains movies of his work on limb development, which identified the AER and the ZPA as two of the major signaling centers in limb formation. His transplantation studies provided the framework for the molecular characterization of the mechanisms of limb formation.

Sex Determination

14

HOW AN INDIVIDUAL'S SEX IS DETERMINED has been one of the great questions of natural philosophy since antiquity. Aristotle claimed that sex was determined by the heat of the male partner during intercourse (Aristotle, ca. 335 BCE). The more heated the passion, said the sage, the greater the probability of male offspring. He also counseled elderly men to conceive in the summer if they wished to have male heirs.

Aristotle's hypothesis of sex determination was straightforward: women were men whose development was arrested too early. The female was "a mutilated male" whose development was forestalled when the coldness of the mother's womb overcame the heat of the father's semen. Women were therefore colder and more passive than men, and female sex organs did not mature to the point where they could provide active seeds. These Aristotelian views were accepted both by the Christian church and by the Greco-Roman physician Galen,* whose anatomy texts were the standard for more than a thousand years.

Indeed, until the twentieth century, the environment—temperature and nutrition, in particular—was believed to be important in determining sex. In 1890, Geddes and Thomson summarized all available data on sex determination and came to the conclusion that the "constitution, age, nutrition, and environment of the parents must be especially considered." They argued that factors favoring the storage of energy and nutrients predisposed one to have female offspring, whereas factors favoring the utilization of energy and nutrients influenced one to have male offspring.

This environmental view of sex determination remained the only major scientific theory until the identification (derived primarily from studies on insects) of the X and Y chromosomes and the correlation of the female sex with an XX karyotype (chromosome complement) and the male sex with either XY or XO karyotypes (Stevens 1905; Wilson 1905; see Gilbert 1978). This correlation suggested strongly that a specific nuclear component was responsible for directing the development of the sexual phenotype. Evidence that sex determination occurs by nuclear inheritance rather than by environmental happenstance continued to accumulate.

Today we know that both internal and environmental mechanisms of sex determination operate in different species. We will first discuss the chromoso-

---

*The imperial physician to the Roman emperors Marcus Aurelius and Commodus, Galen first achieved fame as a physician to gladiators, from whose wounds and corpses he undoubtedly learned much anatomy.

Sexual reproduction is ... the masterpiece of nature.

ERASMUS DARWIN (1791)

It is quaint to notice that the number of speculations connected with the nature of sex have well-nigh doubled since Drelincourt, in the eighteenth century, brought together two hundred and sixty-two "groundless hypotheses," and since Blumenbach caustically remarked that nothing was more certain than that Drelincourt's own theory formed the two hundred and sixty-third.

J. A. THOMSON (1926)

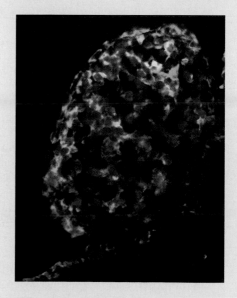

mal mechanisms of sex determination and then consider the ways in which the environment can regulate sexual phenotype.

See WEBSITE 14.1 Social critique of sex determination research

CHROMOSOMAL SEX DETERMINATION

There are several ways chromosomes can determine the sex of an embryo. In mammals, the presence of either a second X chromosome or a Y chromosome determines whether the embryo is to be female (XX) or male (XY). In birds, the situation is reversed (Smith and Sinclair 2001): the male has the two similar sex chromosomes (ZZ), and the female has the unmatched pair (ZW). In flies, the Y chromosome plays no role in sex determination, but the number of X chromosomes appears to determine the sexual phenotype. In other insects (especially hymenopterans such as bees, wasps, and ants), fertilized, diploid eggs develop into females, while the unfertilized, haploid eggs become males (Beukeboom 1995; Gempe et al. 2009). This chapter will discuss only two of the many chromosomal modes of sex determination: sex determination in placental mammals and sex determination in *Drosophila*.

The Mammalian Pattern: Primary and Secondary Sex Determination

Primary sex determination is the determination of the gonads—the egg-forming ovaries or sperm-forming testes. In mammals, primary sex determination is chromosomal and is not usually influenced by the environment. The formation both of ovaries and of testes is an active, gene-directed process. Moreover, as we shall see, both the male and female gonads diverge from a common precursor, the **bipotential**, or **indifferent**, **gonad**.

In most cases, the female's karyotype is XX and the male's is XY (Figure 14.1A). Every individual must carry at least one X chromosome. Since the female is XX, each of her haploid eggs has a single X chromosome. The male, being XY, generates two populations of haploid sperm: half will bear an X chromosome, half a Y. If at fertilization the egg receives a second X chromosome from the sperm, the resulting individual is XX, forms ovaries, and is female; if the egg receives a Y chromosome from the sperm, the individual is XY, forms testes, and is male.

The Y chromosome is a crucial factor for determining male sex in mammals. The Y chromosome carries a gene that encodes a **testis-determining factor**, which organizes the gonad into a testis rather than an ovary. A person with five X chromosomes and one Y chromosome (XXXXXY) would be male. Furthermore, an individual with a single X chromosome and no second X or Y (i.e., XO) develops as

a female and begins making ovaries (although the ovarian follicles cannot be maintained; for a complete ovary, a second X chromosome is needed).

Secondary sex determination involves the sexual phenotype outside the gonads. This includes the male or female duct systems and external genitalia. A male mammal has a penis, scrotum (testicle sac), seminal vesicles, and prostate gland. A female mammal has a vagina, clitoris, labia, cervix, uterus, oviducts, and mammary glands.* In many species, each sex has a sex-specific body size, vocal cartilage, and musculature. Secondary sex characteristics are usually determined by hormones and paracrine factors secreted from the gonads. In the absence of gonads, however, the female phenotype is generated. When Jost (1947, 1953) removed fetal rabbit gonads before they had differentiated, the resulting rabbits had a female phenotype, regardless of whether their genotype was XX or XY.

The general scheme of mammalian sex determination is shown in **Figure 14.1B**. If the embryonic cells have two X chromosomes (and no Y chromosome), the gonadal primordia develop into ovaries. The ovaries produce **estrogen**, a hormone that enables the development of the **Müllerian duct** into the uterus, oviducts, cervix, and upper end of the vagina (Fisher et al. 1998; Couse et al. 1999; Couse and Korach 2001). If embryonic cells contain both an X and

---

*Linnaeus named the mammals—our particular class of vertebrates—after this female secondary sexual trait. The politics of this decision is discussed in Schiebinger 1993.

FIGURE 14.1 Sex determination in mammals. (A) Mammalian chromosomal sex determination results in approximately equal numbers of male and female offspring. (B) Postulated cascades leading to male and female phenotypes in mammals. The conversion of the genital ridge into the bipotential gonad requires, among others, the *LHX9*, *SF1*, and *WT1* genes, since mice lacking any of these genes lack gonads. The bipotential gonad appears to be moved into the female pathway (ovary development) by the *WNT4*, *RSPO1*, and (perhaps) *DAX1* genes and into the male pathway (testis development) by the *SRY* gene (on the Y chromosome) in conjunction with autosomal genes such as *SOX9*. (Lower levels of Dax1 and Wnt4 are also present in the male gonad.) The ovary makes thecal cells and granulosa cells, which together are capable of synthesizing estrogen. Under the influence of estrogen (first from the mother, then from the fetal gonads), the Müllerian duct differentiates into the female reproductive tract, the internal and external genitalia develop, and the offspring develops the secondary sex characteristics of a female. The testis makes two major compounds involved in sex determination. The first, anti-Müllerian factor (AMF), causes the Müllerian duct to regress. The second, testosterone, causes differentiation of the Wolffian duct into the male internal genitalia. In the urogenital region, testosterone is converted into dihydrotestosterone (DHT), and this hormone causes the morphogenesis of the penis and prostate gland. (B after Marx 1995; Birk et al. 2000.)

a Y chromosome, testes form and secrete two major factors. The first is a TGF-β family paracrine factor called **anti-Müllerian factor** (**AMF**), traditionally called anti-Müllerian hormone, or AMH. It destroys the Müllerian duct, preventing formation of the uterus and oviducts. The second factor is the steroid hormone **testosterone**, which masculinizes the fetus, stimulating the formation of the penis, male duct system, scrotum, and other portions of the male anatomy, as well as inhibiting development of the breast primordia. We will now take a more detailed look at these events.

Primary Sex Determination in Mammals

The developing gonads

Mammalian gonads embody a unique embryological situation. All other organ rudiments normally can differentiate into only one type of organ. A lung rudiment can only become a lung, and a liver rudiment can develop only into a liver. The gonadal rudiment, however, has two options: it can develop into either an ovary or a testis. The path of differentiation taken by this rudiment is dictated by the genotype and determines the future sexual development of the organism (Lillie 1917). But before this decision is made, the mammalian gonad first develops through a **bipotential**, or **indifferent**, **stage**, during which time it has neither female nor male characteristics.

In humans, the bipotential gonadal rudiments appear during week 4 and remain sexually indifferent until week 7. The gonadal rudiments are paired regions of the intermediate mesoderm; they form adjacent to the developing kidneys. The ventral portions of these rudiments comprise the **genital ridge epithelium**. During the indifferent stage, the genital ridge epithelium proliferates (**Figure 14.2A,B**). These epithelial layers will form the somatic (i.e., non-germ cell) component of the gonads. The germ cells migrate into the gonad during week 6 and are surrounded by the somatic cells.

(A)

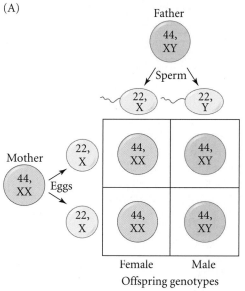

(B)

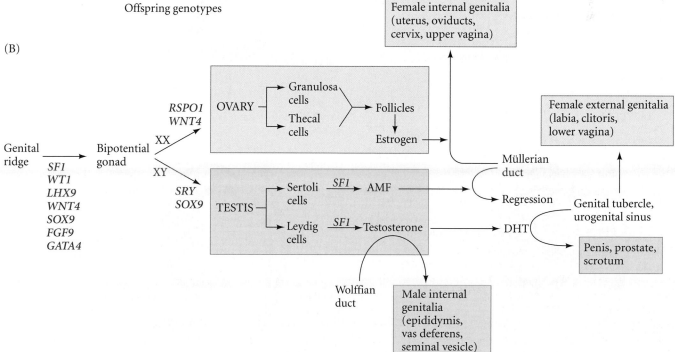

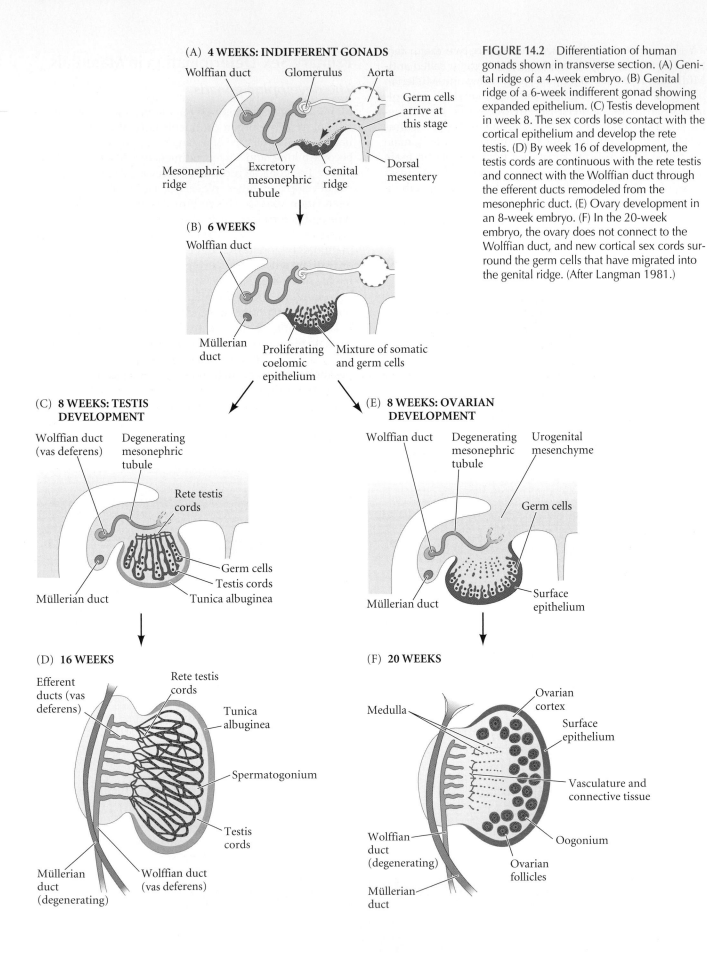

(A) 4 WEEKS: INDIFFERENT GONADS

Wolffian duct   Glomerulus   Aorta

Germ cells arrive at this stage

Mesonephric ridge   Excretory mesonephric tubule   Genital ridge   Dorsal mesentery

(B) 6 WEEKS

Wolffian duct

Müllerian duct   Proliferating coelomic epithelium   Mixture of somatic and germ cells

(C) 8 WEEKS: TESTIS DEVELOPMENT

Wolffian duct (vas deferens)   Degenerating mesonephric tubule

Rete testis cords

Germ cells
Testis cords
Tunica albuginea

Müllerian duct

(D) 16 WEEKS

Efferent ducts (vas deferens)   Rete testis cords

Tunica albuginea

Spermatogonium

Testis cords

Müllerian duct (degenerating)   Wolffian duct (vas deferens)

(E) 8 WEEKS: OVARIAN DEVELOPMENT

Wolffian duct   Degenerating mesonephric tubule   Urogenital mesenchyme

Germ cells

Müllerian duct   Surface epithelium

(F) 20 WEEKS

Medulla   Ovarian cortex

Surface epithelium

Vasculature and connective tissue

Oogonium

Wolffian duct (degenerating)   Ovarian follicles

Müllerian duct

FIGURE 14.2   Differentiation of human gonads shown in transverse section. (A) Genital ridge of a 4-week embryo. (B) Genital ridge of a 6-week indifferent gonad showing expanded epithelium. (C) Testis development in week 8. The sex cords lose contact with the cortical epithelium and develop the rete testis. (D) By week 16 of development, the testis cords are continuous with the rete testis and connect with the Wolffian duct through the efferent ducts remodeled from the mesonephric duct. (E) Ovary development in an 8-week embryo. (F) In the 20-week embryo, the ovary does not connect to the Wolffian duct, and new cortical sex cords surround the germ cells that have migrated into the genital ridge. (After Langman 1981.)

MALE GONADAL DEVELOPMENT If the fetus is XY, the somatic cells continue to proliferate through the eighth week, and then initiate their differentiation into **Sertoli cells**. During week 8, the developing Sertoli cells surround the incoming germ cells and organize themselves into the **testis cords**. These cords form loops in the medullary (central) region of the developing testis and are connected to a network of thin canals, called the **rete testis**, located near the mesonephric duct (Figure 14.2C,D). Eventually, the region containing the testis cords and germ cells becomes enclosed by a thick extracellular matrix, the **tunica albuginea**. Thus, when the germ cells enter the male gonads, they will develop within the testis cords, inside the organ.

The Sertoli cells of the fetal testis cords secrete the anti-Müllerian factor that blocks development of the female ducts, and these epithelial cells will later support the development of sperm throughout the lifetime of the male mammal. Meanwhile, during fetal development, the interstitial mesenchyme cells of the testes differentiate into **Leydig cells**, which make testosterone.

Later in development (at puberty in humans; shortly after birth in mice, which procreate much faster), the testis cords hollow out to form the **seminiferous tubules**. The germ cells migrate to the periphery of these tubules, where they establish the spermatogonial stem cell population that produces sperm throughout the lifetime of the male (see Figure 16.27). In the mature seminiferous tubule, sperm are transported from the inside of the testis through the rete testis, which joins the **efferent ducts**. These efferent ducts are the remodeled tubules of the mesonephric kidney. They link the seminiferous tubules to the **Wolffian duct** (also called the **nephric duct**), which used to be the collecting tube of the mesonephric kidney* (see Chapter 11). During male development, the Wolffian duct differentiates to become the **epididymis** (adjacent to the testis) and the **vas deferens**, the tube through which sperm pass into the urethra and out of the body.

FEMALE GONADAL DEVELOPMENT In females, the germ cells accumulate near the outer surface of the gonad, interspersed with the gonadal somatic cells. Near the time of birth, each individual germ cell is surrounded by somatic cells (Figure 14.2E,F). The germ cells will become the ova, and the surrounding cortical epithelial cells will differentiate into **granulosa cells**. The mesenchyme cells of the ovary differentiate into **thecal cells**. Together, the thecal and granulosa cells form **follicles** that envelop the germ cells and secrete steroid hormones. Each follicle will contain a single germ cell—an **oogonium** (egg precursor)—which will enter meiosis at this time. These germ cells are required for the gonadal cells to complete their differentiation into ovar-

ian tissue[†] (McLaren 1991). In females, the Müllerian duct remains intact and differentiates into the oviducts, uterus, cervix, and upper vagina. In the absence of adequate testosterone, the Wolffian duct degenerates. A summary of the development of mammalian reproductive systems is shown in Figure 14.3.

Mechanisms of mammalian primary sex determination: Making decisions

Several human genes have been identified whose function is necessary for normal sexual differentiation. Since the phenotype of mutations in sex-determining genes is often sterility, clinical infertility studies have been useful in identifying those genes that are active in determining whether humans become male or female. Experimental manipulations to confirm the functions of these genes can then be done in mice. Although the story unfolded in the following paragraphs demonstrates the remarkable progress that has been made in recent years, we still do not fully understand how all these gonad-determining genes interact. The problem of primary sex determination remains (as it has since prehistory) one of biology's great mysteries.

The story starts in the genital ridge, which can become either type of gonad. Here, the genes encoding Wt1, Sox9, Wnt4, Lhx9, Fgf9, GATA4, and Sf1 are expressed, and the loss of function of any one of them will prevent the normal development of any gonad (see Figure 14.1). If no Y chromosome is present, these factors activate further expression of Wnt4 (already expressed at low levels in the genital ridge) and of a small soluble protein called **R-spondin1 (Rspo1)**. Rspo1 binds to its cell membrane receptor and further stimulates the Disheveled protein of the Wnt pathway, making the Wnt pathway more efficient at producing β-catenin. One of the functions of β-catenin in the gonadal cells is to further activate Rspo1 and Wnt4, creating a positive feedback loop between these two proteins. A second role of β-catenin is to initiate the ovarian pathway of development; a third role is to prevent the accumulation of Sox9, a protein crucial for testis determination (Maatouk et al. 2008).

However, if each embryonic cell nucleus has a Y chromosome present, the same set of factors in the genital ridge activates the *SRY* gene (the sex-determining gene on the Y chromosome). Sry protein binds to the *SOX9* promoter and

---

*As discussed in Chapter 11, the mesonephric kidney is one of the three major kidney stages seen during mammalian development, but it does not function as a kidney in most mammals.

[†]There is a reciprocal relationship between the germ cells and the gonadal somatic cells. The germ cells are originally bipotential and can become either sperm or eggs. Once in the male or female sex cords, however, they are instructed to either begin (and remain in) meiosis and become eggs, or to remain mitotically dormant and become spermatogonia (McLaren 1995; Brennan and Capel 2004). In XX gonads, germ cells are essential for the maintenance of ovarian follicles. Without germ cells, the follicles degenerate into cord-like structures and express male-specific markers. In XY gonads, the germ cells help support the differentiation of Sertoli cells, although testis cords will form without the germ cells, albeit a bit later.

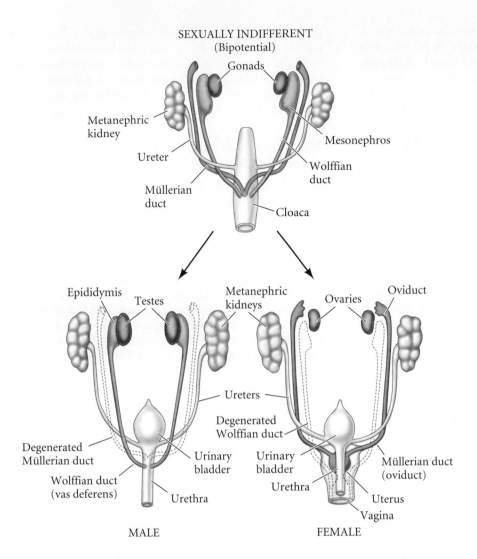

SEXUALLY INDIFFERENT
(Bipotential)

Gonads

Metanephric
kidney

Mesonephros

Ureter

Wolffian
duct

Müllerian
duct

Cloaca

Epididymis

Testes

Metanephric
kidneys

Ovaries

Oviduct

Ureters

Degenerated
Wolffian duct

Degenerated
Müllerian duct

Urinary
bladder

Urinary
bladder

Müllerian duct
(oviduct)

Wolffian duct
(vas deferens)

Urethra

Urethra

Uterus

Vagina

MALE

FEMALE

GONADS		
Gonadal type	Testis	Ovary
Germ cell location	Inside testis cords (in medulla of testis)	Inside follicles of ovarian cortex
DUCTS		
Remaining duct	Wolffian	Müllerian
Duct differentiation	Vas deferens, epididymis, seminal vesicle	Oviduct, uterus, cervix, upper portion of vagina

FIGURE 14.3 Development of the gonads and their ducts in mammals. Note that both the Wolffian and Müllerian ducts are present at the indifferent gonad stage.

elevates expression of this key gene in the testis-determining pathway (Bradford et al. 2009b; Sekido and Lovell-Badge 2009). Sox9 and Fgf9 (and possibly Sry) act to block the ovary-forming pathway, probably by blocking the function of β-catenin (Bernard et al. 2008; Lau and Li 2008). One

possible model of how sex determination can be initiated is shown in **Figure 14.4**.

Here we see an important rule of animal development: A pathway for organogenesis often has two components. One branch of the pathway says "Make A"; the other branch says "and don't make B." So the male pathway says "Make testes and don't make ovaries," while the female pathway says, "Make ovaries and don't make testes."

FIGURE 14.4 Possible mechanism for the initiation of primary sex determination in mammals. While we do not know the specific interactions involved, this model attempts to organize the data into a coherent sequence. In this model (others are possible), the interactions between paracrine and transcription factors in the developing genital ridge activate Wnt4 and Rspo1. Wnt4 activates the canonical Wnt pathway, which is made more efficient by Rspo1. The Wnt pathway causes the accumulation of β-catenin, and the large accumulation of β-catenin stimulates further Wnt4 activity. This continual production of β-catenin both induces the transcription of ovary-producing genes and blocks the testes-determining pathway by interfering with Sox9 activity. If Sry is present, it may block β-catenin signaling (thus halting ovary generation) and, along with Sf1, activate the gene for Sox9. Sox9 activates Fgf9 synthesis, which stimulates testis development and promotes further Sox9 synthesis. Sox9 also prevents β-catenin's activation of ovary-producing genes. In summary, then, a Wnt4/β-catenin loop specifies the ovaries, while a Sox9/Fgf9 loop specifies the testes. (After Sekido and Lovell-Badge 2009).

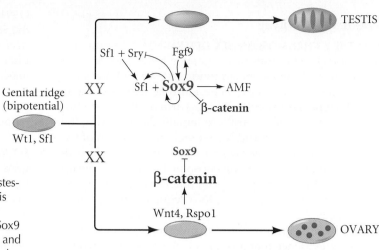

The ovary pathway: Wnt4 and R-spondin1

In mice, Wnt4 is expressed in the genital ridges of both sexes, but its expression later becomes undetectable in XY gonads (which become testes), whereas it is maintained in XX gonads as they begin to form ovaries. In transgenic XX mice that lack the *Wnt4* gene, the ovary fails to form properly, and the cells transiently express testis-specific markers, including Sox9, testosterone-producing enzymes, and AMF (Vainio et al. 1999; Heikkila et al. 2005). Thus, Wnt4 appears to be an important factor in ovary formation, although it is not the only determining factor.

R-spondin1 is also critical in ovary formation, since in human case studies several XX individuals with *RSPO1* gene mutations became males (Parma et al. 2006). Rspo1 acts in synergy with Wnt4 to produce β-catenin, and this β-catenin appears to be critical both in activating further ovarian development and in blocking the synthesis of Sox9. Sox9 (as we will see below) is necessary in testis determination. In XY individuals with a duplication of the region on chromosome 1 that contains both the *WNT4* and *RSPO1* genes, the pathways that make β-catenin override the male pathway and cause a male-to-female sex reversal. Similarly, if XY mice are induced to overexpress β-catenin in their gonadal rudiments, they too will form ovaries rather than testes. Indeed, β-catenin appears to be a key "pro-ovarian" and "anti-testis" signaling molecule in all vertebrate groups, as it is seen in the female gonads (but not in the male gonads) of birds, mammals, and turtles. These groups have very different modes of sex determination, yet Rspo1 and β-catenin are made in the ovaries of each of them (**Figure 14.5**; Maatouk et al. 2008; Cool and Capel 2009; Smith et al. 2009).

Certain transcription factors that appear to be activated by β-catenin are found exclusively in the ovaries. One

possible target for β-catenin is the gene encoding TAFII 105 (Freiman et al. 2002). This subunit of the TATA-binding protein for RNA polymerase binding is seen only in ovarian follicle cells. Female mice lacking this subunit have small ovaries with few, if any, mature follicles. *FoxL2* is another gene that is strongly upregulated in ovaries, and XX mice homozygous for mutant *FoxL2* alleles develop malelike gonad structure and upregulate Sox9 expression and testosterone production. Estrogen receptors (transcription factors that become active by binding estrogen) are expressed in the developing gonads of both sexes but become prevalent in the developing ovaries.

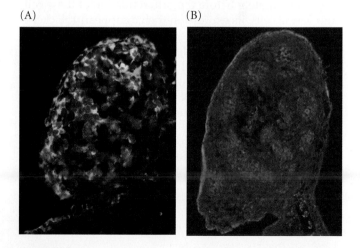

FIGURE 14.5 Localization of Rspo1 protein in embryonic day 14.5 mouse gonads. Immunofluorescent probes were used to identify Rspo1 (green) and the meiotic germ cell marker, Scp3 (red). (A) Rspo1 was found on somatic cells (arrowhead) and at the germ cell surface (arrow) of the ovaries. (B) These antibodies revealed neither Rspo1 nor Scp3 in the developing testes. (The germ cells in the male gonads have not entered meiosis at this point in development, whereas the ovarian germ cells have.) (From Smith et al. 2008; photograph courtesy of C. Smith.)

The testis pathway

SRY: THE Y CHROMOSOME SEX DETERMINANT In humans, the major gene for testis determination resides on the short arm of the Y chromosome. Individuals born with the short but not the long arm of the Y chromosome are male, whereas individuals born with the long arm of the Y chromosome but not the short arm are female. By analyzing the DNA of rare XX men and XY women, the position of the testis-determining gene was narrowed down to a 35,000-base-pair region of the Y chromosome located near the tip of the short arm. In this region, Sinclair and colleagues (1990) found a male-specific DNA sequence that encodes a peptide of 223 amino acids. This gene is called **SRY** (**sex-determining region of the Y chromosome**), and there is extensive evidence that it is indeed the gene that encodes the human testis-determining factor.

SRY is found in normal XY males and also in the rare XX males; it is absent from normal XX females and from many XY females. Approximately 15% of XY females have the SRY gene, but their copies of the gene contain point or frameshift mutations that prevent the Sry protein from binding to DNA (Pontiggia et al. 1994; Werner et al. 1995). If the SRY gene actually does encode the major testis-determining factor, one would expect that it would act in the genital ridge immediately before or during testis differentiation. This prediction has been found to be the case in studies of the homologous gene in mice. The mouse gene (Sry) also correlates with the presence of testes; it is present in XX males and absent in XY females (Gubbay et al. 1990; Koopman et al. 1990). The Sry gene is expressed in the somatic cells of the bipotential mouse gonads of XY mice immediately before the differentiation of these cells into Sertoli cells; its expression then disappears (Hacker et al. 1995; Sekido et al. 2004).

The most impressive evidence for Sry being the gene for testis-determining factor comes from transgenic mice. If Sry induces testis formation, then inserting Sry DNA into the genome of a normal XX mouse zygote should cause that XX mouse to form testes. Koopman and colleagues

See WEBSITE 14.2
Finding the male-determining genes

(1991) took the 14-kilobase region of DNA that includes the Sry gene (and presumably its regulatory elements) and microinjected this sequence into the pronuclei of newly fertilized mouse zygotes. In several instances, XX embryos injected with this sequence developed testes, male accessory organs, and a penis* (**Figure 14.6**). There are thus good reasons to think that Sry/SRY is the major gene on the Y chromosome for testis determination in mammals.

SOX9: AN AUTOSOMAL TESTIS-DETERMINING GENE The SRY gene is probably active for only a few hours. During this time, it synthesizes the Sry transcription factor whose primary (and perhaps only) role is to activate the SOX9 gene (see Sekido and Lovell-Badge 2008). SOX9 is an autosomal gene involved in several developmental processes, most notably bone formation. In the genital ridge, however, SOX9 induces testis formation. XX humans who have an extra copy of SOX9 develop as males, even if they have no SRY gene, and XX mice transgenic for Sox9 develop testes (**Figure 14.7**; Huang et al. 1999; Qin and Bishop 2005). Knocking out the Sox9 genes in the gonads of XY mice causes complete sex reversal (Barrionuevo et al. 2006). Thus, even if Sry is present, mouse gonads cannot form testes if Sox9 is absent—so it appears that Sox9 can replace Sry in testis formation. This is not altogether surprising: although the Sry gene is found specifically in mammals, Sox9 is found throughout the vertebrate phyla.

Indeed, Sox9 appears to be the central male-determining gene in vertebrates. In mammals, it is activated by Sry; in birds, frogs, and fish, it appears to be activated by the dosage of Dmrt1 (see Sidelights & Speculations, p. 521); and in those vertebrates with temperature-dependent sex determination, it is often activated (directly or indirectly) by the male-producing temperature. Sox9 may thus be the older and more central sex determination gene (Pask and

---

*These embryos did not form functional sperm—but they were not expected to. The presence of two X chromosomes prevents sperm formation in XXY mice and men, and the transgenic mice lacked the rest of the Y chromosome, which contains genes needed for spermatogenesis. There is biochemical evidence (not yet tested in living mice) that Sry may have a second function, namely, to block β-catenin-mediated transcription (Bernard et al. 2008; Lau et al. 2009).

(A)

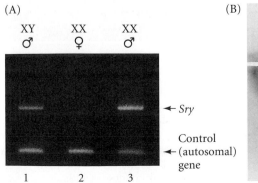

(B)

FIGURE 14.6 An XX mouse transgenic for Sry is male. (A) Polymerase chain reaction followed by electrophoresis shows the presence of the Sry gene in normal XY males and in a transgenic XX/Sry mouse. The gene is absent in a female XX littermate. (B) The external genitalia of the transgenic mouse are male (right) and are essentially the same as those in an XY male (left). (From Koopman et al. 1991; photographs courtesy of the authors.)

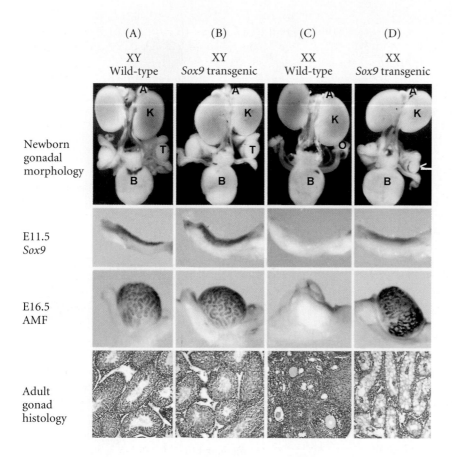

(A)	(B)	(C)	(D)
XY Wild-type	XY *Sox9* transgenic	XX Wild-type	XX *Sox9* transgenic

Newborn gonadal morphology

E11.5 *Sox9*

E16.5 AMF

Adult gonad histology

FIGURE 14.7   Ability of *Sox9* to generate testes. (A) A wild-type XY mouse embryo express-es *Sox9* in the genital ridge at 11.5 days, anti-Müllerian factor (AMF) in the embryonic gonad Sertoli cells at 16.5 days, and eventually forms descended testes (T) with seminiferous tubules. (B) An XY embryo with the *Sox9* transgene (a control for the effects of the transgene) also shows *Sox9* expression, AMF expression, and descended testes with seminiferous tubules. (C) The wild-type XX embryo shows neither *Sox9* expression nor AMF. It constructs ovaries with mature follicle cells. (D) An XX embryo with the *Sox9* transgene expresses the *Sox9* gene and has AMF in its 16.5-day Sertoli cells. It has descended testes, but the seminif-erous tubules lack sperm (because of the presence of two X chromosomes in the Sertoli cells). K, kidneys; A, adrenal glands; B, bladder; T, testis; O, ovary. (From Vidal et al. 2001; photographs courtesy of A. Schedl.)

Graves 1999). Expression of the *Sox9* gene is specifically upregulated by the combined expression of Sry and Sf1 proteins in Sertoli cell precursors (Pask and Graves 1999; Sekido et al. 2004; Sekido and Lovell-Badge 2008). Thus, Sry may act merely as a "switch" operating during a very short time to activate *Sox9*, and the Sox9 protein may ini-tiate the conserved evolutionary pathway to testis forma-tion. So, borrowing Eric Idle's phrase, Sekido and Lovell-Badge (2009) propose that Sry initiates testis formation by "a wink and a nudge."

Sox9 protein has several functions. First, it appears to be able to activate its own promoter, thereby allowing it to tran-scribe for long periods of time. Second, it blocks the ability of β-catenin to induce ovary formation, possibly by block-ing the *FoxL2* gene (Wilhelm et al. 2009). Third, it binds to *cis*-regulatory regions of numerous genes necessary for testis production (Bradford et al. 2009). Fourth, Sox9 binds to the

promoter site on the gene for the anti-Müllerian factor, pro-viding a critical link in the pathway toward a male pheno-type (Arango et al. 1999; de Santa Barbara et al. 2000). Fifth, Sox9 promotes the expression of the gene encoding Fgf9, a paracrine factor critical for establishing testis development. Fgf9 is essential for maintaining *Sox9* transcription, thereby establishing a positive feedforward loop driving the male pathway (Kim et al. 2007).

FIBROBLAST GROWTH FACTOR 9   When the gene for fibrob-last growth factor 9 (*Fgf9*) is knocked out in mice, the homozygous mutants are almost all female. Fgf9, whose expression is activated by Sox9 (Capel et al. 1999; Colvin et al. 2001), plays several roles. First, it causes proliferation of the Sertoli cell precursors and stimulates their differen-tiation (Schmahl et al. 2004; Willerton et al. 2004). Second, Fgf9 activates the migration of endothelial cells from the

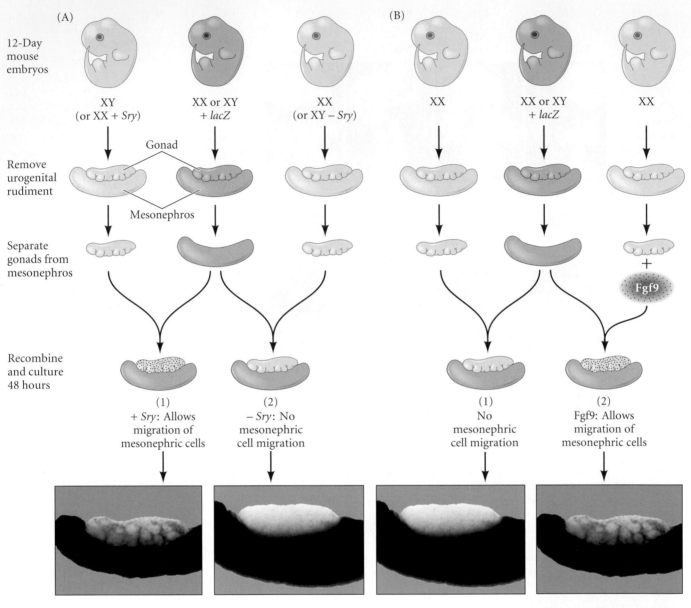

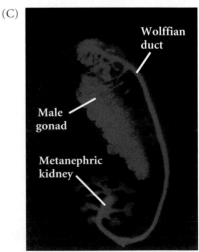

FIGURE 14.8   Migration of mesonephric endothelial cells into *Sry*⁺ gonadal rudiments. In the experiment diagrammed, urogenital ridges (containing both the mesonephric kidneys and bipotential gonadal rudiments) were collected from 12-day embryonic mice. Some of the mice were marked with a β-galactosidase transgene (*lacZ*) that is active in every cell. Thus, every cell of these mice turned blue when stained for β-galactosidase. The gonad and mesonephros were separated and recombined, using gonadal tissue from unlabeled mice and mesonephros from labeled mice. (A) Migration of mesonephric cells into the gonad was seen (1) when the gonadal cells were XY or when they were XX with a *Sry* transgene. No migration of mesonephric tissue into the gonad was seen (2) when the gonad contained either XX cells or XY cells in which the Y chromosome had a deletion in the *Sry* gene. The sex chromosomes of the mesonephros did not affect the migration. (B) Gonadal rudiments for XX mice could induce mesonephric cell migration if these rudiments had been incubated with Fgf9. (C) Intimate relation between the mesonephric duct and the developing gonad in a 16-day male mouse embryo. The duct tissue will form the efferent ducts of the testes and has been stained for cytokeratin-8. (A,B after Capel et al. 1999, photographs courtesy of B. Capel; C from Sariola and Saarma 1999, courtesy of H. Sariola.)

adjacent mesonephros into the XY gonad. While this is normally a male-specific process, incubating XX gonads in Fgf9 leads to the migration of endothelial cells into XX gonads (**Figure 14.8**). These endothelial cells form the major artery of the testis and play an instructive role in inducing the Sertoli cell precursors to form the testis cords; in their absence, these cords do not form (Brennan et al. 2002; Combes et al. 2009).

SF1: A CRITICAL LINK BETWEEN SRY AND THE MALE DEVELOPMENTAL PATHWAYS The transcription factor **steroidogenic factor 1 (Sf1)** is necessary to make the bipotential gonad. But whereas Sf1 levels decline in the genital ridge of XX mouse embryos, they remain high in the developing testis. It is thought that Sry (either directly or indirectly) maintains Sf1 expression. Sf1 protein appears to be active in masculinizing both the Leydig and the Sertoli cells. In the Sertoli cells, Sf1 works in collaboration with Sry to activate Sox9 (Sekido and Lovell-Badge 2008) and, working with Sox9, elevates levels of anti-Müllerian factor transcription (Shen et al. 1994; Arango et al. 1999). In the Leydig cells, Sf1 activates genes encoding the enzymes that make testosterone.

The right time and the right place

Having the right genes doesn't necessarily mean you'll get the organ you expect. Studies of mice have shown that the *Sry* gene of some strains of mice failed to produce testes when bred onto a different genetic background (Eicher and Washburn 1983; Washburn and Eicher 1989; Eicher et al. 1996). This is attributed to a delay in Sry expression or to the failure of the protein to accumulate to the critical threshold level required to trigger Sox9 expression and launch the male pathway. By the time *Sox9* gets turned on, it is too late—the gonad is already following the path to become an ovary (Bullejos and Koopman 2005; Wilhelm et al. 2009). This provides an important clue to how primary sex determination may take place. Timing is critical. There may be a brief window during which the testis-forming genes can function. If these genes are not turned on, the ovary-forming pathway is activated.

Hermaphrodites are individuals in which both ovarian and testicular tissues exist; they have either ovotestes (gonads containing both ovarian and testicular tissue) or an ovary on one side and a testis on the other.* In humans, hermaphroditism is a very rare condition that can result when a Y chromosome is translocated to an X chromosome. In those tissues where the translocated Y is on the active X chromosome, the Y chromosome will be active and the *SRY* gene will be transcribed; in cells where the Y chromosome is on the inactive X chromosome, the Y chromosome will also be inactive (Berkovitz et al. 1992; Margarit et al. 2000). A gonadal mosaic for expression of *SRY* can develop into a testis, an ovary, or an ovotestis, depending on the percentage of cells expressing *SRY* in the Sertoli cell precursors (see Brennan and Capel 2004).

---

*Hermaphroditos was the young man in Greek mythology whose beauty inflamed the ardor of the water nymph Salmacis. She wished to be united with him forever, and the gods, in their literal fashion, granted her wish. The language used to group these conditions is being debated. Hermaphroditism has often been called an "intersex" phenotype; but some activists, physicians, and parents wish to eliminate the term "intersex" to avoid confusion of these anatomical conditions with identity issues such as homosexuality. They prefer to call these conditions "disorders of sex development." In contrast, other activists do not want to medicalize this condition and find the "disorder" category offensive to individuals who do not feel there is anything wrong with their health. For a more detailed analysis of intersexuality, see Gilbert et al. 2005 or the websites listed at the end of this chapter.

SIDELIGHTS & SPECULATIONS

Mysteries of Gonad Differentiation

There are some genes that are definitely involved in primary sex determination, but we don't know how they act or what their normal functions are.

DAX1: A potential testis-suppressing gene on the X chromosome

One of the most puzzling loci in the catalog of sex-determining genes is *DAX1*, located on the X chromosome. In XY humans having a duplication of the *WNT4* region, Dax1 protein is overproduced and the gonads develop into ovaries despite the presence of the Y chromosome (Jordan et al. 2001). Moreover, in XY individuals having a duplication of the *DAX1* gene on their X chromosome, the *SRY* signal is inhibited (**Figure 14.9**; Bernstein et al. 1980; Bardoni et al. 1994). In XY embryos, *DAX1* normally would be suppressed; having two active copies of the gene, however, overrides this suppression.

The mouse homologue, *Dax1*, was cloned and shown to encode a member of the nuclear hormone receptor family (Muscatelli et al. 1994; Zanaria 1994). In mice, *Dax1* is initially expressed in the genital ridges of both male and female embryos prior to the

(Continued on next page)

SIDELIGHTS & SPECULATIONS (Continued)

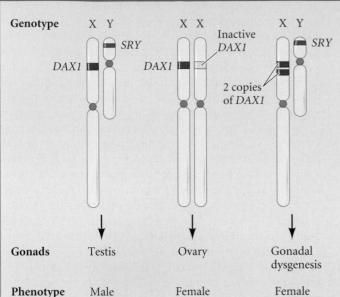

Genotype

Gonads   Testis   Ovary   Gonadal dysgenesis

Phenotype   Male   Female   Female

Figure 14.9 Phenotypic sex reversal in humans having two copies of the *DAX1* locus. *DAX1* (on the X chromosome) plus *SRY* (on the Y chromosome) produces testes. *DAX1* without *SRY* (since the other *DAX1* locus is on the inactive X chromosome) produces ovaries. Two active copies of *DAX1* (on the active X chromosome) plus *SRY* (on the Y chromosome) lead to a poorly formed gonad. Since the gonad makes neither anti-Müllerian factor nor testosterone, the phenotype is female. (From Genetics Review Group 1995.)

time that sex determination occurs, but gradually its expression becomes female-specific. As predicted from the data in humans, overexpression of *Dax1* in male mice appears to antagonize the function of *Sry* and *Sox9* and downregulates *Sf1* expression (Nachtigal et al. 1998; Swain et al. 1998; Iyer and McCabe 2004). However, unexpectedly, the *loss* of *Dax1* does not disrupt ovary development, but instead impairs testis formation (Meeks et al. 2003; Bauma et al. 2005). The apparent contradictory roles of this protein are still not understood.

Dmrt genes: Highly conserved testis-determining genes

The *Dmrt* (doublesex and mab-3-related transcription factor) genes are involved in male sex determination throughout the animal kingdom (Raymond et al. 1998). The male-specific Doublesex (DsxᴹM) protein in *Drosophila* (which will be discussed later in the chapter) has a region of amino acids similar to the male-specific regulatory protein Mab-3 in *Caenorhabditis elegans*. Both proteins positively regulate male-specific development. Moreover, if the male-specific *doublesex* transcript from *Drosophila* is inserted into *C. elegans* males lacking *mab-3*, the newly introduced *Drosophila* message will restore male-specific neurogenesis to the mutant *C. elegans* strain at levels

the same as the introduction of *mab-3*. The female-specific *doublesex* transcript will not work.

Dmrt genes play a critical role in the primary sex determination of Medaka fish, where sex is chromosomally specified by the Y chromosome (Matsuda et al. 2002). Here a *Dmrt* gene on the Y chromosomes encodes a protein that is made in the somatic cells of the developing testes (and not ovaries) and whose function is critical for testis formation. It probably binds to and activates the *Sox9* gene (Bagheri-Fam et al. 2009). In the chicken (and other birds), the male is the homogametic sex (ZZ), and the female is the heterogametic sex (ZW). Smith and colleagues (2009) have shown that the *Dmrt1* gene on the Z chromosome is the major male sex-determining gene, and that a double dose of this gene is needed for the gonad to become a testis. *Dmrt1*, too, appears to activate (either directly or indirectly) the *Sox9* gene.

In mammals (including humans), *Dmrt1*, which is autosomal, appears to function later in testis development. It is no longer involved in specification of the testis, rather it is important in testis differentiation. Kim and colleagues (2007; Hong et al. 2007) showed that *Dmrt1* in mice is not critical for sex determination but is crucial for Sertoli cell differentiation and germ cell survival.

Sex-specific microRNAs

There is evidence that sexual distinctions exist even at the blastocyst stage. Tavakoli and colleagues (2009) have found that XX embryonic stem cells appear to have greater differentiation potential (and slower growth rates) than XY stem cells. This difference may be the result of different populations of microRNAs found in the XY and XX embryonic stem cells. Ciaudo and colleagues (2009) have found that XX and XY human embryonic stem cells transcribe different populations of microRNAs, whose functions, though overlapping, may differentially regulate genes involved in pluripotency and cell cycling.

The apparent relationship between cell division rates and sex determination is provocative. Indeed, the rapid proliferation of Sertoli cell precursors has been hypothesized to be the unifying feature in testis determination (Mittwoch 1986; Schmahl and Capel 2003). In its most extreme form, this hypothesis predicts that anything that increases the division rate of somatic cell precursors in the gonad will direct gonadogenesis in the male direction. While we do not know if this is true under all conditions, we do know that in mice, one of the first distinguishing features of testis development is an increase in cell proliferation seen immediately after *Sry* expression. There appears to be a threshold number of Sertoli cells needed for mammalian testis development (Schmahl et al. 2000). In turtles and alligators, rapid division of Sertoli cell precursors is initiated during the critical stage for forming the males (Mittwoch 1986; Schmahl et al. 2003). From this perspective, anything that increases the division rate of Sertoli cell precursors might direct gonadogenesis in the male direction.

Secondary Sex Determination in Mammals: Hormonal Regulation of the Sexual Phenotype

Primary sex determination is the formation of either an ovary or a testis from the bipotential gonad. This process, however, does not give the complete sexual phenotype. In mammals, secondary sex determination is the development of the female and male phenotypes in response to hormones secreted by the ovaries and testes. Both female and male secondary sex determination have two major temporal phases. The first phase occurs within the embryo during organogenesis; the second occurs at puberty.

During embryonic development, hormones and paracrine coordinate the development of the gonads with the development of secondary sexual organs. In females, the Müllerian ducts persist and, through the actions of estrogen, will later differentiate to become the uterus, cervix, oviducts, and upper vagina (see Figure 14.3). The **genital tubercle** becomes differentiated into the clitoris, and the **labioscrotal folds** become the labia. The Wolffian ducts atrophy in females, since they need testosterone to persist, and become the male accessory organs. As mentioned earlier, if the bipotential gonads are removed from an embryonic mammal, the female phenotype is realized: the Müllerian duct develops and the Wolffian ducts degenerate. This pattern is also seen in certain humans who are born without functional gonads.

The coordination of the male phenotype involves the secretion of two testicular factors. The first of these is AMF, a BMP-like paracrine factor made by the Sertoli cells that causes the degeneration of the Müllerian duct. The second

is the steroid hormone testosterone, which is secreted from the fetal Leydig cells. Testosterone causes the Wolffian ducts to differentiate into the epididymis, vas deferens, and seminal vesicles, and it causes the genital tubercle to develop into the penis and the labioscrotal folds to develop into the scrotum.

The mechanism by which testosterone (and, as we shall see, its more powerful derivative, dihydrotestosterone) masculinizes the external genitalia is thought to involve its interaction with the Wnt pathway (**Figure 14.10**). The Wnt pathway, which in the genital *ridge* activates the female trajectory, acts in the genital *tubercle* to activate male development. In one recent model, the mesenchyme in the XX urogenital swellings is seen to make inhibitors of the Wnt pathway (such as Dickkopf). This prevents the activity of Wnt in the mesenchyme and leads to the feminization of the genital tubercle by estrogens (Holderegger and Keefer 1986; Miyagawa et al. 2009). In this case, the genital tubercle becomes the clitoris and the labioscrotal folds become the labia majora. In males, however, testosterone and dihydrotestosterone bind to the androgen (testosterone) receptor in the mesenchyme and prevent expression of the Wnt inhibitors. Thus, when Wnt expression is permitted in the mesenchyme, male urogenital swellings are converted into the penis and the scrotum.

The genetic analysis of secondary sex determination

The existence of separate and independent AMF and testosterone pathways of masculinization is demonstrated by people with **androgen insensitivity syndrome**.

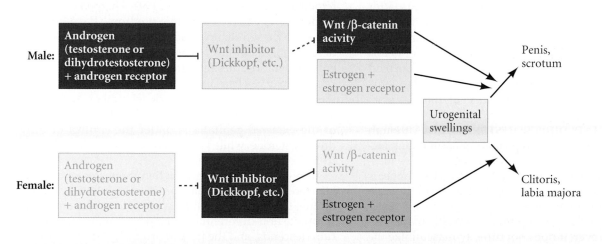

FIGURE 14.10   Model for the formation of external genitalia. In this schema, the mesenchyme in the urogenital swellings secretes inhibitors of Wnt signaling. In the absence of Wnt signaling, estrogen modifies the genital tubercle into the clitoris and the labioscrotal folds into the labia majora surrounding the vagina. In males, however, androgens (such as testosterone and dihy-

drotestosterone) bind to the androgen receptor in the mesenchymal cells and prevent the synthesis of the Wnt inhibitors. Wnt signaling is permitted, and it causes the genital tubercle to become the penis and the labioscrotal folds to become the scrotum. (After Miyagawa et al. 2009.)

These XY individuals have the *SRY* gene and thus have testes that make testosterone and AMF. However, they have a mutation in the gene encoding the androgen receptor protein that binds testosterone and brings it into the nucleus. Therefore these individuals cannot respond to the testosterone made by their testes (Meyer et al. 1975). They can, however, respond to the estrogen made by their adrenal glands (which is normal for both XX and XY individuals), so they develop female external sex characteristics (Figure 14.11). Despite their distinctly female appearance, these individuals have testes, and even though they cannot respond to testosterone, they produce and respond to AMF. Thus, their Müllerian ducts degenerate. Persons with androgen insensitivity syndrome develop as normal-appearing but sterile women, lacking a uterus and oviducts and having internal testes in the abdomen.

Such phenotypes, in which male and female traits are seen in the same individual, are called **intersex** conditions. Indeed, although most people have a reasonably good correlation of their genetic and anatomical sexual phenotypes, about 0.4–1.7% of the population differs from the strictly dimorphic condition (Blackless et al. 2000; Hull 2003). Androgen insensitivity syndrome is one of several intersex conditions that have traditionally been labeled **pseudohermaphroditism**. In pseudohermaphrodites, there is only one type of gonad, but the secondary sex characteristics differ from what would be expected from the gonadal sex. In humans, male pseudohermaphroditism (wherein the gonadal sex is male and the secondary sex characteristics are female) can be caused by mutations in the androgen (testosterone) receptor or by mutations affecting testosterone synthesis (Geissler et al. 1994).

Female pseudohermaphroditism (in which the gonadal sex is female but the person is outwardly male) can be caused by the overproduction of androgens in the ovary or adrenal gland. The most common cause of this latter condition is **congenital adrenal hyperplasia**, in which there is a genetic deficiency of an enzyme that metabolizes cortisol steroids in the adrenal gland. In the absence of this enzyme, testosterone-like steroids accumulate and can bind to the androgen receptor, thus masculinizing the fetus (Migeon and Wisniewski 2001; Merke et al. 2002).

TESTOSTERONE AND DIHYDROTESTOSTERONE Although testosterone is one of the two primary masculinizing factors, there is evidence that it might not be the active masculinizing hormone in certain tissues. Testosterone appears to be responsible for promoting the formation of the male reproductive structures (the epididymis, seminal vesicles, and vas deferens) that develop from the Wolffian duct primordium. However, it does not directly masculinize the male urethra, prostate, penis, or scrotum. These latter functions are controlled by **5α-dihydrotestosterone**, or **DHT** (Figure 14.12). Siiteri and Wilson (1974) showed that testosterone is converted to DHT in the urogenital sinus and swellings, but not in the Wolffian duct. DHT appears to be a more potent hormone than testosterone.

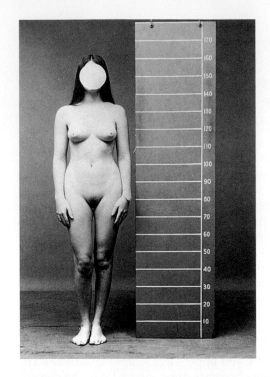

FIGURE 14.11 Androgen insensitivity syndrome. Despite having the XY karyotype, individuals with this syndrome appear female. They cannot respond to testosterone but can respond to estrogen, so they develop female secondary sex characteristics (i.e., labia and a clitoris rather than a scrotum and a penis). Internally, they lack the Müllerian duct derivatives and have undescended testes. (Courtesy of C. B. Hammond.)

The importance of DHT was demonstrated by Imperato-McGinley and her colleagues (1974). They found a small community in the Dominican Republic in which several inhabitants lacked a functional gene for the enzyme 5α-ketosteroid reductase 2—the enzyme that converts testosterone to DHT (Andersson et al. 1991; Thigpen et al. 1992). Although XY children with this syndrome have functional testes, these remain inside the abdomen and do not descend. Because they have a blind vaginal pouch and an enlarged clitoris, these children appear to be girls, and are raised as such. Their internal anatomy, however, is male: they have testes, Wolffian duct development, and Müllerian duct degeneration. Thus it appears that the formation of the external genitalia is under the control of dihydrotestosterone, while Wolffian duct differentiation is controlled by testosterone itself.

Interestingly, at puberty, when the testes of children affected with this syndrome produce high levels of testosterone, their external genitalia are able to respond to the hormone and differentiate. The penis enlarges, the scrotum descends, and the person originally believed to be a girl is revealed to be a young man.

See WEBSITE 14.3
Dihydrotestosterone in adult men

See WEBSITE 14.4
Insulin-like hormone 3

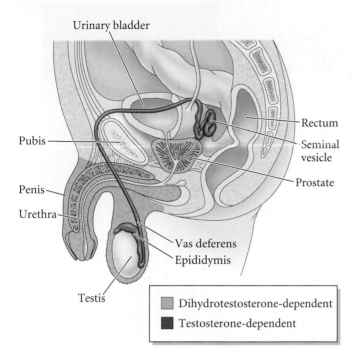

FIGURE 14.12 Testosterone- and dihydrotestosterone-dependent regions of the human male genital system. (After Imperato-McGinley et al. 1974.)

ANTI-MÜLLERIAN FACTOR Anti-Müllerian factor, or AMF, is a member of the TGF-β family of growth and differentiation factors. It is secreted from the fetal Sertoli cells and causes the degeneration of the Müllerian duct (Tran et al. 1977; Cate et al. 1986). AMF is thought to bind to the mesenchyme cells surrounding the Müllerian duct, causing these cells to secrete a paracrine factor that induces apoptosis in the duct's epithelium and breaks down the basal lamina surrounding the duct (Trelstad et al. 1982; Roberts et al. 1999).

ESTROGEN The steroid hormone estrogen is needed for complete postnatal development of both the Müllerian and the Wolffian ducts, and is necessary for fertility in both males and females. In females, estrogen induces the differentiation of the Müllerian duct into the uterus, oviducts, cervix, and upper vagina. In female mice whose genes for estrogen receptors are knocked out, the germ cells die in the adult, and the granulosa cells that had enveloped them start developing into Sertoli-like cells (Couse et al. 1999). Male mice with knockouts of estrogen receptor genes produce few sperm. One of the functions of the male efferent duct (vas efferens) cells is to absorb most of the water from the lumen of the rete testis. This absorption of water, which is regulated by estrogen, concentrates the sperm, giving them a longer life span and providing more sperm per ejaculate. If estrogen or its receptor is absent in male mice, water is not absorbed and the mouse is sterile (Hess et al. 1997). Although, in general, blood concentrations of estrogen are higher in females than in males, the concentration of estrogen in the rete testis is even higher than in female blood.

Brain sex: Secondary sex determination through another pathway?

We have known for a long time that the brain, like other tissues, is responsive to the steroid hormones produced by the gonads. Now, recent evidence suggests that sex differences in the brain can be observed even before the gonads mature, and that the brain may experience direct regulation by the X and Y chromosomes (Arnold and Burgoyne 2004).

The first indication that something besides the gonadal hormones testosterone and estrogen was important in forming sexually different structures in the brain came from studies on Parkinson disease, during which embryonic rat brains were dissected before the gonads matured. These studies indicated that brains from XX embryos had more epinephrine-secreting neurons than XY embryonic brains (Beyer et al. 1991). Later studies, using microarrays and PCR, demonstrated that more than 50 genes in the mouse brain are expressed in sexually dimorphic patterns *before gonad differentiation has occurred* (Dewing et al. 2003). Moreover, the mouse *Sry* gene, in addition to being expressed in the embryonic testes, is also expressed in the fetal and adult brain. The human *SRY* gene appears to be expressed in the adult brain as well (Lahr et al. 1995; Mayer et al. 1998, 2000). *SRY* is specifically active in the substantia nigra of the male hypothalamus, where it helps regulate the gene for tyrosine hydoxylase, an enzyme that is critical for the production of the neurotransmitter dopamine (Dewing et al. 2006).

Stunning demonstrations that sexual dimorphism in the brain can be caused before gonadal hormone synthesis come from natural and experimental conditions in birds. One big difference between male and female finches is that large regions of the male brain are devoted to producing songs. Male finches sing; the females do not. While testosterone is important in the formation of the song centers in finches (and, when added experimentally, can cause female birds to sing), blocking those hormones in males does not prevent normal development of the song centers or singing. Genetically male birds form these brain regions even without male hormones (Mathews and Arnold 1990).

A natural experiment presented itself in the form of a bird that was half male and half female, divided down the middle (**Figure 14.13**). Such animals, where some body parts are male and others female, are called **gynandromorphs** (Greek *gynos*, "female"; *andros*, "male"; *morphos*, "form"). Agate and colleagues (2002) showed that the gynandromorph finch had ZZ (male) sex chromosomes on its right side and ZW (female) sex chromosomes on its left. Its testes produced testosterone, and the bird sang like a male and copulated with females. However, although many brain structures were similar on both sides, some brain regions differed between the male and female halves. The song circuits on the right side had a more masculine phenotype than similar structures on the left, showing that both intrinsic and hormonal influences were important.

Gahr (2003) generated his own avian sexual chimeras by surgically switching the forebrain regions (which control adult sexual behaviors) between ZZ and ZW quail embryos before their gonads had matured. If hormones were all that mattered, the brains of the resulting birds would be appropriate to the gonad that developed. For the females that received male forebrains, this was indeed the case: they looked and behaved like normal female quail. However, male birds given female forebrains did not act normally; they did not crow to attract mates, nor did they attempt copulation. Moreover, their testes failed to develop normally, suggesting that (in quail, at least) a chromosomally male brain is needed to complete development of the testes.

Thus, although brain sex is usually correlated with gonadal sex, it seems probable that this harmony is created both by intrinsic, cell-autonomous differences as well as hormonal regulation from outside the cell.*

---

*The sexual characteristics of the tammar wallaby (*Macropus eugenii*) appear to be another case where secondary sex traits are controlled intrinsically by chromosomal genes rather than extrinsically by circulating hormones. In this marsupial, the pouch and its mammary glands are found only in the female, while the scrotum is made only in males. However, the marsupial pouch, mammary glands, and scrotum are each made before the gonads have matured and are producing hormones (O et al. 1988; Glickman et al. 2005). Renfree and Short (1988) showed that the number of X chromosomes determines whether the wallaby develops a pouch and mammary glands or develops a scrotum for its still undeveloped testes.

Right (♂)   Left (♀)

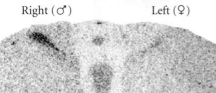

FIGURE 14.13   Gynandromorph finch with ZZ (male) cells on its right side and ZW (female) cells on its left side. Since plumage is controlled by genes on the sex chromosomes, the adult finch has male plumage on its right and female plumage on its left. Micrographs show the difference in brain regions between the right and left sides, indicated by staining of the neurons of the HVC nucleus (a neuron cluster involved in bird song production). (From Agate et al. 2002.)

SIDELIGHTS & SPECULATIONS

Brain Sex: Sex Determinants and Behaviors in Mice and Rats

Does prenatal (or neonatal) exposure to particular steroid hormones impose permanent sex-specific changes on the central nervous system? Such sex-specific neural changes have been shown in regions of the brain that regulate involuntary sexual physiology. The cyclic secretion of luteinizing hormone by the pituitary in adult female rats, for example, is dependent on the lack of testosterone during the first week of the animal's life. The luteinizing hormone secretion of female rats can be made noncyclical by giving them testosterone 4 days after birth. Conversely, the luteinizing hormone secretion of males can be made cyclical by removing their testes within a day of birth (Barraclough and Gorski 1962).

α-Fetoprotein and the organization/activation hypothesis

It is believed that sex hormones act during the fetal or neonatal stage of a mammal's life to organize the nervous system in a sex-specific manner; and that during adult life, the same hormones may have transitory motivational (or "activational") effects. This model of the hormonal basis of sex-specific brain development and behavior is called the **organization/activation hypothesis**. Ironically, the hormone chiefly responsible for determining the male neural pattern is **estradiol**, a type of estrogen. Testosterone from fetal or neonatal blood can be converted into estradiol by the enzyme **aromatase** (**Figure 14.14A**).* This conversion

*The terms *estrogen* and *estradiol* are often used interchangeably. However, estrogen refers to a class of steroid hormones responsible (among other functions) for establishing and maintaining specific female characteristics. Estradiol is one of these hormones, and in most mammals (including humans) it is the most potent of the estrogens. The enzyme's name, *aromatase*, has nothing to do with aroma (although aromas are certainly crucial to rodent sex), but refers to the destabilization of hydrogen bonds in the steroid ring structure.

SIDELIGHTS & SPECULATIONS (Continued)

(A) Aromatase converts androgens to estrogens

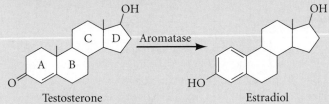

Testosterone   Estradiol

(B)

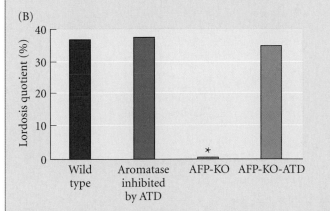

Figure 14.14 Organization of brain development by hormones. (A) The enzyme aromatase converts androgens (such as testosterone) into estrogens (such as estradiol). The name of the enzyme comes from its ability to aromatize the six-carbon ring by reducing the ring-stabilizing keto group (=O) to a hydroxyl group (—OH). This biochemical change allows the hormones to bind to different receptors (dormant transcription factors) and activate different genes. (B) Female lordosis behavior (in which the female mouse bends her spine so the male can readily mate with her) remains present in mice administered an inhibitor of aromatase; the behavior is abolished when the genes for α-fetoprotein are knocked out (AF–KO). However, when aromatase is blocked in female mice without α-fetoprotein (AFP–KO–ATD), lordosis behavior is restored. (After Bakker and Baum 2008.)

occurs in the hypothalamus and limbic system—two areas of the brain known to regulate hormone secretion and reproductive behavior (Reddy et al. 1974; McEwen et al. 1977). Thus, testosterone exerts its effects on the nervous system by being converted into estradiol in the brain.

But the fetal environment is rich in estrogens from the gonads and placenta. What stops these estrogens from masculinizing the nervous system of a female fetus? In both male and female rats, fetal estrogen is bound by **α-fetoprotein**, which binds and inactivates estrogen, but not testosterone. Human fetuses, by contrast, do not make a strong estrogen-binding protein and have a much higher level of free estrogen than do rodent embryos (see Nagel and vom Saal 2003). So although the organization/activation hypothesis explains many of the hormonal effects on rodent development, one of its fundamental assumptions—that α-fetoprotein strongly binds estrogens during prenatal development—does not in fact hold true for humans.

Relationships among estradiol, aromatase, and α-fetoprotein have been analyzed by observing sexual behaviors in mice that have loss-of-function mutations for aromatase and α-fetoprotein. The brain and the behaviors of

mice lacking α-fetoprotein have been defeminized, showing that α-fetoprotein prevents the female brain from receiving circulating estrogens.

Indeed, female mice whose α-fetoprotein genes have been knocked out are sterile because the brain genes controlling ovulation (such as those for gonadotropin-releasing hormone) are downregulated. However, this lack of ovulation can be cured (and the normal female pattern of gene expression established) if such mice are also given drugs that block aromatase. Similarly, the amount of lordosis (a swayback posture taken by female rodents that permits males to mate with them) is almost completely abrogated in female mice lacking functional α-fetoprotein genes. This behavior, too, can be restored by treating the mice prenatally with aromatase inhibitors (**Figure 14.14B**; Bakker et al. 2006; De Mees et al. 2006; Bakker and Baum 2008).

While the *prenatal* lack of estrogen and testosterone may be critical for the formation of female brains, the feminization of the rodent brain may require estrogens *after* birth. This is suggested by the behavioral phenotypes of mice whose aromatase genes have been knocked out. Their female-specific behaviors (lordosis; the ability

to discriminate male pheromones) are also impaired (Bakker and Baum 2008).

Pheromones

Pheromones appear to play a major role in sexual behaviors in rodents. If the vomeronasal organ (responsible for sensing pheromones) or the genes involved in pheromone recognition are removed from male mice, they fail to discriminate between males and females and attempt to mate with both. If this pheromone recognition system is removed from female mice, they lack certain female behaviors and acquire the full set of male courtship behaviors (including mounting, pelvic thrusting, and solicitation of females). In females, pheromones act to repress male behavior patterns and promote female-specific actions. Thus, it appears that the neural circuitry for both male and female behaviors exists in each mouse brain, but the interpretation of pheromone signals is what distinguishes male from female brains. In females, the "feminine" pattern of behavior is activated (sexual receptivity to males, lactating behavior toward pups), while the "masculine" pattern of behavior (fight if male, mount if female) is repressed. In males, the pheromones activate this "masculine" pattern, while the "feminine" pathway is suppressed (Kimchi et al. 2007). The interpretation of pheromone signals is thought to take place in the medial preoptic area/anterior hypothalamus region of the brain, and we know this region to be sexually dimorphic as a result of prenatal estrogen exposure. Thus, the organizational abilities of

SIDELIGHTS & SPECULATIONS (Continued)

testosterone may act largely to effect changes in this small area of the brain, and once this region is organized, it will interpret the pheromone signals to activate either the male or the female sets of neurons (Baum 2009).

The role of testosterone

If testosterone's conversion to estrogen is involved in brain masculinization, then what exactly is going on? It appears that many of testosterone's orders are carried out by **prostaglandin E2 (PGE2)**. PGE2 is made from arachidonic acid by the enzyme **cyclooxygenase-2 (COX2)**; COX2 is induced by estrogen in the brain. So estrogen acts to produce more PGE2 (**Figure 14.15A**).

Studies by Amateau and McCarthy (2004) demonstrated that PGE2 is stimulated by estradiol in the newborn rat brain, and that PGE2 induces the growth and differentiation of those regions of the brain involved in male sexual behaviors. They found that PGE2 was as effective as estrogen in masculinizing these brain regions, and that PGE2 induces male-specific morphology and behaviors when injected into female rat brains. Moreover, when male rats were given COX2 inhibitors, their brains became similar to those of females. Not only did their brain anatomy change, but so did their behaviors. The PGE2-treated females acted like males, attempting to mount and copulate with other females, while male mice treated with COX2 inhibitors lost their sexual drive and did not exhibit male behaviors (**Figure 14.15B**). Studies comparing the structure of brain regions responding to pheromones in mice treated with PGE2 or estrogens indicate that the processing of pheromone signals is a sexually dimorphic phenomenon regulated by early exposure to PGE2 and estradiol.

The human element

Extrapolating from rodents to humans is a very risky business. No sex-specific behavior has yet been identified in humans, humans do not use α-fetoprotein to bind circulating estradiol, and humans do not use pheromones as a primary sexual attractant (sight and touch being far more critical). No "gay

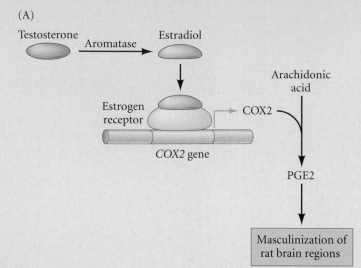

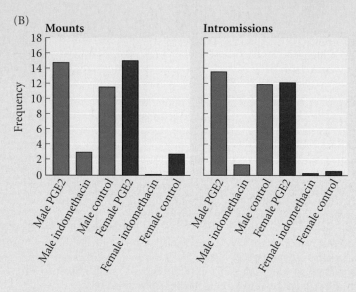

Figure 14.15 Masculinization of the brain by hormones. (A) Testosterone mediates its effect on the rodent brain by being converted to estradiol by aromatase. Estradiol activates the gene for cyclooxygenase, which converts arachidonic acid to PGE2. (B) PGE2 promotes male behavior, and its inhibition by COX2 inhibitors prevents that behavior. Males treated with indomethacin (COX2 inhibitor) did not display the mounting behavior expected of males, nor did they attempt intercourse ($P < 0.001$ between control and treated males). Conversely, females given PGE2 acted like control males in both cases and had significantly more mounting and mating behavior than control females ($P < 0.001$). (After Amateau and McCarthy 2004.)

gene" has been discovered, and the concordance of gender identity between identical twins is only 30%— far from the 100% expected if sexual orientation were strictly genetic (Bailey et al. 2000; CRC 2006). Moreover, behaviors that are seen as "masculine" in one culture may be considered "feminine" in another, and vice versa

(see Jacklin 1981; Bleier 1984; Fausto-Sterling 1992; Kandel et al. 1995). How humans acquire their gendered behavior appears to involve a remarkably complex set of interactions between genes and environment. Like many other behavioral phenotypes, how individuals acquire sexual behaviors has yet to be determined.

Chromosomal Sex Determination in *Drosophila*

Although both mammals and fruit flies produce XX females and XY males, their chromosomes achieve these ends using very different means. In mammals, the Y chromosome plays a pivotal role in determining the male sex. Thus, XO mammals are females, with ovaries, a uterus, and oviducts (but usually very few, if any, ova).

In *Drosophila*, the Y chromosome is not involved in determining sex. A fruit fly's sex is determined predominantly by the number of X chromosomes in each cell. If there is only one X chromosome in a diploid cell, the fly is male. If there are two X chromosomes in a diploid cell, the fly is female. Should a fly have two X chromosomes and three sets of autosomes, it is a mosaic, where some of the cells are male and some of the cells are female. In flies, the Y chromosome appears to be important in sperm cell differentiation, but it plays no role in sex determination. Rather, it seems to be a collection of genes that are active in forming sperm in adults. Thus, XO *Drosophila* are sterile males.

In *Drosophila*, and in insects in general, one can observe gynandromorphs—animals in which certain regions of the body are male and other regions are female (**Figure 14.16**; see also Figure 14.13). Gynandromorph fruit flies result when an X chromosome is lost from one embryonic nucleus. The cells descended from that cell, instead of being XX (female), are XO (male). The XO cells display male characteristics, whereas the XX cells display female traits, suggesting that, in *Drosophila*, each cell makes its own sexual "decision." Indeed, in their classic discussion of gynandromorphs, Morgan and Bridges (1919) concluded, "Male and female parts and their sex-linked characters are strictly self-

determining, each developing according to its own aspiration," and each sexual decision is "not interfered with by the aspirations of its neighbors, nor is it overruled by the action of the gonads."

Although there are organs that are exceptions to this rule (notably the external genitalia), it remains a good general principle of *Drosophila* sexual development. Moreover, molecular data show that differential RNA splicing to create male- and female-specific transcription factors is critical in determining fly sex and can explain how XX individuals become females and XY individuals become males.

The Sex-lethal *gene*

Although it had long been thought that a fruit fly's sex was determined by the X-to-autosome (X:A) ratio (Bridges 1925), this assessment was based largely on the fact that flies have aberrant numbers of chromosomes. Recent molecular analyses suggest that X chromosome number alone is the primary sex determinant in normal diploid insects (Erickson and Quintero 2007). The main basis for this assertion is the fact that the X chromosome contains genes encoding transcription factors that activate the critical gene in *Drosophila* sex determination, the autosomal locus *Sex-lethal* (*Sxl*).

ACTIVATING *SEX-LETHAL* The *Sxl* gene has two promoters. The early promoter is active only in XX cells; the later one is active in both XX and XY cells. The X chromosome appears to encode four protein factors that activate the early promoter of *Sxl*. Three of these proteins are transcription factors—SisA, Scute, and Runt—which bind to the early promoter to activate transcription (**Figure 14.17**). The fourth protein, Unpaired, is a secreted factor that reinforces the

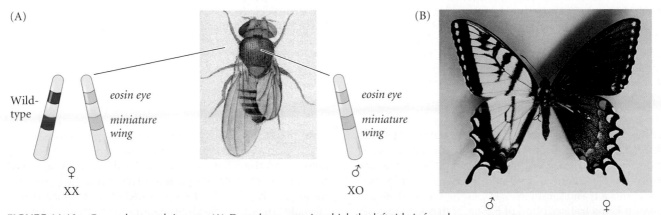

FIGURE 14.16   Gynandromorph insects. (A) *D. melanogaster* in which the left side is female (XX) and the right side is male (XO). The male side has lost an X chromosome bearing the wild-type alleles of eye color and wing shape, thereby allowing expression of the recessive alleles *eosin eye* and *miniature wing* on the remaining X chromosome. (B) Tiger swallowtail butterfly *Paplio glaucus*. The left half (yellow) is male, while the right half (blue-black) is female. (A, drawing by Edith Wallace from Morgan and Bridges 1919; B, photograph courtesy of J. Adams © 2005.)

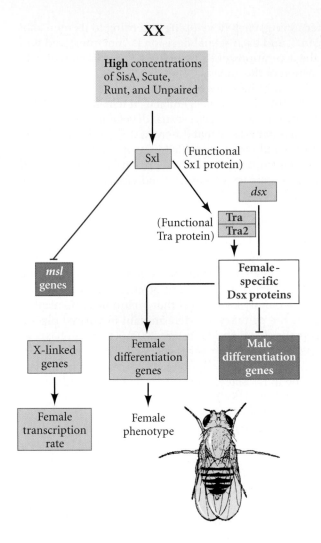

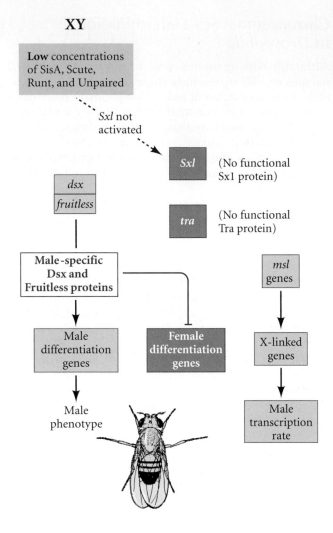

FIGURE 14.17 Proposed regulatory cascade for *Drosophila* somatic sex determination. Transcription factors from the X chromosomes and autosomes compete to activate or repress the *Sxl* gene, which becomes active in females (XX) and inactive in males (XY). The Sex-lethal protein performs three main functions. First, it activates its own transcription, ensuring further Sxl production. Second, it represses the translation of *msl2* mRNA, a factor that facilitates transcription from the X chromosome. This equalizes the amount of transcription from the two X chromosomes in females with that of the single X chromosome in males. Third, Sxl activates the *transformer* (*tra*) genes. The Tra proteins process *doublesex* pre-mRNA in a female-specific manner that provides most of the female body with its sexual fate. They also process the *fruitless* pre-mRNA in a female-specific manner, giving the fly female-specific behavior. In the absence of Sxl (and thus the Tra proteins), *dsx* and *fruitless* pre-mRNAs are processed in the male-specific manner. (After Baker et al. 1987.)

other three proteins through the JAK-STAT pathway (Sefton et al. 2000; Avila and Erickson 2007). If these factors accumulate so they are present in amounts above a certain threshold, the *Sxl* gene is activated through its early promoter (Erickson and Quintero 2007; Gonzales et al. 2008). The result is the transcription of *Sxl* early in XX embryos.

The *Sxl* pre-RNA transcribed from the early promoter lacks exon 3, which contains a stop codon. Thus, Sxl protein that is made early is spliced in a manner such that exon 3 is absent and a complete protein can be made. This functional Sxl binds to its own late promoter to keep the *Sxl* gene active.

In males, the early promoter of *Sxl* is not active. However, later in development, the late promoter becomes active, and the *Sxl* gene is transcribed in both males and females. In XX cells, Sxl protein made from the early promoter can bind to its own pre-mRNA and splice it in a "female" direction, producing more functional Sxl (Bell et al. 1988; Keyes et al. 1992). In XY cells, however, there is no early Sxl protein, and the "male" pre-mRNA is spliced in a manner that yields eight exons, and the termination codon is in exon 3. Protein synthesis thus ends at the third exon, and the protein is nonfunctional. In females, RNA processing yields seven exons, with the male-specific exon 3 being

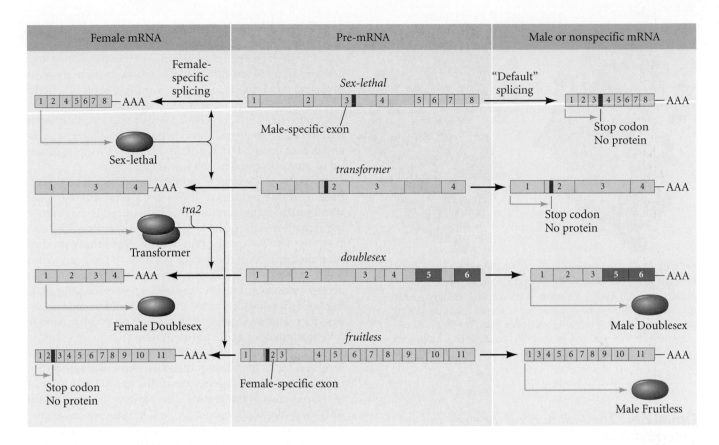

FIGURE 14.18 Sex-specific RNA splicing in four major *Drosophila* sex-determining genes. The pre-mRNAs (shown in the center of diagram) are identical in both male and female nuclei. In each case, the female-specific transcript is shown at the left, while the default transcript (whether male or nonspecific) is shown to the right. Exons are numbered, and the positions of termination codons are marked. *Sex-lethal*, *transformer*, and *doublesex* are all part of the genetic cascade of primary sex determination. The transcription pattern of *fruitless* determines the secondary characteristic of courtship behavior (see pp. 533–534). (After Baker 1989; Baker et al. 2001.)

spliced out as part of a large intron. Thus, the female-specific mRNA lacks the early-termination codon and encodes a protein of 354 amino acids, whereas the male-specific *Sxl* transcript contains a translation termination codon (UGA) after amino acid 48 (**Figure 14.18**, top).

TARGETS OF SEX-LETHAL The protein made by the female-specific *Sxl* transcript contains regions that are important for binding to RNA. There appear to be three major RNA targets to which the female-specific *Sxl* transcript binds. One of these, as mentioned above, is the pre-mRNA of *Sxl* itself. Another target is the *msl2* gene that controls dosage compensation (see below). Indeed, If the *Sxl* gene is nonfunctional in a cell with two X chromosomes, the dosage compensation system will not work, and the result is cell death (hence the gene's macabre name). The third target is

the pre-mRNA of *transformer* (*tra*)—the next gene in the cascade (Bell et al. 1988; Nagoshi et al. 1988).

The pre-mRNA of *transformer* (so named because loss-of-function mutations turn females into males) is spliced into a functional mRNA by Sxl protein. The *tra* pre-mRNA is made in both male and female cells; however, in the presence of Sxl protein, the *tra* transcript is alternatively spliced to create a female-specific mRNA, as well as a nonspecific mRNA that is found in both females and males. Like the male *Sxl* message, the nonspecific *tra* mRNA contains a termination codon early in the message that renders the protein nonfunctional (Boggs et al. 1987). In *tra*, the second exon of the nonspecific mRNA contains the termination codon and is not utilized in the female-specific message (see Figures 14.17 and 14.18).

How is it that females make a different transcript than males? The female-specific Sxl protein activates a 3′ splice site that causes *tra* pre-mRNA to be processed in a way that splices out the second exon. To do this, Sxl protein blocks the binding of splicing factor U2AF to the nonspecific splice site of the *tra* message by specifically binding to the polypyrimidine tract adjacent to it (**Figure 14.19**; Handa et al. 1999). This causes U2AF to bind to the lower-affinity (female-specific) 3′ splice site and generate a female-specific mRNA (Valcárcel et al. 1993). The female-specific Tra protein works in concert with the product of the *transformer-2* (*tra2*) gene to help generate the female phenotype by splicing the *doublesex* gene in a female-specific manner.

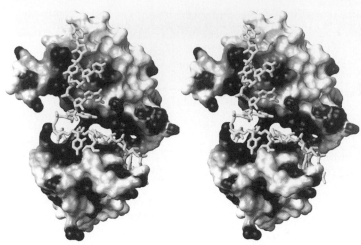

FIGURE 14.19 Stereogram showing binding of *tra* pre-mRNA by the cleft of the Sxl protein. The bound 12-nucleotide RNA (GUUGUUUUUUUU) is shown in yellow. The strongly positive regions are shown in blue, while the scattered negative regions are in red. It is worth crossing your eyes to get the three-dimensional effect. (From Handa et al. 1999, courtesy of S. Yokoyama.)

Doublesex: *The switch gene for sex determination*

The **doublesex** (*dsx*) gene is active in both males and females, but its primary transcript is processed in a sex-specific manner (Baker et al. 1987). This alternative RNA processing is the result of the action of the *tra* and *tra2* gene products on the *dsx* gene (see Figures 14.17 and 14.18; see also Figure 2.28). If the Tra2 and female-specific Tra pro-teins are both present, the *dsx* transcript is processed in a female-specific manner (Ryner and Baker 1991). The female splicing pattern produces a female-specific protein that activates female-specific genes (such as those of the yolk proteins) and inhibits male development. If no functional Tra is produced, a male-specific transcript of *dsx* is made. The male transcript encodes an active transcription factor that inhibits female traits and promotes male traits. In the embryonic gonad, Dsx regulates all known aspects of sexually dimorphic gonad cell fate (**Figure 14.20**).

In XX flies, the female Doublesex protein (Dsx^F) combines with the product of the *intersex* gene (Ix) to make a transcription factor complex that is responsible for promoting female-specific traits. This "Doublesex complex" activates the *Wingless* gene (*Wg*), whose Wnt-family product promotes growth of the female portions of the genital disc. It also represses the *Fgf* genes responsible for making male accessory organs, activates the genes responsible for making yolk proteins, promotes the growth of the sperm storage duct, and modifies *bricabrac* (*bab*) gene expression to give the female-specific pigmentation profile. In contrast, the male Doublesex protein (Dsx^M) acts directly as a transcription factor and directs the expression of male-specific traits. It causes the male region of the genital disc to grow at the expense of the female disc regions. It activates the BMP homologue *Decapentaplegic* (*Dpp*), as well as stimulating *Fgf* genes to produce the male genital disc and accessory structures. Dsx^M also converts certain cuticular structures into claspers and modifies the *bricabrac* gene to produce the male pigmentation pattern (Ahmad and Baker 2002; Christiansen et al. 2002).

According to this model, the result of the sex determination cascade summarized in Figure 14.17 comes down to the type of mRNA processed from the *doublesex* tran-

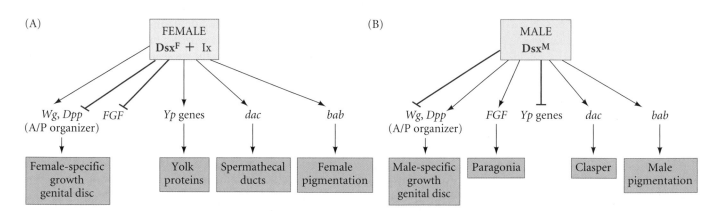

FIGURE 14.20 Roles of Dsx^M and Dsx^F proteins in *Drosophila* sexual development. (A) Dsx^F functions with Intersex (Ix) to promote female-specific expression of those genes that control the growth of the genital disc, the synthesis of yolk proteins, the formation of spermathecal ducts (which keep sperm stored after mating), and pigment patterning. (B) Conversely, Dsx^M acts as a transcription factor to promote the male-specific growth of the genital disc, the formation of male genitalia, the conversion of cuticle into claspers, and the male-specific pigmentation pattern. In addition, Dsx^F represses certain genes involved in specifying male-specific traits (such as the paragonia), and Dsx^M represses certain genes involved in synthesizing female-specific proteins such as yolk protein. (After Christiansen et al. 2002.)

(A)

(B)

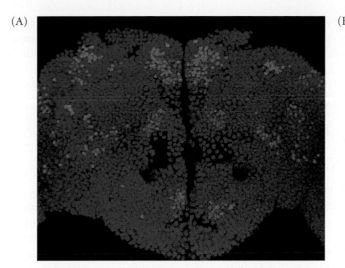

FIGURE 14.21   Subsets of neurons expressing the male-specific form of *fruitless*. (A) Central nervous system of a 48-hour pupa (the time at which sex determination of the brain is established), showing male-specific *fruitless* mRNA in particular cells (pink) of the anterior brain that are involved in courtship behaviors. (B) Some peripheral nervous system cells express the male-specific *fruitless* message. Fruitless protein is stained green in these taste neurons in the labella (the spongy part of the fly mouth). (From Stockinger et al. 2005, courtesy of B. J. Dickson.)

script. If there are two X chromosomes, the transcription factors activating the early promoter of *Sxl* reach a critical concentration, and *Sxl* makes a splicing factor that causes the *transformer* gene transcript to be spliced in a female-specific manner. This female-specific protein interacts with the *tra2* splicing factor, causing *dsx* pre-mRNA to be spliced in a female-specific manner. If the *dsx* transcript is not acted on in this way, it is processed in a "default" manner to make the male-specific message.

Brain Sex in Drosophila

Our discussion of sexual dimorphism in *Drosophila* has so far been limited to nonbehavioral aspects of development. However, as in mammals, there appears to be a separate "brain-sex" pathway in *Drosophila* that provides individuals of each sex with the appropriate set of courtship and aggression behaviors. Among *Drosophila*, there are no parents or other conspecifics to teach "proper" mating behavior, and mating takes place soon after the flies emerge from their pupal cases. So the behaviors must be "hard-wired" into the insect genome.*

Among male *Drosophila*, there is one very simple rule of behavior: If the other fly is male, fight it; if the other fly is female, court it (Certel et al. 2007; Billeter et al. 2008; Wang and Anderson 2010). Although the outcomes of this algorithm are simple, the behaviors of *Drosophila* courtship and mating are quite complicated. A courting male must first confirm that the individual he is approaching is a female. Once this is established, he must orient his body toward the female and follow a specific series of move-

*This is not to say that flies don't learn; indeed, one thing they do learn is to avoid bad sexual encounters. A male who has been brushed off (quite literally) by a female because she has recently mated hesitates before starting to court another female (Siegel and Hall 1979; MacBride et al. 1999).

ments that include following the female, tapping the female, playing a species-specific courtship song by vibrating his wings, licking the female, and finally, curling his abdomen so that he is in a position to mate. Each of these sex-specific courtship behaviors appear to be regulated by the products of *fruitless*, a gene expressed in certain sets of neurons involved with male sexual behaviors (**Figure 14.21**). These include subsets of neurons involved in taste, hearing, smell, and touch, and in total they represent about 2% of all the neurons in the adult male (Lee et al. 2000; Billeter and Goodwin 2004; Stockinger et al. 2005). Fruitless also retains certain male-specific neural circuits; the neurons in these circuits die during female development (and in *fruitless* mutants; see Kimura et al. 2005).

As with *doublesex* pre-mRNA, the Tra and Tra2 proteins splice *fruitless* pre-mRNA into a female-specific message; the default splicing pattern is male (see Figure 14.18). So the female makes Tra protein and processes the *fruitless* pre-mRNA in one way, whereas the male, lacking the Tra protein, processes the *fruitless* message in another way. However, female *fruitless* mRNA includes a termination sequence in an early exon; therefore the female does not make functional Fruitless protein. The male, however, makes an mRNA that does not contain the stop codon (Heinrichs et al. 1998), and the protein it transcribes is a zinc-finger transcription factor. Using homologous recombination to force the transcription of particular splicing forms, Demir and Dickson (2005) showed that it is Fruitless, and not the flies' anatomy, that controls their sexual behavior. When female flies were induced to make the male-specific Fruitless protein, they performed the entire male courtship ritual and tried to mate with other females.

In normal females, the courtship ritual is not as involved as in males. However, females have the ability to be receptive to a male's entreaties or to rebuff them. The product of the *retained* gene (*rtn*) is critical in this female mating behavior. Both sexes express this gene, since it is also involved in axon pathfinding. However, female flies with a loss-of-function allele of *rtn* resist male courtship and are thus rendered sterile by their own behavior (Ditch et al. 2005).

The splicing of the *fruitless* transcripts not only regulates sex-specific courtship patterns, it also regulates sex-specific aggression patterns as well. Female flies having a male Fruitless protein not only tend to court females, they also will fight males and try to establish themselves at the top of a dominance hierarchy. Male flies having a mutant *fruitless* allele will show female-specific aggression against other females (Vrontou et al. 2006).

Dosage Compensation

In animals whose sex is determined by sex chromosomes, there has to be some mechanism by which the amount of X chromosome gene expression is equalized for males and females. This mechanism is known as **dosage compensation**. In Chapter 2 we discussed mammalian X-chromosome inactivation, whereby one of the X chromosomes is inactivated so that the transcription product level is the same in both XX cells and XY cells. In the worm *Caenorhabditis elegans*, dosage compensation occurs by lowering the transcription rates of *both* X chromosomes so that product levels are the same as those of XO individuals.

In *Drosophila*, the female X chromosomes are not suppressed; rather, the male's single X chromosome is hyperactivated. This "hypertranscription" is accomplished at the level of translation, and it is mediated by the Sxl protein. Sxl protein (which you will recall is made by the female cells) binds to the 5′ leader sequence and the 3′ untranslated regions (UTRs) of the *mls2* message. The bound Sxl inhibits the attachment of *msl2* mRNA to the ribosome and prevents the ribosome from getting to the mRNA's coding region (Beckman et al. 2005). The result is that female cells do not produce Msl2 protein (see Figure 14.17). However, Msl2 is made in male cells, in which Sxl is not present. Mls2 is part of a protein-mRNA complex that targets the X chromosome and loosens its chromatin structure by acetylating

histone 4 (see Figure 2.4). In this way, transcription factors gain access to the X chromosome at a much higher frequency in males than in females—hence, "hypertranscription."

ENVIRONMENTAL SEX DETERMINATION

Temperature-Dependent Sex Determination in Reptiles

While the sex of most snakes and lizards is determined by sex chromosomes at the time of fertilization, the sex of most turtles and all species of crocodilians is determined *after* fertilization, by the embryonic environment. In these reptiles, the temperature of the eggs during a certain period of development is the deciding factor in determining sex, and small changes in temperature can cause dramatic changes in the sex ratio (Bull 1980; Crews 2003). Often, eggs incubated at low temperatures produce one sex, whereas eggs incubated at higher temperatures produce the other. There is only a small range of temperatures that permits both males and females to hatch from the same brood of eggs.*

Figure 14.22 shows the abrupt temperature-induced change in sex ratios for the red-eared slider turtle. If a brood of eggs is incubated at a temperature below 28°C, all the turtles hatching from the eggs will be male. Above 31°C, every egg gives rise to a female. At temperatures in between, the brood will give rise to individuals of both sexes. Variations on this theme also exist. The eggs of the

---

*The evolutionary advantages and disadvantages of temperature-dependent sex determination are discussed in Chapter 19.

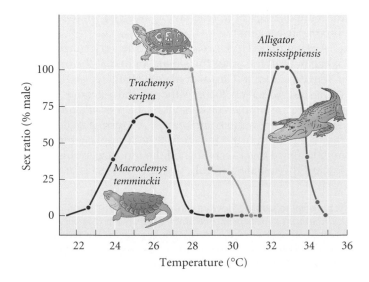

FIGURE 14.22 Temperature-dependent sex determination in three species of reptiles: the American alligator (*Alligator mississippiensis*), red-eared slider turtle (*Trachemys scripta elegans*), and alligator snapping turtle (*Macroclemys temminckii*). (After Crain and Guillette 1998.)

snapping turtle *Macroclemys*, for instance, become female at either cool (22°C or lower) or hot (28°C or above) temperatures. Between these extremes, males predominate.

As we will see in Chapter 18, there can be multiple pathways of sex determination in the same individual. Under normal temperature conditions, the sex of the lizard *Bassiana duperreyi* is determined by sex chromosomes (XY males; XX females). However, at low temperatures, the environmental component overrides the genetic sex-determining mechanism, and all the offspring in cool nests are male (even if their chromosomes are XX; Radder et al. 2008).

One of the best-studied reptiles is the European pond turtle, *Emys obicularis*. In laboratory studies, incubating *Emys* eggs at temperatures above 30°C produces all females, while temperatures below 25°C produce all-male broods. The threshold temperature (at which the sex ratio is even) is 28.5°C (Pieau et al. 1994). The developmental "window" during which sex determination occurs can be discovered by incubating eggs at the male-producing temperature for a certain amount of time and then shifting them to an incubator at the female-producing temperature (and vice versa). In *Emys*, the middle third of development appears to be the most critical for sex determination, and it is believed that the turtles cannot reverse their sex after this period.

The aromatase hypothesis for environmental sex determination

The enzyme aromatase, which converts testosterone into estrogen (see Sidelights & Speculations, p. 528), appears to be a particularly important target for environmental triggers. Unlike the situation in mammals, whose primary sex determination is a function of the X and Y chromosomes, primary sex determination in reptiles and birds is influenced by hormones, and estrogen is essential if ovaries are to develop. In reptiles, estrogen can override temperature, inducing ovarian differentiation even at masculinizing temperatures. Similarly, experimentally exposing eggs to inhibitors of estrogen synthesis produces male offspring, even if the eggs are incubated at temperatures that usually produce females (Dorizzi et al. 1994; Rhen and Lang 1994). The sensitive time for the effects of estrogens and their inhibitors coincides with the time when sex determination usually occurs (Bull et al. 1988; Gutzke and Chymiy 1988).

The estrogen-synthesis inhibitors used in the experiments mentioned above worked by blocking aromatase action, showing that experimentally low aromatase levels yield male offspring.* This correlation appears to hold under natural conditions as well. The aromatase activity

---

*One remarkable finding is that the injection of an aromatase inhibitor into the eggs of an all-female, parthenogenetic species of lizards causes the formation of males (Wibbels and Crews 1994).

of *Emys* is very low at the male-promoting temperature of 25°C. At the female-promoting temperature of 30°C, aromatase activity increases dramatically during the critical period for sex determination (Desvages et al. 1993; Pieau et al. 1994). Temperature-dependent aromatase activity is also seen in diamondback terrapins, and its inhibition masculinizes their gonads (Jeyasuria et al. 1994). Aromatase appears to be involved in the temperature-dependent differentiation of lizards and salamanders as well (Sakata et al. 2005). It is possible that aromatase expression is activated differently in different species. In some species, the aromatase *protein* itself may be temperature sensitive. In other species, the expression of the aromatase *gene* may be differentially activated at high temperatures (see Murdock and Wibbels 2006; Shoemaker et al. 2007).

When turtle gonads are taken out of the embryos and placed in culture, they become testes or ovaries depending on the temperature at which they are incubated (Moreno-Mendoza et al. 2001; Porter et al. 2005). This shows that sex determination is a local activity of the gonadal primordia and not a global activity directed by the pituitary or the brain. By analyzing the timing and regulation of sex-determining genes in such cultured gonads, Shoemaker and colleagues (2005, 2007) proposed that female temperatures not only activate aromatase (which would cause ovary formation), but also activate the *Wnt4* gene (which suppresses the formation of testes in mammals). Conversely, male temperatures appear to activate *Sox9* to a higher level in the male than in the female, promoting testis development (Moreno-Mendoza et al. 2001).

Estrogens, aromatase, sex reversal, and conservation biology

Over the last two decades, data has emerged showing that several of the polychlorinated biphenyl compounds (PCBs), a class of widespread pollutants introduced into the environment by humans, can act as estrogens (e.g., see Bergeron et al. 1994, 1999). PCBs can reverse the sex of turtles raised at "male" temperatures. This knowledge has important consequences in environmental conservation efforts to protect endangered species (such as turtles, amphibians, and crocodiles) in which hormones can effect changes in primary sex determination. Indeed, some reptile conservation biologists advocate using hormonal treatments to elevate the percentage of females in endangered species (www.reptileconservation.org).

The survival of some amphibian species may be at risk from herbicides that promote or destroy estrogens. One such case involves the development of hermaphroditic and demasculinized frogs after exposure to extremely low doses of the weed killer atrazine, the most widely used herbicide in the world (**Figure 14.23**). Hayes and colleagues (2002a) found that exposing tadpoles to atrazine concentrations as low as 0.1 part per billion (ppb) produced gonadal and other sexual anomalies in male frogs. At 0.1 ppb and higher, many male tadpoles developed ovaries in

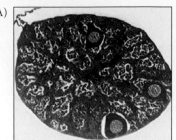

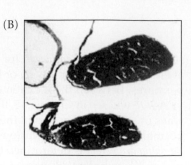

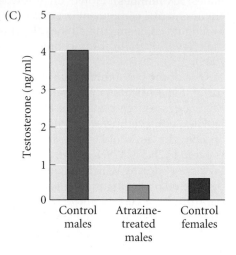

FIGURE 14.23 Demasculinization of frogs by low amounts of atrazine. (A) Testis of a frog from a natural site having 0.5 parts per billion (ppb) atrazine. The testis contains three lobules that are developing both sperm and an oocyte. (B) Two testes of a frog from a natural site containing 0.8 ppb atrazine. These organs show severe testicular dysgenesis, which characterized 28% of the frogs found at that site. (C) Effect of a 46-day exposure to 25 ppb atrazine on plasma testosterone levels in sexually mature male *Xenopus*. Levels in control males were some tenfold higher than in control females; atrazine-treated males had plasma testosterone levels at or below those of control females. (A,B after Hayes et al. 2003, photographs courtesy of T. Hayes; C after Hayes et al. 2002a.)

addition to testes. At 1 ppb atrazine, the vocal sacs (which a male frog must have in order to signal and obtain a potential mate) failed to develop properly.

Atrazine induces aromatase, and aromatase can convert testosterone into estrogen (Crain et al.1997; Fan et al. 2007). In laboratory experiments, the testosterone levels of adult male frogs were reduced by 90% (to levels of control females) by 46 days of exposure to 25 ppb atrazine—an ecologically relevant dose, since the allowable amount of atrazine in U.S. drinking water is 3 ppb, and atrazine levels can reach 224 ppb in streams of the midwestern United States (Battaglin et al. 2000; Barbash et al. 2001).

Given the amount of atrazine in the water supply and the sensitivity of frogs to this compound, the situation could be devastating to wild populations. In a field study, Hayes and his colleagues collected leopard frogs and water at eight sites across the central United States (Hayes et al. 2002b, 2003). They sent the water samples to two separate laboratories for the determination of atrazine, coding the frog specimens so that the technicians dissecting the gonads did not know which site the animals came from. The results showed that all but one site contained atrazine—and this was the only site from which the frogs had no gonadal abnormalities. At concentrations as low as 0.1 ppb, leopard frogs displayed testicular dysgenesis (stunted growth of the testes) or conversion to ovaries. In many examples, oocytes were found in the testes (see Figure 14.23A).

Concern over atrazine's apparent ability to disrupt sex hormones in both wildlife and humans has resulted in bans on the use of this herbicide by France, Germany, Italy, Norway, Sweden, and Switzerland (Dalton 2002). Many geographical and social concerns mediate the amount of atrazine use (**Figure 14.24**), and the company making atrazine has lobbied against the work of independent researchers whose research indicates that it might cause reproductive malfunctions or cancers in wildlife and humans (see Blumenstyk 2003). Endocrine disruptors will be discussed in more detail in Chapter 17.

Location-Dependent Sex Determination

Environmental factors other than temperature can also be sex determinants in some species. For example, it has been known since the nineteenth century that the sex of the echiuroid worm *Bonellia viridis* depends on where a larva settles (Baltzer 1914). If a *Bonellia* larva lands on the ocean floor, it develops into a 10-cm-long female. If the larva is attracted to a female's proboscis, it travels along the tube until it enters the female's body. There it differentiates into a minute (1–3 mm long) male that is essentially a sperm-producing symbiont of the female (**Figure 14.25**).

Another species in which sex determination is affected by the location of the organism is the slipper snail *Crepidula fornicata*. In this species, individuals pile up on top of one another in a great mound. Young individuals are always male, but this phase is followed by the degeneration of the male reproductive system and a period of lability. The next phase can be either male or female, depending on the animal's position in the mound. If the snail is attached to a female, it will become male. If such a snail is removed from its attachment, it will become female. Similarly, the presence of large numbers of males causes some of the males to become females. However, once an individual becomes female, it will not revert to being male (Coe 1936; Collin 1995; Warner et al. 1996).

Many fish change sex based on social interactions; such changes are mediated by the neuroendocrine system (Godwin et al. 2003, 2009). Interestingly, although the trigger of the sex change may be stress hormones (e.g., cortisol) that induce sex-specific neuropeptides, the effector of the change may be (once again) aromatase. There are two vari-

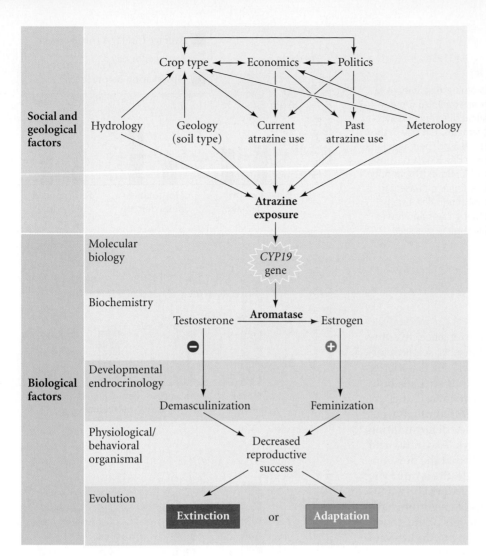

Social and geological factors

Crop type ⟷ Economics ⟷ Politics

Hydrology   Geology (soil type)   Current atrazine use   Past atrazine use   Meterology

Atrazine exposure

Biological factors

Molecular biology

CYP19 gene

Biochemistry

Testosterone — **Aromatase** → Estrogen

⊖   ⊕

Developmental endrocrinology

Demasculinization   Feminization

Physiological/ behavioral organismal

Decreased reproductive success

Evolution

Extinction   or   **Adaptation**

FIGURE 14.24 Possible chain of causation leading to the feminization of male frogs and the decline of frog populations in regions where atrazine has been used to control weed populations. Social, geological, and biological agents are shown. *CYP19* is the gene encoding aromatase, and it has been shown that transcription of the human *CYP19* gene is induced by atrazine (Sanderson et al. 2000). (From Hayes 2005.)

ants of aromatase in many animals, one expressed in the brain and one in the gonads. Black and colleagues (2005) showed a striking correlation between changes in brain aromatase levels and changes in sexual behavior during the sexual transitions of gobies, in which the school typically contains only one male and multiple females. The removal of the male from a stable group caused a rapid increase (>200%) in the aggressive behavior of the largest female, which then became a male in about a week's time. This transformation may have resulted from an increase in brain testosterone levels, since within hours upon removal of the male, these dominant females developed a lower brain aromatase than the other females (**Figure 14.26**). The *gonadal* aromatase levels, however, stayed the same, and gonadal sex change came later. In porgy fish, aromatase inhibitors can block the natural sex change and induce male development (Lee et al. 2002). Thus, changes in the social group, perceived by the nervous system, became expressed by the hormonal system within hours, thereby changing the behavioral phenotype of the female fish. Interestingly, when it comes to behaviors in fish, sex is in the brain before it is in the gonads.

See WEBSITE 14.5 Forms of hermaphroditism

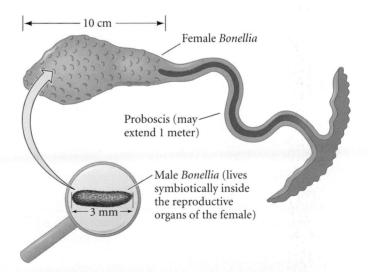

← 10 cm →

Female *Bonellia*

Proboscis (may extend 1 meter)

Male *Bonellia* (lives symbiotically inside the reproductive organs of the female)

← 3 mm →

FIGURE 14.25 Sex determination in *Bonellia viridis*. Larvae that settle on the ocean floor become female. The mature female's body is about 10 cm long, with a 1-meter-long proboscis that emits chemicals which attract other *B. viridis* larvae. Larvae that land on the proboscis are taken into the female's body, where they develop and live symbiotically as tiny (1–3 mm long) males.

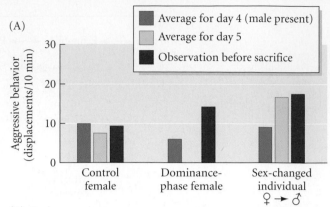

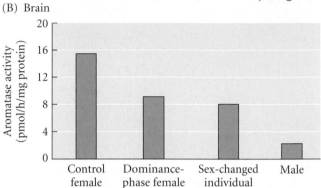

FIGURE 14.26 Aggressive behavior and aromatase activity (AA) in the brain and gonads of the goby *Lythrypnus dalli*. (A) On day 4 (prior to male removal), there was no statistical difference in average daily displacements among the largest females. On day 5, the male was removed and dominant females increased their aggressive behavior. (Dominance-phase fish have no day 5 data because they were sacrificed on day 5 or just after.) (B,C) Brain (B) but not gonadal (C) AA was significantly lower in dominance-phase and sex-changed individuals compared with control females. Established males had lower brain AA than all other groups and lower gonadal AA than all groups except dominance-phase females. (After Black et al. 2005.)

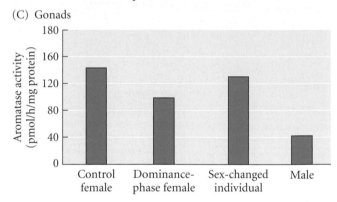

Coda

Nature has many variations on her masterpiece. In some species, including most mammals and insects, sex is determined by chromosomes; in other species, sex is a matter of environmental conditions. In yet other species, both environmental and genotypic sex determination function, often in different geographical areas. Different environmental or genetic stimuli may trigger sex determination through a series of conserved pathways. As Crews and Bull (2009) have reflected, "it is possible that the developmental decision of male versus female does not flow through a single gene but is instead determined by a 'parliamentary' system involving networks of genes that have simultaneous inputs to several components of the downstream cascade." We are finally beginning to understand the mechanisms by which this "masterpiece of nature" is created.

 Snapshot Summary: *Sex Determination*

1. In mammals, primary sex determination (the determination of gonadal sex) is a function of the sex chromosomes. XX individuals are females, XY individuals are males.

2. The Y chromosome plays a key role in male sex determination. XY and XX mammals both have a bipotential gonad. In XY animals, Sertoli cells differentiate and enclose the germ cells within testis cords. The interstitial mesenchyme becomes the Leydig cells.

3. In XX individuals, the germ cells become surrounded by follicle cells in the cortex of the gonadal rudiment. The epithelium of the follicles becomes the granulosa cells; the mesenchyme becomes the thecal cells.

4. In humans, the *SRY* gene is the testis-determining factor on the Y chromosome. It synthesizes a nucleic acid-binding protein that may function as either a transcription factor or as an RNA splicing factor. It activates the evolutionarily conserved *SOX9* gene.

5. The *SOX9* gene product can also initiate testes formation. Functioning as a transcription factor, it binds to the gene encoding anti-Müllerian factor is responsible for activating *FGF9*. Fgf9 and Sox9 proteins have a positive feedback loop that activates testicular development and suppresses ovarian development.

6. Wnt4 and Rspo-1 are involved in ovary formation. These proteins upregulate production of β-catenin; the functions of β-catenin include promoting the ovarian pathway of development while blocking the testicular pathway of development

7. Secondary sex determination in mammals involves the factors produced by the developing gonads. In male mammals, the Müllerian duct is destroyed by the AMF produced by the Sertoli cells, while testosterone produced by the Leydig cells enables the Wolffian duct to differentiate into the vas deferens and seminal vesicle. In female mammals, the Wolffian duct degenerates with the lack of testosterone, whereas the Müllerian duct persists and is differentiated by estrogen into the oviducts, uterus, cervix, and upper portion of the vagina.

8. The conversion of testosterone to dihydrotestosterone in the genital rudiment and prostate gland precursor enables the differentiation of the penis, scrotum, and prostate gland.

9. Individuals with mutations of these hormones or their receptors may have a discordance between their primary and secondary sex characteristics.

10. In *Drosophila*, sex is determined by the number of X chromosomes in the cell; the Y chromosome does not play a role in sex determination. There are no sex hormones, so each cell makes a sex-determination "decision." However, paracrine factors play important roles in forming the genital structures.

11. The *Drosophila Sex-lethal* gene is activated in females (by the accumulation of proteins encoded on the X chromosomes) but repressed in males. Sxl protein acts as an RNA splicing factor to splice an inhibitory exon from the *transformer* (*tra*) transcript. Therefore, female flies have an active Tra protein, while males do not.

12. The Tra protein also acts as an RNA splicing factor to splice exons from the *doublesex* (*dsx*) transcript. The *dsx* gene is transcribed in both XX and XY cells, but its pre-mRNA is processed to form different mRNAs, depending on whether Tra protein is present. The proteins translated from both *dsx* messages are active, and they activate or inhibit transcription of a set of genes involved in producing the sexually dimorphic traits of the fly.

13. Sex determination of the brain may have different downstream agents than in other regions of the body. *Drosophila* Tra proteins also activate the *fruitless* gene in males (but not in females); in mammals, the Y chromosome may activate brain sexual differentiation independently from the hormonal pathways.

14. Dosage compensation is critical for the regulation of gene expression in the embryo. With the same number of autosomes, the transcription from the X chromosome must be equalized for XX females and XY males. In mammals, one X chromosome of XX females is inactivated. In *Drosophila*, the single X chromosome of XY males is hyperactivated.

15. In turtles and alligators, sex is often determined by the temperature experienced by the embryo during the time of gonad determination. Because estrogen is necessary for ovary development, it is possible that differing levels of aromatase (an enzyme that can convert testosterone into estrogen) distinguish male from female patterns of gonadal differentiation.

16. Aromatase may be activated by environmental compounds, causing demasculinization of the male gonads in those animals where primary sex determination can be effected by hormones.

17. In some species, such as *Bonellia* and *Crepidula*, sex is determined by the position of the individual with regard to other individuals of the same species. In fish, numerous environmental factors—especially temperature and the number of males already present in the population—can determine the sex of an individual.

For Further Reading

Complete bibliographical citations for all literature cited in this chapter can be found at the free-access website **www.devbio.com**

Arnold, A. P. and P. S. Burgoyne. 2004. Are XX and XY brain cells intrinsically different? *Trends Endocr. Metab.* 15: 6–11.

Bell, L. R., J. I. Horabin, P. Schedl and T. W. Cline. 1991. Positive autoregulation of *Sex-lethal* by alternative splicing maintains the female determined state in *Drosophila*. *Cell* 65: 229–239.

Bullejos, M. and P. Koopman. 2005. Delayed *Sry* and *Sox9* expression in developing mouse gonads underlie B6-YDOM sex reversal. *Dev. Biol.* 278: 473–481.

Erickson, J. W. and J. J. Quintero. 2007. Indirect effects of ploidy suggest X chromosome dose, not the X:A ratio, signals sex in *Drosophila*. *PLoS Biol.* Dec;5(12):e332.

Hayes, T. B., A. Collins, M. Lee, M. Mendoza, N. Noriega, A. Stuart and A. Vonk. 2002. Hermaphroditic, demasculinized frogs after exposure to the herbicide atrazine at low ecologically relevant doses. *Proc. Natl. Acad. Sci. USA* 99: 5476–5480.

Imperato-McGinley, J., L. Guerrero, T. Gautier and R. E. Peterson. 1974.

Steroid 5α-reductase deficiency in man: An inherited form of male pseudohermaphroditism. *Science* 186: 1213–1215.

Koopman, P., J. Gubbay, N. Vivian, P. Goodfellow and R. Lovell-Badge. 1991. Male development of chromosomally female mice transgenic for *Sry*. *Nature* 351: 117–121.

Maatouk, D. M., L. DiNapoli, A. Alvers, K. L. Parker, M. M. Taketo and B. Capel. 2008. Stabilization of β-catenin in XY gonads causes male-to-female sex-reversal. *Hum. Mol Genet.* 17: 2949–2955.

Miyamoto,Y., H. Taniguchi, F. Hamel, D. W. Silversides and R. S. Viger. 2008. GATA4/WT1 cooperation regulates transcription of genes required for mammalian sex determination and differentiation. *BMC Mol. Biol.* 29: 9–44.

Pilon, N., I. Daneau, V. Paradis, F. Hamel, J. G. Lussier, R. S. Viger and D.

W. Silversides. 2003. Porcine SRY promoter is a target for steroidogenic factor 1. *Biol. Reprod.* 68: 1098–1106.

Sekido, R. and R. Lovell-Badge 2008. Sex determination involves synergistic action of Sry and Sf1 on a specific *Sox9* enhancer. *Nature* 453: 930–934.

Sekido, R. and R. Lovell-Badge. 2009. Sex determination and SRY: Down to a wink and a nudge? *Trends Genet.* 25: 19–29.

Stockinger, P., D. Kvitsiani, S. Rotkopf, L. Tirián and B. J. Dickson. 2005. Neural circuitry that governs *Drosophila* male courtship behavior. *Cell* 121: 795–807.

Go Online

WEBSITE 14.1 Social critique of sex determination research. In numerous cultures, women are seen as the "default state" and men are seen as having "something extra." Historians and biologists show that, until recently, such biases characterized the scientific study of human sex determination.

WEBSITE 14.2 Finding the male-determining genes. The mapping of the testis-determining factor to the SRY region took scientists more than 50 years to accomplish. Moreover, other testis-forming genes that act downstream of SRY have been found on autosomes.

WEBSITE 14.3 Dihydrotestosterone in adult men. The drug finasteride, which inhibits the conversion of testosterone to dihydrotestosterone, is being used to treat prostate growth and male pattern baldness.

WEBSITE 14.4 Insulin-like hormone 3. In addition to testosterone, the Leydig cells secrete insulin-like hormone 3 (INSL3). This hormone is required for the descent of the gonads into the scrotum. Males lacking INSL3 are infertile because the testes do not descend. In females, lack of this hormone deregulates the menstrual cycle.

WEBSITE 14.5 Forms of hermaphroditism. In *C. elegans* and many other invertebrates, hermaphroditism is the general rule. These animals may be born with both ovaries and testes, or they may develop one set of gonads first and the other later (sequential hermaphroditism). In some fish, sequential hermaphroditism is seen, with an individual fish being female in some seasons and male in others.

Outside Sites

For discussions of disorders of sexual development, including intersex conditions, see the American Academy of Pediatrics site at **http://www.pediatriccareonline.org/pco/ub/view/Point-of-Care-Quick-Reference/397188/all/disorders_of_sexual_development**.

The Intersex Society of North America website, **http://isna.org**, is an excellent resource for information on human intersex conditions. The Wikipedia entry on intersexuality, **http://en.wikipedia.org/wiki/Intersexuality**, has a good discussion about the language used to describe these conditions.

Postembryonic Development
Metamorphosis, Regeneration, and Aging

DEVELOPMENT NEVER CEASES. Throughout life, we continuously generate new blood cells, lymphocytes, keratinocytes, and digestive tract epithelium from stem cells. In addition to these continuous daily changes, there are instances in which postembryonic development is obvious—sometimes even startling. One of these instances is metamorphosis, the transition from a larval stage to an adult stage. In many species that undergo metamorphosis, a large proportion of the animal's structure changes, and the larva and the adult are unrecognizable as being the same individual (see Figure 1.4). Another startling type of postembryonic development is regeneration, the creation of a new organ after the original one has been removed from an adult animal. Some adult salamanders, for instance, can regrow limbs and tails after these appendages have been amputated.

The third category of postnatal developmental change encompasses those alterations of form and function associated with aging in adult organisms. This area is more controversial, with some scientists believing that the processes of age-associated degeneration are not properly part of the study of developmental biology. In this view, aging involves the random decay of normative processes. Others scientists claim that the genetically determined, species-specific patterns of aging are an important part of the life cycle and believe that **gerontology**—the scientific study of aging—is rightly part of developmental biology. As Peter Medawar (1957) noted, "That which we call 'development' when looked at from the birth and the life end becomes 'senescence' when looked at from its close." Whatever their relationship to embryonic development, metamorphosis, regeneration, and aging are critical topics for the biology of the twenty-first century.

The earth-bound early stages built enormous digestive tracts and hauled them around on caterpillar treads. Later in the life-history these assets could be liquidated and reinvested in the construction of an entirely new organism—a flying-machine devoted to sex.

CARROLL M. WILLIAMS (1958)

I'd give my right arm to know the secret of regeneration.

OSCAR E. SCHOTTÉ (1950)

METAMORPHOSIS: THE HORMONAL REACTIVATION OF DEVELOPMENT

Animals (including humans) whose young are basically smaller versions of the adult are referred to as **direct developers**. In most animal species, however, embryonic development includes a larval stage with characteristics very different from those of the adult organism, which emerges only after a period of metamorphosis; these animals are **indirect developers**.

Very often, larval forms are specialized for some function such as growth or dispersal, while the adult is specialized for reproduction. *Cecropia* moths, for example, hatch from eggs and develop as wingless juveniles—caterpillars—for several months. After metamorphosis, the insects spend a day or so as fully

TABLE 15.1 Summary of some metamorphic changes in anurans

System	Larva	Adult
Locomotory	Aquatic; tail fins	Terrestrial; tailless tetrapod
Respiratory	Gills, skin, lungs; larval hemoglobins	Skin, lungs; adult hemoglobins
Circulatory	Aortic arches; aorta; anterior, posterior, and common jugular veins	Carotid arch; systemic arch; cardinal veins
Nutritional	Herbivorous: long spiral gut; intestinal symbionts; small mouth, horny jaws, labial teeth	Carnivorous: short gut; proteases; large mouth with long tongue
Nervous	Lack of nictitating membrane; porphyropsin, lateral line system, Mauthner's neurons	Development of ocular muscles, nictitating membrane, rhodopsin; loss of lateral line system, degeneration of Mauthner's neurons; tympanic membrane
Excretory	Largely ammonia, some urea (ammonotelic)	Largely urea; high activity of enzymes of ornithine-urea cycle (ureotelic)
Integumental	Thin, bilayered epidermis with thin dermis; no mucous glands or granular glands	Stratified squamous epidermis with adult keratins; well-developed dermis contains mucous glands and granular glands secreting antimicrobial peptides

Source: Data from Turner and Bagnara 1976 and Reilly et al. 1994.

developed winged moths and must mate quickly before they die. The adults never eat, and in fact have no mouth-parts during this brief reproductive phase of the life cycle. As might be expected, the juvenile and adult forms often live in different environments. During metamorphosis, developmental processes are reactivated by specific hormones, and the entire organism changes morphologically, physiologically, and behaviorally to prepare itself for its new mode of existence.

Among indirect developers, there are two major types of larvae.* **Secondary larvae** are found among those animals whose larvae and adults possess the same basic body plan. Thus, despite the obvious differences between the caterpillar and the butterfly, these two life stages retain the same axes and develop by deleting and modifying old parts while adding new structures into a pre-existing framework. Similarly, the frog tadpole, although specialized for an aquatic environment, is a secondary larva, organized on the same pattern as the adult will be (Jagersten 1972; Raff and Raff 2009).

Larvae that represent dramatically different body plans than the adult form and that are morphologically distinct from the adult are called **primary larvae**. Sea urchin larvae, for instance, are bilaterally symmetrical organisms that float among and collect food in the plankton of the open ocean. The sea urchin adult is pentameral (organized on fivefold symmetry) and feeds by scraping algae from rocks on the seafloor. There is no trace of the adult form in the body plan of the juvenile (see Figure 5.22).

Amphibian Metamorphosis

Amphibians are named for their ability to undergo metamorphosis, their appellation coming from the Greek *amphi* ("double") and *bios* ("life"). Amphibian metamorphosis is associated with morphological changes that prepare an aquatic organism for a primarily terrestrial existence. In **urodeles** (salamanders), these changes include the resorption of the tail fin, the destruction of the external gills, and a change in skin structure. In **anurans** (frogs and toads), the metamorphic changes are more dramatic, with almost every organ subject to modification (Table 15.1; see also Figure 1.1). The changes in amphibian metamorphosis are initiated by thyroid hormones such as **thyroxine (T4)** and **tri-iodothyronine (T3)** that travel through the blood to reach all the organs of the larva. When the larval organs encounter these thyroid hormones, they can respond in any of four ways: growth, death, remodeling, and respecification.

Morphological changes associated with amphibian metamorphosis

GROWTH OF NEW STRUCTURES The hormone tri-iodothyronine induces certain adult-specific organs to form. The limbs of the adult frog emerge from specific sites on the metamorphosing tadpole, and in the eye, both nictitating membranes and eyelids emerge. Moreover, T_3 induces the proliferation and differentiation of new neurons to serve

---

*Although there is controversy on the subject, larvae probably evolved after the adult form had been established. In other words, animals evolved through direct development, and larval forms came about as specializations for feeding or dispersal during the early part of the life cycle (Jenner 2000; Rouse 2000; Raff and Raff 2009). Even so, the biphasic life cycle may be a trait characteristic of metazoans (see Degnan and Degnan 2010).

these organs. As the limbs grow out from the body axis, new neurons proliferate and differentiate in the spinal cord. These neurons send axons to the newly formed limb musculature (Marsh-Armstrong et al. 2004). Blocking T_3 activity prevents these neurons from forming and causes paralysis of the limbs.

One readily observed consequence of anuran metamorphosis is the movement of the eyes to the front of the head from their originally lateral position (**Figure 15.1**).* The lateral eyes of the tadpole are typical of preyed-upon herbivores, whereas the frontally located eyes of the frog befit its more predatory lifestyle. To catch its prey, the frog needs to see in three dimensions. That is, it has to acquire a *binocular field of vision*, where inputs from both eyes converge in the brain (see Chapter 10). In the tadpole, the right eye innervates the left side of the brain, and vice versa; there are no ipsilateral (same-side) projections of the retinal neurons. During metamorphosis, however, ipsilateral pathways emerge, enabling input from both eyes to reach the same area of the brain (Currie and Cowan 1974; Hoskins and Grobstein 1985a).

In *Xenopus*, these new neuronal pathways result not from the remodeling of existing neurons, but from the formation of new neurons that differentiate in response to thyroid hormones (Hoskins and Grobstein 1985a,b). The ability of these axons to project ipsilaterally results from the induction of ephrin B in the optic chiasm by the thyroid hormones (Nakagawa et al. 2000). Ephrin B is also found in the optic chiasm of mammals (which have ipsilateral projections throughout life) but not in the chiasm of fish and birds (which have only contralateral projections). As shown in Chapter 10, ephrins can repel certain neurons, causing them to project in one direction rather than in another.

CELL DEATH DURING METAMORPHOSIS The hormone T_3 also induces certain larval-specific structures to die. Thus, T_3 causes the degeneration of the paddle-like tail and the oxygen-procuring gills that were important for larval (but not adult) movement and respiration. While it is obvious that the tadpole's tail muscles and skin die, is this death murder or suicide? In other words, is T_3 telling the cells to kill themselves, or is T_3 telling something else to kill the cells? Recent evidence suggests that the first part of tail resorption is caused by suicide, but that the last remnants of the tadpole tail must be killed off by other means. When tadpole muscle cells were injected with a dominant negative T_3 receptor (and therefore could not respond to T_3), the muscle cells survived, indicating that T_3 told them to kill themselves by apoptosis (Nakajima and Yaoita 2003; Naka-

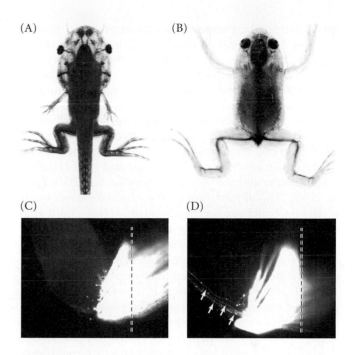

FIGURE 15.1 Eye migration and associated neuronal changes during metamorphosis of the *Xenopus laevis* tadpole. (A) The eyes of the tadpole are laterally placed, so there is relatively little binocular field of vision. (B) The eyes migrate dorsally and rostrally during metamorphosis, creating a large binocular field for the adult frog. (C,D) Retinal projections of metamorphosing tadpole. The dye DiI was placed on a cut stump of the optic nerve to label the retinal projection. (C) In early and middle stages of metamorphosis, axons project across the midline (dashed line) from one side of the brain to the other. (D) In late metamorphosis, ephrin B is produced in the optic chiasm as certain neurons (arrows) are formed that project ipsilaterally. (A,B from Hoskins and Grobstein 1984, courtesy of P. Grobstein; C,D from Nakagawa et al. 2000, courtesy of C. E. Holt.)

jima et al. 2005). This was confirmed by the demonstration that the apoptosis-inducing enzyme caspase-9 is important in causing cell death in the tadpole muscle cells (Rowe et al. 2005). However, later in metamorphosis, the tail muscles are destroyed by phagocytosis, perhaps because the extracellular matrix that supported the muscle cells has been digested by proteases.

Death also comes to the tadpole's red blood cells. During metamorphosis, tadpole hemoglobin is changed into adult hemoglobin, which binds oxygen more slowly and releases it more rapidly (McCutcheon 1936; Riggs 1951). The red blood cells carrying the tadpole hemoglobin have a different shape than the adult red blood cells, and these larval red blood cells are specifically digested—"eaten," if you will—by macrophages in the liver and spleen (Hasebe et al. 1999).

REMODELING DURING METAMORPHOSIS Among frogs and toads, certain larval structures are remodeled for adult needs. Thus, the larval intestine, with its numerous coils

---

*One of the most spectacular movements of eyes during metamorphosis occurs in flatfish such as flounder. Originally, a flounder's eyes, like the lateral eyes of other fish species, are on opposite sides of its face. However, during metamorphosis, one of the eyes migrates across the head to meet the eye on the other side (see Figure 19.32A; Hashimoto et al. 2002; Bao et al. 2005). This allows the flatfish to dwell on the ocean bottom, looking upward.

for digesting plant material, is converted into a shorter intestine for a carnivorous diet. Schrieber and his colleagues (2005) have demonstrated that the new cells of the adult intestine are derived from functioning cells of the larval intestine (instead of there being a subpopulation of stem cells that give rise to the adult intestine). The formation and differentiation of this new intestinal epithelium are probably triggered by the digestion of the old extracellular matrix by the metalloproteinase stromelysin-3, and by the new transcription of the *bmp4* and *sonic hedgehog* genes (Stolow and Shi 1995; Ishizuya-Oka et al. 2001; Fu et al. 2005). The elimination of the original extracellular matrix probably causes the apoptosis of those epithelial cells that were attached to it.* Therefore, the regional remodeling of the organs formed during metamorphosis may be generated by the reappearance of some of the same paracrine factors that modeled those organs in the embryo.

Much of the nervous system is remodeled as neurons grow and innervate new targets. The change in the optic nerve pathway was described earlier. Other larval neurons, such as certain motor neurons in the tadpole jaw, switch their allegiances from larval muscle to newly formed adult muscle (Alley and Barnes 1983). Still others, such as the cells innervating the tongue muscle (a newly formed muscle not present in the larva), have lain dormant during the tadpole stage and form their first synapses during metamorphosis (Grobstein 1987). The lateral line system of the tadpole (which allows the tadpole to sense water movement and helps it to hear) degenerates, and the ears undergo further differentiation (see Fritzsch et al. 1988). The middle ear develops, as does the tympanic membrane characteristic of frog and toad outer ears. Tadpoles experience a brief period of deafness as the neurons change targets (Boatright-Horowitz and Simmons 1997). Thus, the anu-

ran nervous system undergoes enormous restructuring as some neurons die, others are born, and others change their specificity.

The shape of the anuran skull also changes significantly as practically every structural component of the head is remodeled (Trueb and Hanken 1992; Berry et al. 1998). The most obvious change is that new bone is being made. The tadpole skull is primarily neural crest-derived cartilage; the adult skull is primarily neural crest-derived bone (Gross and Hanken 2005; **Figure 15.2**). Another outstanding change is the formation of the lower jaw. Here, Meckel's cartilage elongates to nearly double its original length, and dermal bone forms around it. While Meckel's cartilage is growing, the gills and pharyngeal arch cartilage (which were necessary for aquatic respiration in the tadpole) degenerate. Other cartilage, such as the ceratohyal cartilage (which will anchor the tongue), is extensively remodeled. Thus, as in the nervous system, some skeletal elements proliferate, some die, and some are remolded.

BIOCHEMICAL RESPECIFICATION In addition to the obvious morphological changes, important biochemical transformations occur during metamorphosis as T_3 induces a new set of proteins in existing cells. One of the most dramatic biochemical changes occurs in the liver. Tadpoles, like most freshwater fish, are ammonotelic—that is, they excrete

---

*Many epithelial cells are dependent on their attachment to the extracellular matrix to prevent apoptosis. The rapid apoptosis that occurs with the loss of extracellular matrix attachment has a special designation, *anoikis* (Frisch and Screaton 2001; see Chapter 3).

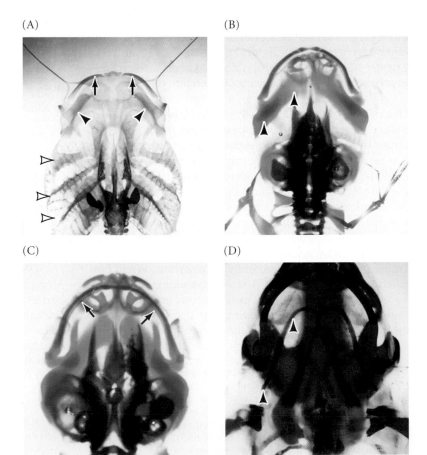

(A)   (B)

(C)   (D)

FIGURE 15.2 Changes in the *Xenopus* skull during metamorphosis. Whole mounts were stained with alcian blue to stain cartilage and alizarin red to stain bone. (A) Prior to metamorphosis, the pharyngeal (branchial) arch cartilage (open arrowheads) is prominent, Meckel's cartilage (arrows) is at the tip of the head, and the ceratohyal cartilage (arrowheads) is relatively wide and anteriorly placed. (B–D) As metamorphosis ensues, the pharyngeal arch cartilage disappears, Meckel's cartilage elongates, the mandible (lower jawbone) forms around Meckel's cartilage, and the ceratohyal cartilage narrows and becomes more posteriorly located. (From Berry et al. 1998, courtesy of D. D. Brown.)

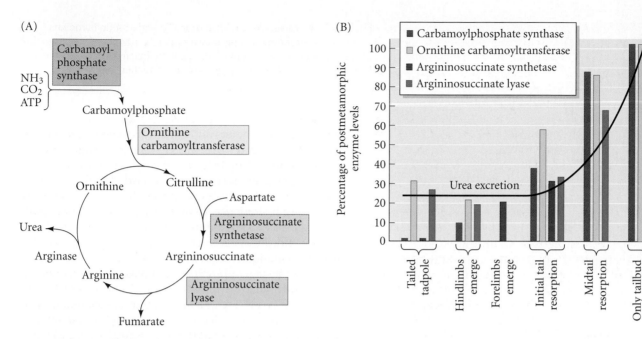

FIGURE 15.3 Development of the urea cycle during anuran metamorphosis. (A) Major features of the urea cycle, by which nitrogenous wastes are detoxified and excreted with minimal water loss. (B) The emergence of urea-cycle enzyme activities correlates with metamorphic changes in the frog *Rana catesbeiana*. (After Cohen 1970.)

ammonia. Like most terrestrial vertebrates, many adult frogs (such as the genus *Rana*, although not the more aquatic *Xenopus*) are ureotelic: they excrete urea, which requires less water than ammonia excretion. During metamorphosis, the liver begins to synthesize the enzymes necessary to create urea from carbon dioxide and ammonia (**Figure 15.3**). T_3 may regulate this change by inducing a set of transcription factors that specifically activates expression of the urea-cycle genes while suppressing the genes responsible for ammonia synthesis (Cohen 1970; Atkinson et al. 1996, 1998).

Hormonal control of amphibian metamorphosis

The control of metamorphosis by thyroid hormones was first demonstrated in 1912 by Gudernatsch, who discovered that tadpoles metamorphosed prematurely when fed powdered horse thyroid glands. In a complementary study, Allen (1916) found that when he removed or destroyed the thyroid rudiment of early tadpoles (thyroidectomy), the larvae never metamorphosed but instead grew into giant tadpoles. Subsequent studies showed that the sequential steps of anuran metamorphosis are regulated by increasing amounts of thyroid hormone (see Saxén et al. 1957; Kollros 1961; Hanken and Hall 1988). Some events (such as the development of limbs) occur early, when the concentration of thyroid hormones is low; other events (such as the resorption of the tail and remodeling of the intes-

tine) occur later, after the hormones have reached higher concentrations. These observations gave rise to a **threshold model**, wherein the different events of metamorphosis are triggered by different concentrations of thyroid hormones. Although the threshold model remains useful, molecular studies have shown that the timing of the events of amphibian metamorphosis is more complex than just increasing hormone concentrations.

The metamorphic changes of frog development are brought about by (1) the secretion of the hormone thyroxine (T_4) into the blood by the thyroid gland; (2) the conversion of T_4 into the more active hormone, tri-iodothyronine (T_3) by the target tissues; and (3) the degradation of T_3 in the target tissues (**Figure 15.4**). T_3 binds to the nuclear **thyroid hormone receptors** (**TRs**) with much higher affinity than does T_4, and causes these receptors to become transcriptional activators of gene expression. Thus, the levels of both T_3 and TRs in the target tissues are essential for producing the metamorphic response in each tissue (Kistler et al. 1977; Robinson et al. 1977; Becker et al. 1997).

The concentration of T_3 in each tissue is regulated by the concentration of T_4 in the blood and by two critical intracellular enzymes that remove iodine atoms from T_4 and T_3. **Type II deiodinase** removes an iodine atom from the outer ring of the precursor hormone (T_4) to convert it into the more active hormone T_3. **Type III deiodinase** removes an iodine atom from the inner ring of T_3 to convert it into an inactive compound that will eventually be metabolized to tyrosine (Becker et al. 1997). Tadpoles that are genetically modified to overexpress type III deiodinase in their target tissues never complete metamorphosis (Huang et al. 1999).

There are two types of thyroid hormone receptors. In *Xenopus*, **thyroid hormone receptor α** (**TRα**) is widely distributed throughout all tissues and is present even before the organism has a thyroid gland. **Thyroid hormone recep-**

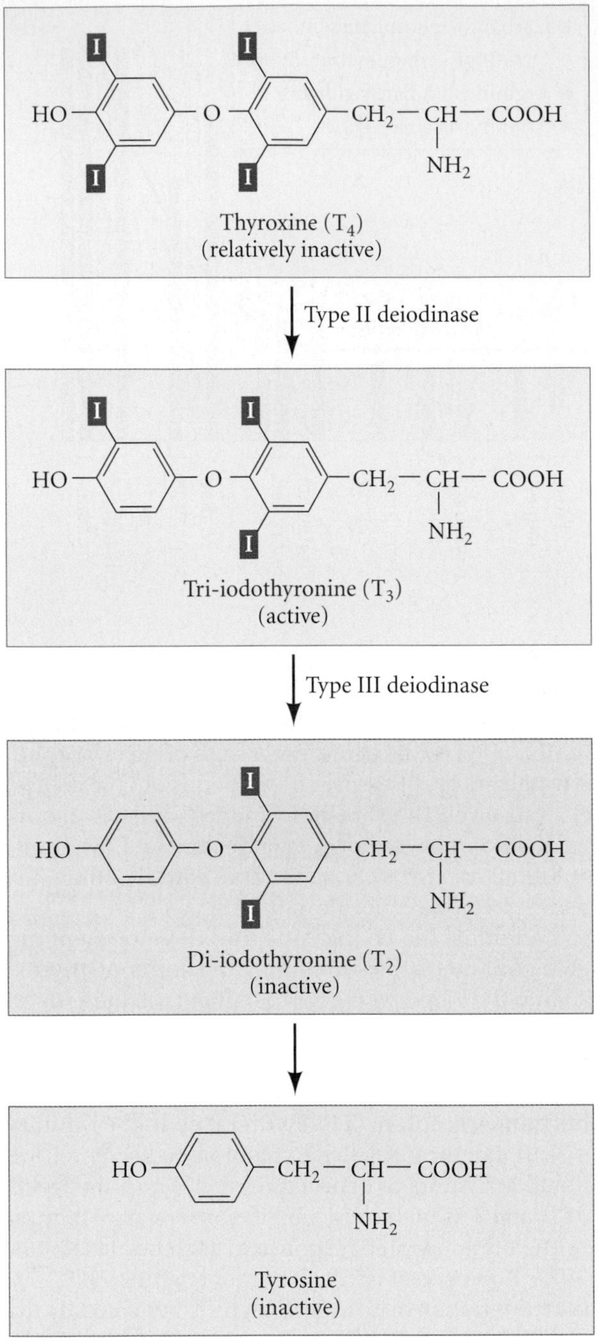

FIGURE 15.4 Metabolism of thyroxine (T_4) and tri-iodothyronine (T_3). T_4 serves as a prohormone. It is converted in the peripheral tissues to the active hormone T_3 by deiodinase II. T_3 can be inactivated by deiodinase III, which converts T_3 into di-iodothyronine and then to tyrosine.

appropriate promoters and enhancers even before it binds T_3. In its unbound state, the TR-RXR is a transcriptional repressor, recruiting histone deacetylases to the region of these genes. However, when T_3 is added to the complex, the T_3-TR-RXR complex activates those same genes by recruiting histone acetyltransferases (Sachs et al. 2001; Buchholz et al. 2003; Havis et al. 2003; Paul and Shi 2003).

Metamorphosis is often divided into stages based on the concentration of thyroid hormones in circulation. During the first stage, **premetamorphosis**, the thyroid gland has begun to mature and is secreting low levels of T_4 (and very low levels of T_3). The initiation of T_4 secretion may be brought about by corticotropin releasing hormone (CRH, which in mammals initiates the stress response). CRH may act directly on the frog pituitary, instructing it to release thyroid stimulating hormone (TSH), or it may generally to make the body cells responsive to low amounts of T_3 (Denver 1993, 2003).

The tissues that respond earliest to the thyroid hormones are those that express high levels of deiodinase II, and can thereby convert T_4 directly into T_3 (Cai and Brown 2004). For instance, the limb rudiments, which have high levels of both deiodinase II and TRα, can convert T_4 into T_3 and use it immediately through the TRα receptor. Thus, during the early stage of metamorphosis, the limb rudiments are able to receive thyroid hormone and use it to start leg growth (Becker et al. 1997; Huang et al. 2001; Schreiber et al. 2001).

As the thyroid matures to the stage of **prometamorphosis**, it secretes more thyroid hormones. However, many major changes (such as tail resorption, gill resorption, and intestinal remodeling) must wait until the **metamorphic climax** stage. At that time, the concentration of T_4 rises dramatically, and TRβ levels peak inside the cells. Since one of the target genes of T_3 is the TRβ gene, TRβ may be the principal receptor that mediates the metamorphic climax. In the tail, there is only a small amount of TRα during premetamorphosis, and deiodinase II is not detectable then. However, during prometamorphosis, the rising levels of thyroid hormones induce higher levels of TRβ. At metamorphic climax, deiodinase II is expressed, and the tail begins to be resorbed. In this way, the tail undergoes absorption only *after* the legs are functional (otherwise, the poor amphibian would have no means of locomotion). The wisdom of the frog is simple: never get rid of your tail before your legs are working.

Some tissues do not seem to be responsive to thyroid hormones. For instance, thyroid hormones instruct the *ventral* retina to express ephrin B and to generate the ipsilateral neurons shown in Figure 15.1D. The *dorsal* retina, how-

tor β (**TRβ**), however, is the product of a gene that is directly activated by thyroid hormones. TRβ levels are very low before the advent of metamorphosis; as the levels of thyroid hormone increase during metamorphosis, so do the intracellular levels of TRβ (Yaoita and Brown 1990; Eliceiri and Brown 1994).

The TRs do not work alone, however, but form dimers with the retinoid receptor, RXR. These dimers bind thyroid hormones and can effect transcription (Mangelsdorf and Evans 1995; Wong and Shi 1995; Wolffe and Shi 1999). The TR-RXR complex appears to be physically associated with

FIGURE 15.5 Regional specificity during frog metamorphosis. (A) Tail tips regress even when transplanted into the trunk. (B) Eye cups remain intact even when transplanted into the regressing tail. (After Schwind 1933.)

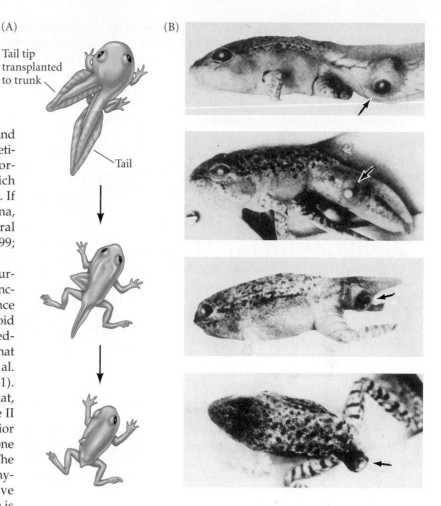

(A)

Tail tip transplanted to trunk

Tail

(B)

ever, is not responsive to thyroid hormones and does not generate new neurons. The dorsal retina appears to be insulated from thyroid hormones by expressing deiodinase III, which degrades the T_3 produced by deiodinase II. If deiodinase III is activated in the ventral retina, neurons will not proliferate and no ipsilateral axons will be formed (Kawahara et al. 1999; Marsh-Armstrong et al. 1999).

The frog brain also undergoes changes during metamorphosis, and one of the brain's functions is to downregulate metamorphosis once metamorphic climax has been reached. Thyroid hormones eventually induce a negative feedback loop, shutting down the pituitary cells that instruct the thyroid to secrete them (Saxén et al. 1957; Kollros 1961; White and Nicoll 1981). Huang and colleagues (2001) have shown that, at the climax of metamorphosis, deiodinase II expression is seen in those cells of the anterior pituitary that secrete thyrotropin, the hormone that activates thyroid hormone expression. The resulting T_3 suppresses transcription of the thyrotropin gene, thereby initiating the negative feedback loop so that less thyroid hormone is made.

See VADE MECUM
Amphibian metamorphosis and frog calls

Regionally specific developmental programs

By regulating the amount of T_3 and TRs in their cells, the different regions of the body can respond to thyroid hormones at different times. The type of response (proliferation, apoptosis, differentiation, migration) is determined by other factors already present in the different tissues. The same stimulus causes some tissues to degenerate while stimulating others to develop and differentiate, as exemplified by the process of tail degeneration. Thus, thyroid hormone instructs the limb bud muscles to grow (they die without thyroxine) while instructing the tail muscles to undergo apoptosis (Cai et al. 2007).

The resorption of the tadpole's tail structures is brought about by apoptosis and is relatively rapid, since the bony skeleton does not extend to the tail (Wassersug 1989). After apoptosis has taken place, macrophages collect in the tail region and digest the debris with their enzymes, especially collagenases and metalloproteinases. The result is that the tail becomes a large sac of proteolytic enzymes*

(Kaltenbach et al. 1979; Oofusa and Yoshizato 1991; Patterson et al. 1995). The tail epidermis acts differently than the head or trunk epidermis. During metamorphic climax, the larval skin is instructed to undergo apoptosis. The tadpole head and body are able to generate a new epidermis from epithelial stem cells. The tail epidermis, however, lacks these stem cells and fails to generate new skin (Suzuki et al. 2002).

Organ-specific responses to thyroid hormones have been dramatically demonstrated by transplanting a tail tip to the trunk region and by placing an eye cup in the tail (Schwind 1933; Geigy 1941). Tail tip tissue placed in the trunk is not protected from degeneration, but the eye cup retains its integrity despite the fact that it lies within the degenerating tail (**Figure 15.5**). Thus, the degeneration of the tail represents an organ-specific programmed cell death response, and only specific tissues die when the signal is given. Such programmed cell deaths are important in molding the body.

---

*Interestingly, the degeneration of the human tail during week 4 of gestation resembles the resorption of the tadpole tail (see Fallon and Simandl 1978).

The metamorphosis of tadpoles into frogs is one of the most rapid and accessible examples of development, obvious even to the eyes of children. Yet it still presents an enormous set of enigmas. As Don Brown and Liquan Cai (2007), have asked, "What will encourage the modern generation of scientists to study the wonderful biological problems presented by amphibian metamorphosis?" Humans may have medically centered interests in preventing aging and regenerating amputated parts; undergoing metamorphosis is not high on our to-do lists. However, the sheer wonder of these phenomena, the ability of computer-aided sequence analysis to obtain the data, and the desire to preserve these species should be enough motivation to study these phenomena.

SIDELIGHTS & SPECULATIONS

Variations on the Theme of Amphibian Metamorphosis

Many amphibians have altered their life cycle by modifying the duration of their larval stage. This phenomenon, whereby animals change the relative time of appearance and rate of development of characters present in their ancestors, is called **heterochrony**. Here we will describe three extreme types of heterochrony:

1. *Neoteny* refers to the retention of the juvenile form as a result of retarded body development relative to the development of the germ cells and gonads (which achieve maturity at the normal time).

2. *Progenesis* also involves the retention of the juvenile form, but in this case, the gonads and germ cells develop at a faster rate than normal, be-

coming sexually mature while the rest of the body is still in a juvenile phase.

3. In *direct development*, the embryo abandons the stages of larval development entirely and proceeds to construct a small adult.

Neoteny

In certain salamanders, the reproductive system and germ cells mature while the rest of the body retains its juvenile form throughout life. In most such species, metamorphosis fails to occur and sexual maturity takes place in a "larval" body.

The Mexican axolotl, *Ambystoma mexicanum*, does not undergo metamorphosis in nature because its pituitary gland does not release the thyrotropin (thyroid-stimulating hormone)

that would activate T_4 synthesis (Prahlad and DeLanney 1965; Norris et al. 1973; Taurog et al. 1974). The axolotl does synthesize functional thyroid hormone receptors, however, and when investigators administered either thyroid hormones or thyrotropin, they found that the salamander metamorphosed into an adult form not seen in nature (**Figure 15.6**; Huxley 1920; Safi et al. 2004).

Other species of *Ambystoma*, such as *A. tigrinum*, metamorphose only in response to cues from the environment. In parts of its range, *A. tigrinum* is neotenic: its gonads and germ cells mature and the salamander mates successfully while the rest of the body retains its aquatic larval form. However, in other regions of its range, the larval form is transitory, leading to the

(A)

(B)

Figure 15.6   Metamorphosis in *Ambystoma*. (A) Normal adult *Ambystoma*, with prominent gills and broad tail. (B) Metamorphosed *Ambystoma* not seen in natural populations. This individual was grown in water supplemented with thyroxine. Its gills have regressed, and its skin has changed significantly. (A © Mark Boulton/Alamy; B courtesy of K. Crawford.)

SIDELIGHTS & SPECULATIONS (Continued)

land-dwelling adult tiger salamander. The ability to remain aquatic is highly adaptive in locations where the terrestrial environment is too dry to sustain the adult form of this salamander (Duellman and Trueb 1986).

Some salamanders are permanently neotenic, even in the laboratory. Whereas T_3 is able to produce the long-lost adult form of *A. mexicanum*, the neotenic species of *Necturus* and *Siren* remain unresponsive to thyroid hormones (Frieden 1981). Strangely, *Necturus* was recently found to have functional thyroid hormone receptors. It appears that these receptors do not bind to those genes that initiate and promote metamorphosis (Safi et al. 2006; Vlaeminck-Guillem et al. 2006).

De Beer (1940) and Gould (1977) have speculated that neoteny is a major factor in the evolution of more complex taxa. By retarding the development of somatic tissues, neoteny may give natural selection a flexible substrate. According to Gould (1977, p. 283), neoteny may "provide an escape from specialization. Animals can relinquish their highly specialized adult forms, return to the lability of youth, and prepare themselves for new evolutionary directions."

Progenesis

In progenesis, gonadal maturation is accelerated while the rest of the body develops normally to a certain stage.

Progenesis has enabled some salamander species to find new ecological niches. *Bolitoglossa occidentalis* is a tropical salamander that, unlike other members of its genus, lives in trees. This salamander's webbed feet and small body size suit it for arboreal existence, the webbed feet producing suction for climbing and the small body making such traction efficient. Alberch and Alberch (1981) showed that *B. occidentalis* resembles juveniles of the related species *B. subpalmata* and *B. rostrata* (whose young are small, with digits that have not yet grown past their webbing). *B. occidentalis* reaches sexual maturity at a much smaller size than its relatives, and this appears to have given it a phenotype that made tree-dwelling possible.

Direct development

While some animals have extended their larval life stage, others have "accelerated" their development by abandoning their larval form for direct development. Thus there are frog species that lack tadpoles and sea urchins that have no pluteus larvae.

Elinson and his colleagues (del Pino and Elinson 1983; Elinson 1987) have studied *Eleutherodactylus coqui*, a small frog that is one of the most abundant vertebrates on the island of Puerto Rico. Unlike the eggs of *Rana* and *Xenopus*, the eggs of *E. coqui* are

fertilized while they are still in the female's body. Each egg is about 3.5 mm in diameter (roughly 20 times the volume of a *Xenopus* egg). After the eggs are laid, the male gently sits on the developing embryos, protecting them from predators and desiccation (Taigen et al. 1984).

Early *E. coqui* development is like that of most frogs. Cleavage is holoblastic, gastrulation is initiated at a subequatorial position, and the neural folds become elevated from the surface. However, shortly after the neural tube closes, limb buds appear on the surface (**Figure 15.7A,B**). This early emergence of limb buds is the first indication that this animal will not pass through the usual limbless tadpole stage. Moreover, the development of *E. coqui* is modified such that the modeling of most of its features—including its limbs—does not depend on thyroid hormones. Its thyroid gland does develop, however, and thyroid hormones appear to be critical for the eventual resorption of the tail (which is used as a respiratory rather than a locomotor organ) and of the primitive kidney (Lynn and Peadon 1955). It appears that the thyroid-dependent phase has been pushed back into embryonic growth (Hanken et al. 1992; Callery et al. 2001). What emerges from the egg jelly 3 weeks after fertilization is not a tadpole but a tiny frog (**Figure 15.7C**).

(Continued on next page)

Figure 15.7 Direct development of the frog *Eleutherodactylus coqui*. (A) Limb buds are seen as the embryo develops on the yolk. (B) As the yolk is used up, the limb buds are easily seen. (C) Three weeks after fertilization, tiny froglets hatch. They are seen here in a petri dish and on a Canadian dime. (Courtesy of R. P. Elinson.)

(A)

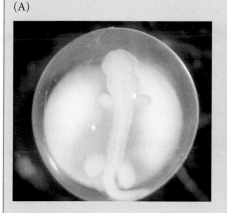

(B)

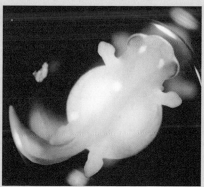

(C)

SIDELIGHTS & SPECULATIONS (Continued)

Direct-developing frogs do not need ponds for their larval stages and can therefore colonize habitats that are inaccessible to other frogs. Direct development also occurs in other phyla, in which it is also correlated with a large egg. It seems that if nutrition can be provided in the egg, the life cycle need not have a food-gathering larval stage.

Tadpole-rearing behaviors

Most temperate-zone frogs do not invest time or energy in providing for their tadpoles. However, among tropical frogs, there are numerous species in which adult frogs take painstaking care of their tadpoles. An example is the poison arrow frog *Dendrobates*, found in the rain forests of Central and South America. Most of the time, these highly toxic frogs live in the leaf litter of the forest floor. After the eggs are laid in a damp leaf, a parent (sometimes the male, sometimes the female, according to the species) stands guard over the eggs. If the ground gets too dry, the frog will urinate on the eggs to keep them moist. When the eggs mature into tadpoles, the guarding parent allows them to wriggle onto its back (**Figure 15.8A**). The parent then climbs into the canopy until it finds a bromeliad plant with a small pool of water in its leaf base. Here it deposits one of its tadpoles, then goes back for another, and so on until the entire brood has been placed into numerous small pools. The female returns each day to these pools and deposits a small number of unfertilized eggs into them, thus replenishing the tadpoles' food supply until they complete metamorphosis (Mitchell 1988; van Wijngaarden and Bolanos 1992; Brust 1993). It is not known how the female frog remem-

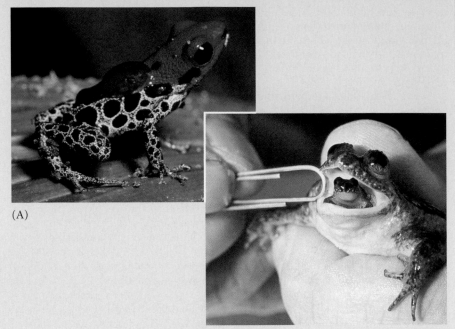

(A)

(B)

bers—or is informed about—where the tadpoles have been deposited.

Brooding frogs carry their developing eggs in depressions in their skin. Some species brood their tadpoles in their mouth and spit out their progeny when their tadpoles undergo metamorphosis. Even more impressive, the gastric-brooding frogs of Australia, *Rheobatrachus silus* and *R. vitellinus*, eat their eggs. The eggs develop into larvae, and the larvae undergo metamorphosis in the mother's stomach. About 8 weeks after being swallowed alive, about two dozen small frogs emerge from the female's mouth (**Figure 15.8B**; Corben et al. 1974; Tyler 1983). What stops the *Rheobatrachus* eggs from being digested or excreted? It appears that the eggs secrete prostaglandins that stop acid secretion and prevent peristaltic contractions in

Figure 15.8 Parental care of tadpoles. (A) Tadpoles of the poison arrow frog *Dendrobates* are carried on their parent's back to small pools of water in the Peruvian rain forest canopy. (B) This female *Rheobatrachus* of Australia brooded more than a dozen tadpoles in her stomach. They emerged after completing metamorphosis. Unfortunately, the last time anyone saw a *Rheobatrachus* frog alive was in 1985. (A © Michael Doolittle/Alamy; B courtesy of M. Tyler.)

the stomach (Tyler et al. 1983). During this time, the stomach is fundamentally a uterus, and the frog does not eat. After the oral birth, the parent's stomach morphology and function return to normal. Unfortunately, both of these remarkable frog species are now feared extinct. No member of either *Rheobatrachus* species has been seen since the mid-1980s.

Metamorphosis in Insects

Whereas amphibian metamorphosis is largely characterized by the remodeling of existing tissues, insect metamorphosis primarily involves the destruction of larval tissues and their replacement by an entirely different population of cells. Insects grow by molting—shedding their cuticle—and forming a new cuticle as their size increases. There are three

major patterns of insect development. A few insects, such as springtails, have no larval stage and undergo direct development. These are called the **ametabolous** insects (**Figure 15.9A**). Immediately after they hatch, these insects have a **pronymph** stage bearing the structures that enabled it to get out of the egg. But after this transitory stage, the insect looks like a small adult; it grows larger after each molt but is unchanged in form (Truman and Riddiford 1999).

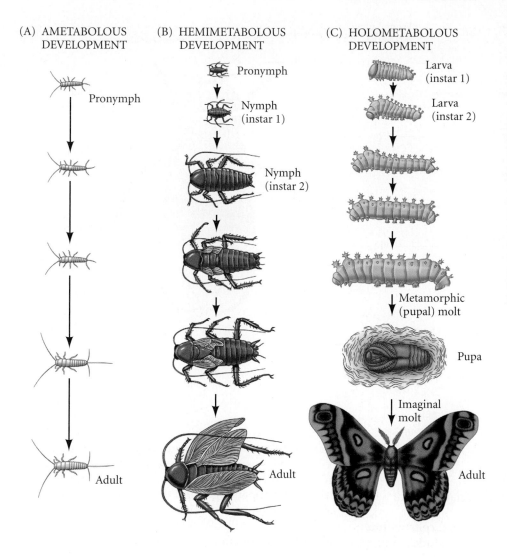

(A) AMETABOLOUS DEVELOPMENT

Pronymph

Adult

(B) HEMIMETABOLOUS DEVELOPMENT

Pronymph

Nymph (instar 1)

Nymph (instar 2)

Adult

(C) HOLOMETABOLOUS DEVELOPMENT

Larva (instar 1)

Larva (instar 2)

Metamorphic (pupal) molt

Pupa

Imaginal molt

Adult

FIGURE 15.9 Modes of insect development. Molts are represented as arrows. (A) Ametabolous (direct) development in a silverfish. After a brief pronymph stage, the insect looks like a small adult. (B) Hemimetabolous (gradual) metamorphosis in a cockroach. After a very brief pronymph phase, the insect becomes a nymph. After each molt, the next nymphal instar looks more like an adult, gradually growing wings and genital organs. (C) Holometabolous (complete) metamorphosis in a moth. After hatching as a larva, the insect undergoes successive larval molts until a metamorphic molt causes it to enter the pupal stage. Then an imaginal molt turns it into an adult.

Other insects, notably grasshoppers and bugs, undergo a gradual, **hemimetabolous** metamorphosis (Figure 15.9B). After spending a very brief period of time as a pronymph (whose cuticle is often shed as the insect hatches), the insect looks like an immature adult and is called a **nymph**. The rudiments of the wings, genital organs, and other adult structures are present and become progressively more mature with each molt. At the final molt, the emerging insect is a winged and sexually mature adult, or **imago**.

In the **holometabolous** insects such as flies, beetles, moths, and butterflies, there is no pronymph stage (Figure 15.9C). The juvenile form that hatches from the egg is called a **larva**. The larva (a caterpillar, grub, or maggot) undergoes a series of molts as it becomes larger. The stages between these larval molts are called **instars**. The number of larval molts before becoming an adult is characteristic of a species, although environmental factors can increase or decrease the number. The larval instars grow in a stepwise fashion, each instar being larger than the previous one. Finally, there is a dramatic and sudden transformation between the larval and adult stages: after the final instar, the larva undergoes a **metamorphic molt** to become

a **pupa**. The pupa does not feed, and its energy must come from those foods it ingested as a larva. During pupation, adult structures form and replace the larval structures. Eventually, an **imaginal molt** enables the adult (imago) to shed its pupal case and emerge. While the larva is said to *hatch* from an egg, the imago is said to *eclose* from the pupa.

Imaginal discs

In holometabolous insects, the transformation from juvenile into adult occurs within the pupal cuticle. Most of the larval body is systematically destroyed by programmed cell death, while new adult organs develop from relatively undifferentiated nests of **imaginal cells**. Thus, within any larva there are two distinct populations of cells: the larval cells, which are used for the functions of the juvenile insect; and thousands of imaginal cells, which lie within the larva in clusters, awaiting the signal to differentiate.

There are three main types of imaginal cells:

1. The cells of **imaginal discs** will form the cuticular structures of the adult, including the wings, legs, antennae, eyes, head, thorax, and genitalia (**Figure 15.10**).

Discs for:

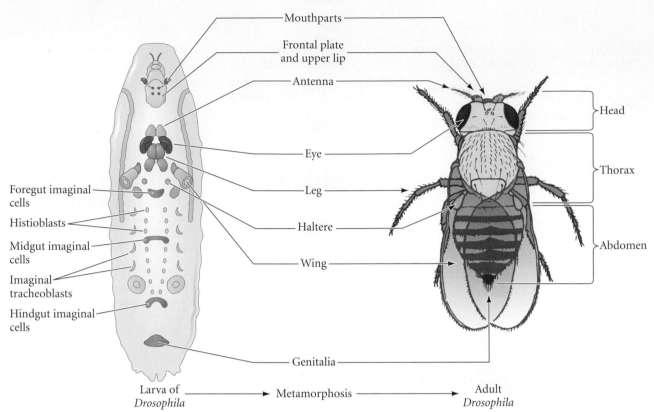

Foregut imaginal cells

Histioblasts

Midgut imaginal cells

Imaginal tracheoblasts

Hindgut imaginal cells

Mouthparts

Frontal plate and upper lip

Antenna

Eye

Leg

Haltere

Wing

Genitalia

Head

Thorax

Abdomen

Larva of *Drosophila* → Metamorphosis → Adult *Drosophila*

FIGURE 15.10 Locations and developmental fates of imaginal discs and imaginal tissues in the third instar larva of *Drosophila melanogaster*. (After Kalm et al. 1995.)

2. **Histoblast nests** are clusters of imaginal cells that will form the adult abdomen.
3. In addition there are clusters of imaginal cells within each organ that will proliferate to form the adult organ as the larval organ degenerates.

The imaginal discs can be seen in the newly hatched larva as local thickenings of the epidermis. Whereas most larval cells have a very limited mitotic capacity, imaginal discs divide rapidly at specific characteristic times. As their cells proliferate, the discs form a tubular epithelium that folds in on itself in a compact spiral (**Figure 15.11A**). At metamorphosis, these cells proliferate even further as they differentiate, and elongate (**Figure 15.11B**).

The fate map and elongation sequence of one of the six *Drosophila* leg discs is shown in **Figure 15.12**. At the end of the third instar, just before pupation, the leg disc is an epithelial sac connected by a thin stalk to the larval epidermis. On one side of the sac, the epithelium is coiled into a series of concentric folds "reminiscent of a Danish pastry" (Kalm et al. 1995). As pupation begins, the cells at the cen-

ter of the disc telescope out to become the most distal portions of the leg—the claws and the tarsus. The outer cells become the proximal structures—the coxa and the adjoining epidermis (Schubiger 1968). After differentiating, the cells of the appendages and epidermis secrete a cuticle appropriate for each specific region. Although the disc is composed primarily of epidermal cells, a small number of **adepithelial cells** migrate into the disc early in develop-

(A)

(B)

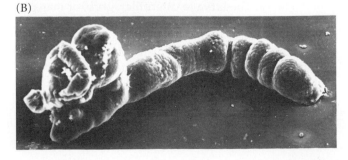

FIGURE 15.11 Imaginal disc elongation. Scanning electron micrograph of *Drosophila* third instar leg disc (A) before and (B) after elongation. (From Fristrom et al. 1977; courtesy of D. Fristrom.)

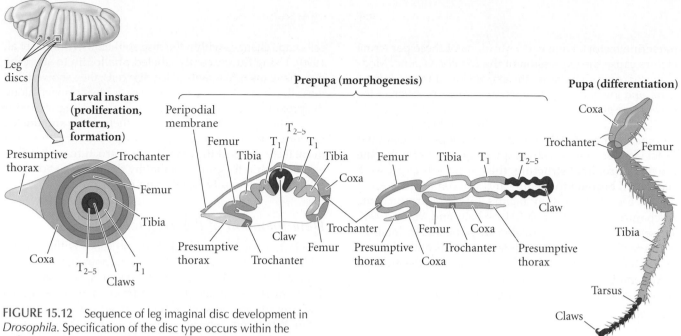

Figure (top illustration labels):

Embryo (specification)

Leg discs

Larval instars (proliferation, pattern, formation)

Presumptive thorax — Trochanter — Femur — Tibia — Coxa — T_{2-5} — T_1 — Claws

Prepupa (morphogenesis)

Peripodial membrane — Femur — Tibia — T_{2-5} — T_1 — T_1 — Tibia — Coxa — Claw — Trochanter — Femur — Presumptive thorax — Trochanter — Femur — Tibia — T_1 — T_{2-5} — Claw — Coxa — Trochanter — Presumptive thorax — Presumptive thorax — Coxa

Pupa (differentiation)

Coxa — Trochanter — Femur — Tibia — Tarsus — Claws

FIGURE 15.12 Sequence of leg imaginal disc development in *Drosophila*. Specification of the disc type occurs within the embryo. Proliferation of the disc cells and specification as to the type of leg cell each will produce are accomplished in the larval stages. Elongation of the disc takes place in the early pupal ("prepupa") stage, and differentiation of the leg tissues occurs while the insect is a pupa. T_1, basitarsus; T_{2-5}, tarsal segments 2–5. (After Fristrom and Fristrom 1975; Kalm et al. 1995.)

ment. During the pupal stage, these cells give rise to the muscles and nerves that serve the leg.

SPECIFICATION AND PROLIFERATION Specification of the general cell fates (i.e., that the disc is to be a leg disc and not a wing disc) occurs in the embryo. The more specific cell fates are specified in the larval stages, as the cells proliferate (Kalm et al. 1995). The type of leg structure

(claw, femur, etc.) generated is determined by the interactions between several genes in the imaginal disc. **Figure 15.13** shows the expression of three genes involved in determining the proximal-distal axis of the fly leg. In the third instar leg disc, the center of the disc secretes the highest concentration of two morphogens, Wingless (Wg; a Wnt paracrine factor) and Decapentaplegic (Dpp; a BMP

FIGURE 15.13 The fates of the imaginal disc cells are directed by transcription factors found in different regions. (A–D) Expression of transcription factor genes in the *Drosophila* leg disc. At the periphery, the *homothorax* gene (purple) establishes the boundary for the coxa. The expression of the *dachshund* gene (green) locates the femur and proximal tibia. The most distal structures, the claw and lower tarsal segments, arise from the expression domain of *Distal-less* (red) in the center of the imaginal disc. The overlap of *dachshund* and *Distal-less* appears yellow and specifies the distal tibia and upper tarsal segments. (A–C) Gene expression at successively later stages of pupal development. (D) Localization of expression domains of the genes onto a leg immediately prior to eclosion. The areas where there is overlap between expression domains are shown in yellow, aqua, and orange. (From Abu-Shaar and Mann 1998, courtesy of R. S. Mann.)

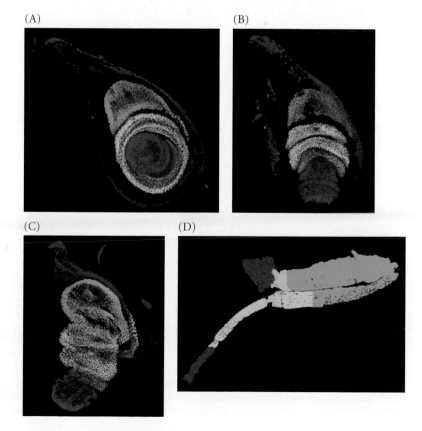

(A)   (B)

(C)   (D)

paracrine factor). High concentrations of these paracrine factors cause the expression of the *Distal-less* gene. Moderate concentrations cause the expression of the *dachshund* gene, and lower concentrations cause the expression of the *homothorax* gene.

Those cells expressing *Distal-less* telescope out to become the most distal structures of the leg—the claw and distal tarsal segments. Those expressing *homothorax* become the most proximal structure, the coxa. Cells expressing *dachshund* become the femur and proximal tibia. Areas where the transcription factors overlap produce the trochanter and distal tibia (Abu-Shaar and Mann 1998). These regions of gene expression are stabilized by inhibitory interactions between the protein products of these genes and of the neighboring genes. In this manner, the gradient of Wg and Dpp proteins is converted into discrete domains of gene expression that specify the different regions of the *Drosophila* leg.

EVERSION AND DIFFERENTIATION The mature leg disc in the third instar of *Drosophila* does not look anything like the adult structure. It is determined but not yet differentiated; its differentiation requires a signal, in the form of a set of pulses of the "molting" hormone **20-hydroxyecdysone (20E**; see Figure 15.17). The first pulse, occurring in the late larval stages, initiates formation of the pupa, arrests cell division in the disc, and initiates the cell shape changes that drive the eversion of the leg. Studies by Condic and her colleagues have demonstrated that the elongation of imaginal discs occurs without cell division and is due primarily to

cell shape changes within the disc epithelium (Condic et al. 1990). Using fluorescently labeled phalloidin to stain the peripheral microfilaments of leg disc cells, they showed that the cells of early third instar discs are tightly arranged along the proximal-distal axis. When the hormonal signal to differentiate is given, the cells change their shape and the leg is everted, the central cells of the disc becoming the most distal (claw) cells of the limb. The leg structures will differentiate within the pupa, so that by the time the adult fly ecloses, they are fully formed and functional.

Determination of the wing imaginal discs

The largest of *Drosophila*'s imaginal discs is that of the wing, containing some 60,000 cells (in contrast, the leg and haltere discs contain about 10,000 cells each; Fristrom 1972). The wing discs are distinguished from the other imaginal discs by the expression of the *vestigial* gene (Kim et al. 1996). When this gene is expressed in any other imaginal disc, wing tissue emerges.

ANTERIOR AND POSTERIOR COMPARTMENTS The axes of the wing are specified by gene expression patterns that divide the embryo into discrete but interacting compartments (Figure 15.14A; Meinhardt 1980; Causo et al. 1993; Tabata et al. 1995). In the first instar, expression of the *engrailed* gene distinguishes the posterior compartment of the wing from the anterior compartment. The Engrailed transcription factor is expressed only in the posterior compartment, and in those cells, it activates the gene for the BMP-like paracrine

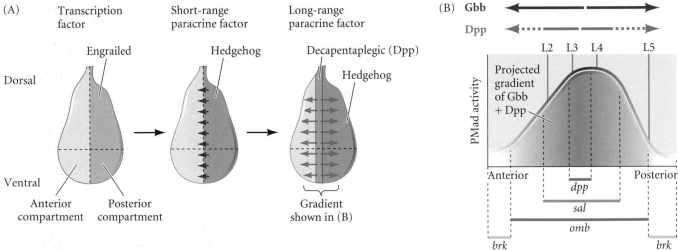

FIGURE 15.14 Compartmentalization and anterior-posterior patterning in the wing imaginal disc. (A) In the first instar larva, the anterior-posterior axis has been formed and can be recognized by the expression of the *engrailed* gene in the posterior compartment. Engrailed, a transcription factor, activates the *hedgehog* gene. Hedgehog acts as a short-range paracrine factor to activate *decapentaplegic* (*dpp*) in the anterior cells adjacent to the posterior compartment, where Dpp and a related protein, Glass-bottom boat (Gbb), act over a longer range. (B) Dpp and Gbb proteins

create a concentration gradient of BMP-like signaling, measured by the phosphorylation of Mad (pMad). High concentrations of Dpp plus Gbb near the source activate both the *spalt* (*sal*) and *optomotor blind* (*omb*) genes. Lower concentrations (near the periphery) activate *omb* but not *sal*. When Dpp plus Gbb levels drop below a certain threshold, *brinker* (*brk*) is no longer repressed. L2–L5 mark the longitudinal wing veins, with L2 being the most anterior. (After Bangi and Wharton 2006.)

(A) (B)

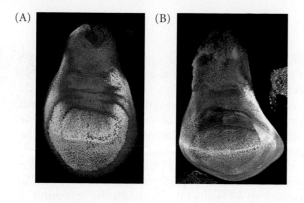

FIGURE 15.15 Determining the dorsal-ventral axis. (A) The prospective ventral surface of the wing is stained by antibodies to Vestigial protein (green), while the prospective dorsal surface is stained by antibodies to Apterous protein (red). The region of yellow illustrates where the two proteins overlap in the margin. (B) Wingless protein (purple) synthesized at the marginal juncture organizes the wing disc along the dorsal-ventral axis. The expression of Vestigial (green) is seen in cells close to those expressing Wingless. (C) The dorsal and ventral portions of the wing disc telescope out to form the two-layered wing. Gene expression patterns are indicated on the double-layered wing. (A,B courtesy of S. Carroll and S. Paddock.)

(C)

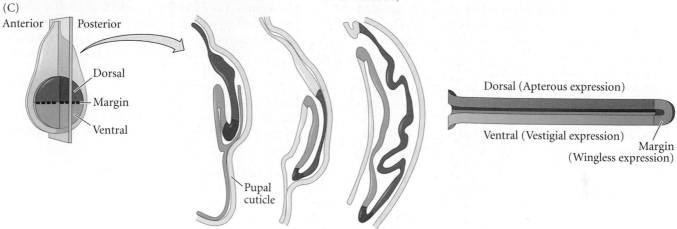

factor Hedgehog. Hedgehog functions only when cells have the receptor (Patched) to receive it. In a complex manner, the diffusion of Hedgehog activates the gene encoding Decapentaplegic (Dpp) in a narrow stripe of cells in the anterior region of the wing disc (Ho et al. 2005).

Dpp and a co-expressed BMP called Glass-bottom boat (Gbb) act to establish a gradient of BMP signaling activity. BMPs activate the Mad transcription factor (a Smad protein) by phosphorylating it, so this gradient can be measured by the phosphorylation of Mad. Dpp is a short-range paracrine factor, while Gbb exhibits a much longer range of diffusion to create a gradient (**Figure 15.14B**; Bangi and Wharton 2006). This signaling gradient regulates the amount of cell proliferation in the wing regions and also specifies cell fates (Rogulja and Irvine 2005). Several transcription factor genes respond differently to activated Mad. At high levels, the *spalt* (*sal*) and *optomotor blind* (*omb*) genes are activated, whereas at low levels (where Gbb provides the primary signal), only *omb* is activated. Below a particular level of phosphorylated Mad activity, the *brinker* (*brk*) gene is no longer inhibited; thus *brk* is expressed outside the signaling domain. Specific cell fates of the wing are specified in response to the action of these transcription factors. (For example, the fifth longitudinal vein of the wing is formed at the border of *optomotor blind* and *brinker*; see Figure 15.14B).

DORSAL-VENTRAL AND PROXIMAL-DISTAL AXES The dorsal-ventral axis of the wing is formed at the second instar stage by the expression of the *apterous* gene in the prospective dorsal cells of the wing disc (Blair 1993; Diaz-Benjumea

and Cohen 1993). Here, the upper layer of the wing is distinguished from the lower layer of the wing blade (Bryant 1970; Garcia-Bellido et al. 1973). The *vestigial* gene remains "on" in the ventral portion of the wing disc (**Figure 15.15A**). The dorsal portion of the wing synthesizes transmembrane proteins that prevent the intermixing of the dorsal and ventral cells (Milán et al. 2005). At the boundary between the dorsal and ventral compartments, the Apterous and Vestigial transcription factors interact to activate the gene encoding the Wnt paracrine factor Wingless (**Figure 15.15B**). Neumann and Cohen (1996) showed that Wingless protein acts as a growth factor to promote the cell proliferation that extends the wing.* Wingless also helps establish the proximal-distal axis of the wing: high levels of Wingless activate the *Distal-less* gene, which specifies the most distal regions of the wing (Neumann and Cohen 1996, 1997; Zecca et al. 1996). This is in the central region of the disc and "telescopes" outward as the distal margin of the wing blade (**Figure 15.15C**).

See WEBSITE 15.1
The molecular biology of wing formation

See WEBSITE 15.2
Homologous specification

---

*The diffusion of paracrine factors such as Wingless and Hedgehog is facilitated when these factors cluster on lipid spheres that can travel between cells without getting caught in the extracellular matrix (Glise et al. 2005; Gorfinkiel et al. 2005; Panáková et al. 2005).

Hormonal control of insect metamorphosis

Although the details of insect metamorphosis differ among species, the general pattern of hormonal action is very similar. Like amphibian metamorphosis, the metamorphosis of insects is regulated by systemic hormonal signals, which are controlled by neurohormones from the brain (for reviews, see Gilbert and Goodman 1981; Riddiford 1996). Insect molting and metamorphosis are controlled by two effector hormones: the steroid 20-hydroxyecdysone (20E) and the lipid **juvenile hormone (JH)** (**Figure 15.16A**). 20-Hydroxyecdysone[†] initiates and coordinates each molt and regulates the changes in gene expression that occur during metamorphosis. Juvenile hormone prevents the ecdysone-induced changes in gene expression that are necessary for metamorphosis. Thus, its presence during a molt ensures that the result of that molt is another larval instar, not a pupa or an adult.

The molting process (**Figure 15.16B**) is initiated in the brain, where neurosecretory cells release **prothoracicotropic hormone (PTTH)** in response to neural, hormonal, or environmental signals. PTTH is a peptide hormone with a molecular weight of approximately 40,000, and it stimulates the production of **ecdysone** by the **prothoracic gland** by activating the RTK pathway in those cells (Rewitz et al. 2009). Ecdysone is modified in peripheral tissues to become the active molting hormone 20E. Each molt is initiated by one or more pulses of 20E. For a larval molt, the first pulse produces a small rise in the 20E concentration in the larval hemolymph (blood) and elicits a change in cellular commitment in the epidermis. A second, larger pulse of 20E initiates the differentiation events associated with molting. These pulses of 20E commit and stimulate the epidermal cells to synthesize enzymes that digest the old cuticle and synthesize a new one.

Juvenile hormone is secreted by the **corpora allata**. The secretory cells of the corpora allata are active during larval molts but inactive during the metamorphic molt and the imaginal molt. As long as JH is present, the 20E-stimulated molts result in a new larval instar. In the last larval instar, however, the medial nerve from the brain to the corpora allata inhibits these glands from producing JH, and there is a simultaneous increase in the body's ability to degrade existing JH (Safranek and Williams 1989). Both these mechanisms cause JH levels to drop below a critical threshold value, triggering the release of PTTH from the brain (Nijhout and Williams 1974; Rountree and Bollenbacher 1986). PTTH, in turn, stimulates the prothoracic gland to secrete a small amount of ecdysone. The resulting pulse of 20E, in the absence of high levels of JH, commits the epidermal cells to pupal development. Larva-spe-

cific mRNAs are not replaced, and new mRNAs are synthesized whose protein products inhibit the transcription of the larval messages.

There are two major pulses of 20E during *Drosophila* metamorphosis. The first pulse occurs in the third instar larva and triggers the initiation of ("prepupal") morphogenesis of the leg and wing imaginal discs (as well as the death of the larval hindgut). The larva stops eating and migrates to find a site to begin pupation. The second 20E pulse occurs 10–12 hours later and tells the "prepupa" to become a pupa. The head inverts and the salivary glands degenerate (Riddiford 1982; Nijhout 1994). It appears, then, that the first ecdysone pulse during the last larval instar triggers the processes that inactivate the larva-specific genes and initiates the morphogenesis of imaginal disc structures. The second pulse transcribes pupa-specific genes and initiates the molt (Nijhout 1994). At the imaginal molt, when 20E acts in the absence of juvenile hormone, the imaginal discs fully differentiate and the molt gives rise to an adult.

See WEBSITE 15.3 Insect metamorphosis

The molecular biology of 20-hydroxyecdysone activity

ECDYSONE RECEPTORS 20-Hydroxyecdysone cannot bind to DNA by itself. Like amphibian thyroid hormones, 20E first binds to nuclear receptors. These proteins, called ecdysone receptors (EcRs), are almost identical in structure to the thyroid hormone receptors of amphibians. An EcR protein forms an active molecule by pairing with an Ultraspiracle (Usp) protein, the homologue of the amphibian RXR that helps form the active thyroid hormone receptor (Koelle et al. 1991; Yao et al. 1992; Thomas et al. 1993). In the absence of the hormone-bound EcR, the Usp protein binds to the ecdysone-responsive genes and inhibits their transcription.[*] This inhibition is converted into activation when the ecdysone receptor binds to the Usp (Schubiger and Truman 2000).

Although there is only one gene for EcR, the EcR mRNA transcript can be spliced in at least three different ways to form three distinct proteins. All three EcR proteins have the same domains for 20E and DNA binding, but they differ in their amino-terminal domains. The type of EcR present in a cell may inform that cell how to act when it receives a hormonal signal (Talbot et al. 1993; Truman et al. 1994). All cells appear to have some EcRs of each type, but the strictly larval tissues and neurons that die when exposed to 20E are characterized by their abundance of the EcR-B1 isoform of the ecdysone receptor. Imaginal discs and differentiating neurons, by contrast, show a preponderance of the EcR-A

[†]Since its discovery in 1954, when Butenandt and Karlson isolated 25 mg of ecdysone from 500 kg of silkworm moth pupae, 20-hydroxyecdysone has gone under several names, including ecdysterone, β-ecdysone, and crustecdysone.

[*]The Ultraspiracle protein may be a juvenile hormone receptor, or JHR (see Figure 15.16), suggesting mechanisms whereby JH can block 20E at the level of transcription (Jones et al. 2001; Jones et al. 2001, 2007; Sasorith et al. 2002).

(A) Juvenile hormone (JH)

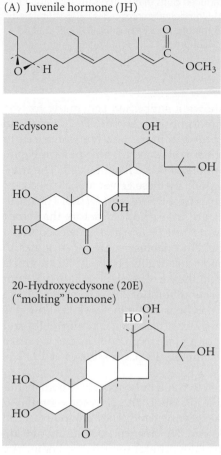

Ecdysone

20-Hydroxyecdysone (20E) ("molting" hormone)

(B)

Brain

Neurosecretory cells of prothoracic gland

Juvenile hormone

Corpora allata

Prothoracicotropic hormone (PTTH)

JH receptor (JHR)

Prothoracic gland

JH-JHR

"Differentiating signal"

"Molting signal"

Ecdysone

20E

Chromosomes

RNA (L)

Protein synthesis

Larval structures

Cuticle

Chromosomes

RNA (P)

Protein synthesis

Pupal structures

Chromosomes

RNA (A)

Protein synthesis

Adult structures

Larva

Pupa

Adult

FIGURE 15.16 Regulation of insect metamorphosis. (A) Structures of juvenile hormone (JH), ecdysone, and the active molting hormone 20-hydroxyecdysone (20E). (B) General pathway of insect metamorphosis. 20E and JH together cause molts that form the next larval instar. When the concentration of JH becomes low enough, the 20E-induced molt produces a pupa instead of an instar. When ecdysone acts in the absence of JH, the imaginal discs differentiate and the molt gives rise to an adult (imago). (After Gilbert and Goodman 1981.)

isoform. Mutations in specific codons that are found in only some of the splicing isoforms indicate that the different forms of EcR play different roles in metamorphosis and that the different receptors activate different sets of genes when they bind 20E (Davis et al. 2005).

BINDING OF 20-HYDROXYECDYSONE TO DNA During molting and metamorphosis, certain regions of the polytene chromosomes of *Drosophila* puff out in the cells of certain organs at certain times (Clever 1966; Ashburner 1972; Ashburner and Berondes 1978). These chromosome puffs are areas where DNA is being actively transcribed. Moreover,

these organ-specific patterns of chromosome puffing can be reproduced by culturing larval tissue and adding hormones to the medium, or by adding 20E to an early-stage larva. When 20E is added to larval salivary glands, certain puffs are produced and others regress (**Figure 15.17**). The puffing is mediated by the binding of 20E at specific places on the chromosomes; fluorescent antibodies against 20E find this hormone localized to the regions of the genome that are sensitive to it (Gronemeyer and Pongs 1980). At these sites, the ecdysone-bound receptor complex recruits a histone methyltransferase that methylates lysine-4 of histone H3, thereby loosening the nucleosomes in that area (Sedkov et al. 2003).

See VADE MECUM Chromosome squash

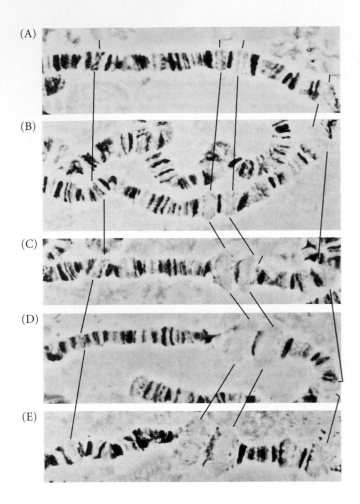

(A)

(B)

(C)

(D)

(E)

FIGURE 15.17 20E-induced puffs in cultured salivary gland cells of *D. melanogaster*. (A) Uninduced control. (B–E) 20E-stimulated chromosomes at (B) 25 minutes, (C) 1 hour, (D) 2 hours, and (E) 4 hours. (Courtesy of M. Ashburner.)

20E-regulated chromosome puffing occurs during the late stages of the third instar *Drosophila* larva, as it prepares to form the pupa. The puffs can be divided into three categories: "early" puffs that 20E induces rapidly; "intermolt" puffs that 20E causes to regress; and "late" puffs that are first seen several hours after 20E stimulation. For example, in the larval salivary gland, about six puffs emerge within a few minutes of hydroxyecdysone treatment. No new protein has to be made in order for these early puffs to be induced. A much larger set of puffs is induced later in development, and these late puffs do need protein synthesis to become transcribed. Ashburner (1974, 1990) hypothesized that the "early puff" genes make a protein product that is essential for the activation of the "late puff" genes and that, moreover, this early regulatory protein itself turns off the transcription of the early genes.* These insights have been confirmed by molecular analyses.

---

*The observation that 20E controlled the transcriptional units of chromosomes was an extremely important and exciting discovery. This was our first real glimpse of gene regulation in eukaryotic organisms. At the time when this discovery was made, the only examples of transcriptional gene regulation were in bacteria.

Figure 15.18A shows a simplified schematic for the framework of metamorphosis in *Drosophila*. 20E binds to the EcR/USP receptor complex. It activates the "early response genes," including *E74* and *E75* (the puffs in Figure 15.17), as well as *Broad* and the *EcR* gene itself. The transcription factors encoded by these genes activate a second series of genes, such as *E75*, *DHR4*, and *DHR3*. The products of these genes are transcription factors that work together. First, they activate *βFTZ-F1*, a gene encoding a transcription factor that enables a new set of genes to respond to the second burst of 20E. Secondly, the products of these genes shut off the early genes so that they do not interfere with the second burst of 20E. Moreover, DHR4 coordinates growth and behavior in the larva. It allows the larva to stop feeding once it reaches a certain weight and to begin searching for a place to glue itself to and form a pupa (Urness and Thummel 1995; Crossgrove et al. 1996; King-Jones et al. 2005).

The effects of these two 20E pulses can be extremely different. One example of this is the ecdysone-mediated changes in the larval salivary gland. The early pulse of 20E activates the *Broad* gene, which encodes a family of transcription factors through differential RNA splicing. The targets of the Broad complex proteins include those genes that encode the salivary gland "glue proteins." The glue proteins allow the larva to adhere to a solid surface, where it becomes a pupa (Guay and Guild 1991). So the first 20E pulse stimulates the function of the larval salivary gland. However, the second pulse of 20E calls for the destruction of this larval organ (Buszczak and Segraves 2000; Jiang et al. 2000). Here, 20E binds to the EcR-A form of the ecdysone receptor (**Figure 15.18B**). When complexed with USP, it activates the transcription of early response genes *E74*, *E75*, and *Broad*. But now a different set of targets is activated. These transcription factors activate the genes encoding the apoptosis-promoting proteins Hid and Reaper, as well as blocking the expression of the *diap2* gene (which would otherwise repress apoptosis). Thus, the first 20E pulse activates the salivary gland, and the second 20E pulse destroys it.

Like the ecdysone receptor gene, the *Broad* gene can generate several different transcription factor proteins through differentially initiated and spliced messages. Moreover, the variants of the ecdysone receptor may induce the synthesis of particular variants of the Broad proteins. Organs such as the larval salivary gland that are destined for death during metamorphosis express the Z1 isoform; imaginal discs destined for differentiation express the Z2 isoform; and the central nervous system (which undergoes marked remodeling during metamorphosis) expresses all isoforms, with Z3 pre-

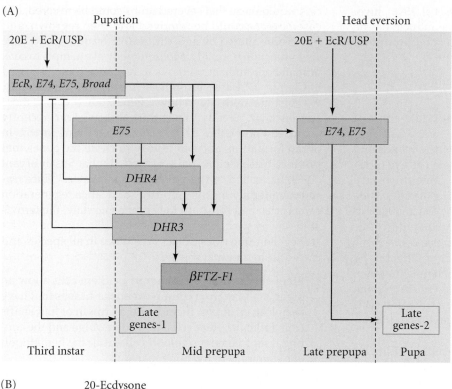

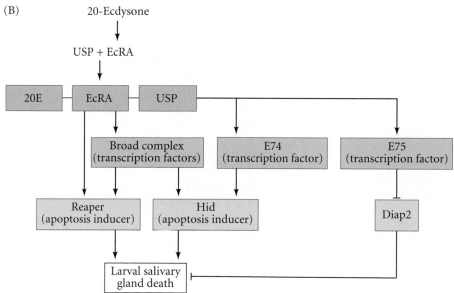

FIGURE 15.18 20-Hydroxyecdysone initiates developmental cascades. (A) Schematic of the major gene expression cascade in *Drosophila* metamorphosis. When 20E binds to the EcR/USP receptor complex, it activates the early response genes, including *E74*, *E75*, and *Broad*. Their products activate the "late genes." The activated EcR/USP complex also activates a series of genes whose products are transcription factors and which activate the *βFTZ-F1* gene. The βFTZ-F1 protein modifies the chromatin so that the next 20E pulse activates a different set of late genes. The products of these genes also inhibit the early-expressed genes, including the EcR receptor. (B) Postulated cascade leading from ecdysone reception to death of the larval salivary gland. Ecdysone binds to the EcR-A isoform of the ecdysone receptor. After complexing with USP, the activated transcription factor complex stimulates transcription of the early response genes *E74A*, *E75B*, and the *Broad* complex. These make transcription factors that promote apoptosis in the salivary gland cells. (A after King-Jones et al. 2005; B after Buszczak and Segraves 2000.)

dominating (Emery et al. 1994; Crossgrove et al. 1996). Juvenile hormone may act to prevent ecdysone-inducible gene expression by interfering with the Broad complex of proteins (Riddiford 1972; Restifo and White 1991).

See WEBSITE 15.4 Precocenes and synthetic JH

COORDINATION OF RECEPTOR AND LIGAND Like those of amphibian metamorphosis, the stories of insect metamorphosis involve complex interactions between ligands and receptors. The "target tissues" are not mere passive recipients of hormonal signals. Rather, they become responsive to hormones only at particular times. For example, when there is a pulse of 20E at the middle of the fourth instar of the tobacco hornworm moth *Manduca*, the epidermis is able to respond because this tissue is expressing ecdysone receptors. The wing discs, however, are unaffected by ecdysone until the prepupal stage, at which time they synthesize ecdysone receptors, grow, and differentiate (Nijhout 1999). Thus, the timing of metamorphic events in insects can be controlled by the synthesis of receptors in the target tissues.

Metamorphosis remains one of the most striking of developmental phenomena, yet we know only an outline of the molecular bases of metamorphosis, and only for a handful of species.

REGENERATION

Regeneration is the reactivation of development in postembryonic life to restore missing tissues. The ability to regenerate amputated body parts or nonfunctioning organs is so "unhuman" that it has been a source of fascination to humans since the beginnings of biological science. It is difficult to behold the phenomenon of limb regeneration in newts or starfish without wondering why we cannot grow back our own arms and legs. What gives salamanders this ability we so sorely lack? In fact, experimental biology was born of the efforts of eighteenth-century naturalists to answer this question. The regeneration experiments of Tremblay* (hydra), Réaumur (crustaceans), and Spallanzani (salamanders) set the standard for experimental research and for the intelligent discussion of one's data (see Dinsmore 1991).

More than two centuries later, we are beginning to find answers to the great questions of regeneration, and at some point we may be able to alter the human body so as to permit our own limbs, nerves, and organs to regenerate. Success would mean that severed limbs could be restored, diseased organs could be removed and then regrown, and nerve cells altered by age, disease, or trauma could once again function normally. Modern medical attempts to coax human bone and neural tissue to regenerate are discussed in Chapter 17, but to bring these treatments to humanity, we must first understand how regeneration occurs in those species that already have this ability.[†] Our recently acquired knowledge of the roles of paracrine factors in organ formation, and our ability to clone the genes that produce those factors, have propelled what Susan Bryant (1999) has called "a regeneration renaissance." Since *renaissance* literally means "rebirth," and since regeneration can be seen as a return to the embryonic state, the term is apt in many ways.

Regeneration does in fact take place in all species and can occur in four major ways:

1. **Stem-cell mediated regeneration**. Stem cells allow an organism to regrow certain organs or tissues that have been lost; examples include the regrowth of hair shafts from follicular stem cells in the hair bulge and the continual replacement of blood cells from the hematopoietic stem cells in the bone marrow.
2. **Epimorphosis**. In some species, adult structures can undergo *de*differentiation to form a relatively undifferentiated mass of cells that then redifferentiates to form the new structure. Such epimorphosis is characteristic of planarian flatworm regeneration and also of regenerating amphibian limbs.
3. **Morphallaxis**. Here, regeneration occurs through the repatterning of existing tissues, and there is little new growth. Such regeneration is seen in *Hydra* (a cnidarian).
4. **Compensatory regeneration**. Here, the differentiated cells divide but maintain their differentiated functions. The new cells do not come from stem cells, nor do they come from the dedifferentiation of the adult cells. Each cell produces cells similar to itself; no mass of undifferentiated tissue forms. This type of regeneration is characteristic of the mammalian liver.

Numerous examples of stem-cell mediated regeneration have been discussed throughout this book. In this chapter we will concentrate on epimorphosis in the salamander limb, morphallaxis in *Hydra*, and compensatory regeneration in the mammalian liver.

---

*Tremblay's advice to researchers who would enter this new field is pertinent even today: he advises us to go directly to nature and to avoid the prejudices that our education has given us. Moreover, "one should not become disheartened by want of success, but should try anew whatever has failed. It is even good to repeat successful experiments a number of times. All that is possible to see is not discovered, and often cannot be discovered, the first time" (quoted in Dinsmore 1991).

[†]Mammals do have a small amount of regenerational ability. In addition to regenerating body parts continuously through adult stem cells, rodents and humans can regenerate the tips of their digits if the animal is young enough. This ability has been correlated with the expression of the homeodomain transcription factor MSX1 (Han et al. 2003; Kumar et al. 2004). Apparently, amputated human fetal digit tips express MSX1 in the migrating epidermis and subjacent mesenchyme, just as regenerating amphibian limbs do (Allan et al. 2005).

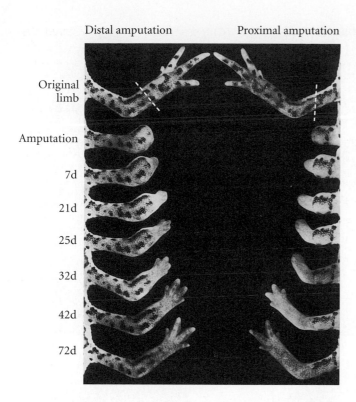

Distal amputation   Proximal amputation

Original limb

Amputation

7d

21d

25d

32d

42d

72d

FIGURE 15.19 Regeneration of a salamander forelimb. The amputation shown on the left was made below the elbow; the amputation shown on the right cut through the humerus. In both instances, the correct positional information was re-specified and a normal limb was regenerated within 72 days. (From Goss 1969, courtesy of R. J. Goss.)

Epimorphic Regeneration of Salamander Limbs

When an adult salamander limb is amputated, the remaining limb cells are able to reconstruct a complete new limb, with all its differentiated cells arranged in the proper order. In other words, the new cells construct only the missing structures and no more. For example, when a wrist is amputated, the salamander forms a new wrist and not a new elbow (**Figure 15.19**). In some way, the salamander limb "knows" where the proximal-distal axis has been severed and is able to regenerate from that point on. Salamanders accomplish epimorphic regeneration by cell dedifferentiation to form a **regeneration blastema**—an aggregation of relatively dedifferentiated cells derived from the originally differentiated tissue—which then proliferates and redifferentiates into the new limb parts (see Brockes and Kumar 2002; Gardiner et al. 2002). Bone, dermis, and cartilage just beneath the site of amputation contribute to the regeneration blastema, as do satellite cells from nearby muscles (Morrison et al. 2006).

Formation of the apical ectodermal cap and regeneration blastema

When a salamander limb is amputated, a plasma clot forms; within 6–12 hours, epidermal cells from the remaining stump migrate to cover the wound surface, forming the **wound epidermis**. In contrast to wound healing in mammals, no scar forms, and the dermis does not move with the epidermis to cover the site of amputation. The nerves innervating the limb degenerate for a short distance proximal to the plane of amputation (see Chernoff and Stocum 1995).

During the next 4 days, the extracellular matrices of the tissues beneath the wound epidermis is degraded by proteases, liberating single cells that undergo dramatic dedifferentiation: bone cells, cartilage cells, fibroblasts, and myocytes all lose their differentiated characteristics. Genes that are expressed in differentiated tissues (such as the *mrf4* and *myf5* genes expressed in muscle cells) are downregulated, while there is a dramatic increase in the expression of genes such as *msx1* that are associated with the proliferating progress zone mesenchyme of the embryonic limb (Simon et al. 1995). This cell mass is the regeneration blastema, and these are the cells that will continue to proliferate, and which will eventually redifferentiate to form the new structures of the limb (**Figure 15.20**; Butler 1935). Moreover, during this time, the wound epidermis thickens to form the **apical epidermal cap** (**AEC**), which acts similarly to the apical ectodermal ridge during normal limb development (Han et al. 2001).

Thus, the previously well-structured limb region at the cut edge of the stump forms a proliferating mass of indistinguishable cells just beneath the apical ectodermal cap. One of the major questions of regeneration has been: do the cells keep a "memory" of what they had been? In other words, do new muscles arise from old muscle cells, or can any cell of the blastema become a muscle? Kragl and colleagues (2009) found that the blastema is not a collection of homogeneous, fully dedifferentiated cells. Rather, in the regenerating limbs of the axolotl salamander, muscle cells arise only from old muscle cells, dermal cells come only from old dermal cells, and cartilage can arise only from old cartilage or old dermal cells. Thus, the blastema is not a collection of unspecified multipotential progenitor cells. Rather, the cells retain their specification, and the blastema is a heterogeneous assortment of *restricted* progenitor cells.

Kragl and colleagues performed an experiment in which they transplanted limb tissue from a salamander whose cells expressed green fluorescent protein (GFP) into different regions of limbs of normal salamanders that did not have the GFP transgene (**Figure 15.21**). If they transplanted the GFP-expressing limb cartilage into a salamander limb that did not contain the GFP transgene, the GFP-expressing cartilage would integrate normally into the limb skeleton. They later amputated the limb through the region containing GFP-marked cartilage cells. The blastema was

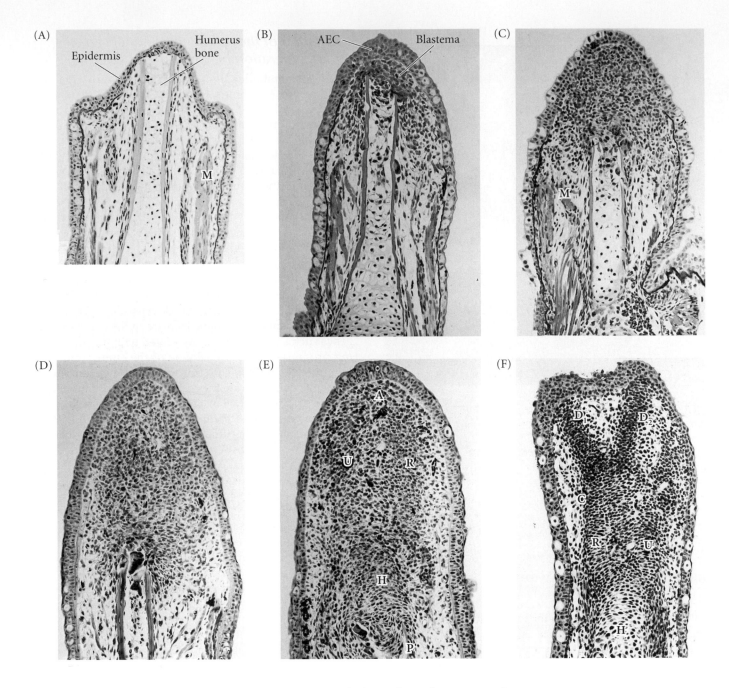

FIGURE 15.20 Regeneration in the larval forelimb of the spotted salamander *Ambystoma maculatum*. (A) Longitudinal section of the upper arm, 2 days after amputation. The skin and muscle (M) have retracted from the tip of the humerus. (B) At 5 days after amputation, a thin accumulation of blastema cells is seen beneath the thickened epidermis, where the apical ectodermal cap (AEC) forms. (C) At 7 days, a large population of mitotically active blastema cells lies distal to the humerus. (D) At 8 days, the blastema elongates by mitotic activity; much dedifferentiation has occurred. (E) At 9 days, early redifferentiation can be seen. Chondrogenesis has begun in the proximal part of the regenerating humerus, H. The letter A marks the apical mesenchyme of the blastema, and U and R are the precartilaginous condensations that will form the ulna and radius, respectively. P represents the stump where the amputation was made. (F) At 10 days after amputation, the precartilaginous condensations for the carpal bones (ankle, C) and the first two digits (D_1, D_2) can also be seen. (From Stocum 1979, courtesy of D. L. Stocum.)

(A)

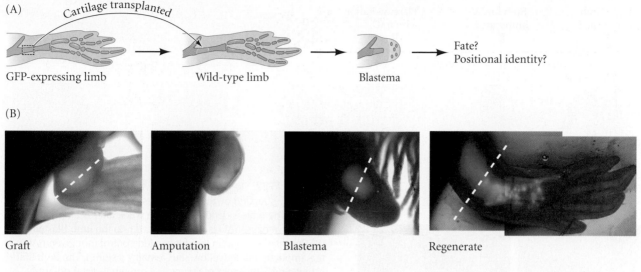

Cartilage transplanted

GFP-expressing limb   Wild-type limb   Blastema

Fate?
Positional identity?

(B)

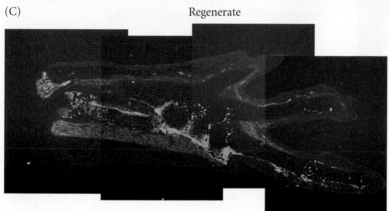

Graft   Amputation   Blastema   Regenerate

(C)   Regenerate

FIGURE 15.21   Blastema cells retain their specification, even though they dedifferentiate. (A,B) Schematic of the procedure, wherein a particular tissue (in this case, cartilage) is transplanted from a salamander expressing a green fluorescent protein (GFP) transgene into a wild-type salamander limb. Later, the limb is amputated through the region of the limb containing GFP expression, and a blastema is formed containing GFP-expressing cells that had been cartilage precursors. The regenerated limb is then studied to see if GFP is found only in the regenerated cartilage tissues or in other tissues. (C) Longitudinal section of a regenerated limb 30 days after amputation. The muscle cells are stained red, and nuclei are stained blue. The majority of GFP cells (green) were found in the regenerated cartilage; no GFP was seen in the muscle. (After Kragl et al. 2009, courtesy of E. Tanaka.)

found to contain GFP-expressing cells, and when the blastema differentiated, the only GFP-expressing cells found were in the limb cartilage. Similarly, GFP-marked muscle cells gave rise only to muscle, and GFP-marked epidermal cells only produced the epidermis of the regenerated limb.

Proliferation of the blastema cells: The requirement for nerves and the AEC

The growth of the regeneration blastema depends on the presence of both the apical ectodermal cap and nerves. The AEC stimulates the growth of the blastema by secreting Fgf8 (just as the apical ectodermal ridge does in normal limb development), but the effect of the AEC is only possible if nerves are present (Mullen et al. 1996). Singer (1954) demonstrated that a minimum number of nerve fibers must be present for regeneration to take place. The neurons are also believed to release factors necessary for the proliferation of the blastema cells (Singer and Caston 1972; Mescher and Tassava 1975). There have been many candidates for such a nerve-derived blastema mitogen, but the one that is probably the best candidate is **newt anterior**

gradient protein (nAG). This protein can cause blastema cells to proliferate in culture, and it permits normal regeneration in limbs that have been denervated (**Figure 15.22**; Kumar et al. 2007a). If activated nAG genes are electoporated into the dedifferentiating tissues of limbs that have been denervated, the limbs are able to regenerate. If nAG is not administered, the limbs remain as stumps. Moreover, nAG is only minimally expressed in normal limbs, but it is induced in the Schwann cells that surround the neurons within 5 days of amputation.

The creation of the amphibian regeneration blastema may also depend on the maintenance of ion currents driven through the stump: if this electric field is suppressed, the regeneration blastema fails to form (Altizer et al. 2002). Such fields have been shown to be necessary for the regeneration of tails in the frog *Xenopus laevis* (an anuran amphibian). The *Xenopus* tadpole regenerates its tail, and the notochord, muscles, and spinal cord each regenerate from the corresponding tissue in the stump (Deuchar 1975; Slack et al. 2004). In this frog, the V-ATPase proton pump is activated within 6 hours after tail amputation, changing

(A)

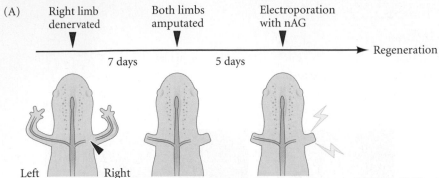

Right limb denervated   Both limbs amputated   Electroporation with nAG

Regeneration

7 days   5 days

Left   Right

(B)

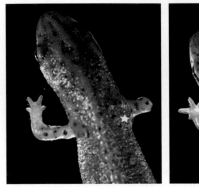

Control   nAG administered

FIGURE 15.22 Regeneration of newt limbs depends on nAG (normally supplied by the limb nerves). (A) Schematic of the procedure. The limb is denervated and a week later is amputated. After 5 five days, nAG is electroporated into the limb blastema. (B) Results show that in the denervated control (not given nAG), the amputated limb (yellow star) remains a stump. The limb that is given nAG regenerates tissues and proximal-distal polarity. (After Yin and Poss 2008, courtesy of K. Poss.)

the membrane voltage and establishing flow of protons through the blastema (Adams et al. 2007). If this proton pump is inactivated either by mutation or by drugs, depolarization of the blastema cells fails to occur and there is no regeneration.

See WEBSITE 15.5   Regeneration in annelid worms

SIDELIGHTS & *SPECULATIONS*

How Do the Blastema Cells Know Their Proximal and Distal Levels?

The regeneration blastema resembles the progress zone mesenchyme of the developing limb in many ways (see Chapter 13), and the genes involved in patterning the limb are often re-expressed appropriately in the regenerating limb. There are some differences, but the similarities are striking. The anterior-posterior axis is established by Sonic hedgehog activation and the graded expression of the HoxD series of genes, while the proximal-distal axis is reestablished using retinoic acid (RA), HoxA genes, and fibroblast growth factors, just as in the developing limbs (Yakushiji et al. 2009).

The limb blastema appears to be specified early, and labeling cells of the blastema with fluorescent dyes shows, for instance, that cells at the distal tip of the blastema are fated to become autopod structures (Echeverri and Tanaka 2005). By transplanting regenerating limb blastemas onto developing limb buds, Muneoka and Bryant (1982) showed that the blastema cells could respond to limb bud signals and contribute to the developing limb. At the molecular level, just as Sonic hedgehog is seen in the posterior region of the developing limb progress zone mesenchyme, it is seen in the early posterior regen-

eration blastema (Imokawa and Yoshizato 1997; Torok et al. 1999).

The initial pattern of Hox gene expression in regenerating limbs is not the same as that in developing limbs. However, the nested pattern of *Hoxa* and *Hoxd* gene expression characteristic of limb development is established as the limb regenerates (Torok et al. 1998). Hox gene expression may be controlled by RA. If regenerating animals are treated with sufficient concentrations of RA (or other retinoids), their regenerated limbs have duplications along the proximal-distal axis (**Figure 15.23**; Niazi and Saxena 1978; Maden 1982). This response is dose-dependent and at maximal dosage can result in a complete new limb regenerating (starting at the most proximal girdle), regardless of the original level of amputation. Dosages higher than this maximum result in the *inhibition* of regeneration. It appears that the RA causes the cells to be respecified to a more proximal position (**Figure 15.24**; Crawford and Stocum 1988b; Pecorino et al. 1996).

Retinoic acid is synthesized in the wound epidermis of the regenerating limb and forms a gradient along the proximal-distal and anterior-posterior axes of the blastema (Brockes 1992; Scadding and Maden 1994; Viviano et al. 1995). This RA gradient is thought to facilitate three major processes that might inform cells of their position along that axis in the limb. First, RA can activate the *Hoxa* genes differentially across the blastema, resulting in the specification of pattern in the regenerating limb. Gardiner and colleagues (1995) have shown that the expression pattern of certain *Hoxa* genes in the distal cells of the regeneration blastema is changed by exogenous RA into an expression pattern characteristic of more proximal cells. It is probable that during normal regeneration, the wound epidermis/apical ectodermal cap secretes RA, which activates the genes needed for cell proliferation, downregulates the genes that are specific for differentiated cells, and activates a

set of Hox genes that tells the cells where they are in the limb and how much they need to grow.

The mechanisms by which the Hox genes do this are not known, but one possibility is that they regulate changes in blastema cell-cell adhesion similar to those postulated to mediate the actions of Hox genes in the developing limb (see Chapter 13; Bryant

and Gardiner 1992). A proximodistal gradient of blastema cell adhesivity has been demonstrated in the regenerating urodele limb (Nardi and Stocum 1983; Stocum and Crawford 1987). It is possible that one of the critical changes in the cell surface mediated by *Hoxa* genes is the upregulation of the cell adhesion protein CD59 (sometimes called Prod1) in response

(Continued on next page)

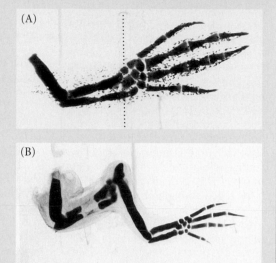

Figure 15.23 Effects of vitamin A (a retinoid) on regenerating salamander limbs. (A) Normal regenerated *Ambystoma mexicanum* limb (9×) with humerus, paired radius and ulna, carpals, and digits. The dotted line shows the plane of amputation through the carpal area. (B) Regeneration after amputation at the same location (5×), but after the regenerating animal had been placed in retinol palmitate (vitamin A) for 15 days. A new humerus, ulna, radius, carpal set, and digit set have emerged. (From Maden 1982, courtesy of M. Maden.)

Figure 15.24 Proximalization of blastema respecification by retinoic acid. (A) When a wrist blastema from a recently cut axolotl forelimb is placed on a host hindlimb cut at the mid-thigh level, it will generate only the wrist. The host (whose own leg was removed) will fill the gap and regenerate the limb up to the wrist. However, if the donor animal is treated with RA, the wrist blastema will regenerate a complete limb and, when grafted, will fail to cause the host to fill the gap. (B) A wrist blastema from a darkly pigmented axolotl was treated with RA and placed on the amputated mid-thigh region of a golden axolotl. The treated blastema regenerated a complete limb. (After Crawford and Stocum 1998a,b; photograph courtesy of K. Crawford.)

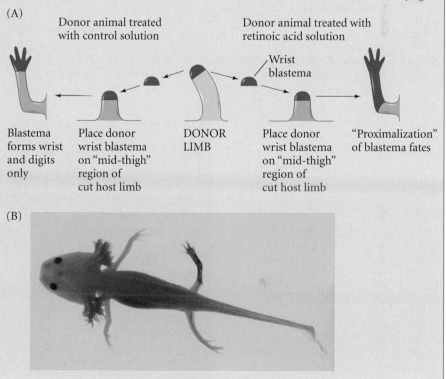

(A) Donor animal treated with control solution

Wrist blastema

Donor animal treated with retinoic acid solution

Blastema forms wrist and digits only — Place donor wrist blastema on "mid-thigh" region of cut host limb — DONOR LIMB — Place donor wrist blastema on "mid-thigh" region of cut host limb — "Proximalization" of blastema fates

(B)

to RA (Morais Da Silva et al. 2002; Echeverri and Tanaka 2005). Although cells at the distal tip of the blastema usually became autopod cells, when CD59 was overexpressed in the distal-most blastema cells (by electroporating plasmids encoding activated CD59 into them), these cells were found in proximal positions (i.e., in limb structures closer to the shoulder or pelvic girdle). When a large percentage of blastema cells overexpressed CD59 protein, the patterning of the regenerating limb was severely disrupted and the distal structures—fingers, ulna, and radius—were frequently missing. Moreover, CD59 appears to be activated by nAG from the glial cells, thereby linking proliferation and patterning together (Kumar et al. 2007b).

See WEBSITE 15.6
Polar coordinate and boundary models

Retinoic acid may also act through a second pathway. The *Meis1* and *Meis2* genes encode homeodomain proteins associated with the proximal (stylopod) portion of the developing limb. Originally, RA establishes a domain of *Meis* gene expression across the entire limb bud. However, FGFs secreted by the apical ectodermal cap suppress *Meis* gene activation, limiting *Meis* gene products to the proximal region of the limbs. During regeneration of the salamander limb, the *Meis* genes are activated by RA and also appear to be associated with the proximal identities of the limb bones. Overexpression of *Meis* genes in the distal blastema cells causes these cells to relocalize in proximal locations, whereas antisense oligonucleotides that block *Meis* expression inhibit RA from proximalizing the regenerating limbs (Mercader et al. 2000, 2005). *Hoxa* and *Meis* genes appear to be the targets of RA's ability to specify proximal cell fate during regeneration.

Morphallactic Regeneration in *Hydra*

*Hydra** is a genus of freshwater cnidarians. Most hydras are tiny—about 0.5 cm long. A hydra has a tubular body, with a "head" at its distal end and a "foot" at its proximal end. The "foot," or **basal disc**, enables the animal to stick to rocks or the undersides of pond plants. The "head" consists of a conical **hypostome** region (containing the mouth) and a ring of tentacles (which catch food) beneath it. *Hydra*, like all cnidarians, has only ectoderm and endoderm; these animals lack a true mesoderm. Hydras can reproduce sexually, but they do so only under adverse conditions (such as crowding or cold weather). They usually multiply asexually, by budding off a new individual (**Figure 15.25A**). The buds form about two-thirds of the way down the animal's body axis.

A hydra's body is not particularly "stable." In humans and flies, for instance, a skin cell in the body's trunk is not expected to migrate and eventually be sloughed off from the face or foot. But that is precisely what does happen in *Hydra*. The cells of the body column are constantly under-

---

*The Hydra is another character from Greek mythology. Whenever one of this serpent's many heads was chopped off, it regenerated two new ones. Hercules finally defeated the Hydra by cauterizing the stumps of its heads with fire. Hercules seems to have had a significant interest in regeneration—he also finally freed the bound Prometheus, thus stopping his daily hepatectomies (see p. 570).

(A)

(B)

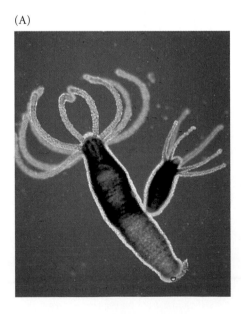
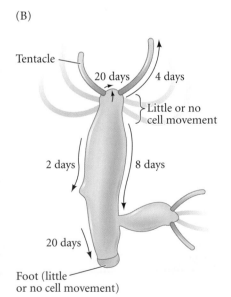

FIGURE 15.25 Budding in *Hydra*. (A) A new individual buds about two-thirds down the side of an adult hydra. (B) Cell movements in *Hydra* were traced by following the migration of labeled tissues. The arrows indicate the starting and leaving positions of the labeled cells. The bracket indicates regions in which no net cell movement took place. Cell division takes place throughout the body column except at the tentacles and foot. (A © Biophoto/Photo Researchers Inc.; B after Steele 2002.)

Tentacle

20 days   4 days

Little or no cell movement

2 days   8 days

20 days

Foot (little or no cell movement)

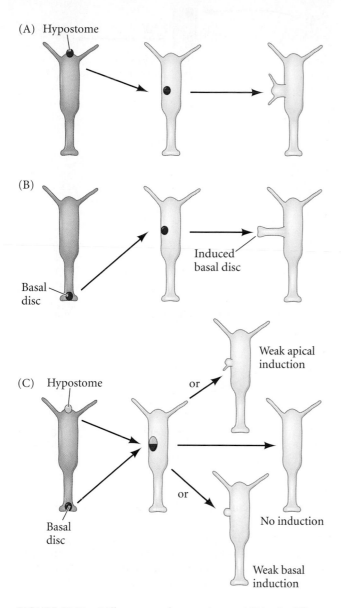

(A) Hypostome

(B)

Basal disc

Induced basal disc

(C) Hypostome

or

Weak apical induction

or

No induction

Basal disc

Weak basal induction

FIGURE 15.26 Different morphogenetic capabilities in different regions of the *Hydra* apical-basal axis. (A) Hypostome tissue grafted onto a host trunk induces a secondary axis with an extended hypostome. (B) Basal disc tissue grafted onto a host trunk induces a secondary axis with an extended basal disc. (C) If hypostome and basal disc tissues are transplanted together, only weak (if any) inductions are seen. (After Newman 1974.)

going mitosis and are eventually displaced to the extremities of the column, from which they are shed (**Figure 15.25B**; Campbell 1967a,b). Thus, each cell plays several roles, depending on how old it is; and the signals specifying cell fate must be active all the time. In a sense, a hydra's body is always regenerating.

See VADE MECUM
Planarian regeneration

If a hydra's body column is cut into several pieces, each piece will regenerate a head at its original apical end and a foot at its original basal end. No cell division is required for this to happen, and the result is a small hydra. Since each cell retains its plasticity, each piece can re-form a smaller organism; this is the basis of morphallaxis.

The head activation gradient

Every portion of the hydra body column along the apical-basal axis is potentially able to form both a head and a foot. However, the polarity of the hydra is coordinated by a series of morphogenetic gradients that permit the head to form only at one place and the basal disc to form only at another. Evidence for such gradients was first obtained from grafting experiments begun by Ethel Browne in the early 1900s. When hypostome tissue from one hydra is transplanted into the middle of another hydra, the transplanted tissue forms a new apical-basal axis, with the hypostome extending outward (**Figure 15.26A**). When a basal disc is grafted to the middle of a host hydra, a new axis also forms, but with the opposite polarity, extending a basal disc (**Figure 15.26B**). When tissues from both ends are transplanted simultaneously into the middle of a host, no new axis is formed, or the new axis has little polarity (**Figure 15.26C**; Browne 1909; Newman 1974). These experiments have been interpreted to indicate the existence of a **head activation gradient** (highest at the hypostome) and a **foot activation gradient** (highest at the basal disc).

The head activation gradient can be measured by implanting rings of tissue from various levels of a donor hydra into a particular region of the host trunk (Wilby and Webster 1970; Herlands and Bode 1974; MacWilliams 1983b). The higher the level of head activator in the donor tissue, the greater the percentage of implants that will induce the formation of new heads. The head activation factor is concentrated in the head and decreases linearly toward the basal disc. Three peptides have been associated with this head activation gradient. Two of them, Heady and Head Activator, are critical for head formation and the initiation of the bud. The other, Hym301, regulates the number of tentacles formed (Takahashi et al. 2005).

The head inhibition gradient

If the tissue of the *Hydra* body column is capable of forming a head, why doesn't it do so? In 1926, Rand and colleagues showed that the normal regeneration of the hypostome is inhibited when an intact hypostome is grafted adjacent to the amputation site. Moreover, if a graft of subhypostomal tissue (from the region just below the hypostome, where there is a relatively high concentration of head activator) is placed in the same region of a host hydra, no secondary axis forms (**Figure 15.27A**). The host head appears to make an inhibitor that prevents the grafted tissue from forming a head and secondary axis. However, if one grafts subhypostomal tissue to a decapitated host

FIGURE 15.27 Grafting experiments providing evidence for a head inhibition gradient. (A) Subhypostomal tissue does not generate a new head when placed close to an existing host head. (B) Subhypostomal tissue generates a head if the existing host head is removed. A head also forms at the site where the host's head was amputated. (C) Subhypostomal tissue generates a new head when placed far away from an existing host head. (After Newman 1974.)

(A) Intact host: No secondary axis induced

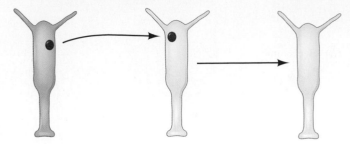

(B) Host's head removed: Secondary axis induced

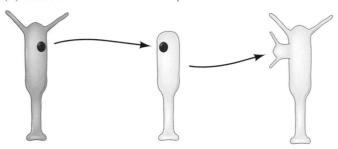

(C) Intact host: Graft away from head region induces secondary axis

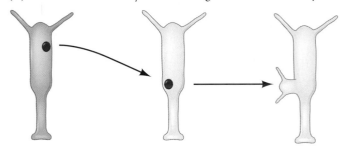

hydra, a second axis does form (**Figure 15.27B**). A gradient of this inhibitor appears to extend from the head down the body column, and can be measured by grafting subhypostomal tissue into various regions along the trunks of host hydras. This tissue will not produce a head when implanted into the apical area of an intact host hydra, but it will form a head if placed lower on the host (**Figure 15.27C**). Thus, there is a gradient of head inhibitor as well as head activator (Wilby and Webster 1970; MacWilliams 1983a).

The hypostome as an "organizer"

Ethel Browne (1909; Lenhoff 1991) noted that the hypostome acted as an "organizer" of the hydra. This notion has been confirmed by Broun and Bode (2002), who demonstrated that (1) when transplanted, the hypostome can induce host tissue to form a second body axis, (2) the hypostome produces both the head activation and head inhibition signals, (3) the hypostome is the only "self-differentiating" region of the hydra, and (4) the head inhibition signal is actually a signal to inhibit the formation of new organizing centers.

See WEBSITE 15.7 Ethel Browne and the organizer

By inserting small pieces of hypostome tissue into a host hydra whose cells were labeled with India ink (colloidal carbon), Broun and Bode found that the hypostome induced a new body axis and that almost all of the result-

ing head tissue came from *host* tissue, not from the differentiation of donor tissue (**Figure 15.28A**). In contrast, when tissues from other regions (such as the subhypostomal region) were grafted into a host trunk, the head and apical trunk of the new hydra were made from the grafted *donor* tissue (**Figure 15.28B**). In other words, only the hypos-

FIGURE 15.28 Formation of secondary axes following transplantation of head regions into the trunk of a hydra. The host endoderm was stained with India ink. (A) Hypostome tissue grafted onto the trunk induces the host's own trunk tissue to become tentacles and head. (B) Subhypostomal donor tissue placed on the host trunk self-differentiates into a head and upper trunk. (From Broun and Bode 2002, courtesy of H. R. Bode.)

(A)

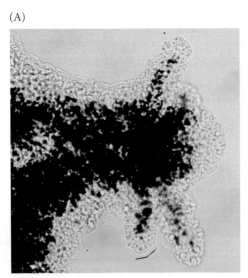

(B)

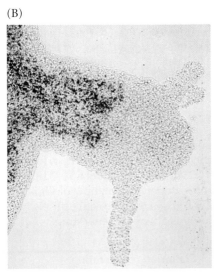

(A)   (B)   (C)   (D)

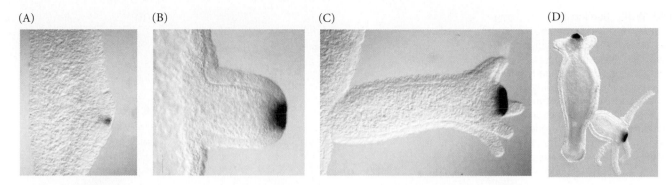

FIGURE 15.29   Expression of a *Wnt* gene during *Hydra* budding. (A) Early bud. (B) Mid-stage bud. (C) Bud with early tentacles. (D) Adult with late bud. (From Hobmayer et al. 2000, courtesy of T. W. Holstein and B. Hobmeyer.)

tome region could alter the fates of the trunk cells and cause them to become head cells. Broun and Bode also found that the signal did not have to emanate from a permanent graft. Even transient contact with the hypostome region was sufficient to induce a new axis from a host hydra. In these cases, *all* the tissue of the new axis came from the host. The head inhibitor appears to repress the effect of the inducing signal from the donor hypostome, and it normally functions to prevent other portions of the hydra from having such organizing abilities.

At least three genes are known to be active in the hypostome organizer area, and their expression in this cnidarian structure suggests an evolutionarily conserved set of signals that function as organizers throughout the animal kingdom. First, a set of *Hydra* Wnt proteins is seen in the apical end of the early bud, defining the hypostome region as the bud elongates (**Figure 15.29**). These proteins act to form the head organizer: signaling through the canonical Wnt pathway, they inhibit GSK3 to stabilize β-catenin in the cell nucleus* (Hobmayer et al. 2000; Broun et al. 2005; Lengfeld et al. 2009; also see Bode 2009). If GSK3 is inhibited throughout the body axis, ectopic tentacles form at all levels, and each piece of the trunk has the ability to stimulate the outgrowth of new buds. The expression of another vertebrate organizer molecule, Goosecoid, is restricted to the *Hydra* hypostome region. Moreover, when the hypostome is brought into contact with the trunk of an adult hydra, it induces expression of the *Brachyury* gene, just as vertebrate organizers do—even though hydras lack mesoderm (Broun et al. 1999; Broun and Bode 2002).

---

*Wnts and β-catenins also appear to regulate anterior-posterior polarity during planarian (flatworm) regeneration (Gurley et al. 2008; Petersen and Reddien 2008). Here, β-catenin is expressed (via Wnts) in the posterior-facing blastema (which generates tails), and if RNA interference against β-catenin eliminates this protein, the posterior-facing blastema forms a head. A frizzled-like repressor of Wnt signaling forms in the anterior-facing blastema (which forms heads).

The basal disc activation and inhibition gradients

Certain properties of the basal disc suggest that it is the source of both a foot inhibition and a foot activation gradient (MacWilliams et al. 1970; Hicklin and Wolpert 1973; Schmidt and Schaller 1976; Meinhardt 1993; Grens et al. 1999). The inhibition gradients for the head and the foot may be important in determining where and when a bud can form. In young adult hydras, the gradients of head and foot inhibitors appear to block bud formation. However, as the hydra grows, the sources of these labile substances grow farther apart, creating a region of tissue about two-thirds down the trunk where levels of both inhibitors are minimal. This region is where the bud forms (**Figure 15.30A**; Shostak 1974; Bode and Bode 1984; Schiliro et al. 1999).

Certain mutants of *Hydra* have defects in their ability to form buds, and these defects can be explained by alterations of the inhibition gradients. The *L4* mutant of *Hydra magnipapillata*, for instance, forms buds very slowly, and only after reaching a size about twice as long as wild-type individuals. The amount of head inhibitor in these mutants was found to be much greater than in wild-type *Hydra* (Takano and Sugiyama 1983).

Several small peptides have been found to activate foot formation, and researchers are just beginning to sort out the mechanisms by which these proteins arise and function (see Harafuji et al. 2001; Siebert et al. 2005). However, the specification of cells as they migrate from the basal region through the body column may be mediated by a gradient of tyrosine kinase. The product of the *shinguard* gene is a tyrosine kinase that extends in a gradient from the ectoderm just above the basal disc through the lower region of the trunk. Buds appear to form where this gradient fades (**Figure 15.30B**). The *shinguard* gene appears to be activated through the product of the *manacle* gene, a putative transcription factor that is expressed earlier in the basal disc ectoderm.

The inhibition and activation gradients also inform the hydra "which end is up" and specify positional values along the apical-basal axis. When the head is removed, the head inhibitor no longer is made, causing the head activa-

FIGURE 15.30 Bud location as a function of head and foot inhibition gradients. (A) Head inhibition (blue) and foot inhibition (red) gradients in newly dropped buds, young adults, and budding adults. (B) Expression of the Shinguard protein in a graded fashion in a budding hydra. (A after Bode and Bode 1984; B from Bridge et al. 2000.)

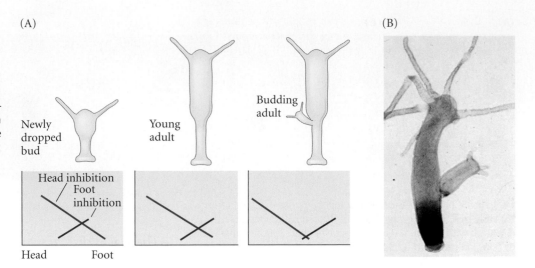

tor to induce a new head. The region with the most head activator (i.e., those cells directly beneath the amputation site) will form the head. Once the head is made, it generates the head inhibitor, and thus equilibrium is restored.

Compensatory Regeneration in the Mammalian Liver

Compensatory regeneration—wherein differentiated cells divide to recover the structure and function of an injured organ—has been demonstrated in the mammalian liver and in the zebrafish heart (Poss et al. 2002).

According to Greek mythology, Prometheus's punishment for bringing the gift of fire to humans was to be chained to a rock and to have an eagle tear out and eat a portion of his liver each day. His liver then regenerated each night, providing a continuous food supply for the eagle and eternal punishment for Prometheus. Today the standard assay for liver regeneration is a **partial hepatectomy**. In this procedure, specific lobes of the liver are removed (after administering anesthesia), leaving the other hepatic lobes intact. Although the removed lobe does not grow back, the remaining lobes enlarge to compensate for the loss of the missing tissue (Higgins and Anderson 1931). The amount of liver regenerated is equivalent to the amount of liver removed.

The human liver regenerates by the proliferation of existing tissue. Surprisingly, the regenerating liver cells do not fully dedifferentiate when they re-enter the cell cycle. No regeneration blastema is formed. Rather, the five types of liver cells—hepatocytes, duct cells, fat-storing (Ito) cells, endothelial cells, and Kupffer macrophages—all begin dividing to produce more of themselves. Each cell type retains its cellular identity, and the liver retains its ability to synthesize the liver-specific enzymes necessary for glucose regulation, toxin degradation, bile synthesis, albumin production, and other hepatic functions (Michalopoulos and DeFrances 1997).

The removal or injury of the liver is sensed through the bloodstream, as some liver-specific factors are lost while others (such as bile acids and gut lipopolysaccharides) increase. These lipopolysaccharides activate two of the non-hepatocyte cells to secrete paracrine factors that allow the hepatocytes to re-enter the cell cycle. The Kupffer cell secretes interleukin 6 (IL6) and tumor necrosis factor-α (which are usually involved with activating the adult immune system), while the stellate cells secrete **hepatocyte growth factor** (**HGF**, or **scatter factor**) and TGF-β. However, hepatocytes that are still connected to one another in an epithelium cannot respond to HGF. The hepatocytes activate cMet, the receptor for HGF, within an hour of partial hepatectomy, and the blocking of cMet (by RNA interference or knockout) blocks liver regeneration (Borowiak et al. 2004; Huh et al. 2004; Paranjpe et al. 2007).

The trauma of partial hepatectomy may activate metalloproteinases that digest the extracellular matrix and permit the hepatocytes to separate and proliferate. These enzymes also may cleave HGF to its active form (Mars et al. 1995). Together, the factors produced by the Kupffer and stellate cells allow the hepatocytes to divide by preventing apoptosis, activating cyclins D and E, and repressing cyclin inhibitors such as p27 (see Taub 2004). The liver stops growing at the appropriate size; the mechanism for how this is achieved is not yet known. One clue comes from parabiosis experiments, wherein the circulatory systems of two rats are linked together. Here, partial hepatectomy in one rat will cause liver enlargement in the other (Moolten and Bucher 1967). Therefore, it seems like there is some factor or factors in the blood that are establishing the size of the liver. Huang and colleagues (2006) have proposed that these factors are bile acids produced by the liver and that stimulate hepatocyte growth. Partial hepatectomy stimulates the release of bile acids into the blood, and these bile acids are received by the hepatocytes and activate the Fxr transcription factor. The Fxr protein promotes cell division, and mice without this protein cannot regenerate their livers. Therefore, these bile acids (a relatively small percentage of secreted liver products) appear to regulate the size of the liver and keep it at a particular volume of cells.

Because human livers have the power to regenerate, a patient's diseased liver can be replaced by compatible liver tissue from a living donor (usually a relative). The donor's liver has always grown back. Human livers regenerate more slowly than those of mice, but function is restored quickly (Pascher et al. 2002; Olthoff 2003). In addition, mammalian livers possess a "second line" of regenerative ability. If the hepatocytes are unable to regenerate the liver sufficiently within a certain amount of time, the **oval cells** divide to form new hepatocytes. Oval cells are a small progenitor cell population that can produce hepatocytes and bile duct cells. They appear to be kept in reserve and to be used only *after* the hepatocytes have attempted to heal the liver (Fausto and Campbell 2005; Knight et al. 2005). The molecular mechanisms by which these factors interact and by which the liver is first told to begin regenerating and then to stop regenerating after reaching the appropriate size remain to be discovered.

AGING: THE BIOLOGY OF SENESCENCE

Entropy always wins. A multicellular organism is able to develop and maintain its identity for only so long before deterioration prevails over synthesis, and the organism ages. **Aging** can be defined as the time-related deterioration of the physiological functions necessary for survival and fertility. The characteristics of aging—as distinguished from diseases of aging, such as cancer and heart disease—affect all the individuals of a species. The aging process has two major facets. The first is simply how long an organism lives; the second concerns the physiological deterioration, or **senescence**, that characterizes old age. These topics are often viewed as being interrelated.

Aging and senescence have both genetic and environmental components. The interplay between mutations, environmental factors, and random epigenetic change makes these phenomena both fascinating and frustrating to study. Moreover, in recent years, new molecular approaches and stem cell technologies have allowed new approaches to these age-old questions (see Stocum 2006; Carlson 2007).

Genes and Aging

Genetic factors play roles both between species and within species. The **maximum lifespan**, which is the maximum number of years any member of a given species has been known to survive, is characteristic of a species. The maximum human life span is estimated to be 121 years (Arking 1998). The life spans of some tortoises and lake trout are

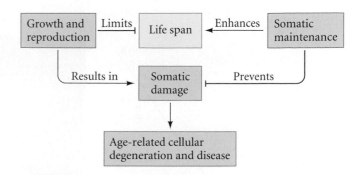

FIGURE 15.31 Conceptual framework, first proposed by Kirkwood (1977), that organisms must effect a compromise between the energy allocated to reproduction and growth and the energy allocated to the maintenance and repair of bodily tissues. (After Vijg and Campisi 2008.)

both unknown but are estimated to be more than 150 years. The maximum life span of a domestic dog is about 20 years, and that of a laboratory mouse is 4.5 years. If a *Drosophila* fruit fly survives to eclose (in the wild, more than 90% die as larvae), it has a maximum life span of 3 months.

The species-specific life span appears to be determined by genes that effect a trade-off between early growth and reproduction and somatic maintenance (**Figure 15.31**). In other words, aging results from natural selection operating more on early survival and reproduction than on having a vigorous post-reproductive life. If longevity is a selectable trait, one should expect to find hereditable variation within populations. Recently, long-term studies of wild populations of animals have provided convincing data that there is hereditable variation within a species for aging (Wilson et al. 2007).

Molecular evidence (see Kenyon 2001; Vijg and Campisi 2008) indicates that certain genetic components of longevity are conserved between species: flies, worms, mammals, and even yeast all appear to use the same set of genes to promote survival and longevity. There are two sets of genes that are well known to be involved in aging and its prevention, and both sets appear to be conserved between phyla and even kingdoms of organisms. These are the genes encoding DNA repair enzymes and the genes encoding proteins involved in the insulin signaling pathway.

Genes encoding DNA repair proteins

DNA repair and synthesis may be important in preventing senescence. Individuals of species whose cells have more efficient DNA repair enzymes live longer (**Figure 15.32A**; Hart and Setlow 1974). Certain premature aging syndromes (**progerias**) in humans appear to be caused by mutations in such DNA repair enzymes (Sun et al. 1998; Shen and Loeb 2001). In humans, Hutchinson-Gilford progeria is a rapid-aging syndrome; children born with this condition age rapidly, dying (usually of heart failure) as early

(A)

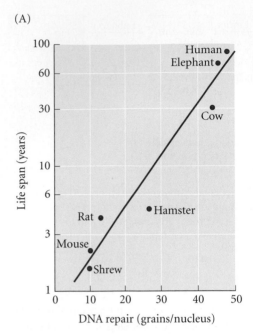

(B)

FIGURE 15.32 Life span and the aging phenotype. (A) Correlation between life span and the ability of fibroblasts to repair DNA in various mammalian species. Repair capacity is represented in autoradiography by the number of grains from radioactive thymidine per cell nucleus. Note that the *y* axis (life span) is a logarithmic scale. (B) Hutchinson-Gilford progeria. Although they are not yet 8 years old, these children have a phenotype similar to that of an aged person. The hair loss, fat distribution, and skin transparency are characteristic of the normal aging pattern as seen in elderly adults, and may result from mutations in DNA repair enzymes (A after Hart and Setlow 1974; B © Associated Press.)

as 12 years of age (**Figure 15.32B**). Symptoms include thin skin with age spots, resorbed bone mass, hair loss, and arteriosclerosis—all characteristics of the human senescent phenotype. Hutchinson-Gilford progeria is the result of a dominant mutation in the gene that encodes lamin A, a nuclear membrane protein, and these same mutations can be seen in age-related senescence (Scaffidi and Misteli 2006).

Another intriguing progeria has been documented in mice and is caused by loss-of-function mutations of the *Klotho** gene (Kuro-o et al. 1997). Conversely, the same gene's gain-of-function phenotype (causing its overexpression) has been known to prolong a mouse's life by 30% (Kurosu et al. 2005). *Klotho* appears to encode a hormone that downregulates insulin signaling. As we shall soon see, the suppression of signaling by insulin and insulin-like

growth factor 1 (IGF-1) is one of the ways life span can be extended in many species.

The protein **p53** also appears to be important in aging. This transcription factor, one of the most important regulators of cell division, has been called the "guardian of the genome" because of its ability to block cancer in several ways. It can stop the cell cycle, cause cellular senescence in rapidly dividing cells, instruct the *Bax* genes to initiate cellular apoptosis, and activate DNA repair enzymes. In most cells, p53 is bound to another protein that keeps p53 inactive. However, ultraviolet radiation, oxidative stress, and other factors that cause DNA damage will also separate and activate p53. The induction of apoptosis by p53 can be beneficial (when destroying cancer cells) or deleterious (when destroying, say, neurons). It is possible that animals with high levels of p53 have increased protection against cancer, but they may also age more rapidly (Tyner et al. 2002). Indeed, p53 can be activated by the absence of lamin A (Varela et al. 2005), thereby suggesting a mechanism for Hutchinson-Gifford progeria.

Another set of genes important in aging are the **sirtuin genes**, which encode histone deacetylation (chromatin-silencing) enzymes. The sirtuin proteins guard the genome, preventing genes from being expressed at the wrong times and places, and blocking chromosomal rearrangements. They are usually found in regions of chromatin (especially repetitive DNA sequences) where such mistaken chromosomal rearrangements can occur. However, when DNA strands break (as inevitably happens as the body ages), sirtuin proteins are called on to fix them and cannot attend to their usual functions. Thus genes that are usually silenced become active as the cells age. Sirtuin proteins have been found to prevent aging throughout the eukaryotic kingdoms, including in yeasts and mammals (Howitz et al. 2003; Oberdoerffer et al. 2008).

---

*The gene is named after one of the Fates of Greek mythology—Klotho, who spun the thread of life. The other two Fates, Lachesis and Atropos, measured out and cut life's thread, respectively.

Aging and the insulin signaling cascade

One of the criticisms of the idea of genetic "programs" for aging asks how evolution could have selected for them. Once the organism has passed maturity and raised its offspring, it is "an excrescence" on the tree of life (Rostand 1962); natural selection presumably cannot act on traits that affect an organism only *after* it has reproduced. But "How can evolution select for a way to degenerate?" may be the wrong question. Evolution probably can't select for such traits. The right question may be "How can evolution select for phenotypes that can postpone reproduction or sexual maturity?" There is often a trade-off between reproduction and maintenance, and in many species reproduction and senescence are closely linked.

Recent studies of mice, *Caenorhabditis elegans*, and *Drosophila* suggest that there is a conserved genetic pathway that regulates aging, and that it can be selected for during evolution. This pathway involves the response to insulin or insulin-like growth factors. In *C. elegans*, a larva proceeds through four larval stages, after which it becomes an adult. If the nematodes are overcrowded or if there is insufficient food, however, the larva can enter a metabolically dormant **dauer larva** stage, a nonfeeding state of **diapause** during which development and aging are suspended. The nematode can remain in the dauer larva stage for up to 6 months, rather than becoming an adult that lives only a few weeks. In this diapausal state, the nematode has increased resistance to oxygen radicals that can crosslink proteins and destroy DNA. The pathway that regulates both dauer larva formation and longevity has been identified as the insulin signaling pathway (Kimura et al. 1997; Guarente and Kenyon 2000; Gerisch et al. 2001; Pierce et al. 2001). Favorable environments signal the activation of the insulin receptor homologue DAF-2, and this receptor stimulates the onset of adulthood (**Figure 15.33A**). Poor environments fail to activate the DAF-2 receptor, and dauer formation ensues. While severe loss-of-function alleles in this pathway cause the formation of dauer larvae in any environment, weak mutations in the insulin signaling pathway enable the animals to reach adulthood and live four times longer than wild-type animals.

Downregulation of the insulin signaling pathway also has several other functions. First, it appears to influence metabolism, decreasing mitochondrial electron transport. When the DAF-2 receptor is not active, organisms have decreased sensitivity to *reactive oxygen species* (ROS), metabolic by-products that can damage cell membranes and proteins and even destroy DNA (Feng et al. 2001; Scott et al. 2002; we discuss these in more detail on p. 577). Second, downregulating the insulin pathway increases the production of enzymes that prevent oxidative damage, as well as DNA repair enzymes (Honda and Honda 1999; Tran et al. 2002). Third, this lack of insulin signaling decreases fertility (Gems et al. 1998). This increase in DNA synthetic

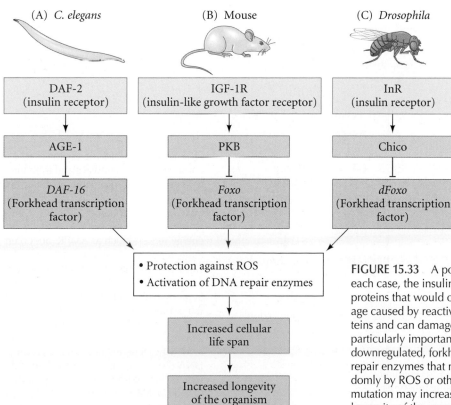

FIGURE 15.33 A possible pathway for regulating longevity. In each case, the insulin signaling pathway inhibits the synthesis of proteins that would otherwise protect cells against oxidative damage caused by reactive oxygen species (ROS) that crosslink proteins and can damage DNA. These protective proteins may be particularly important in mitochondria. When insulin signaling is downregulated, forkhead transcription factors may activate DNA repair enzymes that may protect against mutations caused randomly by ROS or other agents. Such protection against ROS and mutation may increase the functional life span of the cells and the longevity of the organism.

enzymes and in enzymes that protect against ROS is due to the DAF-16 transcription factor. This forkhead-type transcription factor is inhibited by the insulin receptor (DAF-2) signal. When that signal is absent, DAF-16 can function, and this factor appears to activate the genes encoding several enzymes (such as catalase and superoxide dismutase) that are involved in reducing ROS, several enzymes that increase protein and lipid turnover, and several stress proteins (Lee et al. 2003; Libina et al. 2003; Murphy et al. 2003).

It is possible that this system also operates in mammals, but the mammalian insulin and insulin-like growth factor pathways are so integrated with embryonic development and adult metabolism that mutations often have numerous and deleterious effects (such as diabetes or Donahue syndrome). However, there is some evidence that the insulin signaling pathway does affect life span in mammals (**Figure 15.33B**). Dog breeds with low levels of insulin-like growth factor 1 (IGF-1) live longer than breeds with higher levels of this factor. Mice with loss-of-function mutations of the insulin signaling pathway live longer than their wild-type littermates (see Partridge and Gems 2002; Blüher et al. 2004). Holzenberger and colleagues (2003) found that mice heterozygous for the insulin-like growth factor 1 receptor (IGF-1R) not only lived about 30% longer than their wild-type littermates, they also had greater resistance to oxidative stress. In addition, mice lacking one copy of their IGF-1R gene lived about 25% longer than wild-type

mice (and had higher ROS resistance, but otherwise normal physiology and fertility).

The insulin signaling pathway also appears to regulate life span in *Drosophila* (**Figure 15.33C**). Flies with weak loss-of-function mutations of the insulin receptor gene or genes in the insulin signaling pathway (such as *chico*) live nearly 85% longer than wild-type flies (Clancy et al. 2001; Tatar et al. 2001). These long-lived mutants are sterile, and their metabolism resembles that of flies that are in diapause (Kenyon 2001). The insulin receptor in *Drosophila* is thought to regulate a Forkhead transcription factor (dFoxo) similar to the DAF-16 protein of *C. elegans*. When the *Drosophila dFoxo* gene is activated in the fat body, it can lengthen the fly's life span (Giannakou et al. 2004; Hwangbo et al. 2004). Although it has not been demonstrated that dFoxo regulates the enzymes that protect against ROS, independent studies have shown that when these enzymes (such as superoxide dismutase) are downregulated by mutation or by RNA interference, the resulting flies die early, have increased oxidative stress, and display higher levels of DNA damage (Kirby et al. 2002; Woodruff et al. 2004). Conversely, overexpression of superoxide dismutase genes can lengthen the *Drosophila* life span (Parkes et al. 1998). While some evidence points to a correlation between longer life span, lower insulin signaling, and elevated ROS protection in *Drosophila* (Broughton et al. 2005), other studies suggest that some flies and other insects can obtain longer life spans without increasing the enzymes known to protect

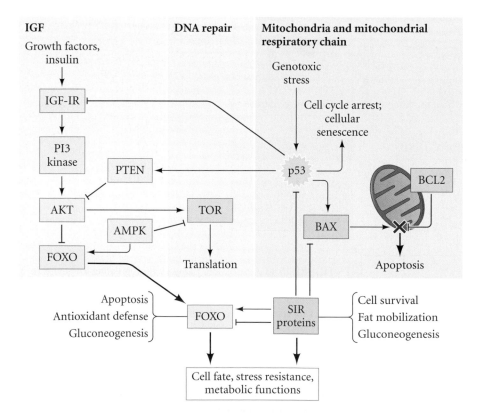

FIGURE 15.34 Possible evolutionarily conserved aging pathways and their interactions. The insulin signaling pathway and mitochondrial pathway for apoptosis are indicated. The pro-aging activities of these pathways are conserved across phyla, and energy sensors, such as AMPK, are potentially important integrators of these pathways. Many longevity signals converge on members of the FOXO and sirtuin protein families, which can interact. Note that sirtuin proteins can both activate and repress FOXO, depending on the context. TOR is a positive regulator of general translational activity. Here, insulin-like factors and DNA-disrupting ("genotoxic") chemicals initiate these pathways. (After Vijg and Campisi 2008.)

against oxidative stress (Le Bourg and Fournier 2004; Parker et al. 2004).

From an evolutionary point of view, the insulin pathway may mediate a trade-off between reproduction and survival/maintenance. Many (although not all) of the long-lived mutants have reduced fertility. Thus, it is interesting that another longevity signal originates in the gonad. When the germline cells are removed from *C. elegans*, the animals live longer. It is thought that the germline stem cells produce a substance that blocks the effects of a longevity-inducing steroid hormone (Hsin and Kenyon 1999; Gerisch et al. 2001; Arantes-Oliviera 2002). Conversely, ROS appears to promote germline development at the expense of somatic development in *C. elegans*. The oxidation of certain lipids accelerates germ cell development, while those same lipids, in their unoxidized form, prevent germ cell proliferation (Shibata et al. 2003).

Integrating the conserved aging pathways

The proteins involved in the insulin signaling pathway and the DNA repair pathway interact with one another (**Figure 15.34**). The p53 factor that induces cell cycle arrest also blocks the activity of the receptor for insulin-like growth factor 1. And sirtuin proteins, in addition to activating Foxo proteins, can also block p53. In some cases, the same protein is involved in both the DNA repair and insulin signaling pathways (Niedernhofer et al. 2006). This is the case in the protein encoded by the *XPF-ERCC1* gene in humans. Most people with mutations of this gene have xeroderma

pigmentosum, a defect in DNA repair that makes them susceptible to cancers, especially melanomas. However, if the mutation occurs in a different part of the same gene, the affected individuals have a premature aging syndrome in which the genes involved in the insulin signaling pathway are downregulated. It is possible that the enzyme encoded by this gene has two functions. Initially it may be used for DNA repair, but later it might act to prolong life by downregulating the insulin pathway.

Environmental and Epigenetic Causes of Aging

Most people cannot expect to live 121 years, and most mice in the wild do not live to celebrate even their first birthday. **Life expectancy**—the length of time an average individual of a given species can expect to live—is not characteristic of species, but of populations. It is usually defined as the age at which half the population still survives. A baby born in England during the 1780s could expect to live to be 35 years old. In Massachusetts during that same time, life expectancy was 28 years. These ages represent the normal range of human life expectancy for most of the human race throughout recorded history (Arking 1998). Even today, in some areas of the world (Cambodia, Togo, Afghanistan, and several other countries) life expectancy is less than 40 years. In the United States, a male born in 1986 can expect to live 74 years, while females have a life expectancy of around 80 years.*

Given that in most times and places people did not live much past the age of 40, our awareness of human aging is relatively new. In 1900, 50% of Americans were dead before the age of 60; in 1950, the comparable age was 72; by 2000, this "median survival" age had climbed to 80 years (**Figure 15.35**). A 70-year-old person was exceptional in 1900 but is commonplace today. People in 1900 did not have the "luxury" of dying from heart attacks or cancers, because these conditions are most likely to affect people over 50. Rather, people died (as they are still dying in many parts of the world) from microbial and viral infections (Arking 1998). Thus, the phenomena of senescence and the diseases of aging are much more common today than they were a century ago. Until recently, relatively few people exhibited the general human senescent phenotype: gray hair, sagging and wrinkling skin, stiff joints, osteoporosis (loss of bone calcium), loss of muscle fibers and muscular strength, memory loss, eyesight deterioration, and slowed sexual responsiveness. As the melancholy Jacques notes in Shakespeare's *As*

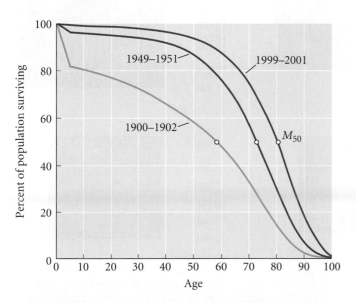

FIGURE 15.35 Survival curves for the United States population for the periods 1900–1902, 1949–1951, and 1999–2001. The circles represent M_{50}, the age at which 50% of individuals of that age survived (From CDC/NCHS, National Vital Statistics Reports 8/5/08.)

*You can see why the funding of Social Security is problematic in the United States. When it was created in 1935, the average working citizen died before age 65. Thus, he (and it usually was a he) was not expected to get back as much as he had paid into the system. Similarly, marriage "until death do us part" was easier to achieve when death occurred in the third or fourth decade of life. Before antibiotics, the death rate of young women due to infections associated with childbirth was high throughout the world.

You Like It, those who did survive to senescence left the world "*sans* teeth, *sans* eyes, *sans* taste, *sans* everything."

The general senescent phenotype is characteristic of each species. But what causes it? This question can be asked at many levels. Here we will look primarily at the cellular level of organization. While there is not yet a consensus on what causes aging (even at the cellular level), a

*There is a popular proposal that the shortening of telomeres—repeated DNA sequences at the ends of chromosomes—is responsible for senescence. Telomere shortening has been connected to a decrease in the ability of cells to divide. However, no correlation between telomere length and the life span of an animal (humans have much shorter telomeres than mice) has been found, nor is there a correlation between human telomere length and a person's age (Cristofalo et al. 1998; Rudolph et al. 1999; Karlseder et al. 2002). Nematodes can have mutations that extend or shorten longevity, and the length of the telomere does not correlate with the age in these roundworms (Raices et al. 2005). Telomeres appear to be critical in stem cell maintenance, and the telomere-dependent inhibition of cell division might serve primarily as a defense against cancer (see Blasco 2005; Flores et al. 2005).

theory is emerging that includes oxidative stress, hormones, and DNA damage.*

WEAR-AND-TEAR AND GENETIC INSTABILITY "Wear-and-tear" theories of aging are among the oldest hypotheses proposed to account for the human senescent phenotype (Weismann 1891; Szilard 1959). As one gets older, small traumas to the body and its genome build up. At the molecular level, the number of point mutations increases, and the efficiency of the enzymes encoded by our genes decreases. If mutations occur in a part of the protein synthetic apparatus, the cell produces a large percentage of faulty proteins (Orgel 1963). If mutations were to arise in the DNA-synthesizing enzymes, the overall rate of mutation in the organism would be expected to increase markedly; Murray and Holliday (1981) have documented such faulty DNA polymerases in senescent cells.

A new variant of this idea is the hypothesis of **random epigenetic drift**. Given that appropriate methylation is essential for normal development, one can immediately see that diseases would result as a consequence of inappropriate epi-

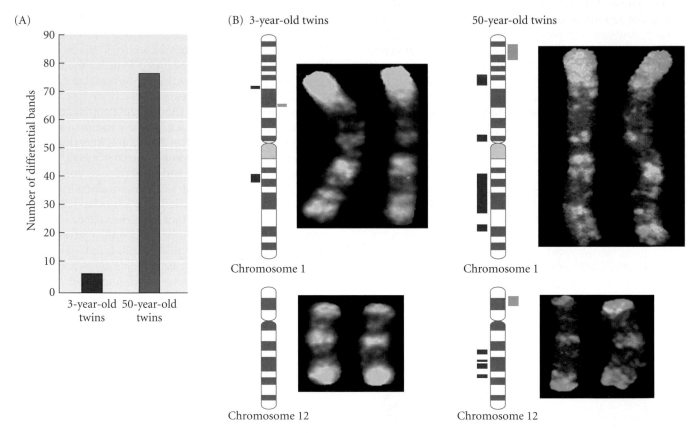

FIGURE 15.36 Differential DNA methylation patterns in aging twins. (A) In bisulfite sequence mapping, regions of DNA that are unmethylated will be cut by restriction enzymes (because bisulfite converts unmethylated cytosine to uracil), but methylated sites will not be cut. The histogram summarizes the number of differences in the resulting restriction maps of 3-year-old and 50-year-old twins. (B) A more recent technique of revealing methylation differences and similarities between twins is to mark the DNA from one twin with a red dye and that from the other with a green dye. One can then collect only the nonmethylated DNA and bind it to metaphase chromosomes. If the bands are red or green, it means the DNA from one twin bound but the DNA from the other twin did not. If the region is yellow, it means the red and green DNAs bound equally. (From Fraga et al. 2005, courtesy of M. Esteller.)

genetic methylation. Recent studies have confirmed that inappropriate methylation can be the critical factor in aging and cancers. Some of the evidence for this hypothesis comes from identical twins. Most "identical" twins start life with very few differences in appearance or behaviors, but accumulate these differences with age. Experience counts, and both random events and lifestyles may be reflected in phenotypes. Fraga and colleagues (2005) found that twin pairs were nearly indistinguishable in methylation patterns when young, but older monozygous twins exhibited very different patterns of methylation. This affected their gene expression patterns, such that older twin pairs had different patterns of DNA expression, while younger twin pairs had very similar expression patterns. **Figure 15.36** shows that monozygotic twin pairs start off with identical amounts of methylated DNA and acetylation of histones H3 and H4 (three epigenetic markers). As the twins age, however, both methylation and acetylation increase, but to different extents and at different chromosomal locations in each twin.

The idea that random epigenetic drift inactivates important genes without any particular environmental cue gives rise to an entirely new hypothesis of aging. Instead of randomly accumulated mutations—which might be due to specific mutagens—we are at the mercy of chance accumulations of errors made by the DNA methylating and demethylating enzymes. Indeed, our DNA methylating enzymes, unlike the DNA polymerases, are prone to errors. DNA methyltransferases are not the most fastidious of enzymes. At each round of DNA replication, they must methylate the appropriate cytosine residues and leave the others unmethylated. This is not always done properly, and such errors accumulate as we age.

For instance, the methylation of the promoter region of estrogen receptors is known to increase with age (Issa et al. 1994). **Figure 15.37** shows a linear relationship between the methylation of a promoter region in an estrogen receptor gene and increased age. The methylation of the promoters of the genes for the alpha and beta estrogen receptors increases with age, resulting in the inactivation of this gene in the smooth muscles of the circulatory system. Moreover, methylation of the estrogen receptor genes is even more prominent in the atherosclerotic plaques (thickened artery walls) that occlude the blood vessels. The atherosclerotic plaques show more methylation of the estrogen receptor genes than does the tissue around it (Post et al. 1999; Kim et al. 2007). Thus, DNA methylation-associated inactivation of the estrogen receptor genes in vascular tissue may play a role in atherogenesis and aging of the vascular system. This potentially reversible defect may provide a new target for intervention in heart disease.

OXIDATIVE DAMAGE One major theory views metabolism as the cause of aging. According to this theory, aging is a result of metabolism and its by-products, **reactive oxygen species** (**ROS**). The ROS produced by normal metabolism can oxidize and damage cell membranes, proteins, and nucleic acids. Some 2–3% of the oxygen atoms taken up by our mitochon-

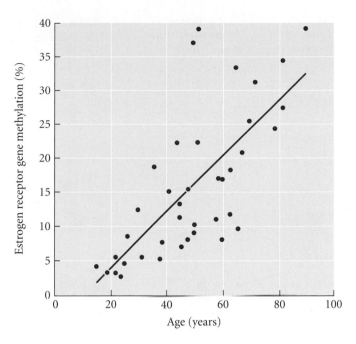

FIGURE 15.37 Methylation of the estrogen receptor gene occurs as a function of normal aging. (After Issa et al. 1994.)

dria are reduced insufficiently and form ROS: superoxide ions, hydroxyl ("free") radicals, and hydrogen peroxide. Evidence that ROS molecules are critical in the aging process includes the observation that fruit flies overexpressing the enzymes that destroy ROS (catalase and superoxide dismutase) live 30–40% longer than do control flies (Orr and Sohal 1994; Parkes et al. 1998; Sun and Tower 1999). Moreover, flies with mutations in the *methuselah* gene (named after the biblical fellow said to have lived 969 years) live 35% longer than wild-type flies. These mutants have enhanced resistance to ROS (Lin et al. 1998). In *C. elegans*, too, individuals with mutations that result in either the degradation of ROS or the prevention of ROS formation live much longer than wild-type nematodes (Larsen 1993; Vanfleteren and De Vreese 1996; Feng et al. 2001). These findings not only suggest that aging is under genetic control, but also provide evidence for the role of ROS in the aging process.

DIET Calorie restriction is one of the few known ways of extending mammalian longevity (again, at the expense of fertility), and it may do so through several routes. First, restricting calorie intake may reduce levels of IGF-1 and of circulating insulin (Kenyon 2001; Roth et al. 2002; Holzenberger et al. 2003). This association of increased longevity with the downregulation of the insulin pathway through diet is seen in yeast, flies, nematodes, and mice. Dietary restriction may also work through the sirtuin proteins (see above; Lamming et al. 2005), thereby uniting the insulin metabolic pathway with the genomic protection hypotheses. The insulin pathway in mammals also negatively regulates *Foxo4*, the gene for a transcription factor that acti-

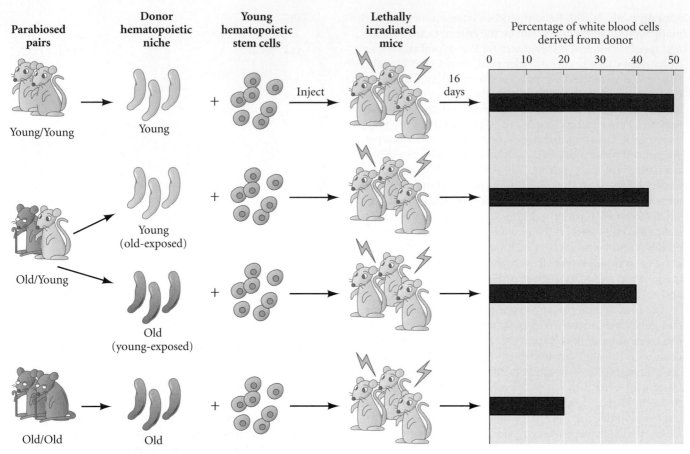

| Parabiosed pairs | Donor hematopoietic niche | Young hematopoietic stem cells | Lethally irradiated mice | Percentage of white blood cells derived from donor |

FIGURE 15.38 Circulating factors "rejuvenate" hematopoietic niche cells. Hematopoietic niche cells were isolated from parabiosed (surgically conjoined) mouse pairs. The pairs were either same age/young (2 months); same age/old (21 months); or one young and one old mouse. Niche cells were cultured with young hematopoietic stem cells and injected into lethally irradiated mice (i.e., mice whose own stem cells had been destroyed by radiation). When the white blood cells produced by the injected stem cells were analyzed 16 weeks later, stem cells residing in young niches with young blood produced the most white blood cells; those residing in old niches bathed in old blood had the worst reconstitution. However, old-niche cells that developed in contact with young blood (i.e., those from the "mixed" parabiosed pair) reconstituted the host white cell population almost as well as the "ever-young" cells. (After Takahashi et al. 2000; Mayack et al. 2010.)

vates ROS-protective enzymes (see Figure 15.33B; Essers et al. 2004). And finally, calorie restriction represses a ribosomal activator whose absence is associated with increased longevity (Selman et al. 2009). So, although calorie restriction does seem to be able to retard aging, the mechanisms by which it does so are still controversial.

YOUNG BLOOD: SERUM FACTORS AND PROGENITOR CELLS
One of the hallmarks of aging is the declining ability of stem cells and progenitor cells to restore damaged or nonfunctioning tissues. A decline in muscle progenitor (satellite) cell activity when Notch signaling is lost results in a significant decrease in the ability to maintain muscle function. Similarly, an age-dependent decline in liver progenitor cell division impairs liver regeneration due to a decline in transcription factor cEBP-α.

The problem, however, may not be in the stem cells themselves as much as in their environment. If an aged and a young mouse are *parabiosed* (i.e., their circulatory systems are surgically joined so that the two mice share one blood supply), the stem cells of the old mouse are exposed to factors in young blood serum (and vice versa). This heterochronic parabiosis has been seen to restore the activity of old stem cells. Notch signaling of the muscle stem cells regained its youthful levels, and muscle cell regeneration was restored. Similarly, liver progenitor cells regained "young" levels of cEBP-α—and their ability to regenerate (Conboy et al. 2005). When young hematopoietic stem cells were placed into "old" stem cell niches that had been exposed to old blood, they did not make many new cells when injected into lethally irradiated mice. However, when these young stem cells were placed into "old" niches that had been exposed by parabiosis to "young" blood, they produced almost as many new blood cells as the young stem cells that had seen young niches exposed only to young blood (**Figure 15.38**).

Exceptions to the Aging Rule

There are a few species in which aging seems to be optional, and these may hold some important clues to how animals can live longer and retain their health. Turtles, for instance, are a symbol of longevity in many cultures. Many turtle species not only live a long time, but they do not undergo the typical aging syndrome. In these species, older females lay as many eggs (if not more) as their younger counterparts. Miller (2001) showed that a 60-year-old female three-toed box turtle (*Terrapene carolina triunguis*) lays as many eggs annually as she ever did. Interestingly, turtles have special adaptations against oxygen deprivation, and these enzymes also protect against ROS (Congdon et al. 2003; Lutz et al. 2003).

In monarch butterflies (*Danaus plexippus*), adults that migrate to wintering grounds in the mountains of central Mexico live several months (August–March), whereas their summer counterparts live only about 2 months (May–July). The regulation of this difference appears to be juvenile hormone (Herman and Tatar 2001). The migrating butterflies are sterile because of suppressed synthesis of JH. If migrants are given JH in the laboratory, they regain fertility but lose their longevity. Conversely, when summer monarchs have their corpora allata removed (so they no longer make JH), their longevity increases 100%. Mutations in the insulin signaling pathway of *Drosophila* likewise decrease JH synthesis (Tu et al. 2005). This decrease in JH makes the flies small, sterile, and long-lived, adding to whatever longevity-producing effect protection against ROS might have.

Finally, there may be organisms that have actually cheated death. The hydrozoan cnidarian *Turritopsis nutricula* may be such an immortal animal. Most hydrozoans have a complex life cycle in which a colonial (polyp) stage asexually buds off the sexually mature, solitary, adult medusa (usually called a *jellyfish*). Eggs and sperm from the medusa develop into an embryo and then a planula larva. Planula larvae then form a colonial polyp stage. Medusae, like the polyps, have a limited life span, and in most hydrozoans they die shortly after releasing their gametes (Martin 1997). *Turritopsis*, however, has evolved a remarkable variation on this theme. The solitary medusa of this species can revert to its polyp stage *after* becoming sexually mature (Bavestrello et al. 1992; Piraino et al. 1996). In the laboratory, 100% of *Turritopsis* medusae undergo this change.

How does the jellyfish accomplish this feat? Apparently, it can alter the differentiated state of a cell, transforming it into another cell type. Such a phenomenon is called **transdifferentiation**, and is usually seen only when parts of an organ regenerate. However, it appears to occur normally in the *Turritopsis* life cycle. In the transdifferentiation process, the medusa is transformed into the stolons and polyps of a hydroid colony. These polyps feed on zooplankton and soon are budding off new medusae. Thus, it is possible that organismic death does not occur in this species.

The factor involved in aging the niche appears to be none other than IGF-1. This factor appears to be produced locally in the niche and is regulated by factors in the blood (Mayack et al. 2010). Thus, although the quality of the progenitor cells themselves does not appear to decline, age-dependent changes in serum factors may produce an environment that is less supportive of progenitor proliferation and proper cellular differentiation.

Promoting longevity

Several interacting agents may promote longevity. These include calorie restriction, protection against oxidative stress, and the factors activated by a suppressed insulin pathway. It is not yet known how these factors interact—whether they are part of a single "longevity pathway," or if they act separately. Moreover, genetics and diet do not appear to be the full answer to aging. Chance, it seems, still plays a role. When clonally identical *C. elegans* are fed an identical diet, some organisms still live longer than others, and different organs deteriorate more rapidly in different individuals (Herndon et al. 2002). Mutations are randomly occurring events, and they may play a role in the aging process.

As advances in our ability to prevent and cure disease lead to increased human life expectancy, we are still left with a general aging syndrome that is characteristic of our species. Unless attention is paid to this general aging syndrome, we risk ending up like Tithonios, the miserable wretch of Greek mythology to whom the gods awarded eternal life, but not eternal youth. However, our new knowledge of regeneration is being put to use by medicine, and we may soon be able to ameliorate some of the symptoms of aging. The potentially far-reaching consequences of this interaction of developmental biology and medicine will be discussed in Chapter 17.

Snapshot Summary: *Metamorphosis, Regeneration, and Aging*

1. Amphibian metamorphosis includes both morphological and biochemical changes. Some structures are remodeled, some are replaced, and some new structures are formed.

2. The hormone responsible for amphibian metamorphosis is tri-iodothyronine (T_3). The synthesis of T_3 from thyroxine (T_4) and the degradation of T_3 by deiodinases can regulate metamorphosis in different tissues. T_3 binds to thyroid hormone receptors and acts predominantly at the transcriptional level.

3. Many changes during amphibian metamorphosis are regionally specific. The tail muscles degenerate; the trunk muscles persist. An eye will persist even if transplanted into a degenerating tail.

4. Heterochrony involves changes in the relative rates of development of different parts of the animal. In neoteny, the larval form is retained while the gonads and germ cells mature at their normal rate. In progenesis, the gonads and germ cells mature rapidly while the rest of the body matures normally. In both instances, the animal can mate while retaining its larval or juvenile form.

5. Animals with direct development do not have a larval stage. Primary larvae (such as those of sea urchins) specify their body axes differently than the adult, whereas secondary larvae (such as those of insects and amphibians) have body axes that are the same as adults of the species.

6. Ametabolous insects undergo direct development. Hemimetabolous insects pass through nymph stages wherein the immature organism is usually a smaller version of the adult.

7. In holometabolous insects, there is a dramatic metamorphosis from larva to pupa to sexually mature adult. In the stages between larval molts, the larva is called an instar. After the last instar, the larva undergoes a metamorphic molt to become a pupa. The pupa undergoes an imaginal molt to become an adult.

8. During the pupal stage, the imaginal discs and histoblasts grow and differentiate to produce the structures of the adult body.

9. The anterior-posterior, dorsal-ventral, and proximal-distal axes are sequentially specified by interactions between different compartments in the imaginal discs. The disc "telescopes out" during development, its central regions becoming distal.

10. Molting is caused by the hormone 20-hydroxyecdysone (20E). In the presence of high levels of juvenile hormone, the molt gives rise to another larval instar. In low concentrations of juvenile hormone, the molt produces a pupa; if no juvenile hormone is present, the molt is an imaginal molt.

11. The ecdysone receptor gene produces a nuclear RNA that can form at least three different proteins. The types of ecdysone receptors in a cell may influence the response of that cell to 20E. The ecdysone receptors bind to DNA to activate or repress transcription.

12. There are four major types of regeneration. In stem-cell mediated regeneration, new cells are routinely produced to replace the ones that die. In epimorphosis (such as regenerating limbs), tissues form into a regeneration blastema, divide, and redifferentiate into the new structure. In morphallaxis (characteristic of *Hydra*), there is a repatterning of existing tissue with little or no growth. In compensatory regeneration (such as in the mammalian liver), cells divide but retain their differentiated state.

13. In regenerating limb blastemas, cells do not become multipotent. Rather, the cells retain their specification, such that neurons come from pre-existing neurons and muscles come from pre-existing muscle cells (that have become mononucleate). The mitogens, such as nAG, are provided by the AEC and the glial surrounding the limb axons.

14. Salamander limb regeneration appears to use the same pattern formation system as the developing limb.

15. In *Hydra*, there appears to be a head activation gradient, a head inhibition gradient, a foot activation gradient, and a foot inhibition gradient. Budding occurs where these gradients are minimal.

16. The hypostome region of *Hydra* appears to be an organizer region that secretes paracrine factors to alter the fates of surrounding tissue.

17. The maximum life span of a species is the longest time an individual of that species has been observed to survive. Life expectancy is usually defined as the age at which approximately 50% of the members of a given population still survive.

18. Aging is the time-related deterioration of the physiological functions necessary for survival and reproduction. The phenotypic changes of senescence (which affect all members of a species) are not to be confused with diseases of senescence, such as cancer and heart disease (which affect some individuals but not others).

19. Reactive oxygen species (ROS) can damage cell membranes, inactivate proteins, and mutate DNA. Mutations that alter the ability to make or degrade ROS can change the life span.

20. Proteins that regulate DNA repair and cell division (such as p53 and sirtuins) may be important regulators of aging.

21. An insulin signaling pathway, involving a receptor for insulin and insulin-like proteins, may be an important component of genetically limited life spans.

For Further Reading

Complete bibliographical citations for all literature cited in this chapter can be found at the free access website **www.devbio.com**

Becker, K. B., K. C. Stephens, J. C. Davey, M. J. Schneider and V. A. Galton. 1997. The type 2 and type 3 iodothyronine deiodinases play important roles in coordinating development in *Rana catesbeiana* tadpoles. *Endocrinology* 138: 2989–2997.

Broun, M., L. Gee, B. Reinhardt, and H. R. Bode. 2005. Formation of the head organizer in hydra involves the canonical Wnt pathway. *Development* 132: 2907.

Cai, L. and D. D. Brown. 2004. Expression of type II iodothyronine deiodinase marks the time that a tissue responds to thyroid hormone-induced metamorphosis in *Xenopus laevis. Dev. Biol.* 266: 87–95.

Carlson, B. M. 2007. *Principles of Regenerative Biology.* Academic Press, New York.

Fraga, M. F. and 20 others. 2005. Epigenetic differences arise during the lifetime of monozygotic twins. *Proc. Natl. Acad. Sci. USA* 102: 10604–10609.

Jiang, C., A. F. Lamblin, H. Steller and C. S. Thummel. 2000. A steroid-triggered transcriptional hierarchy controls salivary gland cell death during *Drosophila* metamorphosis. *Mol. Cell* 5: 445–455.

Kragl, M. D. Knapp, E. Nacu. S. Khattak, M. Maden, H. H. Epperlein and E. M. Tanaka. 2009. Cells keep a memory of their tissue origin during axolotl limb regeneration. *Nature* 460: 60–65.

Kumar, A., J. W. Godwin, P. B. Gates, A. A. Garza-Garcia and J. P. Brockes. 2007. Molecular basis for the nerve dependence of limb regeneration in an adult vertebrate. *Science* 318: 772–777.

Schubiger, M. and J. W. Truman. 2000. The RXR ortholog USP suppresses early metamorphic processes in *Drosophila* in the absence of ecdysteroids. *Development* 127: 1151–1159.

Stocum, D. L. 2006. *Regenerative Biology and Medicine.* Academic Press, New York.

Taub, R. 2004. Liver regeneration: From myth to mechanism. *Nature Rev. Mol. Cell Biol.* 5: 836–847.

Vijg, J. and J. Campisi. 2008. Puzzles, promises and a cure for ageing. *Nature* 454: 1065–1071.

Go Online

WEBSITE 15.1 The molecular biology of wing formation. Formation of the *Drosophila* wing involves the interaction of more than 200 genes. This site discusses some of these gene interactions.

WEBSITE 15.2 Homologous specification. If a group of cells in one imaginal disc are mutated such that they give rise to a structure characteristic of another imaginal disc (for instance, cells from a leg disc giving rise to antennal structures), the regional specification of those structures will be in accordance with their position in the original disc.

WEBSITE 15.3 Insect metamorphosis. The four links on this website discuss (1) the experiments of Wigglesworth and others who identified the hormones of metamorphosis and the glands producing them; (2) the variations that

Drosophila and other insects play on the general theme of metamorphosis; (3) the remodeling of the insect nervous system during metamorphosis; and (4) a microarray analysis of *Drosophila* metamorphosis wherein several thousand genes are simultaneously screened.

WEBSITE 15.4 Precocenes and synthetic JH. Given the voracity of insect larvae, it is amazing that any plants survive. However, many plants get revenge on their predators by making compounds that alter insect metamorphosis, thus preventing the animals from developing or reproducing.

WEBSITE 15.5 Regeneration in annelid worms. An easy laboratory exercise can discover the rules by which worms regenerate their segments. This website details some of those experiments.

WEBSITE 15.6 Polar coordinate and boundary models. The phenomena of epimorphic regeneration can be seen formally as events that reestablish continuity among tissues that the amputation has severed. The polar coordinate and boundary models attempt to explain the numerous phenomena of limb regeneration.

WEBSITE 15.7 Ethel Browne and the organizer. As detailed in Chapter 7, Spemann and Mangold's work with amphibians brought the concept of the Organizer into embryology, and Spemann's laboratory helped make the idea a unifying principle of embryology. However, it has been argued that the concept actually had its origins in Ethel Browne's experiments on *Hydra*.

Vade Mecum

Amphibian metamorphosis and frog calls. For photographs of amphibian metamorphosis (and for the sounds of the adult frogs), check out the metamorphosis and frog call sections of the Amphibian segment.

Chromosome squash. The Fruit Fly segments contains a sequence showing how to do a chromosome squash using the *Drosophila* larval salivary gland.

Planarian regeneration. Freshwater planarians can reproduce by splitting their body in half. The front can regenerate a back, and the back can regenerate a front. This provides a laboratory exercise in stem cells and potency that is easy to perform.

The Saga of the Germ Line

16

WE ARE ABOUT TO COME FULL CIRCLE. We began our analysis of animal development by discussing fertilization, and we will finish our studies of individual development by investigating **gametogenesis**, the processes by which the sperm and the egg are formed. In addition to forming its own body, an individual animal must set aside cells that will provide the material and instructions for initiating bodies in the *next* generation. Germ cells provide the continuity of life between generations, and the mitotic ancestors of our own germ cells once resided in the gonads of reptiles, amphibians, fish, and invertebrates.

In many animals, including insects, roundworms, and vertebrates, there is a clear and early separation of germ cells from somatic cell types. In several other animal phyla (and throughout the entire plant kingdom), this division is not as well established. In these animal species (which include cnidarians, flatworms, and tunicates), somatic cells readily form new organisms. The zooids, buds, and polyps of many invertebrate phyla testify to the ability of somatic cells to give rise to new individuals (Liu and Berrill 1948; Buss 1987).

In those organisms in which there is an established germ line that separates from the somatic cells early in development, the germ cells often do not arise within the gonad itself. Rather, the gamete progenitor cells—the **primordial germ cells** (**PGCs**)—arise elsewhere and migrate into the developing gonads. The first step in gametogenesis, then, involves forming the PGCs and getting them into the genital ridge as the gonad is forming. Therefore, our discussion of gametogenesis will include:

- Formation of the germ plasm and the determination of the primordial germ cells
- Migration of the PGCs into the developing gonads
- The process of meiosis and the modifications of meiosis for forming sperm and eggs
- Differentiation of the sperm and egg cells (gametogenesis)
- Hormonal control of gamete maturation and ovulation

Germ Plasm and the Determination of the Primordial Germ Cells

All sexually reproducing animals arise from the fusion of gametes—sperm and eggs. All gametes arise from primordial germ cells. In most laboratory model organisms (including frogs, nematodes, and flies), the primordial germ cells are specified autonomously by cytoplasmic determinants in the egg that are parceled

And the end of all our exploring
Will be to arrive where we started
And know the place for the first time.

T.S. ELLIOT (1942)

When the spermatozoon enters the
egg, it enters a cell system which has
already achieved a certain degree
of organization.

ERNST HADORN (1955)

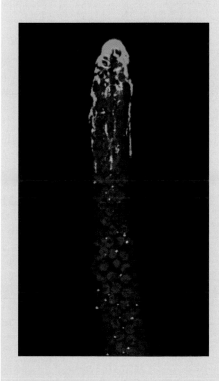

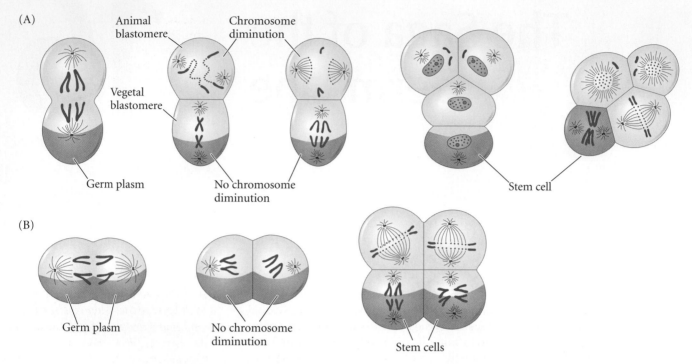

(A)

Animal blastomere

Chromosome diminution

Vegetal blastomere

Germ plasm

No chromosome diminution

Stem cell

(B)

Germ plasm

No chromosome diminution

Stem cells

FIGURE 16.1 Distribution of germ plasm during cleavage of normal and centrifuged zygotes of *Parascaris*. (A) In normal cleavage, the germ plasm is localized in the vegetalmost blastomere, as shown by the lack of chromosomal diminution in that particular cell. Thus, at the 4-cell stage, the embryo has a single stem cell for its gametes. (B) When centrifugation is used to displace the first cleavage by 90 degrees, both of the resulting cells have vegetal germ plasm, and neither cell undergoes chromosome diminution. After the second cleavage, both of these two cells give rise to germinal stem cells. (After Waddington 1966.)

out to specific cells during cleavage. However, evidence suggests that in the majority of species (including salamanders and mammals), the germ cells are specified by interactions among neighboring cells (Extravour and Akam 2003). In those species in which determination of the primordial germ cells is brought about by the autonomous localization of specific proteins and mRNAs, these cytoplasmic components are collectively referred to as the **germ plasm**.

Germ cell determination in nematodes

BOVERI'S EXPERIMENTS ON *PARASCARIS* Theodor Boveri (1862–1915) was the first person to observe an organism's chromosomes throughout its development. In so doing, he discovered a fascinating feature in the development of the roundworm *Parascaris aequorum* (formerly known as *Ascaris megalocephala*). This nematode worm has only two chromosomes per haploid cell, thereby allowing detailed observations of its individual chromosomes. The cleavage plane of the first embryonic division is unusual in that it is equatorial, separating the animal half from the vegetal half of the zygote (**Figure 16.1A**). More bizarre, however, is the behavior of the chromosomes in the subsequent division of these first two blastomeres. The chromosomes in the animal blastomere fragment into dozens of pieces just before this cell divides. This phenomenon is called **chromosome diminution**, because only a portion of the original chromosome survives. Numerous genes are lost when the chromosomes fragment, and these genes are not included in the newly formed nuclei (Tobler et al. 1972; Müller et al. 1996).

Meanwhile, in the vegetal blastomere, the chromosomes remain normal. During second cleavage, the animal cell splits meridionally while the vegetal cell again divides equatorially. Both vegetally derived cells have normal chromosomes. However, the chromosomes of the more animally located of these two vegetal blastomeres fragment before the third cleavage. Thus, at the 4-cell stage, only one cell—the most vegetal—contains a full set of genes. At successive cleavages, nuclei with diminished chromosomes are given off from this vegetalmost line until the 16-cell stage, when there are only two cells with undiminished chromosomes. One of these two blastomeres gives rise to the germ cells; the other eventually undergoes chromosome diminution and forms more somatic cells. The chromosomes are kept intact only in those cells destined to form the germ line. If this were not the case, the genetic information would degenerate from one generation to the next. The cells that have undergone chromosome diminution generate the somatic cells.

Boveri has been called the last of the great observers of embryology and the first of the great experimenters. Not content with observing the retention of the full chromo-

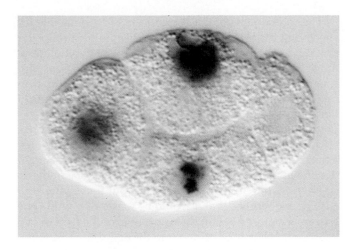

FIGURE 16.2 Inhibition of transcription in germ cell precursors of *Caenorhabditis elegans*. The photograph shows in situ hybridization to β-galactosidase mRNA expressed under control of the *pes-10* promoter. The *pes-10* gene is one of the earliest genes expressed in *C. elegans*. The P-blastomere that gives rise to the germ cells (far right) does not transcribe the gene. (From Seydoux and Fire 1994, courtesy of G. Seydoux.)

some complement by the germ cell precursors, he set out to test whether a specific region of cytoplasm protects the nuclei within it from diminution. If so, any nucleus happening to reside in this region should remain undiminished. In 1910, Boveri tested this hypothesis by centrifuging *Parascaris* eggs shortly before their first cleavage. This treatment shifted the orientation of the mitotic spindle. When the spindle forms perpendicular to its normal orientation, both resulting blastomeres contain some of the vegetal cytoplasm (**Figure 16.1B**). Boveri found that after the first division, neither nucleus underwent chromosomal diminution. However, the next division was equatorial along the animal-vegetal axis. Here the resulting animal blastomeres both underwent diminution, whereas the two vegetal cells did not. Boveri concluded that the vegetal cytoplasm contains a factor (or factors) that protects nuclei from chromosomal diminution and determines germ cells.

C. ELEGANS In the nematode *Caenorhabditis elegans*, the germline precursor cell is the P4 blastomere. The **P-granules** that enter this cell are critical for instructing it to become the germline precursor (see Figure 5.45). The P-granule protein repertoire includes several transcriptional inhibitors and RNA-binding proteins, including homologues of *Drosophila* Vasa, Piwi, and Nanos, whose functions we will discuss below (Kawasaki et al. 1998; Seydoux and Strome 1999; Subramanian and Seydoux 1999). In addition, as discussed in Chapter 5, the *C. elegans* germ plasm contains the PIE-1 protein, which prevents the phosphorylation of RNA polymerase II, thereby preventing transcription in the germ cell lineage (Ghosh and Seydoux 2008). This is critical for preventing the germ line from dif-

ferentiating into somatic cells, and germ cell differentiation cannot commence until the disappearance of PIE-1 in later embryonic stages. Until that time, the germline nuclei are silenced (**Figure 16.2**). It is possible that these germ plasm proteins can aggregate into P-granules only in the P1–P4 cells. The MEX-5 and PAR-1 proteins inhibit P-granule stability in the remaining somatic cells (Brangwynne et al. 2009).

> **See WEBSITE 16.1**
> **Germline sex determination in *C. elegans***

> **See WEBSITE 16.2**
> **Mechanisms of chromosome diminution**

Germ cell determination in insects

In *Drosophila*, PGCs form as a group of **pole cells** at the posterior pole of the cellularizing blastoderm. These nuclei migrate into the posterior region at the ninth nuclear division and become surrounded by the **pole plasm**, a complex collection of mitochondria, fibrils, and **polar granules** (**Figure 16.3**; Mahowald 1971a,b; Schubiger and Wood 1977). If the pole cell nuclei are prevented from reaching the pole plasm, no germ cells will be made (Mahowald et al. 1979). The germ cells are responsible for forming the **germline stem cells**, each of which divides asymmetrically to produce another stem cell and a differentiated daughter cell called a **cystoblast**. Cystoblasts undergo four mitotic divisions with incomplete cytokinesis to form a cluster of 16 cells interconnected by cytoplasmic bridges called

(A)

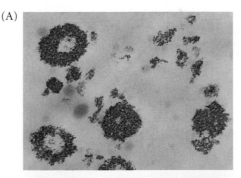

(B)

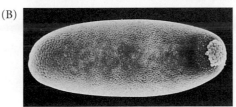

FIGURE 16.3 Pole plasm of *Drosophila*. (A) Electron micrograph of polar granules from particulate fraction of *Drosophila* pole cells. (B) Scanning electron micrograph of a *Drosophila* embryo just prior to completion of cleavage. The pole cells can be seen at the right of the photograph. (Photographs courtesy of A. P. Mahowald.)

ring canals. Only those two cells having four interconnections are capable of developing into oocytes, and of those two, only one becomes the egg (the other begins meiosis but does not complete it). Thus, only one of the 16 cystocytes becomes an ovum; the remaining 15 cells become **nurse cells** (Figure 16.4).

As it turns out, the cell destined to become the oocyte is that cell residing at the most posterior tip of the egg chamber, or **ovariole**, enclosing the 16-cell clone. However, since the nurse cells are connected to the oocyte by the ring canals, the entire complex can be seen as one egg-producing unit. The nurse cells produce numerous RNAs and proteins that ultimately are transported into the oocyte through the ring canals.

The components of the germ plasm are responsible for specifying these cells to be germ cells and for inhibiting somatic gene expression in these cells. These two functions might be interrelated, since the inhibition of gene transcription appears to be essential for germ cell determination (see Santos and Lehmann 2004). One of the components of the pole plasm is the mRNA of the **germ cell-less** (**gcl**) gene. This gene was discovered by Jongens and his colleagues (1992) when they mutated *Drosophila* and screened for females that did not have "grandoffspring." They assumed that if a female did not place functional pole plasm in her eggs, she could still have offspring—but those offspring would lack germ cells and would be sterile. The wild-type *gcl* gene is transcribed in the nurse cells of the fly's ovary, and its mRNA is transported into the egg. Once inside the egg, it is transported to the posteriormost portion and

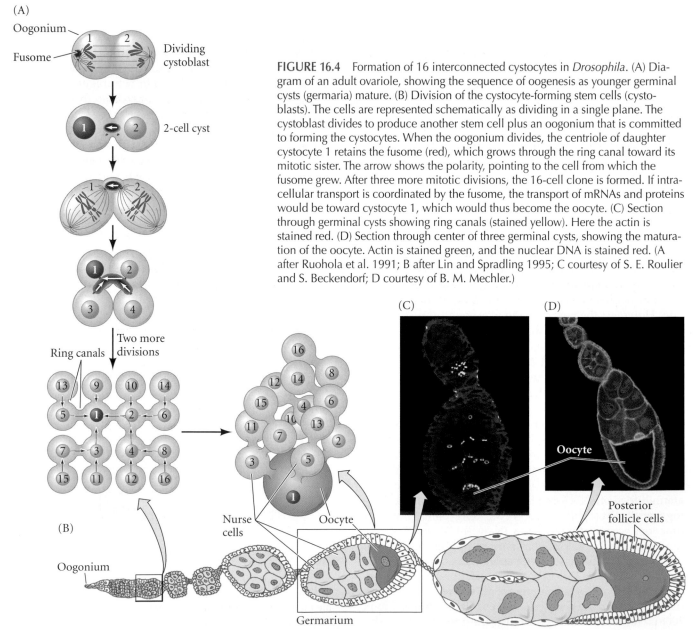

FIGURE 16.4 Formation of 16 interconnected cystocytes in *Drosophila*. (A) Diagram of an adult ovariole, showing the sequence of oogenesis as younger germinal cysts (germaria) mature. (B) Division of the cystocyte-forming stem cells (cystoblasts). The cells are represented schematically as dividing in a single plane. The cystoblast divides to produce another stem cell plus an oogonium that is committed to forming the cystocytes. When the oogonium divides, the centriole of daughter cystocyte 1 retains the fusome (red), which grows through the ring canal toward its mitotic sister. The arrow shows the polarity, pointing to the cell from which the fusome grew. After three more mitotic divisions, the 16-cell clone is formed. If intracellular transport is coordinated by the fusome, the transport of mRNAs and proteins would be toward cystocyte 1, which would thus become the oocyte. (C) Section through germinal cysts showing ring canals (stained yellow). Here the actin is stained red. (D) Section through center of three germinal cysts, showing the maturation of the oocyte. Actin is stained green, and the nuclear DNA is stained red. (A after Ruohola et al. 1991; B after Lin and Spradling 1995; C courtesy of S. E. Roulier and S. Beckendorf; D courtesy of B. M. Mechler.)

Wild type

(A)

gcl
mRNA

Gcl
protein

Mutant

(B)

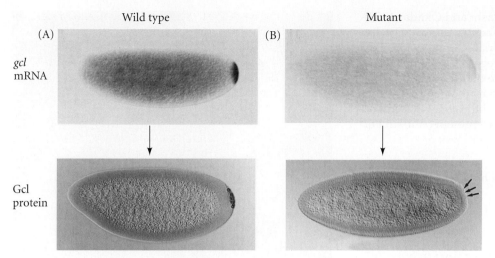

FIGURE 16.5 Localization of *germ cell-less* gene products in the posterior of the *Drosophila* egg and embryo. The *gcl* mRNA can be seen in the posterior pole of early-cleavage embryos produced by wild-type females (A), but not in embryos produced by *gcl*-deficient mutant females (B). The protein encoded by the *gcl* gene can be detected in the germ cells at the cellular blastoderm stage of embryos produced by wild-type females, but not in embryos from mutant females (arrows). (From Jongens et al. 1992, courtesy of T. A. Jongens.)

resides in what will become the pole plasm. This message is translated into protein during the early stages of cleavage (**Figure 16.5A**). The *gcl*-encoded protein appears to enter the nucleus, and it is essential for pole cell production. Flies with mutations of this gene lack germ cells (**Figure 16.5B**). The *gcl* gene encodes a nuclear envelope protein that prevents gene transcription and is critical for specifying the pole cells (Leatherman et al. 2002). Homologues of the *gcl* gene have been found in the germ cells of mice and humans, and the mouse *Gcl* gene also represses transcription. Human males with mutant *GCL* genes have defective spermatozoa and are often sterile (Nili et al. 2001; Kleiman et al. 2003).

A second protein found in *Drosophila* pole plasm (and one that becomes localized in the polar granules) is the **polar granule component** (**Pgc**). *Pgc* mRNA is part of the original pole plasm and becomes translated there. The Pgc protein inhibits transcription, and it does so (like PIE-1 in *C. elegans*) by preventing the phosphorylation of RNA polymerase II (Martinho et al. 2004; Hanyu-Nakamura et al. 2008; see Figure 2.7). Without this phosphorylation, the RNA polymerase cannot transcribe any genes. If the maternal *Pgc* gene is mutated, the germ cells begin expressing the genes characteristic of their neighboring somatic cells.

A third set of pole plasm components are the **posterior group determinants**. The Oskar protein (see Chapter 6) appears to be the critical member of this group, since expression of *oskar* mRNA in ectopic sites will cause the nuclei in those areas to form germ cells. The genes that restrict Oskar to the posterior pole are also necessary for germ cell formation (Ephrussi and Lehmann 1992; Newmark et al. 1997; Riechmann et al. 2002). Moreover, Oskar appears to be the limiting step of germ cell formation, since adding more *oskar* message to the oocyte causes more germ cells to form. Oskar functions by localizing the proteins and RNAs necessary for germ cell formation (such as *germ cell-less*) to the posterior pole (Ephrussi and Lehmann 1992; Snee and Macdonald 2004).

One of the mRNAs localized by Oskar is the *Nanos* message, whose product is essential for posterior segment formation and germ cell specification. Pole cells lacking Nanos do not migrate into the gonads and fail to become gametes. While Gcl and Pgc appear to be critical in regulating transcription, Nanos appears to be essential for inhibiting the translation of certain messages. In embryos lacking Nanos, the germline cells usually die; but if inhibited from dying, these germline cells can become somatic cells (see Chapter 6). Nanos thus prevents the pole cells from activating the pathway that would lead to the formation of somatic cells (Hayashi et al. 2004).

Another of the posterior mRNAs encodes **Vasa**, an RNA-binding protein. The mRNAs for this protein are seen in the germ plasm of many species, and Vasa is critical for initiating germ cell differentiation and meiosis (Ghabrial and Schüpbach 1999). Two other nucleic acid-binding proteins, **Piwi***** and its relative **Aubergine**, are also found in the pole plasm. They, too, have the ability to repress transcription. Piwi will later become critical in establishing the germ cell as a stem cell in the gonad (Cox et al. 1998; Megosh et al. 2006).

There are numerous components of the pole plasm that we know little about (see Santos and Lehmann 2004). For instance, mitochondrial ribosomes are seen transiently in

---

*Piwi appears to be required for stem cell maintenance and proliferation throughout the eukaryotic kingdoms. In addition to being present in germ stem cells, *Piwi* genes have also been found expressed in the totipotent stem cells of planaria and regenerating annelids. Inhibiting *Piwi* gene expression in the adult flatworm blocks the worm's regeneration (Reddien 2004). *Piwi* is also expressed in the somatic stem cells of jellyfish and is upregulated immediately before transdifferentiation. The continuous low expression of Piwi in differentiated cells of jellyfish may underlie their ability to remodel their bodies so profoundly (Seipel et al. 2004). *Piwi* may even be responsible for stem cell maintenance across kingdoms: two *Piwi* genes in *Arabidopsis* are crucial for maintaining meristem proliferation at the root and shoot of the plant (Bohmert et al. 1998; Moussian et al. 1998).

the *Drosophila* pole plasm. Kobayashi and Okada (1989) demonstrated ribosomal component's importance by showing that injecting mitochondrial RNA into embryos formed from ultraviolet-irradiated eggs restored their ability to form pole cells. It is possible that some of the pole plasm mRNAs are being translated by these mitochondrial ribosomes. Inhibiting protein synthesis by these ribosomes impairs production of the Gcl protein (Amikura et al. 2005).

VADE MECUM
Germ cells in the *Drosophila* embryo

See WEBSITE 16.3
The insect germ plasm

Germ cell determination in frogs and fish

FROGS Cytoplasmic localization of germ cell determinants has also been observed in vertebrate embryos. Bounoure (1934) showed that the vegetal region of fertilized frog eggs contains material with staining properties similar to those of *Drosophila* pole plasm. He was able to trace this cortical cytoplasm into the few cells in the presumptive endoderm that would normally migrate into the genital ridge. By transplanting genetically marked cells from one embryo into another of a differently marked strain, Blackler (1962) showed that these cells are the primordial germ cell precursors.

The germ plasm of amphibians consists of germinal granules and a matrix around them. It contains many of the same RNAs and proteins (including the large and small mitochondrial ribosomal RNAs) as the pole plasm of *Drosophila*, and they appear to repress transcription and translation (Kloc et al. 2002). The early movements of amphibian germ plasm have been analyzed in detail by Savage and Danilchik (1993), who labeled the germ plasm with a fluorescent dye. They found that the germ plasm of unfertilized eggs consists of tiny "islands" that appear to be tethered to the yolk mass near the vegetal cortex. These islands move with the vegetal yolk mass during the cortical rotation just after fertilization. After this rotation, the islands are released from the yolk mass and begin fusing together and migrating to the vegetal pole. Their aggregation depends on microtubules, and their movement to the vegetal pole depends on a kinesin-like protein that may act as the motor for germ plasm movement (Robb et al. 1996; Quaas and Wylie 2002). Savage and Danilchik (1993) found that UV light prevents vegetal surface contractions and inhibits the migration of germ plasm to the vegetal pole. Furthermore, the *Xenopus* homologues of *Nanos* and *Vasa* messages are specifically localized to the vegetal region (**Figure 16.6**; Forristall et al. 1995; Ikenishi et al. 1996; Zhou and King 1996).

ZEBRAFISH In zebrafish, the germ plasm forms a dense structure characterized by polar granules, mitochondria,

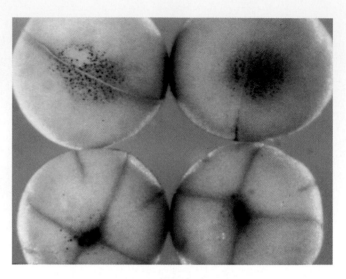

FIGURE 16.6 Germ plasm at the vegetal pole of frog embryos. In situ hybridization to the mRNA for *Xcat2* (the *Xenopus* homologue of *Nanos*) localizes the message in the vegetal cortex of first-cleavage (upper) and fourth-cleavage (lower) embryos. (After Kloc et al. 1998, courtesy of L. Etkin.)

and concentrated mRNAs. Two of these mRNAs are *Vasa* and *Nanos*. These messages are maternally supplied, and they appear to be associated with the cleavage furrows of the early dividing egg (Yoon et al. 1997). *Vasa* mRNA and other components of the germ plasm form a compact structure that is inherited by only one of the two daughter cells at each division. Thus, at late cleavage (around 1000 cells), only four cells have the germ plasm. However, after this stage, the germ plasm is distributed evenly at cell division, creating four clusters of primordial germ cells (see Figure 16.12).

Germ cell determination in mammals

In insects, frogs, nematodes, and flies, the germ cells are determined by material in the egg cytoplasm. However, in mammals, there is no obvious germ plasm. Rather, germ cells are *induced* in the embryo (Wakahara 1996; Hayashi et al. 2007).

In mice, the germ cells form at the posterior region of the epiblast, at the junction of the extraembryonic ectoderm, epiblast, primitive streak, and allantois (**Figure 16.7A,B**). This is called the posterior proximal epiblast because it is close (proximal) to the extraembryonic ectoderm, and it will be at the posterior of the embryo. Thus, the cells that become the PGCs in mice are not intrinsically different from the other cells of the epiblast and contain no specific germ plasm. Rather, the posterior epiblast cells are induced by the extraembryonic tissue. Wnts from the visceral endoderm are probably responsible for giving the posterior proximal epiblast cells the competence to

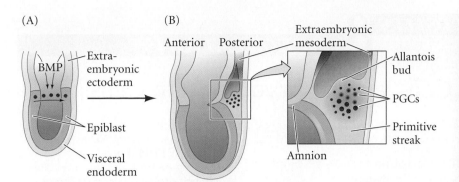

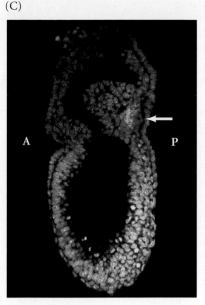

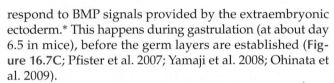

FIGURE 16.7 Specification and migration of mammalian primordial germ cells. (A) In the mouse embryo, BMP signals (blue) from the extraembryonic ectoderm induce neighboring epiblast cells (purple circles) to become precursors of PGCs and extraembryonic mesoderm. During gastrulation (arrow), these cells come to reside in the posterior epiblast. (B) On embryonic day 7, these cells emerge from the posterior primitive streak. The PGC precursors also express *Prdm14* and *blimp1* (red). (C) Late-streak stage (approximately embryonic day 7.0). Nuclei are stained white (with DAPI), and the expression of the *Prdm14* gene (having been fused to the mRNA of a green fluorescent protein) is seen by the green fluorescence. Arrowhead shows expression in the PGC in the extraembryonic mesoderm. A, anterior; P, posterior. (A,B after Hogan 2002; C from Yamaji et al. 2008, courtesy of M. Saitou.)

respond to BMP signals provided by the extraembryonic ectoderm.* This happens during gastrulation (at about day 6.5 in mice), before the germ layers are established (**Figure 16.7C**; Pfister et al. 2007; Yamaji et al. 2008; Ohinata et al. 2009).

The BMPs induce the expression of *blimp1* and *Prdm14* in a small cluster of cells (about six cells in the 6.5-day mouse embryo). Blimp1 is a transcriptional regulator that represses somatic-type gene expression, while activating those genes (such as *Sox2* and *Nanog*) associated with pluripotency. Blimp1 also activates the germline determinant *Nanos3*, which protects the germ cells against apoptosis during their migration (Tsuda et al. 2003). Prdm14 helps establish pluripotency by also activating Sox2, and it is critical for the chromatin modifications that will later silence the genome of the germ cells (Yamaji et al. 2008). Cells that express *blimp1* and *Prdm14* are restricted to the germ cell fate (Saitou et al. 2002; Ohinata et al. 2005).

The requirement for germ cell induction was shown by transplanting clumps of tissue from the distal portions of the epiblast to the proximal posterior portion of the epiblast. These cells then gave rise to PGCs (Tam and Zhou 1996). Moreover, cultured epiblast cells exposed to Wnt

signals and BMP4 gave rise to PGCs. When such PGCs from cultured male epiblast cells were transferred into testicular tubules, they produced viable sperm that could fertilize mouse eggs (Ohinata et al. 2009).

The inert genome hypothesis

As indicated above, one of the critical events in specifying germ cells appears to be the global repression of gene expression According to this hypothesis, the cells become germ cells because they are forbidden to become any other type of cell (Nieuwkoop and Sutasurya 1981; Wylie 1999; Cinalli et al. 2008). This suppression of transcription is seen in the germ cells of several species, including mammals, flies, frogs, and nematodes (see Figure 16.2). In mice, the germ cells undergo extensive chromatin modification (Seki et al. 2007), causing them to become transcriptionally inert at embryonic day 8.5 (as they begin migrating). Many of the components in the germ plasm (such as Gcl, Pgc, Piwi, and Nanos in *Drosophila* and PIE-1 in *C. elegans*) act by inhibiting either transcription or translation (Leatherman et al. 2002; de las Heras et al. 2009). Many such proteins are found throughout the animal kingdom. It is interesting that when animal germ cells are separated from the somatic cells—whether in chicks, mice, or flies—the germ cells are often specified *outside* the developing body proper. Perhaps this exile into an extraembryonic "enclave" insulates the primordial germ cells from paracrine signaling taking place within the somatic cells of the growing embryo (Dickson 1994). Once the repression of somatic gene expression is accomplished, the germ cells can return to the embryo and travel to the gonads. This germ cell migration will be our next topic.

---

*This induction can occur only in the posteriormost region of the epiblast; BMP antagonists prevent it from occurring in the trunk and anterior.

SIDELIGHTS & SPECULATIONS

Pluripotency, Germ Cells, and Embryonic Stem Cells

Primordial germ cells and embryonic stem cells are both characterized by their ability to generate any cell type in the embryo. Embryonic stem (ES) cells are derived from the inner cell masses of mammalian blastocysts and are believed to be the functional equivalent of the inner cell mass (ICM) blastomeres (see Chapter 8). One of the best pieces of evidence for this equivalence is that when ES cells are injected into the ICMs of mouse blastocysts, they behave like mouse blastocyst cells and contribute cells to the embryo. One of the interesting species-specific differences between human and mouse ES cells is that human ES cells appear to

contribute to the trophoblast, whereas mouse ES cells do not (Xu et al. 2002).

Transcription factors associated with totipotency

In mammals, the retention of totipotency or pluripotency has been correlated with the expression of three nuclear transcription factors: Oct4, Stat3, and Nanog (see Chapter 8). Oct4 is a homeodomain transcription factor expressed in all early-cleavage blastomere nuclei, but its expression becomes restricted to the ICM. During gastrulation, Oct4 becomes expressed solely in those posterior epiblast cells thought to give rise to the primordial germ cells. After that, Oct4 is seen

only in the primordial germ cells and, later, in oocytes (**Figure 16.8**; see also Figure 8.18). Oct4 is not seen in the developing sperm after the germ cells reach the testes and become committed to sperm production (Yeom et al. 1996; Pesce et al. 1998). Nanog is another homeodomain transcription factor found in the pluripotential cells of the mouse blastocyst, as well as in ES cells and germline tumors. Nanog expression is high in the PGCs of certain mouse embryos (Hatano et al. 2005; Yamaguchi et al. 2005). Knockout experiments indicate that Nanog is critical in maintaining the pluripotency of stem cells, and overexpression experiments demonstrate that elevated

(A)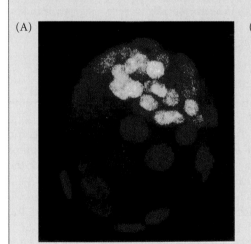

(B)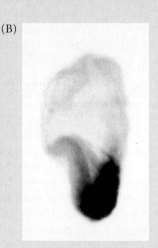

(C)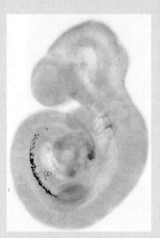

(D)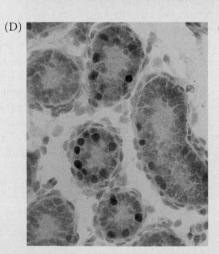

(E)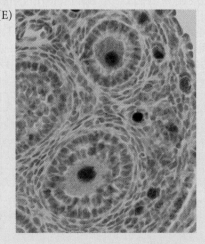

Figure 16.8 Expression of *Oct4* mRNA correlates with totipotency and ability to form germ cells in mammals. (A) Oct4 transcription factor is stained green with a fluorescent antibody, while all cell nuclei are stained red with propidium iodide. The overlap (yellow) shows that Oct4 is found only in the inner cell mass. (B,C) An *Oct4/lacZ* transgene driven by the *Oct4* promoter region shows its expression (dark color) (B) in the posterior epiblast of the 8.5-day mouse embryo and (C) in migrating PGCs in the 10.5-day embryo. (D,E) Labeled antibody (brown) staining shows Oct4 protein in the nuclei of (D) spermatogonia in postnatal testes and (E) oogonia in postnatal ovaries. (A–C from Yeom et al. 1996; D,E from Pesce et al. 1998; courtesy of H. R. Schöler.)

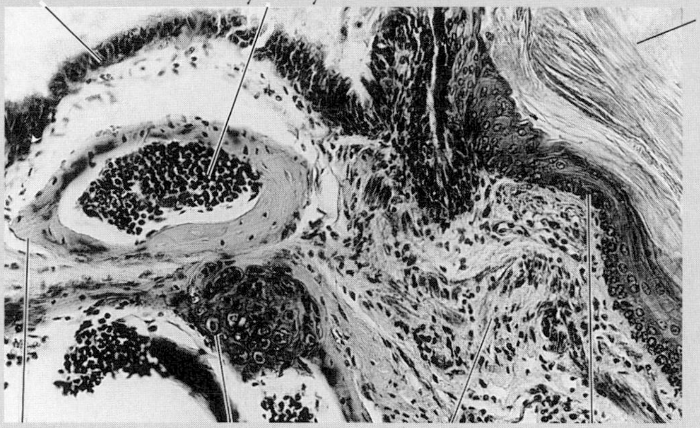

Figure 16.9 Photomicrograph of a section through a mouse teratocarcinoma, showing numerous differentiated cell types. (From Gardner 1982; photograph by C. Graham, courtesy of R. L. Gardner.)

Nanog negates the need for Stat3 and by itself maintains Oct4 transcription in ES cells (Chambers et al. 2003; Mitsui et al. 2003).

The same pluripotency and transcription factor expression pattern is seen not only in the PGCs but in two derivatives of the PGCs: cultured embryonic germ cells, and tumorous germ cells called *teratocarcinomas*.

Embryonic germ cells

When PGCs are first placed into culture, they resemble ES cells. Stem cell factor increases the proliferation of migrating mouse primordial germ cells in culture, and this proliferation can be further increased by adding another growth factor, leukemia inhibition factor (LIF). However, the life span of these PGCs is short, and the cells soon die. But if an additional mitotic regulator—basic fibroblast growth factor, Fgf2—is added, a remarkable change takes place. The cells continue to proliferate, producing pluripotent embryonic stem cells with characteristics resembling those of the inner cell mass (Matsui et al. 1992; Resnick et al. 1992; Rohwedel et al. 1996). These PGC-derived cells are called **embryonic germ (EG) cells**, and they have the potential to differentiate into all the cell types of the body.

In 1998, researchers in John Gerhart's laboratory cultured human EG cells (Shamblott et al. 1998). These cells were able to generate differentiated cells from all three primary germ layers, so they are presumably pluripotent. Such cells could be used medically to create neural or hematopoietic stem cells, which might be used to regenerate damaged neural or blood tissues. EG cells are often considered ES cells, and the distinction of their origin is ignored.

Embryonal carcinoma cells

What happens if a PGC became malignant? In one type of tumor, the germ cells become embryonic stem cells, much like the Fgf2-treated PGCs in the experiment above. This type of tumor is called a **teratocarcinoma**. Whether spontaneous or experimentally produced, teratocarcinomas contain an undifferentiated stem cell population that has biochemical and developmental properties remarkably similar to those of the inner cell mass (Graham 1977; see Parson 2004). Moreover, these stem cells not only divide but can also differentiate into a wide variety of tissues, including gut and respiratory epithelia, muscle, nerve, cartilage, and bone (**Figure 16.9**). These undifferentiated pluripotential stem cells are called **embryonal carcinoma (EC) cells**. Once differentiated, these cells no longer divide, and are therefore no longer malignant. Such tumors can give rise to most of the tissue types

in the body (Stevens and Little 1954; Kleinsmith and Pierce 1964; Kahan and Ephrussi 1970). Thus, the teratocarcinoma stem cells mimic early mammalian development, but the tumor they form is characterized by random, haphazard development.

In 1981, Stewart and Mintz formed a mouse from cells derived in part from a teratocarcinoma stem cell. Stem cells that had arisen in a teratocarcinoma of an agouti (yellow-tipped) strain of mice were cultured for several cell generations and were seen to maintain the characteristic chromosome complement of the parental mouse. Individual stem cells descended from the tumor were injected into the blastocysts of black-furred mice. The blastocysts were then transferred to the uterus of a foster mother, and live mice were born. Some of these mice had coats of two colors, indicating that the tumor cell had integrated itself into the embryo. This, in itself, is a remarkable demonstration that the tissue context is critical for the phenotype of a cell—a malignant cell was made nonmalignant.

But the story does not end here. When these chimeric mice were mated to mice carrying alleles recessive to those of the original tumor cell, the alleles of the tumor cell were expressed in many of the offspring. This means that the originally malignant tumor cell had produced many, if not all, types of normal somatic cells, and had even produced normal, functional germ cells! When such mice (being heterozygous for tumor cell genes) were mated with each other, the resulting litter contained mice that were homozygous for a large number of genes from the tumor cell (**Figure 16.10**). Thus, germ cell tumors can retain their pluripotency.

Germ cells and stem cells: Possible interactions

One idea emerging from this study is that some descendants of the pluripotent cells (such as the teratocarcinoma or ES cells) form PGCs that can undergo meiosis to form sperm or eggs. Indeed, there is evidence that ES cells

(Continued on next page)

can develop into oogonia that enter meiosis and recruit adjacent cells into follicle-like structures (Hübner et al. 2003). There is also parallel evidence that mouse ES cells can be made to differentiate into spermatocytes that can become functional sperm when transplanted into testes (Toyooka et al. 2003). These studies need to be extensively confirmed and extended to humans if they are to provide a new way of curing infertility.

It is even possible that ES, EG, and EC cells have a common origin in the presumptive PGC cells. Zwaka and Thomson (2005) hypothesize that the ES cells are actually the equivalent of PGCs and not of the inner cell mass. Not every inner cell mass blastomere can become an ES cell, and Zwaka and Thomson suggest that perhaps the successful stem cells are those that have been positioned next to trophoblast at the future posterior proximal region of the embryo. In other words, the blastomeres that become the ES cells might actually be the presumptive PGCs. While this idea remains hypothetical, it would relate these four pluripotent cell types.

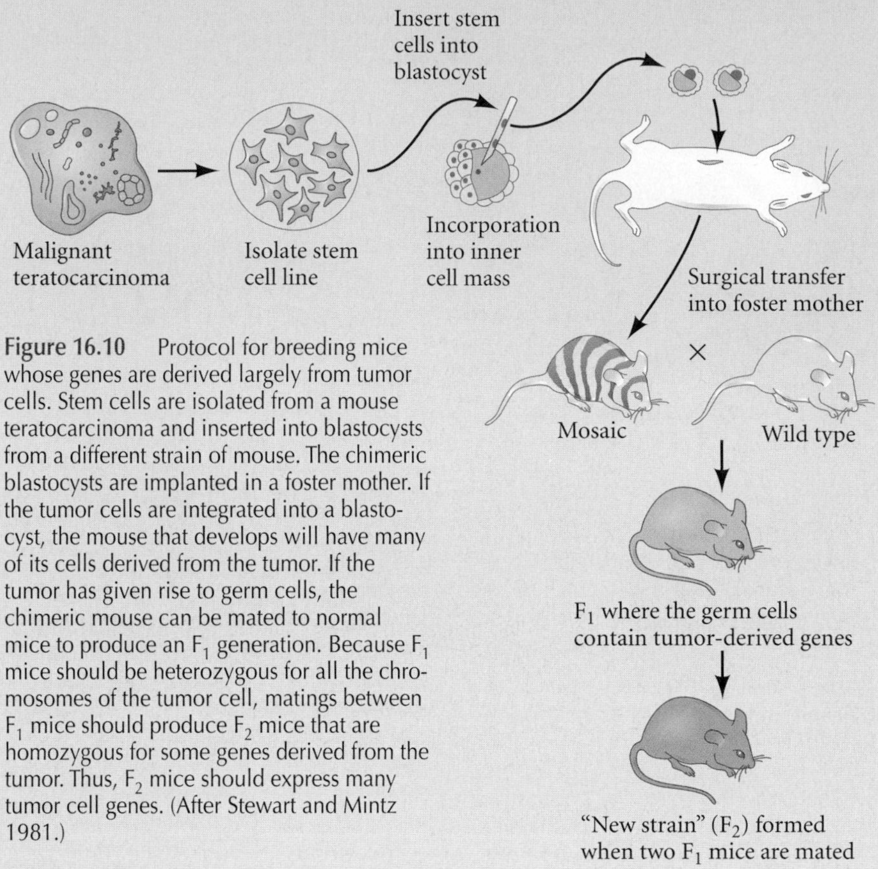

Figure 16.10 Protocol for breeding mice whose genes are derived largely from tumor cells. Stem cells are isolated from a mouse teratocarcinoma and inserted into blastocysts from a different strain of mouse. The chimeric blastocysts are implanted in a foster mother. If the tumor cells are integrated into a blastocyst, the mouse that develops will have many of its cells derived from the tumor. If the tumor has given rise to germ cells, the chimeric mouse can be mated to normal mice to produce an F_1 generation. Because F_1 mice should be heterozygous for all the chromosomes of the tumor cell, matings between F_1 mice should produce F_2 mice that are homozygous for some genes derived from the tumor. Thus, F_2 mice should express many tumor cell genes. (After Stewart and Mintz 1981.)

Germ Cell Migration

Germ cell migration in Drosophila

During *Drosophila* embryogenesis, the primordial germ cells move from the posterior pole to the gonads. The first step in this migration is a passive one, wherein the 30–40 pole cells are displaced into the posterior midgut by the movements of gastrulation (Figure 16.11A,B). The germ cells are actively prevented from migrating during this stage (Jaglarz and Howard 1994; Li et al. 2003). In the second step, the gut endoderm triggers the germ cells to actively migrate by diapedesis (i.e., squeezing amoebically) through the blind end of the posterior midgut (Kunwar et al. 2003). The germ cells migrate from the endoderm into the visceral mesoderm. In the third step, the PGCs split into two groups, each of which will become associated with a developing gonad primordium.

In the fourth step, the germ cells migrate to the gonads, which are derived from the lateral mesoderm of parasegments 10–12 (Warrior 1994; Jaglarz and Howard 1995; Broihier et al. 1998). This step involves both attraction and repulsion. The products of the *wunen* genes appear to be responsible for directing the migration of the primordial germ cells from the endoderm into the mesoderm and their division into two streams (Figure 16.11C–E). This protein is expressed in the endoderm immediately before PGC migration and in many other tissues that the germ cells avoid, and it appears to be repelling the germ cells. In loss-of-function mutants of this gene, the PGCs wander randomly (Zhang et al. 1997; Hanyu-Nakamura et al. 2004; Sano et al. 2005).

HMG-CoA reductase, the product of the *columbus* gene, appears to be critical for attracting the *Drosophila* PGCs to the gonads (Van Doren et al. 1998). This protein is made in the mesodermal cells of the gonads and probably acts as part of a biosynthetic pathway required to produce lipids that either modulate the activity of a germ cell attractant or act directly to attract PGCs (Ricardo and Lehmann 2009). In loss-of-function mutants of this gene, the PGCs wander randomly from the endoderm, and if the *columbus* gene is expressed in other tissues (such as the nerve cord), those tissues will attract the PGCs. In the last step, the gonad coalesces around the germ cells, allowing the germ cells to

(A) *Vasa* probe labeling the pole plasm

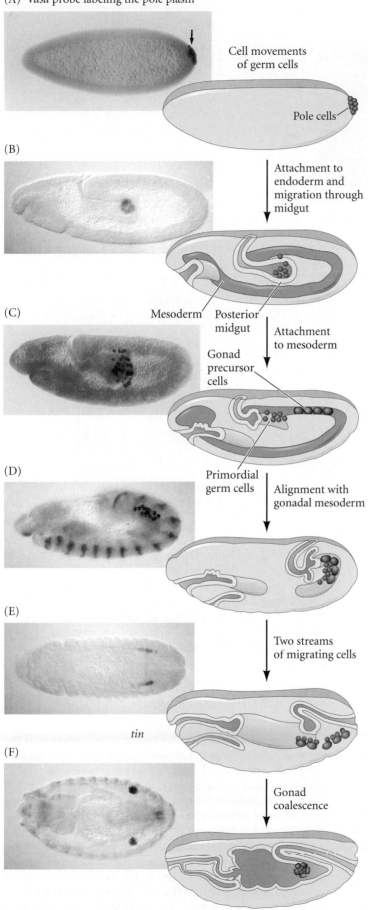

Cell movements
of germ cells

Pole cells

(B)

Attachment to
endoderm and
migration through
midgut

Mesoderm Posterior
midgut

(C)

Attachment
to mesoderm

Gonad
precursor
cells

Primordial
germ cells

(D)

Alignment with
gonadal mesoderm

(E)

Two streams
of migrating cells

tin

(F)

Gonad
coalescence

divide and mature into gametes (**Figure 16.11F**). This step requires E-cadherin (Jenkins et al. 2003).

Neither the gonads nor the germ cells differentiate until metamorphosis. During the larval stages, both the PGCs and the somatic gonadal cells divide, but they remain relatively undifferentiated. At the larval-pupal transition, gonadal morphogenesis occurs (Godt and Laski 1995; King 1970). During this transition, those PGCs in the anterior region of the gonad become the germline stem cells (Asaoka and Lin 2004), which divide asymmetrically to produce both another stem cell and a cystoblast. The cystoblasts eventually develop into an egg chamber (King 1970; Zhu and Xie 2003; see Chapter 6).

We are just beginning to understand how the germline stem cells retain their stem cell properties in the gonad (Gilboa and Lehmann 2004). As mentioned in Part Opener III, stem cells must be in a "niche" that supports their proliferation and inhibits their differentiation. Daughter cells that travel outside this niche begin differentiating. In ovaries, the germline stem cells are attached to the stromal cap, where they are maintained by the BMP4-like factor Decapentaplegic (Dpp). Dpp protein represses the gene encoding a transcription factor (Bag-of-marbles) that initiates oogenesis; Dpp cannot reach cells that leave the stromal cap. Without Dpp to repress the *bag-of marbles* gene, the germline cell begins the developmental cascade that produces the 15 nurse cells and the single oocyte (Chen and McKearin 2003; Decotto and Spradling 2005).

The stem cells of the male germ line are connected to "hub" cells that create a stem cell microenvironment by secreting BMP signals as well as the Unpaired protein. Unpaired activates the JAK-STAT pathway in the germline stem cells (see Figure 3.23). If the JAK-STAT signaling pathway is disrupted, the germline stem cells differentiate into spermatogo-

FIGURE 16.11 Migration of germ cells in the *Drosophila* embryo. The left column shows the germ plasm as stained by antibodies to Vasa, a protein component of the germ plasm (D has been counterstained with antibodies to Engrailed protein to show the segmentation; E and F are dorsal views). The right column diagrams the movements of the germ cells. (A) Germ cells originate from the pole plasm at the posterior end of the egg. (B) Passive movements carry the PGCs into the posterior midgut. (C) PGCs move through the endoderm and into the caudal visceral mesoderm by diapedesis. The *wunen* gene product expressed in the endoderm expels the PGCs, while the product of the *columbus* gene expressed in the caudal mesoderm attracts them. (D–F) Movements of the mesoderm bring the PGCs into the region of parasegments 10–12, where the mesoderm coalesces around them to form the gonads. (Photographs from Warrior et al. 1994, courtesy of R. Warrior; diagrams after Howard 1998.)

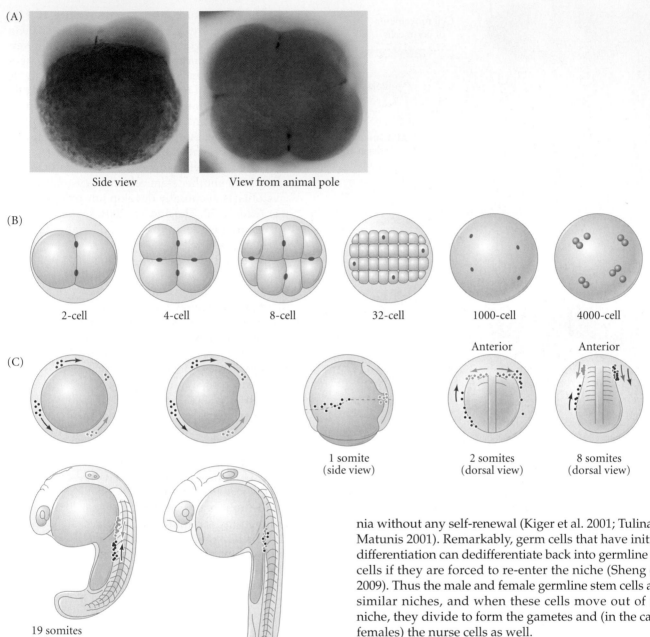

(A)

Side view   View from animal pole

(B)

2-cell   4-cell   8-cell   32-cell   1000-cell   4000-cell

(C)

1 somite
(side view)

Anterior   Anterior

2 somites
(dorsal view)

8 somites
(dorsal view)

19 somites
(side view)

1 day
(side view)

FIGURE 16.12 Specification and migration of germ cells in zebrafish. (A) In situ hybridization of *Vasa* mRNA, showing the accumulation of these messages and germ plasm along the cleavage planes during the first two divisions of the *Danio rerio* embryo. (B,C) Germ plasm movement during zebrafish cleavage (B) and further development (C). The earlier embryos are viewed from the animal pole; other view orientations are noted. The germ plasm remains in four clusters (colored dots). Movement of the clusters is probably caused by both attractive and repulsive chemical interactions. The route to the gonad is directed by the Sdf1 chemoattractant. (A courtesy of N. Hopkins; B,C after Yoon 1997.)

nia without any self-renewal (Kiger et al. 2001; Tulina and Matunis 2001). Remarkably, germ cells that have initiated differentiation can dedifferentiate back into germline stem cells if they are forced to re-enter the niche (Sheng et al. 2009). Thus the male and female germline stem cells are in similar niches, and when these cells move out of their niche, they divide to form the gametes and (in the case of females) the nurse cells as well.

Germ cell migration in vertebrates

ZEBRAFISH Whereas *Drosophila* PGC migration is motivated by both chemoattractants and chemorepellents of the germ cell precursors, zebrafish PGCs arrive at the gonads via chemoattraction. Using the *Vasa* message as a marker, Weidinger and colleagues (1999) detailed the migration of the four clusters of zebrafish PGCs (**Figure 16.12**). These PGC clusters follow different routes, but by the end of the first day of development (at the 1-somite stage), the PGCs are found in two discrete clusters along the border of the trunk mesoderm. From there, they migrate posteriorly into the developing gonad. In zebrafish, the primordial germ cells follow a gradient of the Sdf1 protein that is secreted by the developing gonad. The receptor for this protein is

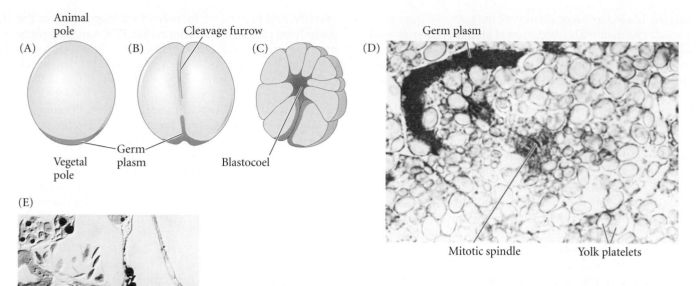

FIGURE 16.13 Migration of *Xenopus* germ plasm. (A–C) Changes in the position of the germ plasm (color) in an early frog embryo. Originally located near the vegetal pole of the uncleaved egg (A), the germ plasm advances along the cleavage furrow (B) until it becomes localized at the floor of the blastocoel (C). (D) A germ plasm-containing cell in the endodermal region of a blastula in mitotic anaphase. Note the germ plasm entering into only one of the two yolk-laden daughter cells. (E) Migration of two primordial germ cells (arrows) along the dorsal mesentery connecting the gut region to the gonadal mesoderm. (A–C after Bounoure 1934; D courtesy of A. Blackler; E from Heasman et al. 1977, courtesy of the authors.)

the CXCR4 protein on the PGC surface (Doitsidou et al. 2002; Knaut et al. 2003). This Sdf1/CXCR4 chemotactic guidance system is known to be important in the migration of lymphocytes and hematopoietic progenitor cells. Loss of either CXCR4 from the PGCs or Sdf1 from the somatic cells results in random migration of the zebrafish primordial germ cells.

FROGS The germ plasm of anuran amphibians (frogs and toads) collects around the vegetal pole in the zygote (see Figure 16.6). During cleavage, this material is brought upward through the yolky cytoplasm. Periodic contractions of the vegetal cell surface appear to push it along the cleavage furrows of the newly formed blastomeres. Germ plasm eventually becomes associated with the endodermal cells lining the floor of the blastocoel (Figure 16.13; Bounoure 1934; Ressom and Dixon 1988; Kloc et al. 1993). The PGCs become concentrated in the posterior region of the larval gut, and as the abdominal cavity forms, they migrate along the dorsal side of the gut, first along the dorsal mesentery (which connects the gut to the region where the mesodermal organs are forming; see Figure 16.13E) and then along the abdominal wall and into the genital ridges. They migrate up this tissue until they reach the developing gonads.

Xenopus PGCs move by extruding a single filopodium and then streaming their yolky cytoplasm into that filopodium while retracting their "tail." Contact guidance

in this migration seems likely, as both the PGCs and the extracellular matrix over which they migrate are oriented in the direction of the migration (Wylie et al. 1979). Furthermore, PGC adhesion and migration can be inhibited if the mesentery is treated with antibodies against *Xenopus* fibronectin (Heasman et al. 1981). Thus, the pathway for germ cell migration in these frogs appears to be composed of an oriented fibronectin-containing extracellular matrix. The fibrils over which the PGCs travel lose this polarity soon after migration has ended. As they migrate, *Xenopus* PGCs divide about three times, so that approximately 30 PGCs will colonize the gonads (Whitington and Dixon 1975; Wylie and Heasman 1993). These cells will divide to form the germ cells. The mechanism by which the *Xenopus* PGCs are directed to the gonad involves a CXCR4 protein on the PGC responding to a Sdf1 ligand along the migration path (Nishiumi et al. 2005; Takeuchi et al. 2010). Knocking out the the CXCR4 mRNA with morpholinos results in fewer PGCs reaching the gonads, and ectopically expressing Sdf1 will misdirect the PGCs into other areas.

MAMMALS Based on differential staining of fixed tissue, it had long been thought that the mouse germ cell precursors migrated from the posterior epiblast into the extraembryonic mesoderm and then back again into the embryo by way of the allantois (see Chiquoine 1954; Mintz 1957). However, the ability to label mouse primordial germ cells with green fluorescent protein and to watch these living cells migrate has led to a reevaluation of the germ cell migration pathway in mammals (Anderson et al. 2000; Molyneaux et al. 2001; Tanaka et al. 2005). First, it appears that mammalian PGCs forming in the posterior epiblast migrate directly into the endoderm from the posterior region of the primitive streak. (Those cells that are seen to enter the allantois are believed to die.) These cells find themselves in the hindgut (**Figure 16.14A**). Although they move actively, they cannot get out of the gut until about embryonic day 9. At that time, the PGCs exit the gut but do not yet migrate toward the genital ridges. By the following day, however, PGCs are seen migrating into the genital ridges (**Figure 16.14B,C**). By embryonic day 11.5, the PGCs enter the developing gonads. During this trek, they have proliferated from an initial population of 10–100 cells to the 2500–5000 PGCs present in the gonads by day 12.

Like the PGCs of *Xenopus*, mammalian PGCs appear to be closely associated with the cells over which they migrate, and they move by extending filopodia over the underlying cell surfaces. Mammalian PGCs are also capable of penetrating cell monolayers and migrating through cell sheets (Stott and Wylie 1986). The mechanism by which these cells know the route of their journey is still unknown. Fibronectin is likely to be an important substrate for PGC migration (ffrench-Constant et al. 1991), and germ cells lacking the integrin receptor for such extracellular matrix proteins cannot migrate to the gonads (Anderson et al. 1999). During the time from their specification to their entrance into the genital ridges, the PGCs are surrounded by cells secreting stem cell factor (SCF). SCF is necessary for PGC motility and survival. Moreover, the cluster of SCF-secreting cells appears to migrate with the PGCs, forming a "traveling niche" of cells that support the persistence and movement of the PGCs (Gu et al. 2009). However, the mechanism by which directionality is provided for migrating toward the gonads remains controversial (see Ara et al. 2003; Molyneaux et al. 2003; Farini et al. 2007; Saga 2008).

BIRDS AND REPTILES In birds and reptiles, the primordial germ cells are derived from epiblast cells that migrate from the central region of the area pellucida to a crescent-shaped zone in the hypoblast at the anterior border of the area pellucida (**Figure 16.15**; Eyal-Giladi et al. 1981; Ginsburg and Eyal-Giladi 1987). This extraembryonic region is called the **germinal crescent**, and the PGCs multiply there.

Unlike those of amphibians and mammals, the PGCs of birds and reptiles migrate to the gonads primarily by means of the bloodstream (**Figure 16.16**). When blood vessels form in the germinal crescent (anterior to the future head region), the PGCs enter those vessels and are carried

(A) Migration of PGCs to endoderm

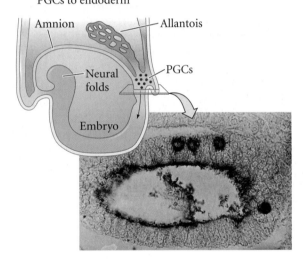

(B) Migration of PGCs into gonad

(C)

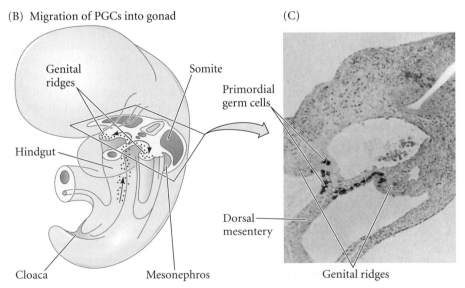

FIGURE 16.14 Primordial germ cell migration in the mouse. (A) On day 8, the PGCs established in the posterior epiblast (see Figure 16.7) migrate into the definitive endoderm of the embryo. The photo shows four large PGCs (stained for alkaline phosphatase) in the hindgut of a mouse embryo. (B) The PGCs migrate through the gut and, dorsally, into the genital ridges. (C) Alkaline phosphatase-staining cells are seen entering the genital ridges around embryonic day 11. (A from Heath 1978; C from Mintz 1957, courtesy of the author.)

(A)

Germinal
crescent

(B)

Germ
cells

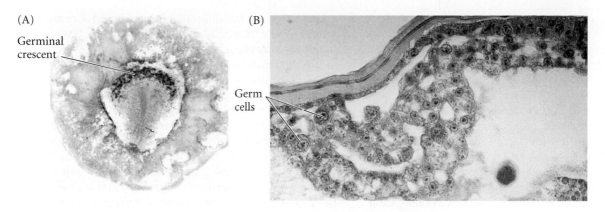

FIGURE 16.15 Germinal crescent of the chick embryo. (A) Germ cells of a stage 4 (definitive primitive streak stage, roughly 18 hours) chick embryo, stained purple for the chick Vasa homologue. The stained cells are confined to the germinal crescent. (B) Higher magnification of the stage 4 germinal crescent region, showing germ cells (stained brown) in the thickened epiblast. (From Tsunekawa et al. 2000, courtesy of N. Tsunekawa.)

by the circulation to the intermediate mesoderm. Here they leave the circulation and migrate into the genital ridges (Swift 1914; Nakamura et al. 2007).

The PGCs of the germinal crescent appear to enter the blood vessels by diapedesis, a type of amoeboid movement common to lymphocytes and macrophages that enables cells to squeeze between the endothelial cells of small blood vessels. In some as-yet-undiscovered way, the PGCs are instructed to exit the blood vessels and enter the gonads (Pasteels 1953; Dubois 1969; Nakamura et al. 2007). Evidence for chemotaxis comes from studies in which circulating chick PGCs were isolated from the blood and cultured between gonadal rudiments and other embryonic tissues (Kuwana et al. 1986). During a 3-hour incubation, the PGCs migrated specifically into the gonadal rudiments.

The molecules that chick PGCs use for chemotaxis may be the same Sdf1/CXCR4 chemotactic system seen in zebrafish. Like mammals, chicks only use chemotaxis during the latter stages of migration. Thus, after they leave the blood vessels, chick PGCs appear to utilize Sdf1 gradients to reach the gonads (Stebler et al. 2004). Indeed, if Sdf1-secreting cells are transplanted into late-stage chick embryos, the PGCs will be attracted to them.

(A)

Primordial
germ cell
(PGC)

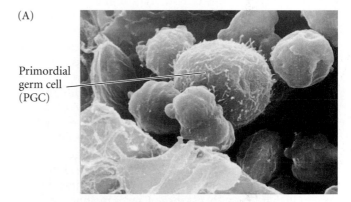

(B)

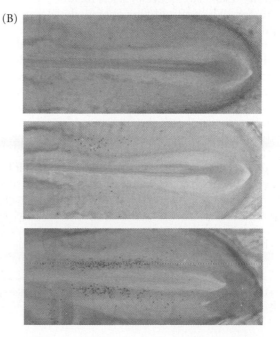

FIGURE 16.16 Migration of primordial germ cells in the chick embryo. (A) Scanning electron micrograph of a chick PGC in a capillary of a gastrulating embryo. Note the larger size of the PGC, as well as the microvilli on its surface. (B) After leaving the blood vessels, PGCs migrate into the intermediate mesodermal region that forms the gonad. These whole mounts show chick PGCs (stained with antibodies against the Vasa protein) in the posterior region of stage 14, 15, and 17 embryos. (A from Kuwana 1993, courtesy of T. Kuwana; B from Nakamura et al. 2007, courtesy of T. Takahiro.)

Meiosis

Meiosis is perhaps the most revolutionary invention of eukaryotes. It is difficult now to appreciate how startling this concept was for biologists at the end of the nineteenth century. Rather than just being a list of Greek names for the different stages of the germ cell cycle, the discovery of meiosis signaled the critical breakthrough for the investigation of inheritance. Van Beneden's 1883 observations that the divisions of germ cells caused the resulting gametes to contain half the diploid number of chromosomes "demonstrated that the chromosomes of the offspring are derived in equal numbers from the nuclei of the two conjugating germ-cells and hence equally from the two parents" (Wilson 1924). All subsequent theories of heredity, including the Sutton-Boveri model that united Mendelism with cell biology, are based on meiosis as the mechanism for sexual reproduction and the transmission of genes from one generation to the next. So let's return to the primordial germ cells that have migrated to the gonads.

Once in the gonad, the PGCs continue to divide mitotically, often producing millions of potential gamete precursors. The germ cells of both male and female gonads are then faced with the necessity of reducing their chromosomes from the diploid to the haploid condition. In the haploid condition, each chromosome is represented by only one copy, whereas diploid cells have two copies of each chromosome. To accomplish this reduction, the germ cells undergo **meiosis** (see Figure 1.5). Meiosis differs from mitosis in that (1) meiotic cells undergo two cell divisions without an intervening period of DNA replication, and (2) homologous chromosomes (each consisting of two sister chromatids joined at a kinetochore) pair together and recombine genetic material. Meiosis is therefore at the center of sexual reproduction. Villeneuve and Hillers (2001) conclude that "the very essence of sex is meiotic recombination." Yet for all its centrality in genetics, development, and evolution, we know surprisingly little about meiosis.

After the germ cell's last mitotic division, a period of DNA synthesis occurs, so that the cell initiating meiosis doubles the amount of DNA in its nucleus. In this state, each chromosome consists of two sister **chromatids** attached at a common kinetochore.* (In other words, the diploid nucleus contains four copies of each chromosome.) Meiosis entails two cell divisions. In the first division, homologous chromosomes (for example, the two copies of chromosome 3 in the diploid cell) come together and are then separated into different cells. Hence, the first meiotic division *splits two homologous chromosomes* between two daughter cells such that each daughter cell has only one copy of the chromosome. But each of the chromosomes has already replicated (i.e., each has two chromatids). The second meiotic division

---

*Although the terms *centromere* and *kinetochore* are often used interchangeably, the *kinetochore* is the complex protein structure that assembles on a sequence of DNA known as the *centromere*.

separates the two sister chromatids from each other. Consequently, each of the four cells produced by meiosis has a single (haploid) copy of each chromosome.

The first meiotic division begins with a long prophase, which is subdivided into five stages. During the **leptotene** (Greek, "thin thread") stage, the chromatin of the chromatids is stretched out very thinly, and it is not possible to identify individual chromosomes. DNA replication has already occurred, however, and each chromosome consists of two parallel chromatids. At the **zygotene** (Greek, "yoked threads") stage, homologous chromosomes pair side by side. This pairing is called **synapsis**, and it is characteristic of meiosis; such pairing does not occur during mitotic divisions. Although the mechanism whereby each chromosome recognizes its homologue is not known (see Barzel and Kupiec 2008), synapsis seems to require the presence of the nuclear membrane and the formation of a proteinaceous ribbon called the **synaptonemal complex**. This complex is a ladder-like structure with a central element and two lateral bars (von Wettstein 1984; Schmekel and Daneholt 1995). The chromatin becomes associated with the two lateral bars, and the chromosomes are thus joined together (**Figure 16.17A,B**).

Examinations of meiotic cell nuclei with the electron microscope (Moses 1968; Moens 1969) suggest that paired chromosomes are bound to the nuclear membrane, and Comings (1968) has suggested that the nuclear envelope helps bring together the homologous chromosomes. The configuration formed by the four chromatids and the synaptonemal complex is referred to as a **tetrad** or a **bivalent**.

During the next stage of meiotic prophase, **pachytene** (Greek, "thick thread"), the chromatids thicken and shorten. Individual chromatids can now be distinguished under the light microscope, and crossing over may occur. **Crossing over** represents an exchange of genetic material whereby genes from one chromatid are exchanged with homologous genes from another. Crossing over may continue into the next stage, **diplotene** (Greek, "double threads"). Here, the synaptonemal complex breaks down, and the two homologous chromosomes start to separate. Usually, however, they remain attached at various points called **chiasmata**, which are thought to represent regions where crossing over is occurring (**Figure 16.17C**). The diplotene stage is characterized by a high level of gene transcription. In some species, the chromosomes of both male and female germ cells take on the "lampbrush" appearance characteristic of chromosomes that are actively making RNA (see below).

During the next stage, **diakinesis** (Greek, "moving apart"), the kinetochores move away from each other, and the chromosomes remain joined only at the tips of the chromatids. This last stage of meiotic prophase ends with the breakdown of the nuclear membrane and the migration of the chromosomes to the **metaphase plate**. Anaphase of meiosis I does not commence until the chromosomes are properly aligned on the mitotic spindle fibers. This alignment is accomplished by proteins that prevent cyclin B

(A)

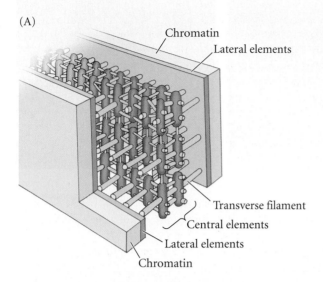

Chromatin
Lateral elements
Transverse filament
Central elements
Lateral elements
Chromatin

(B)

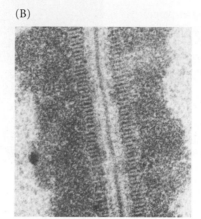

(C)

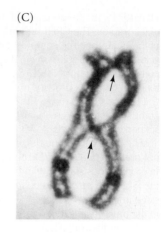

FIGURE 16.17   The synaptonemal complex and recombination. (A) Interpretive diagram of the ladderlike synaptonemal complex structure. (B) Homologous chromosomes held together in a synaptonemal complex during the zygotene phase of the first meiotic prophase in a *Neottiella* (mushroom) oocyte. (C) Chiasmata in diplotene bivalent chromosomes of salamander oocytes. Kinetochores are visible as darkly stained circles; the arrows point to the two chiasmata. (A after Schmekel and Daneholt 1995; B from von Wettstein 1971, courtesy of D. von Wettstein; C courtesy of J. Kezer.)

from being degraded until after all the chromosomes are securely fastened to microtubules. If these proteins are deficient, aneuploidies such as Down syndrome can occur (Homer et al. 2005; Steuerwald et al. 2005).

During anaphase I, the homologous chromosomes are separated from each other in an independent fashion. This stage leads to telophase I, during which two daughter cells are formed, each cell containing one partner of each homologous chromosome pair. After a brief **interkinesis**, the second division of meiosis takes place. During this division, the kinetochore of each chromosome divides during anaphase so that each of the new cells gets one of the two chromatids, the final result being the creation of four haploid cells. Note that meiosis has also reassorted the chromosomes into new groupings. First, each of the four haploid cells has a different assortment of chromosomes. Humans have 23 different chromosome pairs; thus 2^{23} (nearly 10 million) different haploid cells can be formed from the genome of a single person. In addition, the crossing-over that occurs during the pachytene and diplotene stages of prophase I further increases genetic diversity and makes the number of potential different gametes incalculably large.

See WEBSITE 16.4
Human meiosis

Meiosis is the core of sexual reproduction, and it seems to have arisen in the ancestor of fungi, plants, and animals. Indeed, nearly all the genes and proteins used in yeast (fungus) meiosis also function in mammalian meiosis. This observation has allowed the identification of a "core meiotic recombination complex" used by plants, fungi, and animals. This meiotic recombination complex is built on rings of the **cohesin proteins**, which encircle the sister chromatids. The rings of cohesin resist the pulling forces of the spindle microtubules and thereby keep the sister chromatids attached together (Haering et al. 2008; Brar et

al. 2009). This complex recruits another set of proteins that help promote pairing between the homologous chromosomes and allow recombination to occur (Pelttari et al. 2001; Villeneuve and Hillers 2001). These recombination-inducing proteins are involved in making and repairing double-stranded DNA breaks. The cohesin proteins will be degraded in the second meiotic division.

Although the relationship between the synaptonemal complex and the cohesin-recruited recombination complex is not clear, it appears in mammals that the synaptonemal complex stabilizes the associations initiated by the recombination complex, giving a morphological scaffolding to the tenuous protein connections (Pelttari et al. 2001). If the synaptonemal complex fails to form, the germ cells arrest at the pachytene stage and their chromosomes fragment (Figure 16.18). If the murine synaptonemal complex forms but lacks certain proteins, chiasmata formation fails, and the germ cells are often aneuploid (having multiple copies of one or more chromosomes) (Tay and Richter 2001; Yuan et al. 2002). The events of meiosis appear to be coordinated through cytoplasmic connections between the dividing cells. Whereas the daughter cells formed by mitosis routinely separate from each other, the products of the meiotic cell divisions remain coupled to each other by **cytoplasmic bridges**.

Animals can modify meiosis in different ways. Some of the most interesting modifications have occurred in those animal species that have no males. In these species, meiosis is modified such that the resulting gamete is diploid

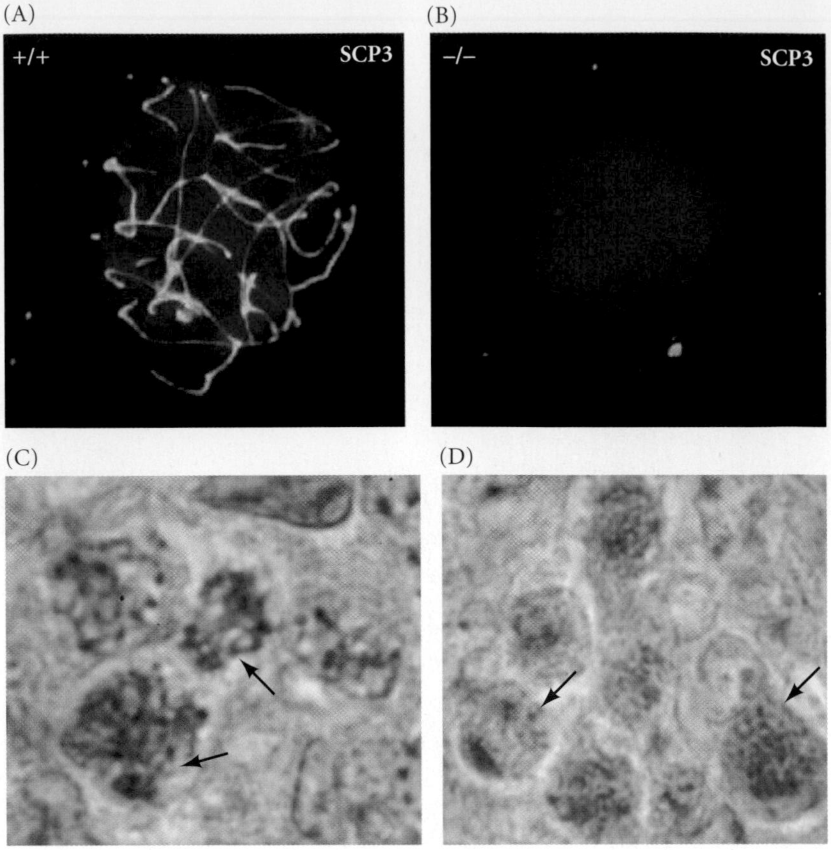

FIGURE 16.18 Importance of the synaptonemal complex. The CPEB protein is a constituent of the synaptonemal complex, and when this gene is knocked out in mice, the synaptonemal complex fails to form. (A,B) Staining for synaptonemal complex proteins in wild-type (A) and CPEB-deficient (B) mice. The synaptonemal complex is stained green, and the DNA is stained blue. The synaptonemal complex is absent in the mutant nuclei, and their DNA is not organized into discrete chromosomes. (C,D) In preparations stained with hematoxylin-eosin, the DNA can be seen to be in the pachytene stage in the wild-type cells (C) and fragmented in the mutant cells (D). (From Tay and Richter 2001, courtesy of J. D. Richter.)

and need not be fertilized to develop. Such animals are said to be **parthenogenetic** (Greek, "virgin birth"). In the fly *Drosophila mangabeirai*, one of the polar bodies (a meiotic cell having very little cytoplasm) acts as a sperm and "fertilizes" the oocyte after the second meiotic division. In some other insects and in the lizard *Cnemidophorus uniparens*, the oogonia further double their chromosome number before meiosis, so that the halving of the chromosomes restores the diploid number. The germ cells of the grasshopper *Pycnoscelus surinamensis* dispense with meiosis altogether, forming diploid ova by two mitotic divisions (Swanson et al. 1981). All of these species consist entirely of females. In other species, haploid parthenogenesis is widely used not only as a means of

reproduction but also as a mechanism of sex determination. In the Hymenoptera (bees, wasps, and ants), unfertilized haploid eggs develop into males, whereas fertilized eggs are diploid and develop into females. The haploid males are able to produce sperm by abandoning the first meiotic division, thereby forming two sperm cells through second meiosis.

SIDELIGHTS & SPECULATIONS

Big Decisions: Mitosis or Meiosis? Sperm or Egg?

In many species, the germ cells migrating into the gonad are bipotential and can differentiate into either sperm or eggs, depending on their gonadal environment. When the ovaries of salamanders are experimentally transformed into testes, the resident germ cells cease their oogenic differentiation and begin developing as sperm (Burns 1930; Humphrey 1931). Similarly, in the housefly and mouse, the gonad is able to direct the differentiation of the germ cells (McLaren

1983; Inoue and Hiroyoshi 1986). Thus, in most organisms, the sex of the gonad and that of its germ cells is the same.

But what about hermaphroditic animals, where the change from sperm production to egg production is a naturally occurring physiological event? How is the same animal capable of producing sperm during one part of its life and oocytes during another part? Using *Caenorhabditis elegans*, Kimble and her colleagues

identified two "decisions" that presumptive germ cells have to make. The first is whether to enter meiosis or to remain a mitotically dividing stem cell. The second is whether to become an egg or a sperm.

There is evidence that these decisions are intimately linked. The mitosis/meiosis decision in *C. elegans* is controlled by a single nondividing cell—the **distal tip cell**—located at the end of each gonad. The germ cell precursors near this cell divide mitotical-

SIDELIGHTS & SPECULATIONS (Continued)

(A)

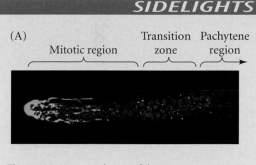

Figure 16.19 Regulation of the mitosis/meiosis decision in the adult germline of *C. elegans*. (A) The distal tip cell and its long projections comprise the stem cell niche. The distal tip of this adult worm is stained green (by attaching the *GFP* gene to the promoter of the *lag2* gene). Germline cells are stained red, and the mitotic region, transition zone, and pachytene region are shown. (B) The balance of mitotic and meiotic regulator proteins determines whether a cell remains in mitosis or enters meiosis. Vertical yellow bars represent the relative number of mitotic nuclei at each position in the mitotic region. The transition from mitosis to meiosis begins where 60% of nuclei have the crescent-shaped morphology typical of early meiotic prophase (red crescents in A). Green bars represent the percentage of nuclei in meiotic prophase at a given position along the distal-proximal axis. Levels of GLP-1 and FBF mitotic regulators are high throughout the mitotic region and decrease dramatically as the germ cells enter meiosis (horizontal yellow bands). Conversely, levels of most meiotic regulators (horizontal green bands) gradually increase in the proximal part of the mitotic region, reaching high levels as germ cells enter meiosis. (One exception is NOS-3, which is distributed uniformly throughout the germ line.) (C) Simplified summary of a network controlling the mitosis/meiosis decision. Notch signaling activates FBF-2. FBF-1 and FBF-2 are very similar proteins whose negative feedback loop may specify the size of the mitotic region, as it negatively regulates levels of GLD-1 and GLD-3, inhibiting meiosis in the distal region of the germ line. (After Kimble and Crittenden 2005; A courtesy of S. Crittenden and J. Kimble.)

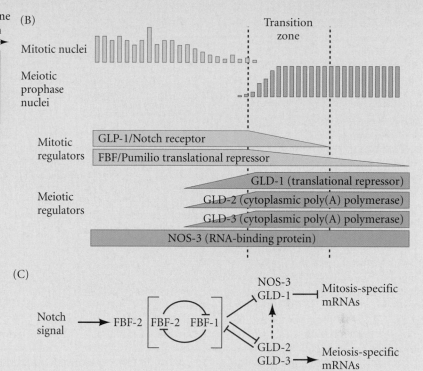

ly, forming the pool of germ cells; but as these cells get farther away from the distal tip cell, they enter meiosis. If the distal tip cell is destroyed by a focused laser beam, all the germ cells enter meiosis; and if the distal tip cell is placed in a different location in the gonad, germline stem cells are gener-

ated near its new position (Kimble 1981; Kimble and White 1981). The distal tip cell extends long filaments that touch the distal germ cells (**Figure 16.19A**). The extensions contain in their cell membranes the LAG-2 protein, a *C. elegans* homologue of Delta (Henderson et al. 1994; Tax et al. 1994; Hall et al. 1999). LAG-2 maintains the germ cells in mitosis and inhibits their meiotic differentiation.

Austin and Kimble (1987) isolated a mutation that mimics the phenotype obtained when the distal tip cell is removed. It is not surprising that this mutation involves the gene encoding GLP-1, the *C. elegans* homologue of Notch—the receptor for Delta. All the germ cell precursors of nematodes homozygous for the recessive mutation of *glp-1* initiate meiosis, leaving no mitotic population. Instead of the 1500 germ cells usually found in the fourth larval stage of hermaphroditic development, these mutants produce only 5–8 sperm cells. When genetic chimeras are made in which wild-type germ cell precursors are found in a mutant larva, the wild-type cells are able to respond to the distal tip cells and undergo mitosis. However, when

mutant germ cell precursors are found in wild-type larvae, they all enter meiosis. Thus, the *glp-1* gene appears to be responsible for enabling the germ cells to respond to the distal tip cell's signal.*

As is usual in development, the binary decision entails both a push and a pull (**Figure 16.19B**). The decision to enter meiosis must be amplified by a decision to end mitosis. This appears to be accomplished by the FBF (*fem-3* mRNA-binding factor) proteins, similar to the *Drosophila* Pumilio RNA-binding protein mentioned in Chapter 6. Notch appears to activate FBFs, which are translational repressors of the GLD (germline development) proteins. GLD-1 (in combination with a Nanos protein) suppresses the translation of mitosis-specific mes-

*The *glp-1* gene appears to be involved in a number of inductive interactions in *C. elegans*. You may recall that GLP-1 is also needed by the AB blastomere for it to receive inductive signals from the EMS blastomere to form pharyngeal muscles (see Chapter 5).

(Continued on next page)

SIDELIGHTS & SPECULATIONS (Continued)

sages. This includes suppressing the translation of *glp-1* mRNA (Eckmann et al. 2002, 2004; Marin and Evans 2003; Kimble and Crittenden 2005). FBF also represses the translation of GLD-2 and GLD-3, two proteins necessary for polyadenylating meiosis-specific mRNAs, allowing them to be translated. Thus, the Notch signal, acting through FBF, simultaneously promotes mitosis and blocks meiosis (**Figure 16.19C**).

After the germ cells begin their meiotic divisions, they still must become either sperm or ova. Generally, in each hermaphrodite gonad (called an ovotestis), the most proximal germ cells produce sperm, while the most distal (near the tip) become eggs (Hirsh et al. 1976). This means that the germ cells entering meiosis early become sperm, and those entering meiosis later become eggs. (It also means that, unlike the situation in vertebrates, the germ cells form in the gonads.) The genetics of this switch are currently being analyzed. The laboratories of Hodgkin (1985) and Kimble (Kimble et al. 1986) have isolated several genes needed for germ cell pathway selection, but the switch appears to involve the activity or inactivity of *fem-3* mRNA. **Figure 16.20** presents a scheme for how these genes might function.

During early development, the *fem* genes, especially *fem-3*, are critical for the specification of sperm cells. Loss-of-function mutations of these genes convert XX *C. elegans* into

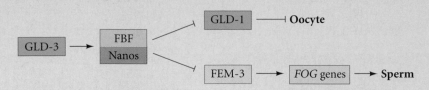

Figure 16.20   Model of sex determination switch in the germ line of *C. elegans* hermaphrodites. FBF and Nanos simultaneously promote oogenesis (by blocking an inhibitor) and inhibit spermatogenesis (by blocking an activator). Expression of the *fbf* and *nanos* genes appears to be regulated by a GLD-3 protein. As long as GLD-3 is made, sperm are produced. After GLD-3 production stops, the germ cells become oocytes. (After Eckmann et al. 2004.)

females (i.e., spermless hermaphrodites). As long as the FEM proteins are made in the germ cells, sperm are produced. FEM protein is thought to activate the *fog* genes (whose loss-of-function mutations cause feminization of the germ line and eliminate spermatogenesis). The *fog* gene products activate the genes involved in transforming the germ cell into sperm and also inhibit those genes that would otherwise direct the germ cells to initiate oogenesis.

Oogenesis can begin only when FEM activity is suppressed. This suppression appears to act at the level of RNA translation. The 3' untranslated region of *fem-3* mRNA contains a sequence that binds a repressor protein during normal development. If

this region is mutated such that the repressor cannot bind, the *fem-3* mRNA remains translatable and oogenesis never takes place. The result is a hermaphrodite body that produces only sperm (Ahringer and Kimble 1991; Ahringer et al. 1992). The *trans*-acting repressor of the *fem-3* message is a combination FBF with the Nanos and Pumilio proteins (the same combination that represses *hunchback* message translation in *Drosophila*).

As in the meiosis/mitosis decision, there are pushes and pulls in the sex-determining pathways. The same Nanos and FBF signal that inhibits the sperm-producing *fem-3* message also inhibits the oocyte-inhibiting *gld-1* message. Thus, the Nanos/FBF signal simultaneously blocks sperm production (by inhibiting an activator) while promoting oocyte production (by inhibiting the inhibitor). The use of the same proteins in both the mitosis/meiosis decision and the sperm/egg decision has allowed Eckmann and colleagues (2004) to speculate that these two pathways evolved from a single original pathway whose function was to regulate the balance between cell growth and cell differentiation.

Gamete Maturation

The regulation of meiosis can differ dramatically between males and females. The egg is usually a nonmotile cell that has conserved its cytoplasm and has stored the ribosomes, mitochondria, and mRNAs needed to initiate development. The sperm is usually a smaller, motile cell that has eliminated most of its cytoplasm to become a nucleus attached to a propulsion system. As we will soon see, there are often large differences between **oogenesis**, the production of eggs, and **spermatogenesis**, the production of sperm. Thus, gametogenesis is more than making the nucleus haploid. The formation of the sperm involves constructing the flagellum and the acrosome. Constructing the

egg involves building the organelles involved in fertilization, synthesizing and positioning the mRNAs and proteins used in early development, and accumulating energy sources and energy-producing organelles (ribosomes, yolk, and mitochondria) in the cytoplasm. A partial catalogue of the materials stored in the oocyte cytoplasm of a frog is shown in **Table 16.1**; a partial list of stored mRNAs found in several organisms was shown in Table 2.2.

The mechanisms of oogenesis vary among species more than those of spermatogenesis. This variation should not be surprising, since patterns of reproduction vary so greatly among species. In some species, such as sea urchins and frogs, the female routinely produces hundreds or thousands of eggs at a time, whereas in other species, such as

TABLE 16.1 Cellular components stored in the mature oocyte of *Xenopus laevis*

Component	Approximate excess over amount in larval cells
Mitochondria	100,000
RNA polymerases	60,000–100,000
DNA polymerases	100,000
Ribosomes	200,000
tRNA	10,000
Histones	15,000
Deoxyribonucleoside triphosphates	2,500

Source: After Laskey 1979.

humans and most other mammals, only a few eggs are produced during an individual's lifetime. In those species that produce thousands of ova each breeding season, the female PGCs produce **oogonia**, self-renewing stem cells that endure for the lifetime of the organism. In those species that produce fewer eggs, the oogonia divide to form a limited number of egg precursor cells.

Maturation of the oocytes in frogs

The eggs of sea urchins, fish, and amphibians are derived from an oogonial stem cell population that can generate a new cohort of oocytes each year. In the frog *Rana pipiens*, oogenesis takes 3 years. During the first 2 years, the oocyte increases its size very gradually. During the third year, however, the rapid accumulation of yolk in the oocyte causes the egg to swell to its characteristically large size (**Figure 16.21**). Eggs mature in yearly batches, with the first cohort maturing shortly after metamorphosis; the next group matures a year later.

VITELLOGENESIS Vitellogenesis—the accumulation of yolk proteins—occurs when the oocyte reaches the diplotene stage of meiotic prophase. Yolk is not a single substance, but a mixture of materials for embryonic nutrition. The major yolk component in frog eggs is a 470-kDa protein called **vitellogenin**. It is not made in the frog oocyte (as are the major yolk proteins of organisms such as annelids and crayfish), but is synthesized in the liver and carried by the bloodstream to the ovary (Flickinger and Rounds 1956; Danilchik and Gerhart 1987).

FIGURE 16.21 Growth of oocytes in the frog. During the first 3 years of life, three cohorts of oocytes are produced. The drawings follow the growth of the first-generation oocytes. (After Grant 1953.)

COMPLETION OF AMPHIBIAN MEIOSIS: PROGESTERONE AND FERTILIZATION Amphibian primary oocytes can remain in the diplotene stage of meiotic prophase for years. This state resembles the G_2 phase of the mitotic cell division cycle. Resumption of meiosis in the amphibian oocyte is thought to require progesterone. This hormone is secreted by the follicle cells in response to gonadotropic hormones secreted by the pituitary gland. Within 6 hours of progesterone stimulation, **germinal vesicle breakdown** (**GVBD**) occurs, the microvilli retract, the nucleoli disintegrate, and the chromosomes contract and migrate to the animal pole to begin division. Soon afterward, the first meiotic division occurs, and the mature ovum is released from the ovary by a process called **ovulation**. The ovulated egg is in second meiotic metaphase when it is released (**Figure 16.22**).

How does progesterone enable the egg to break its dormancy and resume meiosis? To understand the mechanisms by which this activation is accomplished, it is necessary to briefly review the model for early blastomere division (see Chapter 5). Entry into the mitotic (M) phase of the cell cycle (in both meiosis and mitosis) is regulated by **mitosis-promoting factor**, or **MPF** (originally called maturation-promoting factor, after its meiotic function). MPF contains two subunits, **cyclin B** and the **p34** protein. The p34 protein is a cyclin-dependent kinase—its activity is dependent on the presence of cyclin. Since all the components of MPF are present in the amphibian oocyte, it is generally thought that progesterone somehow converts a pre-MPF complex into active MPF.

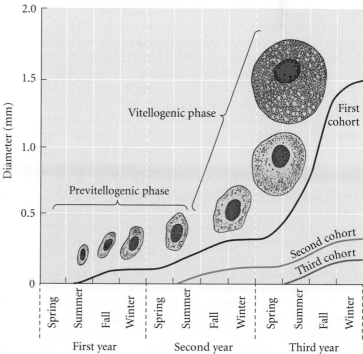

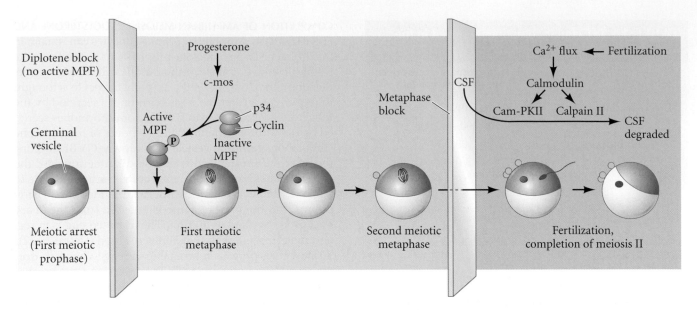

FIGURE 16.22 Schematic representation of *Xenopus* oocyte maturation, showing the regulation of meiotic cell division by progesterone and fertilization. Oocyte maturation is arrested at the diplotene stage of first meiotic prophase by the lack of active MPF. Progesterone activates the production of the c-mos protein. This protein initiates a cascade of phosphorylation that eventually phosphorylates the p34 subunit of MPF, allowing the MPF to become active. The MPF drives the cell cycle through the first meiotic division, but further division is blocked by CSF, a compound containing c-mos, cyclin-dependent kinase 2, and Erp1. CSF inhibits the anaphase-promoting complex from degrading cyclin. Upon fertilization, calcium ions released into the cytoplasm are bound by calmodulin and are used to activate two enzymes, calmodulin-dependent protein kinase II and calpain II, which inactivate and degrade CSF. Second meiosis is completed, and the two haploid pronuclei can fuse. At this time, cyclin B is resynthesized, allowing the first cell cycle of cleavage to begin.

The mediator of the progesterone signal is the **c-mos** protein. Progesterone reinitiates meiosis by causing the egg to polyadenylate the maternal *c-mos* mRNA that has been stored in its cytoplasm (Sagata et al. 1988; Sheets et al. 1995; Mendez et al. 2000). This message is translated into a 39-kDa phosphoprotein. This c-mos protein is detectable only during oocyte maturation and is destroyed quickly upon fertilization. Yet during its brief lifetime, it plays a major role in releasing the egg from its dormancy. The c-mos protein activates a phosphorylation cascade that phosphorylates and activates the p34 subunit of MPF (Ferrell and Machleder 1998; Ferrell 1999). The active MPF allows the germinal vesicle to break down and the chromosomes to divide. If the translation of *c-mos* is inhibited by injecting *c-mos* antisense mRNA into the oocyte, germinal vesicle breakdown and the resumption of oocyte maturation do not occur.

However, oocyte maturation then encounters a second block. MPF can take the chromosomes only through the first meiotic division and prophase of the second meiotic division. The oocyte is arrested once again in the metaphase of the second meiotic division. This metaphase block is caused by **cytostatic factor** (**CSF**; Matsui 1974). CSF is a complex of proteins that includes c-mos, cyclin-dependent kinase 2 (cdk2), MAP kinase, and Erp1 (Gabrielli et al. 1993; Inoue et al. 2007; Nishiyama et al. 2007). Erp1 is the active protein, and it is synthesized immediately after the first meiotic division. The proteins of the CSF complex interact, eventually activating Erp1 by phosphorylating it. Phosphorylated Erp1 blocks the degradation of cyclin by the anaphase-promoting complex (**Figure 16.23**).

This metaphase block is broken by fertilization. The calcium ion flux attending fertilization activates the calcium-binding protein **calmodulin**, and calmodulin, in turn, can activate two enzymes that inactivate CSF. These enzymes are calmodulin-dependent protein kinase II, which inactivates cdk2, and calpain II, a calcium-dependent protease that degrades c-mos (Watanabe et al. 1989; Lorca et al. 1993). This action promotes cell division in two ways. First, without CSF, cyclin can be degraded, and the meiotic division can be completed. Second, calcium-dependent protein kinase II also allows the centrosome to duplicate, thus forming the poles of the meiotic spindle (Matsumoto and Maller 2002). In 1911, Frank Lillie wrote, "The nature of the inhibition that causes the need for fertilization is a most fundamental problem." The solution to that problem appears to be oocyte-derived CSF and the sperm-induced wave of calcium ions.

Gene transcription in amphibian oocytes

The amphibian oocyte has certain periods of very active RNA synthesis. During the diplotene stage, certain chromosomes stretch out large loops of DNA, causing them to

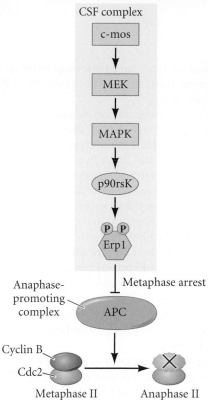

FIGURE 16.23   The main pathway leading to metaphase arrest in the second meiotic division. The CSF protein complex consists of c-mos, three transducer kinases, and the effector protein Erp1. Activation of c-mos activates the kinases, which eventually phosphorylate Erp1. Phophorylated Erp1 binds to and inhibits the anaphase-promoting complex, thus blocking the degradation of cyclin B that would allow the cell to enter anaphase. (After Inoue et al. 2007.)

resemble a lampbrush (which was a handy instrument for cleaning test tubes in the days before microfuges). In situ hybridization reveals these **lampbrush chromosomes** to be sites of RNA synthesis. Oocyte chromosomes can be incubated with a radioactive RNA probe and autoradiography used to visualize the precise locations where genes are being transcribed (**Figure 16.24A**). Electron micrographs of gene transcripts from lampbrush chromosomes also enable one to see chains of mRNA coming off each gene as it is transcribed (**Figure 16.24B**; also see Hill and MacGregor 1980).

In addition to mRNA synthesis, ribosomal RNA and transfer RNA are also transcribed during oogenesis. **Figure 16.25A** shows the pattern of rRNA and tRNA synthesis during *Xenopus* oogenesis. Transcription appears to begin in early (stage I, 25–40 μm) oocytes, during the diplotene stage of meiosis. At this time, all the rRNAs and tRNAs needed for protein synthesis until the mid-blastula stage are made, and all the maternal mRNAs needed for early development are transcribed. This stage lasts for months in *Xenopus*. The rate of rRNA production is prodigious. The *Xenopus* oocyte genome has over 1800 genes encoding 18S and 28S rRNA (the two large RNAs that form the ribosomes), and these genes are selectively amplified such that there are over 500,000 genes making rRNA in the oocyte (**Figure 16.25B**; Brown and Dawid 1968). When the mature (stage VI) oocyte reaches a certain size, its chromosomes condense, and the rRNA genes are no longer tran-

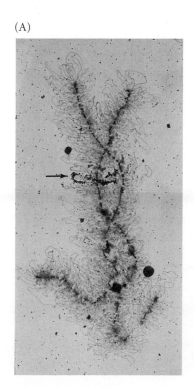

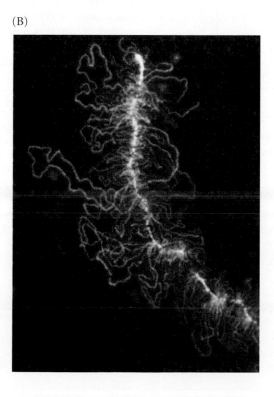

FIGURE 16.24   In amphibian oocytes, lampbrush chromosomes are active in the diplotene germinal vesicle during first meiotic prophase. (A) Autoradiograph of chromosome I of the newt *Triturus cristatus* after in situ hybridization with radioactive histone mRNA. A histone gene (or set of histone genes) is being transcribed (arrow) on one of the loops of this lampbrush chromosome. (B) Lampbrush chromosome of the salamander *Notophthalmus viridescens*. Extended DNA (white) loops out and is transcribed into RNA (red). (A from Old et al. 1977, courtesy of H. G. Callan; B courtesy of M. B. Roth and J. Gall.)

scribed. This "mature oocyte" condition can also last for months. Upon hormonal stimulation, the oocyte completes its first meiotic division and is ovulated. The mRNAs stored by the oocyte now join with the ribosomes to initiate protein synthesis. Within hours, the second meiotic division has begun, and the egg is fertilized in second meiotic metaphase. The embryo's genes do not begin active transcription until the mid-blastula transition (Newport and Kirschner 1982).

As we saw in Chapter 2, the oocytes of several species make two classes of mRNAs—those for immediate use in the oocyte, and those that are stored for use during early development. In frogs, the translation of stored oocyte messages (maternal mRNAs) is initiated by progesterone as the egg is about to be ovulated. One of the results of the MPF activity induced by progesterone may be the phosphorylation of proteins on the 3' UTR of stored oocyte mRNAs. The phosphorylation of these factors is associated with the lengthening of the polyA tails of the stored messages and their subsequent translation (Paris et al. 1991).

Meroistic oogenesis in insects

There are several types of oogenesis in insects, but most studies have focused on those insects (including *Drosophi-*

la and moths) that undergo **meroistic oogenesis**, in which cytoplasmic connections remain between the cells produced by the oogonium.

The oocytes of meroistic insects do not pass through a transcriptionally active stage, nor do they have lampbrush chromosomes. Rather, RNA synthesis is largely confined to the nurse cells, and the RNA made by those cells is actively transported into the oocyte cytoplasm (see Figure 6.7). Oogenesis takes place in only 12 days, so the nurse cells are metabolically very active during this time. Nurse cells are aided in their transcriptional efficiency by becoming polytene—instead of having two copies of each chromosome, they replicate their chromosomes until they have produced 512 copies. The 15 nurse cells pass ribosomal and messenger RNAs as well as proteins into the oocyte cytoplasm, and entire ribosomes may be transported as well. The mRNAs do not associate with polysomes, and they are not immediately active in protein synthesis (Paglia et al. 1976; Telfer et al. 1981).

The meroistic ovary confronts us with some interesting problems. If all 16 cystocytes derived from the PGC are connected so that proteins and RNAs can shuttle freely among them, how do 15 cystocytes become RNA-producing nurse cells while one cell is fated to become the oocyte? Why is the flow of protein and RNA in one direction only?

As the cystocytes divide, a large, spectrin-rich structure called the **fusome** forms and spans the ring canals between the cells (see Figure 16.4A). It is constructed asymmetrically, as it always grows from the spindle pole that remains in one of the cells after the first division (Lin and Spradling 1995; de Cuevas and Spradling 1998). The cell that retains the greater part of the fusome during the first division becomes the oocyte. It is not yet known if the fusome contains oogenic determinants, or if it directs the traffic of materials into this particular cell.

(A)

FIGURE 16.25 Ribosomal and transfer RNA production in *Xenopus* oocytes. (A) Relative rates of DNA, tRNA, and rRNA synthesis in amphibian oogenesis during the last 3 months before ovulation. (B) Transcription of the large RNA precursor of the 28S, 18S, and 5.8S ribosomal RNAs. These units are tandemly linked together, with some 450 per haploid genome. (A after Gurdon 1976; B courtesy of O. L. Miller, Jr.)

(B)

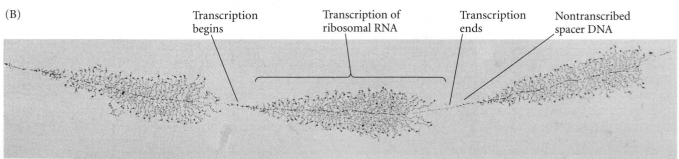

Transcription begins

Transcription of ribosomal RNA

Transcription ends

Nontranscribed spacer DNA

TABLE 16.2 Sexual dimorphism in mammalian meioses

Female oogenesis	Male spermatogenesis
Meiosis initiated once in a finite population of cells	Meiosis initiated continuously in a mitotically dividing stem cell population
One gamete produced per meiosis	Four gametes produced per meiosis
Completion of meiosis delayed for months or years	Meiosis completed in days or weeks
Meiosis arrested at first meiotic prophase and reinitiated in a smaller population of cells	Meiosis and differentiation proceed continuously without cell cycle arrest
Differentiation of gamete occurs while diploid, in first meiotic prophase	Differentiation of gamete occurs while haploid, after meiosis ends
All chromosomes exhibit equivalent transcription and recombination during meiotic prophase	Sex chromosomes excluded from recombination and transcription during first meiotic prophase

Source: Handel and Eppig 1998.

Once the patterns of transport are established, the cytoskeleton becomes actively involved in transporting mRNAs from the nurse cells into the oocyte cytoplasm (Cooley and Theurkauf 1994). An array of microtubules that extends through the ring canals (see Figure 16.4C) is critical for oocyte determination. In the nurse cells, the Exuperantia protein binds *bicoid* message to the microtubules and transports it to the anterior of the oocyte (Cha et al. 2001; see Chapter 6). If the microtubular array is disrupted (either chemically or by mutations such as *bicaudal-D* or *egalitarian*), the nurse cell gene products are transmitted in all directions and all 16 cells differentiate into nurse cells (Gutzeit 1986; Theurkauf et al. 1992, 1993; Spradling 1993).

The Bicaudal-D and Egalitarian proteins are probably core components of a dynein motor system that transports mRNAs and proteins throughout the oocyte (Bullock and Ish-Horowicz 2001). It is possible that some compounds transported from the nurse cells into the oocyte become associated with transport proteins such as dynein and kinesin, which would enable them to travel along the tracks of microtubules extending through the ring canals (Theurkauf et al. 1992; Sun and Wyman 1993). The *oskar* message, for instance, is linked to kinesin through the Barentsz protein, and kinesin can transport the *oskar* message to the posterior of the oocyte (van Eeden et al. 2001; see Figure 6.7).

Actin may become important for maintaining the polarity of transport during later stages of oogenesis. Mutations that prevent actin microfilaments from lining the ring canals prevent the transport of mRNAs from the nurse cells to the oocyte, and disruption of the actin microfilaments randomizes the distribution of mRNA (Cooley et al. 1992; Watson et al. 1993). Thus, the cytoskeleton controls the movement of organelles and RNAs between nurse cells and oocyte such that developmental cues are exchanged only in the appropriate direction.

Gametogenesis in Mammals

As outlined in **Table 16.2**, there are profound differences between spermatogenesis and oogenesis in mammals. One of the fundamental differences concerns the timing of meiosis onset. In females, meiosis begins in the embryonic gonads; in males, meiosis is not initiated until puberty. This critical difference in timing is due to retinoic acid (RA) produced by the mesonephric kidneys (**Figure 16.26**). This RA stimulates the germ cells to undergo a new round of DNA replication and initiate meiosis (Baltus et al. 2006; Bowles et al. 2006; Lin et al. 2008). In males, however, the embryonic testes secrete the RA-degrading enzyme Cyp26b1. This prevents RA from promoting meiosis. Later, Nanos2 will be expressed in the male germ cells, and this will also prevent meiosis and ensure that the cells follow the pathway to become sperm (Koubova et al. 2006; Suzuki and Saga 2008).

VADE MECUM Gametogenesis in mammals

Spermatogenesis

Once mammalian PGCs arrive at the genital ridge of a male embryo, they are called **gonocytes** and become incorporated into the sex cords (Culty 2009). They remain there until maturity, at which time the sex cords hollow out to form the **seminiferous tubules**. The epithelium of the tubules differentiates into the **Sertoli cells** that will nourish and protect the developing sperm cells. The gonocytes differentiate into a population of stem cells that have recently been named the **undifferentiated type A spermatogonia** (Yoshida et al. 2007). These cells can reestablish spermatogenesis when transferred into mice whose sperm production was eliminated by toxic chemicals. They appear to reside in stem cell niches created by the junction of Sertoli cells, interstitial (testosterone-producing) cells, and blood vessels.

FIGURE 16.26 Retinoic acid (RA) determines the timing of meiosis and sexual differentiation of mammalian germ cells. (A) In female mouse embryos, RA secreted from the mesonephros reaches the gonad and triggers meiotic initiation via the induction of Stra8 transcription factor in female germ cells (beige). However, if activated *Nanos2* genes are added to female germ cells, they suppress Stra8 expression, leading the germ cells into a male pathway (gray). (B) In embryonic testes, Cyp26b1 blocks RA signaling, thereby preventing male germ cells from initiating meiosis until embryonic day 13.5 (left panel). After embryonic day 13.5, when Cyp26b1 expression is decreased, Nanos2 is expressed and prevents meiotic initiation by blocking Stra8 expression. This induces male-type differentiation in the germ cells (right panel). (C,D) Day 12 mouse embryos stained for mRNAs encoding the RA-synthesizing enzyme Aldh1a2 (C) and the RA-degrading enzyme Cyp26b1 (D). The RA-synthesizing enzyme is seen in the mesonephros of both the male and female; the RA-degrading enzyme is seen only in the male gonad. (A,B from Saga 2008; C,D from Bowles et al. 2006, courtesy of P. Koopman.)

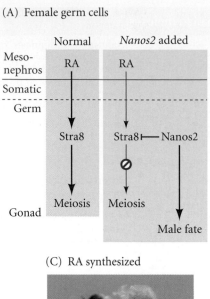

(A) Female germ cells

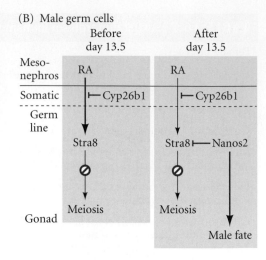

(B) Male germ cells

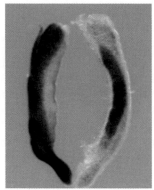

(C) RA synthesized

Male   Female

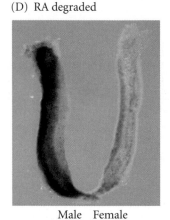

(D) RA degraded

Male   Female

The decision to proliferate or differentiate may involve interactions between the Wnt and BMP pathways. Wnt signaling appears to promote proliferation of stem cells, and the spermatogonia appear to have receptors for both Wnts and BMPs (Golestaneh et al. 2009). The initiation of spermatogenesis during puberty is probably regulated by the synthesis of BMPs by the spermatogenic germ cells, the spermatogonia. When BMP8b reaches a critical concentration, the germ cells begin to differentiate. The differentiating cells produce high levels of BMP8b, which can then further stimulate their differentiation. Mice lacking BMP8b do not initiate spermatogenesis at puberty (Zhao et al. 1996).

The spermatogenic germ cells are bound to the Sertoli cells by N-cadherin molecules on the surfaces of both cell types, and by galactosyltransferase molecules on the spermatogenic cells that bind a carbohydrate receptor on the Sertoli cells (Newton et al. 1993; Pratt et al. 1993). Spermatogenesis—the developmental pathway from germ cell to mature sperm—occurs in the recesses between the Sertoli cells (**Figure 16.27**).

FORMING THE HAPLOID SPERMATID The undifferentiated type A_1 spermatogonia (sometimes called the dense type

A spermatogonia) are found adjacent to the outer basal lamina of the sex cords. They are stem cells, and upon reaching maturity are thought to divide to make another type A_1 spermatogonium as well as a second, paler type of cell, the type A_2 spermatogonia. The A_2 spermatogonia divide to produce type A_3 spermatogonia, which then beget type A_4 spermatogonia. The A_4 spermatogonia are thought to differentiate into the first committed stem cell type, the **intermediate spermatogonia**. Intermediate spermatogonia are committed to becoming spermatozoa, and they divide mitotically once to form **type B spermatogonia** (see Figure 16.27). These cells are the precursors of the spermatocytes and are the last cells of the line that undergo mitosis. They divide once to generate the **primary spermatocytes**—the cells that enter meiosis.

The transition between spermatogonia and spermatocytes appears to be mediated by the opposing influences of glial cell line-derived neurotrophic factor (GDNF) and stem cell factor (SCF), both of which are secreted by the Sertoli cells. GDNF levels determine whether the dividing spermatogonia remain spermatogonia or enter the pathway to become spermatocytes. Low levels of GDNF favor the differentiation of the spermatogonia, whereas high levels favor self-renewal of the stem cells (Meng et al. 2000).

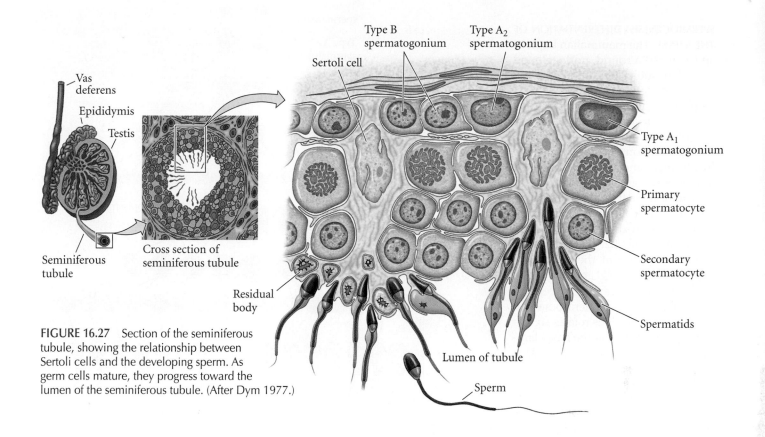

Type B
spermatogonium

Type A$_2$
spermatogonium

Sertoli cell

Type A$_1$
spermatogonium

Primary
spermatocyte

Secondary
spermatocyte

Spermatids

Lumen of tubule

Sperm

Residual
body

Vas
deferens

Epididymis

Testis

Seminiferous
tubule

Cross section of
seminiferous tubule

FIGURE 16.27 Section of the seminiferous tubule, showing the relationship between Sertoli cells and the developing sperm. As germ cells mature, they progress toward the lumen of the seminiferous tubule. (After Dym 1977.)

SCF promotes the transition to spermatogenesis (Rossi et al. 2000). Since both GDNF and SCF are upregulated by follicle-stimulating hormone (FSH), these two factors may serve as a link between the Sertoli cells and the endocrine system, and they provide a mechanism for FSH to instruct the testes to produce more sperm (Tadokoro et al. 2002). Keeping the stem cells in equilibrium—producing neither too many undifferentiated cells nor too many differentiated cells—is not easy. Mice with the *luxoid* mutation are sterile because they lack a transcription factor that regulates this division. All their spermatogonia become sperm at once, leaving the testes devoid of stem cells (Buaas et al. 2004; Costoya et al. 2004).

Looking at **Figure 16.28**, we find that during the spermatogonial divisions, cytokinesis is not complete. Rather, the cells form a syncytium in which each cell communicates with the others via cytoplasmic bridges about 1 μm in diameter (Dym and Fawcett 1971). The successive divisions produce clones of interconnected cells, and because ions and molecules readily pass through these cytoplasmic bridges, each cohort matures synchronously. During this time, the spermatocyte nucleus often transcribes genes whose products will be used later to form the axoneme and acrosome.

Each primary spermatocyte undergoes the first meiotic division to yield a pair of **secondary spermatocytes**, which complete the second division of meiosis. The haploid cells thus formed are called **spermatids**, and they are still connected to one another through their cytoplasmic bridges. The spermatids that are connected in this manner have haploid nuclei but are functionally diploid, since a gene product made in one cell can readily diffuse into the cytoplasm of its neighbors (Braun et al. 1989).

During the divisions from type A$_1$ spermatogonia to spermatids, the cells move farther and farther away from the basal lamina of the seminiferous tubule and closer to its lumen (see Figure 16.27; Siu and Cheng 2004). Thus, each type of cell can be found in a particular layer of the tubule. The spermatids are located at the border of the lumen, and here they lose their cytoplasmic connections and differentiate into spermatozoa. In humans, the progression from spermatogonial stem cell to mature spermatozoa takes 65 days (Dym 1994).

The processes of spermatogenesis require a very specialized network of gene expression (Sassone-Corsi 2002). Not only are histones substantially remodeled and replaced by sperm-specific variants (see below), but even the basal RNA polymerase II transcription factors are exchanged for sperm-specific variants. The TFIID complex, which contains the TATA-binding protein and 14 TAFs, functions in the recognition of RNA polymerase. One of these TAFs, TAF4b, is a sperm-specific TAF required for mouse spermatogenesis (Falender et al. 2005). Without this factor, the spermatogonial stem cells fail to make Ret (the receptor for GDNF) or the luxoid transcription factor, and spermatogenesis fails to occur.

SPERMIOGENESIS: DIFFERENTIATION OF THE SPERM The mammalian haploid spermatid is a round, unflagellated cell that looks nothing like the mature vertebrate sperm. The next step in sperm maturation, then, is **spermiogenesis** (or **spermateliosis**), the differentiation of the sperm cell. For fertilization to occur, the sperm has to meet and bind with an egg, and spermiogenesis prepares the sperm for these functions of motility and interaction. The process of mammalian sperm differentiation was shown in Figure 4.2. The first step is the construction of the acrosomal vesicle from the Golgi apparatus. The acrosome forms a cap that covers the sperm nucleus. As the acrosomal cap is formed, the nucleus rotates so that the cap will be facing the basal lamina of the seminiferous tubule. This rotation is necessary because the flagellum, which is beginning to form from the centriole on the other side of the nucleus, will extend into the lumen. During the last stage of spermiogenesis, the nucleus flattens and condenses, the remaining cytoplasm (the residual body, or "cytoplasmic droplet") is jettisoned, and the mitochondria form a ring around the base of the flagellum.

During spermiogenesis, the histones of the spermatogonia are often replaced by histone variants, and widespread nucleosome dissociation takes place. This remodeling of nucleosomes might also be the point at which the PGC pattern of methylation is removed and the male genome-specific pattern of methylation is established on the sperm DNA (see Wilkins 2005). As spermiogenesis ends, the histones of the haploid nucleus are eventually replaced by protamines.* This replacement results in the complete shutdown of transcription in the nucleus and facilitates the nucleus assuming an almost crystalline structure (Govin et al. 2004). The resulting sperm then enter the lumen of the seminiferous tubule.

See WEBSITE 16.5 The Nebenkern

---

*Protamines are relatively small proteins that are over 60% arginine. Transcription of the genes for protamines is seen in the early haploid spermatids, although translation is delayed for several days (Peschon et al. 1987). The replacement, however, is not complete, and "activating" nucleosomes, having trimethylated H3K4, cluster around developmentally significant loci, including Hox gene promoters, certain microRNAs, and imprinted loci that are paternally expressed (Hammoud et al. 2009).

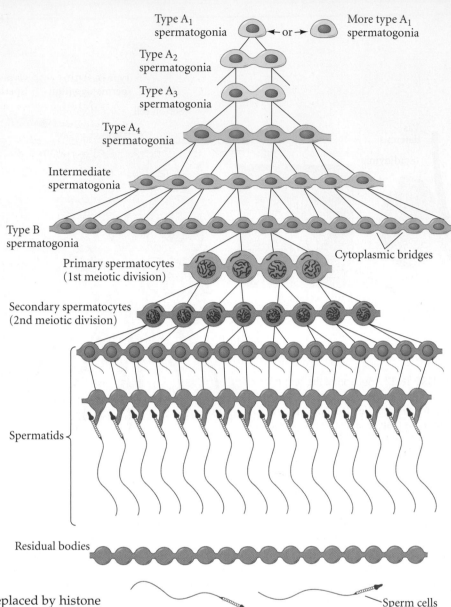

FIGURE 16.28 Formation of syncytial clones of human male germ cells. (After Bloom and Fawcett 1975.)

In the mouse, development from stem cell to spermatozoon takes 34.5 days: the spermatogonial stages last 8 days, meiosis lasts 13 days, and spermiogenesis takes another 13.5 days. Human sperm development takes nearly twice as long. Because the original type A spermatogonia are stem cells, spermatogenesis can occur continuously. Each day, some 100 million sperm are made in each human testicle, and each ejaculation releases 200 million sperm. Unused sperm are either resorbed or passed out of the body in urine. During his lifetime, a human male can produce 10^{12} to 10^{13} sperm (Reijo et al. 1995).

Oogenesis

In the human embryo, the thousand or so oogonia divide rapidly from the second to the seventh month of gestation

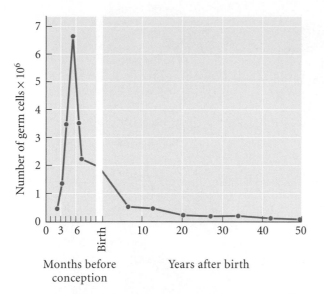

FIGURE 16.29 The number of germ cells in the human ovary changes over the life span. (After Baker 1970.)

to form roughly 7 million germ cells (**Figure 16.29**). After the seventh month of embryonic development, however, the number of germ cells drops precipitously. Most oogonia die during this period, while the remaining oogonia enter the first meiotic division (Pinkerton et al. 1961). These latter cells, called **primary oocytes**, progress through the first meiotic prophase until the diplotene stage, at which point they are maintained until the female matures. With the onset of puberty, groups of oocytes periodically resume meiosis. Thus, in the human female, the first part of meiosis begins in the embryo, and the signal to resume meiosis is not given until roughly 12 years later. In fact, some oocytes are maintained in meiotic prophase for nearly 50 years. As Figure 16.29 illustrates, primary oocytes continue to die. Of the millions of primary oocytes present at her birth, only about 400 mature during a woman's lifetime.*

See **WEBSITE 16.6 Synthesizing oocyte ribosomes**

OOGENIC MEIOSIS Oogenic meiosis differs from spermatogenic meiosis in its placement of the metaphase plate. When the primary oocyte divides, its nucleus, called the

*It has long been thought that the number of oocytes in a female mammal (including humans) is established during embryogenesis and never increases (Zuckerman 1951). Radioactive labeling of oocyte nuclei also supported the view that the number of oocytes is fixed during embryonic life (see Telfer 2004). Recently, however, Zou and colleagues (2008) have claimed to find female germline stem cells in the ovaries of adult mice. Such a finding would mean that the loss of female fertility in mammals might not be due solely to the aging of oocytes but also to the depletion of the PGC stem cells (Oktem and Oktay 2009). Thus, the ovary may have some regenerative abilities that remain unexplored.

germinal vesicle, breaks down, and the metaphase spindle migrates to the periphery of the cell. At telophase, one of the two daughter cells contains hardly any cytoplasm, whereas the other cell retains nearly the entire volume of cellular constituents (**Figure 16.30**). The smaller cell is called the **first polar body**, and the larger cell is referred to as the **secondary oocyte**. During the second division of meiosis, a similar unequal cytokinesis takes place. Most of the cytoplasm is retained by the mature egg (the ovum), and a second polar body receives little more than a haploid nucleus. (The first polar body usually does not divide.) Thus, oogenic meiosis conserves the volume of oocyte cytoplasm in a single cell rather than splitting it equally among four progeny.

MATURATION OF THE MAMMALIAN OOCYTE Ovulation in mammals follows one of two patterns, depending on the species. One type of ovulation is stimulated by the act of copulation. Physical stimulation of the cervix triggers the release of gonadotropins from the pituitary. These gonadotropins signal the egg to resume meiosis and initiate the events that will expel it from the ovary. This mechanism ensures that most copulations will result in fertil-

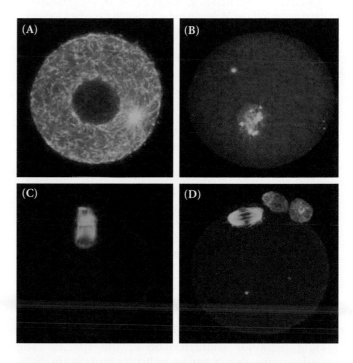

FIGURE 16.30 Meiosis in the mouse oocyte. The tubulin of the microtubules is stained green; the DNA is stained blue. (A) Mouse oocyte in meiotic prophase. The large haploid nucleus (the germinal vesicle) is still intact. (B) The nuclear envelope of the germinal vesicle breaks down as metaphase begins. (C) Meiotic anaphase I, wherein the spindle migrates to the periphery of the egg and releases a small polar body. (D) Meiotic metaphase II, wherein the second polar body is given off (the first polar body has also divided). (From De Vos 2002, courtesy of L. De Vos.)

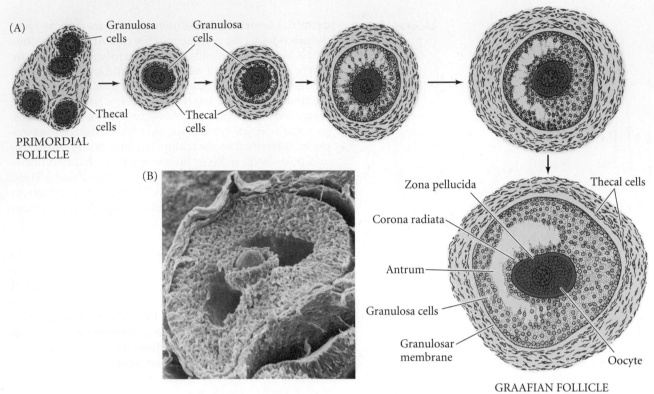

(A) Granulosa cells / Granulosa cells / Thecal cells / Thecal cells / PRIMORDIAL FOLLICLE

(B)

Zona pellucida / Thecal cells / Corona radiata / Antrum / Granulosa cells / Granulosar membrane / Oocyte / GRAAFIAN FOLLICLE

FIGURE 16.31 The ovarian follicle of mammals. (A) Maturation of the ovarian follicle. When mature, it is often called a Graafian follicle. (B) Scanning electron micrograph of a mature follicle in the rat. The oocyte (center) is surrounded by the smaller granulosa cells that will make up the cumulus. (A after Carlson 1981, B courtesy of P. Bagavandoss.)

ized ova, and animals that utilize this method of ovulation—such as rabbits and minks—have a reputation for procreative success.

Most mammals, however, have a periodic ovulation pattern, in which the female ovulates only at specific times of the year. This ovulatory period is called **estrus** (or its English equivalent, *heat*). In these animals, environmental cues (most notably the amount and type of light during the day) stimulate the hypothalamus to release **gonadotropin-releasing hormone (GRH)**. GRH stimulates the pituitary to release the gonadotropins—**follicle-stimulating hormone (FSH)** and **luteinizing hormone (LH)**—that cause the ovarian follicle cells to proliferate and secrete estrogen. Estrogen enters certain neurons and evokes the pattern of mating behavior characteristic of the species. The gonadotropins also stimulate follicular growth and initiate ovulation. Thus, mating behavior and ovulation occur close together.

Humans have a variation on the theme of periodic ovulation. Although human females have cyclical ovulation (averaging about once every 29.5 days) and no definitive yearly estrus, most of human reproductive physiology is shared with other primates. The characteristic primate periodicity in maturing and releasing ova is called the **men-**

strual cycle because it entails the periodic shedding of blood and endothelial tissue from the uterus at monthly intervals.* The menstrual cycle represents the integration of three very different cycles:

1. The **ovarian cycle**, the function of which is to mature and release an oocyte.
2. The **uterine cycle**, the function of which is to provide the appropriate environment for the developing blastocyst.
3. The **cervical cycle**, the function of which is to allow sperm to enter the female reproductive tract only at the appropriate time.

These three functions are integrated through the hormones of the pituitary, hypothalamus, and ovary.

---

*The periodic shedding of the uterine lining is a controversial topic. Some scientists speculate that menstruation is an active process, with adaptive significance in evolution. Profet (1993) and Howes (2010) have argued that menstruation is a crucial immunological adaptation, protecting the uterus against infections contracted from semen or other environmental agents. Strassmann (1996) suggested that cyclicity of the endometrium is an energy-saving adaptation in times of poor nutrition, and that vaginal bleeding would be a side effect of this adaptive process. Finn (1998) claimed that menstruation has no adaptive value but is necessitated by the immunological crises that are a consequence of bringing two genetically dissimilar organisms together in the uterus. Martin (1992) pointed out that it might be wrong to think of there being a single function of menstruation—its role might change during a woman's life cycle.

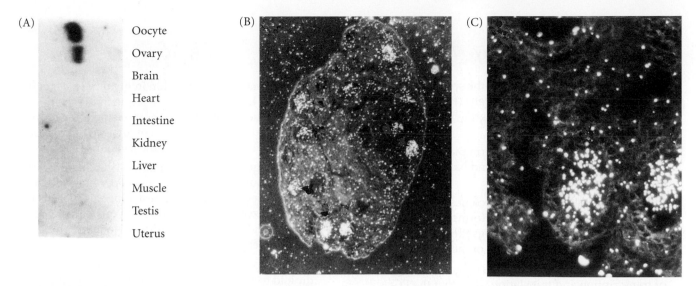

FIGURE 16.32 Expression of the *ZP3* gene in the developing mouse oocyte. (A) Northern blot of *ZP3* mRNA accumulation in the tissues of a 13-day mouse embryo. A radioactive probe to the *ZP3* message found it expressed only in the ovary, and specifically in the oocytes. (B) When the luciferinase reporter gene is placed onto the *ZP3* promoter and inserted into the mouse genome, *luciferinase* message is seen only in the developing oocytes of the ovary. (C) Higher magnification of a section of (B), showing two of the ovarian follicles containing maturing oocytes. (A from Roller et al. 1989; B,C from Lira et al. 1990; photographs courtesy of P. Wassarman.)

The majority of the oocytes in the adult human ovary are maintained in the diplotene stage of the first meiotic prophase, often referred to as the **dictyate state**. Each oocyte is enveloped by a primordial follicle consisting of a single layer of epithelial granulosa cells and a less organized layer of mesenchymal thecal cells (**Figure 16.31**). Periodically, a group of primordial follicles enters a stage of follicular growth. During this time, the oocyte undergoes a 500-fold increase in volume (corresponding to an increase in oocyte diameter from 10 μm in a primordial follicle to 80 μm in a fully developed follicle).

See WEBSITE 16.7
Hormones and mammalian egg maturation

See WEBSITE 16.8
The reinitiation of mammalian meiosis

Concomitant with oocyte growth is an increase in the number of **granulosa cells**, which form concentric layers around the oocyte. This proliferation of granulosa cells is mediated by a paracrine factor, GDF9, a member of the TGF-β family (Dong et al. 1996). Throughout this growth period, the oocyte remains in the dictyate stage. The fully grown follicle thus contains a large oocyte surrounded by several layers of granulosa cells. The innermost of these cells will stay with the ovulated egg, forming the **cumulus**, which surrounds the egg in the oviduct. In addition, during the growth of the follicle, an **antrum** (cavity) forms

and becomes filled with a complex mixture of proteins, hormones, and other molecules.

Just as the maturing oocyte synthesizes paracrine factors that allow the follicle cells to proliferate, the follicle cells secrete growth and differentiation factors (TGF-β2, VEGF, leptin, Fgf2) that allow the oocyte to grow and bring blood vessels into the follicular region (Antczak et al. 1997). The oocytes are maintained in the dictyate stage by the ovarian follicle cells. The release from this dictyate stage and the reinitiation of meiosis are driven by lutenizing hormone from the pituitary. The pituitary hormone is received by the granulosa cells of the ovary, and the granulosa cells send a paracrine or juxtacrine signal that induces activation of mitosis-promoting factor (MPF) in the oocyte (Eppig et al. 2004; Mehlmann et al. 2004).

Meanwhile, the growing oocyte is actively transcribing genes whose products are necessary for cell metabolism, for oocyte-specific processes, and for early development before the zygote-derived nuclei begin to function. In mice, for instance, the growing diplotene oocyte is actively transcribing the genes for zona pellucida proteins ZP1, ZP2, and ZP3 (see Figure 4.31). Moreover, these genes are transcribed only in the oocyte and not in any other cell type, as the proteins essential for fertilization are being synthesized (**Figure 16.32**; Roller et al. 1989; Lira et al. 1990; Epifano et al. 1995).

The fertilizable mammalian oocyte is arrested in second meiotic metaphase by MPF. As in amphibian oocytes, a cytostatic factor stabilizes the cyclin B-Cdc2 MPF dimers. Shoji and colleagues (2006) showed that loss of the cytostatic factor occurred upon egg activation.

Coda

We are now back where we began: the stage is set for fertilization to take place. The egg and the sperm will both die if they do not meet. As F. R. Lillie recognized in 1919, "The elements that unite are single cells, each on the point of death; but by their union a rejuvenated individual is formed, which constitutes a link in the eternal process of Life."

Snapshot Summary: *The Saga of the Germ Line*

1. The precursors of the gametes—the sperm and eggs—are the primordial germ cells (PGCs). They form outside the gonads and migrate into the gonads during development.

2. In many species, a distinctive germ plasm exists. It often contains the Oskar, Vasa, and Nanos proteins or the mRNAs encoding them.

3. In *Drosophila*, the germ plasm becomes localized in the posterior of the embryo and forms pole cells, the precursors of the gametes. In frogs, the germ plasm originates in the vegetal portion of the oocyte.

4. The germ plasm in many species contains inhibitors of transcription and translation, such that the PGCs derived from them are thought to be both translationally and transcriptionally silent.

5. In amphibians, the germ cells migrate on fibronectin matrices from the posterior larval gut to the gonads. In mammals, a similar migration is seen, and fibronectin pathways may also be used. Stem cell factor (SCF) is critical in this migration, and the germ cells proliferate as they travel.

6. In birds, the germ plasm is first seen in the germinal crescent. The germ cells migrate through the blood, then leave the blood vessels and migrate into the genital ridges.

7. In zebrafish, the germ cell determinants enter specific cells that are attracted to the gonad by a gradient of chemoattractants such as the Sdf1 protein.

8. Germ cell migration in *Drosophila* occurs in several steps involving transepithelial migration, repulsion from the endoderm, and attraction to the gonads.

9. Once the germ cells reach the gonads, they may initiate meiosis. The timing and details of this process depend on the species and sex of the organism. In humans and mice, germ cells entering ovaries initiate meiosis while in the embryo; germ cells entering testes do not initiate meiosis until puberty.

10. Before meiosis, the DNA is replicated and the resulting sister chromatids remain bound at the kinetochore. Homologous chromosomes are connected through the synaptonemal complex.

11. The first division of meiosis separates the homologous chromosomes. The second division of meiosis splits the kinetochore and separates the chromatids.

12. The meiosis/mitosis decision in nematodes is regulated by a Delta protein homologue in the membrane of the distal tip cell. The decision for a germ cell to become either a sperm or an egg is regulated at the level of translation of the *fem-3* message.

13. Spermatogenic meiosis in mammals is characterized by the production of four gametes per meiosis and by the absence of meiotic arrest. Oogenic meiosis is characterized by the production of one gamete per meiosis and by an arrest at first meiotic prophase to allow the egg to grow.

14. In some species, meiosis is modified such that a diploid egg is formed. Such species can produce a new generation parthenogenetically, without fertilization.

15. The egg not only synthesizes numerous compounds, but also absorbs material produced by other cells. Moreover, it localizes many proteins and messages to specific regions of the cytoplasm, often tethering them to the cytoskeleton.

16. The *Xenopus* oocyte transcribes actively from lampbrush chromosomes during the first meiotic prophase.

17. In *Drosophila*, nurse cells make mRNAs that enter the developing oocyte. Which of the cells derived from the primordial germ cell becomes the oocyte and which become nurse cells is determined by the fusome and the pattern of divisions.

18. In mammals, retinoic acid from the mesonephros initiates germ cell meiosis in the ovaries. In the testes, however, retinoic acid is degraded and meiosis is blocked until puberty.

19. In male mammals, the PGCs generate stem cells that last for the life of the organism. PGCs do not become stem cells in female mammals (although in many other animal groups, PGCs do become germ stem cells in the ovaries).

20. In female mammals, germ cells initiate meiosis and are retained in the first meiotic prophase (dictyate stage) until ovulation. In this stage, they synthesize mRNAs and proteins that will be used for gamete recognition and early development of the fertilized egg.

For Further Reading

Complete bibliographical citations for all literature cited in this chapter can be found at the free-access website **www.devbio.com**

Bowles, J. and 11 others. 2006. Retinoid signaling determines germ cell fate in mice. *Science* 312: 596–600.

Decotto, E. and A. C. Spradling. 2005. The *Drosophila* ovarian and testis stem cell niches: similar somatic stem cells and signals. *Dev. Cell* 9: 501–510.

Doitsidou, M. and 8 others. 2002. Guidance of primordial germ cell migration by the chemokine SDF-1. *Cell* 111: 647–659.

Ephrussi, A. and R. Lehmann. 1992. Induction of germ cell formation by oskar. *Nature* 358: 387–392.

Extravour, C. G. and M. Akam. 2003. Mechanisms of germ cell specification across the metazoans: epigenesis and preformation. *Development* 130: 5869–5884.

Hayashi, Y., M. Hayashi and S. Kobayashi. 2004. Nanos suppresses somatic cell fate in *Drosophila* germ line. *Proc. Natl. Acad. Sci. USA* 101: 10338–10342.

Knaut, H., C. Werz, R. Geisler, The Tübingen 2000 screen consortium, and C. Nüsslein-Volhard. 2003. A zebrafish homologue of the chemokine receptor Cxcr4 is a germ-cell guidance receptor. *Nature* 421: 279–282.

Molyneaux, K. A., J. Stallock, K. Schaible and C. Wylie. 2001. Time-lapse analysis of living mouse germ cell migration. *Dev. Biol.* 240: 488–498.

Ohinata, Y. and 11 others. 2005. Blimp1 is a critical determinant of the germ cell lineage in mice. *Nature* 436: 207–213.

Saga, Y. 2008. Mouse germ cell development during embryogenesis. *Curr. Opin. Genet. Dev.* 18: 337–341.

Seydoux, G. and S. Strome. 1999. Launching the germline in *Caenorhabditis elegans*: Regulation of gene expression in early germ cells. *Development* 126: 3275–3283.

Stewart, T. A. and B. Mintz. 1981. Successful generations of mice produced from an established culture line of euploid teratocarcinoma cells. *Proc. Natl. Acad. Sci. USA* 78: 6314–6318.

Weidinger, G., U. Wolke, M. Koprunner, M. Klinger and E. Raz. 1999. Identification of tissues and patterning events required for distinct steps in early migration of zebrafish primordial germ cells. *Development* 126: 5295–5307.

Go Online

WEBSITE 16.1 Germline sex determination in *C. elegans*. The establishment of whether a germ cell is to become a sperm or an egg involves multiple levels of inhibition. Translational regulation is seen in several of these steps.

WEBSITE 16.2 Mechanisms of chromosome diminution. The somatic cells do not lose DNA randomly. Rather, specific regions of DNA are lost during chromosome diminution.

WEBSITE 16.3 The insect germ plasm. The insect germinal cytoplasm was discovered as early as 1911, when Hegner found that removing the posterior pole cytoplasm of beetle eggs caused sterility in the resulting adults.

WEBSITE 16.4 Human meiosis. Nondisjunction—the failure of chromosomes to sort properly during meiosis—is not uncommon in humans. Its frequency increases with maternal age.

WEBSITE 16.5 The Nebenkern. Sperm mitochondria are often highly modified to fit the streamlined cell. The mitochondria of flies fuse together to form a structure called the Nebenkern; this fusion is controlled by the *fuzzy onions* gene.

WEBSITE 16.6 Synthesizing oocyte ribosomes. Ribosomes are almost a "differentiated product" of the oocyte, and the *Xenopus* oocyte contains 20,000 times as many ribosomes as somatic cells do. Gene repetition and gene amplification are both used to transcribe these enormous amounts of rRNA.

WEBSITE 16.7 Hormones and mammalian egg maturation. To survive, the follicle and its oocyte have to "catch the wave" of gonadotropic hormone release. The hormones of the menstrual cycle synchronize egg maturation with the anatomical changes of the uterus and cervix.

WEBSITE 16.8 The reinitiation of mammalian meiosis. The hormone-mediated disruption of communication between the oocyte and its surrounding follicle cells may be critical in the resumption of meiosis in female mammals.

Vade Mecum

Germ cells in the *Drosophila* embryo. In the Fruit Fly segment, a view of gametogenesis follows the primordial germ cells of the living *Drosophila* embryo from their formation as pole cells through gastrulation as they move from the posterior end of the embryo into the region of the developing gonad.

Gametogenesis in mammals. Stained sections of testis and ovary illustrate the process of gametogenesis, the streamlining of developing sperm, and the remarkable growth of the egg as it stores nutrients for its long journey. You can see this in movies and labeled photographs that take you at each step deeper into the mammalian gonad.

PART IV

SYSTEMS BIOLOGY

Expanding Developmental Biology to Medicine, Ecology, and Evolution

Systems biology is an attempt to redefine the priorities of biology as a discipline, focusing on complexity and integration. Throughout this book, we have focused on the physical, looking at development in terms of DNA sequences, cells, and organs. Systems theory provides a complementary, if more abstract, way of looking at development: it views development as the flow of information. This information is embodied in the genes, cells, organs, and external environment. Indeed, if information flow is the process unifying these disparate entities, one might be able to establish equations and models for developmental phenomena. Such modeling is one of the goals of many systems developmental biologists (see Van Speybroeck et al. 2005; Madar et al. 2009; Edelman 2010).

Different streams of systems biology emphasize different principles (see De Backer et al. 2010). Some are extensions of molecular biology and aim to model the molecular networks underlying phenotypic diversity within and between species (see the discussion of micromere specification in Chapter 5 or the discussion of *Drosophila* segmentation genes in Chapter 6). In such approaches, models are proposed based on bioinformatics (genome sequences, protein composition, etc.), and experiments are set up confirm or deny the model—and, as in all science, to predict further properties that can be verified (or not) by further experimentation. Thus, there is a continuing cycle of laboratory experimentation and computer modeling.

Another approach to systems biology is embodied in the "systems theory approach," which attempts to understand the principles behind biological organization and to relate the different levels of organization (gene, cell, tissue, organ, organism, ecosystem, etc.) to one another. Modern systems theory approach can be said to have had its start in developmental biology, which was one of the first sciences to utilize these principles of causation, integration, and context dependency.

The demonstration of embryonic regulation, where one blastomere of a 4-cell sea urchin embryo could become an entire larva and where the fate of a newt blastomere depended on its neighbors, caused a crisis in the field of experimental embryology. In response to the vitalism of Driesch (who held that there must be some non-physical wisdom directing development) and the reductionism of Roux (who held that the embryo was basically a machine), Oskar Hertwig (1894) proposed a type of materialist philosophy called **wholist organicism**. This philosophy embraced the views that, first, the properties of the whole cannot be predicted solely from the properties of its component parts; and second, that the properties of the parts are informed by their relationship to the whole. Subsequently, several developmental biologists—notably Paul Weiss (1926) and Ludwig von Bertalanffy (1928, 1932)—formalized these notions into a systems approach to development (Brauckmann 2000; Drack et al. 2007).

Both the reductionist and organicist approaches are materialist in that they do not invoke any extramaterial agent (entelechy; the soul; *Bildungstrieb*) as directing development. However, while reductionism claims that all complex entities (including proteins, cells, organisms, and ecosystems) can be completely explained by the properties of their component parts, organicism claims that complex wholes are inherently greater than the sum of their parts in the sense that the properties of each part are dependent on the context of that part within the whole in which it operates. Thus, when we try to explain how the whole system behaves, we cannot get away with speaking only about the parts. These explanations are no less materialistic than reductionism. As Denis Noble (2006) has stated,

> Systems biology ... is about putting together rather than taking apart, integration rather than reduction. It requires that we develop ways of thinking about integration that are as rigorous as our reductionist programmes, but different. ...It means changing our philosophy, in the full sense of the term.

Although the emphasis differs widely among systems biologists, the theoretical systems approach can be characterized as having five principles that apply directly to the study of development.

- **Context-dependent properties**. The meaning or role of an individual component of a system is dependent on its context—on what is before and after, above and below it. What BMP4 does, for instance, depends on the history and context of the cell that receives it. At one time and place, BMP4 may signal bone formation; at another time and place, it may signal apoptosis; at yet another, it may specify epidermis.

- **Level-specific properties and emergence**. The properties of a system at any given level of organization cannot be totally explained by those of levels "below" it. Thus, temperature is not a property of an atom, but a property that "emerges" from an aggregate of atoms. Similarly, voltage potential is a property of a biological membrane but not of any of its components. Higher-level properties result from lower-level activities, but they must be understood in the context of the whole.

- **Heterogeneous causation and integration**. In biology, causation is seen as being both "upward," from the genes to the environment, and "downward," from the environment to the genes. What a cell is depends both on its genes and on the other cells surrounding it (i.e., on the input it receives from both internal and external sources). Systems biology demands the integration of different types of explanations. It also calls for the integration of analysis (taking things apart) and synthesis (putting things together).

- **Modularity and robustness**. The organism develops as a system of modules. Raff (1996) characterized developmental modules as having discrete genetic specification, hierarchical organization, interactions with other modules, a particular physical location within a developing organism, and the ability to undergo transformations on both developmental and evolutionary time scales. **Robustness** (sometimes called **canalization**) refers to the ability of an organism to develop the same phenotype despite perturbations from the environment or from mutations. This is a function of interactions within and between modules.

We will look briefly at each of these in light of animal development.

Context-Dependent Properties

Language provides an analogy by which organicism can be understood (Collier 1985). Certain combinations of letters form words, and certain organizations of words form sentences. The meaning of a sentence obviously depends on the meaning of each component word (the parts define the whole). But most words have multiple and distinct meanings, and it is the context of the sentence that determines which specific meaning of a word is most appropriate (the whole defines the parts). Consider the following three sentences:

1. The party leaders were split on the platform.

2. The disc jockey was a black rock star.

3. The pitcher was driven home on a sacrifice fly.

In these sentences, each word's meaning is suspended until the sentence is complete; the words and sentence mutually define each other. In other words, when put into relation with each other, specific meanings of words are singled out, resulting in a series of words with a particular ("emergent") meaning (i.e., a story is formed). Parts determine the whole, and the whole determines each of its parts.

In embryology, we are constantly aware of the parts being determined by their context within the whole (Needham 1943; Haraway 1976; Hamburger 1988; Gilbert and Sarkar 2000). Indeed, in vertebrate embryos, the fate of a cell is specified by its position. Similarly, as said earlier, the function of a protein is determined by its context. BMP can induce apoptosis or bone formation; δ-crystallin can be a transparent lens protein or a liver enzyme; β-catenin can be a cofactor of transcription factors or adhesion molecules.

Emergence

In our linguistic metaphor, emergent properties can be seen in the relationship of letters to words. Standing alone, most letters usually have no intrinsic meaning. But when letters are grouped together in certain arrangements and according to certain rules, meaningful words emerge. In developmental biology, one often encounters these emergent properties.

For example, in renal development, the nephron is formed by interactions between the ureteric bud and the metanephrogenic mesenchyme. If one cultures these two tissues separately, neither develops any portion of the kidney. However, if you place these tissues together, the mesenchyme cells form the 10 cell types characteristic of the renal filtration apparatus, and the ureteric bud tissue branches just as it would in the intact organism. Thus, 10 new cell types "emerge" from the interactions between 2 cell types, neither of which had any of the specific properties of the proximal convoluted tubule cells, juxtaglomerular cells, or Bowman's capsule cells. Moreover, the shape of the filtration apparatus, with the tubules descending into Henle's loop before ascending into the glomerulus; the connection between the nephron and the collecting duct; and the histotypic arrays of the nephrons and their relationship to the collecting ducts are all higher-order properties that are reproducibly seen from kidney to kidney. Yet none is a property predicted from the properties of the isolated metanephrogenic mesenchyme cells or the ureteric bud.

Level-Specific Rules

Parts are organized into wholes, and these wholes are often components of larger wholes. Moreover, at each biological level there are appropriate rules, and one cannot necessarily "reduce" all the properties of body tissues to atomic phenomena. As mentioned above, temperature is not an appropriate concept for an atom, nor is semipermeability a proper concept for a protein. When you have an entity as complex as the cell, the fact that quarks have certain spins is irrelevant. This is not to say, however, that each level is independent of those "below" it. To the contrary, laws at one level may be almost deterministically dependent on those at lower levels; but they may also be dependent on levels "above." This notion of level-specific interactive modules forms the basis of many new computer programs. Indeed, the importance of level-specific rules to our conception of reality in general has been emphasized by philosophers and has been summarized by Dyke (1988) and Wimsatt (1995).

The noted embryologist R. G. Harrison was firm on this point: that there were "integrative levels" of organization in the embryo and that one could not homogenize these levels. As Joseph Needham (1943) wrote,

> The deadlock [between mechanism and vitalism] is overcome when it is realized that every level of organization has its own regularities and principles, not reducible to those appropriate to lower levels of organization, nor applicable to higher levels, but at the same time in no way inscrutable or immune from scientific analysis and comprehension.

Heterogeneous Causation

What is the "cause" of a cell's fate or behavior? Often we depict this as a one-way arrow, from the nucleus to the cytoplasm, or from the "bottom up." But as we have seen, paracrine factors and extracellular matrix (i.e., "higher levels") also play roles in determining which genes are expressed. In the next three chapters, we will see that "top-down" causation—that is, from the environment to the genome—is critical in medicine, ecology, and evolution. We will see that a variety of substances present in the environment, both anthropogenic and natural, can adversely affect animal development. We will see that symbionts (organisms that live in and on a second, host, organism) are often crucial to the host's development. And we will see how environmental forces interacting with developing organisms can drive the processes of biological evolution.

To say that there is a "gene for" something is often poor shorthand for a large variety of factors (see Tauber and Sarkar 1992; Moss 2003). As we have seen, gene products work together, and the genetic approach usually is able to determine only that a particular gene is necessary to make a particular cell type or tissue. A gene's loss of function means that an organ doesn't form or function. In terms of the whole organism, this is analogous to saying that a "gas pedal gene" is responsible for automobiles, since cars do not function without one. Even when one has gain-of-function studies (such as those showing that three transcription factors will convert an exocrine pancreatic cell into an insulin-secreting beta cell), the inserted genes are working with a host of transcription factors, methylating enzymes, and other components that are the cell's heritage.

That causation occurs from below and from above is becoming an important concept in medicine and conservation biology. The specification of sex in turtles and many other reptiles is dependent not only on the genome (which provides the information that sets the threshold temperature separating males and females) but also on the immediate temperature of the organism's environment. This knowledge is important for biologists attempting to protect endangered turtle species. The link between an organism's development and the symbionts that naturally infect that organism is often so tight that development cannot take place without the symbionts being present. Such linkage is critical in development and evolution, and even in medicine. One new way of treating infections of parasitic worms, for example, is to give patients antibiotics. These drugs do not destroy the worms directly; rather, they wipe out the prokaryote symbionts that provide factors essential for the worm's development (Hoerauf et al. 2003; Coulibaly et al. 2009). Thinking of the host and symbiont as a developing system allows us to see development in a new way and allows new questions to be asked.

Integration

One of the chief components of systems biology is its insistence on synthesis as well as analysis. This is also one of the biggest sources of heterogeneity among those who consider themselves systems biologists. True to the notion of level-specific rules, there can be different systems biologies at different levels, but each of them will be synthetic.

For instance, on a single level—that of gene transcription—Davidson and colleagues (see Chapter 5) have attempted to map all the gene inputs involved in the specification of the early sea urchin embryo. This systems approach creates a "wiring diagram" of all the inputs and outputs of each gene involved in cell specification, and finds the rules (such as the "double-negative gate"; see Figure 5.12B) that are the "logic" of development. But this approach can be expanded by defining the "system" not as the cell or the organism itself, but as the cell within an organism within an ecosystem. This is done routinely in the subspecialty of obstetrics and gynecology called maternal-fetal medicine. Here, neither the mother nor the fetus is seen as an entity separate from the other. As we will see, the field of teratology (birth defects) concerns the modulation of cell behavior during development (differential gene transcription, cell specification and adhesion, etc.) by environmental agents mediated through the mother. Similarly, cancer research now sees many epithelial tumors as arising from defects in the mesenchymal cells surrounding the epithelium, and these defects can be caused by environmental agents (Soto et al. 2008, 2009). In these cases, the explanation of cell behavior is as much a function of "top-down causation" from the environment as it is "bottom-up causation" from the genome.

In eukaryotic embryos, there are so many levels of complexity that it will be difficult to get an inventory of every component involved (as well as an inventory of potential interactions between the components). The compilation of such an inventory is being undertaken by the "-omics ": genomics (the inventory of genes), proteomics (the inventory of proteins), transcriptomics (the inventory of RNAs), and metabalomics (the inventory of the small molecule end-products of gene expression). Comparisons between the genomes of different species, and between the transcriptomes and proteomes of the same cell type in different species, have become a necessary precondition for seeing how organisms can change as they develop and evolve (Ge et al. 2003; Bonneau et al. 2007; Vidal 2009). Indeed, we will see numerous examples (such as the effects of symbionts or different environmental conditions on cell transcription) where comparing inventories of gene products reveals that the outside agent is actually changing gene expression patterns. However, the lack of technology to measure the exact (yet often fluctuating) quantity of these components, and to map their context sensitivity and their often short and rapid reactions in real time, remains a hurdle to this endeavor.

Modules and Robustness

Systems biology views the embryo as a complex adaptive system (Gell-Mann 1964; Edelman et al. 2010). It is "complex" because of the diversity of parts and the multiple interacting modules. It is "adaptive" because the component modules can modify their behaviors in response to environmental or genetic perturbations.

Modularity is one of the most important concepts in all biology, and it is fundamental to any discussion of development and evolution. According to Bolker (2000), discussions of modularity in development have emphasized (1) that a module is a biological entity (e.g., a structure, a *cis*-regulatory region, a cell lineage, a morphogenetic field) characterized by more internal than external integration (Wagner 1996; Wagner and Altenberg 1996; von Dassow and Munro 1999); and (2) that modules are biological entities that can be delineated from their surroundings or context, and whose behavior or function reflects the integration of their parts, not simply the arithmetical sum.

Living beings are organized according to what philosopher Chuck Dyke (1988) calls "level-interactive modular arrays." This concept implies a nest of modular structures. Each entity is an organized array of constituent modular parts, and is at the same time the constituent of a larger module. The components of each module (which are themselves modules) interact to form the coherent (larger) module. Moreover, the modules interact with the levels above and below them. Just as each organelle must function to make a coherent cell, so each cell and cell type must function to make a coherent tissue. Thus, the tissue architecture of the liver can regulate elements in the modules below it (the liver extracellular matrix and intercellular interactions within the tissue regulating gene expression within the cell); and the lower-level modules such as those of the cell obviously determine the function of the higher-level modules such as the tissue.

Development depends on modularity. Suppose (as Dyke does) that you bought a bicycle online, and at the bottom of the product description a note said "Some assembly required." Since you're good with your hands and the price is reasonable, you order it. A week later a truck delivers to your door a basket of iron ore, a large sack of limestone, a box of bauxite, a bag of charcoal, and a container full of sap from a rubber tree. The packing slip says "One bicycle." The synthesis of a bicycle (and certainly the development of an animal) is a path-dependent process comprising a large number of steps, and the path involves the synthesis of modular intermediates (Dyke 1988). Without the production of these intermediates, there is no way that an organism can come into being. Tissues must be made from pre-assembled cells, and cells are made from pre-assembled components such as proteins. Proteins must be made from pre-assembled amino acids, and so forth.

Simon (1962) argues for the importance of modularity, which he offers both as a prescription for human designers and as a description of the complex systems we see in nature. To make the latter point, he offers the parable of the two watchmakers. Tempus and Hora are both expert watchmakers who make complicated watches from numerous parts. However, they are both interrupted frequently in their work. Tempus does not design his watches as modular systems. He puts every piece in one at a time, so every time he is interrupted and forced to set aside his work, the unfinished assembly falls apart. By contrast, Hora first builds stable subassemblies that he can then put together in hierarchic fashion into larger, stable subassemblies. Thus, when Hora is interrupted, only the last unfinished subassembly falls apart. In a selective environment, such developmental stability would be rewarded with evolutionary survival (Simon 1962).

What makes Tempus's unfinished watches so unstable is not the number of parts involved. Rather, it is the parts' interdependency. In a non-modular system, the successful operation of any given part is likely to depend on the characteristics of many other components throughout the system. So when such a system has missing or defective parts, the whole ceases to function. In a modular system, by contrast, the proper working of a given part depends primarily on the characteristics of other parts within its subassembly, although it depends to some degree on the characteristics of parts outside that subassembly. As a result, a modular system is usually able to function to some degree even if some subsystems are damaged or incomplete. If one adds an editing function—such as redundancy or the ability to replace a defective module—then one has assurance that the entire system will function.

As we will see, modularity allows anatomical change to occur in one module without necessarily affecting another module. This change during development is predicated on the modularity of enhancer sequences, mentioned in Chapter 2. Modularity also allows for robustness: the ability to retain the phenotype despite environmental or genetic changes. Robustness is a characteristic of living systems (Kitano 2004) and is responsible for the normal phenotype developing so typically, despite the millions of interactions that have to occur sequentially and at the right place during embryogenesis. This robustness is due to redundancy in control and to feedback loops between modules (von Dassow et al. 2000).

Thus, Spemann (1927) spoke about "double assurance" (which he likened to wearing both a belt and suspenders) on the anatomical level. The lens could form in more than one way, and if one tissue didn't induce the lens, the other tissue would. If the operculum did not open in its normal manner during frog metamorphosis, another system kicked in and opened that flap of skin. More recently, redundancy has been found at the level of genes. Muscles will form when the muscle-forming *MyoD* gene is knocked out of a mouse embryo, even if the details of the muscle development are not perfect (Wang et al. 1996). This is because *MyoD* usually downregulates the synthesis of *Myf5*, a *MyoD*-like gene that can also direct muscle development. In the absence of *MyoD*, *Myf5* is synthesized and can perform most of the functions of the absent molecule. The segment polarity genes of *Drosophila* are interconnected by a series of regulatory loops such that changing any one of them still gave proper segmentation 90% of the time.

As C. H. Waddington (1957) pointed out, however, studying systems biology in development is going to be more difficult than studying systems biology in physiology. Physiologists attempt to understand the interactions among parts that exist throughout the life of the organism. But Waddington pointed out that the task of the developmental biologist is not to understand homeostasis—how the organism maintains itself. Rather, the task of the developmental biologist is to understand **homeorhesis**—how the organism stabilizes its different cell lineages while it is still constructing itself.

Donna Haraway points out that modularity depends on integrated *processes*, not factors. "Relationships," she writes (2008, p. 228) "are the smallest possible pattern for analysis." This way of thinking may make systems biology different from what has come before, and new insights are beginning to emerge. Waddington, who made substantial contributions to both developmental biology and systems biology, viewed development as a mountainous landscape and the cell types as basins. Like a ball rolling down a hill, a cell eventually finds itself in one or another of these basins. Basins are constructed through the interactions of cells during development. Kauffman (1969, 1987, 1993) and others (Huang 2009a,b) see these basins as "attractor states," where a stable and robust configuration of gene expression becomes possi-

FIGURE IV.1 An "epigenetic land-scape" for a complex network wherein *N* genes can interact with one another. The overall slope (from back to front) represents the progression of development over time. Normal developmental trajectories (blue lines) lead to attractors representing distinct cell types and are prevented from entering "abnormal attractors" (red dashed circle) by epigenetic barriers (lavender hill). Mutations or methylation differences can lower this barrier, opening access to attractors that encode an abnormal, immature phenotype, including cancerous cell types (red arrow). (After Huang et al. 2009b.)

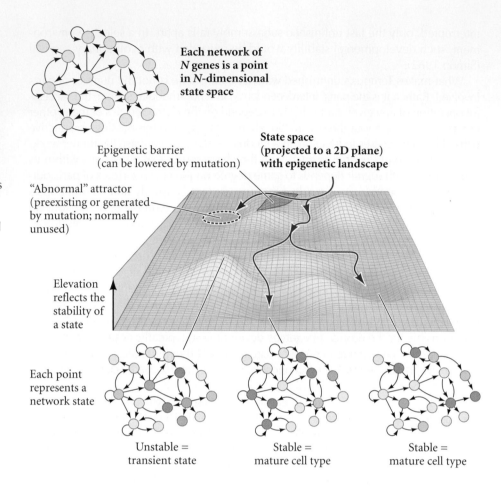

Each network of ***N* genes is a point in *N*-dimensional state space**

State space (projected to a 2D plane) with epigenetic landscape

Epigenetic barrier (can be lowered by mutation)

"Abnormal" attractor (preexisting or generated by mutation; normally unused)

Elevation reflects the stability of a state

Each point represents a network state

Unstable = transient state

Stable = mature cell type

Stable = mature cell type

ble (**Figure IV.1**). The ability of a set of transcription factors to rearrange transcriptional networks to convert exocrine pancreas or liver cells into endocrine pancreas cells can be seen as an example of such an attractor state (Horb et al. 2003; Zhou et al. 2008). Most combinations of transcription factors and enhancers are expected to be unstable; those combinations that are stable, however, can become cell types. Huang and colleagues (2005, 2009a) have provided evidence based on systems analysis that the neutrophil (a white blood cell) is such a stable cell type. Their computer analysis connecting over 2750 genes into networks found that the neutrophil expression pattern is the most stable pattern, and that the only "rival" to it is that of neutrophilic leukemia. The formations of new cell types (such as the neural crest cells of vertebrates) can be looked at as constructions of new stable attraction states over the course of evolution. The stem cell state of cell division so important for cancer may be another stable attractor state (Huang et al. 2009b).

Denis Noble (2006) stated that the "ultimate goal of systems biology" is "to reconnect physiology and developmental models to theories of evolution." He also noted that we have just begun this enterprise. In the closing chapters of this book, we will examine the roles of development in disease, ecosystem dynamics, and evolution, keeping in mind the properties of systems.

Medical Aspects of Developmental Biology

17

OUR UNDERSTANDING OF MAMMALIAN DEVELOPMENT promises to expand our medical capacities in this next century as much as microbiology allowed us to regulate common infectious diseases in the past century. Organ regeneration, cancer therapies, and even the prolongation of healthy lives may be possible through stem cell technologies and our new knowledge of paracrine factors and transcription factors. However, our technological society is constantly producing new chemical compounds that can produce developmental anomalies in humans and wildlife. In both cases, developmental biology is now critical to issues of medicine and public health. Developmental biology has been the basis of numerous breakthroughs in medicine, including in vitro fertilization to circumvent infertility; bone marrow transplantation, which uses hematopoietic stem cells to treat anemia; and the administration of erythropoietin (a paracrine factor that stimulates red blood cell production) to patients whose cancer chemotherapy has led to anemia.

This chapter will first focus on four major developmental aspects of disease:

- *Genetic syndromes*, wherein mutations alter development in deleterious ways
- *Teratogenesis*, wherein environmental substances cause birth defects
- *Endocrine disruption*, wherein environmental substances that alter the development of the endocrine system result in pathologies later in life
- *Cancer*, wherein cells escape the normal regulation of their neighbors and regain embryonic properties such as cell division, migration, and invasiveness

The chapter will conclude by looking at how insights from developmental biology are being used to cure and repair injured tissues.* These developmental therapies include:

- *Anti-angiogenesis treatment* for cancer, wherein tumor cells are prevented from receiving necessary blood vessels
- *Induced pluripotent stem cells*, wherein normal adult body cells can be converted into pluripotent cells resembling embryonic stem cells that can generate any cell type in the body
- *Regenerative biology*, wherein adult stem cells and paracrine factors can restore damaged or diseased organs

---

*When one discusses the medical aspects of any science, one must take social issues as well as purely scientific concerns into account. Discussions concerning how genetic diseases, birth defects, cancers, or infertility might be controlled also raise issues concerning social, political, and economic motives and opportunities (see Gilbert et al. 2005). While this chapter seeks to provide general information about medical topics, it is not intended to provide medical advice for specific persons or disorders.

The amazing thing about mammalian development is not that it sometimes goes wrong, but that it ever succeeds.

VERONICA VAN HEYNINGEN (2000)

The future is already here. It's just not evenly distributed yet.

WILLIAM GIBSON (1999)

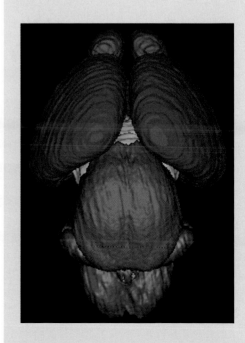

DISEASES OF DEVELOPMENT

Genetic Errors of Human Development

If you think it is amazing that any one of us survives to be born, you are correct. It is estimated that one-half to two-thirds of all human conceptions do not develop successfully to term (**Figure 17.1**). Many of these embryos express their abnormality so early that they fail to implant in the uterus. Others implant but fail to establish a successful pregnancy. Thus, most embryos are spontaneously aborted, often before a woman even knows she is pregnant (Boué et al. 1985). Using a sensitive immunological test that detects the presence of human chorionic gonadotropin (hCG) as early as 8 or 9 days after fertilization, Edmonds and co-workers (1982) monitored 112 pregnancies in normal women. Of these hCG-determined pregnancies, 67 (about 59%) were not maintained.

Most early embryonic and fetal demise is probably due to chromosomal abnormalities that interfere with developmental processes. There are also defects that are not deleterious to the fetus (which does not depend on organs such as the brain, kidneys, and lungs while inside the mother) but that can threaten life once the baby is born. Winter (1996; Epstein 2008) has estimated that approximately 2.5% of newborns have a recognizable birth defect. Congenital ("present at birth") abnormalities and losses of the fetus prior to birth have both intrinsic and extrinsic causes. Those abnormalities caused by genetic events may result from mutations, aneuploidies (improper chromosome number), and translocations (Opitz 1987).

Even an extra copy of the tiny chromosome 21 disrupts numerous developmental functions. This **trisomy 21** causes a set of anomalies—among them facial muscle changes, heart and gut abnormalities, and cognitive problems—collectively known as **Down syndrome** (Figure 17.2). Cer-

tain genes on chromosome 21 are thought to encode transcription factors, and the extra copy of chromosome 21 probably causes an overproduction of these regulatory proteins. Such overproduction would cause the misregulation of genes necessary for heart, muscle, and nerve formation (Chang and Min 2009; Korbel et al. 2009). Down syndrome is one of the few trisomies mild enough to allow the fetus to be born and to survive beyond the first weeks of infancy.*

Until recently, the molecular study of human genetics focused almost exclusively on inborn errors in metabolic and structural proteins. Thus, diseases of enzymes, collagens, and globins predominated. But these proteins are the final products of differentiated cells. Errors in the proteins involved with development—transcription factors, paracrine factors, and elements of signal transduction pathways—were little understood. In the past decade, however, we have made great gains in our knowledge of human development, and we now know that many of these malformations are caused by mutations of genes encoding transcription factors and signal transduction proteins (Table 17.1). Many of these mutations have been used to discover the protein responsible for the deficient reaction, and these have been mentioned throughout this book. Indeed, Charles Epstein (2008) wrote that "Science has finally caught up with human malformations," and his co-edited 1600-page book documents that claim.

See WEBSITE 17.1
Human embryology and genetics

---

*With proper medical care, some infants born with trisomies 13 and 18 (Patau syndrome and Edward syndrome, respectively) can live for years, although they usually suffer from lung and intestinal defects and heart malformations.

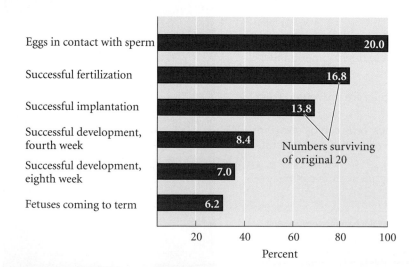

FIGURE 17.1   Fate of 20 hypothetical human eggs in the United States and western Europe. Under normal conditions, only 6.2 (or fewer) of the original 20 eggs would be expected to develop successfully to term. (After Volpe 1987.)

(A)

(B)

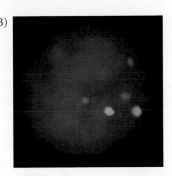

FIGURE 17.2   Down syndrome. (A) Down syndrome, caused by a third copy of chromosome 21, is characterized by a particular facial pattern and by cognitive deficiencies, the absence of a nasal bone, and often heart and gastrointestinal defects. (B) The procedure shown here tests for chromosome number using fluorescently labeled probes that bind to DNA on chromosomes 21 (pink) and 13 (blue). This person has Down syndrome (trisomy 21) but has the normal two copies of chromosome 13. (A © Mood-Board/Alamy; B courtesy of Vysis, Inc.)

TABLE 17.1 Some genes encoding human transcription factors and phenotypes resulting from their mutation

Gene	Mutation phenotype
Androgen receptor	Androgen insensitivity syndrome
AZF1	Azoospermia
CBFA1	Cleidocranial dysplasia
CSX	Heart defects
EMX2	Schizencephaly
Estrogen receptor	Growth regulation problems, sterility
Forkhead-like 15	Thyroid agenesis, cleft palate
GLI3	Grieg syndrome
HOXA13	Hand-foot-genital syndrome
HOXD13	Polysyndactyly
LMX1B	Nail-patella syndrome
MITF	Waardenburg syndrome type 2
PAX2	Renal-coloboma syndrome
PAX3	Waardenburg syndrome type 1
PAX6	Aniridia
PTX2	Reiger syndrome
PITX3	Congenital cataracts
POU3F4	Deafness and dystonia
SOX9	Campomelic dysplasia, male sex reversal
SRY	Male sex reversal
TBX3	Schinzel syndrome (ulna-mammary syndrome)
TBX5	Holt-Oram syndrome
TCOF	Treacher-Collins syndrome
TWIST	Seathre-Chotzen syndrome
WT1	Urogenital anomalies

The Nature of Human Syndromes

Human birth defects, which range from life-threatening to relatively benign, are often linked into **syndromes**, where there are several abnormalities occurring together. The word *syndrome* comes from the Greek *syndromos*, "running together." Genetically based syndromes are caused either by (1) a chromosomal event (such as aneuploidies) where several genes are deleted or added, or (2) by one gene having many effects. The production of several effects by one gene or pair of genes is called **pleiotropy** (see Grüneberg 1938; Hadorn 1955). Pleiotropic effects are called **mosaic pleiotropy** when the effects are produced independently as a result of the gene being critical in different parts of the body (**Figure 17.3A**). For instance, the *KIT* gene is expressed in blood stem cells, pigment stem cells, and germ stem cells, where it is needed for their proliferation (see p. 27). When this gene is defective, the resulting syndrome of anemia (lack of red blood cells), sterility (lack of germ cells), and albinism (lack of pigment cells) is evidence of mosaic pleiotropy.

Syndromes are said to have **relational pleiotropy** when a defective gene in one part of the embryo causes a defect in another part, even though the gene is not expressed in

(A) Mosaic pleiotropy   (B) Relational pleiotropy

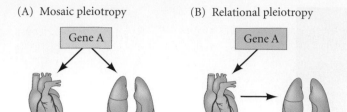

FIGURE 17.3 Mosaic and relational pleiotropy. (A) In mosaic pleiotropy, a gene is independently expressed in several tissues. Each tissue needs the gene product and develops abnormally in its absence. (B) In relational pleiotropy, a gene product is needed by only one particular tissue. However, a second tissue needs a signal from the first tissue in order to develop properly. If the first tissue develops abnormally, the signal is not given, so the second tissue also develops abnormally.

the second tissue (**Figure 17.3B**). The failure of *MITF* expression in the pigmented retina prevents this structure from fully differentiating. This failure of pigmented retina growth, in turn, causes a malformation of the choroid fissure of the eye, resulting in the drainage of vitreous humor. Without this fluid, the eye fails to enlarge (hence *microphthalmia*, or "small eye"). The lenses and corneas are smaller, even though they themselves do not express *MITF*.

Genetic Heterogeneity and Phenotypic Heterogeneity

In pleiotropy, the same gene can produce different effects in different tissues. However, the opposite phenomenon is an equally important feature of genetic syndromes: mutations in *different genes* can produce the *same phenotype*. If several genes are part of the same signal transduction pathway, a mutation in any of them often produces a similar phenotypic result. This production of similar phenotypes by mutations in different genes is called **genetic heterogeneity**. The syndrome of sterility, anemia, and albinism caused by the absence of Kit protein (discussed above) can also be caused by the absence of its paracrine ligand, stem cell factor (SCF). Another example is cyclopia, a phenotype that can be produced by mutations in the *sonic hedgehog* gene (see Chapter 3) or by mutations in the genes *activated* by hedgehog or in the genes controlling cholesterol synthesis (since cholesterol is essential for hedgehog signaling).

Not only can different mutations produce the same phenotype, but the same mutation can produce a different phenotype in different individuals (Wolf 1995, 1997; Nijhout and Paulsen 1997). This phenomenon, called **phenotypic heterogeneity**, is discussed at length in Chapter

18. Phenotypic heterogeneity comes about because genes are not autonomous agents. Rather, genes interact with other genes and gene products, becoming integrated into complex pathways and networks. Bellus and colleagues (1996), for instance, analyzed the phenotypes derived from the same mutation in the *FGFR3* gene in 10 different, unrelated human families. These phenotypes ranged from relatively mild anomalies to potentially lethal malformations. Similarly, Freire-Maia (1975) reported that within one family, the homozygous state of a mutant gene affecting limb development caused phenotypes ranging from severe phocomelia (lack of limb development) to a mild abnormality of the thumb. The severity of a mutant gene's effect often depends on the *other* genes in the pathway as well as on environmental factors.

Phenotypic heterogeneity in women can also be caused by statistical variation in X chromosome inactivation (see Sidelights & Speculations, p. 501). One such case involved identical twins, each of whom carried one normal and one mutant allele for an X-linked blood clotting factor. One twin's random inactivation pattern resulted in the loss of a large percentage of the X chromosomes that carried the normal allele, and she had severe hemophilia (inability of the blood to clot after injury). The other twin, with a lower percentage of her normal X chromosomes inactivated, did not have the disease (Tiberio 1994; Valleix et al. 2002).

Teratogenesis: Environmental Assaults on Human Development

In addition to genetic mutations, we now know that numerous environmental factors can disrupt development. The summer of 1962 brought two portentious discoveries. The first was the disclosure by Rachel Carson (1962) that the pesticide DDT was destroying bird eggs and preventing reproduction in several species (see Chapter 18). The second (Lenz 1962) was the discovery that thalidomide, a sedative used to help manage pregnancies, could cause limb and ear abnormalities in the fetus (see Chapter 1). These two discoveries showed that the embryo was vulnerable to environmental agents.* This was underscored in 1964, when an epidemic of rubella (German measles) spread across the United States. Adults showed relatively mild symptoms when infected by this virus, but over 20,000 fetuses infected by rubella were born blind, deaf, or both. Many of these infants were also born with heart defects and/or mental retardation (CDC 2002).

---

*Indeed, Rachel Carson realized the connection, commenting that "It is all of a piece, thalidomide and pesticides. They represent our willingness to rush ahead and use something without knowing what the results will be" (Carson 1962).

Prenatal Diagnosis and Preimplantation Genetics

One of the consequences of in vitro fertilization (IVF) and the ability to detect genetic mutations early in development is a new area of medicine called **preimplantation genetic diagnosis (PGD)**. Preimplantation genetics seeks to test for genetic disease *before* the embryo enters the uterus. After that, many genetic diseases can still be diagnosed before a baby is born. This **prenatal diagnosis** can be done by chorionic villus sampling at 8–10 weeks of gestation, or by amniocentesis around the fourth or fifth month of pregnancy.

Figure 17.4   Preimplantation genetics is performed on one or two blastomeres (seen here in the pipette) taken from an early blastocyst. The polymerase chain reaction is then used to determine whether certain genes in these cells are present, absent, or mutant. (Courtesy of The Institute for Reproductive Medicine and Science of St. Barnabas, Livingston, NJ.)

Chorionic villus sampling and amniocentesis

Chorionic villus sampling involves taking a sample of the placenta, whereas **amniocentesis** involves taking a sample of the amnionic fluid. In both cases, fetal cells from the sample are grown and then analyzed for the presence or absence of certain chromosomes, genes, or enzymes.

However useful these procedures have been in detecting genetic disease, they have brought with them a serious ethical concern: if a fetus is found to have a genetic disease, the only means of prevention presently available is to abort the pregnancy. The need to make such a choice can be overwhelming to prospective parents. Indeed, the waiting time between the knowledge of being pregnant and the results from amniocentesis or chorionic villus sampling has created a new phenomenon, the "tentative pregnancy." Many couples do not announce their pregnancy during this stressful period for fear that it might have to be terminated (Rothman et al. 1995).

By using IVF and PGD, one can consider implanting only those embryos that are most likely to be healthy, as opposed to aborting those fetuses that are most likely to produce malformed or nonviable children. This can be achieved by screening embryonic cells before the embryo is implanted in the womb. While the embryos are still in the petri dish (at the 6- to 8-cell stage), a small hole is made in the zona pellucida and two blastomeres are removed from the embryo (**Figure 17.4**). Since the mammalian egg undergoes regulative development, the removal of these blastomeres does not endanger the embryo, and the isolated blastomeres are tested immediately. **Fluorescent in situ hybridization**, or **FISH**, is used to determine whether the normal numbers and types of chromosomes are present (see Figure 17.2B), and the polymerase chain reaction technique (PCR) can be used to determine the presence or absence of certain genes (Kanavakis and Traeger-Synodinos 2002; Miny et al. 2002). Results are often available within 2 days. Presumptive normal embryos can be implanted into the uterus, while embryos presumed to have deleterious mutations are discarded.

Sex selection and sperm selection

The same procedures that enable preimplantation genetics also enable the physician to know the sex of the embryo. Sometimes parents wish to have this information; sometimes they do not. However, knowing the sex of an embryo prior to its implantation raises the possibility that parents could decide to have only embryos of the desired sex implanted. Sex selection using PGD is seen by many as a beneficial way of preventing X-linked diseases, but in fact it is often used to choose an offspring's sex. Opponents of sex selection point to its possible use as a method of preventing the birth of girls in cultures where women are not as highly valued as men (see Gilbert et al. 2005; Zhu and Hesketh 2009). Different countries and even different hospitals have different policies on the use of preimplantation genetic diagnosis solely for the purpose of sex determination.

Another way to accomplish sex selection is through sperm selection. The X chromosome is substantially larger than the Y chromosome; therefore, human sperm cells containing an X chromosome contain nearly 3% more total DNA than sperm cells containing a Y chromosome. This DNA difference can be measured, and the X- and Y-bearing sperm cells separated based on their size/mass ratio, using a flow cytometer. The separated sperm can then be used for artificial insemination or IVF. Recent studies have shown that this technique is about 90% reliable for sorting X-bearing sperm, and about 78% reliable for sorting Y-bearing sperm (Stern et al. 2002).

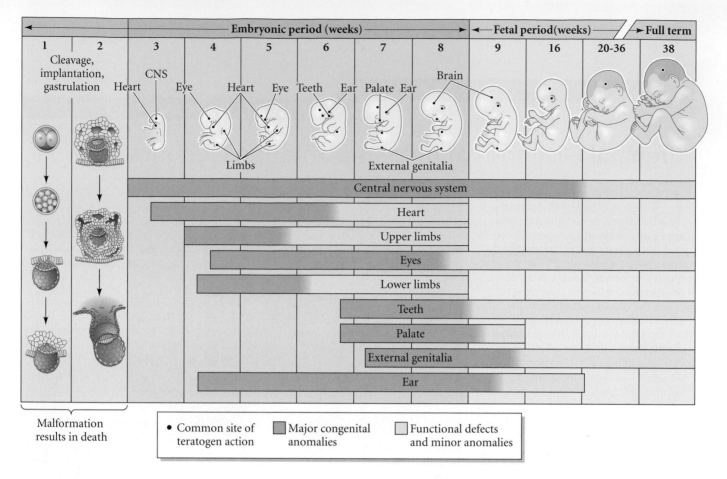

FIGURE 17.5 Weeks of gestation and sensitivity of embryonic organs to teratogens. (After Moore and Persaud 1993.)

Exogenous agents that cause birth defects are called **teratogens.*** Most teratogens produce their effects during certain critical periods of development. Human development is usually divided into two periods, the **embryonic period** (to the end of week 8) and the **fetal period** (the remaining time in utero). It is during the embryonic period that most of the organ systems form; the fetal period is generally one of growth and modeling. Figure 17.5 indicates the times at which various organs are most susceptible to teratogens.

The period of maximum susceptibility to teratogens is between weeks 3 and 8, since that is when most organs are forming. The nervous system, however, is constantly forming and remains susceptible throughout development. Prior to week 3, exposure to teratogens does not usually produce congenital anomalies because a teratogen encountered at this time either damages most or all of the cells of an embryo, resulting in its death, or kills only a few cells, allowing the embryo to fully recover.

Different agents are teratogenic in different organisms (see Gilbert and Epel 2009). The largest class of teratogens includes drugs and chemicals, but viruses, radiation, hyperthermia, and metabolic conditions in the mother can also act as teratogens. A partial list of agents that are teratogenic in humans is given in Table 17.2.

Some chemicals that are naturally found in the environment can cause birth defects.[†] For example, jervine and cyclopamine are chemical products of the plant *Veratrum californicum* that block cholesterol synthesis. As mentioned above and in Chapter 3, blocking cholesterol synthesis can

---

*In some cases, the same condition can be caused either by an exogenous agent or by a genetic mutation. Chondrodysplasia punctata is a congenital defect of bone and cartilage, characterized by abnormal bone mineralization, underdevelopment of nasal cartilage, and shortened fingers. It is caused by a defective gene on the X chromosome. An identical phenotype is produced by exposure of fetuses to the anticoagulant (blood-thinning) compound warfarin. It appears that the defective gene normally produces an arylsulfatase protein (CDPX2) necessary for cartilage growth. The warfarin compound inhibits this same enzyme (Franco et al. 1995).

[†]The widely used natural substance caffeine has not been proved to cause congenital anomalies (see Browne et al. 2007). There is, however, evidence that nicotine, another natural substance and an important component of cigarette smoke, may damage the brain and lungs during development. Dwyer and colleagues (2008) showed that nicotine induces abnormalities of synapse formation and cell survival in the developing brain. Nicotine can also induce lung cells to have an altered metabolism that ages them more rapidly; and prenatal exposure to cigarette smoke is associated with increased risks of impaired lung functions later in life (Maritz 2008; Wang and Pinkerton 2008). Smoking also significantly lowers the number, quality, and motility of sperm in the semen of males who smoke at least four cigarettes a day (Kulikauskas et al. 1985; Mak et al. 2000; Shi et al. 2001).

TABLE 17.2 Some agents thought to cause disruptions in human fetal development[a]

DRUGS AND CHEMICALS	IONIZING RADIATION (X-RAYS)
Alcohol	
Aminoglycosides (Gentamycin)	HYPERTHERMIA (FEVER)
Aminopterin	INFECTIOUS MICROORGANISMS
Antithyroid agents (PTU)	Coxsackie virus
Bromine	Cytomegalovirus
Cortisone	Herpes simplex
Diethylstilbesterol (DES)	Parvovirus
Diphenylhydantoin	Rubella (German measles)
Heroin	*Toxoplasma gondii* (toxoplasmosis)
Lead	*Treponema pallidum* (syphilis)
Methylmercury	
Penicillamine	METABOLIC CONDITIONS IN THE MOTHER
Retinoic acid (Isotretinoin, Accutane)	Autoimmune disease (including Rh incompatibility)
Streptomycin	
Tetracycline	Diabetes
Thalidomide	Dietary deficiencies, malnutrition
Trimethadione	
Valproic acid	Phenylketonuria
Warfarin	

Source: Adapted from Opitz 1991.

[a]This list includes known and possible teratogenic agents and is not exhaustive.

block Sonic hedgehog function and lead to cyclopia (see Figure 3.26).

See WEBSITE 17.2 Thalidomide as a teratogen

See VADE MECUM Somites and thalidomide

Alcohol as a teratogen

In terms of the frequency of its effects and its cost to society, the most devastating teratogen is undoubtedly alcohol (ethanol). In 1968, Lemoine and colleagues in France noticed a syndrome of birth defects in the children of alcoholic mothers, and **fetal alcohol syndrome (FAS)** was confirmed in the United States by Jones and Smith (1973). Babies with FAS are characterized by their small head size, indistinct philtrum (the pair of ridges that runs between the nose and mouth above the center of the upper lip), narrow vermillion border on the upper lip, and low nose bridge. The brain of such a child may be dramatically smaller than normal and often shows defects in neuronal and glial migration, resulting in lack of brain development (**Figure 17.6**; Clarren 1986). FAS is the most prevalent type of congenital mental retardation syndrome, occurring in

approximately 1 out of every 650 children born in the United States (May and Gossage 2001). Although the IQs of children with FAS vary substantially, the mean is about 68 (Streissguth and LaDue 1987). FAS patients with a mean chronological age of 16.5 years have been found to have the functional vocabulary of 6.5-year-olds and the mathematical abilities of fourth-graders. Most adults and adolescents with FAS cannot handle money and have difficulty learning from past experiences.*

FAS represents only a portion of a range of defects caused by prenatal alcohol exposure. The term *fetal alcohol spectrum disorder* (FASD) has been coined to encompass all of the alcohol-induced malformations and functional deficits that occur. In many FASD manifestations, behavioral abnormalities exist without any gross physical changes in head size or notable reductions in IQ. It is estimated that for every case of FAS, there are at least three times as many instances of FASD (CDC 2009).

As with other teratogens, the amount and timing of fetal exposure to alcohol, as well as the genetic background of the fetus, contribute to the developmental outcome. Variability in the metabolism of alcohol by the mother may also account for some outcome differences (Warren and Li 2005). While FASD is most strongly associated with high levels of alcohol consumption, the results of animal studies suggest that even a single episode of consuming the equivalent of two alcoholic drinks during

*For remarkable accounts of raising children with fetal alcohol syndrome, read Michael Dorris's *The Broken Cord* (1989) and Liz and Jodee Kulp's *The Best I Can Be* (2000). For an excellent account of the debates within the medical profession about this syndrome, see Janet Golden's *Message in a Bottle* (2005).

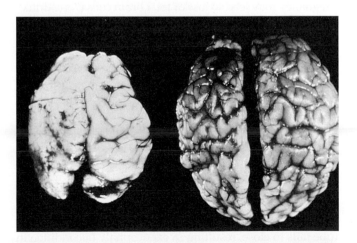

FIGURE 17.6   Comparison of a brain from an infant with fetal alcohol syndrome (FAS) with a brain from a normal infant of the same age. The brain from the infant with FAS (left) is smaller, and the pattern of convolutions is obscured by glial cells that have migrated over the top of the brain. (Courtesy of S. Clarren.)

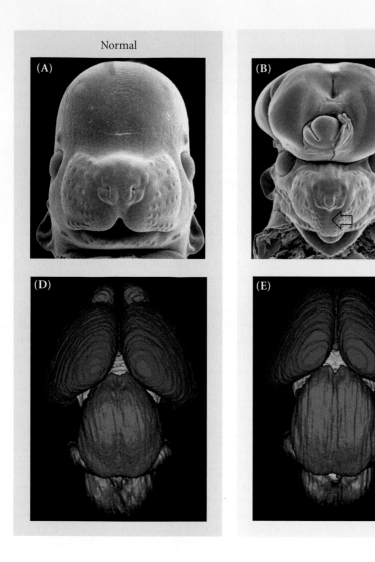

Normal

Alcohol-exposed

(A)

(B)

(C)

(D)

(E)

FIGURE 17.7 Alcohol-induced craniofacial and brain abnormalities in mice. (A–C) Normal (A) and abnormal (B,C) day-14 embryonic mice. In (B), the anterior neural tube failed to close, resulting in exencephaly, a condition in which the brain tissue is exposed to the exterior. Later in development, the exposed brain tissue will erode away, resulting in anencephaly. (B,C) Prenatal alcohol exposure can also affect facial development, resulting in a small nose and an abnormal upper lip (open arrow). These facial features are present in fetal alcohol syndrome. (D,E) Three-dimensional reconstructions prepared from magnetic resonance images of the brains of normal (D) and alcohol-exposed (E) 17-day embryonic mice. In the alcohol-exposed specimen, the olfactory bulbs (pink) are absent and the cerebral hemispheres (red) are abnormally united in the midline. Light green, diencephalon; magenta, mesencephalon; teal, cerebellum; dark green, pons and medulla. (Courtesy of K. Sulik.)

pregnancy may lead to loss of fetal brain cells ("one drink" is defined as 12 oz. of beer, 5 oz. of wine, or 1.5 oz. of "hard" liquor). Importantly, alcohol can cause permanent damage to an unborn child at a time before most women even realize they are pregnant.

A mouse model system has been used to examine the developmental stage specificity of alcohol-induced birth defects and to study the effects of alcohol on the developing face and nervous system (Sulik 2005). When mice are exposed to alcohol at the time of gastrulation, it induces defects of the face that are comparable to those in humans with FAS (**Figure 17.7**). Most affected is the upper midface, with deficiencies being particularly evident in the nose and upper lip. The brain is concurrently affected. Malformations are subtle in some affected mice, while in others the neural tube fails to close, resulting in exencephaly (a condition in which brain tissue is subsequently eroded, resulting in anencephaly). In some of the affected mice, the forebrain lacks median tissue and is holoprosencephalic (see Figure 9.31). When mice are exposed to alcohol at other stages of development, different patterns of abnormality result.

Alcohol-induced interference with cell migration, proliferation, adhesion, and survival have all been illustrated in model systems. Hoffman and Kulyk (1999) showed that instead of migrating and dividing, alcohol-treated neural crest cells prematurely initiate their differentiation into facial cartilage. Among the numerous genes that are misregulated following maternal alcohol exposure in mice are several involved in the cytoskeletal reorganization that enables cell movement (Green et al. 2007). In addition, the death of some cell populations is readily observed as early as 12 hours following exposure of mouse embryos to high alcohol concentrations. When the time of alcohol exposure corresponds to the third and fourth weeks of human development, cells that should form the median portion of the forebrain, upper midface, and cranial nerves are killed. One reason for this cell death is that alcohol treatment results in the generation of superoxide radicals that can

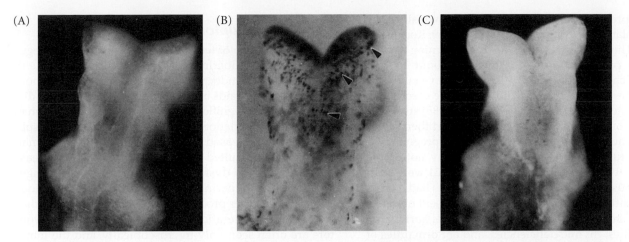

FIGURE 17.8 Possible mechanisms producing fetal alcohol syndrome. (A–C) Cell death caused by alcohol-induced superoxide radicals. Staining with Nile blue sulfate shows areas of cell death. (A) Head region of control day-9 mouse embryo. (B) Head region of alcohol-treated embryo, showing areas of cell death. (C) Head region of embryo treated with both alcohol and superoxide dismutase, an inhibitor of superoxide radicals. The enzyme prevents the alcohol-induced cell death. (D) Inhibition of L1-mediated cell adhesion by alcohol. (A–C from Kotch et al. 1995, courtesy of K. Sulik; D after Ramanathan et al. 1996.)

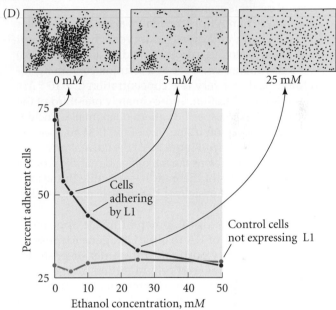

damage cell membranes (**Figure 17.8A–C**; Davis et al. 1990; Kotch et al. 1995; Sulik 2005).

In model systems, antioxidants have been effective in reducing both the cell death and the malformations caused by alcohol (Chen et al. 2004). Abnormal signaling may also underlie excessive cell death. In alcohol-exposed embryos, expression of Sonic hedgehog (which is important in establishing the facial midline structures; see Chapter 9) is downregulated. While the mechanism for this downregulation remains incompletely understood, the finding that Shh-secreting cells placed into the head mesenchyme can prevent the alcohol-induced death of cranial neural crest cells highlights the significance of Shh pathway perturbation for alcohol's teratogenesis (Ahlgren et al. 2002; Chrisman et al. 2004). Another mechanism that may be involved in alcohol's teratogenesis is interference with the ability of the cell adhesion molecule L1 to hold cells together. Ramanathan and colleagues (1996) have shown that at levels as low as 7 mM, an alcohol concentration produced in the blood or brain with a single drink, alcohol can block the adhesive function of the L1 protein in vitro (**Figure 17.8D**). Moreover, mutations in the human *L1* gene cause a syndrome of mental retardation and malformations similar to that seen in severe FAS cases.

Retinoic acid as a teratogen

In some instances, even a compound involved in normal metabolism can have deleterious effects if it is present in large enough amounts and/or at particular times. As we have seen throughout this book, retinoic acid (RA) is a vitamin A derivative that is important in specifying the anterior-posterior axis and in forming the jaws and heart of the mammalian embryo (see Chapters 8, 10, and 12). In normal development, RA forms gradients whereby the relative amount of RA informs the cells of their respective positions along the axis. RA is secreted from particular cells, diffuses, and is then degraded by other cells. However, if RA is present in large amounts, the gradient is perturbed, and cells that normally would not receive high concentrations of this molecule are exposed to it and will respond to it.

In its pharmaceutical form, 13-*cis*-retinoic acid (also called isotretinoin and sold under the trademark Accutane) has been useful in treating severe cystic acne and has been available for this purpose since 1982. The deleterious effects of administering large amounts of vitamin A or its analogues to pregnant animals have been known since the 1950s (Cohlan 1953; Giroud and Martinet 1959; Kochhar et al. 1984). The drug now carries a strong warning against its use by pregnant women. However, about 160,000 women of childbearing age (15–45 years) have taken the drug since it was introduced,* and some have used it during pregnancy. Lammer and co-workers (1985) studied a group of women who inadvertently exposed themselves to retinoic acid and who elected to remain pregnant. Of their 59 fetuses, 26 were born without any noticeable anomalies, 12 aborted spontaneously, and 21 were born with obvious anomalies. The affected infants had a characteristic pattern of anomalies, including absent or defective ears, absent or small jaws, cleft palate, aortic arch abnormalities, thymic deficiencies, and abnormalities of the central nervous system.

This pattern of multiple congenital anomalies is similar to that seen in rat and mouse embryos whose pregnant mothers were given these drugs. Goulding and Pratt (1986) placed 8-day mouse embryos in a solution containing 13-cis-retinoic acid at a very low concentration (2×10^{-6} M). Even at this concentration, approximately one-third of the embryos developed a specific pattern of anomalies, including dramatic reduction in the size of the first and second pharyngeal arches (see Figure 8.33). In normal mice, the first arch eventually forms the maxilla and mandible of the jaw and two ossicles of the middle ear, while the second arch forms the third ossicle of the middle ear as well as other facial bones.

The basis for this developmental insult appears to reside in the ability of RA to alter Hox gene expression and thereby respecify portions of the anterior-posterior axis and inhibit neural crest cell migration from the cranial region of the neural tube (Moroni et al. 1994; Studer et al. 1994). Radioactively labeled RA binds to the cranial neural crest cells and arrests both their proliferation and their migration (Johnston et al. 1985; Goulding and Pratt 1986). The teratogenic period during which cranial neural crest cells are affected occurs on days 20–35 in humans (days 8–10 in mice).

---

*Retinoic acid is a critical public health concern precisely because there is significant overlap between the population using acne medicine and the population of women of childbearing age, and because it is estimated that half of the pregnancies in the U.S. are unplanned (Nulman et al. 1997). Vitamin A is itself teratogenic in megadose amounts. Rothman and colleagues (1995) found that pregnant women who took more than 10,000 international units of vitamin A per day (in the form of vitamin supplements) had a 2% chance of having a baby born with disruptions similar to those produced by RA. According to the new rules of the U.S. Food and Drug Administration, each patient using isotretinoin, each physician prescribing it, and each pharmacy selling it must sign a registry. Moreover, each woman using this drug is expected to take a pregnancy test within 7 days before filling her prescription, agree to use two methods of birth control, and adhere to pregnancy testing on a monthly basis.

Endocrine disruptors

A specialized area of teratogenesis involves the misregulation of the endocrine system during development. According to the endocrine disruptor hypothesis, hormonally active compounds in the environment—**endocrine disruptors**—have a significant impact on the health of human and wildlife populations. The vast amounts of these substances being released into the environment by human activities have resulted in increased incidences of disease and physiological dysfunction. The phenotypic changes produced by endocrine disruptors are not the obvious anatomical changes produced by classic teratogens. Rather, the anatomical alterations induced by endocrine disruptors are often seen only microscopically, and the major changes are physiological. These functional changes are more subtle than those produced by the teratogens, but they can be extremely important phenotypic alterations. Moreover, they may persist in generations after the exposure to the disruptor.

The term *endocrine disruptor* was coined by Theo Colborn, Frederic vom Saal, and Ana Soto in 1993, but these compounds are also known by many other names: hormone mimics, environmental signal modulators, environmental estrogens, or hormonally active agents. Endocrine disruptors are exogenous (coming from outside the body) chemicals that interfere with the normal functions of hormones, and consequently disrupt development (Colborn et al. 1993, 1997). These chemicals can interfere with hormonal functions in many ways.

1. They can be agonists, mimicking the effect of a natural hormone and binding to its receptors. An example is the paradigmatic endocrine disruptor diethylstilbestrol (DES), which binds to the estrogen receptor and mimics the sex hormone estradiol, a common form of estrogen.
2. They can act as antagonists and inhibit the binding of a hormone to its receptor or block the synthesis of a hormone. DDE, a metabolic product of the insecticide DDT, can act as an anti-testosterone, binding to the androgen receptor and preventing normal testosterone from functioning properly.
3. They can affect the synthesis, elimination, or transportation of a hormone in the body. One of the ways that polychlorinated biphenyls (PBCs) disrupt the endocrine system is by interfering with the elimination and degradation of thyroid hormones. The herbicide atrazine elevates the synthesis of estrogen.
4. Some endocrine disruptors can "prime" the organism to be more sensitive to hormones later in life. For instance, bisphenol A experienced in utero causes the embryonic mammary gland to make more estrogen receptors. These extra receptors cause the mammary gland to have altered growth responses to natural estrogen later in life, predisposing it to form cancers (see Wadia et al. 2007).

Given the paradigm of teratogens, it had been thought that there were only a few "bad" agents, and that the only people who received these were pregnant women who inadvertently exposed their embryos to high doses of these chemicals. However, we now recognize that hormone disruptors are everywhere in our technological society (and even in rural areas where pesticides and herbicides are abundant), and that low-dose exposure can be enough to produce significant disabilities later in life. Endocrine disruptors include chemicals that line baby bottles and the brightly colored plastic containers from which we drink our water; chemicals used in cosmetics, sunblocks, and hair dyes; and chemicals used to give cars their "new car smell" and to prevent clothing from being highly flammable.

DIETHYLSTIBESTROL One of the most potent environmental estrogens is **diethylstilbestrol**, or **DES**. This drug was thought to ease pregnancy and prevent miscarriages, and it is estimated that in the United States over 1 million pregnant women and fetuses were exposed to DES between 1947 and 1971. This is probably a small fraction of exposures worldwide.* Research from the 1950s showed that DES actually had no beneficial effects on pregnancy, but it was still prescribed until the FDA banned it in 1971. The ban was imposed when a specific type of tumor (clear-cell adenocarcinoma) was discovered in the reproductive tracts of women whose mothers took DES during pregnancy.

DES interferes with sexual and gonadal development by causing cell type changes in the female reproductive tract (the derivatives of the Müllerian duct, which forms the upper portion of the vagina, cervix, uterus, and oviducts; see Chapter 14). In many cases, DES causes the boundary between the oviduct and the uterus (the uterotubal junction) to be lost, resulting in

---

*In addition to the DES that was administered to pregnant women individually, biologically relevant levels of DES were found in meat; it had been fed to cattle to accelerate their growth (see Knight 1980).

infertility or subfertility (Robboy et al. 1982; Newbold et al. 1983). Moreover, the distal Müllerian ducts often fail to come together to form a single cervical canal (**Figure 17.9**).

Symptoms similar to the human DES syndrome occur in mice exposed to DES in utero, allowing the mechanisms of this endocrine disruptor to be uncovered. Normally, the regions of the female reproductive tract are specified by the HoxA genes, which are expressed in a nested fashion throughout the Müllerian duct (**Figure 17.10**). Ma and colleagues (1998) showed that the effects of DES on the female mouse reproductive tract could be explained as the result of altered *Hoxa10* expression in the Müllerian duct. DES was injected under the skin of pregnant mice, and the fetuses were allowed to develop almost to birth. When the fetuses from the DES-injected mothers were compared with fetuses from mothers that had not received DES, it was seen that DES almost completely repressed the expression

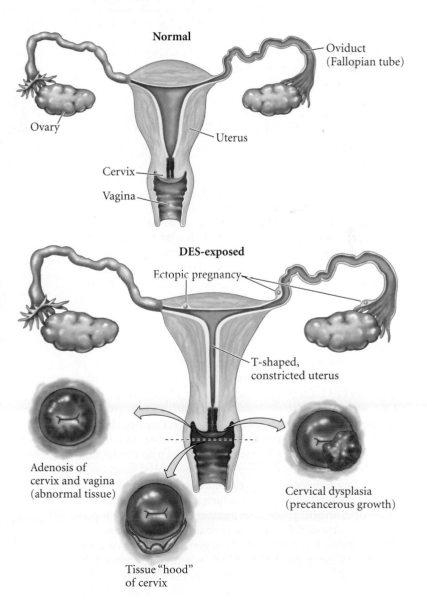

FIGURE 17.9 Genital anomalies can occur in women exposed to DES in utero. In these "DES daughters," the cervical tissue (red) is often displaced into the vagina. Such individuals may have a uterus that is T-shaped and constricted, as well as adenosis of the cervix and vagina (where the lining differentiates into mucosal cells), precancerous cells, ectopic pregnancies, adenocarcinomas, and other effects.

(A)

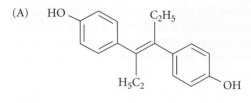

(B)

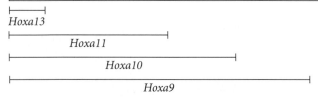

Hoxa11
high expression
border at uterus/
oviduct boundry

Hoxa9
expression border
throughout most
of oviduct

Hoxa13
expression
Cervix

Hoxa10
expression border
at uterus/oviduct boundary

├─┤ *Hoxa13*

├────────┤ *Hoxa11*

├───────────┤ *Hoxa10*

├──────────────┤ *Hoxa9*

FIGURE 17.10 Effects of DES exposure on the female reproductive system. (A) The chemical structure of DES. (B) 5_ HoxA gene expression in the reproductive system of a normal 16.5-day embryonic female mouse. A whole-mount in situ hybridization of the *Hoxa13* probe is shown (red) along with a probe for *Hoxa10* (purple). *Hoxa9* expression extends throughout the uterus and about through much of the presumptive oviduct. *Hoxa10* expression has a sharp anterior border at the transition between the presumptive uterus and the oviduct. *Hoxa11* has the same anterior border as *Hoxa10*, but its expression diminishes closer to the cervix. *Hoxa13* expression is found only in the cervix and upper vagina. (After Ma et al. 1998.)

of *Hoxa10* in the Müllerian duct (**Figure 17.11**). This repression was most pronounced in the stroma (mesenchyme) of the duct, where experimental embryologists had localized the effect of DES (Boutin and Cunha 1997). The case for DES acting through repression of *Hoxa10* is strengthened by the phenotype of the *Hoxa10* knockout mouse (Benson et al. 1996; Ma et al. 1998), in which there is a transformation of the proximal quarter of the uterus into oviduct tissue, as well as abnormalities of the uterotubal junction.

The link between Hox gene expression and uterine morphology is the Wnt proteins, which are associated with cell proliferation and protection against apoptosis. The reproductive tracts of DES-exposed female mice resemble those of *Wnt7a* knockout mice. The Hox genes and the Wnt genes communicate to keep each other activated. Moreover, the Hox and Wnt proteins are involved in the specification and morphogenesis of the reproductive tissues (**Figure 17.12**).

However, DES, acting through the estrogen receptor, represses the *Wnt7a* gene. This repression prevents the maintenance of the Hox gene expression pattern, and also prevents the activation of another Wnt gene, *Wnt5a*, which encodes a protein necessary for cell proliferation (Miller et al. 1998; Carta and Sassoon 2004).

The DES tragedy is a complex story of public policy, medicine, and developmental biology (Bell 1986; Palmlund 1996). However, according to biologists Frederick vom Saal, Ana Soto, and others, the public and the pharmaceutical industry have learned little from the DES tragedy, and we may be experiencing the same endocrine disruption today, only on a much larger scale. These researchers claim that some of the major constituents of plastics are estrogenic compounds, and that they are present in doses large enough to have profound effects on sexual development and behavior. Two of these products are bisphenol A and nonylphenol.

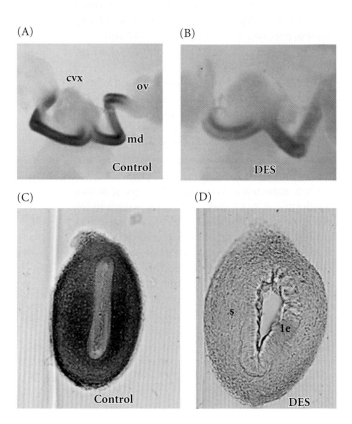

FIGURE 17.11 In situ hybridization of a *Hoxa10* probe shows that DES exposure represses *Hoxa10*. (A) Normal 16.5-day embryonic female mice show *Hoxa10* expression from the boundary of the cervix through the uterus primordium and most of the oviduct (cvx, cervix; md, Müllerian duct; ov, ovary). (B) In mice exposed prenatally to DES, this expression is severely repressed. (C) In control female mice at 5 days after birth (when reproductive tissues are still forming), a section through the uterus shows abundant expression of *Hoxa10* in the uterine mesenchyme. (D) In female mice that are given high doses of DES 5 days after birth, *Hoxa10* expression in the mesenchyme is almost completely suppressed (le, luminal epithelium; s, stroma). (After Ma et al. 1998.)

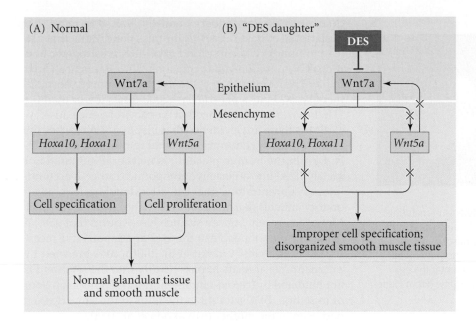

(A) Normal

(B) "DES daughter"

FIGURE 17.12 Misregulation of Müllerian duct morphogenesis by DES. (A) During normal morphogenesis, the *Hoxa10* and *Hoxa11* genes in the mesenchyme are activated and maintained by Wnt7a from the epithelium. Wnt7a also induces Wnt5a in the mesenchyme, and Wnt5a protein both maintains *Wnt7a* expression and causes mesenchymal cell proliferation. Together, these factors specify and order the morphogenesis of the uterus. (B) DES, acting through the estrogen receptor, blocks *Wnt7a* expression. Proper activation of the Hox genes and *Wnt5a* in the mesenchyme does not occur, leading to a radically altered morphology of the female genitalia. (After Kitajewsky and Sassoon 2000.)

PLASTICS AND PLASTICITY: NONYLPHENOL Estrogenic compounds are in the food we eat and in the plastic wrapping that surrounds them. The discovery of the estrogenic effect of plastic stabilizers was made in a particularly alarming way. Investigators at Tufts University Medical School had been studying estrogen-responsive tumor cells, which require estrogen in order to proliferate. Their studies were going well until 1987, when the experiments suddenly went awry. Their control cells began to show high growth rates, suggesting stimulation comparable to that of the estrogen-treated cells. It was as if someone had contaminated the control culture medium by adding estrogen to it. What was the source of contamination? After spending 4 months testing all the components of their experimental system, the researchers discovered that the source of estrogen was the plastic containers that held their water and serum. The company that made the containers refused to describe its new process for stabilizing the polystyrene plastic, so the scientists had to discover it themselves.

The culprit turned out to be *p*-nonylphenol, a compound that is also used to harden the plastic of the pipes that bring us water and to stabilize the polystyrene plastics that hold water, milk, orange juice, and other common liquid food products (Soto et al. 1991; Colborn et al. 1996). This compound is also the degradation product of detergents, household cleaners, and contraceptive creams. Nonyphenol has been shown to alter reproductive physiology in female mice and to disrupt sperm function. It is also correlated with developmental anomalies in wildlife (Fairchild et al. 1999; Hill et al. 2002; Kim et al. 2002; Adeoya-Osiguwa et al. 2003; Kurihara et al. 2007).

BISPHENOL A Bisphenol A (BPA) was actually synthesized as an estrogenic compound in the 1930s. In the early years

of hormone research, the steroid hormones were very difficult to isolate, so chemists manufactured synthetic analogues that would accomplish the same tasks. BPA was synthesized by Dodds, who demonstrated it to be estrogenic in 1936. (Two years later, Dodds synthesized DES.) Later, polymer chemists realized that BPA could be used in plastic production, and today it is one of the top 50 chemicals in production. Four corporations in the United States make almost 2 billion pounds of it each year for use in the resin lining in most cans, the polycarbonate plastic in baby bottles and children's toys, and dental sealant. BPA is also used in making the brightly colored polycarbonate water bottles. Its modified form, tetrabromo-bisphenol A, is the major flame retardant on the world's fabrics.

However, as several laboratories have shown (Krishnan et al. 1993; vom Saal 2000; Howdeshell et al. 2003), BPA is not fixed in plastic forever. If you let water sit in an old polycarbonate rat cage at room temperature for a week, you can measure around 300 µg per liter of BPA in the water. That is a biologically active amount—a concentration that will reverse the sex of a frog and cause weight changes in the uterus of a young mouse. It also can cause chromosome anomalies. When a laboratory technician mistakenly rinsed some polycarbonate cages in an alkaline detergent, the female mice housed in the cages had meiotic abnormalities in 40% of their oocytes (the normal amount is about 1.5%). When BPA was administered to pregnant mice under controlled circumstances, Hunt and her colleagues (2003) showed that a short, low-dose exposure to BPA was sufficient to cause meiotic defects in maturing mouse oocytes (**Figure 17.13**).

BPA at environmentally relevant concentrations can cause abnormalities in fetal gonads, low sperm counts, and behavioral changes when these fetuses become adults

(A)   (B)

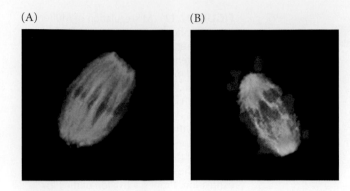

FIGURE 17.13 Bisphenol A causes meiotic defects in maturing mouse oocytes. (A) Chromosomes (red) normally line up at the center of the spindle during first meiotic metaphase. (B) Short exposures to BPA cause chromosomes to align randomly on the spindle. Different numbers of chromosomes then enter the egg and polar body, resulting in aneuploidy and infertility. (From Hunt et al. 2003, courtesy of P. Hunt.)

(vom Saal et al. 1998; Palanza et al. 2002; Kubo et al. 2003). Indeed, recent research has implicated BPA (and PCBs) in a suite of trends, including the lowering of human sperm counts, an increase in prostate enlargement, and a lowering of the age of female sexual maturation (vom Saal 2000; vom Saal and Hall 2005).

Research from several laboratories showed that female mice exposed in utero to low doses of BPA underwent sexual maturity faster than unexposed mice. They also showed altered mammary development at puberty, alterations in the organization of their breast tissue and ovaries, and altered estrous cyclicity as adults (Howdeshell et al. 1999, 2000; Markey et al. 2003). Each mammary gland produced more terminal buds and was more sensitive to estrogen. This may predispose these mice toward breast cancer as adults (Muñoz-de-Toro et al. 2005). What is remarkable in these studies is that the dosage of BPA was 2000 times lower than the dosage set as "safe" by the U.S. government.

See WEBSITE 17.3 Our stolen future

BPA also appears to make breast tissue more sensitive to estrogens later in life, predisposing women exposed to BPA in utero to breast cancer. Fetal exposure to BPA caused the development of early-stage cancer in the mammary glands of one-third of the mice exposed to environmentally relevant doses of BPA (Figure 17.14; Muñoz-de-Toro et al. 2005; Murray et al. 2007). None of the control mice developed such tumors. Fetal exposure to BPA also increased the number of "preneoplastic lesions" (areas of rapid cell growth in the ducts) three- to fourfold. Furthermore, daily gestational exposure to as little as 25 ng/kg body weight BPA followed at puberty by a "subcarcinogenic dose" of a chemical mutagen resulted in the formation of tumors only in the animals exposed to BPA (Duran-

do et al. 2007; Vandenberg et al. 2007). It is thought that gestational exposure to BPA induces conditions that can lead to tumors when a second exposure of estrogenic hormones or mutation-producing agents is experienced later in life. The lowest dose of BPA to alter mammary development was 25 parts per *trillion*—2000 times lower than the U.S. Environmental Protection Agency reference dose.

But is there any evidence that bisphenol A reaches the human fetus in concentrations that matter? Unfortunately, BPA in the human placenta is neither eliminated nor metabolized into an inactive compound; rather, it accumulates to concentrations that can alter development in laboratory animals (Ikezuki et al. 2002; Schönfelder et al. 2002). Moreover, recent survey studies have found that 95% of urine samples of people in the U.S. and Japan have measurable BPA levels (Calafat et al. 2005). BPA ingested by pregnant rats appears to pass readily into the fetus and is not hindered by steroid-binding hormones. Within an hour of ingestion, BPA is found in the fetus at the same doses it had been in the mother (Miyakoda et al. 1990).

Females are not the only ones affected by BPA. As mentioned above, BPA has been implicated (along with other estrogenic compounds) in the decline in sperm quality and number and in the increase in prostate enlargement seen in men. When vom Saal and colleagues (1997) gave pregnant mice 2 parts per billion BPA—that is, 2 nanograms per gram of body weight—for the 7 days at the end of pregnancy (equivalent to the period when human reproductive organs are developing), treated males showed an increase in prostate size of about 30% (Wetherill 2002; Timms et al. 2005).

The plastics industry has countered these claims, saying the evidence is meager and that the experiments cannot be repeated (Cagen et al. 1999; see also Lamb 2002). However, a review of the industry's own studies (claiming that mice exposed in utero to BPA do not have enlarged prostates or low sperm counts) points out that the positive control of the industry-sponsored research did not produce the expected effects. Indeed, when one study (Lernath et al. 2008) showed that BPA disrupted monkey brain development (at concentrations lower than what the U.S. EPA considers safe), the American Chemical Council replied that "there is no direct evidence that exposure to bisphenol-A adversely affects human reproduction or development" (see Layton and Lee 2008; Gilbert and Epel 2009).* Reviewing the literature, vom Saal and Hughes (2005; Chapel Hill Consensus 2007) conclude that BPA is one of the most dangerous chemicals known and that governments should consider banning its use in products containing liquids that humans and animals might drink.

---

*The catch is that "direct evidence" would mean testing the drugs in known concentrations on human fetuses. Interestingly, in the absence of government regulation, Nalgene and Wal-Mart voluntarily stopped making and selling BPA-containing bottles.

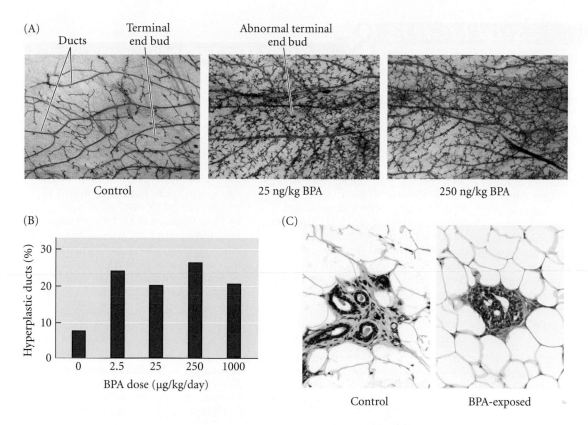

FIGURE 17.14 Bisphenol A induces mammary lesions in mice. (A) Photomicrographs of mammary gland tissue show profound differences between a control animal and animals exposed perinatally to nanogram levels of BPA. The increase in terminal end buds and branching in the BPA-exposed mammary glands predisposes the animals to cancer later in life. (B) The percentage of mammary glands showing intraductal hyperplasia (indicating a cancer-prone state) is significantly increased at postnatal day 50 in BPA-treated animals. (C) Later in life, BPA-exposed animals display significant cancer in situ. (From Murray et al. 2007 and Soto et al. 2008, courtesy of Ana Soto.)

Other teratogenic agents

In addition to natural chemicals, hundreds of new artificial compounds come into general use each year in our industrial society. Pesticides and organic mercury compounds have caused neurological and behavioral abnormalities in infants whose mothers have ingested them during pregnancy. Moreover, drugs that are used to control diseases in adults may have deleterious effects on fetuses. Such drugs include cortisone, warfarin, tetracycline, and valproic acid (see Table 17.2).

Over 87,000 artificial chemicals are currently licensed for use in the United States, and around 500 new compounds are being made each year. The Toxic Substances Control Act of the United States assumes chemicals are safe unless demonstrated to be otherwise, and only around 8,000 chemicals have been tested for their potential health effects (Johnson 1980; EPA 2008). Although teratogenic compounds have always been with us, the risks increase as more and more untested compounds enter our environ-

ment. Most industrial chemicals have not been screened for their teratogenic effects. Standard screening protocols are expensive, long, and subject to interspecies differences in metabolism. There is still no consensus on how to test a substance's teratogenicity for human embryos.

HEAVY METALS Heavy metals such as zinc, lead, and mercury are powerful teratogens. Industrial pollution has resulted in high concentrations of heavy metals in the environment in many places (see Gilbert and Epel 2009). In the former Soviet Union, the unregulated "industrial production at all costs" approach left behind a legacy of soaring birth defect rates. In some regions of Kazakhstan, heavy metals are found in high concentrations in drinking water, vegetables, and the air. In such locations, nearly half the people tested have extensive chromosome breakage, and in some areas the incidence of birth defects has doubled since 1980 (Edwards 1994). In the United States, lax enforcement of antipollution laws has led to the contamination of most lakes in the country by heavy metals. This

SIDELIGHTS & SPECULATIONS

Testicular Dysgenesis Syndrome and the Transgenerational Effects of Endocrine Disruptors

As noted earlier, our technological environment has increasingly become an estrogenic environment. As such, one would expect not only females but also males to have reproductive problems, and they do. Laboratories from all over the world have reported cases of a "testicular dysgenesis* syndrome" in men. This syndrome encompasses low sperm count, poorly formed testes and penises, and testicular tumors (**Figure 17.15**).

Dysgenesis is from the Greek for "bad beginning" and denotes defects in development.

In the past 3 decades, there has been an increase in testicular cancers and a decrease in sperm concentration throughout the industrialized world (Carlsen et al. 1992; Aitken et al. 2004). The sperm count (number of sperm per ml) has dropped precipitously throughout much of Europe and the Americas. Skakkebaek and his team at Copenhagen University reviewed 61 international studies

done between 1938 and 1992, involving 14,947 men. They found the average sperm count had fallen from 113 million per ml in 1940 to 66 million in 1990 (Carlsen et al. 1992). In addition, the number of "normal" sperm fell from 60 million per ml to 20 million in the same period.

Subsequent studies have confirmed and extended Skakkebaek's findings (Merzenich et al. 2010). A survey of

Figure 17.15   Developmental estrogen syndrome is manifest in climbing rates of breast cancer and testicular dysgenesis. The rise of testicular cancer (A) parallels the rise of breast cancer (B) and anomalies of penis development (C) such as hypospadias (failure to completely close the penis). Sperm counts (D) among North American males have declined nearly 50% within the past century, and the decline has even been steeper among European men. (After Sharpe and Irvine 2004.)

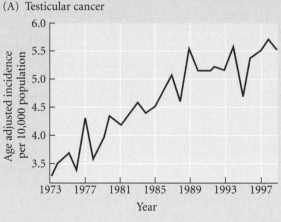

(A) Testicular cancer

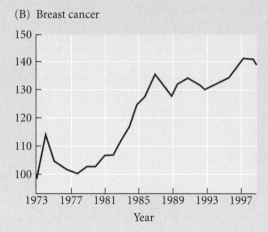

(B) Breast cancer

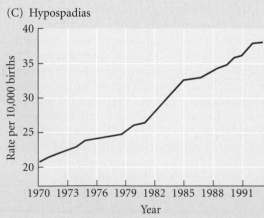

(C) Hypospadias

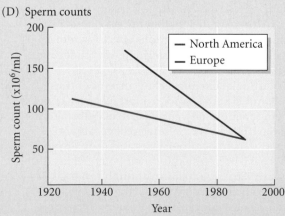

(D) Sperm counts

SIDELIGHTS & SPECULATIONS (Continued)

(A)

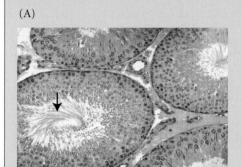

(B)

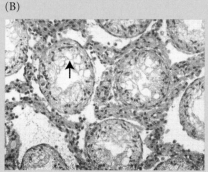

Figure 17.16   Cross section of seminiferous tubules from the testes of (A) a control rat and (B) a rat whose grandfather was born from a mother that had been injected with vinclozolin. This rat was infertile. The arrow in (A) shows the tails of the sperm. The arrow in (B) shows the lack of germ cells in the much smaller tubule. (From Anway et al. 2005, courtesy of M. K. Skinner.)

1,350 sperm donors in Paris found that sperm counts declined by around 2% each year from the 1970s through the mid-1990s, with younger men having the poorest quality semen (Auger et al. 1995). (Average sperm count was 90 million sperm per ml in 1973 and only 60 million per ml in 1992.) A study from Finland showed a similar decline, and also showed that the average weight of the testes had decreased, while the proportion of fibrous testicular tissue had increased at the expense of the sperm-producing seminiferous tubules (Pajarinen et al. 1997). A Scottish study showed that men born in the 1970s were producing some 24% fewer motile sperm in their ejaculate than men born in the 1950s (Irvine et al. 1996). In addition to the drop in sperm count documented in these studies, there has also been an increase in testicular cancers over recent decades (see Figure 17.15A,C).

Sharpe (1994) has suggested that testicular dysgenesis syndrome may be due in large part to endocrine disruptors. While a chain of causation has not been completely established (see Sharpe and Irvine 2004), there is evidence that the pathologies of this syndrome can be caused by environmentally relevant concentrations of endocrine disruptors. Indeed, all the developmental anomalies (but not the testicular tumors) can be induced by administering phthalate derivatives to pregnant rats (Fisher et al. 2003). Among male mice exposed in utero to dibutyl phthalate, more than 60% exhibited cryptorchidism (undescended testicles), hypospadias (misplaced urinary aperture), low sperm count,

and testis abnormalities—very similar to conditions found in human testicular dysgenesis syndrome.

Phthalates are ubiquitous in industrialized society and are widely used in plastics and in cosmetics (that great "new car smell" consists largely of volatilizing phthalates). Evidence obtained from newborns showed that male babies exposed in utero to relatively high phthalate levels had some morphological changes in their testes (Duty et al. 2003).

Other endocrine disruptors that adversely affect sperm are dioxins, nonylphenol, bisphenol A, acrylamide, and certain pesticides and herbicides* (see Aitken et al. 2004; Newbold et al. 2006). The sunscreen 4-MBC, a camphor derivative, has been found to decrease the size of the testes and prostate glands, and it can delay male puberty in rats (Schlumpf et al. 2004). Pesticides may be critically important in impairing male fertility. The link between pesticides and infertility has been known for a long time (Carson 1962; Colborn et al. 1996).

*Genistein, the estrogenic compound found in soy (and soy products such as tofu) is also being scrutinized (Newbold et al. 2001; Cederroth et al. 2010). For people eating an omnivorous diet with soy supplementation, genistein should not be a concern (indeed, this compound may protect against certain cancers). However, researchers at the National Institute of Environmental Health Science are worried that infants born to vegan or vegetarian mothers and who are fed solely soy-based formula may develop abnormalities of the reproductive system and thyroid glands.

One of the most important endocrine disrupting pesticides is DDT. In humans, DDT has been linked to preterm births and immature babies, and it is banned in the United States (Longnecker et al. 2001). DDT is probably not the active compound, though. It breaks down into DDE (dichlorodiphenyldichloroethylene), and DDE binds to the androgen receptor, preventing testosterone binding (Xu et al. 2006).

The fungicide vinclozolin (used extensively in grape farming) also works as an anti-androgen, inserting itself into the androgen receptor and preventing testosterone from binding there (Grey et al. 1999; Monosson et al. 1999). Vinclozolin administered to rats around the time of birth causes penis malformations, absent sex accessory glands, and very low (around 20% of normal) sperm production (Figure 17.16). Male rats born to mothers injected with vinclozolin late in pregnancy are sterile. This impaired fertility affects not only the generation exposed in utero but has been observed in males for at least four generations afterward. The sons of rats injected with vinclozolin during mid-pregnancy were able to reproduce, but their testis cells underwent apoptosis more than usual, their sperm count dropped 20%, and the sperm that remained had significantly lowered motility. When affected males were mated with normal females, the male offspring also had this testicular dysgenesis syndrome. Some of the offspring were sterile, and some had reduced fertility. The

(Continued on next page)

study (Anway et al. 2005) ended after the fourth generation of males continued to show low sperm count, low sperm motility, prostate disease, and high testicular cell apoptosis. This transgenerational effect is thought to be the result of methylation of genes involved in spermatogenesis (Chang et al. 2006).

Some scientists argue that these claims are exaggerated and that their tests on mice indicate that litter size, sperm concentration, and development are not affected by environmentally relevant concentrations of environmental estrogens. However, investigations by Spearow and colleagues (1999) have shown a remarkable genetic difference in sensitivity to estrogen among different mouse strains. The strain that was used for testing environmental estrogens, the CD-1 strain of laboratory mice, is at least 16 times more resistant to endocrine disruption than the most sensitive strains, such as B6. When estrogen-containing pellets were implanted beneath the skin of young male CD-1 mice, very little happened. However, when the same pellets were placed beneath the skin of B6 mice, their testes shrank and the number of sperm seen in the seminiferous tubules dropped dramatically (**Figure 17.17**). This widespread range of sensitivities has important consequences for determining safety limits for humans.

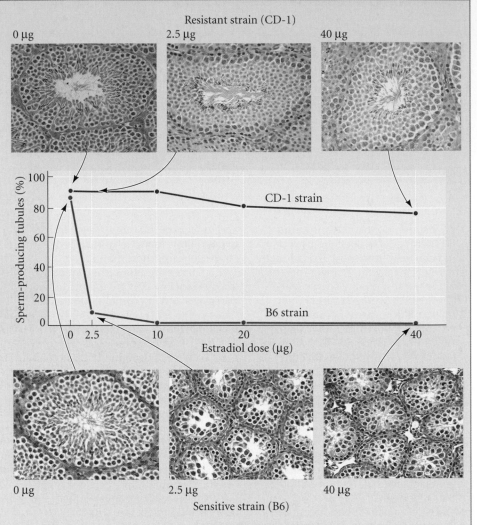

Figure 17.17 Effects of estrogen implants on different strains of mice. The graph shows the percentage of seminiferous tubules containing elongated spermatozoa. (The mean +/– standard error is for an average of six individuals.) The micrographs show cross sections of the testicles and are all at the same magnification. Forty µg of estradiol did not affect spermatogenesis in the CD-1 strain, but as little as 2.5 µg almost completely abolished spermatogenesis in the B6 strain. (After Spearow et al. 1999, photographs courtesy of J. L. Spearow.)

contamination is especially prevalent where mining interests have been allowed to disgorge metal-containing slag into streams feeding the lakes (USGS 2009). An International Joint Commission of the U.S. and Canada had warned that pregnant women should not eat fish caught in the Great Lakes.

Lead and mercury can damage the developing nervous system (Bellinger 2005). The polluting of Minamata Bay, Japan, with mercury in 1956 produced brain and eye deficiencies both by transmission of the mercury across the placenta and by its transmission through mother's milk.

Mercury is selectively absorbed by regions of the developing cerebral cortex (Eto 2000; Kondo 2000; Eto et al. 2001). When pregnant mice are given mercury on day 9 of gestation, nearly half the pups are born with small brains or small eyes* (O'Hara et al. 2002).

---

*At the same time in America, the lead industry lobbied to assure consumers that lead was safe to use in paints. The U.S. banned leaded paint in 1977—52 years after it was banned in Europe (see Steingraber 2003).

PATHOGENS Another class of teratogens includes viruses and other pathogens. Gregg (1941) first documented the fact that women who contracted rubella (German measles) during the first trimester of their pregnancy had a 1 in 6 chance of giving birth to an infant with eye cataracts, heart malformations, or deafness. This study provided the first evidence that the mother could not fully protect the fetus from the outside environment. The rubella virus is able to enter many cell types, where it produces a protein that prevents mitosis by blocking kinases that allow the cell cycle to progress (Atreya et al. 2004). Thus, numerous organs are affected, and the earlier in pregnancy the rubella infection occurs, the greater the risk that the embryo will be malformed. The first 6 weeks of development appear to be the most critical, because that is when the heart, eyes, and ears are formed (see Figure 17.5). The rubella epidemic of 1963–1965 probably resulted in over 10,000 fetal deaths and 20,000 infants with birth defects in the United States (CDC 2002b). Two other viruses, cytomegalovirus and the herpes simplex virus, are also teratogenic. Cytomegalovirus infection of early embryos is nearly always fatal; infection of later embryos can lead to blindness, deafness, cerebral palsy, and mental retardation.

Bacteria and protists are rarely teratogenic, but some of them are known to damage human embryos. *Toxoplasma gondii*, a protist carried by rabbits and cats (and their feces), can cross the placenta and cause brain and eye defects in the fetus. *Treponema pallidum*, the bacterium that causes syphilis, can kill early fetuses and produce congenital deafness and facial damage in older fetuses.

Cancer as a Disease of Development

Many endocrine disruptors are known to cause tumors as well as developmental abnormalities, and cancer is increasingly being studied as a disease of development. Although the genetics-centered approach to malignancy, which sees cancers as resulting from an accumulation of mutations, explains the formation of numerous tumor types, it is not the whole story. As Folkman and colleagues (2000) noted, the genetic approach to cancer therapy must be complemented by a developmental perspective seeing that "epigenetic, cell-cell, and extracellular interactions are also pivotal in tumor progression."

Indeed, both carcinogenesis and teratogenesis can be seen as diseases of tissue organization and intercellular communication. Some defects in signaling pathways can cause cancers, while other defects in the same pathways (even the same protein, but at different sites) can cause malformations (see Cohen 2003). There are many reasons to view malignancy and metastasis in terms of development. Four of them will be discussed here:

- Context-dependent tumor formation
- Deficient stem cell regulation in tumor formation
- Reactivation of embryonic migration pathways
- Epigenetic reprogramming of cancer cells

Context-dependent tumors

Many tumor cells have normal genomes, and whether or not these tumors are malignant depends on their environment. The most remarkable of these cases is the **teratocarcinoma**, which is a tumor of germ cells or stem cells (see Chapter 16; Illmensee and Mintz 1976; Stewart and Mintz 1981). Teratocarcinomas are malignant growths of cells that resemble the inner cell mass of the mammalian blastocyst, and they can kill the organism. However, if a teratocarcinoma cell is placed on the inner cell mass of a mouse blastocyst, it will integrate into the blastocyst, lose its malignancy, and divide normally. Its cellular progeny can become part of numerous embryonic organs. Should its progeny form part of the germ line, sperm or egg cells formed from the tumor cell will transmit the tumor genome to the next generation (see Figure 16.10). Thus, whether the cell becomes a tumor or part of the embryo can depend on its surrounding cells.

DEFECTS IN CELL-CELL COMMUNICATION This brings us to the idea that cancer can be caused by miscommunication between cells. In many cases, tissue interactions are required to prevent cells from dividing. Thus, tumors can arise through defects in tissue architecture, and the surroundings of a cell are critical in determining malignancy (Sonnenschein and Soto 1999, 2000; Bissell et al. 2002). Studies have shown that tumors can be caused by altering the structure of the tissue, and that these tumors can be suppressed by restoring an appropriate tissue environment (Coleman et al. 1997; Weaver et al. 1997; Sternlicht et al. 1999). In particular, whereas 80% of human tumors are from epithelial cells, these cells do not always appear to be the site of the cancer-causing lesion. Rather, epithelial cell cancers are often caused by defects in the mesenchymal stromal cells that surround and sustain the epithelia. When Maffini and colleagues (2004) recombined normal and carcinogen-treated epithelia and mesenchyme in rat mammary glands, tumorous growth of mammary *epithelial* cells occurred not in carcinogen-treated epithelia, but only in epithelia placed in combination with mammary mesenchyme that had been exposed to the carcinogen. Thus, the carcinogen caused defects in the mesenchymal stroma of the mammary gland, and apparently the treated mammary cells could not hold back the epithelium from dividing.

DEFECTS IN PARACRINE PATHWAYS This brings us to the next notion: tumors can occur by disruptions of paracrine signaling between cells. In some instances, tumor cells reactivate paracrine pathways that were used during development. Indeed, this finding links to the importance of stromal tissue just mentioned. Many tumors, for instance, secrete the paracrine factor Sonic hedgehog. Shh does not act on the tumor cells themselves but on the stromal cells, causing the stromal cells to produce factors that support the tumor cells. If the Shh pathway is blocked, the tumor regresses (Yauch et al. 2008, 2009).

In addition, Shh is normally required for the maintenance of hematopoietic stem cells, and inhibitors of the Shh pathway can reverse certain leukemias (Zhao et al. 2009). Cyclopamine, a teratogen that blocks Shh, can prevent certain tumors from growing (Berman et al. 2002, 2003; Thayer et al. 2003).

DIFFERENTIATION THERAPY And this brings us to a possibility for treatment: differentiation therapy. In 1978, Pierce and his colleagues noted that cancer cells were in many ways reversions to embryonic cells, and they hypothesized that cancer cells should revert to normalcy if they were made to differentiate. That same year, Sachs (1978) discovered that certain leukemias could be controlled by making their cells differentiate rather than proliferate. One of these leukemias, acute promyelocytic leukemia (APL), is caused by a somatic recombination creating a new transcription factor, one of whose parts is a retinoic acid receptor. The expression of this transcription factor in neutrophil progenitors causes the cell to become malignant (Miller et al. 1992; Grignani et al. 1998). Treatment of APL patients with all-*trans* retinoic acid causes remission of APL in more than 90% of cases, since the additional retinoic acid is able to effect the differentiation of the leukemic cells into normal neutrophils (Hansen et al. 2000; Fontana and Rishi 2002).

The cancer stem cell hypothesis: Cancer as a disease of stem cell regulation

Another aspect of viewing cancers as diseases of development is that the properties of cancer cells resemble those of adult stem cells. Indeed, it is possible that tumor cells are stem cells that have mutated such that they can exist outside their niche. Most stem cells die when they leave their niche, but many tumor cells have amplified the genes that block apoptosis. Melanomas for instance, are tumors of the pigment cells. Recent studies have demonstrated that the MITF transcription factor, in addition to activating tyrosinase and other melanin-forming genes, also activates the anti-apoptosis gene *BLC2*.* MITF is important in the self-renewal of melanocyte stem cells and (through the activation of *BCL2*) in the survival of these cells (Nishimura et al. 2005). That is probably why melanomas are so resistant to treatment. Moreover, melanomas have amplified *MITF*-containing regions on their chromosomes (i.e., they have several copies of the *MITF* gene; McGill et al. 2002; Garraway et al. 2005; Hornyak et al. 2009). It is possible that melanomas are not differentiated melanocytes that have reverted to a primitive stage, but are melanocyte stem cells that have amplified their *MITF* genes and thus are able to survive outside their niche.

Such thinking has led to the **cancer stem cell hypothesis** (Figure 17.18), which postulates that the malignant part of a tumor actually arises from an adult stem cell (Reya et al. 2001; Dean et al. 2005). In numerous cases, including glioblastomas (the major brain tumor), prostate cancer, melanomas, and myeloid leukemias, there is a rapidly

---

*It is probable that many cancer cells survive only if their anti-apoptosis genes or anti-senescence genes are suppressed (see Sharpless and DePinho 2005).

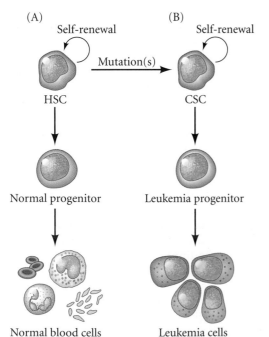

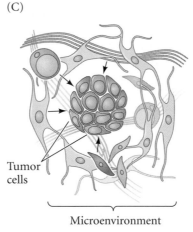

FIGURE 17.18 Model of cancer stem cell production, using leukemia (a white blood cell tumor) as an example. (A) A hematopoietic stem cell (HSC) usually gives rise to normal blood progenitor cells that can become mature white blood cells. (B) If the HSC undergoes mutations or epigenetic changes involving gene activation, it can become a cancer stem cell (CSC) that can divide to produce more of itself plus other relatively differentiated (leukemic) cells. As in normal blood development, the CSC retains the ability for self-renewal and thereby becomes the malignant portion of the cancer. (C) The CSC may also be produced by changes in the microenvironment, which allows certain cells to display a stem cell phenotype that they would not otherwise possess. (After Rosen and Jordan 2009.)

dividing stem cell population that gives rise to more cancer stem cells and to populations of relatively slowly dividing differentiated cells (Lapidot et al. 1994; Bonnet and Dick 1997; Singh et al. 2004; Schatton 2008). Whether the tumor is initiated by an adult stem cell "gone bad" or by a more differentiated cell that has regained stem cell abilities is a matter of controversy (Gupta et al. 2009; Rosen and Jordan 2009; Schatton et al. 2009). In some tumors, such as prostate cancer, the origin of the tumor is most likely a normal adult stem cell that has escaped the control of its niche (Wang et al. 2009).

Cancer as a return to embryonic invasiveness: Migration reactivated

Another crucial point in considering cancers as diseases of disrupted development involves **metastasis**, the invasion of the malignant cell into other tissues. Like embryonic cells, tumor cells do not usually stay put—they migrate and form colonies. Chapter 3 discussed the roles of cadherin proteins in the sorting-out of cells to form tissues during development, and how cells form boundaries and segregate into tissues by altering the strengths of their attachments. In cancer metastasis, this property is lost: cadherin levels are downregulated, and the strength of attachment to the extracellular matrix and other types of cells becomes greater than the cohesive force binding the tissue together. As a result, the cells become able to spread into other tissues (Foty and Steinberg 1997, 2004).

Another phase of metastasis involves the digestion of extracellular matrices by **metalloproteinases**. These enzymes are used by migrating embryonic cells to digest a path to their destination. They are commonly secreted by trophoblast cells, axon growth cones, sperm cells, and somitic cells. Metalloproteinases can be reactivated in malignant cancer cells, allowing the cancer to invade other tissues. The presence of these enzymes is a marker that the tumor is particularly dangerous (see Gu et al. 2005).

Cancer and epigenetic gene regulation

In Chapter 15, we saw evidence that the methylation patterns of mammalian genes change with age. We specifically looked at genes that might cause elements of the aging phenotype. But what would happen if the random, age-dependent patterns of gene methylation altered the transcription of the genes regulating cell division?

Two types of genes control cell division. The first are **oncogenes**, which promote cell division, reduce cell adhesion, and prevent cell death. These are the genes that can promote tumor formation and metastasis. The second set of regulatory genes are the **tumor suppressor genes**. These genes usually put the brakes on cell division and increase the adhesion between cells; they can also induce apoptosis of rapidly dividing cells. The interplay of oncogenes and tumor suppressor genes has to be very finely regulat-

ed. Thus, one might get cancer if faulty methylation either inappropriately methylated the tumor suppressor genes (turning them off) or inappropriately demethylated the oncogenes (turning them on; **Figure 17.19**).

Some genes may be oncogenes in one set of cells and tumor suppressor genes in another set of cells. In the breast, for instance, estrogen receptors can act as oncogenes for estrogen-dependent breast cancer. In the colon, however, estrogen stops the proliferation of cells, and estrogen receptors function as tumor-suppressor genes. Issa and colleagues (1994) showed that in addition to the age-associated methylation of estrogen receptors, there was a much higher level of DNA methylation in the estrogen receptor genes in colon cancers. Even the smallest colon cancers had nearly 100% methylation of the cytosines in the promoter of the estrogen receptor gene.

The epigenetic causation of cancer does not exclude a genetic cause. Indeed, several studies indicate that these mechanisms augment one another. Numerous mutations occur in each cancer cell, and recent evidence suggests that as many as 14 significant tumor-promoting mutations are found in each cancer cell (Sjöblom et al. 2006). Jacinco and Esteller (2007) have presented evidence that the large number of mutations that accumulate in cancer cells may have an epigenetic cause. DNA has several means of protecting itself from mutations. One is the editing subunits on DNA polymerase; these "proofreaders" get rid of mismatched bases and insert the correct ones. Another mechanism is the set of enzymes that repair DNA when the DNA has been damaged by light or by cellular compounds that are products of metabolism. In cancer cells, the genes encoding these DNA repair enzymes appear to be susceptible to inactivation by methylation. Once DNA repair enzymes have been downregulated, the number of mutations increases.

It is therefore possible that aging and cancer may be linked by the common denominator of aberrant DNA methylation. If metabolically or structurally important genes (such as the estrogen receptors) become heavily methylated, they don't produce enough receptor proteins, and our body function suffers. If tumor suppressor genes or the genes encoding DNA repair enzymes are heavily methylated, tumors can arise.

Tumors can be generated by a combination of genetic and epigenetic means. Changes in DNA methylation can activate oncogenes and repress tumor-suppressor genes, thereby initiating tumor formation. Conversely, oncogenes can cause the methylation of tumor suppressor genes, which also aids tumorigenesis. Moreover, the tissue environment of the cell may be critical in regulating these processes. The complexities of tumors, including their multiple somatic mutations and their resistance to agents that induce apoptotic cell death, may best be explained by a combination of genetic and epigenetic factors rather than just by the basis of mutations. Knowledge of the epigenetic causes of cancer can provide the basis for new methods of cancer therapy.

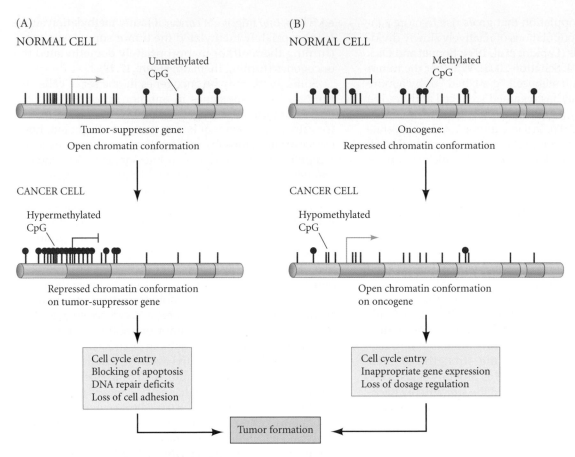

(A)
NORMAL CELL

Unmethylated
CpG

Tumor-suppressor gene:
Open chromatin conformation

CANCER CELL

Hypermethylated
CpG

Repressed chromatin conformation
on tumor-suppressor gene

Cell cycle entry
Blocking of apoptosis
DNA repair deficits
Loss of cell adhesion

(B)
NORMAL CELL

Methylated
CpG

Oncogene:
Repressed chromatin conformation

CANCER CELL

Hypomethylated
CpG

Open chromatin conformation
on oncogene

Cell cycle entry
Inappropriate gene expression
Loss of dosage regulation

Tumor formation

FIGURE 17.19 Cancer can arise (A) if tumor-suppressor genes are inappropriately turned off by DNA methylation or (B) if oncogenes are inappropriately demethylated (and thereby activated). (After Esteller 2007.)

SIDELIGHTS & SPECULATIONS

The Embryonic Origins of Adult-Onset Illnesses

Teratogenesis is usually associated with *congenital* disease (i.e., a condition appearing at birth) and is also associated with disruptions of organogenesis during the embryonic period. However, D. J. P. Barker and colleagues (1994a,b) have offered evidence that certain *adult-onset* diseases may also result from conditions in the uterus prior to birth. Based on epidemiological evidence, they hypothesize that there are critical periods of development during which certain physiological insults or stimuli can cause specific changes in the body.

The "Barker hypothesis" postulates that certain anatomical and physiological parameters get "programmed" during embryonic and fetal development, and that deficits in nutrition during this time can produce permanent changes in the pattern of metabolic activity—changes that can predispose the adult to particular diseases.

Specifically, Barker and colleagues showed that infants whose mother experienced protein deprivation (because of wars, famines, or migrations) during certain months of pregnancy were at high risk for having certain diseases as adults. Undernutrition during a fetus's first trimester could lead to hypertension and strokes in adult life, while those fetuses experiencing undernutrition during the second trimester had a high risk of developing heart disease and diabetes as adults. Those fetuses experiencing undernutrition during the third trimester were prone to blood clotting defects as adults.

Recent studies have tried to determine whether there are physiological or anatomical reasons for these correlations (Gluckman and Hanson 2004,

Figure 17.20 Anatomical changes associated with hypertension. (A) In age-matched individuals, the kidneys of men with hypertension had about half the number of nephrons as the kidneys of men with normal blood pressure. (B) The glomeruli of the nephrons in hypertensive kidneys were much larger than the glomeruli in control subjects. (After Keller et al. 2003, photographs courtesy of G. Keller.)

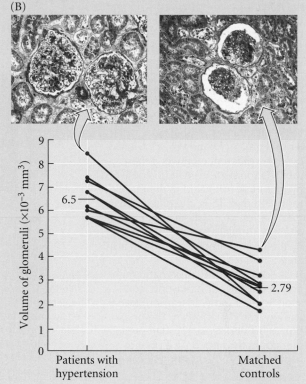

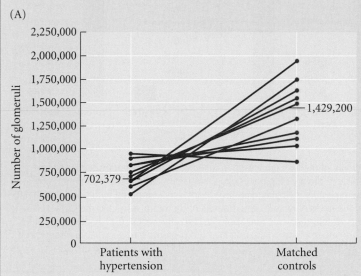

2005; Lau and Rogers 2005). Anatomically, undernutrition can change the number of cells produced during a critical time of organ formation. When pregnant rats are fed low-protein diets at certain times during their pregnancy, the resulting offspring are at high risk for hypertension as adult. The poor diet appears to cause low nephron numbers in the adult kidney (see Moritz et al. 2003). In humans, the number of nephrons present in the kidneys of men with hypertension was only about half the number found in men without hypertension (**Figure 17.20A**; Keller et al. 2003). In addition, the glomeruli (the blood-filtering unit of the nephron) of hypertensive men were larger than those in control subjects (**Figure 17.20B**).

Similar trends have been reported for non-insulin dependent (Type II) diabetes and glucose intolerance (Hales et al. 1991; Hales and Barker 1992). Here, poor nutrition reduces the number of β cells in the pancreas and hence the ability to synthesize insulin. Moreover, the pancreas isn't the only organ involved. Undernutrition in rats changes the histological

architecture in the liver as well. A low-protein diet during gestation appeared to *increase* the amount of periportal cells that produce the glucose-synthesizing enzyme phosphoenolpyruvate carboxykinase while *decreasing* the number of perivenous cells that synthesize the glucose-degrading enzyme glucokinase in the offpsring (Burns et al. 1997). These changes may be coordinated by glucocorticoid hormones that are stimulated by malnutrition and which act to conserve resources, even though such actions might make the person prone to hypertension later in life (see Fowden and Forhead 2004). (Since, historically, most humans died before age 50, this would not be a detrimental evolutionary trade-off.)

Hales and Barker (2001) have proposed a "thrifty phenotype" hypothesis wherein the malnourished fetus is "programmed" to expect an energy-deficient environment. The developing

fetus sets its biochemical parameters to conserve energy and store fat.* Resulting adults who do indeed meet with the expected poor environment are ready for it and can survive better than individuals whose metabolisms were set to utilize energy and not store it as efficiently. However, if such a "deprivationally developed" person lives in an energy- and protein-rich environment, their cells store more fats and their heart and kidneys have developed to survive more stringent conditions. Both these developments put the person at risk for several later-onset diseases.

How can conditions experienced in the uterus create anatomical and biochemical conditions that will be maintained throughout adulthood? One place to look is DNA methylation. Lillycrop and colleagues (2005) have shown that rats born to mothers having a low-protein diet had a different pattern of liver gene methylation than did

*In other words, the embryo has phenotypic plasticity—the ability to modulate its phenotype depending on the environment; this plasticity will be discussed further in Chapter 18.

the offspring of mothers fed a normal diet. These differences in methylation changed the metabolic profile of the rats' livers. For instance, the methylation of the promoter region of the PPARα gene (which is critical in the regulation of carbohydrate and lipid metabolism) is 20% lower in the offspring of protein-restricted rats, and the gene's transcriptional activity is tenfold greater (**Figure 17.21**). Moreover, the difference between these methylation patterns can be abolished by including folic acid in the protein-restricted diet. Thus, the difference in methylation probably results from changes in folate metabolism caused by the limited amount of protein available to the fetus.

It does appear that prenatal nutrition can induce long-lasting, gene-specific alterations in transcriptional activity and metabolism. The prevention of adult disease through prenatal diet could thus become a public health issue in the coming decades.

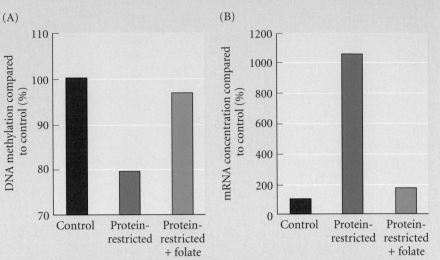

Figure 17.21 Activity of the liver gene for peroxisomal proliferator-activated receptor (PPARα) is susceptible to dietary differences. (A) DNA methylation pattern of the PPARα promoter region, showing highly methylated control promoters compared with poorly methylated promoters from the livers of mice whose mothers had protein-restricted diets ($p < 0.001$). Adding folate to the protein-restricted diet abolished this difference. (B) Levels of mRNA for the PPARα gene were much higher in the mice fed the protein-restricted diet ($p < 0.0001$). (After Lillycrop et al. 2005.)

DEVELOPMENTAL THERAPIES

Knowledge gained from research in the field of developmental biology is now being focused on several diseases. The ability to block paracrine factors, to use stem cells to regenerate body parts, and to induce nearly any cell in the body to become a pluripotential stem cell may enable us to block the spread of cancer, repair bodily injuries, and even to ameliorate genetic disease.

Anti-Angiogenesis

Judah Folkman (1974) has estimated that as many as 350 billion mitoses occur in the human body every day. With each cell division comes the chance that the resulting cells will be malignant. Indeed, autopsies have shown that every person over 50 years old has microscopic tumors in their thyroid glands, although less than 1 in 1000 persons have thyroid cancer (Folkman and Kalluri 2004). Folkman suggested that cells capable of forming tumors develop at a certain frequency, but that most never form observable tumors. The reason is that a solid tumor, like any other rapidly dividing tissue, needs oxygen and nutrients to survive. Without a blood supply, potential tumors either die or remain as dormant "microtumors," stable cell populations wherein dying cells are replaced by new cells. Thus one important area in which knowledge of development

can contribute to cancer therapies is the inhibition of angiogenesis (blood vessel formation).

The critical point at which a node of cancerous cells becomes a rapidly growing tumor occurs when the node becomes vascularized. A microtumor can expand to 16,000 times its original volume in the 2 weeks following vascularization (Folkman 1974; Ausprunk and Folkman 1977). To achieve vascularization, the microtumor secretes substances called **tumor angiogenesis factors**, which often include the same factors that engender blood vessel growth in the embryo—VEGFs, Fgf2, placenta-like growth factor, and others. Tumor angiogenesis factors stimulate mitosis in endothelial cells and direct the cell differentiation into blood vessels in the direction of the tumor.

Tumor angiogenesis can be demonstrated by implanting a piece of tumor tissue within the layers of a rabbit or mouse cornea. The cornea itself is not vascularized, but it is surrounded by a vascular border, or limbus. The tumor tissue induces blood vessels to form and grow toward the tumor (**Figure 17.22**; Muthukkaruppan and Auerbach 1979). Once the blood vessels enter the tumor, the tumor cells undergo explosive growth, eventually bursting the eye. Other adult solid tissues do not induce blood vessels to form. It might therefore be possible to block tumor development by blocking angiogenesis. Numerous chemicals are being tested as natural and artificial angiogenesis inhibitors. These compounds act by preventing endothe-

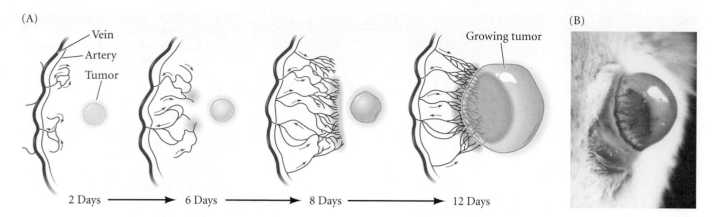

FIGURE 17.22   New blood vessel growth to the site of a mammary tumor transplanted into the cornea of an albino mouse. (A) Sequence of events leading to vascularization of the tumor on days 2, 6, 8, and 12. Both the veins and the arteries in the limbus surrounding the cornea supply blood vessels to the tumor. (B) Photograph of living cornea of an albino mouse, with new blood vessels from the limbus approaching the tumor graft. (From Muthukkaruppan and Auerbach 1979; B courtesy of R. Auerbach.)

lial cells from responding to the angiogenetic signal of the tumor.* One of the advantages of these compounds is that the tumor cells are unlikely to evolve resistance to them, since the tumor cell itself is not the target of these agents (Boehm et al. 1997; Kerbel and Folkman 2002).

In one set of clinical trials, an antibody against VEGF-A was found to be successful against colon cancer but not against breast cancer. This is probably because colon cancer is more dependent on VEGF-induced angiogenesis than are mammary tumors (Whisenant and Bergsland 2005). Antibodies against a placental form of VEGF were able to inhibit the growth of tumors without affecting healthy blood vessels (Fischer et al. 2007). Blocking the VEGF receptor VEGFR3 prevents the angiogenic sprouting needed for new blood vessels (Tammela et al. 2008) and stops tumors from getting blood-borne nutrients and oxygen. Interestingly, thalidomide, the teratogen responsible for birth defects in the 1960s, is now being used to block tumor-induced blood vessels. Thalidomide has been found to be a potent anti-angiogenesis factor that can reduce the growth of cancers in rats and mice (D'Amato et al. 1994; Dredge et al. 2002).

Cancer and congenital malformations are opposite sides of the same coin. Both involve disruptions of normal development. Thus, as we have seen, agents that have been known to cause congenital malformations—thalidomide, retinoic acid, and cyclopamine—can be used as drugs to

---

*This is the flip side of differentiation therapy discussed above. In differentiation therapy, paracrine factors or hormones are added to promote differentiation. Here, the substances being administered block paracrine signals in order to prevent tissue (i.e., blood vessel) formation.

prevent cancers. Just as they disrupt normal development, these substances can disrupt the caricature of development that is caused by tumor cells. When angiogenesis is blocked, the tumor cells can be starved.

Stem Cells and Tissue Regeneration

Embryonic stem cells

As we discussed in Chapter 8, the inner cell mass of the mammalian blastocyst generates the entire embryo. Identical twins and chimeric individuals show that each of the cells of the inner cell mass are pluripotent. When cultured, the cells of the inner cell mass blastomeres can become **embryonic stem (ES) cells**, which remain pluripotent. Currently, pluripotent stem cells are obtained by two major techniques (**Figure 17.23A**). They can be derived from the inner cell mass of blastocysts, such as those left over from in vitro fertilization (Thomson et al. 1998), and they can also be generated from germ cells derived from spontaneously aborted fetuses. The latter are generally referred to as **embryonic germ (EG) cells** (Gearhart 1998). Some experimental evidence (Strelchenko et al. 2004) suggests that it may also be possible to derive embryonic stem cells from late morulae, before they form blastocysts. While adult stem cells are rare and do not usually remain undifferentiated in culture, embryonic stem cells can be readily harvested and retain their undifferentiated state for years of culturing.

The importance of pluripotent stem cells in medicine is potentially enormous. The hope is that human ES cells can be used to produce new neurons for people with degenerative brain disorders (such as Alzheimer and Parkinson disease) or spinal cord injuries, and new pancreatic β cells for people with diabetes. People with deteriorating hearts might be able to have damaged tissue replaced with new heart cells, and those suffering from immune deficiencies might be able to replenish their failing immune systems. Such therapies have already worked in mice. Murine ES cells have been cultured under conditions causing them to form insulin-secreting cells, muscle stem cells, glial stem cells, and

(A)

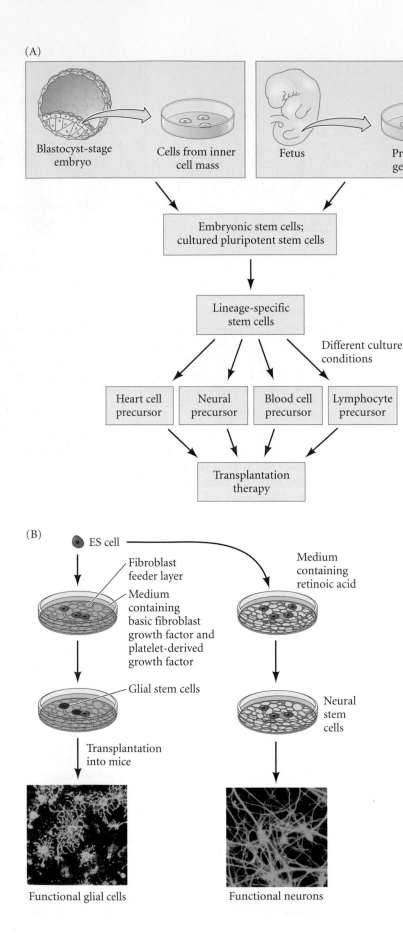

Blastocyst-stage
embryo

Cells from inner
cell mass

Fetus

Primordial
germ mass

Embryonic stem cells;
cultured pluripotent stem cells

Lineage-specific
stem cells

Different culture
conditions

Heart cell
precursor

Neural
precursor

Blood cell
precursor

Lymphocyte
precursor

Transplantation
therapy

(B)

● ES cell

Fibroblast
feeder layer

Medium
containing
basic fibroblast
growth factor and
platelet-derived
growth factor

Medium
containing
retinoic acid

Glial stem cells

Neural
stem
cells

Transplantation
into mice

Functional glial cells

Functional neurons

FIGURE 17.23 Embryonic stem cell therapeutics. (A) ES cells can be obtained from the inner cell mass of the blastocyst or from primordial germ cells and cultured in different ways to produce lineage-specific stem cells. These cells (or the precursor cells derived from them) can then be transplanted into a host. (B) Differentiation of mouse ES cells into lineage-restricted (neuronal and glial) stem cells can be accomplished by altering the media in which the ES cells grow. (A after Gearhart 1998; B, photographs from Brüstle et al. 1999 and Wickelgren 1999, courtesy of O. Brüstle and J. W. McDonald.)

neural stem cells (**Figure 17.23B**; Brüstle et al. 1999; McDonald et al. 1999). Dopaminergic neurons derived from ES cells have been shown to significantly reduce the symptoms of Parkinson disease in rodents (Bjorklund et al. 2002; Kim et al. 2002). When neural stem cells derived from germ cell-derived ES cells were transplanted into the injured brains of mice, the neural stem cells replaced neurons and glial cells in the forebrains of the newborn mice (Mueller et al. 2005). However, in most cases when human ES cells have been transplanted into the brains of immunosuppressed animals, the results have been less encouraging. Typically, transplantation of thousands of ES cell-generated neurons into animals results in very few surviving dopaminergic neurons and a high frequency of tumors (Li et al. 2008). For such transplantations to work, one must be able to find the neural stem cells made by the ES cells and to prevent undifferentiated ES cells from being transplanted.

Human embryonic stem cells differ from their murine counterparts in their growth requirements. In most ways, however, they are very similar, and have a similar, if not identical, pluripotency. Like mouse ES cells, human ES cells can be directed down specific developmental paths. For example, Kaufman and his colleagues (2001) directed human ES cells to become blood-forming stem cells by placing them on mouse bone marrow or endothelial cells. These ES-derived hematopoietic stem cells could further differentiate into numerous types of blood cells (**Figure 17.24**). Human embryonic germ cells were able to cure virus-induced paraplegia in rats. These stem cells appear to do this both by differentiating into new neurons and by producing paracrine factors (BDNF and TGF-α) that prevent the death of the existing neurons (Kerr et al. 2003). Similarly, ES cells from monkey blastocysts have been able to cure a Parkinson-like condition in adult

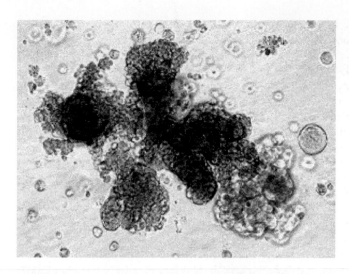

FIGURE 17.24   Differentiated blood cells developing from human ES cells cultured on mouse bone marrow. (Courtesy of The University of Wisconsin.)

monkeys whose dopaminergic neurons had been destroyed (Tagaki et al. 2005).

Research is now being done to find ways of directing the differentiation of ES cells by using small molecules (rather than paracrine factors). Chen and colleagues (2009), for instance, have found that indolactam V molecules can induce *Pdx1* expression in ES cells, directing them into the pancreatic lineage. Thus, ES cells may be able to provide a reusable and readily available source of cells to heal damaged tissue in adult men and women.

See **WEBSITE 17.4   Therapeutic cloning**

Induced pluripotent stem cells

The mammalian inner cell mass is known to be characterized by certain transcription factors, including Nanog, Oct4, and Sox2. Knocking out these genes in mice abolishes the pluripotency and self-renewal of the inner mass blastomeres, eventually leading to the demise of the embryo (see Boyer et al. 2006; Niwa 2007). Sox2 and Oct4 can dimerize to form a transcription factor complex that activates their own genes (*Oct4* and *Sox2*) as well as the *Nanog* gene. It appears that Oct4 not only promotes the synthesis of Sox2 and Klf4, but it also blocks the production of the *miR145* microRNA that would otherwise prevent the translation of *Oct4*, *Sox2*, and *Klf4* messages (Chivukula and Mendell 2009; Xu et al. 2009). These transcription factors (and microRNAs) initiate a transcription network in the inner cell mass and ES cells that is essential for pluripotency (see Welstead et al. 2008). This first set of transcription factors probably acts by activating other transcription factors, such as Nanog and Fbx15, which are critical in maintaining the pluripotent and dividing state.

The newfound knowledge of the transcription factors needed to maintain pluripotency has illuminated a startlingly easy way to generate embryonic stem cells that have the exact genotype of the patient.* In 2006, Kazutoshi Takahashi and Shinya Yamanaka of Kyoto University demonstrated that by inserting activated copies of four genes that encoded some of these critical transcription factors, nearly any cell in the adult mouse body could be made into a cell with the pluripotency of embryonic stem cells. Such cells are called **induced pluripotent stem (iPS) cells**.

Using a strategy very similar to the one used to identify the active components of Spemann's organizer (see Chapter 7), Takahashi and Yamanaka obtained mRNA from mouse ES cells and made these into cDNAs, which they placed onto active viral promoters. They transfected sets of 24 of these recombinant viruses into cultured fibroblasts that had a neomycin-resistance gene placed onto an *Fbx15* regulatory region. The *Fbx15* gene is usually turned on in ES cells, so if the transfected genes activated the *Fbx15* gene, then the neomycin-resistance gene would be activated and the cells would survive in neomycin-containing culture medium (**Figure 17.25A**). If a cohort of 24 active genes was found to activate the *Fbx15* promoter (as shown by the cells' growing in neomycin-containing medium), then the group of cloned genes was further split. Eventually, Takahashi and Yamanaka discovered that only four active genes were needed to turn on the *Fbx15* gene. These genes encoded the transcription factors Oct4, Sox2, Klf4, and c-Myc (**Figure 17.25B**). When the researchers cultured those cells whose *Fbx15* promoter was activated, they found that in many cases the cells had become pluripotent, as demonstrated in a series of tests:

- When the cells were aggregated together, they formed a teratoma—a tumorlike amalgam containing cell types of all three germ layers.
- When injected onto the blastocyst of a normal mouse, the induced cells showed that they could contribute to the production of cells in each of the three germ layers (**Figure 17.25C**; Maherali et al. 2007; Okita et al. 2007; Wernig et al. 2007).
- When the inner cell masses of mouse embryos were completely composed of induced pluripotential cells, normal mice were generated (Boland et al. 2009; Zhao et al. 2009).
- The transcription and DNA methylation pattern of the induced pluripotential cells was found to be almost identical to that of actual mouse embryonic stem cells. The genes for Nanos, Sox2, and other ES-cell transcription factors were hypomethylated, as they are in ES cells. Interestingly, the methylation problems that plague

*In order to be a successful therapeutic agent, embryonic stem cells have to be the same genotype as the patient, so that their differentiated products will not be rejected. In fact, however, they are not quite exact, because the viruses used to induce pluripotency are integrated into the chromosomes, as we'll see next.

(A)

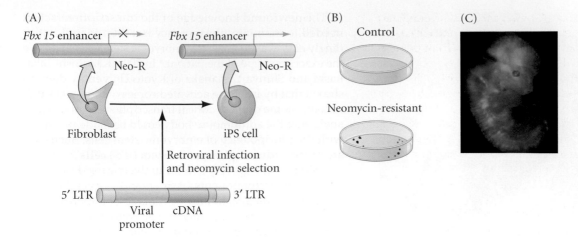

(B)

Control

Neomycin-resistant

(C)

FIGURE 17.25 Production of induced pluripotent stem cells by expression of four transcription factors in adult mouse tail fibroblasts. (A) A mouse was produced wherein the 5' regulatory region of the *Fbx15* gene was attached to a neomycin-resistance (Neo-R) gene. When *Fbx15* gene is activated (as it is in ES cells), the neomycin-resistance gene is activated and the cells are able to grow in medium containing neomycin. Various combinations of genes encoding mRNAs found in ES cells were added to the fibroblasts, and the fibroblasts grown in neomycin to see if any combination of genes activated the *Fbx15* promoter. (B) Some neomycin-resistant colonies survived. These resistant colonies were found to have genes encoding Sox2, Oct4, Klf4, and c-Myc. (C) The pluripotency of these cells was shown by their insertion into the inner cell mass of a mouse blastocyst. GFP-labeled iPS cells were subsequently found in all organs of the embryo. (After Takahashi and Yamanaka 2006.)

cloned animals do not appear to be a problem for embryonic stem cells derived by somatic cell nuclear transfer. It appears that the nuclei in the small population of ES cells that survive in culture have had their methylation patterns erased, thus enabling them to redifferentiate (Jaenisch 2004; Rugg-Gunn et al. 2005).

This technique does nothing less than transform any cell in the body into an embryonic stem cell. By 2007, the Yamanaka laboratory (Takahashi et al. 2007) had used the same set of four transcription factors to induce pluripotentiality in adult *human* fibroblast cells. These induced cells formed teratomas containing cells of all three germ layers, and they had the same transcription profile as the human ES cells.

Yamanaka (2007) proposed that c-Myc helped "immortalize" the cell, keeping it in an undifferentiated state of proliferation, and that Klf4 prevented apoptosis and senescence. But this situation would have created a tumor, were it not for Oct4 and Sox2, which redirected growth and gave the cell properties similar to those of germ cells. The need for c-Myc was disconcerting, however, since this is a well-known oncogene, capable of initiating tumor formation in

normal cells. However, Nakagawa and colleagues (2007) showed that by altering the culturing techniques, one could circumvent the need for c-Myc. In other words, one should be able to obtain induced pluripotent stem cells by adding merely three active genes to a patient's cell. Another way of "getting rid of c-Myc" was to place the genes for the key ES transcription factors onto episomal viral vectors (Kaji et al. 2009; Yu et al. 2009). Episomal vectors are derived from viruses (such as the Epstein-Barr virus that causes mononucleosis) that do not insert themselves into host DNA. In this manner, the genes from the vector generated the transcription factors that converted human fibroblasts into iPS cells. Once the iPS cells were produced, the culture media could be changed so that the vector would be eliminated. This technique generated human iPS cells that did not have a virus inserted in it. Thus, not only was the problem of c-Myc circumvented, but so was the problem of the viral insertion capable of causing a gene mutation.

Recently, adult stem cells have been found that are already "part of the way" to becoming pluripotent and do not need as many factors to make them so. Giorgetti and colleagues (2009) showed that blood stem cells from the human umbilical cord can be converted into pluripotential stem cells merely by adding activated *OCT4* and *SOX2* genes, and Kim and colleagues (2009) found that they could transform human neural stem cells into iPS cells by adding only a single activated gene, *OCT4*.

The therapeutic potential of iPS cells was demonstrated by the ability of iPS-derived hematopoietic stem cells to correct a sickle-cell anemia phenotype in mice (**Figure 17.26**; Hanna et al. 2007). Here, tail-tip fibroblasts from a mouse with sickle-cell hemoglobin were made into iPS cells by being infected in culture by viruses containing Oct4, Sox2, Klf4, and c-Myc. The iPS cells were selected, then electroporated with DNA containing wild-type globin genes. These genetically corrected iPS cells were cultured in media that promoted the production of hematopoietic stem cells, and the hematopoietic stem cells were injected back into the mice with sickle-cell anemia. By 2 months, the anemia had been cured.

The combination of iPS cells and genetic engineering may be able to cure certain genetic diseases. Raya and colleagues (2009) have cultured dermal fibroblast cells from patients with the genetic disease Fanconi anemia and have added to them a good copy of the defective gene. These cells were then converted into iPS cells by the addition of retroviruses bearing the activated forms of *OCT4*, *SOX2*, *KLF4*, and *c-MYC*. Moreover, these iPS cells could be directed to become the blood stem cells needed by the patients. However, we still do not know if these cells can become normal parts of the patients and not produce tumors themselves.

Adult stem cells and regeneration therapy

In the introduction to Part III, we mentioned the potential medical uses of adult stem cells, especially the mesenchymal stem cells. We will now provide details about some of these new therapies.

CARDIAC REGENERATION When mesenchymal stem cells from the bone marrow of heart attack patients are injected into their own hearts, these cells can differentiate into heart and vessel cells and can significantly improve the patients' outcomes (Yousef et al. 2009). Interestingly, in many instances the stem cells do not create new structures themselves to circumvent the blockage. Rather, they secrete

paracrine factors that appear to activate the heart's own stem cells to repair the damage (Cho et al. 2007; Mirotsou et al. 2007).

BONE REGENERATION Several stem cell therapies can be seen in an incredibly important area of regenerative medicine: forming new adult bone. While fractured bones can heal, bone cells in adults usually do not regrow to bridge wide gaps. The finding that the same paracrine and endocrine factors involved in endochondral ossification are also involved in fracture repair (Vortkamp et al. 1998) raises the possibility that new bone could grow if the proper paracrine factors and extracellular environment were provided. Several methods are now being tried to develop new functional bone in patients with severely fractured or broken bones.

One solution to the problem of delivery was devised by Bonadio and his colleagues (1999), who developed a collagen gel containing plasmids carrying the human parathyroid hormone gene. The plasmid-impregnated gel was placed in the gap between the ends of a broken dog tibia or femur. As cells migrated into the collagen matrix, they incorporated the plasmid and made parathyroid hormone. A dose-dependent increase in new bone formation was seen in about a month (**Figure 17.27**). This type of treatment has the potential to help people with large bone fractures as well as those with osteoporosis.

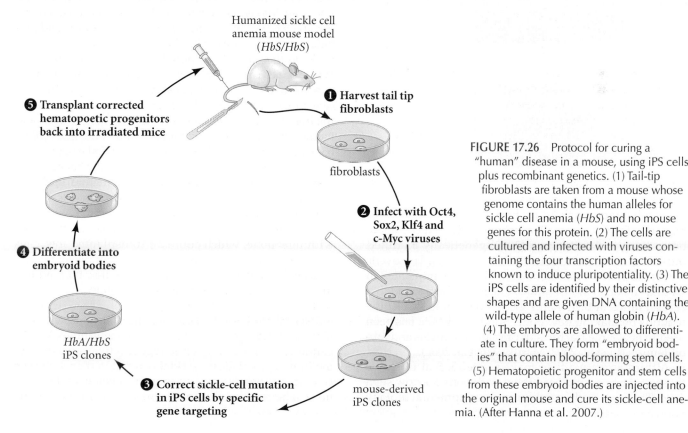

FIGURE 17.26 Protocol for curing a "human" disease in a mouse, using iPS cells plus recombinant genetics. (1) Tail-tip fibroblasts are taken from a mouse whose genome contains the human alleles for sickle cell anemia (*HbS*) and no mouse genes for this protein. (2) The cells are cultured and infected with viruses containing the four transcription factors known to induce pluripotentiality. (3) The iPS cells are identified by their distinctive shapes and are given DNA containing the wild-type allele of human globin (*HbA*). (4) The embryos are allowed to differentiate in culture. They form "embryoid bodies" that contain blood-forming stem cells. (5) Hematopoietic progenitor and stem cells from these embryoid bodies are injected into the original mouse and cure its sickle-cell anemia. (After Hanna et al. 2007.)

(A) Treated fracture

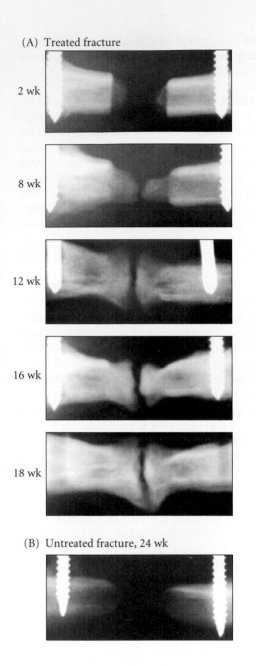

2 wk

8 wk

12 wk

16 wk

18 wk

(B) Untreated fracture, 24 wk

(C) Whole bone, 53 wk

FIGURE 17.27 Bone formation from collagen matrix containing plasmids bearing the human parathyroid hormone. (A) A 1.6-cm gap was made in a dog femur and stabilized with screws. Plasmid-containing gel was placed on the edges of the break. Radiographs of the area at 2, 8, 12, 16, and 18 weeks after the surgery show the formation of bone bridging the gap at 18 weeks. (B) Control fracture (no plasmid in the gel) at 24 weeks. (C) Whole bone a year after surgery, showing repaired region. (From Bonadio et al. 1999, courtesy of J. Bonadio.)

and mechanical engineering, is called **tissue engineering**. Li and colleagues (2005) made scaffolds of material that resembles normal extracellular matrix and which can be molded to form the shape of bone needed. The bone marrow stem cells can be placed in these scaffolds and placed into the existing bone. These stem cells are told how to differentiate by the local conditions, and histological and gene expression studies have shown that these cells form bone with the appropriate amounts of osteocytes and chondrocytes. Similarly, when cells are removed from the tracheae of recent cadavers, they can be reseeded with bone marrow stem cells of patients with damaged tracheae. The stem cells differentiate into chondrocytes, and the newly reseeded tracheae can be substituted for the damaged tracheae and provide a normal airway (Macchiarini et al. 2008).

A fourth approach has been to let the body do the work. While bones can't regenerate if wide gaps appear, many bones can undergo normal healing of small wounds. Here, cells in the periosteum (the sheath of cells that surrounds the bone) will differentiate into new cartilage, bone, and ligaments to fill the crack. Stevens and colleagues (2005) have used this normal healing process to make new bones. They injected saline solutions between the rabbit tibial bone and its periosteum, mimicking a fracture, and kept this space open by adding a gel containing calcium (to push the periosteal cells to differentiate into bone rather than cartilage). Within a few weeks, these cavities filled with new bone, which could be transplanted into sites where bone had been damaged. This technique might provide a relatively painless way to produce new bone for fusing vertebrae and other surgical techniques.

NEURONAL REGENERATION While the central nervous system is characterized by its ability to change and make new connections, it has very little regenerative capacity. The motor neurons of the peripheral nervous system, however, have significant regenerative powers, even in adult mammals. The regeneration of motor neurons involves the

Another approach is to find the right mixture of paracrine factors to recruit stem cells and produce normal bone. Peng and colleagues (2002) genetically modified muscle stem cells to secrete BMP4 or VEGF-A. These cells were placed in gel matrix discs, which were implanted in wounds made in mouse skulls. The researchers found that certain ratios of BMP4 and VEGFs were able to heal the wounds by making new bone. Similarly, BMP2 has been used to heal large fractures of primate mandibles and rabbit femurs (Li et al. 2002; Marukawa et al. 2002).

A third approach is to make scaffolds that resemble those of the bone and seed them with bone marrow stem cells. This approach, combining developmental biology

regrowth of a severed axon, not the replacement of a missing or diseased cell body. If the cell body of a motor neuron is destroyed, it cannot be replaced.

The myelin sheath that covers the axon of a motor neuron is necessary for its regeneration. This sheath is made by the Schwann cells, a type of glial cell in the peripheral nervous system (see Chapter 9). When an axon is severed, the Schwann cells divide to form a pathway along which the axon can regrow from the proximal stump. This proliferation of the Schwann cells is critical for directing the regenerating axon to the original Schwann cell basal lamina. If the regrowing axon can find that basal lamina, it can be guided to its target and restore the original connection. In turn, the regenerating neuron secretes mitogens that allow the Schwann cells to divide. Some of these mitogens are specific to the developing or regenerating nervous system (Livesey et al. 1997).

The neurons of the central nervous system cannot regenerate their axons under normal conditions. Thus, spinal cord injuries can cause permanent paralysis. One strategy to get around this block is to find ways of enlarging the population of adult neural stem cells and to direct their development in ways that circumvent the lesions caused by disease or trauma. The neural stem cells found in adult mammals may be very similar to embryonic neural stem cells and may respond to the same growth factors (Johe et al. 1996; Johansson et al. 1999; Kerr et al. 2003).

Another strategy for CNS neural regeneration is to create environments that encourage axonal growth. Unlike the Schwann cells of the peripheral nervous system, the myelinating glial cells of the central nervous system, the oligodendrocytes, produce substances that inhibit axon regeneration (Schwab and Caroni 1988). Schwann cells transplanted from the peripheral nervous system into a CNS lesion are able to encourage the growth of CNS axons to their targets (Keirstead et al. 1999; Weidner et al. 1999). Three substances that inhibit axonal outgrowth have been isolated from oligodendrocyte myelin: myelin-associated glycoprotein, Nogo-1, and oligodendrocyte-myelin glycoprotein (Mukhopadyay et al. 1994; Chen et al. 2000; Grand-Pré et al. 2000; Wang et al. 2002). Each of these substances binds to the Nogo receptor (NgR). Thus, NgR may be the critical target for therapies allowing regeneration. Chen and colleagues (2009) used RNA interference to block the synthesis of NgR and found that this helped rat optic neurons to regenerate. The axons regenerated even better when this therapy was combined with nutrients and positive growth factors.

As mentioned in the case with bone, there is a close relationship between wound healing and regeneration. Patients with multiple sclerosis suffer from the demyelination of axons in the brain and spinal cord. Several transcription factors are needed to reinitiate the pathway to myelination, and these transcription factors appear to be repressed by high levels of Wnt signaling (Arnett et al. 2004; Fancy et al. 2009). Therefore, the use of Wnt inhibitors (such as Frzb and Dickkopf) may be a mechanism for treating this disease. Research into CNS axon regeneration may become one of the most important contributions of developmental biology to medicine.

Direct transdifferentiation

One of the newest developmental therapies involves using transcription factors to convert one cell type into another without the intermediary of stem cells, a procedure known as **transdifferentiation**. In Chapter 2, we discussed the transdifferentiation of exocrine pancreatic cells into insulin-secreting pancreatic β cells by three transcription factors (Ngn3, Mafa1, and Pdx1). More recently, Kajiyama and colleagues (2010) have transfected one of these factors, Pdx1, into mouse adipose-derived stem cells and converted the stem cells into insulin-producing pancreatic cells. These induced pancreatic cells were able to reduce the hyperglycemia in diabetic mice.

Vierbuchen and colleagues (2010) found that the viral insertion of three active transcription factor genes (Ascl1, Brn2, and Myt1l) sufficed to efficiently convert mouse dermal fibroblasts into functional neurons in vitro. These induced neuronal cells expressed several neuron-specific proteins, were able to generate action potentials, and formed functional synapses (resembling excitatory neurons of the forebrain). It remains to be seen if this transdifferentiation is stable and whether it will ameliorate disease in vivo. However, the ability to generate pancreatic β cells and neural cells that are genetically identical to a patient holds the promise of significant therapies for patients with diabetes and degenerative neural diseases.

Coda

Developmental biology is gaining increasing importance in modern medicine. First, preventive medicine, public health, and conservation biology demand that we learn more about the mechanisms by which industrial chemicals and drugs can damage embryos. The ability to effectively and inexpensively assay compounds for potential harm is critical. Second, developmental biology is providing us new ways of preventing and curing cancers. Third, developmental biology is providing the explanations for how mutated genes and aneuploidies cause their aberrant phenotypes. Fourth, the field of regenerative medicine is using developmental biology to provide induced pluripotent stem cells as well as the knowledge of paracrine factors needed to form new cells, tissues, and organs in adults.

Snapshot Summary: *Medical Aspects of Developmental Biology*

1. Pleiotropy occurs when several different effects are produced by a single gene. In mosaic pleiotropy, each effect is caused independently by the expression of the same gene in different tissues. In relational pleiotropy, abnormal gene expression in one tissue influences other tissues, even though those other tissues do not express that gene.

2. Genetic heterogeneity occurs when mutations in more than one gene can produce the same phenotype.

3. Phenotypic heterogeneity arises when the same gene can produce different defects (or differing severities of the same defect) in different individuals.

4. Preimplantation genetics involves testing for genetic abnormalities in early embryos in vitro, and implanting only those embryos that may develop normally. Selection for sex is also possible using preimplantation genetics.

5. Teratogenic agents include certain chemicals such as alcohol and retinoic acid, as well as heavy metals, certain pathogens, and ionizing radiation. These agents adversely affect normal development, yielding malformations and functional deficits.

6. Fetal alcohol syndrome is completely preventable. There may be multiple effects of alcohol on cells and tissues that result in this syndrome of cognitive and physical abnormalities.

7. Endocrine disruptors can bind to or block hormone receptors or block the synthesis, transport, or excretion of hormones. DES is a powerful endocrine disruptor. Presently, bisphenol A and other compounds are being considered as possible agents of low sperm counts in men and a predisposition to breast cancer in women.

8. Environmental estrogens can cause reproductive system anomalies by suppressing Hox gene expression and Wnt pathways.

9. In some instances, endocrine disruptors methylate DNA, and these patterns of methylation can be inherited from one generation to the next. Such methylation can alter metabolism and development by turning genes off.

10. Cancer can be seen as a disease of altered development. Some tumors revert to non-malignancy when placed in environments that fail to support rapid cell division.

11. Cancers can arise from errors in cell-cell communication. These errors include alterations of paracrine factor synthesis.

12. Cancers metastasize in manners similar to embryonic cell movement.

13. In many instances, tumors have a rapidly dividing stem cell population which produces more stem cells as well as more quiescent and differentiated cells.

14. The methylation pattern of cancer cells is often aberrant, and these methylation differences can cause cancer by inappropriately inactivating tumor suppressor genes or activating oncogenes.

15. Disrupting tumor-induced angiogenesis may become an important means of stopping tumor progression.

16. Skin fibroblasts, and perhaps any normal adult cell, can be induced to form pluripotent stem cells by the incorporation of activated genes encoding certain transcription factors. These pluripotent stem cells would not be rejected by the patient from whom they were formed.

17. By altering conditions to resemble those in the embryo and by providing surfaces on which adult stem cells might grow, some adult stem cells might be directed to form numerous cell types.

19. Neurons in the central nervous system can be aided in regenerating by blocking the glial-derived paracrine factors that stabilize neurons and prevent their growth.

20. Transdifferentiation is the viral insertion of transcription factors (or transcription factor activation by small molecules), which can convert one stable cell type into another.

For Further Reading

Complete bibliographical citations for all literature cited in this chapter can be found at the free-access website **www.devbio.com**

Anway, M. D., A. S. Cupp, M. Uzumcu and M. K. Skipper. 2005. Epigenetic transgeneration effects of endocrine disruptors and male fertility. *Science* 308: 1466–1469.

Baksh, D., L. Song and R. S. Tuan. 2004. Adult mesenchymal stem cells: Characterization, differentiation, and application in cell and gene therapy. *J. Cell Mol. Med.* 8: 301–316.

Bissell, M. J., D. C. Radisky, A. Rizki, V. M. Weaver and O. W. Petersen. 2002. The organizing principle: Microenvironmental influences in the normal and malignant breast. *Differentiation* 70: 537–546.

Foty, R. A. and M. S. Steinberg. 1997. Measurement of tumor cell cohesion and suppression of invasion by E- and P-cadherin. *Cancer Res.* 57: 5033–5036.

Gilbert, S. F. and Epel, D. 2009. *Ecological Developmental Biology: Integrating Epigenetics, Medicine, and Evolution.* Sinauer Associates. Sunderland, MA.

Gluckman, P. D. and M. A. Hanson. 2004. Living with the past: Evolution, development, and patterns of disease. *Science* 305: 1733–1739.

Gupta, P. B., C. L. Chaffer and R. A. Weinberg. 2009. Cancer stem cells: Mirage or reality? *Nature Medicine* 15: 1010–1012.

Howdeshell, K. L., A. K. Hotchkiss, K. A. Thayer, J. G. Vandenbergh and F. S. vom Saal. 1999. Plastic bisphenol A speeds growth and puberty. *Nature* 401: 762–764.

Huang, G. T., S. Gronthos and S. Shi. 2009. Mesenchymal stem cells derived from dental tissues vs. those from other sources: Their biology and role in regenerative medicine. *J. Dental Res.* 88:792–806.

Keller, G., G. Zimmer, G. Mall, E. Ritz and K. Amann. 2003. Nephron number in patients with primary hypertension. *New Engl. J. Med.* 348: 101–108.

Lammer, E. J. and 11 others. 1985. Retinoic acid embryopathy. *N. Engl. J. Med.* 313: 837–841.

Lillycrop, K.A., E. S. Phillips, A. A. Jackson, M. A. Hanson and G. C. Burdge. 2005. Dietary protein restriction of pregnant rats induces and folic acid supplementation prevents epigenetic modification of hepatic gene expression in the offspring. *J. Nutrition* 135: 1382–1386.

Macchiarini, P. and 14 others. 2008. Clinical transplantation of a tissue-engineered airway. *Lancet* 372: 2023–2030.

Maffini, M. V., A. M. Soto, J. M. Calabro, A. A. Ucci and C. Sonnenschein. 2004. The stroma as a crucial target in mammary gland carcinogenesis. *J. Cell Sci.* 117: 1495–1502.

Nakayama, A., M. T. Nguyen, C. C. Chen, K. Opdecamp, C. A. Hodgkinson and H. Arnheiter. 1998. Mutations in microphthalmia, the mouse homolog of the human deafness gene MITF, affect neuroepithelial and neural crest-derived melanocytes differently. *Mech Dev.* 70: 155–166.

Sulik, K. K. 2005. Genesis of alcohol-induced craniofacial dysmorphism. *Exp. Biol. Med.* 230: 366–375.

Takahashi, K. and S. Yamanaka. 2006. Induction of pluripotent stem cells from mouse embryonic and adult fibroblast cultures by defined factors. *Cell* 126: 663–676.

Tammela, T. and 20 others. 2008. Blocking VEGFR-3 suppresses angiogenic sprouting and vascular network formation. *Nature* 454: 656–660.

Go Online

WEBSITE 17.1 Human embryology and genetics. This links to other websites that will connect you to tutorials in human development, as well as to the Online Mendelian Inheritance in Man (OMIM), which details all human genetic conditions.

WEBSITE 17.2 Thalidomide as a teratogen. The drug thalidomide caused thousands of babies to be born with malformed arms and legs, and it provided the first major evidence that drugs could induce congenital anomalies. The mechanism of its action is still hotly debated.

WEBSITE 17.3 Our stolen future. This website monitors the environmental effects of endocrine disruptors. It is a political and consumer action site as well as a scientific clearinghouse for endocrine disruption. Run by the authors of the book *Our Stolen Future*, it also provides links to the websites of people who disagree with them.

WEBSITE 17.4 Therapeutic cloning. Prior to induced pluripotent stem cells, another technique, therapeutic cloning, was seen as a way around the immune rejection of stem cell-derived differentiated cells.

Vade Mecum

Somites and thalidomide. These movies are from the laboratory of Jay Lash, whose insightful work on cartilage formation resulted in some of our first insights into the mechanisms by which the drug thalidomide halts limb growth.

Outside Sites

Association of Reproductive Health Professionals: **www.arhp.org/topics/enviro-repro-health** contains resources for health care providers and their clients.

Endocrine disruptor exchange: **www.endocrinedisruption.com**

Reproductive Toxicology website: **www.reprotox.org** has summaries of over 5000 agents, exposure levels, and their effects on development and reproduction.

Stem cell basics: The NIH website on stem cell education. **http://stemcells.nih.gov/info/basics**

"All things stem cell." An informative stem cell blog: **http://www.allthingsstemcell.com**

For information on fetal alcohol syndrome: **http://www.cdc.gov/ncbddd/fasd/data.html** .

NIH, FASD: **http://www.niaaa.nih.gov/AboutNIAAA/Interagency/AboutFAS.htm**

Substance Abuse and Mental Health Services Administration (SAMHSA): **http://www.fascenter.samhsa.gov**

Developmental Plasticity and Symbiosis

18

IT WAS LONG THOUGHT THAT THE ENVIRONMENT played only a minor role in development. Nearly all developmental phenomena were believed to be a "read-out" of nuclear genes, and those organisms whose development *was* significantly controlled by the environment were considered interesting oddities. When environmental agents played roles in development, they appeared to be destructive, such as the roles played by teratogens and endocrine disruptors (see Chapter 17). However, recent studies have shown that the environmental context plays significant roles in the *normal* development of almost all species, and that animal genomes have evolved to respond to environmental conditions. Moreover, symbiotic associations, wherein the genes of one organism are regulated by the products of another organism, appear to be the rule rather than the exception.

One reason developmental biologists have largely ignored the environment's effects is that a criterion for selecting which animals to study has been their ability to develop in the laboratory (Bolker 1995). Given adequate nutrition and temperature, all "model organisms"—*C. elegans*, *Drosophila*, zebrafish, *Xenopus*, chicks, and laboratory mice—develop independently of their particular environment, leaving us with the erroneous impression that everything needed to form the embryo is within the fertilized egg. Today, with new concerns about the loss of organismal diversity and the effects of environmental pollutants, there is renewed interest in the regulation of development by the environment (see van der Weele 1999; Gilbert and Epel 2009).

The Environment as a Normal Agent in Producing Phenotypes

Although the nucleus and cytoplasm of the zygote contribute a majority of phenotypic instructions, everything needed for producing a particular phenotype is not pre-packaged in the fertilized egg. Rather, crucial parts of phenotypic determination are regulated by environmental factors outside the organism. **Phenotypic plasticity** is the ability of an organism to react to an environmental input with a change in form, state, movement, or rate of activity (West-Eberhard 2003). When seen in embryonic or larval stages of animals or plants, this ability to change phenotype is often called **developmental plasticity**.

We have already encountered several examples of developmental plasticity in this book. When we discussed environmental sex determination in turtles, fish, and echiuroid worms (see Chapter 14), we were aware that the sexual phenotype was being instructed not by the genome but by the environment. When we discussed in Chapter 12 the ability of shear stress to activate gene expression in capillary, heart, and bone tissue, we similarly were studying the effect of an

We may now turn to consider adaptations towards the external environment; and firstly the direct adaptations ... in which an animal, during its development, becomes modified by external factors in such a way as to increase its efficiency in dealing with them.

C. H. WADDINGTON (1957)

Honor thy symbionts.

JIAN XU AND JEFFREY I. GORDON (2003)

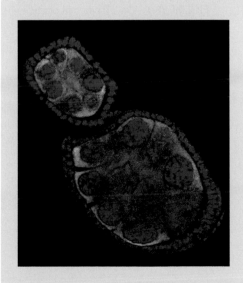

(A)

(B)

(C)

(D)

Spring morph among catkins

Summer morph on twig

(E)

FIGURE 18.1 Developmental plasticity in insects. (A,B) Density-induced polyphenism in the desert (or "plague") locust *Schistocerca gregaria*. (A) The low-density morph has green pigmentation and miniature wings. (B) The high-density morph has deep pigmentation and wings and legs suitable for migration. (C,D) *Nemoria arizonaria* caterpillars. (C) Caterpillars that hatch in the spring eat young oak leaves and develop a cuticle that resembles the oak's flowers (catkins). (D) Caterpillars that hatch in the summer, after the catkins are gone, eat mature oak leaves and develop a cuticle that resembles a young twig. (E) Gyne (reproductive queen) and worker of the ant *Pheidologeton*. This picture shows the remarkable dimorphism between the large, fertile queen and the small, sterile worker (seen near the queen's antennae). The difference between these two sisters is the result of larval feeding. (A,B from Tawfik et al. 1999, courtesy of S. Tanaka; C,D courtesy of E. Greene; E © Mark W. Moffett/National Geographic Society.)

environmental agent on phenotype. While studies of phenotypic plasticity play a central role in plant developmental biology, the mechanisms of plasticity have only recently been studied in animals. These studies are showing that the integration of animals into ecological communities is accomplished largely through developmental interactions. Indeed, these studies show that we do not develop as autonomous entities, but rather that we are co-constructed by other organisms and that we respond to abiotic agents within our local communities.

Two main types of phenotypic plasticity are currently recognized: reaction norms and polyphenisms (Woltereck 1909; Schmalhausen 1949; Stearns et al. 1991). In a **reaction norm,** the genome encodes the potential for a *continuous range* of potential phenotypes, and the environment the individual encounters determines the phenotype (usually the most adaptive one). For instance, our muscle phenotype is determined by the amount of exercise our body is exposed to (even though there is a genetically defined limit to how much muscular hypertrophy is possible). Similarly, the length of a male's horn in some dung beetle species is determined by the quantity and quality of food (i.e., the dung) the larva eats before metamorphosis (see the next section). The upper and lower limits of a reaction norm are also a property of the genome that can be selected. The different phenotypes produced by environmental conditions are called **morphs** (or, occasionally, *ecomorphs*).

The second type of phenotypic plasticity, **polyphenism,** refers to *discontinuous* ("either/or") phenotypes elicited by

(A)   Horned male   Hornless male

(B)

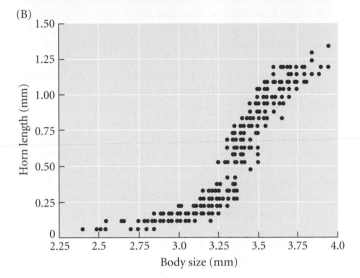

FIGURE 18.2   Diet and *Onthophagus* horn phenotype. (A) Horned and hornless males of the dung beetle *Onthophagus acuminatus* (horns have been artificially colored). Whether a male is horned or hornless is determined by the titer of juvenile hormone at the last molt, which in turn depends on the size of the larva. (B) There is a sharp threshold of body size under which horns fail to form and above which horn growth is linear with the size of the beetle. This threshold effect produces males with no horns and males with large horns, but very few with horns of intermediate size. (After Emlen 2000; photographs courtesy of D. Emlen.)

phenotypes in genetically identical organisms. Diet is also largely responsible for the formation of fertile "queens" in ant, wasp, and bee colonies. Among these insects, each larva has the genetic potential to become either a worker or a queen; only those larvae fed adequately become queens (**Figure 18.1E**). In honeybees, we know that extra nutrition results in the demethylation of particular genes associated with ovary growth and general metabolic rate (Maleszka 2008; Elango et al. 2009; Foret et al. 2009).

See WEBSITE 18.1
Inducible caste determination in ant colonies

the environment. One obvious example is sex determination in turtles, where one range of temperatures will induce female development in the embryo, and another set of temperatures will induce male development. Between these sets of temperatures is a small band of temperatures that will produce different proportions of males and females, but they do not induce intersexual animals. Another important example of polyphenism is the migratory locust *Schistocerca gregaria*. These grasshoppers exist either as a short-winged, green, solitary morph or as a long-winged, brown, gregarious morph (**Figure 18.1A,B**). Cues in the environment determine which morphology a larva will develop upon molting (Rogers et al. 2003; Simpson and Sword 2008).

Diet-induced polyphenisms

The effects of diet in development can be seen in the caterpillar of *Nemoria arizonaria*. When it hatches on oak trees in the spring, it has a form that blends remarkably with the young oak flowers (*catkins*). But those larvae hatching from their eggs in the summer would be very obvious if they still looked like oak flowers. Instead, they resemble newly formed twigs. Here, it is the diet (young versus old oak leaves) that determines the phenotype (**Figure 18.1C,D**; Greene 1989).

Diets having different amounts of proteins or different concentrations of methyl donors have also been found to cause different genes to be expressed in mammalian embryos. Different diets can lead to remarkably distinct

WHEN DUNG REALLY MATTERS   For the male dung beetle (*Onthophagus*), what really matters in life is the amount and quality of the dung he eats as a larva. The hornless female dung beetle digs tunnels, then gathers balls of dung and buries them in these tunnels. She then lays a single egg on each dung ball; when the larvae hatch, they eat the dung. Metamorphosis occurs when the dung ball is finished, and the anatomical and behavioral phenotypes of the male dung beetle are determined by the quality and quantity of this maternally provided food (Emlen 1997; Moczek and Emlen 2000). The amount of food determines the size of the larva at metamorphosis; the size of the larva at metamorphosis determines the titre of juvenile hormone during its last molt; and the titre of juvenile hormone regulates the growth of the imaginal discs that make the horns (**Figure 18.2A**; Emlen and Nijhout 1999; Moczek 2005). If juvenile hormone is added to tiny *O. taurus* males during the sensitive period of their last molt, the cuticle in their heads expands to produce horns. Thus, whether a male is horned or hornless depends not on the male's genes but on the food his mother left for him.

Horns do not grow until the male beetle reaches a certain size. After this threshold body size, horn growth is very rapid.* Thus, although body size has a normal distribution, there is a bimodal distribution of horn sizes: about half the males have no horns, while the other half have horns of considerable length (**Figure 18.2B**).

---

*Interestingly, the threshold size at which the phenotype changes from hornless to horned *is* genetically transmitted and can change when conditions favor one morph over the other (Emlen 1996, 2000).

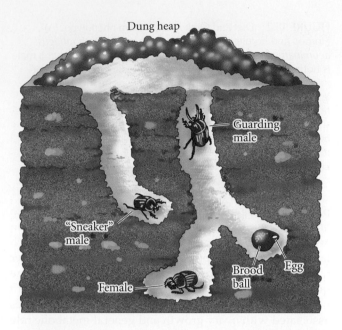

FIGURE 18.3 The presence or absence of horns determines the male reproductive strategy in some dung beetle species. Females dig tunnels in the soil beneath a pile of dung and bring dung fragments into the tunnels. These will be the food supply of the larvae. Horned males guard the tunnel entrances and mate repeatedly with the females. They fight to prevent other males from entering the tunnels, and thee males with long horns usually win such contests. Smaller, hornless males do not guard tunnels, but dig their own to connect with those of females. They can then mate and exit, unchallenged by the guarding male. (After Emlen 2000.)

The size of the horns determines a male's behavior and chances for reproductive success. Horned males guard the females' tunnels and use their horns to prevent other males from mating with the female; the male with the biggest horns wins such contests. But what about the males with no horns? Hornless males do not fight with the horned males for mates. Since they, like the females, lack horns, they are able to dig their own tunnels. These "sneaker males" dig tunnels that intersect those of the females and mate with the females while the horned male stands guard at the tunnel entrance (**Figure 18.3**; Emlen 2000; Moczek and Emlen 2000). Indeed, about half the fertilized eggs in most populations are from hornless males. The ability to produce a horn is inherited; but whether to produce a horn and how big to make it is regulated by the environment.

DIET AND DNA METHYLATION Diet can also directly influence the DNA. As mentioned earlier, honeybee caste (queen or worker) is determined by diet-induced changes in DNA methylation patterns. Dietary alterations can also produce changes in mammalian DNA methylation, and these methylation changes can affect the phenotype. Waterland and Jirtle (2003) demonstrated this by using mice containing the *viable-yellow* allele of *Agouti*. *Agouti* is a dominant gene

that gives mice yellowish hair color; it also affects lipid metabolism such that the mice become fatter. The *viable-yellow* allele of *Agouti* has a transposable element inserted into its *cis*-regulatory regions. These transposon insertion sites are very interesting for regulation: whereas most regions of the adult genome have hardly any intraspecies variation in CpG methylation, there are large DNA methylation differences between individuals at the sites of transposon insertion. Such CpG methylation can block gene transcription. When the promoter of the *Agouti* gene is methylated, the gene is not transcribed. The mouse's fur remains black, and lipid metabolism is not altered.

Waterland and Jirtle fed pregnant *viable-yellow Agouti* mice methyl donor supplements, including folate, choline, and betain. They found that the more methyl supplementation, the greater the methylation of the transposon insertion site in the fetus' genome, and the darker the pigmentation of the offspring. Although the mice in **Figure 18.4** are genetically identical, their mothers were fed different diets during pregnancy. The mouse whose mother did not receive methyl donor supplementation is fat and yellow—the *Agouti* gene promoter was unmethylated, and the gene was active. The mouse born to the mother that was given supplements is sleek and dark; the methylated *Agouti* gene was not transcribed.

As we saw in Chapter 17, such differential gene methylation has been linked to human health problems. Dietary restrictions during a woman's pregnancy may show up as heart or kidney problems in her adult children. Moreover, studies in rats showed that differences in protein and methyl donor concentration in the mother's prenatal diet affected metabolism in the pup's livers (Lillycrop et al. 2005).

FIGURE 18.4 Maternal diet can affect phenotype. These two mice are genetically identical; both contain the *viable-yellow* allele of the *Agouti* gene, whose protein product converts brown pigment to yellow and accelerates fat storage. The obese yellow mouse is the offspring of a mother whose diet was not supplemented with methyl donors (e.g., folic acid) during her pregnancy. The embryo's *Agouti* gene was not methylated, and Agouti protein was made. The sleek brown mouse was born of a mother whose prenatal diet was supplemented with methyl donors. The *Agouti* gene was turned off, and no Agouti protein was made. (After Waterland and Jirtle 2003, photograph courtesy of R. L. Jirtle.)

Predator-induced polyphenisms

Imagine an animal who is frequently confronted by a particular predator. One could then imagine an individual who could recognize soluble molecules secreted by that predator and could use those molecules to activate the development of structures that would make this individual less palatable to the predator. This ability to modulate development in the presence of predators is called predator-induced defense, or **predator-induced polyphenism**.

To demonstrate predator-induced polyphenism, one has to show that the phenotypic modification is caused by the presence of the predator, and that the modification increases the fitness of its bearers when the predator is present (Adler and Harvell 1990; Tollrian and Harvell 1999). Figure 18.5A shows both the typical and predator-induced morphs for several species. In each case, the induced morph is more successful at surviving the predator, and

soluble filtrate from water surrounding the predator is able to induce the changes. Chemicals that are released by a predator and can induce defenses in its prey are called **kairomones**.

Several rotifer species will alter their morphology when they develop in pond water in which their predators were cultured (Dodson 1989; Adler and Harvell 1990). The predatory rotifer *Asplanchna* releases a soluble compound that induces the eggs of a prey rotifer species, *Keratella slacki*, to develop into individuals with slightly larger bodies and anterior spines 130% longer than they otherwise would be, making the prey more difficult to eat. When exposed to the effluent of the crab species that preys on it, the snail *Thais lamellosa* develops a thickened shell and a "tooth" in its aperture. In a mixed snail population, crabs will not attack the thicker-shelled snails until more than half of the typical-morph snails are devoured (Palmer 1985).

(A)

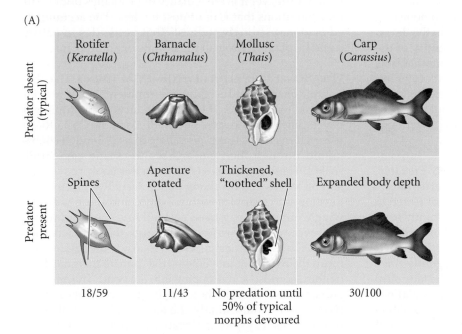

(B)

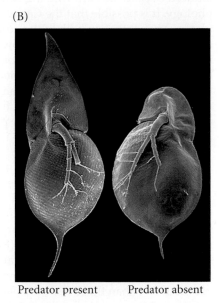

Predator present　　　Predator absent

(C) Predator present

(D) Predator absent

FIGURE 18.5 Predator-induced defenses. (A) Typical (upper row) and predator-induced (lower row) morphs of various organisms. The numbers beneath each column represent the percentages of organisms surviving predation when both induced and uninduced individuals were presented with predators (in various assays). (B) Scanning electron micrographs show predator-induced (left) and typical (right) morphs of genetically identical individuals of the water flea *Daphnia*. In the presence of chemical signals from a predator, *Daphnia* grows a protective "helmet." (C,D) Tadpole phenotypes. (C) Tadpoles of the tree frog *Hyla chrysoscelis* developing in the presence of cues from a predator's larvae develop strong trunk muscles and a red coloration. (D) When predator cues are absent, the tadpoles grow sleeker, which helps them compete for food. (A after Adler and Harvell 1990 and references cited therein; B courtesy of A. A. Agrawal; C,D courtesy of T. Johnson/USGS.)

One of the more interesting mechanisms of predator-induced polyphenism is that of certain echinoderm larvae. When exposed to the mucus of their fish predator, sand dollar plutei clone themselves, budding off small groups of cells that quickly become larvae themselves. The small plutei are below the visual detection of the fish, and thereby escape being eaten (Vaughn and Strathmann 2008; Vaughn 2009).

DAPHNIA AND THEIR KIN The predator-induced polyphenism of the parthenogenetic water flea *Daphnia* is beneficial not only to itself but also to its offspring. When *D. cucullata* encounter the predatory larvae of the fly *Chaeoborus*, their "helmets" grow to twice the normal size (**Figure 18.5B**). This increase lessens the chances that *Daphnia* will be eaten by the fly larvae. This same helmet induction occurs if the *Daphnia* are exposed to extracts of water in which the fly larvae had been swimming. Agrawal and colleagues (1999) have shown that the offspring of such an induced *Daphnia* are born with this same altered head morphology. It is possible that the *Chaeoborus* kairomone regulates gene expression both in the adult and in the developing embryo. We still do not know the identity of the kairomone, the identity of its receptor, or the mechanisms by which the binding of the kairomone to the receptor initiates the adaptive morphological changes. There are trade-offs, however; the induced *Daphnia*, having put resources into making protective structures, produce fewer eggs (Tollrian 1995).

AMPHIBIAN PHENOTYPES INDUCED BY PREDATORS Predator-induced polyphenism is not limited to invertebrates.* Among amphibians, tadpoles found in ponds or reared in the presence of other species may differ significantly from tadpoles reared by themselves in aquaria. For instance, newly hatched wood frog tadpoles (*Rana sylvetica*) reared in tanks containing the predatory larval dragonfly *Anax* (confined in mesh cages so they cannot eat the tadpoles) grow smaller than those reared in similar tanks without predators. Moreover, their tail musculature deepens, allowing faster turning and swimming speeds (Van Buskirk and Relyea 1998). The addition of more predators to the tank causes a continuously deeper tail fin and tail musculature, and in fact what initially appeared to be a polyphenism may be a reaction norm that can assess the number (and type) of predators.

McCollum and Van Buskirk (1996) have shown that in the presence of its predators, the tail fin of the tadpole of the tree frog *Hyla chrysoscelis* grows larger and becomes bright red (**Figure 18.5C,D**). This phenotype allows the tadpole to swim away faster and to deflect predator strikes toward the tail region. The trade-off is that noninduced tadpoles grow more slowly and survive better in predator-free environments. In some species, phenotypic plasticity is reversible, and removing the predators can restore the non-induced phenotype (Relyea 2003a).

The metabolism of predator-induced morphs may differ significantly from that of the uninduced morphs, and this has important consequences. Relyea (2003b, 2004) has found that in the presence of the chemical cues emitted by predators, the toxicity of pesticides such as carbaryl (Sevin®) can become up to 46 times more lethal than it is without the predator cues. Bullfrog and green frog tadpoles were especially sensitive to carbaryl when exposed to predator chemicals. Relyea has related these findings to the global decline of amphibian populations, saying that governments should test the toxicity of the chemicals under more natural conditions, including that of predator stress. He concludes (Relyea 2003b) that "ignoring the relevant ecology can cause incorrect estimates of a pesticide's lethality in nature, yet it is the lethality of pesticides under natural conditions that is of utmost interest. The accumulated evidence strongly suggests that pesticides in nature could be playing a role in the decline of amphibians."

VIBRATIONAL CUES ALTER DEVELOPMENTAL TIMING The phenotypic changes induced by environmental cues are not confined to structure. They can also include the timing of developmental processes. The embryos of the Costa Rican red-eyed treefrog (*Agalychnis callidryas*) use vibrations transmitted through their egg masses to escape egg-eating snakes. These egg masses (shown on this book's cover) are laid on leaves that overhang ponds. Usually, the embryos develop into tadpoles within 7 days, and these tadpoles wiggle out of the egg mass and fall into the pond water. However, when snakes feed on the eggs, the vibrations they produce cue the remaining embryos inside the egg mass to begin the twitching movements that initiate their hatching (within seconds!) and dropping into the pond. The embryos are competent to begin these hatching movements at day 5 (**Figure 18.6**). Interestingly, the embryos have evolved to respond this way only to vibrations given at a certain frequency and interval (Warkentin et al. 2005, 2006). Up to 80% of the remaining embryos can escape snake predation in this way, and research has shown that these vibrations alone (and not smell or sight) cue these hatching movements in the embryos. There is a trade-off here, too. Although these embryos have escaped their snake predators, they are now at greater risk from waterborne predators than are fully developed embryos, because the musculature of the early hatchers has not developed fully.

Temperature as an environmental agent

TEMPERATURE AND SEX There are many species in which temperature does control whether testes or ovaries develop. Indeed, among the cold-blooded vertebrates such as fish, turtles, and alligators there are many species in which

---

*Indeed, when viewed biologically rather than medically, the vertebrate immune system is a wonderful example of predator-induced polyphenism. Here, our immune cells utilize chemicals from our predators (viruses and bacteria) to change our phenotype so that we can better resist them (see Frost 1999).

(A)

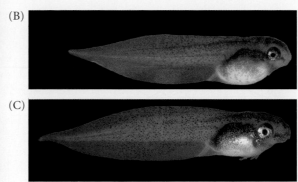

FIGURE 18.6   Predator-induced polyphenism in the red-eyed tree frog *Agalychnis callidryas*. (A) When a snake eats a clutch of *Agalychnis* eggs, most of the remaining embryos inside the egg mass respond to the vibrations by hatching prematurely (arrow) and falling into the water. (B) Immature tadpole, induced to hatch at day 5. (C) Normal tadpoles hatch at day 7 and have better-developed musculature. (Courtesy of K. Warkentin.)

the environment determines whether an individual is male or female (Crews and Bull 2009). This type of environmental sex determination has advantages and disadvantages. One advantage is that it probably gives the species the benefits of sexual reproduction without tying the species to a 1:1 sex ratio. In crocodiles, in which extreme temperatures produce females while moderate temperatures produce males, the sex ratio may be as great as 10 females to each male (Woodward and Murray 1993). In such instances, where the number of females limits the population size, this ratio is better for survival than the 1:1 ratio demanded by genotypic sex determination.

The major disadvantage of temperature-dependent sex determination may be its narrowing of the temperature limits within which a species can persist. Thus thermal pollution (either locally or due to global warming) could conceivably eliminate a species in a given area (Janzen and Paukstis 1991). Researchers (Ferguson and Joanen 1982; Miller et al. 2004) have speculated that dinosaurs may have had temperature-dependent sex determination and that their sudden demise may have been caused by a slight change in temperature creating conditions wherein only males or only females hatched. (Unlike many turtle species, whose members have long reproductive lives, can hibernate for years, and whose females can store sperm, dinosaurs may have had a relatively narrow time to reproduce and no ability to hibernate through prolonged bad times.)

See **WEBSITE 18.2** *Volvox*: **When heat brings out sex**

Charnov and Bull (1977) argued that environmental sex determination would be adaptive in those habitats characterized by patchiness—that is, a habitat having some regions where it is more advantageous to be male and other regions where it is more advantageous to be female. Conover and Heins (1987) provided evidence for this hypothesis. In certain fish species, females benefit from being larger, since larger size translates into higher fecundity. If you are a female Atlantic silverside fish (*Menidia menidia*), it is advantageous to be born early in the breeding season, because you have a longer feeding season and thus can grow larger. (The size of males in this species doesn't influence mating success or outcomes.) In the southern range of *Menidia*, females are indeed born early in the breeding season, and temperature appears to play a major role in this pattern. However, in the northern reaches of its range, the species shows no environmental sex determination. Rather, a 1:1 sex ratio is generated at all temperatures. Conover and Heins speculated that the more northern populations have a very short feeding season, so there is no advantage for females in being born earlier. Thus, this fish has environmental sex determination in those regions where it is adaptive and genotypic sex determination in those regions where it is not.

BUTTERFLY WINGS   In tropical parts of the world, there is often a hot wet season and a cooler dry season. In Africa, a polyphenism of the dimorphic Malawian butterfly (*Bicyclus anynana*) is adaptive to seasonal changes. The dry (cool) season morph is a mottled brown butterfly that survives by hiding in dead leaves on the forest floor. In contrast, the wet (hot) season morph, which routinely flies, has prominent ventral eyespots that deflect attacks from predatory birds and lizards (**Figure 18.7**).

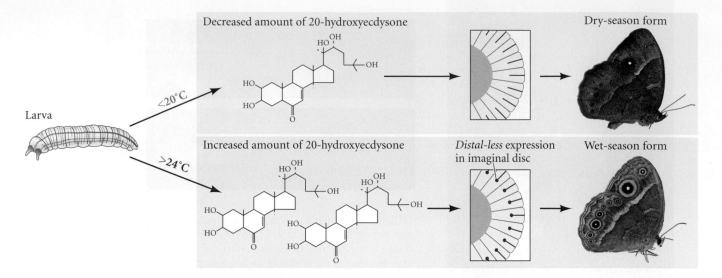

Decreased amount of 20-hydroxyecdysone
Larva
<20°C
>24°C
Increased amount of 20-hydroxyecdysone
Distal-less expression in imaginal disc
Dry-season form
Wet-season form

FIGURE 18.7 Phenotypic plasticity in *Bicyclus anynana* is regulated by temperature during pupation. High temperature (either in the wild or in controlled laboratory conditions) allows the accumulation of 20-hydroxyecdysone (20E), a hormone that is able to sustain *Distal-less* expression in the pupal imaginal disc. The region of *Distal-less* expression becomes the focus of each eyespot. In cooler weather, 20E is not formed, *Distal-less* expression in the imaginal disc begins but is not sustained, and eyespots fail to form. (Courtesy of S. Carroll and P. Brakefield.)

The factor determining the seasonal pigmentation of *B. anynana* is not diet, but the temperature during pupation. Low temperatures produce the dry-season morph; higher temperatures produce the wet-season morph (Brakefield and Reitsma 1991). The mechanism by which temperature regulates the *Bicyclus* phenotype is becoming known. In the late larval stages, transcription of the *distal-less* gene in the wing imaginal discs is restricted to a set of cells that will become the signaling center of each eyespot. In the early pupa, higher temperatures elevate the formation of 20-hydroxyecdysone (20E; see Chapter 15). This hormone sustains and expands the expression of *distal-less* in those regions of the wing imaginal disc, resulting in prominent eyespots. In dry season, the cooler temperatures prevent the accumulation of 20E in the pupa, and the foci of Distal-less signaling are not sustained. In the absence of the Distal-less signal, the eyespots do not form (Brakefield et al. 1996; Koch et al. 1996). Distal-less protein is believed to be the activating signal that determines the size of the eyespot (see Figure 18.7).

The importance of hormones such as 20E for mediating environmental signals controlling wing phenotypes has been documented in the *Araschnia* butterfly (**Figure 18.8**). *Araschnia* develops alternative phenotypes depending on whether the fourth and fifth instars experience a photoperiod (hours of daylight) that is longer or shorter than a particular critical day length. Below this critical day length, ecdysone levels are low and the butterfly has the orange wings characteristic of spring butterflies. Above the critical point, ecdysone is made and the summer pigmentation forms. The summer form can be induced in spring pupae by injecting 20E into the pupae. Moreover, by altering the timing of 20E injections, one can generate a series of intermediate forms not seen in the wild (Koch and Bückmann 1987; Nijhout 2003).

FIGURE 18.8 Environmentally induced morphs of the European map butterfly (*Araschnia levana*). The orange morph (bottom) forms in the spring, when levels of ecdysone in the larva are low. The dark morph with a white stripe (top) forms in summer, when higher temperatures and longer photoperiods induce greater ecdysone production in the larva. Linnaeus classified the two morphs as different species. (Courtesy of H. F. Nijhout.)

Environmental Induction of Behavioral Phenotypes

In many instances, the morphological phenotype is accompanied by a behavioral phenotype. This is obvious in the environmental determination of sex, where an individual's sexual behavior generally matches the gonads and genitalia. This is also seen in the cases of butterfly wings (fliers vs. crawlers) and dung beetle horns (fighters vs. "sneakers"). Sometimes, however, the behavior is the major developmental phenotype induced by the environment.

Adult anxiety and environmentally regulated DNA methylation

Environmentally derived methylation differences near birth may lead to important behavioral differences in adults. In rats, behavioral differences in the response to stressful situations have been correlated with the number of glucorticoid receptors in the brain's hippocampus. The more glucocorticoid receptors, the better the adult rat is able to downregulate these adrenal hormones and deal with stress. The number of glucocorticoid receptors appears to depend on the quality of grooming and licking the rat pup experiences during the first week after birth.

How is the adult phenotype regulated by these perinatal (near the time of birth) experiences? Weaver and colleagues (2004) have shown that the behavioral difference involves the methylation of a particular site in the enhancer region on the glucocorticoid receptor gene. Before birth, there is no methylation at this site; 1 day after birth, this site is methylated in all rat pups. However, in those pups that experience intensive grooming and licking during the first week after birth, this site *loses* its methylation; but methylation is retained in those rats that do not have such extensive care. Moreover, this methylation difference is not seen at other sites in or near the gene (**Figure 18.9**).

By switching pups and parents, Weaver and colleagues demonstrated that this methylation difference was dependent on the mother's care, and was not the result of differences in the pups themselves. When unmethylated, this enhancer site binds the Egr1 transcription factor and is associated with "active" acetylated nucleosomes. The transcription factor does not bind to the methylated site, and the chromatin in such cases is not activated. These chromatin differences, established during the first week after birth, are retained throughout the life of the rat. Thus, adult rats that received extensive perinatal grooming have more glucocorticoid receptors and are able to deal with stress better than rats that received less care.* Just *how* grooming can alter DNA methylation patterns, however, remains to be discovered.

---

*Does this relate to humans? Using an "extreme" set of cases, McGowan and colleagues (2009) showed that the *cis*-regulatory region of the hippocampus-specific glucocorticoid receptor is more highly methylated in the brains of suicide victims with a history of childhood abuse than in suicide victims with no childhood abuse, or in controls.

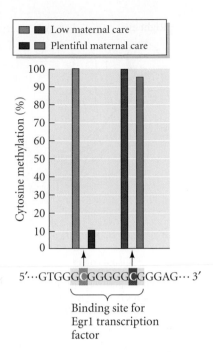

FIGURE 18.9 Differential DNA methylation due to behavioral differences in parental care. A portion of an enhancer sequence of the rat glucocorticoid receptor (*GR*) gene is shown, indicating the binding site for the Egr1 transcription factor. Two cytosine residues within this binding site have the potential to be methylated. The cytosine at the 5′ end is completely methylated in the brains of pups that did not receive extensive licking and grooming from their mothers (red bar). The Egr1 transcription factor did not bind these methylated sites, and thus the *GR* gene remained inactive. If the pups received proper maternal care, this same site was largely unmethylated (orange bar), and the gene was transcribed in the brain. The cytosine at the 3′ end of the enhancer (blue bars) was always methylated and had no effect on Egr1 binding. (After Weaver et al. 2004.)

Learning: The Developmentally Plastic Nervous System

Learning provides remarkable examples of phenotypic plasticity. Since neurons, once formed, do not divide, the "birthday" of a neuron can be identified by treating the organism with radioactive thymidine. Normally, very little radioactive thymidine is taken up into the DNA of a neuron that has already been formed. However, if a neural precursor cell divides during the treatment, it will incorporate radioactive thymidine into its DNA.

Such new neurons are seen to be generated when male songbirds first learn their songs. Juvenile zebra finches memorize a model song and then learn the pattern of muscle contractions necessary to sing a particular phrase. In this learning and repetition process, new neurons are generated in the hyperstriatum of the finch's brain. Many of these new neurons send axons to the archistriatum, which is responsible for controlling the vocal musculature

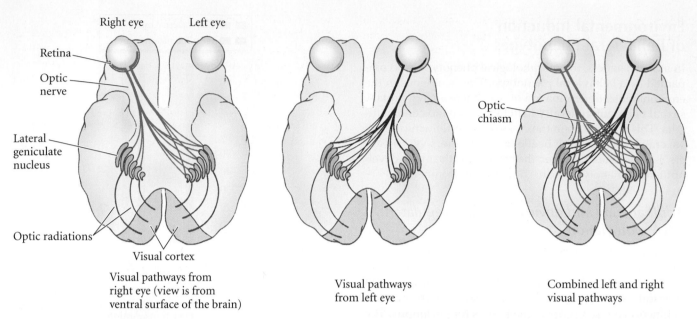

Visual pathways from right eye (view is from ventral surface of the brain)

Visual pathways from left eye

Combined left and right visual pathways

FIGURE 18.10 Major pathways of the mammalian visual system. In mammals, the optic nerve from each eye branches, sending nerve fibers to a lateral geniculate nucleus on each side of the brain. On the ipsilateral side, a particular part of the retina projects to a particular part of the lateral geniculate nucleus. On the contralateral side, the lateral geniculate nucleus receives input from all parts of the retina. Neurons from each lateral geniculate nucleus innervate the visual cortex on the same side.

(Nordeen and Nordeen 1988). These changes are not seen in males that are too old to learn the song, nor are they seen in juvenile females (which do not sing these phrases). In white-crowned sparrows (*Zonotrichia leucophrys*), whose song is regulated by photoperiod and hormones, exposing adult males to long hours of light and to testosterone induces over 50,000 new neurons in their vocal centers (Tramontin et al. 2000). The neural circuitry of these birds' brains shows seasonal plasticity. Testosterone is believed to increase the level of brain-derived neurotropic factor (BDNF) in the song-producing vocal centers. If female birds are given BDNF, they also produce more neurons there (Rasika et al. 1999).

The cerebral cortices of young rats reared in stimulating environments are packed with more neurons, synapses, and dendrites than are found in rats reared in isolation (Turner and Greenough 1983). Even the adult brain continues to develop in response to new experiences. When adult canaries learn new songs, they generate new neurons whose axons project from one vocal region of the brain to another (Alvarez-Buylla et al. 1990). Studies on adult rats and mice indicate that environmental stimulation can increase the number of new neurons in the dentate gyrus of the hippocampus (Kempermann et al. 1997a,b; Gould et al. 1999; van Praag et al. 1999). Similarly, when adult rats learn to keep their balance on dowels, their cerebellar Purkinje neurons develop new synapses (Black et al. 1990).

In addition to inducing the formation of new neurons, learning and new experiences can also remodel old neurons into new patterns of connections. When students were taught the classic three-ball cascade juggling routine (which takes months to get right), the neurons in a specific area of the temporal lobe of the brain took on a new pattern—a pattern not seen in students who were not taught this skill

(Draganski et al. 2004). Similarly, mice reared in cages experienced changes in their neural circuitry when they were placed into more natural environments (Polley et al. 2004). Thus, the pattern of neuronal connections is a product of inherited patterning and patterning produced by experiences. This interplay between innate and experiential development has been detailed most dramatically in studies on mammalian vision.

Experiential changes in mammalian visual pathways

Some of the most interesting research on mammalian neuronal patterning concerns the effects of sensory deprivation on the developing visual system in kittens and monkeys. The paths by which electric impulses pass from the retina to the brain in mammals are shown in **Figure 18.10**. Axons from the retinal ganglion cells form the two optic nerves, which meet at the optic chiasm. As in *Xenopus* tadpoles, some axons go to the opposite (contralateral) side of the brain, but unlike in most other vertebrates, mammalian retinal ganglion cells also send inputs into the same (ipsilateral) side of the brain (see Chapter 10). These axons end at the two lateral geniculate nuclei. Here the input from each eye is kept separate, with the uppermost and anterior layers receiving the axons from the contralateral eye, and the middle of the layers receiving input from the ipsilateral eye. The situation becomes even more complex as neurons from the lateral geniculate nuclei connect with the neurons of the visual cortex. Over 80% of the neural cells in the visual cortex receive input from both eyes. The result is binocular vision and depth perception.

A remarkable finding is that the retinocortical projection pattern is the same for both eyes. If a certain cortical neuron is stimulated by light flashing across a region of the left eye 5° above and 1° to the left of the fovea,* it will also be stimulated by a light flashing across a region of the *right* eye 5° above and 1° to the left of the fovea. Moreover, the response evoked in the cortical neuron when both eyes are stimulated is greater than the response when either retina is stimulated alone.

Hubel, Wiesel, and their co-workers demonstrated that the development of the nervous system depends to some degree on the experience of the individual during a critical period of development (see Hubel 1967). In other words, not all neuronal development is encoded in the genome; some is the result of learning. Experience appears to strengthen or stabilize some neuronal connections that are already present at birth and to weaken or eliminate others. These conclusions come from studies of partial sensory deprivation. Hubel and Wiesel (1962, 1963) sewed shut the right eyelids of newborn kittens and left them closed

---

*The *fovea* is a depression in the center of the retina where only cones are present (rods and blood vessels are absent). In this instance, it serves as a convenient landmark.

for 3 months. After this time, they unsewed the right eyelids. The cortical neurons of these kittens could not be stimulated by shining light into the right eye. Almost all the inputs into the visual cortex came from the left eye only. The behavior of the kittens revealed the inadequacy of their right eyes; when the left eyes of these kittens were covered, they became functionally blind. Because the lateral geniculate neurons appeared to be stimulated by input from both right and left eyes, the physiological defect appeared to be in the connections between the lateral geniculate nuclei and the visual cortex. Similar phenomena have been observed in rhesus monkeys, where the defect has been correlated with a lack of protein synthesis in the lateral geniculate neurons innervated by the covered eye (Kennedy et al. 1981).

Although it would be tempting to conclude that the blindness resulting from these experiments was the result of failure to form the proper visual connections, this is not the case. Rather, when a kitten or monkey is born, axons from lateral geniculate neurons receiving input from each eye overlap extensively in the visual cortex (Hubel and Wiesel 1963; Crair et al. 1998). However, when one eye is covered early in the animal's life, its connections in the visual cortex are taken over by those of the other eye (Figure 18.11). The axons compete for connections, and experience plays a role in strengthening and stabilizing the connections that are made. Thus, when *both* eyes of a kitten are sewn shut for 3 months, most cortical neurons can still be stimulated by appropriate illumination of one eye or the other. The critical time in kitten development for this

(A)

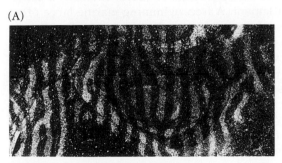

(B)

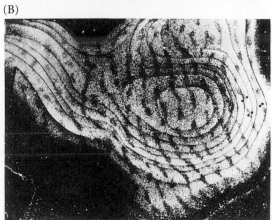

FIGURE 18.11 Experience alters neural connections. (A,B) Dark-field autoradiographs of monkey striate (visual) cortex 2 weeks after one eye was injected with [³H]proline in the vitreous humor. Each retinal neuron takes up the radioactive label and transfers it to the cells with which it forms synapses. (A) Normal labeling pattern. The white stripes indicate that roughly half the columns took up the label, while the other half did not—a pattern reflecting that half the cells were innervated by the labeled eye and half by the unlabeled eye. (B) Labeling pattern when the unlabeled eye was sutured shut for 18 months. Axonal projections from the normal (labeled) eye have taken over the regions that would normally have been innervated by the sutured eye. (C,D) Drawings of axons from the lateral geniculate nuclei of kittens in which one eye was occluded for 33 days. The terminal branching of axons receiving input from the occluded eye (C) was far less extensive than that of axons receiving input from the nonoccluded eye (D). (A,B from Wiesel 1982, courtesy of T. Wiesel; C,D after Antonini and Stryker 1993.)

(C)   (D)

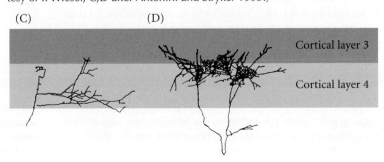

Cortical layer 3

Cortical layer 4

validation of neuronal connections begins between 4 and 6 weeks after birth. Monocular deprivation up to the fourth week produces little or no physiological deficit, but through the sixth week it produces all the characteristic neuronal changes. If a kitten has had normal visual experience for the first 3 months, any subsequent monocular deprivation (even for a year or more) has no effect. At that point, the synapses have been stabilized.

Two principles, then, can be seen in the patterning of the mammalian visual system. First, the neuronal connections involved in vision are present even before the animal sees. Second, experience plays an important role in determining whether or not certain connections persist.[*] Just as experience refines the original neuromuscular connections, experience plays a role in refining and improving the visual connections. It is possible, then, that adult functions such as learning and memory arise from the establishment and/or strengthening of different synapses by experience. As Purves and Lichtman (1985) remark:

> The interaction of individual animals and their world continues to shape the nervous system throughout life in ways that could never have been programmed. Modification of the nervous system by experience is thus the last and most subtle developmental strategy.

Life Cycles and Polyphenisms

Diapause

Throughout nature, organisms are able to respond to the environment by changing their developmental strategy. One of these strategies is to delay development to a better time, and many species of insects and mammals have evolved a developmental strategy called diapause to survive periodically harsh conditions. **Diapause** is a suspension of development that can occur at the embryonic, larval, pupal, or adult stage, depending on the species (see Chapter 15). Diapause is not a physiological response brought about by harsh conditions. Rather, it is induced by stimuli (such as changes in the duration of daylight) that *presage* a change in the environment—cues beginning before the severe conditions actually arise. Diapause is especially important for temperate-zone insects, enabling them to survive the winter. The overwintering eggs of the hickory aphid provide an example of this strategy. The development in the egg is suspended over the winter, so the larvae do not hatch when food is unavailable. In this case, diapause occurs during early development. The silkworm moth *Bombyx mori* similarly overwinters as an

embryo, entering diapause just before segmentation. The gypsy moth *Lymantria dispar* initiates diapause as a larva, and needs an extended period of cold weather to end diapause, which is why this pest is not found in the southern regions of Europe or the United States.

Over 100 mammalian species undergo diapause. The two most common mammalian strategies are delayed fertilization (the sperm are stored for later use) and delayed implantation (the blastocyst remains unimplanted in the uterus, and the rate of cell divisions diminishes or vanishes). Some species have *seasonal* diapause, so embryos conceived in autumn will be born in spring rather than winter; in other species, diapause is induced by the presence of a newborn that is still getting milk. In the tammar kangaroo (*Macropus eugenii*), diapause can be a response to suckling-induced prolactin release, but it can also be induced by prolactin synthesized in response to changes in day length. In both cases, progesterone seems to be the signal that restores implantation and embryonic growth. Different groups of mammals use different hormones to induce or break diapause, but the result is the same: diapause lengthens the gestation period, allowing mating to occur and young to be born at times and seasons appropriate to the habitat of that species (Renfree and Shaw 2000).

See WEBSITE 18.3 Mechanisms of diapause

Larval settlement

The ability of marine larvae to suspend development until they experience a particular environmental cue is called larval settlement. A free-swimming marine larva often needs to settle near a source of food or on a firm substrate on which it can metamorphose. Among the molluscs, there are often very specific cues for settlement (Hadfield 1977; Hadfield and Paul 2001; Zardus et al. 2008). In some cases, the prey supply the cues, while in other cases the substrate itself gives off molecules used by the larvae to initiate settlement. These cues may not be constant, but they need to be part of the environment if further development is to occur[†] (Pechenik et al. 1998).

---

[*]Studies have shown that differences in neurotransmitter release result in changes in synaptic adhesivity and cause the withdrawal of the axon providing the weaker stimulation (Colman et al. 1997). Studies in mice suggest that BDNF is crucial during the critical period (Huang et al. 1999; Katz 1999; Waterhouse and Xu 2009; Cowansage et al. 2010).

[†]The importance of substrates for larval settlement and metamorphosis was first demonstrated in 1880, when William Keith Brooks, an embryologist at Johns Hopkins University, was asked to help the ailing oyster industry of Chesapeake Bay. For decades, oysters had been dredged from the bay, and there had always been a new crop to take their place. But by 1880, each year brought fewer oysters. What was responsible for the decline? Experimenting with larval oysters, Brooks discovered that the American oyster (unlike its better-studied European relative) needs a hard substrate on which to metamorphose. For years, oystermen had thrown the shells back into the sea, but with the advent of suburban sidewalks, they started selling the shells to cement factories. Brooks's solution: throw the shells back into the bay. The oyster population responded, and the Baltimore wharves still sell their descendants.

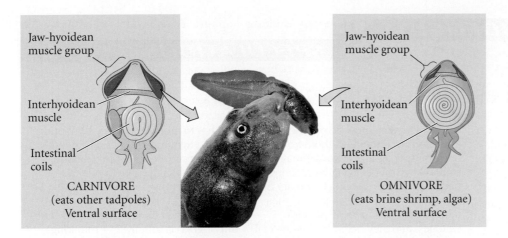

FIGURE 18.12   Polyphenism in tadpoles of the spadefoot toad (*Scaphiopus couchii*). The typical morph (right) is an omnivore, feeding on arthropods and algae. When ponds are drying out quickly, however, a carnivorous (cannibalistic) morph forms (left). It develops a wider mouth, larger jaw muscles, and an intestine modified for a carnivorous diet. The center photograph shows a cannibalistic tadpole eating a smaller pondmate. (Photograph © Thomas Wiewandt; drawings courtesy of R. Ruibel.)

In many marine invertebrate species, larval settlement is regulated by mats of bacteria called **biofilms** that help determine the distribution of the invertebrate populations. Humans are changing this distribution, however, with our desire to place large objects in the oceans. Such objects readily acquire biofilms and the resultant marine fauna that attach to them. As early as 1854, Charles Darwin speculated that barnacles were transported to new locales when their larvae settled on the hulls of ships. Indeed, the ability of biofilms to aid invertebrate larval settlement and colony formation explains the ability of barnacles and tubeworms ("biofouling invertebrates") to accumulate on ship keels, clog sewer pipes, and deteriorate underwater structures (Zardus et al. 2008).

The spadefoot toad: A hard life

The spadefoot toad (*Scaphiopus couchii*) has a remarkable strategy for coping with a harsh environment. These toads are called out from hibernation by the thunder that accompanies the first spring storm in the Sonoran desert. (Unfortunately, motorcycles produce the same sounds, causing the toads to come out of hibernation only to die in the scorching Arizona sun.) The toads breed in temporary ponds formed by the rain, and the embryos develop quickly into larvae. After the larvae metamorphose, the young toads return to the desert, burrowing into the sand until the next year's storms bring them out.

Desert ponds are ephemeral pools that can either dry up quickly or persist, depending on their initial depth and the frequency of rainfall. One might envision two alternative scenarios confronting a tadpole in such a pond: either (1) the pond persists until you have time to fully metamor-

phose, and you live; or (2) the pond dries up before your metamorphosis is complete, and you die. *S. couchii* (and several other amphibians), however, have evolved a third alternative. The timing of their metamorphosis is controlled by the pond. If the pond persists at a viable level, development continues at its normal rate, and the algae-eating tadpoles develop into juvenile toads. However, if the pond is drying out and getting smaller, some of the tadpoles embark on an alternative developmental pathway. They develop a wider mouth and powerful jaw muscles, which enables them to eat (among other things) other *Scaphiopus* tadpoles (**Figure 18.12**). These carnivorous tadpoles metamorphose quickly, albeit into a smaller version of the juvenile spadefoot toad. But they survive while other *Scaphiopus* tadpoles perish from desiccation (Newman 1989, 1992).

The signal for accelerated metamorphosis appears to be the change in water volume. In the laboratory, *Scaphiopus* tadpoles are able to sense the removal of water from aquaria, and their acceleration of metamorphosis depends on the rate at which the water is removed. The stress-induced corticotropin-releasing hormone signaling system appears to modulate this effect (Denver et al. 1998, 2009). This increase in brain corticotropin-releasing hormone is thought to be responsible for the subsequent elevation of the thyroid hormones that initiate metamorphosis (Boorse and Denver 2003). As in many other cases of polyphenism, the developmental changes are mediated through the endocrine system. Sensory organs send a neural signal to regulate hormone release. The hormones then can alter gene expression in a coordinated and relatively rapid fashion.

See WEBSITE 18.4
Pressure as a developmental agent

Life Cycle Choices: *Dictyostelium*

Of the many developmental changes triggered by the environment, few are as profound as the effect of nutrient starvation on *Dictyostelium discoideum*.* *Dictyostelium* is a multicellular organism only when environmental conditions demand it. The life cycle of this fascinating organism is illustrated in **Figure 18.13**. In its asexual cycle, solitary haploid amoe-

*Although colloquially called a "cellular slime mold," *Dictyostelium* is not a mold, nor is it consistently slimy. It is perhaps best to think of *Dictyostelium* as a social amoeba.

bae (called myxamoebae or "social amoebae" to distinguish them from amoeba species that always remain solitary) live on decaying logs, eating bacteria and reproducing by binary fission. When they have exhausted their food supply, their gene expression changes, and they begin expressing cell adhesion molecules. Moreover, they start migrating to a common place. Tens of thousands of these myxamoebae join together to form moving streams of cells that converge at a central point. Here they synthesize new adhesion molecules and pile atop one

another to produce a conical mound called a tight aggregate. Subsequently, a tip arises at the top of this mound, and the tight aggregate bends over to produce the migrating slug (with the tip at the front). The **slug** (often given the more dignified title of **pseudoplasmodium** or **grex**) is usually 2–4 mm long and is encased in a slimy sheath. The grex begins to migrate (if the environment is dark and moist) with its anterior tip slightly raised. When it reaches an illuminated area, migration ceases, and the culmination stages of the life cycle take place as the grex differentiates into

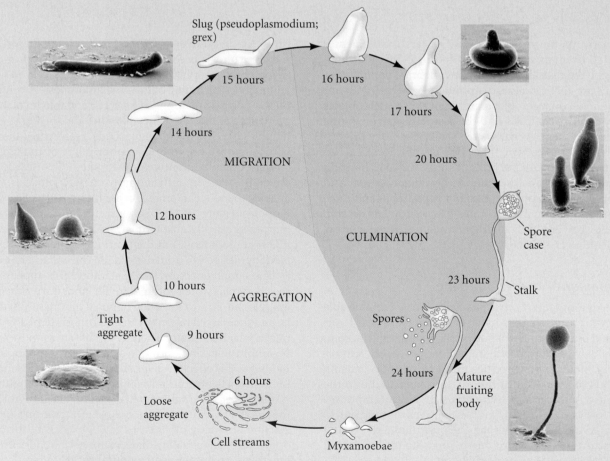

FIGURE 18.13 Life cycle of *Dictyostelium discoideum*. Haploid spores give rise to myxamoebae, which can reproduce asexually to form more haploid myxamoebae. As the food supply diminishes, aggregation occurs and a migrating slug is formed. The slug culminates in a fruiting body that releases more spores. Times refer to hours since the onset of nutrient starvation. Prestalk cells are indicated in yellow. (Photographs courtesy of R. Blanton and M. Grimson.)

SIDELIGHTS & SPECULATIONS (Continued)

(A) **cAMP**

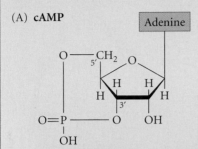

(B)

(C)

(D)

FIGURE 18.14 Chemotaxis of *Dictyostelium* myxamoebae is a result of spiral waves of cAMP. (A) Chemical structure of cAMP. (B) Visualization of several cAMP "waves." Central cells secrete cAMP at regular intervals, and each pulse diffuses outward as a concentric wave. The waves were charted by saturating filter paper with radioactive cAMP and placing it on an aggregating colony. The cAMP from the secreting cells dilutes the radioactive cAMP. When the radioactivity on the paper is recorded (by placing it over X-ray film), the regions of high cAMP concentration in the culture appear lighter than those of low cAMP concentration. (C) Spiral waves of myxamoebae moving toward the initial source of cAMP. Because moving and nonmoving cells scatter light differently, the photograph reflects cell movement. The bright bands are composed of elongated migrating cells; the dark bands are cells that have stopped moving and have rounded up. As cells form streams, the spiral of movement can still be seen moving toward the center. (D) Computer simulation of cAMP wave spreading across migrating *Dictyostelium* cells. The model takes into account the reception and release of cAMP, and changes in cell density due to the movement of the cells. The cAMP wave is plotted in dark blue. The population of myxamoebae goes from green (low) to red (high). Compare with the actual culture shown in (C). (B from Tomchick and Devreotes 1981; C from Siegert and Weijer 1989; D from Dallon and Othmer 1997.)

(1) a fruiting body composed of spore cells and (2) a stalk that supports the fruiting body (Raper 1940; Bonner 1957). The anterior cells, representing 15–20% of the entire cellular population, form the tubed stalk. This process begins as some of the central anterior cells, the **prestalk** cells, begin secreting an extracellular cellulose coat and extending a tube through the grex. As the prestalk cells differentiate, they form vacuoles and enlarge, lifting up the mass of **prespore** cells that made up the posterior four-fifths of the grex

(Jermyn and Williams 1991). The stalk cells die, but the prespore cells, elevated above the stalk, become spore cells. These spore cells disperse, each one becoming a new myxamoeba.

Aggregation of *Dictyostelium* cells

What causes the myxamoebae to aggregate? Time-lapse videomicroscopy has shown that no directed movement occurs during the first 4–5 hours following nutrient starvation. During the next 5 hours, however, the

cells can be seen moving at about 20 mm/min for 100 seconds. This movement ceases for about 4 minutes, then resumes. Although the movement is directed toward a central point, it is not a simple radial movement. Rather, cells join with one another to form streams; the streams converge into larger streams, and eventually all streams merge at the center. Bonner (1947) and Shaffer (1953) showed that this movement is a result of **chemotaxis**: the cells are guided to aggregation centers by a soluble substance. This substance was later identified as **cyclic adenosine 3′,5′-monophosphate (cAMP**; Konijn et al. 1967; Bonner et al. 1969), the chemical structure of which is shown in **Figure 18.14A**.

See **WEBSITE 18.5**
The *Dictyostelium* life cycle:
Variations within variations

See **VADE MECUM**
Slime mold life cycle

(Continued on next page)

SIDELIGHTS & SPECULATIONS (Continued)

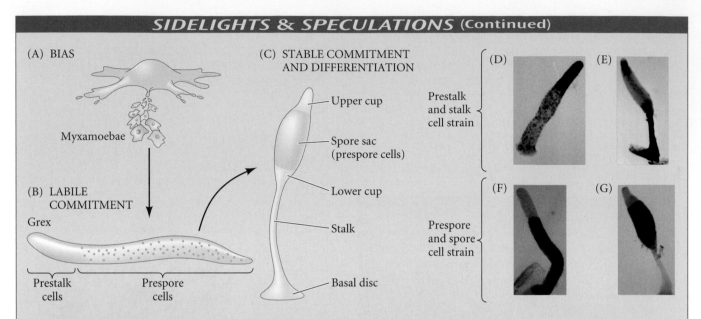

(A) BIAS

Myxamoebae

(B) LABILE COMMITMENT

Grex

Prestalk cells   Prespore cells

(C) STABLE COMMITMENT AND DIFFERENTIATION

Upper cup

Spore sac (prespore cells)

Lower cup

Stalk

Basal disc

(D)   (E)   Prestalk and stalk cell strain

(F)   (G)   Prespore and spore cell strain

FIGURE 18.15 Alternative cell fates in *Dictyostelium discoideum*. (A–C) Progressive commitment of cells to become either spore or stalk cells. (A) Myxamoebae may have biases toward stalk or spore formation due to the stage of the cell cycle they were in when starved. (B) As the grex migrates, most prestalk cells are in the anterior third of the grex, while most of the posterior consists of prespore cells. Some prestalk cells are also seen in the posterior, and these cells will contribute to the cups of the spore sac and to the basal disc at the bottom of the stalk. The cell fates are not yet fixed, however, and if the stalk-forming anterior is cut off, the anteriormost cells remaining will convert from stem to stalk. (C) At culmination, the spore-forming cells are massed together in the spore sac. The stalk cells form the cups of the spore sac, as well as the stalk and basal disc. (D,E) Grex and culminant stained with dye that recognizes the extracellular matrix of the prestalk and stalk cells. (F,G) Grex and culminant stained with dye that recognizes the extracellular matrix of prespore and spore cells. (After Escalante and Vicente 2000; photographs courtesy of R. Escalante.)

Aggregation is initiated as each of the myxamoebae begins to synthesize cAMP. There are no dominant cells that begin the secretion or control the others. Rather, the sites of aggregation are determined by the distribution of the myxamoebae (Keller and Segal 1970; Tyson and Murray 1989). Neighboring cells respond to cAMP in two ways: they initiate a movement toward the cAMP pulse for about a minute, and they release cAMP of their own (Robertson et al. 1972; Shaffer 1975). The movement of each myxamoeba is caused by the change in cytoskeletal polarity brought about by the cAMP (Parent et al. 1998; Iijima et al. 2002). After this happens, the cell is unresponsive to further cAMP pulses for several minutes. During this time, an extracellular membrane-associated phosphodiesterase then cleaves the remaining cAMP from the environment, allowing the receptors to get ready to receive another pulse. The result is a rotating spiral wave of cAMP that is propagated throughout the population of cells (**Figure 18.14B–D**). As each wave arrives, the cells take another step toward the center.

The differentiation of individual myxamoebae into either stalk (somatic) or spore (reproductive) cells is a complex matter. In *Dictyostelium*, differentiation involves a dichotomous decision. There appears to be a pro-

gressive commitment to one of the two alternative pathways (**Figure 18.15**). At first there is a *bias* toward one path or another. For instance, cells starved in the S and early G2 phases of the cell cycle have relatively high levels of calcium and display a tendency to become stalk cells, while those starved in mid or late G2 have lower calcium levels and tend to become spore cells (Nanjundiah 1997; Azhar et al. 2001). Then there is a *labile specification*, a time when the cell will normally become either a spore cell or a stalk cell, but when it can still change its fate if placed in a different position in the organism. Due to the biases in these cells, cAMP is used in different ways by the prespore and prestalk cells (see Kimmel and Firtel 2004). In the prespore cells of the grex, extracellular cAMP initiates the

expression of spore-specific mRNAs. It does this by inducing β-catenin, which enters the nucleus to activate certain spore-specific genes (Ginsburg and Kimmel 1997; Plyte et al. 1999; Kim et al. 2002). In the prestalk cells that are in the anterior tip of the grex, cAMP suppresses this pathway and causes these cells to become prestalk cells. Another group of prestalk cells are formed by a secreted chlorinated lipid, DIF-1, which is made by the prespore cells (Fukuzawa et al. 2003; Thompson et al. 2004). The third and fourth stages are a *determination* to a specific fate, followed by the cell's *differentiation* into a particular cell type, either a stalk cell or a spore cell.

Several proteins, including Trishanku and spore differentiation factors SDF1 and SDF2, appear to be important in the final differentiation of the prespore cells into encapsulated spores (Anjard et al. 1998a,b; Mujumdar et al. 2009). SDF1 and Trishanku are important in initiating culmination, while SDF2 seems to cause the prespore cells (but not prestalk cells) to become spores. The prespore cells appear to have a receptor that enables them to respond to SDF2, while the prestalk cells lack this receptor (Wang et al. 1999). Culmination is also brought about by declining ammonia concentrations (Follstaedt et al. 2003). Ammonia is released preferentially in the anterior portion of the grex, and it appears to help regulate chemotaxis of the pre-

stalk cells as well as aid in the production of spore cells (Oyama and Blumberg 1986; Feit et al. 2001). The formation of stalk cells from prestalk cells is similarly complicated and may involve several factors working synergistically (Early 1999). Indeed, prestalk cells from different parts of the grex pass through different intermediary cell types before reaching the final stage of stalk cell. Thus, the stalk cells that cover the spores have a slightly different history than those stalk cells that hold the ball of spores above the ground. The differentiation of stalk cells appears to need a signal from the intracellular enzyme PKA,

and at least one type of stalk cell is induced by the DIF-1 lipid (Thompson and Kay 2000; Fukuzawa et al. 2001).

So there are many environmental factors triggering alternative developmental pathways in *Dictyostelium*. First, nutrient depletion causes a wholesale change in gene expression and cell adhesivity; second, cAMP orients the migration of individual cells toward a common center; third, the neighboring cells determine whether any particular cell is to be stalk or spore; and fourth, molecules such as ammonia can help complete the processes of differentiation.

Developmental Symbioses

Contrary to the popular use of the term to mean a mutually beneficial relationship, the word **symbiosis** (Greek, *sym*, "together"; *bios*, "life") can refer to any close association between organisms of different species (see Sapp 1994). In many symbiotic relationships, one of the organisms involved is much larger than the other, and the smaller organism may live on the surface or inside the body of the larger. In such relationships, the larger organism is referred to as the **host** and the smaller as the **symbiont**. There are two important categories of symbiosis:

- **Parasitism** occurs when one partner benefits at the expense of the other. An example of a parasitic relationship is that of a tapeworm living in the human digestive tract, wherein the tapeworm steals nutrients from its host.

- **Mutualism** is a relationship that benefits both partners. A striking example of this type of symbiosis can be found in the partnership between the Egyptian plover (*Pluvianus aegyptius*) and the Nile crocodile (*Crocodylus niloticus*). Although it regards most birds as lunch, the crocodile allows the plover to roam its body, feeding on the harmful parasites there. In this mutually beneficial relationship, the bird obtains food while the crocodile is rid of parasites.

A third type of symbiosis, known as **commensalism**, is defined as a relationship that is beneficial to one partner and neither beneficial nor harmful to the other partner. Although many symbiotic relationships may appear on the surface to be commensal, the more we learn, the more we

are discovering that few symbiotic relationships are truly neutral with respect to either party.

In addition, the term **endosymbiosis** ("living inside") is widely used to describe the situation in which one cell lives inside another cell, a circumstance thought to account for the evolution of the organelles of the eukaryotic cell (see Margulis 1971), and one that describes the *Wolbachia* developmental symbioses discussed at length later in this chapter.

Symbiosis, and especially mutualism, is the basis for life on Earth. The symbiosis between *Rhyzobium* bacteria and the roots of legume plants is responsible for converting atmospheric nitrogen into a usable form for generating amino acids, and is therefore essential for life. Symbioses between fungi and plants are ubiquitous, and are often necessary for plant development (see Gilbert and Epel 2009; Pringle 2009). Orchid seeds, for example, contain no energy reserves, so a developing orchid plant must acquire carbon from mycorrhizal fungi. (This is why orchids grow best in moist tropical environments, where fungi are plentiful.)

In some cases, the development of one individual is brought about by signals from organisms of a different species. In some organisms, this relationship has become symbiotic—the symbionts have become so tightly integrated into the host organism that the host cannot develop without them (Sapp 1994). Indeed, recent evidence indicates that developmental symbioses appear be the rule rather than the exception (McFall-Ngai 2002, 2008a).

See WEBSITE 18.6
Developmental symbiosis and parasitism

Mechanisms of developmental symbiosis: Getting the partners together

All symbiotic associations must meet the challenge of maintaining their partnerships over successive generations. In the partnerships that are the main subject here, in which microbes are crucial to the development of their animal hosts, the task of transmission is usually accomplished in one of two ways: by vertical or horizontal transmission.

VERTICAL TRANSMISSION Vertical transmission refers to the transfer of symbionts from one generation to the next through the germ cells, usually the eggs (Krueger et al. 1996). There are several ways by which embryos can become infected by their mothers, but one of the most common is for the symbiont to be transmitted through the oocyte.

Bacteria of the genus *Wolbachia* reside in the egg cytoplasm of invertebrates and provide important signals for the development of the individuals produced by those eggs. As we shall see, many species of invertebrates have "outsourced" important developmental signals to *Wolbachia* bacteria, which are transmitted like mitochondria—that is, in the oocyte cytoplasm. Feree and colleagues (2005) have shown that in *Drosophila* development, *Wolbachia* utilize the host's nurse cell microtubule system and dynein motors to travel from the nurse cells into the developing oocyte (**Figure 18.16A**). In other words, the bacteria use the same cytoskeletal pathway as mitochondria, ribosomes, and *bicoid* mRNA (see Chapter 6). Once in the oocyte, the bacteria enter every cell, becoming endosymbionts. In numerous *Drosophila* species, *Wolbachia* provide resistance against viruses (Teixeira et al. 2008; Osborne et al. 2009).

HORIZONTAL TRANSMISSION *Wolbachia* also can be transmitted horizontally. In horizontal transmission, the metazoan host is born free of symbionts but subsequently becomes infected, either by its environment or by other members of the species. In pill bugs such as *Amadillidium vulgare*, genetically male insects infected with *Wolbachia* are transformed by the bacteria into females (**Figure 18.16B**). As females, the pill bugs can then transmit the *Wolbachia* symbionts to the next generation (Cordaux et al. 2004).

A different type of horizontal transmission involves aquatic eggs that attract photosynthetic algae. Clutches of amphibian and snail eggs, for example, are packed together in tight masses. The supply of oxygen limits the rate of their development, and there is a steep gradient of oxygen from the outside of the cluster to deep within it; thus embryos on the inside of the cluster develop more slowly than those near the surface (Strathmann and Strathmann 1995). The embryos seem to get around this problem by coating themselves with a thin film of photosynthetic algae, which they obtain from the pond water. In clutches of amphibian and snail eggs, photosynthesis from this algal "fouling" enables net oxygen production in the light, while respiration exceeds photosynthesis in the dark (Bachmann et al. 1986; Pinder and Friet

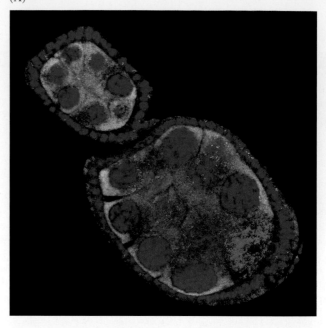

(A)

(B)

FIGURE 18.16 Vertical and horizontal transmission of *Wolbachia* bacteria. (A) In *Drosophila*, *Wolbachia* are transmitted vertically through the female germ cells. In the germinarium, 15 nurse cells transport proteins, RNAs, and organelles to the distal-most oocyte cell. The symbiotic bacterium (stained red) is also transported by these microtubules into the oocyte. Cytoplasm of the ovary is green, and blue indicates DNA. (B) Male and female *Armadillidium vulgare*. Genetically male pill bugs (right) can be transformed into phenotypic egg-producing females (left) by infection (i.e., horizontal transmission) of *Wolbachia* bacteria. (A after Ferree et al. 2005, courtesy of H. M. Frydman and E. Wieschaus; B © David McIntyre.)

1994; Cohen and Strathmann 1996). Thus, the symbiotic algae "rescue" the eggs by their photosynthesis.

Horizontal transmission is crucial for the symbiotic gut bacteria found in many animals, including humans. As we will see later in this chapter, mammalian gut bacteria are critical in forming the blood vessels of the intestine, and possibly in regulating stem cell proliferation (Pull et al. 2005; Liu et al. 2010). Human infants usually acquire these symbionts as they travel through the birth canal. Before then, development is aseptic; but once the amnion breaks, the microbiota of the mother's reproductive tract can colonize the infant's skin and gut. This is supplemented by bacteria from the parents' skin. Although each baby starts with a unique bacterial profile, within a year the types and proportions of bacteria have converged to the adult human profile that characterizes the human digestive tract (Palmer et al. 2007).

The Euprymna-Vibrio *symbiosis*

Horizontal transmission also plays a major role in one of the best-studied examples of developmental symbiosis: that between the squid *Euprymna scolopes* and the luminescent bacterium *Vibrio fischeri* (McFall-Ngai and Ruby 1991; Montgomery and McFall-Ngai 1995). The adult *Euprymna* is equipped with a light organ composed of sacs filled with these bacteria (**Figure 18.17A**). The newly hatched squid, however, does not contain these light-emitting symbionts, nor does it have the light organ to house them. The juvenile squid acquires *V. fischeri* from the seawater pumped through its mantle cavity (Nyholm et al. 2000). The bacte-

ria bind to a ciliated epithelium that extends into this cavity; the epithelium only binds *V. fischeri*, allowing other bacteria to pass through (**Figure 18.17B**). The bacteria then induce hundreds of genes in the epithelium, leading to the apoptotic death of these epithelial cells, their replacement by a nonciliated epithelium, the differentiation of the surrounding cells into storage sacs for the bacteria, and the expression of genes encoding opsins and other visual proteins in the light organ (**Figure 18.17C**; Chun et al. 2008; McFall-Ngai 2008b; Tong et al. 2009).

The substance *V. fischeri* secretes to effect these changes turns out to be fragments of the bacterial cell wall, and the active agents are tracheal cytotoxin and lipopolysaccharide (Koropatnick et al. 2004). This finding was surprising, because these two agents have been long known to cause inflammation and disease. Indeed, tracheal cytotoxin is responsible for the tissue damage in both whooping cough and gonorrheal infections. The destruction and replacement of ciliated tissue in the respiratory tract and oviduct are due to these bacterial compounds. After the symbiotic bacteria have induced the morphological changes in the host, the host secretes a peptide into the *Vibrio*-containing crypts which neutralizes the bacterial toxin (Troll et al. 2010). Both organisms change their gene expression patterns, and both benefit from their developmental association. The bacteria get a home and express their light-generating enzymes, and the squid develops a light organ that allows it to swim at night in shallow waters without casting a shadow.

Obligate developmental mutualism

In an **obligate mutualism**, the species involved are interdependent with one another to such an extent that neither could survive without the other. The most common example of obligate mutualism is the lichens, in which fungal and algal species are joined in a relationship that results in an essentially new species. More and more examples of obligate mutualism are being described, and most of these have important consequences for medicine and conservation biology.

(A)

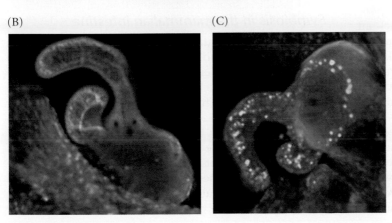

(B)   (C)

FIGURE 18.17 The *Euprymna scolopes-Vibrio fischeri* symbiosis. (A) An adult Hawaiian bobtail squid (*E. scolopes*) is about 2 inches long. The symbionts are housed in a two-lobed light organ on the squid's underside. (B) The light organ of a juvenile squid is poised to receive *V. fischeri*. Ciliary currents and mucus secretions create an environment (diffuse yellow stain) that attracts seaborne Gram-negative bacteria, including *V. fischeri*, to the organ. Over time all bacteria except *V. fischeri* will be eliminated by mechanisms yet to be exactly elucidated. (C) Once *V. fischeri* are established in the crypts of the light organ, they induce apoptosis of the epithelial cells (yellow dots) and shut down production of the mucosal secretions that attracted other bacteria. (Courtesy of M. McFall-Ngai.)

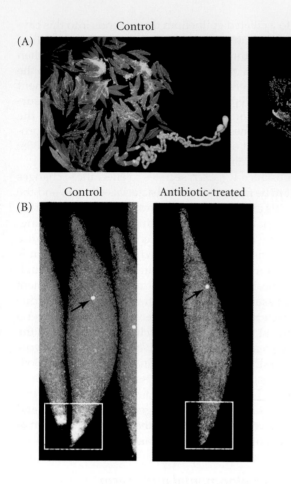

(A) Control   Antibiotic-treated

(B) Control   Antibiotic-treated

FIGURE 18.18 Comparison of ovaries and oocytes of the wasp *Asobara tabida* from control females and females treated with rifampicin antibiotic to remove *Wolbachia*. (A) The ovaries of control females had an average of 228 oocytes, while those of rifampicin-treated females had an average of 36 oocytes. (B) When DNA in the oocytes was stained, oocytes from control females had a nucleus (arrow) as well as a mass of *Wolbachia* at one end (boxed area). Oocytes from rifampicin-treated females had a nucleus but no *Wolbachia*; these eggs were sterile. (From Dedeine et al. 2001.)

One example of obligate developmental mutualism has been described in both the parasitic wasp *Asobara tabida* and the leafhopper *Euscelis incisus*. In these insects, symbiotic bacteria are found in the egg cytoplasm and, as in the *Wolbachia* example discussed above, are vertically transferred through the female germ plasm. In the leafhopper, these bacteria have become so specialized that they can multiply only inside the leafhopper's cytoplasm, and the host has become so dependent on the bacteria that it cannot complete embryogenesis without them; the bacteria appear to be essential for the formation of the embryonic gut. If the bacterial symbionts are removed from the eggs surgically or by feeding antibiotics to larvae or adults, the symbiont-free oocytes develop into embryos that lack an abdomen (Sander 1968; Schwemmler 1974, 1989). In *Asobara*, the bacteria enable the wasp to complete yolk production and egg maturation (Dedeine et al. 2001; Pannebakker et al. 2007). If the symbionts are removed, the ovaries undergo apoptosis and no eggs are produced (Figure 18.18).

In obligate developmental mutualisms, the death of the host can result from killing the symbiont. Atrazine was mentioned in Chapter 14 for its ability to induce aromatase and cause sex-determination anomalies in amphibians. But the major use and effect of atrazine is to kill plant life; it is a potent nonspecific herbicide. Once applied, atrazine can remain active in the soil for more than 6 months, and it can be carried by wind and rainwater to new sites. However, the egg masses of many amphibian and snail species depend on algal symbionts to provide oxygen to the eggs deepest in the clutch. The spotted salamander (*Ambystoma maculatum*) lays eggs that recruit a green algal symbiont so specific that its name is *Oophilia amblystomatis* ("lover of ambysoma eggs"). Concentrations of atrazine as low as 50 μg/L completely eliminate this algae from the eggs, and the amphibian's hatching success is greatly lowered (Figure 18.19A; Gilbert 1944; Mills and Barnhart 1999; Olivier and Moon 2010).

An example of using a symbiont as a means to eradicate an unwanted host is that of the filariasis worm, a human parasite that causes elephantiasis and other debilitating conditions. Most of these worms have *Wolbachia* bacteria as endosymbionts; in this case, *Wolbachia* produce chemicals that enable the worm to molt, and without them, the worm dies. Many species of filariasis worms have become resistant to the drugs traditionally used to kill these parasites in humans. A new strategy has been to employ antibacterial antibiotics (such as doxycycline) against the *symbionts* rather than the hosts (Figure 18.19B; Hoerauf et al. 2003; Coulibaly et al. 2009). Once the antibiotic destroys the symbiont, the worms cannot develop further; they die, and the patient is no longer infected.

Symbiosis in the mammalian intestine

Even mammals maintain developmental symbioses with bacteria. Using the polymerase chain reaction (PCR) and high-frequency sequencing techniques, researchers have recently been able to identify many anaerobic bacterial species present in the human gut (see Qin et al. 2010). Their presence was not realized earlier because these species cannot yet be cultured in the laboratory.

These studies have revealed particular distributions of the bacterial symbionts in our bodies. The hundreds of different bacterial species of the human colon are stratified into specific regions along the length and diameter of the gut tube, where they can attain densities of 10^{11} cells per

(A)

(B)

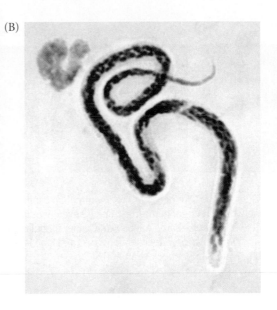

FIGURE 18.19 Obligate developmental symbionts. (A) Spotted salamander (*Ambystoma maculata*) eggs at the center of the cluster cannot survive the lack of oxygen when their algal symbiont is eliminated by herbicides. (B) Filariasis worms such as *Mansonella ozzardi* cannot complete molting when their *Wolbachia* symbiont is eliminated by antibiotics. (A © Gustav Verderber/OSF/Photolibrary.com; B courtesy of Mae Melvin/CDC.)

milliliter (Hooper et al. 1998; Xu and Gordon 2003). Indeed, by cell count, 90% of the cells in our body are microbial. We never lack these microbial components; we pick them up from the reproductive tract of our mother as soon as the amnion bursts. We have coevolved to share our space with them, and we have even co-developed such that our cells are primed to bind to them, and the bacteria induce gene expression in the intestinal epithelial cells (Bry et al. 1996; Hooper et al. 2001).

Bacteria-induced expression of mammalian genes was first demonstrated in the mouse gut (**Figure 18.20**). Umesaki (1984) noticed that a particular fucosyl transferase enzyme characteristic of mouse intestinal villi was induced by bacteria, and more recent studies (Hooper et al. 1998) have shown that the intestines of germ-free mice can initiate, but not complete, their differentiation. For complete development, the microbial symbionts of the gut are needed. Normally occurring gut bacteria can upregulate the

FIGURE 18.20 Induction of mammalian genes by symbiotic microbes. Mice raised in "germ-free" environments were either left alone or inoculated with one or more types of bacteria. After 10 days, their intestinal mRNAs were isolated and tested on microarrays. Mice grown in germ-free conditions had very little expression of the genes encoding colipase, angiogenin-4, or Sprr2a. Several different bacteria—*Bacteroides thetaiotaomicron*, *Escherichia coli*, *Bifidobacterium infantis*, and an assortment of gut bacteria harvested from conventionally raised mice—induced the genes for colipase and angiogenin-4 *B. thetaiotaomicron* appeared to be totally responsible for the 50-fold increase in *Sprr2a* expression over that of germ-free animals. This ecological relationship between the gut microbes and the host cells could not have been discovered without the molecular biological techniques of polymerase chain reaction and microarray analysis. (After Hooper et al. 2001.)

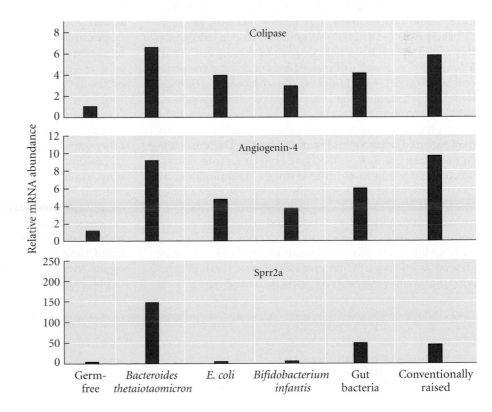

(A)   (B)   (C)

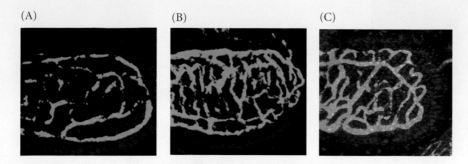

FIGURE 18.21 Gut microbes are necessary for mammalian capillary development. (A) The capillary network (green) of germ-free mice is severely reduced compared with (B) the capillary network in those same mice 10 days after inoculation with normal gut bacteria. (C) The addition of *Bacteroides thetaiotaomicron* alone is sufficient to complete capillary formation. (From Stappenbeck et al. 2002.

transcription of several mouse genes, including those encoding colipase, which is important in nutrient absorption; angiogenin-4, which helps form blood vessels; and Sprr2a, a small, proline-rich protein that is thought to fortify matrices that line the intestine (see Figure 18.20; Hooper et al. 2001). Stappenbeck and colleagues (2002) have demonstrated that in the absence of particular intestinal microbes, the capillaries of the small intestinal villi fail to develop their complete vascular networks (**Figure 18.21**).

Intestinal microbes also appear to be critical for the maturation of the mammalian gut-associated lymphoid tissue (GALT). GALT mediates mucosal immunity and oral immune tolerance, allowing us to eat food without making an immune response to it (see Rook and Stanford 1998; Cebra 1999; Steidler 2001). When introduced into germ-free rabbit appendices, neither *Bacillus fragilis* nor *B. subtilis* alone was capable of consistently inducing the proper formation of GALT. However, the combination of these two common mammalian gut bacteria consistently induced GALT (Rhee et al. 2004). The major inducer appears to be the protein bacterial polysaccharide-A (PSA), especially that encoded by the genome of *B. fragilis*. The PSA-deficient mutant of *B. fragilis* is not able to restore normal immune function to germ-free mice (Mazmanian et al. 2005). Thus, a bacterial compound appears to play a major role in inducing the host's immune system.

In short, mammals have coevolved with bacteria to the point that our bodily phenotypes do not fully develop without them. The microbial community of our gut can be viewed as an "organ" that provides us with certain functions that we haven't evolved (such as the ability to process plant polysaccharides). And, like our developing organs, microbes induce changes in neighboring tissues. As Mazmanian and colleagues (2005) have concluded, "The most impressive feature of this relationship may be that the host not only tolerates but has evolved to require colonization by commensal microorganisms for its own development and health."

Coda

Phenotype is not merely the expression of one's inherited genome. Rather, there are interactions between an organism's genotype and environment that elicit a particular phenotype from a genetically controlled repertoire of possible phenotypes. Environmental factors such as temperature, diet, physical stress, the presence of predators, and crowding can generate a phenotype that is suited for that particular environment. Environment is therefore considered to play a role in the *generation* of phenotypes in addition to its well-established role in the *selection* of phenotypes. The fact that we co-develop with other organisms is an important concept for developmental biology. It also calls into question the notions of autonomy and independent development. Moreover, if we are not truly "individuals," but have a phenotype based on community interactions, what exactly is natural selection selecting? Can natural selection select teams or relationships? (See Gilbert and Epel 2009; Gilbert et al. 2010.) The ramifications of developmental plasticity and developmental symbiosis on the rest of biology are just beginning to become appreciated.

Snapshot Summary: *Developmental Plasticity and Symbiosis*

1. The environment plays critical roles during normal development. These agents include temperature, diet, crowding, and the presence of predators.

2. Developmental plasticity makes it possible for environmental circumstances to elicit different phenotypes from the same genotype. The genome encodes a repertoire of possible phenotypes. The environment often selects which of those phenotypes will become expressed.

3. Reaction norms are phenotypes that quantitatively respond to environmental conditions, such that the phenotype reflects small differences in the environmental conditions.

4. Polyphenisms represent "either/or" phenotypes; that is, one set of conditions elicits one phenotype while another set of conditions elicits another.

5. Seasonal cues such as photoperiod, temperature, or type of food can alter development in ways that make the organism more fit under the conditions it encounters. Changes in temperature also are responsible for determining sex in several organisms, including many reptiles and fish.

6. Predator-induced polyphenisms have evolved such that prey species can respond morphologically to the presence of a specific predator. In some instances, this induced adaptation can be transmitted to the progeny of the prey.

7. There are several routes through which gene expression can be influenced by the environment. Environmental factors can methylate genes differentially; they can induce gene expression in surrounding cells; and they can be monitored by the nervous system, which then produces hormones that affect gene expression.

8. Behavioral phenotypes can also be induced by the environment. Learning results from changes in the nervous system brought about by experiences with the environment.

9. Organisms usually develop with symbiotic organisms, and signals from the symbionts can be critical for normal development.

10. Symbionts can be acquired horizontally (through infection) or vertically (through the oocyte).

11. In an obligate mutalism, both partners are needed for the survival of the other; in an obligate developmental mutualism, at least one partner is needed for the proper development of another.

12. The mammalian gut contains symbionts that actively regulate intestinal gene expression to generate proteins that are normal physiological components of intestinal development and function. Without these symbionts, the intestinal blood vessels and gut-associated lymphoid tissue of some mammalian species fail to form properly.

For Further Reading

Complete bibliographical citations for all literature cited in this chapter can be found at the free access website **www.devbio.com**

Agrawal, A. A., C. Laforsch and R. Tollrian. 1999. Transgenerational induction of defenses in animals and plants. *Nature* 401: 60–63.

Brakefield, P. M. and N. Reitsma. 1991. Phenotypic plasticity, seasonal climate, and the population biology of *Bicyclus* butterflies (Satyridae) in Malawi. *Ecol. Entomol.* 16: 291–303.

Gilbert, S. F. and D. Epel. 2009. *Ecological Developmental Biology: Integrating Epigenetics, Medicine, and Evolution.* Sinauer Associates, Sunderland, MA.

Hooper, L. V., M. H. Wong, A. Thelin, L. Hansson, P. G. Falk and J. I. Gordon. 2001. Molecular analysis of commensal host-microbial relationships in the intestine. *Science* 291: 881–884.

McFall-Ngai, M. J. 2002. Unseen forces: The influence of bacteria on animal development. *Dev. Biol.* 242: 1–14.

Moczek, A. P. 2005. The evolution of development of novel traits, or how beetles got their horns. *BioScience* 55: 937–951.

Relyea, R. A. and N. Mills. 2001. Predator-induced stress makes the pesticide carbaryl more deadly to grey treefrog tadpoles (*Hyla versicolor*). *Proc. Natl. Acad. Sci. USA* 2491–2496.

Stappenbeck, T. S., L. V. Hooper and J. I. Gordon. 2002. Developmental regulation of intestinal angiogenesis by indigenous microbes via Paneth cells. *Proc. Natl. Acad. Sci. USA* 99: 15451–15455.

Waterland, R. A. and R. L. Jirtle. 2003. Transposable elements: Targets for early nutritional effects of epigenetic gene regulation. *Mol. Cell. Biol.* 23: 5293–5300.

Go Online

WEBSITE 18.1 Inducible caste determination in ant colonies. In some species of ants, the loss of soldier ants creates conditions that induce more workers to become soldiers.

WEBSITE 18.2 *Volvox*: When heat brings out sex. During most of its life cycle, *Volvox* (a "green algae") reproduces asexually. However, when the ponds *Volvox* inhabits become hot—a signal that they may soon dry up—the next generation becomes sexual and produces sperm and eggs that unite in zygotes that can survive desiccation.

WEBSITE 18.3 Mechanisms of diapause. Both light and temperature are critical for the induction and maintenance of diapause. Different species use different signals for this event.

WEBSITE 18.4 Pressure as a developmental agent. Mechanical stress is critical for gene expression in numerous tissues, including bone, heart, and muscle. Without physical force, we would not develop our kneecaps.

WEBSITE 18.5 The *Dictyostelium* life cycle: Variations within variations *Dictyostelium* has some fascinating variations on its life cycle, including the ability to become sexual.

WEBSITE 18.6 Developmental symbioses and parasitism. Some embryos acquire protection and nutrients by forming symbiotic associations with other organisms. The mechanisms by which these associations form are now being elucidated. In other situations, one species uses material from another to support its development. Blood-sucking mosquitoes are examples of such parasites.

Vade Mecum

Slime mold life cycle. The life cycle of this fascinating organism is shown in movies. There is also original footage of some of John Tyler Bonner's experiments, along with an interview with Dr. Bonner, whose pioneering work demonstrated the major principles of *Dictyostelium* development.

Outside Sites

Karen Warkentin's Boston University website includes an amazing movie of the snake-induced rapid development of red-eyed tree frogs. (**http://people.bu.edu/kwarken/KWvideoSMALLER.html**)

Developmental Mechanisms of Evolutionary Change

WHEN WILHELM ROUX ANNOUNCED the creation of experimental embryology in 1894, he broke many of the ties that linked embryology to evolutionary biology. However, he promised that embryology would someday return to evolutionary biology, bringing with it new knowledge of how animals are generated and how evolutionary changes might occur. He predicted that once we knew how bodies are formed, we would be able to analyze alterations in development that lead to new body plans. A little more than a century later, we are at the point of fulfilling Roux's prophecy.

Developmental biology is returning to evolutionary biology, forging a new discipline, **evolutionary developmental biology** ("evo-devo"), and producing a new model of evolution that integrates developmental genetics and population genetics to explain and define the diversity of life on Earth (Raff 1996; Hall 1999; Arthur 2004; Carroll et al. 2005; Kirschner and Gerhart 2005; Gilbert and Epel 2009). Contemporary evolutionary developmental biology is analyzing how changes in development can create the diverse variation that natural selection can act on. Rather than concentrating on the "survival of the fittest," evolutionary developmental biology gives us new insights into the "arrival of the fittest" (Gilbert and Epel 2009).

"Unity of Type" and "Conditions of Existence": Charles Darwin's Synthesis

In the nineteenth century, debates over the origin of species pitted two ways of viewing nature against each other. One view, championed by Georges Cuvier and Charles Bell, focused on the *differences* among species that allowed each species to adapt to its environment. Thus, the hand of the human, the flipper of the seal, and the wings of birds and bats were seen as marvelous contrivances, each fashioned by the Creator, to allow these animals to adapt to their "conditions of existence." The other view, championed by Étienne Geoffroy Saint-Hilaire and Richard Owen, was that "unity of type" (the *similarities* among organisms, which Owen called "homologies") was critical. The human hand, the seal's flipper, and the wings of bats and birds were all modifications of the same basic plan (see Figure 1.19). In discovering that plan, one could find the form upon which the Creator designed these animals. The adaptations were secondary.

Darwin acknowledged his debt to these earlier debates when he wrote in 1859, "It is generally acknowledged that all organic beings have been formed on two great laws—Unity of Type, and Conditions of Existence." Darwin went on to explain that his theory would explain unity of type by descent from a common ancestor. The changes creating the marvelous adaptations to the conditions of

How does newness come into the world? How is it born? Of what fusions, translations, conjoinings is it made? How does it survive, extreme and dangerous as it is? What compromises, what deals, what betrayals of its secret nature must it make to stave off the wrecking crew, the exterminating angel, the guillotine?

SALMAN RUSHDIE (1988)

A study of the effects of genes during development is as essential for an understanding of evolution as are the study of mutation and that of selection.

JULIAN HUXLEY (1942)

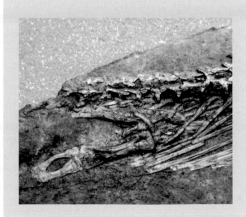

existence would be explained by natural selection. Darwin called this concept **descent with modification**. Darwin noted that the homologies between the embryonic and larval structures of different phyla provided excellent evidence for descent with modification. He also argued that adaptations that depart from the "type" and allow an organism to survive in the "conditions" of its particular environment develop late in the embryo. Thus, Darwin recognized two ways of looking at descent with modification. One could emphasize *common descent* in the embryonic homologies between two or more groups of animals, or one could emphasize the *modifications* by showing how development was altered to produce diverse adaptive structures (Gilbert 2003). Or, as Darwin's friend Thomas Huxley aptly remarked, "Evolution is not a speculation but a fact; and it takes place by epigenesis" (Huxley 1893, p. 202).

See WEBSITE 19.1
Relating evolution to development in the nineteenth century

Preconditions for Evolution through Developmental Change

Darwin was greatly perplexed about variation. After all, if natural selection could only operate on existing variants, where was all that variation coming from? He concluded (in his next books) that it would have to come from changes in development; but how could the development of an embryo change when development is so finely tuned and complex? How could such change occur without destroying the entire organism? Darwin's German colleague, Ernst Haeckel, proposed that the major way to evolve was to add a step to the *end* of embryonic development. But there turned out to be so many exceptions to that rule that it fell into disrepute.

When the molecular biology of protein synthesis became understood, the problem did not go away. If a protein-encoding gene were mutated, the abnormal protein would be made in all the places where the protein was normally expressed. There was no way the mutation could cause the protein to be made in one place and not another. The matter lay unsolved until evolutionary developmental biologists demonstrated that large morphological changes could be made during development because of two conditions in the ways that organisms develop: **modularity** and **molecular parsimony**.

Modularity: Divergence through dissociation

We now know that even early stages of development can be altered to produce evolutionary novelties. Such changes can occur because development occurs through a series of discrete and interacting **modules** (Riedl 1978; Bonner 1988; Kuratani 2009). Examples of developmental modules include morphogenetic fields (for example, those described

for the limb or eye), signal transduction pathways (described throughout this book), imaginal discs, cell lineages (such as the inner cell mass or trophoblast), insect parasegments, and vertebrate organ rudiments (Gilbert et al. 1996; Raff 1996; Wagner 1996; Schlosser and Wagner 2004). The ability of one module to develop differently from other modules (a phenomenon sometimes called **dissociation**) was well known to early experimental embryologists. For instance, when Victor Twitty (Twitty and Schwind 1931; Twitty and Elliott 1934) grafted the limb bud from the early larva of a large salamander onto the embryonic trunk of a small salamander larva, the limb grew to its normal large dimensions within the small larva, indicating that the limb field module was independent from the global growth patterning of the embryo. The same independence was seen for the eye field. Modular units allow certain parts of the body to change without interfering with the functions of other parts.

One of the most important discoveries of evolutionary developmental biology is that not only are the *anatomical* units modular (such that one part of the body can develop differently than the others), but that the DNA regions that form the *enhancers* of genes are modular. This was shown in Figure 2.10. The modularity of enhancer elements allows particular sets of genes to be activated together and permits a particular gene to become expressed in several discrete places. Thus, if a particular gene loses or gains a modular enhancer element, the organism containing that particular allele will express that gene in different places or at different times than those organisms retaining the original allele. This mutability can result in the development of different anatomical and physiological morphologies (Sucena and Stern 2000; Shapiro et al. 2004), and major morphological changes can proceed through a mutation in a DNA regulatory region. Thus, the modularity of enhancers can be critical in providing selectable variation.

DUFFY BLOOD GROUP SUBSTANCE Enhancer modularity has been known from studies of selectable traits in human populations. *Plasmodium vivax* is a protozoan parasite that causes about 75 million cases of malaria each year. This form of malaria is not as lethal as *falciparum* malaria, but it can be incapacitating, causing severe pain, diarrhea, and fever. Some African populations are immune to *P. vivax* because their red blood cells lack a protein, the Duffy glycoprotein, that *P. vivax* needs to attach itself to the host's red blood cells. The Duffy glycoprotein is probably one of several receptors for interleukin 8 (IL8, a paracrine factor involved in the circulatory system and in the cerebellum), and it is found on Purkinje neurons, veins, and erythrocytes (red blood cells). People who lack Duffy glycoprotein on their red blood cells still have it on their veins and Purkinje neurons. So if one asks, "Why don't these people have Duffy glycoprotein on their red blood cells?" the ultimate answer is probably that the lack of the Duffy glycoprotein was selected in these populations because it gives these people resistance to *vivax* malaria, given that

Plasmodium uses the protein to infect the red blood cells. The proximate answer is that the lack of Duffy glycoprotein is caused by a mutation in the erythrocyte enhancer, a C-to-G substitution at position –36, which prevents the binding of the GATA1 transcription factor present in red blood cell precursors (Tournamille et al. 1995). Thus, the mutation blocks one of the enhancers (the one for expression in red blood cells) from functioning, but allows the enhancers that permit the gene's expression in veins and Purkinje neurons to function.

PITX1 AND STICKLEBACK EVOLUTION The importance of enhancer modularity has been dramatically demonstrated by the analysis of evolution in the threespine stickleback fish (*Gasterosteus aculeatus*). Freshwater sticklebacks evolved from marine sticklebacks about 12,000 years ago, when marine populations colonized the newly formed freshwater lakes at the end of the last ice age. Marine sticklebacks (**Figure 19.1A**) have pelvic spines that serve as protection against predation, lacerating the mouths of predatory fish that try to eat the stickleback. (Indeed, the scientific name of the fish translates as "bony stomach with spines.") Freshwater sticklebacks, however, do not have pelvic spines (**Figure 19.1B**). This may be because the freshwater fish lack the piscine predators that the marine fish face, but instead must deal with invertebrate predators that can easily capture them by grasping onto such spines. Thus, a pelvis without spines was selected in freshwater populations of this species.

To determine which genes might be involved in this pelvic difference, researchers mated individuals from marine (with spines) and freshwater (no spines) populations. The resulting offspring were bred to each other and produced numerous progeny, some of which had pelvic spines and some of which didn't. Using molecular markers to identify specific regions of the parental chromosomes, Shapiro and co-workers (2004) found that the major gene for pelvic spine development mapped to the distal end of chromosome 7. That is to say, nearly all the fish with pelvic spines inherited this "hindfin-encoding" chromosomal region from the marine parent, while fish lacking pelvic spines obtained this region from the freshwater parent. The researchers then tested numerous candidate genes (genes known to be present in the hindlimb structures of mice, for instance) and found that the gene encoding transcription factor Pitx1 was located on this region of chromosome 7.

When they compared the amino acid sequences of the Pitx1 protein between marine and freshwater sticklebacks, there were no differences. However, there was a critically important difference when they compared the *expression patterns* of *Pitx1* between these populations. In both populations, *Pitx1* was expressed in the precursors of the thymus, nose, and sensory neurons. In the marine populations, *Pitx1* was also expressed in the pelvic region. But in the freshwater populations, the pelvic expression of *Pitx1* was absent or severely reduced (**Figure 19.1C**). Since the coding region of *Pitx1* was not mutated (and since the gene involved in the pelvic spine differences maps to the site of the *Pitx1* gene, and the difference between the freshwater and marine populations involves the expression of this gene at a particular site), it was reasonable to conclude that the *enhancer region* allowing expression of *Pitx1* in the pelvic area (i.e., the pelvic spine enhancer) no longer functions in the freshwater populations.

FIGURE 19.1 Modularity of development: enhancers. Loss of *Pitx1* gene expression in the pelvic region of freshwater populations of the threespine stickleback (*Gasterosteus aculeatus*). Bony plates and pelvic spines characterize marine populations of this species (A). In freshwater populations (B), the pelvic spines are absent, as is much of the bony armor. In magnified ventral views of embryos (inset photos), in situ hybridization reveals *Pitx1* expression (purple) in the pelvic area (as well as in sensory neurons, thymic cells, and nasal regions) of the marine population. The staining in the pelvic region is absent in freshwater populations, although it is still seen in the other areas (the arrow points to one of several expression points). (C) Model for the evolution of pelvic spine loss. Four enhancer regions are postulated to reside near the *Pitx1* coding region. These enhancers direct the expression of this gene in the thymus, pelvic spines, sensory neurons, and nasal pit, respectively. In freshwater populations of threespine sticklebacks, the pelvic spine (hindlimb) enhancer module has been mutated and the *Pitx1* gene fails to function there. (After Shapiro et al. 2004; photographs courtesy of D. M. Kingsley.)

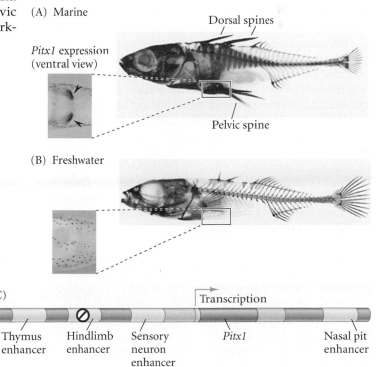

(A) Marine

Pitx1 expression (ventral view)

Dorsal spines

Pelvic spine

(B) Freshwater

(C)

Transcription

Thymus enhancer

Hindlimb enhancer

Sensory neuron enhancer

Pitx1

Nasal pit enhancer

This conclusion was confirmed when high-resolution genetic mapping showed that the DNA of the "hindlimb" enhancer of *Pitx1* differed between sticklebacks with pelvic spines and those without pelvic spines* (Chan et al. 2010). When this 2.5-kb DNA fragment from marine (spined) fish was fused to a gene for green fluorescent protein (GFP) and inserted into fertilized freshwater stickleback eggs, it caused the GFP protein to be expressed in the pelvis. Moreover, when this region from the marine sticklebacks was placed next to the *Pitx1*-coding sequence from the freshwater (spine-deficient) fish and then injected into fertilized eggs of the spine-deficient fish, pelvic spines were properly formed in the freshwater fish.

See WEBSITE 19.2
Correlated progression

Molecular parsimony: Gene duplication and divergence

The second precondition for macroevolution through developmental change is molecular parsimony, sometimes called the "small toolkit." In other words, although development differs enormously from lineage to lineage, development within all lineages uses the same types of molecules. The transcription factors, paracrine factors, adhesion molecules, and signal transduction cascades are remarkably similar from one phylum to another. Certain transcription factors such as Pax and Hox genes are found in all animal phyla, including cnidarians, insects, and primates. In fact, some "toolkit genes" appear to play the same roles in all animal lineages. Thus, Pax6 appears to be involved in specifying light-sensing organs, irrespective of whether the eye is that of a mollusc, an insect, or a primate (**Figure 19.2**; Gehring 2005).

Hox genes appear to specify the anterior-posterior body axis, even though the way these genes are activated differs enormously among nematodes, flies, and birds. Similarly, homologues of *Otx* specify head formation in both vertebrates and invertebrates; and, though insect and vertebrate hearts are very different, both are specified by *tinman/Nkx2–5* (see Erwin 1999). The Wnt and TGF-β paracrine factors and their signaling pathways can be found in the primitive cnidarians as well as in primitive bilaterians such as flatworms (Finnerty et al. 2004; Carroll et al. 2005). There is descent with modification for the signal transduction pathways and transcription factors, just as there is descent with modification of whole organisms.

FIGURE 19.2 The *Pax6* gene for eye development is an example of a gene ancestral to both protostomes and deuterostomes. The micrograph shows ommatidia emerging in the leg of a fruit fly (a protostome) in which mouse (deuterostome) *Pax6* cDNA was expressed in the leg disc. (From Halder et al. 1995, courtesy of W. J. Gehring and G. Halder.)

One theme that echoes through this book is that Hox gene expression provides the basis for anterior-posterior axis specification throughout the animals. This means that the enormous variation of morphological form among animals is underlain by a common set of instructions. Indeed, Hox genes provide one of the most remarkable pieces of evidence for deep evolutionary homologies among all the animals of the world. The similarity of all the Hox genes is best explained by descent from a common ancestor. First, the different Hox genes in the homeotic gene cluster each originated from duplications of an ancestral gene. This would mean that in *Drosophila*, the *Deformed*, *Ultrabithorax*, and *Antennapedia* genes all emerged as duplications of an original gene. The sequence patterns of these three genes (especially in the homeodomain region) are extremely well conserved. Such tandem gene duplications are thought to be the result of errors in DNA replication, and they are very common. Many genomes have at least two genes (usually located close to each other) that resulted from a replication error. Once replicated, the gene copies can diverge by random mutations in their coding sequences and enhancers, developing different expression patterns and new functions (Lynch and Conery 2000; Damen 2002; Locascio et al. 2002).

This scenario of **duplication and divergence** is seen in the Hox genes, the globin genes, the collagen genes, the *Distal-less* genes, and in many paracrine factor families (e.g., the Wnt genes). Each member of such a gene family is homologous to the others (that is, their sequence similarities are due to descent from a common ancestor and are not the result of convergence for a particular function), and they are called **paralogues**. Thus, the *Antennapedia* gene of *Drosophila* is a paralogue of *Drosophila Ultrabithorax*. Susumu Ohno (1970), one of the founders of the gene fam-

---

*Interestingly, the loss of the pelvic spines in several stickleback populations appears to have been the result of independent losses of this *Pitx1* expression domain. This finding suggests that if the loss of *Pitx1* expression in the pelvis occurs, this trait can be readily selected (Colosimo et al. 2004). Here we see that by combining population genetics approaches and developmental genetics approaches, one can determine the mechanisms by which evolution can occur.

ily concept, likened gene duplication to a method used by a sneaky criminal to circumvent surveillance. While the "police force" of natural selection makes certain that there is a "good" gene properly performing its function, that gene's duplicate, unencumbered by the constraints of selection, can mutate and undertake new functions.

One of the most significant fates for duplicated genes is the subdivision of the ancestral expression and function (Force et al. 2005). Thus, the original *Distal-less* gene was probably expressed in many places (as it still is in insects, which have only one *Distal-less* gene); but in vertebrates, which have numerous *Distal-less* genes, the original domains have become subdivided so that the different *Distal-less* paralogues regulate different functions. *Distal-less-3*, for instance, is necessary for epidermal differentiation, while *Distal-less-5* helps specify neural crest cells (Panganiban and Rubenstein 2002). Such subfunctionalization has since been shown to be the case for many genes, including Hox genes.

Each Hox gene in *Drosophila* has a homologue in vertebrates. In some cases, the homologies go very deep and can also be seen in the gene's functions. Not only is the vertebrate *Hoxb4* gene similar in sequence to its *Drosophila* homologue, *Deformed* (*Dfd*), but the human *HOXB4* gene can perform the functions of *Dfd* when introduced into *Dfd*-deficient *Drosophila* embryos (Malicki et al. 1992). As mentioned in Chapter 8, the Hox genes in insects and humans are not just homologous—they occur in the same order on their respective chromosomes. Their expression patterns are also remarkably similar: the more 3' Hox genes have more anterior expression boundaries* (see Figure 8.30). Thus they are homologous genes between species (as opposed to members of a gene family being homologous within a species). Genes that are homologous between species are called **orthologues**.

All multicellular organisms—animals, plants, and fungi—have Hox-like genes, so it is most likely that there was an ancestral Hox gene that encoded a basic helix-loop-helix transcription factor in protozoans. In the earliest animal groups, this gene became replicated. One of the two Hox genes present in some extant cnidarians (e.g., jellyfish) corresponds to the anterior set of vertebrate Hox genes (and is expressed in the anterior portion of the cnidarian larva), while sequences in the other cnidarian gene are a posterior class Hox gene (**Figure 19.3**; Yanze et al. 2001; Hill et al. 2003; Finnerty et al. 2004). Perhaps even the ancient cnidarians used Hox genes to distinguish their anterior and posterior tissues. In bilateral phyla, the cen-

(A)   (B)

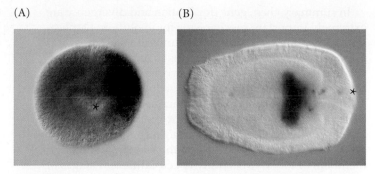

FIGURE 19.3   Evidence of the evolutionary conservation of regulatory genes. (A) The cnidarian homologue of the vertebrate *Bmp4* and *Drosophila Decapentaplegic* genes is expressed asymmetrically at the edge of the blastopore (marked with an asterisk) in the embryo of the sea anemone *Nematostella*. This gene represents an ancestral form of the protostome and deuterostome forms of the gene. (B) The Hox gene *Anthox6*, a cnidarian member of the paralogue 1 group of Hox genes, is expressed at the blastopore side of the larval sea anemone. (From Finnerty et al. 2004, courtesy of M. Martindale.)

tral Hox genes emerged as a duplication from one of the earlier genes (de Rosa et al. 1999).

In the deuterostome chordate lineage, two large-scale duplications of the entire Hox cluster took place, such that vertebrates have four Hox clusters per haploid genome instead of just one. Thus, instead of having a single *Hox4* gene (orthologous to *Deformed* in *Drosophila*), vertebrates have *Hoxa4*, *Hoxb4*, *Hoxc4*, and *Hoxd4*. This constitutes the *Hox4* **paralogue group** in vertebrates. Such large-scale duplications have had several consequences. First, as seen in Chapters 11 and 13, these duplications create much redundancy. It is difficult to obtain a loss-of-function mutant phenotype, since to do so all copies of these paralogue group genes must be deleted or made nonfunctional (Wellik and Capecchi 2003). However, in some instances, the genes have become specialized. *Hoxd11*, for instance, plays an important role in the mammalian limb bud but not in the reproductive system. Mammalian *Hoxa11*, however, plays roles in both the limb (where it is critical in specifying the zeugopod) and in the female reproductive tract (where it helps construct the uterus; Lynch et al. 2004; Wong et al. 2004).

The second duplication event of the Hox genes within the vertebrates took place during fish evolution. In the ancestral group that gave rise to the teleost fishes (such as zebrafish), the entire genome was duplicated. After this duplication, some clusters were lost. But most teleosts still have six or seven Hox clusters, whereas the other vertebrates have four. Again, there is much redundancy, causing a stabilization of phenotype. But the fish have also used some of these genes for different functions. For instance, the expression pattern of one zebrafish *Hoxc13* paralogue, *Hoxc13a*, is different from the second paralogue, *Hoxc13b*, indicating their roles diverged after the second duplication (Thummel et al. 2004).

---

*The conservation of Hox genes and their colinearity demands an explanation. One recent proposal (Kmita et al. 2000, 2002) contends that the Hox genes "compete" for a remote enhancer that recognizes the Hox genes in a polar fashion. This enhancer most efficiently activates Hox genes at the 5' end. If the positions of the genes are changed by recombination or deletion, then different genes are activated in different regions of the body, and morphology changes.

In summary, then, gene duplication and divergence are extremely important mechanisms for evolution. Duplication allows the formation of redundant structures, and divergence allows these structures to assume new roles. While one gene copy maintains its original role, the other copies are free to mutate and diverge functionally. Numerous transcription factors and paracrine factors are members of such paralogue families. Hox genes are used to pattern the body and limb axes, *Distal-less* genes are used to extend appendages and to pattern the vertebrate skull, and members of the MyoD family specify different stages of muscle development. These genes, each family derived from a single ancestral gene, are active in different tissues and provide instructions for the formation of different cell types.* Thus one of the most important differences between the genome of a fruit fly and that of a human "is not that the human has new genes but that where the fly only has one gene, our species has multigene families" (Morange 2001, p. 33).

Deep Homology

One of the most exciting findings of evolutionary developmental biology has been the discovery not only of homologous regulatory genes, but also of homologous signal transduction pathways, many of which have been mentioned earlier in this book. In different organisms, these pathways are composed of homologous proteins arranged in a homologous manner (Zuckerkandl 1994; Gilbert 1996; Gilbert et al. 1996). This shows a level of parsimony even deeper than that of the individual genes.

In some instances, homologous pathways made of homologous parts are used for the same function in both protostomes and deuterostomes. This has been called **deep homology** (Shubin et al. 1997, 2009). Conserved similarities in both the pathway and its function over millions of years of phylogenetic divergence are considered to be evidence of deep homology between these structures (Shubin et al. 1997). One example is the chordin/BMP4 pathway discussed in Chapter 7. In both vertebrates and invertebrates, chordin/Short-gastrulation (Sog) inhibits the lateralizing effects of Bmp4/Decapentaplegic (Dpp), thereby allowing the ectoderm protected by chordin/Sog to become the neurogenic ectoderm. These reactions are so similar that *Drosophila* Dpp protein can induce ventral fates in *Xenopus* and can substitute for Sog* (**Figure 19.4**; Holley et al. 1995).

According to this scheme, the central nervous system (CNS) only originated once, and its ventral and dorsal position was a later occurrence. This idea has been supported by evidence that annelid worms and cephalochordates also

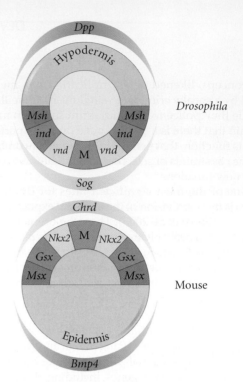

FIGURE 19.4 The same set of instructions forms the nervous systems of both protostomes and deuterostomes. In the fruit fly (protostome), the TGF-β family member *Dpp* (*Decapentaplegic*) is expressed dorsally and is opposed by *Sog* ventrally. In the mouse (deuterostome), the TGF-β family member *Bmp4* is expressed ventrally and is countered dorsally by *chordin* (*Chrd*). The highest concentration of chordin/Sog becomes the midline (M). The midline is dorsal in vertebrates and ventral in insects, and the concentration gradient of the TGF-β family protein (Bmp4 or Dpp) activates the genes specifying the regions of the nervous system in the same order in both groups: *vnd/Nkx2*, followed by *ind/Gsx*, and finally *Msh/Msx*. These genes have also been seen to be expressed in a similar fashion in cnidarians. (After Ball et al. 2004.)

use the inhibition of the BMP pathway to make their CNS (Denes et al. 2007; Yu et al. 2007).

Thus the protostome and deuterostome nervous systems, despite their obvious differences, seem to be formed by the same set of instructions. The plan for specifying the animal nervous system may have been laid down only once.

See WEBSITE 19.3
The search for the Urbilaterian ancestor

---

*The duplication-and-divergence scheme for generating paralogous and orthologous homologies was first described in the 1840s by Sir Richard Owen, a friend (and later rival) of Charles Darwin (see Gilbert 1980). The relationship between the number of genes or gene families and complexity is not simple. Horseshoe crabs and teleost fishes have more Hox clusters than humans.

*In addition to this central inhibitory reaction, there are other reactions that add to the deep homology of the instructions for forming the protostome and deuterostome neural tube. For instance, the spread of Dpp in *Drosophila* is aided by Tolloid, a metalloprotease that degrades Sog. The gradient of Dpp concentration from dorsal to ventral is created by the opposing actions of Tolloid (increasing Dpp) and Sog (decreasing it) (Marqués et al. 1997). In *Xenopus* and zebrafish, the homologues of Tolloid (Xolloid and BMP1, respectively) have the same function: they degrade chordin. The gradient of BMP4 from ventral to dorsal is established by the antagonistic interactions of Xolloid or BMP1 (increasing BMP4) and chordin (decreasing BMP4) (Blader et al. 1997; Piccolo et al. 1997). Even the regulators of BMP stability are conserved between these two groups of animals and function in the same way (Larrain et al. 2001).

Mechanisms of Evolutionary Change

In 1940, Richard Goldschmidt wrote that the accumulation of small genetic changes was not sufficient to generate evolutionarily novel structures such as the neural crest, teeth, turtle shells, feathers, or cnidocysts. He claimed that such evolution could occur only through inheritable changes in the genes that regulated development. This idea was brought up again in 1975, when Mary-Claire King and Alan Wilson published a paper titled "Evolution at Two Levels in Humans and Chimpanzees." This study showed that despite the large anatomical differences between chimpanzees and humans, their DNA was almost identical. The differences would be found in the regulatory genes that acted during development:

> The organismal differences between chimpanzees and humans would ... result chiefly from genetic changes in a few regulatory systems, while amino acid substitutions in general would rarely be a key factor in major adaptive shifts.

In other words, the allelic substitutions of the genes that encode protein sequences—which seem to be pretty much the same for chimpanzees and humans—were not seen as being important. The important differences are where, when, and how much the genes are activated. In 1977, the idea that change within regulatory genes is critical to evolution was extended by François Jacob, the Nobel laureate who helped establish the operon model of gene regulation. First, Jacob said, evolution works with what it has: it combines existing parts in new ways rather than creating new parts. Second, he predicted that such "tinkering" would be most likely to occur in those genes that construct the embryo, not in the genes that function in adults (Jacob 1977).

Wallace Arthur (2004) has catalogued four ways in which Jacob's "tinkering" can take place at the level of gene expression to generate phenotypic variation available for natural selection:

- Heterotopy (change in location)
- Heterochrony (change in time)
- Heterometry (change in amount)
- Heterotypy (change in kind)

These changes can only be accomplished if the gene expression patterns are modular, that is, if they are controlled by different enhancer elements. The modularity of development allows one part of the organism to change without necessarily affecting the other parts.*

---

*This chapter concentrates on *transcriptional-level* changes that can generate new morphological forms, but morphological changes can be instigated at these levels as well. Abzhanov and Kaufman (1999), for instance, have shown that translational differences in the *Sex combs reduced* gene are critical in converting legs into maxillipeds in the crustacean pill bug *Porcellio scaber*.

Heterotopy

One important way of creating new structures is to alter the *location* where a transcription factor or paracrine factor is expressed. This spatial alteration of gene expression is called **heterotopy** (Greek, "different place"). Heterotopy allows different cells to take on a new identity (as sea urchin micromeres did when they recruited the genes for the skeleton formation; see Chapter 5) or to activate or inhibit a paracrine factor-mediated process in a new area of the body (as when Gremlin inhibits BMP-mediated apoptosis in the webbing between digits; see Figure 13.27). There are many other examples, some of which we will describe next.

HOW THE BAT GOT ITS WINGS AND THE TURTLE GOT ITS SHELL
In Chapter 1, we mentioned that the bat evolved its wing by changing the development of the forelimb such that the cells in the interdigital webbing did not die. It turns out that the bat retains its forelimb webbing in a manner very similar to how the duck embryo retains its hindlimb webbing—by blocking the BMPs that would otherwise cause the interdigital cells to undergo apoptosis (see Figures 1.20 and 13.28). Both Gremlin and FGF signaling appears to block BMP functions in the bat wing. Unlike other mammals, bats express Fgf8 in their interdigital webbing, and this protein is critical for maintaining the cells there. If FGF signaling is inhibited (by drugs such as SU5402), BMPs can induce apoptosis of the forelimb webbing, just as in other mammals (Laufer et al. 1997; Weatherbee et al. 2006). The Fgf8 in the webbing is probably also responsible for providing the mitotic signal that extends the digits of the bat, thereby expanding its wing (Hockman et al. 2008).

The formation of the turtle shell also uses BMPs and FGFs, but in different ways. What distinguishes turtles from other vertebrates are their ribs—they migrate laterally into the dermis instead of forming a rib cage (Figure 19.5). Certain regions of the turtle dermis attract rib precursor cells, and these dermal regions differ from those of other vertebrates because they synthesize Fgf10. Fgf10 seems to attract the ribs, since the ribs do not enter the dermis if the Fgf10 signal is blocked (Burke 1989; Cebra-Thomas et al. 2005). The lateral growth of the ribs causes some muscles to establish new attachment sites and causes the scapula (shoulder blades) to reside inside the ribs. This phenomenon is seen only in the turtles (Nagashima et al. 2009). Once inside the dermis, the rib cells do what rib cells are expected to do—they undergo endochondral ossification wherein the cartilage cells are replaced by bone. To do this, BMPs are made. But the rib is embedded in dermis, and the dermal cells can also respond to the BMPs by becoming bone (Cebra-Thomas et al. 2005). In this way, each of the newly positioned ribs instructs the dermis around it to become bone, and the turtle gets its shell.

HOW BIRDS GOT THEIR FEATHERS Although it had long been accepted that feathers emerged as modified reptilian scales

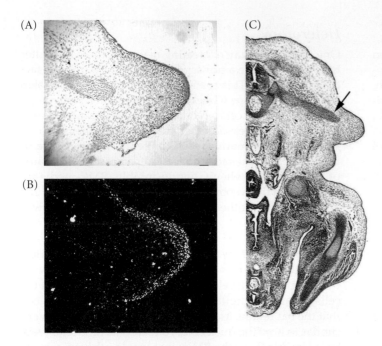

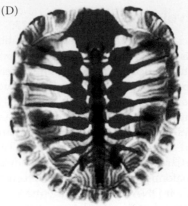

(A)

(B)

(C)

(D)

FIGURE 19.5   Heterotopy on several levels in turtle development. The carapace (dorsal shell) of the turtle is formed through sequential layers of heterotopies. *Fgf10* expression in certain regions of the dermis impels rib precursor cells to migrate laterally into the dermis instead of forming a rib cage. (A,B) Cross section of early turtle embryo as the rib enters the dermis (A, brightfield; B, autoradiograph staining for *Fgf10*). (C) Half cross section of a slightly later turtle embryo, showing a rib extending from the vertebra into the region of the dermis that will expand to form the shell. (D) Hatchling turtle stained with alizarin to show bones. Bones can be seen in the dermis around the ribs that entered into it. Heterotopies include *Fgf10* expression, rib placement, and bone location. (After Loredo et al. 2001.)

(see Maderson 1972; Prum et al. 1999; Maderson and Alibardi 2000), the mechanism that produces feathers has remained elusive. Harris and his colleagues (2002) have provided a developmental mechanism for feather evolution, showing that the feather most likely evolved from the archosaurian (dinosaur/bird ancestor) scale through an alteration of the expression pattern of the Sonic hedgehog (Shh) and BMP2 proteins.

Scales and feathers start off the same way, with *Bmp2* and *Shh* expressed in separate domains. However, in the feather, both expression domains shift to the distal region

of the appendage. This feather-specific pattern is repeated serially around the proximal-distal axis. The interaction between BMP2 and Shh proteins then causes each of these regions to form its own axis—the barbs of the feather (Figure 19.6). Moreover, when this serially repeated pattern was

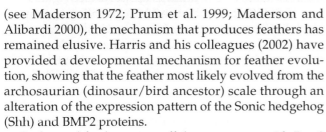

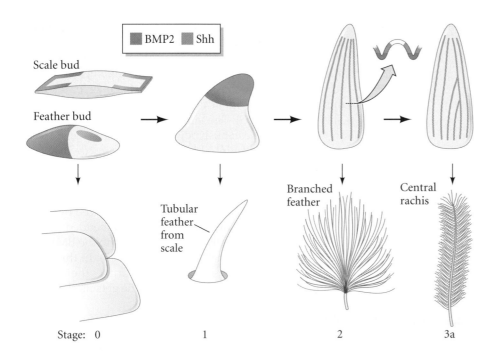

Stage: 0   1   2   3a

FIGURE 19.6   Model for the evolution of the feather by changes in the pattern of *Bmp2* and *Shh* expression. Stage 0 shows the *Shh* and *Bmp2* expression in the scale (above) and feather bud (below). Stage 1 represents a tubular feather as evolved from an archosaurian scale. The *Shh* and *Bmp2* expression patterns are postulated to be at the tip. Stage 2 represents the emergence of a branched feather evolved by further changing the expression patterns of *Bmp2* and *Shh* to form rows along the proximal-distal axis. In stage 3a, changes in feather morphology evolved by altering the pattern to produce a central rachis. (After Harris et al. 2002.)

experimentally modified, the feather pattern was modified in a predictable manner (Harris et al. 2002; Yu et al. 2002).

HOW SNAKES LOST THEIR LIMBS One of the most radical alterations of the vertebrate body plan is seen in snakes. Snakes evolved from four-limbed reptiles; they appear to have lost their legs in a two-step process. Both paleontological and embryological evidence support the view that snakes lost their forelimbs first and later lost their hindlimbs (Caldwell and Lee 1997; Graham and McGonnell 1999; Chipman 2009). Fossil snakes with hindlimbs, but no forelimbs, have been found (Figure 19.7A; Tchernov et al. 2000). Moreover, while the most derived snakes (such as vipers) are completely limbless, the more primitive snakes (such as boas and pythons) have pelvic girdles and rudimentary femurs.

The missing forelimbs can be explained by the Hox expression pattern in the anterior portion of the snake. As described in Chapter 8, the expression pattern of Hox genes in vertebrates determines the type of vertebral structure formed. Thoracic (chest) vertebrae, for instance, have ribs, while cervical (neck) vertebrae and lumbar (lower back) vertebrae do not. The type of vertebra produced is specified by the Hox genes expressed in the somite. In most vertebrates, the forelimbs form just anterior to the most anterior expression domain of *Hoxc6* (Gaunt 1994, 2000; Burke et al. 1995). Caudal to that point, *Hoxc6*, in combination with *Hoxc8*, helps specify vertebrae to be thoracic. During early

python development, *Hoxc6* is not expressed in the absence of *Hoxc8*, so the forelimbs do not form. Rather, the combination of *Hoxc6* and *Hoxc8* is expressed for most of the length of the organism, telling the vertebrae to form ribs throughout most of the body (Figure 19.7B–D; Cohn and Tickle 1999; see also Woltering et al. 2009).

The loss of hindlimbs apparently occurred by a different mechanism. Hindlimb buds do begin to form in some snakes, such as pythons, but do not produce anything more than a femur. This appears to be due to the lack of *Sonic hedgehog* expression by the limb bud mesenchyme. Sonic hedgehog is needed both for limb polarity and for maintenance of the apical ectodermal ridge (AER; see Chapter 13). Python hindlimb buds lack the AER, and the phenotype of the python hindlimb resembles that of mouse embryos with loss-of-function mutations of *Sonic hedgehog* (Chiang et al. 1996).

Heterochrony

Heterochrony (Greek, "different time") is a shift in the relative timing of two developmental processes. Heterochrony can be seen at any level of development, from gene regulation to adult animal behaviors (West-Eberhard 2003). In heterochrony, one module changes its time of expression or growth rate relative to the other modules of the embryo. One sees heterochronic development throughout the animal kingdom. As Darwin (1859, p. 209) noted,

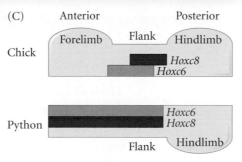

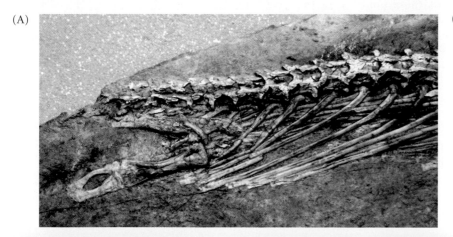

FIGURE 19.7 Loss of limbs in snakes. (A) *Haasiophis terrasanctus*, a fossil snake from the upper Cretaceous carbonates near Jerusalem, Israel. The stylopod and zeugopod can readily be seen. (B) Skeleton of the garter snake, *Thamnophis*, stained with alcian blue. Ribbed vertebrae are seen from the head to the tail. (C,D) Hox expression patterns in chick (C) and python (D). (A photograph by S. Gilbert; B courtesy of A. C. Burke; C,D after Cohn and Tickle 1999.)

(C) Chick

Anterior · · · · · · · · · · · · · Posterior
Forelimb · · · Flank · · · Hindlimb
Hoxc8
Hoxc6

Python

Hoxc6
Hoxc8
Flank · · · Hindlimb

"we may confidently believe that many modifications, wholly due to the laws of growth, and at first in no way advantageous to a species, have been subsequently taken advantage of by the still further modified descendants of this species."

Heterochronies are quite common in vertebrate evolution. We have already discussed the extended growth of the human skull and the heterochronies of amphibian metamorphosis. Another example is found in marsupials, whose mouth and forelimbs develop at a faster rate than do those of placental mammals, allowing the marsupial to climb into the maternal pouch and suckle (Smith 2003; Sears 2004). The enormous number of ribs formed in embryonic snakes (more than 500 in some species; see Figure 19.7B) is likewise due to heterochrony. The segmentation reactions cycle nearly four times faster relative to tissue growth in snake embryos than they do in related vertebrate embryos (Gomez et al. 2008).

In some instances we can determine the heterochronic expression of certain genes. The elongated fingers in the dolphin flipper appear to be the result of the heterochronic expression of *Fgf8*, which as we saw in Chapter 13 encodes a major paracrine factor for limb outgrowth (**Figure 19.8**; Richardson and Oelschläger 2002; Cooper 2010). Another "digital" example of molecular heterochrony occurs in the lizard genus *Hemiergis*, which includes species with three, four, or five digits per limb. The number of digits is regulated by the length of time the *Sonic hedgehog* gene remains active in the limb bud's zone of polarizing activity. The shorter the duration of *Shh* expression, the fewer the number of digits (Shapiro et al. 2003). In primates, there is a heterochronic shift

in the transcription of a set of cerebral mRNAs, such that the expression pattern in adult humans resembles that seen in juvenile chimpanzees (Somel et al. 2009).

Heterometry

Heterometry is a change in the *amount* of a gene product or structure. We have already mentioned such heterometric changes in Chapter 9, where we discussed the evolution of the Mexican blind cavefish. Here, we saw that an overabundance of Sonic hedgehog production from the midline prechordal plate downregulates the *Pax6* gene, preventing eye formation. But this overexpression of *Shh* has other consequences as well. Not only does it cause the degeneration of the eyes, it also causes the jaw size and number of taste buds to increase (Yamamoto et al. 2009). Since the cavefish have no light to see in, an expansion of their jaw and gustatory sense at the expense of sight can be selected. Heterometry can also be seen in the human response to parasitic worms: a mutation causing overproduction of interleukin 4 has been (and is being) selected in populations where worm parasites are endemic (Rockman et al. 2003).

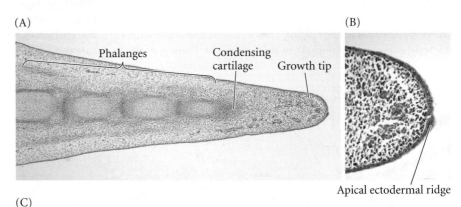

(A) Phalanges Condensing cartilage Growth tip

(B)

Apical ectodermal ridge

FIGURE 19.8 Heterochrony in flipper development of the spotted dolphin (*Stenella attenuata*). (A) Dolphins show "hyperphalangy" of digits 2 and 3: these two digits continue growing long after the other digits, and they keep adding new cartilaginous regions. (B) The mechanism for this appears to be the retention of the apical ectodermal ridge which secretes growth factors necessary for digit growth. (C) A heterochronic hypothesis for cetacean limb development. When limb growth terminates early, loss of distal structures is seen. This leads to rudimentary appendages in many whales. Archaeocetes (the fossilized progenitors of whales) appear to have had a normal amount of limb growth. In toothed whales (such as dolphins), limb development terminates late, giving rise to extra long digits. (After Richardson and Oelschläger 2002.)

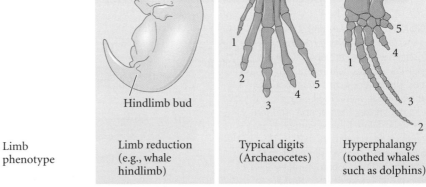

(C)

Duration of limb outgrowth | Short | Normal | Long

Hindlimb bud

Limb phenotype | Limb reduction (e.g., whale hindlimb) | Typical digits (Archaeocetes) | Hyperphalangy (toothed whales such as dolphins)

Bmp4 expression

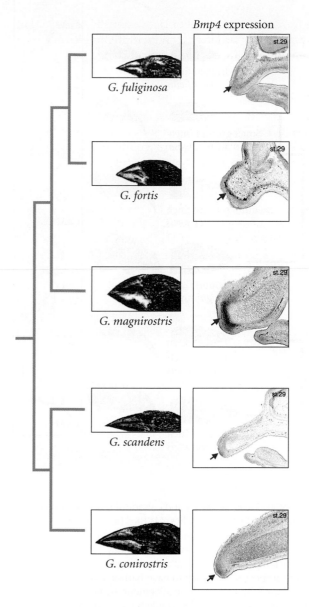

G. fuliginosa

G. fortis

G. magnirostris

G. scandens

G. conirostris

FIGURE 19.9   Correlation between beak shape and the expression of *Bmp4* in Darwin's finches. In the genus *Geospiza*, the ground finches (represented by *G. fuliginosa*, *G. fortis*, and *G. magnirostris*) diverged from the cactus finches (represented by *G. scandens* and *G. conirostris*). The differences in beak morphology correlate to heterochronic and heterometric changes in *Bmp4* expression in the beak. *Bmp4* is expressed earlier and at higher levels in the seed-crushing ground finches. This gene expression difference provides one explanation for the role of natural selection on these birds. (After Abzhanov et al. 2004.)

DARWIN'S FINCHES   One of the best examples of heterometry involves Darwin's celebrated finches. Darwin's finches are a set of 15 closely related birds collected by Charles Darwin and his shipmates during their visit to the Galápagos and Cocos islands in 1835. These birds helped Darwin frame his evolutionary theory of descent with modification, and they still serve as one of the best examples of adaptive radiation and natural selection (see Weiner 1994; Grant and Grant 2008). Systematists have shown that these finch species evolved in a particular manner, with a major speciation event being the split between the cactus finches and the ground finches (**Figure 19.9**). The ground finches evolved deep, broad beaks that enable them to crack seeds open, whereas the cactus finches evolved narrow, pointed beaks that allow them to probe cactus flowers and fruits for insects and flower parts. Earlier research (Schnei-

der and Helms 2003) had shown that species differences in the beak pattern were caused by changes in the growth of the neural crest-derived mesenchyme of the frontonasal process (i.e., those cells that form the facial bones). Abzhanov and his colleagues (2004) found a remarkable correlation between the beak shape of the finches and the timing and amount of *Bmp4* expression. No other paracrine factor showed such differences. The expression of *Bmp4* in ground finches started earlier and was much greater than *Bmp4* expression in cactus finches. In all cases, the *Bmp4* expression pattern correlated with the breadth and depth of the beak.

The importance of these expression differences was confirmed experimentally by changing the *Bmp4* expression pattern in chick embryos (Abzhanov et al. 2004; Wu et al. 2004). When *Bmp4* expression was enhanced in the frontonasal process mesenchyme, the chick developed a broad beak reminiscent of the beaks of the ground finches. Conversely, when BMP signaling was inhibited in this region (by Noggin, a BMP inhibitor), the beak lost depth and width. Thus, enhancers controlling the amount of beak-specific BMP4 synthesis may have been critically important in the evolution of Darwin's finches.

But this was only the beginning of the story. Gene chip technology showed that the level of *calmodulin* gene expression in the beak primordia of the sharp-beaked cactus finches was 15-fold greater than in the beak primordia of the blunt-beaked ground finches. Calmodulin is a protein that combines with many enzymes to make their activity dependent on calcium ions. In situ hybridization and other techniques demonstrated that the *calmodulin* gene is expressed at higher levels in the embryonic beaks of cactus finches than in the embryonic beaks of ground finches (**Figure 19.10**). When calmodulin was upregulated in the embryonic chicken beak, it too became long and pointed.

BMP4 and calmodulin represent two targets for natural selection (**Figure 19.11**; Abzhanov et al. 2006). BMP4 is regulated in heterochronic and heterometric ways, while calmodulin is regulated heterometrically. These studies demonstrate the role for modularity in evolution with morphological variation regulated along two independent axes. While natural selection will allow certain morphologies to survive, the generation of those morphologies depends on variations of developmental regulatory genes such as *BMP4* and *calmodulin*.

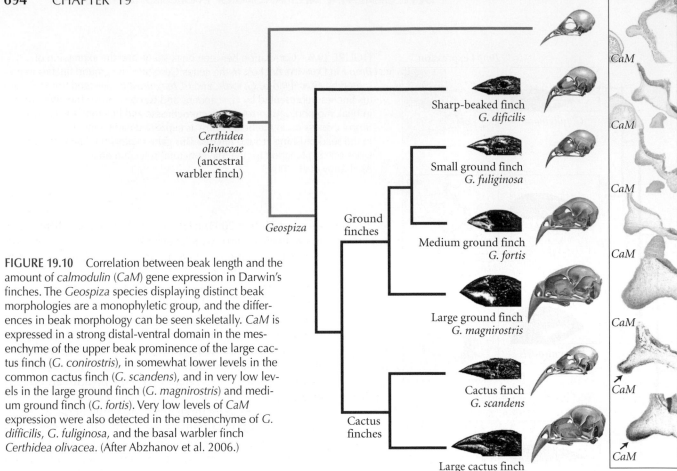

FIGURE 19.10 Correlation between beak length and the amount of *calmodulin* (*CaM*) gene expression in Darwin's finches. The *Geospiza* species displaying distinct beak morphologies are a monophyletic group, and the differences in beak morphology can be seen skeletally. *CaM* is expressed in a strong distal-ventral domain in the mesenchyme of the upper beak prominence of the large cactus finch (*G. conirostris*), in somewhat lower levels in the common cactus finch (*G. scandens*), and in very low levels in the large ground finch (*G. magnirostris*) and medium ground finch (*G. fortis*). Very low levels of *CaM* expression were also detected in the mesenchyme of *G. difficilis*, *G. fuliginosa*, and the basal warbler finch *Certhidea olivacea*. (After Abzhanov et al. 2006.)

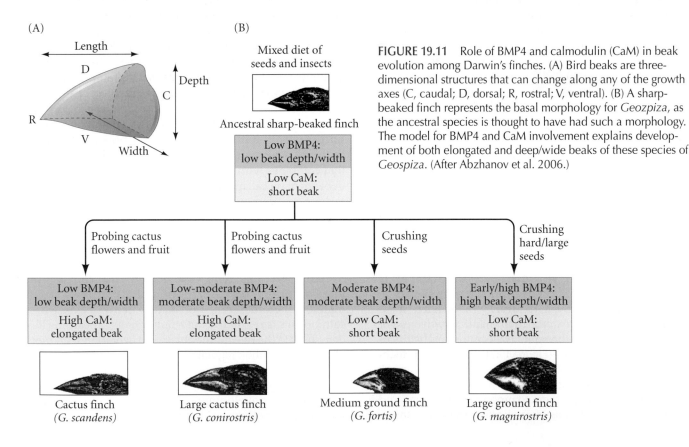

FIGURE 19.11 Role of BMP4 and calmodulin (CaM) in beak evolution among Darwin's finches. (A) Bird beaks are three-dimensional structures that can change along any of the growth axes (C, caudal; D, dorsal; R, rostral; V, ventral). (B) A sharp-beaked finch represents the basal morphology for *Geozpiza*, as the ancestral species is thought to have had such a morphology. The model for BMP4 and CaM involvement explains development of both elongated and deep/wide beaks of these species of *Geospiza*. (After Abzhanov et al. 2006.)

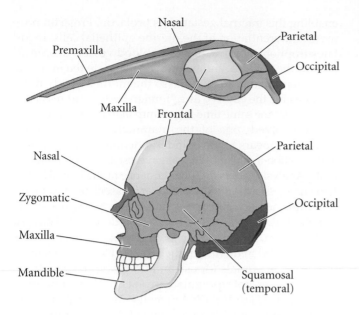

FIGURE 19.12   Allometric growth in the whale head. An adult human skull is shown for comparison. The whale's upper jaw (maxilla) has pushed forward, causing the nose to move to the top of the skull. The mandible is not shown. (The premaxilla is present in the early human fetus, but it fuses with the maxilla by the end of the third month of gestation. The human premaxilla was discovered by Goethe, among others, in 1786.) (After Slijper 1962.)

ALLOMETRY   Another consequence of modularity associated with heterometry is allometry—changes that occur when different parts of an organism grow at different rates (Huxley and Teissier 1936; Gayon 2000). Allometry can be very important in forming diverse body plans within a lineage. Such differential changes in growth rate can involve altering a target cell's sensitivity to growth factors, or altering the amounts of growth factors produced. Again, the vertebrate limb provides a useful illustration. Local differences among chondrocytes caused the central toes of embryonic mammals to grow at a rate 1.4 times that of the lateral toes (Wolpert 1983). As horses grew larger over evolutionary time, this regional difference in chondrocytes resulted in the one-toed state seen in modern horses.

A particularly dramatic example of allometry in evolution comes from vertebrate skull development. In the very young (4 to 5 mm long) whale embryo, the nose is in the usual mammalian position. However, the enormous growth of the maxilla and premaxilla (upper jaw) pushes over the frontal bone and forces the nose to the top of the skull (**Figure 19.12**). This new position turns the mammalian nose into the cetacean blowhole, allowing the whale to have a large and highly specialized jaw apparatus and (not incidentally) to breathe while swimming at the water's surface (Slijper 1962).

Allometry can also generate evolutionary novelty by small, incremental changes that eventually cross some developmental threshold (sometimes called a **bifurcation point**). When such a threshold is crossed, a change in quantity eventually becomes a change in quality. It has been postulated that this type of mechanism produced the unique cheek pouches of pocket gophers and kangaroo rats that live in deserts. These external pouches differ from internal ones in that they are fur-lined and have no internal connection to the mouth. They allow these animals to store seeds outside their mouths, thus minimizing water loss that would occur to seeds held inside the mouth.

Brylski and Hall (1988) dissected pocket gopher and kangaroo rat embryos to examine the way the external cheek pouch is constructed. They found that these external pouches and the internal cheek pouches of animals such as hamsters are actually formed in a very similar manner. Both pouch types are formed by outpocketings of the cheek (buccal) epithelium into the facial mesenchyme. In animals with internal cheek pouches, these evaginations stay within the cheek. However, in animals that form external pouches, the elongation of the snout draws the outpocketings up into the region of the lip. As the lip epithelium rolls out of the oral cavity, so do the outpockets. What had been internal becomes external. The fur lining is probably derived from the external pouches coming into contact with dermal mesenchyme, which can induce hair to form in epithelia (see Chapter 9). The molecular bases for these growth rate changes have not yet been identified.

The transition from internal to external pouch is one of threshold. The placement of the evaginations—anteriorly or posteriorly—determines whether the pouch is internal or external. There is no "transitional stage" displaying two openings, one internal and one external.* One could envision this externalization occurring by a chance mutation or concatenation of alleles that shifted the outpocketings to a slightly more anterior location. Such a trait would be selected for in desert environments, where desiccation is a constant risk. As Leigh Van Valen reflected (1976), evolution can be defined as "the control of development by ecology."

---

*The lack of such transitional forms is often cited by creationists as evidence against evolution. For instance, in the transition from reptiles to mammals, three of the bones of the reptilian jaw became the incus and malleus, leaving only one bone (the dentary) in the lower jaw (see Chapter 1) Gish (1973), a creationist, says that this is an impossible situation, since no fossil has been discovered showing two or three jaw bones and two or three ear ossicles. Such an animal, he claims, would have dragged its jaw on the ground. However, such a specific transitional form (and there are over a dozen documented transitional forms between reptilian and mammalian skulls) need never have existed. Hopson (1966) has shown on embryological grounds how the bones of the jaw could have divided and been used for different functions, and Romer (1970) has found reptilian fossils wherein the new jaw articulation was already functional while the older bones were becoming useless. Several species of therapsid reptiles had two jaw articulations, with the stapes brought into close proximity with the upper portion of the quadrate bone (which would become the incus).

Heterotypy

In heterochrony, heterotopy, and heterometry, mutations affect the regulatory regions of the gene. The gene's product—the protein—remains the same, although it may be synthesized in a new place, at a different time, or in different amounts. The changes of **heterotypy** affect the actual coding region of the gene, and thus can change the functional properties of the protein being synthesized.

WHY INSECTS HAVE ONLY SIX LEGS Insects have six legs, but most other arthropod groups (think of spiders, millipedes, centipedes, and shrimp) have many more. How is it that the insects came to form legs only in their three thoracic segments? The answer seems to reside in the relationship, mentioned earlier, between Ultrabithorax protein and the *Distal-less* gene.

In most arthropod groups, Ubx protein does not inhibit the *Distal-less* gene. However, in the insect lineage, a mutation occurred in the *Ubx* gene wherein the original 3′ end of the protein-coding region was replaced by a group of nucleotides encoding a stretch of about 10 alanine residues (**Figure 19.13**; Galant and Carroll 2002; Ronshaugen et al. 2002). This polyalanine region represses *Distal-less* transcription. When a shrimp *Ubx* gene is experimentally modified to encode this polyalanine region, it too represses the *Distal-less* gene. The ability of insect Ubx to inhibit *Distal-less* thus appears to be the result of a gain-of-function mutation that characterizes the insect lineage.

HOW PREGNANCY MAY HAVE EVOLVED IN MAMMALS One of the most amazing features of mammals is the female uterus, a structure that can hold, nourish, and protect a developing fetus within its mother's body. One of the key proteins

enabling this internal gestation is prolactin. Prolactin promotes differentiation of the uterine epithelial cells, regulates trophoblast growth, allows blood vessels to spread toward the embryo, and helps downregulate the immune and inflammatory responses (so the mother's body does not perceive the embryo as a "foreign body" and reject it).

At about the same time the mammalian uterus and pregnancy evolved, one of the mammalian Hox genes—*Hoxa11*—appears to have undergone intensive mutation and selection in the lineage that gave rise to placental mammals. Analysis shows that the sequence of the Hoxa11 protein changed in mammals such that it associates and interacts with another transcription factor, Foxo1A (**Figure 19.14**; Lynch et al. 2004, 2008). This association with Foxo1A enables Hoxa11 to upregulate prolactin expression from the enhancer used in uterine epithelial cells. Hoxa11 from non-eutherian mammals (opossum and platypus) and from chickens does not upregulate prolactin. If morpholinos knock out the *Hoxa11* mRNA in mouse uterine cells, no prolactin is expressed. Therefore, one of the most important evolutionary changes in the lineage leading to mammals involved the heterotypic alteration of the Hoxa11 sequence.

HETEROTOPY AND THE EMERGENCE OF CORN Heterotypy is also responsible for one of the most important regulatory mutations in agriculture—the exposure of the kernels on the ears of corn (*Zea mays*). The gene *Tga1* (*teosinte glume architecture-1*) encodes a transcription factor that reduces the area of the glume (the sheath of the ear of corn), enabling the kernels to be exposed and harvested (Wang et al. 2005).

As in the other cases, the new phenotype cannot be produced by a single genetic change. One model that has been proposed numerous times (but which is now receiving experimental support from examples such as those above) is that one particular mutation (i.e., *Tga1* or the hindlimb enhancer allele of *Pitx1*) makes a big change and then other mutations fine-tune the change. This type of evolution has been seen in bacteria (see Chouard 2010). By looking at changes in gene expression, evolutionary developmental biology is now able to explore the origin of numerous

FIGURE 19.13 Changes in Ubx protein associated with the insect clade in the evolution of arthropods. Of all arthropods, only the insects have Ubx protein that is able to repress *Distal-less* gene expression and thereby inhibit abdominal legs. This ability to repress *Distal-less* is due to a mutation that is seen only in the insect *Ubx* gene. (After Galant and Carroll 2002; Ronshaugen et al. 2002.)

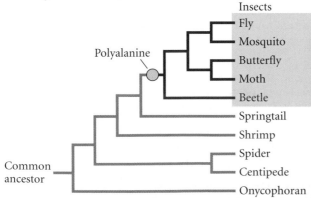
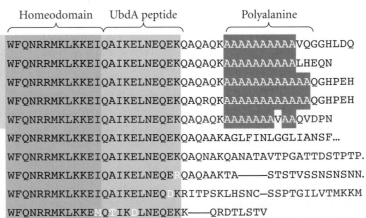

FIGURE 19.14 Ability of the mammalian Hoxa11 protein, in combination with Foxo1A, to promote expression of the uterine prolactin enhancer. The activated luciferinase reporter gene (*d332/luc3*), activated human *Hoxa11* gene (HsaHoxA11), and activated human *Foxo1A* (HsaFoxo1A) each failed to activate the prolactin gene from the enhancer transcription. Mammalian (but not opossum, platypus, or chicken) Hoxa11 increased transcription from this enhancer, but only in the presence of Foxo1A. "Eutherian A11" indicates generalized *HoxA11* from placental mammals. "Therian A11" indicates the consensus *HoxA11* sequence from all mammals. (After Lynch et al. 2008.)

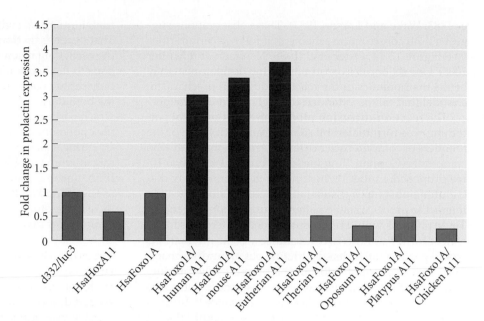

Developmental Constraints on Evolution

There are only about three dozen major animal lineages, and they encompass all the different body plans seen in the animal kingdom. One can easily envision other body plans by imagining animals that do not exist (science fiction writers do it all the time). So why don't we see more body plans among the living animals? To answer this, we have to consider the constraints that development imposes on evolution.

The number and forms of possible phenotypes that can be created are limited by the interactions that are possible among molecules and between modules.* These interactions also allow change to occur in certain directions more easily than in others. Collectively, the restraints on phenotype production are called **developmental constraints**. Constraints on evolution fall into three major categories: physical, morphogenetic, and phyletic (see Richardson and Chipman 2003).

> **See WEBSITE 19.4**
> **How the chordates got a head**

Physical constraints

The laws of diffusion, hydraulics, and physical support are immutable and will permit only certain physical phenotypes to arise. For example, blood cannot circulate to a rotating organ; thus a vertebrate on wheeled appendages (of the sort that Dorothy saw in Oz) cannot exist, and this entire evolutionary avenue is closed off. Similarly, structural parameters and fluid dynamics would prohibit the existence of 6-foot-tall mosquitoes or 25-foot-long leeches.

The elasticity and tensile strength of tissues are also physical constraints. In *Drosophila* sperm, for example, the type of tubulin that can be used in the axoneme is constrained by the need for certain physical properties in the exceptionally long flagellum (Nielsen and Raff 2002). The six cell behaviors used in morphogenesis (division, growth, shape change, migration, death, and matrix secretion) are each limited by physical parameters, and thereby provide limits on what structures animals can form. Interactions between different sets of tissues involve coordinating the behaviors of sheets, rods, and tubes of cells in a limited number of ways (Larsen 1992).

Morphogenetic constraints: The reaction-diffusion model

Bateson (1894) and Alberch (1989) noted that when organisms depart from their normal development, they do so in only a limited number of ways. Although there have been many modifications of the vertebrate limb over 300 million years, some modifications (such as a middle digit shorter than its surrounding digits) are never seen (Hold-

---

*G. W. Leibniz, who may have been the philosopher who most influenced Darwin, noted that existence must be limited not only to the possible but to the *compossible*—that is, whereas numerous things *can* come into existence, only those that are mutually compatible *will* actually exist (see Lovejoy 1964). So, even though many developmental changes are possible, only those that can integrate into the rest of the organism (or that can cause a compensatory change in the rest of the organism) will be seen. Developmental biologists see constraints as limiting the appearance of certain phenotypes, whereas population geneticists see constraints as limiting "ideal" adaptation (such as constraints on optimal foraging) (Amundson 1994, 2005).

---

adaptive variations. One of Darwin's greatest puzzles—the origins of selectable variation, or "the arrival of the fittest"—is now capable of being solved.

er 1983; Wake and Larson 1987). These observations suggest a limb construction scheme that follows certain rules (see Figure 13.14; Oster et al. 1988; Newman and Müller 2005). One set of rules constraining limb development may be the mathematics of the reaction-diffusion mechanism, a model that can be extended throughout development.

The **reaction-diffusion model** for developmental patterning was formulated by Alan Turing (1952), one of the founders of computer science (and the mathematician who cracked the German "Enigma" code during World War II). He proposed a model wherein two homogeneously distributed substances ("substance P" and "substance S") interact to produce stable patterns during morphogenesis. These patterns would represent regional differences in the concentrations of the two substances. Their interactions would produce an ordered structure out of random chaos.

In Turing's model, substance P promotes the production of more substance P as well as substance S. Substance S, however, inhibits the production of substance P. Turing's mathematics show that if S diffuses more readily than P, sharp waves of concentration differences will be generated for substance P (**Figure 19.15**). The reaction-diffusion model predicts alternating areas of high and low concentrations of some substance. When the concentration of the substance is above a certain threshold level, a cell (or group of cells) can be instructed to differentiate in a certain way. An important feature of Turing's model is that particular chemical wavelengths will be amplified while all others will be suppressed. As local concentrations of P increase, the values of S form a peak centering on the P peak, but becoming broader and shallower because substance S diffuses more rapidly. These S peaks inhibit other P peaks

from forming. But which of the many P peaks will survive? That depends on the size and shape of the tissues in which the oscillating reaction is occurring. This pattern is analogous to the harmonics of vibrating strings, as in a guitar: only certain resonance vibrations are permitted, based on the boundaries of the string. The mathematics describing which particular wavelengths are selected consist of complex polynomial equations.

Turing's mathematics have been used to model the spiral patterning of slime molds, the polar organization of the limb, and the pigment patterns of mammals, fish, and snails (Kondo and Asai 1995; Meinhardt 2004). One of the most successful of these attempts applied the model to the evolution of mammalian teeth.

THE EVOLUTION OF MAMMALIAN TEETH Stephen J. Gould (1989) once joked that paleontologists believe mammalian evolution occurs when two teeth mate to produce slightly altered descendant teeth. Since tooth enamel is far more durable than ordinary bone, teeth often remain after all the other bones have decayed, and the study of tooth morphology has been critical to mammalian systematics and ecology. Changes in the cusp pattern of molars are regarded as especially important in allowing the radiation of mammals into new ecological niches. What mechanism allows molars to change their form so rapidly?

Jukka Jernvall and colleagues (Jernvall et al. 2000; Salazar-Ciudad 2008; Salazar-Ciudad and Jernvall 2002, 2004) pioneered a computer-based approach to phenotype production using Geographic Information Systems (GIS) to map gene expression patterns in incipient tooth buds. Their studies showed that gene expression patterns

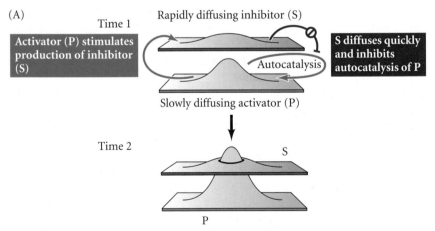

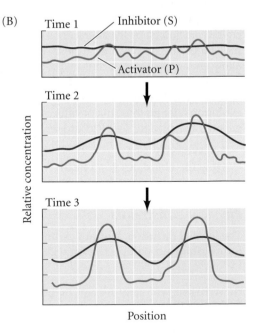

FIGURE 19.15 Reaction-diffusion (Turing) model of pattern generation. (A) Generation of periodic spatial heterogeneity can come about spontaneously when two reactants, S and P, are mixed together under the conditions that S inhibits P; P catalyzes production of both S and P; and S diffuses faster than P. These conditions yield a peak of P and a lower peak of S at the same place. (B) The distribution of the reactants is initially random, and their concentrations fluctuate over a given average. As P increases locally, it produces more S, which diffuses to inhibit more peaks of P from forming in the vicinity of its production. The result is a series of P peaks ("standing waves") at regular intervals.

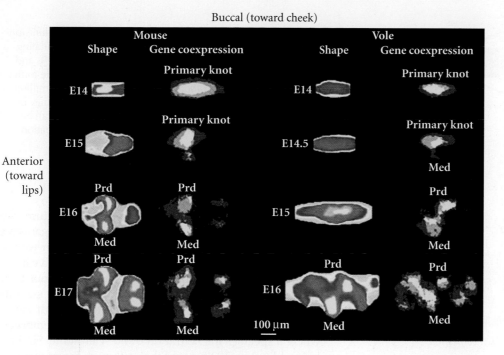

FIGURE 19.16 GIS analysis of gene activity in the formation of the first set of cusps in mouse and vole molars. (The first two cusps are the proconid, labeled Prd, and the metaconid, labeled Med). For both species, GIS mapping of molar shape is shown on the left, and the expression of *Fgf4* and *Shh* (two genes expressed from the enamel knots) is shown at the right. The gene expression pattern on embryonic day 15 predicts the formation of the new cusps seen on day 16; the gene expression pattern on day 16 predicts the formation of cusps in those areas on day 17. Similarly, in the vole molar, whose cusps are diagonal to one another, gene expression predicts cusp formation. (After Jernvall et al. 2000, courtesy of J. Jernvall.)

forecast the exact location of the tooth cusps in mice and voles based on differences in gene expression patterns (**Figure 19.16**).

Salazar-Ciudad and Jernvall (2004, 2010) also showed how reaction-diffusion models can explain the differences in gene expression between mice and voles. The signaling center of the tooth, the enamel knot, secretes BMPs, FGFs, and Shh (see Chapter 10). Shh and FGFs inhibit BMP production, while BMPs stimulate both the production of more BMPs and the synthesis of their own inhibitors. BMPs also induce epithelial differentiation, while FGFs induce epithelial growth. For this study, two additions to the classic reaction-diffusion equation involved (1) chang-

ing the "constants" of diffusion as development progresses, and (2) changing the amount of elasticity associated with the extracellular matrix as development progresses. (This is because the extracellular matrix changes as the cells differentiate, altering tissue shape and the diffusion of paracrine factors.) The result is a pattern of gene activity that changes as the shape of the tooth changes, and vice versa (**Figure 19.17**). Under this model, the large differences between mouse and vole molars can be generated by small changes in the binding constants and diffusion rates of the BMP and Shh proteins. A small increase in the diffusion rate of BMP4 and a stronger binding constant of

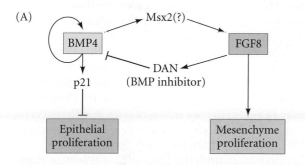

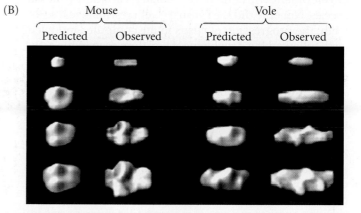

FIGURE 19.17 Basic model for cusp development in mice and voles. (A) Experimentally derived gene network wherein BMPs activate their own production as well as the production of their inhibitors, Shh and FGFs. The FGFs and Shh stimulate cell proliferation; the BMPs inhibit it. (B) Predicted and observed results from this model. The model can generate the final and intermediate forms of molar development in mice and voles, and the difference between mouse and vole molars can be reproduced by slight alterations in the rate of BMP diffusion and binding to inhibitors. (After Salazar-Ciudad and Jernvall 2002; photographs courtesy of J. Jernvall.)

(A)

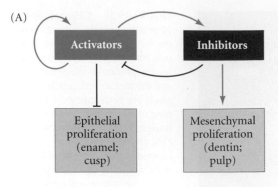

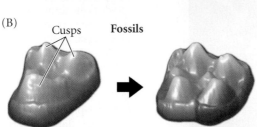

(B)

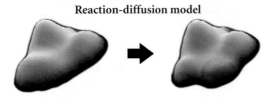

FIGURE 19.18 Mammalian tooth cusp pattern modeled by reaction-diffusion equations. (A) The reaction-diffusion mechanism serves as a motor regulating the genes responsible for slowing the growth of the enamel-forming cells and promoting the proliferation of the dentin pulp-forming cells. (B) Possible formation of the four-cusped tooth of *Hyracotherium* (a fossil horse from 55 million years ago) from the three-cusped tooth of *Loxolophus*, a mammal that may be an earlier member of the horse lineage. This transition in tooth shape can be achieved by modifying a single parameter of the reaction-diffusion equation. (Courtesy of J. Jernvall.)

its inhibitor is sufficient to change the vole pattern of tooth growth into that of the mouse.

The work on mouse and vole molars showed how large morphological changes can result from very small changes in initial conditions. The set of equations that emerged also modeled the observed tooth shape variation seen in natural populations of seals, thereby relating small changes in development with microevolutionary variation within a species. Not only did the "virtual" teeth resemble real seals' teeth; the progression of cusps in the computer-modeled teeth followed the developmental pattern seen in the actual teeth. Moreover, by altering the parameter of epithelial growth, the equations modeled interspecies variation of jaw dentition. Another conclusion is that all the cells can start off with the same basic set of instructions, and specific instructions will emerge as the cells interact.

The Turing model also predicts that some types of teeth are much more likely to evolve in certain ways than in others (**Figure 19.18**). Moreover, the ecological context of tooth use (herbivory versus carnivory, for instance) would select certain variants and not others, and this model can predict both the number and size of molars under different ecological conditions (Kavanagh et al. 2007; Polly 2007). The predictions conform to what paleontologists have observed about mammalian evolution. These studies show the power of mathematical modeling to integrate development, cell biology, and genetics into a predictive model for evolution.

Phyletic constraints

Phyletic constraints on the evolution of new structures are historical restrictions based on the genetics of an organism's development (Gould and Lewontin 1979). In other words, once a structure comes to be established by inductive interactions, it is difficult to start over again. The notochord, for example, is functional in adult protochordates such as *Amphioxus* (Berrill 1987) but degenerates in adult vertebrates. Yet it is transiently necessary in vertebrate embryos, where it specifies the neural tube and sclerotome. Similarly, Waddington (1938) noted that, although the pronephric kidney of the chick embryo is considered vestigial (since it has no ability to concentrate urine), it is the source of the ureteric bud that induces the formation of a functional kidney during chick development (see Chapter 11).

One fascinating example of a phyletic constraint is the lack of variation among marsupial limbs. Although eutherian limbs show a dramatic range of diversity (claws, wings, paddles, flippers, hands), limbs among marsupial species are all pretty much the same. Sears (2004) has documented that the necessity for the marsupial fetus to crawl into its mother's pouch has constrained limb development such that the limb musculature and cartilage must develop very early into a structure that can grasp and crawl. Any variation in this trait has effectively been eliminated.

As genes acquire new functions during the course of evolution, they may become active in more than one module, making change difficult. Galis and colleagues provide evidence that the reason the segment polarity gene network is conserved in all types of insects is that these genes play roles in several different pathways (Galis et al. 2002). Such pleiotropy constrains the possibilities for alternative mechanisms, since it makes change difficult. **Pleiotropy**, the ability of a gene to play different roles in different cells, is the "opposite" of modularity, involving the connections between parts rather than their independence.

Pleiotropies may underlie the constraints seen in mammalian development. Galis speculates that mammals have only seven cervical vertebrae (while birds may have dozens) because the Hox genes that specify these vertebrae have become linked to stem cell proliferation in mammals (Galis 1999; Galis and Metz 2001; Abramovich et al. 2002; Schiedlmeier et al. 2007). Thus, changes in Hox gene expression that might facilitate evolutionary changes in the skeleton might also *mis*regulate cell proliferation and lead to cancers. Galis supports this speculation with epidemiological

(A)

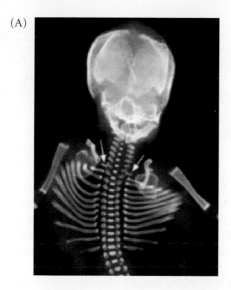

(B)

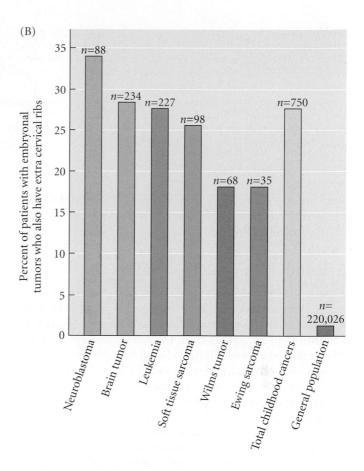

FIGURE 19.19 Extra cervical ribs are associated with childhood cancers. (A) Radiogram showing an extra cervical rib. (B) Nearly 80% of fetuses with extra cervical ribs die before birth. Those surviving often develop cancers very early in life. This indicates strong selection against changes in the number of mammalian cervical ribs. (After Galis et al. 2006, courtesy of F. Galis.)

evidence showing that changes in skeletal morphology correlate with childhood cancer. The intraembryonic selection against having more or fewer than seven cervical ribs appears to be remarkably strong. At least 78% of human embryos with an extra anterior rib (i.e., six cervical vertebrae) die before birth, and 83% die by the end of the first year. These deaths appear to be caused by multiple congenital anomalies or cancers (**Figure 19.19**; Galis et al. 2006).

Until recently, it was thought that the earliest stages of development would be the most resistant to change, because altering them would either destroy the embryo or generate a radically new phenotype. But recent work (and the reappraisal of older work; see Raff et al. 1991) has shown that certain alterations can be made to early cleavage without upsetting the organism's final form. Evolutionary modifications of cytoplasmic determinants in mollusc embryos can give rise to new types of larvae that still metamorphose into molluscs, and changes in sea urchin cytoplasmic determinants can generate sea urchins that develop directly (with no larval stage) but still become sea urchins.

The earliest stages of development, then, appear to be extremely plastic. The later stages are very different from species to species, as the different phenotypes of adult vertebrates amply demonstrate. There is something in the middle of development, however, that appears to be invariant. Sander (1983) and Raff (1994) have argued that the formation of new body plans (*Baupläne*) is inhibited by the global consequences of induction during the **phylotypic stage**—the stage that typifies a phylum. For instance, the late neurula, also known as the **pharyngula**, is the phylotypic stage that appears to be critical for all vertebrates (see

Figure 1.10; Slack et al. 1993). And in fact, while all the vertebrates arrive at the pharyngula stage, they do so by very different means. Birds, reptiles, and fish arrive there after meroblastic cleavages of different sorts; amphibians get there by way of radial holoblastic cleavage; and mammals reach the same stage after constructing a blastocyst, chorion, and amnion (see Chapters 7 and 8).

Before the vertebrate pharyngula stage, there are few inductive events, and most of them are on global scales (involving axis specification). In these early stages of development, there is a great deal of regulation, so small changes in morphogen distributions or the position of cleavage planes can be accommodated (Henry et al. 1989). After the pharyngula stage, there are a great many inductive events, but almost all of them occur within discrete modules (**Figure 19.20**). The lens induces the cornea, but if it fails to do so, only the eye is affected. But during the phylotypic pharyngula stage, the modules interact. Failure to have the heart in a certain place can affect the induction of eyes. Failure to induce the mesoderm in a certain region leads to malformations of the kidneys, limbs, and tail. By searching the literature on congenital anomalies, Galis and Metz (2001) have documented that the pharyngula is much more vulnerable than any other stage. Moreover, based on patterns of multiple organ anomalies in the same person, they concluded that the multiple malformations were due to the interactivity of the modules at this stage. Thus, this phylotypic stage that typifies the vertebrate phylum

FIGURE 19.20 Mechanism for the bottleneck at the pharyngula stage of vertebrate development. (A) In the cleaving embryo, global interactions exist, but there are very few of them (mainly to specify the axes of the organism). (B) At the neurula to pharyngula stages, there are many global inductive interactions. (C) After the pharyngula stage, there are even more inductive interactions, but they are primarily local in effect, confined to their own modules. (After Raff 1994.)

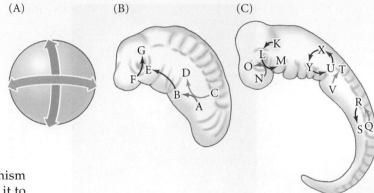

appears to constrain its evolution. Once an organism becomes a vertebrate, it is probably impossible for it to evolve into anything else.

Selectable Epigenetic Variation

Changes in development provide the raw material of variation. But we have seen earlier in the book (especially in Chapter 18) that developmental signals can come from the environment as well as from the nuclei and cytoplasm. Might this environmentally induced variation be inherited and selectable? This idea smacks of Lamarckism, wherein environmentally induced traits could be inherited through the germ line. We now know that Lamarck was wrong in thinking that phenotypes acquired by use or disuse could be transmitted. Children of weight lifters don't inherit their parents' physiques, and accident victims who have lost limbs can rest assured that their children will be born with normal arms and legs.

But what if an environmental agent were to cause changes not only in the somatic DNA but also in the germline DNA? Then the effect might be able to be transmitted from one generation to the next. There are dozens of such examples where the environmentally induced phenotype can be inherited from one generation to the next in the absence of an environmental inducer (Jablonka and Raz 2009). So this is another example of developmental changes generating variation. The question becomes: "How important is phenotypic plasticity for evolution?" Some scientists say it is of minor concern, valid for a few special cases. Other scientists say this is a critical agent in the evolution of many, if not most, species on the planet. Evolutionary developmental biology is still a very young science. It will take time to find out.

Transgenerational inheritance of environmentally induced traits

So far we have discussed the roles of genetic variation in producing variations that can be tested through natural selection. But developmental plasticity also plays a role in producing variation. In some instances, this epigenetic variation is stable between generations, and the processes creating such transmissible epigenetic inheritance have been called **epigenetic inheritance systems**. These epigenetic inheritance systems function in parallel with the genetic inheritance systems of the nucleus and mitochondria.

Early in Chapter 18, we discussed the polyphenisms of *Daphnia* and locusts. Here, environmental conditions experienced by the juveniles (predators in *Daphnia*; population density in locusts) alter the adult phenotype. But in both cases, the altered phenotype is transmitted to the offspring, even in the absence of the environmental cues that initiated it. The mechanism by which predator-induced polyphenisms in *Daphnia* are transmitted to their progeny is not known (Agrawal et al. 1999); the case of the locust, however, is better understood.

The gregarious brown, long-winged migratory morph of the locust *Schistocerca gregaria* is retained for several generations after the crowding stimulus initiates the transformation from the solitary morph. This transgenerational effect is now known to be mediated during oviposition (egg laying) by a chemical agent introduced into the foam surrounding the eggs. Part of the gregarious female's phenotype involves the production of a particular chemical that she introduces into the foam surrounding her eggs. This chemical agent (which appears to be a modified form of the neurotransmitter L-dopa) is thought to be synthesized in the accessory glands of the female reproductive tract and to act at the time of egg laying (Hägele and Simpson 2000; Miller et al. 2008). If foam is transferred from egg masses laid by gregarious females to egg masses produced by solitary females, the solitary eggs turn into gregarious locusts. If the foam is washed off, the gregarious state reverts to the solitary phenotype after a few molts (**Figure 19.21**; McCaffery and Simpson 1998; Simpson and Miller 2007).

These examples show that environmentally induced phenotypes can be inherited. There are two major epigenetic inheritance systems known: epialleles and symbionts. A third process, genetic assimilation, shows that some environmentally induced traits, when continually selected, are stabilized genetically so that the trait is inherited without having to be induced in each generation.

EPIALLELES While the *alleles* that are the basis of the genetic inheritance system are variants of the DNA sequence, the **epialleles** of epigenetic inheritance systems are variants of

FIGURE 19.21   Summary of experiments demonstrating that the source of the transgenerational gregarizing agent in *Schistocerca* is a substance found in the egg froth produced by gregarious females. (After Simpson and Miller 2007; Simpson and Sword 2008.)

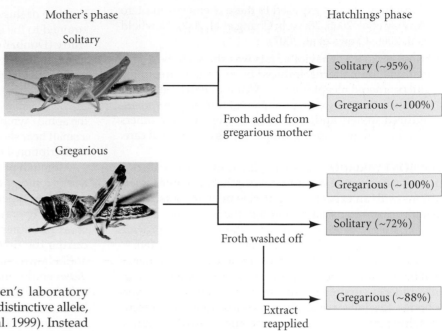

Mother's phase

Solitary

Froth added from gregarious mother

Solitary (~95%)

Gregarious (~100%)

Gregarious

Froth washed off

Extract reapplied

Hatchlings' phase

Gregarious (~100%)

Solitary (~72%)

Gregarious (~88%)

chromatin structure that can be inherited between generations. In most known cases, epialleles are differences in DNA methylation patterning that are able to affect the germ line and thereby be transmitted to offspring. The asymmetrical *peloria* variant of the toadflax plant (*Linaria vulgaris*; **Figure 19.22**) was first described by Linnaeus in 1742 as a stably inherited form. Coen's laboratory showed that this variant was due not to a distinctive allele, but rather to a stable epiallele (Cubas et al. 1999). Instead of carrying a mutation in the *cycloidea* gene, the *peloria* form of this gene was hypermethylated. It does not matter to the developing system whether a gene has been inactivated by a mutation or by an altered chromatin configuration. The effect is the same.

There are dozens of examples of epiallelic inheritance (Gilbert and Epel 2009; Jablonka and Raz 2009). We have discussed some epialleles in their developmental context:

- In the viable *Agouti* phenotype in mice, methylation differences affect coat color and obesity. When a pregnant female is fed a diet high in methyl donors, the specific methylation pattern at the *Agouti* locus is transmitted not only to the progeny developing in utero, but also to

the progeny of those mice and to their progeny (Jirtle and Skinner 2007).

- Enzymatic and metabolic phenotypes are established in utero by protein-restricted diets in rats when protein restriction during a grandmother rat's pregnancy leads to a specific methylation pattern in her pups and grandpups (Burdge et al. 2007).

- The endocrine disruptors vinclozolin, methoxychlor, and bisphenol A have the ability to alter DNA methylation patterns in the germ line, thereby causing developmental anomalies and predispositions to diseases in the

(A)

(B)

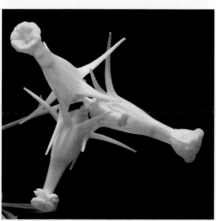

FIGURE 19.22   Epigenetic forms of toadflax. (A) Typical *Linaria*, with a relatively unmethylated *cycloidea* gene. (B) The *cycloidea* gene of the *peloria* variant is relatively heavily methylated. The epialleles that create the different phenotypes of this species are stably inherited. (Courtesy of R. Grant-Downton.)

grandpups of mice exposed to these chemicals in utero (Anway et al. 2005, 2006a,b; Chang et al. 2006; Newbold et al. 2006; Crews et al. 2007).

- Stress-resistant behavior of rats was shown to be due to methylation patterns, induced by maternal care, in the glucocorticoid receptor genes. Meaney (2001) found that rats that received extensive maternal care had less stress-induced anxiety and, if female, developed into mothers that gave their offspring similar levels of maternal care.

SYMBIONT VARIATION As we explored in Chapter 18, one important aspect of phenotypic plasticity involves interactions with an expected population of symbionts. When symbionts are transmitted through the germ line (as *Wolbachia* bacteria are in many insects), the symbionts provide a second system of inheritance (Gilbert and Epel 2009).

Most symbiotic relationships involve microorganisms that have fast growth rates and can thus change more rapidly under environmental stresses than invertebrates or vertebrates. Rosenberg et al. (2007) describe four mechanisms by which microorganisms may confer greater adaptive potential to the whole organism than can the host genome alone. First, the relative abundance of microorganisms associated with the host can be changed due to environmental pressure. Second, adaptive variation can result from the introduction of a new symbiont to the community. Third, changes to the microbial genome can occur through recombination or random mutation, and these changes can occur in a microbial symbiont more rapidly than in the host. And fourth, there is the possibility of horizontal gene transfer between members of the symbiotic community.

One example of symbionts conferring variation to the entire organism involves pea aphids and their symbiotic bacteria. The pea aphid *Acrythosiphon pisum* and its bacterial symbiont *Buchnera aphidicola* have a mutually obligate symbiosis. That is, neither the aphids nor the bacteria will flourish without their partner. Pea aphids rely on *Buchnera* to provide essential amino acids that are absent from the pea aphids' phloem sap diet (Baumann 2005). In exchange, the pea aphids supply nutrients and intracellular niches that permit the *Buchnera* to reproduce (Sabeter et al. 2001).

Because of this interdependence, aphids are highly constrained to the ecological tolerances of *Buchnera*. A recent study (Dunbar et al. 2007) showed that heat tolerance of pea aphids and *Buchnera* could be destroyed with a single nucleotide deletion in a heat shock gene promoter. They discovered that a single-base deletion in the promoter of the *Buchneria ibpA* gene can lower the thermotolerance of the aphid/symbiont organism. This microbial gene encodes a small heat shock protein, and the deletion eliminates the transcriptional response of *ibpA* to heat.

Although pea aphids harboring *Buchnera* with the short *ibpA* promoter allele suffer from decreased thermotolerance, they experience increased reproductive rates under cooler temperatures (15–20°C). Aphid lines containing the short-promoter *Buchnera* produce more nymphs per day during the first 6 days of reproduction compared with aphid lines containing long-allele *Buchnera*. This trade-off between thermotolerance and fecundity allows the pea aphids and *Buchnera* to diversify. As Rosenberg et al. (2007) have pointed out, advantageous mutations will spread more quickly in bacterial genomes than in host genomes because of the rapid reproductive rates of bacteria. This may be especially important in species such as aphids that are produced parthenogenetically (without males) and therefore are essentially a clonal population. The symbionts can be responsible for the variation needed to survive in different environments. Such partnerships, where thermotolerance is provided by the symbiont, are also seen in coral and Christmas cactuses (see Gilbert et al. 2010).

In addition to providing selectable epigenetic variation for *homeostatic* functions (i.e., thermotolerance), experiments on mice indicate there may be instances where different symbiotic bacterial populations alter the development of an organism. When mice with mutations in their *leptin* genes become obese, their guts contain a 50% higher proportion of Firmicutes bacteria and a 50% reduction in Bacteroidetes bacteria than wild-type mouse guts. Moreover, when the gut symbionts from the obese mice were transplanted into genetically wild-type germ-free mice, these mice gained 20% more weight than those germ-free mice receiving gut microbes from wild-type mice (**Figure 19.23**). Thus, there may be an interaction between the genotype of the host and the types of microbial symbionts that are selected by that host environment (Ley et al. 2005; Turnbaugh et al. 2006). Together, they generate a particular phenotype—in this case, obesity.

FIGURE 19.23 Mice with mutations in the *leptin* genes are genetically obese. Their gut microbes are also different from those of wild-type mice, having 50% more Firmicutes and 50% less Bacteroidetes. The mix of microbes in the obese mouse's intestine is more effective at releasing calories from food. When the bacteria from obese mice are transferred into wild-type mice, the wild-type mice also become obese. (Photograph © Science VU Visuals Unlimited.)

Genetic assimilation

In the early 1900s, some evolutionary biologists speculated that the environment could select one of a variety of environmentally induced phenotypes, and that this phenotype would then become the dominant one for the species. In other words, the environment could both induce and select a phenotype. But these scientists had no theory of development or genetics to provide mechanisms for their hypotheses. This idea was revisited in the middle of the twentieth century, and at that time several models were proposed to explain how constant selection could "fix" a particular environmentally induced phenotype in a population.

One of the most important hypotheses of such plasticity-driven adaptation schemes is the concept of **genetic assimilation**, defined as the process by which a phenotypic character initially produced only in response to some environmental influence becomes, through a process of selection, taken over by the genotype so that it is formed even in the absence of the environmental influence that had first been necessary (King and Stanfield 1985). The idea of genetic assimilation was introduced independently by Waddington (1942, 1953, 1961) and Schmalhausen (1949) to explain the remarkable outcomes of artificial selection experiments in which an environmentally induced phenotype became expressed even in the *absence* of the external stimulus that was initially necessary to induce it.

GENETIC ASSIMILATION IN THE LABORATORY Genetic assimilation is readily demonstrated in the laboratory, as Waddington's many experiments proved. For example, when pupae from a laboratory population of wild-type *Drosophila melanogaster* were exposed to a heat shock of 40°C, some of the emerging adults exhibited in their wings a gap in the posterior crossveins. This gap is not normally present in untreated flies (Waddington 1952, 1953). Two selection regimens were followed, one in which only the aberrant flies (termed *crossveinless*) were bred to one another, and another in which only non-aberrant flies were mated. After some generations of selection, when only the individuals showing this gap were allowed to breed, the proportion of adults

with broken crossveins induced by heat shock at the pupal stage rose to above 90%. Moreover (and significantly), by generation 14 of such inbreeding, a small proportion of individuals were *crossveinless* even among flies of this line that had not been exposed to temperature shock. When Waddington extended this artificial selection by breeding the adults that had developed the abnormality without heat shock, the frequency of *crossveinless* individuals among untreated flies became very high, reaching 100% in some lines. The phenotypically induced trait had become "genetically assimilated" into the population.*

In addition to finding the *crossveinless* phenotype on exposure to heat shock, Waddington showed that his laboratory strains of *Drosophila* had a particular reaction norm in their response to ether. Embryos exposed to ether at a particular stage developed a phenotype similar to the *bithorax* mutation and had four wings instead of two. The flies' halteres—balancing structures on the third thoracic segment—were transformed into wings. Generation after generation was exposed to ether, and individuals showing the four-winged state were selectively bred each time. After 20 generations, the mated *Drosophila* produced the mutant phenotype even when no ether was applied (**Figure 19.24**; Waddington 1953, 1956).

Subsequent experiments have borne out Waddington's findings (see Bateman 1959a,b; Matsuda 1982; Ho et al. 1983). In 1996, Gibson and Hogness repeated Waddington's *bithorax* experiments and got similar results (Gibson 1996). Moreover, they found that four alleles of the *Ultrabithorax* (*Ubx*) gene had existed in the population and were

FIGURE 19.24 Phenocopy of the *bithorax* mutation. (A) A *bithorax* (four-winged) phenotype produced after treatment of the embryo with ether. The forewings have been removed to show the aberrant metathorax. This particular individual is actually from the "assimilated" stock that produced this phenotype without being exposed to ether. (B) Selection experiments for or against the *bithorax*-like response to ether treatment. Two experiments are shown (red and blue lines). In both cases, one group was selected for the trait and the other group was selected against the trait. (After Waddington 1956.)

*Note that in these artificial selection experiments, the original phenotype induced by the environment was not intrinsically adaptive. Only the hand of the experimenter choosing which flies mated made it so.

(A)

(B)

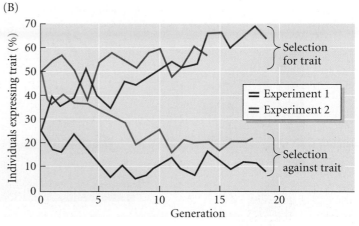

Hsp90 and Genetic Assimilation

How can the results of genetic assimilation be explained? Waddington proposed that the nuclei were buffered such that the same phenotype arose despite minor genetic or environmental perturbations. Schmalhausen suggested that the environmental perturbation unmasked genetic heterogeneity for modifier genes that already existed in the population.

There does indeed appear to be a lot of "buffering" in the cell, so potentially deleterious mutations do not always express deleterious phenotypes. Somehow, the mutant proteins fold correctly (or at least correctly enough to do their jobs.) In 1998, Suzanne Rutherford and Susan Lindquist showed that a major agent responsible for this buffering was the "heat shock protein" Hsp90. Hsp90 binds to a set of signal transduction molecules that are inherently unstable. When it binds to them, it stabilizes their tertiary structure so they can respond to upstream signaling molecules. Heat shock, however, causes other proteins in the cell to become unstable, and Hsp90 is diverted from its normal function (i.e., stabilizing signal transduction proteins) to the more general function of stabilizing any of the now partially denatured peptides in the cell (Jakob et al. 1995; Nathan et al. 1997). Since Hsp90 was known to be involved with inherently unstable proteins and could be diverted by stress, the researchers suspected that it might be involved in buffering developmental pathways against environmental contingencies.

Evidence for the role of Hsp90 as a developmental buffer first came from mutations of *Hsp83*, the gene for Hsp90. Homozygous mutations of *Hsp83* are lethal in *Drosophila*. Heterozygous mutations increase the proportion of developmental abnormalities; in *Drosophila* populations heterozygous for mutant *Hsp83*, deformed eyes, bristle duplications, and abnormalities of legs and wings appeared (**Figure 19.25**). When different mutant alleles of *Hsp83* were brought together in the same flies, both the incidence and severity of the abnormalities increased. Abnormalities were also seen when a specific inhibitor of Hsp90 (geldanamycin) was added to the food of wild-type flies, and the types of defects seen differed among different stocks of flies. The abnormalities observed did not show simple Mendelian inheritance, but were the outcome of the interactions of several gene products. Selective breeding of the flies with the abnormalities led over a few generations to populations in which 80–90% of the progeny had the mutant phenotype. But not all of the mutant progeny carried the *Hsp83* mutation. In other words, once a mutation in *Hsp83* had allowed cryptic mutations to be expressed, selective matings could retain the abnormal phenotype even in the absence of abnormal Hsp90.

Thus, Hsp90 is probably a major component of a buffering system that enables normalization of potentially mutant phenotypes. It provides one way to resist phenotype fluctuations that would otherwise result from slight mutations or slight environmental changes. Hsp90 might also be responsible for allowing mutations to accumulate, but keeping them from being expressed until the environment changes. No individual mutation would change the phenotype, but mating would allow these mutations to be "collected" by members of the population. An environmental change (anything that might stress the cells) might thereby release the hidden phenotypic possibilities of the population. In other words, transient decreases in Hsp90 (resulting from its aiding stress-damaged proteins) would uncover preexisting genetic variations that would produce morphological variations. Most of these morphological variations would probably be deleterious, but some might be selected for in the new environment. Such release of hidden morphological variation may be responsible for the many examples of rapid speciation found in the fossil record.

Figure 19.25 Hsp90 buffers development. (A,B) Developmental abnormalities in *Drosophila* associated with mutations in the *Hsp83* gene include (A) deformed eyes and (B) thickened wing veins. (C) The deformed eye trait seen in (A) was selected by breeding only those individuals expressing the trait. This abnormality was not observed in the original stock, but it can be seen in a high proportion of the descendants of individuals that were mated to heterozygous *Hsp83* flies. The strong response to selection showed that even though the population was small, it contained a large amount of hidden genetic variation. (After Rutherford and Lindquist 1998; photographs courtesy of the authors.)

(A)

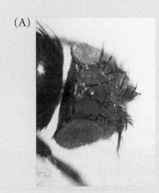

(B)

(C)

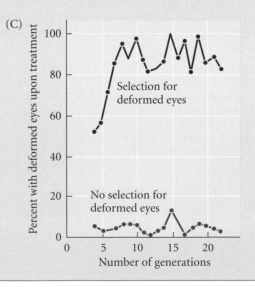

(A)

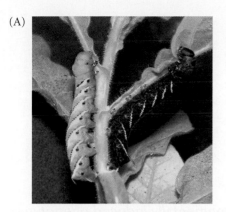

FIGURE 19.26   Effect of selection on temperature-mediated larval color change in the black mutant of the moth *Manduca sexta*. (A) The two color morphs of *Manduca sexta*. (B) Changes in the coloration of heat-shocked larvae in response to selection for increased (black) and decreased (green) color response to heat shock treatments, compared with no selection (red). The color score indicates the relative amount of colored regions in the larvae. (C) Reaction norm for generation-13 flies reared at constant temperatures between 20°C and 33°C, and heat shocked at 42°C. Note the steep polyphenism at around 28°C. (After Suzuki and Nijhout 2006; photograph courtesy of Fred Nijhout.)

(B)

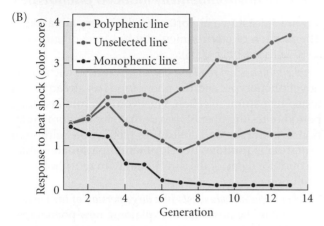

(C)

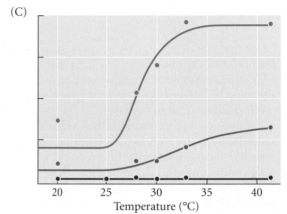

critical for the genetic assimilation of the ether-induced *bithorax* phenotype. "Waddington's experiment showed some fruit flies were more sensitive to ether-induced phenocopies than others, but he had no idea why," Gibson said. "In our experiment, we show that differences in the *Ubx* gene are the cause of these morphological changes."

Genetic assimilation also has also been demonstrated in Lepidoptera. Brakefield and colleagues (1996) were able to genetically assimilate the different morphs of the adaptive polyphenism in *Bicyclus* butterflies (see Figure 18.7), and Suzuki and Nijhout (2006) have shown genetic assimilation in the larvae of the tobacco hornworm moth *Manduca sexta* (**Figure 19.26**). By judicious selection protocols, Suzuki and Nijhout were able to breed lines in which the environmentally induced phenotype (larval color) was selected for and was eventually produced without the environmental agent (temperature shock). The underlying genetic differences concerned the rise of juvenile hormone titres in the larvae. Therefore, at least in the laboratory, genetic assimilation can be shown to work.

GENETIC ASSIMILATION IN NATURAL ENVIRONMENTS
Although it is difficult to document genetic assimilation in nature, there are at least two instances where it appears that phenotypic variation due to developmental plasticity was later fixed by genes. The first involves pigment variations in butterflies. As early as the 1890s, scientists used

heat shock to disrupt the pattern of butterfly wing pigmentation. In some instances, the color patterns that develop after temperature shock mimic the normal genetically controlled patterns of races (or related species) living at different temperatures. A race (subspecies) whose phenotype is characteristic of the species in a particular geographic area is called an **ecotype** (Turesson 1922). Standfuss (1896) demonstrated that a heat-shocked phenocopy of the Swiss subspecies of *Iphiclides podalirius* resembled the normal form of the Sicilian subspecies of that butterfly, and Richard Goldschmidt (1938) observed that heat-shocked specimens of the central European subspecies of *Aglais urticae* produced wing patterns that resembled those of the Sardinian subspecies (**Figure 19.27**). Conversely, cold-shocked individuals of the central European ecotype of *Aglais* developed the wing patterns of the subspecies from northern Scandinavia.

Further observations by Shapiro (1976) on the mourning cloak butterfly (*Nymphalis antiopa*) and by Nijhout (1984) on the buckeye butterfly (*Precis coenia*) have confirmed the view that temperature shock can produce phenocopies that mimic genetically controlled patterns of related races or species existing in colder or warmer conditions. Chilling the pupa of *Pieris occidentalis* will cause it to have the short-day phenotype (Shapiro 1982), which is similar to that of the northern subspecies of pierids. Even "instinctive" behavioral phenotypes associated with these color changes (such as mating

(A)   (B)   (C)

FIGURE 19.27   Temperature shocking *Aglais urticae* produces phenocopies of geographic variants. (A) Usual central European variant. (B) Heat-shock phenocopy resembling Sardinian form. (C) Sardinian form of the species. (From Goldschmidt 1938.)

and flying) are phenocopied (see Burnet et al. 1973; Chow and Chan 1999). Thus, an *environmentally* induced phenotype might become the standard *genetically* induced phenotype in one part of the range of that organism.

The genetic assimilation of environmentally induced phenotypes has also been postulated in the evolution of asymmetric organisms. Throughout the animal kingdom, numerous species exhibit profound asymmetries. These include the direction of snail coiling (discussed in detail in Chapter 5; see Figure 5.25). Other examples are the eyes of flatfish that migrate to one side of the animal's body and the crushing claws of crabs and lobsters (**Figure 19.28**). Palmer (2004, 2009) has documented that most genetically asymmetrical animals evolved from organisms whose asymmetry had been random. Moreover, he and others (Govind 1992; Huber et al. 2007) have shown that the environment can induce the asymmetry in those species where the asymmetry is not genetic. Juvenile lobsters, for instance, will develop a crushing claw only if they have hard objects to manipulate, and whether the right or the left claw becomes the crusher is random. However, if a lobster is forced to preferentially use one claw, that claw will become the crusher (Govind 1992). And although modern flatfish display species-specific laterality, they probably arose from flatfish whose eye-side was determined either by chance or by random environmental cues (Friedman et al. 2008). Only later did the genes for laterality become genetically based. Thus, asymmetry appears to display the phenotype-precedes-genotype mode of evolution that would be expected of genetic assimilation.

Fixation of environmentally induced phenotypes

There are at least two important evolutionary advantages to the fixation of environmentally induced phenotypes (West-Eberhard 1989, 2003):

- *The phenotype is not random.* The environment elicited the novel phenotype, and the phenotype has already been tested by natural selection. This would eliminate a long period of testing phenotypes derived by random mutations. As Garson and colleagues (2003) note, although mutation is random, developmental parameters may account for some of the directionality in morphological evolution.

- *The phenotype already exists in a large portion of the population.* One of the problems of explaining new phenotypes is that the bearers of such phenotypes are "monsters" compared with the wild type. How would such muta-

(A)

(B)

FIGURE 19.28   Asymmetric species. (A) Adult flounder have eyes on one side of the head. In flounder larvae, the eyes are normally separated; during metamorphosis one eye routinely migrates across the fish's skull to the same side as the other eye. Which side gets the eyes is subject to both genetic and environmental influences. (B) Among fiddler crabs (*Uca* spp.), environmental influences shape development of the crusher claw. (A © Marevision/AGE Fotostock; B © OSF/Photolibrary.com.)

tions, perhaps present only in one individual or one family, become established and eventually take over a population? The developmental model solves this problem: this phenotype has been around for a long while, and the capacity to express it is widespread in the population; it merely needs to be genetically stabilized by modifier genes that already exist in the population.

Given these two strong advantages, the genetic assimilation of morphs originally produced through developmental plasticity may contribute significantly to the origin of new species. Ecologist Mary Jane West-Eberhard has noted, "Contrary to popular belief, environmentally initiated novelties may have greater evolutionary potential than mutationally induced ones. Therefore, the genetics of speciation can profit from studies in changes of gene expression as well as changes in gene frequency and genetic isolation."

A Developmental Account of Evolution

As mentioned at the start of this chapter, experimental embryology separated itself from evolutionary biology to mature on its own. However, Roux promised that once it did mature, embryology would return to evolutionary biology with powerful mechanisms for how evolution takes place. Evolution is a theory of change, and population genetics can identify and quantify the dynamics of such change (Amundson 2005). However, Roux realized that evolutionary biology also needed a theory of body construction that would show how any specific mutation becomes manifest as a selectable phenotype.

This is what developmental biology is doing. First, it has provided the underpinning of variation—modularity, molecular parsimony, gene duplication—that enables a change in development to occur without destroying the organism. Second, it has uncovered four modes of genetic changes that can produce new and large variations: heterotopy, heterochrony, heterometry, and heterotypy. And last, it has shown that epigenetic inheritance—epialleles, symbionts, and genetic assimilation—provide selectable variations and aid their propagation through a population.

In 1922, Walter Garstang declared that **ontogeny** (an individual's development) does not recapitulate **phyloge-**ny (evolutionary history); rather, it creates phylogeny. Evolution is generated by heritable changes in development. "The first bird," said Garstang, "was hatched from a reptile's egg." The developmental genetic model has been formulated to account for both the homologies seen in evolution and the differences. We are still approaching evolution in the two ways that Darwin recognized, and descent with modification remains central. We are now at the point where we can answer evolutionary questions using both population genetics and developmental genetics. Thus, when confronted with the question of how the arthropod body plan arose, Hughes and Kaufman (2002) begin their study by saying:

> To answer this question by invoking natural selection is correct—but insufficient. The fangs of a centipede … and the claws of a lobster accord these organisms a fitness advantage. However, the crux of the mystery is this: From what developmental genetic changes did these novelties arise in the first place?

By integrating population genetics with developmental genetics and embryology, we can now begin to explain the construction and evolution of biodiversity.

In his review of evolution, J. B. S. Haldane (1953) expressed his thoughts about evolution with the following developmental analogy:

> The current instar of the evolutionary theory may be defined by such books as those of Huxley, Simpson, Dobzhansky, Mayr, and Stebbins [the founders of the Modern Synthesis of evolution with genetics]. We are certainly not ready for a new moult, but signs of new organs are perhaps visible.

This recognition of developmental ideas "points forward to a broader synthesis in the future." We have finally broken through the old pupal integument, and a new, broader, developmentally inclusive evolutionary synthesis is taking wing.

See WEBSITE 19.5
"Intelligent design" and evolutionary developmental biology

Snapshot Summary: *Evolutionary Developmental Biology*

1. Evolution is the result of inherited changes in development. Modifications of embryonic or larval development can create new phenotypes that can then be selected.

2. Darwin's concept of "descent with modification" explained both homologies and adaptations. The similarities of structure are due to common ancestry (homology), while the modifications are due to natural selection (adaptation to the environmental circumstances).

3. Homology means that similarity between organisms or genes can be traced to descent from a common ancestor. In some instances, certain genes specify the same traits thoughout the animal phyla.

4. The ways of effecting evolutionary change through development at the level of gene expression are: change in location (heterotopy), change in timing (heterochrony), change in amount (heterometry), and change in kind (heterotypy).

5. Changes in their sequence can give Hox genes new properties that may have significant developmental effects. The constraint on insect anatomy of having only six legs is one example; the evolution of the uterus is another.

6. Evolution can occur through the "tinkering" of existing genes. Changes in the location of gene expression can explain the origins of novel structures such as feathers, turtle shells, and duck feet.

7. Changes in Hox gene expression are correlated with the limbless phenotypes of snakes.

8. Changes in Hox gene number may allow Hox genes to take on new functions. Large changes in the numbers of Hox genes correlate with major transitions in evolution.

9. Duplications of genes may enable these genes to become expressed in new places. The formation of new cell types may result from duplicated genes whose regulation has diverged.

10. In addition to structures and genes being homologous, signal transduction pathways can be homologous. In these cases, homologous proteins are organized in homologous ways. These pathways can be used for different developmental processes both in different organisms and within the same organism.

11. The modularity of development allows parts of the embryo to change without affecting other parts. The modularity of enhancers controls this modularity and has allowed one part of the embryo to change without changing other parts.

12. The dissociation of one module from another is shown by heterochrony (a shift in the timing of the development of one region with respect to another) and by allometry (a shift in the growth rates of different parts of the organism relative to one another).

13. Duplication and divergence are important mechanisms of evolution. On the genetic level, the Hox genes and many other gene families started as single genes that were duplicated. The divergent members of such a gene family can assume different functions.

14. Changes in the location of gene expression during development appear to account for the evolution of the turtle shell, the loss of limbs in snakes, the emer-

gence of feathers, and the evolution of differently shaped molars. Changes in the timing of gene expression have been important in the formation of whale flippers and limbs throughout the animal kingdom.

15. Changes in the amount of gene expression can account for the development of beak phenotypes in Darwin's finches, while changes in the developmental regulatory gene itself has been critical in the evolution of insect leg pattern, the mammalian uterus, and the corn glume.

16. Co-option (recruitment) of existing genes and pathways for new functions is a fundamental mechanism for creating new phenotypes. One such case is the use of the limb development signaling pathway to form eyespots in butterfly wings.

17. Developmental constraints prevent certain phenotypes from arising. Such constraints may be physical (no rotating limbs), morphogenetic (no middle digit smaller than its neighbors), or phyletic (no neural tube without a notochord).

18. Developmental constrains can be modeled mathematically. Models such as reaction-diffusion allow predictions of which evolutionary changes are more likely than others to occur.

19. The merging of the population genetic model with the developmental genetic model of evolution is creating a new evolutionary synthesis that can account for macroevolutionary as well as microevolutionary phenomena.

20. Epigenetic inheritance systems include epialleles, wherein inherited patterns of DNA methylation can regulate gene expression. A heavily methylated gene can be as nonfunctional as a genetically mutant allele.

21. Symbiotic organisms are often needed for development to occur, and variants of these organisms may cause different modes of development. Different bacterial strains, for instance, appear to bias a mouse for obesity.

22. Genetic assimilation, wherein a phenotypic character initially induced by the environment becomes, though a process of selection, produced by the genotype in all permissive environments, has been well documented in the laboratory.

23. Evolutionary developmental biology is able to show how small genetic or epigenetic changes can generate large phenotypic changes and enable the production of new anatomical structures.

For Further Reading

Complete bibliographical citations for all literature cited in this chapter can be found at the free access website **www.devbio.com**

Abzhanov, A., W. P. Kuo, C. Hartmann, P. R. Grant, R. Grant and C. Tabin. 2006. The calmodulin pathway and evolution of beak morphology in Darwin's finches. *Nature* 442: 563–567.

Amundson, R. 2005. *The Changing Role of the Embryo in Evolutionary Thought: Roots of Evo-Devo.* Cambridge University Press, New York.

Carroll, S. B. 2006. *Endless Forms Most Beautiful: The New Science of Evo-Devo.* Norton, New York.

Cohn, M. J. and C. Tickle. 1999. Developmental basis of limblessness and axial patterning in snakes. *Nature* 399: 474–479.

Gilbert, S. F. and D. Epel. 2009. *Ecological Developmental Biology: Integrating*

Epigenetics, Medicine, and Evolution. Sinauer Associates, Sunderland, MA.

Kuratani, S. 2009. Modularity, comparative embryology and evo-devo: Developmental dissection of body plans. *Dev. Biol.* 332: 61–69.

Lynch, V. J., A. Tanzer, Y. Wang, F. C. Leung, B. Gelelrsen, D. Emera and G. P. Wagner. 2008. Adaptive changes in the transcription factor HoxA-11 are essential for the evolution of pregnancy in mammals. *Proc. Natl. Acad. Sci. USA.* 105: 14928–14933.

Merino, R., J. Rodríguez-Leon, D. Macias, Y. Ganan, A. N. Economides and J. M. Hurle. 1999. The BMP antagonist Gremlin regulates outgrowth, chondrogenesis and programmed cell death

in the developing limb. *Development* 126: 5515–5522.

Rockman, M. V., M. W. Hahn, N. Soranzo, D. B. Goldstein and G. A. Wray. 2003. Positive selection on a human-specific transcription factor binding site regulating IL4 expression. *Curr. Biol.* 13: 2118–2123.

Shapiro, M. D. and 7 others. 2004. Genetic and developmental basis of evolutionary pelvic reduction in three-spine sticklebacks. *Nature* 428: 717–723.

Shubin, N., C. Tabin and S. B. Carroll. 2009. Deep homology and the origins of evolutionary novelty. *Nature* 457: 818–823.

Smith, K. 2003. Time's arrow: Heterochrony and the evolution of development. *Int. J. Dev. Biol.* 47: 613–621.

Go Online

WEBSITE 19.1 Relating evolution to development in the nineteenth century. Immediately after Darwin's *Origin of Species* biologists attempted to relate evolution to changes in development. Here the attempts of three such scientists—Frank Lillie, Edmund B. Wilson, and Ernst Haeckel—are highlighted.

WEBSITE 19.2 Correlated progression. In many cases, modules must co-evolve. The upper and lower jaws, for instance, have to fit together properly. If one changes, so must the other. If the sperm-binding proteins on the egg change, then so must the egg-binding proteins on the sperm. This site looks at correlated changes during evolution.

WEBSITE 19.3 The seach for the Urbilaterian ancestor. Homologous genes specifying the formation of the eye, heart, body axis, and nervous system enable biologists to intuit an "Urbilaterian ancestor"—an organism that may

have been the precursor of both protostomes and deuterostomes. This organism may have resembled the planula larvae of contemporary cnidarians.

WEBSITE 19.4 How the chordates got a head. The neural crest is responsible for forming the heads of chordates. But how did the neural crest come into existence? It is probable that ancestral deuterostomes had all the requisite genes, but only in the chordates did these genes become linked together into the network that became the neural crest cell.

WEBSITE 19.5 "Intelligent design" and evolutionary developmental biology. Evolutionary developmental biology explains many of the "problems" (such as the evolution of the vertebrate eye and the evolution of turtle shells) that proponents of "intelligent design" and other creationists claimed were impossible to explain by evolution.

Glossary

A

Achondroplasic dwarfism Condition wherein chondrocytes stop proliferating earlier than usual, resulting in short limbs. Often caused by mutations that activate *FgfR3* prematurely.

Acron The terminal portion of the *Drosophila* head; includes the brain.

Acrosome (acrosomal vesicle) Cap-like organelle containing proteolytic enzymes that, together with the sperm nucleus, forms the sperm head.

Acrosome reaction The Ca^{2+}-dependent fusion of the acrosome with the sperm cell membrane, resulting in exocytosis and release of proteolytic enzymes that allow the sperm to penetrate the egg extracellular matrix and fertilize the egg.

Adepithelial cells Cells that migrate into the insect imaginal disc early in development. During the pupal stage, these cells give rise to the muscles and nerves that serve the leg.

Adhesion Attachment between cells or between a cell and its extracellular substrate. The latter provides a surface for migrating cells to travel along.

Adult stem cells Stem cells found in the tissues of organs after the organ has matured. Adult stem cells are usually involved in replacing and repairing tissues of that particular organ, and can form only a subset of cell types.

Afferent neurons Neurons that carry information from sensory receptor cells (e.g., sound waves from the ear, light signals from the retina, touch or pain sensations from the skin) to the central nervous system (spinal cord and brain).

Aging The time-related deterioration of the physiological functions necessary for survival and fertility.

Allantois In amniote species, extraembryonic membrane that stores urinary wastes and helps mediate gas exchange. It is derived from splanchnopleure at the caudal end of the primitive streak. In mammals, the size of the allantois depends on how well nitrogenous wastes can be removed by the chorionic placenta. In reptiles and birds, the allantois becomes a large sac, as there is no other way to keep the toxic by-products of metabolism away from the developing embryo.

Allometry Developmental changes that occur when different parts of an organism grow at different rates.

Alternative splicing A means of producing multiple different proteins encoded by a single gene by splicing together different sets of exons to generate different types of mRNAs.

Ametabolous A pattern of insect development in which there is no larval stage and the insect undergoes direct development to a small adult form following a transitory *pronymph* stage.

Amniocentesis Medical procedure that removes a sample of amnionic fluid around the fourth or fifth month of pregnancy. Fetal cells from the fluid are cultured and analyzed for the presence or absence of certain chromosomes, genes, or enzymes.

Amnion "Water sac." A membrane enclosing and protecting the embryo and its surrounding amnionic fluid. Defining the "amniote vertebrates," this epithelium is derived from somatopleure. Ectodermal tissue supplies epithelial cells, and the mesoderm generates the essential blood supply.

Amnionic fluid A secretion that serves as a shock absorber for the developing embryo while preventing it from drying out.

Amniotes Vertebrates whose embryos form an amnion: the reptiles, birds, and mammals.

Ampulla Latin, "flask." The segment of the mammalian oviduct, distal to the uterus and near the ovary, where fertilization takes place.

Analogous Structures and/or their respective components whose similarity arises from their performing a similar function rather than their arising from a common ancestor (e.g., the wing of a butterfly vs. the wing of a bird). Compare with **homologous**.

Anchor cell The cell connecting the overlying gonad to the vulval precursor cells in *C. elegans*. If the anchor cell is destroyed, the VPCs will not form a vulva, but instead become part of the hypodermis (skin).

Androgen insensitivity syndrome Intersex condition in which an XY individual has a mutation in the gene encoding the androgen receptor protein that binds testosterone. This results in a female external phenotype, lack of a uterus and oviducts and presence of abdominal testes.

Anencephaly A lethal congenital defect resulting from failure to close the anterior neuropore. The forebrain remains in contact with the amnionic fluid and subsequently degenerates, so the vault of the skull fails to form.

Angioblasts From *angio*, blood vessel; and *blast*, a rapidly dividing cell (usually a stem cell). The progenitor cells of blood vessels.

Angiogenesis Process by which the primary network of blood vessels created by vasculogenesis is remodeled and pruned into a distinct capillary bed, arteries, and veins.

Angiopoietins Paracrine factors that mediate the interaction between endothelial cells and pericytes.

Animal cap In amphibians, the roof of the blastocoel (in the animal hemisphere).

Animal hemisphere The non-yolk-containing (upper) half of the amphibian egg. During embryogenesis, cells in the animal hemisphere divide rapidly and become actively mobile ("animated").

Anoikis Rapid apoptosis that occurs when epithelial cells lose their attachment to the extracellular matrix.

Antennapedia complex A region of *Drosophila* chromosome 3 containing the homeotic genes *labial* (*lab*), *Antennapedia* (*Antp*), *sex combs reduced* (*scr*), *deformed* (*dfd*), and *proboscipedia* (*pb*), which specify head and thoracic segment identities.

Anterior heart field Cells of the heart field forming the outflow tract (conus and truncus arteriosus, right ventricle).

Anterior intestinal portal (AIP) The posterior opening of the developing foregut region of the primitive gut tube; it opens into the future midgut region which is contiguous with the yolk sac at this stage.

Anterior visceral endoderm (AVE) Mammalian equivalent to the chick hypoblast and similar to the head portion of the amphibian organizer, it creates an anterior region by secreting antagonists of Nodal.

Anterior-posterior (anteroposterior) axis The body axis extending from head to tail (or mouth to anus in those organisms that lack a head and tail). When referring to the limb, this refers to the thumb (anterior)-pinkie (posterior) axis.

Anti-Müllerian factor (AMF) TGF-β family paracrine factor secreted by the embryonic testes that induces apoptosis of the epithelium and destruction of the basal lamina of the Müllerian duct, preventing formation of the uterus and oviducts. Formerly known as anti-Müllerian hormone, or AMH.

Anurans Frogs and toads.

Aorta-gonad-mesonephros region (AGM) A mesnchymal area in the lateral plate splanchnopleure near the aorta that produces the controversial putative definitive hematopoietic stem cells of the fetus and adult.

Aortic arches These begin as symmetrically arranged, paired vessels that develop within the paired pharyngeal arches and link the ascending and descending/dorsal paired aortae. Some arches degenerate. Left aortic arch IV becomes the Arch of the Aorta and aortic arch VI is the ductus arteriosus on the left.

Apical ectodermal ridge (AER) A ridge along the distal margin of the limb bud that will become a major signaling center for the developing limb. Its roles include (1) maintaining the mesenchyme beneath it in a plastic, proliferating state that enables the linear (proximal-distal) growth of the limb; (2) maintaining the expression of those molecules that generate the anterior-posterior axis; and (3) interacting with the proteins specifying the anterior-posterior and dorsal-ventral axes so that each cell is given instructions on how to differentiate.

Apical epidermal cap (AEC) Forms in the wound epidermis of an amputated salamander limb and acts similarly to the apical ectodermal ridge during normal limb development.

Apoptosis Programmed cell death. Apoptosis is an active process that prunes unneeded structures (e.g., frog tails, male mammary tissue), controls the number of cells in particular tissues, and sculpts complex organs (e.g., palate, retina, digits, and heart). Not to be confused with *necrosis*, pathological cell death, caused by external factors such as inflammation or toxic injury. See also **anoikis**.

Archenteron The primitive gut of the sea urchin blastula, formed by invagination of the vegetal plate into the blastocoel.

Area opaca The peripheral ring of avian blastoderm cells that have not shed their deep cells.

Area pellucida A 1-cell-thick area in the center of the avian blastoderm (following shedding of most of the deep cells) that forms most of the actual embryo.

Aromatase Enzyme that converts testosterone to estradiol. Excess aromatase in the environment is linked to herbicides and other chemicals and is believed to contribute to reproductive disorders (demasculization and feminization, particularly in male amphibians).

Arthrotome Mesenchymal cells in the center of the somite that contribute to the sclerotome, becoming the vertebral joints, the intervertebral discs, and those portions of the ribs closest to the vertebrae.

Autocrine regulation The same cells that secrete paracrine factors also respond to them.

Autonomous specification A mode of cell commitment in which the blastomere inherits a determinant, usually a set of transcription factors from the egg cytoplasm, and these transcription factors regulate gene expression to direct the cell into a particular path of development.

Autopod The distal limb bones of any vertebrate limb, consisting of carpals and metacarpals (forelimb), tarsals and metatarsals (hindlimb) and the phalanges (fingers and toes).

Axon Continuous extension of the nerve cell body. Transmits signals (action potentials) to targets in the central and peripheral nervous systems. Axonal migration is crucial to development of the vertebrate nervous system.

B

Basal disc The "foot" of a hydra, which enables the animal to stick to rocks or the undersides of pond plants.

Basal lamina Specialized, closely knit sheets of extracellular matrix that underlie epithelia, composed of largely of laminin and type IV collagen. Epithelial cells adhere to the basal lamina in part via binding between integrins and laminin. Sometimes called the basement membrane.

Basal layer (stratum germinativum) The inner layer of both the embryonic and adult epidermis, it contains epidermal stem cells attached to a basement membrane that they help to make.

β-Catenin A protein that can act as an anchor for cadherins or as a transcription factor (induced by the Wnt pathway). It is important in the specification of germ layers throughout the animal phyla.

bHLH proteins The basic helix-loop-helix family of transcription factors. It includes such proteins as scleraxis, the MRFs (MyoD, Myf5, and myogenin), and c-Myc.

Bicoid Anterior morphogen critical for establishing anterior-posterior polarity in the *Drosophila* embryo. Functions as a transcription factor to activate anterior-specific gap genes and as a translational repressor to suppress posterior-specific gap genes.

Bilaminar germ disc An amniote embryo prior to gastrulation; consists of epiblast and hypoblast layers.

Bilateral holoblastic cleavage Cleavage pattern, found primarily in tunicates, in which the first cleavage plane establishes the right-left axis of symmetry in the embryo and each successive division orients itself to this plane of symmetry. Thus the half-embryo formed on one side of the first cleavage plane is the mirror image of the other side.

Biofilm Mats of bacteria that regulate larval settlement of many marine invertebrate species.

Bipotential gonad (indifferent gonad) A common precursor, derived from the genital ridge, from which both the male and female gonads diverge.

Bisphenol A (BPA) Synthetic estrogenic chemical compound used in plastics and flame retardants. BPA has been associated with meiotic defects, reproductive abnormalities, and precancerous conditions in rodents.

Bithorax complex The second region of *Drosophila* chromosome 3 containing the homeotic genes *Ultrabithorax* (*Ubx*), which is required for the identity of the third thoracic segment; and the *abdominal A* (*abdA*) and *Abdominal B* (*AbdB*) genes, which are responsible for the segmental identities of the abdominal segments.

Blastocoel A fluid-filled cavity that forms in the animal hemisphere of early amphibian and echinoderm embryos, or between the epiblast and hypoblast of avian, reptilian and mammalian blastoderm-stage embryos.

Blastocyst A mammalian blastula. The blastocoel is expanded and the inner cell mass is positioned on one side of the ring of trophoblast cells.

Blastodisc Small region at the animal pole of the telolecithal eggs of fish and chicks, containing the yolk-free cytoplasm where cleavage can occur and that gives rise to the embryo. Following cleavage, the blastodisc becomes the blastoderm.

Blastomere A cleavage-stage cell resulting from mitosis.

Blastopore The invagination point where gastrulation begins. In deuterostomes, this marks the site of the anus. In protostomes, this marks the site of the mouth.

Blastula Early-stage embryo consisting of a sphere of cells surrounding an inner fluid-filled cavity, the blastocoel.

Blood islands Aggregations of hemangioblasts in the splanchnic mesoderm. It is generally thought that the inner cells of these blood islands become blood progenitor cells, while the outer cells become angioblasts.

Bone marrow-derived stem cells (BMDCs) *See* Mesenchymal stem cells.

Bone morphogenetic proteins (BMPs) Members of the TGF-β superfamily. Originally identified by their ability to induce bone formation, they are extremely multifunctional, having been found to regulate cell division, apoptosis, cell migration, and differentiation.

Bottle cells Invaginating cells during amphibian gastrulation, the main body of each cell is displaced toward the inside of the embryo while maintaining contact with the outside surface by way of a slender neck.

Brain-derived neurotrophic factor (BDNF) A paracrine factor that regulates neural activity and appears to be critical for synapse formation by inducing local translation of neural messages in the dendrites. BDNF is required for the survival of a particular subset of neurons in the striatum (a region of the brain involved in movement).

Bulge A region of the hair follicle that serves as a niche for adult stem cells.

C

Cadherins *Ca*lcium-*de*pendent ad*he*sion molecules. Transmembrane proteins that interact with other cadherins on adjacent cells and are critical for establishing and maintaining intercellular connections, spatial segregation of cell types, and the organization of animal form.

Calorie restriction Dietary restriction as a means of extending mammalian longevity (at the expense of fertility).

Cancer stem cell hypothesis The hypothesis that the malignant part of a tumor is either an adult stem cell that has escaped the control of its niche or a more differentiated cell that has regained stem cell properties.

Cap sequence A 5′ cap is added to nuclear RNA before it leaves the nucleus and is necessary for the binding of mRNA to the ribosome and for subsequent translation.

Capacitation The set of physiological changes by which sperm become capable of fertilizing an egg.

Cardia bifida A condition in which two separate hearts form, resulting from manipulation of the embryo or genetic defects that prevent fusion of the two endocardial tubes.

Cardiac neural crest Subregion of the cranial neural crest that extends from the otic (ear) placodes to the third somites. Cardiac neural crest cells develop into melanocytes, neurons, cartilage, and connective tissue. Cardiac neural crest also ciontributes to the muscular-connective tissue wall of the large arteries (the "outflow tracts") of the heart, as well as contributing to the septum that separates pulmonary circulation from the aorta.

Cardiogenic mesoderm (heart field) Two groups of cardiac cells in the lateral plate mesoderm, at the level of the node. The cardiac cells of the heart field migrate through the primitive streak during gastrulation such that the medial-lateral arrangement of these early cells will become the ante-rior-posterior (rostral-caudal) axis of the developing heart tube.

Cardiomyocytes Cardiac cells derived from cardiogenic mesoderm that form the muscular layers of the heart and its inflow and outflow tracts.

Catenins A complex of proteins that anchor cadherins inside the cell. The cadherin-catenin complex forms the classic adherens junctions that help hold epithelial cells together and, by binding to the actin (microfilament) cytoskeleton of the cell, integrate the epithelial cells into a mechanical unit. One of them, **β-catenin**, can also be a transcription factor.

Caudal intestinal portal (CIP) The anterior opening of the developing hindgut region of the primitive gut tube; it opens into the future midgut region which is contiguous with the yolk sac at this stage.

Caudal Referring to the tail.

Cavitation A process whereby the trophoblast cells secrete fluid into the morula to create a blastocoel. The membranes of trophoblast cells pump Na ions into the central cavity, which draw in water osmotically, thus creating and enlarging the blastocoel.

Cell lineage The series of cell types starting from an undifferentiated, pluripotent stem cell through stages of increasing differentiation, to the terminally differentiated cell type.

Cellular blastoderm Stage of *Drosophila* development in which all the cells are arranged in a single-layered jacket around the yolky core of the egg

Centrolecithal Type of egg, such as those of insects, which have yolk in the center and undergo superficial cleavage.

Centromere A region of DNA where chromatids are attached to each other by the kinetochore.

Centrosome-attracting body (CAB) Cellular structure that, in some invertebrate blastomeres, positions the centrosomes asymmetrically and recruits particular mRNAs so that the resulting daughter cells have different properties.

Cephalic furrow A transverse furrow formed during gastrulation in *Drosophila* that separates the future head region (procephalon) from the germ band, which will form the thorax and abdomen.

Cephalic Referring to the head.

Chemoattractant A biochemical that causes cells to move toward it.

Chemotaxis Movement of a cell down a chemical gradient, such as sperm following a chemical (chemoattractant) secreted by the egg.

Chiasmata Points of attachment between homologous chromosomes during meiosis which are thought to represent regions where crossing-over is occurring.

Chimera an organism consisting of a mixture of cells from two individuals.

Chimeric embryo Embryo made from tissues of more than one genetic source.

Chondrocyte-like osteoblasts Cranial neural crest cells undergoing early stages of intramembranous ossification. These cells downregulate Runx2 and begin expressing the *osteopontin* gene, giving them a phenotype similar to a developing chondrocyte.

Chondrocytes Cartilage cells.

Chordamesoderm Axial mesoderm that produces the notochord and head process.

Chordate An animal that has, at some stage of its life cycle, a notochord and a dorsal nerve cord or neural tube.

Chordin A paracrine factor with organizer activity. Chordin binds directly to BMP4 and BMP2 and prevents their complexing with their receptors, thus inducing dorsal ectoderm to form neural tissue.

Chorioallantoic membrane Forms in some amniote species, such as chickens, by fusion of the mesodermal layer of the allantoic membrane with the mesodermal layer of the chorion. This extremely vascular envelope is crucial for chick development and is responsible for transporting calcium from the eggshell into the embryo for bone production.

Chorion An extraembryonic membrane essential for gas exchange in amniote embryos. It is generated from trophoblast and extraembryonic mesoderm (somatopleure). The chorion adheres to the shell in birds and reptiles, allowing the exchange of gases between the egg and the environment. It forms the embryonic/fetal portion of the placenta in mammals.

Chorionic villus sampling Taking a sample from the placenta at 8–10 weeks of gestation to grow fetal cells to be analyzed for the presence or absence of certain chromosomes, genes, or enzymes.

Chromatid Half of a mitotic prophase chromosome, which consists of duplicate "sister" chromatids that are attached to each other by the kinetochore.

Chromatin The complex of DNA and protein in which eukaryotic genes are contained.

Chromosome diminution The fragmentation of chromosomes just prior to cell division, resulting in cells in which only a portion of the original chromosome survives. Chromosome dimunution occurs during cleavage in *Parascaris aequorum* in the cells that will generate the somatic cells while the future germ cells are protected from this phenomenon and maintain an intact genome.

Ciliary body A vascular structure at the junction between the neural retina and the iris that secretes the aqueous humor.

***cis*-regulatory elements** Regulatory elements (promoters and enhancers) that reside on the same strand of DNA.

Cleavage furrow A groove formed in the cell membrane in a dividing cell due to tightening of the microfilamentous ring.

Cleavage A series of rapid mitotic cell divisions following fertilization in many early embryos; cleavage divides the embryo without increasing its mass.

Cloaca Latin, "sewer." An endodermally lined chamber at the caudal end of the embryo that will become the waste receptacle for both the intestine and the kidney. Amphibians, reptiles, and birds retain this organ and use it to void both liquid and solid wastes. In mammals, the cloaca becomes divided by a septum into the urogenital sinus and the rectum.

Cloacal membrane Caudal end of the primitive streak formed by closely apposed endoderm and ectoderm, future site of the anus.

Cloning *See* Somatic cell nuclear transfer.

Coelom Space between the somatic mesoderm and splanchnic mesoderm that becomes the body cavity. In mammals, the coelom becomes subdivided into the pleural, pericardial, and peritoneal cavities, enveloping the thorax, heart, and abdomen, respectively.

Coherence Scientific evidence that fits into a system of other findings and is therefore more readily accepted.

Cohesin proteins Protein rings that encircle the sister chromatids during meiosis, provide a scaffold for the assembly of the meiotic recombination complex, resist the pulling forces of the spindle microtubules, and thereby keep the sister chromatids attached together and promote recombination.

Combinatorial association In developmental genetics, the principle that enhancers contain regions of DNA that bind transcription factors, and it is this combination of transcription factors that activates the gene.

Commensalism A symbiotic relationship that is beneficial to one partner and neither beneficial nor harmful to the other partner.

Commitment Describes a state in which a cell's developmental fate has become restricted even though it is not yet displaying overt changes in cellular biochemistry and function.

Committed stem cells Describes the multipotent and unipotent stem cells, which have the potential to become relatively few cell types.

Compaction A unique feature of mammalian cleavage, mediated by the cell adhesion molecule E-cadherin. The cells in the early (around eight-cell) embryo change their adhesive properties and become tightly opposed to each other.

Comparative embryology Study of how anatomy changes during the development of different organisms.

Compensatory regeneration Form of regeneration in which the differentiated cells divide but maintain their differentiated functions (e.g., mammalian liver).

Competence The ability to respond to a specific inductive signal.

Conditional specification The ability of cells to achieve their respective fates by interactions with other cells. What a cell becomes is in large measure specified by paracrine factors secreted by its neighbors.

Congenital adrenal hyperplasia A condition causing female pseudohermaphroditism due to the presence of excess testosterone.

Congenital defect Any defect that an animal or person is born with. Congenital defects can be heriditary or they may have an environmental cause (e.g., exposure to teratogenic plants, drugs, chemicals, radiation, etc.). They may also be "idiopathic" (we don't know the cause).

Consensus sequence Located at the 5′ and 3′ ends of the introns that signal the "splice sites" of the intron.

Contact inhibition The mechanism for directional cell movement wherein cells are prohibited from moving "backwards" due to interactions with the cell membranes of other migrating cells.

Context-dependent properties A principle of the theoretical systems approach: The meaning or role of an individual component of a system is dependent on its context.

Conus arteriosus Cardiac outflow tract; precursor of both ventricles.

Convergent extension A phenomenon wherein cells intercalate to narrow the tissue and at the same time move it forward. Mechanism used for elongation of the archenteron in the sea urchin embryo, notochord of the tunicate embryo, and involuting mesoderm of the amphibian. This movement is reminiscent of traffic on a highway when several lanes must merge to form a single lane.

Cornified layer (stratum corneum) The outer layer of the epidermis, consisting of keratinocytes that are now dead, flattened sacs of keratin protein with their nuclei pushed to one edge of the cell. These cells are continually shed throughout life and are replaced by new cells.

Corona radiata The innermost layer of cumulus cells, immediately adjacent to the zona pellucida.

Corpora allata Insect glands that secrete juvenile hormone during larval molts.

Correlative evidence Evidence based on the association of events.

Cortex An outer structure (in contrast with medulla, an inner structure).

Cortical cytoplasm A thin layer of gel-like cytoplasm lying immediately beneath the cell membrane of most eggs. The cortex contains high concentrations of globular actin molecules that will polymerize to form microfilaments and microvilli during fertilization.

Cortical granule reaction The basis of the slow block to polyspermy in many animal species, including sea urchins and most mammals. A mechanical block to polyspermy that becomes active about a minute after successful sperm-egg fusion, in which enzymes from the egg's cortical granules contribute to the formation of a fertilization envelope that blocks further sperm entry.

Cortical granules Membrane-bound, Golgi-derived structures located in the egg's cortex that contain proteolytic enzymes and are thus homologous to the acrosomal vesicle of the sperm.

CpG islands Regions of DNA rich in cytosine and guanosine that flank many promoter TATA box regions.

Cranial (cephalic) neural crest cells Neural crest cells in the future head region that migrate to produce the craniofacial mesenchyme, which differentiates into the cartilage, bone, cranial neurons, glia, and connective tissues of the face. These cells also enter the pharyngeal arches and pouches to give rise to thymic cells, the odontoblasts of the tooth primordia, and the bones of the middle ear and jaw.

Cranial ectodermal placodes Epidermal thickenings that form neurons and sensory epithelia.

Cranium The vertebrate skull, composed of the neurocranium (skull vault and base) and the viscerocranium (jaws and other pharyngeal arch derivatives).

Crossing over The exchange of genetic material during meiosis, whereby genes from one chromatid are exchanged with homologous genes from another.

Crystallins Transparent, lens-specific proteins.

Cumulus oophorus A layer of cells surrounding the mammalian egg, which is made up of the innermost layer of ovarian follicular (granulosa) cells that were nurturing the egg at the time of its release from the ovary.

Cutaneous appendages Species-specific epidermal modifications that include hairs, scales, feathers, hooves, claws, and horns.

Cyclic adenosine 3′,5′-monophosphate (cAMP) An important component of several intracellular signaling cascades and the soluble chemotactic substance that directs the aggregation of the myxamoebae of *Dictyostelium* to form a grex.

Cyclin B The larger subunit of mitosis-promoting factor, shows the cyclical behavior that is key to mitotic regulation, accumulating during S and being degraded after the cells have reached M. Cyclin B regulates the small subunit of MPF, the cyclin-dependent kinase.

Cyclin-dependent kinase Small subunit of MPF, activates mitosis by phosphorylating several target proteins, including histones, the nuclear envelope lamin proteins, and the regulatory subunit of cytoplasmic myosin resulting in chromatin condensation, nuclear envelope depolymerization, and the organization of the mitotic spindle. Requires cyclin B to function.

Cyclooxygenase-2 (COX2) An enzyme that generates prostaglandins from the fatty acid, arachidonic acid.

Cyclopia Congenital defect characterized by a single eye, caused by mutations in genes that encode either Sonic hedgehog or the enzymes that synthesize cholesterol and can be induced by certain chemicals that interfere with the cholesterol biosynthetic enzymes

Cystoblasts/cystocytes Derived from the asymmetric division of the germline stem cells of *Drosophila*, a cystoblast undergoes four mitotic divisions with incomplete cytokinesis to form a cluster of 16 cystocytes (one ovum and 15 nurse cells) interconnected by ring canals.

Cytokines Paracrine factors that are collected and concentrated by the extracellular matrix of the stromal (mesenchymal) cells at the sites of hematopoiesis and are involved in blood cell and lymphocyte formation.

Cytokinesis The division of the cell cytoplasm into two daughter cells. The mechanical agent of cytokinesis is a contractile ring of microfilaments made of actin. Each daughter cell receives one of the nuclei produced by nuclear division (karyokinesis).

Cytoplasmic bridges Continuity between adjacent cells that results from incomplete cytokinesis, e.g., during gametogenesis.

Cytotrophoblast Mammalian extraembryonic epithelium composed of the original trophoblast cells, it adheres to the endometrium through adhesion molecules and, in species with invasive placentation such as the mouse and human, secretes proteolytic enzymes that enables the cytotrophoblast to enter the uterine wall and remodel the uterine blood vessels so that the maternal blood bathes fetal blood vessels.

D

Dauer larva A metabolically dormant larval stage in *C. elegans*. See also **diapause**.

Decidua The maternal portion of the placenta, made from the endometrium of the uterus.

Deep cells A population of cells in the zebrafish blastula between the EVL and the YSL that give rise to the embryo proper.

Deep homology Signal transduction pathways composed of homologous proteins arranged in a homologous manner that are used for the same function in both protostomes and deuterostomes.

Delamination The splitting of one cellular sheet into two more or less parallel sheets.

Delta protein Cell surface ligand for Notch, participates in juxtacrine interaction and activation of the Notch pathway.

Dendrites The fine, branching extensions of the neuron that are used to pick up electric impulses from other cells.

Dermal bone Bone that forms in the dermis of the skin, such as most of the bones of the skull and face. They can be derived from head mesoderm or cranial neural crest-derived mesenchymal cells.

Dermal papilla   A component of mesenchymal-epithelial induction during hair formation; A small node formed by dermal fibroblasts beneath the epidermal hair germ that stimulates proliferation of the overlying epidermal basal stem cells.

Dermamyotome   Dorsolateral portion of the somite that contains skeletal muscle progenitor cells, (including those that migrate into the limbs) and the cells that generate the dermis of the back.

Dermatome   The central portion of the dermamyotome that produces the precursors of the dermis of the back and a third population of muscle cells.

Descent with modification   Darwin's theory to explain unity of type by descent from a common ancestor with changes creating the adaptations to the conditions of particular environments.

Determination   The second, and irreversible, stage of cell or tissue commitment in which the cell or tissue is capable of differentiating autonomously even when placed into a non-neutral environment.

Deuterostomes   In the deuterostome animal phyla (echinoderms, tunicates, cephalochordates, and vertebrates), the first opening (i.e., the blastopore) becomes the anus while the second opening becomes the mouth (hence, *deutero stoma*, "mouth second"). Compare with **protostomes**.

Development   The process of progressive and continuous change that generates a complex multicellular organism from a single cell. Development occurs throughout embryogenesis, maturation to the adult form, and continues into senescence.

Developmental constraints on evolution   The number and forms of possible phenotypes that can be created are limited by the interactions that are possible among molecules and between modules.

Developmental plasticity   The ability of an embryo or larva to react to an environmental input with a change in form, state, movement, or rate of activity (i.e., phenotypic change).

Diacylglycerol (DAG)   Second messenger generated in the IP_3 pathway from membrane phospholipid phosphatidylinositol 4,5-bisphosphate (PIP2), along with IP_3. DAG activates protein kinase C, which in turn activates a protein that exchanges sodium ions for hydrogen ions, raising the pH.

Diapause   A metabolically dormant, nonfeeding stage, during which development and aging are suspended, that can occur at the embryonic, larval, pupal, or adult stage.

Dickkopf   German, "thick head," "stubborn." A protein that interacts directly with the Wnt receptors, preventing Wnt signaling.

Diencephalon   The caudal subdivision of the prosencephalon, will form the optic vesicles, retinas, and the thalamic and hypothalamic brain regions, which receive neural input from the retina.

Diethylstilbestrol (DES)   A potent environmental estrogen. DES administration to pregnant women interferes with sexual and gonadal development in their female offspring resulting in infertility, subfertility, ectopic pregnancies, adenocarcinomas, and other effects.

Differential adhesion hypothesis   A model explaining patterns of cell sorting based on thermodynamic principles. Cells interact so as to form an aggregate with the smallest interfacial free energy and therefore, the most thermodynamically stable pattern.

Differential expression   A basic principle of developmental genetics: In spite of the fact that all the cells of an individual body contain the same genome, the specific proteins expressed by the different cell types are widely diverse. Differential gene expression, differential nRNA processing, differential mRNA translation, and differential protein modification all work to allow the extensive differentiation of cell types.

Differentiation   The process by which an unspecialized cell becomes specialized into one of the many cell types that make up the body.

Digestive tube   The primitive gut of the embryo, which extends the length of the body from the pharynx to the cloaca. Buds from the digestive tube form the liver, gallbladder, and pancreas.

5α-Dihydrotestosterone (DHT)   A steroid hormone derived from testosterone by the action of the enzyme 5α-ketosteroid reductase 2. DHT is required for masculinization of the male urethra, prostate, penis, or scrotum.

Diploblastic   Refers to "two-layer" animals of certain phyla, such as the poriferans (sponges) and ctenophores (comb jellies), that lack a true mesoderm, in contrast to triploblastic animals.

Direct development   Embryogenesis characterized by the lack of a larval stage, where the embryo proceeds to construct a small adult.

Discoidal   Meroblastic cleavage pattern for telolecithal eggs, in which the cell divisions occur only in the small blastodisc, as in birds, reptiles, and fish.

Disruption   Abnormality or congenital defect caused by exogenous agents (teratogens) such as plants, chemicals, viruses, radiation, or hyperthermia.

Dissociation   The ability of one module to develop differently from other modules.

Distal tip cell   A single nondividing cell located at the end of each gonad in *C. elegans* which maintains the nearest germ cells in mitosis by inhibiting their meiotic differentiation.

Dizygotic   "Two eggs." Describes twins that result from two separate fertilization events (fraternal twins). Compare with **monozygotic**.

DNA-binding domain   Transcription factor domain that recognizes a particular DNA sequence.

DNA methylation   A method of controlling the level of gene transcription in vertebrates by the enzymatic methylation of the promoters of inactive genes. Certain cytosine residues that are followed by guanosine residues are methylated and the resulting methylcytosine stabilizes nucleosomes and prevents transcription factors from binding. Important in X chromosome inactivation and DNA imprinting.

Dorsal blastopore lip   Location of the involuting marginal zone cells of amphibian gastrulation. Migrating marginal cells sequentially become the dorsal lip of the blastopore, turn inward and travel along the inner surface of the outer animal hemisphere cells (i.e., the blastocoel roof).

Dorsal closure   A process that brings together the two sides of the epidermis of the *Drosophila* embryo together at the dorsal surface.

Dorsal root ganglia (DRG)   Sensory spinal ganglia derived from the neural crest lying laterally paired and dorsally to the spinal cord. Sensory neurons of the DRG connect centrally with neurons in the dorsal horn of the spinal cord.

Dorsal-ventral (dorsoventral) axis   The line extending from back (dorsum) to belly (ventrum). When referring to the limb, this axis refers to the knuckles (dorsal) and palms (ventral).

Dorsolateral hinge points (DLHPs)   Two hinge regions, besides the MHP, that form furrows near the connection of the avian and mammalian neural plate with the remainder of the ectoderm.

Dosage compensation   Equalization of expression of X chromosome-encoded gene products in male and female cells. Achieved by either (1) doubling the transcription rate of the male X chromosomes (*Drosophila*), (2) partially repressing both X chromosomes (*C. elegans*), or (3) inactivating one X chromosome in each female cell (mammals).

Double-negative gate   A mechanism whereby a repressor locks the genes of specification, and these genes can be unlocked by the repressor of that repressor. (In other words, activation by the repression of a repressor.)

Ductus arteriosus   A vessel that forms from left aortic arch VI, serves as a shunt between the embryonic/fetal pulmonary artery and the descending aorta. It normally closes at birth (if not, a pathological condition results called patent ductus arteriosus).

Duplication and divergence   Tandem gene duplications resulting from replication errors. Once replicated, the gene copies can diverge by random mutations developing different expression patterns and new functions.

Dynein   The protein attached to the axoneme microtubules that provides the force for sperm propulsion. Dynein is an ATPase, an enzyme that hydrolyzes ATP, converting the released chemical energy into mechanical energy to allow the active sliding of the outer doublet microtubules, causing the flagellum to bend.

Dysgenesis   Greek, "bad beginning." Defective development.

E

Early allocation and progenitor expansion model   An alternative to the progress zone model of proximal-distal specification of the limb, wherein the cells of the entire early limb bud are already specified; subsequent cell divisions simply expand these cell populations.

Ecdysone   Insect steroid hormone, secreted by the prothoracic glands, that is modified in peripheral tissues to become the active molting hormone 20-hydroxyecdysone. Crucial to insect metamorphosis.

Ectoderm   Greek *ektos*, "outside." The cells that remain on either the outside (amphibian) or dorsal (avian, mammalian) surface of the embryo following gastrulation. Of the 3 germ layers, the ectoderm is the one that forms the nervous system from the neural tube and neural crest and also generates the epidermis covering the embryo.

Ectodysplasin (EDA) cascade   A gene cascade specific for cutaneous appendage formation. Vertebrates with dysfunctional EDA proteins exhibit a syndrome called anhidrotic ectodermal dysplasia characterized by absent or malformed cutaneous

appendages (hair, teeth, and sweat glands).

Efferent ducts   Ducts that link the rete testis to the Wolffian duct, formed from remodeled tubules of the mesonephric kidney.

Efferent neurons   Neurons (most often motor neurons) that carry information away from the central nervous system (brain and spinal cord) to be acted on by the peripheral nervous system (muscles). Compare with **afferent neurons**.

Egg chamber   The ovary in which the *Drosophila* oocyte will develop, containing 15 interconnected nurse cells and a single oocyte.

Egg jelly   A glycoprotein meshwork outside the vitelline envelope in many species, most commonly it is used to attract and/or to activate sperm.

Embryo   A developing organism prior to birth or hatching. In mammals, the term embryo generally refers to the early stages of development, starting with the fertilized egg until the end of organogenesis (see embryonic period.). After this, the developing mammal is called a fetus until birth, at which time it becomes a neonate.

Embryology   The study of animal development from fertilization to hatching or birth.

Embryonal carcinoma (EC)   *See* Teratocarcinoma.

Embryonic axis   Any of the positional axes in an embryo; includes anterior-posterior (head-tail), dorsal-ventral (back-belly), and right-left.

Embryonic epiblast   Separates from the epiblast cells that line the amnionic cavity and will form the embryo proper.

Embryonic germ (EG) cells   Pluripotent embryonic cells with characteristics of the inner cell mass derived from PGCs that have been treated particular paracrine factors to maintain cell proliferation.

Embryonic period   Up until the eighth week of gestation in humans, the period that includes organogenesis and is a time of maximum sensitivity to teratogens. Compare with **fetus**.

Embryonic shield   A localized thickening on the future dorsal side of the fish embryo; functionally equivalent to the dorsal blastopore lip of amphibians.

Embryonic stem cells (ES cells)   Pluripotent stem cells derived from inner cell mass blastomeres that are cultured in a manner that lets them retain their pluripotency.

Enamel knot   The signaling center for tooth development, a group of cells induced in the epithelium by the neu-

ral crest-derived mesenchyme that secretes paracrine factors that pattern the cusp of the tooth.

Endocardium   The internal layer of the heart chambers, derived from cardiogenic mesoderm.

Endocardial cushions   Form from the endocardium and divide the tube into right and left atrioventricular channels. The atrioventricular valves are also derived from endocardial cells.

Endochondral ossification   Bone formation in which mesodermal mesenchyme becomes cartilage and the cartilage is replaced by bone. It characterizes the bones of the trunk and limbs.

Endocrine disruptors   Hormonally-active compounds in the environment.

Endocrine factors   Hormones that travel through the blood to exert their effects.

Endoderm   Greek *endon*, "within." The innermost germ layer; forms the epithelial lining of the respiratory tract, the gastrointestinal tract, and the accessory organs (e.g., liver, pancreas) of the digestive tract. In the amphibian embryo, the yolk-containing cells of the vegetal hemisphere become endoderm. In mammalian and avian embryos, the endoderm is the most ventral of the three germ layers, continuous with the yolk sack epithelium.

Endometrium   The epithelial lining of the uterus.

Endosteal osteoblasts   Osteoblasts that line the bone marrow and are responsible for providing the niche that attracts HSCs, prevents apoptosis, and keeps the HSCs in a state of plasticity.

Endosymbiosis   Greek, "living within." Describes the situation in which one cell lives inside another cell one organism lives within another.

Endothelial   Refers to the lining of the blood vessels.

Endothelins   Small peptides secreted by blood vessels that have a role in vasoconstriction and can direct the extension of certain sympathetic axons that have endothelin receptors, e.g. targeting of neurons from the superior cervical ganglia to the carotid artery.

Energids   In *Drosophila*, the nuclei at the periphery of the syncytial blastoderm and their associated cytoplasmic islands of cytoskeletal proteins.

Enhancer   A DNA sequence that controls the efficiency and rate of transcription from a specific promoter. Enhancers bind specific transcription factors that activate the gene by (1) recruiting enzymes (such as histone acetyltransferases) that break up the nucleosomes in the area or (2) stabilizing the transcription initiation complex.

Enhancer modularity   The principle that having multiple enhancers allows a protein to be expressed in several different tissues while not being expressed at all in others, according to the combination of transcription factor proteins the enhancers bind.

Enveloping layer (EVL)   A cell population in the zebrafish embryo at the mid-blastula transition made up of the most superficial cells from the blastoderm, which form an epithelial sheet a single cell layer thick. The EVL is an extraembryonic protective covering that is sloughed off during later development.

Environmental integration   Describes the influence of cues from the environment surrounding the embryo, fetus, or larva on their development.

Eph receptors   Receptor for ephrin ligands, involved in juxtacrine signaling.

Ephrin ligands   Juxtacrine factors, binding between an ephrin on one cell and the eph receptor on an adjacent cell results in signals sent to each of the two cells. These signals are often those of either attraction or repulsion, and ephrins are often seen where cells are being told where to migrate or where boundaries are forming. Ephrins and the eph receptors function in the formation of blood vessels, neurons, and somites and direct neural crest cell migration.

Epialleles   Variants of chromatin structure that can be inherited between generations. In most known cases, epialleles are differences in DNA methylation patterning that are able to affect the germ line and thereby be transmitted to offspring.

Epiblast   The outer layer of the thickened margin of the epibolizing blastoderm in the gastrulating zebrafish embryo or the upper layer of the bilaminar gastrulating embryonic disc in birds and mammals. The epiblast contains ectoderm precursors in fish and all three germ layer precursors of the embryo proper (plus the amnion) in amniotes.

Epiboly   The movement of epithelial sheets (usually of ectodermal cells) that spread as a unit (rather than individually) to enclose the deeper layers of the embryo. Epiboly can occur by the cells dividing, by the cells changing their shape, or by several layers of cells intercalating into fewer layers. Often, all three mechanisms are used.

Epicardium   The outer surface of the heart that forms the coronary blood vessels that feed the heart, derived from cardiogenic mesoderm.

Epidermis   Outer layer of skin.

Epididymis   Derived from the Wolffian duct, the tube adjacent to the testis that links the efferent tubules to the ductus deferens.

Epigenesis   The view supported by Aristotle and William Harvey that the organs of the embryo are formed de novo ("from scratch") at each generation.

Epigenetics   The study of genetic mechanisms that act on the phenotype without changing the nucleotide sequence of the DNA. Specifically, these changes work by altering gene *expression* rather than altering the gene sequence as mutation does. Epigenetic changes can sometimes be transmitted to future generations, a phenomenon referred to as epigenetic inheritance.

Epimorphosis   Form of regeneration observed when adult structures undergo *de*differentiation to form a relatively undifferentiated mass of cells that then redifferentiates to form the new structure (e.g., amphibian limb regeneration).

Epiphyseal growth plates   Cartilaginous areas at the ends of the long bones that allow continued bone growth.

Episomal vectors   Vehicles for gene delivery derived from viruses that do not insert themselves into host DNA.

Epithelial-mesenchymal interactions   Induction involving interactions of sheets of epithelial cells with adjacent mesenchymal cells. Properties of these interactions include regional specificity (when placed together, the same epithelium develops different structures according to the region from which the mesenchyme was taken), genetic specificity (the genome of the epithelium limits its ability to respond to signals from the mesencyme, i.e., the response is species-specific).

Epithelial-mesenchymal transition (EMT)   An orderly series of events whereby epithelial cells are transformed into mesenchymal cells. In this transition, a polarized stationary epithelial cell, which normally interacts with basement membrane through its basal surface, becomes a migratory mesenchymal cell that can invade tissues and form organs in new places.

Epithelium   Epithelial cells tightly linked together on a basement membrane to form a sheet or tube with little extracellular matrix.

Equatorial region   In mammalian sperm, the junction between this inner acrosomal membrane and the mammalian sperm cell membrane, where membrane fusion between sperm and egg begins.

Erythroblast   Cell that matures from the proerythroblast and synthesizes enormous amounts of hemoglobin.

Erythrocyte   The mature red blood cell that enters the circulation where it delivers oxygen to the tissues. It is incapable of division, RNA synthesis, or protein synthesis. Amphibians, fish, and birds retain the functionless nucleus; mammals extrude it from the cell.

Erythroid progenitor cell (BFU-E)   A committed stem cell that can form only red blood cells.

Erythropoietin   A hormone that acts on erythroid progenitor cells to produce proerythroblasts, which will generate red blood cells.

Estradiol   An estrogenic steroid hormone. The most active in humans is 17β-estradiol.

Estrogen   A group of steroid hormones (including estradiol) needed for complete postnatal development of both the Müllerian and the Wolffian ducts and necessary for fertility in both males and females

Estrus   Greek *oistros*, "frenzy"; also called "heat." The estrogen-dominated stage of the ovarian cycle in female mammals that are spontaneous or periodic ovulators, characterized by the display of behaviors consistent with receptivity to mating.

Eukaryotic initiation factor-4E (eIF4E)   A protein that is important for the initiation of translation by binding to the 5′ cap of mRNAs and contributing to the protein complex that mediates RNA unwinding and brings the 3′ end of the message next to the 5′ end, allowing the messenger RNA to bind to and be recognized by the ribosome.

Eukaryotic initiation factor-4G (eIF4G)   A scaffold protein that allows the mRNA to bind to the ribosome through its interaction with eIF4E.

Evolutionary developmental biology (evo-devo)   A model of evolution that integrates developmental genetics and population genetics to explain and the origin of biodiversity.

External granule layer   A germinal zone of cerebellar neuroblasts that migrate from the germinal neuroepithelium to the outer surface of the developing cerebellum.

Extracellular matrix (ECM)   Macromolecules secreted by cells into their immediate environment, forming a region of noncellular material in the interstices between the cells. Extracellular matrices are made up of collagen, proteoglycans, and a variety of specialized glycoprotein molecules such as fibronectin and laminin.

Extraembryonic endoderm   Formed by delamination of the hypoblast cells from the inner cell mass to line the blastocoel cavity; will form the yolk sac.

F

Fast block to polyspermy   Mechanism by which additional sperm are prevented from fusing with a fertilized sea urchin egg by changing the electric potential to a more positive level. Has not been demonstrated in mammals.

Fate map   Diagrams that follow cell lineages from specific regions of the embryo in order to "map" larval or adult structures onto the region of the embryo from which they arose. The superimposition of a map of "what is to be" onto a structure that has yet to develop into these organs.

Female pronucleus   The haploid nucleus of the egg.

Fertilization   Fusion of male and female gametes followed by fusion of the haploid gamete nuclei to restore the full complement of chromosomes characteristic of the species and initiation in the egg cytoplasm of those reactions that permit development to proceed.

Fertilization envelope   Forms from the vitelline envelope of the sea urchin egg following cortical granule release. Glycosaminoglycans released by the cortical granules absorb water to expand the space between the cell membrane and fertilization envelope.

Fetal alcohol syndrome (FAS)   Condition of babies born to alcoholic mothers, characterized by small head size, specific facial features, and small brain that often shows defects in neuronal and glial migration. FAS is the most prevalent ongenital mental retardation syndrome.

α-Fetoprotein   A protein that binds and inactivates fetal estrogen, but not testosterone, in both male and female rats and is critical for normal sexual differentiation of the rat brain.

Fetus   Refers to the developing human from the ninth week of gestation to birth, a period characterized by growth and modeling.

Fibroblast growth factors (FGFs)   A family of paracrine factors. Genes (*Fgf*) and their encoded proteins (FGF) that regulate cell proliferation and differentiation.

Fibroblast growth factor receptors (FGFRs)   A set of receptor tyrosine kinases that are activated by FGFs, resulting in activation of the dormant kinase and phosphorylation of certain proteins (including other FGF receptors) within the responding cell.

Fibronectin   A very large (460 kDa) glycoprotein dimer synthesized by numerous cell types and secreted into the extracellular matrix. Functions as a general adhesive molecule, linking cells to one another and to other substrates such as collagen and proteoglycans, and provides a substrate for cell migration.

Filopodia   Long, thin processes produced by migrating mesenchymal cells.

Flagellum   The tail of the sperm, containing the central axoneme, supporting structures called outer dense fibers and the fibrous sheath, and the mitochondria-containing midpiece.

Floor plate   Ventral region of the neural tube important in the establishment of dorsal-ventral polarity. Floor plate forms in the ventral neural tube due to induction by Sonic hedgehog secreted from the adjacent notochord. It becomes a secondary signaling center that also secretes Sonic hedgehog, establishing a gradient that is highest ventrally.

Fluorescent dye   Compounds, such as fluorescein and green fluorescent protein (GFP), that emit bright light at a specific wavelength when excited with ultraviolet light.

Fluorescent in situ hybridization (FISH)   Fluorescent labeling using chromosome-specific probes. Used to determine whether the normal numbers and types of chromosomes are present.

Focal adhesions   Where the cell membrane contacts the extracellular matrix in migrating cells, mediated by connections between actin, integrin and the extracellular matrix.

Follicle   Composed of a single oogonium surrounded by granulosa cells and thecal cells.

Follicle-stimulating hormone (FSH)   A peptide hormone secreted by the mammalian pituitary that promotes ovarian follicle development and spermatogenesis.

Follicular stem cell   A multipotent adult stem cell that resides in the bulge niche of the hair follicle. It gives rise to the hair shaft, sheath, and sebaceous gland.

Follistatin   A paracrine factor with organizer activity, an inhibitor of both activin and BMPs, causes ectoderm to become neural tissue.

Foramen ovale   An opening in the septum separating the right and left atria.

Forkhead transcription factors   (Foxa1, Foxa2, HNF4α) Transcription factors. Especially important in the endoderm that will form liver, where they displace nucleosomes from the regulatory regions surrounding liver-specific genes.

Frizzled   Transmembrane receptor for Wnt family of paracrine factors.

Frontonasal process   Cranial prominence formed by neural crest cells from the midbrain and rhombomeres 1 and 2 of the hindbrain that forms the forehead, the middle of the nose, and the primary palate. Thus, the cranial neural crest cells generate the facial skeleton

Fusome   A large, spectrin-rich structure that spans the ring canals between the cystocytes of meroistic oogenesis in *Drosophila* and other insects.

G

Gain-of-function evidence   A strong type of evidence, wherein the initiation of the first event causes the second event to happen even in instances where or when neither event usually occurs.

Gamete   A specialized reproductive cell through which sexually reproducing parents pass chromosomes to their offspring; a sperm or an egg.

Gametogenesis   The production of gametes.

Ganglia   Clusters of neuronal cell bodies whose axons form a nerve.

Gap genes   *Drosophila* zygotic genes expressed in broad (about three segments wide), partially overlapping domains. Gap mutants lacked large regions of the body (several contiguous segments).

Gastrula   A stage of the embryo following gastrulation that contains the three germ layers that will interact to generate the organs of the body.

Gastrulation   A process involving movement of the blastomeres of the embryo relative to one another resulting in the formation of the three germ layers of the embryo.

Genetic assimilation   The process by which a phenotypic character initially produced only in response to some environmental influence becomes, through a process of selection, taken over by the genotype so that it is formed even in the absence of the environmental influence that had first been necessary.

Genetic heterogeneity   The production of similar phenotypes by mutations in different genes.

Genital disc   Region of the *Drosophila* larva that will generate male or female genitalia. Male and female genitalia are derived from separate cell populations of the genital disc, as induced by paracrine factors.

Genital ridge   A thickening of the splanchnic mesoderm (**germinal epithelium**) and of the underlying intermediate mesodermal mesenchyme on the medial edge of the mesonephros; it forms the testis or ovary. Also called the germinal ridge.

Genital tubercle   A structure cranial to the cloacal membrane during the indifferent stage of differentiation of the mammalian external genitalia. It will form either the clitoris in the female fetus or the glans penis in the male.

Genome   The complete DNA sequence of an individual organism.

Genomic equivalence   The theory that every cell of an organism has the same genome as every other cell.

Genomic imprinting   A phenomenon in mammals whereby only the sperm-derived or only the egg-derived allele of the gene is expressed, sometimes due to inactivation of one allele by DNA methylation during spermatogenesis or oogenesis.

Germ band   A collection of cells along the ventral midline of the *Drosophila* embryo that forms during gastrulation by convergence and extension of the surface ectoderm and will form the trunk of the embryo and the thorax and abdomen of the adult.

Germ cells   A group of cells set aside from the somatic cells that form the rest of the embryo for reproductive function. Consists of the cells of the gonads (ovary and testis) that undergo meiotic cell divisions to generate the gametes.

Germinal crescent   A region in the anterior portion of the avian and reptilian blastoderm area pellucida containing the hypoblasts displaced by migrating endodermal cells. It contains the precursors of the germ cells, which later migrate through the blood vessels to the gonads.

Germinal epithelium   Epithelium of the bipotential gonad, derived from splanchnic mesoderm, that will form the somatic (i.e., non-germ cell) component of the gonads.

Germinal neuroepithelium   A layer of rapidly dividing neural stem cells one cell layer thick that constitute the original neural tube.

Germinal vesicle breakdown (GVBD)   Disintegration of the oocyte nuclear membrane (germinal vesicle) upon resumption of meiosis during oogenesis.

Germ layer   One of the three layers of the vertebrate embryo, ectoderm, mesoderm, and endoderm, generated by the process of gastrulation, that will form all of the tissues of the body except for the germ cells.

Germline stem cells   The *Drosophila* pole cell derivatives that divide asymmetrically to produce another stem cell and a differentiated daughter cell called a cystoblast which in turn produces a single ovum and 15 nurse cells.

Germ plasm   Cytoplasmic determinants (mRNA and proteins) in the eggs of some species, including frogs, nematodes, and flies, that autonomously specify the primordial germ cells.

Germ plasm theory   A model of cell specification proposed by Weismann, in which each cell of the embryo would develop autonomously. Instead of dividing equally, the chromosomes were hypothesized to divide in such a way that different chromosomal determinants entered different cells. Only the nuclei in those cells destined to become germ cells (gametes) were postulated to contain all the different types of determinants. The nuclei of all other cells would have only a subset of the original determinants.

Glia   Supportive cells of the central nervous system, derived from neuroepithelial cells.

Glial-derived neurotrophic factor (GDNF)   A paracrine factor that binds to the Ret receptor tyrosine kinase. It is produced by the gut mesenchyme that attracts vagal and sacral neural crest cells, and it is produced by the metanephrogenic mesenchyme to inducethe formation and branching of the ureteric buds.

Glial guidance   A mechanism important for positioning young neurons in the developing mammalian brain (e.g., the granule neuron precursors travel on the long processes of the Bergmann glia in the cerebellum).

Glycogen synthase kinase 3 (GSK3)   Targets β-catenin for destruction.

Glycosaminoglycans (GAGs)   Complex acidic polysaccharides consisting of unbranched chains assembled from many repeats of a two-sugar unit. The carbohydrate component of proteoglycans.

Gonadotropin-releasing hormone (GRH; GnRH)   Peptide hormone released from the hypothalamus that stimulates the pituitary to release the gonadotropins follicle-stimulating hormone and luteinizing hormone, which are required for mammalian gametogenesis and steroidogenesis.

Gonocytes   Mammalian PGCs that have arrived at the genital ridge of a male embryo and have become incorporated into the sex cords.

G protein   A protein that binds GTP and is activated or inactivated by GTP

modifying enzymes (such as GTPases.) They play important roles in the RTK pathway and in cytoskeletal maintenance.

Granule neurons   Derived from neuroblasts of the external granule layer of the developing cerebellum. Granule neurons migrate back toward the ventricular (ependymal) zone, where they produce a region called the internal granule layer.

Granulosa cells   Cortical epithelial cells of the fetal ovary, granulosa cells surround individual germ cells that will become the ova and will form, with thecal cells, the follicles that envelop the germ cells and secrete steroid hormones. The number of granulosa cells increase and form concentric layers around the oocyte as the oocyte matures prior to ovulation.

Gray crescent   A band of inner gray cytoplasm that appears following a rotation of the cortical cytoplasm with respect to the internal cytoplasm in the marginal region of the 1-cell amphibian embryo. Gastrulation starts in this location.

Green fluorescent protein (GFP)   A protein that occurs naturally in certain jellyfish. It emits bright green fluorescence when exposed to ultraviolet light. The *GFP* gene is widely used as a transgenic label for cells in developmental and other research, since cells that express GFP are easily identified by a bright green glow.

Growth and differentiation factors (GDFs)   *See* Paracrine factors.

Growth cone   The motile tip of a neuronal axon; leads nerve outgrowth.

Growth factor   A secreted protein that binds to a receptor and initiates signals to promote or retard cell division and growth.

Growth plate closure   Causes the cessation of bone growth at the end of puberty. High levels of estrogen induce apoptosis in the hypertrophic chondrocytes and stimulate the invasion of bone-forming osteoblasts into the growth plate.

Gynandromorph   Greek *gynos*, "female"; *andros*, "male." An animal in which some body parts are male and others are female. Compare with **hermaphrodite**.

H

Halteres   A pair of balancers on the third thoracic segment in *Drosophila*.

Haptotaxis   Migration on preferred substrates.

Hatched blastula   Free-swimming sea urchin embryo, after the cells of the animal hemisphere synthesize and secrete a hatching enzyme that digests the fertilization envelope.

Head activation gradient   A morphogenetic gradient in *Hydra* that is highest at the hypostome and permits the head only to form in one place.

Head mesoderm   Mesoderm located anterior to the trunk mesoderm, consisting of the unsegmented paraxial mesoderm and prechordal mesoderm. This region provides the head mesenchyme that forms much of the connective tissues and musculature of the face and eyes.

Head process   *See* Chordamesoderm.

Heart field   *See* Cardiogenic mesoderm.

Hedgehog   A family of paracrine factors used by the embryo to induce particular cell types and to create boundaries between tissues. Hedgehog proteins must become complexed with a molecule of cholesterol in order to function. Vertebrates have at least three homologues of the *Drosophila hedgehog* gene: *sonic hedgehog* (*shh*), *desert hedgehog* (*dhh*), and *indian hedgehog* (*ihh*).

Hedgehog pathway   Proteins activated by the binding of a Hedgehog protein to the Patched receptor. When Hedgehog binds to Patched, the Patched protein's shape is altered such that it no longer inhibits Smoothened. Smoothened acts to release the Ci protein from the microtubules and to prevent its being cleaved. The intact Ci protein can now enter the nucleus, where it acts as a transcriptional *activator* of the same genes it used to repress.

Hemangioblasts   Greek "blood vessel" plus "blast," a rapidly dividing cell, usually a stem cell. These are the stem cells that form blood vessels and blood cells.

Hematopoesis   The generation of blood cells.

Hematopoietic inductive microenvironments (HIMs)   Cell regions that induce different sets of transcription factors in multipipotent hematopoietic stem cells, and these transcription factors specify the developmental path taken by the descendents of those cells.

Hematopoietic stem cell (HSC)   A multipotent stem cell type that generates a series of intermediate progenitor cells whose potency is restricted to certain blood cell lineages. These lineages are then capable of producing all the blood cells and lymphocytes of the body.

Hemimetabolous   A form of insect metamorphosis that includes pronymph, nymph, and imago stages.

Hemogenic endothelial cell   Primary endothelial cells of the dorsal aorta, especially those in the ventral area, that are derived from angioblasts that migrated from the sclerotome. They give rise to the hematopoietic stem cells that migrate to the liver and bone marrow and become the adult hematopoietic stem cells.

Hensen's node (primitive knot)   A regional thickening of cells at the anterior end of the primitive streak. The center of Hensen's node contains a funnel-shaped depression (sometimes called the primitive pit) through which cells can enter the embryo to form the notochord and prechordal plate. Hensen's node is the functional equivalent of the dorsal lip of the amphibian blastopore (i.e., Spemann's organizer) and the fish embryonic shield.

Hepatic diverticulum   The liver precursor, a bud of endoderm that extends out from the foregut into the surrounding mesenchyme.

Hepatocyte growth factor (HGF)   A paracrine factor secreted by the stellate cells of the liver that allows the hepatocytes to re-enter the cell cycle during compensatory regeneration. Also called scatter factor.

Hermaphrodite   An individual in which both ovarian and testicular tissues exists, having either ovotestes (gonads containing both ovarian and testicular tissue) or an ovary on one side and a testis on the other. Compare with **gynandromorph**.

Heterochromatin   Chromatin that remains condensed throughout most of the cell cycle and replicates later than most of the other chromatin. Usually transcriptionally inactive.

Heterochrony   The phenomenon wherein animals change the relative time of appearance and rate of development of characters present in their ancestors.

Heterogeneous causation   A principle of the theoretical systems approach: Causation is seen as being both "upward," from the genes to the environment, and "downward," from the environment to the genes.

Heterochrony   Greek, "different time." A shift in the relative timing of two developmental processes as a mechanism to generate phenotypic variation available for natural selection. One module changes its time of expression or growth rate relative to the other modules of the embryo.

Heterometry   Greek, "different measure." A change in the amount of a gene product as a mechanism to generate phenotypic variation available for natural selection.

Heterotopy   Greek, "different place." The spatial alteration of gene expression (e.g., transcription factors or paracrine factors) as a mechanism to generate phenotypic variation available for natural selection.

Heterotypy   Greek, "different kind." The alteration of the actual coding region of the gene, changing the functional properties of the protein being synthesized, as a mechanism to generate phenotypic variation available for natural selection.

Histoblast nests   Clusters of imaginal cells that will form the adult abdomen in holometabolous insects.

Histone   Positively charged proteins that are the major protein component of chromatin. See also **nucleosome**.

Histone acetylation   The addition of negatively charged acetyl groups to histones which neutralizes the basic charge of lysine and loosens the histones, and thus activates transcription.

Histone acetyltransferases   Enzymes that place acetyl groups on histones (especially on lysines in histones H3 and H4), destabilizing the nucleosomes so that they come apart easily, thus facilitating transcription.

Histone deacetylases   Enzymes that remove acetyl groups, stabilize the nucleosomes, and prevent transcription.

Histone methylation   The addition of methyl groups to histones. Can either activate or further repress transcription, depending on the amino acid that is methylated and the presence of other methyl or acetyl groups in the vicinity.

Histone methyltransferases   Enzymes that add methyl groups to histones and either activate or repress thanscription.

Holoblastic   Greek *holos*, "complete." Refers to a cell division (cleavage) pattern in the embryo in which the entire egg is divided into smaller cells, as it is in frogs and mammals.

Holometabolous   The type of insect metamorphosis found in flies, beetles, moths, and butterflies. There is no pronymph stage. The insect hatches as a larva (a caterpillar, grub, or maggot) and progresses through instar stages as it gets bigger between larval molts, a metamorphic molt to become a pupa, an imaginal molt and finally the emergence (eclose) of the adult (imago).

Homeorhesis   How the organism stabilizes its different cell lineages while it is still constructing itself.

Homeostasis   Maintenance of a stable physiological state by means of feedback responses.

Homeotic complex (Hom-C)   The region of *Drosophila* chromosome 3 containing both the Antennapedia complex and the bithorax complex.

Homeotic selector genes   A class of *Drosophila* genes regulated by the protein products of the gap, pair-rule, and segment polarity genes whose transcription determines the developmental fate of each segment.

Homeotic mutants   Result from mutations of homeotic selector genes, in which one structure is replaced by another (as where an antenna is replaced by a leg).

Homodimer   Two identical protein molecules bound together.

Homologous   Structures and/or their respective components whose similarity arises from their being derived from a common ancestral structure. For example, the wing of a bird and the forelimb of a human. Compare with **analogous**.

Horizontal transmission   When a host that is born free of symbionts but subsequently becomes infected, either by its environment or by other members of the species. Can also refer to the transfer of genes from one bacterium to another. Compare with **vertical transmission**.

Host   The larger organism in a symbiotic relationship in which one of the organisms involved is much larger than the other, and the smaller organism may live on the surface or inside the body of the larger.

Hox genes   Abbreviation of "homeobox" genes. Large family of related genes that dictate (at least in part) regional identity in the embryo, particularly along the anterior-posterior axis. Hox genes encode transcription factors that regulate the expression of other genes. All known mammalian genomes contain four copies of the Hox complex per haploid set, located on four different chromosomes (*Hoxa* through *Hoxd* in the mouse, *HOXA* through *HOXD* in humans). The mammalian Hox/HOX genes are numbered from 1 to 13, starting from that end of each complex that is expressed most anteriorly.

Hub   A regulatory microenvironment in *Drosophila* testes where the stem cells for sperm reside.

Hyaline layer   A coating around the sea urchin egg formed by the cortical granule protein hyalin. The hyaline layer provides support for the blastomeres during cleavage.

Hydatidiform mole   A human tumor which resembles placental tissue, arise when a haploid sperm fertilizes an egg in which the female pronucleus is absent and the entire genome is derived from the sperm which pre-

cludes normal development and is cited as evidence for genomic imprinting.

20-hydroxyecdysone (20E)   An insect hormone that initiates and coordinates each molt, regulates the changes in gene expression that occur during metamorphosis, and signals imaginal disc differentiation.

Hyperactivation   Describes the type of motility displayed by capacitated sperm of some mammalian species. Hyperactivation has been proposed to help detach capacitated sperm from the oviductal epithelium, allow sperm to travel more effectively through viscous oviductal fluids, and facilitate penetration of the extracellular matrix of the cumulus cells.

Hypertrophic chondrocytes   Formed during the fourth phase of endochondral ossification, when the chondrocytes, under the influence of the transcription factor Runx2, stop dividing and increase their volume dramatically.

Hypoblast   The inner layer of the thickened margin of the epibolizing blastoderm in the gastrulating zebrafish embryo or the lower layer of the bilaminar embryonic blastoderm in birds and mammals. The hypoblast in fish (but not in birds and mammals) contains the precursors of the endoderm and mesoderm.

Hypoblast islands (primary hypoblast)   Derived from area pellucida cells of the avian blastoderm that migrate individually into the subgerminal cavity to form individual disconnected clusters containing 5–20 cells each. Does not contribute to the embryo proper.

Hypostome   A conical region of the "head" of a hydra that contains the mouth.

I

Imaginal discs   Clusters of relatively undifferentiated cells set aside to produce adult structures. Imaginal discs will form the cuticular structures of the adult, including the wings, legs, antennae, eyes, head, thorax, and genitalia in holometabolous insects.

Imaginal rudiment   Develops from the left coelomic sac of the pluteus larva and will form many of the structures of the adult sea urchin.

Imago   A winged and sexually mature adult insect.

Indirect developers   Animals for which embryonic development includes a larval stage with characteristics very different from those of the adult organism, which emerges only after a period of metamorphosis.

Induced pluripotent stem (iPS) cells   Adult mouse or human cells that have been converted to cells with the pluripotency of embryonic stem cells. Usually this is accomplished by the activation of certain transcription factors.

Inducer   Tissue that produces a signal (or signals) that induces a cellular behavior in some other tissue.

Induction   The process by which one cell population influences the development of neighboring cells via interactions at close range.

Ingression   Migration of individual cells from the surface layer into the interior of the embryo. The cells become mesenchymal (i.e., they separate from one another) and migrate independently.

Inner cell mass (ICM)   A small group of internal cells within a mammalian morula or blastocyst that will eventually develop into the embryo proper and its associated yolk sac, allantois, and amnion.

Inositol 1,4,5-trisphosphate (IP$_3$)   A second messenger generated by the phospholipase C enzyme that releases intracellular Ca^{2+} stores. Important in the initiation of both cortical granule release and sea urchin development.

Instructive interaction   A mode of inductive interaction in which a signal from the inducing cell is necessary for initiating new gene expression in the responding cell.

Insulin-like growth factors (IGFs)   Growth factors that initiate an FGF-like signal transduction cascade that interferes with the signal transduction pathways of both BMPs and Wnts. IGFs are required for the formation of the anterior neural tube, including the brain and sensory placodes of amphibians.

Integration   A principle of the theoretical systems approach: How the parts are put together and how they interact to form the whole.

Integrins   A family of receptor proteins so named because they *integrate* the extracellular and intracellular scaffolds, allowing them to work together. On the extracellular side, integrins bind to the sequence arginine-glycine-aspartate (RGD), found in several adhesive proteins in extracellular matrices, including fibronectin, vitronectin (found in the basal lamina of the eye), and laminin. On the cytoplasmic side, integrins bind to talin and α-actinin, two proteins that connect to actin microfilaments. This dual binding enables the cell to move by contracting the actin microfilaments against the fixed extracellular matrix.

Intermediate mesoderm   Mesoderm immediately lateral to the paraxial mesoderm. It forms the outer (cortical) portion of the adrenal gland and the urogenital system, consisting of the kidneys, gonads, and their associated ducts.

Intermediate progenitor cells (IPCs)   Neuroblast precursor cells of the subventricular zone derived from radial glial cells.

Intermediate spermatogonia   The first committed stem cell type of the mammalian testis, they are committed to becoming spermatozoa.

Intersex   A condition in which male and female traits are observed in the same individual.

Intracytoplasmic sperm injection (ICSI)   A method of assisted reproduction for the treatment of infertility. A sperm is directly injected into the oocyte cytoplasm, bypassing any interaction with the egg plasma membrane.

Intramembranous ossification   Bone formation directly from mesenchyme. There are three main types of intramembranous bone: sesamoid bone and periostal bone, which come from mesoderm, and dermal bone which originate from cranial neural crest-derived mesenchymal cells.

Introns   Non-protein-coding regions of DNA within a gene.

Invagination   The infolding of a region of cells, much like the indenting of a soft rubber ball when it is poked.

Involuting marginal zone (IMZ)   Cells that involute during *Xenopus* gastrulation, includes precursors of the pharyngeal endoderm, head mesoderm, notochord, somites, heart, kidney, and ventral mesoderms.

Involution   Inturning or inward movement of an expanding outer layer so that it spreads over the internal surface of the remaining external cells.

Ionophore   A compound that allows the diffusion of ions such as Ca^{2+} across lipid membranes, permitting them to traverse otherwise impermeable barriers.

Iris   A pigmented ring of muscular tissue in the eye that controls the size of the pupil and determines eye color.

Isolecithal   Greek, "equal yolk." Describes eggs with sparse, equally distributed yolk particles, as in sea urchins, mammals, and snails.

Isthmus   The narrow segment of the mammalian oviduct adjacent to the uterus.

J

Jagged protein   Ligand for Notch, participates in juxtacrine interaction and activation of the Notch pathway.

JAK   *J*anus *k*inase proteins. Linked to FGF receptors in the JAK-STAT cascade.

JAK-STAT cascade   A pathway activated by paracrine factors binding to receptors that are linked to JAK (Janus kinase) proteins. The binding of ligand to the receptor phosphorylates the STAT (signal *t*ransducers and *a*ctivators of *t*ranscription) family of transcription factors.

Juvenile hormone (JH)   A lipid hormone in insects that prevents the ecdysone-induced changes in gene expression that are necessary for metamorphosis. Thus, its presence during a molt ensures that the result of that molt is another larval instar, not a pupa or an adult.

Juxtacrine interactions   When cell membrane proteins on one cell surface interact with receptor proteins on adjacent (juxtaposed) cell surfaces.

K

Kairomones   Chemicals that are released by a predator and can induce defenses in its prey.

Karyokinesis   The mitotic division of the cell's nucleus. The mechanical agent of karyokinesis is the mitotic spindle.

Keratinocytes   Differentiated epidermal cells that are bound tightly together and produce a water-impermeable seal of lipid and protein.

Kit RTK   Cell-surface receptor tyrosine kinase for *stem cell factor* ligand. Binding of stem cell factor to kit inhibits apoptosis and promotes proliferation.

Koller's sickle   *See* Primitive streak.

Kupffer's vesicle   Transient fluid-filled organ housing the cilia that control left-right asymmetry in zebrafish.

L

Labioscrotal folds   Folds surrounding the cloacal membrane in the indifferent stage of differentiation of mammalian external genitalia. They will form the labia in the female and the scrotum in the male. Also called urethral folds or genital swellings.

***lacZ* gene**   The *E. coli* gene for β-galactosidase; commonly used as a reporter gene.

Laminin   A large glycoprotein and major component of the basal lamina, plays a role in assembling the extracellular matrix, promoting cell adhesion and growth, changing cell shape, and permitting cell migration.

Lampbrush chromosomes   Chromosomes during the diplotene stage of amphibian oocyte meiosis that stretch out large loops of DNA that represent sites of upregulated RNA synthesis.

Large micromeres   A tier of cells produced by the fifth cleavage in the sea urchin embryo when the micromeres divide.

Larva   The sexually immature stage of an organism, often of significantly different appearance than the adult and frequently the stage that lives the longest and is used for feeding or dispersal.

Laryngotracheal groove   An outpouching of endodermal epithelium in the center of the pharyngeal floor, between the fourth pair of pharyngeal pouches, that extends ventrally. The laryngotracheal groove then bifurcates into the branches that form the paired bronchi and lungs.

Lateral plate mesoderm   Mesodermal sheet lateral to the intermediate mesoderm. Gives rise to appendicular bones, connective tissues of the limb buds, circulatory system (heart, blood vessels, and blood cells), muscles and connective tissues of the digestive and respiratory tracts, and lining of coelom and its derivatives. It also helps form a series of extraembryonic membranes that are important for transporting nutrients to the embryo.

Lateral somitic frontier   The boundary between the primaxial and abaxial muscles and between the somite-derived and lateral plate-derived dermis.

Lens placode   Paired epidermal thickenings induced by the underlying optic cups that invaginates to form the transparent lens that allows light to impinge on the retina.

Level-specific properties and emergence   A principle of the theoretical systems approach: The properties of a system at any given level of organization cannot be totally explained by those of levels "below" it.

Leydig cells   Testis cells derived from the interstitial mesenchyme cells surrounding the testis cords that make the testosterone required for secondary sex determination and, in the adult, required to support spermatogenesis.

Life expectancy   The length of time an average individual of a given species can expect to live which is characteristic populations, not of species. It is usually defined as the age at which half the population still survives.

Limb bud   A circular bulge that will form the future limb. The limb bud is formed by the proliferation of mesenchyme cells from the somatic layer of the limb field lateral plate mesoderm (the limb *skeletal* precursor cells) and from the somites (the limb *muscle* precursor cells).

Limb field   An area of the embryo containing all of the cells capable of forming a limb.

Lineage-restricted stem cells   Stem cells derived from multipotent stem cells, and which can now generate only a particular cell type or set of cell types.

Loss-of-function evidence (negative inference evidence)   The absence of the postulated cause is associated with the absence of the postulated effect. While stronger than correlative evidence, loss-of-function evidence still does not exclude other explanations.

Luteinizing hormone (LH)   A peptide hormone secreted by the mammalian pituitary that stimulates the production of steroid hormones, such as estrogen from the ovarian follicle cells and testosterone from the testicular leydig cells. A surge in LH levels causes the primary oocyte to complete meiosis I and prepares the follicle for ovulation.

M

Macromeres   Larger cells generated by asymmetrical cleavage, e.g., the four large cells generated by the fourth cleavage when the vegetal tier of the sea urchin embryo undergoes an unequal equatorial cleavage.

Male pronucleus   The haploid nucleus of the sperm.

Malformation   Abnormalities caused by genetic events such as gene mutations, chromosomal aneuploidies, and translocations.

Mantle (intermediate) zone   Second layer of the developing spinal cord and medulla that forms around the original neural tube. Because it contains neuronal cell bodies and has a grayish appearance grossly, it will form the gray matter.

Marginal zone   (1) The third and outer zone of the developing spinal cord and medulla composed of a cell-poor region composed of axons extending from neurons residing in the mantle zone. Will form the white matter as glial cells cover the axons with myelin sheaths, which have a whitish appearance. (2) In amphibians: Where gastrulation begins, the region surrounding the equator of the blastula, where the animal and vegetal hemispheres meet. (3) In birds and reptiles (= marginal belt), a thin layer of cells between the area pellucida and the area opaca is a thin layer of cells, important in determining cell fate during early chick development.

Maternal effect genes   Encode messenger RNAs that are placed in different regions of the *Drosophila* egg.

Medial hinge point (MHP)   Derived from the portion of the avian and mammalian neural plate just anterior to Hensen's node and from the anterior midline of Hensen's node. MHP cells become anchored to the notochord beneath them and form a hinge, which forms a furrow at the dorsal midline.

Medullary cord   Forms by condensation of mesenchyme cells and then mesenchymal-to-epithelial transition in the caudal region of the embryo during the process of secondary neurulation. It will then cavitate to form the caudal section of the neural tube.

Meiosis   A unique division process occurring only in germ cells, to reduce the number of chromosomes to a haploid complement. All other cells divide by mitosis. Meiosis differs from mitosis in that (1) meiotic cells undergo two cell divisions without an intervening period of DNA replication, and (2) homologous chromosomes (each consisting of two sister chromatids joined at a kinetochore) pair together and recombine genetic material.

Melanocytes   Cells containing the pigment melanin. Derived from neural crest cells and undergo extensive migration to all regions of the epidermis.

Melanocyte stem cell   Adult stem cell derived from melanocyte trunk neural crest cells that resides in the bulge niche of the hair or feather follicle and which gives rise to the pigment of the skin, hair, and feathers.

Meroblastic   Greek *meros*, "part." Refers to the cell division (cleavage) pattern in zygotes containing large amounts of yolk, wherein only a portion of the cytoplasm is cleaved. The cleavage furrow does not penetrate the yolky portion of the cytoplasm because the yolk platelets impede membrane formation there. Only part of the egg is destined to become the embryo, while the other portion—the yolk—serves as nutrition for the embryo, as in insects, fish, reptiles, and birds.

Meroistic oogenesis   Type of oogenesis found in certain insects (including *Drosophila* and moths), in which cytoplasmic connections remain between the cells produced by the oogonium.

Mesencephalon   The midbrain, the middle vesicle of the developing vertebrate brain; major derivatives include optic tectum and tegmentum. Its lumen becomes the cerebral aqueduct.

Mesenchymal stem cells (MSCs)   Also called bone marrow-derived stem cells, or BMDCs. Multipotent stem cells that originate in the bone marrow, MSCs are able to give rise to numerous bone, cartilage, muscle, and fat lineages.

Mesenchyme   Loosely organized embryonic connective tissue consisting of scattered fibroblast-like and sometimes migratory mesenchymal cells separated by large amounts of extracellular matrix.

Mesentoblasts   In snail embryos, the cells derived from the 4d blastomere that give rise to both the mesodermal (heart) and endodermal (intestine) organs.

Mesoderm   Greek *mesos*, "between." The middle of the three embryonic germ layers, lying between the ectoderm and the endoderm. The mesoderm gives rise to muscles and skeleton; connective tissue; the reproductive organs; and to kidneys, blood, and most of the cardiovascular tissue.

Mesodermal mantle   The cells that involute through the ventral and lateral blastopore lips during amphibian gastrulation and will form the heart, kidneys, bones, and parts of several other organs.

Mesomeres   The eight cells generated in the sea urchin embryo by the fourth cleavage when the four cells of the animal tier divide meridionally into eight blastomeres, each with the same volume.

Mesonephros   The second kidney of the amniote embryo, induced in the adjacent mesenchyme by the middle portion of the nephric duct. It functions briefly in urine filtration in some mammalian species and mesonephric tubules form the tubes that transport the sperm from the testes to the urethra (the epididymis and vas deferens).

Mesonephric duct   *See* Wolffian duct.

Metalloproteinases   Matrix metalloproteinases (MMP) Enzymes that digest extracellular matrices and are important in many types of tissue remodeling in disease and development, including metastasis, branching morphogenesis of epithelial organs, placental detachment at birth, and arthritis.

Metamorphosis   Changing from one form to another, such as the transformation of an insect larva to a sexually mature adult or a tadpole to a frog.

Metanephrogenic mesenchyme   An area of mesenchyme, derived from posterior regions of the intermediate mesoderm, involved in mesenchymal-epithelial interactions that generate the metanephric kidney and will form the secretory nephrons.

Metanephros/metanephric kidney The third kidney of the embryo and the permanent kidney of amniotes.

Metaphase plate A structure present during mitosis or meiosis in which the chromosomes are attached via their kinetochores to the microtubule spindle and are lined up midway between the two poles of the cell.

Metastasis The invasion of a malignant cell into other tissues.

Metencephalon The anterior subdivision of the rhombencephalon; gives rise to the cerebellum, which coordinates movements, posture, and balance.

Microfilaments Long cables of polymerized actin necessary for cytokinesis and also formed during fertilization in the egg's cortex where they are used to form microvilli.

Micromeres Small cells created by asymmetrical cleavage, e.g., four small cells generated by the fourth cleavage at the vegetal pole when the vegetal tier of the sea urchin embryo undergoes an unequal equatorial cleavage.

Micropyle The only place where *Drosophila* sperm can enter the egg, at the future dorsal anterior region of the embryo, a tunnel in the chorion (eggshell) that allows sperm to pass through it one at a time.

MicroRNA (miRNA) A small (about 22 nucleotide) RNA complementary to a portion of a particular mRNA that regulates translation of a specific message. MicroRNAs often bind to the 3′UTR of mRNAs.

Microspikes Essential for neuronal pathfinding, microfilament-containing pointed filopodia of the growth cone that elongate and contract to allow axonal migration. Microspikes also sample the microenvironment and sends signals back to the soma.

Microvilli Small projections that extend from the egg surface during fertilization that may aid sperm entry into the cell.

Mid-blastula transition The transition from the early rapid biphasic (only M and S phases) mitoses of the embryo to a stage characterized by (1) mitoses that include the "gap" stages (G1 and G2) of the cell cycle, (2) loss of syncronicity of cell division, and (3) transcription of new (zygotic) mRNAs needed for gastrulation and cell specification.

Midpiece Section of sperm flagellum near the head that contains rings of mitochondria that provide the ATP needed to fuel the dynein ATPases and support sperm motility.

Mitosis-promoting factor (MPF) Consists of cyclin B and p34/Cdk1 cyclin, required to initiate entry into the mitotic (M) phase of the cell cycle in both meiosis and mitosis.

mRNA cytoplasmic localization The spatial regulation of mRNA translation, mediated by (1) diffusion and local anchoring, (2) localized protection, and (3) active transport along the cytoskeleton.

Model systems Species that are easily studied in the laboratory and have special properties that allow their mechanisms of development to be readily observed (e.g., sea urchins, snails, ascidians, and *C. elegans*).

Modularity A principle of the theoretical systems approach. The organism develops as a system of discrete and interacting modules.

Module A biological entity characterized by more internal than external integration.

Molecular parsimony (the "small toolkit") Development within all lineages uses the same types of molecules: transcription factors, paracrine factors, adhesion molecules, and signal transduction cascades are remarkably similar from one phylum to another.

Monospermy Only one sperm enters the egg, and a haploid sperm nucleus and a haploid egg nucleus combine to form the diploid nucleus of the fertilized egg (zygote), thus restoring the chromosome number appropriate for the species.

Monozygotic Greek, "one-egg." Describes twins that form from a single embryo whose cells become dissociated from one another, either by the separation of early blastomeres, or by the separation of the inner cell mass into two regions within the same blastocyst. "Identical" twins. Compare with **dizygotic**.

Morph One of several different potential phenotypes produced by environmental conditions.

Morphallaxis Type of regeneration that occurs through the repatterning of existing tissues with little new growth (e.g., *Hydra*).

Morphogens Greek, "form-givers." Substances that, by their differing concentrations, differentially specify cell fates. Morphogens are made in specific sites in the embryo, diffuse over long distances, and form concentration gradients where the highest concentration is at the point of synthesis, becoming lower as the morphogen diffuses away from its source and degrades over time.

Morphogenesis The organization of the cells of the body into functional structures via coordinated cell growth, cell migration, and cell death.

Morphogenetic determinants Transcription factors or their mRNAs that will influence the cell's development.

Morpholino An antisense oligonucleotide against an mRNA used to inhibit protein expression.

Morula Latin, "mulberry." Vertebrate embryo of 16–64 cells;, precedes the blastula or blastocyst stage. Mammalian morula occurs at the 16-cell stage, consists of a small group of internal cells (the inner cell mass) surrounded by a larger group of external (trophoblast) cells.

Mosaic embryos Embryos in which most of the cells are determined by autonomous specification, with each cell receiving its instructions independently and without cell-cell interaction.

Mosaic pleiotropy A syndrome characterized by the expression of multiple, independently produced effects resulting from a gene being critical in different parts of the body.

Müllerian duct (paramesonephric duct) Duct running lateral to the mesonephric duct in both male and female mammalian embryos. These ducts regress in the male fetus, but form the oviducts, uterus, cervix, and upper part of the vagina in the female fetus. See also **Wolffian duct**.

Multipotent cardiac progenitor cells Progenitor cells of the heart field that form cardiomyocytes, endocardium, epicardium, and the Purkinje neural fibers of the heart.

Multipotent stem cells Adult stem cells whose commitment is limited to a relatively small subset of all the possible cells of the body.

Mutualism A form of symbiosis in which the relationship benefits both partners.

Myelencephalon The posterior subdivision of the rhombencephalon; becomes the medulla oblongata.

Myelin sheath Modified oligodendrocyte or schwann cell plasma membrane that surrounds nerve axons, providing insulation that confines electrical impulses transmitted along axons, and is necessary for normal neural transmission.

Myoblast Muscle precursor cell.

Myocardium Heart muscles.

Myogenic regulatory factors (MRFs) bHLH transcription factors (such as MyoD and Myf5, that are critical regulators of muscle development.

Myostatin (Greek, "muscle stopper") A member of the TGF-β family of paracrine factors that negatively regulates muscle development. Genetic defects in the gene or its negative regulatory miRNA cause huge muscles in some mammals, including humans.

Myotome   Portion of the somite that gives rise to skeletal muscles. The myotome has two components: the primaxial component, closest to the neural tube, which forms the musculature of the back and rib cage, and the abaxial component, away from the neural tube, which forms the muscles of the ventral body wall.

N

NAD⁺ kinase   Activated during the early response of the sea urchin egg to the sperm, converts NAD⁺ to NADP⁺ which can be used as a coenzyme for lipid biosynthesis and may be important in the construction of the many new cell membranes required during cleavage. NADP⁺ is also used to make NAADP.

Nanos   Protein critical for the establishment of anterior-posterior polarity of the *Drosophila* embryo. *Nanos* mRNA is localized to the posterior pole.

Negative inference evidence   *See* Loss-of-function evidence.

Neoteny   Retention of the juvenile form as a result of retarded body development relative to the development of the germ cells and gonads (which achieve maturity at the normal time). See also **progenesis**.

Nephric duct   *See* Wolffian duct.

Nephron   Functional unit of the kidney.

Nerve growth factor (NGF)   Neurotrophin released from potential target tissues that works at short ranges as either a chemotactic factor or chemorepulsive factor for axonal guidance. Also important in the selective survival of different subsets of neurons.

Netrin-1, Netrin-2   Paracrine factors found in a gradient that guide axonal growth cones. They are important in commissural axon migration and retinal axon migration.

Neural crest   A transient band of cells, arising from the lateral edges of the neural plate, that joins the neural tube to the epidermis. It gives rise to a cell population—the neural crest cells—that detach during formation of the neural tube and migrate to form a variety of cell types and structures, including sensory neurons, enteric neurons, glia, pigment cells, and (in the head) bone and cartilage.

Neural folds   Thickened edges of the neural plate that move upward during neurulation and migrate toward the midline and eventually fuse to form the neural tube.

Neural groove   U-shaped groove that forms in the center of the neural plate during primary neurulation.

Neural plate   The region of the dorsal ectoderm that is specified to become neural ectoderm. The cells of this region have a columnar appearance.

Neural restrictive silencer element (NRSE)   A regulatory DNA sequence found in several mouse genes which prevents their expression in adult neurons.

Neural restrictive silencer factor (NRSF)   A zinc finger transcription factor that binds the NRSE and is expressed in every cell that is *not* a mature neuron.

Neural retina   Derived from the inner layer of the optic cup, composed of a layered array of cells that include the light- and color-sensitive photoreceptor cells (rods and cones); the cell bodies of the ganglion cells; bipolar interneurons that transmit electric stimuli from the rods and cones to the ganglion cells, Müller glial cells that maintain its integrity, amacrine neurons (which lack large axons), and horizontal neurons that transmit electric impulses in the plane of the retina.

Neural tube   The embryonic precursor to the central nervous system (brain and spinal cord).

Neuroblast   A neural precursor cell.

Neuron   Nerve cell; neurons are derived from neuroepithelial cells.

Neuropore   The two open ends (anterior neuropore and posterior neuropore) of the neural tube that later close.

Neurotransmitters   Chemical messengers molecules secreted at the ends of axons that cross the synaptic cleft and are received by the adjacent neuron, thus relaying the neural signal.

Neurotrophin   Greek *trophikos*, "nourish." Neurotropins supply factors (usually growth factors) that keep the neuron alive.

Neurotropin   Greek *tropikos*, "turn." Chemoattractant of neurons.

Neurula   Refers to an embryo during neurulation (i.e., while the neural tube is forming).

Neurulation   Process of folding of the neural plate and closing of the cranial and caudal neuropores to form the neural tube.

Neural crest effectors   Transcription factors activated by neural crest specifiers that give the neural crest cells their migratory properties and some of their differentiated properties.

Neural crest specifiers   A set of transcription factors (e.g., FoxD3, Sox9, Id, Twist, and Snail) induced by the border-specifying transcription factors,

that specify the cells that are to become the neural crest.

Neural plate border specifiers   A set of transcription factors induced by the neural plate inductive signals. These factors, including Distalless-5, Pax3, and Pax7, collectively prevent the border region from becoming either neural plate or epidermis.

Neural plate inductive signals   BMPs and Wnts secreted from the ventral ectoderm and paraxial mesoderm interact to specify the boundaries between neural and non-neural ectoderm during chick gastrulation. Relative concentrations of the same signals also specify neural crest and cranial placode cells.

Nieuwkoop center   The dorsalmost vegetal blastomeres of the amphibian blastula, formed as a consequence of the cortical rotation initiated by the sperm entry; an important signaling center on the dorsal side of the embryo. One of its main functions is to induce Spemann's Organizer.

Nodal   A paracrine factor and member of the transforming growth factor-β (TGF-β) family involved in establishing left-right asymmetry in vertebrates and invertebrates.

Nodal vesicular parcels (NVPs)   Small, membrane-bound particles in the mammalian embryo that contain Sonic hedgehog protein and retinoic acid. Secreted from the node cells under the influence of FGF signals, Ciliary flow carries the NVPs to the left side of the body, delivering paracrine factors for the establishment of left-right asymmetry.

Node   The mammalian homologue of Hensen's node.

Noggin   A soluble BMP antagonist that blocks BMP signaling.

Noninvoluting marginal zone (NIMZ)   Region of cells on the exterior of the gastrulating amphibian embryo that expand by epiboly to cover the entire embryo, eventually forming the surface ectoderm.

Non-skeletogenic mesenchyme   Formed from the veg2 layer of the 60-cell sea urchin embryo, it generates pigment cells, immunocytes, and muscle cells. Also called secondary mesenchyme.

Notch protein   Receptor for Delta, Jagged, or Serrate, participants in juxtacrine interactions. Ligand binding causes Notch to undergo a conformational change that enables a part of its cytoplasmic domain to be cut off by the Presenilin-1 protease. The cleaved portion enters the nucleus and binds to a dormant transcription factor of the CSL family. When bound to the Notch protein, the CSL transcription factors activate their target.

Notochord A transient mesodermal rod in the most dorsal portion of the embryo that plays an important role in inducing and patterning the nervous system.

Nuclear RNA (nRNA) The original transcription product, sometimes called *heterogeneous nuclear RNA* (hnRNA) or *pre-messenger RNA* (pre-mRNA): contains the cap sequence, the 5′ UTR, exons, introns, and the 3′ UTR.

Nuclear RNA selection Means of controlling gene expression by processing specific subsets of the nRNA population into mRNA in different types of cells.

Nucleus (1) An organized cluster of neurons in the brain with specific functions and connections. (2) The membrane-enclosed organelle housing the eukaryotic chromosomes.

Nucleosome The basic unit of chromatin structure, composed of an octamer of histone proteins (two molecules each of histones H2A, H2B, H3, and H4) wrapped with two loops containing approximately 147 base pairs of DNA.

Nurse cells Fifteen interconnected cells that generate mRNAs and proteins that are transported to the single developing oocyte during *Drosophila* oogenesis.

Nymph Insect larval stage that resembles an immature adult of the species. Becomes progressively more mature though a series of molts.

O

Obligate mutualism Symbiosis in which the species involved are interdependent with one another to such an extent that at least one partner could not survive without the other.

Olfactory placodes Paired epidermal thickenings that form the nasal epithelium (smell receptors) as well as the ganglia for the olfactory nerves.

Omphalomesenteric (umbilical) veins The veins that form from yolk sack blood islands, and that bring nutrients to the mammalian embryo and transport gases to and from the sites of respiratory exchange.

Oncogenes Regulatory genes which promote cell division, reduce cell adhesion, and prevent cell death. Can promote tumor formation and metastasis. Cancer can result from either mutations or inappropriate methylations that activate oncogenes.

Oocyte The developing egg (prior to reaching the stage of meiosis at which it is fertilized).

Oogonium A single female germ cell that will form an oocyte.

Optic cups Double-walled chambers formed by the invagination of the optic vesicles.

Optic vesicle Extend from the diencephalon and activate the head ectoderm's latent lens-forming ability.

Oral plate (stomodeum) A region of ectoderm that blocks the oral end of the primitive gut.

Organization/activation hypothesis The theory that sex hormones act during the fetal or neonatal stage of a mammal's life to organize the nervous system in a sex-specific manner; and that during adult life, the same hormones may have transitory motivational (or "activational") effects.

Organizer *See* Spemann's organizer.

Organogenesis Interactions between, and rearrangement of, cells of the three germ layers to produce tissues and organs.

Orthologues Genes that are homologous between species.

Osteoblast A committed bone precursor cell.

Osteoclasts Multinucleated cells derived from a blood cell lineage that enter the bone through the blood vessels and destroy bone tissue during remodeling.

Osteocytes Bone cells, derived from osteoblasts that become embedded in the calcified osteoid matrix.

Osteogenesis Bone formation.

Osteoid matrix A collagen-proteoglycan secreted by osteoblasts that is able to bind calcium.

Otic placodes Paired epidermal thickenings that invaginate to form the inner ear labyrinth, whose neurons form the acoustic ganglia that enable us to hear.

Ovariole The *Drosophila* egg chamber.

Ovulation Release of the oocyte from the ovarian follicle.

Ovum The mature egg (at the stage of meiosis at which it is fertilized).

P

p53 A transcription factor that can stop the cell cycle, cause cellular senescence in rapidly dividing cells, instruct the initiation of apoptosis, and activate DNA repair enzymes. One of the most important regulators of cell division.

Pair-rule genes *Drosophila* zygotic genes regulated by gap gene proteins which divide the embryo into periodic units, resulting in a striped pattern of seven transverse bands perpendicular to the anterior-posterior axis. Pair-rule mutants lacked portions of every other segment.

Parabiosis Procedure that surgically joins the circulatory systems of two experimental animals so that they share one blood supply.

Paracrine factor A secreted, diffusible protein that provides a signal that interacts with and changes the cellular behavior of neighboring cells and tissues.

Paralogues Each member of a gene family, which is homologous to the others (that is, their sequence similarities are due to descent from a common ancestor and are not the result of convergence for a particular function).

Parasegment A "transegmental" unit in *Drosophila* that includes the posterior compartment of one segment and the anterior compartment of the immediately posterior segment; appears to be the fundamental unit of embryonic gene expression.

Parasympathetic (enteric) ganglia Neural ganglia of the gut that are required for peristaltic movement of the bowels.

Paraxial (somitic) mesoderm Thick bands of embryonic mesoderm immediately adjacent to the neural tube and notochord. In the trunk, paraxial mesoderm gives rise to somites, in the head it gives rise to the connective tissues and musculature of the face.

Paraxial protocadherin Adhesion protein expressed specifically in the paraxial (somite-forming) mesoderm during amphibian gastrulation; essential for convergent extension.

Parthenogenesis Greek, "virgin birth." When an oocyte is activated in the absence of sperm. Normal development can proceed in many invertebrates and some vertebrates.

Pathway selection The first step in the specification of axonal connection, wherein the axons travel along a route that leads them to a particular region of the embryo.

Pattern formation The set of processes by which embryonic cells form ordered spatial arrangements of differentiated tissues.

Pericardial cavity The division of the coelom that surrounds the heart in mammals.

Pericytes Smooth muscle-like cells the endothelial cells recruit to cover them during vasculogenesis.

Periderm A temporary epidermis-like covering in the embryo that is shed once the inner layer differentiates to form a true epidermis.

Periosteal bone Bone which adds thickness to long bones and is derived from mesoderm via intramembranous ossification.

Periosteum   A fibrous sheath containing connective tissue, capillaries, and bone progenitor cells and that covers the developing and adult bone.

Permissive interaction   Inductive interaction in which the responding tissue has already been specified, and needs only an environment that allows the expression of these traits.

P granules   Ribonucleoprotein complexes containing translation regulators that specify the germ cells in *C. elegans* and are localized by PAR ("partitioning") proteins.

Pharyngeal arches   Also called a branchial arches, these are bars of mesenchymal tissue derived from paraxial mesoderm, lateral plate mesoderm, and neural crest cells. Found in the pharyngeal region (near the pharynx) of the vertebrate embryo, the arches will form gill supports in fishes and many skeletal and connective tissue structures in the face, jaw, mouth, and larynx in other vertebrates.

Pharyngeal clefts   Clefts (invaginations) of external ectoderm that separate the pharyngeal arches. There are four pharyngeal clefts in the early embryo, but only the first becomes a structure (the external auditory meatus).

Pharyngeal pouches   Inside the pharynx, these are where the pharyngeal epithelium pushes out laterally to form 4 pairs of pouches between the pharyngeal arches. These give rise to the auditory tube, wall of the tonsil, thymus gland, parathyroids and thyroid.

Pharyngula   Term often applied to the late neurula stage of vertebrate embryos.

Pharynx   The region of the digestive tube anterior to the point at which the respiratory tube branches off.

Phenotypic heterogeneity   Refers to the same mutation producing different phenotypes in different individuals.

Phenotypic plasticity   The ability of an organism to react to an environmental input with a change in form, state, movement, or rate of activity.

Pheromones   Vaporized chemicals emitted by an individual that results in communication with another individual. Pheromones are recognized by the vomeronasal organ of many mammalian species and play a major role in sexual behavior.

Phospholipase C (PLC)   Enzyme in the IP_3 pathway that splits membrane phospholipid phosphatidylinositol 4,5-bisphosphate (PIP2) to yield IP_3 and diacylglycerol (DAG).

Phylotypic stage   The stage that typifies a phylum, such as the late neurula or pharyngula of vertebrates, and which appears to be relatively invariant and to constrain its evolution.

Pioneer nerve fibers   Axons that go ahead of other axons and serve as guides for them.

Pioneer transcription factors   Transcription factors that can penetrate repressed chromatin and bind to their enhancer DNA sequences, a step critical to establishing certain cell lineages. Examples: FoxA1, Pax7, Pbx.

Placenta   The organ that serves as the interface between fetal and maternal circulations and has endocrine, immune, nutritive and respiratory functions. It consists of a maternal portion (the uterine endometrium, or decidua, which is modified during pregnancy) and a fetal component (the chorion).

Placodes   Precursors of cutaneous appendages, such as hair, that are formed via inductive interactions between the dermal mesenchyme and the ectodermal epithelium.

PLCξ   Phospholipase C-zeta, a soluble activator of Ca^{2+} release, initiating Ca^{2+} oscillations, stored in the mammalian sperm head and delivered to the egg by gamete fusion.

Pleiotropy   The production of several effects by one gene or pair of genes.

Pluripotent   Latin, "capable of many things." A single pluripotent stem cell has the ability to give rise to different types of cells that develop from the three germ layers (mesoderm, endoderm, ectoderm) from which all the cells of the body arise. The cells of the mammalian inner cell mass are pluripotent, as are embryonic stem cells. Each of these cells can generate any cell type in the body, but because the distinction between ICM and trophoblast has been established, it is thought that ICM cells are not able to form the trophoblast. Germ cells and germ cell tumors (such as teratocarcinomas) can also form pluripotent stem cells. Compare with **totipotent**.

Polar body   The smaller cell, containing hardly any cytoplasm, generated during the asymmetrical meiotic division of the mammalian oocyte. The first polar body is diploid and results from the first meiotic division and the secondary polar body is haploid and results from the second meiotic division.

Polar granules   Particles containing factors important for germ line specification that are localized to the pole plasm and pole cells of *Drosophila*.

Polar granule component (PGC)   A protein important for germ line specification and localized to *Drosophila* polar granules. Pgc inhibits transcription of somatic cell-determining genes by preventing the phosphorylation of RNA polymerase II.

Polarization   The first stage of cell migration, wherein a cell defines its front and its back, directed by diffusing signals (such as a chemotactic protein) or by signals from the extracellular matrix. These signals will reorganize the cytoskeleton such that the front part of the cell will form *lamellipodia* (or *filopodia*) with newly polymerized actin.

Polar lobe   An anucleate bulb of cytoplasm extruded immediately before first cleavage in certain spirally cleaving embryos (mostly in the mollusc and annelid phyla). It contains the determinants for the proper cleavage rhythm and the cleavage orientation of the D blastomere.

Pole cells   About five nuclei in the *Drosophila* embryo that reach the surface of the posterior pole during the ninth division cycle and become enclosed by cell membranes. The pole cells give rise to the gametes of the adult.

Pole plasm   Cytoplasm at the posterior pole of the *Drosophila* oocyte which contains the determinants for producing the abdomen and the germ cells.

Polyadenylation   The insertion of a "tail" of some 200–300 adenylate residues on the RNA transcript, about 20 bases downstream of the AAUAAA sequence. This polyA tail (1) confers stability on the mRNA, (2) allows the mRNA to exit the nucleus, and (3) permits the mRNA to be translated into protein.

PolyA tail   A series of adenine (A) residues that are added by enzymes to the 3′ terminus of the mRNA transcript in the nucleus. The polyA tail may stabilize the message by protecting it from exonucleases that would otherwise digest it.

Polycomb proteins   Family of proteins that bind to condensed nucleosomes, keeping the genes in an inactive state.

Polyphenism   Refers to *discontinuous* ("either/or") phenotypes elicited by the environment. Compare with **reaction norm**.

Polyspermy   The entrance of more than one sperm during fertilization resulting in aneuploidy (abnormal chromosome number) and either death or abnormal development.

Posterior group determinants   *Drosophila* pole plasm component that includes the Oskar protein which localizes the proteins and RNAs necessary for germ cell formation to the posterior pole. This group also includes the RNA-binding protein Vasa, and the transcription repressors, Piwi and Aubergine.

Posterior marginal zone (PMZ) The end of the chick blastoderm where primitive streak formation begins and acts as the equivalent of the amphibian Nieuwkoop center. The cells of the PMZ initiate gastrulation and prevent other regions of the margin from forming their own primitive streaks.

Posttranslational regulation Protein modifications that determine whether or not a protein will be active. These modifications can include cleaving an inhibitory peptide, sequestration and targeting to specific cell regions, assembly with other proteins in order to form a functional unit, binding an ion (such as Ca^{2+}), or modification by the covalent addition of a phosphate or acetate group.

Prechordal plate mesoderm Cells that migrate anteriorly through Hensen's node, between the endoderm and the epiblast, and form the precursor of the head mesoderm.

Predator-induced polyphenism The ability to modulate development in the presence of predators in order to express a more defensive phenotype.

Preformation The view, supported by the early microscopist, Marcello Malpighi, that the organs of the embryo are already present, in miniature form, within the egg (or sperm). A corollary, *embôitment* (encapsulation), stated that the next generation already existed in a prefigured state within the germ cells of the first prefigured generation, thus ensuring that the species would remain constant.

Preimplantation genetics Testing for genetic diseases using blastomeres from embryos produced by in vitro fertilization before implanting the embryo in the uterus.

Prenatal diagnosis The use of chorionic villus sampling or amniocentesis to diagnose many genetic diseases before a baby is born.

Primary capillary plexus A network of capillaries formed by endothelial cells during the third phase of vasculogenesis.

Primary embryonic induction The process whereby the dorsal axis and central nervous system forms through interactions with the underlying mesoderm, derived from the dorsal lip of the blastopore.

Primary heart field The cells of the heart field that will form the inflow tract (left ventricle and the atria).

Primary larvae Larvae that represent dramatically different body plans than the adult form and that are morphologically distinct from the adult; the plutei of sea urchins are such larvae. Compare with **secondary larvae**.

Primary neurulation The process that forms the anterior portion of the neural tube. The cells surrounding the neural plate direct the neural plate cells to proliferate, invaginate, and pinch off from the surface to form a hollow tube.

Primary oocytes Formed in the mammalian fetus from oogonia that have entered the first meiotic division and are arrested in first meiotic prophase until puberty.

Primary sex determination The determination of the gonads to form either the egg-forming ovaries or sperm-forming testes. Primary sex determination is chromosomal and is not usually influenced by the environment in mammals, but can be affected by the environment in other vertebrates.

Primary spermatocytes Derived from mitotic division of the type B spermatogonia, these are the cells that enter meiosis.

Primaxial muscles The intercostal musculature between the ribs and the deep muscles of the back, formed from those myoblasts in the myotome closest to the neural tube.

Primitive groove A depression that forms within the primitive streak that serves as an opening through which migrating cells pass into the deep layers of the embryo.

Primitive knot/pit *See* Hensen's node.

Primitive streak the first morphological sign of gastrulation in amniotes, it first arises from a local thickening of the epiblast at the posterior edge of the area pellucida, called Koller's sickle. Homologous to the amphibian blastopore.

Primordial germ cells (PGCs) Gamete progenitor cells, which typically arise elsewhere and migrate into the developing gonads.

Proacrosin A mammalian sperm protein that adheres to the inner acrosomal membrane and binds to sulfated carbohydrate groups on the zona pellucida glycoproteins.

Proerythroblast A red blood cell precursor.

Progenesis Condition in which the gonads and germ cells develop at a faster rate than than the rest of the body, becoming sexually mature while the rest of the body is still in a juvenile phase (see **neoteny**).

Progenitor cells Relatively undifferentiated cells that have the capacity to divide a few times before differentiating and, unlike stem cells, are not capable of unlimited self-renewal. They are sometimes called *transit-amplifying cells*, since they divide while migrating. Both unipotent stem cells and progenitor cells have been called *lineage-restricted cells*, but progenitor cells are usually more differentiated than stem cells.

Progesterone A steroid hormone important in the maintenance of pregnancy in mammals. Progesterone secreted from the cumulus cells may act as a chemotactic factor for sperm.

Programmed cell death *See* Apoptosis.

Progress zone The limb bud mesenchyme directly beneath the apical ectodermal ridge. The proximal-distal growth and differentiation of the limb bud are made possible by a series of interactions between the AER and the progress zone.

Progress zone model Model for specification of proximal-distal specification of the limb which postulates that each mesoderm cell is specified by the amount of time it spends dividing in the progress zone. The longer a cell spends in the progress zone, the more mitoses it achieves and the more distal its specification becomes.

Prometamorphosis The second stage of metamorphosis, during which the thyroid matures and secretes more thyroid hormones.

Promoter Region of a gene where RNA polymerase binds to the DNA to initiate transcription. Most contain the sequence TATA (the TATA box), to which RNA polymerase binds and are usually located immediately upstream from the site where the RNA polymerase initiates transcription.

Pronephric duct The pronephric duct arises in the intermediate mesoderm, migrates caudally, and induces the adjacent mesenchyme to form the pronephros, or tubules of the initial kidney of the embryo. The pronephric tubules form functioning kidneys in fish and in amphibian larvae but are not believed to be active in amniotes. In mammals, the caudal pronephric tubules become the Wolffian duct.

Prosencephalon The forebrain; the most anterior vesicle of the developing vertebrate brain, Will form two secondary brain vesicles: the telencephalon and the diencephalon.

Protamines Basic proteins, tightly compacted through disulfide bonds, that package the DNA of the sperm nucleus.

Proteoglycans Large extracellular matrix molecules consisting of core proteins (such as syndecan) with covalently attached glycosaminoglycan polysaccharide side chains. Two of the most widespread proteoglycans are heparan sulfate and chondroiton sulfate.

Proteome   The number and type of proteins encoded by the genome.

Prothoracicotropic hormone (PTTH)   A peptide hormone that initiates the molting process in insects when it is released by neurosecretory cells in the brain in response to neural, hormonal, or environmental signals. PTTH stimulates the production of ecdysone by the prothoracic gland.

Protocadherins   A class of cadherins that lack the attachment to the actin skeleton through catenins. They are an important means of keeping migrating epithelia together, and they are important in separating the notochord from surrounding tissues during its formation.

Protostomes   Greek, "mouth first." Animals that form their mouth regions from the blastopore, such as molluscs. Compare with **deuterostomes**.

Proximal-distal axis   The close-far axis of the limb, e.g., shoulder-finger or hip-toe.

Pseudohermaphroditism   Intersex conditions in which the secondary sex characteristics differ from what would be expected from the gonadal sex. Male pseudohermaphroditism (e.g., androgen insensitivity syndrome) describes conditions wherein the gonadal sex is male and the secondary sex characteristics are female while Female pseudohermaphroditism describes the reverse situation (e.g., congenital adrenal hyperplasia).

R

Radial glial cells (RGCs)   Neural progenitor cells found in the ventricular zone (VZ) of the developing brain. At each division, they generate another VZ cell and a more committed cell type that leaves the VZ to differentiate.

Radial holoblastic cleavage   Cleavage pattern in echinoderms. The cleavage planes are parallel and perpendicular to the animal-vegetal axis of the egg.

Random epigenetic drift   The hypothesis that the chance accumulation of inappropriate epigenetic methylation due to errors made by the DNA methylating and demethylating enzymes could be the critical factor in aging and cancers.

Ras   A G-protein in the RTK pathway. Mutations in the *RAS* gene account for a large proportion of cancerous human tumors.

Reaction-diffusion model   Model for developmental patterning wherein two homogeneously distributed substances (an activator that activates itself as well as forming its own, faster-diffusing inhibitor) interact to produce stable patterns during morphogenesis. These patterns would represent regional differences in the concentrations of the two substances.

Reaction norm   A type of phenotypic plasticity in which the genome encodes the potential for a *continuous range* of potential phenotypes, and the environment the individual encounters determines the phenotype. Compare with **polyphenism**.

Reactive oxygen species (ROS)   Metabolic by-products that can damage cell membranes and proteins and destroy DNA. ROS are generated by mitochondria due to insufficient reduction of oxygen atoms and include superoxide ions, hydroxyl ("free") radicals, and hydrogen peroxide.

Receptor tyrosine kinase   A receptor that spans the cell membrane and has an extracellular region, a transmembrane region, and a cytoplasmic region. Ligand (paracrine factor) binding to the extracellular domain causes a conformational change in the receptor's cytoplasmic domains, activating kinase activity that uses ATP to phosphorylate specific tyrosine residues of particular proteins.

Reciprocal inductions   A common sequential feature of induction: An induced tissue in turn induces other tissues, including the original inducer.

Regeneration   The ability to reform an adult structure or organ that has been damaged or destroyed by trauma or disease.

Regeneration blastema   An aggregation of relatively dedifferentiated cells derived from the originally differentiated tissue, which then proliferates and redifferentiates into the new limb parts during epimorphic regeneration.

Regulation   The ability to respecify cells so that the removal of cells destined to become a particular structure can be compensated for by other cells producing that structure. This is seen when an entire embryo is produced by cells thatwould have contributed only certain parts to the original embryo. It is also seen in the ability of two or more early embryos to form one chimeric individual rather than twins, triplets, or a multiheaded individual.

Relational pleiotropy   Describes syndromes in which a defective gene in one part of the embryo causes a defect in another part, even though the gene is not expressed in the second tissue.

Reporter gene   A gene with a product that is readily identifiable and not usually made in the cells of interest. Can be fused to regulatory elements from a gene of interest, inserted into embryos, and then monitored for reporter gene expression If the sequence contains an enhancer, the reporter gene should become active at particular times and places. The genes for green fluorescent protein (*GFP*) and β-galactosidase (*lacZ*) are common examples.

Resact   A 14-amino-acid peptide that has been isolated from the egg jelly of the sea urchin *Arbacia punctulata* that acts as a chemotactic factor and sperm-activating peptide for *Arbacia* sperm.

Resegmentation   Occurs during formation of the vertebrae from sclerotomes; the rostral segment of each sclerotome recombines with the caudal segment of the next anterior sclerotome to form the vertebral rudiment and this enables the muscles to coordinate the movement of the skeleton, permitting the body to move laterally.

Respiratory tube   The future respiratory tract, which forms as an epithelial outpocketing of the pharynx, and eventually bifurcates into the two lungs.

Responder   During induction, the tissue being induced. Cells of the responding tissue must have a receptor protein for the paracrine factor and have the competence to respond to it.

Rete testis   A network of thin canals located near, and eventually connecting with, the mesonephric duct that connects with the testis cords of the developing testis and conveys spermatozoa from the seminiferous tubules to the mesonephric duct derivatives, the epididymis and ductus deferens (vas deferens) in the adult.

Reticulocyte   Cell that derived from the mammalian erythroblast that has expelled its nucleus. Although reticulocytes, lacking a nucleus, can no longer synthesize globin mRNA, they can translate *existing* messages into globins.

Retina   *See* Neural retina.

Retinal homeobox (*Rx*)   Gene expressed in the central bulge region of the diencephalon; required for optic cup and retina formation.

Retinoic acid (RA)   A derivative of vitamin A and morphogen involved in anterior-posterior axis formation.

Retinoic acid-4-hydroxylase   An enzyme that degrades retinoic acid.

Retinotectal projection   The map of retinal connections to the optic tectum. Point-for-point correspondence between the cells of the retina and the cells of the tectum that enables the animal to see an unbroken image.

Rhombencephalon   The hindbrain, the most caudal vesicle of the developing vertebrate brain; will form two secondary brain vesicles, the metencephalon and myelencephalon.

Rhombomeres   Periodic swellings that divide the rhombencephalon into smaller compartments, each with a different fate and different associated cranial nerve ganglia.

Right-left axis   A line between the two lateral sides of the body.

Ring canals   The cytoplasmic interconnections between the cystocytes that become the ovum and nurse cells in *Drosophila.*

RNA-induced silencing complex (RISC)   A complex containing several proteins and a microRNA which can then bind to the 3′ UTR of messages and inhibit their translation.

Robustness   Also called canalization; a principle of the theoretical systems approach: Refers to the ability of an organism to develop the same phenotype despite perturbations from the environment or from mutations.

Roof plate   Dorsal region of the neural tube important in the establishment of dorsal-ventral polarity. The adjacent epidermis induces a secondary signaling center and BMP4 expression in the roof plate cells of the neural tube that in turn induces a cascade of TGF-β proteins in adjacent cells.

Rostral-caudal   Latin, "beak-tail." An anterior-posterior positional axis, especially when referring to chick embryos or the head and brain of mammals.

Rotational cleavage   The cleavage pattern for mammalian and nematode embryos. In mammals, the first cleavage is a normal meridional division while in the second cleavage, one of the two blastomeres divides meridionally and the other divides equatorially. In *C. elegans*, each asymmetrical division produces one founder cell that produces differentiated descendants; and one stem cell. The stem cell lineage always undergoes meridional division to produce (1) an anterior founder cell and (2) a posterior cell that will continue the stem cell lineage.

R-spondin1 (Rspo1)   Small, soluble protein that upregulates the Wnt pathway and is critical for ovary formation.

RTK pathway   The receptor tyrosine kinase (RTK) is dimerized by ligand, which causes autophosphorylation of the receptor. An adaptor protein recognizes the phosphorylated tyrosines on the RTK and activates an intermediate protein, GNRP, which activates the Ras G protein by allowing the phosphorylation of the GDP-bound Ras. At the same time, the GAP protein stimulates the hydrolysis of this phosphate bond, returning Ras to its inactive state. The active Ras activates the Raf protein kinase C (PKC), which in turn phosphorylates a series of kinases. Eventually, an activated kinase alters gene expression in the nucleus of the responding cell by phosphorylating certain transcription factors (which can then enter the nucleus to change the types of genes transcribed) and certain translation factors (which alter the level of protein synthesis). In many cases, this pathway is reinforced by the release of Ca^{2+}.

S

Sacral neural crest   Neural crest cells that lie posterior to chick somite 28 that generate the parasympathetic (enteric) ganglia of the gut that are required for peristaltic movement in the bowels.

Satellite cell   Putative muscle stem cell found within the basal lamina of mature myofibers that can respond to injury or exercise by proliferating into myogenic cells that fuse and form new muscle fibers.

Sclerotomes   Blocks of mesodermal cells in the ventromedial half of each somite that will differentiate into the vertebral bodies and intervertebral discs of the spine and ribs. They are also critical in patterning the neural crest and motor neurons.

Secondary hypoblast   Underlies the epiblast in the bilaminar amniotic blastoderm. A sheet of cells derived from deep yolky cells at the posterior margin of the blastoderm that migrates anteriorly, displacing the hypoblast islands (primary hypoblast). Hypoblast cells do not contribute to the primary germ layers of the embryo proper, but instead form portions of the external membranes and provide chemical signals that specify the migration of epiblast cells. Also called endoblast.

Secondary larvae   Larvae that possess the same basic body plan as the adult; caterpillars and tadpoles are examples. Compare with **primary larvae**.

Secondary neurulation   The process that forms the posterior portion of the neural tube by the coalescence of mesenchyme cells into a solid cord that subsequently forms cavities that coalesce to create a hollow tube.

Secondary oocyte   The diploid oocyte following the first meiotic division, which also generates the first polar body.

Secondary sex determination   Developmental events, directed by hormones or factors produced by the gonads, that affect the phenotype outside the gonads. This includes the male or female duct systems and external genitalia, and, in many species, sex-specific body size, vocal cartilage, and musculature.

Secondary spermatocytes   A pair of cells derived from the first meiotic division of a primary spermatocyte, which then complete the second division of meiosis to generate the four haploid spermatids.

Segmental plate   The bands of paraxial mesoderm in chick embryos.

Segment polarity genes   *Drosophila* zygotic genes activated by the proteins encoded by the pair-rule genes whose mRNA and protein products divide the embryo into 14-segment-wide units, establishing the periodicity of the embryo. Segment polarity mutants showed defects (deletions, duplications, polarity reversals) in every segment.

Selective affinity   Principle that explains why disaggregated cells reaggregate to reflect their embryonic positions. Specifically, the inner surface of the ectoderm has a positive affinity for mesodermal cells and a negative affinity for the endoderm, while the mesoderm has positive affinities for both ectodermal and endodermal cells.

Semaphorins   Extracellular matrix proteins that repel migrating neural crest cells and axonal growth cones.

Seminiferous tubules   Form at puberty from the testis cords, containing the sertoli cells and the gonocytes/spermatogonial stem cells, by the initiation of spermatogenesis, proliferation and differentiation of the spermatogenic epithelium, and formation of a fluid-filled lumen.

Senescence   The physiological deterioration that characterizes old age.

Septum   A partition that divides a chamber, such as the atrial septa that split the developing atrium into left and right atria.

Sertoli cells   Large secretory support cells in the seminiferous tubules of the testes involved in spermatogenesis in the adult through their role in nourishing and maintaining the developing sperm cells. They secrete AMF in the fetus and provide a niche for the incoming germ cells. They are derived from somatic cells, which are in turn derived from the genital ridge epithelium.

Sesamoid bone   Small bones at joints that form as a result of mechanical stress (such as the patella). They are derived from mesoderm via intramembranous ossification.

Shield   *See* Embryonic shield.

Short neural precursors (SNPs) Neuroblast precursor cells of the ventricular zone derived from radial glial cells.

Signal transduction cascades Pathways of response whereby paracrine factors bind to a receptor that initiates a series of enzymatic reactions within the cell which in turn have as their end point either the regulation of transcription factors (such that different genes are expressed in the cells reacting to these paracrine factors) or the regulation of the cytoskeleton (such that the cells responding to the paracrine factors alter their shape or are permitted to migrate).

Silencer A DNA regulatory element that binds transcription factors that actively repress the transcription of a particular gene.

Sinus venosus The posterior region of the developing heart, where the two major veins fuse. Inflow tract to the atrial area of the heart.

Sirtuin genes Encode histone deacetylation (chromatin-silencing) enzymes that guard the genome, preventing genes from being expressed at the wrong times and places, and blocking chromosomal rearrangements. They may be important defenses against premature aging.

Skeletogenic mesenchyme Also called primary mesenchyme, formed from the first tier of micromeres (the large micromeres) of the 60-cell sea urchin embryo, forms the larval skeleton.

Slow block to polyspermy *See* Cortical granule reaction.

SMAD pathway The pathway activated by members of the TGF-β superfamily. The TGF-β ligand binds to a type II TGF-β receptor, which allows that receptor to bind to a type I TGF-β receptor. Once the two receptors are in close contact, the type II receptor phosphorylates a serine or threonine on the type I receptor, thereby activating it. The activated type I receptor can now phosphorylate the Smad proteins. Smads 1 and 5 are activated by the BMP family of TGF-β factors, while the receptors binding activin, Nodal, and the TGF-β family phosphorylate Smads 2 and 3. These phosphorylated Smads bind to Smad4 and form the transcription factor complex that will enter the nucleus.

Small-interfering RNA (siRNA) Similar to miRNA, is the basis of RNA interference technique. siRNA is made from double-stranded RNA and is also packaged into RISCs.

Small micromeres A cluster of cells produced by the fifth cleavage at the vegetal pole in the sea urchin embryo when the micromeres divide.

Somatic cells Greek *soma*, body. Cells that form the body; all cells in the organism that are not germ cells.

Somatic cell nuclear transfer (SCNT) Better but less accurately known as "cloning," this is the procedure by which a cell nucleus is transferred into an activated enucleated egg and directs the development of a complete organism with the same genome as the donor cell.

Somatic mesoderm (parietal mesoderm) Derived from lateral mesoderm closest to the ectoderm (dorsal) and separated from other component of lateral mesoderm (splanchnic, near endoderm, ventral) by the intraembryonic coelom. It forms the somatopleure with the overlying ectoderm, and it will form the body wall, and the lining of the coelom.

Somite Segmental block or ball of mesoderm formed from paraxial mesoderm adjacent to notochord (the axial mesoderm). Differentiates to form, initially, sclerotome and dermamyotome; the latter goes on to form dermotome and myotome. Somites will form the axial skeleton (vertebrae, ribs), all skeletal muscle, and the dorsal dermis, tendons, joints, and dorsal aortic cells.

Somitic mesoderm *See* Paraxial mesoderm.

Somitogenesis the process of segmentation of the paraxial mesoderm to form somites, beginning cranially and extending caudally. Its components are (1) periodicity, (2) fissure formation (to separate the somites), (3) epithelialization, (4) specification, and (5) differentiation.

Somitomeres Early pre-somites, consisting of paraxial mesoderm cells organized into whorls of cells.

Sonic hedgehog (SHH) The major hedgehog family paracrine factor. SHH has distinct functions in different tissues of the embryo. For example, it is secreted by the notochord and ventralizes the neural tube. It is also involved in the establishment of left-right asymmetry, primitive gut tube differentiation, proper placement of feather formation in chicks, and in the differentiation of the sclerotome and the epaxial myotome.

Specification The first stage of commitment of cell or tissue fate during which the cell or tissue is capable of differentiating autonomously (i.e., by itself) when placed in an environment that is neutral with respect to the developmental pathway. At the stage of specification, cell commitment is still capable of being reversed.

Spemann's Organizer More correctly, the Spemann-Mangold Organizer. In amphibians, the dorsal lip cells of the blastopore and their derivatives (notochord and head endomesoderm). Functionally equivalent to Hensen's node in chick, the node in mammals, and the shield in fish. Organizer action establishes the basic body plan of the early embryo.

Spermatids Haploid sperm cells generated by meiosis that are still connected to one another through their cytoplasmic bridges, and are therefore functionally diploid due to the diffusion of gene products across the cytoplasmic bridges.

Spermatogenesis The production of sperm.

Spermatozoa The male gamete or mature sperm cell.

Sperm head Consists of the nucleus, acrosome, and minimal cytoplasm.

Spermiogenesis The differentiation of the mature spermatozoa from the haploid round spermatid.

Sperm-surface galactosyltransferase A mammalian sperm surface protein that recognizes carbohydrate residues on ZP3 and may contribute to sperm-zona binding.

Spina bifida A congenital defect resulting from failure to close the posterior neuropore, the severity of which depends on how much of the spinal cord remains exposed.

Spiral holoblastic cleavage Characteristic of several animal groups, including annelid worms, some flatworms, and most molluscs. Cleavage is at oblique angles, forming a "spiral" arrangement of daughter blastomeres, the cells touch one another at more places than do those of radially cleaving embryos, assuming the most thermodynamically stable packing orientation.

Splanchnic mesoderm Also called the visceral mesoderm; derived from lateral mesoderm closest to the endoderm (ventral) and separated from other component of lateral mesoderm (somatic, near ectoderm, dorsal) by the intraembryonic coelom. It forms the splanchnopleure together with the underlying endoderm. It will form the heart, capillaries, gonads, the visceral mesothelial and serous membranes that cover the organs, the mesenteries, and blood cells.

Spliceosome A complex, made up of small nuclear RNAs (snRNAs) and splicing factors, that bind to splice sites and mediates the splicing of nRNA.

Splicing enhancer A *cis*-acting sequence on nRNA that promotes the assembly of spliceosomes at RNA cleavage sites.

Splicing factors Proteins that bind to splice sites or to the areas adjacent to them.

Splicing isoforms Different proteins encoded by the same gene and generated by alternative splicing.

Splicing silencer A *cis*-acting sequence on nRNA that acts to exclude exons from an mRNA sequence.

SRY Sex-determining *region* of the *Y* chromosome. Gene encoding the mammalian testis-determining factor. The *SRY* gene is probably active for only a few hours in the genital ridge, during which time, it synthesizes the Sry transcription factor, whose primary role is to activate the *SOX9* gene required for testis formation.

STAT Signal *t*ransducers and *a*ctivators of *t*ranscription. A family of transcription factors, part of the JAK-STAT pathway. Important in the regulation of human fetal bone growth.

Stem cell A relatively undifferentiated cell from the embryo, fetus, or adult that, when it divides, produces (1) one cell that retains its undifferentiated character and remains in the stem cell niche; and (2) a second cell that leaves the niche and can undergo one or more paths of differentiation.

Stem cell mediated regeneration Stem cells allow an organism to regrow certain organs or tissues (e.g., hair, blood cells) that have been lost.

Stem cell niche An environment (regulatory microenvironment) that provides a milieu of extracellular matrices and paracrine factors that allows cells residing within it to remain relatively undifferentiated. Regulates stem cell proliferation and differentiation.

Stereoblastulae Blastulae produced by spiral cleavage; have no blastocoel.

Stomodeum *See* Oral plate.

Stratum corneum *See* Cornified layer.

Stratum germinativum *See* Basal layer.

Stylopod The proximal limb bones of any vertebrate limb, adjacent to the body wall, consisting of either humerus (forelimb) or femur (hindlimb).

Subendothelial cells A minor cell population in the bone marrow that may be critical for forming the hematopoietic inductive microenvironment.

Subgerminal cavity A space between the blastoderm and the yolk of avian eggs which is created when the blastoderm cells absorb water from the albumen ("egg white") and secrete the fluid between themselves and the yolk.

Sulcus limitans A longitudinal groove that divides the developing spinal cord and medulla into dorsal (receives sensory input) and ventral (initiates motor functions) halves.

Superficial cleavage The divisions of the cytoplasm of centrolecithal zygotes occur only in the rim of cytoplasm around the periphery of the cell due to the presence of a large amount of centrally-located yolk, as in insects.

Surfactant A secretion of specific proteins and phospholipids such as sphingomyelin and lecithin produced by the type II alveolar cells of the lungs very late in gestation. The surfactant enables the alveolar cells to touch one another without sticking together.

Symbiont The smaller organism in a symbiotic relationship in which one of the organisms involved is much larger than the other, and the smaller organism may live on the surface or inside the body of the larger.

Symbiosis Greek, "living together." Refers to any close association between organisms of different species.

Syncytial blastoderm Describes the *Drosophila* embryo since no cell membranes exist other than that of the egg itself.

Syncytial specification The interactions of nuclei and transcription factors, which eventually result in cell specification, take place in a common cytoplasm.

Syncytiotrophoblast A population of cells from the murine and primate trophoblast that undergoes mitosis without cytokinesis resulting in multinucleate cells. The syncytiotrophoblast tissue is thought to further the progression of the embryo into the uterine wall by digesting uterine tissue.

Syncytium Many nuclei reside in a common cytoplasm, results either from karyokinesis without cytokinesis or from cell fusion.

Syndetome Greek *syn*, "connected." Derived from the most dorsal sclerotome cells, which express the *scleraxis* gene and generate the tendons.

Syndrome Greek, "happening together." Several malformations or pathologies that occur concurrently. Genetically based syndromes are caused either by (1) a chromosomal event (such as trisomy 21) where several genes are deleted or added, or (2) by one gene having many effects.

T

T-box (*Tbx*) A specific DNA-binding domain found in certain transcription factors, including the T (*Brachyury*) gene, Tbx4 and Tbx5. Tbx4 and Tbx5 help specify hindlimbs and forelimbs, respectively.

Target selection The second step in the specification of axonal connection, wherein the axons, once they reach the correct area, recognize and bind to a set of cells with which they may form stable connections.

TATA box The sequence TATA within the promoter, to which RNA polymerase will be bound. Usually occurs about 30 base pairs upstream from the first transcribed base

Telencephalon The anterior subdivision of the prosencephalon, will eventually form the cerebral hemispheres.

Telolecithal Describes the eggs of birds and fish which have only one small area of the egg that is free of yolk.

Telomeres Repeated DNA sequences at the ends of chromosomes.

Telson Tail of *Drosophila* and other dipteran insects.

Teratocarcinoma A tumor derived from malignant primordial germ cells and containing an undifferentiated stem cell population (embryonal carcinoma, or EC cells) that has biochemical and developmental properties similar to those of the inner cell mass. EC cells can differentiate into a wide variety of tissues, including gut and respiratory epithelia, muscle, nerve, cartilage, and bone.

Teratogens Greek, "monster-formers." Exogenous agents that cause disruptions in development resulting in teratogenesis, the formation of congenital defects. Teratology is the study of birth defects and of how environmental agents disrupt normal development.

Testis cords Loops in the medullary (central) region of the developing testis formed by the developing Sertoli cells and the incoming germ cells. Will become the seminiferous tubules and site of spermatogenesis.

Testis-determining factor A protein encoded by the SRY (sex-determining region on Y) gene on the mammalian Y chromosome that organizes the gonad into a testis rather than an ovary.

Testosterone A steroid hormone secreted by the fetal testes which masculinizes the fetus, stimulating the formation of the penis, male duct system, scrotum, and other portions of the male anatomy, as well as inhibiting development of the breast primordia.

TGF-β superfamily *Transforming growth factor,* More than 30 structurally related members of a group of paracrine factors. The proteins encoded by TGF-β superfamily genes are processed such that the carboxy-terminal region contains the mature peptide. These peptides are dimerized into homodimers (with themselves) or heterodimers (with other TGF-β peptides) and are secreted from the cell. The TGF-β superfamily includes the TGF-β family, the activin family, the bone morphogenetic proteins (BMPs), the Vg1 family, and other proteins, including glial-derived neurotrophic factor (GDNF; necessary for kidney and enteric neuron differentiation) and Müllerian inhibitory factor (which is involved in mammalian sex determination).

Thecal cells Steroid hormone-secreting ovarian cells surrounding the follicle that differentiate from mesenchyme cells of the ovary.

Thermotaxis Migration towards an area of higher temperature. Capacitated sperm of some mammalian species may sense the 2°C temperature difference between the isthmus of the oviduct and the warmer ampullary region and preferentially swim from cooler to warmer sites.

Threshold model A model of development wherein the different events of are triggered by different concentrations of morphogens or hormones.

Tip cells Certain endothelial cells that can respond to the VEGF signal and begin "sprouting" to form a new vessel during angiogenesis.

Tissue engineering A regenerative medicine approach whereby a scaffold is generated from material that resembles extracellular matrix or decellularized extracellular matrix from a donor, is seeded with stem cells, and is used to replace an organ or part of an organ.

Totipotent Latin, "capable of all." Describes the earliest mammalian blastomeres (such as each blastomere of an 8-cell embryo), which can form both trophoblast cells and the embryo precursor cells. Compare with **pluripotent**.

***trans*-activating domain** The transcription factor domain that activates or suppresses the transcription of the gene whose promoter or enhancer it has bound, usually by enabling the transcription factor to interact with the proteins involved in binding RNA polymerase or with enzymes that modify histones.

***trans*-regulatory elements** Soluble molecules whose genes are located elsewhere in the genome and which bind to the *cis*-regulatory elements. They are usually transcription factors.

Transcription-associated factors (TAFs) Proteins required to facilitate efficient binding of eukaryotic RNA polymerases to the TATA sequence within the promoter.

Transcription factor A protein that binds to enhancer DNA to alter gene expression.

Transcription factor domains The three major domains are a DNA-binding domain, a trans-activating domain and a protein-protein interaction domain.

Transcription initiation site Nucleotide sequence of at the 5′ end signaling the start of gene transcription. Also called the cap sequence because it encodes the 5′ end of the RNA, which will receive a "cap" of modified nucleotides soon after it is transcribed.

Transcription termination sequence Transcription continues beyond the AATAAA site for about 1000 nucleotides before being terminated.

Transdifferentiation The use of transcription factors to directly transform one differentiated cell type into another.

Transforming growth factor *See* TGF-β.

Transgene Exogenous DNA or gene introduced through experimental manipulation into a cell's genome.

Trefoil stage A three-lobed stage of certain spirally cleaving embryos, resulting from the extrusion of a particularly large polar lobe at first cleavage.

Triploblastic "three-layer" animals containing tissues derived from the three germ layers, ectoderm, mesoderm, and endoderm, as opposed to diploblastic animals that lack mesoderm tissue.

Trophoblast The external cells of the early mammalian embryo (i.e., the morula and the blastocyst) that bind to the uterus. Trophoblast cells form the chorion (the embryonic portion of the placenta). Also called trophectoderm.

Truncus arteriosus Cardiac outflow tract precursor that will form the roots and proximal portionof the aorta and pulmonary artery.

Trunk neural crest Neural crest cells migrating from this region become the dorsal root ganglia containing the sensory neurons, the adrenal medulla, the nerve clusters surrounding the aorta, and Schwann cells if they migrate along a ventral pathway, and they generate melanocytes of the dorsum and belly if they migrate along a dorsolateral pathway.

Tubulin A dimeric protein that polymerizes to form the microtubules that are the basis of the sperm flagellar axoneme and the mitotic spindle.

Tumor angiogenesis factors Factors secreted by microtumors, including VEGFs, Fgf2, placenta-like growth factor, and others, which stimulate mitosis in endothelial cells and direct the cell differentiation into blood vessels in the direction of the tumor.

Tumor suppressor genes Regulatory genes that inhibit cell division and increase the adhesion between cells; they can also induce apoptosis of rapidly dividing cells. Cancer can result from either mutations or inappropriate methylations that inactivate tumor suppressor genes.

Tunica albuginea A thick, whitish capsule of extracellular matrix that encases the testis.

Type B spermatogonia Precursors of the spermatocytes and the last cells to undergo mitosis, they are derived from the mitotic division of intermediate spermatogonia.

U

Umbilical cord Connecting cord derived from the allantois that brings the embryonic blood circulation to the uterine vessels of the mother.

Unipotent stem cells Stem cells from particular tissues that are involved in regenerating a particular type of cell.

Unsegmented mesoderm The bands of paraxial mesoderm in non-avian vertebrate embryos.

3′ Untranslated region (3′ UTR) A region of a eukaryotic gene and RNA following the translation termination codon that, although transcribed, is not translated into protein. It often contains regulatory regions that control the rate of translation.

5′ Untranslated region (5′ UTR) A region of a eukaryotic gene or RNA: the sequence of base pairs intervening between the initiation points of transcription and translation. These are not translated into protein, but may contain important regulatory sequences.

Upstream promoter elements DNA sequences near the TATA box and usually upstream from it that bind to the transcription-associated factors (TAFs).

Ureteric buds Paired epithelial branches induced by the metanephrogenic mesenchyme to branch from each of the paired nephric ducts. Ureteric buds will form the collecting ducts, renal pelvis, and ureters that take the urine to the bladder.

Urodeles Salamanders.

Uterine cycle A component of the menstrual cycle, the function of the uterine cycle is to provide the appropriate environment for the developing blastocyst.

V

Vagal neural crest Neural crest cells from the neck region, which overlaps the cranial/trunk crest boundary. Generates the parasympathetic (enteric) ganglia of the gut, which are required for peristaltic movement of the bowels.

Vascular endothelial growth factors (VEGFs) A family of proteins involved in vasculogenesis which includes several VEGFs, as well as placental growth factor. Each VEGF appears to enable the differentiation of the angioblasts and their multiplication to form endothelial tubes.

Vasculogenesis The de novo creation of a network of blood vessels from the lateral plate mesoderm.

Vas (ductus) deferens Derived from the Wolffian duct, the tube through which sperm pass from the epididymis to the urethra.

VegT pathway Involved in dorsal-ventral polarity and specification of the organizer cells in the amphibian embryo. The VegT pathway activates the expression of Vg1 and other Nodal-related paracrine factors, which in turn activate the Smad2/4 transcription factor in the mesodermal cells above them, activating genes that give these cells their "organizer" properties.

Vegetal hemisphere The bottom portion of the amphibian egg, containing yolk, which serves as food for the developing embryo. The yolk-filled cells are divide more slowly and undergo less movement during embryogenesis (and hence are like plants, or "vegetal").

Vegetal plate Area of thickened cells at the vegetal pole of the sea urchin blastula.

Vegetal pole The yolk containing portion of the egg or embryo.

Vegetal rotation During frog gastrulation, internal cell rearrangements place the prospective pharyngeal endoderm cells adjacent to the blastocoel and immediately above the involuting mesoderm.

Vellus Short and silky hair of the fetus and neonate that remains on many parts of the human body that are usually considered hairless, such as the forehead and eyelids. In other areas of the body, vellus hair gives way to longer and thicker "terminal" hair.

Ventral furrow Invagination of the prospective mesoderm, about 1000 cells constituting the ventral midline of the embryo, at the onset of gastrulation in *Drosophila.*

Ventricular zone (VZ) Inner layer of the developing spinal cord and brain medulla. Forms from the germinal neuroepithelium of the original neural tube and will eventually form the ependyma.

Vertical transmission Transfer from one generation to the next through the germ cells. Can refer to the transfer of symbionts via the germ cells, usually the eggs.

Vital dyes Stains used to label living cells without killing them. When applied to embryos, vital dyes have been used to follow cell migration during development and generate fate maps of specific regions of the embryo.

Vitelline envelope In invertebrates, the extracellular matrix that forms a fibrous mat around the egg outside the cell membrane and is often involved in sperm-egg recognition and is essential for the species-specific binding of sperm. The vitelline envelope contains several different glycoproteins. It is supplemented by extensions of membrane glycoproteins from the cell membrane and by proteinaceous "posts" that adhere the vitelline envelope to the membrane.

Vitelline veins The veins, continuous with the endocardium, that carry nutrients from the yolk sac into the sinus venosus of the developing heart. These veins form from yolk sac blood islands, and bring nutrients to the avian embryo and transport gases to and from the sites of respiratory exchange.

Vitellogenesis The accumulation of yolk proteins in telolecithal and mesolecithal eggs.

Vulval precursor cells (VPCs) Six cells in the larval stage of *C. elegans* that will form the vulva via inductive signals.

W

Wholist organicism Philosophical notion stating that the properties of the whole cannot be predicted solely from the properties of its component parts, and that the properties of the parts are informed by their relationship to the whole. It was very influential in the construction of developmental biology.

Wnts A gene family of cysteine-rich glycoprotein paracrine factors. Their name is a fusion of the name of the *Drosophila* segment polarity gene *wingless* with the name of one of its vertebrate homologues, *integrated.* Wnt proteins are critical in establishing the polarity of insect and vertebrate limbs; promoting the proliferation of stem cells; and in several steps of urogenital system development.

Wnt pathway (canonical) Binding of Wnt proteins to Frizzled causes Frizzled to activate the Disheveled protein. Once Disheveled is activated, it inhibits the activity of the glycogen synthase kinase 3 (GSK3) enzyme. GSK3, if it were active, would prevent the dissociation of the β-catenin protein from the APC protein, which targets β-catenin for degradation. However, when the Wnt signal is present and GSK3 is inhibited, β-catenin can dissociate from the APC protein and enter the nucleus. Once inside the nucleus, it can form a heterodimer with another DNA-binding protein, becoming a transcription factor. This complex binds to and activates the Wnt-responsive genes.

Wnt pathway (non-canonical) When Wnt activates Disheveled, the Disheveled protein can interact with a Rho GTPase. This GTPase can activate the kinases that phosphorylate cytoskeletal proteins and thereby alter cell shape, cell polarity (where the upper and lower portions of the cell differ), and motility. A third Wnt pathway diverges earlier than Disheveled. Here, the Frizzled receptor protein activates a phospholipase (PLC) that synthesizes a compound that releases calcium ions from the endoplasmic reticulum. The released calcium can activate enzymes, transcription factors, and translation factors.

Wolffian duct (nephric duct) Derived from the caudal portions of the pronephric duct. In mammals, it is the central component of the excretory system throughout development. Degenerates in females; in males, becomes the epididymis and vas deferens.

X

X chromosome inactivation Mechanism of dosage compensation in mammals; the irreversible conversion of the chromatin of one X chromosome in each female (XX) cell into highly condensed heterochromatin (Barr bodies), thus preventing excessive transcription of X-chromosome genes. See also **dosage compensation.**

Y

Yellow crescent Region of the tunicate zygote cytoplasm containing yellow lipid inclusions that will become mesoderm.

Yolk cell The uncleaved, yolk-containing portion of the telolecithal egg that underlies the blastoderm.

Yolk plug The large endodermal cells that remain exposed on the vegetal surface surrounded by the blastopore of the amphibian gastrulating embryo.

Yolk sac The first extraembryonic membrane to form, derived from splanchnopleure that grows over the yolk to enclose it. The yolk sac mediates nutrition in developing birds and reptiles. It is connected to the midgut by the yolk duct (vitelline duct), so that the walls of the yolk sac and the walls of the gut are continuous.

Yolk syncytial layer (YSL) A cell population in the zebrafish cleavage stage embryo formed at the ninth or tenth cell cycle, when the cells at the vegetal edge of the blastoderm fuse with the underlying yolk cell producing a ring of nuclei in the part of the yolk cell cytoplasm that sits just beneath the blastoderm. Important for directing some of the cell movements of gastrulation.

Z

Zeugopod The middle limb bones of any vertebrate limb, consisting of either radius and ulna (forelimb) or tibia and fibula (hindlimb).

Zona pellucida Glycoprotein coat (extracellular matrix) around the mammalian egg, synthesized and secreted by the growing oocyte.

Zona proteins (ZP1, ZP2, ZP3) The three major glycoproteins of the mouse zona pellucida. ZP3 binds the sperm and, once the sperm is bound, initiates the acrosome reaction. ZP2 binds to acrosome-reacted sperm. ZP1 holds the zona pellucida together.

Zone of polarizing activity (ZPA) A small block of mesodermal tissue near the posterior junction of the young limb bud and the body wall that specifies the anterior-posterior axis of the developing limb through the action of the paracrine factor, *sonic hedgehog*.

Zygote A fertilized egg with a diploid chromosomal complement in its zygote nucleus generated by fusion of the haploid male and female pronuclei.

Chapter-Opening Source Credits

Part Opening Art

The original cartoons that open Parts I, II, III, and IV are by Dave Granlund, www.davegranlund.com.

Chapter 1

Franklin Mall, ca. 1890. Cited in L. M. Morgan. 2009. *Icons of Life: A Cultural History of Human Embryos*. University of California Press, Berkeley, p. 72.

Jane M. Oppenheimer, 1955. Analysis of development: Problems, concepts, and their history. In *Analysis of Development*, B. H. Willier, P. A. Weiss, and V. Hamburger (eds.). Saunders, Philadelphia, pp. 1–24.

Tunicate fate map, Figure 1.12A. Adapted from M. T. Veeman, Y. Nakatani, C. Hendrickson, V. Ericson, C. Lin and W. C. Smith. 2008. *chongmague* reveals an essential role for laminin-mediated boundary formation in chordate convergence and extension movements. *Development* 135: 33–41. Courtesy of Michael Veeman, University of California, Santa Barbara.

Chapter 2

C. H. Waddington, 1956. *Principles of Embryology*. Macmillan, New York, p. 5.

Albert Claude, 1974. The coming of age of the cell. Nobel lecture, reprinted in *Science* 189: 433–435.

Model of nucleosome structure, Figure 2.3A. Model by David McIntyre, after C. A. Davey, D. F. Sargent, K. Luger, A. W. Maeder and T. J. Richmond. 2002. Solvent-mediated interactions in the structure of the nucleosome core particle at 1.9 Å resolution. *J. Mol. Biol.* 319: 1097–1113.

Chapter 3

Octavia Butler, 1998. *Parable of the Talents*. Warner Books, New York, p. 3.

Jonathan Bard, 1997. Explaining development. *BioEssays* 20: 598–599.

Cross-section through *Drosophila* embryo, Figure 3.9B. Micrograph courtesy of Verena Kölsch and Maria Leptin, University of Cologne.

Chapter 4

Walt Whitman, 1855. "Song of Myself." In *Leaves of Grass and Selected Prose*. S. Bradley (ed.), 1949. Holt, Rinehart & Winston, New York, p. 25.

Arthur Schopenhauer, quoted in C. Darwin, 1871. *The Descent of Man*. Murray, London, p. 893.

Sea urchin sperm and egg, Figure 4.15A. Scanning electron micrograph © Mia Tegner/SPL/Photo Researchers Inc.

Chapter 5

E. E. Just, 1939. *The Biology of the Cell Surface*. Blakiston, Philadelphia, p. 288.

Lewis Wolpert, 1986. *From Egg to Embryo: Determinative Events in Early Development*. Cambridge University Press, Cambridge, p. 1.

Unequal cleavage in the 16-cell sea urchin embryo, Figure 5.6B. Confocal micrograph courtesy of George von Dassow and the Center for Cell Dynamics.

Chapter 6

Jack Schultz, 1935. Aspects of the relation between genes and development in *Drosophila*. *American Naturalist* 69: 30–54.

Robert E. Kohler, 1944. *Lords of the Fly*: Drosophila *Genetics and the Experimental Life*. University of Chicago Press, Chicago, p. 33.

GFP-labeled transgenic *Drosophila*, Figure 6.6B. Photograph courtesy of Ansgar Klebes, Freie Universität Berlin.

Chapter 7

Hans Spemann, 1943. *Forschung und Leben*. Quoted in T. J. Horder, J. A. Witkowski and C. C. Wylie, 1986, *A History of Embryology*. Cambridge University Press, Cambridge, p. 219.

Jean Rostand, 1960. *Carnets d'un Biologiste*. Librairie Stock, Paris.

Sagittal section of late-gastrula *Xenopus*, Figure 7.12B. From M. Marsden and D. W. DeSimone. 2001. Regulation of cell polarity, radial intercalation, and epiboly in *Xenopus*. *Development* 128: 3635–3647, courtesy of the authors.

Chapter 8

Arthur Conan Doyle, 1891. "A Case of Identity." In *The Adventures of Sherlock Holmes*. Reprinted in *The Complete Sherlock Holmes Treasury*, 1976. Crown, New York, p. 31.

Miroslav Holub, 1990. "From the Intimate Life of Nude Mice." In *The Dimension of the Present Moment*. Trans. D. Habova and D. Young. Faber and Faber, London, p. 38.

Axial skeletons of chick and mouse at comparable stages of development, Figure 8.31A. From M. Kmita and D. Duboule. 2003. Organizing axes in time and space: 25 years of colinear tinkering. *Science* 301: 331–333, courtesy of the authors.

Chapter 9

Lewis Thomas, 1979. "On Embryology." In *The Medusa and the Snail*, Viking Press, New York, p. 157.

The neural tube and neural crest, Figure 10.1A. Micrograph courtesy of James Briscoe, National Institute for Medical Research, London.

Chapter 10

Alfred North Whitehead, 1934. *Nature and Life.* Cambridge University Press, Cambridge, p. 41.

Santiago Ramón y Cajal, 1937. *Recollections of My Life.* Trans. E. H. Craigie and J. Cano. MIT Press, Cambridge, MA, pp. 36–37.

"Brainbow" visualizing individual neurons and their axons, Figure 10.32B. Courtesy of Jeff Lichtman, Harvard University.

Chapter 11

Walt Whitman, 1867. "Inscriptions." In *Leaves of Grass and Selected Prose.* S. Bradley (ed.), 1949. Holt, Rinehart & Winston, New York, p. 1.

Natalie Angier, 1994. *The New York Times,* November 1.

Neural tube with surrounding somites and paraxial mesoderm, Figure 11.4A. Scanning electron micrograph courtesy of Kathryn Tosney, University of Miami.

Chapter 12

William Harvey, 1628. *Exercitio Anatomica de Motu Cordis et Sanguinis Animalibus.* Reprinted in 1928, C. C. Thomas, Baltimore, p. A2.

Johann Wolfgang von Goethe, 1805. *Faust,* Part I. Trans. R. Jarrell, 1976.

Capillary network in a 9.5-day mouse embryo, Figure 12.17B. Micrograph courtesy of Kari Alitalo, University of Helsinki.

Chapter 13

Charles Darwin, 1859. *On the Origin of Species.* Reprinted by New American Library, New York, p. 403

Debra Niehoff, 2005. *The Language of Life: How Cells Communicate in Health and Disease.* Joseph Henry Press, Chevy Chase, MD.

Gene expression in the apical ectodermal ridge of a 3-day chick embryo, Figure 13.10C. Micrograph courtesy of John F. Fallon and A. López-Martínez, University of Wisconsin.

Chapter 14

Erasmus Darwin, 1791. Quoted in M. T. Ghiselin, 1974, *The Economy of Nature and the Evolution of Sex,* University of California Press, Berkeley, p. 49.

J. A. Thomson, 1926. *Heredity.* Putnam, New York, p. 477.

Rspo1 protein in embryonic mouse gonads, Figure 14.5A. From C. A. Smith, C. M. Shoemaker, K. N. Roeszler, J. Queen, D. Crews and A. H. Sinclair. 2008. Cloning and expression of *R-Spondin1* in different vertebrates suggests a conserved role in ovarian development. *BMC Dev. Biol.* 8: 72, courtesy of C. A. Smith.

Chapter 15

Carroll M. Williams, 1959. Hormonal regulation of insect metamorphosis. In *The Chemical Basis of Development,* W. D. McElroy and B. Glass (eds.). Johns Hopkins University Press, Baltimore, p. 794.

Oscar E. Schotte, 1950. Quoted in R. J. Goss, The natural history (and mystery) of regeneration. In C. E. Dinsmore (ed.), *A History of Regeneration Research,* 1991. Cambridge University Press, Cambridge, p. 12.

Parental care in *Dendrobates* frogs, Figure 15.8A. Photograph © Michael Doolittle/Alamy.

Chapter 16

T. S. Elliot, 1942. "Little Gidding." In *Four Quartets.* Harcourt, Brace and Company, New York. 1943. p. 39. Copyright © T. S. Elliot. All rights reserved.

Ernst Hadorn, 1955. *Developmental Genetics and Lethal Factors.* London 1961, p. 105.

Distal tip cell of *C. elegans,* Figure 16.19A. Micrograph courtesy of Sarah Crittenden and Judith Kimble, University of Wisconsin.

Chapter 17

Veronica van Heyningen, 2000. Gene games of the future. *Nature* 408: 769–771.

William Gibson, 1999. Quoted in M. Payser, "The Home of the Gay," *Newsweek* March 1, 1999, p. 50.

Reconstruction of 17-day embryonic mouse brain based on MRI, Figure 17.7E. Courtesy of Kathleen Sulik, University of North Carolina.

Chapter 18

C. H. Waddington, 1957. *The Strategy of the Genes.* Allen & Unwin, London, pp. 154–155.

Jian Xu and Jeffrey I. Gordon. 2003. Honor thy symbionts. *Proc. Natl. Acad. Sci. USA* 100: 10452–10459.

Drosophila germinarium infected with *Wolbachia* symbionts, Figure 18.16A. Confocal micrograph courtesy of Horacio Frydman and Eric Weischaus, Princeton University.

Chapter 19

Salman Rushdie, 1989. *The Satanic Verses.* Viking, New York, p. 8.

Julian Huxley, 1942. *Evolution: The Modern Synthesis.* Allen & Unwin, London. Reprint edition 2010, MIT Press, Cambridge, MA.

Haasiophis terrasanctus fossil laid down in the upper Cretaceous, with visible stylopod and zeugopod, Figure. 19.7A. Photograph by Scott F. Gilbert.

Author Index

Subject Index

Italic type indicates the information will be found in an illustration.

The designation "n" indicates the information will be found in a footnote.

A

α4β1 Integrin, 380
α5β1 Integrin, 427
A23187, 138
ABa cell, 199
Abaxial, 423n
Abaxial muscles, 423, 426
Abaxial myotome, 422
Abaxial somitic bud, 422
AbdB protein, 236
abdominal A gene, 234, 235
abdominal B gene, 234, 235, 236
Abdominal B protein, 236
Abortion, spontaneous, 626
ABp cell, 199
Accutane, 631, 633–634
N-Acetylglucosamine, 153
N-Acetylglucosaminidase, 153
Achondroplasia, 90, 508, 509
Acidic FGF, 85
Acila, 185
Acne medication, 633–634
Acron, 225
Acrosomal cap, 610
Acrosomal process, 123
 in the acrosome reaction, 130, 131
 bindin and, 131, 132
 species-specific binding to the egg surface, 131, 132
 sperm-egg fusion and, 133, 134
Acrosomal vesicle, 122–123, 151, 610
Acrosome, 122–123, 610
Acrosome reaction
 in external fertilization, 130–131
 induction by ZP3, 150–151
Acrythosiphon pisum, 704
actin gene, 190
Actin microfilaments
 cadherins and cell adhesion, 73, 74

of the contractile ring, 161
of the egg cortex, 127
left-right axis in *Drosophila* and, 217
RNA transport in meroistic oogeneis and, 607
Activin, 261, 262
 distance of action, 84
 function, 95
 left-right axis specification in birds, 298
 suppression of nephric duct cells, 438
Acute premyelocytic leukemia (APL), 644
ADAM2, 150
ADAM3, 150
Adepithelial cells, 552–553
Adherens junctions, 73
Adhesion, in cell migration, 79
Adipose tissue, 425
ADMP, 265, 266
Adolescents, brain development, 358
Adrenal cortex, 382
Adrenal gland, 382
Adrenal medulla, 380–381
Adrenergic neurons, 381
Adrenomedullary cells, 381
Adult anxiety, 667
Adult-onset illnesses, 646–648
Adult stem cell niches, 328–330
Adult stem cells
 defined, 326
 mulitpotent, 330–331
 neural, 354–355, 655
 overview, 327
 regeneration therapy and, 653–655
Aequorin, 137
Afferent neurons, 380n
Agalychnis callidryas, 664, 665
Aggregations, thermodynamic model of, 72–73
Aging
 defined, 571
 environmental and epigenetic causes, 575–579
 genetic factors, 571–575
 insulin signaling pathway, 573–575

integrating the pathways in, 575
Aglais urticae, 707, 708
Agouti phenotype, 662, 703
Agrican, 429
Air sacs, 478
AIx1, 176
Albumen, 288
Albumin proteins, 147
Alcohol, teratogenic effects, 631–633
Algal symbionts, 676, 678
Allantoic artery, 455
Allantois, 480, 482
Alligator mississippiensis, 534
Allometry, 695
Alternative nRNA splicing, 54–56
Alternative splicing factor, 54
Alveoli, 478
Alx1 gene, 170, 171
Alzheimer's disease, 401
Amacrine neurons, 363
Amadillidium vulgare, 676
Ambystoma, limb bud formation, 487
Ambystoma jeffersonianum, 139n
Ambystoma maculatum, 486, 562, 678, 679
Ambystoma mexicanum, 15, 548
Ambystoma platineum, 139n
Ambystoma tigrinum, 548–549
American Chemical Council, 638
American oyster, 670n
Ametabolous development, 550, 551
Aminoglycosides, 631
Aminopterin, 631
Amniocentesis, 629
Amnion, 307, 310, 480, 481
Amnionic cavity, 305, 306
Amnionic sac, 310
Amnioserosa, 207, 208, 215
Amniotes
 defined, 287
 extraembryonic membranes, 480–482
Amniotic cavity, 307
Amniotic fluid, 12, 305, 481
Amphibian axis formation
 left-right axis, 272
 model of, 272

About the Book

Publisher: Andrew D. Sinauer

Editor: Carol Wigg

Copy Editor: Elizabeth C. Pierson

Production Manager: Christopher Small

Book and Cover Design: Janice Holabird

Electronic Bookbuilders: Janice Holabird and Jefferson Johnson

Illustration Program: Dragonfly Media Group

Indexer: Grant Hackett

Book and Cover Manufacture: Courier Companies, Inc.